New Trends in Cross-Coupling
Theory and Applications

RSC Catalysis Series

Editor-in-Chief:
Professor James J Spivey, *Louisiana State University, Baton Rouge, USA*

Series Editors:
Professor Chris Hardacre, *Queen's University Belfast, Northern Ireland*
Professor Zinfer Ismagilov, *Boreskov Institute of Catalysis, Novosibirsk, Russia*
Professor Umit Ozkan, *Ohio State University, USA*

Titles in the Series:

How to obtain future titles on publication:
A standing order plan is available for this series. A standing order will bring
delivery of each new volume immediately on publication.

For further information please contact:
Book Sales Department, Royal Society of Chemistry, Thomas Graham House,
Science Park, Milton Road, Cambridge, CB4 0WF, UK
Telephone: +44 (0)1223 420066, Fax: +44 (0)1223 420247
Email: booksales@rsc.org

Visit our website at www.rsc.org/books

New Trends in Cross-Coupling
Theory and Applications

Edited by

Thomas J Colacot
*Johnson Matthey Catalysis and Chiral Technologies and
Johnson Matthey, West Deptford, New Jersey, USA
Email: thomas.colacot@jmusa.com*

THE QUEEN'S AWARDS
FOR ENTERPRISE:
INTERNATIONAL TRADE
2013

RSC Catalysis Series No. 21

Print ISBN: 978-1-84973-896-5
PDF eISBN: 978-1-78262-025-9
ISSN: 1757-6725

A catalogue record for this book is available from the British Library

Published by The Royal Society of Chemistry,
Thomas Graham House, Science Park, Milton Road,
Cambridge CB4 0WF, UK

Registered Charity Number 207890

For further information see our web site at www.rsc.org

Printed in the United Kingdom

Foreword

Transition metal-catalyzed cross-coupling involving the use of catalysts containing Pd, Ni, Cu, and other d-block transition metals has emerged as a collectively highly useful, selective, and widely applicable method for synthesizing a wide range of organic compounds through highly selective cross-coupling *via* C–C bond formation. Aside from earlier brief reviews published in the 1970s and early 1980s[1] a massive compilation of this important subject was first published in 2002.[2] Just a few years later, this publication was updated and significantly supplemented by de Meijere and Diederich in 2004,[3] and again in 2014 by de Meijere and his associates.[4] These facts simply point to the extraordinary fundamental significance of cross-coupling in organic synthesis.

We now have a welcome addition to the above-mentioned impressive and massive collection entitled *"New Trends in Cross-Coupling: Theory and Applications"*, edited by T. J. Colacot at Johnson Matthey Catalysis & Chiral Technologies, USA. He and his associates have also written the first three fundamentally important chapters.

As indicated by the title, it focuses its attention on new trends in cross-coupling, thereby distinguishing itself from the others mentioned above.

It consists of 16 chapters. Most impressively, well over 3000 references mostly on new subjects for cross-coupling are cited, making this compilation a "must" for anyone interested in learning about and using newer trends in cross-coupling. Specifically, the following topics are ably discussed: Introduction to New Trends in Cross-Coupling (Chapter 1), Prominent Ligand Types in Modern Cross-Coupling (Chapter 2), Pd-Phosphine Precatalysts (Chapter 3), Use of Pd N-Heterocyclic Carbene (Pd-NHC) Complexes (Chapter 4), Ancillary Ligand Design for Challenging Selective Monoarylation (Chapter 5), Transition Metal-Catalyzed Formation of C–O and C–S Bonds (Chapter 6), Pd(0)-Catalyzed Carboiodination (Chapter 7), Boron

RSC Catalysis Series No. 21
New Trends in Cross-Coupling: Theory and Applications
Edited by Thomas J Colacot
© The Royal Society of Chemistry 2015
Published by the Royal Society of Chemistry, www.rsc.org

Reagents in Suzuki–Miyaura Coupling (Chapter 8), The Modern Heck Reactions (Chapter 9), Palladium-Catalyzed Carbonylative-Coupling (Chapter 10), Stereospecific and Stereoselective Suzuki–Miyaura Cross-Coupling Reactions (Chapter 11), Direct Arylation *via* C–H Activation (Chapter 12), Cross-Coupling Chemistry in Continuous Flow (Chapter 13), Green Approaches to Cross-Coupling (Chapter 14), Recent Large-Scale Applications of Transition Metal-Catalyzed Couplings for the Synthesis of Pharmaceuticals (Chapter 15), and Palladium Detection Technologies for Active Pharmaceutical Ingredients Prepared *via* Cross-Couplings (Chapter 16).

Inasmuch as this compilation focuses its attention on newer trends in cross-coupling, it would be useful to explicitly remind the readers of earlier compilations, such as references 1–4.

Ei-ichi Negishi

References

1. E. Negishi, *Acc. Chem. Res.*, 1982, **15**, 340–348.
2. *Handbook of Organopalladium Chemistry for Organic Synthesis*, ed. E. Negishi and A. de Meijere, John Wiley & Sons, Inc., New York, 2002, vol. 1 and 2, pp. 1–3279.
3. *Metal-Catalyzed Cross-Coupling Reactions*, ed. A. de Meijere and F. Diederich, Wiley-VCH Verlag GmbH & Co. KGaA, 2nd edn, 2004, pp. 1–916.
4. *Metal-Catalyzed Cross-Coupling Reactions and More*, ed. A. de Meijere, S. Bräse and M. Oestreich, Wiley-VCH Verlag GmbH & Co. KGaA, 2014, pp. 1–1511.

About Ei-ichi Negishi

Professor Negishi was awarded the 2010 Nobel Prize in Chemistry "for palladium catalyzed cross couplings in organic synthesis" jointly with Professors Richard F. Heck and Akira Suzuki. He is a distinguished Professor of Organic Chemistry & Teijin Limited Director of the Negishi-Brown Institute at Purdue University and the inventor of Negishi Coupling.

Foreword

Helping to address the problems of society ranging from electronics to medicine is a major goal of science. What differentiates a chemist from other scientists is the ability to design a structure that one feels is desirable regardless of whether such structures exist since the chemist can go into the lab and synthesize the structure. A major obstacle is doing so in a time-efficient manner. Thus, there is a need to have a synthetic toolbox that will address such a problem. Herein is the need to improve our synthetic reactions to permit the timely synthesis of any structure regardless of its molecular complexity.

This monograph relates to the explicit role that a subset of Pd-catalyzed reactions, notably with homogeneous catalysts, is having on meeting the above goal. It is instructive to examine how we got here. To begin, let's examine a brief overview of the early days in the discovery of precious metals. At the start, individuals involved with isolation of precious metals frequently spanned a broad range of aspects of the science, from the initial discovery all the way to creating a market. Let me make this point with respect to palladium, the topic of this work. Historically, around the year 1700, miners in Brazil were aware of a metal which they referred to as "ouro podre" or worthless gold, a native alloy of Pd and Au, today known as one of the forms of white gold. It was, however, through the mining of Pt that ultimately led to the actual refining of Pd. By the eighteenth century Pt had found numerous uses, and Percival Norton Johnson, son of assayer John Johnson, played a major role through his company which he co-founded in 1817. In 1838, George Matthey, a banker, joined and led to what is known today as Johnson Matthey. It was the purification of Pt by William Wollaston that led him to find a way to remove a pesky impurity. He ultimately isolated the "impurity" which he believed was a new metal. He called this new material "palladium" after a new asteroid named "Pallas", named so after the Goddess of

RSC Catalysis Series No. 21
New Trends in Cross-Coupling: Theory and Applications
Edited by Thomas J Colacot
© The Royal Society of Chemistry 2015
Published by the Royal Society of Chemistry, www.rsc.org

Wisdom. Incidentally, he also discovered rhodium. Believing there was commercial value in palladium as a metal, he anonymously announced the properties of this new metal and made it available for sale but refused to reveal the process for making it until shortly before his death in 1826. Since Pd was obtained as a by-product of Pt production, its quantities increased but, unfortunately with no market for the growing supplies. It wasn't until the 1930's that a German company developed and patented alloys of palladium and gold or silver for use in dentistry.

In 1959, the use of the "noble" metals in catalysis was virtually only taught with respect to heterogeneous hydrogenation. The first industrial process other than hydrogenation involving Pd was the Wacker process which came about in 1956 with the conversion of ethylene to acetaldehyde. Such discoveries stimulated studies into how ligands bind to metals and effect their chemical behavior, thus creating a steep growth spurt. With respect to Pd, the Wacker oxidation allowed the conversion of cheap, readily available hydrocarbons to higher value added "oxidation" products wherein useful oxygen functionality in the form of simple reagents like water and acetic acid was installed via olefinic bonds. This type of Pd process constitutes one of the important fundamental transformations in catalysis which has morphed into a host of synthetic reactions that elaborate olefins into many types of higher value products. Indeed, it was the study of the mechanism of this process that led Richard Heck, then at Hercules Chemical Co. in Wilmington, DE to invent by design what we now call the Heck reaction. In a tour-de-force, Heck disclosed these studies in a series of seven papers of which he was the sole author published back to back in the *Journal of the American Chemical Society* in 1968. Heck's reports revealed a type of chemical reactivity that heretofore did not exist, the direct addition of a carbon–metal bond to a non-activated carbon–carbon π bond. The only synthetic problem for its use was that the reaction required stoichiometric amounts of palladium. In 1968, Fitton discovered that tetrakis(triphenylphosphine)palladium underwent a stoichiometric reaction with iodobenzene to form a stable phenylpalladium iodide complex. Armed with this information, Heck published a catalytic version of his reaction about 4 years later in 1972 and the rest is history – his sharing of the Nobel Prize in 2010.

During this time, another type of reactivity of Pd(+2) species was reported by Professor Arthur Cope at MIT, best known for reactions like the Cope elimination and Cope rearrangement, who also played with the organic chemistry of Pd. In 1965, he demonstrated what I believe is the first example of an unactivated C–H insertion by a Pd(+2) salt upon reaction with azobenzene to give an isolable organometallic. This process revealed that appropriate coordination to an organic molecule can direct the resultant complex to facilitate the insertion of the Pd into a proximal C–H bond, a type of reactivity that has proven invaluable on the types of Pd catalyzed reactions described in this monograph.

In 1976, Ishikawa noted that the palladium complex of Fitton catalyzed the cross-coupling of aryl Grignard reagents with aryl iodides. A major drawback of this method and related nickel catalyzed reactions was the lack of chemoselectivity associated with the use of Grignard reagents. That same year, Negishi began examining the prospect of using more chemoselective nucleophilic partners such as organoalanes and organozirconium complexes in palladium (as well as nickel) catalyzed vinyl-vinyl cross-coupling processes. Shortly thereafter Negishi noted that *in situ* generated organozinc compounds participated in chemoselective cross coupling and gave higher yields than Grignard reagents or organoalanes. It is interesting to note that Negishi reports the failure of organoboranes in such processes. In 1979, the landscape changed when Suzuki reported conditions that allowed organoboranes to be used. The grandfather of metal-promoted coupling, the Ullman reaction, morphed from a very limited process stoichiometric in metal into a widely divergent carbon–carbon bond forming process that had the characteristics of nearly perfect selectivity. Correspondingly, palladium has moved from being an esoteric metal of no known use to one of being among the most versatile type of transition metal homogeneous catalyst of any metal known to date. Thus, Negishi and Suzuki joined Heck in the recognition of these pioneers by their receipt of the Noble Prize in 2010.

In a period of slightly more than 30 years, palladium catalysts literally changed the way complex molecules could be made. This monograph vividly illustrates the enormity of the invention. The book opens with an account of the key parameters and mechanistic characteristics of metal-catalyzed reactions in Chapter 1. Chapters 2 to 5 note the remarkable influence of ligands on the chemistry of palladium complexes. Indeed, the ability to tune any specific reaction at one's will by appropriate choice of ligand environment is both a power of this synthetic tool as well as a complication that must be realized. Key to extending the cross-coupling reaction beyond Grignard reagents is understanding how the transmetalation process works, a topic dealt with in Chapter 8. Chapters 5 and 6 demonstrate the breadth of the concept beyond C–C bond forming events to carbon-heteroatom bond forming events, with nitrogen being the most notable. The status of the Heck reaction today is dealt with in Chapter 9. An important component of selectivity in complex molecule synthesis, stereochemical control both relative and absolute, is the topic of Chapter 11. Carbon–carbon bond forming reactions extrapolating from the core Heck–Negishi–Suzuki type are illustrated in a so-called carboiodination pathway as an alternative to the Heck process in Chapter 7 and the ability to intercept an organopalladium intermediate by carbon monoxide to generate the extremely important carbonyl containing products is covered in Chapter 10. Making the reaction more atom economic by effecting such reactions by direct C–H activation is examined in Chapter 12 whereas Chapter 14 deals more broadly with making these palladium catalyzed processes "greener". A new dimension in homogeneous catalytic processes is performing them under flow conditions. The benefits of these techniques are examined in Chapter 13. In developing new synthetic tools, a critical question is the ability to scale up the processes. Chapter 15 examines this component.

This monograph vividly illustrates that these methods have rapidly gained immense impact on making truly complicated structures. But we must be careful in coming to any conclusion about what is left to be discovered. The trite saying "you don't know what you don't know" is especially true in synthetic chemistry. Indeed many of these chapters show that many un-imaginable processes have become real. Given the enormity of the variables, it is impossible to guess how much more can be done. It is perhaps why palladium has been referred to as the metal of the twenty-first century. This monograph is only one small step on the path toward being able to make everything imaginable. This book is invaluable to anyone involved in synthesis of organic compounds for any purpose.

Barry M. Trost

About Barry M. Trost

Born in Philadelphia, Pennsylvania, he obtained a BA degree from the University of Pennsylvania in 1962 and PhD degree just three years later at the Massachusetts Institute of Technology (1965). He directly moved to the University of Wisconsin where he was promoted to Professor of Chemistry in 1969 and subsequently became the Vilas Research Professor in 1982. He joined the faculty at Stanford as Professor of Chemistry in 1987 and became Tamaki Professor of Humanities and Sciences in 1990. In addition, he has been Visiting Professor in Germany (Universities of Marburg, Hamburg, Munich and Heidelberg), Denmark (University of Copenhagen), France (Universities of Paris VI and Paris-Sud), Italy (University of Pisa) and Spain (University of Barcelona). He received honorary degrees from the Université Claude-Bernard (Lyon I), France (1994), and the Technion, Haifa, Israel (1997). In recognition of his many contributions, Professor Trost has received a large number of awards, a few among which are the ACS Award in Pure Chemistry (1977), the Dr Paul Janssen Prize (1990), the ASSU Graduate Teaching Award (1991), Bing Teaching Award (1993), the ACS Roger Adams Award (1995), the Presidential Green Chemistry Challenge Award (1998), the Belgian Organic Synthesis Symposium Elsevier Award (2000), the ACS Nobel Laureate Signature Award for Graduate Education in Chemistry (2002), the ACS Cope Award (2004), the Nagoya Medal (2008), the Ryoji Noyori Prize (2013), the International Precious Metals Institute's Tanaka Distinguished Achievement Award (2014), and the German Chemical Society's August-Wilhelm-von-Hofmann Denkmuenze (2014). Professor Trost has been elected a fellow of the American Academy of Sciences (1992) and a member of the U.S. National Academy of Sciences (1990).

Foreword

The relative difficulty of achieving the direct formation of C_{sp}^2–C_{sp}^2 bonds through "conventional" organic chemistry was well-appreciated at the time that pioneering studies of cross-coupling reactions began to emerge about 40 years ago. Nevertheless, the chemistry community was surprisingly slow to embrace the seminal discoveries of Heck, Negishi, and Suzuki (along with so many other important early contributors).

During the past 15 years, that situation has changed dramatically, as evidenced by the recognition of the field with the Nobel Prize in Chemistry in 2010. Thus, more versatile and active catalysts have been developed for the classic bond-forming processes of Heck, Negishi, Suzuki, and others, and these have now been applied across many disciplines (*e.g.*, biology, chemistry, and materials science) and in large-scale manufacturing. Equally importantly, others areas of investigation have emerged: couplings that achieve the formation of C–N (Buchwald–Hartwig reaction) and other C–heteroatom bonds, as well as cross-couplings of alkyl electrophiles, direct arylations of C–H bonds...the list goes on.

Books such as the present one, *New Trends in Cross-Coupling: Theory and Applications*, can play a critical role by assessing where the field currently stands and by pointing to the unsolved challenges that represent the future of the field. In this monograph, Dr Colacot has assembled leaders who do exactly that, describing not only the remarkable progress that has been achieved to date, but also the wealth of exciting opportunities that lie ahead.

Gregory Fu

RSC Catalysis Series No. 21
New Trends in Cross-Coupling: Theory and Applications
Edited by Thomas J Colacot
© The Royal Society of Chemistry 2015
Published by the Royal Society of Chemistry, www.rsc.org

About Greg Fu

After earning a PhD from Harvard in 1991 under the guidance of Prof. David A. Evans, and post-doctoral fellowship with Prof. Robert H. Grubbs at Caltech, in 1993, Greg Fu joined MIT, as a faculty member. In 2012, he was appointed the Altair Professor of Chemistry at the California Institute of Technology.

Greg received the Corey Award of the American Chemical Society in 2004, the Mukaiyama Award of the Society of Synthetic Organic Chemistry of Japan in 2006, and the Award for Creative Work in Synthetic Organic Chemistry of the American Chemical Society in 2012. He is a fellow of the Royal Society of Chemistry, the American Academy of Arts and Sciences, and the National Academy of Sciences. Greg serves as an associate editor for the *Journal of the American Chemical Society*. His current research interests include metal-catalyzed coupling reactions and the design of chiral catalysts. His work on "bulky, electron-rich" in 1998 was an important milestone the area of cross-coupling.

Preface

Although the major seminal discoveries in cross-coupling occurred during the 1970s, the importance of this technology has been realized only during the last two decades. Since then, the field has experienced rapid growth, ultimately leading to the award of the 2010 Nobel Prize in Chemistry.[1] The original pioneers may not have anticipated during that time that this area would become such a force in industry and academia, revolutionizing the way we think about organic synthesis, be it for a small molecule or a complex drug molecule or natural product. In fact, Prof. Heck in a BBC interview after he shared the 2010 Nobel Prize with Prof. Suzuki and Prof. Negishi, commented that he "did not make a dime out of this technology", while Prof. Negishi commented that he was "lucky enough to be alive" to receive the award. I heard Prof. Suzuki speaking at a conference that he had difficulties in publishing his pioneering work in top-rated journals as the reviewers were "not so nice". However, in the above-mentioned BBC interview, he joked that his hypertension is well controlled by one of the *sartans* developed using the Suzuki–Miyaura coupling technology – an ultimate satisfaction for any inventor to reach the climax.

I got into the area of cross-coupling accidentally when I joined Johnson Matthey in late 1995. I originally worked in the area of process chemistry and non-precious metal-based new product development, and in my spare time I started looking at the dppf ligand based on an inquiry. Although we developed a good process, the project died. Quite frustrated, I started to read about this ligand in a book titled *Ferrocenes* by Prof. Togni and Prof. Hayashi, where the importance of the bite angle was mentioned with respect to a Kumada–Corriu coupling involving a *sec*-alkyl Grignard reagent.[2] Although we could not sell even gram quantities for 1–2 years, we pushed dppfPdCl$_2$ as a "magic catalyst" based on the description by Gan and Hor,[2] in terms of its air stability and activity/selectivity in comparison with the air-sensitive

RSC Catalysis Series No. 21
New Trends in Cross-Coupling: Theory and Applications
Edited by Thomas J Colacot
© The Royal Society of Chemistry 2015
Published by the Royal Society of Chemistry, www.rsc.org

$Pd(PPh_3)_4$. This was during the time that Prof. Hartwig started publishing several amination reactions, where the use of the dppf ligand in conjunction with $Pd(dba)_2$ or even the preformed catalyst was highlighted, while Prof. Buchwald was using a $BINAP/Pd_2dba_3$ combination for the same technology. We were the first group to develop a good process for both dppf and $dppfPdCl_2$ for multi-kilogram quantities for the industry. This helped us gain an early understanding of the importance of preformed catalysts *versus in situ* in terms of selectivity, activity, scalability and ease of handling, which has become a main theme today in the area of cross-coupling. Very soon, the field began to develop rapidly – 1998 was marked as an exciting year for cross-coupling. The first book on this topic was published by Prof. Diederich and Prof. Stang.[3] Prof. Buchwald's work[4] using a novel electron-rich phosphine and Prof. Fu's work[5] on the use of the known "bulky electron-rich" $P(t\text{-}Bu)_3$ ligand, in conjunction with Pd, were other significant events in the new trends in cross-coupling with respect to C–C and C–N coupling, although Dr. Koie[6] also reported related studies for C–N coupling. Continuing on the same theme, in subsequent years many modern ligands were developed by Prof. Buchwald (biaryl ligands – 1998),[7] Prof. Nolan (*N*-heterocyclic carbenes – 1999),[8] Prof. Hartwig (Q-Phos – 2000)[9] and Prof. Beller (adamantyl-based ligands – 2000)[10] to solve many problems in cross-coupling (see Chapters 2 and 4).

Making the catalyst technology accessible to fine chemical and pharmaceutical industries all over the world and also academia to meet their needs has been my primary concern in concert with understanding the structure–activity relationships of the catalysts and substrates. Several new preformed catalysts have been developed involving the following classes of compounds: L_2PdX_2, $L_2Pd(0)$, $(L–L)PdX_2$ and precursors to $LPd(0)$-based catalysts, which indeed changed the landscape (Chapter 3).[11] New versions of the catalysts have emerged from time to time just as in the case of the iPhone™ technology, where the ever-increasing requirements of the customers such as air stability, low loading/high activity and selectivity, operating under mild reaction conditions, ease of handling, scalability of the catalyst and the coupling reactions, broader substrate scope and waste minimization have been addressed.

In spite of the new developments, demands continue to increase, broadening the scope of this technology. This was a significant driving force when I agreed to do this book, following an invitation from Dr Merlin Fox of the Royal Society of Chemistry. The initial feeling was overwhelming given my other responsibilities. Fortunately, I was very much aware of the academic push and industry expectations, and hence decided to create this book by providing the reader with a basic understanding about coupling, while discussing the modern trends from a technology and applied point of view. On behalf of the contributors, I hope that this book will serve as a textbook-*cum*-reference guide for both undergraduate and graduate students and also those who are experts in the area, irrespective of their academic or industrial background. It was designed to have contributions from both junior and senior authors, who are world leaders in their field, covering a wide range of topics in 16 chapters; including the logic behind choosing the ligand and catalyst, new reactions such as carboiodination, metal detection in the APIs, flow chemistry, API synthesis, reaction mechanisms, green chemistry, *etc.*

Because of my industry background, quality was one of my prime concerns, considering the size of the book. I was fortunate enough to be able to implement a peer review process for each chapter by at least two or three subject experts. With the reputation of Royal Society of Chemistry, I promised to maintain a high quality standard in terms of the book production.

I knew how busy each author was, considering their day-to-day responsibilities. A few had to go through some personal, unforeseen difficulties after they accepted the offer. I sincerely thank them all for putting their trust in me and completing the chapters with utmost sincerity. I am sure that their contributions in this book will stimulate the area further to grow at an advanced pace. I also personally acknowledge Johnson Matthey Catalysis and Chiral Technologies, the Royal Society of Chemistry and all the reviewers for their support and help in different capacities to make this project a success. My co-workers, Dr Carin Johansson Seechurn and Dr Andrew DeAngelis, are also thanked for their assistance, and William Carole is acknowledged for assistance in the design of the cover graphic.

Thomas J. Colacot

References

1. C. C. C. Johanson Seechurn, M. O. Kitching, T. J. Colacot and V. Snieckus, *Angew. Chem. Int. Ed.*, 2011, **51**, 5062, and references therein.
2. A. Togni and T. Hayashi (eds), *Ferrocenes: Homogeneous Catalysis, Organic Synthesis, Materials Science*, VCH, Weinheimk, 1995.
3. F. Diederich and P. J. Stang (eds), *Metal-Catalyzed Cross-Coupling Reactions*. Wiley–VCH, Weinheim, 1998.
4. D. W. Old, J. P. Wolfe and S. L. Buchwald, *J. Am. Chem. Soc.*, 1998, **120**, 9722.
5. A. F. Littke and G. C. Fu, *Angew. Chem. Int. Ed.*, 1998, **37**, 3387.
6. (a) M. Nishiyama, T. Yamamoto and Y. Koie, *Tetrahedron Lett.*, 1998, **39**, 617; (b) T. Yamamoto, M. Nishiyama and Y. Koie, *Tetrahedron Lett.*, 1998, **39**, 2367.
7. (a) R. Martin and S. L. Buchwald, *Acc. Chem. Res.*, 2008, **41**, 1461; (b) D. S. Surry and S. L. Buchwald, *Angew. Chem. Int. Ed.*, 2008, **47**, 6338.
8. S. Díez-González, N. Marion and S. P. Nolan, *Chem. Rev.*, 2009, **109**, 3612.
9. N. Kataoka, Q. Shelby, J. P. Stambuli and J. F. Hartwig, *J. Org. Chem.*, 2002, **67**, 5553.
10. A. Zapf, A. Ehrentraut and M. Beller, *Angew. Chem. Int. Ed.*, 2000, **39**, 4153.
11. H. Li, C. C. C. Johansson Seechurn and T. J. Colacot, *ACS Catal*, 2012, **2**, 1147.

Acknowledgements

Each chapter has been reviewed meticulously by 2–3 subject experts, to maintain the quality of the book. They are acknowledged for giving their valuable time considering their hectic schedules.

Prof. Gary Molander, University of Pennsylvania, PA, USA
Dr Izzat Raheem, Merck Research Laboratories, PA, USA
Prof. John Hartwig, University of California, Berkeley, CA, USA
Dr Joseph Martinelli, Eli Lilly and Company, IN, USA
Prof. Bruce Lipshutz, University of California, Santa Barbara, CA, USA
Prof. Brian Frost, University of Nevada, NV, USA
Dr Jason Mulder, Boehringer Ingelheim Pharmaceuticals, Inc., CT, USA
Prof. Richmond Sarpong, University of California, Berkeley, CA, USA
Dr Christopher Barnard, Johnson Matthey Technology Center,
    Sonning, UK
Prof. Mahesh K. Lakshman, The City College and The City University of
    New York, NY, USA
Dr Jinkun Haung, Chengdu R&D Center, Jiangsu Hengrui Medicine, China
Dr Jann Pesti, Bristol Meyers Squibb (past), Organic Process Research and
    Development (Associate Editor), American Chemical Society, USA
Prof. Oliver Kappe, Karl-Franzens-University Graz Heinrichstrasse, Austria
Prof. Narayan Hosmane, Northern Illinois University, IL, USA
Prof. Martin Burke, University of Illinois at Urbana-Champaign, IL, USA
Prof. Kevin Shaughnessy, University of Alabama at Tuscaloosa, AL, USA
Prof. Guy Lloyd-Jones, University of Edinburgh, UK
Prof. Steve Nolan, University of St Andrews, UK
Prof. Joseph Fox, University of Delaware, DE, USA
Prof. T. V. Rajanbabu, The Ohio State University, OH, USA
Prof. Mark Lautens, University of Toronto, Canada

RSC Catalysis Series No. 21
New Trends in Cross-Coupling: Theory and Applications
Edited by Thomas J Colacot
© The Royal Society of Chemistry 2015
Published by the Royal Society of Chemistry, www.rsc.org

Prof. Pierre Dixneuff, Institute of Chemical Sciences, University of
    Rennes, France
Prof. Lee Silverberg, Pennsylvania State University, PA, USA
Dr Thomas Colacot, Johnson Matthey Catalysis and Chiral Technologies,
    NJ, USA
Dr Christopher Welch, Merck, NJ, USA

This book is dedicated to the following people:

My father, the late Colacot K. John, a teacher who inspired me with the story of the Nobel Laureate, Prof. C. V. Raman even when I was a 2nd grader in a school in Kerala, India.

My mother, the late Achamma John, who taught me the importance of faith, humility, self-confidence and higher education.

My brother-in-law, the late Dr Roy Joseph, a young communications professor, whose passion to become a great writer was ended tragically by a hit and run accident while I was finishing this book.

Biographies

Thomas J. Colacot

Dr Thomas John Colacot was born in Central Kerala, India. After finishing his PhD in Chemistry at IIT Madras with Prof. M.N.S. Rao in 1989, he moved to the University of Alabama at Birmingham for a combined teaching and post-doctoral research position in the area of Group III-V Chemistry with Prof. L. K. Krannich. In 1992, he became an Asst. Professor at Florida A & M University, while collaborating with Prof. Will Rees at Florida State University. Due to family commitments, he moved to Southern Methodist University, Dallas in 1993 to work with Professor N.S. Hosmane on an Advanced Technology Program/AMCO project. Although Prof. Hosmane and Dr Colacot did not have any previous experience in catalysis, Dr Hosmane gave him complete freedom to explore catalysis research using mixed carborane–cyclopentadiene complexes of early transition metals for the manufacture of plastics (olefin polymerization).

In 1995, Dr Colacot began his career at Johnson Matthey, USA, directing the homogeneous catalysis research group. With an extensive background in ligand technology obtained during his time at IIT Madras, Chennai and UAB, in conjunction with his studies in catalysis from SMU, Dr Colacot focused his research in the area of precious metal catalysis and has become an industrial expert in cross coupling.

Through his research, Dr Colacot and his group have generated highly active, practical palladium based cross coupling catalysts for applications in

RSC Catalysis Series No. 21
New Trends in Cross-Coupling: Theory and Applications
Edited by Thomas J Colacot
© The Royal Society of Chemistry 2015
Published by the Royal Society of Chemistry, www.rsc.org

pharmaceutical, fine chemical and academic labs. Currently, Dr Colacot is the R&D Global Manager of Homogeneous Catalysis. He has given numerous invited and plenary lectures in many international conferences and acts as an external PhD thesis examiner to IITs and universities and a visiting professor in many universities. He has contributed several publications, patents, book chapters and reviews. He is a Fellow of the Royal Society of Chemistry and has obtained a MBA from Pennsylvania State University. He received Royal Society's 2012 RSC Applied Catalysis Award and Medal for his "exceptional contributions" to Homogeneous Catalysis, particularly for making this technology accessible to industry and academia, and recently he has received the 2015 American Chemical Society Award for Industrial Chemistry in "recognition of his contributions and leadership in the development and commercialization of ligands and precatalysts for metal-catalyzed organic synthesis, particularly cross-couplings, for industrial and academic use and applications". Dr Colacot has coauthored Chapters 1–3 of this book.

Christopher F. J. Barnard

Dr Christopher F. J. Barnard has worked for Johnson Matthey for 35 years. His principal contributions have been to the development of platinum compounds as the active ingredients in anti-cancer drugs and in developing catalytic chemistry for synthesis in pharmaceutical applications. Palladium-based chemistries such as coupling and carbonylation have been areas of particular interest in recent years.

Irina P. Beletskaya

Irina P. Beletskaya received her Diploma degree in 1955, her PhD degree in 1958 and her Doctor of Chemistry degree in 1963 from Moscow State University. The subject of the last degree was Electrophilic Substitution at Saturated Carbon. She became a Full Professor at Moscow State University in 1970, and in 1974 she became a Corresponding Member of the Academy of Sciences (USSR), of which she became a full member (Academician) in 1992. She is currently Head of the Laboratory of Organoelement Compounds, Department of Chemistry, Moscow State University. Irina Beletskaya is Chief Editor of the *Russian Journal of Organic Chemistry*. She was President of the Organic Chemistry Division of IUPAC from 1989 to 1991. She was a recipient of the Lomonosov Prize (1979), the Mendeleev Prize (1982), The Nesmeyanov Prize (1991), the Demidov Prize (2003) and the State Prize (2004). She is the author of more 600 articles and four monographs. Her current scientific interests are manifold, including transition metal catalysis and organocatalysis in organic synthesis, organometallic derivatives of lanthanides, carbanions and nucleophilic aromatic substitution, supramolecular chemistry, and many others.

Anthony Chartoire

Anthony Chartoire received his MSc and PhD in organic and organometallic chemistry from the University of Nancy, where he worked under the supervision of Prof. Yves Fort until 2010. In 2011, he joined the group of Prof. Steven P. Nolan at the University of St Andrews, where he worked as a Postdoctoral Research Fellow until August 2013. He is currently a research scientist at Econic Technologies. His research interests include organic and organometallic chemistry and homogeneous catalysis.

Andrei V. Cheprakov

Andrei V. Cheprakov graduated from the Department of Chemistry of Moscow State University in 1983 and joined Prof. I. Beletskaya's Laboratory of Organoelement Compounds. His PhD thesis was devoted to the oxidative halogenation of aromatic compounds and was defended in 1989. Currently he is a Docent at the Chair of Organic Chemistry of Moscow University. His research interests include the methodology of transition metal-catalyzed reactions, reactions in non-conventional aqueous microheterogeneous media, chemistry and applications of brassinosteroids, chemistry of π-extended porphyrinoids and fluorogenic oligopyrrolic ligands.

Cathleen M. Crudden

Cathleen Crudden obtained her BSc and MSc from the University of Toronto with Dr Mark Lautens, and her PhD jointly with Dr Howard Alper at the University of Ottawa and Dr Shinji Murai at Osaka University, Japan. Following this, she took up an NSERC PDF with Prof. Scott Denmark at the University of Illinois at Urbana Champaign. She then became an Assistant Professor at the University of New Brunswick, where she was awarded a Research and Innovation Award and a UNB Merit Award, and received the first university Research

Professorship. In 2002, she took up the position of Queen's National Scholar at Queen's University, Kingston, ON, Canada, where she was awarded a Premier's Research and Excellence Award and a Chancellor's Research Award. One of her publications was in the top 10 most cited papers of the year for all of science in Canada for the year 2006. In 2012 she was appointed Research Professor at the Institute of Transformative Bio-Molecules in Nagoya, Japan. She has won numerous research awards, including the Clara Benson Award for the top female chemist in Canada and an NSERC Accelerator Award. Her diverse research program includes materials chemistry, organic chemistry, catalysis and chirality. She has been a Visiting Professor at the Research Center for Materials Science in Nagoya University, in the laboratories of Prof. Ryoji Noyori (joint recipient of the Nobel Prize in Chemistry, 2001). In 2008, she was awarded a Global Center of Excellence Professorship at Kyoto University. She was also awarded a Visiting Professorship by the Catalan Government in Tarragona, Spain, in 2007. Cathleen was President of the Canadian Society for Chemistry (CSC) in 2012–13. Prior to becoming CSC President, she served on the Board of Directors for two terms representing the Catalysis Division. She also served on the Editorial Advisory Board for ACCN for 10 years. She has been a member of the organizing committee of Pacifichem for the past 7 years and served as area coordinator for materials and inorganic chemistry.

Andrew DeAngelis

Andrew DeAngelis received his BS in Chemistry from Moravian College, Bethleham, PA, in 2005, where he conducted undergraduate research with Prof. Carl Salter. He then went on to earn his PhD in Chemistry and Biochemistry from the University of Delaware in 2011, where he worked on the development of new catalytic reactions with Rh-carbenoids under the guidance of Prof. Joseph M. Fox. He subsequently pursued postdoctoral studies with Prof. Stephen L. Buchwald at Massachusetts Institute of Technology, where he worked on the development of new cross-coupling processes in continuous flow. In 2012, he joined Johnson Matthey as a research scientist working in the Catalysis and Chiral Technologies Division in West Deptford, NJ, where he is engaged in the development and applications of novel homogeneous metal catalysts and also process development for scale-up.

Joshua R. Dunetz

Joshua R. Dunetz graduated from Haverford College, Haverford, PA in 2000 with a BA in Chemistry after undergraduate research with Prof. Karin Åkerfeldt. He received his PhD in Organic Chemistry from MIT in 2005 under the guidance of Prof. Rick Danheiser, and then completed postdoctoral studies with Prof. William Roush at the Scripps Research Institute, Jupiter, FL. In early 2008, Joshua joined Pfizer Chemical R&D, where he developed processes for the GMP manufacture of small molecules on the gram to multi-kilogram scale. After six wonderful years with Pfizer, Joshua relocated to California to join the Process Chemistry team at Gilead Sciences.

Ben W. Glasspoole

Ben Glasspoole was raised in Montreal, Quebec and completed his BSc at McGill University in 2006. He then completed his PhD under the guidance of Prof. Cathleen Crudden at Queen's University, Kingston, ON, Canada in 2011, working in the field of Pd catalysis. During his PhD term he carried out a research exchange with Prof. Varinder Aggarwal. After performing postdoctoral work with Prof. Michael Krische at the University of Texas at Austin, Ben joined Sigma-Aldrich in 2013.

Volker Hessel

Volker Hessel studied chemistry at Mainz University, and from 1994 he was an employee of the Institut für Mikrotechnik Mainz, where in 1999 he was appointed Head of the Microreaction Technology Department. In 2007, Prof. Hessel was appointed Director R&D at IMM. In 2011, he was appointed as full Professor of the Chair of Micro Flow Chemistry and Process Technology at Eindhoven University of Technology. Prof. Hessel received the AIChE award for Excellence in Process Development Research in 2007 and in 2010 an ERC Advanced Grant for Novel Process Windows. He is currently Editor-in-Chief of the journal *Green Processing and Synthesis*.

Eric C. Keske

Eric Keske was born in London, ON, Canada. He obtained his BSc at the University of Western Ontario in 2009 and then moved to Queen's University to pursue a PhD in organometallic chemistry with Prof. Cathleen Crudden. During this time he enjoyed short research stays with Dr Eric Fillion at the University of Waterloo and Dr Martin Albrecht at University College Dublin. Eric's research interests involve late transition metal catalysis and reaction discovery. Eric considers himself a bourbon connoisseur, and can easily be distracted from the laboratory by loud music, whiskey and judo practice.

Kazunori Koide

Professor Koide obtained his bachelor's and master's degrees from the Department of Pharmaceutical Sciences at the University of Tokyo, Japan, where he developed new glycosylation methods in the Ohno group. He then moved to the United States and received a PhD in chemistry at the University of California, San Diego, under the guidance of Professor K. C. Nicolaou. As a Merck Fellow of Cancer Research Fund of the Damon Runyon-Walter Winchell Foundation, he conducted postdoctoral research with Professor Gregory L. Verdine in the Department of Chemistry and Chemical Biology at Harvard University. He then began his independent academic career in the Department of Chemistry at the University of Pittsburgh. His current research interests include development of new synthetic methods, total synthesis of biologically important natural products, chemical biology and medicinal chemistry of natural products, and development and applications of fluorescent chemosensors and chemodosimeters.

Mark Lautens

Mark Lautens was born in Ontario, Canada. He obtained his BSc degree from the University of Guelph, where he graduated with distinction. He conducted his doctoral studies at the University of Wisconsin-Madison under the direction of Barry M. Trost, where he discovered Mo-catalyzed allylic alkylation and Pd enyne cycloisomerization. In 1985, he moved to Harvard University where he conducted his NSERC PDF with David A. Evans on studies directed toward the synthesis of bryostatin, a potent anti-cancer agent. He joined the University of Toronto in 1987 as an NSERC University Research Fellow and Assistant Professor. He was promoted to Associate Professor in 1992 and Professor in 1995. Since 1998, he has held an Endowed Chair, the AstraZeneca Professor of Organic Synthesis, and from 2003–13 he was named an NSERC/Merck Frosst Industrial Research Chair in New Medicinal Agents via Catalytic Reactions. In 2012, he was appointed as University Professor, the highest rank at the University of Toronto.

Hongbo Li

Hongbo Li completed his both BS and MS in Chemistry at Peking University, China before he moved to the USA in 2000. He completed his PhD at Northwestern University under the supervision of Prof. Tobin Marks in 2005. His PhD work focused on the development of novel olefin polymerization catalysts. After working briefly at IBM Almaden Research Center, in 2006 he joined Los Alamos National Laboratory as Research Associate, where he worked on the development of immobilized catalysts. In 2008, he joined the Catalysis and Chiral Technologies Division of Johnson Matthey in West Deptford, NJ, where he is working on the development of novel homogeneous organometallic catalysts.

Christine M. Le

Christine Le was born in 1989 in Canada and completed her BSc degree at Western University. She then moved to the University of Toronto and obtained her Master's degree in the group of Prof. Vy Dong, where she carried out research in the field of Rh-catalyzed hydroacylation. She is currently pursuing her PhD at the University of Toronto under the supervision of Prof. Mark Lautens. Her research involves asymmetric transformations enabled by Rh and Pd catalysis.

Alastair Lennox

Alastair Lennox is a graduate of Manchester University (2008, 1st class, MChem), where he conducted a final year research project with Dr Ian Watt and spent a year studying at the University of California, Los Angeles. He obtained his PhD in 2012 at the University of Bristol, where he worked under the supervision of Prof. Guy Lloyd-Jones and studied the reactivity of potassium organotrifluoroborate salts in Suzuki–Miyaura couplings.

Guy Lloyd-Jones

Guy Lloyd-Jones studied at Huddersfield Polytechnic (BSc, 1989) and Oxford University (DPhil with John M. Brown FRS, 1992) before tenure of a Royal Society postdoctoral fellowship at the University of Basel with Andreas Pfaltz. He joined the University of Bristol in 1996 and was promoted to full Professor in 2003. In 2013 he was elected to the Royal Society (FRS) and moved to Edinburgh University as the Forbes Professor of Organic Chemistry.

Javier Magano

Javier Magano was born in Madrid, Spain. He received a BS degree in organic chemistry from Complutense University in Madrid and an MS degree in chemistry from the University of Michigan. After working for the oil industry in Spain for several years, he obtained an MS degree in rubber and polymer science at the Center for Advanced Scientific Research in Madrid. After moving back to the USA, he joined the early process chemistry group at Pfizer in 1998 in Ann Arbor, MI, where he spent 9 years developing scalable processes for the preparation of drug candidates. In 2007, he moved to Groton, CT to continue his work as a process chemist and, during this period, he also worked in the area of biologics for $1\frac{1}{2}$ years on the preparation of linkers for bioconjugation processes. Javier currently holds a position in the Chemical Technology group at Pfizer, where he is involved in the applications of high-throughput screening to transition metal-catalyzed cross-couplings. His research interests also include the development of catalytic processes that employ non-precious metals in cross-coupling reactions.

Debabrata Maiti

Debabrata Maiti received his PhD from Johns Hopkins University (USA) in 2008 under the supervision of Prof. Kenneth D. Karlin. After postdoctoral studies at Massachusetts Institute of Technology (MIT) with Prof. Stephen L. Buchwald (2008–10), he joined the Department of Chemistry of IIT Bombay as an Assistant Professor in 2011. His research interests are focused on the development of new and sustainable synthetic methodologies.

Soham Maity

Soham Maity was born in 1988 in West Bengal (India). He studied chemistry at St Xavier's College, Kolkata, and received his BSc degree in 2009. After completing his MSc degree at the University of Calcutta he joined Prof. D. Maiti's group in 2011, where he is currently a third-year PhD student.

Arun Maji

Arun Maji was born in 1989 in Durgapur, West Bengal (India). He completed his graduation from Presidency College (Kolkata) and received his BSc degree in 2010. After completing his MSc degree at the Indian Institute of Technology, Kanpur, he joined Prof. Maiti's group in 2012, where he has completed his first year of PhD studies.

Atanu Modak

Atanu Modak was born in 1989 in West Bengal (India). He graduated from Ramakrishna Mission Vidyamandira, Belur, with a BSc degree in chemistry from the University of Calcutta in 2009. He then pursued an MSc degree in chemistry at the Indian Institute of Technology, Kanpur, in 2011. From January 2012 he has been working as a PhD student in Prof. D. Maiti's laboratories at IIT Bombay.

Stephen G. Newman

Stephen G. Newman was born in Newfound-land, Canada in 1985. He obtained a BSc degree with a major in chemistry from Dalhousie University in 2008. In 2012, he graduated with a PhD degree from the University of Toronto under the supervision of Prof. Mark Lautens. He is currently an NSERC Postdoctoral Fellow in the laboratory of Prof. Klavs F. Jensen, and will joining the faculty at the University of Ottawa in summer 2014 as an Assistant Professor of Chemistry.

Timothy Noël

Timothy Noël was born in Aalst, Belgium, and received an MSc degree (Industrial Chemical Engineering) from the KaHo Sint-Lieven in 2004. In 2009, he received his PhD at the University of Ghent with Prof. Johan Van der Eycken (Department of Organic Chemistry). He then moved to Massachusetts Institute of Technology as a Fulbright Postdoctoral Fellow with Prof. Stephen L. Buchwald (Department of Chemistry), where he worked on flow chemistry (MIT–Novartis Center for Continuous Manufacturing). In 2012, he accepted a position as Assistant Professor at Eindhoven University of Technology, The Netherlands. In 2011, Dr Noël received an Incentive Award for Young Researchers from the Comité de Gestion du Bulletin des Sociétés Chimiques Belges. In 2012, he received a prestigious VENI Award from the Dutch Government (NWO). He is also Associate Editor of the *Journal of Flow Chemistry*. His research interests are focused on the combination of flow chemistry, organic synthetic chemistry and catalysis.

Steven P. Nolan

Steven P. Nolan received his BSc in Chemistry from the University of West Florida and his PhD from the University of Miami where he worked under the supervision of Prof. Carl D. Hoff. After a postdoctoral stay with Prof. Tobin J. Marks at Northwestern University, he joined the Department of Chemistry of the University of New Orleans in 1990. In 2006 he joined the Institute of Chemical Research of Catalonia (ICIQ) as Group leader and ICREA Research Professor. In early 2009, he joined the School of Chemistry at the University of St Andrews where he holds the Chair in Inorganic Chemistry. His research interest include organometallic chemistry and homogeneous catalysis.

David A. Petrone

David A. Petrone was born in Ontario, Canada, in 1989. He obtained his BSc degree with Honors in Chemistry in 2011 from the University of Guelph under the supervision of Prof. Kathryn Preuss and Prof. William Tam. He is currently a PhD student in Mark Lautens' research laboratory at the University of Toronto, where he is studying Pd(0)-catalyzed carbohalogenation reactions.

Carin C. C. Johansson Seechurn

Carin Johansson Seechurn completed her MChem with French degree at UMIST (University of Manchester Institute of Science and Technology) in 2003. This included a year abroad at CPE (l'Ecole Supérieure Chimie, Physique, Electronique de Lyon) in France. From 2003 until 2007 she carried out PhD studies at the University of Cambridge, under the supervision of Prof. Matthew Gaunt. Her PhD work focused on the development of novel organocatalytic methodology. After the completion of the PhD studies, she continued to work for Prof. Gaunt, now in the area of palladium catalysis. In 2008 she joined the Catalysis and Chiral Technologies Division of Johnson Matthey in Royston, UK, where she is working on the development of novel homogeneous metal catalysts.

Upendra Sharma

Upendra Sharma studied chemistry at Guru NankDev University, Amritsar (India) until 2005, and subsequently obtained a PhD degree in organic chemistry with Dr Bikram Singh, Chief Scientist, CSIR–IHBT, Palampur. In the fall of 2012, he joined Prof. D. Maiti at IIT Bombay as Young Scientist. His research interests involve metal-catalyzed selective C–H activation and natural product synthesis.

Kevin H. Shaughnessy

Kevin H. Shaughnessy grew up in the small town of Trumbull, Nebraska. He earned his BS degree from the University of Nebraska–Lincoln in 1992, where he worked with Reuben D. Rieke. He then joined the group of Robert M. Waymouth at Stanford University, where he developed zirconocene-catalyzed carbometallation reactions. Upon completing his PhD degree in 1998, he moved to Yale University to work with John F. Hartwig on palladium-catalyzed enolate arylations and the development of high-throughput screening assays. In 1999, he joined the Chemistry Faculty at the University of Alabama, where he is currently Professor and Chair of Chemistry. His research interests are focused on the development of catalytic methodologies in alternative solvents, design of new ligands, and mechanistic studies of ligand structure–activity relationships.

James P. Stambuli

James Stambuli was born in Paterson, NJ and received his BA in Chemistry with Honors from Rutgers University in Newark, NJ in 1998. He obtained his PhD from Yale University in the laboratory of Prof. John Hartwig in 2003 and was an NIH Postdoctoral Fellow with Prof. Barry Trost at Stanford University from 2003 to 2006. He began an independent research career at the Ohio State University in 2006 and recently joined the PR&D division of AbbVie. Prof. Stambuli's research is in the fields of organometallic and organic chemistry with particular emphasis on transition metal catalyst design to assist in the invention of new technologies for chemical transformations.

Mark Stradiotto

Mark Stradiotto received his BSc (Hons) degree in Applied Chemistry in 1995 and a PhD in Organometallic Chemistry in 1999 from McMaster University, Canada, the latter under the supervision of Prof. Michael A. Brook and Prof. Michael J. McGlinchey. After conducting research as an NSERC Postdoctoral Fellow at the University of California at Berkeley (USA) in the research group of Prof. T. Don Tilley (1999–2001), Mark moved to the Department of Chemistry at Dalhousie University, where he is now the Alexander McLeod Professor of Chemistry. Mark has served as a member of the Editorial Advisory Board of *Organometallics*, has been named a *Synlett/Synthesis* Promising Young Professor Journal Awardee, and was awarded the Canadian Society for Chemistry 2012 Strem Chemicals Award for Pure or Applied Inorganic Chemistry.

Xiao-Feng Wu

Xiao-Feng Wu was born in 1985 in China. He studied chemistry in Zhejiang Sci-Tech University (ZSTU), where he obtained his BSc degree in science in 2007. In the same year, he moved to Rennes 1 University in France and worked with Prof C. Darcel on iron-catalyzed reactions. After earning his Master's degree in 2009, he joined Matthias Beller's group in the Leibniz Institute for Catalysis (LIKAT) in Germany, where he completed his PhD thesis in January 2012. He subsequently began independent research at ZSTU and LIKAT. His research interests include carbonylation reactions, synthesis of heterocycles and the catalytic application of cheap metals. He was also a Fellow of the Max-Buchner-Forschungsstiftung.

Contents

RSC Catalysis Series No. 21
New Trends in Cross-Coupling: Theory and Applications
Edited by Thomas J Colacot
© The Royal Society of Chemistry 2015
Published by the Royal Society of Chemistry, www.rsc.org

Chapter 15 Recent Large-Scale Applications of Transition Metal-Catalyzed Couplings for the Synthesis of Pharmaceuticals 697
Javier Magano and Joshua R. Dunetz

Chapter 16 Palladium Detection Techniques for Active Pharmaceutical Ingredients Prepared *via* Cross-Couplings 779
Kazunori Koide

CHAPTER 1

Introduction to New Trends in Cross-Coupling

CARIN C. C. JOHANSSON SEECHURN,[a] ANDREW DEANGELIS[b] AND THOMAS J. COLACOT*[b]

[a] Johnson Matthey Catalysis and Chiral Technologies, Orchard Road, Royston, SG8 5HE, UK; [b] Johnson Matthey Catalysis and Chiral Technologies, 2001 Nolte Drive, West Deptford, NJ 08066, USA
*Email: thomas.colacot@jmusa.com

1.1 Importance of Cross-Coupling in Homogeneous Catalysis

Transition metal-catalyzed reactions play a vital role in the production of many industrially important chemicals, where homogeneous catalysis (reactions that take place in the same phase as the catalyst) is rapidly growing, as evidenced by the awarding of three distinct Nobel Prizes in Chemistry during the last decade – chiral catalysis (2001; Noyori, Sharpless and Knowles),[1] olefin metathesis (2005; Grubbs, Chauvin and Schrock)[2] and cross-coupling [2010; Heck (Figure 1.1), Suzuki (Figure 1.2) and Negishi (Figure 1.3)].[3] The field of cross-coupling, well dominated by homogeneous catalysis, has undoubtedly turned into an area appreciated by all synthetic chemists, irrespective of their prominence in academia or industry.

Recently, heterogeneous catalysis (reactions that take place in a different phase than the catalyst) has also been used for simple cross-coupling reactions, relying on metal leaching to mediate the desired reaction. However, the leached metal must subsequently be readsorbed in order not to

RSC Catalysis Series No. 21
New Trends in Cross-Coupling: Theory and Applications
Edited by Thomas J Colacot

Figure 1.1 Prof. Heck giving a lecture at Queens University, Canada, in 2006 using a
transparency projector.
Courtesy of Prof. Snieckus.

Figure 1.2 Prof. Suzuki (right) with the Editor, Dr Colacot (left), and Prof. Fu (middle).
Courtesy of Prof. Dixneuf.

contaminate the final product ("release-and-catch" strategy).[4] This is
not always ideal, depending on the target use of the product and from a
reproducibility point of view. In addition, with reactions catalyzed hetero-
geneously, it is difficult to carry out reactions with high selectivity, in terms
of stereo-, regio- or, in some cases, chemoselectivity. Pd-catalyzed cross-
coupling has enriched the area of homogeneous catalysis, where rapid
growth has been taking place in the past several years, as evidenced by the
growing total number of publications/patents[†] in the area (Table 1.1).

[†]Based on SciFinder searches, May 2014.

Figure 1.3   Prof. Negishi at Stockholm during the 2010 Nobel Prize ceremony. Courtesy of Prof. Negishi.

Table 1.1   Growth in the total number of publications and patents on cross-coupling reactions through April 2014.

	Pre-1990	Through 2000	Through 2010	Through April 2014
Suzuki–Miyaura	20	824	10175	15883
Heck	56	768	4029	5816
Sonogashira	0	156	3623	5689
Stille	3	470	2380	3537
Negishi	0	13	429	737
Buchwald–Hartwig	0	7	253	498
Kumada–Corriu	1	8	136	298
Hiyama	0	1	91	172
Carbonyl α-arylation	13	33	113	193

Thus, Pd-catalyzed cross-coupling reactions comprise one of the most important classes of synthetic transformations in modern organic chemistry, providing chemists with an exceptionally powerful tool for the construction of carbon–carbon (C–C) and carbon–heteroatom (C–X) bonds. These and many related transformations have become ubiquitous in both industry and academia. Indeed, as mentioned above, the 2010 Nobel Prize in Chemistry was a monumental accomplishment for the assiduous

Figure 1.4   'It all started with Grignard' – Negishi.
Photograph courtesy of the Nobel Foundation.

contributions of Professors Richard F. Heck (University of Delaware), Akira Suzuki (University of Hokkaidō) and Ei-ichi Negishi (Purdue University) for their achievements within the area of Pd-catalyzed C–C bond-forming reactions.[5] According to Negishi, the roots of cross-coupling can be traced all the way back to Victor Grignard (Figure 1.4). These early studies laid the foundation of what would become one of the most important and most studied classes of catalytic reactions. Intense research efforts would soon spawn several new C–C coupling reactions in addition to C–X coupling reactions as the chemistry evolved into what it has become today.

The aim of this book is to serve both academia and industry. In the following sections, some key parameters and basic concepts are introduced.

1.2   Definition of Some Key Parameters

1.2.1   Turnover Number (TON)

The turnover number is defined as the absolute number of passes through the catalytic cycle before the catalyst becomes deactivated. In general, industrial chemists are interested in both TON and turnover frequency (TOF) (see the next section). A large TON (*e.g.*, 10^6–10^{10}) indicates a stable, very long-lived catalyst. The TON can be calculated by dividing the amount of reactant (moles) by the amount of catalyst (moles):

$$\text{TON} = \frac{\text{number of moles (equivalents) of reactant}}{\text{number of moles (equivalents) of catalyst}}$$

This assumes a yield of the product of 100%, which is most often not the case. To calculate the true number of turnovers, the yield obtained needs to be taken into account. For example, if 10 mol of reactant and 2.5 mol of catalyst are used, then the TON becomes

$$\text{TON} = \frac{10}{2.5} = 4$$

If the yield of the product is 94%, then the actual number of turnovers is

$$\text{Actual TON} = 4 \times 0.94 = 3.76$$

Authors often report mole % of catalyst used. This refers to the fraction of catalyst used relative to the amount of limiting reactant present.

1.2.2 Turnover Frequency (TOF)

Turnover frequency is defined as the number of passes through the catalytic cycle per unit time (typically seconds, minutes or hours). This number is usually determined by dividing the TON by the time required to produce the given amount of product.

However, as with the TON, the actual yield of the product also needs to be taken into account. Continuing the example above, if the reaction in question was run for 7 h to obtain the 94% yield, the TOF is

$$\text{TOF} = \frac{3.76\ \text{turnovers}}{7\,\text{h}} = 0.54\,\text{h}^{-1}$$

1.3 General Elementary Steps

The generally accepted simplified catalytic cycle for cross-coupling reactions is shown in Scheme 1.1, where $L_n\text{Pd}(0)$, the active catalytic species, acts as a "matchmaker". In Japanese language, "catalyst" is pronounced *shoku bai* and in Chinese it is *chu mei* (the same character as for matchmaker).[6]

In C–C bond-forming cross-coupling, there are two coupling partners: an aryl/vinyl halide or pseudohalide and an organometallic reagent such as a Grignard reagent. There are three basic steps in the catalytic cycle: *oxidative addition, transmetallation* and *reductive elimination*.

Here is an analogy: one of the partners with a family member or friend (R–X) establishes a connection with the matchmaker $[L_n\text{Pd}(0)]$ with the profile of "R". This is called the *oxidative addition* of an organic halide/

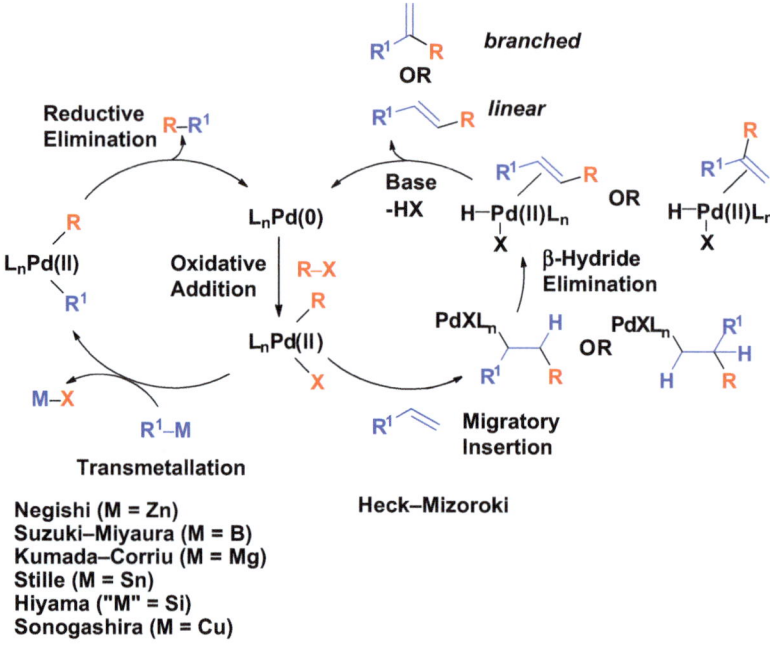

Scheme 1.1 General mechanism of cross-coupling reactions and the Heck–Mizoroki reaction.

pseudohalide, R–X, to $L_nPd(0)$ to generate an R–Pd(Ar)(X)(II) intermediate. In the second step, the other partner (R^1) in the form of R^1–M also forms a connection with the matchmaker so that R and R^1 can communicate with each other through the Pd. This is the second step, called *transmetallation*, where M (a friend or family member of R^1) forms a "bond" with X. In the third step, R and R^1 are united and detach from the matchmaker (Pd catalyst) in an event called *reductive elimination*. The success of a matchmaker depends on how many challenging coupling partners are successfully coupled (get married) without any deleterious incidents, within a short time frame. This is related to the TON and TOF of the catalyst. Although Heck shared the 2010 Nobel Prize for Pd-catalyzed cross-coupling reactions with Suzuki and Negishi, some argue that the Heck–Mizoroki reaction (often shortened to the Heck reaction) is not a true cross-coupling reaction as it does not involve a transmetallation step. In the Heck reaction, the Pd(II)–R species undergoes a migratory insertion with the alkene substrate, followed by a *syn*-periplanar β-hydride elimination event to give the product. This step was well established by the work of Fu and Hartwig.[7] Base is necessary to turn over palladium catalyst by inducing the reductive elimination of HX in the last step. Depending on the nature of substituents on the olefin, linear or branched coupled products are obtained, as these olefin substituents can influence the *regioselectivity* of the product. The general rule of thumb is that electron-withdrawing groups on the olefin favor linear products with neutral Pd complexes. Bidentate ligands such as dppf

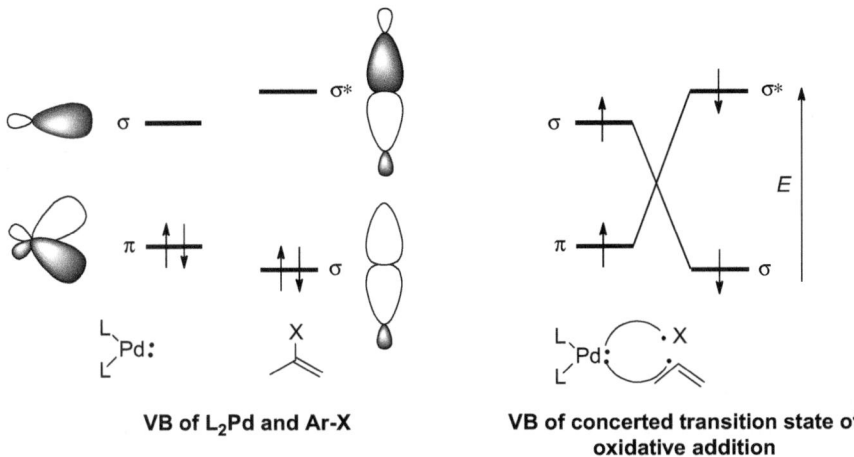

VB of L₂Pd and Ar-X   **VB of concerted transition state of oxidative addition**

Figure 1.5 Valence bond (VB) representations for the two components L$_2$Pd and Ar–X and for a concerted transition state of the oxidative addition process. Reproduced from Ref. 14.

[1,1'-bis(diphenylphosphino)ferrocene] under cationic conditions[8] and dnpf [1,1'-bis(dinaphthylphosphino)ferrocene] in presence of a polar solvent and TBAC (tetrabutylammonium chloride) additive[9] produce branched products for electron-rich and electron-neutral olefins.

Since the original discoveries of cross-coupling reactions, there has been a great deal of effort in this area to better understand the reaction mechanism, where the role of the ligand is important. The electronic and steric nature of the ligand (L) and the coordination number of Pd can significantly influence two important steps of the cycle; oxidative addition and reductive elimination (Figure 1.5).[10] The role of ligands in the transmetallation step is not as well understood; however, the groups of Hartwig, Amatore and Lloyd-Jones have carried out some impressive work in the area of Suzuki–Miyaura coupling.[11] The groups of Beletskaya[12] and Buchwald[13] have shown that more electron-deficient ligands can increase the rate of C–N cross-coupling reactions involving ureas and amides, respectively, likely reflecting an increased rate of "transmetallation" (amide binding).[13] Oxidative addition was considered to be the rate-limiting step, where the choice of the ligand is important. For example, it is proposed that electron-rich ligands make the Pd basic enough to do the oxidative addition of challenging aryl chlorides, while with aryl iodides and bromides oxidative addition is relatively facile, even with less electron-rich ligands such as Ph$_3$P. Figure 1.5 shows the valence bond (VB) representations for the two components L$_2$Pd and Ar–X and for a concerted, three-centered transition state of the oxidative addition process.[14] The energy (ΔG) required to excite one electron into the antibonding (σ^*) orbital of the Ar–X bond decreases in the series Ar–Cl > Ar–Br > Ar–I.

The low reactivity of more challenging substrates such as unactivated aryl chlorides was often attributed to the large bond dissociation energy of the C–Cl bond (95 kcal mol^{-1}) in comparison with Ar–Br (79 kcal mol^{-1}) or

Ar–I (64 kcal mol^{-1}),[15] which highlights the difficulty for an aryl chloride to add oxidatively to a less electron-rich L$_n$Pd(0) species.[16]

Interestingly, in the transmetallation step, recent evidence suggests that the trend is the opposite: chloride complexes are transmetallated faster than those of bromides and iodides.[13,17] The size of the ligand also plays an important role in the reductive elimination,[10] in addition to stabilizing the coordinatively unsaturated L$_n$Pd(0).

1.4   Brief Historical Notes on Cross-Coupling Reactions and the Contents of This Book

The intent of this chapter is not to provide an exhaustive review of the history of cross-coupling reactions,[18] but to identify the most notable milestones (Figure 1.6) and the genesis of some of the topics of the chapters presented here.

Some argue that the history of the use of metals as catalysts to accomplish organic transformations was initiated by Fittig, who recorded sodium-mediated alkylations of halogenated arenes in 1862.[19] In the early 1900s, Ullmann and Goldberg carried out extensive studies on copper-catalyzed C–C, C–O and C–N bond-forming reactions.[20] Noteworthy is that the first person to combine successfully organometallic reagents with catalysis, in this case NiCl$_2$, was the French chemist André Job.[21] He reported that PhMgBr, in the presence of NiCl$_2$, was able to absorb CO, NO, C$_2$H$_4$, C$_2$H$_2$ and H$_2$. Since Job's underappreciated revolutionary discovery, nickel has been overshadowed by palladium in similar transformations. Since this early discovery, carbonylation has become an industrially important process and its modern version, carbonylative cross-coupling, is reviewed in detail by Xiao-Feng Wu and Christopher Barnard in Chapter 10.

Following Job's discoveries, the next notable milestone would be the reports by Kharasch on the metal-catalyzed homo-couplings of organo-magnesium reagents.[22] More specifically, he employed catalytic amounts of CoCl$_2$, MnCl$_2$, FeCl$_3$ or NiCl$_2$ in the presence of Grignard reagents and organic halides to affect this homo-coupling reaction (Scheme 1.2, top).

The use of vinyl bromide in place of bromobenzene, under the same conditions, resulted not in the expected homo-coupling of the Grignard reagent, but in the first-ever reported catalytic cross-coupling reaction (Scheme 1.2, bottom).[23] These findings, to some extent, make Kharasch (Figure 1.7) the "grandfather" of cross-coupling reactions.

More than 20 years later came the next breakthrough in the Pd-catalyzed cross-coupling area. Heck reported in 1968 that arylations of alkenes could be achieved by using an organomercury arylating reagent and a palladium catalyst (Scheme 1.3).[24]

A modification of this Pd-catalyzed reaction was subsequently published by Moritani and Fujiwara. They disclosed the direct coupling between arenes and alkenes, first in the presence of stoichiometric amounts of palladium compounds[25] and later using catalytic amounts (Scheme 1.4).[26] This finding

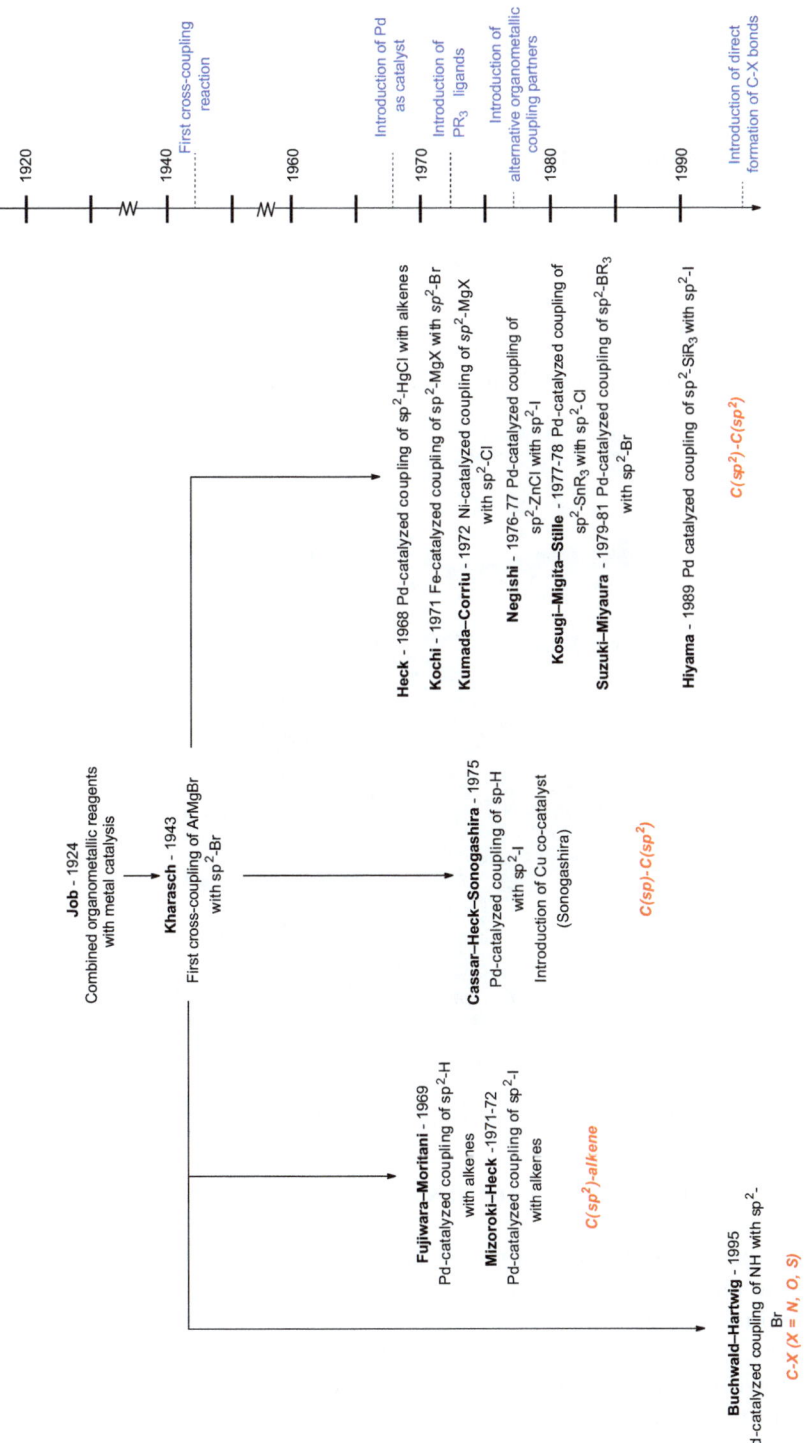

Figure 1.6 Simplified time-line of events in the early development of cross-coupling reactions.

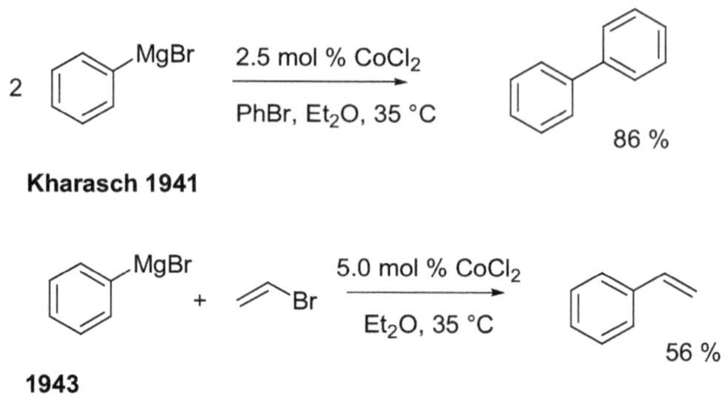

Scheme 1.2 Cobalt-catalyzed homo- and cross-coupling of Grignard reagents.

Figure 1.7 Morris Kharasch. "The grandfather of metal-catalyzed cross-coupling reactions."
Photograph courtesy of http://www.ashoftruth.org/history.

can be classified as one of the first direct C–C bond formations *via* C–H activation chemistry.

Most of the early developments involved the use of prefunctionalized coupling partners in terms of organometallic reagents as nucleophiles and aryl halides as electrophiles. An alternative attractive approach would be (as Fujiwara and Moritani showed) the direct functionalization of arene C–H bonds, without the need for prefunctionalization. In addition to Fujiwara and Moritani's disclosure, a few examples of C–H activation were reported in the 1980s by Ames[27] (intramolecular) and Ohta[28] (intermolecular). During the past two decades, the development of palladium-catalyzed direct

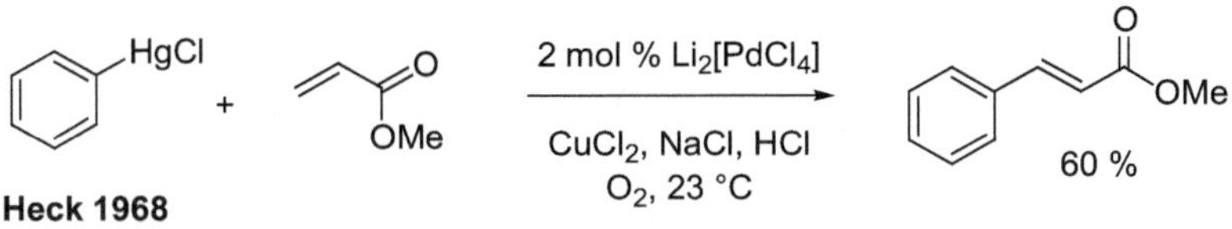

Heck 1968

Scheme 1.3 Palladium-catalyzed arylation of alkenes using an organomercury reagent.

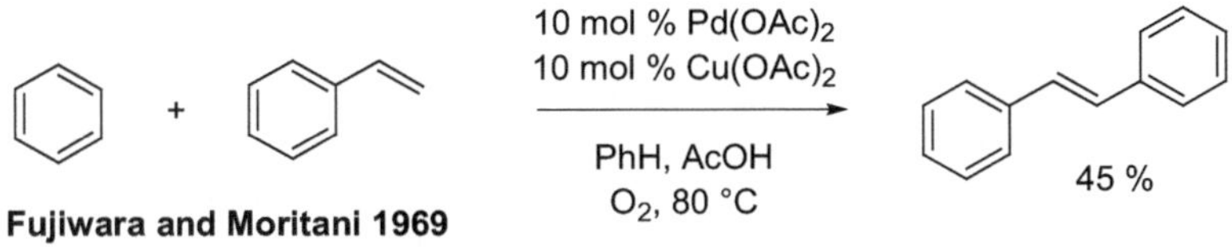

Fujiwara and Moritani 1969

Scheme 1.4 Palladium-catalyzed direct coupling of arenes with alkenes.

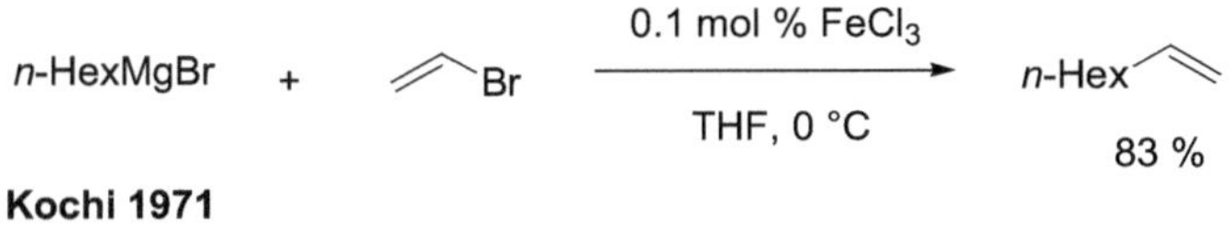

Kochi 1971

Scheme 1.5 Cross-coupling of Grignard reagents with $C(sp^2)$-bromides.

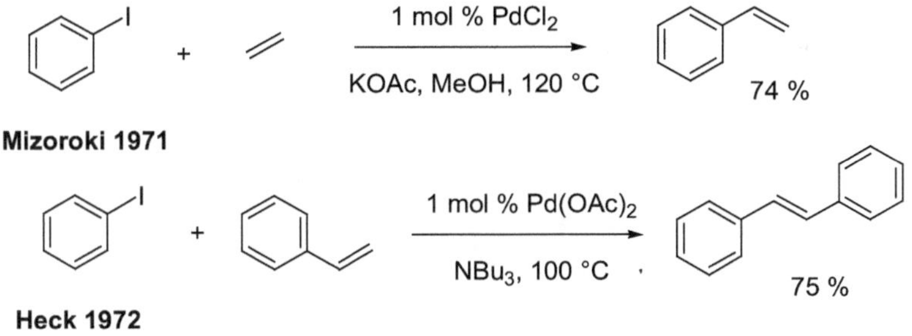

Mizoroki 1971

Heck 1972

Scheme 1.6 Pd-catalyzed coupling of alkenes with aryl iodides.

arylations has progressed enormously and these advances are discussed by Upendra Sharma, Atanu Modak, Soham Maity, Arun Maji and Debabrata Maiti in Chapter 12.

Building on Kharasch's cobalt-catalyzed cross-coupling reaction, Kochi accomplished an iron-catalyzed reaction between $C(sp^2)$-Br electrophiles and Grignard reagents (Scheme 1.5).[29]

In the same year, Mizoroki and co-workers presented a related reaction to the one reported by Heck in 1968 that importantly did not require the use of toxic arylmercury, -tin or -lead reagents. The C–C bond formation between ethylene or monosubstituted alkenes and iodobenzene could be achieved using catalytic amounts of $PdCl_2$ or heterogeneous Pd black (Scheme 1.6).[30]

Concurrently, Heck demonstrated independently the Pd-catalyzed reaction of aryl halides with alkenes in the presence of a hindered amine base.[31] Heck's work on aryl and vinyl halide substrates led to the second most practiced reaction in cross-coupling, namely the Mizoroki–Heck reaction.[32] Irina Beletskaya and Andrei Cheprakov in Chapter 9 discuss the role of modern Heck reactions in organic synthesis.

So far, only simple metal salts had been employed as catalysts. Corriu and Masse[33] and Tamao and Kumada[34] independently described the nickel-catalyzed coupling reaction of Grignard reagents with aryl halides. Tamao and Kumada (Figure 1.8) thereby pioneered the area of cross-coupling by showing the effects of adding phosphine ligands to the catalysts.[33–35]

The benefit of using phosphine ligands was particularly striking in reactions with less reactive aryl chlorides. Chapter 2, authored by Andrew DeAngelis and Thomas Colacot, covers the emergence of the development and use of ligands in Pd-catalyzed cross-coupling reactions in detail, with some theoretical background in choosing the right ligands for specific reaction types. Only during the past 10–15 years has the importance of the steric and electronic properties of the ligands used been fully recognized and evaluated.

A few years after Tamao and Kumada's disclosure of the importance of added phosphine ligands, four groups made independent reports on the cross-coupling reactions between alkynes and $C(sp^2)$–halide reagents.[36] Cassar made use of $Pd(PPh_3)_4$ as catalyst, whereas Heck and Sonogashira (Figure 1.9) used $Pd(PPh_3)_2(OAc)_2$ and $Pd(PPh_3)_2Cl_2$,

Figure 1.8 Prof. Makoto Kumada.
Photograph courtesy of Kyoto University.

Figure 1.9 Prof. Sonogashira (Pd/Cu-catalyzed sp–sp^2 coupling).

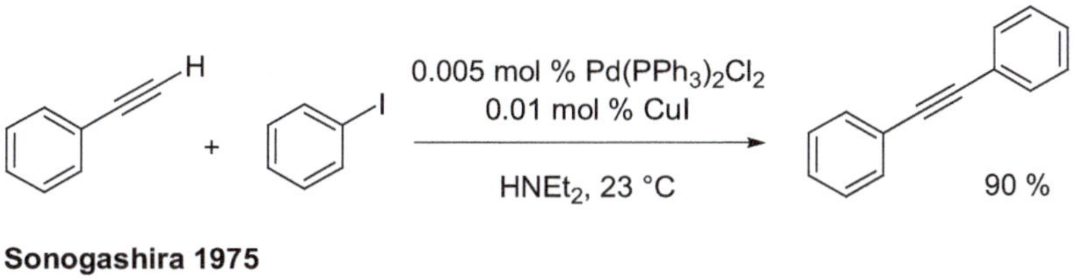

Sonogashira 1975

Scheme 1.7 Cu-mediated Pd-catalyzed coupling of alkenes with aryl iodides.

respectively. In addition, Sonogashira's studies showed that the presence of catalytic amounts of CuI enabled the reaction to proceed efficiently under significantly milder reaction conditions (Scheme 1.7). Recent work from a few laboratories, including ours, demonstrated that Cu is not required; in fact, it is detrimental for alkynylation reactions of aryl chlorides and unactivated aryl bromides where electron-rich Pd catalysts are required.[37] An insight into this phenomenon has been proposed.[37a]

Until 1976, the development of cross-coupling reactions had focused on the use of magnesium-based Grignard reagents as nucleophiles, with the exceptions of the discoveries of Mizoroki, Heck and Sonogashira. However, in 1976, Negishi showed that other organometallic reagents could be used as coupling partners.[38] First, Ni- or Pd-catalyzed cross-coupling reactions with organoaluminium reagents as nucleophiles were disclosed,[39] followed by the use of arylzinc reagents (Negishi reaction) (Scheme 1.8).[40]

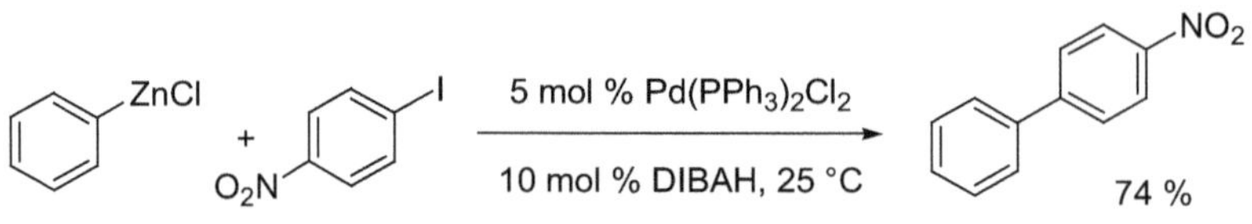

Negishi 1977

Scheme 1.8   Pd-catalyzed coupling of organozinc reagents with aryl iodides.

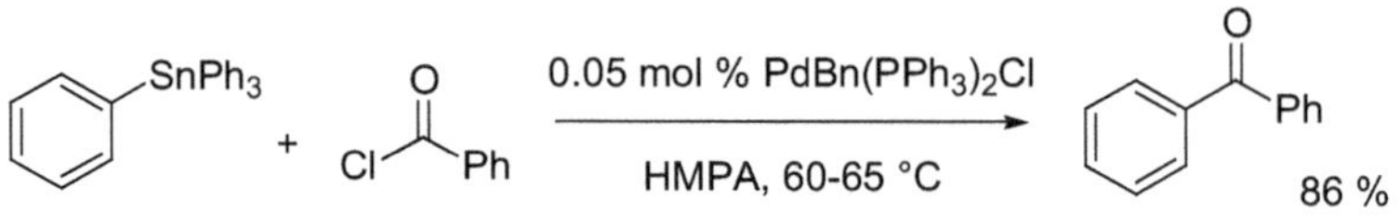

Stille 1978

Scheme 1.9   Pd-catalyzed coupling of organotin reagents with aroyl chlorides.

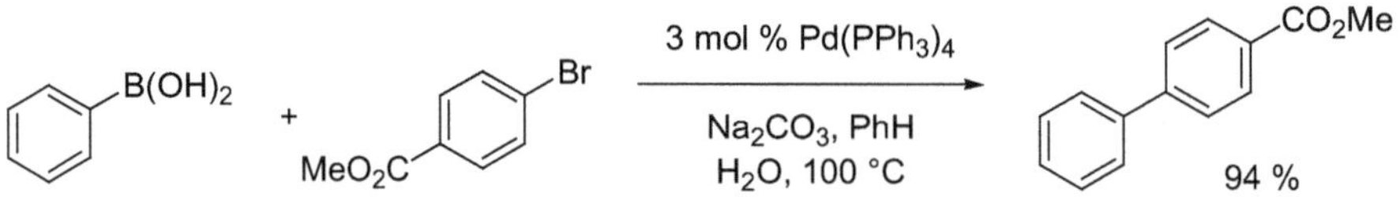

Suzuki 1981

Scheme 1.10   Pd-catalyzed coupling of organoboron reagents with aryl bromides.

Continuing the investigations of the use of alternative nucleophiles, Kosugi and Migita,[41] followed by Stille,[42] reported cross-coupling reactions involving organotin reagents (Scheme 1.9).

Extending the range of possible organometallic coupling partners, following the report by Murahashi's on a transmetallation strategy for using trialkylboranes,[43] Suzuki and Miyaura disclosed their findings on the beneficial effect of added bases in cross-coupling reactions of alkenylboranes with aryl halides catalyzed by palladium (Suzuki–Miyaura reaction) (Scheme 1.10).[44]

Since Suzuki and Miyaura's seminal report, this cross-coupling reaction has been extensively developed and its mechanism studied in great detail. Chapter 8, authored by Alastair Lennox and Guy Lloyd-Jones, describes the discovery, development and deployment of boron reagents in cross-coupling reactions, including mechanistic analysis. In Chapter 11, the topic of stereospecific and stereoselective Suzuki–Miyaura coupling reactions is discussed by Ben Glasspoole, Eric Keske and Cathleen Crudden.

Subsequent to the introduction of organoboron reagents as cross-coupling partners, Hiyama reported efficient cross-coupling reactions of arylsilanes by the use of fluoride additives (Scheme 1.11).[45]

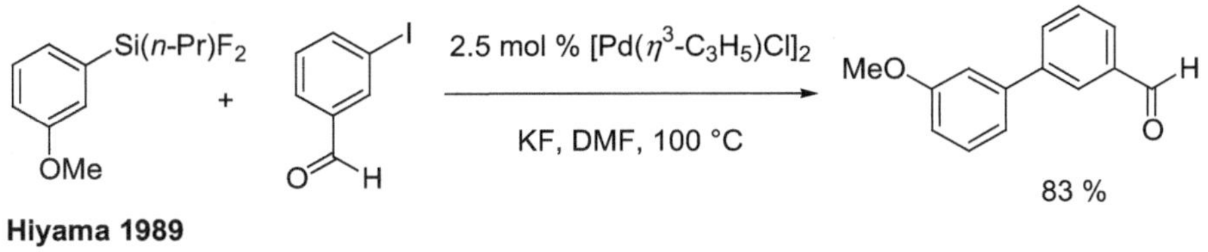

Hiyama 1989

Scheme 1.11 Pd-catalyzed coupling of organosilicon reagents with aryl iodides.

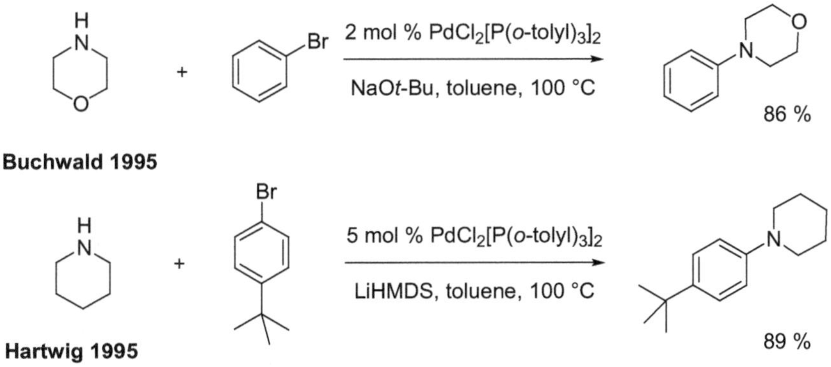

Buchwald 1995

Hartwig 1995

Scheme 1.12 Pd-catalyzed direct coupling of amines with aryl bromides.

By 1989, it was already possible to achieve C–C bond formation through cross-coupling reactions using a wide variety of organometallic reagents. The next breakthrough came with the independent discoveries by Buchwald and Hartwig that C–N bond formation could also be efficiently accomplished through palladium-catalyzed cross-coupling reactions of free amines with aryl halides (Buchwald–Hartwig amination) (Scheme 1.12).[46] This methodology was also extended to cover C–O and C–S bond formation. Chapter 6, authored by James Stambuli, provides a specific discussion of the development of C–O and C–S bond-forming reactions using cross-coupling methodologies.

At this stage, it may seem like most of the monumental discoveries had already been made; however, there is by no means an end to the advancement of cross-coupling reactions. Reports continue to appear in the literature regarding novel reactivity, enhanced substrate compatibility, new trends and solutions to previously problematic transformations.

Following the seminal discoveries made before 2000, the first decade of the new millennium would see the expansion of cross-coupling reactions to include the use of the previously very challenging aryl chlorides as coupling partners. This was achieved by the groups of Buchwald,[47] Fu[48] and Koie[49] by means of using bulky, electron-rich monophosphines as ligands. In addition, enormous progress has been made to be able to employ C(sp^3)-based coupling partners, such as alkyl–BR$_2$ or alkyl halide reagents.[50]

The intent of the chapters in this book is to highlight and discuss the most recent advances and discoveries within the field of cross-coupling.

Chapter 3 by Carin Johansson Seechurn, Hongbo Li and Thomas Colacot discusses the advantages of the use of well-defined preformed Pd catalysts in cross-coupling reactions. This is an area that has only been recognized for its importance in the past 10 years.[51]

Although Chapter 2 briefly touches the subject of ligands other than phosphine, Chapter 4, by Anthony Chartoire and Steven Nolan, details the advances in C–X coupling that has been made using NHC (*N*-heterocyclic carbene) ligands in conjunction with palladium.

Specific examples of the importance of the structural design of the ligands and its impact on the outcome of cross-coupling reactions are demonstrated in Chapter 5 by Mark Stradiotto. Here, selective monoarylation reactions of amines and carbonyl compounds are discussed.

In addition to more thorough mechanistic investigations and realizations and the continuous expansion of substrate scope for the named cross-coupling reactions, there are also examples of novel reactions appearing in the literature. An example is highlighted and discussed in Chapter 7 by David Petrone, Christine Le, Stephen Newman and Mark Lautens, where the recently disclosed Pd(0)-catalyzed carboiodination reaction, a reaction fundamentally rooted in the Heck reaction, is also discussed.

The importance of cross-coupling reactions is reflected in their frequent use within the chemical industry. To meet the criteria of the economic demands for cheaper and more efficient processes, the challenges within industry are somewhat different from those met within the academic community. In recent years, continuous flow technology has provided an attractive alternative to the more traditional batch preparations usually carried out in process chemistry. This technology has also reached the field of cross-coupling, an area which is discussed by Timothy Noël and Volker Hessel in Chapter 13. The chemical industry is constantly faced with stricter regulations regarding the environmental impact of the processes being carried out. Kevin Shaughnessy reviews the area of green chemistry *via* cross-coupling reactions in Chapter 14. Javier Magano and Joshua Dunetz demonstrate in Chapter 15 some recent applications of cross-coupling reactions in the large-scale synthesis of pharmaceuticals and showcase the problems encountered in a process chemistry environment. Chapter 16, authored by Kazunori Koide, highlights the problem with residual Pd in final APIs (active pharmaceutical ingredients) and discusses detection methods.

In addition to all the improvements developed to expand the field of cross-coupling involving the introduction of novel ligands, precatalysts, chemical engineering designs and ever-more in-depth understanding of reaction mechanisms, the area continues to grow as challenges in terms of incorporating problematic nucleophilic and electrophilic reaction partners are met with innovative solutions. As an example, shown in Scheme 1.13, Feringa and co-workers recently demonstrated the previously difficult[52]

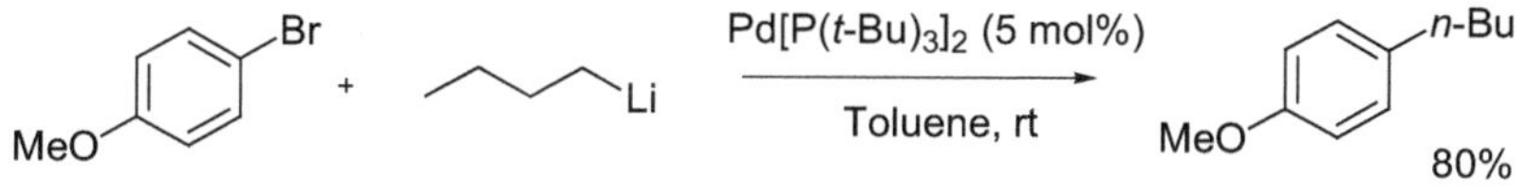

Feringa 2013

Scheme 1.13 Direct coupling of organolithium reagents with aryl bromides.

direct coupling reaction of organolithium reagents with aryl halides (Murahashi coupling),[53] demonstrating how important problems can be solved through the utilization of the ever-growing technology made available by the numerous research groups across the globe who are active within the field of cross-coupling.

Substantial discoveries within the field of cross-coupling have not reached an end; new reactions are waiting to be discovered, as chemists will never run out of problems to solve. Even fundamental questions such as why Pd is significantly better than Ni or Pt in most cases are still not fully understood.

References

1. Nobelprize.org, *The Nobel Prize in Chemistry 2001*, Nobel Media AB, 2013, http://www.nobelprize.org/nobel_prizes/chemistry/laureates/2001/, accessed 16 April 2014.
2. Nobelprize.org, *The Nobel Prize in Chemistry 2005*, Nobel Media AB, 2013, http://www.nobelprize.org/nobel_prizes/chemistry/laureates/2005/, accessed 17 April 2014.
3. Nobelprize.org, *The Nobel Prize in Chemistry 2010*, Nobel Media AB, 2013, http://www.nobelprize.org/nobel_prizes/chemistry/laureates/2010/, accessed 16 April 2014.
4. I. W. Davies, L. Matty, D. L. Hughes and P. J. Reider, *J. Am. Chem. Soc.*, 2001, **123**, 10139.
5. For Nobel lectures see: (a) E. Negishi, *Angew. Chem. Int. Ed.*, 2011, **50**, 6738; (b) A. Suzuki, *Angew. Chem. Int. Ed.*, 2011, **50**, 6722.
6. A. Behr and P. Neubert, *Applied Homogeneous Catalysis*, Wiley-VCH, Weinheim, 2012.
7. (a) I. D. Hills and G. C. Fu, *J. Am. Chem. Soc.*, 2004, **126**, 13178; (b) F. Barrios-Landeros, B. P. Carrow and J. F. Hartwig, *J. Am. Chem. Soc.*, 2008, **130**, 5842.
8. R. J. Deeth, A. Smith and J. M. Brown, *J. Am. Chem. Soc.*, 2004, **126**, 7144.
9. L. Qin, H. Hirao and J. Zhou, *Chem. Commun.*, 2013, **49**, 10236.
10. A. de Mejiere and F. Diederich (eds), *Metal Catalyzed Cross-Coupling Reactions*, 2nd edn, Wiley, New York, 2004.
11. (a) B. P. Carrow and J. F. Hartwig, *J. Am. Chem. Soc.*, 2011, **133**, 2116; (b) M. Butters, J. Harvey, J. Jover, A. Lennox, G. Lloyd-Jones and P. Murray, *Angew. Chem. Int. Ed.*, 2010, **49**, 5156; (c) C. Amatore, A. Jutand and G. Le Duc, *Chem. Eur. J.*, 2011, **17**, 2492.

12. A. G. Sergeev, G. A. Artamkina and I. P. Beletskaya, *Tetrahedron Lett.*, 2003, **44**, 4719.
13. J. D. Hicks, A. M. Hyde, A. M. Cuezva and S. L. Buchwald, *J. Am. Chem. Soc.*, 2009, **131**, 16720.
14. S. Shaik, *Phys. Chem. Chem. Phys.*, 2010, **12**, 8706.
15. Y.-R. Luo, *Handbook of Bond Dissociation Energies in Organic Compounds*, CRC Press, Boca Raton, FL, 2003.
16. F. Barrios-Landeros, B. P. Carrow and J. F. Hartwig, *J. Am. Chem. Soc.*, 2009, **131**, 8141.
17. (a) B. P. Fors and S. L. Buchwald, *J. Am. Chem. Soc.*, 2009, **131**, 12898; (b) T. Kinzel, Y. Zhang and S. L. Buchwald, *J. Am. Chem. Soc.*, 2010, **132**, 14073; (c) E. V. Vinogradova, B. P. Fors and S. L. Buchwald, *J. Am. Chem. Soc.*, 2012, **134**, 11132.
18. For reviews on the history of cross-coupling reactions, see: (a) C. C. C. Johansson Seechurn, M. O. Kitching, T. J. Colacot and V. Snieckus, *Angew. Chem. Int. Ed.*, 2012, **51**, 5062; (b) F. Bellina and R. Rossi, *Chem Rev.*, 2010, **110**, 1082.
19. (a) B. Tollens and R. Fittig, *Justus Liebigs Ann. Chem.*, 1864, **131**, 303; (b) R. Fittig, *Justus Liebigs Ann. Chem.*, 1862, **121**, 361.
20. (a) F. Ullmann and J. Bielecki, *Chem. Ber.*, 1901, **34**, 2174; (b) F. Ullmann, *Chem. Ber.*, 1903, **36**, 2382; (c) F. Ullmann and P. Sponagel, *Chem. Ber.*, 1905, **38**, 2211; (d) I. Goldberg, *Chem. Ber.*, 1906, **39**, 1691.
21. A. Job and R. Reich, *C. R. Acad. Sci. Paris*, 1924, **179**, 330.
22. (a) M. S. Kharasch and E. K. Fields, *J. Am. Chem. Soc.*, 1941, **63**, 2316; (b) M. S. Kharasch and O. Reinmuth, *Grignard Reagents of Nonmetallic Substances*, Prentice-Hall, New York, 1954.
23. M. S. Kharasch and C. F. Fuchs, *J. Am. Chem. Soc.*, 1943, **65**, 504.
24. R. F. Heck, *J. Am. Chem. Soc.*, 1968, **90**, 5518.
25. (a) I. Moritani and Y. Fujiwara, *Tetrahedron Lett.*, 1967, **8**, 1119; (b) Y. Fujiwara, I. Moritani and M. Matsuda, *Tetrahedron*, 1968, **24**, 4819.
26. Y. Fujiwara, I. Moritani, S. Danno, R. Asano and S. Teranishi, *J. Am. Chem. Soc.*, 1969, **91**, 7166.
27. (a) D. E. Ames and D. Bull, *Tetrahedron*, 1982, **38**, 383; (b) D. E. Ames and A. Opalko, *Synthesis*, 1983, 234; (c) D. E. Ames and A. Opalko, *Tetrahedron*, 1984, **40**, 1919.
28. (a) Y. Akita, A. Inoue, K. Yamamoto, A. Ohta, T. Kurihara and M. Shimizu, *Heterocycles*, 1985, **23**, 2327; (b) Y. Akita, Y. Itagaki, S. Takizawa and A. Ohta, *Chem. Pharm. Bull.*, 1989, **37**, 1477.
29. (a) M. Tamura and J. Kochi, *J. Am. Chem. Soc.*, 1971, **93**, 1487; (b) M. Tamura and J. Kochi, *Synthesis*, 1971, 303.
30. (a) T. Mizoroki, K. Mori and A. Ozaki, *Bull. Chem. Soc. Jpn.*, 1971, **44**, 581; (b) K. Mori, T. Mizoroki and A. Ozaki, *Bull. Chem. Soc. Jpn.*, 1973, **46**, 1505.
31. R. F. Heck and J. P. Nolley, *J. Org. Chem.*, 1972, **37**, 2320.
32. M. Oestreich (ed.) *The Mizoroki–Heck Reaction*, Wiley, Chichester, 2009.
33. R. J. P. Corriu and J. P. Masse, *J. Chem. Soc., Chem. Commun.*, 1972, 144.

34. K. Tamao, K. Sumitani and M. Kumada, *J. Am. Chem. Soc.*, 1972, **94**, 4374.
35. For a review, see: M. Kumada, *Pure Appl. Chem.*, 1980, **52**, 669.
36. (a) L. Cassar, *J. Organomet. Chem.*, 1975, **93**, 253; (b) M. Yamamura, I. Moritani and S.-I. Murahashi, *J. Organomet. Chem.*, 1975, **91**, C39; (c) H. A. Dieck and R. F. Heck, *J. Organomet. Chem.*, 1975, **93**, 259; (d) K. Sonogashira, Y. Tohda and N. Hagihara, *Tetrahedron Lett.*, 1975, **50**, 4467.
37. For examples, see: (a) X. Pu, H. Li and T. J. Colacot, *J. Org. Chem.*, 2013, **78**, 568; (b) D. Gelman and S. L. Buchwald, *Angew. Chem. Int. Ed.*, 2003, **42**, 5993; (c) C. Torborg, J. Huang, T. Schulz, B. Schäffner, A. Zapf, A. Spannenberg, A. Börner and M. Beller, *Chem. Eur. J.*, 2009, **15**, 1329.
38. E.-i. Negishi, *Acc. Chem. Res.*, 1982, **15**, 340.
39. (a) E.-i. Negishi and S. Baba, *J. Chem. Soc., Chem. Commun.*, 1976, 596; (b) S. Baba and E.-i. Negishi, *J. Am. Chem. Soc.*, 1976, **98**, 6729.
40. A. O. King, N. Okukado and E.-i. Negishi, *J. Chem. Soc., Chem. Commun.*, 1977, **19**, 683.
41. (a) M. Kosugi, K. Sasazawa, Y. Shimizu and T. Migita, *Chem. Lett.*, 1977, 301; (b) M. Kosugi, Y. Shimizu and T. Migita, *Chem. Lett.*, 1977, 1423.
42. D. Milstein and J. K. Stille, *J. Am. Chem. Soc.*, 1978, **100**, 3636.
43. K. Kondo and S.-I. Murahashi, *Tetrahedron Lett.*, 1979, **20**, 1237.
44. (a) N. Miyaura, K. Yamada and A. Suzuki, *Tetrahedron Lett.*, 1979, **20**, 3437; (b) N. Miyaura, T. Yanagi and A. Suzuki, *Synth. Commun.*, 1981, **11**, 513.
45. Y. Hatanaka, S. Fukushima and T. Hiyama, *Chem. Lett.*, 1989, 1711.
46. (a) A. S. Guram, R. A. Rennels and S. L. Buchwald, *Angew. Chem. Int. Ed.*, 1995, **34**, 1348; (b) J. Louie and J. F. Hartwig, *Tetrahedron Lett.*, 1995, **36**, 3609.
47. D. W. Old, J. P. Wolfe and S. L. Buchwald, *J. Am. Chem. Soc.*, 1998, **120**, 9722.
48. A. F. Littke and G. C. Fu, *Angew. Chem. Int. Ed.*, 1998, **37**, 3387.
49. (a) M. Nishiyama, T. Yamamoto and Y. Koie, *Tetrahedron Lett.*, 1998, **39**, 617; (b) T. Yamamoto, M. Nishiyama and Y. Koie, *Tetrahedron Lett.*, 1998, **39**, 2367.
50. For a review of this area, see: R. Jana, T. P. Pathak and M. S. Sigman, *Chem. Rev.*, 2011, **111**, 1417.
51. H. Li, C. C. C. Johansson Seechurn and T. J. Colacot, *ACS Catal.*, 2012, **2**, 1147.
52. (a) S. Murahashi, M. Yamamura, K. Yanagisawa, N. Mita and K. Kondo, *J. Org. Chem.*, 1979, **44**, 2408; (b) For a continuous flow-based approach, see: A. Nagaki, A. Kenmoku, Y. Moriwaki, A. Hayashi and J.-i. Yoshida, *Angew. Chem. Int. Ed.*, 2010, **49**, 7543.
53. (a) C. Vila, M. Giannerini, V. Hornillos, M. Fañanás-Mastral and B. L. Feringa, *Chem. Sci.*, 2014, **5**, 1361; (b) V. Hornillos, M. Giannerini, C. Vila, M. Fañanás-Mastral and B. L. Feringa, *Org. Lett.*, 2013, **15**, 5114; (c) M. Giannerini, M. Fañanás-Mastral and B. L. Feringa, *Nat. Chem.*, 2013, **5**, 667; (d) M. Giannerini, V. Hornillos, C. Vila, M. Fañanás-Mastral and B. L. Feringa, *Angew. Chem. Int. Ed.*, 2013, **52**, 13329.

CHAPTER 2

Prominent Ligand Types in Modern Cross-Coupling Reactions

ANDREW DEANGELIS AND THOMAS J. COLACOT*

Johnson Matthey Catalysis and Chiral Technologies, 2001 Nolte Drive, West Deptford, NJ 08066, USA
*Email: thomas.colacot@jmusa.com

2.1   Introduction

Intense research efforts have been ongoing in the area of ligand development since the late 1990s, which significantly accelerated the advancement of Pd-catalyzed cross-coupling to reach the premium status that it has achieved today. Several excellent reviews[1] and books[2] encompass the historical perspectives of the development of the state-of-the-art technology in cross-coupling, wherein the importance of ligands has been highlighted.

Although cross-coupling emerged in large part from Glaser's seminal work in the 1800s for copper-mediated processes, such as alkyne dimerization,[3] followed by the use of Ni-, Fe- and Co-based catalysts,[2a] palladium has become the metal of choice in the majority of cross-coupling reactions as it has proven to be unparalleled in terms of reaction scope and functional group compatibility. The reactive intermediates formed along the catalytic cycle feature an excellent balance of reactivity and stability when palladium is employed relative to other metals. Furthermore, phosphine ligands have become the ligands of choice in conjunction with Pd owing to their high activity, ease of synthesis and fine tuning, although many useful catalyst

RSC Catalysis Series No. 21
New Trends in Cross-Coupling: Theory and Applications
Edited by Thomas J Colacot

systems have been developed utilizing other ligand platforms such as *N*-heterocyclic carbenes (NHCs).[4] This chapter focuses primarily on phosphine-based systems as Chapter 4 is devoted to cross-coupling reactions with NHC-based catalysts.

The soft nature of the phosphine is an ideal match for soft low-valent metals including palladium. According to Crabtree, "An important part of the art of organometallic chemistry is to pick suitable spectator ligand sets to facilitate certain types of reactions",[5] underlining the importance of ligand attributes to catalysis. Here, we aim to cover the prominent classes of ligands that have emerged to generate highly active catalysts in modern Pd-catalyzed cross-coupling and related processes for broad applications.

2.2 Evolving from Traditional Ligands: the Significance of Ligand Properties

Owing to its wide availability and ease of handling, early studies on the development of Pd-catalyzed cross-coupling reactions utilized PPh_3 as the major supporting ligand. However, in the early 1980s, Heck discovered that a palladium complex of $P(o\text{-}tol)_3$ (tri-*o*-tolylphosphine), a ligand more sterically demanding than PPh_3, was more active in vinylation reactions in comparison with Pd–Ph_3P complexes, for example, $Pd(PPh_3)_4$.[6] Furthermore, an even earlier report in 1979 by Kumada's group revealed a significant improvement in catalytic activity in the cross-coupling reactions of organobromides with *sec*-butylmagnesium chloride when a bidentate phosphine ligand, dppf [1,1'-bis(diphenylphosphino)ferrocene], was utilized relative to a control reaction with PPh_3.[7] Soon, in an effort to understand these effects and to design new and improved catalyst systems, chemists began to probe the effects of ligands in all of the three major steps in cross-coupling reactions, namely oxidative addition, transmetallation and reductive elimination. Several parameters have since been established to quantify these properties in order to map their effects on the individual steps of the catalytic cycle.

Today, the role of the ligand is well understood in terms of steric and electronic effects, which are often correlated with "cone angle" and basicity/nucleophilicity, respectively. In bidentate ligands, steric properties are compounded by another effect called the "bite angle" effect. The following sections explain these effects in detail with some theoretical understanding on the choice of the ligand.

2.2.1 Steric Considerations

Early studies regarding the effects of ligand properties on cross-coupling reactions were focused primarily on investigating steric considerations. The steric demand of supporting ligands is significant, as increased bulk promotes ligand dissociation to generate a 12 electron-based monoligated

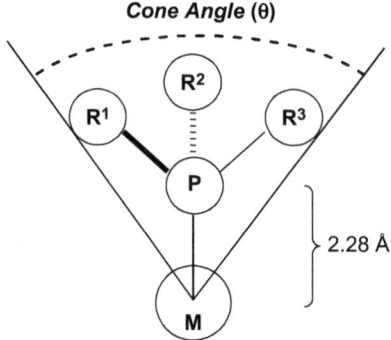

Figure 2.1   Definition of cone angle.

palladium intermediate (L_1Pd), which is generally proposed to be the catalytically active species in most modern cross-coupling processes, with some exceptions.[8] Steric properties of ligands are often described in terms of "cone angle" (θ), a parameter introduced by Tolman to measure the physical space that a ligand occupies (Figure 2.1).[9] The value of θ for symmetrically substituted ligands (cases where for PR_3 all R groups are equivalent) is defined by the angle that is created by a cylindrical cone extending 2.28 Å (typical Ni– P bond length) from the metal (or other atom center) to the outermost atoms of the R groups on the ligand using CPK models (space-filling models, named after the scientists who pioneered the use: Robert Corey, Linus Pauling and Walter Koltun). For unsymmetrically substituted ligands, Tolman also described a model for the determination of θ by the summation of the partial cone angles.[9b] Thus, an individual partial cone angle (θ_i) is determined for each of the R^n substituents ($n = 1$–3) and the total cone angle is calculated according to eqn (2.1). Even today, this method is still widely used for the estimation of ligand bulkiness (Table 2.1).

$$\theta = \frac{2}{3} \sum_{i=1}^{3} \frac{\theta_i}{2} \tag{2.1}$$

As a representative example of a calculation of θ for an unsymmetrical ligand, the cone angle θ of $PPh_2(t\text{-Bu})$ is calculated as $(2/3)(145°/2 + 145°/2 + 182°/2) = 157°$.

However, the CPK-based cone angle parameter to quantify the steric demand imposed by a supporting ligand is not without limitations. Most ligands do not actually form perfect cones. In addition, when multiple ligands are coordinated to a single metal center, there can be inter-ligand meshing, resulting in decreased steric demand relative to what one would predict based on the sum of the individual partial cone angles. Tolman's model uses several rough approximations: the 2.28 Å M–L bond length assumption is not necessarily accurate for non-phosphine ligands and different metal centers, and an idealized tetrahedral geometry (R–P–R bond

Table 2.1 Common Tolman cone angles θ and exact ligand cone angles $\theta°$.

Entry	Ligand	Cone angle θ (°)[9c]	$\theta°$ (Pd) Min. (°)[10]	Max. (°)[10]
1	PH$_3$	87		
2	PMe$_3$	118	120	
3	P(CF$_3$)$_3$	137		
4	PPhMe$_2$	122	149	
5	PPh$_2$Me	136	151	
6	PPh$_3$	145	170	
7	P(p-tol)$_3$	145	171	
8	P(C$_6$F$_5$)$_3$	184		
9	P(o-tol)$_3$	194	176	208
10	P(mesityl)$_3$	212		
11	PCy$_2$Ph	159[15]		
12	PPh$_2$(t-Bu)	157	167	
13	PPh(t-Bu)$_2$	170	187	
14	PEt$_3$	132	136	169
15	P(n-Bu)$_3$	132	136	169
16	P(i-Pr)$_3$	160	169	177
17	PCy$_3$	170		
18	P(i-Bu)$_3$	143		214
19	P(s-Bu)$_3$	160		
20	P(t-Bu)$_3$	182[9] (194[12b])	188	
21	P(1-adamantyl)$_2$(n-Bu)	176[15]		
22	P(t-Bu)$_2$(neopentyl)	198[12b]		
23	P(t-Bu)(neopentyl)$_2$	210[12b]		
24	P(neopentyl)$_3$	~180[9] (227[12b])		
25	P(OMe)$_3$	107		
26	P(OPh)$_3$	128		
27	2-(Di-*tert*-butylphosphino)biphenyl (JohnPhos)	246[15]		
28	2-Dicyclohexylphosphino-2′,6′-dimethoxybiphenyl (SPhos)	240[15]		
29	2-Dicyclohexylphosphino-2′,4′,6′-triisopropylbiphenyl (XPhos)	256[15]		

angle $= 109.5°$) of the phosphorus atom is not always valid. Additionally, the substituents are folded back to make a minimum cone when flexibility in the ligand is present, which may not always be a reasonable assumption. To address these limitations, Allen and co-workers recently developed an innovative method for determining "exact ligand cone angles" utilizing a mathematical approach.[10] They define the exact ligand cone angle ($\theta°$) as the angle of the most acute right circular cone which contains all of the atoms of the ligand (described by spheres with the atom's corresponding van der Waals radius) and is tangent to up to three of the atoms (Figure 2.2). The models are constructed from either X-ray data or DFT-optimized structures and, importantly, no approximations are used. Thus, only the Cartesian coordinates are required to determine $\theta°$. For ligands that can adopt multiple conformations, $\theta°$ is determined as a range between a maximum $\theta°$ and a minimum $\theta°$. In certain cases, the value of $\theta°$ is vastly different from

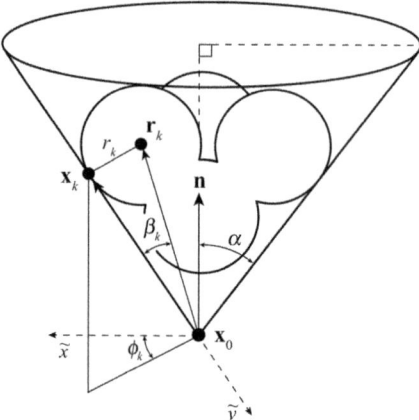

Figure 2.2 Definition of exact ligand cone angle θ°. $\theta^\circ = 2\alpha$, $x_0 =$ metal center, $r_k =$ ligand atom center, $x_k =$ tangent point, $r_k =$ van der Waals radius of ligand atom.
Reproduced with permission from J. A. Bilbrey, A. H. Kazez, J. Locklin and W. D. Allen, *J. Comput. Chem.*, 2013, **34**, 1189. Copyright 2013 Wiley Periodicals, Inc.

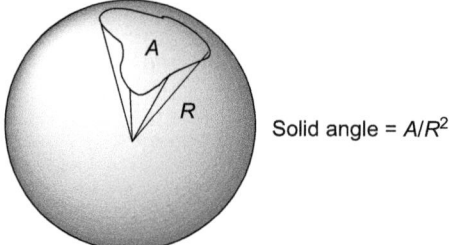

Figure 2.3 Definition of solid angle.

the traditional Tolman cone angle θ (see Table 2.1 for θ° values and for comparison with θ).

White *et al.* introduced the method of measuring "solid angles" to characterize ligand steric demand.[11] The solid angle is defined as the normalized area A of a theoretical shadow cast by the ligand on to a sphere of radius R which encompasses the complex, with the source of light at the origin (Figure 2.3).

Shaughnessy and co-workers found that although ligands with increased steric demand typically give rise to more active catalysts, a definitive correlation could not be drawn when comparing catalyst activities with Tolman cone angles, especially with neopentyl-based phosphines.[12] Other methods have since been developed for calculating steric parameters based on X-ray and computational data. Using a computational approach developed by Taverner,[13] Shaughnessy and co-workers found that cone angles calculated

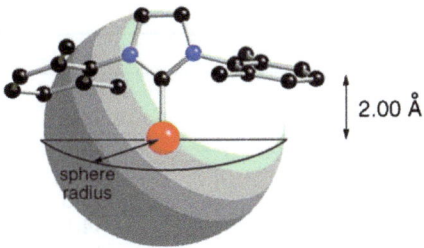

Figure 2.4   Definition of $\%V_{\mathrm{bur}}$.
Reproduced from Ref. 15.

from DFT-optimized structures can be larger than those calculated from the CPK models by Tolman.[12] For instance, the cone angle of P(t-Bu)$_3$ using the method of Tolman is 182°,[9] whereas an angle of 194° is determined by the computational method[12b] (Table 2.1, Entry 20). Moreover, as noted by Nolan and co-workers, the Tolman calculation is not valid for *N*-heterocyclic carbene ligands as the steric bulk of these ligands is directed *towards* the metal center, not away as in the case of tertiary phosphines.[14] Nolan and co-workers therefore introduced the "percentage buried volume" ($\%V_{\mathrm{bur}}$) to quantify the steric demand of NHC ligands and also others that are misrepresented by conventional Tolman cone angle calculations.[15] The $\%V_{\mathrm{bur}}$ is defined as the percentage of the total volume occupied by a ligand of a sphere with a defined radius (2.00 and 2.28 Å) and a metal at the core (Figure 2.4). Using crystallographic data, the $\%V_{\mathrm{bur}}$ can be calculated using the web-based Samb*V*ca software (Salerno molecular buried volume calculation), which is available free of charge.[16]

A good correlation was found between $\%V_{\mathrm{bur}}$ and θ for tertiary phosphines,[17] and it was also demonstrated that cone angles could be estimated from the $\%V_{\mathrm{bur}}$ by linear regression based on this correlation. This method provided a means not only to evaluate the steric properties of NHC ligands, but also to allow for direct comparison with phosphines. Table 2.2 presents a list of $\%V_{\mathrm{bur}}$ values.

2.2.2   Electronic Considerations

As illustrated in Figure 2.5, the phosphorus–palladium bond is a dative interaction which arises primarily from σ-donation from the phosphorus electron lone pair to the unfilled d-orbitals of Pd (Lewis acid–base interaction). Tertiary phosphines also act as π-acceptors (π-acids), wherein the filled metal d-orbitals overlap primarily with P–R σ*-orbitals (Figure 2.5).[18]

The electron-donating ability of the phosphine is significant in terms of increasing the electron density on the Pd, thereby helping a more facile oxidative addition. Various methods have been used to characterize the electronic contributions from ancillary ligands; however, the separation of the σ-donor *versus* π-backbonding effects has proven particularly challenging. One of the most straightforward methods to quantify σ-donation is

Table 2.2 %V_{bur} for LAuCl complexes.

Entry	Ligand	%V_{bur} (2.00 Å)	%V_{bur} (2.28 Å)
1	PMe$_3$	27.3	23.3
2	PPhMe$_2$	30.5	26.2
3	PPh$_3$	34.8	29.9
4	P(p-tol)$_3$	34.2	29.3
5	P(C$_6$F$_5$)$_3$	42.6	37.3
6	P(o-tol)$_3$	44.8	39.4
7	P(mesityl)$_3$	50.5	45.0
8	PCy$_2$Ph	38.0	32.7
9	PEt$_3$	31.7	27.1
10	P(n-Bu)$_3$	30.4^a	25.9^a
11	P(i-Pr)$_3$	39.1	24.0
12	PCy$_3$	38.8	33.4
13	P(t-Bu)$_3$	43.9	38.1
14	P(1-adamantyl)$_2$(n-Bu)	41.9	36.3
15	P(OMe)$_3$	30.8	26.4
16	P(OPh)$_3$	36.5	31.9
17	2-(Di-*tert*-butylphosphino)biphenyl (JohnPhos)	55.5	50.9
18	2-Dicyclohexylphosphino-2′,6′-dimethoxybiphenyl (SPhos)	53.7	49.7
19	2-Dicyclohexylphosphino-2′,4′,6′-triisopropylbiphenyl (XPhos)	57.4	53.1

^{a}Calculated from the crystal structure of the free phosphine.
Adapted from Ref. 15.

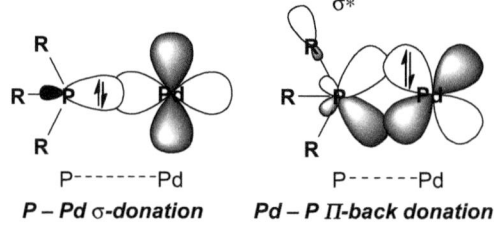

Figure 2.5 Phosphine–palladium orbital interactions.

to examine the pK_a of the corresponding phosphonium salt (see Table 2.3). Despite the fact that the nature of the interaction between H$^+$ and a phosphine is hard–soft and that of a phosphine metal bond is soft–soft, remarkably good correlations have been found.[19] Hard–soft acid–base theory can also be invoked to explain why phosphines (soft Lewis bases) are significantly better ligands for palladium (soft Lewis acid) than amines (hard Lewis bases) in spite of their similar basicity.

The classic method for evaluating electronic contributions from donor ligands was developed by Tolman.[9c] This method of quantifying "nucleophilicity" utilizes IR spectroscopy to measure the vibrational frequency of the A$_1$ carbonyl stretching mode (v_{CO}) of [Ni(CO)$_3$L] complexes. The v_{CO} value is related to the degree of backbonding between Ni and CO, which is related to the electron density on the Ni. This, in turn, is influenced by the ligand's

Table 2.3 Phosphonium pK_a values.

Entry	Ligand	pK_a (phosphonium)
1	PMe$_3$	8.65[20a]
2	P(Et)$_3$	8.69[21]
3	PPhMe$_2$	6.50[20a]
4	PPh$_2$Me	4.57[20b]
5	PPh$_3$	2.73[20a]
6	P(p-OMePh)$_3$	4.46[20a]
7	P(p-ClPh)$_3$	1.03[20b]
8	P(p-FPh)$_3$	1.97[20c]
9	P(o-tol)$_3$	3.08[20c]
10	PCy$_3$	9.7[20a]
11	P(i-Bu)$_3$	7.97[21]
12	P(n-Bu)$_3$	8.43[21]
13	P(t-Bu)$_3$	11.4[20c]

Table 2.4 ν_{CO} values for [Ni(CO)$_3$L].

Entry	Ligand (L)	ν_{CO} (cm^{-1})[22]
1	P(t-Bu)$_3$	2056
2	PCy$_3$	2056
3	P(i-Pr)$_3$	2059
4	P(n-Bu)$_3$	2060
5	PEt$_3$	2062
6	PMe$_3$	2064
7	PPhMe$_2$	2065
8	P(4-OMeC$_6$H$_4$)$_3$	2066
9	P(o-tol)$_3$	2067
10	P(p-tol)$_3$	2067
11	PPh$_2$Me	2067
12	PPh$_3$	2069
13	P(p-FPh)$_3$	2071
14	P(p-ClPh)$_3$	2073
15	P(OMe)$_3$	2079
16	P(OPh)$_3$	2085
17	P(C$_6$F$_5$)$_3$	2091
18	PCl$_3$	2097
19	PF$_3$	2111

Adapted with permission from C. A. Tolman, *J. Am. Chem. Soc.*, 1970, **92**, 2953. Copyright 1970 American Chemical Society.

electron-donating ability. Thus, a lower ν_{CO} stretching frequency correlates with a stronger σ-donor ability of the ligand. See Table 2.4.

These methods do have some limitations. For instance, contributions from π-backbonding are underrepresented. A computational method using natural bond orbital (NBO) analysis[23] to measure the relative π-accepting abilities of different ligands was reported by Leyssens *et al.*[24] This scale sets the π-accepting ability of CO at 100; several phosphine and amine ligands have been evaluated. Electron-rich donor ligands feature increased σ-donation, whereas electron-deficient donor ligands increase the degree of π-backbonding (Table 2.5).

Table 2.5 π-Acceptor indices.

Entry	Ligand	π-Acceptor index
1	NMe_3	2
2	NH_3	4
3	Pyridine	15
4	PPh_3	16
5	PH_3	17
6	PMe_3	18
7	$P(OMe)_3$	22
8	NF_3	38
9	PF_3	38
10	PCl_3	48
11	CO	100

Adapted with permission from T. Leyssens, D. Peeters, A. G. Orpen and J. N. Harvey, *Organometallics*, 2007, **26**, 2637. Copyright 2007 American Chemical Society.

Experimental studies aimed at measuring the π-accepting ability of ligands have produced mixed results. Particularly, several studies using photoelectron spectroscopy (PES) indicated that alkylphosphines are poor π-acids;[25] however, more recent studies provide evidence that alkylphosphines can act as π-acceptors with early transition metals in low oxidation states (electron-rich metals), *e.g.*, $TiMe_2(dmpe)_2$ [dmpe = 1,2-bis(dimethylphosphino)ethane].[26]

2.2.3 Bite Angle Considerations (Chelating Ligands)

As mentioned previously, the seminal work of Kumada's group highlighted the beneficial effect of a bidentate ligand, dppf, in cross-coupling reactions of organobromides with secondary alkyl Grignard reagents relative to the PPh_3 system.[7] In fact, many catalytic applications utilize chelating ligands due to the increased stability, unique reactivity and selectivity that they provide.[27] The P–M–P angle created by the metal chelate is commonly referred to as the "bite angle" (Figure 2.6).

Casey and Whiteker introduced the concept of "natural bite angle" to quantify the strain imposed by chelating bisphosphine ligands.[30] In this model, molecular mechanics calculations are used to determine the preferred bite angle (P–M–P angle) imposed by ligand backbone constraints – not by metal valence effects. In this simulation, a dummy "metal" atom is used at a distance of 2.315 Å (M–P distances of similar complexes determined crystallographically), the purpose of which is to maintain bidentate coordination during the calculation. The P–M–P force constant is set to zero to exclude any contributions from the metal center (Figure 2.7).

Ligand bite angle effects can be classified as electronic or steric in nature. Electronic bite angle effects are described as electronic changes at the metal center imposed by the natural bite angle of the ligand. Dierkes and van Leeuwen introduced the concept of "metal-preferred bite angle", which they

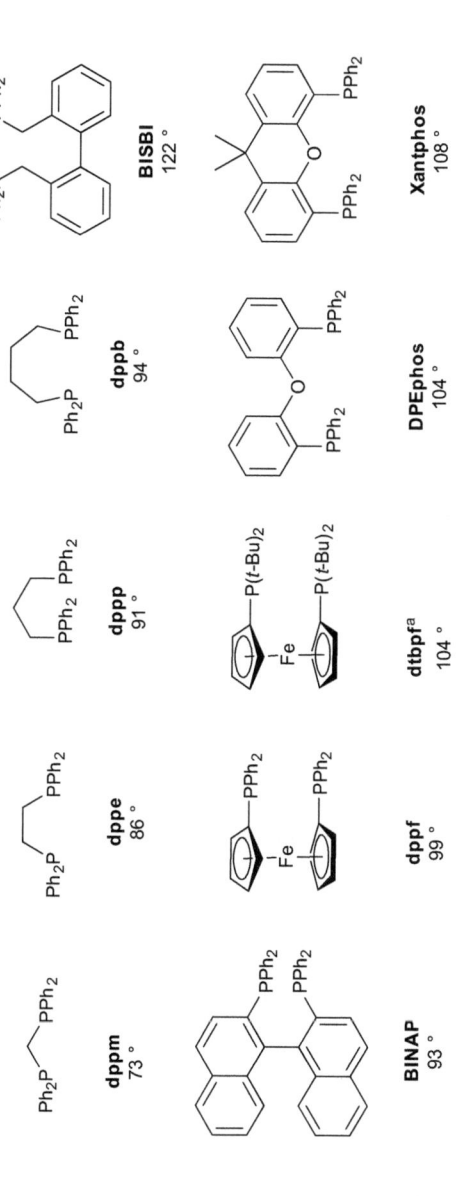

Figure 2.6 Common chelating ligand bite angles.[27a] Abbreviations: dppm = 1,1-bis(diphenylphosphino)methane; dppe = 1,2-bis(diphenyl-phosphino)ethane; dppp = 1,3-bis(diphenylphosphino)propane; dppb = 1,4-bis(diphenylphosphino)butane; BISBI = 2,2′-bis[(diphenylphosphino)methyl]-1,1′-biphenyl; BINAP = 2,2′-bis(diphenylphosphino)-1,1′-binaphthalene; dppf = 1,1′-bis(diphenyl-phosphino)ferrocene; dtbpf = 1,1′-bis(di-*tert*-butylphosphino)ferrocene; DPEPhos = (oxydi-2,1-phenylene)bis(diphenylphosphine); Xantphos = 4,5-bis(diphenylphosphino)-9,9-dimethylxanthene.
[a]Data for dtbpf taken from Refs 28 and 29.

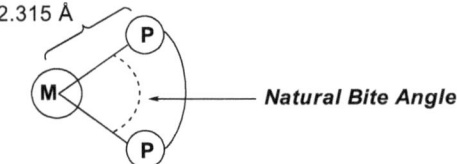

Figure 2.7   Natural bite angle.

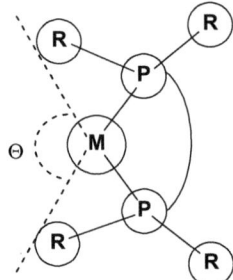

Figure 2.8   Definition of pocket angle.

defined as the lowest energy P–M–P angle of the metal complex when steric effects are not taken into consideration.[31] In a rough approximation, the metal-preferred bite angle for square-planar complexes is 90°, for tetrahedral complexes 109° and for trigonal bipyramidal complexes 120°. The more closely the ligand natural bite angle matches the metal-preferred bite angle, the better the ligand can *stabilize* the complex. Conversely, larger differences between them lead to increased *destabilization* and, hence, differences in reactivity. Steric *bite angle* effects are those which change the steric environment around the metal center by modification of the ligand backbone while the substituents on the donor phosphorus atoms are held constant. In this regard, the steric bite angle effect is independent of the steric effect imposed by the groups on phosphorus (or other donor atom). The concept of "pocket angle" (Θ) was introduced by Barron and co-workers as a method to quantify steric bite angle effects.[32] As shown in Figure 2.8, the pocket angle is the interior cone angle created by ligand at the metal center and provides a measure of the interior space of the metal complex (Table 2.6).

   Collectively, these ligand properties have served as the basis for the design and synthesis of new ligands which continue to drive new reactivity in homogeneous catalysis, especially in the area of Pd-catalyzed cross-coupling reactions.

2.3   Modern Ligands in Cross-Coupling Reactions

Several classes of prominent supporting ligands have emerged that give rise to highly active and selective catalysts that collectively make up the state of the art in modern cross-coupling reactions. The following sections provide

Table 2.6 Pocket angles (Θ).

Entry	Ligand	Θ (°)
1	dppe	132
2	dppp	108
3	dppb	93
4	dmpe	141
5	dcpe	106

Abbreviations: dmpe = 1,2-bis(dimethylphosphino)ethane; dcpe = 1,2-bis(dicyclohexylphosphino)ethane.
Adapted with permission from Y. Koide, S. G. Bott and A. R. Barron, *Organometallics*, 1996, **15**, 2213. Copyright 1996 American Chemical Society.

the details of the various types of ligands developed and utilized in various cross-coupling reactions.

2.3.1 Bulky Trialkylphosphines – The Origins of Modern Ligands for Cross-Coupling

Until the 1990s, studies on cross-coupling reactions were almost exclusively focused on the coupling of aryl iodides, aryl bromides or aryl triflates, despite the fact that aryl chlorides are often more attractive substrates owing to their lower cost and wider availability. Aryl chlorides, however, had been particularly problematic for existing catalyst systems, owing to the strong C–halogen bond (BDE[33] = 95 kcal mol^{-1}) relative to those of aryl bromides (BDE = 80 kcal mol^{-1}) and aryl iodides (BDE = 65 kcal mol^{-1}).[34] For this reason, their use in cross-coupling reactions had been largely unexplored with the exception of a few activated aryl and heteroaryl systems.[35] In 1989, the groups of Osborn[36] and Milstein[37] independently reported Pd-catalyzed carbonylation reactions of chlorobenzenes. Key to the success of these transformations was the use of either a Pd(PCy$_3$)$_2$Cl$_2$[36] (Cy = cyclohexyl) or a Pd(OAc)$_2$/dippb[37] [dippb = bis(diisopropylphosphino)butane] system. Both groups attributed this new reactivity to the electron richness and steric demand of these ligand systems. Osborn noted that "significant catalytic activity is found only with phosphines which are both strongly basic (pK_a >6.5) and with well-defined steric volume, that is, the cone angle must exceed 160°."[36] These findings began to shed light upon the importance of ligand properties, which set the stage for further research in this area, ultimately leading to the constant development of more active catalysts in the area of cross-coupling.

Koie and co-workers recognized the potential of bulky trialkylphosphines as supporting ligands in Buchwald–Hartwig amination reactions and reported that a Pd/P(t-Bu)$_3$ catalyst system was highly effective in the *N*-arylation of piperazine with primarily aryl and heteroaryl bromides[38] and in the formation of triarylamines from aryl halides and diarylamines (Figure 2.9).[39] Noted in both studies was the reduced efficiency when similar

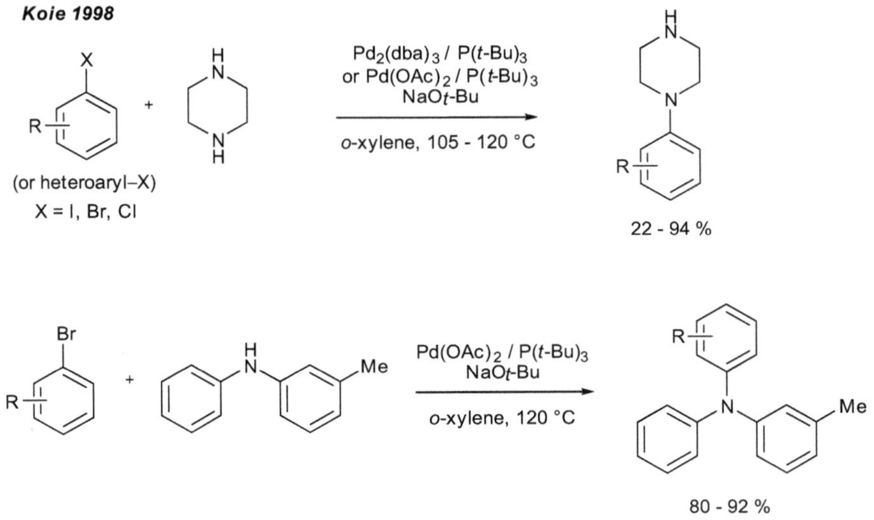

Koie 1998

Figure 2.9   Koie's amination reactions with Pd/P(*t*-Bu)$_3$.

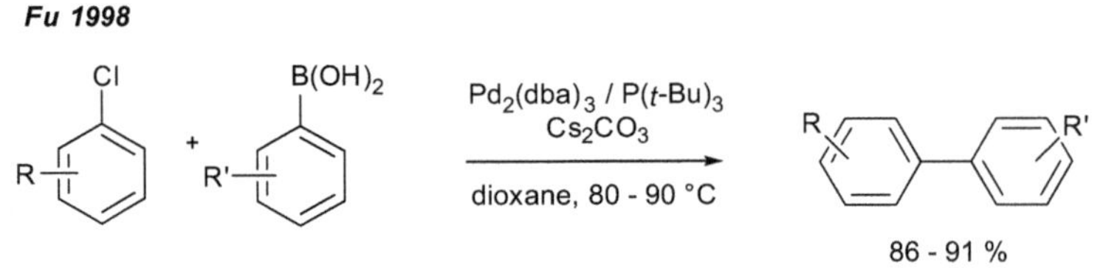

Fu 1998

Figure 2.10   Suzuki–Miyaura reactions of aryl chlorides using Pd/P(*t*-Bu)$_3$.

ligands such as P(mesityl)$_3$ or P(*n*-Bu)$_3$ were employed, highlighting the significance of the electron richness in concert with the steric bulk of the *tert*-butyl groups to make the ligand exceptionally unique.

Concurrently, Littke and Fu reported that both Pd/PCy$_3$ and Pd/P(*t*-Bu)$_3$ combinations provided highly active catalysts for Suzuki–Miyaura cross-coupling reactions, with the latter being marginally more active (Figure 2.10).[40] This breakthrough represented one of the first general SMC reactions with unactivated aryl chlorides. The authors found a Pd:ligand ratio of near 1:1 to be optimal, hence they suggested the active catalytic species to be a mono-ligated Pd species. Furthermore, they observed that no coupling took place when less electron-rich triarylphosphines were employed as ligands. This led to the assumption that "the steric bulk and the electron-richness of P(*t*-Bu)$_3$ are critical for this unprecedented reactivity."[40] This study sparked significant interest in the use of P(*t*-Bu)$_3$ (and also other highly electron-rich and sterically demanding phosphines) as a ligand in coupling reactions. Even today, Pd in combination with P(*t*-Bu)$_3$ remains the most widely used system in cross-coupling.[1k]

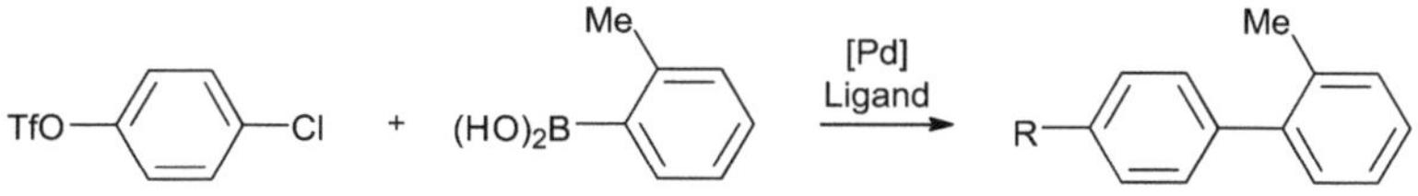

Ligand	R	Yield
P(*t*-Bu)$_3$	OTf	95%
PCy$_3$	Cl	87%

Figure 2.11 Ligand-controlled chemoselective Suzuki–Miyaura couplings with *p*-chlorophenyltriflate.

Fu's group further expanded the scope and versatility of this and related catalyst systems in SMC reactions.[41] Notably, they described conditions for the room temperature coupling of aryl bromides, activated and heterocyclic aryl chlorides and aryl triflates. Although P(*t*-Bu)$_3$ proved to be an efficient ligand for promoting SMC reactions of aryl chlorides and bromides, the analogous reactions utilizing aryl triflates showed diminished activity with this ligand. They speculated that the increased steric demand of triflates relative to bromides and chlorides may be the cause of the precipitous decrease in reactivity. They discovered, however, that with the use of the less bulky PCy$_3$ ligand, aryl triflates could be coupled in high yield at room temperature. Further expanding on this catalyst-controlled selectivity observation, they further elegantly demonstrated that a Pd catalyst system based on P(*t*-Bu)$_3$ could be used to couple an aryl chloride chemoselectively in the presence of an aryl triflate – an unexpected selectivity switch (Figure 2.11). With the use of a catalyst system based on PCy$_3$, the "conventional" pattern of reactivity was observed (ArOTf > ArCl).

A theoretical mechanistic investigation by Schoenebeck and Houk identified the origins of this observed ligand-controlled chemoselectivity: the monoligated phosphine complex Pd[P(*t*-Bu)$_3$] was the active catalyst in the selective oxidative addition to the aryl chloride, whereas the bis-ligated complex Pd(PCy$_3$)$_2$ was implicated in the chemoselective aryl triflate reactivity.[42] More recently, Proutiere and Schoenebeck demonstrated that Pd/P(*t*-Bu)$_3$ can engage in chemoselective Ar–OTf activation in polar solvents.[43] A "bis-ligated" anionic Pd species – [PdP(*t*-Bu)$_3$X]⁻ in which a salt anion is acting as a ligand – was proposed as the active catalytic species in polar solvent on the basis of experimental and DFT studies. Fu and co-workers have expanded their early work into several facets of cross-coupling, demonstrating that Pd/P(*t*-Bu)$_3$-based catalysts are similarly highly reactive towards aryl chlorides in Stille,[44] Negishi[45] and Heck–Mizoroki[46] reactions.

In an early study on ketone enolate α-arylation, Hartwig's group found that an efficient catalyst was generated using the bisphosphine ligand dtbpf (1,1'-*bis*(di-*tert*-butylphosphino)ferrocene).[47,48] Although the ligand in this study is a chelating bisphosphine (see Section 2.3.2), through ³¹P NMR studies of a transmetallation complex (dtbpf)Pd(Ar)enolate generated in a stoichiometric reaction, it was observed that only one of the phosphines was coordinated to

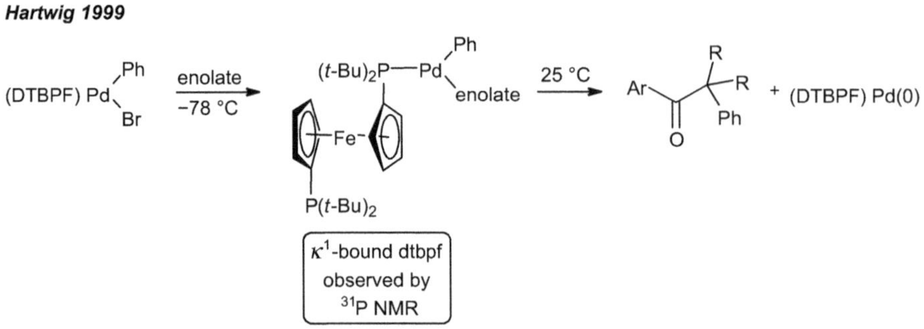

Hartwig 1999

Figure 2.12 Stoichiometric enolate arylation with Pd(dtbpf).

the Pd center, leading the authors to speculate that electron-rich, sterically demanding *mono*phosphines should be suitable for ketone α-arylation (Figure 2.12).

Indeed, they found that ketone α-arylation, and also the arylation of malonic esters, employing sterically demanding monophosphines [P(*t*-Bu)$_3$ and PCy$_3$] as supporting ligands, were highly efficient for coupling of aryl bromides and aryl chlorides and the Pd/P(*t*-Bu)$_3$ system was similarly effective in room temperature aryl amination reactions of aryl bromides, whereas some chlorides required higher temperatures (r.t.–70 °C).[49] They also identified hindered trialkylphosphines to be efficient at promoting Heck–Mizoroki reactions[50] and the arylation of ethyl cyanoacetate[51] through a series of fluorescence-based high-throughput screening experiments. This catalyst system was also used for the *N*-arylation of indoles and carbamates. In continuation, Hartwig's group also expanded the use of P(*t*-Bu)$_3$-based catalysts in α-arylation reactions of ester enolates using either Pd(dba)$_2$/P(*t*-Bu)$_3$ mixtures or a single-component pre-ligated Pd(I) dimer, {[P(*t*-Bu)$_3$]PdBr}$_2$[48,52] (Figure 2.13) (for a detailed discussion on preformed complexes as precatalysts, see Chapter 3).

This dimeric Pd(I) precatalyst, although air sensitive, has also been demonstrated by Hartwig and co-workers to achieve aryl amination and Suzuki–Miyaura couplings with aryl chlorides and bromides with exceptionally high rates.[53,54]

In 2000, Beller's group introduced another new type of bulky trialkylphosphine[55] for Suzuki–Miyaura coupling reactions: di-(1-adamantyl)-*n*-butylphosphine [Ad$_2$P(*n*-Bu)].[56] The catalyst based on this ligand was reported to be active for the coupling of chloroarenes with arylboronic acids with high turnover numbers. Electron-rich, electron-deficient and heterocyclic chloroarenes were all successfully coupled (Figure 2.14).

An entire class of this family of ligands has since been developed and evaluated by the same group (Figure 2.15, top).[57] The synthesis of these diadamantylphosphines involves treating di(1-adamantyl)phosphine with an alkyl iodide or bromide to give a quaternary phosphonium salt, which is then liberated by NEt$_3$ to give the corresponding free ligand (Figure 2.15, bottom).

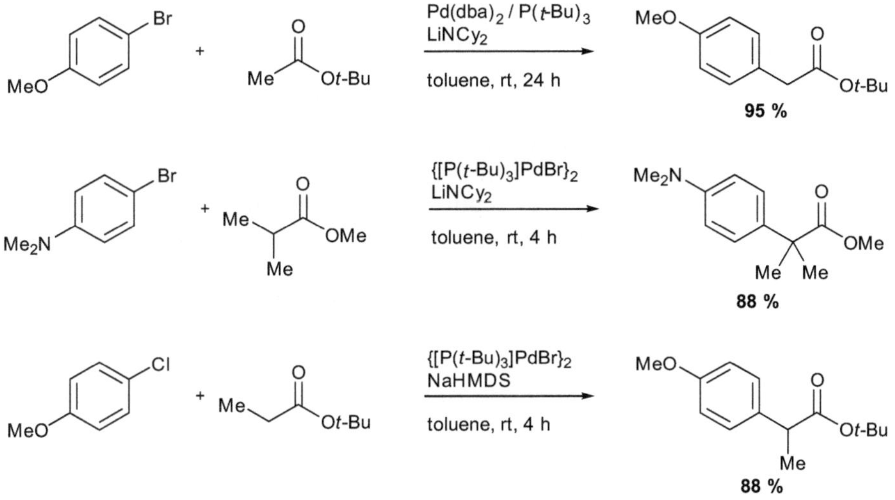

Figure 2.13 Ester enolate arylation using P(t-Bu)₃ complexes.

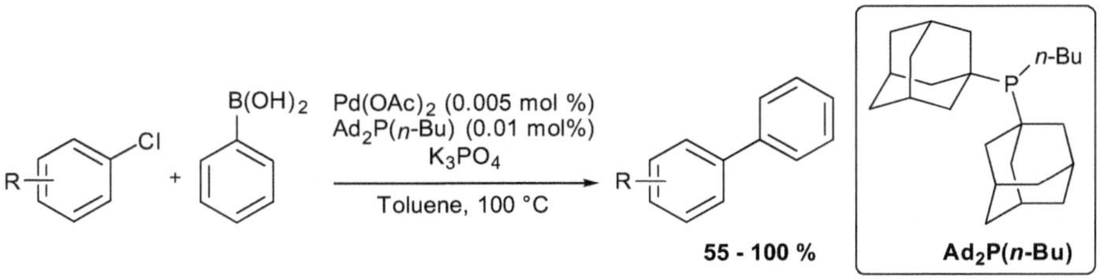

Figure 2.14 Suzuki–Miyaura reactions using Ad₂P(n-Bu).

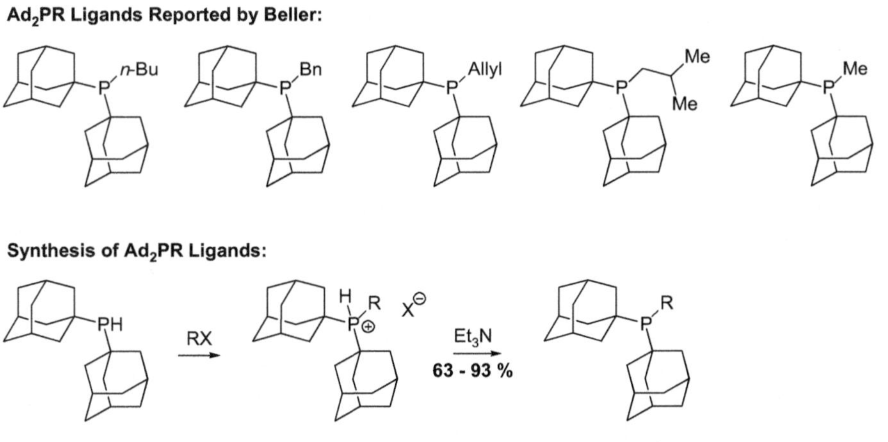

Ad₂PR Ligands Reported by Beller:

Synthesis of Ad₂PR Ligands:

Figure 2.15 Ad₂PR ligand family (top) and synthesis (bottom).

Catalysts based on diadamantylalkylphosphines have also been demonstrated to be useful in several related cross-coupling reactions. For example, Molander and Gormisky used $Ad_2P(n\text{-}Bu)$ in SMC reactions of aryl and heteroaryl chlorides with cyclopropyl- and cyclobutyltrifluoroborate potassium salts.[58] Buchwald–Hartwig aryl amination reactions are successful with a broad scope of electronically and sterically varied aryl chlorides with $Ad_2P(n\text{-}Bu)$/Pd systems as reported by Beller and co-workers.[59] They additionally showed that the arylation of ketone enolates using the Pd/$Ad_2P(n\text{-}Bu)$ system is successful with a usefully broad substrate scope,[60] while Hartwig's group utilized the related ligand $Ad_2P(t\text{-}Bu)$ in α-arylation reactions of aryl bromides with aza-γ-lactones towards the synthesis of quaternary amino acid derivatives.[61]

Plenio and co-workers were the first to develop general conditions for Heck alkynylation reactions (sometimes referred to as the Cu-free Sonogashira reaction) of aryl chlorides using Ad_2PBn,[62] although improved systems have been developed subsequently.[63] Reductive carbonylation (formylation) reactions of aryl bromides with Pd/$Ad_2P(n\text{-}Bu)$ was also reported by Beller's group.[64–66] with some useful mechanistic studies.[66] Specifically, through a detailed mechanistic study, Beller and co-workers proposed that the cluster complex $(CO)_mPd_nL_n$ and the hydrobromide complex $PdL_2(H)(Br)$ act as "reservoirs" of Pd and that the highly active monoligated PdL complex is slowly released during the course of the reaction. The low level of catalyst concentration maintained throughout the catalytic cycle allows oxidative addition to outcompete palladium(0) aggregation with eventual formation of palladium black, a process which is second or higher order in palladium[67] (Figure 2.16).

In 2006, Shaughnessy and co-workers introduced a new type of trialkylphosphine for aryl amination reactions of aryl bromides and chlorides: neopentyl-substituted phosphines (Figure 2.17).[12b] These ligands had been known in the context of coordination chemistry since the 1970s,[68] but their use in catalysis was essentially non-existent.[69] A series of neopentyl-based ligands, such as (di-*tert*-butyl)neopentylphosphine (DTBNpP), *tert*-butyldineopentylphosphine (TBDNpP) and trineopentylphosphine (TNpP) have been studied, with the motivation being their larger cone angles reflecting the increased steric demand of the neopentyl-substituted phosphines.

Cone angles calculated from DFT-optimized structures revealed that DTBNpP ($\theta = 198°$) is more sterically encumbered than $P(t\text{-}Bu)_3$ ($\theta = 194°$). Substitution of *tert*-butyl groups for additional neopentyl groups gives rise to even more sterically demanding ligands (TBDNpP, $\theta = 210°$; TNpP, $\theta = 227°$). This ligand series was further evaluated for electron donation by measuring the CO stretching frequencies of the $ClRh(PR_3)_2(CO)$ complexes; not unexpectedly, DTBNpP is less electron rich than $P(t\text{-}Bu)_3$ as its CO stretch at 1946 cm^{-1} is 18 cm^{-1} higher than that of the $P(t\text{-}Bu)_3$ complex. Additional substitution of neopentyl groups for *t*-Bu groups further increases the CO stretch, but to a lesser extent with each substitution (TBDNpP, $\nu = 1953$ cm^{-1}; TNpP, $\nu = 1957$ cm^{-1}). The neopentylphosphines were compared along with $P(t\text{-}Bu)_3$ in the aminations of deactivated

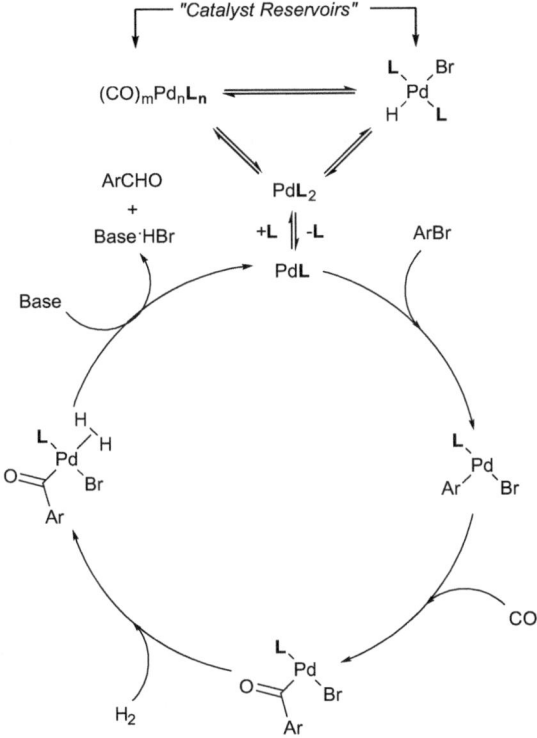

Figure 2.16 Mechanistic rationale for catalyst stability.
Adapted with permission from A. G. Sergeev, A. Spannenberg and M. Beller, *J. Am. Chem. Soc.*, 2008, **130**, 15549. Copyright 2008 American Chemical Society.

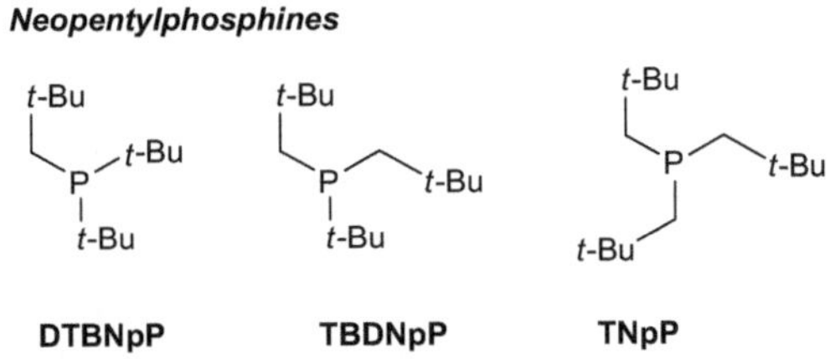

Figure 2.17 Neopentylphosphines studied by Shaughnessy and co-workers.

4-bromoanisole with aniline and morpholine, and DTBNpP outperformed the other ligands, including P(*t*-Bu)₃ (Figure 2.18).

The Pd/DTBNpP catalyst system was demonstrated to be effective for a broad scope of amination reactions of aryl bromides with primary and secondary amines under mild conditions (r.t.–50 °C) with catalyst loadings of 0.5–1.0 mol%. Aryl chlorides could also be successfully coupled, although higher temperatures (50–140 °C) with increased catalyst loadings were required (0.5–5 mol%) for deactivated substrates such as 4-chloroanisole.

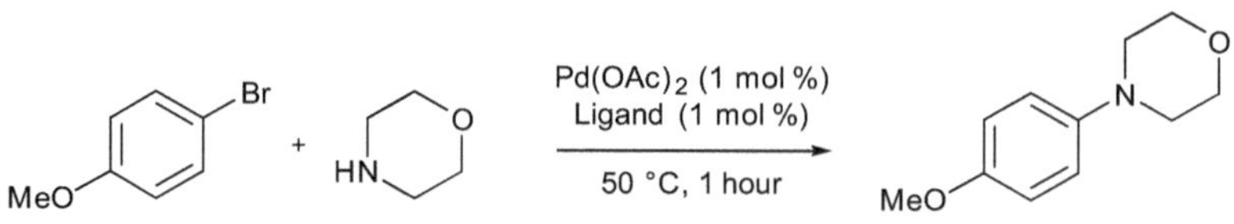

Ligand	Conversion
P(t-Bu)$_3$	36 %
DTBNpP	>99 %
TBDNpP	87 %
TNpP	0 %

Figure 2.18 Comparison of P(t-Bu)$_3$ and neopentylphosphines in the amination of 4-bromoanisole.

The forcing conditions required to convert aryl chlorides were attributed to the decreased electron donation of neopentylphosphines relative to P(t-Bu)$_3$. With aryl bromides, oxidative addition is more facile and the increased steric demand gives rise to a highly active catalyst under mild conditions. Owing to the high activity observed with DTBNpP relative to TBDNpP and TNpP and previously reported data on Buchwald–Hartwig aminations using less sterically demanding ligands (*i.e.*, PCy$_3$) with cone angles <190°, Shaughnessy and co-workers concluded that the optimal cone angle for the reactions under the conditions studied existed between 190 and 200°. Shaughnessy and co-workers also described Pd/DTBNpP to be highly active in Suzuki–Miyaura, Sonogashira–Heck and Heck–Mizoroki reactions of aryl bromides and chlorides.[70] They further showed that Pd complexes of DTBNpP are highly active in Buchwald–Hartwig aminations and arylation of propiophenone enolate.[71] Subsequent work by Colacot's group revealed that the preformed (π-allyl)Pd(DTBNpP)Cl is more efficient than the *in situ* system.[72] The details of the study are discussed in Chapter 3. Recently, Shaughnessy and co-workers reported that TNpP is a useful ligand in Buchwald–Hartwig amination reactions of substrates with high steric demand and also in Suzuki–Miyaura coupling reactions with moderately hindered reaction partners (Figure 2.19).[73] It was proposed the conformational flexibility of this ligand is crucial to its ability to accommodate hindered substrates.

2.3.2 Chelating Bisphosphine Ligands

As mentioned earlier, the beneficial effects of the chelating bisphosphine ligand dppf [1,1′-bis(diphenylphosphino)ferrocene] was originally observed by Kumada's group in 1979 while studying the cross-coupling of aryl bromides with β-branched alkyl Grignard reagents.[7] The more facile

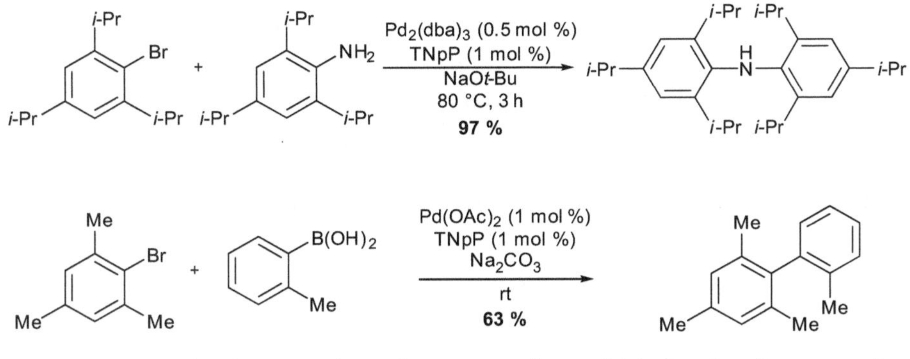

Figure 2.19 Buchwald–Hartwig and SMC reactions of hindered substrates using trineopentylphosphine.

reductive elimination induced with the use of dppf allowed for selective cross-coupling to proceed over other pathways occurring *via* β-hydride elimination. Kumada and co-workers established a correlation between increasing P–Pd–P angles (bite angles) with increasing catalyst activity while studying Kumada–Tamao–Corriu and Negishi reactions of alkyl nucleophiles.[74] (dppf)PdCl$_2$ (bite angle 99.1°) significantly outperformed (dppp)PdCl$_2$ (bite angle 90.6°) and (dppe)PdCl$_2$ (bite angle 85.8°). Driver and Hartwig also reported the beneficial effects of bidentate (dppf)PdCl$_2$ in aryl amination reactions involving primary alkylamines and rationalized that the high selectivity stemmed from chelation and bite angle effects.[75] The intermediate four-coordinate (dppf)Pd(Ar)(amido) complexes are devoid of a vacant coordination site, thus shutting down the β-hydride elimination pathway.[76] A review by van Leeuwen *et al.* further covered the significance of the ligand bite angle.[77] Indeed, it has now been well established that the use of bidentate ligands in Pd-catalyzed cross-coupling reactions has an accelerating effect on reductive elimination.

Building on these original findings, countless studies emerged using dppf as a supporting ligand. Even today, dppf remains as a useful ligand for Pd-catalyzed cross-coupling reactions in the fine chemical and pharmaceutical industries. Johnson Matthey was the first company to commercialize dppf and also (dppf)PdCl$_2$ and provide these compounds on an industrial scale.[78] The history of the use of dppf in homogeneous catalysis applications has been summarized in other work.[79] The aim of this section is to provide a review of the modern prominent bidentate phosphine ligands in Pd-catalyzed processes. Ferrocene provides an excellent scaffold for ligand synthesis owing to the relative ease with which it can be functionalized. The methods of bis-lithiation with 2 equiv. of *n*-BuLi in the presence of TMEDA (TMEDA = tetramethylethylenediamine) followed by electrophile capture with ClPR$_2$ or, alternatively, treatment of 1,1′-bis(dichlorophosphino)ferrocene with organolithium or Grignard reagents have been used to generate a substantial library of 1,1′-bis(phosphino)ferrocenes (Figure 2.20).[80]

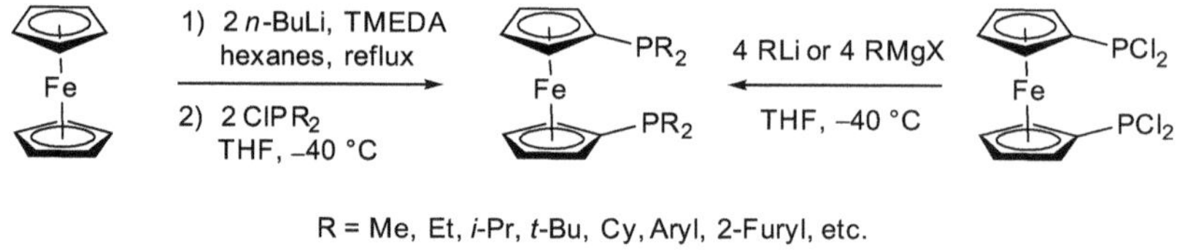

R = Me, Et, *i*-Pr, *t*-Bu, Cy, Aryl, 2-Furyl, etc.

Figure 2.20   Synthesis of symmetrical 1,1′-bis(phosphino)ferrocenes.

Table 2.7   Superior activity of (dtbpf)PdCl$_2$ in Suzuki–Miyaura coupling.

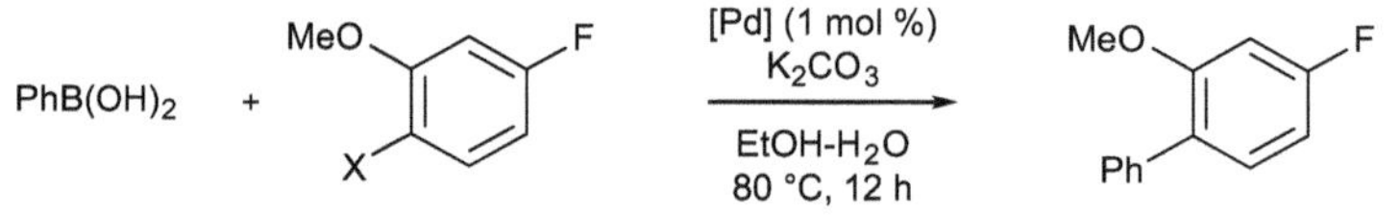

Entry	Catalyst	X	Yield (%)
1	(PPh$_3$)$_2$PdCl$_2$	Br	93
2	(PCy$_3$)$_2$PdCl$_2$	Br	100
3	(dppe)PdCl$_2$	Br	100
4	(dppf)PdCl$_2$	Br	100
5	(dippf)PdCl$_2$	Br	100
6	(PPh$_3$)$_2$PdCl$_2$	Cl	2
7	(PCy$_3$)$_2$PdCl$_2$	Cl	2
8	(dppe)$_2$PdCl$_2$	Cl	5
9	(dppf)PdCl$_2$	Cl	4
10	(dippf)PdCl$_2$	Cl	9
11	**(dtbpf)PdCl$_2$**	**Cl**	**65**
12^a	**(dtbpf)PdCl$_2$**	**Cl**	**100**

^{a}Run in DMF at 120 °C. dppe = bis(diphenylphosphino)ethane.
Adapted with permission from T. J. Colacot and H. A. Shea, *Org. Lett.*, 2004, **6**, 3731. Copyright 2008 American Chemical Society.

In 2004, Colacot and Shea demonstrated for the first time that 1,1′-bis(di-*tert*-butylphosphino)ferrocenepalladium dichloride [(dtbpf)PdCl$_2$][48] is very active in Suzuki–Miyaura reactions utilizing unactivated aryl bromides and chlorides.[81] Importantly, this air-stable complex seemed to be uniquely reactive compared with similar catalysts including 1,1′-bis(diisopropylphosphino)ferrocenepalladium [(dippf)PdCl$_2$], indicating that the balance of the electronic and steric nature of the ligand in combination with the large bite angle (104.2°)[29] is integral to the catalyst activity (Table 2.7). By comparison, the bite angle of (dippf)PdCl$_2$ (103.6°)[80d] nearly matches that of the *tert*-butyl analog; however, the isopropyl groups are both less electron-donating and sterically demanding.

The superior activity of (dtbpf)PdCl$_2$ relative to other similar catalysts was also observed by Itoh *et al.* in Suzuki–Miyaura reactions to generate 2-aryl-purines with unprotected NH$_2$ groups.[82]

As mentioned above, Hartwig and co-workers identified dtbpf as a competent ligand for α-arylation reactions.[51] Regardless of the mode of ligand

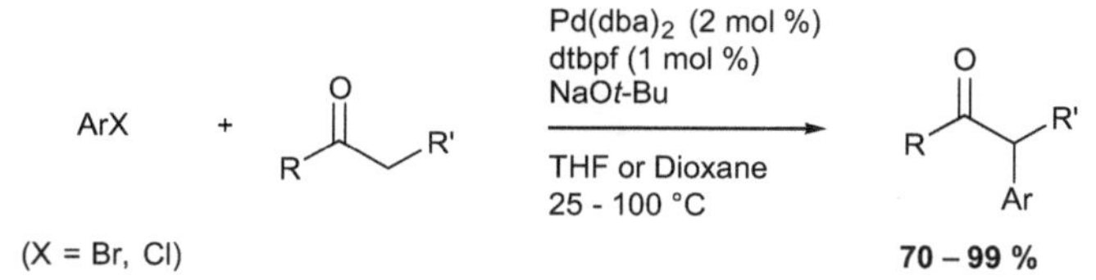

Figure 2.21   α-Arylation of ketones using Pd/dtbpf.

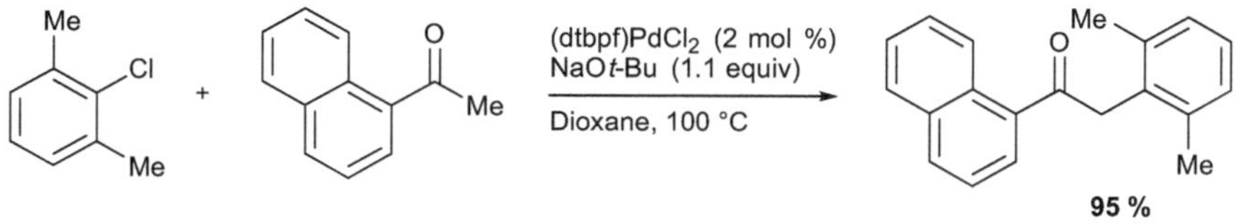

Figure 2.22   Arylation of ketone enolates using preformed (dtbpf)PdCl$_2$.

binding (κ^1 *versus* κ^2), an array of ketone enolates and malonates were monoarylated with bromo- and chloroarenes (Figure 2.21).[47] It was hypothesized that the steric bulk of the ligand promotes the formation of the active low-coordinate Pd(0) species and that the alkyl groups are more resistant to migration relative to aryl groups, thus leading to the observed higher turnover numbers.

Grasa and Colacot later demonstrated that the preformed (dtbpf)PdCl$_2$[48,83] complex was successful in the α-arylation of ketone enolates.[29,84] Using this precatalyst, α-arylation reactions of various ketones were achieved with a broad range of aryl bromides and chlorides (Figure 2.22).

Lu and co-workers developed a modification of the Larock indole synthesis utilizing o-bromo- and o-chloroanilines and internal alkynes (Figure 2.23).[85] Several ligands, including PCy$_3$, P(t-Bu)$_3$, dppf, and biaryldialkylphosphines (Buchwald ligands), were evaluated in this process and dtbpf was optimal in terms of both yield and regioselectivity as 2,3-disubstituted indoles were formed in 60–97% yield.

Hamman and Hartwig examined the effects of systematic variation of bidentate ligands in aryl amination (Buchwald–Hartwig) reactions, with particular attention to dppf analogs.[86] Interestingly, they observed that electron-rich, bulky ligands with *small* bite angles ($\sim 90°$) gave the best selectivities over hydrodehalogenation of the aryl halides. Additionally, they found a similar trend with respect to selectivity for monoarylation of primary amines; bulky, electron-rich ligands promoted monoarylation. A subsequent study identified dtbpf as a highly active ligand in amination reactions of aryl halides and the first amination of an aryl tosylate.[87] Also described in this paper was the use of Solvias's Josiphos-based chelating ligands in aryl amination reactions, a family of ligands initially developed for asymmetric reduction reactions (Figure 2.24).[88]

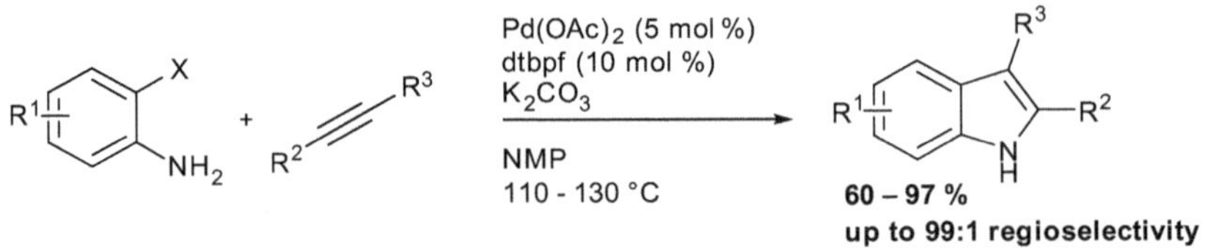

Figure 2.23   Regioselective Pd-catalyzed indolization.

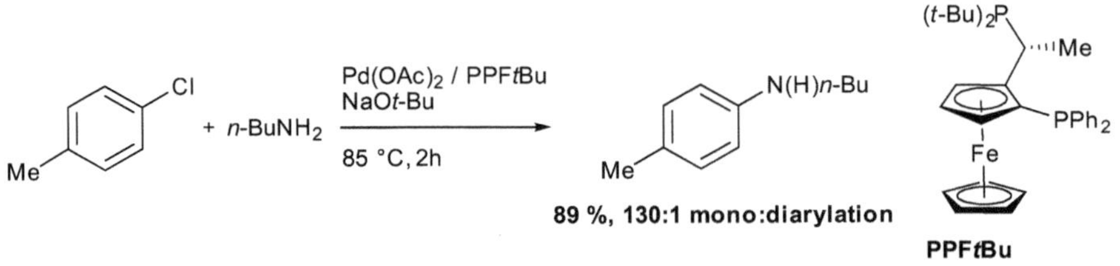

Figure 2.24   Aryl amination using Josiphos PPF*t*Bu.

Hartwig later revisited this class of ligands with the aim of identifying a catalyst that is both highly active and selective for arylation of primary amines and thiols, including the use of heteroaryl halides.[89] The air stability and chelating ability of CyPF*t*Bu, which contains a dicyclohexylphosphino group in addition to a di-*tert*-butylphosphino group, was attractive for this purpose as it has two electron-donating PR_2 groups with considerable steric bulk, although the ligand synthesis is not straightforward.[90] Furthermore, it was surmised that the chelating ability (six-membered chelate) of this ligand should give rise to a more stable catalyst that is robust towards ligand displacement by amines or heterocycles.[91] Studies on amination reactions demonstrated this to be true. In the chloropyridine series, the CyPF*t*Bu/Pd catalyst system was shown to be effective for the coupling of octylamine with yields ranging from 83 to 93% and selectivity for monoarylation with exceptionally low catalyst loadings (0.001–0.01% Pd), as shown in Figure 2.25.[92] The high selectivity for monoarylation, however, makes the arylation of secondary amines challenging with this ligand class.[92b]

This system was also successfully applied to a variety of nitrogen-containing heterocyclic aryl chloride derivatives and chloroarenes utilizing primary alkylamines, benzophenone imine and benzophenone hydrazone as nucleophiles. Aryl bromides underwent effective amination with primary alkylamines with catalyst loadings as low as 0.0005% (5 ppm) Pd. More interestingly, the Pd/CyPF*t*Bu system was found to be an efficient catalyst for the amination of aryl/heteroaryl iodides, substrates which are typically more challenging in aryl amination reactions.[93] A variety of aryl and heteroaryl iodides underwent amination with primary amines, although high catalyst loadings (up to 0.5 mol% Pd) were required in comparison with the analogous reactions involving aryl bromides and chlorides. Shen and Hartwig also utilized the CyPF*t*Bu-based catalyst to develop the first aryl amination

Figure 2.25 Aryl aminations of chloropyridines using Josiphos CyPFtBu.

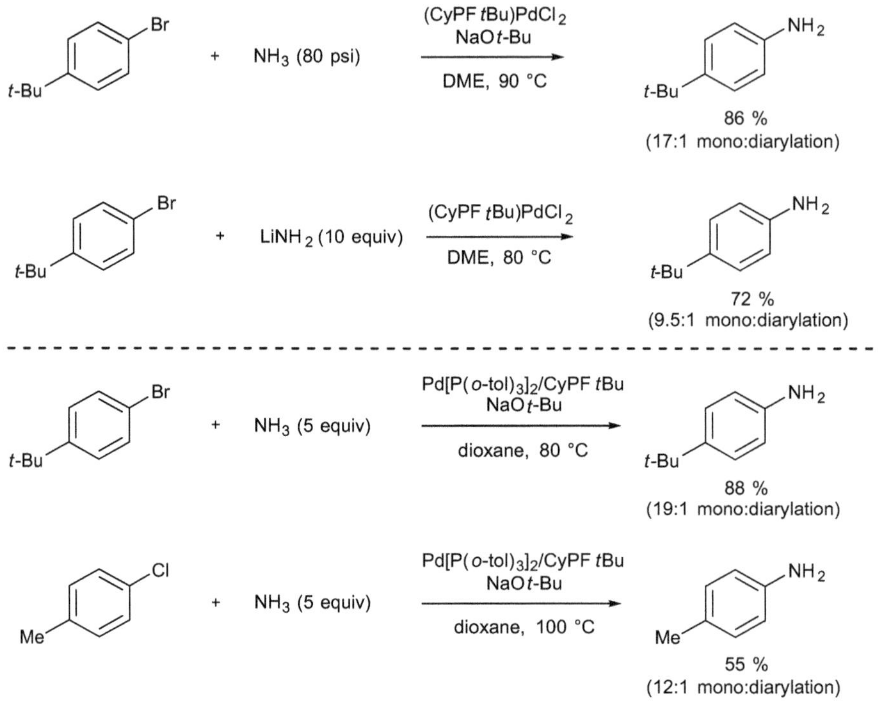

Figure 2.26 Cross-coupling with NH_3 and $LiNH_2$ using Josiphos CyPF*t*Bu (top) and improved ammonia arylation using Pd[P(*o*-tol)$_3$]$_2$/CyPF*t*Bu (bottom).

reactions using ammonia or lithium amide to generate primary anilines (Figure 2.26, top).[94] This reaction required a large excess of ammonia (80 psi) or $LiNH_2$ (10 equiv.) to achieve high selectivity for monoarylation. Aryl chlorides, bromides and sulfonates and bromoisoquinolines were converted to the corresponding anilines in high yield (69–94%) with selectivities ranging from 17:1 to >50:1 (monoarylation:diarylation) using ammonia; slightly lower selectivities were observed when lithium amide was employed (8:1 was the lowest selectivity reported). A subsequent improvement to this method was reported in 2009 to expand the substrate scope in addition to obviate the high pressure requirement (Figure 2.26, bottom).[95] The highest yields and selectivities were observed using a 1:1 ratio of CyPF*t*Bu ligand to Pd[P(*o*-tol)$_3$]$_2$[48,96] as a palladium source, whereas lower conversions and/or decreased selectivities for monoarylation were reported using other catalyst precursors such as Pd(OAc)$_2$, Pd$_2$(dba)$_3$ and (CyPF*t*Bu)PdCl$_2$.

Another reaction that Hartwig and co-workers investigated using CyPF*t*Bu as a supporting ligand was the coupling of aryl halides with thiols to generate aryl sulfides.[97] The strong electron-donating and chelating ability of the ligand was rationalized to generate a catalyst that is not only highly active, but also resistant to thiol poisoning. Aryl chlorides and sulfonates were successfully coupled with a range of alkyl and aryl sulfides at very low

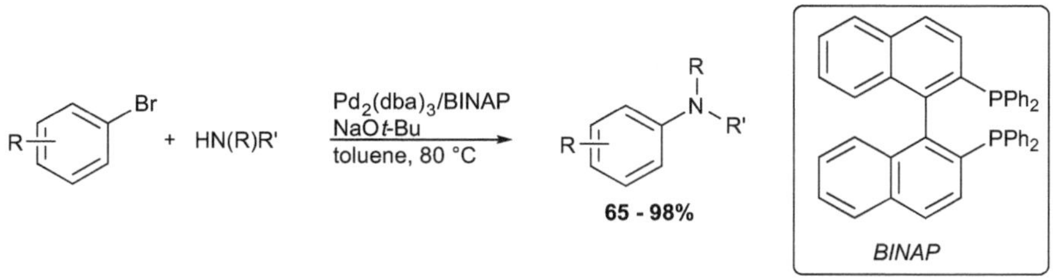

Figure 2.27   Amination of aryl bromides using Pd/BINAP.

catalyst loadings (many examples at 0.05 mol% Pd), including the use of triisopropylsilanethiol [TIPS-SH] as an H_2S equivalent. This methodology was then later extended to include aryl bromides and iodides.[98]

In a 1996 report, Buchwald and co-workers described the use of both racemic and non-racemic BINAP [2,2'-bis(diphenylphosphino)-1,1'-binaphthyl][99] in successful C–N cross-coupling reactions of aryl bromides with primary and secondary amines (Figure 2.27).[100] Catalyst loadings as low as 0.05% Pd could be employed in some cases. It was also demonstrated that this catalyst combination was more active and selective for monoarylation of a primary amine than other chelating bisphosphine ligands such as dppe and dppf.

A more detailed study of Pd/BINAP-catalyzed amination of aryl bromides followed a few years later,[101] and it was proposed that the smaller bite angle of BINAP (92.7°)[102] in comparison with dppf (99.1°)[74] coupled with the rigid nature of the ligand backbone provides a tight chelate, which disfavors the ligand from adopting a monodentate coordination mode and accessing the β-hydride elimination pathway. The high selectivity for monoarylation of primary amines was attributed to the steric demand of BINAP. Buchwald and co-workers also reported the first amination of a heteroaryl chloride using the Pd/BINAP system. Similarly, the same catalyst system has also been successfully used for amination reactions of aryl triflates,[103] iodides[104] and chloropyridines[91b] by the same group, and benzophenone imine,[105] benzophenone hydrazone[106] and α-chiral amines[107] have been used as nitrogen nucleophiles. Importantly, for intermolecular *N*-arylation of optically active α-chiral amines, the use of (±)-BINAP as a supporting ligand was essential in providing perfect stereochemical fidelity; analogous reactions using other ligands such as P(*o*-tol)$_3$ led to the erosion of optical purity. The loss of enantiopurity with monodentate phosphines was determined to be due to an off-cycle β-hydride elimination/reinsertion pathway, a pathway which was shut down by the tight chelating ability of the BINAP (Figure 2.28).[107]

BINAP has also been shown to be a very useful ligand for many other C–N cross-coupling reactions[108] and also others including C–O cross-coupling[109] and α-arylation reactions of amides,[110] ketone enolates[111] and aldehydes.[112] However, a major limitation with the use of BINAP is the general inability to utilize aryl chloride substrates owing to the more difficult oxidative addition.

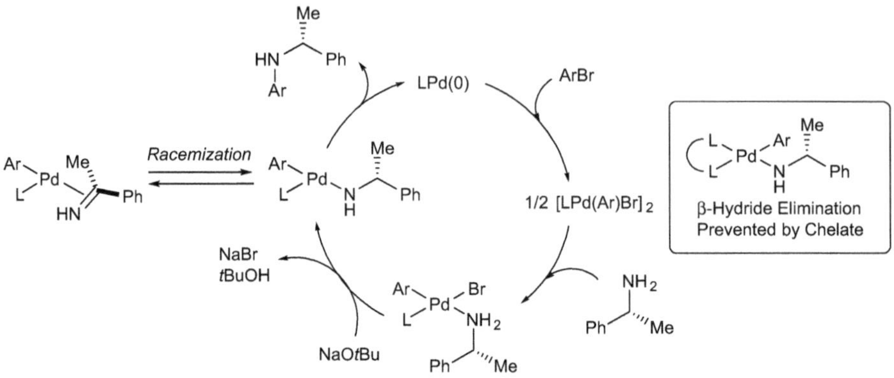

Figure 2.28   Mechanistic rationale for the racemization of α-chiral amines in Pd-
catalyzed amination reactions.
Adapted from Ref. 107.

The mechanism of aryl amination catalyzed by Pd/BINAP has been a
subject of intense discussion. On the basis of kinetic studies, in 2000
Hartwig and co-workers proposed a mechanism in which Pd(BINAP)$_2$ in the
catalytic cycle underwent a ligand dissociation to generate Pd(BINAP) as the
turnover-limiting step and ligand reassociation was required to release
the product (Figure 2.29, left),[113] This was largely based on the fact that the
amination reaction appeared to be zeroth order in [ArBr]. Subsequently, in
2002 Buchwald and co-workers proposed a different mechanism based on
kinetic studies of the catalytic amination of bromobenzene with both
primary and secondary amines using reaction calorimetry.[114] A positive
order in both [ArBr] and [amine] was observed, but only after an induction
period. This was attributed to slow catalyst activation. These data were used
to rationalize the apparent zeroth-order rate dependence on [ArBr] observed
by Hartwig and co-workers.[113] It was further proposed that oxidative addition of the aryl bromide occurs more rapidly to the amine-bound Pd–BINAP
complex than to Pd(BINAP) based on kinetic modeling (Figure 2.29, right).
    However, in 2006 Hartwig's group studied the rate of the stoichiometric
oxidative addition of bromobenzene to Pd(BINAP)$_2$ in the presence and absence of various concentrations of primary and secondary amines.[115] It was
determined that oxidative addition with (amine)Pd(BINAP) is not faster
than that with Pd(BINAP), and even in the presence of an amine oxidative
addition still occurs with Pd(BINAP). The mechanism was subsequently
reexamined and revised in 2006 by Hartwig, Buchwald, Blackmond and
co-workers.[116] New kinetic data were obtained which revealed that the
catalytic rate of the reactions is zeroth order in [amine], first order in [aryl
bromide] and inverse first order in [ligand]. A new mechanistic proposal was
set forth based on these data: oxidative addition occurs to Pd(BINAP), amine
and base react with the oxidative addition complex to generate the corresponding amido complex and reductive elimination releases the arylamine.

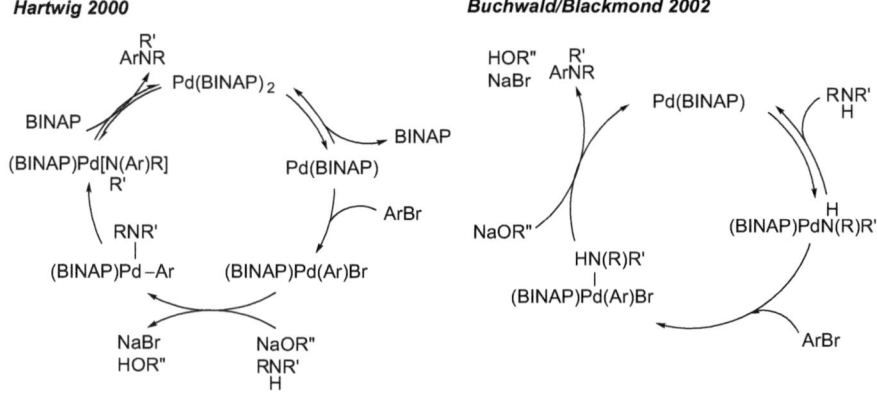

Figure 2.29 Original mechanisms for aryl amination catalyzed by Pd/BINAP proposed by the groups of Hartwig (left) and Buchwald (right).

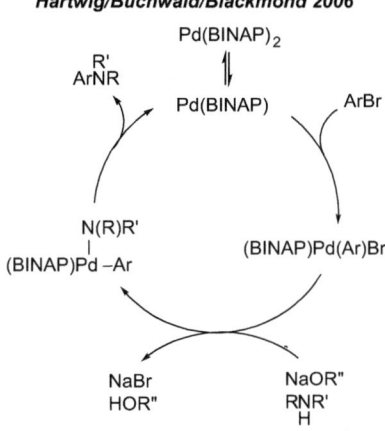

Figure 2.30 Revised mechanism of Pd/BINAP-catalyzed aryl amination.

Pd(BINAP) is generated by ligand dissociation of Pd(BINAP)$_2$, which lies off the catalytic cycle (Figure 2.30).

van Leeuwen and co-workers introduced the use of the chelating ligand 9,9-dimethyl-4,6-bis(diphenylphosphino)xanthene (Xantphos) for the Pd-catalyzed amination of aryl bromides (Figure 2.31).[117] Xantphos, originally developed by van Leeuwen's group for Rh-catalyzed hydroformylation processes,[118] features a very large bite angle[119] (108.8° for the (Xantphos)Pd(tetracyanoethylene) complex and 111.7° calculated natural bite angle). This ligand, along with a family of similar chelating bisphosphines, was easily synthesized from 9,9-dimethylxanthene *via* selective bis-lithiation with *sec*-butyllithium/TMEDA and then reacted with diphenylchlorophosphine to give the ligand in 75% yield.

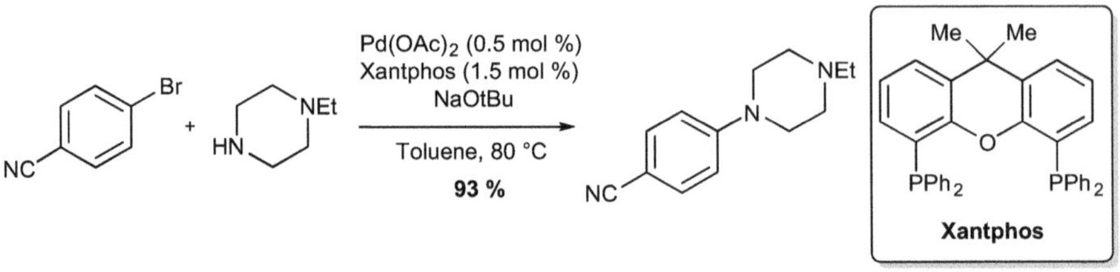

Figure 2.31 Arylation of *N*-ethylpiperazine using Pd/Xantphos.

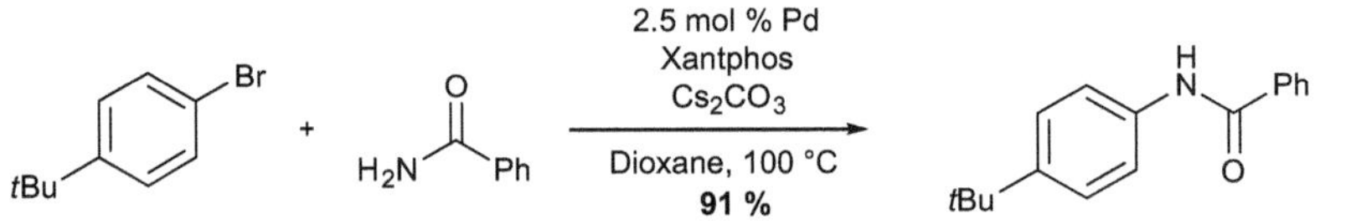

Figure 2.32 Aryl amidation catalyzed by Pd/Xantphos.

In 2000, Yin and Buchwald found that the Pd/Xantphos combination was generally effective in its ability to catalyze amidation reactions of activated and electronically neutral aryl halides/triflates with reasonably broad scope (Figure 2.32).[120]

In 2002, they undertook a detailed study of this transformation with the aim of addressing some of the limitations encountered in their previous studies.[121] For instance, electron-rich or sterically encumbered aryl halides were not productive substrates, as also were not sulfonamides or secondary acyclic amides. Interestingly, it was observed that in the reactions of these more challenging substrates, aryl group exchange[86,122] became significant and more rapid catalyst decomposition was observed when the catalyst loading was increased. Ultimately, they determined that both catalyst loading and reaction concentration must be carefully controlled for the success of reactions involving challenging substrates; Pd loadings of 1–4 mol% and concentrations ranging from 0.125 to 0.5 M were required compared with 1–2 mol% Pd and 1 M concentration for most of the less challenging examples. During their mechanistic studies on this process, a Pd–Xantphos oxidative addition complex was prepared by stirring 4-bromobenzonitrile, Pd$_2$dba$_3$ and Xantphos in benzene at room temperature (Figure 2.33). Unexpectedly, the ^{31}P NMR spectrum of this complex contained only one singlet at 9.3 ppm, compared with the pair of doublets observed for the analogous *cis*-chelated (BINAP)Pd(4-cyanophenyl)Br oxidative addition complex, providing evidence for a *trans*-chelated complex. X-ray crystallography was used to confirm the rare *trans*-chelation of the bisphosphine.[123]

The most striking structural feature of this complex is its extremely large bite angle (150.7°), which is significantly larger than the bite angle of *cis*-(Xantphos)Pd(tetracyanoethylene) (104.6°),[119] but closer to that of *trans*-(Xantphos)Pd(Me)Cl (153°) measured by van Leeuwen and co-workers.[123]

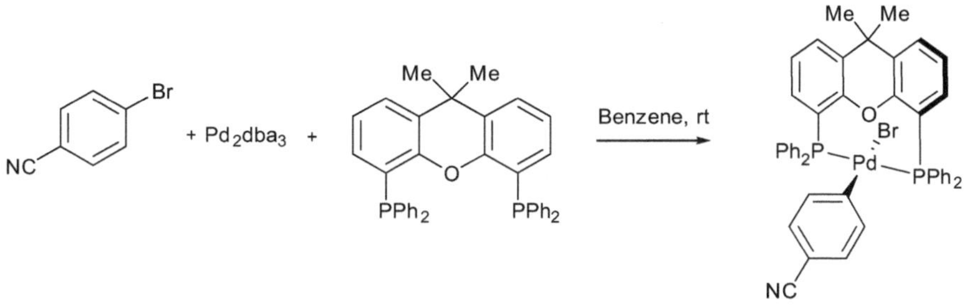

trans-(Xantphos)Pd(Ar)Br

Figure 2.33 Synthesis of *trans*-(Xantphos)Pd(Ar)Br.

Related observations of *trans*-chelating bisphosphines in *cationic* Pd complexes have been made by Sato *et al.*[124] and Hartwig and co-workers[28] with (dppf)Pd(PPh$_3$)(BF$_4$)$_2$ and (dtbpf)Pd(Ar)(BF$_4$), respectively, both of which feature a Pd–Fe interaction. It was proposed that a weak interaction between Pd and the oxygen atom (2.697 Å) could be a contributing factor to the extremely large bite angle.[119] The *trans*-(Xantphos)Pd(Ar)Br complex also proved to be catalytically competent in aryl amidation.[121] Like BINAP, the use of Xantphos as a ligand has widespread applications in many C–N bond-forming processes such as the synthesis of alkyldiarylamines,[125] arylation of heteroarylamines,[126] regioselective aminations of polyhalopyridines,[127] amidation of enol triflates[128] and several others.[129]

Another prominent ligand from the family of chelating bisphosphines which was first synthesized by van Leeuwen's group is (oxydi-2,1-phenylene)bis(diphenylphosphine) (DPEPhos).[118] DPEPhos[130] is structurally similar to Xantphos, but lacks only the bridge junction. They then studied the Kumada–Tamao–Corriu coupling of *sec*-butylmagnesium chloride and bromobenzene (the same reaction studied earlier by Hayashi *et al.* when they observed the beneficial effect of dppf with respect to activity and selectivity over β-hydride elimination[7]) and observed that the analogous reaction using DPEPhos was significantly more active and selective (Figure 2.34).[131] A ratio of 98 : 1 : 1 of branched coupling product (**A**):linear product (**B**):homocoupling (**C**) was observed with DPEPhos, whereas the Pd/dppf-catalyzed reaction gave 95 : 2 : 3. The increased activity using DPEPhos was attributed to its slightly larger bite angle relative to dppf (102.9° *versus* 99.1°), hence it was able to promote reductive elimination better. However, the corresponding Pd/Xantphos-catalyzed reaction gave poor selectivity with only 24% conversion, implying that a bite angle of ∼100° is optimal and ligands with smaller or larger bite angles gave diminished reactivity/selectivity.

The first use of DPEPhos in aryl amination was reported by Buchwald and co-workers in 1998 in a study to identify a more cost-efficient alternative to *rac*-BINAP, which was "moderately expensive" at that time.[132] The DPEPhos-based catalyst system was demonstrated to be at least as active as the BINAP systems, but significantly more active than Pd/dppf in the amination of aryl bromides with anilines. Preliminary studies indicated that this system was

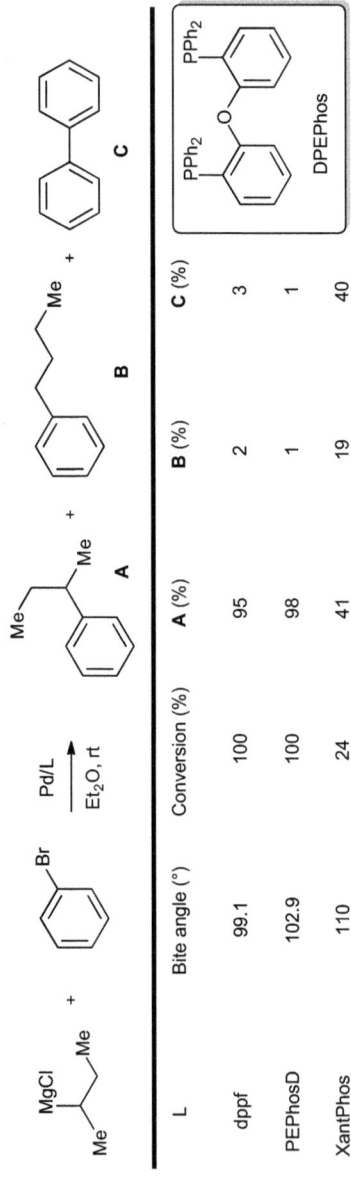

L	Bite angle (°)	Conversion (%)	A (%)	B (%)	C (%)
dppf	99.1	100	95	2	3
PEPhosD	102.9	100	98	1	1
XantPhos	110	24	41	19	40

Figure 2.34 Kumada–Tamao–Corriu coupling catalyzed by Pd/DPEPhos with high selectivity for the branched product.

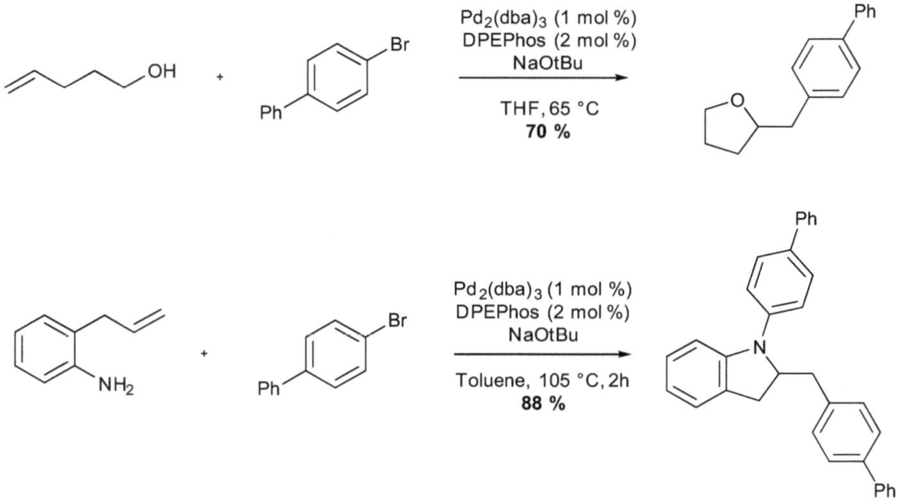

Figure 2.35 Carboetherification (top) and carboamination (bottom) reactions promoted by Pd/DPEPhos.

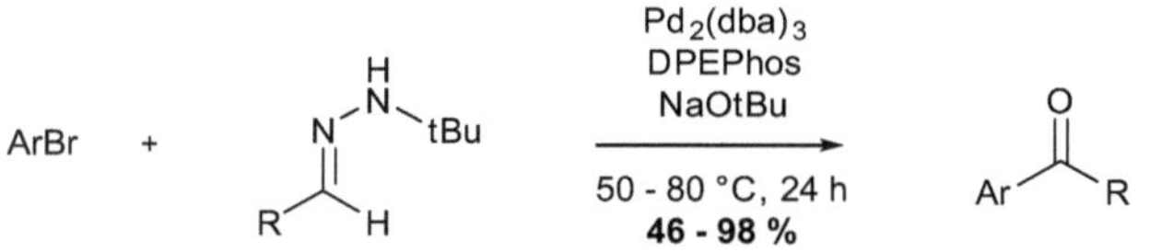

Figure 2.36 Coupling of aryl bromides with *N-tert*-butylhydrazone as an acyl anion equivalent.

not suitable for coupling aliphatic amines or *N*-alkylanilines as large amounts of hydrodebromination products were observed. DPEPhos has proven to be a very useful ligand in many amination reactions.[133] It was also effective in promoting Negishi reactions,[134] regioselective Suzuki–Miyaura reactions of dihalogenated electrophiles,[135] thioetherification of (hetero)aryl iodides with (hetero)aryl thiols[136] and a few examples of ketone enolate α-arylations.[137] Wolfe and co-workers developed a series of interesting carboetherification[138] and carboamination[139] reactions in which Pd/DPEPhos is often the catalyst of choice (Figure 2.35).[140]

Takemiya and Hartwig demonstrated that Pd/DPEPhos is uniquely suited for coupling reactions of aryl bromides and *N-tert*-butylhydrazone as an acyl anion equivalent (Figure 2.36).[141]

2.3.3 Biaryl Monophosphines

In 1998, Buchwald and co-workers reported the synthesis and use of a novel electron-rich phosphine ligand [2-dicyclohexylphosphino-2'-(*N,N*-dimethyl-amino)biphenyl (DavePhos[142])] for Suzuki–Miyaura, ketone α-arylation and

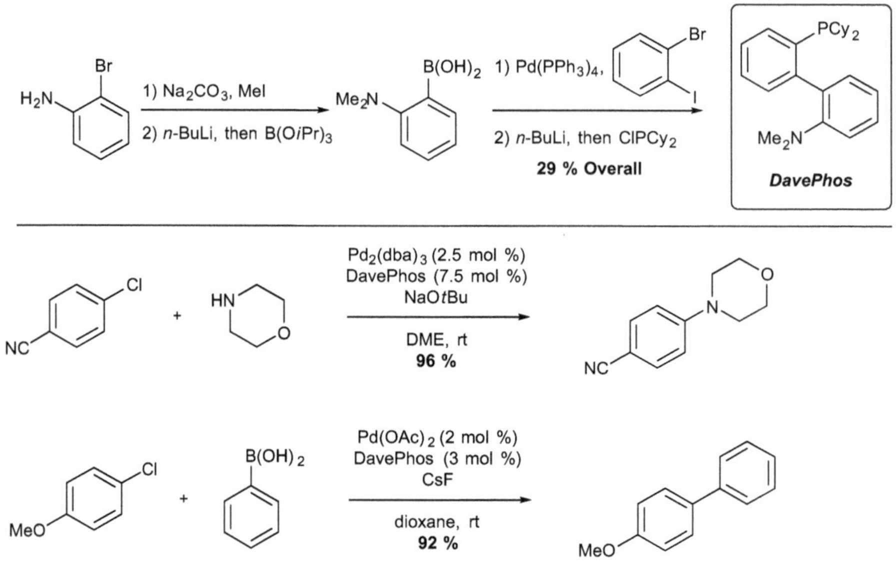

Figure 2.37　Synthesis (top) and applications (bottom) of DavePhos.

aryl amination reactions of unactivated aryl chlorides.[143] Noteworthy is that this work *preceded* Littke and Fu's studies using trialkylphosphines.[40] DavePhos was originally prepared in four steps from commercially available 2-bromoaniline in 29% overall yield (Figure 2.37). This ligand was inspired by the authors' "reasonable success" in the amination of a chloroarene using electron-rich 2,2'-bis(dicyclohexylphosphino)-1,1'-binaphthyl coupled with their previous success using Hayashi-type "bidentate" monophosphine PPFA and PPF-OMe ligands,[144,145] prompting the installation of the dicyclohexylphosphino group and dimethylamino group, respectively, on the biaryl backbone (Figure 2.38). The resulting Pd/DavePhos-based catalyst was highly active as it promoted room temperature amination reactions of aryl bromides and also the unprecedented room temperature amination of aryl chlorides, and it was the first general catalyst system for room temperature Suzuki–Miyaura couplings of aryl chlorides. It was believed at that time that coordination of the dimethylamino group to palladium was important to the catalysis as other electron-rich monodentate phosphines (Cy_3P, $PhPCy_2$) gave significantly diminished reactivity.

In subsequent studies of Suzuki–Miyaura reactions,[146] aryl amination reactions[146a,147] and C–O cross-coupling reactions[148] with phenols, it was found in many instances that the desamino derivatives of DavePhos (JohnPhos, CyJohnPhos) were equally efficient as supporting ligands for many transformations with loadings as low as 0.000001% Pd (Figure 2.39). They were prepared in good yield in one step from commercial 2-bromobiphenyl by a Grignard reaction (JohnPhos) or lithium–halogen exchange (CyJohnPhos) and subsequent reaction with the appropriate chlorodialkylphosphine

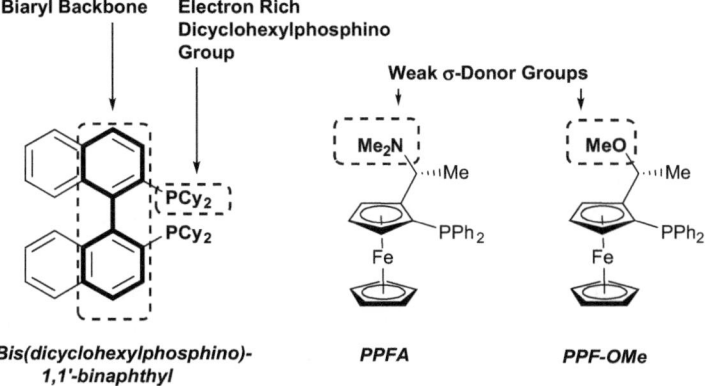

Figure 2.38 Structural elements inspiring the design of DavePhos.

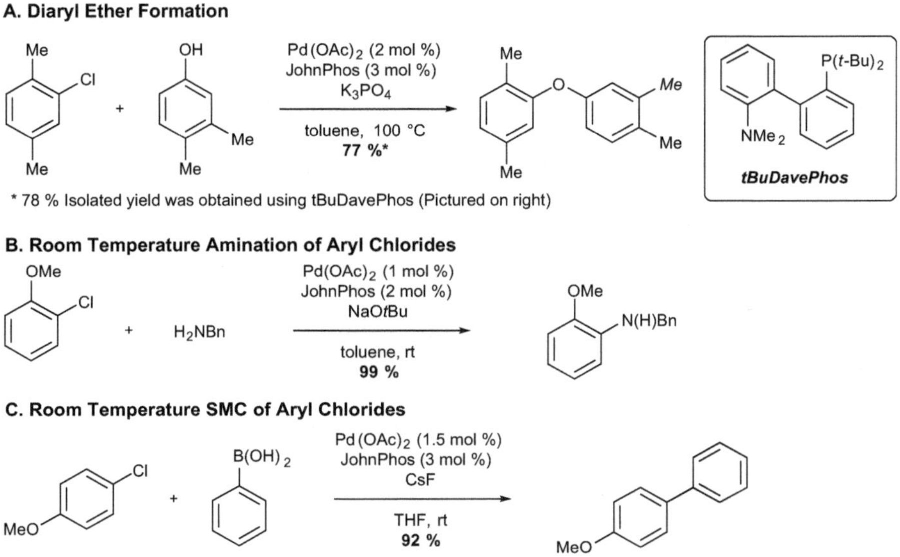

A. Diaryl Ether Formation

* 78 % Isolated yield was obtained using tBuDavePhos (Pictured on right)

B. Room Temperature Amination of Aryl Chlorides

C. Room Temperature SMC of Aryl Chlorides

Figure 2.39 Pd/JohnPhos-catalyzed reactions: (a) diaryl ether formation; (b) room temperature aryl chloride amination; (c) room temperature Suzuki–Miyaura reactions of aryl chlorides.

(Figure 2.40).[146b] It was postulated that the high activity of catalysts based on JohnPhos and CyJohnPhos was due to the electron richness (facilitates oxidative addition with tight binding to the Pd) and the steric bulkiness (promotes monoligated Pd and reductive elimination) and a possible stabilizing interaction[149] between the metal and the non-phosphine arene (*ipso*-C; see below) of the ligand.[146a] Interestingly, despite the electron-rich nature of JohnPhos, it is stable to air and moisture (as are all members of this ligand class).[150]

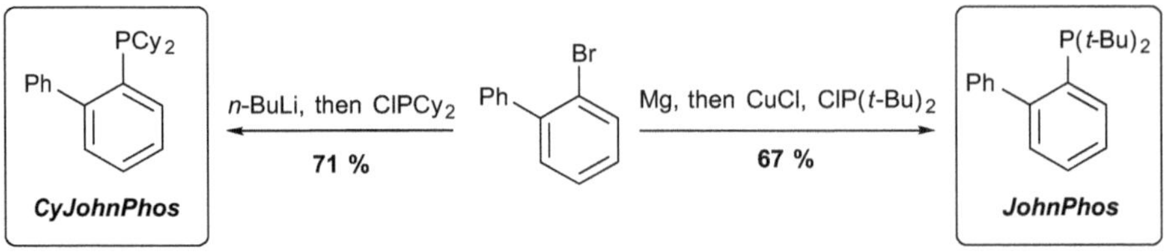

Figure 2.40 Synthesis of CyJohnPhos (left) and JohnPhos (right).

In related studies on ketone enolate α-arylation, it was found that analogs of DavePhos with alkyl substituents in the 2′-position on the biaryl backbone (R = Me, i-Pr) were particularly effective (Figure 2.41).[151] These ligands showed similar or better activities than DavePhos. Similar earlier findings with JohnPhos derivatives in other reactions[144–148] prompted the authors to suggest that binding of the dimethylamino group in DavePhos is not important for catalysis.

In fact, Kočovský and co-workers had earlier reported a Pd–C interaction in closely related Pd complexes of MAP[152] and MOP[153] through NMR and X-ray crystallographic studies (Figure 2.42).[154] The MAP complex was highly active in amination and Suzuki–Miyaura reactions, suggesting that the Pd–C interaction should be considered in catalysis.

It became apparent that subtle changes to the ligand backbone can lead to dramatic differences in reactivity. In some instances (*e.g.*, Suzuki–Miyaura reaction of hindered substrates,[146,147] ketone α-arylation[151]), the substituted biarylphosphine ligands cannot be replaced with the more easily prepared JohnPhos ligands. However, as mentioned above, the syntheses of the substituted biarylphosphine ligands required multiple steps and only modest yields were obtained. Accordingly, a serious effort was undertaken to streamline the syntheses of the ligands and provide a means of practical modification of the ligand scaffold. An improved route was developed in which aryl Grignard reagents were added to benzyne (generated *in situ* from 1-bromo-2-chlorobenzene), generating functionalized *o*-metallated biaryls which were then reacted directly with ClPCy$_2$ or ClP(t-Bu)$_2$ in the presence of CuCl (Figure 2.43).[155] Using this convenient one-pot procedure, functionalized biarylphosphines were synthesized in modest to moderate yields (18–53%) on several gram scales from readily available starting materials without chromatography.

This modular synthesis[156] allowed the facile modification and steady improvement/fine tuning of this ligand class over the next several years, leading to an entire library of ligands, many of which are now commercially available in multi-kilogram quantities.[48] The ability to tailor specific ligand attributes towards different applications has made this ligand class exceptionally powerful in many cross-coupling processes, especially C–N,[157] C–C[158] and C–O[159] bond-forming reactions. The high reactivity and selectivity of biarylphosphines can be traced to specific structural elements that they contain (Figure 2.44). It has been shown that oxidative addition is faster

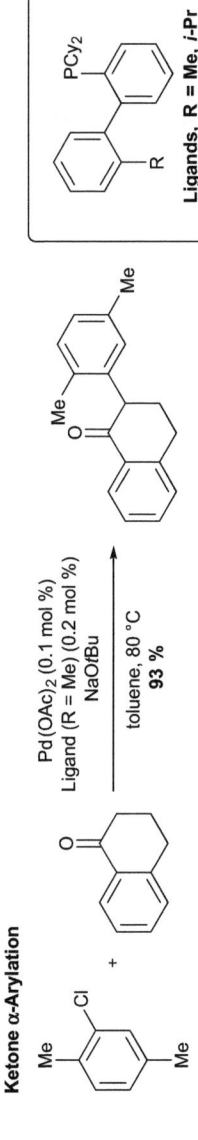

Figure 2.41 α-Arylation of ketones using 2′-substituted biarylphosphines.

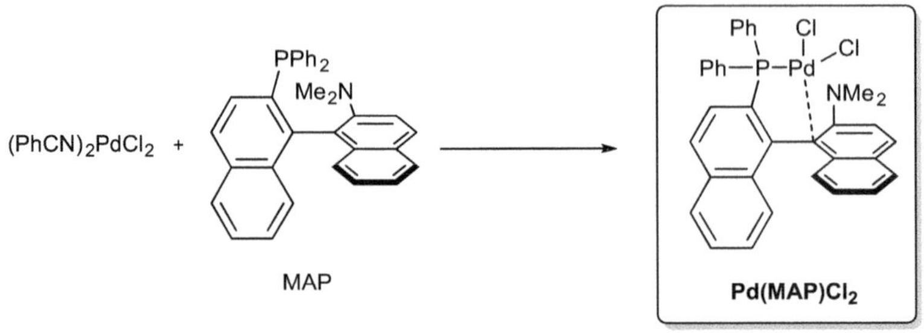

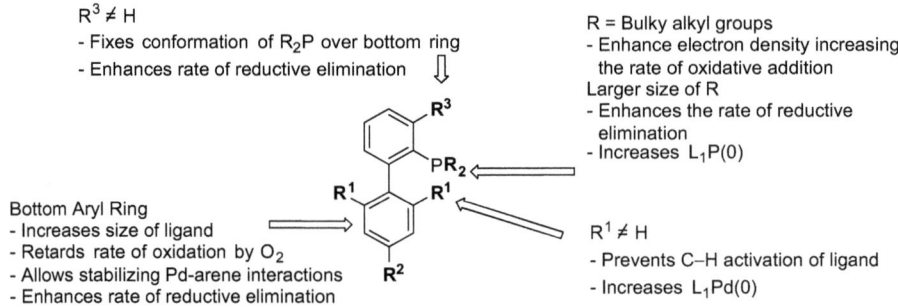

Figure 2.42   Synthesis of Pd(MAP)Cl$_2$ showing the Pd–*ipso*-C interaction.

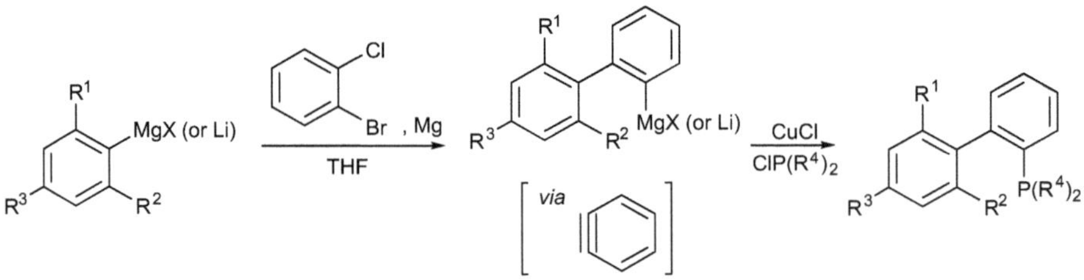

Figure 2.43   The improved "benzyne route" to substituted biarylphosphines.

R^3 ≠ H
- Fixes conformation of R$_2$P over bottom ring
- Enhances rate of reductive elimination

R = Bulky alkyl groups
- Enhance electron density increasing
  the rate of oxidative addition
Larger size of R
- Enhances the rate of reductive
  elimination
- Increases L$_1$P(0)

Bottom Aryl Ring
- Increases size of ligand
- Retards rate of oxidation by O$_2$
- Allows stabilizing Pd-arene interactions
- Enhances rate of reductive elimination

R^1 ≠ H
- Prevents C–H activation of ligand
- Increases L$_1$Pd(0)

Figure 2.44   Structural features of biarylphosphines.
             Adapted from Ref. 1m.

with L$_1$Pd(0) than L$_2$Pd(0) for steric reasons.[160] The large size of these ligands shifts the equilibrium in favor of the active L$_1$Pd species over L$_2$Pd, thus increasing the rate of oxidative addition. The rate of oxidative addition is further increased by the presence of electron-donating alkyl groups on phosphorus. The rate of transmetallation is presumed to be similarly faster with monoligated palladium species for steric reasons,[158f] and reductive elimination has also previously been shown to be faster from L$_1$Pd species than from the corresponding L$_2$ complexes.[161] The bulky nature of the alkyl groups on phosphorus further enhances the rate of reductive elimination. Theoretical studies have identified catalytic intermediates which feature important Pd–arene interactions involving the non-phosphine-containing aryl ring of the ligand.[161b,162] Substitution in the *ortho*-position on the non-phosphine-containing aryl ring also prevents cyclometallation *via* C–H activation of the lower ring, a process shown to occur with non-*ortho*-substituted (lower aryl ring) biarylphosphines.[163]

The ligand library that has resulted is vast, and catalysts based on many of these ligands promote several challenging transformations. Several of the most prominent and widely used are shown in Figure 2.45.

XPhos, introduced in 2003, featured significant advances over the existing technology.[164] The scope of aryl amination was broadened, the first amidation of aryl sulfonates was achieved and amination reactions were conducted in water with no cosolvent for the first time (Figure 2.46).

XPhos would prove not only to be useful in C–N bond-forming processes,[165] but also would become one of the most versatile ligands in the biarylphosphine family. Examples include challenging Suzuki–Miyaura,[158a,c–f,166] Negishi,[158b] Kumada–Corriu,[167] Stille,[168] α-arylation[169] and Miyaura borylation[170] reactions. In the following year, SPhos was introduced as a new ligand which allowed for unprecedented reactivity in Suzuki–Miyaura reactions to generate highly hindered biaryls with low Pd loadings, as low as 0.0005 mol%, even at room temperature (Figure 2.47).[171]

An X-ray structure of an (SPhos)Pd(dba) complex was obtained,[171b] and an η^1 interaction[172] between the Pd and *ipso*-carbon of the non-phosphine-containing ring was observed. It was again postulated that this interaction may be important to the catalyst stability. With the aid of detailed computational studies, Pd–arene interactions were identified in catalytic intermediates, thereby further underlining their importance to catalyst stability/longevity (Figure 2.48).[161b,162]

SPhos-based catalysts have been widely used for various C–C[173] and C–N[174] bond-forming reactions. Additional key breakthrough ligands have been reported over the past several years. For example, RuPhos, originally developed for challenging Negishi couplings,[175] has been shown to be an extremely useful ligand for aryl aminations with secondary amines,[157a] and BrettPhos[157e] is highly selective for monoarylation of primary amines,[157a,e] including methylamine.[157e] Taken together, catalysts based on either RuPhos or BrettPhos cover an extremely broad scope of aryl amination reactions.[157a] Furthermore, these ligands can operate in concert with one another.[157b] Starting with either a RuPhos precatalyst[165a] and additional BrettPhos or *vice versa*, a catalyst system equivalent to (or better than) *both* ligands was observed (for a more detailed discussion of palladacycle precatalysts, see Chapter 3). This was demonstrated by the "multiligand" reactions of 3-bromoanisole with morpholine or octylamine as shown in Figure 2.49.

Thus, primary and secondary amines can both be coupled efficiently from a single catalyst system based on both ligands. Mechanistic evidence was provided that showed ligand exchange to be facile under the reaction conditions (>80 °C). Ligand exchange was shown to occur at both the Pd(0)

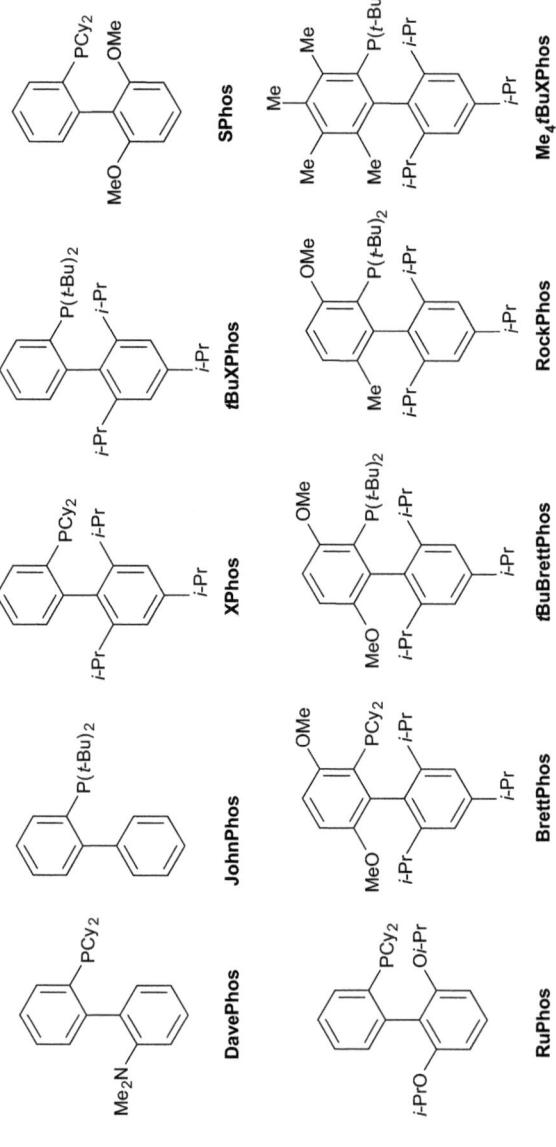

Figure 2.45 Prominent biarylmonophosphine ligands used in various cross-coupling reactions.

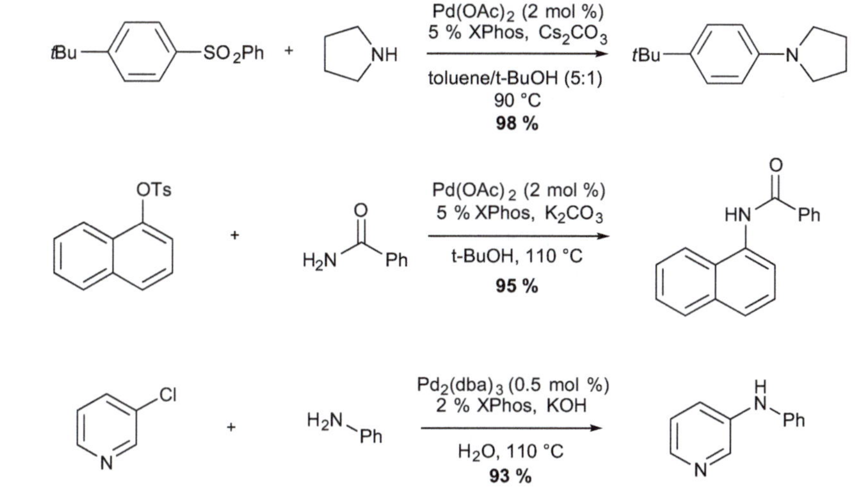

Figure 2.46 Challenging C–N cross-coupling reactions using Pd/XPhos.

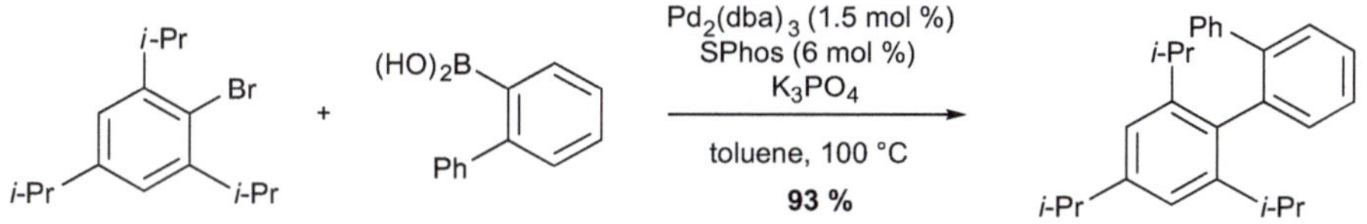

Figure 2.47 Highly hindered biaryl formation *via* Suzuki–Miyaura coupling using Pd/SPhos.

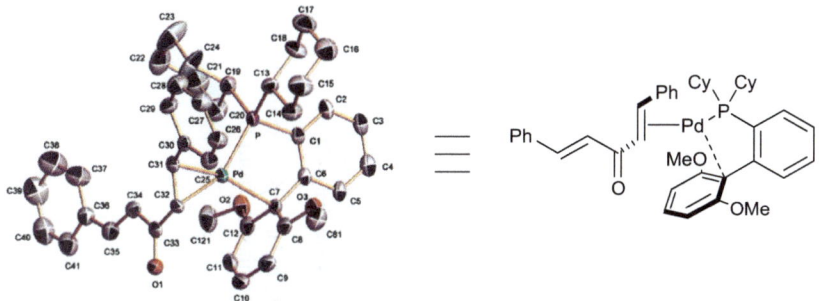

Figure 2.48 X-ray structure of (SPhos)Pd(dba) with hydrogens removed for clarity (dba = dibenzylideneacetone).
Reprinted with permission from T. E. Barder, S. D. Walker, J. R. Martinelli and S. L. Buchwald, *J. Am. Chem. Soc.*, 2005, **127**, 4685. Copyright 2005 American Chemical Society.

and Pd(II) oxidative addition intermediates, but not at the amine-bound Pd(II) intermediate. The following mechanism was proposed in which the palladium is being transferred between the two cycles *via* ligand exchange at the above-mentioned intermediates (Figure 2.50).

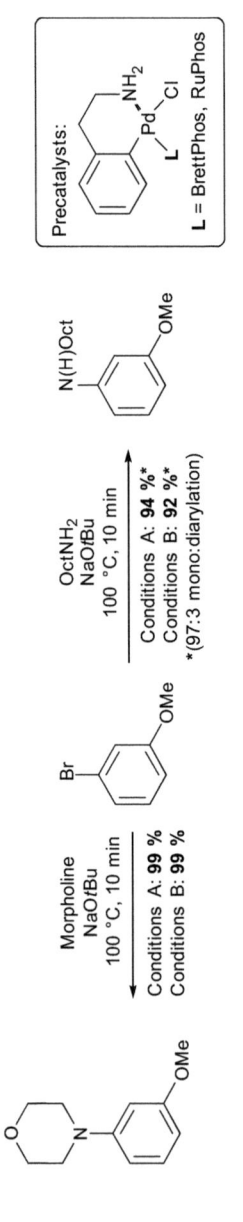

Figure 2.49 Multiligand system for aryl amination based on BrettPhos and RuPhos. Adapted with permission from B. P. Fors and S. L. Buchwald, *J. Am. Chem. Soc.*, 2010, **132**, 15914. Copyright 2010 American Chemical Society.

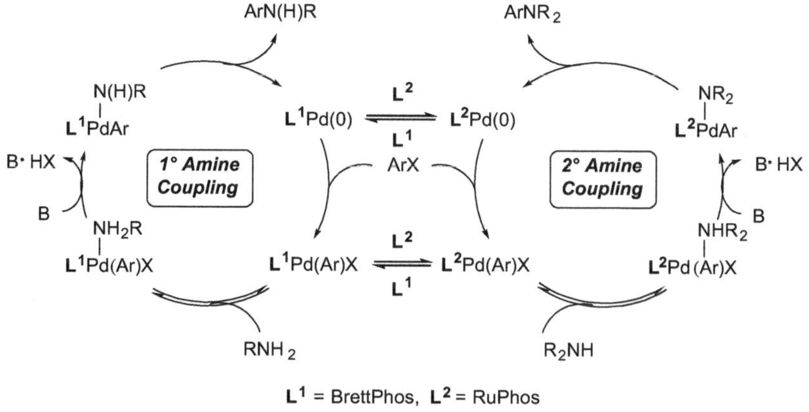

Figure 2.50 Proposed mechanism for the BrettPhos/RuPhos multiligand system for aryl amination.
Adapted with permission from B. P. Fors and S. L. Buchwald, *J. Am. Chem. Soc.*, 2010, **132**, 15914. Copyright 2010 American Chemical Society.

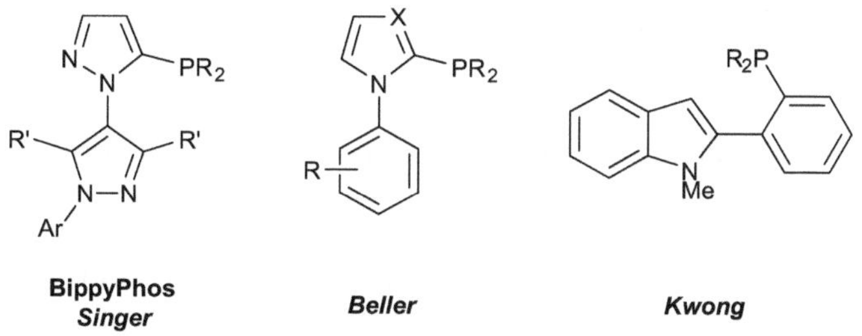

Figure 2.51 Heterobiarylphosphine ligands.

Additionally, Pd catalysts based on the BrettPhos subset of ligands have been shown to be uniquely effective in several challenging couplings, including the formation of Ar–CF$_3$,[176] Ar–F,[177] Ar–SCF$_3$[178] and others.[179]

Numerous variations on the biarylphosphine theme have been reported by several groups.[180] Notable are the bipyrazole "BippyPhos" ligands reported by Singer and co-workers,[180a,b] which show high activity in C–N[180a,b,d,e] and C–O[180c] coupling reactions, the heterocyclic biaryl ligands developed by Beller and co-workers for amination reactions[180f,g] and the indole-based biarylphosphines developed by Kwong and co-workers for amination of less reactive aryl mesylates (Figure 2.51).[180h]

2.3.4 Other Aryl Dialkyl-Based Monophosphines

In the course of a study on aryl ether formation using di-*tert*-butylphosphinoferrocene [PFc(*t*-Bu)$_2$ (Fc = 1-ferrocenyl)], Hartwig's group reported

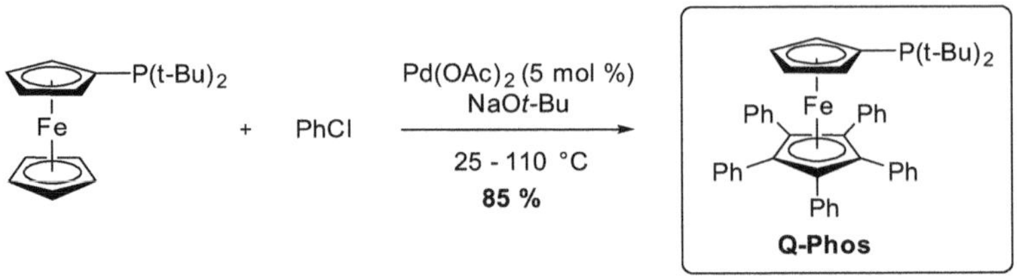

Figure 2.52   Q-Phos synthesis *via* perphenylation of FcP(*t*-Bu)₂.

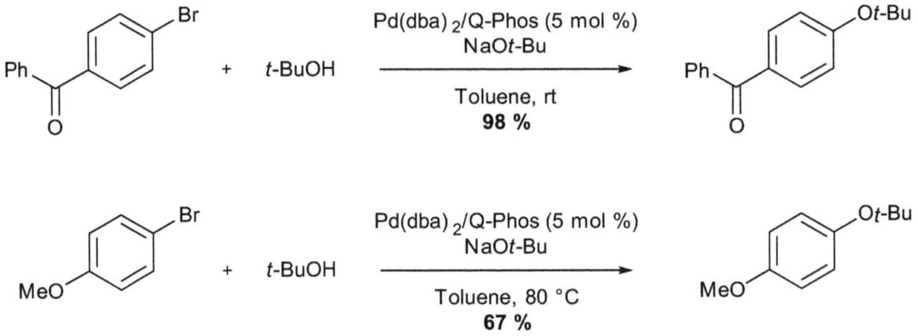

Figure 2.53   Synthesis of aryl ethers using Pd/Q-Phos.

some unusual observations: (1) whereas the catalytic reactions of aryl halides with PhONa proceeded in high yield with Pd(dba)₂ and PFc(*t*-Bu)₂, stoichiometric reductive elimination reactions gave much lower yield and (2) in the catalytic reactions a significant induction period was observed and aryl halide was initially consumed without the generation of aryl ether product.[181] Mechanistic studies were carried out to find an explanation for these phenomena and it was ultimately established that the FcP(*t*-Bu)₂ ligand was being catalytically perphenylated[182] on the unsubstituted cyclopentadienyl ring, thus forming a significantly more active catalyst system (Figure 2.52).

The combination of Pd(dba)₂ and the modified ligand, named Q-Phos,[48,183] was demonstrated to be an efficient catalyst for the mild (r.t.–80 °C) formation of aryl ethers from aryl chlorides and bromides, as shown in Figure 2.53.

Q-Phos has also been shown by Hartwig and co-workers to be an efficient ligand in various Pd-catalyzed C–C, C–N and C–O cross-coupling reactions,[183] including the arylation of malonates and cyanoacetate[184] and the α-arylation of azalactones,[61] zinc amide enolates[185] and aldehydes.[186] Colacot and co-workers showed that (π-crotyl)Pd(Q-Phos)Cl is an effective and convenient precatalyst for a variety of C–C and C–N cross-coupling reactions.[72] Lautens and co-workers described an interesting carbohalogenation reaction using Pd(Q-Phos)₂[48] as a precatalyst (Figure 2.54).[187] The key step in these transformations is an unusual C–I reductive elimination. A more detailed account of this transformation can be found in Chapter 7.

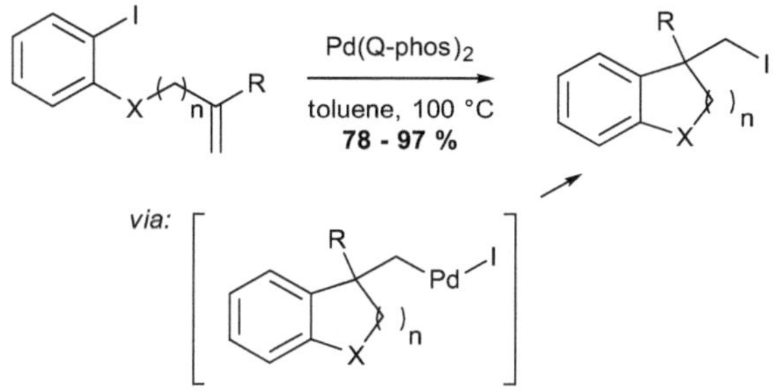

Figure 2.54 Carboiodination reactions promoted by Pd(Q-Phos)$_2$.

N,N-Dimethyl-4-(di-*tert*-butylphosphino)aniline[188] (A^{ta}-phos, also referred to as AmPhos[189])[48] is an aryl dialkylmonophosphine ligand originally developed by Guram *et al.* at Amgen for SMC reactions of heteroaryl chlorides.[190] Using Pd(AmPhos)$_2$Cl$_2$[148] (Pd-132), a variety of heteroaryl chlorides underwent efficient coupling with arylboronic acids with 1 mol% of catalyst (Figure 2.55). They later expanded the scope and included the use of aryl-boronate esters and heteroarylboronic acids.[191] The success of this ligand *versus* the unsubstituted phenyl derivative highlights the importance of the *electronic* properties of the ligand as PhP(*t*-Bu)$_2$ and AmPhos are isosteric.

Lipshutz and co-workers showed that Pd(AmPhos)$_2$Cl$_2$ (Pd-132) is an excellent catalyst for micellar Negishi-type couplings in or "on" water using the phase-transfer surfactant polyoxyethanyl α-tocopheryl sebacate (Figure 2.56).[192]

Stereospecific Suzuki–Miyaura reactions of arylboronic acids and alkyl α-cyanohydrin triflates occur in high yield and with nearly complete *inversion* of absolute configuration using Pd-132, as shown by He and Falck (Figure 2.57).[193]

Additionally, Pd-132 has been used in a kilogram-scale Suzuki–Miyaura reaction in the process route to a pyrazolopyridinone-based p38 MAP kinase inhibitor by the Amgen process group (Figure 2.58).[194]

Recently, Colacot and co-workers reported a highly efficient Heck alkynylation protocol using Pd-132 or the analogous Pd(0) complex Pd(AmPhos)$_2$ (Pd-149). Reactions of both aryl and heteroaryl bromides and chlorides were described (Figure 2.59).[63f]

Stradiotto and co-workers introduced a new type of aryldialkylmonophosphine ligands which are structurally related to AmPhos. The design plan was to find a middle ground between bisphosphine ligands which feature two strong P–Pd interactions and biarylphosphines which feature one strong P–Pd interaction and one weak C–Pd interaction (from the non-phosphine-containing aryl ring; see above). The result was the discovery of a class of ligands which feature a κ^2-P,N chelate; a strong P–Pd interaction and a moderate N–Pd interaction, referred to as the DalPhos ligands[195] (Figure 2.60)

This class of ligands, containing an *ortho*-amino group and bulky 1-adamantyl groups on the phosphine, has been used to generate highly active catalysts that are effective in aryl amination reactions,[196] including mono-arylation of ammonia[195,196a,c,f] and carbonyl α-arylation[197] including the

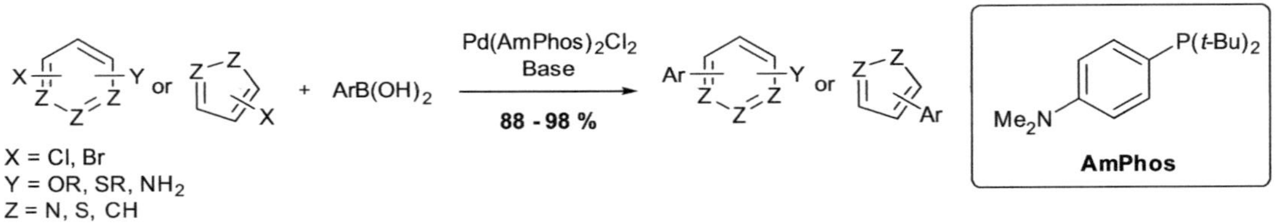

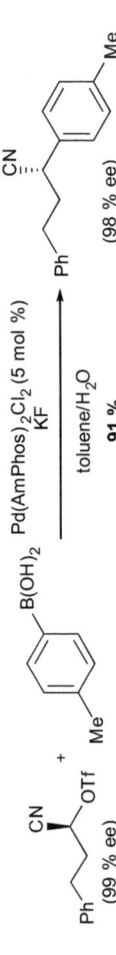

Figure 2.55 Suzuki–Miyaura reactions of heteroaryl halides catalyzed by Pd(AmPhos)$_2$Cl$_2$.

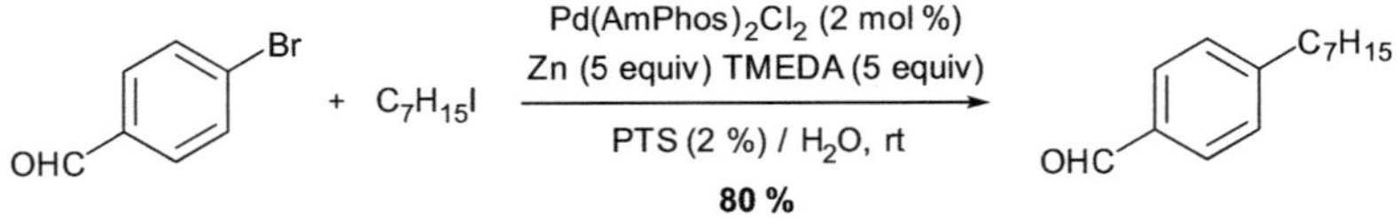

Figure 2.56 Micellar Negishi coupling using Pd(AmPhos)$_2$Cl$_2$ and a phase-transfer surfactant.

Figure 2.57 Stereospecific Suzuki–Miyaura couplings catalyzed by Pd(AmPhos)$_2$Cl$_2$.

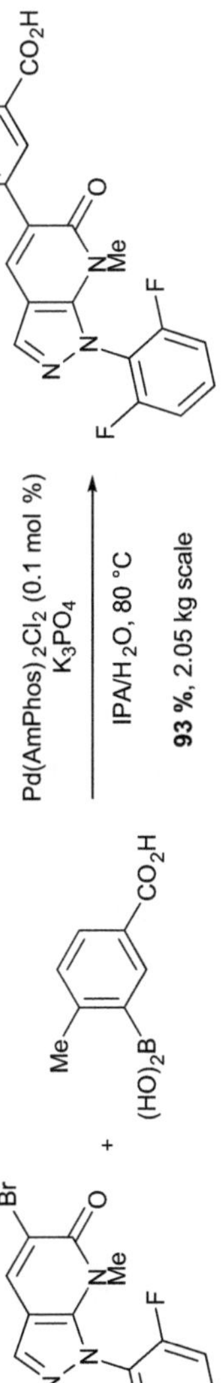

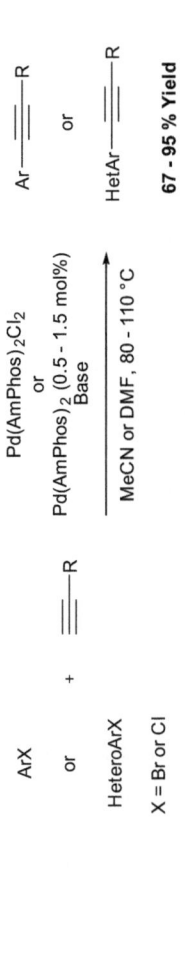

Figure 2.58 Kilogram-scale Suzuki–Miyaura reaction using Pd(AmPhos)₂Cl₂.

Figure 2.59 Heck alkynylation reactions of aryl and heteroaryl halides using Pd/AmPhos complexes.

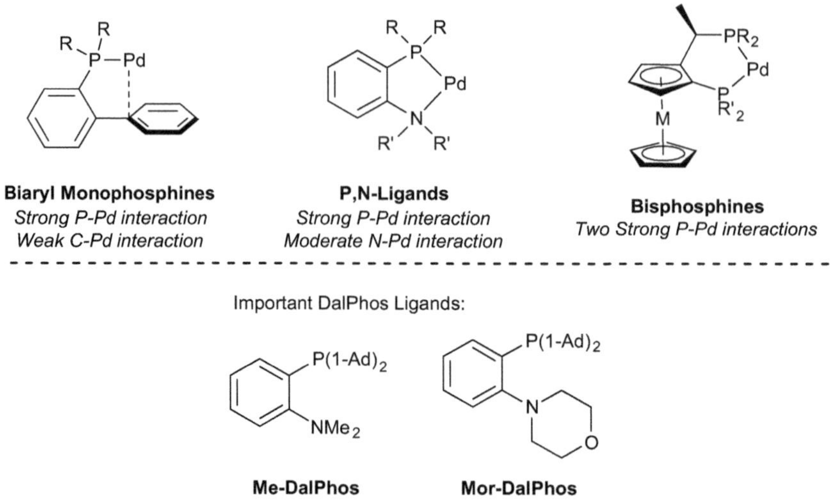

Biaryl Monophosphines
Strong P-Pd interaction
Weak C-Pd interaction

P,N-Ligands
Strong P-Pd interaction
Moderate N-Pd interaction

Bisphosphines
Two Strong P-P interactions

Important DalPhos Ligands:

Me-DalPhos   **Mor-DalPhos**

Figure 2.60 Pd–ligand interactions in biarylphosphines, P,N ligands and bisphosphines (top) and commonly used DalPhos ligands (bottom).

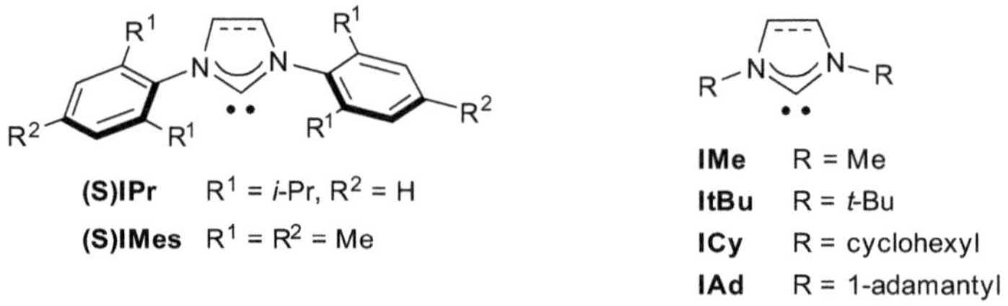

(S)IPr R^1 = *i*-Pr, R^2 = H
(S)IMes R^1 = R^2 = Me

IMe   R = Me
ItBu R = *t*-Bu
ICy   R = cyclohexyl
IAd   R = 1-adamantyl

Figure 2.61 Commonly employed *N*-heterocyclic carbene ligands in cross-coupling reactions.

monoarylation of acetone.[197a] A detailed discussion of this class of ligands and their applications is presented in Chapter 5.[198]

2.3.5 *N*-Heterocyclic Carbene (NHC) Ligands

N-Heterocyclic carbenes, first isolated by Arduengo *et al.* in 1991,[199] have been introduced as an alternative to phosphines as supporting ligands in several transition metal-catalyzed reactions.[4d] The first applications in Pd-catalyzed coupling processes were Heck reactions, SMC reactions and alkynylations reported by Herrmann.[4,122c] Since these early reports, the area of Pd-catalyzed cross-coupling reactions utilizing NHCs as supporting ligands has grown immensely.[4] The work of the groups of Nolan,[4a,c-e] Glorius[4b] and Organ[200] has taken the area significantly forward. The high activity of NHCs can be traced to their electronic and steric properties. NHCs are known to be stronger σ-donors than trialkylphosphines and, by some measures, even more sterically demanding.[4b] The most common NHC ligands used in cross-coupling reactions are depicted in Figure 2.61.

A more detailed discussion on NHC ligands in Pd-catalyzed cross-coupling reactions can be found in Chapter 4.

2.3.6 Chiral Ligands Used in Asymmetric Transformations

Compared with the breadth of ligands available for non-asymmetric cross-coupling reactions, the number of chiral ligands commonly used is significantly lower. This is likely because asymmetric transformations are typically limited to the generation of enantioenriched axially chiral biaryl structures, α-arylcarbonyl compounds that contain a quaternary chiral center and products of asymmetric Heck reactions. The pioneering work of Hayashi and co-workers in asymmetric Kumada–Corriu coupling reactions, however, demonstrated the feasibility of this concept.[201] Using a *P,N* ligand containing carbon central chirality [(*S*)-Phephos], axially chiral biaryls were synthesized in high yield (77–87%) with enantiomeric excesses (*ees*) ranging from 84 to 93% from ditriflates (Figure 2.62).

Since this early report, several prominent ligand types have emerged that promote highly enantioselective cross-coupling processes. BINAP (either racemic or non-racemic),[99] which was introduced earlier as a broadly useful ligand for non-asymmetric reactions, is also used as a supporting ligand in several enantioselective processes in its optically active forms. Since the earliest examples of asymmetric Heck reactions were reported independently by the groups of Shibasaki[202] and Overman,[203] Pd/BINAP systems are still the most broadly used in enantioselective Heck reactions.[204] Buchwald and co-workers showed that a Pd/(*S*)-BINAP catalyst can give reasonable enantioselectivity in the α-arylation of ketone enolates,[205,206] and a BINAP/Ni catalyst can give high enantioinduction in α-arylation reactions of γ-butyrolactones (Figure 2.63).[207]

Hartwig and co-workers showed that both Pd and Ni catalysts of BINAP and the BINAP-related difluorphos[208] ligand can give excellent enantioselectivities in the α-arylation of cyclic ketones with aryl/heteroaryl chlorides[209] and aryl triflates (Figure 2.64).[210]

Kwong, Chan and co-workers reported asymmetric arylations of ketone enolates with mainly aryl bromides using an Ni/(*R*)-P-Phos catalyst system,[211] a chiral bisphosphine ligand developed by Chan and co-workers (Figure 2.65).[212]

Asymmetric Suzuki–Miyaura reactions[213] have also been reported using BINAP as a supporting ligand; however, the enantioselectivities were typically modest–moderate.[214] Yin and Buchwald reported one of the earliest processes for asymmetric Suzuki–Miyaura coupling using a monophosphine containing the axially chiral binaphthyl backbone (Figure 2.66).[215] This new ligand, named KenPhos, was used in combination with Pd$_2$(dba)$_3$ to give biaryls in high yield (74–98%) with high enantiocontrol (up to 92% *ee*).

Pd/KenPhos-based catalysts (or similarly related ligands) have since been used by Buchwald and co-workers to improve upon asymmetric arylation[216] and vinylation[217] of ketone enolates, expand the scope of asymmetric

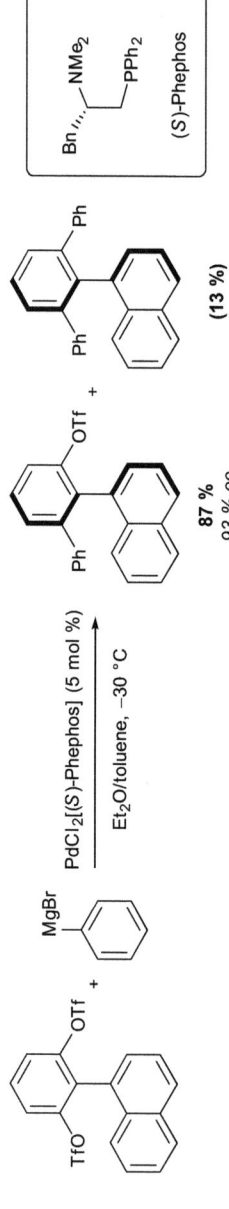

Figure 2.62   Enantioselective biaryl formation with PdCl₂[(*S*)-Phephos].

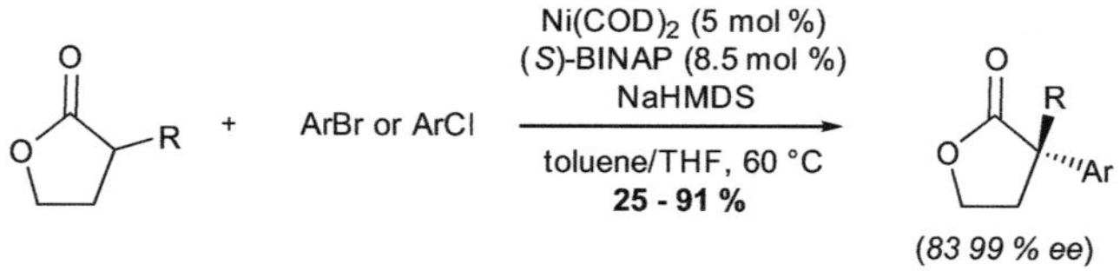

Figure 2.63   Enantioselective α-arylation of lactones using Ni/(*S*)-BINAP.

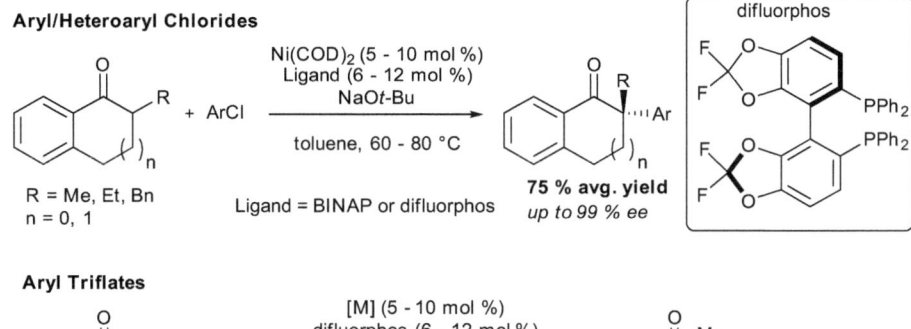

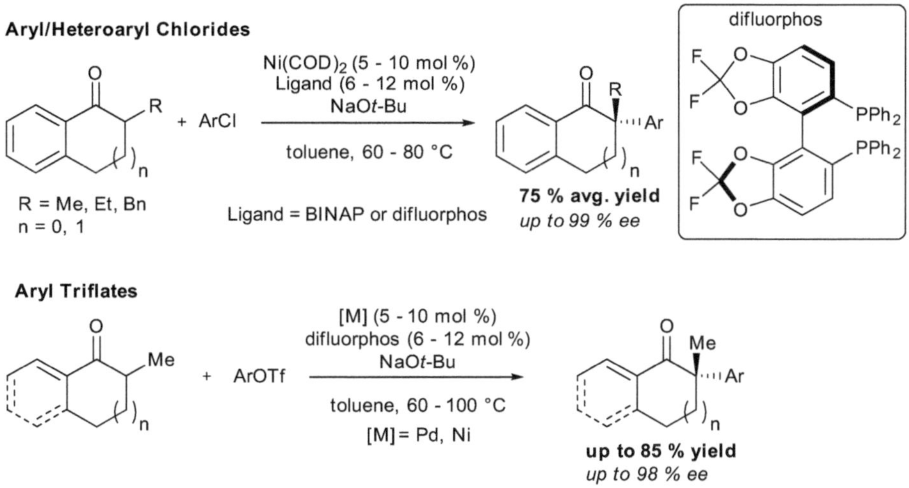

Figure 2.64 Asymmetric α-arylations of ketones with BINAP- or difluorphos-based catalysts.

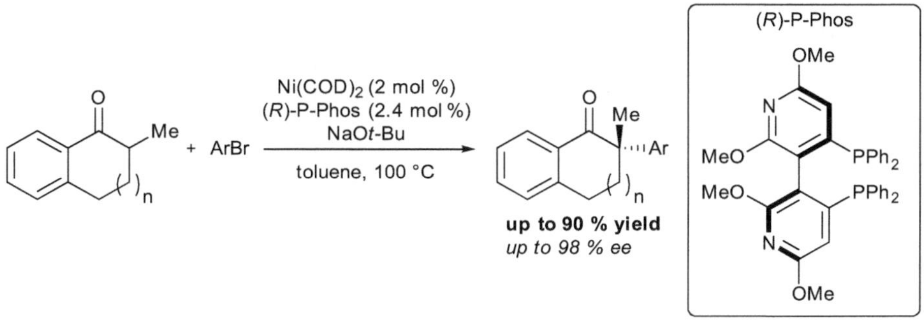

Figure 2.65 Asymmetric α-arylations using an (*R*)-P-Phos-based nickel catalyst.

Suzuki–Miyaura reactions[218] and promote intramolecular asymmetric dearomatization reactions of naphthalenes (Figure 2.67).[219]

Ligands that are stereogenic at phosphorus have also been used in asymmetric cross-coupling reactions. For instance, Hamada and Buchwald developed monophosphine ligands that contain both axial chirality and central chirality at phosphorus (Figure 2.68).[220] It was found that this ligand was effective for highly asymmetric α-vinylation of a ketone enolate, but significantly less so for corresponding α-arylation.

This P-stereogenic ligand, however, was broadly effective in Pd-catalyzed highly enantioselective α-arylation and α-vinylation of oxindoles, as reported by Buchwald and co-workers.[221] Furthermore, the same group also demonstrated that asymmetric dearomative spirocyclization of phenols can be promoted by a Pd catalyst containing this P-stereogenic ligand with good to high enantiocontrol (Figure 2.69).[222]

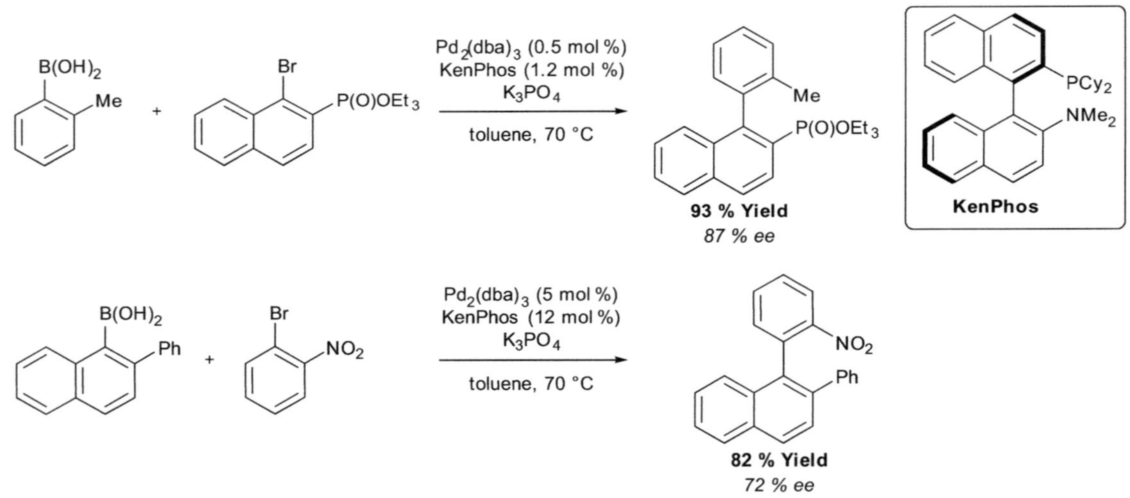

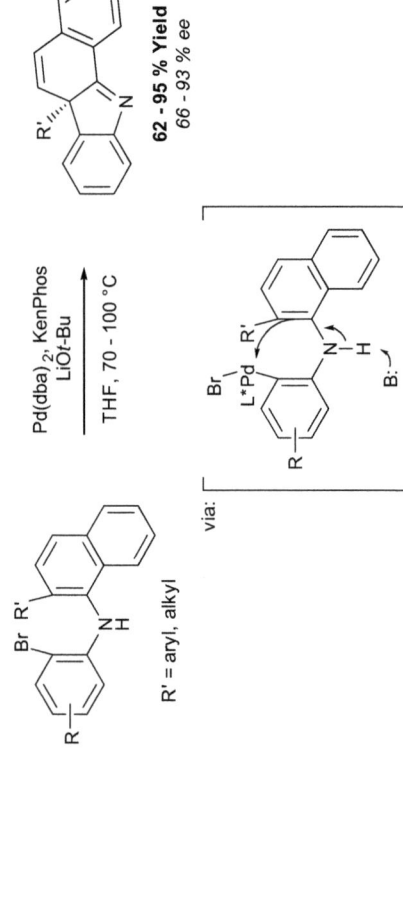

Figure 2.66 Pd-catalyzed axially chiral biaryl synthesis using KenPhos.

Figure 2.67 Pd-catalyzed enantioselective dearomative intramolecular arylation of naphthalenes.

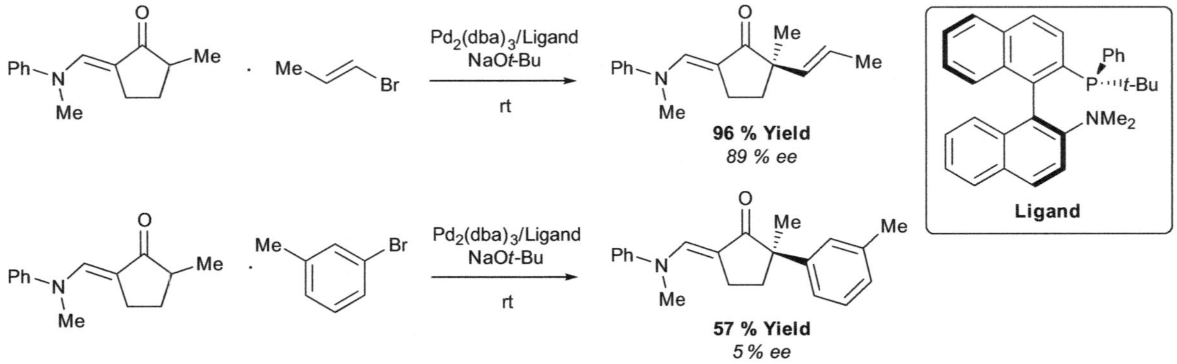

Figure 2.68 Enantioselective α-arylation of cyclic ketones using a P-stereogenic ligand.

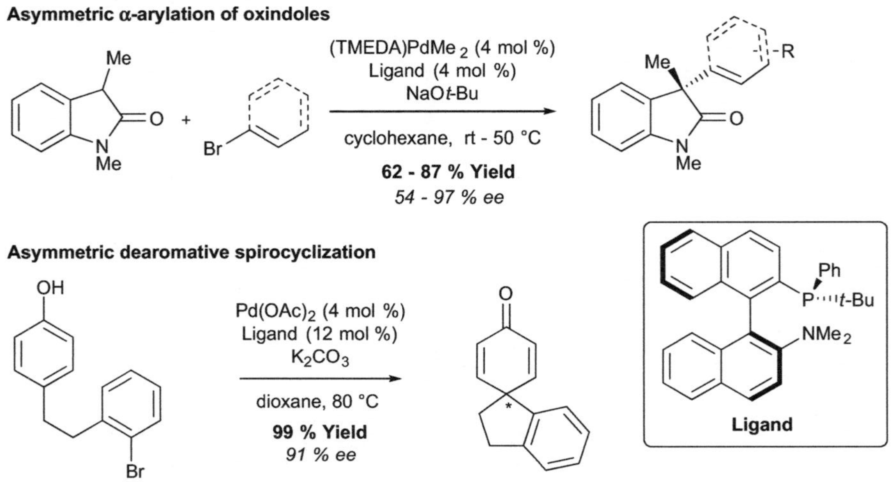

Figure 2.69 Other Pd-catalyzed asymmetric reactions using a P-stereogenic ligand.

Senanayake and co-workers recently developed a biarylmonophosphine ligand with only phosphorus and carbon central chirality and applied it to asymmetric Suzuki–Miyaura reactions.[223] High yields and enantioselectivities were observed in reactions of arylboronic acids and aryl bromides bearing *o*-acyloxazolidinones. They also demonstrated the utility of this transformation using a similar ligand in the syntheses of axially chiral biaryl natural products korupensamine A and B and michellamine B (Figure 2.70).[224]

García-Fortanet and Buchwald accomplished the asymmetric intra-molecular Pd-catalyzed α-arylation of aldehydes[225] using a phox ligand,[226] a ligand family which features only carbon central chirality (Figure 2.71).

The phox family of ligands can also promote highly enantioselective Heck reactions,[204] as demonstrated by Pfaltz and co-workers (Figure 2.72).[227]

Lee and Hartwig investigated the Pd-catalyzed intramolecular α-arylation of amides to form oxindoles.[228] Having observed disappointing enantio-selectivities with available chiral phosphines, they turned to chiral NHC ligands to enhance the enantioselectivities up to 70% (Figure 2.73).

Several groups have since reported efforts in the asymmetric formation of oxindoles.[229] The first highly enantioselective example was reported by Kündig *et al.* using imidazolium iodide ligand precursors (Figure 2.74).[230]

2.4 Outlook

Clearly, the development and understanding of new ligands and how they affect individual steps in catalytic reactions have been crucial to the ad-vancement of the field, leading to the 2010 Nobel Prize-winning work and beyond.[1] The sheer number of transformations that can be accomplished today by modern cross-coupling technology is indeed vast, but not limitless. For instance, many challenging reductive elimination processes limit the scope of coupling partners that can be used. In fact, only recently have Pd-catalyzed trifluoromethylation[176,231] and fluorination[177] been achieved and more challenging processes such as catalytic aryl trifluoromethoxylation have not yet been realized.[232] Additionally, cross-coupling reactions have traditionally been limited to aryl halide/pseudohalide (primarily sulfonate) electrophiles. Efforts are intensifying to expand the scope of the electrophilic partners that can be employed. Accordingly, exciting results have been re-ported recently in which relatively inert C–X (X = N, O) bonds have been activated in cross-coupling reactions, primarily with Ni catalysts.[233] Also of significant interest is the ability to functionalize C–H bonds selectively, a mode of reactivity that obviates the need to use prefunctionalized coupling partners such as aryl halides. Over the past several years, this dream is rapidly becoming a reality and despite the fact that this area is still in its infancy, a wide scope of literature exists on this topic. Beginning with the groundbreaking work of Lkeiman and Dubeck in 1963 up to the ever-advancing state-of-the-art driven by the work of the groups of Fagnou, White, Du Bois, Gaunt, Hartwig, Miura, Sanford, Yu and others, the area is pro-gressing by leaps and bounds and the reader is directed to Chapter 12,

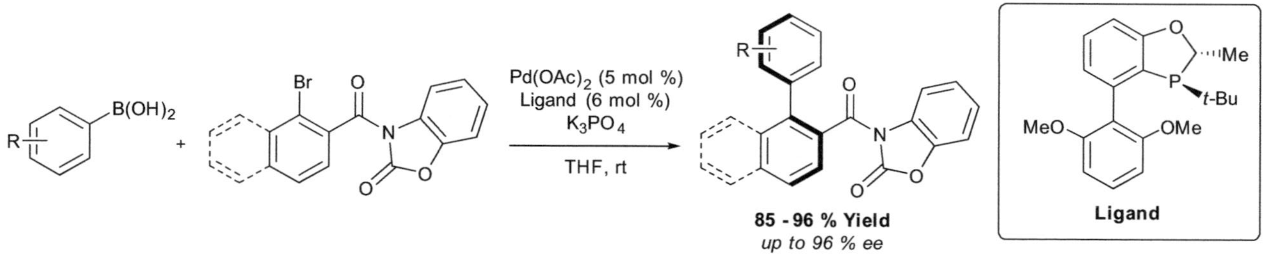

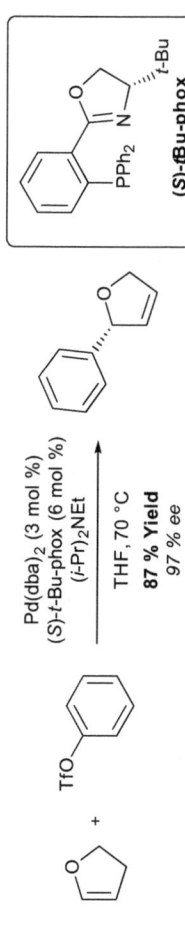

Figure 2.70 Asymmetric Suzuki–Miyaura coupling using a P,C-stereogenic ligand.

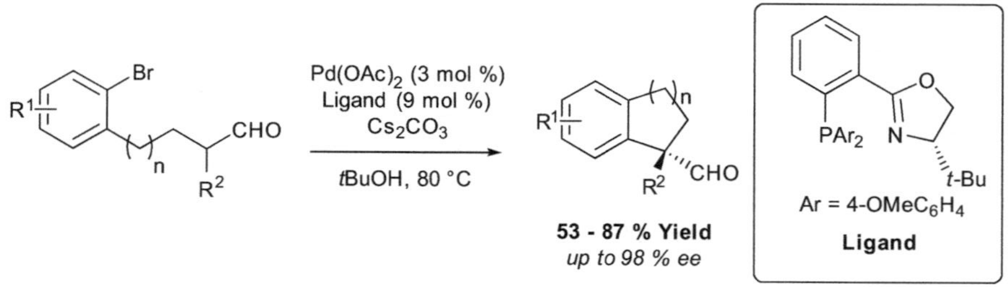

Figure 2.71 Asymmetric intramolecular α-arylation of aldehydes using a phox ligand.

Figure 2.72 Enantioselective Heck reaction employing (*S*)-*t*-Bu-phox.

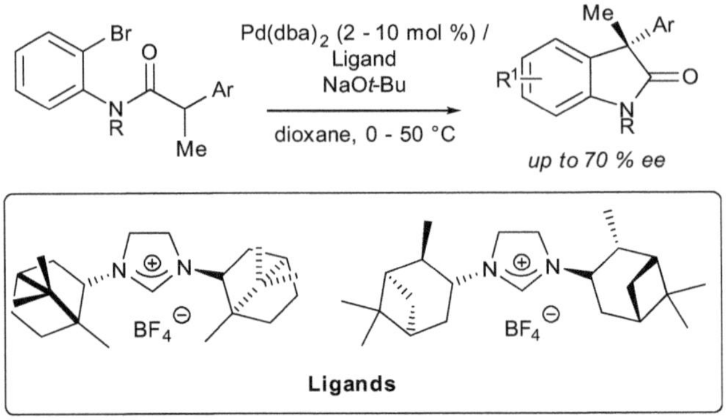

Figure 2.73   Enantioselective intramolecular oxindole synthesis using chiral NHCs.

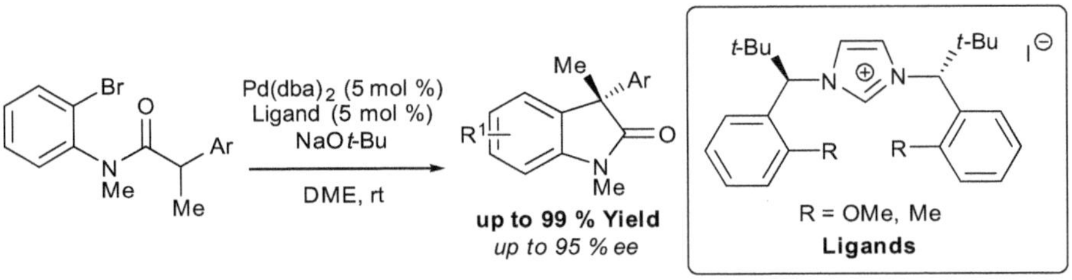

Figure 2.74   Enantioselective intramolecular oxindole synthesis using imidazolium iodides as chiral NHC ligand precursors.

devoted to C–H activation/functionalization, and to some excellent reviews.[234] Additionally, owing to the more difficult oxidative addition of alkyl halides relative to aryl/vinyl halides coupled with the propensity of the resulting alkyl–Pd(II) intermediate to undergo β-hydride elimination, the use of C(sp³)-electrophiles has been slow to develop. Nonetheless, beginning with groundbreaking results from the groups of Suzuki[235] and Knochel[236] and with the pioneering work in large part from Fu and co-workers,[237] this is a rapidly developing area[238] with exciting and challenging transformations yet to be achieved. These challenges will undoubtedly be met with intense research efforts aimed at identifying appropriate metal catalysts and finely tuned ligands. A more applied and practical trend that is emerging is the development and implementation of stable preformed complexes to achieve better efficiency from cost and environmental perspectives. These trends are discussed at length in Chapter 3.

References

1. (a) C. C. C. Johansson Seechurn, M. O. Kitching, T. J. Colacot and V. Snieckus, *Angew. Chem. Int. Ed.*, 2012, **51**, 5062 and references therein; (b) T. J. Colacot, *Platinum Met. Rev.*, 2011, **55**, 84; (c) *Acc. Chem.*

Res., 2008, **41**, 1439 (entire issue); (d) J. F. Hartwig, *Angew. Chem. Int. Ed.*, 1998, **37**, 2046; (dd) J. F. Hartwig, *Acc. Chem. Res.*, 1998, **31**, 852; (e) A. F. Littke and G. C. Fu, *Angew. Chem. Int. Ed.*, 2002, **41**, 4176; (f) V. Farina, *Adv. Synth. Catal.*, 2004, **346**, 1553; (g) U. Christmann and R. Vilar, *Angew. Chem. Int. Ed.*, 2005, **44**, 366; (h) A. Zapf and M. Beller, *Chem. Commun.*, 2005, 431; (i) D. S. Surry and S. L. Buchwald, *Angew. Chem. Int. Ed.*, 2008, **47**, 6338; (j) L. Xue and Z. Lin, *Chem. Soc. Rev.*, 2010, **39**, 1692; (k) C. A. Fleckenstein and H. Plenio, *Chem. Soc. Rev.*, 2010, **39**, 694; (l) C. C. C. Johansson and T. J. Colacot, *Angew. Chem. Int. Ed.*, 2010, **49**, 676; (m) D. S. Surry and S. L. Buchwald, *Chem. Sci.*, 2011, **2**, 27; (n) N. Miyaura and A. Suzuki, *Chem. Rev.*, 1995, **95**, 2457; (o) J. P. Wolfe, S. Wagaw, J.-F. Marcoux and S. L. Buchwald, *Acc. Chem. Res.*, 1998, **31**, 805.

2. (a) A. de Meijere and F. Diederich (eds), *Metal-Catalyzed Cross-Coupling Reactions*, Wiley-VCH, Weinheim, 2004; (b) G. A. Molander (ed.), *Handbook of Reagents for Organic Synthesis, Catalyst Components for Coupling Reactions*, Wiley, Chichester, 2008; (c) J. Tsuji, *Palladium Reagents and Catalysts, Innovations in Organic Synthesis*, Wiley, Chichester, 1995; (d) J. F. Hartwig, *Organotransition Metal Chemistry*, University Science Books, Mill Valley, CA, 2010; (e) M. Beller and C. Bolm (eds), *Transition Metals for Organic Synthesis*, Vol. 1, Wiley-VCH, Weinheim, 2nd edn, 2004.

3. (a) C. Glaser, *Ber. Dtsch. Chem. Ges.*, 1869, **2**, 422; (b) C. Glaser, *Ann. Chem. Pharm.*, 1870, **154**, 137.

4. For reviews, see: (a) N. Marion and S. P. Nolan, *Acc. Chem. Res.*, 2008, **41**, 1440; (b) S. Würtz and F. Glorius, *Acc. Chem. Res.*, 2008, **41**, 1523; (c) G. C. Fortman and S. P. Nolan, *Chem. Soc. Rev.*, 2011, **40**, 5151; (d) S. Díez-González, N. Marion and S. P. Nolan, *Chem. Rev.*, 2009, **109**, 3612; (e) A. C. Hillier, G. A. Grasa, M. S. Viciu, H. M. Lee, C. Yang and S. P. Nolan, *J. Organomet. Chem.*, 2002, **653**, 69.

5. R. H. Crabtree, *The Organometallic Chemistry of the Transition Metals*, Wiley, Hoboken, NJ, 5th edn, 2009.

6. R. F. Heck, *Org. React.*, 1982, **27**, 345.

7. T. Hayashi, M. Konishi and M. Kumada, *Tetrahedron Lett.*, 1979, **20**, 1871.

8. For a review, see Ref. 1g.

9. (a) C. A. Tolman, *J. Am. Chem. Soc.*, 1970, **92**, 2956; (b) C. A. Tolman, W. C. Seidel and L. W. Gosser, *J. Am. Chem. Soc.*, 1974, **96**, 53 for a useful review, see: (c) C. A. Tolman, *Chem. Rev.*, 1977, **77**, 313.

10. J. A. Bilbrey, A. H. Kazez, J. Locklin and W. D. Allen, *J. Comput. Chem.*, 2013, **34**, 1189.

11. (a) D. White, B. C. Taverner, N. J. Coville and P. W. Wade, *J. Organomet. Chem.*, 1995, **495**, 41; (b) D. White, B. C. Taverner, P. G. L. Leach and N. J. Coville, *J. Organomet. Chem.*, 1994, **478**, 205; (c) D. White, B. C. Taverner, P. G. L. Leach and N. J. Coville, *J. Comput. Chem.*, 1993, **14**, 1042; (d) N. J. Coville, M. S. Loonat, D. White and L. Carlton, *Organometallics*, 1992, **11**, 1082.

12. (a) R. B. DeVasher, J. M. Spruell, D. A. Dixon, G. A. Broker, S. T. Griffin, R. D. Rogers and K. H. Shaughnessy, *Organometallics*, 2005, **24**, 962; (b) L. L. Hill, L. R. Moore, R. Huang, R. Craciun, A. J. Vincent, D. A. Dixon, J. Chou, C. J. Woltermann and K. H. Shaughnessy, *J. Org. Chem.*, 2006, **71**, 5117.

13. (a) B. C. Taverner, *J. Comput. Chem.*, 1996, **17**, 1612; (b) B. C. Taverner, *Steric*, http://www.ccl.net/cca/software/SOURCES/C/steric/index.shtml, 1996 (accessed 3 May 2014).

14. J. Huang, H.-J. Schanz, E. D. Stevens and S. P. Nolan, *Organometallics*, 1999, **18**, 2370.

15. H. Clavier and S. P. Nolan, *Chem. Commun.*, 2010, **46**, 841.

16. (a) A. Poater, B. Cosenza, A. Correa, S. Giudice, F. Ragone, V. Scarano and L. Cavallo, *Eur. J. Inorg. Chem.*, 2009, 1759; (b) https://www.molnac.unisa.it/OMtools/sambvca.php (accessed 3 May 2014).

17. It was noted by the authors that this correlation does not apply well to phosphite ligands, as the Tolman cone angles may have been underestimated.

18. (a) S. Xiao, W. C. Trogler, D. E. Ellis and Z. Berkovitch-Yellin, *J. Am. Chem. Soc.*, 1983, **105**, 7033; (b) D. S. Marynick, *J. Am. Chem. Soc.*, 1984, **106**, 4064; (c) J. A. Tossell, J. H. Moore and J. C. Giordan, *Inorg. Chem.*, 1985, **24**, 1100; (d) J. C. Giordan, J. H. Moore and J. A. Tossell, *Acc. Chem. Res.*, 1986, **19**, 281; (e) A. G. Orpen and N. G. Connelly, *Organometallics*, 1990, **9**, 1206.

19. P. C. J. Kamer and P. W. N. M. van Leeuwen (eds), *Phosphorus(III) Ligands in Homogeneous Catalysis, Design and Synthesis*, Wiley, Chichester, 2012.

20. (a) C. A. Streuli, *Anal. Chem.*, 1960, **32**, 985; (b) M. N. Golovin, M. M. Rahman, J. E. Belmonte and W. P. Giering, *Organometallics*, 1985, **4**, 1981; (c) T. Allman and R. G. Goel, *Can. J. Chem.*, 1982, **60**, 716.

21. C. H. Suresh and N. Koga, *Inorg. Chem.*, 2002, **41**, 1573.

22. C. A. Tolman, *J. Am. Chem. Soc.*, 1970, **92**, 2953.

23. (a) F. Weinhold and C. R. Landis, *Valency and Bonding: a Natural Bond Orbital Donor–Acceptor Perspective*, Cambridge University Press, Cambridge, 2005; (b) A. E. Reed, R. B. Weinstock and F. Weinhold, *J. Chem. Phys.*, 1985, **83**, 735; (c) A. E. Reed and F. Weinhold, *J. Chem. Phys.*, 1985, **83**, 1736; (d) A. E. Reed, L. A. Curtiss and F. Weinhold, *Chem. Rev.*, 1988, **88**, 899.

24. T. Leyssens, D. Peeters, A. G. Orpen and J. N. Harvey, *Organometallics*, 2007, **26**, 2637.

25. (a) B. R. Higginson, D. R. Lloyd, J. A. Connor and I. H. Hillier, *J. Chem. Soc., Faraday Trans. II*, 1974, **70**, 1418; (b) L. W. Yarbrough II and M. B. Hall, *Inorg. Chem.*, 1978, **17**, 2269; (c) H. Daaman, A. Oskam and D. J. Stufkens, *Inorg. Chim. Acta*, 1980, **38**, 71; (d) M. Gerloch and R. G. Woolley, *Prog. Inorg. Chem.*, 1983, **31**, 371; (e) B. E. Bursten, D. J. Darensbourg, G. E. Kellogg and D. L. Lichtenberger, *Inorg. Chem.*, 1984, **23**, 4361.

26. R. J. Morris and G. S. Girolami, *Inorg. Chem.*, 1990, **29**, 4167.
27. For reviews, see: (a) M.-N. Birkholz, Z. Freixa and P. W. N. M. van Leeuwen, *Chem. Soc. Rev.*, 2009, **38**, 1099; (b) Z. Freixa and P. W. N. M. van Leeuwen, *J. Chem. Soc., Dalton Trans.*, 2003, 1890; (c) P. C. J. Kamer, P. W. N. M. van Leeuwen and J. N. H. Reek, *Acc. Chem. Res.*, 2001, **34**, 895; (d) P. W. N. M. van Leeuwen, P. C. J. Kamer, J. N. H. Reek and P. Dierkes, *Chem. Rev.*, 2000, **100**, 2741.
28. G. Mann, Q. Shelby, A. H. Roy and J. F. Hartwig, *Organometallics*, 2003, **22**, 2775.
29. G. A. Grasa and T. J. Colacot, *Org. Lett.*, 2007, **9**, 5489.
30. C. P. Casey and G. T. Whiteker, *Isr. J. Chem.*, 1990, **20**, 199.
31. P. Dierkes and P. W. N. M. van Leeuwen, *J. Chem. Soc., Dalton Trans.*, 1999, 1519.
32. Y. Koide, S. G. Bott and A. R. Barron, *Organometallics*, 1996, **15**, 2213.
33. BDE = bond dissociation energy.
34. Y.-R. Luo, *Handbook of Bond Dissociation Energies in Organic Compounds*, CRC Press, Boca Raton, FL, 2003.
35. Selected examples: (a) M. B. Mitchell and P. J. Wallbank, *Tetrahedron Lett.*, 1991, **32**, 2273; (b) N. P. Reddy and M. Tanaka, *Tetrahedron Lett.*, 1997, **38**, 4807; (c) W. Shen, *Tetrahedron Lett.*, 1997, **38**, 5575; (d) N. A. Bumagin and V. V. Bykov, *Tetrahedron*, 1997, **53**, 14437; (e) F. Firooznia, C. Gude, K. Chan and Y. Satoh, *Tetrahedron Lett.*, 1998, **39**, 3985.
36. M. Huser, M.-T. Youinou and J. A. Osborn, *Angew. Chem. Int. Ed. Engl.*, 1989, **28**, 1386.
37. Y. Ben-David, M. Portnoy and D. Milstein, *J. Am. Chem. Soc.*, 1989, **111**, 8742.
38. M. Nishiyama, T. Yamamoto and Y. Koie, *Tetrahedron Lett.*, 1998, **39**, 617.
39. T. Yamamoto, M. Nishiyama and Y. Koie, *Tetrahedron Lett.*, 1998, **39**, 2367.
40. A. F. Littke and G. C. Fu, *Angew. Chem. Int. Ed.*, 1998, **37**, 3387.
41. A. F. Littke, C. Dai and G. C. Fu, *J. Am. Chem. Soc.*, 2000, **122**, 4020.
42. F. Schoenebeck and K. N. Houk, *J. Am. Chem. Soc.*, 2010, **132**, 2496.
43. F. Proutiere and F. Schoenebeck, *Angew. Chem. Int. Ed.*, 2011, **50**, 8192.
44. (a) A. F. Littke and G. C. Fu, *Angew. Chem. Int. Ed.*, 1999, **38**, 2411; (b) A. F. Littke, L. Schwarz and G. C. Fu, *J. Am. Chem. Soc.*, 2002, **124**, 6343.
45. C. Dai and G. C. Fu, *J. Am. Chem. Soc.*, 2001, **123**, 2719.
46. (a) A. F. Littke and G. C. Fu, *J. Org. Chem.*, 1999, **64**, 10; (b) A. F. Littke and G. C. Fu, *J. Am. Chem. Soc.*, 2001, **123**, 6989.
47. M. Kawatsura and J. F. Hartwig, *J. Am. Chem. Soc.*, 1998, **121**, 1473.
48. Available in bulk quantities from Johnson Matthey Catalysis and Chiral Technologies, http://www.jmcct.com (accessed 3 May 2014).
49. J. F. Hartwig, M. Kawatsura, S. I. Hauck, K. H. Shaughnessy and L. M. Alcazar-Roman, *J. Org. Chem.*, 1999, **64**, 5575.

50. (a) K. H. Shaughnessy, P. Kim and J. F. Hartwig, *J. Am. Chem. Soc.*, 1999, **121**, 2123; (b) J. P. Stambuli, S. R. Stauffer, K. H. Shaughnessy and J. F. Hartwig, *J. Am. Chem. Soc.*, 2001, **123**, 2677.
51. S. R. Stauffer, N. A. Beare, J. P. Stambuli and J. F. Hartwig, *J. Am. Chem. Soc.*, 2001, **123**, 4641.
52. (a) M. Jørgensen, S. Lee, X. Liu, J. P. Wolkowski and J. F. Hartwig, *J. Am. Chem. Soc.*, 2002, **124**, 12557; (b) T. Hama and J. F. Hartwig, *Org. Lett.*, 2008, **10**, 1545; (c) T. Hama and J. F. Hartwig, *Org. Lett.*, 2008, **10**, 1549; (d) D. S. Huang, R. J. DeLuca and J. F. Hartwig, *Org. Synth.*, 2011, **88**, 4.
53. J. P. Stambuli, R. Kuwano and J. F. Hartwig, *Angew. Chem. Int. Ed.*, 2002, **41**, 4746.
54. (a) T. J. Colacot, *Platinum Met. Rev.*, 2009, **53**, 183; (b) T. J. Colacot, in *e-EROS Encyclopedia of Reagents for Organic Synthesis*, Wiley, Hoboken, NJ, 2009, DOI: 10.1002/047084289X; (c) T. J. Colacot, M. W. Hooper and G. A. Grasa, *Int. Pat.*, WO 2011012889, 2010.
55. Hartwig and co-workers have also reported the synthesis and use of both 1-adamantyl- and 2-adamantyl-substituted trialkylphosphines in high-throughput screening studies: see Refs 50b and 51.
56. A. Zapf, A. Ehrentraut and M. Beller, *Angew. Chem. Int. Ed.*, 2000, **39**, 4153.
57. A. Tewari, M. Hein, A. Zapf and M. Beller, *Synthesis*, 2004, **6**, 935.
58. G. A. Molander and P. E. Gormisky, *J. Org. Chem.*, 2008, **73**, 7481.
59. (a) A. Ehrentraut, A. Zapf and M. Beller, *J. Mol. Catal. A: Chem.*, 2002, **182–183**, 515; (b) A. Tewari, M. Hein and M. Beller, *Tetrahedron*, 2005, **61**, 9705.
60. A. Ehrentraut, A. Zapf and M. Beller, *Adv. Synth. Catal.*, 2002, **344**, 209.
61. X. Liu and J. F. Hartwig, *Org. Lett.*, 2003, **5**, 1915.
62. A. Köllhofer, T. Pullmann and H. Plenio, *Angew. Chem. Int. Ed.*, 2003, **42**, 1056.
63. Selected examples: (a) D. Gelman and S. L. Buchwald, *Angew. Chem. Int. Ed.*, 2003, **42**, 5993; (b) K. W. Anderson and S. L. Buchwald, *Angew. Chem. Int. Ed.*, 2005, **44**, 6173; (c) C. Yi and R. Hua, *J. Org. Chem.*, 2006, **71**, 2535; (d) C. Torborg, J. Huang, T. Schulz, B. Schäffner, A. Zapf, A. Börner and M. Beller, *Chem. Eur. J.*, 2009, **15**, 1329; (e) W. Shu and S. L. Buchwald, *Chem. Sci.*, 2011, **2**, 2321; (f) X. Pu, H. Li and T. J. Colacot, *J. Org. Chem.*, 2013, **78**, 568.
64. (a) S. Klaus, H. Neumann, A. Zapf, D. Strübing, S. Hübner, J. Almena, T. Riermeier, P. Gross, M. Sarich, W.-R. Krahnert, K. Rossen and M. Beller, *Angew. Chem. Int. Ed.*, 2006, **45**, 154; (b) A. Brennführer, H. Neumann, S. Klaus, T. Riermeier, J. Almena and M. Beller, *Tetrahedron*, 2007, **63**, 6252.
65. A. G. Sergeev, A. Zapf, A. Spannenberg and M. Beller, *Organometallics*, 2008, **27**, 297.
66. A. G. Sergeev, A. Spannenberg and M. Beller, *J. Am. Chem. Soc.*, 2008, **130**, 15549.
67. See Ref. 66 and references therein.

68. (a) R. Mason, M. Textor, N. Al-Salem and B. L. Shaw, *J. Chem. Soc., Chem. Commun.*, 1976, 292; (b) R. B. King, J. C. Cloyd Jr and R. H. Reimann, *J. Org. Chem.*, 1976, **41**, 972; (c) R. B. King, J. C. Cloyd Jr, M. E. Norins and R. H. Reimann, *J. Coord. Chem.*, 1977, 7, 23; (d) L. Dahlenburg and N. Höck, *Inorg. Chim. Acta*, 1995, **104**, L29.

69. For the single attempt at a cross-coupling reaction prior to 2006 using trineopentylphosphine as a supporting ligand, see: K. Ogawa, K. R. Radke, S. D. Rothstein and S. C. Rasmussen, *J. Org. Chem.*, 2001, **66**, 9067

70. L. L. Hill, J. M. Smith, W. S. Brown, L. R. Moore, P. Guevera, E. S. Pair, J. Porter, J. Chou, C. J. Wolterman, R. Craciun, D. A. Dixon and K. H. Shaughnessy, *Tetrahedron*, 2008, **64**, 6920.

71. L. L. Hill, J. L. Crowell, S. L. Tutwiler, N. L. Massie, C. C. Hines, S. T. Griffin, R. D. Rogers, K. H. Shaughnessy, G. A. Grasa, C. C. C. Johansson Seechurn, H. Li, T. J. Colacot, J. Chou and C. J. Woltermann, *J. Org. Chem.*, 2010, **75**, 6477.

72. C. C. C. Johansson Seechurn, S. L. Parisel and T. J. Colacot, *J. Org. Chem.*, 2011, **76**, 7918; this work is described in more detail in Chapter 3.

73. S. M. Raders, J. N. Moore, J. K. Parks, A. D. Miller, T. M. Leissing, S. P. Kelley, R. D. Rogers and K. H. Shaughnessy, *J. Org. Chem.*, 2013, **78**, 4649.

74. T. Hayashi, M. Konishi, Y. Kobori, M. Kumada, T. Higuchi and K. Hirotsu, *J. Am. Chem. Soc.*, 1984, **106**, 158.

75. M. S. Driver and J. F. Hartwig, *J. Am. Chem. Soc.*, 1996, **118**, 7217.

76. G. M. Whitesides, J. F. Gaasch and E. R. Stedronsky, *J. Am. Chem. Soc.*, 1972, **94**, 5258.

77. See Ref. 27d and references therein.

78. (a) T. J. Colacot, H. Qian, R. Cea-Olivares and S. Hernandez-Ortega, *J. Organomet. Chem.*, 2001, **637–639**, 691; (b) www.jmcct.com (accessed 3 May 2014).

79. (a) T. J. Colacot, *Platinum Met. Rev.*, 2001, **45**, 22; (b) S. W. Chien and T. S. A. Hor, in *Ferrocenes: Ligands, Materials and Biomolecules*, ed. P. Štěpnička, Wiley, Chichester, 2008, Ch. 2, pp. 33–116; see also Ref. 1d.

80. (a) M. Kumada, Y. Kiso and M. Umeno, *J. Chem. Soc. D, Chem. Commun.*, 1970, 611; (b) W. R. Cullen, T.-J. Kim, F. W. B. Einstein and T. Jones, *Organometallics*, 1983, **2**, 714; (c) I. R. Butler, W. R. Cullen and T.-J. Kim, *Synth. React. Inorg. Met.-Org. Chem.*, 1985, **15**, 109; (d) A. R. Elsagir, F. Gassner, H. Görls and E. Dinjus, *J. Organomet. Chem.*, 2000, **597**, 139.

81. T. J. Colacot and H. A. Shea, *Org. Lett.*, 2004, **6**, 3731.

82. (a) T. Itoh, K. Sato and T. Mase, *Adv. Synth. Catal.*, 2004, **346**, 1859; (b) T. Itoh and T. Mase, *Tetrahedron Lett.*, 2005, **46**, 3573.

83. I. R. Butler, W. R. Cullen, T.-J. Kim, S. J. Rettig and J. Trotter, *Organometallics*, 1985, **4**, 972.

84. G. A. Grasa and T. J. Colacot, *Org. Process Res. Dev.*, 2008, **12**, 522.

85. M. Shen, G. Li, B. Z. Lu, A. Hossain, F. Roschangar, V. Farina and C. H. Senanayake, *Org. Lett.*, 2004, **6**, 4129.

86. B. C. Hamann and J. F. Hartwig, *J. Am. Chem. Soc.*, 1998, **120**, 3694.

87. B. C. Hamann and J. F. Hartwig, *J. Am. Chem. Soc.*, 1998, **120**, 7369.

88. (a) A. Togni, C. Breutel, M. C. Soares, N. Zanetti, T. Gerfin, V. Gramlich, F. Spindler and G. Rihs, *Inorg. Chim. Acta*, 1994, **222**, 213; (b) R. Imwinkelried, *Chimia*, 1997, **51**, 300; (c) N. C. Zanetti, F. Spindler, J. Spencer, A. Togni and G. Rihs, *Organometallics*, 1996, **15**, 860.

89. For a review, see: J. F. Hartwig, *Acc. Chem. Res.*, 2008, **41**, 1534

90. A. Togni, *Chimia*, 1996, **50**, 86.

91. (a) F. Paul, J. Patt and J. F. Hartwig, *Organometallics*, 1995, **14**, 3030; (b) S. Wagaw and S. L. Buchwald, *J. Org. Chem.*, 1996, **61**, 7240.

92. (a) Q. Shen, S. Shekhar, J. P. Stambuli and J. F. Hartwig, *Angew. Chem. Int. Ed.*, 2005, **44**, 1371; (b) Q. Shen, T. Ogata and J. F. Hartwig, *J. Am. Chem. Soc.*, 2008, **130**, 6586.

93. (a) S. Urgaonkar, J.-H. Xu and J. G. J. Verkade, *J. Org. Chem.*, 2003, **68**, 8416; (b) C. Bolm and J. P. Hildebrand, *J. Org. Chem.*, 2000, **65**, 169; (c) J. P. Wolfe and S. L. Buchwald, *J. Org. Chem.*, 1996, **61**, 1133; (d) M. Kosugi, M. Kameyama and T. Migita, *Chem. Lett.*, 1983, **12**, 927.

94. Q. Shen and J. F. Hartwig, *J. Am. Chem. Soc.*, 2006, **128**, 10028.

95. G. D. Vo and J. F. Hartwig, *J. Am. Chem. Soc.*, 2009, **131**, 11049.

96. H. Li, G. A. Grasa and T. J. Colacot, *Org. Lett.*, 2010, **12**, 3332.

97. (a) M. A. Fernández-Rodríguez, Q. Shen and J. F. Hartwig, *J. Am. Chem. Soc.*, 2006, **128**, 2180; (b) M. A. Fernández-Rodríguez, Q. Shen and J. F. Hartwig, *Chem. Eur. J.*, 2006, **12**, 7782.

98. M. A. Fernández-Rodríguez and J. F. Hartwig, *J. Org. Chem.*, 2009, **74**, 1663.

99. A. Miyashita, A. Yasuda, H. Takaya, K. Toriumi, T. Ito, T. Souchi and R. Noyori, *J. Am. Chem. Soc.*, 1980, **102**, 7932.

100. J. P. Wolfe, S. Wagaw and S. L. Buchwald, *J. Am. Chem. Soc.*, 1996, **118**, 7215.

101. J. P. Wolfe and S. L. Buchwald, *J. Org. Chem.*, 2000, **65**, 1144.

102. F. Ozawa, A. Kubo, Y. Matsumoto, T. Hayashi, E. Nishioka, K. Yanagi and K. Moriguchi, *Organometallics*, 1993, **12**, 4188.

103. J. P. Wolfe and S. L. Buchwald, *J. Org. Chem.*, 1997, **62**, 1264.

104. J. P. Wolfe and S. L. Buchwald, *J. Org. Chem.*, 1997, **62**, 6066.

105. J. P. Wolfe, J. Åhman, J. P. Sadighi, R. A. Singer and S. L. Buchwald, *Tetrahedron Lett.*, 1997, **38**, 6367.

106. (a) S. Wagaw, B. H. Yang and S. L. Buchwald, *J. Am. Chem. Soc.*, 1998, **120**, 6621; (b) S. Wagaw, B. H. Yang and S. L. Buchwald, *J. Am. Chem. Soc.*, 1999, **121**, 10251.

107. S. Wagaw, R. A. Rennels and S. L. Buchwald, *J. Am. Chem. Soc.*, 1997, **119**, 8451.

108. Selected examples: (a) M. P. Wentland, W. Duan, D. J. Cohen and J. M. Bidlack, *J. Med. Chem.*, 2000, **43**, 3558; (b) C. Meier and S. Gräsl, *Synlett*, 2002, 802; (c) M. Hepperle, J. Eckert, D. Gala, L. Shen,

C. A. Evans and A. Goodman, *Tetrahedron Lett.*, 2002, **43**, 3359; (d) J. T. Kuethe, A. Wong and I. W. Davies, *J. Org. Chem.*, 2004, **69**, 7752; (e) C. E. Elmquist, J. S. Stover, Z. Wang and C. J. Rizzo, *J. Am. Chem. Soc.*, 2004, **126**, 11189. For a study on solvent effects in Pd/BINAP-catalyzed amination reactions, see: (f) H. Christensen, S. Kiil and K. Dam-Johansen, *Org. Process Res. Dev.*, 2006, **10**, 762.

109. M. Palucki, J. P. Wolfe and S. L. Buchwald, *J. Am. Chem. Soc.*, 1997, **119**, 3395.
110. K. H. Shaughnessy, B. C. Hamann and J. F. Hartwig, *J. Org. Chem.*, 1998, **63**, 6546.
111. M. P. Palucki and S. L. Buchwald, *J. Am. Chem. Soc.*, 1997, **119**, 11108.
112. R. Martín and S. L. Buchwald, *Angew. Chem. Int. Ed.*, 2007, **46**, 7236.
113. L. M. Alcazar-Roman, J. F. Hartwig, A. L. Rheingold, L. M. Liable-Sands and I. A. Guzei, *J. Am. Chem. Soc.*, 2000, **122**, 4618.
114. U. K. Singh, E. R. Strieter, D. G. Blackmond and S. L. Buchwald, *J. Am. Chem. Soc.*, 2002, **124**, 14104.
115. S. Shekhar, P. Ryberg and J. F. Hartwig, *Org. Lett.*, 2006, **8**, 851.
116. S. Shekhar, P. Ryberg, J. F. Hartwig, J. S. Mathew, D. G. Blackmond, E. R. Strieter and S. L. Buchwald, *J. Am. Chem. Soc.*, 2006, **128**, 3584.
117. Y. Guari, D. S. van Es, J. N. H. Reek, P. C. J. Kamer and P. W. N. M. van Leeuwen, *Tetrahedron Lett.*, 1999, **40**, 3789.
118. M. Kranenburg, Y. E. M. van der Burgt, P. C. J. Kamer, P. W. N. M. van Leeuwen, K. Goubitz and J. Fraanje, *Organometallics*, 1995, **14**, 3081.
119. M. Kranenburg, J. G. P. Delis, P. C. J. Kamer, P. W. N. M. van Leeuwen, K. Vrieze, N. Veldman, A. L. Spek, K. Goubitz and J. Fraanje, *J. Chem. Soc., Dalton Trans.*, 1997, 1839.
120. J. Yin and S. L. Buchwald, *Org. Lett.*, 2000, **2**, 1101.
121. J. Yin and S. L. Buchwald, *J. Am. Chem. Soc.*, 2002, **124**, 6043.
122. (a) K.-C. Kong and C.-H. Cheng, *J. Am. Chem. Soc.*, 1991, **113**, 6313; (b) B. E. Segelstein, T. W. Butler and B. L. Chenard, *J. Org. Chem.*, 1995, **60**, 12; (c) W. A. Herrmann, C. Brossmer, K. Öfele, M. Beller and H. Fischer, *J. Organomet. Chem.*, 1995, **491**, C1.
123. Shortly prior, van Leeuwen and co-workers had reported the X-ray crystallographically determined structure of *trans*-(Xantphos)Pd(Me)Cl: P. C. J. Kamer, P. W. N. M. van Leeuwen and J. N. H. Reek, *Acc. Chem. Res.*, 2001, **34**, 895.
124. M. Sato, H. Shigeta, M. Sekino and S. Akabori, *J. Organomet. Chem.*, 1993, **458**, 199.
125. M. C. Harris, O. Geis and S. L. Buchwald, *J. Org. Chem.*, 1999, **64**, 6019.
126. J. Yin, M. M. Zhao, M. A. Huffman and J. M. McNamara, *Org. Lett.*, 2002, **4**, 3481.
127. J. Ji, T. Li and W. H. Bunnelle, *Org. Lett.*, 2003, **5**, 4611.
128. D. J. Wallace, D. J. Klauber, C.-y. Chen and R. P. Volante, *Org. Lett.*, 2003, **5**, 4749.
129. Selected examples: Ref. 106b and (a) J. M. Harris and A. Padwa, *Org. Lett.*, 2003, **5**, 4195; (b) D. Steinhuebel, M. Palucki, D. Askin and

U. Dolling, *Tetrahedron Lett.*, 2004, **45**, 3305; (c) A. Kuwahara, K. Nakano and K. Nozaki, *J. Org. Chem.*, 2005, **70**, 413; (d) M. C. Willis, R. H. Snell, A. J. Fletcher and R. L. Woodward, *Org. Lett.*, 2006, **8**, 5089; (e) S. Piguel and M. Legraverend, *J. Org. Chem.*, 2007, **72**, 7026; (f) Z. Shen, Y. Hong, X. He, W. Mo, B. Hu, N. Sun and X. Hu, *Org. Lett.*, 2010, **12**, 552; (g) F. Saab, V. Bénéteau, F. Schoentgen, J.-Y. Mérour and S. Routier, *Tetrahedron*, 2010, **66**, 102; (h) M. Vimolratana, J. L. Simard and S. P. Brown, *Tetrahedron Lett.*, 2011, **52**, 1020; (i) G. M. Noonan, A. P. Dishington, J. Pink and A. D. Campbell, *Tetrahedron Lett.*, 2012, **53**, 3038; (j) G. W. Stewart, K. M. J. Brands, S. E. Brewer, C. J. Cowden, A. J. Davies, J. S. Edwards, A. W. Gibson, S. E. Hamilton, J. D. Katz, S. P. Keen, P. R. Mullens, J. P. Scott, D. J. Wallace and C. S. Wise, *Org. Res. Process Dev.*, 2010, **14**, 849; (k) J. Barluenga, C. Valdés, G. Beltrán, M. Escribano and F. Aznar, *Angew. Chem. Int. Ed.*, 2006, **45**, 6893.

130. J. P. Wolfe, in *e-EROS Encyclopedia of Reagents for Organic Synthesis*, Wiley, Hoboken, NJ, 2011, DOI: 10.1002/047084289X.rn00854.pub2.
131. M. Kranenburg, P. C. J. Kamer and P. W. N. M. van Leeuwen, *Eur. J. Inorg. Chem.*, 1998, **1998**, 155.
132. J. P. Sadighi, M. C. Harris and S. L. Buchwald, *Tetrahedron Lett.*, 1998, **39**, 5327.
133. Selected examples: (a) R. A. Singer and S. L. Buchwald, *Tetrahedron Lett.*, 1999, **40**, 1095; (b) B. H. Yang and S. L. Buchwald, *Org. Lett.*, 1999, **1**, 35; (c) A. Y. Lebedev, A. S. Khartulyari and A. Z. Voskoboynikov, *J. Org. Chem.*, 2005, **70**, 596; (d) B. J. Margolis, K. A. Long, D. L. T. Laird, J. C. Ruble and S. R. Pulley, *J. Org. Chem.*, 2007, **72**, 2232.
134. (a) J.-c. Shi, X. Zeng and E.-i. Negishi, *Org. Lett.*, 2003, **5**, 1825; (b) X. Zeng, Q. Hu, M. Qian and E.-i. Negishi, *J. Am. Chem. Soc.*, 2003, **125**, 13636; (c) M. Qian and E.-i. Negishi, *Tetrahedron Lett.*, 2005, **46**, 2927.
135. I. N. Houpis, C. Huang, U. Nettekoven, J. G. Chen, R. Liu and M. Canters, *Org. Lett.*, 2008, **10**, 5601.
136. U. Schopfer and A. Schlapbach, *Tetrahedron*, 2001, **57**, 3069.
137. Selected examples: (a) W. Yin, M. S. Kabir, Z. Wang, S. K. Rallapalli, J. Ma and J. M. Cook, *J. Org. Chem.*, 2010, **75**, 3339; (b) J. Y. L. Chung, D. Steinhuebel, S. W. Krska, F. W. Hartner, C. Cai, J. Rosen, D. E. Mancheno, T. Pei, L. DiMichele, R. G. Ball, C.-y Chen, L. Tan, A. D. Alorati, S. E. Brewer and J. P. Scott, *Org. Process Res. Dev.*, 2012, **16**, 1832.
138. (a) J. P. Wolfe and M. A. Rossi, *J. Am. Chem. Soc.*, 2004, **126**, 1620; (b) M. B. Hay, A. R. Hardin and J. P. Wolfe, *J. Org. Chem.*, 2005, **70**, 3099; (c) M. B. Hay and J. P. Wolfe, *Angew. Chem. Int. Ed.*, 2007, **46**, 6492. Also see Ref. 140 and references therein.
139. (a) R. Lira and J. P. Wolfe, *J. Am. Chem. Soc.*, 2004, **126**, 13906; (b) M. B. Bertrand, J. D. Neukom and J. P. Wolfe, *J. Org. Chem.*, 2008, **73**, 8851; (c) N. C. Giampietro and J. P. Wolfe, *J. Am. Chem. Soc.*, 2008, **130**, 12907. Also see Ref. 140 and references therein.

140. (a) J. P. Wolfe, *Eur. J. Org. Chem.*, 2007, **2007**, 571; (b) J. P. Wolfe, *Synlett*, 2008, 2913.
141. A. Takemiya and J. F. Hartwig, *J. Am. Chem. Soc.*, 2006, **128**, 14800.
142. The common names were later reported with the family of biarylphosphines developed by Buchwald and co-workers.
143. D. W. Old, J. P. Wolfe and S. L. Buchwald, *J. Am. Chem. Soc.*, 1998, **120**, 9722.
144. J.-F. Marcoux, S. Wagaw and S. L. Buchwald, *J. Org. Chem.*, 1997, **62**, 1568.
145. T. Hayashi, T. Mise, M. Fukushima, M. Kagotani, N. Nagashima, Y. Hamada, A. Matsumoto, S. Kawakami, M. Konishi, K. Yamamoto and M. Kumada, *Bull. Chem. Soc. Jpn.*, 1980, **53**, 1138.
146. (a) J. P. Wolfe and S. L. Buchwald, *Angew. Chem. Int. Ed.*, 1999, **38**, 2413; (b) J. P. Wolfe, R. A. Singer, B. H. Yang and S. L. Buchwald, *J. Am. Chem. Soc.*, 1999, **121**, 9550.
147. J. P. Wolfe, H. Tomori, J. P. Sadighi, J. Yin and S. L. Buchwald, *J. Org. Chem.*, 2000, **65**, 1158.
148. A. Aranyos, D. W. Old, A. Kiyomori, J. P. Wolfe, J. P. Sadighi and S. L. Buchwald, *J. Am. Chem. Soc.*, 1999, **121**, 4369.
149. (a) H. Ossor, M. Pfeffer, J. T. B. H. Jastrzebski and C. H. Stam, *Inorg. Chem.*, 1987, **26**, 1169; (b) L. R. Falvello, J. Forniés, R. Navarro, V. Sicilia and M. Tomás, *Angew. Chem. Int. Ed. Engl.*, 1990, **29**, 891; (c) C.-S. Li, C.-H. Cheng, F.-L. Liao and S.-L. Wang, *J. Chem. Soc., Chem. Commun.*, 1991, 710; (d) S. Kannan, A. J. James and P. R. Sharp, *J. Am. Chem. Soc.*, 1998, **120**, 215.
150. The air stability of dialkylbiarylphosphines has since been studied in detail: T. E. Barder and S. L. Buchwald, *J. Am. Chem. Soc.*, 2007, **129**, 5096.
151. J. M. Fox, X. Huang, A. Chieffi and S. L. Buchwald, *J. Am. Chem. Soc.*, 2000, **122**, 1360.
152. S. Vyskočil, M. Smrčina, V. Hanuš, M. Polášek and P. Kočovský, *J. Org. Chem.*, 1998, **63**, 7738.
153. Y. Uozumi and T. Hayashi, *J. Am. Chem. Soc.*, 1991, **113**, 9887.
154. P. Kočovský, S. Vyskočil, I. Císařová, J. Sejbal, I. Tišlerová, M. Smrčina, G. C. Lloyd-Jones, S. C. Stephen, C. P. Butts, M. Murray and V. Langer, *J. Am. Chem. Soc.*, 1999, **121**, 7714.
155. H. Tomori, J. M. Fox and S. L. Buchwald, *J. Org. Chem.*, 2000, **65**, 5334.
156. Improvements have been made to the benzyne synthesis of biarylphosphines since the initial report, see: (a) S. Kaye, J. M. Fox, F. A. Hicks and S. L. Buchwald, *Adv. Synth. Catal.*, **343**, 789; (b) N. Hoshiya and S. L. Buchwald, *Adv. Synth. Catal.*, 2012, **354**, 2031.
157. Recent examples: (a) D. Maiti, B. P. Fors, J. L. Henderson, Y. Nakamura and S. L. Buchwald, *Chem. Sci.*, 2011, **2**, 57; (b) B. P. Fors and S. L. Buchwald, *J. Am. Chem. Soc.*, 2010, **132**, 15914; (c) J. L. Henderson and S. L. Buchwald, *Org. Lett.*, 2010, **12**, 4442; (d) J. L. Henderson, S. M. McDermott and S. L. Buchwald, *Org. Lett.*, 2010, **12**, 4438;

(e) B. P. Fors, D. A. Watson, M. R. Biscoe and S. L. Buchwald, *J. Am. Chem. Soc.*, 2008, **130**, 13552. For reviews, see Refs 1i and 1m.

158. Recent examples: Ref. 63e and (a) M. A. Düfert, K. L. Billingsley and S. L. Buchwald, *J. Am. Chem. Soc.*, 2013, **135**, 12877; (b) Y. Yang, N. J. Oldenhius and S. L. Buchwald, *Angew. Chem. Int. Ed.*, 2013, **52**, 615; (c) M. A. Oberli and S. L. Buchwald, *Org. Lett.*, 2012, **14**, 4606; (d) T. Kinzel, Y. Zhang and S. L. Buchwald, *J. Am. Chem. Soc.*, 2010, **132**, 14073. For a review, see: (e) R. Martin and S. L. Buchwald, *Acc. Chem. Res.*, 2008, **41**, 1461.

159. Recent examples: (a) C. W. Cheung and S. L. Buchwald, *Org. Lett.*, 2013, **15**, 3998; (b) L. Salvi, N. R. Davis, S. Z. Ali and S. L. Buchwald, *Org. Lett.*, 2012, **14**, 170; (c) X. Wu, B. P. Fors and S. L. Buchwald, *Angew. Chem. Int. Ed.*, 2011, **50**, 9943; (d) K. W. Anderson, T. Ikawa, R. E. Tundel and S. L. Buchwald, *J. Am. Chem. Soc.*, 2006, **128**, 10694.

160. (a) F. Barrios-Landeros and J. F. Hartwig, *J. Am. Chem. Soc.*, 2005, **127**, 6944; (b) J. F. Hartwig and F. Paul, *J. Am. Chem. Soc.*, 1995, **117**, 5373.

161. (a) J. F. Hartwig, *Inorg. Chem.*, 2007, **46**, 1936; and references therein; (b) T. E. Barder and S. L. Buchwald, *J. Am. Chem. Soc.*, 2007, **129**, 12003.

162. T. E. Barder, M. R. Biscoe and S. L. Buchwald, *Organometallics*, 2007, **26**, 2183.

163. E. R. Strieter and S. L. Buchwald, *Angew. Chem. Int. Ed.*, 2006, **45**, 925.

164. X. Huang, K. W. Anderson, D. Zim, L. Jiang, A. Klapars and S. L. Buchwald, *J. Am. Chem. Soc.*, 2003, **125**, 6653.

165. Representative examples: Ref. 164 and (a) M. R. Biscoe, B. P. Fors and S. L. Buchwald, *J. Am. Chem. Soc.*, 2008, **130**, 6686; (b) K. W. Anderson, R. E. Tundel, T. Ikawa, R. A. Altman and S. L. Buchwald, *Angew. Chem. Int. Ed.*, 2006, **45**, 6523; (c) M. D. Charles, P. Schultz and S. L. Buchwald, *Org. Lett.*, 2005, 7, 3965.

166. G. R. Dick, E. M. Woerly and M. D. Burke, *Angew. Chem. Int. Ed.*, 2012, **51**, 2667.

167. R. Martin and S. L. Buchwald, *J. Am. Chem. Soc.*, 2007, **129**, 3844.

168. (a) J. R. Naber, B. P. Fors, X. Wu, J. T. Gunn and S. L. Buchwald, *Heterocycles*, 2010, **80**, 1215; (b) R. Selig, M. Goettert, V. Schattel, D. Schollmeyer, W. Albrecht and S. Laufer, *J. Med. Chem.*, 2012, **55**, 8429; (c) J. R. Naber and S. L. Buchwald, *Adv. Synth. Catal.*, 2008, **350**, 957.

169. Representative examples: (a) M. R. Biscoe and S. L. Buchwald, *Org. Lett.*, 2009, **11**, 1773; (b) H. N. Nguyen, X. Huang and S. L. Buchwald, *J. Am. Chem. Soc.*, 2003, **125**, 11818.

170. Representative examples: (a) G. A. Molander, S. L. J. Trice, S. M. Kennedy, S. D. Dreher and M. T. Tudge, *J. Am. Chem. Soc.*, 2012, **134**, 11667; (b) G. A. Molander, S. L. J. Trice and S. M. Kennedy, *Org. Lett.*, 2012, **14**, 4814; (c) G. A. Molander, S. L. J. Trice and S. D. Dreher, *J. Am. Chem. Soc.*, 2010, **132**, 17701; (d) K. L. Billingsley, T. E. Barder and S. L. Buchwald, *Angew. Chem. Int. Ed.*, 2007, **46**, 5359.

171. (a) S. D. Walker, T. E. Barder, J. R. Martinelli and S. L. Buchwald, *Angew. Chem. Int. Ed.*, 2004, **43**, 1871. For a more detailed study, see: (b) T. E. Barder, S. D. Walker, J. R. Martinelli and S. L. Buchwald, *J. Am. Chem. Soc.*, 2005, **127**, 4685.

172. Pd(0)–arene interactions had been previously observed in related complexes: (a) J. Yin, M. P. Rainka, X.-X. Zhang and S. L. Buchwald, *J. Am. Chem. Soc.*, 2002, **124**, 1162; (b) S. M. Reid, R. C. Boyle, J. T. Mague and M. J. Fink, *J. Am. Chem. Soc.*, 2003, **125**, 7816.

173. Representative examples: Refs 158a and 167 and (a) K. L. Billingsley, K. W. Anderson and S. L. Buchwald, *Angew. Chem. Int. Ed.*, 2006, **45**, 3484; (b) A. Metzger, L. Melzig and P. Knochel, *Synthesis*, 2010, 2853; (c) Q. Tang and R. Gianatassio, *Tetrahedron Lett.*, 2010, **51**, 3473. Also see Ref. 158f and references therein.

174. Examples: Ref. 165c and (a) R. C. Hodgkinson, J. Schulz and M. C. Willis, *Tetrahedron*, 2009, **65**, 8940; (b) F. Perez and A. Minatti, *Org. Lett.*, 2011, **13**, 1984.

175. J. E. Milne and S. L. Buchwald, *J. Am. Chem. Soc.*, 2004, **126**, 13028.

176. E. J. Cho, T. D. Senecal, T. Kinzel, Y. Zhang, D. A. Watson and S. L. Buchwald, *Science*, 2010, **328**, 1679.

177. (a) D. A. Watson, M. Su, G. Teverovskiy, Y. Zhang, J. García-Fortanet, T. Kinzel and S. L. Buchwald, *Science*, 2009, **325**, 1661; (b) T. Noël, T. J. Maimone and S. L. Buchwald, *Angew. Chem. Int. Ed.*, 2011, **50**, 8900; (c) T. J. Maimone, P. J. Milner, T. Kinzel, Y. Zhang, M. K. Takase and S. L. Buchwald, *J. Am. Chem. Soc.*, 2011, **133**, 18106; (d) H. G. Lee, P. J. Milner and S. L. Buchwald, *Org. Lett.*, 2013, **15**, 5602. For the use of a teraryl phosphine ligand in Pd-catalyzed aryl fluorination, see (e) H. G. Lee, P. J. Milner and S. L. Buchwald, *J. Am. Chem. Soc.*, 2014, **136**, 3792.

178. G. Teverovskiy, D. S. Surry and S. L. Buchwald, *Angew. Chem. Int. Ed.*, 2011, **50**, 7312.

179. Cyanate: (a) E. V. Vinogradova, B. P. Fors and S. L. Buchwald, *J. Am. Chem. Soc.*, 2012, **134**, 11132; (b) E. V. Vinogradova, N. H. Park, B. P. Fors and S. L. Buchwald, *Org. Lett.*, 2013, **15**, 13942-Aminothiazoles: (c) M. A. McGowan, J. L. Henderson and S. L. Buchwald, *Org. Lett.*, 2012, **14**, 1432. Amidines: (d) M. A. McGowan, C. Z. McAvoy and S. L. Buchwald, *Org. Lett.*, 2012, **14**, 3800. Alcohols: Refs 159a–c Halogens (not F): (e) J. Pan, X. Wang, Y. Zhang and S. L. Buchwald, *Org. Lett.*, 2011, **13**, 4974.

180. (a) R. A. Singer, M. Doré, J. E. Sieser and M. A. Berliner, *Tetrahedron Lett.*, 2006, **47**, 3727; (b) G. J. Withbroe, R. A. Singer and J. E. Sieser, *Org. Process Res. Dev.*, 2008, **12**, 480; (c) S. Gowrisankar, A. G. Sergeev, P. Anbarasan, A. Spannenberg, H. Neumann and M. Beller, *J. Am. Chem. Soc.*, 2010, **132**, 11592; (d) A. Porzelle, M. D. Woodrow and N. C. O. Tomkinson, *Org. Lett.*, 2009, **11**, 233; (e) B. J. Kotecki, D. P. Fernando, A. R. Haight and K. A. Lukin, *Org. Lett.*, 2009, **11**, 947; (f) F. Rataboul, A. Zapf, R. Jackstell, S. Harkal, T. Riermeier,

A. Monsees, U. Dingerdissen and M. Beller, *Chem. Eur. J.*, 2004, **10**, 2983; (g) N. Schwarz, A. Tillack, K. Alex, I. A. Sayyed, R. Jackstell and M. Beller, *Tetrahedron Lett.*, 2007, **48**, 2897; (h) C. M. So, Z. Zhou, C. P. Lau and F. Y. Kwong, *Angew. Chem. Int. Ed.*, 2008, **47**, 6402; (i) R. A. Singer, S. Caron, R. E. McDermott, P. Arpin and N. M. Do, *Synthesis*, 2003, **2003**, 1727; (j) R. A. Singer, N. J. Tom, H. N. Frost and W. M. Simon, *Tetrahedron Lett.*, 2004, **45**, 4715; (k) K. Suzuki, Y. Hori, T. Nishikawa and T. Kobayashi, *Adv. Synth. Catal.*, 2007, **349**, 2089; (l) S. Doherty, J. G. Knight, C. H. Smyth and G. A. Jorgenson, *Adv. Synth. Catal.*, 2008, **350**, 1801; (m) C. A. Fleckenstein and H. Plenio, *Chem. Eur. J.*, 2007, **13**, 2701; (n) Q. Dai, W. Gao, D. Liu, L. M. Kapes and X. Zhang, *J. Org. Chem.*, 2006, **71**, 3928; (o) J. Ruan, L. Shearer, J. Mo, J. Bacsa, A. Zanotti-Gerosa, F. Hancock, X. Wu and J. Xiao, *Org. Biomol. Chem.*, 2009, 7, 3236; (p) R. Pratap, D. Parrish, P. Gunda, D. Venkataraman and M. K. Lakshman, *J. Am. Chem. Soc.*, 2009, **131**, 12240; (q) A. Mukherjee and A. Sarkar, *Tetrahedron Lett.*, 2004, **45**, 9525.

181. Q. Shelby, N. Kataoka, G. Mann and J. Hartwig, *J. Am. Chem. Soc.*, 2000, **122**, 10718.

182. M. Miura, S. Pivsa-Art, G. Dyker, J. Heiermann, T. Satoh and M. Nomura, *Chem. Commun.*, 1998, 1889.

183. N. Kataoka, Q. Shelby, J. P. Stambuli and J. F. Hartwig, *J. Org. Chem.*, 2002, **67**, 5553.

184. N. A. Beare and J. F. Hartwig, *J. Org. Chem.*, 2002, **67**, 541. P(t-Bu)$_3$ was also shown to be effective in several of these transformations.

185. T. Hama, D. A. Culkin and J. F. Hartwig, *J. Am. Chem. Soc.*, 2006, **128**, 4976.

186. G. D. Vo and J. F. Hartwig, *Angew. Chem. Int. Ed.*, 2008, **47**, 2127.

187. (a) S. G. Newman and M. Lautens, *J. Am. Chem. Soc.*, 2011, **133**, 1778; (b) S. G. Newman, J. K. Howell, N. Nicolaus and M. Lautens, *J. Am. Chem. Soc.*, 2011, **133**, 14916; (c) D. A. Petrone, M. Lischka and M. Lautens, *Angew. Chem. Int. Ed.*, 2013, **52**, 10635.

188. (a) R. L. Lundgren, in *e-EROS Encyclopedia of Reagents for Organic Synthesis*, Wiley, Hoboken, NJ, 2011, DOI: 10.1002/04784289X.rn01354; (b) T. J. Colacot, in *e-EROS Encyclopedia of Reagents for Organic Synthesis*, Wiley, Hoboken, NJ, 2011, DOI: 10.1002/047084289X.rn01283.

189. "Amphos" can also refer to the water-soluble (2-di-*tert*-butylphospinoethyl)trimethylammonium chloride ligand developed by Shaughnessy and Booth: K. H. Shaughnessy and R. S. Booth, *Org. Lett.*, 2001, 3, 2757.

190. A. S. Guram, A. O. King, J. G. Allen, X. Wang, L. B. Schenkel, J. Chan, E. E. Bunel, M. M. Faul, R. D. Larsen, M. J. Martinelli and P. J. Reider, *Org. Lett.*, 2006, **8**, 1787.

191. A. S. Guram, X. Wang, E. E. Bunel, M. M. Faul, R. D. Larsen and M. J. Martinelli, *J. Org. Chem.*, 2007, **72**, 5104.

192. (a) A. Krasovskiy, C. Duplais and B. H. Lipshutz, *J. Am. Chem. Soc.*, 2009, **131**, 15592; (b) A. Krasovskiy, C. Duplais and B. H. Lipshutz, *Org. Lett.*,

2010, **12**, 4742; (c) C. Duplais, A. Krasovskiy, A. Wattenburg and B. H. Lipshutz, *Chem. Commun.*, 2010, **46**, 562; (d) C. Duplais, A. Krasovskiy and B. H. Lipshutz, *Organometallics*, 2011, **30**, 6090.

193. A. He and J. R. Falck, *J. Am. Chem. Soc.*, 2010, **132**, 2524.

194. R. R. Milburn, O. R. Thiel, M. Achmatowicz, X. Wang, J. Zigterman, C. Bernard, J. T. Colyer, E. DiVirgilio, R. Crockett, T. L. Correll, K. Nagapudi, K. Ranganathan, S. J. Headley, A. Allgeier and R. D. Larsen, *Org. Process Res. Dev.*, 2011, **15**, 31.

195. R. J. Lundgren, A. Sappong-Kumankumah and M. Stradiotto, *Chem. Eur. J.*, 2010, **16**, 1983.

196. See Ref. 195 and (a) R. J. Lundgren, B. D. Peters, P. G. Alsabeh and M. Stradiotto, *Angew. Chem. Int. Ed.*, 2010, **49**, 4071; (b) R. J. Lundgren and M. Stradiotto, *Angew. Chem. Int. Ed.*, 2010, **49**, 8686; (c) P. G. Alsabeh, R. J. Lundgren, L. E. Longobardi and M. Stradiotto, *Chem. Commun.*, 2011, **47**, 6936; (d) B. J. Tardiff, R. McDonald, M. J. Ferguson and M. Stradiotto, *J. Org. Chem.*, 2011, 77, 1056; (e) B. J. Tardiff and M. Stradiotto, *Eur. J. Org. Chem.*, 2012, 3972; (f) P. G. Alsabeh, R. J. Lundgren, R. McDonald, C. C. C. Johansson Seechurn, T. J. Colacot and M. Stradiotto, *Chem. Eur. J.*, 2013, **19**, 2131.

197. (a) K. D. Hesp, R. J. Lundgren and M. Stradiotto, *J. Am. Chem. Soc.*, 2011, **133**, 5194; (b) S. M. Crawford, P. G. Alsabeh and M. Stradiotto, *Eur. J. Org. Chem.*, 2012, 6042.

198. For a review, see: R. J. Lundgren, K. D. Hesp and M. Stradiotto, *Synlett,* 2011, 2011, **17**, 2443.

199. A. J. Arduengo III, R. L. Harlow and M. Kline, *J. Am. Chem. Soc.*, 1991, **113**, 361. The first NHC–organometallic complexes were isolated 23 years prior: (a) K. Öfele, *J. Organomet. Chem.*, 1968, **12**, P42; (b) H. W. Wanzlick and H.-J. Schönherr, *Angew. Chem. Int. Ed. Engl.*, 1968, **7**, 141.

200. E. A. B. Kantchev, C. J. O'Brien and M. G. Organ, *Angew. Chem. Int. Ed.*, 2007, **46**, 2768.

201. T. Hayashi, S. Niizuma, T. Kamikawa, N. Suzuki and Y. Uozumi, *J. Am. Chem. Soc.*, 1995, **117**, 9101.

202. Y. Sato, M. Sodeoka and M. Shibasaki, *J. Org. Chem.*, 1989, **54**, 4738.

203. N. E. Carpenter, D. J. Kucera and L. E. Overman, *J. Org. Chem.*, 1989, **54**, 5846.

204. (a) A. B. Dounay and L. E. Overman, *Chem. Rev.*, 2003, **103**, 2945 and references therein; (b) D. McCartney and P. J. Guiry, *Chem. Soc. Rev.*, 2011, **40**, 5122 and references therein.

205. J. Åhman, J. P. Wolfe, M. V. Troutman, M. Palucki and S. L. Buchwald, *J. Am. Chem. Soc.*, 1998, **120**, 1918.

206. For a review, see: C. C. C. Johansson and T. J. Colacot, *Angew. Chem. Int. Ed.*, 2010, **49**, 676.

207. (a) D. J. Spielvogel, W. M. Davis and S. L. Buchwald, *Organometallics*, 2002, **21**, 3833; (b) D. J. Spielvogel and S. L. Buchwald, *J. Am. Chem. Soc.*, 2002, **124**, 3500.

208. S. Jeulin, S. D. de Paule, V. Ratovelomanana-Vidal, J.-P. Genêt, N. Champion and P. Dellis, *Angew. Chem. Int. Ed.*, 2004, **43**, 320. For a review, see: J.-P. Genet, T. Ayad and V. Ratovelomanana-Vidal, *Chem. Rev.*, 2014, **114**, 2824

209. S. Ge and J. F. Hartwig, *J. Am. Chem. Soc.*, 2011, **133**, 16330.

210. X. Liao, Z. Weng and J. F. Hartwig, *J. Am. Chem. Soc.*, 2008, **130**, 195.

211. G. Chen, F. Y. Kwong, H. O. Chan, W.-Y. Yu and A. S. C. Chan, *Chem. Commun.*, 2006, 1413.

212. C.-C. Pai, C.-W. Lin, C.-C. Lin, C.-C. Chen and A. S. C. Chan, *J. Am. Chem. Soc.*, 2000, **122**, 11513.

213. The first enantioselective SMC reaction was reported by Cammidge and Crépy: (a) A. N. Cammidge and K. V. L. Crépy, *Chem. Commun.*, 2000, 1723The first diastereoselective SMC reaction using a chiral ligand to control diastereoselectivity was reported by Nicolaou and co-workers: (b) K. C. Nicolaou, H. Li, C. N. C. Boddy, J. M. Ramanjulu, T.-Y. Yue, S. Natarajan, X.-J. Chu, S. Bräse and F. Rübsam, *Chem. Eur. J.*, 1999, **5**, 2584For a review, see: (c) O. Baudoin, *Eur. J. Org. Chem.*, 2005, 4223.

214. (a) A.-S. Castanet, F. Colobert, P.-E. Broutin and M. Obringer, *Tetrahedron: Asymmetry*, 2002, **13**, 659; (b) K. Mikami, T. Miyamoto and M. Hatano, *Chem. Commun.*, 2004, 2082; (c) K. Sawai, R. Tatumi, T. Nakahodo and H. Fujihara, *Angew. Chem. Int. Ed.*, 2008, **47**, 6917.

215. J. Yin and S. L. Buchwald, *J. Am. Chem. Soc.*, 2000, **122**, 12051.

216. T. Hamada, A. Chieffi, J. Åhman and S. L. Buchwald, *J. Am. Chem. Soc.*, 2002, **124**, 1261.

217. A. Chieffi, K. Kamikawa, J. Åhman, J. M. Fox and S. L. Buchwald, *Org. Lett.*, 2001, **3**, 1897.

218. X. Shen, G. O. Jones, D. A. Watson, B. Bhayana and S. L. Buchwald, *J. Am. Chem. Soc.*, 2010, **132**, 11278.

219. J. García-Fortanet, F. Kessler and S. L. Buchwald, *J. Am. Chem. Soc.*, 2009, **131**, 6676.

220. T. Hamada and S. L. Buchwald, *Org. Lett.*, 2002, **4**, 999.

221. A. M. Taylor, R. A. Altman and S. L. Buchwald, *J. Am. Chem. Soc.*, 2009, **131**, 9900.

222. S. Rousseaux, J. García-Fortanet, M. A. D. A. Sanchez and S. L. Buchwald, *J. Am. Chem. Soc.*, 2011, **133**, 9282.

223. W. Tang, N. D. Patel, G. Xu, X. Xu, J. Savoie, S. Ma, M.-H. Hao, S. Keshipeddy, A. G. Capacci, X. Wei, Y. Zhang, J. J. Gao, W. Li, S. Rodriguez, B. Z. Lu, N. K. Lee and C. H. Senanayake, *Org. Lett.*, 2012, **14**, 2258.

224. G. Xu, W. Fu, G. Liu, C. H. Senanayake and W. Tang, *J. Am. Chem. Soc.*, 2014, **136**, 570.

225. J. García-Fortanet and S. L. Buchwald, *Angew. Chem. Int. Ed.*, 2008, **47**, 8108.

226. (a) P. von Matt and A. Pfaltz, *Angew. Chem. Int. Ed. Engl.*, 1993, **32**, 566; (b) G. J. Dawson, C. G. Frost, J. M. J. Williams and S. J. Coote,

Tetrahedron Lett., 1993, **34**, 3149; (c) J. Sprinz and G. Helmchen, *Tetrahedron Lett.*, 1993, **34**, 1769.

227. O. Loiseleur, M. Hayashi, N. Schmees and A. Pfaltz, *Synthesis*, 1997, **1997**, 1338.

228. S. Lee and J. F. Hartwig, *J. Org. Chem.*, 2001, **66**, 3402.

229. (a) T. Arao, K. Kondo and T. Aoyama, *Tetrahedron Lett.*, 2006, **47**, 1417; (b) T. Arao, K. Sato, K. Kondo and T. Aoyama, *Chem. Pharm. Bull.*, 2006, **54**, 1576; (c) F. Glorius, G. Altenhoff, R. Goddard and C. Lehmann, *Chem. Commun.*, 2002, 2704.

230. E. P. Kündig, T. M. Seidel, Y.-x. Jia and G. Bernardinelli, *Angew. Chem. Int. Ed.*, 2007, **46**, 8484.

231. E. J. Cho and S. L. Buchwald, *Org. Lett.*, 2011, **13**, 6552.

232. Ritter and co-workers reported a silver-mediated trifluoromethoxylation of arylstannanes and boronic acids; however, the method is not catalytic: C. Huang, T. Liang, S. Harada, E. Lee and T. Ritter, *J. Am. Chem. Soc.*, 2011, **133**, 13308.

233. For a review of Ni-catalyzed cross-coupling involving C–O electrophiles, see: (a) B. M. Rosen, K. W. Quasdorf, D. A. Wilson, N. Zhang, A.-M. Resmerita, N. K. Garg and V. Percec, *Chem. Rev.*, 2011, **111**, 1346. For selected examples of C–N activation, see: (b) K. W. Quasdorf, A. Antoft-Finch, P. Liu, A. L. Silberstein, A. Komaromi, T. Blackburn, S. D. Ramgren, K. N. Houk, V. Snieckus and N. K. Garg, *J. Am. Chem. Soc.*, 2011, **133**, 6352; (c) S. B. Blakey and D. W. C. MacMillan, *J. Am. Chem. Soc.*, 2003, **125**, 6046; (d) E. Wenkert, A.-L. Han and C.-J. Jenny, *J. Chem. Soc., Chem. Commun.*, 1988, 975.

234. (a) X. Chen, K. M. Engle, D.-H. Wang and J.-Q. Yu, *Angew. Chem. Int. Ed.*, 2009, **48**, 5094; (b) J. Wencel-Delord, T. Dröge, F. Liu and F. Glorius, *Chem. Soc. Rev.*, 2011, **40**, 4740; (c) R. Giri, B.-F. Shi, K. M. Engle, N. Maugel and J.-Q. Yu, *Chem. Soc. Rev.*, 2009, **38**, 3242; (d) T. W. Lyons and M. S. Sanford, *Chem. Rev.*, 2010, **110**, 1147; (e) C. S. Yeung and V. M. Dong, *Chem. Rev.*, 2011, **111**, 1215; (f) M. C. White, *Synlett*, 2012, **23**, 2746; (g) M. Miura and T. Satoh, in *Modern Arylation Methods*, ed. L. Ackermann, Wiley-VCH, Weinheim, 2009, pp. 335–362.

235. T. Ishiyama, S. Abe, N. Miyaura and A. Suzuki, *Chem. Lett.*, 1992, 691.

236. A. Devasagayaraj, T. Stüdemann and P. Knochel, *Angew. Chem. Int. Ed. Engl.*, 1996, **34**, 2723.

237. For early examples, see: (a) M. R. Netherton, C. Dai, K. Neuschütz and G. C. Fu, *J. Am. Chem. Soc.*, 2001, **123**, 10099; (b) J. H. Kirchhoff, M. R. Netherton, I. D Hills and G. C. Fu, *J. Am. Chem. Soc.*, 2002, **124**, 13662.

238. For reviews, see: (a) E. C. Swift and E. R. Jarvo, *Tetrahedron*, 2013, **69**, 5799; (b) R. Jana, T. P. Pathak and M. S. Sigman, *Chem. Rev.*, 2011, **111**, 1417; (c) A. Rudolph and M. Lautens, *Angew. Chem. Int. Ed.*, 2009, **48**, 2656; (d) B. D. Sherry and A. Fürstner, *Acc. Chem. Res.*, 2008, **41**, 1500; (e) B. Liégault, J.-L. Renaud and C. Bruneau, *Chem. Soc. Rev.*, 2008, **37**, 290; (f) J. Terao and N. Kambe, *Bull. Chem. Soc. Jpn.*, 2006,

79, 663; (g) A. C. Frisch and M. Beller, *Angew. Chem. Int. Ed.*, 2005, **44**, 674; (h) M. R. Netherton and G. C. Fu, in *Topics in Organometallic Chemistry: Palladium in Organic Synthesis*, ed. J. Tsuji, Springer, New York, 2005, pp. 85–108; (i) M. R. Netherton and G. C. Fu, *Adv. Synth. Catal.*, 2004, **346**, 1525; (j) D. J. Cárdenas, *Angew. Chem. Int. Ed.*, 2003, **42**, 384; (k) T.-Y. Luh, M.-k. Leung and K.-T. Wong, *Chem. Rev.*, 2000, **100**, 3187; (l) J. Terao and N. Kambe, *Acc. Chem. Res.*, 2008, **41**, 1545.

Pd–Phosphine Precatalysts for Modern Cross-Coupling Reactions

CARIN C. C. JOHANSSON SEECHURN,[a] HONGBO LI[b] AND THOMAS J. COLACOT*[b]

[a] Johnson Matthey Catalysis and Chiral Technologies, Orchard Road, Royston SG8 5HE, UK; [b] Johnson Matthey Catalysis and Chiral Technologies, 2001 Nolte Drive, West Deptford, NJ 08066, USA
*Email: thomas.colacot@jmusa.com

3.1 Introduction

Over the last few decades, the real-world applications of palladium-catalyzed cross-coupling have continuously expanded into many areas of chemistry, such as the synthesis of natural products,[1] active pharmaceutical ingredients (APIs),[2] agrochemicals,[3] monomers for polymers[4] and advanced materials for electronic applications.[5] A historical perspective on the development of cross-coupling reactions in the light of the 2010 Nobel Prize in Chemistry was recently reviewed by our group in collaboration with Snieckus and co-workers.[6] Concurrently, a more focused review on the theory behind the preformed catalyst development was also published by our group.[7] Therefore, in this chapter we have tried our best not to duplicate those efforts, but to discuss the developments and applications of L_2Pd- and LPd-based preformed catalysts from a reactivity and selectivity point of view, while providing the theoretical understanding in making the catalyst selection for a

RSC Catalysis Series No. 21
New Trends in Cross-Coupling: Theory and Applications
Edited by Thomas J Colacot

specific reaction with a concise approach. Whereas the previous chapter describes the importance of ligands in modern cross-coupling reactions, the focus of this chapter is geared towards providing insights into how the active catalytic Pd(0) species is generated from a preformed catalyst for specific name reactions in cross-coupling.

The way in which $L_nPd(0)$ is generated can significantly affect the overall rate of the catalytic cycle. If formed *in situ*, there are a number of deleterious side reactions that can occur, especially when the new-generation bulky electron-rich monodentate or electron-rich bidentate ligands with large bite angles are used. Part of the reasons could be due to the formation of various $L_nPd(0)$ species. For example, dtbpf (di-*tert*-butylphosphinoferrocene), binap [2,2′-bis(diphenylphosphino)-1,1′-binaphthyl], *etc.*, can form thermo-dynamically stable, and therefore less reactive, 18-electron species even if the ratio of Pd to ligand is 1:1.[8,9] When a preformed catalyst is used, the activation pathway to the catalytically active species will depend not only on its structure, but also on the nature of the spectator ligands on the Pd. Many of these pathways are still unknown or speculative; however, some of the recent studies are encouraging in understanding the structure–activity relationship (Figure 3.1).

As discussed in the previous chapter, ligands play a significant role in cross-coupling. Until the mid-90s, Ph_3P was the most widely used ligand for palladium-catalyzed coupling reactions, where the substrates were normally aryl iodides and bromides. Fu and co-workers' pioneering report in 1998 in line with Koie's work on the use of sterically hindered, electron-rich trialkylphosphine ligands such as t-Bu_3P and Cy_3P, along with Buchwald's studies using dialkylbiarylphosphines, demonstrated the scope of using less reactive organic chlorides as coupling partners.[10–12] These findings indeed rejuvenated the entire area of cross-coupling, making significant advances over the past 15 years. Figure 3.2 lists some of the most notable P-based ligands for use in cross-coupling reactions during the last decade.[13] They are all discussed in more detail in Chapter 2.

In spite of these significant developments, most of the catalyst systems described above are formed *in situ*, by mixing a palladium precursor, such as $Pd(dba)_x$ ($x = 1.5, 2$) or $Pd(OAc)_2$, with a ligand. However, dba (dibenzylidene-acetone) ligand has been reported to retard the activity of the catalyst significantly when $Pd(dba)_x$ is used.[14,15] Also, from a practical point of view, many of the highly active electron-rich monophosphine ligands are air sensitive or even pyrophoric (*e.g.*, t-Bu_3P), which leads to further complications, especially during reaction scale-up. The problems associated with

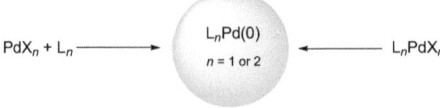

Figure 3.1   Generation of catalytically active species $L_nPd(0)$ *in situ* or from precatalysts.

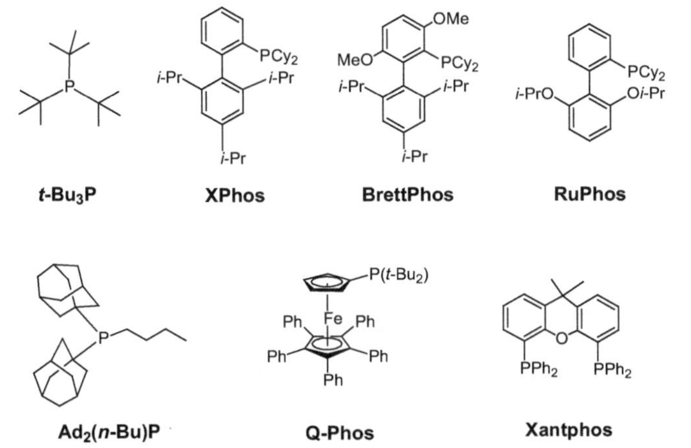

Figure 3.2 Selected electron-rich bulky ligands used in palladium-catalyzed coupling reactions.

in situ-formed catalysts were not anticipated in the early days but were identified during the last 5–10 years. Using the right catalyst at the optimum loading, today one can significantly improve the overall efficiency of a process by reducing environmental waste, process time, and temperature, and thus, take advantage of the process economics.

However, it could be possible to carry out some of the large-scale processes successfully using the *in situ*-generated catalyst either using simple ligands (*e.g.*, Ph$_3$P) or by skillfully transferring the process from laboratory to production with precise engineering control. Sometimes, the reaction solvent used for the coupling reaction may not be suitable to preform the catalyst precisely while prestirring the Pd precursor with a ligand under the coupling conditions, thus leading to decreased catalytic efficiency. The term "precatalyst", however, refers to the isolated and purified L$_n$PdX complex, prepared by a separate process.

3.2 The 18-Electron Rule *Versus* Catalytic Activity and Stability

The stability and reactivity of transition metal complexes can be predicted to a certain extent using the "18-electron rule".[16,17] The valence shells of transition metals consist of nine valence orbitals, which can accommodate 18 electrons. Combining the nine atomic orbitals with the ligand orbitals creates nine molecular orbitals. These orbitals can be either metal–ligand bonding or non-bonding. When the electron count of a metal complex adds up to 18 electrons, it is said to have reached the same electronic configuration as the noble gas in the same period, hence it is assumed to be a stable complex. The 18-electron complexes are viewed as "exchange inert" and ligand dissociation has to occur prior to any other reactions at the metal centre. Complexes with fewer than 18 electrons tend to show enhanced reactivity.

There are two ways of counting electrons in a metal complex: neutral or ionic. In the neutral count, start by locating the central atom in the periodic table and determine the number of valence electrons. To this, add one for every halide or other anionic ligand (X⁻) followed by two electrons for every lone pair binding (*e.g.*: Ph₃P) to the metal. Finally, add one electron for each homoelement bond and one for each negative charge of the complex. Similarly, subtract one for each positive charge. In the ionic count, calculate the number of electrons of the element, taking the oxidation state into account. Then, add two electrons for every halide or any other anionic ligand (X⁻), followed by two for every lone pair bonding to the metal.

For unsaturated ligands, such as alkenes, count the number of carbon atoms binding to the metal (hapticity, represented by the Greek letter η), each of which contributes one electron in the neutral model. For example, allyl is a three-electron donor (η^3), cyclopentadiene (Cp) is a five-electron donor (η^5), whereas in the ionic model one more electron is added for each charge on the anions. For example, for Cp⁻ one more electron is added, making it a six-electron donor.

Both the neutral and the ionic counts are illustrated in Figure 3.3, using (Ph₃P)₄Pd, dppfPdCl₂ and ionic complex dtbpfPdI₂ as examples (dppf = diphenylphosphinoferrocene). Anionic [(Ph₃P)₂PdCl]⁻ is a proposed intermediate in the catalytic cycle, as will be discussed later.

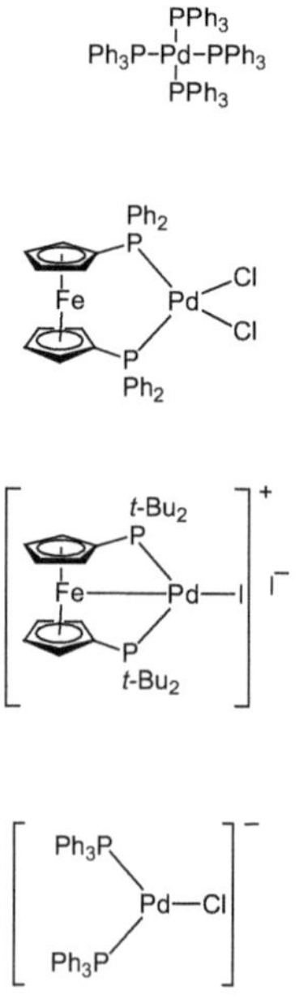

Figure 3.3 Electron-count for (Ph₃P)₄Pd, dppfPdCl₂, dtbpfPdI₂ and [(Ph₃P)₂PdCl]⁻.

There are a number of exceptions to the 18-electron rule, one of which is illustrated in Figure 3.3. The electron count of dppfPdCl$_2$ only adds up to 16, which in theory would make this an unsaturated reactive complex. Instead, it is an air-stable precatalyst. The most common exception to the rule is that these Pd(II) complexes have a low-spin d^8 configuration. They adopt a square-planar conformation where the splitting of the d-orbitals is such that a 16-electron complex is the most stable.

Other exceptions to the rule arise due to the nature of the ligands bonded to the metal center. Very bulky ligands, *e.g.*, *t*-Bu$_3$P, dtbpf, Q-Phos [Q-Phos = 1,2,3,4,5-pentaphenyl-1′-(di-*tert*-butylphosphino)ferrocene)], radical ligands or π-donating ligands are all able to stabilize a low-electron count complex better. Figure 3.4 shows three examples of precatalysts that are very efficient in a number of coupling reactions. The classical coupling catalyst (Ph$_3$P)$_4$Pd is an 18-electron species, yet it is air sensitive as it can easily dissociate two ligands to form a 14-electron species. As expected, (*t*-Bu$_3$P)$_2$Pd is an air-sensitive complex as it has a 14-electron configuration, whereas (Ph$_3$P)$_2$PdCl$_2$ and Q-PhosPd(crotyl)Cl are air-stable complexes with a 16-electron configuration. Interestingly, the dimeric complex [(*t*-Bu$_3$P)Pd(μ-Br)]$_2$, despite its 16-electron configuration, is fairly air sensitive, due to the +1 oxidation state of Pd.

In order to account for all these anomalies, Landis and Weinhold proposed an alternative theoretical treatment, where it is suggested that transition metals do not employ p-orbitals for bonding.[18] They therefore have only a valence shell of one s- and five d-orbitals. These can accommodate up to 12 electrons with two-center–two-electron bonds. If the metal has co-ordination numbers beyond this, the bonds will have to be formed by linear three-center–four-electron bonds, where only one metal orbital is used for

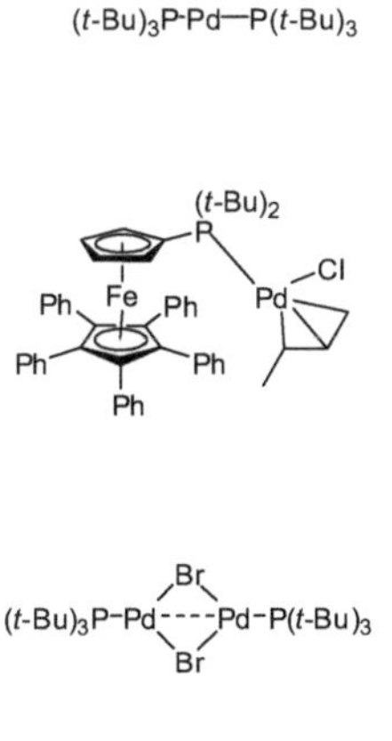

		Neutral Count	Ionic Count
(*t*-Bu)$_3$P-Pd—P(*t*-Bu)$_3$	Pd (0) d^{10}	10	10
	PR$_3$	4	4
		14	14
	Pd (II) d^8	10	8
	PR$_3$	2	2
	Cl	1	2
	η3-allyl	3	4
		16	16
	Pd (I) d^9	10	9
	PR$_3$	2	2
	2 Br	3	4
	Pd–Pd	1	1
		16	16

Figure 3.4   Electron count for (*t*-Bu$_3$P)$_2$Pd, Q-PhosPd(crotyl)Cl and [(*t*-Bu$_3$P)Pd(μ-Br)]$_2$.

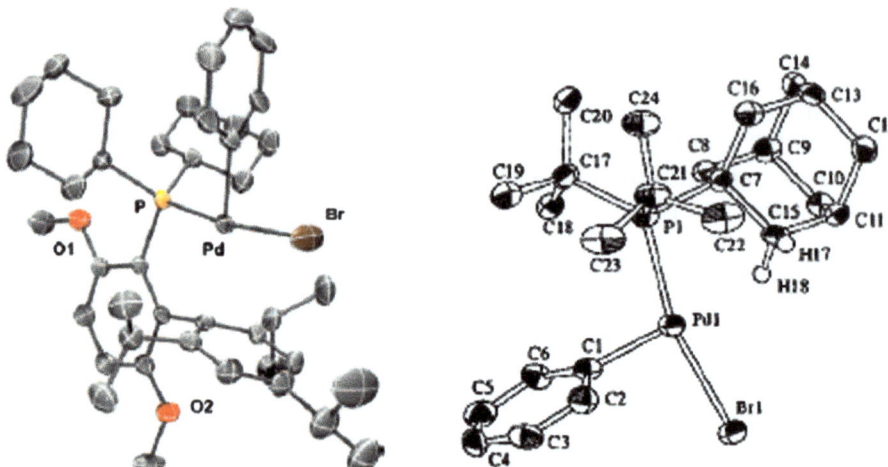

Figure 3.5 Crystal structure of isolated BrettPhosPd(Ar)(Br)[19] and [Ad(*t*-Bu)$_2$P]Pd(Ph)(Br).[20]
Reprinted with permission from the *Journal of the American Chemical Society*. Copyright 2002 and 2008 American Chemical Society.

binding two ligands. This new theory can accommodate previous exceptions to the 18-electron rule, such as square-planar d^8-complexes, and also provide support for the existence of low-ligated Pd(0) in the catalytic cycles.

Therefore, the general rule of thumb is that Pd(II) complexes are air stable whereas Pd(0) and Pd(I) complexes are air sensitive. The 12-electron-based LPd(0) is the catalytically most active species, and hence very sensitive (reactive). As of today, to the best of our knowledge, LPd(0) species has neither been isolated nor detected spectroscopically, although LPd(Ar)(X) has been isolated and characterized (Figure 3.5).[19,20] Since LPd(0) is very active, the catalyst can also become deactivated rapidly in the presence of air and impurities, if the kinetics of the reaction are slow.

3.3 Formation of L$_n$Pd(0) from Precatalysts

The active catalytic species L$_n$Pd(0) can be generated from an array of precatalysts to form LPd(0) or L$_2$Pd(0) as shown in Figure 3.6. L is a monophosphine or, in the case of L$_2$PdX$_2$, it may also be a bisphosphine ligand, where X is an anion, such as chloride, bromide or in some cases mesylate.

Both L$_4$Pd(0) and L$_2$PdX$_2$ are proposed to form L$_2$Pd(0) as the catalytically active species, whereas LPd(allyl)X, [LPdX]$_2$ and LPd(palladacycle)X, with a Pd:L ratio of 1:1, are believed to form the "12-electron" LPd(0). There are different theories on the nature of the catalytically active species formed from L$_2$Pd(0) complexes. Both the electronic and steric properties of the ligand and the type of cross-coupling reaction seem to influence the formation of either L$_2$Pd(0) or LPd(0) species.

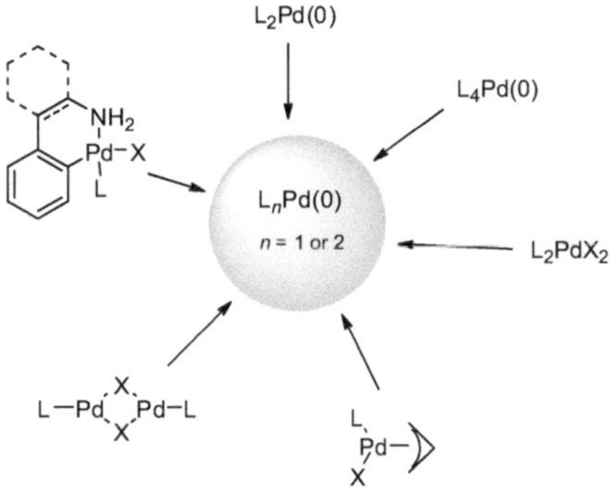

Figure 3.6 Generation of $L_nPd(0)$ from precatalysts.

The proposed mechanism for the formation of the catalytically active species is discussed in detail below for each individual class of precatalyst type: $L_4Pd(0)$, $L_2Pd(0)$, L_2PdX_2, LPd(allyl)X, $[LPdX]_2$ and LPd(palladacycle)X.

3.4 $L_4Pd(0)$

The widely used preformed Pd(0) catalyst $(Ph_3P)_4Pd$, nicknamed "palladium tetrakis", is an air- and somewhat moisture-sensitive crystalline material. It is easy to tell from the appearance of the complex whether or not decomposition has started. If the material is exposed to the air, the color changes from bright yellow to dark yellow and finally to orange–brown (Figure 3.7). The catalyst is, however, produced in multi-kilogram quantities with a reasonably good shelf-life when kept under inert conditions.

The preparation of $(Ph_3P)_4Pd$ was originally reported as involving the use of NH_2NH_2 as a reducing reagent (Scheme 3.1).[21] This method requires harsh reaction conditions such as high temperature (140 °C) in DMSO. In addition, there are toxicity concerns regarding the use of hydrazine as a reducing reagent in the process.

More recently, alternative patented methods were developed to make $(Ph_3P)_4Pd$ in multi-kilogram quantities under milder conditions without the use of hydrazine.[22]

There are a number of early examples where the preformed Pd(0) complex $(Ph_3P)_4Pd$[23,24] has been used as catalyst with excellent results. The solution-phase dissociation of $(Ph_3P)_4Pd$ is shown in Scheme 3.2, where the resulting $(Ph_3P)_3Pd$ is in rapid equilibrium with $(Ph_3P)_2Pd$.[25]

In the Heck reaction, an increase in reaction rate was observed on addition of anions such as chloride in the form of TBAC (tetrabutylammonium chloride), which was explained by the formation of a tricoordinate anionic species $[(Ph_3P)_2PdX]^-$ (16 electrons), which was proposed to be more reactive than the neutral $(Ph_3P)_2Pd$ species.[26] This theory was questioned more recently, based on the evidence that dicoordinate species have been found to be lower in energy than the tricoordinate species.[27] The most likely anionic species present would be the dicoordinate 14-electron complex

| | | |
| 0 h | 24 h | 96 h |

Figure 3.7 Decomposition of $(Ph_3P)_4Pd$ with time.

$$2\ PdCl_2 + 8\ Ph_3P + 5\ NH_2NH_2.H_2O \longrightarrow 2\ (Ph_3P)_4Pd + 4\ NH_2NH_2.HCl + N_2 + 5\ H_2O$$

Scheme 3.1 Reported synthesis of $(Ph_3P)_4Pd$.

$$(Ph_3P)_4Pd \underset{+Ph_3P}{\overset{-Ph_3P}{\rightleftarrows}} (Ph_3P)_3Pd \underset{+Ph_3P}{\overset{-Ph_3P}{\rightleftarrows}} (Ph_3P)_2Pd$$

Scheme 3.2 Solution-phase dissociation of $(Ph_3P)_4Pd$.

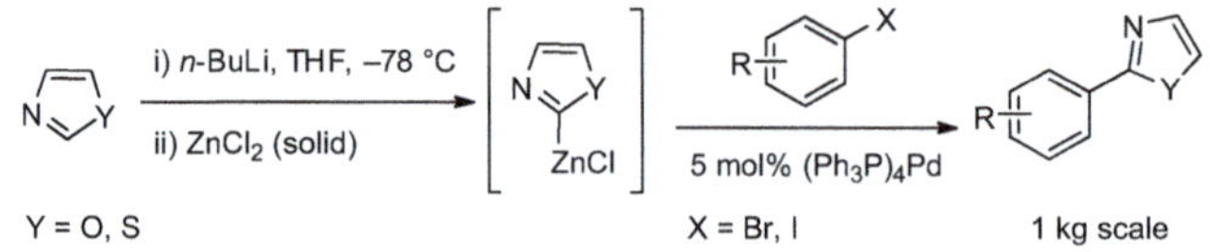

Y = O, S   X = Br, I   1 kg scale

Scheme 3.3 $(Ph_3P)_4Pd$-catalyzed Negishi coupling.

$[(Ph_3P)PdX]^-$, which has been found to be more reactive towards oxidative addition than the 16-electron-based $[(Ph_3P)_2PdX]^-$. The anion effect can therefore be rationalized by an alternative theory, namely that the anionic complexes are more reactive for the formation of the prereactive complexes than their neutral analogs during the oxidative addition step. The transformation from $(Ph_3P)_2Pd$ to $(Ph_3P)Pd(Ph)(I)$ (14–16-electron configuration) requires more energy than going from $[(Ph_3P)PdX]^-$ to $[Pd(PhI)X]^-$ (no change in electronic configuration).[27] The reason for this may be that in the neutral complex, Ph_3P acts as an electron acceptor; therefore, the addition of another electron acceptor such as PhI is then disfavored. On the other hand, in the anionic complex, only an electron-donating ligand (X^-) is coordinated to palladium, hence ligation of the electron acceptor PhI is favored.

Despite the simplicity of the Ph_3P ligand, $(Ph_3P)_4Pd$ is still a highly effective practical catalyst in cross-coupling reactions involving aryl iodides or activated aryl bromides. Many large-scale examples of this can be found even in the recent literature,[28] a few of which are exemplified below. For instance, Reeder *et al.* at Pfizer reported a Negishi process to couple oxazole and thiazole with aryl halides using $(Ph_3P)_4Pd$ as the catalyst (Scheme 3.3).[29] This was demonstrated on a 1 kg scale. The use of a relatively high Pd loading (5 mol%) may have been justified by the lower cost of the precatalyst

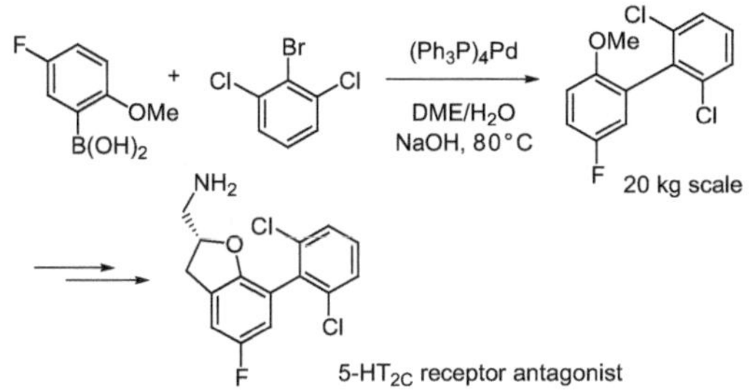

Scheme 3.4   Wyeth's synthesis of a 5-HT$_{2C}$ receptor antagonist using a chemoselective Suzuki–Miyaura coupling.

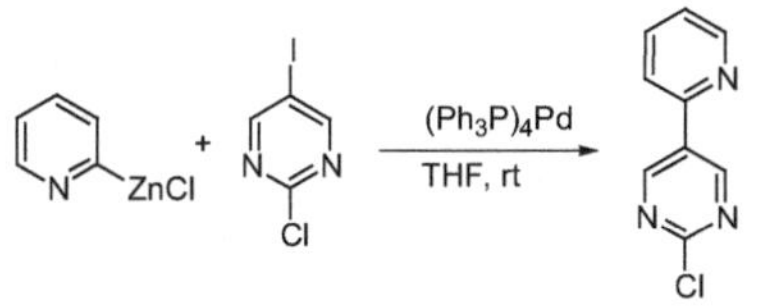

Scheme 3.5   Chemoselective Negishi coupling of an aryl iodide.

compared with the use of more advanced Pd precatalysts with higher reactivities.

Gontcharov *et al.* at Wyeth Research disclosed a process to make a 5-HT$_{2C}$ receptor antagonist in which the first step consisted of a $(Ph_3P)_4Pd$-catalyzed Suzuki–Miyaura coupling reaction carried out on a 20 kg scale (Scheme 3.4).[30] In this case, the chemoselective coupling of a chloride-containing aryl bromide was achieved. This may have proved to be challenging in terms of selectivity while using a more active precatalyst.

In 2007, a Negishi cross-coupling between the *in situ*-prepared 2-pyridylzinc chloride and 5-iodo-2-chloropyrimidine catalyzed by $(Ph_3P)_4Pd$ afforded the product in a single step on a kilogram scale (Scheme 3.5).[31] This chemoselective coupling of a chloride-containing aryl iodide is particularly noteworthy, as the chloride is in the inherently most reactive position for a standard S_NAr reaction.

3.5   L$_2$Pd(0) Complexes

The use of triarylphosphines such as Ph$_3$P as ligands limits the scope of the aryl halide substrate to aryl iodides or activated aryl bromides. Since aryl chlorides would be desirable substrates from a cost point of view as they are industrial feedstocks, substantial research efforts have been focused on overcoming this limitation. The use of trialkyl- and dialkyl-based phosphine ligands came as a breakthrough in this area (see Chapter 2 for details), although some of these ligands are air sensitive or even pyrophoric in nature, posing safety concerns especially during scale-up. Such risks are avoided by using preformed Pd catalysts of these ligands, where the complexes (precatalysts) are easy to handle. In addition, preformed complexes provide superior results in cross-coupling reactions in comparison with the catalysts generated *in situ* using a Pd precursor, such as $Pd_2(dba)_3$ or $Pd(OAc)_2$ (see Section 3.1 for a discussion on this topic). The following sections describe the development of the preformed Pd catalysts containing highly active trialkyl- or aryldialkylphosphine ligands of general formula $L_2Pd(0)$. Their excellent performance in a number of cross-coupling reactions is highlighted by examples from academia and the pharmaceutical industry.

3.5.1 What is the Catalytically Active Species Formed from $L_2Pd(0)$ Complexes?

Generally, in challenging cross-coupling reactions, good results can be achieved by using sterically demanding ligands in conjunction with a Pd(0) precursor. It has been shown that the resting state of a catalyst formed *in situ* from, for example, $Pd_2(dba_3)$ with the bulky ligand t-Bu_3P, is $L_2Pd(0)$.[11] Notably, various deleterious side reactions may take place to produce catalytically inactive species, which diminishes the efficiency of the cross-coupling reaction.[32] Until fairly recently, $L_2Pd(0)$ was also assumed to be the catalytically active species in a cross-coupling cycle, based on conventional knowledge. More recent experimental and computational data, however, support ligand dissociation to LPd(0) and L before oxidative addition of Ar–X (X = Br or Cl; Figure 3.8).[27,33–35] In addition to this new evidence, there is increased support for the recent bonding analysis by Landis and Weinhold[36] (see Section 3.2), suggesting the existence of low-ligated Pd species.

The proposal that the monoligated phosphine LPd(0) species is the true catalytic species has been supported by DFT calculations showing that the transition states of oxidative addition of PhCl and PhBr to $(Ph_3P)_2Pd$ have higher free energies (ΔG_B) than the corresponding ones for $(Ph_3P)Pd$.[37] The transition state for $(t$-$Bu_3P)_2Pd$ does not exist. The free energy barriers for

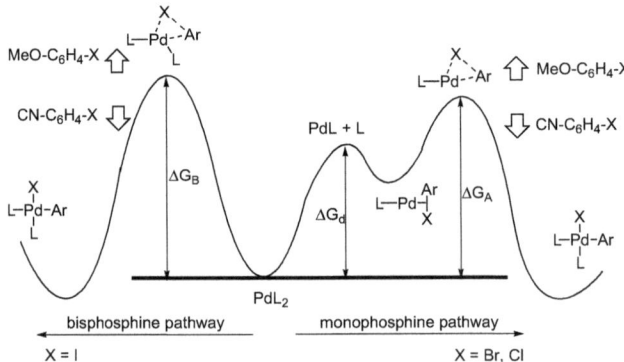

Figure 3.8 Energy barriers in oxidative addition to PdL_2 *versus* PdL. Reproduced from Ref. 35.

oxidative addition on to $(PR_3)Pd$ (ΔG_A) for several different tertiary phosphine ligands were also calculated. It was found that in the R_3P (R = Me, Et, *i*-Pr, *t*-Bu, Ph) series these values did not change significantly. Importantly, this means that the reason for the requirement for $(t\text{-}Bu_3P)Pd$ in reactions with aryl chlorides is not that the oxidative addition is facilitated by the electron-rich ligand. Instead, through calculations, it was demonstrated that the dissociation energy (ΔG_d) of *t*-Bu$_3$P from $[(t\text{-}Bu_3P)_2Pd]$ is significantly lower than those of the other ligands in the series (R = Me, Et, *i*-Pr, Ph). In the case of iodoarenes, however, the oxidative addition is suggested to occur on to a bisphosphine $L_2Pd(0)$ because they are more reactive than the other haloarenes. Typically, C–OTf bonds have similar bond dissociation energy to C–I,[38] and have been proposed to react preferentially with $L_2Pd(0)$.[39] The bromo- and chloroarenes, on the other hand, require the more reactive monophosphine species $[LPd(0)]$ in order to undergo oxidative addition, since the C–Br and C–Cl bonds are stronger and more difficult to cleave.[40]

The proposal that the catalytically active species is the monophosphine (L)-based $LPd(0)$ was further supported by the isolation of monomeric three-coordinate oxidative addition products **1** and **2** by Hartwig and co-workers (Scheme 3.6).[20] It was shown that these complexes are very likely intermediates in the palladium-catalyzed amination of aryl halides.

Interestingly, the corresponding aryl chloride complexes could not be isolated from the direct oxidative addition of ArCl to the $L_2Pd(0)$ catalyst, but had to be prepared *via* anion exchange from the $LPd(Ar)(Br)$ complex.[41]

These results are consistent with the assumption that palladium catalysts bearing sterically demanding phosphine ligands favor the formation of monoligated 12-electron $LPd(0)$ species and are therefore the ligands of choice for most coupling reactions involving the use of unactivated bromo- or chloroarenes. When using iodoarenes, aryl triflates or activated bromoarenes, less bulky ligands such as Ph_3P can be employed since the bisphosphine $L_2Pd(0)$ is reactive enough for the cross-coupling to take place.

The presence of $LPd(0)$ is suggested to be the reason for the enhanced reactivity,[42] but the ligand dissociation step is also proposed to be the reactivity-limiting step. The ligand dissociation can occur by two possible mechanisms: either *via* direct ligand dissociation or *via* substrate-assisted ligand displacement (Scheme 3.7).[42,43] Computational studies suggest that the substrate-assisted ligand displacement pathway would be the favored mechanism.[43]

Based on the above studies, a 12-electron-based $LPd(0)$ species is generated for any bulky phosphine ligand-based Pd complex, although the

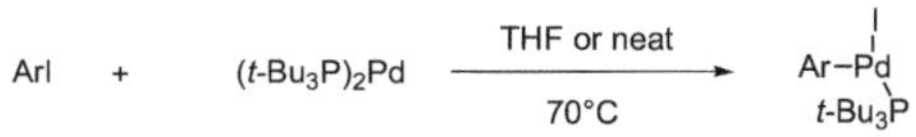

Scheme 3.6 Investigation into catalytically active species using $(t\text{-}Bu_3P)_2Pd$.

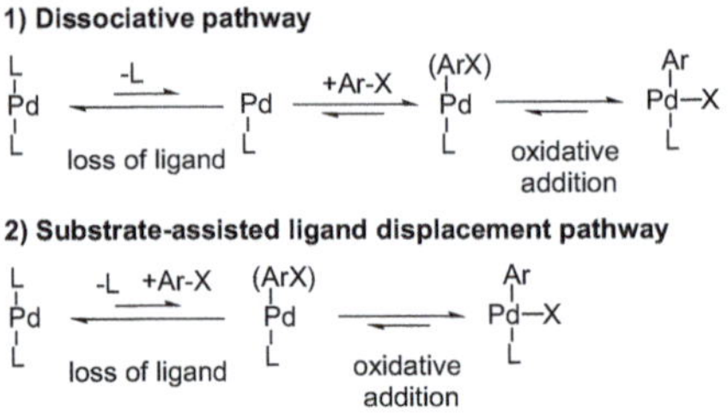

1) Dissociative pathway

Scheme 3.7 Dissociation of L$_2$Pd(0) to LPd(0).

0 h   2 h   24 h   72 h

Figure 3.9 Decomposition of commercially available $(t\text{-Bu}_3\text{P})_2\text{Pd}$ in air.

precatalyst activation pathway may have a profound influence on the outcome of the reaction and rate. However, more experimental studies are required to understand the "active catalytic" species in the cycle. For example, the bulky $(o\text{-tol})_3\text{P}$ is not a good ligand for Ar–Cl coupling, suggesting that the electronic properties play an equally important role.

3.5.2 Synthesis, Properties and Applications of L$_2$Pd(0) Complexes

The first examples of an isolated L$_2$Pd(0) catalyst to be used in cross-coupling reactions were reported by Dai and Fu, who employed $(t\text{-Bu}_3\text{P})_2\text{Pd}$ in Negishi couplings.[44]

The new-generation Pd(0) catalyst $(t\text{-Bu}_3\text{P})_2\text{Pd}$ was introduced as a white, crystalline, air-stable solid by Dai and Fu;[44] however, it was later found to decompose within 1–2 days when kept under an aerobic atmosphere (see Figure 3.9). It is believed to be more stable in pure crystalline form than the widely used $(\text{Ph}_3\text{P})_4\text{Pd}$. This white solid can be stored intact for longer periods under an inert atmosphere. In certain reactions, significant decreases in activities with side reactions or no activities are observed when the orange, grey, or black materials are used.

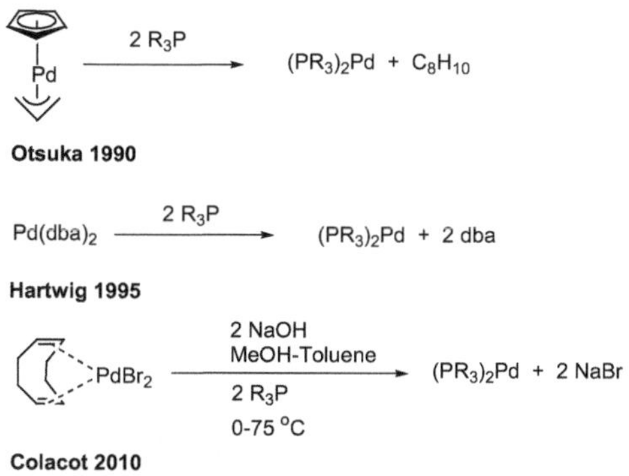

Scheme 3.8 Various synthesis routes for $L_2Pd(0)$.

Until recently, $(t\text{-}Bu_3P)_2Pd$ was the only commercially available $L_2Pd(0)$ catalyst. There were initially two known methods for the preparation of this catalyst. The first employed the highly volatile and unstable precursor $Pd(\eta^3\text{-}C_3H_5)(\eta^5\text{-}C_5H_5)$ (Scheme 3.8).[45] Later, a cinnamyl derivative, $Pd(\eta^3\text{-}1\text{-}PhC_3H_4)(\eta^5\text{-}C_5H_5)$, was reported as a precursor to generate the $L_2Pd(0)$ *in situ*,[46,47] although it has not been demonstrated in a synthesis to isolate the precatalyst. However, the reported synthesis of $Pd(\eta^3\text{-}1\text{-}PhC_3H_4)(\eta^5\text{-}C_5H_5)$ requires extreme cryogenic conditions.

The second method of preparation of $(t\text{-}Bu_3P)_2Pd$ used the readily available $Pd(dba)_2$ as precursor (Scheme 3.8); however, the procedure involved the recrystallization of the final compound with a large amount of solvent under cryogenic conditions in order to remove the black color and dba from the product.[44,48]

Both of these methods seem to not be very practical and are therefore not suited for a facile scale-up to multi-kilogram quantities.

It was not until 15 years later that a general and novel route was developed by Johnson Matthey to make a series of $L_2Pd(0)$ catalysts.[22] Starting from the readily available, inexpensive, air-stable precursor $(COD)PdBr_2$, a stoichiometric amount of phosphine ligand was used in the presence of a Brønsted base (*e.g.*, alkali metal hydroxides) in a protic solvent to generate the desired precatalysts. Employing this method, a number of $L_2Pd(0)$ catalysts could be prepared in near quantitative yields on a large scale (Scheme 3.8).

Of special interest was the use of sterically bulky, electron-rich phosphines such as $t\text{-}Bu_3P$, Cy_3P, $(o\text{-}tol)_3P$, $t\text{-}Bu_2PhP$, $p\text{-}Me_2NC_6H_4(t\text{-}Bu)_2P$ and $t\text{-}Bu_2(C_5H_4FeC_5Ph_5)P$ (Q-Phos), but the method was not limited to these ligands alone (Figure 3.10).

The mechanism for the formation of the $L_2Pd(0)$ complexes was established based on Scheme 3.9, where the intermediates **3** and **4** were isolated and characterized by various methods including X-ray crystallography.

Generally, the $L_2Pd(0)$ (L = *tert*-alkylphosphine) catalysts show higher activities at elevated temperatures, to form presumably LPd(0) as the catalytically active species.

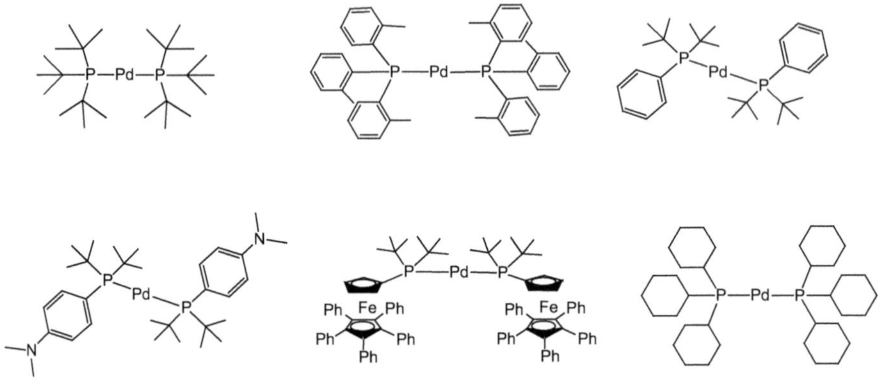

Figure 3.10 Commercially available preformed L$_2$Pd(0) catalysts.

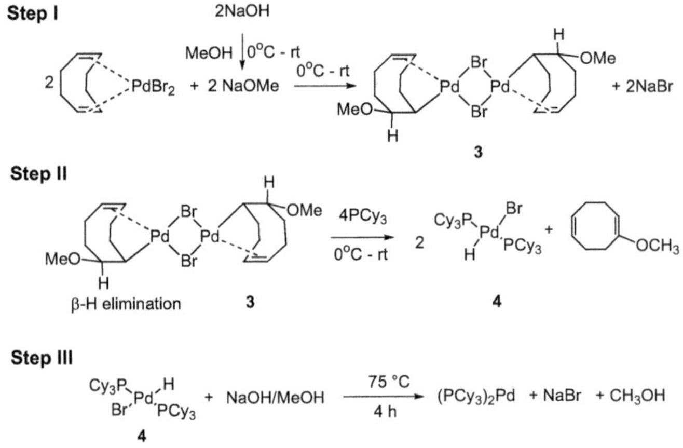

Scheme 3.9 Mechanism of information of L$_2$Pd(0) complexes.[22a]
Reprinted with permission from *Organic Letters*. Copyright 2010 American Chemical Society.

As the 18-electron rule predicts, all the new L$_2$Pd(0) catalysts (14-electron species) are air sensitive and should be stored and handled under an inert atmosphere. They are typically white/off-white in color and slowly turn brown or gray–black when exposed to oxygen. The less bulky species such as (Cy$_3$P)$_2$Pd are more susceptible to degradation and should be stored under an inert atmosphere at low temperatures to extend their shelf-life.

The new Pd(0) precatalysts have been evaluated in a number of different coupling reactions. An analysis of the applications of these catalysts can help identify which catalyst should be used in a specific reaction.

(*t*-Bu$_3$P)$_2$Pd has found widespread use as an effective precatalyst for a number of cross-coupling reactions. Limanto *et al.* made use of a (*t*-Bu$_3$P)$_2$Pd-catalyzed Heck alkynylation (copper-free Sonogashira coupling)

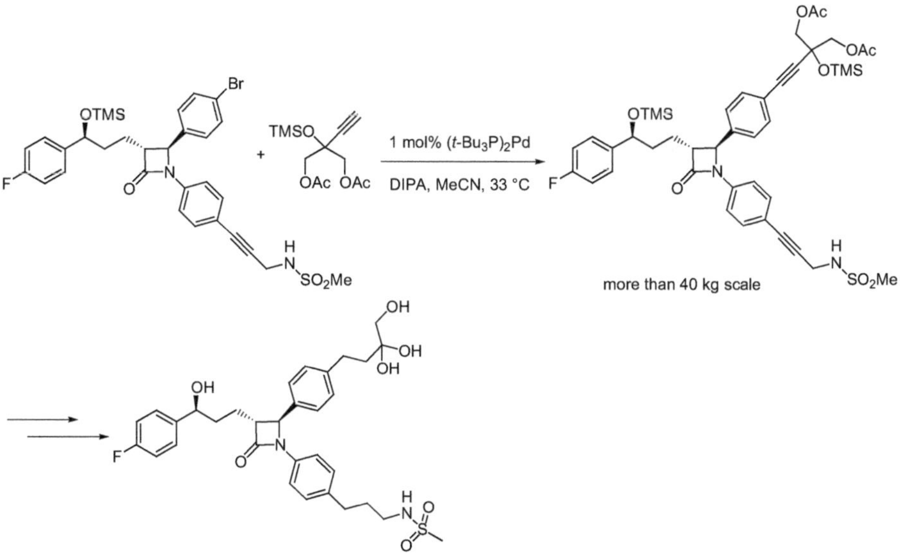

Scheme 3.10 (*t*-Bu₃P)₂Pd-catalyzed Heck alkynylation.

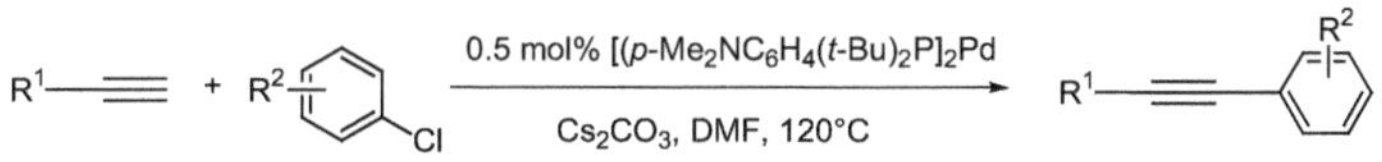

Scheme 3.11 [(*p*-Me₂NC₆H₄(*t*-Bu)₂P]₂Pd-catalyzed Heck alkynylation.

of an aryl bromide as part of a process to prepare antihypercholesterolemic azetidinone compounds (Scheme 3.10).[49] The coupling reaction was carried out on a 40 kg scale.

Notably, our group developed a protocol for the use of aryl chlorides in Heck alkynylation (copper-free Sonogashira) reactions. For this transformation, [(*p*-Me₂NC₆H₄(*t*-Bu)₂P]₂Pd was identified as the best catalyst (Scheme 3.11).[50]

The crystal structure of [(*p*-Me₂NC₆H₄(*t*-Bu)₂P]₂Pd (Figure 3.11) reveals that it has similar Pd–P bond lengths as in the case of the other L₂Pd(0) complexes. However, it has the smallest P–Pd–P angle (174.7°) whereas most other molecules have shown perfectly linear (180°) structures (Table 3.1). It was inferred that the "bent" L₂Pd(0) species tends to be more active than the linear Pd complexes containing monodentate ligands. It is also noted that copper can actually inhibit the coupling reaction.[50]

Continuing with an example from industry, Song *et al.* recently reported a route to a hepatitis C virus inhibitor that involved the use of (*t*-Bu₃P)₂Pd as precatalyst in a Heck reaction (Scheme 3.12).[54] The Heck reaction was followed by hydrogenation of the olefin making both isomers converge to a single product.

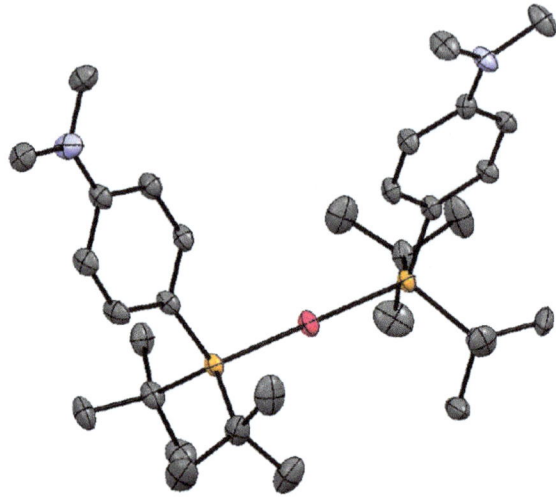

JMCA14/ Pd-149

Figure 3.11   Single-crystal X-ray structure of $[(p\text{-}Me_2NC_6H_4(t\text{-}Bu)_2P]_2Pd.$[50]
Reprinted with permission from *Journal of Organic Chemistry*. Copyright
2013 American Chemical Society.

Table 3.1   P–Pd–P angles *versus* activities of $L_2Pd(0)$ precatalysts.

Entry	Catalyst	Yield (%)[a]	Pd–P bond lengths (Å)	P–Pd–P angle (°)
1	$[p\text{-}Me_2NC_6H_4(t\text{-}Bu)_2P]_2Pd$ (Pd-149)	95	2.299(3); 2.299(2)	174.7(0)
2	$[Ph(t\text{-}Bu)_2P]_2Pd$ (Pd-148)	69	2.282(4); 2.273(4)[51]	176.8(1)
3	$(t\text{-}Bu_3P)_2Pd$ (Pd-116)	48	2.285(3); 2.285(3)[52]	180.0(0)
4	$(Q\text{-}Phos)_2Pd$ (Pd-150)	20	2.276(4); 2.276(4)[53]	180.0(0)
5	$[(p\text{-}tol)_3P]_2Pd$ (Pd-141)	<5	2.276(1); 2.276(1)[48]	180.0(0)

[a]After 7 h.

The choice of ligand can have a profound impact on the chemoselectivity
of certain reactions. Fu and co-workers found that in Suzuki–Miyaura
coupling, different ligands have different relative reactivity with aryl chlorides *versus* triflates. Whereas the catalyst based on the $t\text{-}Bu_3P$ ligand is more
selective towards chloride substrates, the Cy_3P-based catalyst is more selective for triflate substrates.[55] The same chemoselectivity trend has been
observed using preformed $(t\text{-}Bu_3P)_2Pd$ and $(Cy_3P)_2Pd$,[56] which is discussed in
some detail in Chapter 2.

Varying the ligand of the $L_2Pd(0)$ catalyst can also have an impact on the
chemistry of a cross-coupling reaction. For example, Lautens and co-workers

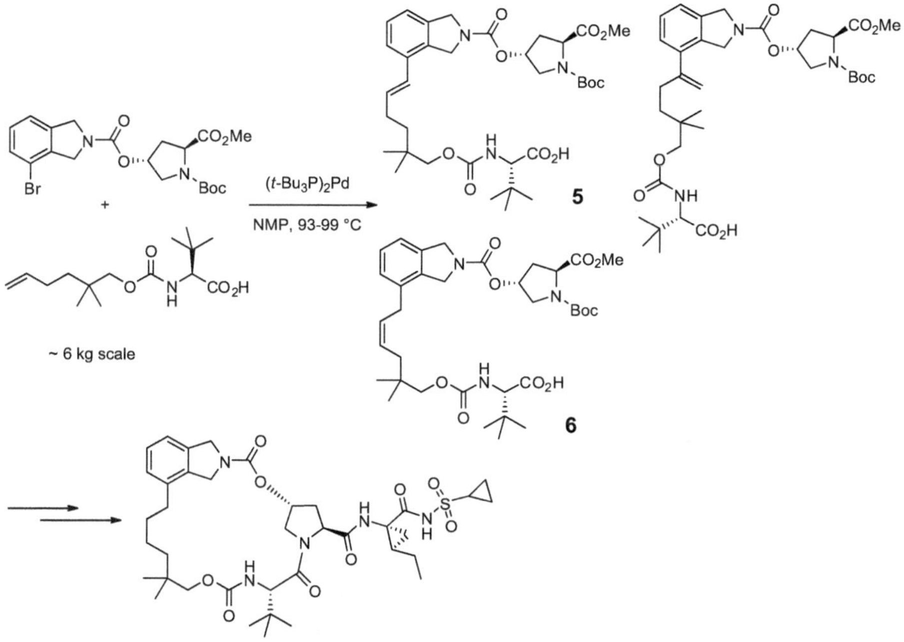

Scheme 3.12 $(t\text{-}Bu_3P)_2Pd$-catalyzed Heck reaction.

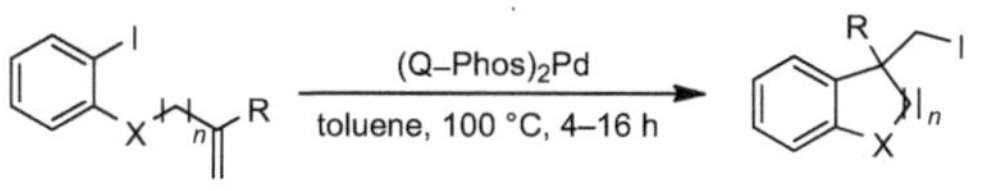

Scheme 3.13 $L_2Pd(0)$-catalyzed carboiodination reaction.

demonstrated the use of $(Q\text{-}Phos)_2Pd$ in a novel carboiodination reaction (Scheme 3.13).[57,58] Other L_2Pd precatalysts gave lower yields of the product. This process could be extended to a number of different substrates in order to form various benzene-fused heterocycles.[59] Interestingly, in a few selected cases $(t\text{-}Bu_3P)_2Pd$ was found to be superior to $(Q\text{-}Phos)_2Pd$.[60,61] Notably, in the carboiodination reaction the LPd-based complex Q-PhosPd(crotyl)(Cl) did not perform well. However, the addition of 1 mol % of Q-Phos ligand to Q-PhosPd(crotyl)(Cl) resulted in very good conversion,[62] suggesting the importance of a 2 : 1 ligand-to-Pd ratio to form $L_2Pd(0)$ for this reaction, under these reaction conditions. An entire chapter is dedicated to this work (see Chapter 7).

Lipshutz and co-workers recently demonstrated that the stereochemical outcome of a coupling reaction depends largely on the nature of the ligand employed.[63] A number of precatalysts, and also *in situ*-generated catalysts, were evaluated in a Suzuki–Miyaura coupling of a (Z)-alkenyl halide with arylboronic acids. Most of the $L_2Pd(0)$-based precatalysts or the well-known Suzuki–Miyaura coupling catalysts gave scrambling of the isomers, whereas

Table 3.2 Effect of precatalysts in stereoselective Suzuki–Miyaura coupling.

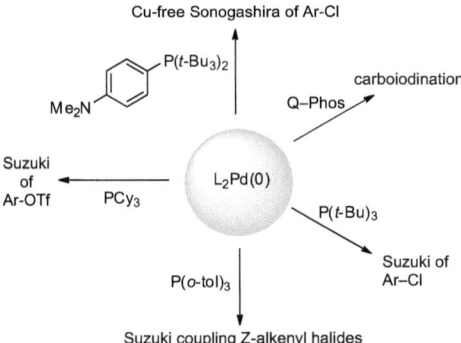

Entry	X	Product	PdL$_n$	Z : E
1	I		(t-Bu$_3$P)$_2$Pd	62 : 38
2	I	n-C$_6$H$_{13}$ ⟍ Ph	[p-Me$_2$NC$_6$H$_4$(t-Bu)$_2$P]$_2$PdCl$_2$	71 : 29
3	I		(o-tol$_3$P)$_2$Pd	99 : 1
4	I		(t-Bu$_3$P)$_2$Pd	9 : 91
5	I	n-C$_6$H$_{13}$	[p-Me$_2$NC$_6$H$_4$(t-Bu)$_2$P]$_2$PdCl$_2$	N/A
6	I		(o-tol$_3$P)$_2$Pd	99 : 1
7	Br		(t-Bu$_3$P)$_2$Pd	58 : 42
8	Br	Ph OBn	[p-Me$_2$NC$_6$H$_4$(t-Bu)$_2$P]$_2$PdCl$_2$	N/A
9	Br		(o-tol$_3$P)$_2$Pd	98 : 2

Figure 3.12 Examples of a wide range of reactions developed using L$_2$Pd(0) precatalysts.

(o-tol$_3$P)$_2$Pd gave exceptionally good stereoretention (Table 3.2) toward the Z-isomer.

Figure 3.12 summarizes how the reactivity, chemoselectivity and stereo-selectivity of specific cross-coupling reactions can be optimized by matching the right L$_2$Pd(0) precatalyst with the right transformation.

3.6 L$_2$PdX$_2$ Complexes

Although L$_2$Pd(0) catalysts are highly active, these compounds are air sensitive both in the solid state and in the solution phase. The class of L$_2$PdX$_2$ compounds is stable to air and moisture even at higher temperatures in solution. These Pd(II) complexes possess square-planar geometry with a 16-electron configuration and therefore belong to the class of stable low-spin d^8 complexes as predicted by one accepted exception to the classical 18-electron rule. Ligand L$_2$ of L$_2$PdX$_2$ can be monodentate or bidentate, as depicted in Figure 3.13. This family of Pd complexes is one of the most active, yet air-stable, frequently employed catalysts in many industrial processes,

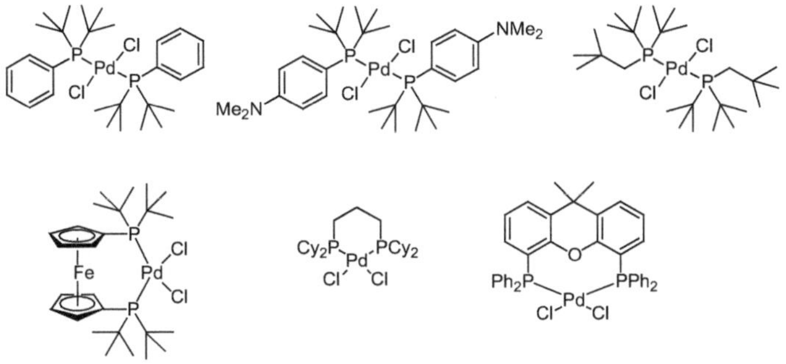

Figure 3.13 Examples of some of the prominent new-generation L_2PdX_2 precatalysts.

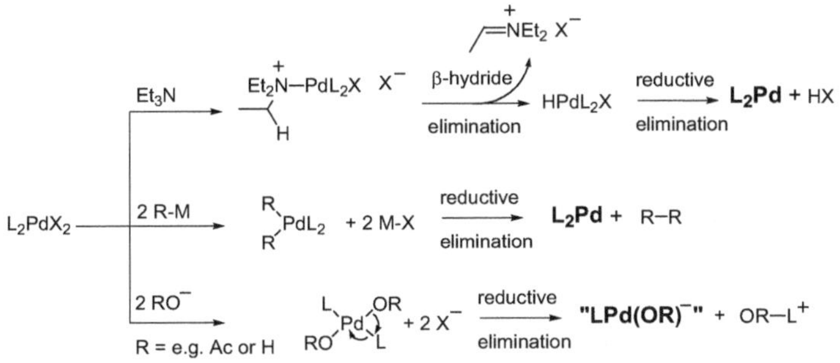

Scheme 3.14 Reduction of L_2PdX_2 to Pd(0).

on both small and large scales. However, these air-stable catalysts become activated to $L_nPd(0)$ in catalysis, hence the coupling reactions have to be conducted under inert conditions to maximize the optimal results, such as low loading and shorter reaction time.

One of the earliest known L_2PdX_2 precatalysts is *trans*-dichlorobis-(triphenyl)phosphine, $(Ph_3P)_2PdCl_2$. As with $(Ph_3P)_4Pd$, this catalyst also generates $(Ph_3P)_2Pd(0)$, but the mechanism of activation is speculative. The reducing agent could be one of the coupling partners (such as boronic acid) or solvent (such as alcohol,[64] *etc.*; Scheme 3.14).

There are a number of early examples where the preformed Pd(II) catalyst, $(Ph_3P)_2PdCl_2$,[65] has been used as a commercial catalyst. Although it is considered to be an old-generation catalyst, it is still very commonly used in many large-scale processes, mainly because of its relatively low cost in comparison with the newly developed advanced catalysts.

The catalyst[26,66,67] is an air-stable Pd(II) d^8 16-electron complex, therefore displaying a square-planar conformation. Its preparation was reported by Miyaura and Suzuki, involving reaction of $PdCl_2$ with Ph_3P ligand in the presence of PhCN (Scheme 3.15),[65] although newer processes are available.

The successful use of $(Ph_3P)_2PdCl_2$ even in recent years can be exemplified by the large-scale preparation of eniluracil (Scheme 3.16).[68] Here, the catalyst

$$\text{PdCl}_2 + 2\text{Ph}_3\text{P} \xrightarrow{\text{PhCN}} (\text{Ph}_3\text{P})_2\text{PdCl}_2$$

Scheme 3.15 Preparation of $(\text{Ph}_3\text{P})_2\text{PdCl}_2$.

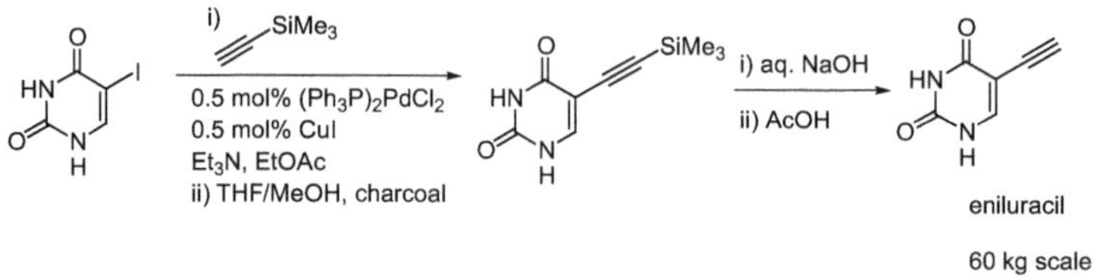

Scheme 3.16 Large-scale Sonogashira coupling catalyzed by $(\text{Ph}_3\text{P})_2\text{PdCl}_2$.

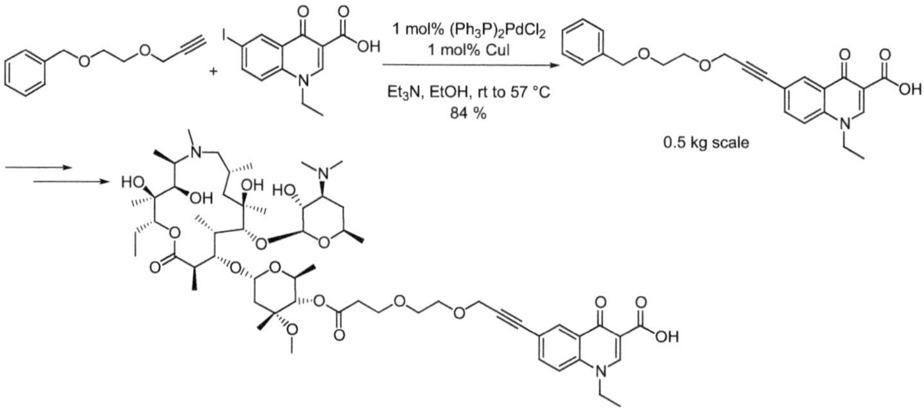

Scheme 3.17 $(\text{Ph}_3\text{P})_2\text{PdCl}_2$ catalyzed Sonogashira coupling.

loading in a Sonogashira reaction could be reduced to 0.5 mol% $(\text{Ph}_3\text{P})_2\text{PdCl}_2$ with 0.5 mol% CuI co-catalyst. The final product was isolated with Pd and Cu levels below 2 and 1 ppm, respectively.

The *in situ* formation of the catalyst by mixing PdCl_2 and Ph_3P was very inefficient (as judged by the almost complete insolubility of PdCl_2 in EtOAc) in comparison with the use of preformed $(\text{Ph}_3\text{P})_2\text{PdCl}_2$.[68] This highlights one of the advantages of using a preformed catalyst over the *in situ*-formed catalyst.

Ŝtimac *et al.* at GlaxoSmithKline in 2010 developed a process to a lead antibacterial compound (Scheme 3.17). One of the initial steps consisted of a $(\text{Ph}_3\text{P})_2\text{PdCl}_2$-catalyzed Sonogashira coupling,[69] as in the previous example.

As discussed in Chapter 2, the properties of bidentate ligands also may have a profound influence on a specific cross-coupling reaction. It was early identified that the larger bite angle (99.1°) of a dppf ligand was very important in facilitating the reductive elimination step in the catalytic cycle (Figure 3.14).

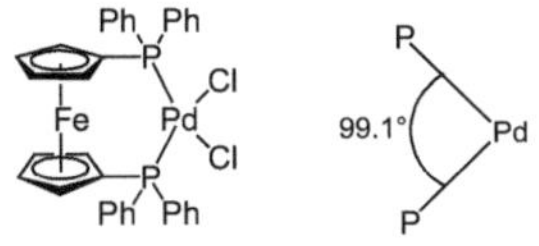

Figure 3.14 Bite angle of the dppf ligand in precatalyst dppfPdCl$_2$.

Scheme 3.18 Preparation of dppfPdCl$_2$.

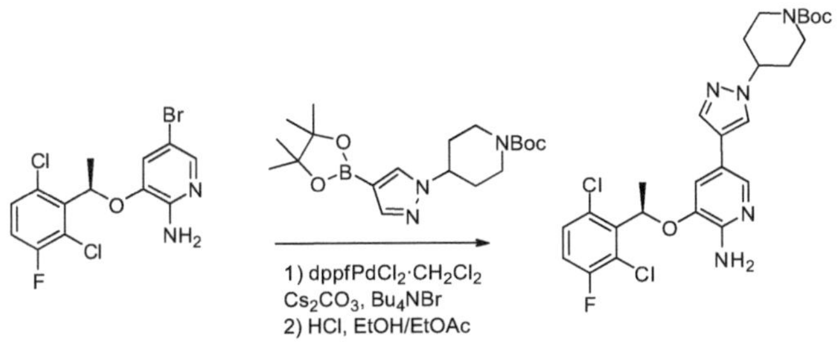

Scheme 3.19 Suzuki–Miyaura coupling as part of the synthesis of crizotinib.

This can help improve reactions that are otherwise problematic, such as Kumada coupling of secondary alkylmagnesium chlorides, where the reductive elimination occurs faster than the β-hydride elimination.[70,71] Numerous examples of the use of the precatalyst dppfPdCl$_2$ in catalysis can be found in the literature.[72] The orange, air-stable catalyst was prepared originally by reacting a suitable Pd precursor with dppf ligand (Scheme 3.18).[70] As in the previous examples, the catalytically active species is proposed to be the dppfPd(0) species, although its isolation is very challenging. This catalyst is commercially available in three forms: dppfPdCl$_2$·CH$_2$Cl$_2$, dppfPdCl$_2$·Me$_2$CO and solvent-free dppfPdCl$_2$.[73] Johnson Matthey was the first company to commercialize both the ligand and catalyst for bulk applications.

Recently, a large-scale Suzuki–Miyaura coupling of ArBr with a boronic ester was reported for the six-step synthesis of crizotinib using dppfPdCl$_2$ catalyst (Scheme 3.19).[74]

In 2010, an efficient synthesis of multi-kilogram quantities of BMS-764459 was reported, which features an efficient palladium-catalyzed cyanation of a 5-chloropyrazinone, as shown in Scheme 3.20.[75]

In general, the dppfPdCl$_2$ precatalyst is particularly useful for borylation of Ar–Br[76] and carbonylation.[77]

Just as in the case of the activity variation of (Ph$_3$P)$_4$Pd *versus* (*t*-Bu$_3$P)$_2$Pd, exchanging the Ph of dppf with a bulkier more electron-rich *t*-Bu group dramatically enhanced the reactivity of the precatalyst to allow the coupling of less reactive aryl halides (Figure 3.15).

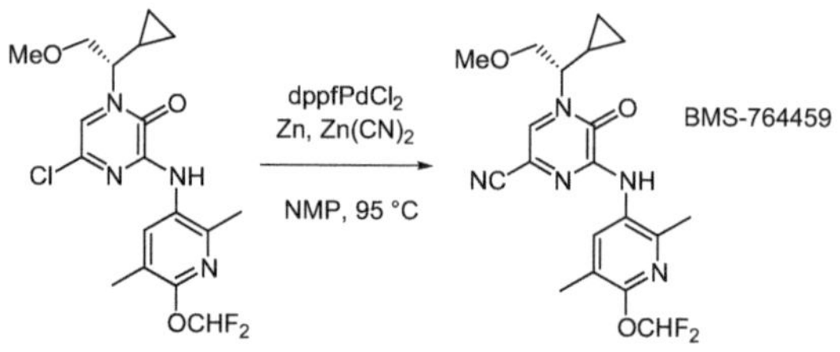

Scheme 3.20 Large-scale cyanation catalyzed by dppfPdCl$_2$.

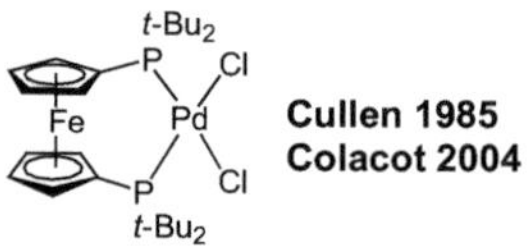

Figure 3.15 Structure of dtbpfPdCl$_2$.

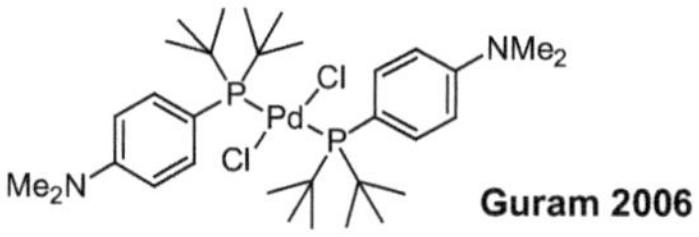

Figure 3.16 Structure of $[p\text{-Me}_2\text{NC}_6\text{H}_4(t\text{-Bu})_2\text{P}]_2\text{PdCl}_2$.

Cullen and co-workers briefly mentioned the first synthesis of the PdL$_2$X$_2$ complex dtbpfPdCl$_2$ in 1985.[78] Colacot and co-workers were the first to develop a scalable route to dtbpfPdCl$_2$;[79] they also reported its X-ray structure, in addition to demonstrating its utility as a highly active air-stable precatalyst in Suzuki–Miyaura[80] and α-arylation reactions.[8,81] Since then, this complex has been used successfully in many challenging cross-coupling reactions and is deemed to be one of the catalysts of choice for a "first-pass" Suzuki–Miyaura reaction[82] and for "first-pass" Heck reactions using aryl bromides.[83]

Another prominent L$_2$PdX$_2$ precatalyst, developed by Guram *et al.*,[84] showed similar activities to dtbpfPdCl$_2$ in many reactions, but it is devoid of Fe (Figure 3.16). The *trans*-dichlorobis[di-*tert*-butyl(4-dimethylaminophenyl)phosphine]palladium(II) catalyst, $[p\text{-Me}_2\text{NC}_6\text{H}_4(t\text{-Bu})_2\text{P}]_2\text{PdCl}_2$, has since been widely employed as a highly effective precatalyst for a number of cross-coupling reactions, such as Suzuki–Miyaura coupling of heteroaryl chlorides[84] and Heck alkynylation (Cu-free Sonogashira) reactions of aryl chlorides.[50,85] Larsen and co-workers at Amgen reported a practical synthesis of a p38 MAP (mitogen-activated protein) kinase inhibitor employing a Pd-catalyzed Suzuki–Miyaura coupling (Scheme 3.21).[86] They found $[p\text{-Me}_2\text{NC}_6\text{H}_4(t\text{-Bu})_2\text{P}]_2\text{PdCl}_2$ to be the best performing catalyst.

Carbonylation of aryl halides, in particular of aryl chlorides, is a challenging C–C bond-forming reaction due to the strong d$_\pi$–p$_\pi$ backbonding of Pd to the CO ligand, when the ligand involved is electron rich. This

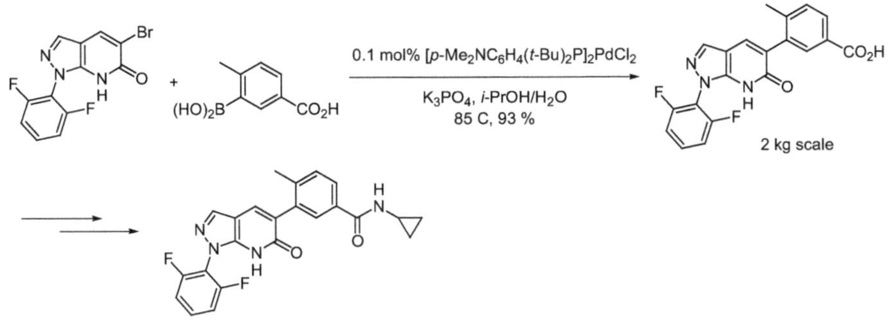

Scheme 3.21 Large-scale [p-Me$_2$NC$_6$H$_4$(t-Bu)$_2$P]$_2$PdCl$_2$-catalyzed Suzuki–Miyaura coupling reaction.

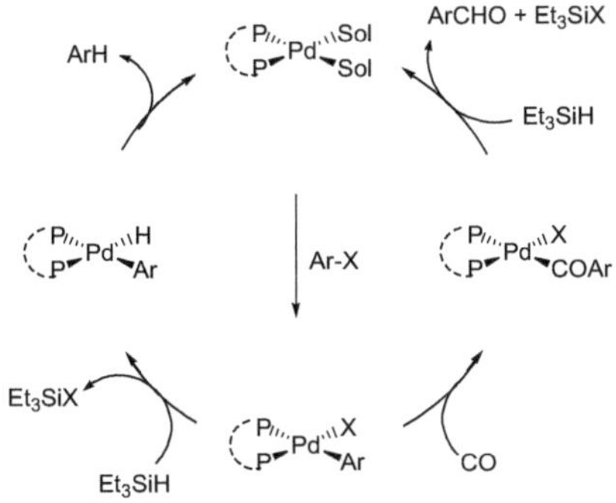

Scheme 3.22 Proposed catalytic cycle for carbonylation using dcppPdCl$_2$.

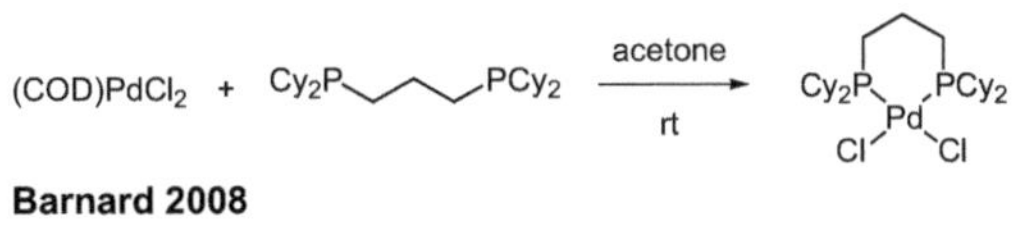

Barnard 2008

Scheme 3.23 Preparation of dcppPdCl$_2$.[89]

strengthens the CO binding to Pd and hence retards its insertion into the Pd–Ar bond after the oxidative addition step (Scheme 3.22). However, to promote the oxidative addition of Ar–Cl, an electron-rich bulky ligand coordinated to the Pd catalyst is needed. By careful choice of a ligand, this transformation can, however, be successfully carried out. Barnard and co-workers employed the L$_2$PdX$_2$ precatalyst dichloro[1,3-bis(dicyclohexylphosphino)propane]palladium(II)[87] (Scheme 3.23) in the carbonylation of a range of less challenging aryl chlorides.[88] This process is discussed in more detail in Chapter 9.

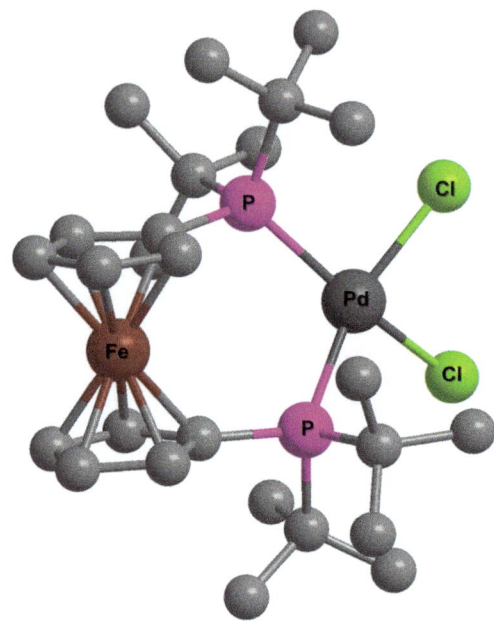

Figure 3.17 Crystal structure of dtbpfPdCl$_2$.[8]

The L$_2$PdCl$_2$ (L = trialkylphosphine) precatalysts can be assumed to follow the same pathway of activation to the Pd(0) species as discussed in the cases where L is a more traditional triarylphosphine ligand. In the case of bidentate phosphine ligands, the bite angle of the phosphine has been suggested to be of great importance for the cross-coupling reaction. The electron-donating *t*-Bu groups in the dtbpf ligand are proposed to facilitate the oxidative addition, whereas the large P–Pd–P angle (104.2°) is said to promote the final reductive elimination step in the catalytic cycle (Figure 3.17).

The importance of the bite angle has been debated, however, since Kawatsura and Hartwig noted that ketone arylation catalyzed by a Pd(dba)$_2$/dtbpf system likely proceeds *via* a Pd(II) intermediate containing the ligand in a monodentate fashion (Scheme 3.24, A; see also Chapter 2).[90] This also relates to the fact that the proposed catalytically active species, in cases with bulky phosphine ligands, is LPd(0) and not L$_2$Pd(0).

α-Arylation reactions catalyzed by the preformed catalyst dtbpfPdCl$_2$, however, seem to proceed *via* a different pathway. In a catalytic reaction monitored by ^{31}P NMR spectroscopy, none of the monocoordinated phosphine intermediate could be observed.[8] Instead, the NMR data indicated that the binding mode in the resting state of the catalyst [dtbpfPd(0)] is bidentate (Scheme 3.24, B). This demonstrates how the reaction mechanism depends not only on the ligands and type of aryl halide, but also on the type of cross-coupling reaction.

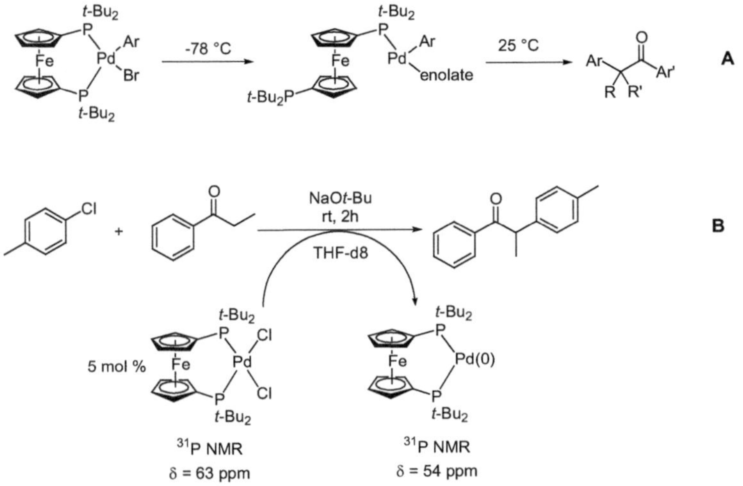

Scheme 3.24 Mono- *versus* bidentate reaction pathway involving dtbpf ligand.

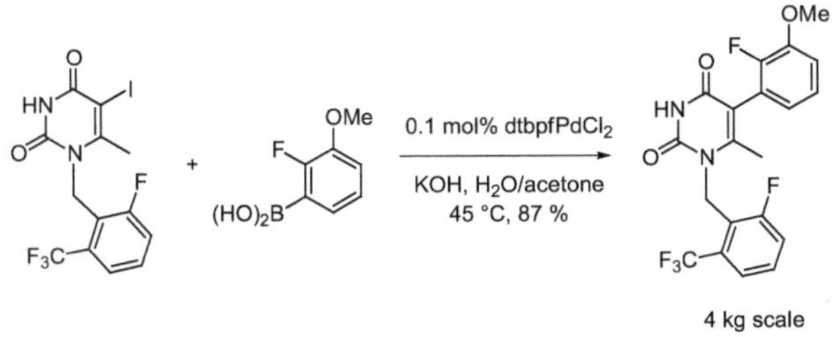

Scheme 3.25 Low-loading Pd-catalyzed Suzuki–Miyaura cross-coupling.

Applications of dtbpfPdCl$_2$ precatalyst in large-scale reactions are notable. Gallagher *et al.* developed a scalable route to uracil derivatives, incorporating a Pd-catalyzed Suzuki–Miyaura coupling (Scheme 3.25).[91] The catalyst of choice was dtbpfPdCl$_2$, which could be used in a loading as low as 0.1 mol% to produce 4 kg of the coupled product.

As part of a route to enantiomerically pure 2-arylpyrrolidines, researchers at GPRD Process Research and Development incorporated a Suzuki–Miyaura coupling catalyzed by 1 mol% dtbpfPdCl$_2$ (Scheme 3.26).[92] The cross-coupling was carried out successfully despite the use of the unstable 2-boronic acid pyrrole coupling partner.

Another noteworthy member of the L$_2$PdCl$_2$ family of catalysts is XantphosPdCl$_2$. This precatalyst is also becoming popular as a second-generation catalyst in industry. A chemoselective Suzuki–Miyaura reaction of bromophenylchloropyrazine was achieved by careful choice of the precatalyst.[93] The use of XantphosPdCl$_2$ favored oxidative addition on the chloropyrazine whereas dtbpfPdCl$_2$ favored reaction with the bromophenyl (Scheme 3.27).

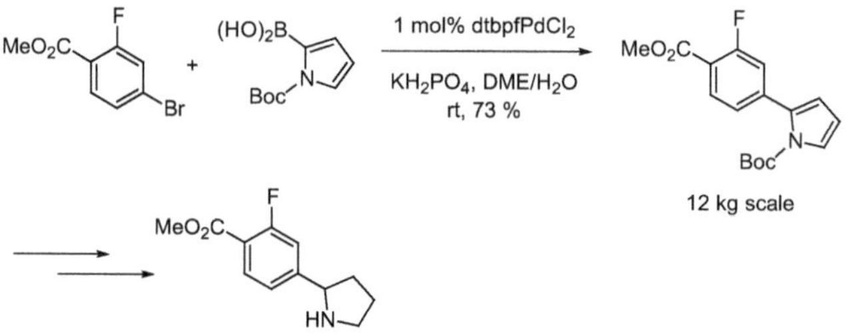

Scheme 3.26 Suzuki–Miyaura coupling using unstable 2-boronic acid N-Boc-pyrrole.

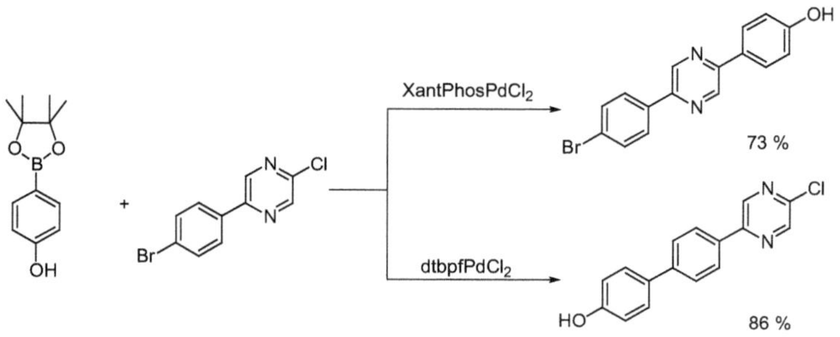

Scheme 3.27 Pd-catalyzed chemoselective Suzuki–Miyaura cross-coupling.

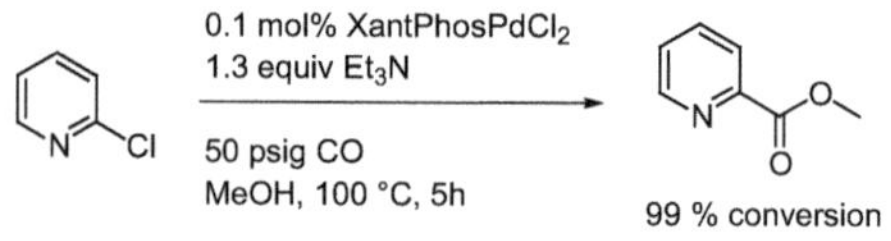

Scheme 3.28 Pd-catalyzed carbonylation of aryl chlorides.

XantphosPdCl$_2$ has also been found to be an effective precatalyst for carbonylation reactions.[94] Despite the challenge of carbonylation of aryl chlorides, this transformation was carried out with excellent conversion using the Xantphos-based catalyst (Scheme 3.28).

3.7 Generation of LPd(0) Catalysts

The metal-to-ligand ratio seems to have a profound influence on certain cross-coupling reactions, although this is not the case in all chemistries. For *in situ*-generated catalysts involving monodentate ligands, a few reports suggest a more active system when the metal-to-ligand ratio is 1:1, rather than with the more traditional approach where excess ligand is used.[95,96] However during the mixing of Pd with ligand, more than one ligand can coordinate with the Pd. Therefore, precatalysts engineered with a precise 1:1 ratio of phosphine (or other ligand) to palladium are of increasing interest. Currently, there are three main classes of precatalysts providing this ratio;

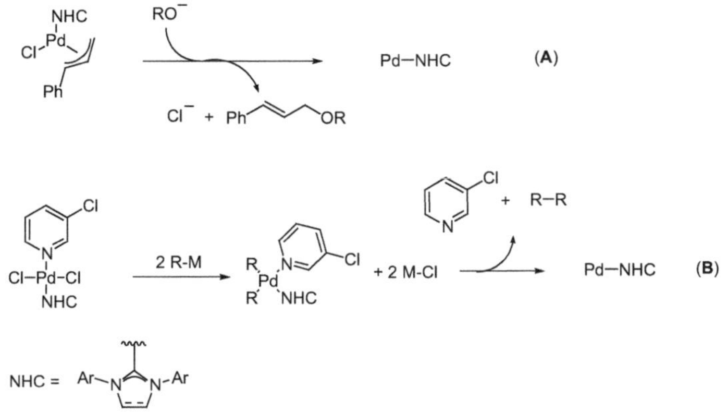

Scheme 3.29 Generation of NHC–Pd(0) from cinnamyl- (A) and PEPPSI (B) precatalysts.

Pd(I) dimer, Pd(II)(R-allyl)Cl and Pd(II) palladacycles. Although the focus of this chapter is on phosphine-containing precatalysts, the NHC (*N*-hetero-cyclic carbene)-based catalysts such as the palladium cinnamyl complexes developed by Nolan and co-workers[97–99] and the PEPPSI complexes introduced by Organ and co-workers[100,101] are noteworthy. Both the cinnamyl and the PEPPSI type of precatalyst also generate the proposed catalytically active species LPd(0). The proposed activation pathways are shown in Scheme 3.29. In the case of the cinnamyl complexes, an added base such as KO*t*-Am (*tert*-amylate) or KOH is suggested to activate the complex by nucleophilic attack on the cinnamyl moiety (A). The PEPPSI complexes are activated by initial transmetallation with an organometallic reagent, followed by reductive elimination of R–R (B). The pyridine ligand is a labile ligand that easily dissociates to generate LPd(0).

However the phosphine-containing Pd(I) dimer, Pd(II)(R-allyl)Cl and Pd(II) palladacycles catalysts all activate by different pathways. This is discussed in more detail in the following sections.

3.7.1   [(*t*-Bu₃P)Pd(*μ*-Br)]₂

The first synthesis of the palladium(I) dimer [(*t*-Bu₃P)Pd(μ-Br)]₂ that was demonstrated by Mingos and co-workers gave a very poor yield,[102] although some improvements were subsequently made by the same group (Scheme 3.30).[103] Colacot *et al.* recently developed a high-yielding patented route to the dimer, and it is now a readily commercially available precatalyst.[104]

Initially, this complex was only investigated in reactions with a selected number of reagents, namely CO, H₂, nitriles, alkenes and alkynes.[103] The complex is air sensitive and it has been shown to react with aerial oxygen to form a second dimeric complex by insertion of O₂ into the Pd(I)–Pd(I) dimer and subsequent C–O bond formation (Scheme 3.31).[105]

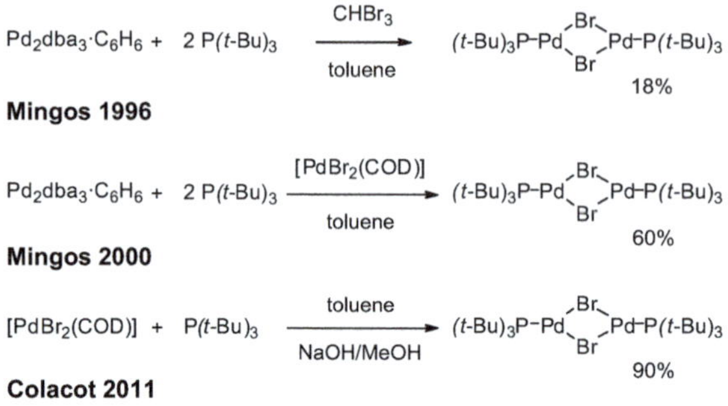

Scheme 3.30 Reported syntheses of [(*t*-Bu$_3$P)Pd(μ-Br)]$_2$.

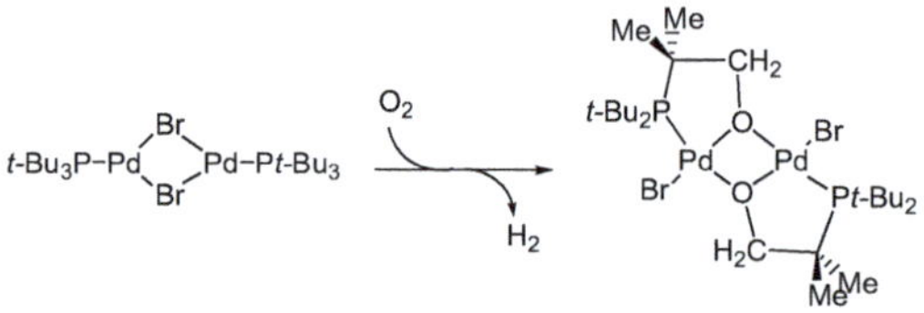

Scheme 3.31 Reaction of [(*t*-Bu$_3$P)Pd(μ-Br)]$_2$ with O$_2$.

Figure 3.18 Gradual decomposition of [(*t*-Bu$_3$P)Pd(μ-Br)]$_2$ in air.

The color change is very indicative of this reaction. The unreacted Pd(I) dimer is a dark-green material; however, on contact with air it slowly reacts with oxygen to give an orange–brown compound (Figure 3.18). This pre-catalyst should be handled and stored under an inert atmosphere, although it can be synthesized and used in multi-kilogram quantities.

[(*t*-Bu$_3$P)Pd(μ-Br)]$_2$ is formally a 16-electron species; however, it is a very reactive complex. The reaction with aerial oxygen results in a coordinatively saturated dimeric Pd(II)-speces..

Schoenebeck and co-workers recently carried out in-depth studies on the likely active catalytic species formed from this Pd(I) dimer and the mechanism of precatalyst activation.[106] Experimental investigations pointed towards the active catalytic species to be consistent with mononuclear Pd(0) catalysis.

Since the reaction rates using Pd(I) dimer as catalyst have been observed to be much higher than those using the corresponding $L_2Pd(0)$ as pre-catalyst, the activation of the Pd(I) precatalyst must be a more facile process than the substrate-assisted ligand displacement process that is presumed to activate $L_2Pd(0)$ to the catalytically active LPd(0) species. Recent computational studies suggested that the precatalyst was converted into the active species *via* a reductive pathway,[106] which is a lower energy pathway than the previously assumed disproportionation mechanism (Scheme 3.32).

The exact mechanism of the reduction to catalytically active LPd(0) species is currently not known. It should be noted that the same LPd(0) species is generated from $(t\text{-}Bu_3P)Pd(\text{crotyl})Cl$, which is discussed in Section 3.7.2. This precatalyst is air stable and convenient to handle, compared with air-sensitive $[(t\text{-}Bu_3P)Pd(\mu\text{-}Br)]_2$.

In addition, it was demonstrated that under Suzuki–Miyaura cross-coupling reaction conditions, the Pd(I) dimer underwent precatalyst deactivation to form $(t\text{-}Bu_3P)_2Pd$ and Pd black.[106] This observation could explain the incomplete conversions when using $[(t\text{-}Bu_3P)Pd(\mu\text{-}Br)]_2$ as pre-catalyst in Suzuki–Miyaura couplings of aryl chlorides, based on the fact that $(t\text{-}Bu_3P)_2Pd$ on its own provided the desired product in very low yield.[106]

Schoenebeck's group further demonstrated the importance of the catalytically active species to the outcome of a chemoselective reaction. Based on the finding that the actual catalyst formed *in situ* from Pd_2dba_3 and $t\text{-}Bu_3P$ could be fine-tuned by the choice of solvent (polar solvents favored the formation of $[LPdX]^-$ and non-polar solvents the formation of neutral PdL),[11,107] $[(t\text{-}Bu_3P)Pd(\mu\text{-}Br)]_2$ was investigated for chemo- and regio-selectivity in Suzuki–Miyaura cross-coupling reactions (Scheme 3.33).[106]

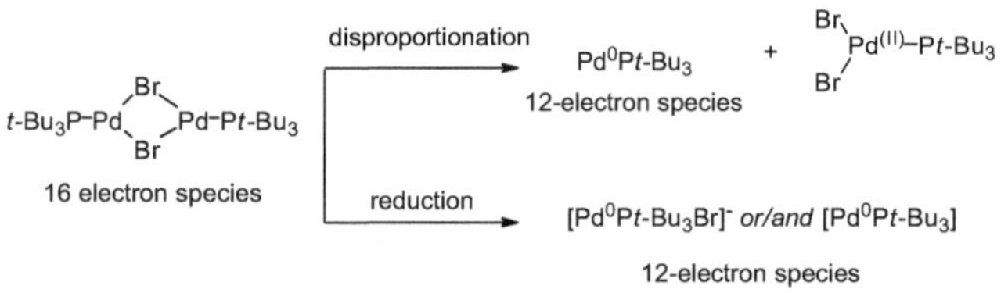

Scheme 3.32 Disproportionation *versus* reduction pathway to LPd(0) from $[(t\text{-}Bu_3P)Pd(\mu\text{-}Br)]_2$.

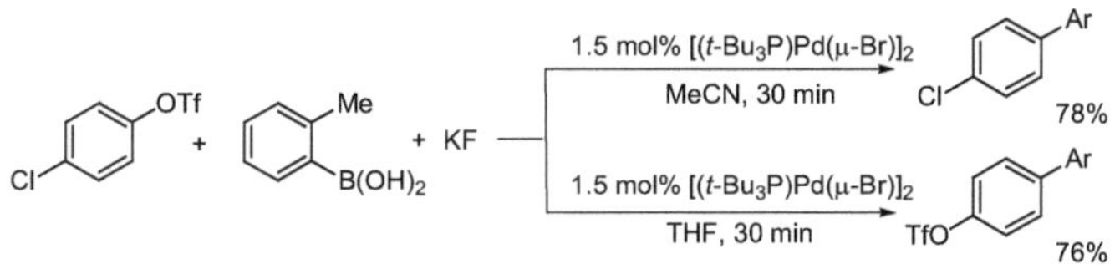

Scheme 3.33 Chemoselective Suzuki–Miyaura cross-coupling reactions using $[(t\text{-}Bu_3P)Pd(\mu\text{-}Br)]_2$.

Identically with the Pd_2dba_3/t-Bu_3P systems, use of the Pd(I) dimer in MeCN resulted in exclusive OTf insertion (Scheme 3.33). In THF, oxidative addition occurred into the C–Cl bond. Interestingly, the reaction rates using the precatalyst $[(t$-$Bu_3P)Pd(\mu$-$Br)]_2$ were much faster than those observed when using the *in situ*-formed catalyst. Similarly enhanced reaction rates were observed in amine arylation (Buchwald–Hartwig) reactions with aryl chlorides and Suzuki–Miyaura reactions using aryl bromides.[108]

The chemoselectivity can be explained by having a closer look at the actual active catalytic species generated under each set of conditions. It is suggested that in a polar solvent, such as acetonitrile, the active species formed is the anionic $[(t$-$Bu_3P)Pd(0)(X)]^-$,[107] resulting in triflate insertion. In a nonpolar solvent such as THF, the neutral species $(t$-$Bu_3P)Pd(0)$ is formed, favoring insertion into the C–Cl bond. This is consistent with their earlier studies to support Fu and co-workers' observation[11] that Cy_3P-based Pd complexes are suitable for OTf coupling.

The precatalyst $[(t$-$Bu_3P)Pd(\mu$-$Br)]_2$ has also been shown to provide regioselective couplings in the case of dihalogenated heterocycles (Scheme 3.34).[106]

The role of the Pd(I) dimer as a precatalyst has been questioned, however. Schoenebeck and co-workers demonstrated that aryl iodides are in fact able to react directly with the dimer and undergo oxidative addition.[109] The second product in this reaction is the corresponding aryl bromide, with the bromide originating from the precatalyst. This reaction can be made catalytic in Pd by the addition of NBu_4Br as a bromide source.

$[(t$-$Bu_3P)Pd(\mu$-$Br)]_2$ has also been used successfully in cross-coupling reactions in industry. Ryberg reported a robust and mild method for the cyanation of indole (Scheme 3.35).[110] Pd(I) dimer (1.25 mol%) was identified as the catalyst of choice and provided high yields of the product even on a 5 kg scale.

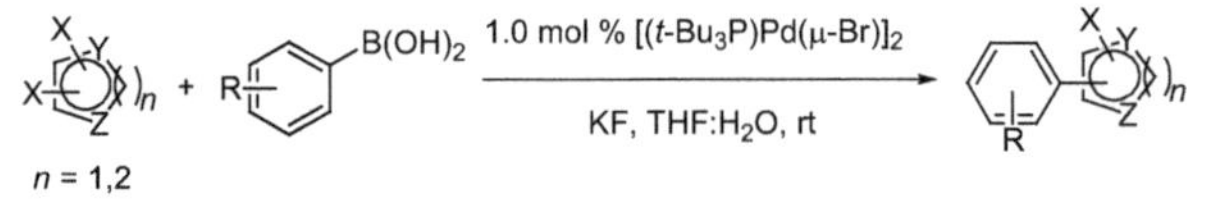

Scheme 3.34   Regioselective   Suzuki–Miyaura   cross-coupling   reaction   using $[(t$-$Bu_3P)Pd(\mu$-$Br)]_2$.

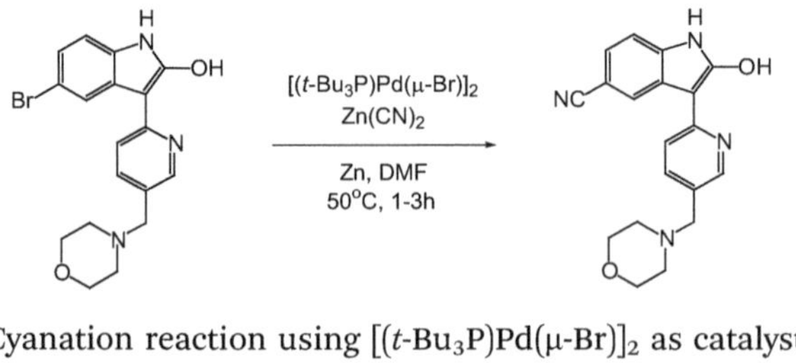

Scheme 3.35   Cyanation reaction using $[(t$-$Bu_3P)Pd(\mu$-$Br)]_2$ as catalyst.

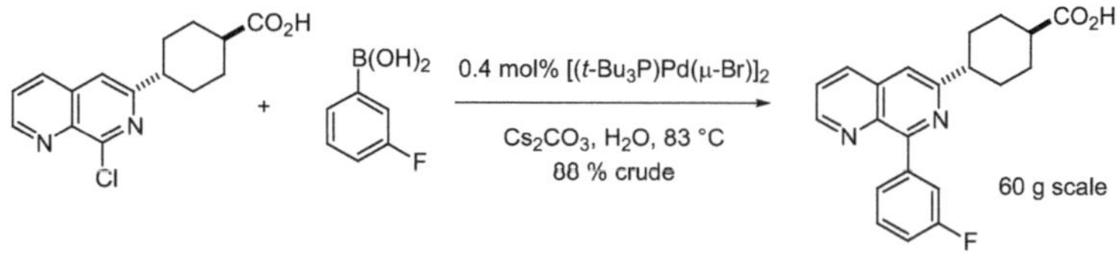

Scheme 3.36 Scalable Suzuki–Miyaura cross-coupling using $[(t\text{-}Bu_3P)Pd(\mu\text{-}Br)]_2$.

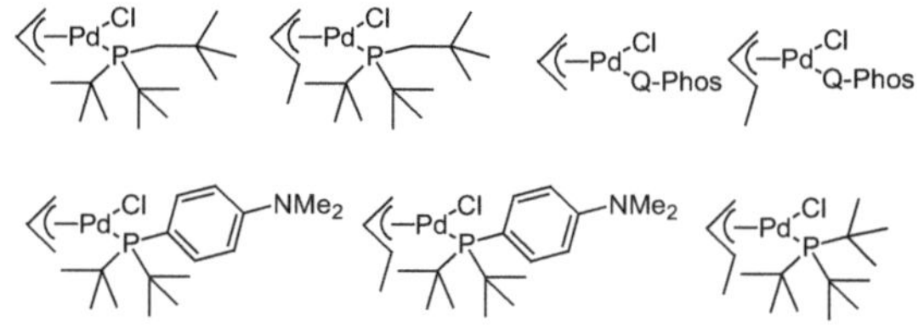

Figure 3.19 LPd(π-allyl)Cl catalysts.

Researchers at Novartis reported a scalable process to make a PDE-4 inhibitor, incorporating a Pd-catalyzed Suzuki–Miyaura coupling as the final step (Scheme 3.36).[111] In order to remove any remaining Pd residues from the active pharmaceutical ingredient (API), the product was treated with the metal scavenger (chelating anion exchanger) Smopex 110.

3.7.2 LPd(R-allyl)(Cl)

Inspired by the π-allyl-based NHC-containing Pd precatalysts developed by Nolan and co-workers,[99] Colacot and co-workers recently reported another P-based class of precatalysts generating catalytically active LPd(0).[112] These LPd(R-allyl)(Cl) (L = PR$_3$) complexes are depicted in Figure 3.19. The (dtbnp)Pd(allyl)Cl (dtbnp = di-*tert*-butylneopentylphosphine) was initially demonstrated as a highly active catalyst in amination and α-arylation reactions.[113] In some cases, amination reactions involving aryl bromides could be carried out at room temperature under air. Aryl chlorides required elevated temperatures under a nitrogen atmosphere.

The preformed catalyst provided results superior to those obtained using the *in situ*-generated catalyst from $Pd_2(dba)_3$ and free dtbnp ligand (Figure 3.20).

The preformed catalyst is stable when stored under air at low temperature. This avoids handling of the pyrophoric free ligand. When stored at ambient temperature, the complex slowly turns from yellow (good catalyst) to green (decomposed).

Nolan and co-workers had previously observed significant effects on the catalysis when the allyl ligand was changed to crotyl or cinnamyl.[99] Subsequent to the initial success of (dtbnp)Pd(allyl)Cl, the role of the π-allyl moiety of the catalyst and the effect of the nature of the tertiary phosphine

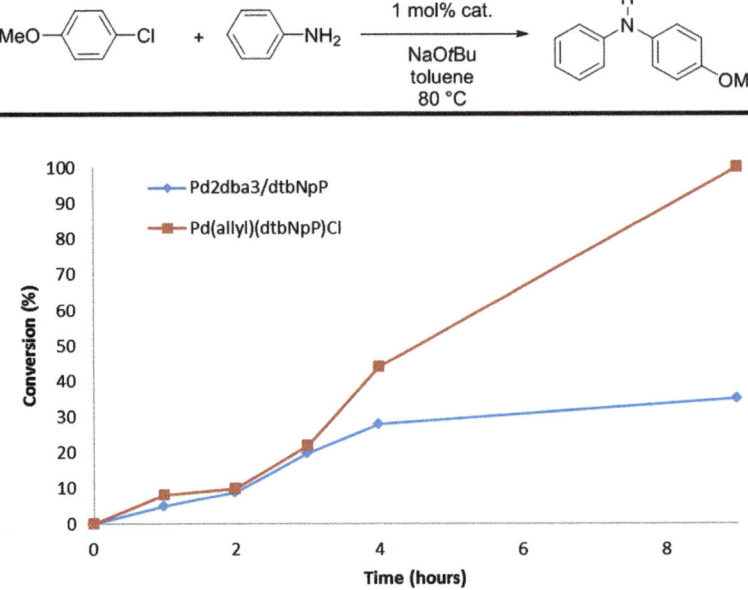

Figure 3.20 (dtbnp)Pd(allyl)Cl *versus in situ*-generated Pd$_2$(dba)$_3$/dtbnp.

Table 3.3 Evaluation of LPd(π-allyl)Cl precatalysts in amination reactions.

Entry	Catalyst (mol%)	Time (h)	Conversion[a] (%)
1	(dtbnp)Pd(allyl)Cl (1.0)	6	99
2	(dtbnp)Pd(crotyl)Cl (1.0)	3	100
3	(Q-Phos)Pd(allyl)Cl (1.0)	6	97
4	(Q-Phos)Pd(crotyl)Cl (1.0)	3	100
5	(Q-Phos)Pd(crotyl)Cl (0.5)	18	99
6	(Q-Phos)Pd(crotyl)Cl (0.1)	18	95
7	[*p*-Me$_2$NC$_6$H$_4$(*t*-Bu)$_2$P]Pd(allyl)Cl (1.0)	22	18
8	[*p*-Me$_2$NC$_6$H$_4$(*t*-Bu)$_2$P]Pd(crotyl)Cl (1.0)	22	95
9	(*t*-Bu$_3$P)Pd(crotyl)Cl (1.0)	3	100

[a]Conversion determined by GC analysis.

ligand were therefore evaluated.[112] A number of additional π-allyl and crotyl P-based precatalysts (Figure 3.19) were prepared and investigated in amination, Suzuki–Miyaura and α-arylation reactions.[112] The general trend observed with amination reactions at ambient temperature was that Pd–crotyl complexes showed higher activities than the corresponding Pd–allyl precatalysts (Table 3.3, Entries 1 *versus* 2, 3 *versus* 4 and 7 *versus* 8). This may indicate that the activation pathway to the catalytically active species is more facile in the case of the crotyl complexes, leading to a more efficient coupling reaction, based on the crystal structure analysis.[112]

The same range of precatalysts were subsequently evaluated in a room temperature Suzuki–Miyaura coupling (Table 3.4). Again, the Pd–crotyl complexes proved superior to the corresponding Pd–allyl precatalysts (Entries 1 *versus* 2 and 3 *versus* 4). Notably, in the case of p-Me$_2$NC$_6$H$_4$(t-Bu)$_2$P-based complexes, both the allyl and crotyl catalysts provided very low conversions to the desired biaryl product (Entries 5 and 6). This demonstrates the importance of matching the right ligand with the right cross-coupling reaction.

Subsequently, a reaction requiring elevated temperature was investigated. Surprisingly, in the chosen α-arylation reaction, the reactivity of the Pd–allyl *versus* the Pd–crotyl complexes was observed to be the reverse of that found in the room temperature amination and Suzuki–Miyaura reactions. The crotyl complexes were outperformed by their allyl counterparts Q-Phos-Pd(allyl)Cl and [p-Me$_2$NC$_6$H$_4$(t-Bu)$_2$P]Pd(allyl)Cl (Table 3.5, Entries 1 *versus* 2 and 4 *versus* 5).

Table 3.4 Room temperature Suzuki–Miyaura cross-coupling reactions.

Entry	Catalyst (0.05 mol%)	Conversion (%)[a]
1	(dtbnp)Pd(allyl)Cl	13
2	(dtbnp)Pd(crotyl)Cl	99
3	(Q-Phos)Pd(allyl)Cl	2
4	(Q-Phos)Pd(crotyl)Cl	100[b]
5	[p-Me$_2$NC$_6$H$_4$(t-Bu)$_2$P]Pd(allyl)Cl	3
6	[p-Me$_2$NC$_6$H$_4$(t-Bu)$_2$P]Pd(crotyl)Cl	7
7	(t-Bu$_3$P)Pd(crotyl)Cl	100

[a]Conversion determined by GC analysis; average of at least two runs.
[b]85% isolated yield.

Table 3.5 α-Arylation of tetralone.

Entry	Catalyst	Time (h)	Conversion (%)[a]
1	(Q-Phos)Pd(allyl)Cl	3	80
2	(Q-Phos)Pd(crotyl)Cl	3	23
3	[p-Me$_2$NC$_6$H$_4$(t-Bu)$_2$P]Pd(allyl)Cl	22	100[b]
4	[p-Me$_2$NC$_6$H$_4$(t-Bu)$_2$P]Pd(allyl)Cl	3	96
5	[p-Me$_2$NC$_6$H$_4$(t-Bu)$_2$P]Pd(crotyl)Cl	3	7

[a]Conversion determined by GC-MS analysis. Average of two runs.
[b]91% isolated yield.

3.7.3 LPd(palladacycle)X

The third, and final, group of LPd(0) precursors in this chapter are the palladacycles, which have shown very promising results in a number of cross-coupling reactions.

The use of P-based palladacycles as precatalysts was demonstrated by Bedford *et al.* in 2003,[114] and corresponding NHC-based palladacycles were concurrently reported by Nolan and co-workers as efficient precatalysts for cross-coupling reactions.[115,116]

More recent examples of P-based palladacycles can be readily prepared from an aliphatic amine, *o*-chlorophenethylamine (Scheme 3.37).[117] When reacted with a base, such as NaO*t*-Bu or K_2CO_3, the catalytically active LPd(0) species is proposed to be generated. These palladacycles have been used in Suzuki–Miyaura coupling and amination reactions with excellent results.

Generally, the catalyst generated *in situ* from $Pd_2(dba)_3$ and XPhos provided significantly lower conversion to the product than the XPhos-based palladacycle 7 in Scheme 3.37 (Table 3.6).[117] Clearly, the active catalytic species LPd(0) is formed much more efficiently from the precatalyst, thereby influencing the subsequent cross-coupling reaction considerably.

A number of palladacycles bearing various different dialkylarylphosphine ligands have been prepared. Notably, the BrettPhos- and RuPhos-containing precatalysts were demonstrated to provide very reliable and general catalysts for amination reactions (Figure 3.21).[118]

Buchwald and co-workers also reported the second-generation palladacycle precatalyst **10**, which can be prepared by a one-pot reaction starting from 2-aminobiphenyl and $Pd(OAc)_2$ (Scheme 3.38). Its activation pathway is identical with that of the first-generation palladacycles (see Scheme 3.37), in

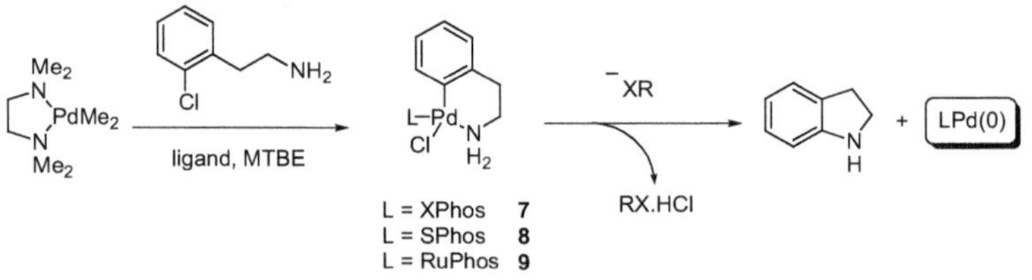

L = XPhos 7
L = SPhos 8
L = RuPhos 9

Scheme 3.37 Palladacycle prepared from a Pd-based aliphatic amine complex and proposed precatalyst activation.

Table 3.6 Preformed *versus in situ* catalysis using XPhos.

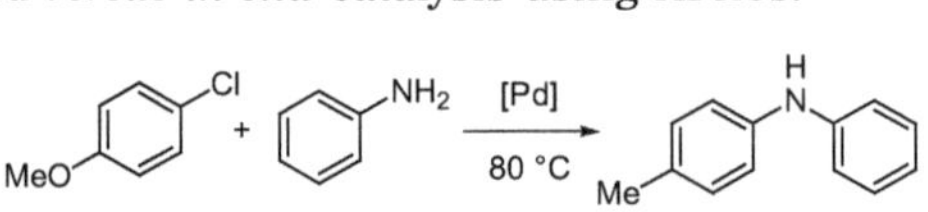

Entry	[Pd]	Time (min)	Conversion (%)
1	0.05 mol% $Pd_2(dba)_3$/XPhos	35	25^a
2	0.1 mol% 7	35	100

^{a}No further conversion was observed.

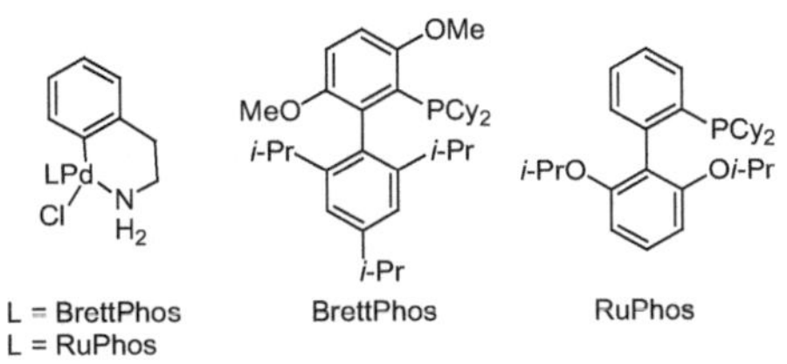

Figure 3.21 Palladacycle catalysts for amination reactions.

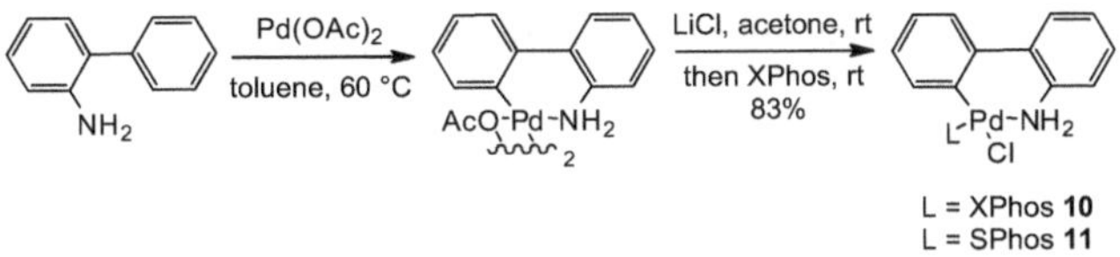

Scheme 3.38 Palladacycle prepared from 2-aminobiphenyl.

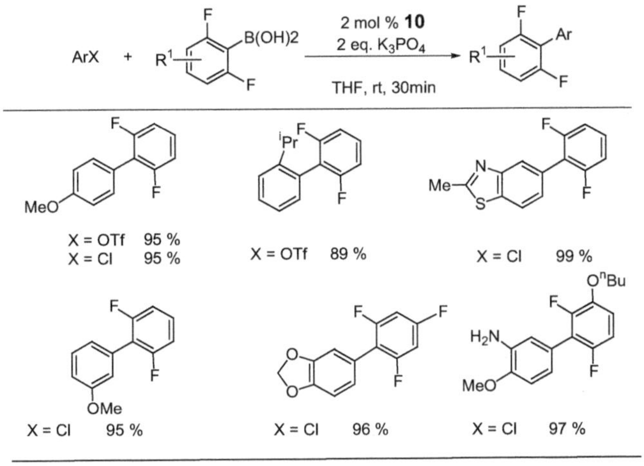

Scheme 3.39 Coupling of polyfluoroboronic acids to aryl chlorides, bromides and triflates.

this case forming carbazole and LPd(0). The-second generation palladacycle was applied in ambient temperature Suzuki–Miyaura coupling reactions.[119] A wide range of (hetero)aryl halides and triflates were used as coupling partners and excellent functional group tolerance was displayed (Scheme 3.39). The otherwise unstable polyfluorophenyl and five-membered 2-heterocyclic boronic acids were coupled with a wide range of aryl (pseudo)halides, thanks to the milder reaction conditions offered by these catalysts. The rate increase is explained by the fast generation of the catalytically active species LPd(0) from the palladacycle **10**.

Another challenging reaction, namely monoarylation of acetate esters and aryl methyl ketones, was also carried out successfully using the first-generation palladacycles.[120] Challenging heteroaryl chloride–heteroaryl boronic acid couplings were carried out using the SPhos-containing palladacycle **11**.[121] It has also been demonstrated that a combination of palladacycle with added free ligand may sometimes improve the reactions compared with the use of the palladacycle alone.[122,123] Despite the potential of the second-generation palladacycles, a couple of drawbacks limit their scope for use in coupling reactions: first, they cannot be generated using extremely bulky ligands, such as *t*-BuBrettPhos; and second, they decompose in solution with extended periods of time.

The most recent addition to the palladacycle class of precatalysts is a third-generation palladacycle that contains an OMs group in place of a chloride (Scheme 3.40).[124] This improved family of catalysts display greater solution stability than the chloride-containing catalysts.

The most recently reported palladacycles containing a mesylate ligand in place of the chloride have been used in a number of applications, such as cyanation of heteroaryl halides,[125] synthesis of aryl ethers,[126,127] monoarylation of ammonia with a wide range of aryl- and heteroaryl halides,[128] arylation of primary amides[127] and Negishi coupling reactions (Scheme 3.41).[129]

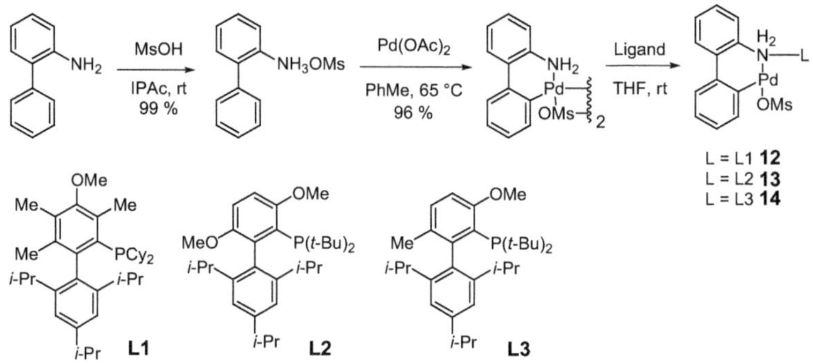

Scheme 3.40 Synthesis of third-generation palladacycles.

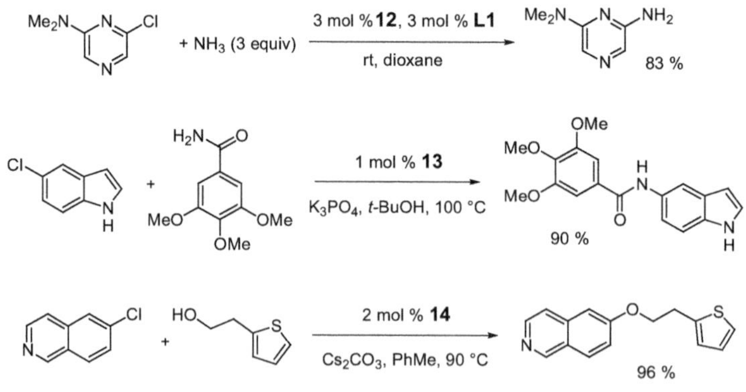

Scheme 3.41 Examples of monoarylation of ammonia, amide arylation and synthesis of aryl ethers using palladacycle precatalysts.

3.8 Effects of the Nature of the Precise Catalyst

Throughout this chapter, it has been highlighted by several examples that the choice of ligand is crucial to the success of a cross-coupling reaction. The second factor that can have a significant impact is the precise nature of the catalyst, more specifically, whether it is generated *in situ* or from a pre-formed catalyst. In turn, the nature of the preformed catalyst may affect the outcome of the reaction, even if the same ligand is employed.

The dtbnp ligand may be used as an illustrative example, where cross-coupling reactions also can be affected by the above-described behavior. The α-arylation of acetophenone with 4-chlorotoluene was investigated using a number of preformed and *in situ*-generated dtbnp-containing catalysts (Table 3.7).[113] Overall, in this reaction, it was noted that catalysts with a Pd-to-ligand ratio of 1 : 2 performed better than catalysts with a 1 : 1 ratio. (dtbnp)$_2$Pd and (dtbnp)$_2$PdCl$_2$ gave quantitative conversion (Entries 3 and 4), whereas (dtbnp)Pd(allyl)Cl and *in situ*-generated LPd(0) provided the product with considerably lower conversion (Entries 1 and 2).

Another example can be extracted from the amination investigations using Q-Phos as ligand. Notably, preformed Q-PhosPd(crotyl)Cl performed better than *in situ*-generated catalyst using Pd(dba)$_x$ ($x = 1.5$ or 2) in conjunction with free Q-Phos ligand (Table 3.8, Entries 1, 2 and 3).[112] Also, use of

Table 3.7 Preformed *versus in situ* catalysis in α-arylation reactions using dtbnp ligand.

Entry	[Pd]	Conversion (%)^a
1	Pd$_2$(dba)$_3$/dtbnp (1 : 1)	42
2	(dtbnp)Pd(allyl)Cl	79
3	(dtbnp)$_2$Pd	99
4	(dtbnp)$_2$PdCl$_2$	99

^{a}Conversion determined by GC analysis.

Table 3.8 Application of Q-Phos-based catalysts in amination reactions.

Entry	Catalyst	Conversion (%)^a
1	0.5 mol% (Q-Phos)Pd(crotyl)Cl	100
2	0.25 mol% Pd$_2$(dba)$_3$/0.5 mol% Q-Phos	35
3	0.5 mol% Pd(dba)$_2$/0.5 mol% Q-Phos	80
4	0.5 mol% (Q-Phos)$_2$Pd	9

^{a}Conversion determined by GC analysis. Average of at least two runs.

Q-PhosPd(crotyl)Cl (Pd : L = 1 : 1) resulted in considerably faster reaction rates than precatalyst (Q-Phos)$_2$Pd (Pd : L = 1 : 2) (Entry 3). This highlights the importance of the precatalyst and the pathway of generation of the catalytically active species, which is here presumed to be LPd(0) in all cases.

Furthermore, the effect of using allyl- *versus* crotyl-based precatalysts containing the same ligand has already been discussed in Section 3.7.2.[111]

3.9 Concluding Remarks

In the past 10 years, great progress has been made in terms of expanding the substrate scope while reducing the palladium loading in cross-coupling reactions. This is partly due to the development of highly active, bulky, electron-rich phosphine and NHC ligand-based catalysts, although phosphines dominate the area. A major factor in the improvement of these reactions is the use of well-defined preformed catalysts. By using precatalysts, it is possible to avoid deleterious side reactions arising from generating the catalyst *in situ*. This avoids the waste of both ligands and Pd.

Substantial research efforts have been focused on elucidating the mechanism of each named coupling reaction by studying in detail the rate-determining step(s). Lately, it has been noted that generation of the catalytically active LPd(0) species is in fact crucial to the success of the subsequent cross-coupling reaction.

By using preformed Pd catalysts, more control can be exerted over the generation of the catalytically active species. Hence, not only is the ligand important in a coupling reaction, but also the nature of the precatalyst [Pd(0), Pd(I) or Pd(II)]. In addition, the ancillary ligands (*e.g.*, π-allyl groups, halides or other counteranions) influence the rate of activation and hence the overall reactivity.

Although research groups have started to study the mechanism of the formation of the catalytically active species from a number of precatalysts, the reaction pathways are still mostly unknown. A more thorough understanding of the activation of preformed catalysts may lead to even more efficient precatalysts and further reduced Pd loadings.

A list of recommended general precatalysts for selected named cross-coupling reactions is presented in Table 3.9.

3.10 Application Table for Second-Generation Preformed Catalysts

Although the precatalysts mentioned in Table 3.10 have been reported as efficient in the specified coupling reactions, it should be kept in mind that each individual cross-coupling may require a different set of conditions to give the optimal result. This could involve using a different base or solvent and sometimes changing the precatalyst. In short, there is no fool-proof way of predicting which catalyst will be the best for any specific set of coupling substrates. This table may, however, provide a good starting point in developing a user guide.

Table 3.9 Recommended precatalysts.

Reaction	Substrate	Recommended precatalyst
α-Arylation	Ketones	PdL_2 or PdL
	Methyl ketones (monoarylation)	PdL
	Esters	PdL
Suzuki–Miyaura	Aryl bromides, chlorides	PdL_2 or PdL
	Aryl triflates	PdL_2
	Heterocycles	PdL_2 or PdL
Heck		PdL_2
Buchwald–Hartwig amination	Ammonia (monoarylation)	PdL
	Amides	PdL
Heck alkynylation/Sonogashira	Aryl bromides	PdL_2
	Aryl chlorides	PdL
Carboiodination		PdL_2
Cyanation		PdL
Etherification		PdL
Direct arylation of heterocycles		PdL_2
Carbonylation		PdL_2
Negishi	Arylzinc chlorides	PdL
	Alkylzinc iodides	PdL_2

Table 3.10 A concise summary of P-based Pd precatalysts used in specific cross-coupling reactions.

Reaction	Precatalyst	Ref.
X = Br, Cl; R' ≠ H	dtbpfPdCl$_2$(p-Me$_2$NC$_6$H$_4$(t-Bu)$_2$P)Pd(allyl)Cl	8, 81, 112
R' = OtBu, Ar; X = Cl	L = XPhos or tBuXPhos	118
	L = tBuXPhos	124
X = Br, Cl	dtbpfPdCl$_2$	80, 82
	[(t-Bu$_3$P)Pd(μ-Br)]$_2$	106
	L = XPhos	124
X = OTf	[p-Me$_2$NC$_6$H$_4$(t-Bu)$_2$P]$_2$PdCl$_2$	130
X = Br	dtbpfPdCl$_2$ (t-Bu$_3$P)$_2$Pd dtbpfPdCl$_2$	83 83 131

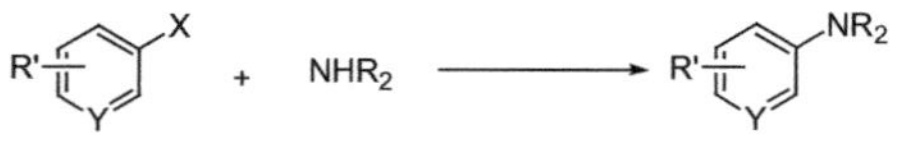

112

118

124

108

132

50

58, 59
60, 61

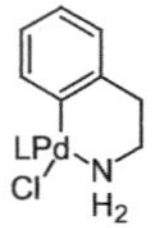

125

121

84

(Q-Phos)Pd(crotyl)Cl

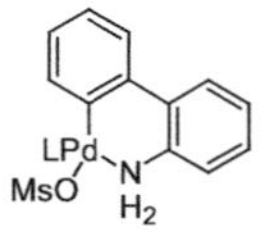

L = BrettPhos or RuPhos

L = BrettPhos or RuPhos

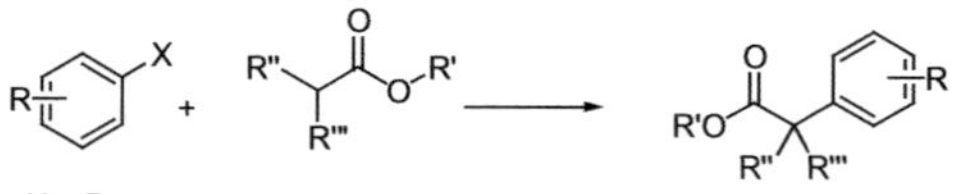

$[(t\text{-}Bu_3P)Pd(\mu\text{-}Br)]_2$

$[(t\text{-}Bu_3P)Pd(\mu\text{-}Br)]_2$

$[(p\text{-}Me_2NC_6H_4(t\text{-}Bu)_2P]_2Pd$

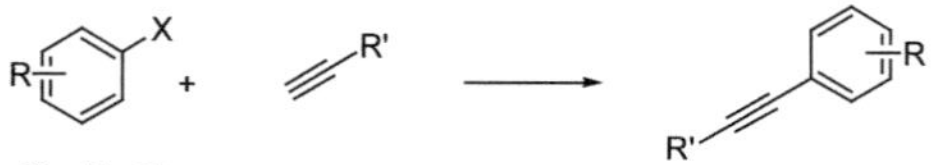

(Q-Phos)$_2$Pd
($t\text{-}Bu_3P)_2$Pd

L = XPhos, tBuXPhos

L = SPhos

$[p\text{-}Me_2NC_6H_4(t\text{-}Bu)_2P]_2PdCl_2$

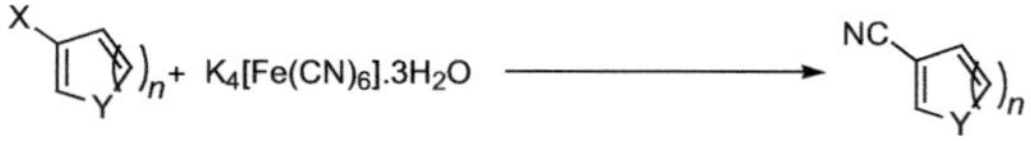

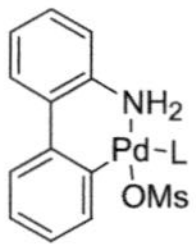

X = Br, Cl

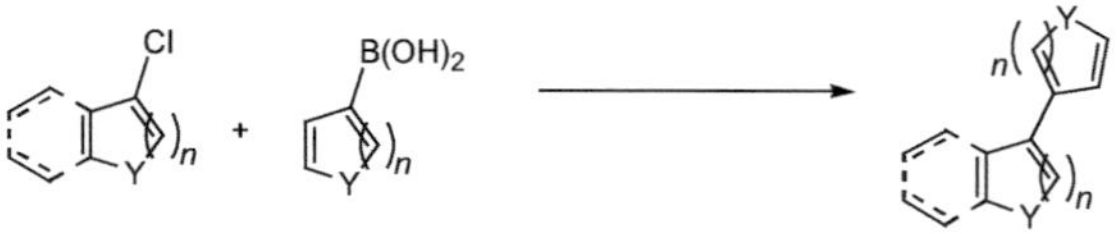

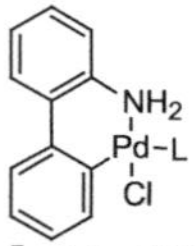

X = Br
R' = alkyl

X = Br, Cl

X = I, Br

+ $K_4[Fe(CN)_6]\cdot 3H_2O$

Table 3.10 (*Continued*)

Reaction	Precatalyst	Ref.
OR + ROH ; X = Cl, Br	L = *t*BuBrettPhos, RockPhos, Cy₃*t*BuBrettPhos	126, 127
NH₂ + NH₃	L = AdBrettPhos, Me₃(OMe)XPhos, Me₃(OMe)PhXPhos	128
+ H₂N–C(O)–R ; X = Cl	L = *t*BuBrettPhos	127
ZnCl + HetArX ; X = I, Br, Cl, OTf	L = XPhos	129
+ R-phenyl ; X = Br, Cl	$(t\text{-Bu}_3\text{P})_2\text{Pd}$	133, 134
+ ZnI	$[(p\text{-Me}_2\text{NC}_6\text{H}_4(t\text{-Bu})_2\text{P})_2]_2\text{Pd}$	135, 136
Cl + CO + ROH or R'₂NH	$(\text{dcpp})_2\text{PdCl}_2$	88

References

1. K. C. Nicolaou and S. A. Snyder, *Classics in Total Synthesis II*, VCH, Weinheim, 2003.
2. J. Magano and J. R. Dunetz, *Chem. Rev.*, 2011, **111**, 2177.
3. C. Torborg and M. Beller, *Adv. Synth. Catal.*, 2009, **351**, 3027.
4. R. A. DeVries, P. C. Vosejpka and M. L. Ash, in *Catalysis of Organic Reactions*, ed. F. E. Herkes, Marcel Dekker, New York, 1998, p. 75.
5. F. Naso, F. Babudri and G. M. Farinola, *Pure Appl. Chem.*, 1999, **71**, 1485.
6. C. C. C. Johansson Seechurn, M. Kitching, T. J. Colacot and V. Snieckus, *Angew. Chem. Int. Ed.*, 2012, **51**, 5062.
7. H. Li, C. C. C. Johansson Seechurn and T. J. Colacot, *ACS Catal.*, 2012, **2**, 1147.
8. G. A. Grasa and T. J. Colacot, *Org. Lett.*, 2007, **9**, 5489.
9. L. M. Klingensmith, E. R. Strieter, T. E. Barder and S. L. Buchwald, *Organometallics*, 2006, **25**, 82.
10. A. F. Littke and G. C. Fu, *Angew. Chem. Int. Ed.*, 1998, **38**, 3387.
11. A. F. Littke, C. Dai and G. C. Fu, *J. Am. Chem. Soc.*, 2000, **122**, 4020.
12. D. W. Old, J. P. Wolfe and S. L. Buchwald, *J. Am. Chem. Soc.*, 1998, **120**, 9722.
13. S. L. Buchwald, *Acc. Chem. Res.*, 2008, **4** special issue.
14. C. Amatore, G. Broeker, A. Jutand and F. Khalil, *J. Am. Chem. Soc.*, 1997, **119**, 5176.
15. I. J. S. Fairlamb, A. R. Kapdi and A. F. Lee, *Org. Lett.*, 2004, **6**, 4435.
16. I. Langmuir, *Science*, 1921, **54**(1386), 59.
17. W. B. Jensen, *J. Chem. Educ.*, 2005, **82**, 28.
18. C. R. Landis and F. Weinhold, *J. Comput. Chem.*, 2007, **28**, 198.
19. B. P. Fors, D. A. Watson, M. R. Biscoe and S. L. Buchwald, *J. Am. Chem. Soc.*, 2008, **130**, 13552.
20. J. P. Stambuli, B. Bühl and J. F. Hartwig, *J. Am. Chem. Soc.*, 2002, **124**, 9346.
21. D. R. Coulson, *Inorg. Synth.*, 1990, **28**, 107.
22. (a) H. Li, G. A. Grasa and T. J. Colacot, *Org. Lett.*, 2010, **12**, 3332; (b) T. J. Colacot, G. A. Grasa and H. Li, *Int. Pat.*, WO 2010/128316, 2010; (c) A. K. Keep, S. Collard, M. W. Hooper and T. J. Colacot, *Int. Pat.*, WO 2007/029031, 2007.
23. M. Yamamura, I. Moritani and S.-I. Murahashi, *J. Organomet. Chem.*, 1975, **91**, C39.
24. S. Murahashi, M. Yamamura, K. Yanagisawa, N. Mita and K. Kondo, *J. Org. Chem.*, 1979, **44**, 2408.
25. (a) J. F. Fauvarque, F. Pflüger and M. Troupel, *J. Organomet. Chem.*, 1981, **208**, 419; (b) E. A. Mitchell and M. C. Baird, *Organometallics*, 2007, **26**, 5230.
26. C. Amatore, M. Azzabi and A. Jutand, *J. Am. Chem. Soc.*, 1991, **113**, 8375.
27. M. Ahlquist, P. Fristrup, D. Tanner and P.-O. Norrby, *Organometallics*, 2006, **25**, 2066.

28. M. Beller and H. U. Blaser (eds), *Top. Orgonomet. Chem.*, 2012, **42** and references therein.

29. M. R. Reeder, H. E. Gleaves, S. A. Hoover, R. J. Imbordino and J. J. Pangborn, *Org. Process Res. Dev.*, 2003, 7, 696.

30. A. Gontcharov, C.-C. Shaw, Q. Yu, S. Tadayon, M. Bernatchez, M. Lankau, M. Cantin, J. Potoski, G. Khafizova, G. Stack, J. Gross and D. Zhou, *Org. Process Res. Dev.*, 2010, **14**, 1438.

31. For an example of the use of (Ph$_3$P)$_4$Pd in a large-scale process, see: C. Perez-Balado, A. Willemsens, D. Ormerod, W. Aelterman and N. Mertens, *Org. Process Res. Dev.*, 2007, **11**, 237.

32. One example of a characterized side product has been demonstrated by Stambuli and co-workers: W. H. Henderson, J. M. Alvarez, C. C. Eichman and J. P. Stambuli, *Organometallics*, 2011, **30**, 5038.

33. S. Kozuch and J. M. L. Martin, *ACS Catal.*, 2011, **1**, 246.

34. C. L. McMullin, B. Ruehle, M. Besora, A. G. Orpen, J. N. Harvey and N. Fey, *J. Mol. Catal. A: Chem.*, 2010, **324**, 48.

35. For a review of L$_2$Pd *versus* LPd active species, see: L. Xue and Z. Lin, *Chem. Soc. Rev.*, 2010, **39**, 1692.

36. F. Weinhold and C Landis, *Valency and Bonding: a Natural Bond Orbital Donor–Acceptor Perspective*, Cambridge University Press, Cambridge, 2005.

37. Z. Li, Y. Fu, Q.-X. Guo and L. Liu, *Organometallics*, 2008, **27**, 4043.

38. (a) D. F. McMillen and D. M. Golden, *Annu. Rev. Phys. Chem.*, 1982, **33**, 493; (b) J. D. Cox and G. Pilcher, *Thermochemistry of Organic and Organometallic Compounds*, Academic Press, London, 1970.

39. F. Schoenebeck and K. N. Houk, *J. Am. Chem. Soc.*, 2010, **132**, 2496.

40. F. Barrios-Landeros, B. P. Carrow and J. F. Hartwig, *J. Am. Chem. Soc.*, 2009, **131**, 8141.

41. J. P. Stambuli, C. D. Incarvito, M. Bühl and J. F. Hartwig, *J. Am. Chem. Soc.*, 2004, **126**, 1184.

42. U. Christmann and R. Vilar, *Angew. Chem. Int. Ed.*, 2005, **44**, 366.

43. C. L. McMullin, J. Jover, J. N. Harvey and N. Fey, *Dalton Trans.*, 2010, **39**, 10833.

44. C. Dai and G. C. Fu, *J. Am. Chem. Soc.*, 2001, **123**, 2719.

45. T. Yoshida and S. Otsuka, *Inorg. Synth.*, 1990, **28**, 113.

46. D. M. Norton, E. A. Mitchell, N. R. Botros, P. G. Jessop and M. C. Baird, *J. Org. Chem.*, 2009, **74**, 6674.

47. A. W. Fraser, B. E. Jaksic, R. Batcup, C. D. Sarsons, M. Woolman and M. C. Baird, *Organometallics*, 2013, **32**, 9.

48. F. Paul, J. Patt and J. F. Hartwig, *Organometallics*, 1995, **14**, 3030.

49. J. Limanto, L. Tan, S. D. Dreher, B. T. Dorner, N. Yoshikawa and S. W. Krska, *PCT Int. Pat. Appl.*, WO 2009054887, 2009.

50. X. Pu, H. Li and T. J. Colacot, *J. Org. Chem.*, 2013, **78**, 568.

51. M. Matsumoto, H. Yoshioka, K. Nakatsu, T. Yoshida and S. Otsuka, *J. Am. Chem. Soc.*, 1974, **96**, 3322.

52. M. Tanaka, *Acta Crystallogr.*, 1992, **C48**, 739.

53. G. Mann, C. Incarvito, A. L. Rheingold and J. F. Hartwig, *J. Am. Chem. Soc.*, 1999, **121**, 3224.
54. Z. J. Song, Y. Wang, L. M. Artino, D. M. Tellers, D. R. Lieberman, *PCT Int. Pat. Appl.*, WO 2011025849, 2011.
55. See Chapter 2 and Ref. 11.
56. H. Li and T. J. Colacot, in-house report, unpublished results.
57. S. G. Newman and M. Lautens, *J. Am. Chem. Soc.*, 2011, **133**, 1778.
58. Y. Lan, P. Liu, S. G. Newman, M. Lautens and K. Houk, *Chem. Sci.*, 2012, **3**, 1987.
59. S. G. Newman, J. K. Howell, N. Nicolaus and M. Lautens, *J. Am. Chem. Soc.*, 2011, **133**, 14916.
60. X. Jia, D. A. Petrone and M. Lautens, *Angew. Chem. Int. Ed.*, 2012, **51**, 9870.
61. D. Petrone, H. A. Malik, A. Clemenceau and M. Lautens, *Org. Lett.*, 2012, **14**, 4806.
62. D. A. Petrone, M. Lischka and M. Lautens, *Angew. Chem. Int. Ed.*, 2013, **40**, 10635.
63. G.-P. Lu, K. R Voigtritter, C. Cai and B. H. Lipshutz, *J. Org. Chem.*, 2012, **77**, 3700.
64. C. Amatore, E. Carré, A. Jutand and M. A. M'Barki, *Organometallics*, 1995, **14**, 1818.
65. N. Miyaura and A. Suzuki, *Org. Synth. Coll. Vol.*, 1993, **8**, 532; N. Miyaura and A. Suzuki, *Org. Synth.*, 1990, **68**, 130.
66. E.-i. Negishi, *J. Chem. Soc., Chem. Commun.*, 1986, 1338.
67. C. Amatore and A. Jutand, *J. Organomet. Chem.*, 1999, **576**, 254.
68. J. W. B. Cooke, R. Bright, M. J. Coleman and K. P. Jenkins, *Org. Process Res. Dev.*, 2001, **5**, 383.
69. V. Ŝtimac, M. M. Ŝkugor, I. P. Jakopović, A. Vinter, M. Ilijaŝ, S. Alihodžić and S. Mutak, *Org. Process Res. Dev.*, 2010, **14**, 1393.
70. T. Hayashi, M. Konishi, Y. Kobori, M. Kumada, T. Higuchi and K. Hirotsu, *J. Am. Chem. Soc.*, 1984, **106**, 158.
71. T. Hayashi, M. Konishi and M. Kumada, *Tetrahedron Lett.*, 1979, 1871.
72. (a) T. J. Colacot, *Platinum Met. Rev.*, 2001, **45**, 22; (b) T. J. Colacot and S. Parisel, Synthesis, coordination chemistry and catalytic use of dppf analogs, in *Ferrocenes: Ligands, Materials and Biomolecules*, ed. P. Ŝtěpnička, Wiley, Chichester, 2008, pp. 117–140.
73. Johnson Matthey, http://jmcct.com (accessed 5 May 2014).
74. P. D. Koning, D. McAndrew, R. Moore and I. B. Moses, *Org. Process Res. Dev.*, 2011, **15**, 1018.
75. D. K. Leahy, J. Li, J. B. Sausker, J. Zhu, M. A. Fitzgerald, C. Lai, F. G. Buono, A. Braem, N. Mas, Z. Manaloto, E. Lo, W. Merkl, B. Su, Q. Gao, A. T. Ng and R. A. Hartz, *Org. Proc. Res. Dev.*, 2010, **14**, 1221.

76. (a) T. Ishiyama, M. Murata and N. Miyaura, *J. Org. Chem.*, 1995, **60**, 7508; (b) M. Murata, S. Watanabe and Y. Masuda, *J. Org. Chem.*, 1997, **62**, 6458.

77. Y. Zhao, L. Jin, P. Li and A. Lei, *J. Am. Chem. Soc.*, 2008, **130**, 9429.

78. I. R. Butler, W. R. Cullen, T. Kim, S. J. Rettig and J. Trotter, *Organometallics*, 1985, **4**, 972.

79. T. J. Colacot, in *e-EROS Encyclopedia of Reagents for Organic Synthesis*, Wiley, Hoboken, NJ, 2009, , DOI: 10.1002/047084289X.rn01062.

80. T. J. Colacot and H. A. Shea, *Org. Lett.*, 2004, **6**, 3731.

81. G. A. Grasa and T. J. Colacot, *Org. Process Res. Dev.*, 2008, **12**, 522.

82. J. D. Moseley, P. M. Murray, E. R. Turp, S. N. G. Tyler and R. T. Burn, *Tetrahedron*, 2012, **68**, 6010.

83. P. M. Murray, J. F. Bower, D. K. Cox, E. K. Galbraith, J. S. Parker and J. B. Sweeney, *Org. Process Res. Dev.*, 2013, **17**, 397.

84. A. S. Guram, A. O. King, J. G. Allen, X. Wang, L. B. Schenkel, J. Chan, E. E. Bunel, M. M. Faul, R. D. Larsen, M. J. Martinelli and P. J. Reider, *Org. Lett.*, 2006, **8**, 1787.

85. T. J. Colacot, in *e-EROS Encyclopedia of Reagents for Organic Synthesis*, Wiley, Hoboken, NJ, 2011, DOI: 10.1002/047084289X.rn01283.

86. R. R. Milburn, O. R. Thiel, M. Achmatowicz, X. Wang, J. Zigterman, C. Bernard, J. T. Colyer, E. DiVirgilio, R. Crockett, T. L. Correll, K. Nagapudi, K. Ranganathan, S. J. Hedley, A. Allgeier and R. D. Larsen, *Org. Process Res. Dev.*, 2011, **15**, 31.

87. J. K. Padia, C. F. J. Barnard and T. J. Colacot, in *e-EROS Encyclopedia of Reagents for Organic Synthesis*, Wiley, Hoboken, NJ, 2009, DOI: 10.1002/047084289X.rd097.pub2.

88. (a) C. F. J. Barnard, *Org. Process Res. Dev.*, 2008, **12**, 566; (b) see also Chapter 2 for a discussion on carbonylation reactions using *in situ*-prepared catalysts and Chapter 9.

89. C. F. J. Barnard and H. Li, *US Pat. Appl.*, 2010/0305349 A1, 2010.

90. M. Kawatsura and J. F. Hartwig, *J. Am. Chem. Soc.*, 1999, **121**, 1473.

91. D. Gallagher, L. Treiber, R. Hughes, O. Campopiano, P. Wang, Y. Zhao, S. Chou, M. Ouellette and D. Hettinger, *PCT Int. Pat. Appl.*, WO 2009062087, 2009.

92. D. M. Barnes, J. Barkalow, Y. Chen, A. Gupta, A. R. Haight, J. E. Hengeveld, F. A. J. Kerdesky, B. J. Kotecki, B. Macri and A. Pal, *Org. Process Res. Dev.*, 2009, **13**, 225.

93. C. P. Ashcroft, S. J. Fussell and K. Wilford, *Tetrahedron Lett.*, 2013, 4529.

94. For carbonylation of aryl chlorides, see: (a) J. Albaneze-Walker, C. Bazaral, T. Leavey, P. G. Dormer and J. A. Murry, *Org. Lett.*, 2004, **6**, 2097. For carbonylation of benzylic C–H bond, see: (b) P. Xie, Y. Xie, B. Qian, H. Zhou, C. Xia and H. Huang, *J. Am. Chem. Soc.*, 2012, **134**, 9902; (c) P. Xie, C. Xia and H. Huang, *Org. Lett.*, 2013, **15**, 3370.

95. J. F. Hartwig, M. Kawatsura, S. I. Hauck, K. H. Shaughnessy and L. M. Alcazar-Roman, *J. Org. Chem.*, 1999, **64**, 55757.

96. A. F. Littke and G. C. Fu, *J. Am. Chem. Soc.*, 2001, **123**, 6989.

97. Selected reference: G. Bastug and S. P. Nolan, *Organometallics*, 2014, **33**, 1253 and references therein.

98. N. Marion and S. P. Nolan, *Acc. Chem. Res.*, 2008, **41**, 1440 and references therein.

99. Selected reference: N. Marion, O. Navarro, J. Mei, E. D. Stevens, N. M. Scott and S. P. Nolan, *J. Am. Chem. Soc.*, 2006, **128**, 4101 and references therein.

100. Selected reference: M. G. Organ, S. Avola, I. Dubovyk, N. Hadei, E. A. B. Kantchev, C. J. O'Brien and C. Valente, *Chem. Eur. J.*, 2006, **12**, 4749.

101. Selected reference: C. J. O'Brien, E. A. B. Kantchev, N. Hadei, C. Valente, G. A. Chass, J. C. Nasielski, A. Lough, A. C. Hopkinson and M. G. Organ, *Chem. Eur. J.*, 2006, **12**, 4743.

102. R. Vilar, D. M. P. Mingos and C. J. Cardin, *J. Chem. Soc., Dalton Trans.*, 1996, 4313.

103. V. Durà-Vilà, D. M. P. Mingos, R. Vilar, A. J. P. White and D. J. Williams, *J. Organomet. Chem.*, 2000, **600**, 198.

104. (a) T. J. Colacot, M. W. Hooper, G. A. Grasa, *Int. Pat.*, WO2011/12889 A1, 2011; (b) T. J. Colacot, in *e-EROS Encyclopedia of Reagents for Organic Synthesis*, Wiley, Hoboken, NJ, 2009, DOI: 10.1002/047084289X. rn01103.

105. V. Durà-Vilà, D. M. P. Mingos, R. Vilar, A. J. P. White and D. J. Williams, *Chem. Commun.*, 2000, 1525.

106. F. Proutiere, M. Aufiero and F. Schoenebeck, *J. Am. Chem. Soc.*, 2012, **134**, 606.

107. F. Proutiere and F. Schoenebeck, *Angew. Chem. Int. Ed.*, 2011, **50**, 8192.

108. J. P. Stambuli, R. Kuwano and J. F. Hartwig, *Angew. Chem. Int. Ed.*, 2002, **41**, 4746.

109. K. J. Bonney, F. Proutiere and F. Schoenebeck, *Chem. Sci.*, 2013, **4**, 4434.

110. P. Ryberg, *Org. Process Res. Dev.*, 2008, **12**, 540.

111. X. Jiang, G. T. Lee, E. B. Villhauer, K. Prasad and M. Prashad, *Org. Process Res. Dev.*, 2010, **14**, 883.

112. C. C. C. Johansson Seechurn, S. L. Parisel and T. J. Colacot, *J. Org. Chem.*, 2011, **76**, 7918.

113. L. L. Hill, J. L. Crowell, S. L. Tutwiler, N. L. Massie, C. C. Hines, S. T. Griffin, R. D. Rogers, K. H. Shaughnessy, G. A. Grasa, C. C. C. Johansson Seechurn, H. Li, T. J. Colacot, J. Chou and C. J. Woltermann, *J. Org. Chem.*, 2010, **75**, 6477.

114. Selected reference: R. B. Bedford, C. S. J. Cazin, S. J. Coles, T. Gelbrich, M. B. Hursthouse and V. J. M. Scordia, *Dalton Trans.*, 2003, 3350.

115. Selected reference: M. S. Viciu, R. A. Kelly III, E. D. Stevens, F. Naud, M. Studer and S. P. Nolan, *Org. Lett.*, 2003, **5**, 1479.

116. Selected reference: J. Broggi, H. Clavier and S. P. Nolan, *Organometallics*, 2008, **27**, 5525.

117. M. R. Biscoe, B. P. Fors and S. L. Buchwald, *J. Am. Chem. Soc.*, 2008, **130**, 6686.

118. D. Maiti, B. P. Fors, J. L. Henderson, Y. Nakamura and S. L. Buchwald, *Chem. Sci.*, 2011, **2**, 57.
119. T. Kinzel, Y. Zhang and S. L. Buchwald, *J. Am. Chem. Soc.*, 2010, **132**, 14073.
120. M. R. Biscoe and S. L. Buchwald, *Org. Lett.*, 2009, **11**, 1773.
121. M. A. Düfert, K. L. Billingsley and S. L. Buchwald, *J. Am. Chem. Soc.*, 2013, **135**, 12877.
122. Selected example: B. P. Fors and S. L. Buchwald, *J. Am. Chem. Soc.*, 2010, **132**, 15914.
123. Selected example: J. L. Henderson and S. L. Buchwald, *Org. Lett.*, 2010, **12**, 4442.
124. N. C. Bruno, M. T. Tudge and S. L. Buchwald, *Chem. Sci.*, 2013, **4**, 916.
125. T. D. Senecal, W. Shu and S. L. Buchwald, *Angew. Chem. Int. Ed.*, 2013, **52**, 10035.
126. C. W. Cheung and S. L. Buchwald, *Org. Lett.*, 2013, **15**, 3998.
127. N. C. Bruno and S. L. Buchwald, *Org. Lett.*, 2013, **15**, 2876.
128. C. W. Cheung, D. S. Surry and S. L. Buchwald, *Org. Lett.*, 2013, **15**, 3734.
129. Y. Yang, N. J. Oldenhuis and S. L. Buchwald, *Angew. Chem. Int. Ed.*, 2013, **52**, 615.
130. A. He and J. R. Falck, *J. Am. Chem. Soc.*, 2010, **132**, 2524.
131. B. H. Lipshutz and B. R. Taft, *Org. Lett.*, 2008, **10**, 1329.
132. T. Hama, S. Ge and J. F. Hartwig, *J. Org. Chem.*, 2013, **78**, 8250.
133. S. Tamba, Y. Okubo, S. Tanaka, D. Monguchi and A. Mori, *J. Org. Chem.*, 2010, **75**, 6998.
134. T. Ohmura, A. Kijima and M. Suginome, *Org. Lett.*, 2011, **13**, 1238.
135. A. Krasovskiy and B. H. Lipshutz, *Org. Lett.*, 2011, **13**, 3822.
136. A. Krasovskiy, C. Duplais and B. H. Lipshutz, *J. Am. Chem. Soc.*, 2009, **131**, 15592.

CHAPTER 4

Advances in C–C and C–X Coupling Using Palladium–N-Heterocyclic Carbene (Pd–NHC) Complexes

ANTHONY CHARTOIRE AND STEVEN P. NOLAN*

EaStCHEM School of Chemistry, University of St Andrews, St Andrews, KY16 9ST, UK
*Email: snolan@st-andrews.ac.uk

4.1 Introduction

Palladium-catalysed cross-coupling reactions are nowadays one of the most powerful methods for the creation of C–C or C–X bonds.[1] The development of these reactions indubitably modified retro-synthetic strategies of organic chemists for the preparation of many important organic molecules, including biologically active compounds[2] and industrially relevant starting materials.[3] The importance of cross-coupling reactions was definitely perceived in 2010, by awarding Akira Suzuki,[4] Ei-ichi Negishi[5] and Richard F. Heck[6] the Nobel Prize in Chemistry. In cross-coupling chemistry, the nature of the palladium catalyst is crucially important. In particular, the nature of the ligand bonded to the metal plays a decisive role in the coupling because it dictates the complex's catalytic properties. Historically, phosphines have been the most widely employed class of ligands in palladium catalysis.[7]

RSC Catalysis Series No. 21
New Trends in Cross-Coupling: Theory and Applications
Edited by Thomas J Colacot

However, *N*-heterocyclic carbenes (NHCs) are nowadays honourable competitors, due to an interesting range of properties including reactivity, stability and tunability.[8]

In this chapter, we focus on the advances made in C–C and C–X coupling using palladium–*N*-heterocyclic carbene (Pd–NHC) complexes, especially since the 2010 Nobel Prize award. After touching upon the fundamental chemistry behind NHCs and Pd–NHC complexes, we survey the use of these fascinating systems in cross-coupling chemistry.[9] The content of each topic varies depending on how much knowledge is already available in the literature. Readers who require more information can consult the cited references.

4.2   *N*-Heterocyclic Carbenes and Pd–NHC Complexes

4.2.1   The Birth of a New Class of Ligands

Until recently, NHCs were long considered as *laboratory curiosities*[10] or *phosphine mimics*[11] and many years of studies were necessary to realise their real potential as an indisputable powerful class of ligands. The first metal complexes bearing NHC ligands were reported separately by Öfele[12] and Wanzlick and Schoenherr[13] in 1968. Öfele prepared the pentacarbonyl(1,3-dihydro-1,3-dimethyl-2*H*-imidazol-2-ylidene)–chromium complex **1** (Figure 4.1) by sublimation of the chromium salt [HIMe][CrH(CO)$_5$]. Wanzlick and Schönherr synthesized bis(1,3-diphenylimidazolo)mercury diperchlorate (**2**) from the carbene salt and [Hg(OAc)$_2$] in DMSO (Figure 4.1).

The seminal breakthrough occurred in 1991, when Arduengo *et al.* reported the preparation of the first isolable free NHC, IAd (Figure 4.2).[14] The discovery that carbenes could be isolated and handled under an inert atmosphere consequently gave rise to considerable interest in their coordination chemistry. Since then, a wide range of NHCs have been prepared and their complexes with nearly all transition metals have been synthesized.[8c] Over the last few decades, NHCs have been modified by varying their skeletal structure and also the nature of the groups attached to it. Despite the fact that many NHCs have been prepared throughout the years, the most efficient and utilized congeners are the five-membered ring imidazolylidenes and imidazolinylidenes illustrated in Figure 4.2.

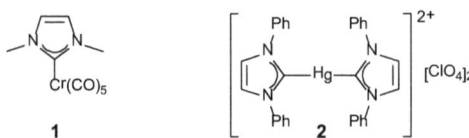

Figure 4.1   First metal–NHC complexes reported by Öfele and Wanzlick and Schoenherr.

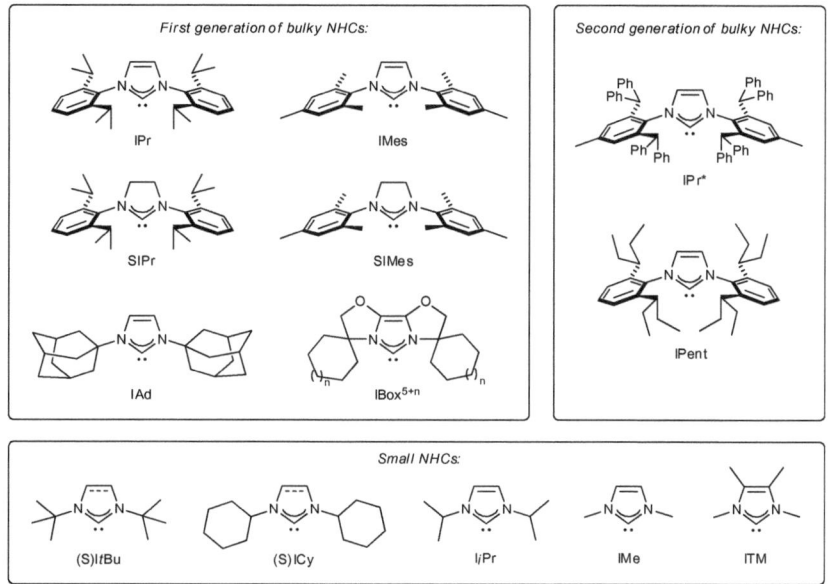

Figure 4.2 The most commonly used NHCs.

4.2.2 Routes to Free Carbenes and to Pd–NHC Complexes

There are several routes for the preparation of free carbenes. The most common approach is *via* the deprotonation of the corresponding NHC salt (Scheme 4.1).[15] However, they can also be synthesized by reduction of the thione moiety with potassium or through the thermal decomposition of a variety of adducts, including chloroform,[16] alcohols[17] and carbon dioxide.[18]

Herrmann and co-workers were the first to report the use of a Pd–NHC complex in catalysis. Complexes of the formula [PdI$_2$(NHC)$_2$] were successfully used in catalysis, notably in Heck coupling.[19] The strength of the Pd–NHC bond was found to be particularly advantageous in these reactions and the catalysts remained active even at high temperatures. Pd–NHC complexes are generally prepared using three main routes (Scheme 4.2).[9d] First is the substitution of a labile ligand using the free carbene [eqn (4.1)]. An inert atmosphere is required in this case as the free carbene is both air and moisture sensitive. The labile ligands can be, for example, dibenzylideneacetone (dba), cyclooctadiene (COD), a phosphine or a bridging halide. The advantages of this method are notably low temperatures and short reaction times. The second method is the *in situ* generation of a free carbene starting from the NHC salt. The ambient atmosphere can often be used in this case as the NHC salt is typically air and moisture stable. The base used to deprotonate the NHC salt can either be external [KOtBu for example, eqn (4.2)] or linked to the palladium source [AcO in Pd(OAc)$_2$ for example, eqn (4.3)]. Finally, transmetallation from Ag complexes can be applied [eqn (4.4)]. [Ag(NHC)X] can be formed from the NHC salt using routes (4.2) or (4.3) and

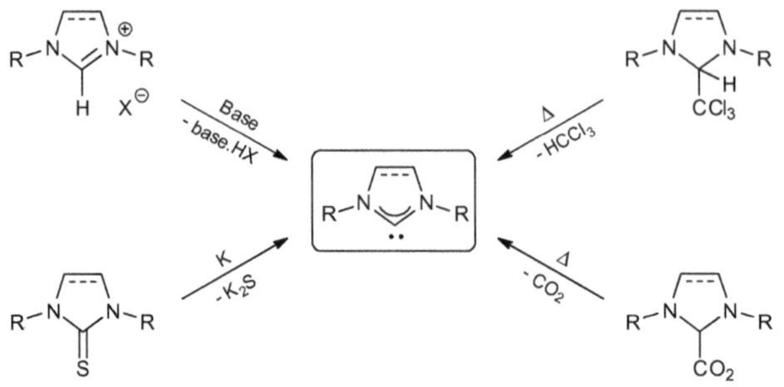

Scheme 4.1   Routes for formation of free NHCs.

$$\text{Pd}\!-\!\text{L} \quad + \quad \text{NHC} \quad \longrightarrow \quad \text{Pd}\!-\!\text{NHC} \quad + \quad \text{L} \qquad\qquad (4.1)$$

L = Labile ligand such as dba, PR$_3$, μ-Cl

$$\text{Pd} \quad + \quad \text{NHC.HX} \quad \xrightarrow[\text{- Base.HX}]{\text{Base}} \quad \text{Pd}\!-\!\text{NHC} \qquad\qquad (4.2)$$

$$\text{Pd}\!-\!\text{R} \quad + \quad \text{NHC.HX} \quad \longrightarrow \quad \text{X}\!-\!\text{Pd}\!-\!\text{NHC} \qquad\qquad (4.3)$$

R = Basic group such as AcO

$$\text{NHC.HX} \quad + \quad \text{1/2 Ag}_2\text{O} \quad \xrightarrow[\text{- 1/2 H}_2\text{O}]{} \quad \text{X}\!-\!\text{Ag}\!-\!\text{NHC} \quad \xrightarrow[\text{- LAgX}]{\text{+ LPd}} \quad \text{Pd}\!-\!\text{NHC} \qquad (4.4)$$

Scheme 4.2   Routes to Pd–NHC.

then transmetallated to Pd in a second step. The disadvantages of the method are the cost and the light sensitivity of the silver complexes. Method (4.4) should be only used when routes (4.1), (4.2) and (4.3) fail or are impossible.

4.2.3   Electronic and Steric Parameters of the NHCs

Electronically, the Pd–NHC bond is composed of three main components: (a) the NHC $\rightarrow$ M σ-donation; (b) the NHC $\rightarrow$ M π-donation; and (c) the M $\rightarrow$ NHC π^*-backdonation.[9d,20] A simplified diagram depicting the bonding molecular orbitals of the M–NHC bond is presented in Figure 4.3. Most of the M–NHC bond is comprised of the NHC $\rightarrow$ M σ-dative bond. NHCs are stronger σ-donors than the most basic phosphines. As a consequence, an M–NHC bond is typically stronger than an M-phosphine bond and the dissociation of the NHC from the metal centre is more difficult. This behaviour is presumed to have a significant positive impact on catalysis. The overall M–NHC π-interaction (NHC $\rightarrow$ M π-donation and M $\rightarrow$ NHC π^*-backdonation) is relatively small compared with the σ-interaction. Nevertheless, it is non-negligible.[9d,20] For the formally d^8 metals such as Pd(II) and the formally d^{10}

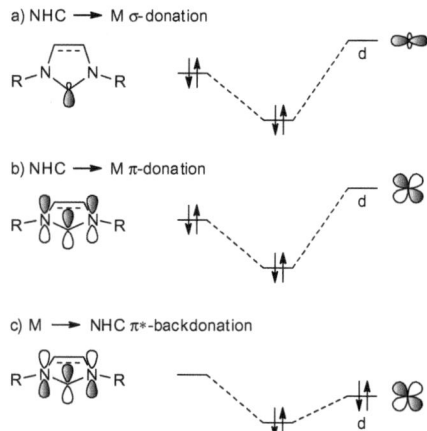

a) NHC → M σ-donation

b) NHC → M π-donation

c) M → NHC π*-backdonation

Figure 4.3   Simplified depiction of the bonding molecular orbitals of the M–NHC bond.

metals such as Pd(0), the contribution of the π-interaction has been estimated at 15 and 20%, respectively.[9d,20] The NHC → M π-donation results from the overlap between the filled π-orbital of the NHC and a d-orbital of the metal. The M → NHC π*-backdonation results from the overlap between a π* orbital of the NHC and the HOMO frontier orbital of the metal. The latter represents >75% of the global M–NHC π-interaction for late transition metals.[9d,20]

Several methods exist to quantify the electronic properties of the NHC ligands,[21] and notably include methodologies based on pK_a or nucleophilicity measurements, IR and NMR spectroscopy and electrochemistry. The most commonly used metric is the Tolman electronic parameter (TEP),[22] which consists of the measurement of the IR stretching frequency of the carbonyl ligand(s) for complexes of formula $[M(CO)_nL]$. The more electron donating the ligand is, the lower the stretching frequency of the carbonyl ligand(s) will be. The TEP is the frequency of the A_1 vibrational frequency and is typically found between 2050 and 2100 cm^{-1}. To measure the TEP, metal complexes of Ni, Rh or Ir can be prepared.[21]

The steric parameters of NHCs are also of crucial importance. Especially in palladium catalysis, *bulky yet flexible*[23] NHCs are believed to have a positive influence on the reactivity of the complexes in catalysis. This is why recent efforts have been devoted towards the development of a second generation of bulky NHC ligands (Figure 4.2).[9e,24] Compared with phosphines, the steric configuration of the NHCs is completely different. Indeed, for the phosphines, the bulk expands away from the metal centre. For the NHCs, the bulk is directed towards the coordination sphere of the metal (Figure 4.4). In order to evaluate and compare the bulk of the ligands, Tolman developed the cone angle for phosphine ligands.[22] Because phosphines and NHCs are geometrically different, the Tolman cone angle cannot be used to quantify

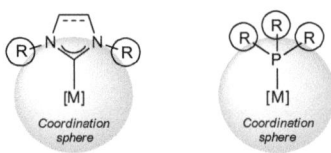

Figure 4.4   Steric environment of NHCs *versus* phosphines.

the steric effect of NHCs. Another tool, developed by Cavallo and co-workers, which allows the assessment of the bulk of an NHC ligand, is the percent buried volume (%V_{bur}).[25] The latter can be calculated from crystal structure data using the software Sam*b*Vca.[26] Recently, Cavallo and co-workers expanded the utility of this software by developing the "steric maps."[24,27] This helps in visualizing the steric environment of an NHC ligand around the metal centre.

4.3   Pd–NHC Complexes in Catalysis

A cross-coupling reaction consists of joining two organic fragments together through the formation of a new bond, using an organometallic catalyst. For decades, palladium has proven to be the metal of choice for such transformations.[1a] Historically, phosphines have been the dominant choice of ligands in palladium-catalysed cross-coupling reactions.[7] However, NHCs are nowadays efficient challengers, owing to the specific properties of the Pd–NHC bond.[8,9] The purpose of this chapter is to discuss the main recent advances in C–C and C–X coupling using Pd–NHC complexes, especially since the awarding of the 2010 Nobel Prize in Chemistry. Each family of cross-coupling is treated separately and the content of each topic is outlined in the corresponding introductions.

4.3.1   Reactions Through Transmetallation Mechanisms

Most cross-coupling reactions require a transmetallating agent, because the use of such reagents very often renders the coupling more efficient and also more selective. The accepted reaction mechanism for these transformations is very similar[9d] and is depicted in Scheme 4.3.

First, the palladium precatalyst is reduced to its active Pd(0) form, then the oxidative addition of the organic halide (X = Cl, Br, I) or pseudohalide (X = OTs, OTf, for example) takes place. In the latter, the difficulty of the coupling increases in the order R–I < R–Br < R–Cl, which is due to the increasing R–X bond dissociation enthalpies.[28] Because organochlorides are more abundant and generally cheaper than organobromides and -iodides, they are the most desirable substrates to use. However, they are also the most difficult to activate, which is why current efforts in cross-coupling chemistry are dedicated to improving their reactivity. The second part of the catalytic cycle is the transmetallation step. This is where the second organic fragment is

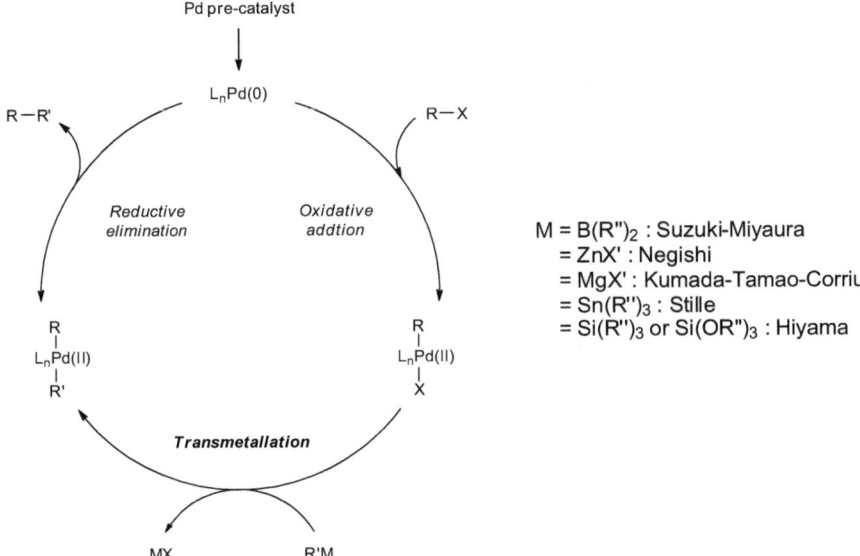

Scheme 4.3 General accepted mechanism for cross-coupling reactions using trans-metallating agents.

delivered on to the palladium centre. This is also the step that differentiates the various types of cross-coupling reactions. Generally, the nature of the transmetallating agents is directly linked to the names of the chemists who discovered or developed their use. Thus, the Suzuki–Miyaura reaction involves boronic acids or esters and the Negishi coupling utilizes organozinc derivatives. The Kumada–Tamao–Corriu reaction uses organomagnesium (Grignard) reagents, the Stille reaction employs organotin derivatives and Hiyama coupling makes use of organosilanes or siloxanes.[29] The last step of the catalytic cycle is reductive elimination, which releases the desired cross-coupling product and regenerates the Pd(0) active species.

In the following sections, we describe the main advances in cross-coupling chemistry, involving transmetallation-type mechanisms and Pd–NHC complexes, divided into subsections tackling each type of the above-mentioned cross-couplings separately.

4.3.1.1 Suzuki–Miyaura Coupling

Suzuki–Miyaura coupling is probably the most commonly employed and also the most convenient protocol for the creation of C–C bonds. This is notably because organoboron reagents offer a wide range of convenient properties. They are generally thermally stable, inert to water and oxygen, highly reactive as transmetallating agents and relatively non-toxic.[30] Suzuki–Miyaura coupling involves the reaction between an organic halide or pseudohalide and an organoboron reagent, generally a boronic acid or ester.

In the literature, a large number of reports and some excellent reviews on this well-known cross-coupling reaction can be found, covering the major achievements before 2010.[9d,30,31] In this section, we mainly focus our attention on recent protocols using Pd–NHC complexes after the award of the 2010 Nobel Prize in Chemistry. In particular, we discuss the recent advances tackling various challenges of the reaction, such as the preparation of tetra-*ortho*-substituted biaryls, the creation of sp³–sp³ or sp³–sp² bonds, the development of new systems promoting the reaction at low catalyst loading, at low temperature and in water; and other topics.

4.3.1.1.1 Preparation of Tetra-*Ortho*-Substituted Biaryls. One of the remaining challenges in the Suzuki–Miyaura reaction is the reaction between sterically hindered halides and boronic acids to yield tetra-*ortho*-substituted biaryls, especially under mild conditions. Ligand design has proven to be important for this kind of reactivity. Indeed, the use of *bulky yet flexible*[23] ligands was shown to be critical for the reaction to take place. In 2002, Buchwald and co-workers were the first to report the preparation of tetra-*ortho*-substituted biaryls using the Suzuki reaction.[32] The reaction used Pd₂(dba)₃, a bulky biarylphosphine and K₃PO₄ as the base at 110 °C in toluene. Other procedures were reported following this discovery, using other bulky tertiary phosphines.[33] All of them suffered from several disadvantages, such as high catalyst loadings, high temperatures and the use of excesses of ligands. The first example of the use of an NHC in this reaction was described in 2004 by Glorius and co-workers.[34] The use of IBox12 (**3**, Figure 4.5) proved to be fairly efficient in promoting the desired reaction; nevertheless, high temperature and high catalyst loading were still required. In 2009, Organ *et al.* were able to decrease the reaction temperature considerably to 65 °C by using [Pd(IPent)(PEPPSI)] (**4**) as the well-defined precatalyst (Figure 4.5);[35] 2 mol% of the latter complex was still necessary to achieve the coupling but the scope of the reaction appeared to be considerably wider than that in previous reports. Schmidt and Rahimi[36] also reported the use of ligand **5** to promote the reaction

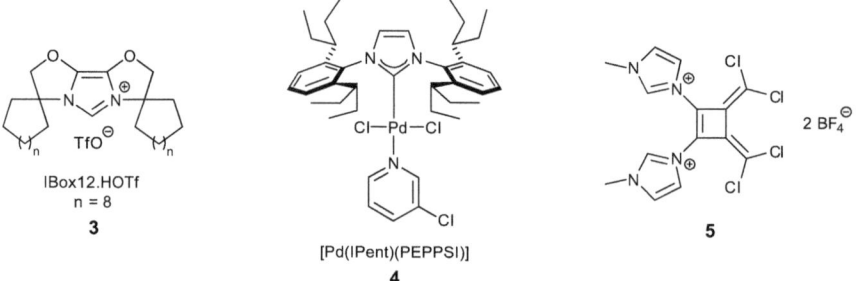

Figure 4.5 Structures of the NHCs and Pd–NHC used for the preparation of tetra-*ortho*-substituted biaryls by Suzuki–Miyaura coupling.

successfully (Figure 4.5). The desired biaryls were obtained in this case using 2 mol% of palladium at temperatures ranging from 60 to 110 °C.

In 2011, Dorta and co-workers introduced [Pd(*anti*-(2,7)-SICyoctNap) (cin)Cl] (**8**) as the first well-defined precatalyst capable of promoting this challenging coupling at room temperature (Figure 4.6).[27c] The reaction used 2 mol% of the complex and KOtBu as the base in toluene. Compared with other very efficient systems described in the literature, complex **8** exhibited astonishing reactivity (Table 4.1) and a wide range of biaryls could be obtained in good yields (Scheme 4.4).

The success of the reaction was attributed to the very bulky nature of the NHC ligand in this system. Indeed, by using steric maps representing the bulk of the ligand around the metal centre, Dorta and co-workers showed that the NHC was twisted around the metal centre. Two faces of the ligand were found to be very bulky, whereas the other two were less hindered. It was then postulated that the less bulky faces were favouring the approach of the substrates during the oxidative addition and transmetallation steps. In contrast, the more bulky faces were postulated to favour the reductive elimination by applying steric pressure and releasing the biaryl. DFT calculations were also performed to support these hypotheses.[27c]

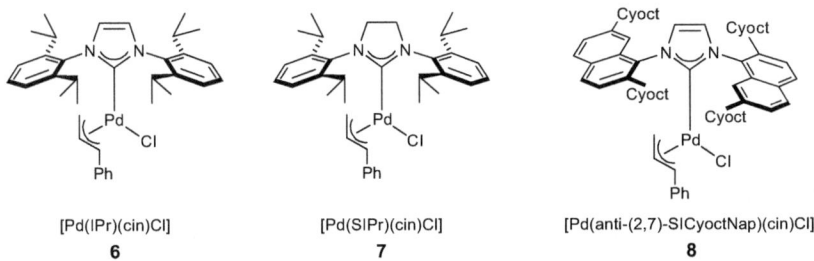

[Pd(IPr)(cin)Cl]
6

[Pd(SIPr)(cin)Cl]
7

[Pd(anti-(2,7)-SICyoctNap)(cin)Cl]
8

Figure 4.6 Structures of some [Pd(NHC)(cin)Cl] precatalysts investigated in Suzuki–Miyaura coupling for the formation of tetra-*ortho*-substituted biaryls.

Table 4.1 [Pd(*anti*-(2,7)-SICyoctNap)(cin)Cl](**8**) *versus* other precatalysts in Suzuki–Miyaura coupling for the preparation of tetra-*ortho*-substituted biaryls.

Entry	Precatalyst Pd–NHC	Yield (%)
1	[Pd(IPr)(cin)Cl] (**6**)	33
2	[Pd(SIPr)(cin)Cl] (**7**)	35
3	[Pd(*anti*-(2,7)-SICyoctNap)(cin)Cl] (**8**)	90
4	[Pd(IPent)(PEPPSI)Cl] (**4**)	29

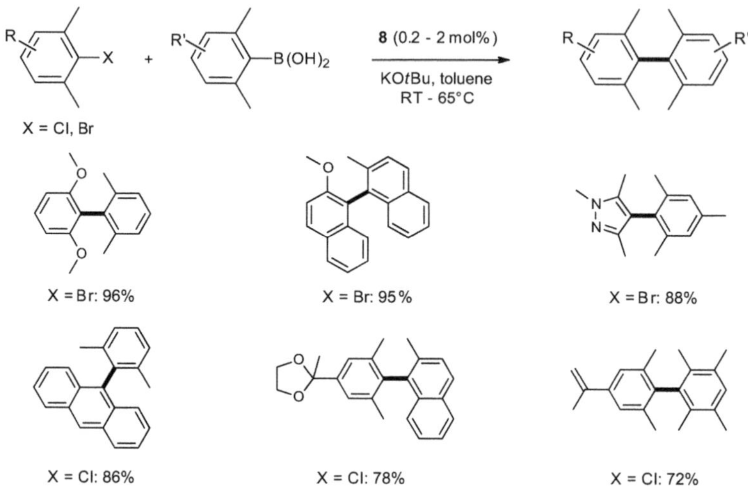

Scheme 4.4   Preparation of tetra-*ortho*-substituted biaryls using [Pd(*anti*-(2,7)-SICyoctNap)(cin)Cl] (**8**).

Table 4.2   Comparison of ligand bulk in [Pd(NHC)(cin)Cl].

Entry	NHC in [Pd(NHC)(cin)Cl]	$\%V_{bur}$
1	IPr	36.7
2	SIPr	37.0
3	*anti*-(2,7)-SICyoctNap	42.0
4	IPr*	44.6

Inspired by the work of Dorta and co-workers and noting that bulky ligands were key to the success of the reaction, Nolan's group next decided to apply [Pd(IPr*)(cin)Cl] (**9**) to this challenging transformation.[24] The IPr* ligand was found to be bigger than the previously reported *anti*-(2,7)-SICyoctNap (Table 4.2) and the reactivity was found to be slightly superior. Notably, 1 mol% of the complex was sufficient to promote the coupling at room temperature. The combination between KOH and DME was found to be the best system in this case, allowing the preparation of a wide range of biaryls at room temperature or using gentle heating (65 °C) (Scheme 4.5). It should be noted that the use of KOH is preferable to KO*t*Bu in this case as KOH is both cheaper and milder than KO*t*Bu. Steric maps of the [Pd(IPr*)(cin)Cl] complex exhibited a similar twist of the ligand around the metal centre.

In 2012, Tu *et al.* investigated the reactivity of robust acenaphthoimidazolylidene PEPPSI-based palladium complexes in the preparation of tetra-*ortho*-substituted biaryls.[37] Complex **10** (Figure 4.7) proved highly efficient in the reaction in comparison with well-established PEPPSI systems reported in the literature. They attributed this higher activity to a stronger σ-donation

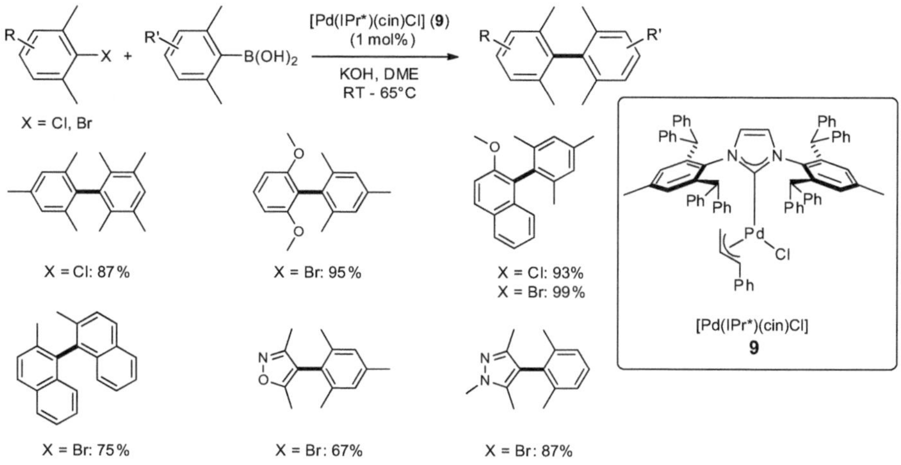

Scheme 4.5   Preparation of tetra-*ortho*-substituted biaryls using [Pd(IPr*)(cin)Cl] (**9**).

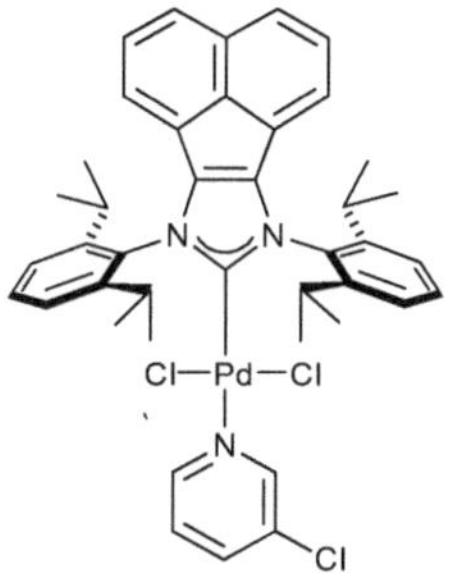

Figure 4.7   Structure of precatalyst **10**.

from the NHC ligand. They concluded that flexible steric bulk was not the only important factor in promoting the coupling and that electronic properties also played a crucial role.

4.3.1.1.2 Formation of sp^3–sp^3 or sp^3–sp^2 Bonds.

Suzuki–Miyaura cross-coupling is a very powerful reaction. Nevertheless, most of the bonds that can be constructed by this methodology are sp^2–sp^2. The development of precatalysts enabling the formation of sp^3–sp^3 or sp^3–sp^2 bonds is therefore highly important.[38] The main issue associated with such reactions is that of β-hydrogen elimination.[39] The first report on the formation of an sp^3–sp^3 bond using an NHC system was reported by Caddick and co-workers in 2004.[40] The system was comprised of Pd(dba)$_2$ (4 mol%) and IPr·HCl (8 mol%) in the presence of KO*t*Bu and AgOTf (4 mol%) in THF at 40 °C. The coupling of various alkyl bromides and alkyl-9-BBN reagents

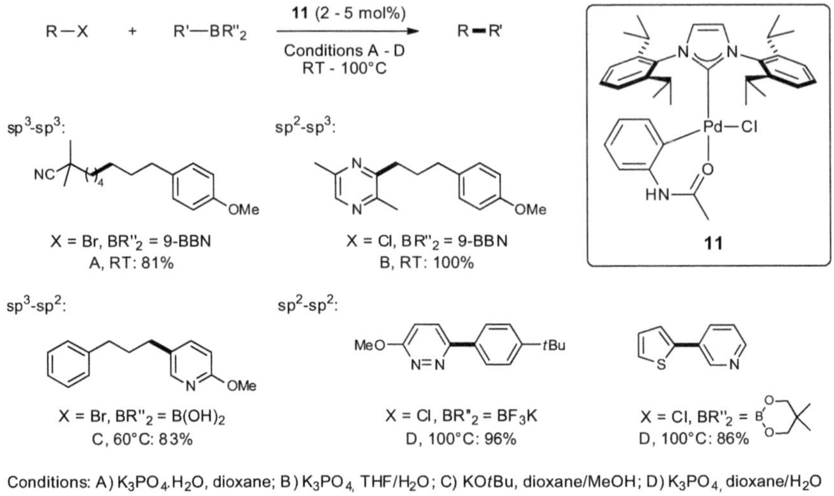

Conditions: A) $K_3PO_4 \cdot H_2O$, dioxane; B) K_3PO_4, THF/H_2O; C) KOtBu, dioxane/MeOH; D) K_3PO_4, dioxane/H_2O

Scheme 4.6   Reactivity of **11** in the Suzuki reaction for the creation of various kinds of bonds.

was then accomplished with moderate yields. A few years later, Organ and co-workers used the well-defined [Pd(IPr)(PEPPSI)] complex (4 mol%) in this reaction to form both sp^3–sp^3 and sp^3–sp^2 bonds.[41] The scope of the reaction was considerably expanded and the desired compounds were obtained in better yields. The reaction conditions were also milder (K_3PO_4 as the base, room temperature).

In 2010, Kantchev and co-workers explored the reactivity of various [Pd(NHC)(palladacycle)] complexes in Suzuki–Miyaura cross-coupling.[42] Among several precatalysts, the IPr-cyclopalladated acetanilide complex **11** was found to be particularly active in the reaction due to its easy activation. Interestingly, **11** allowed the formation of a wide range of C–C bonds including sp^2–sp^2 bonds but also sp^3–sp^3 and sp^3–sp^2 bonds (Scheme 4.6).

Another example of a challenging Suzuki–Miyaura reaction was reported by Organ and co-workers in 2012.[43] [Pd(IPent)(PEPPSI)] (**4**) was used to promote the coupling of allylboronic acids. The challenge in this kind of coupling is to obtain good selectivities between α- and γ-arylated products. [Pd(IPent)(PEPPSI)] showed high selectivity (>97%) for the formation of the α-product and several aryl and heteroaryl halides were successively coupled in the presence of 5 M KOH in THF (Scheme 4.7).

4.3.1.1.3 Asymmetric Suzuki Coupling.   In a rare example of an asymmetric Suzuki reaction, Labande and co-workers prepared a range of new palladium complexes bearing chiral phosphine–NHC ligands with planar chirality (**12**, Figure 4.8).[44] These complexes were then used to prepare chiral binaphthyl compounds. Good yields and moderate enantioselectivities were obtained (*ee* up to 42%, Scheme 4.8), using low catalyst loadings (0.1–0.5 mol%).

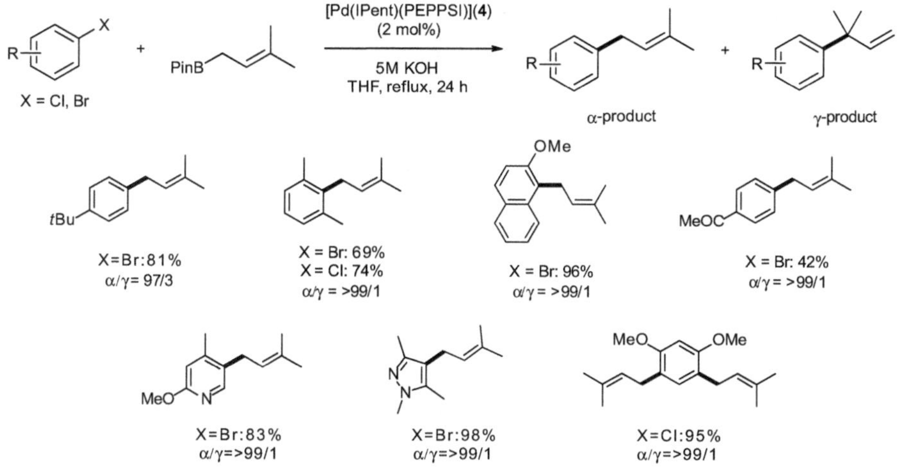

Scheme 4.7 Activity of [Pd(IPent)(PEPPSI)] (**4**) in the selective Suzuki reaction involving allylboronic acids.

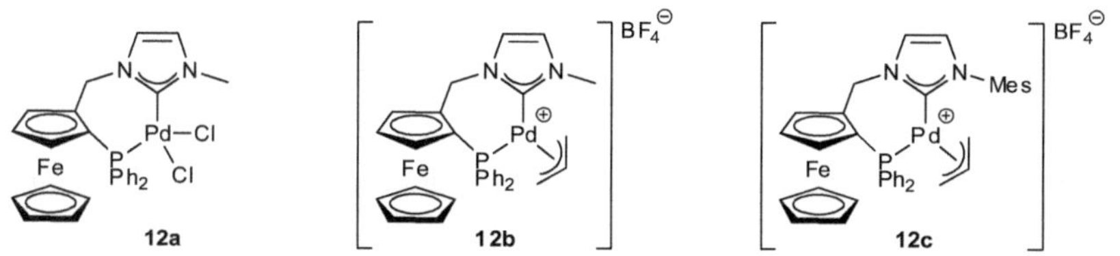

Figure 4.8 Structures of the precatalysts **12a–c** for asymmetric Suzuki coupling.

Yield: 88%
ee = 42%

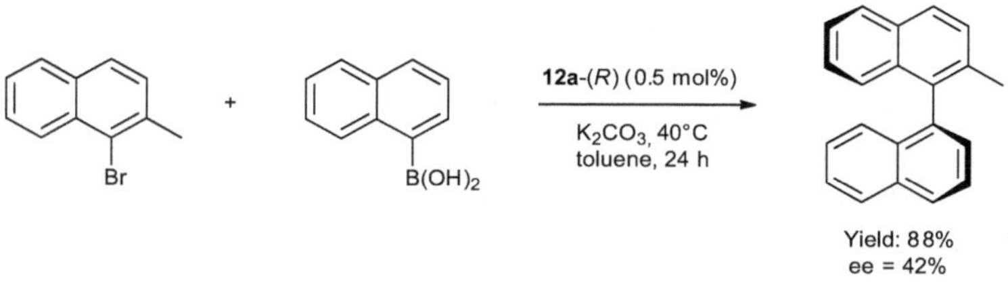

Scheme 4.8 Asymmetric Suzuki coupling for the preparation of chiral binaphthyls.

4.3.1.1.4 Developments of New Well-Defined Precatalysts with New "Throwaway Ligands".

When using a well-defined precatalyst in cross-coupling reactions, one of the keys to success is the ease and the speed with which the precatalyst leads to the formation of the active Pd(0) species in the reaction medium. The development of new labile ligands, also called *throwaway ligands*, is therefore very important. Among the most notable advances, Cazin and co-workers developed a new range of mixed

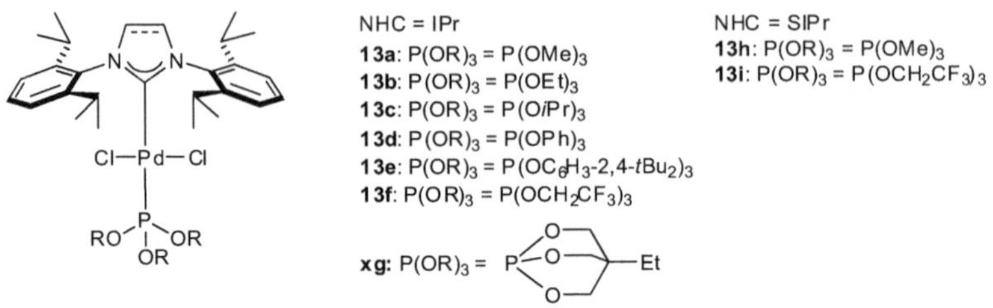

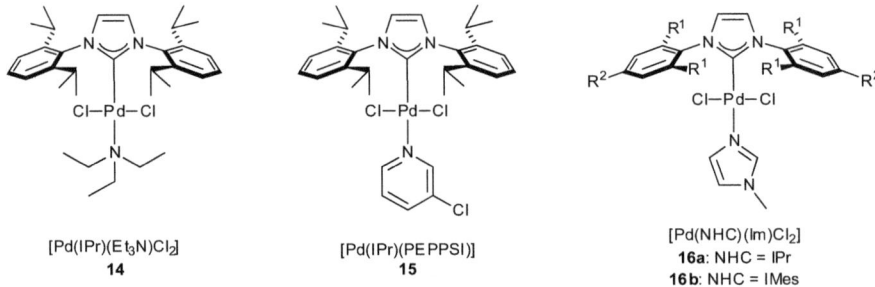

NHC = IPr
13a: P(OR)$_3$ = P(OMe)$_3$
13b: P(OR)$_3$ = P(OEt)$_3$
13c: P(OR)$_3$ = P(O*i*Pr)$_3$
13d: P(OR)$_3$ = P(OPh)$_3$
13e: P(OR)$_3$ = P(OC$_6$H$_3$-2,4-*t*Bu$_2$)$_3$
13f: P(OR)$_3$ = P(OCH$_2$CF$_3$)$_3$

NHC = SIPr
13h: P(OR)$_3$ = P(OMe)$_3$
13i: P(OR)$_3$ = P(OCH$_2$CF$_3$)$_3$

xg: P(OR)$_3$ = [structure] —Et

Figure 4.9   Structure of the mixed phosphite–NHC precatalysts.

[Pd(IPr)(Et$_3$N)Cl$_2$]
14

[Pd(IPr)(PEPPSI)]
15

[Pd(NHC)(Im)Cl$_2$]
16a: NHC = IPr
16b: NHC = IMes

Figure 4.10   Structures of precatalysts active in Suzuki–Miyaura coupling.

phosphite–NHC palladium complexes (**13**, Figure 4.9).[45] Interestingly, in these structures, the NHC was found to modulate its bulkiness depending on the bulkiness of the phosphite co-ligand. The reactivity of the complexes was compared in Suzuki coupling. Compounds **13a** and **13c** exhibited the highest catalytic activities. Alkoxide and hydroxide bases and/or alcohols as solvents were found to be necessary to obtain good catalytic activities. In a similar way, Cazin and co-workers also described the preparation of a range of mixed phosphine–NHC palladium complexes.[46] These complexes were active for Suzuki coupling in a mixture of water and *i*PrOH (9:1) with typically a very low catalyst loading of 0.03 mol%. As previously observed with the phosphites, the NHC was found to adapt its bulkiness according to that of the phosphine co-ligand.

In 2011, Navarro and co-workers utilized Et$_3$N as a cheap and readily available throwaway ligand.[47] The reactivity of the new [Pd(IPr)(Et$_3$N)Cl$_2$] complex (**14**, Figure 4.10) was directly compared with that of the [Pd(IPr)(PEPPSI)] precatalysts (**15**, Figure 4.10) developed by Organ and co-workers[48] (Table 4.3). Complex **14** showed higher catalytic activity, at lower temperature. This was explained by an easier departure of the labile ligand, and/or a higher propensity for the Et$_3$N to recoordinate to the NHC–Pd(0) active species.

Another significant example reported in the literature is the introduction of methylimidazole as a throwaway ligand by Shao and co-workers.[49] Two new complexes of formula [Pd(NHC)(Im)Cl$_2$] were prepared (**16**, Figure 4.10).

Table 4.3 [Pd(IPr)(Et$_3$N)Cl$_2$] *versus* [Pd(IPr)(PEPPSI)] in the Suzuki reaction.

Entry	[Pd(NHC)L$_n$]	T (°C)	Time	Yield (%)^a
1	[Pd(IPr)(PEPPSI)] (**15**)	25	3 h	37
2	[Pd(IPr)(Et$_3$N)Cl$_2$] (**14**)	25	1.5 h	96
3	[Pd(IPr)(PEPPSI)] (**15**)	40	2 h	98
4	[Pd(IPr)(Et$_3$N)Cl$_2$] (**14**)	40	20 min	98

^{a}By GC.

The IPr derivative proved highly active and versatile for various cross-coupling reactions, including the Suzuki reaction. [Pd(IPr)(Im)Cl$_2$] also promoted the coupling in water.[50]

4.3.1.1.5 Development of New NHC Ligands.

The nature of the NHC ligand dictates the catalytic properties of the Pd–NHC catalyst. As a consequence, modifying the nature of the NHC by modifying its steric demand and also its electronic properties is important to achieve different kinds of reactivities. Since 2010, many new NHCs have been developed for Suzuki–Miyaura cross-coupling and we focus on some of the main advances here.

Lee and co-workers developed the amido-*N*-imidazolium salt **17** for the formation of arylheterocycles at extremely low catalyst loadings.[51] The ligand was found to be active with Pd loadings as low as 0.0001 mol%, yielding the 2-phenylpyridine with a turnover number (TON) of 850 000. The use of ligand **17** also allowed the preparation of two pharmaceutical compounds, milrinone and irbesartan (Figure 4.11).

Kuriyama *et al.* also developed an *in situ*-generated system using ligand **18** for the preparation of polyheterocycles.[52] In this case, Cs$_2$CO$_3$ was used as the base for a greater group tolerance. Polyheterocycles were globally prepared in good yields (Scheme 4.9).

The modification of the steric bulk of NHCs has also been widely studied in recent years. Notably, Holland and co-workers studied the effect of increased bulk at the *para*-position of the aryl groups of the NHC ligand.[53] Four new [Pd(NHC)(cin)Cl] complexes were prepared and their reactivities were compared with that of the previously reported [Pd(IPr)(cin)Cl] complex[54] (Figure 4.12, Table 4.4). All of the new complexes were found to be more active for the coupling between 4-chloro(trifluoromethyl)benzene and phenylboronic acid. This suggests that palladium catalysts bearing bulkier ligands are more active, but this does not seem always to be the case. For example, Straub and co-workers showed that [Pd(IPr**)(pyridinyl)Cl$_2$] (**20**, Figure 4.12) was inactive in Suzuki coupling,[55] despite the fact that IPr** is one of the largest known NHCs to date.[56]

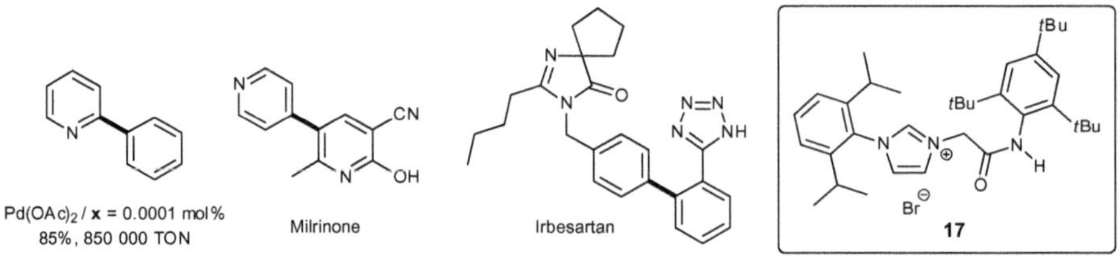

Pd(OAc)₂ / **x** = 0.0001 mol%
85%, 850 000 TON

Milrinone

Irbesartan

17

Figure 4.11 **17** as a ligand for Suzuki coupling: preparation of arylheterocycles at extremely low catalyst loadings and the total synthesis of milrinone and irbesartan.

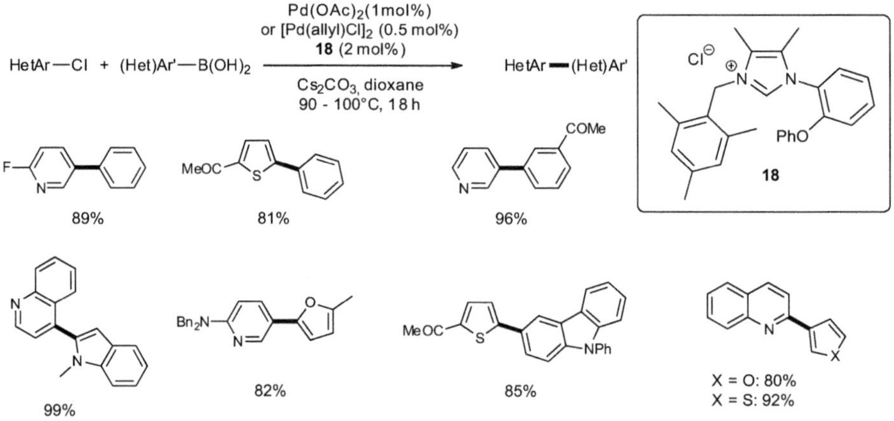

$$\text{HetAr—Cl} + \text{(Het)Ar'—B(OH)}_2 \xrightarrow[\substack{\text{Cs}_2\text{CO}_3, \text{ dioxane} \\ 90 - 100°C, 18 h}]{\substack{\text{Pd(OAc)}_2(1\text{mol}\%) \\ \text{or [Pd(allyl)Cl]}_2 \text{ (0.5 mol\%)} \\ \textbf{18} \text{ (2 mol\%)}}} \text{HetAr—(Het)Ar'}$$

18

89%

81%

96%

99%

82%

85%

X = O: 80%
X = S: 92%

Scheme 4.9 Reactivity of ligand **18** in Suzuki coupling.

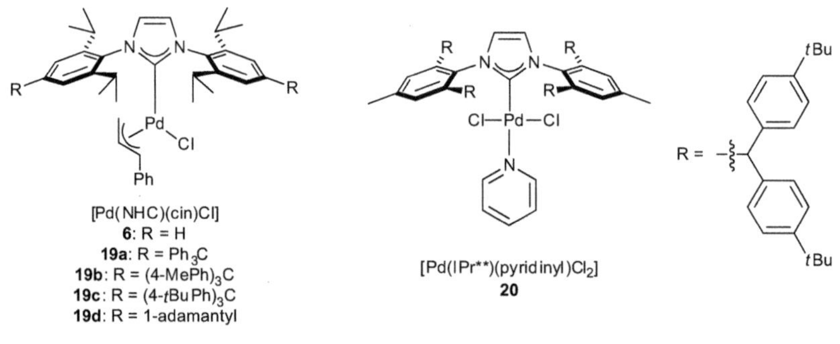

[Pd(NHC)(cin)Cl]
6: R = H
19a: R = Ph₃C
19b: R = (4-MePh)₃C
19c: R = (4-*t*BuPh)₃C
19d: R = 1-adamantyl

[Pd(IPr**)(pyridinyl)Cl₂]
20

R =

Figure 4.12 Structures of the [Pd(NHC)(cin)Cl] complexes **6**, **19a–d** and [Pd(IPr**)-(pyridinyl)Cl₂] (**20**).

Another example showing that the size of the ligand is not an absolutely vital factor was reported by Trzeciak and co-workers in 2013.[57] They showed that small NHCs in complexes of formula [Pd(NHC)₂X₂] (Figure 4.13) gave results comparable to those with IMes for the coupling between 2-bromo-benzene and phenylboronic acid (Table 4.5). The use of mokt-y as the NHC ligand even allowed a TON as high as 760 000 to be reached.

Table 4.4 Comparison of reactivities of various [Pd(NHC)(cin)Cl] complexes.

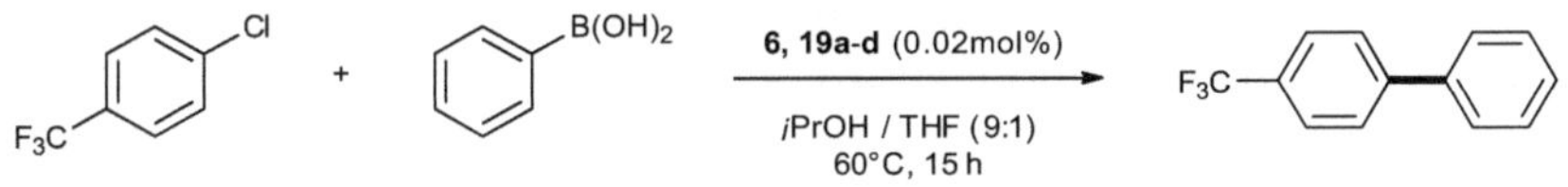
Cl
F3C
B(OH)2
6, 19a-d (0.02mol%)
iPrOH / THF (9:1)
60°C, 15 h
F3C

Entry	[Pd(NHC)(cin)Cl]	Conversion (%)^{a}
1	**6**	55
2	**19a**	87
3	**19b**	86
4	**19c**	85
5	**19d**	69

^{a}By GC.

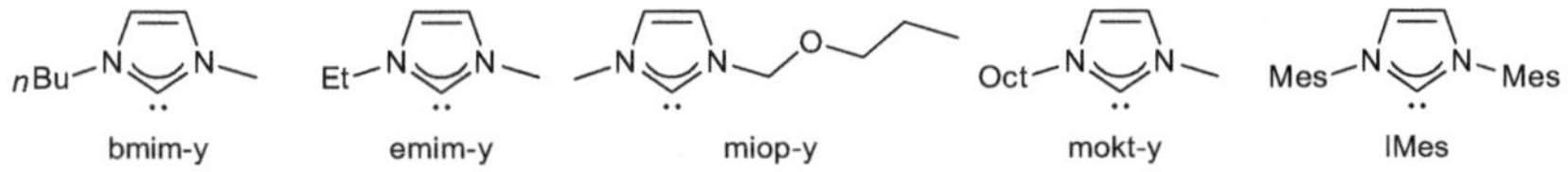
nBu
Et
O
Oct
Mes
Mes
bmim-y
emim-y
miop-y
mokt-y
IMes

Figure 4.13 Small NHCs and IMes used by Trzeciak and co-workers.[57]

Table 4.5 Comparison of reactivities of various [Pd(NHC)$_2$X$_2$] complexes for the Suzuki reaction at very low catalyst loadings.

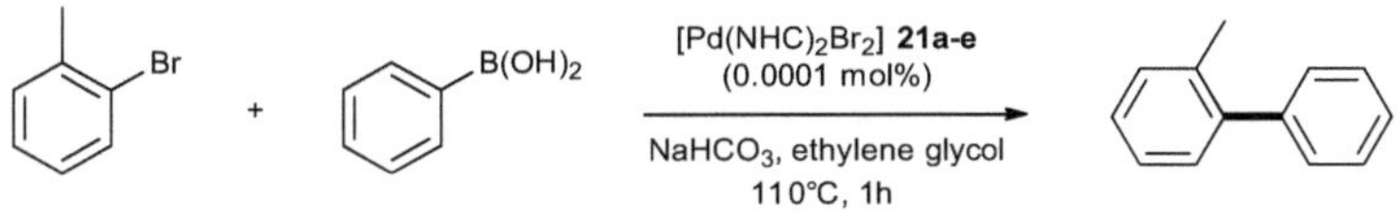
Br
B(OH)2
[Pd(NHC)2Br2] 21a-e
(0.0001 mol%)
NaHCO3, ethylene glycol
110°C, 1h

Entry	[Pd(NHC)$_2$Br$_2$]	Yield (%)	TON
1	[Pd(bmim-y)$_2$Br$_2$] (**21a**)	62	620 000
2	[Pd(emim-y)$_2$Br$_2$] (**21b**)	50	500 000
3	[Pd(miop-y)$_2$Cl$_2$] (**21c**)	2	20 000
4	[Pd(mokt-y)$_2$Cl$_2$] (**21d**)	76	760 000
5	[Pd(IMes)$_2$Cl$_2$] (**21e**)	86	860 000

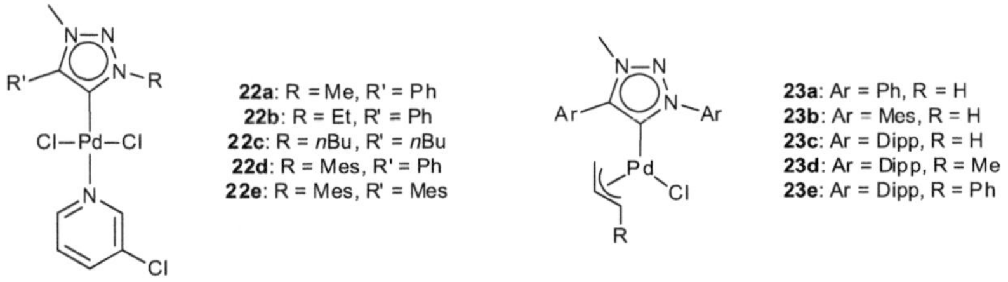
22a: R = Me, R' = Ph
22b: R = Et, R' = Ph
22c: R = nBu, R' = nBu
22d: R = Mes, R' = Ph
22e: R = Mes, R' = Mes
23a: Ar = Ph, R = H
23b: Ar = Mes, R = H
23c: Ar = Dipp, R = H
23d: Ar = Dipp, R = Me
23e: Ar = Dipp, R = Ph

Figure 4.14 Structures of the complexes **22a–e** and **23a–e**.

Some 1,2,3-triazolylidene ligands were also developed by the groups of Albrecht[58] and Fukuzawa[59] as an alternative to more common NHC ligands (Figure 4.14). In the case of the PEPPSI-type precatalysts (**22a–e**), the impact of steric shielding on Suzuki coupling was found to be opposite to that observed with normal NHCs.[58] Indeed, complexes bearing small alkyl groups were more active. In the case of the allyl-type complexes (**23a–e**), the influence of the substitution on the allyl group was studied.[59] The congener bearing a cinnamyl group (**23e**) was found to be the most active, allowing the coupling of a range of aryl chlorides at room temperature. This is in agreement with previous studies carried out by Nolan and co-workers.[54]

4.3.1.1.6 Suzuki Coupling in Water. In recent years, several groups were also interested in developing catalysts active in water. Roy and Plenio notably described the preparation of complex **24** (Figure 4.15),[60] a sulfonated version of the highly active and previously reported [Pd(IPr)(cin)Cl][54] (**6**, Figure 4.6). The new precatalyst was active for the Suzuki reaction in a mixture of water and *n*BuOH (1 : 1) at 100 °C using KOH as the base. Both aryl chlorides and bromides were suitable coupling partners. Godoy and co-workers also used the sulfonate functionality to prepare the bis-NHC precatalyst **25** as a water-soluble complex (Figure 4.15).[61] The use of **25** in pure water at 110 °C in the presence of K_2CO_3 notably allowed the coupling of 4-bromoacetophenone and phenylboronic acid with a TON of 100 000 (Scheme 4.10). Karimi and Akhavan developed the "nanocentipede-like" Pd–NHC polymer **26** (Figure 4.15).[62] This promoted the Suzuki reaction in water at room temperature with catalyst loadings as low as 0.005 mol% for the coupling of aryl bromides and 0.05 mol% for the coupling of aryl chlorides. Interestingly, the catalyst was also recyclable and could be used up to 17 times.

Nechaev and co-workers reported on the use of [Pd(NHC)(cin)Cl] complexes bearing expanded-ring NHCs for the Suzuki coupling of heteroaryl chlorides

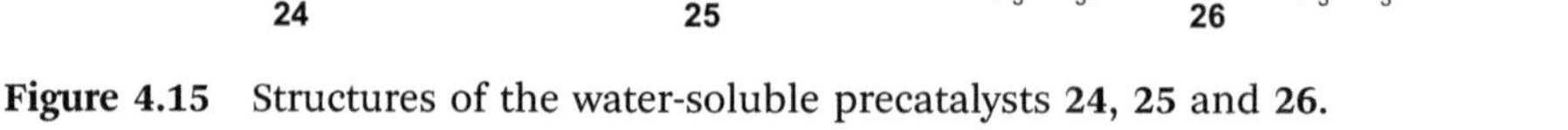

Figure 4.15   Structures of the water-soluble precatalysts **24**, **25** and **26**.

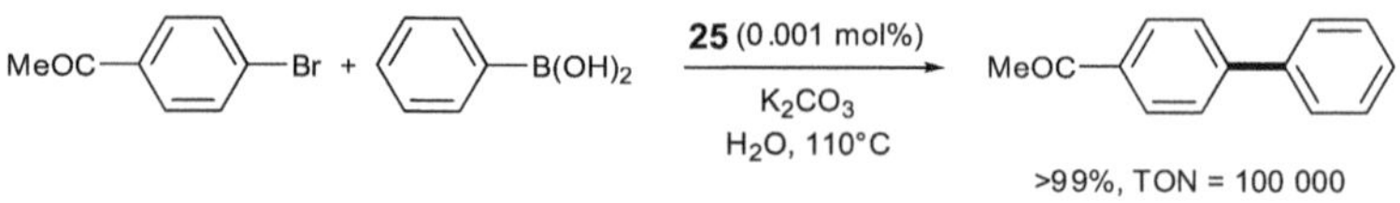

Scheme 4.10   Suzuki coupling using **25** in water.

Table 4.6 Reactivity of [Pd(NHC)(cin)Cl] (**27**) in the Suzuki reaction.

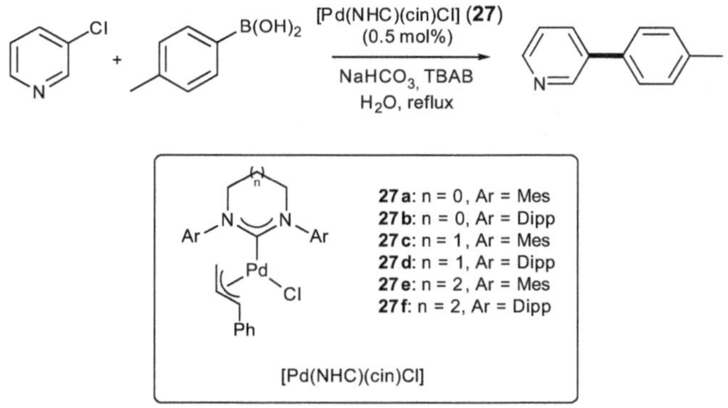

27a: n = 0, Ar = Mes
27b: n = 0, Ar = Dipp
27c: n = 1, Ar = Mes
27d: n = 1, Ar = Dipp
27e: n = 2, Ar = Mes
27f: n = 2, Ar = Dipp

[Pd(NHC)(cin)Cl]

Entry	[Pd(NHC)(cin)Cl] (**27**)	Yield (%)
1	**27a**	3
2	**27b**	33
3	**27c**	15
4	**27d**	84
5	**27e**	86
6	**27f**	25

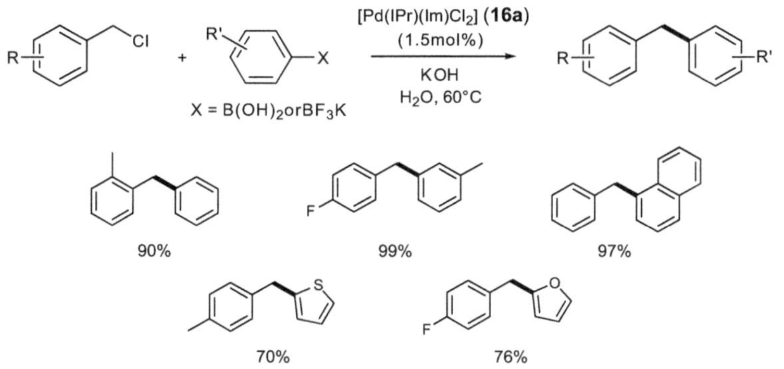

Scheme 4.11 Preparation of diarylmethane derivatives using **16a**.

in water (Table 4.6).[63] Complexes **27d** and **27e** (six- and seven-membered ring, respectively) were found to be the most active congeners for the coupling of 3-chloropyridine and 4-tolylboronic acid. Interestingly, the reaction is successful under air, using TBAB as a transfer agent and NaHCO$_3$ as an environmentally benign base.

In a rare example of coupling involving benzylic chlorides, Lu and coworkers determined the activity of [Pd(IPr)(Im)Cl$_2$] (**16a**) in neat water.[64] The methodology allowed the preparation of an interesting library of diarylmethane derivatives that are important compounds for the pharmaceutical industry (Scheme 4.11).

4.3.1.1.7   Development of Recyclable Catalytic Systems. Another ongoing challenge to solve in cross-coupling chemistry is the development of recyclable systems. One strategy to recycle a catalyst is to anchor it on a solid support.[65] Nevertheless, this is not always an easy task owing to various problems, including catalyst deactivation, catalyst leaching and the fact that anchored catalysts regularly appear to be less active than their non-immobilized congeners. Yang's group have made important contributions in this area. In 2010, they developed mesoporous ethane–silicas functionalized with IMes-type ligands (**28a**, Figure 4.16).[66] The support was able to coordinate Pd(OAc)$_2$ and the supported catalyst was subsequently used in Suzuki–Miyaura cross-coupling. The system could be used eight times without any significant decrease in the catalytic activity. The high recyclability was attributed to the functionalized and stable nanopores that prevented aggregation of the Pd nanoparticles into larger particles. A year later, they extended the utility of the support by anchoring an IPr-type ligand (**28b**, Figure 4.16).[67] The resulting materials were palladated using various palladium sources. Amongst them, [Pd(ES-IPr)(acac)Cl] (**29**) was found to be the most active for Suzuki coupling. In this case, the system could be recycled up to 10 times without any significant decrease in reactivity (Table 4.7). Notably, the supported catalysts are recovered by centrifugation between two catalytic runs. Yang and co-workers also developed three-dimensional cubic mesoporous materials that

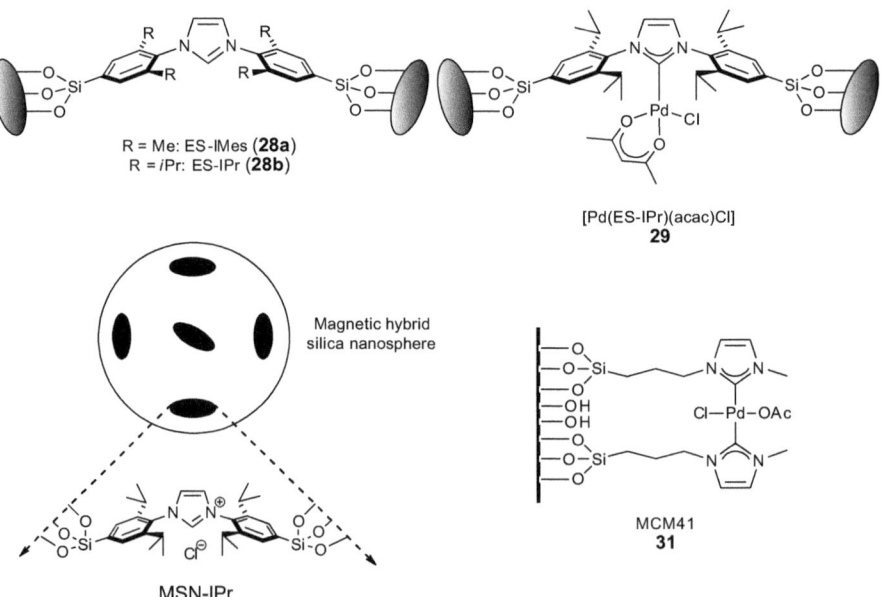

R = Me: ES-IMes (**28a**)
R = *i*Pr: ES-IPr (**28b**)

[Pd(ES-IPr)(acac)Cl]
29

Magnetic hybrid
silica nanosphere

MSN-IPr
30

MCM41
31

Figure 4.16   Structures of some solid supported NHC and Pd–NHC systems used in Suzuki coupling.

Table 4.7 Recyclability of **29**.

	Cycles									
	1	2	3	4	5	6	7	8	9	10
Time (h)	4	4	4	4	4	4	5	5	6	7
Yield (%)	91	93	92	92	93	90	87	89	82	83

proved to be both active and recyclable for the Suzuki reaction,[68] and also magnetic IPr-functionalized silica nanoparticles (**30**, Figure 4.16).[69] After coupling, the latter material could be isolated by applying an external magnetic field and therefore could be reused without loss of activity. Another interesting contribution by Alam and Sarkar highlights the use of the nanostructured mesoporous solid supported catalyst MCM41 (**31**, Figure 4.16).[70] This heterogeneous catalyst was successfully used in the Suzuki coupling of aryl bromides and iodides and could be reused after simple filtration from the reaction mixture.

4.3.1.1.8 Suzuki Coupling Using Pseudohalides. Halides are most often the coupling partner of choice for Suzuki coupling. However, some recent examples involving pseudohalides can be highlighted. For example, in 2010 Luo and co-workers reported the reaction with aryltriazenes using a polymer-supported Pd–NHC catalyst (**32**, Table 4.8).[71] The coupling proceeded smoothly and the biaryls were obtained in good yields at room temperature. Interestingly, iodide and bromide atoms were not affected by the reaction conditions and hence could be tolerated by the system. Of note, the catalyst is also recyclable.

Arylsulfonates were also found to be suitable coupling partners for the first time using a Pd–NHC precatalyst.[72] Both tosylates and mesitylates afforded the expected biaryls using 1 mol% of [Pd(IPr)(Im)Cl$_2$] (**16a**, Figure 4.10), K$_3$PO$_4$·H$_2$O as the base in morpholine as the solvent. However, tosylates were better coupling partners (Scheme 4.12).

4.3.1.2 Negishi Coupling

Despite being one of the first cross-couplings to have been discovered,[73] the Negishi reaction appears to be far less developed than the Suzuki–Miyaura reaction. This is nevertheless a very powerful tool, notably because of its high tolerance to functional groups. The coupling involves the reaction between a halide or pseudohalide and an organozinc derivative. In this section, we focus on the main advances in the field using Pd–NHC systems, before and after the award of the 2010 Nobel Prize in Chemistry.

Table 4.8   Suzuki coupling with aryltriazenes.

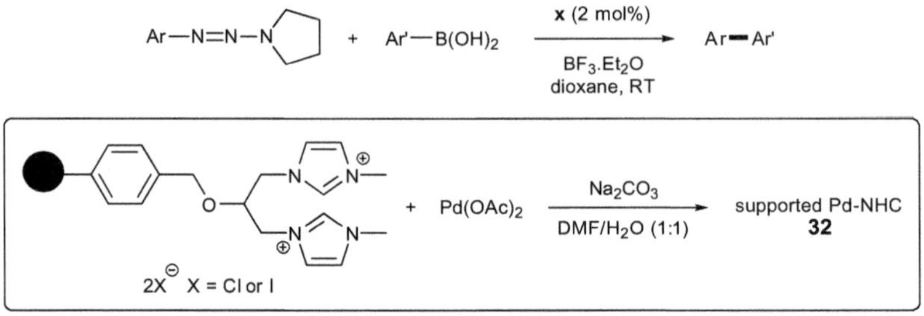

Entry	Ar	Ar'	Product	Yield (%)
1	3-NO$_2$C$_6$H$_4$	Ph	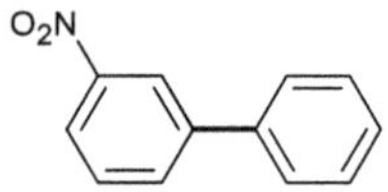	92
2	4-IC$_6$H$_4$	Ph	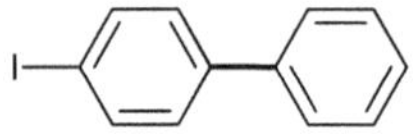	66
3	4-BrC$_6$H$_4$	Ph	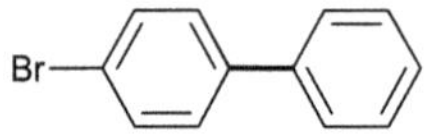	71
4	4-MeC$_6$H$_4$	4-MeOC$_6$H$_4$	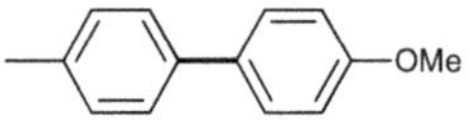	82
5	Ph	3-NO$_2$C$_6$H$_4$	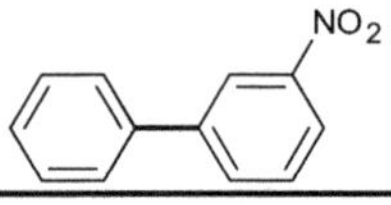	74

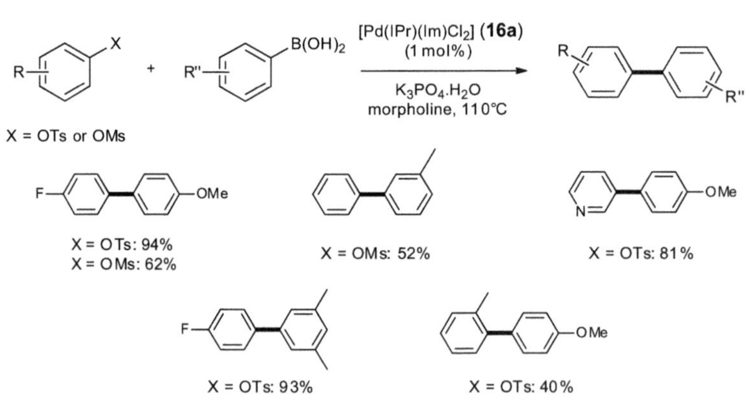

Scheme 4.12   Suzuki reaction using aryl sulfonates.

4.3.1.2.1   **Origins and Advances in Coupling Using Pd–NHC Systems.**
The first publication on the use of a Pd–NHC system in the Negishi reaction appeared in 2005.[74] This involved an *in situ*-generated catalyst composed of Pd$_2$(dba)$_3$ and IPr · HCl (2 and 8 mol%, respectively). The system

was capable of promoting the coupling between primary bromides and alkyl organozinc reagents to create sp^3–sp^3 bonds in high yields, at room temperature, in THF–NMP (2:1) as solvent (Scheme 4.13).

One year later, Organ *et al.* developed the first well-defined palladium precatalyst to achieve the cross-coupling, namely [Pd(IPr)(PEPPSI)] (**15**, Scheme 4.14).[75] The system proved highly efficient in Negishi coupling,

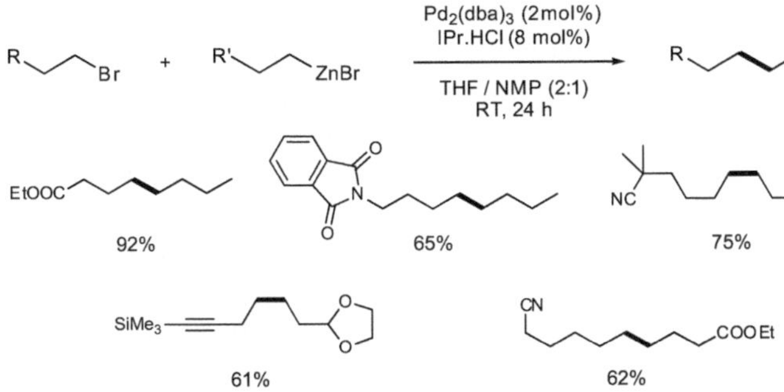

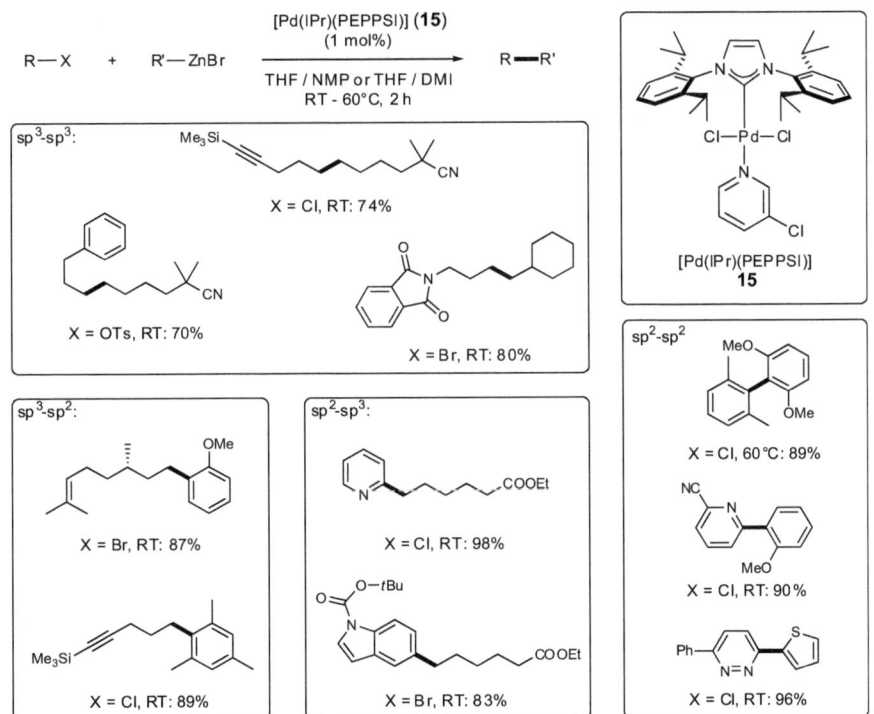

Scheme 4.13   Negishi coupling using an *in situ*-generated catalyst system for the formation of sp^3–sp^3 bonds.

Scheme 4.14   Negishi coupling using the well-defined [Pd(IPr)(PEPPSI)] precatalyst.

allowing the construction of a wide range of bonds: sp^3–sp^3, sp^3–sp^2, sp^2–sp^3 and sp^2–sp^2. The use of the well-defined system in comparison with the older *in situ*-generated system[74] was found to increase substantially the scope and the reliability of the cross-coupling. Most notably, **15** was active at lower catalyst loadings (1 *versus* 2 mol%).

In 2009, Organ and co-workers began to study the mechanism of the alkyl–alkyl Negishi reaction.[76] Using DFT investigations, it was shown that transmetallation was the rate-limiting step and not oxidative addition. Moreover, a Pd–Zn interaction was identified in the mechanism. The latter was postulated to facilitate reductive elimination, in combination with the ligand. This was thought to be possible by the generation of a sterically crowded environment in the coordination sphere of the Pd atom.

Organ's group also investigated the role of additives in Negishi coupling.[77] Bromide salts were found to promote the reaction whereas the counter-cation was observed to have a minor effect. Titration studies also showed that the coupling remained dormant until a minimum ratio of 1:1 between LiBr and *n*BuZnBr was achieved. This suggested that higher-order zincates of formula $Li_mZn(nBu)Br_3^{(2-m)-}$ were likely to be the active transmetallating agents (Scheme 4.15).

To corroborate the postulate that higher-order zincates were involved in the Negishi reaction, Organ and co-workers studied in more detail the addition of lithium bromide to the reaction medium. Mass spectrometry[78] was used to identify the negatively charged species resulting from mixtures of LiBr and *n*BuZnBr in various solvents. DMI and NMP were found to be essential to form and stabilize higher-order zincates such as $nBuZnBr_2^-$ and $nBuZnBr_3^{2-}$. This was also confirmed by NMR studies. Finally, studies involving directly isolated and pure crystalline $RZnBr_3^{2-}$ zincates[79] definitely proved the involvement of these species in the reaction. Reaction occurred in 2 h in THF to yield the desired compounds in excellent yields (Scheme 4.16).

In 2010, Organ and co-workers reported the use of a new complex in the reaction, [Pd(IPent)(PEPPSI)] (**4**).[80] This complex proved to be much more active than the previously reported [Pd(IPr)(PEPPSI)] (**15**), especially for the preparation of highly functionalized tetra-*ortho*-substituted biaryls (Scheme 4.17). As reported previously for Suzuki–Miyaura cross-coupling, the *flexible steric bulk*[23] of the IPent ligand was postulated as the key to success in the more challenging reactions.

[Pd(IPent)(PEPPSI)] (**4**) was also found to be an excellent precatalyst for the coupling of secondary alkylzinc halides with aryl/heteroaryl halides.[81] Most

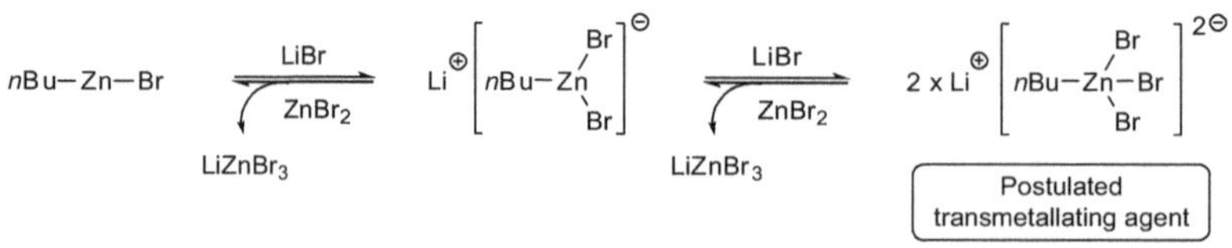

Scheme 4.15   Postulated transmetallating agents in Negishi coupling.

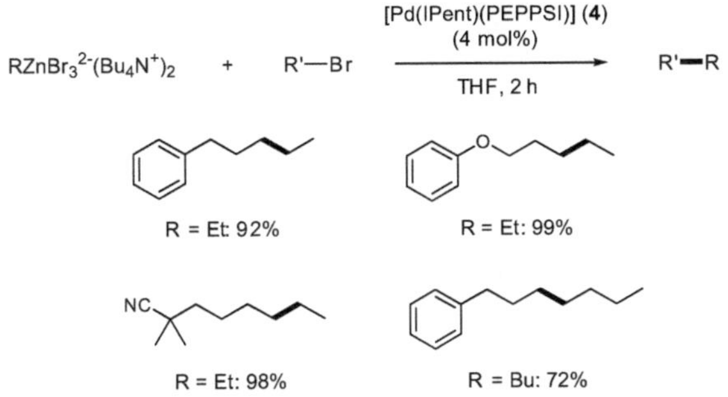

Scheme 4.16 Reactivity of higher order zincates in the Negishi reaction.

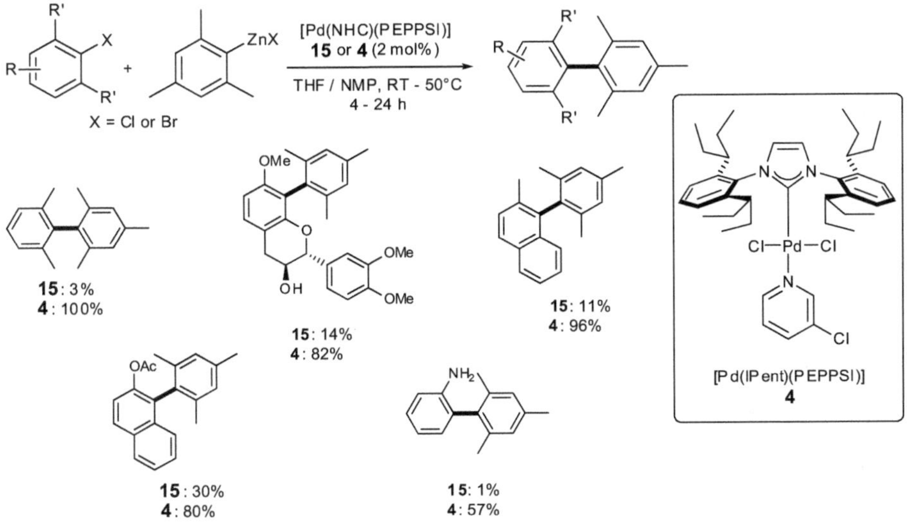

Scheme 4.17 [Pd(IPent)(PEPPSI)] *versus* [Pd(IPr)(PEPPSI)] in Negishi cross-coupling.

importantly, thanks to the IPent ligand, β-hydride elimination/migratory insertion of the zinc reagent was considerably reduced, limiting the production of isomeric coupling product (Table 4.9).

In 2011, Organ and co-workers used [Pd(IPr)(PEPPSI)] (**15**) for the bis-functionalization of chlorobromoalkanes using orthogonal alkyl–alkyl Negishi reactions (Scheme 4.18).[82] The difference in reactivity between the C–Br and the C–Cl bond was triggered by the polarity of the solvent. First, the C–Br bond reacted in a 1:2 DMI–THF mixture. The second coupling on the C–Cl bond was subsequently enhanced by adding the second zinc reagents in DMI, leading to a reaction medium containing a 2:1 DMI–THF mixture. Several functionalized alkanes were then prepared in moderate yields.

In 2011, Larrosa *et al.* noticed an interesting abnormal reactivity of [Pd(IPr)(PEPPSI)] (**15**) in Negishi coupling (Scheme 4.19).[83] When they

Table 4.9 [Pd(IPent)(PEPPSI)] (**4**) *versus* [Pd(IPr)(PEPPSI)] (**15**) for the coupling of secondary alkylzinc derivatives.

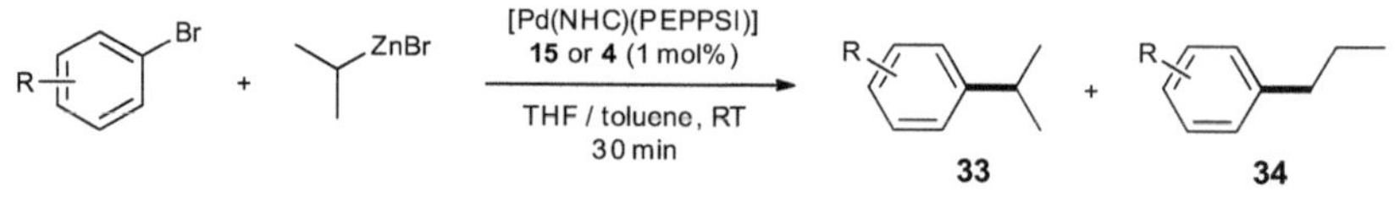

Entry	R	Catalyst	Yield (%)	Selectivity 33:34
1	4-OMe	**15**	89	2.5:1
2	4-OMe	**4**	95	33:1
3	3-CN	**15**	77	1:1.4
4	3-CN	**4**	84	11:1
5	2-CN	**15**	99	1:8
6	2-CN	**4**	80	2.4:1
7	2-OMe	**15**	99	1:9
8	2-OMe	**4**	46	2:1

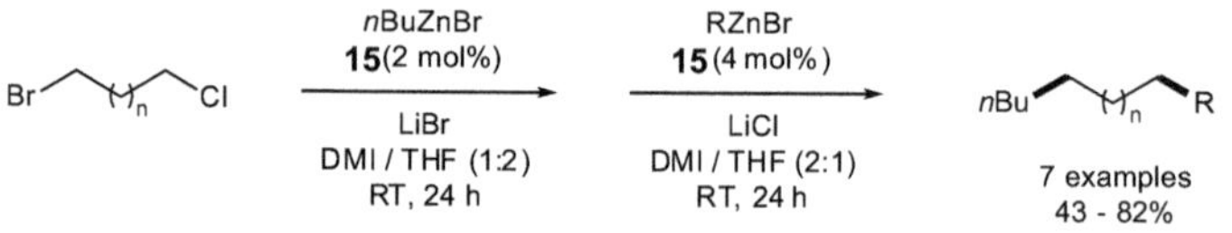

Scheme 4.18 Bisfunctionalization of chlorobromoalkanes.

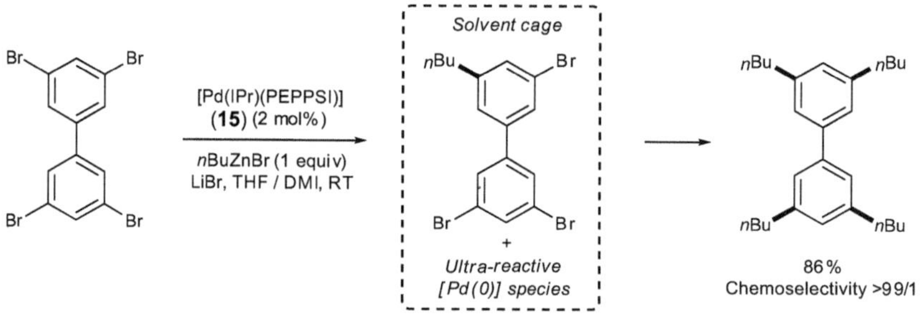

Scheme 4.19 Abnormal reactivity of [Pd(IPr)(PEPPSI)] in the Negishi reaction.

reacted 1 equiv. of *n*BuZnBr to monobutylate polybrominated arenes selectively, they observed the favoured formation of polybutylated rings with high chemoselectivity (up to >99%). They explained this reactivity by the generation of an ultra-reactive Pd(0) species trapped in a solvent cage.

In 2012, Organ and co-workers reported the new [Pd(IPent^{Cl})(PEPPSI)] (**35**) for the selective cross-coupling of secondary organozinc reagents (Scheme 4.20).[84] The new precatalyst exhibited high selectivity towards the desired branched product in comparison with the non-desired linear product (resulting from isomerization in the reaction mixture). Thus a wide range of alkylzincs were coupled to a wide range of functionalized (hetero)aromatic halides. This is an improvement in comparison with the previous protocol, which presented only a limited number of challenging examples.[81]

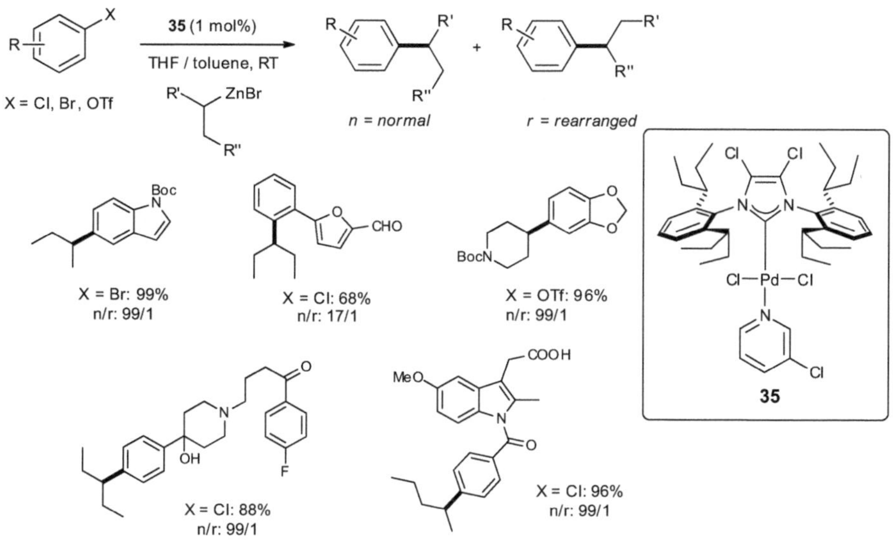

Scheme 4.20 [Pd(IPent^{Cl})(PEPPSI)] (**35**) for the selective cross-coupling of secondary organozinc reagents.

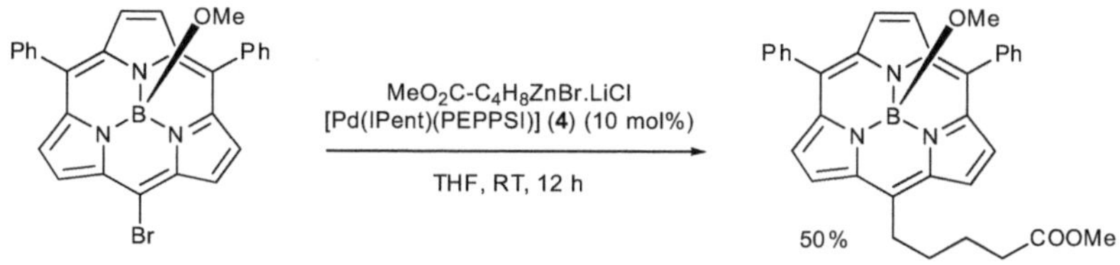

Scheme 4.21 Preparation of a subporphyrin by Negishi coupling using [Pd(IPent)(PEPPSI)] (**4**).

4.3.1.2.2 Applications. The catalysts developed by Organ's group have found a number of applications in the literature in recent years. Notably, Osuka and co-workers reported the use of [Pd(IPent)(PEPPSI)] (**4**) for the *meso* fabrication of a subporphyrin (Scheme 4.21).[85]

Another example was reported by the same group, using [Pd(IPr)(PEPPSI)] (**15**) to prepare a library of 2-aryl-3-trifluoromethylnaphthofurans (Scheme 4.22).[86] These are important compounds found in biologically active molecules and in organic functional materials. SMe groups were used as pseudohalides in the reaction and were coupled to ArZnCl.LiCl derivatives. The use of SMe groups in this case required a large amount of palladium (15 mol%); nevertheless, the reaction proceeded efficiently to yield the desired biaryls in good yields. The synthesized compounds were subsequently studied for their photophysical properties.

Finally, in 2012, Kristensen and co-workers used [Pd(IPr)(PEPPSI)] (**15**) to prepare a key intermediate in the total synthesis of ascididemin.[87] The

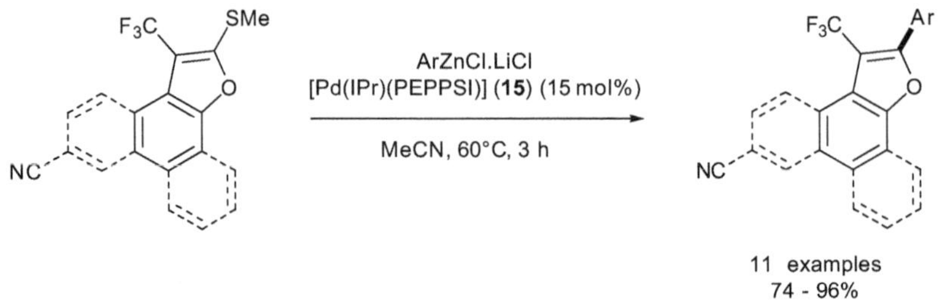

Scheme 4.22 Preparation of a library of 2-aryl-3-trifluoromethylnaphthofurans using [Pd(IPr)(PEPPSI)] (**15**).

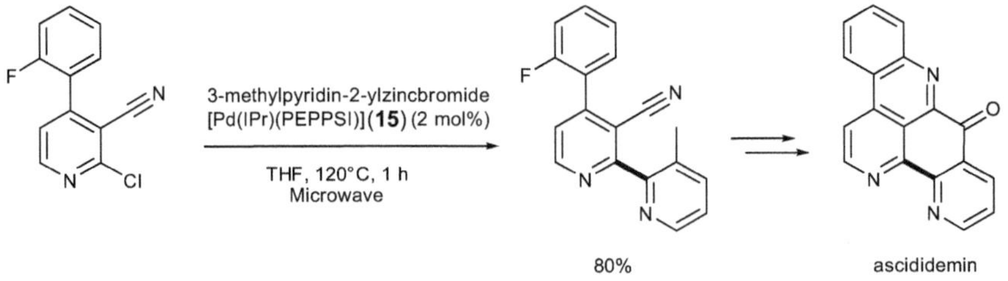

Scheme 4.23 Use of Negishi coupling as a key step in the formation of ascididemin.

reaction proceeded efficiently using 2 mol% of the precatalyst and the target compound was obtained in a good yield (Scheme 4.23).

4.3.1.3 Kumada–Tamao–Corriu Coupling

The Kumada–Tamao–Corriu reaction involves the reaction between an aryl halide and a Grignard reagent (organomagnesium reagent).[88] This chemistry has hardly been investigated using Pd–NHC systems. This section covers most of the protocols reported in the literature.

In 1999, Huang and Nolan investigated for the first time the use of NHCs in the Kumada reaction.[89] The system utilized Pd$_2$(dba)$_3$ (1 mol%) as the palladium source and IPr·HCl (4 mol%) as the ligand. The bulk of IPr was found to be crucial in enhancing catalytic activity in comparison with IMes. The coupling of unactivated chlorides with various Grignard reagents was then achieved efficiently in a mixture of dioxane and THF at 80 °C, yielding the biaryl compounds globally in good yields (Scheme 4.24). Unfortunately, the system did not allow the preparation of tetra-*ortho*-substituted biaryls.

In 2003, Beller and co-workers explored the reactivity of well-defined Pd(0)–NHC catalysts and also *in situ*-generated systems in the reaction between aryl-Grignard reagents and alkyl chlorides.[90] It was shown that well-defined systems were more efficient than the *in situ*-generated systems. In this case, IMes was found to be a better ligand than the bulkier IPr. The reaction proceeded efficiently using 2 mol% of catalyst **36** at room temperature (Scheme 4.25).

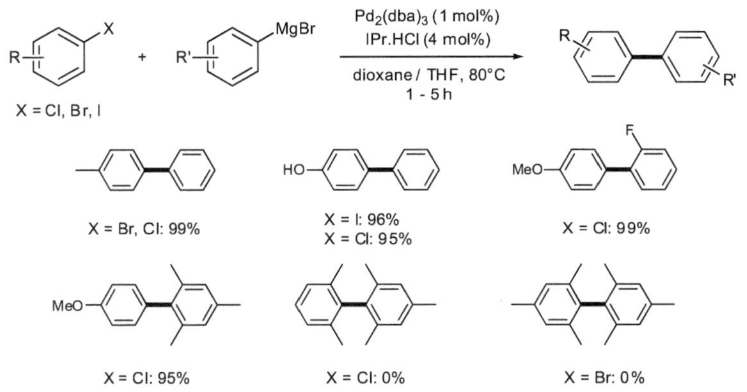

Scheme 4.24 Kumada cross-coupling using an *in situ*-generated system.

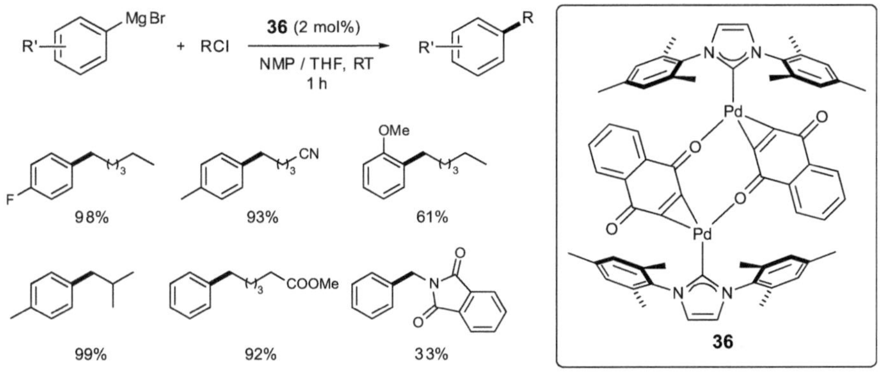

Scheme 4.25 Kumada cross-coupling using the well-defined Pd(0) catalyst **36**.

37a: NHC = SIPr
37b: NHC = IPr
37c: NHC = SIMes
37d: NHC = IMes

Figure 4.17 Structure of the [Pd(NHC)(μ-Cl)Cl]₂ complexes.

In 2009, Nolan and co-workers reported the use of several [Pd(NHC)(μ-Cl)Cl]₂ complexes for Kumada coupling between aryl-Grignard reagents and (hetero)aryl chlorides.[91] Four different NHC ligands were screened and SIPr showed the best catalytic activity (**37**, Figure 4.17). For the first time, the coupling was possible at low catalyst loadings (typically 0.2 mol%), under mild conditions (r.t.–60 °C, Scheme 4.26). The preparation of

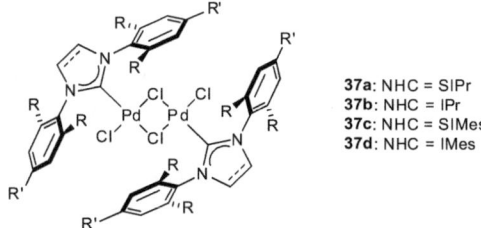

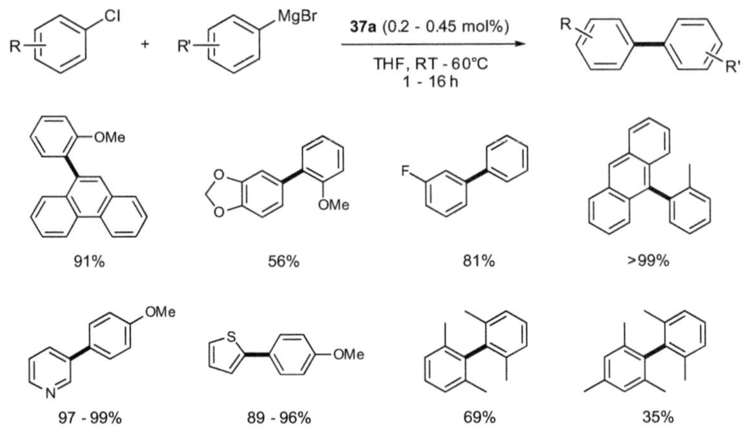

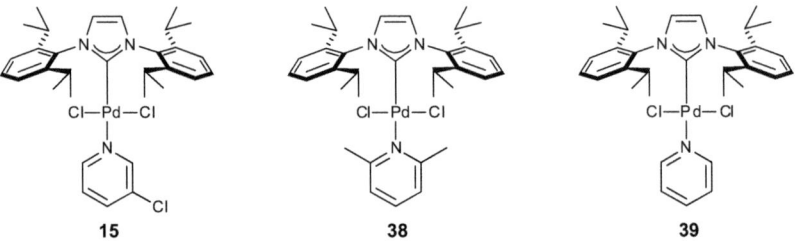

Scheme 4.26 Reactivity of [Pd(SIPr)(μ-Cl)Cl]₂ (**37a**) in Kumada cross-coupling.

Figure 4.18 Three [Pd(NHC)(PEPPSI)] complexes investigated in a structure–activity relationship study of the Kumada reaction.

tetra-*ortho*-substituted biaryls was also reported for the first time using a Pd–NHC precatalyst. A slight increase in the catalyst loading was required (0.45 mol%) but the highly sterically hindered products were obtained in moderate yields (two examples, 35–69%).

Organ and co-workers also studied Kumada coupling using a range of [Pd(NHC)(PEPPSI)] complexes.[92] In a structure–activity relationship study, it was shown that the IPr ligand was crucial for the reaction to occur. Moreover, the effect of the substitution of the pyridine throwaway ligand was also studied (Figure 4.18). Both 3-chloropyridine and pyridine showed good activity, with a slight advantage for the pyridine moiety (Tables 4.10 and 4.11).

Wu and co-workers developed carbene adducts of cyclopalladated ferrocenylimine for the Kumada reaction.[93] The IPr derivative was found to be the most active in comparison with other carbenes (IMes or an abnormal carbene) and some phosphines (PPh₃ and PCy₃). A wide range of biaryls were efficiently prepared under mild conditions (r.t.–60 °C) using complex **40** (0.5 mol%, Scheme 4.27).

Another interesting contribution is that of Jin *et al.*[94] They employed four different [Pd(NHC)(Cp)Cl] complexes in the reaction. In this case, they found the best ligand to be SIMes (**41**, Figure 4.19). Using 1 mol% catalyst loading

Table 4.10 Comparison of the reactivities of **15**, **38** and **39** in the Kumada coupling reaction.

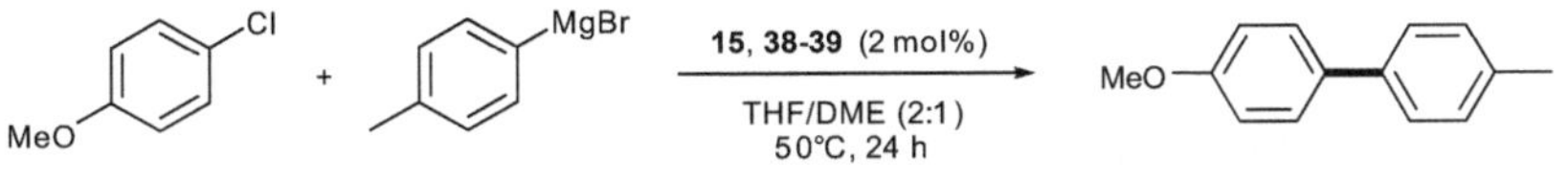

Entry	[Pd(NHC)(PEPPSI)]	Conversion (%)^a
1	**15**	100
2	**38**	85
3	**39**	100

^{a}By GC.

Table 4.11 Comparison of the reactivities of **15**, **38** and **39** in the Kumada coupling reaction.

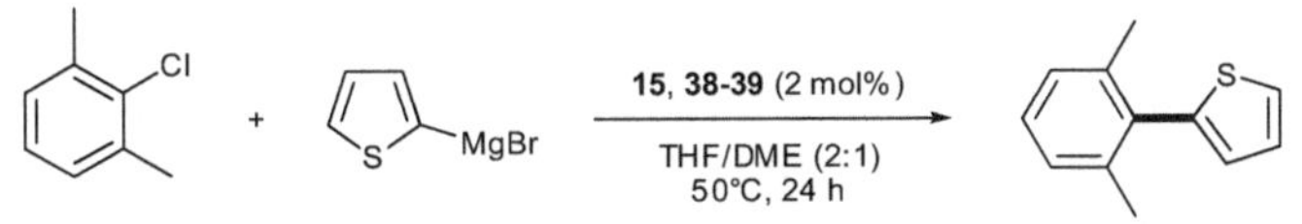

Entry	[Pd(NHC)(PEPPSI)]	Conversion (%)^a
1	**15**	75
2	**38**	88
3	**39**	95

^{a}By GC.

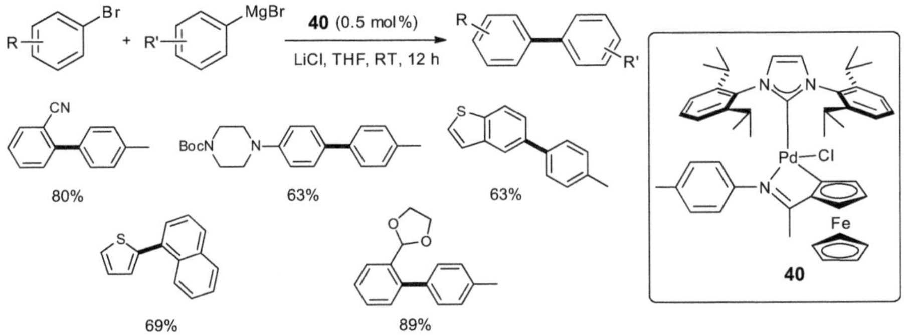

80%   63%   63%

69%   89%

40

Scheme 4.27 Reactivity of **40** in Kumada coupling.

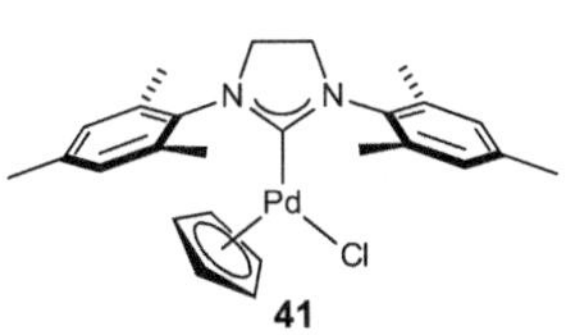

41

Figure 4.19 Structure of [Pd(SIMes)(Cp)Cl] (**41**).

at room temperature, they were able to couple various aryl and heteroaryl chlorides.

4.3.1.4 Stille Coupling

Owing to the toxicity of the stannyl derivatives, the Stille reaction[95] is being used less and less as a cross-coupling reaction and has slowly been replaced by more environmentally benign methods. Nevertheless, this reaction always proved to be very efficient. There are very few reports in the literature of the use of Pd–NHC systems, most of them being prior to the award of the Nobel Prize in Chemistry in 2010.

In 1999, Herrmann and co-workers were the first to investigate the use of well-defined Pd–NHC complexes to promote Stille coupling.[96] The mixed phosphine–NHC complexes 42 were found to combine the stability of the bis(carbene) complexes and the reactivity of the bis(phosphine) complexes. PPh_3 was found to be the best phosphine to promote the coupling of various bromides with tributylphenyltin, affording the corresponding coupling products in good yields (Table 4.12).

In a similar study, Herrmann *et al.* confirmed that PPh_3 was better than other phosphines at promoting the coupling.[97] The presence of an aromatic

Table 4.12 Pioneering work using Pd–NHC precatalysts for the Stille reaction.

Entry	X	Product	PR_3	Yield (%)[a]
1	Br	MeOC—⟨ ⟩—⟨ ⟩	–	11
2	Br	MeOC—⟨ ⟩—⟨ ⟩	PPh_3	>99
3	Br	MeOC—⟨ ⟩—⟨ ⟩	PCy_3	9
4	Br	MeOC—⟨ ⟩—⟨ ⟩	$P(Fur)_3$	68
5	Br	⟨ ⟩—⟨ ⟩	PPh_3	91
6	Br	MeO—⟨ ⟩—⟨ ⟩	PPh_3	82
7	Cl	MeOC—⟨ ⟩—⟨ ⟩	PPh_3	4

[a]By GC.

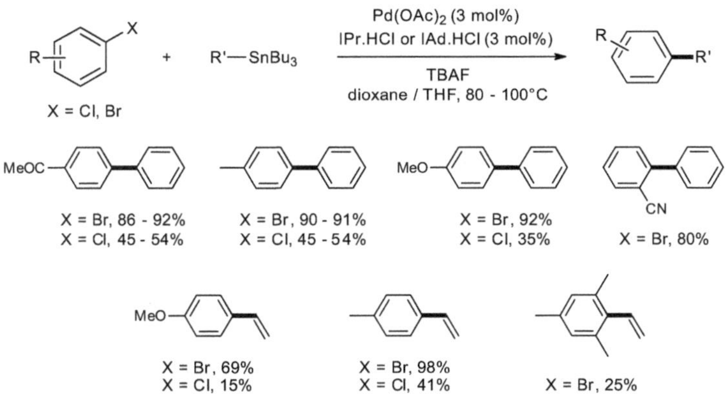

Scheme 4.28 Stille cross-coupling using an *in situ*-generated system.

group on the *N*-substituents of the carbene ligand was also found to be crucial to promote the reaction efficiently.

In 2001, Grasa and Nolan developed an *in situ*-generated system consisting of Pd(OAc)$_2$ and IPr · HCl or IAd · HCl in the presence of TBAF.[98] In this case, TBAF has multiple roles. It acts as a base and a tin transfer reagent but also aids in the removal of the tin by a simple water extraction. Both ligands proved active for the coupling of electron-neutral and -deficient aryl bromides with phenylstannane (Scheme 4.28). IPr · HCl was nevertheless preferred for coupling bromides with vinyltributylstannane.

After the award of the Nobel Prize in Chemistry in 2010, only one report of Pd–NHC-catalysed Stille cross-coupling was reported, by Organ and co-workers in 2010.[99] [Pd(IPent)(PEPPSI)] (**4**) proved particularly active in the reaction, outperforming [Pd(IPr)(PEPPSI)] (**15**). The new system helped solve some of the classical issues of Stille coupling. Reaction could be achieved using aryl chlorides at relatively low temperature (60–80 °C). This lower temperature allowed the toleration of more sensitive heterocycles. As a consequence, a wide and interesting library of heterocycles was prepared (Scheme 4.29). Nevertheless, it is noteworthy that one drawback of this new protocol was the need for high catalyst loadings (4–8 mol%).

4.3.1.5 Hiyama Coupling

Hiyama cross-coupling involves the reaction between a halide and an organosilicon derivative.[100] Despite the non-toxicity and the stability of organosilicon compounds, Hiyama cross-coupling reactions have only been rarely investigated using Pd–NHC catalytic systems. In this section, we cover most of the major achievements in the field prior to and after the award of the 2010 Nobel Prize in Chemistry.

In 2000, Lee and Nolan were the first to examine the effect of NHC ligands on the reaction.[101] The reaction employed an *in situ*-generated system based on Pd(OAc)$_2$ and IPr · HCl in the presence of TBAF as promoter, in

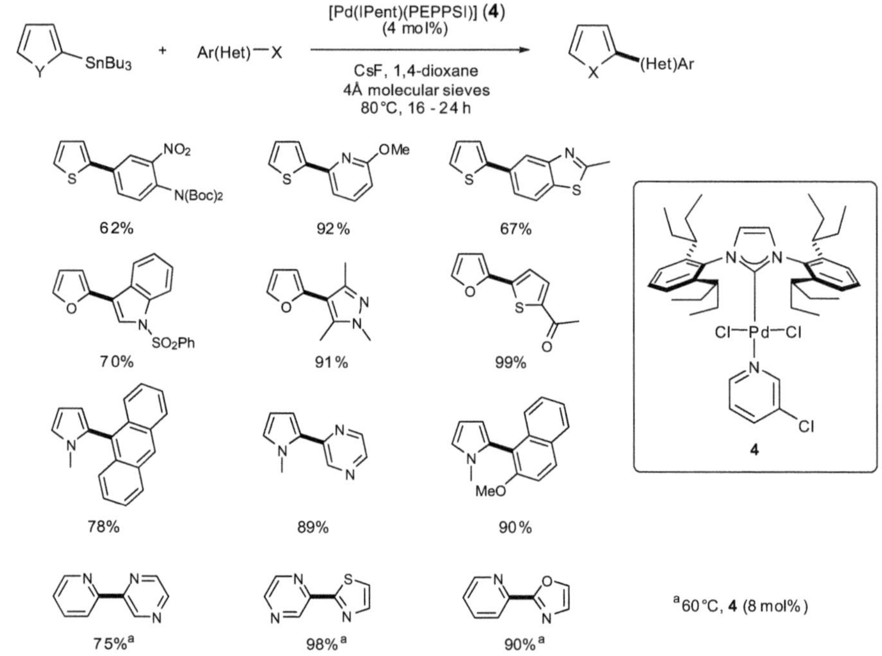

Scheme 4.29 Arylation of heterocycles by Stille cross-coupling using [Pd(IPent)-(PEPPSI)] (**4**).

Table 4.13 Comparison between phosphine and NHC ligands for Hiyama cross-coupling.

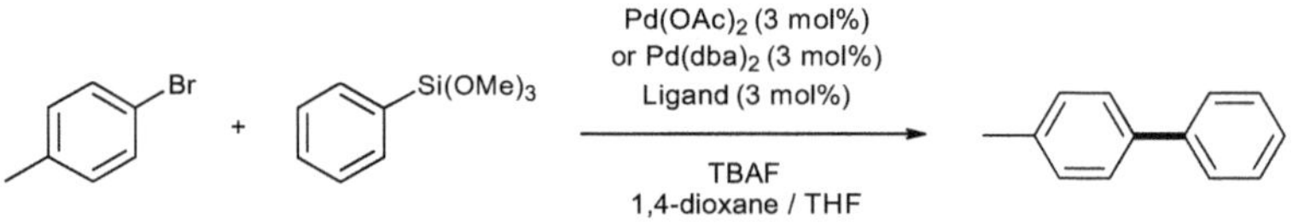

Entry	[Pd]	L	Time (h)	Yield (%)[a]
1	Pd(dba)$_2$	PCy$_3$	1.5	100[b]
2	Pd(dba)$_2$	P(o-tol)$_3$	1	100[b]
3	Pd(OAc)$_2$	IMes · HCl	2	60[c]
4	Pd(OAc)$_2$	IPr · HCl	6	93[c]

[a]By GC.
[b]80 °C, 2 equiv. Ph(SiOMe)$_3$.
[c]60 °C, 3 equiv. Ph(SiOMe)$_3$.

1,4-dioxane–THF as solvent at 80 °C. The system proved particularly efficient for the cross-coupling between aryl bromides and electron-deficient chlorides with phenyl- or vinyltrimethoxysilane (Scheme 4.30). In terms of ligand selection, IPr was found to be better than IMes and as efficient as PCy$_3$ or P(o-tol)$_3$ (Table 4.13).

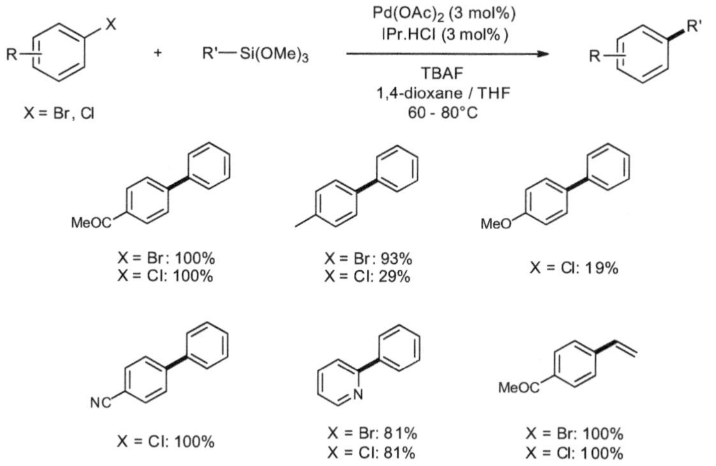

Scheme 4.30 Reactivity of the *in situ*-generated system in Hiyama cross-coupling.

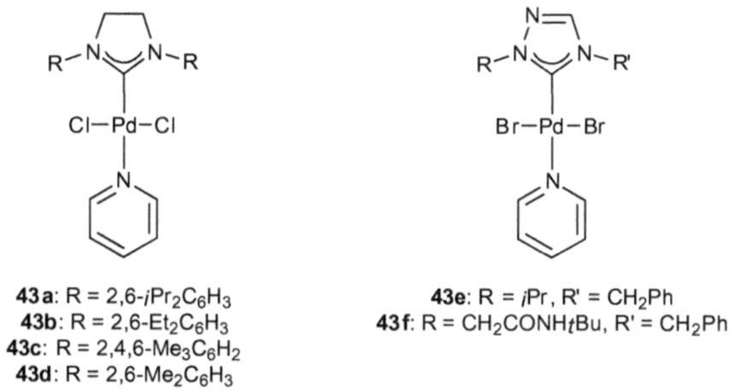

43a: R = 2,6-*i*Pr₂C₆H₃
43b: R = 2,6-Et₂C₆H₃
43c: R = 2,4,6-Me₃C₆H₂
43d: R = 2,6-Me₂C₆H₃

43e: R = *i*Pr, R' = CH₂Ph
43f: R = CH₂CONH*t*Bu, R' = CH₂Ph

Figure 4.20 Structure of the well-defined complexes used in Hiyama coupling.

It was not until 2009 that a second report of Hiyama coupling using a Pd–NHC system appeared, when Ghosh and co-workers reported the use of well-defined PEPPSI-type precatalysts in the reaction (**43**, Figure 4.20).[102] Various improvements to the original protocol were outlined: the use of a fluoride source was not necessary to promote the reaction and the coupling occurred in air and in a mixed aqueous medium (dioxane–water). Because of the similar electronegativities of Si and C, the silane is a poor nucleophile. This explains why Hiyama coupling generally requires fluoride anions in order to proceed. Nevertheless, in this report, NaOH promoted the reaction. Using this system, aryl bromides could be efficiently coupled to phenylsiloxane; however, 4-chloroacetophenone was found to be unreactive and vinyl-trimethoxysilane reacted in a sluggish manner (Table 4.14).

In the same year, Chen and co-workers reported the use of complex **44** in the cross-coupling (Scheme 4.31).[103] The reaction between 4-chloro-acetophenone and phenyltrimethoxysilane was found to be successful in this

Table 4.14 Hiyama coupling using well-defined Pd–NHC complexes.

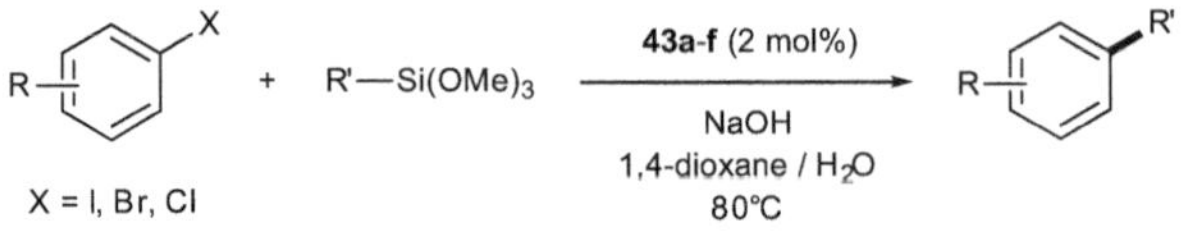

Entry	X	Product	Yield (%)[a]					
			43a	**43b**	**43c**	**43d**	**43e**	**43f**
1	Br	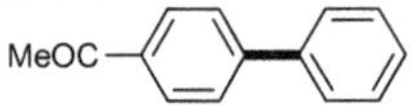 MeOC— (biphenyl)	>99	>99	>99	>99	>99	96
2	Cl	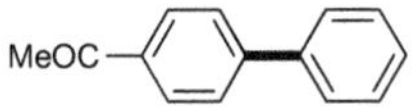 MeOC— (biphenyl)	30	7	14	22	31	7
3	Br	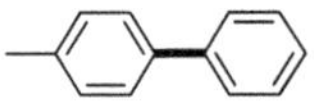 (biphenyl)	>99	80	>99	>99	>99	>99
4	Br	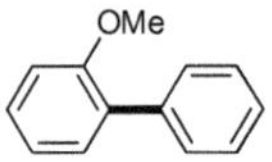 OMe (biphenyl)	61	72	81	89	86	93
5	Br	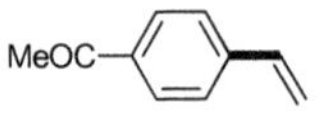 MeOC— (styryl)	15	28	41	45	17	22
6	I	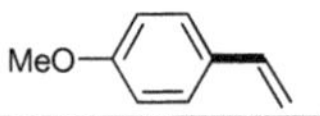 MeO— (styryl)	41	29	27	22	32	26

[a]By GC.

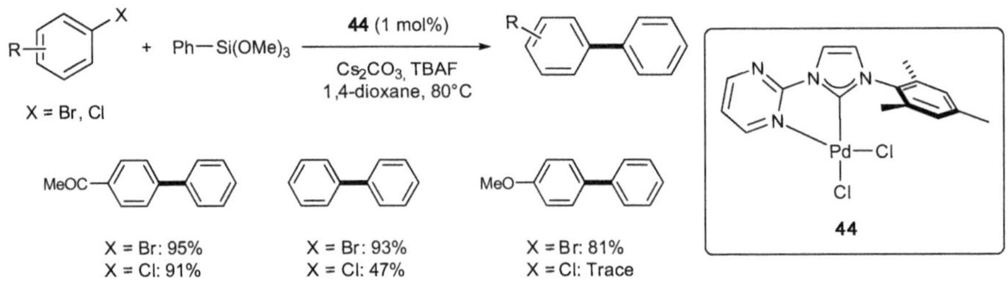

Scheme 4.31 Reactivity of complex **44** in Hiyama cross-coupling.

case in the presence of TBAF. Unfortunately, unactivated and deactivated chlorides were unreactive.

After 2010, Yus and co-workers developed hydroxy-functionalized imidazolium salts for microwave-assisted Hiyama coupling (**45**, Table 4.15).[104] Like Ghosh and co-workers' system, the reaction uses aqueous NaOH instead of TBAF as the reaction promoter. Microwave irradiation considerably accelerated the coupling. Surprisingly, the *in situ*-generated system proved slightly more active than the well-defined complex and the authors preferred to use it to explore the scope of the reaction. A wide range of (hetero)biaryls were thus prepared efficiently using 0.1 mol% of Pd(OAc)$_2$, 0.2 mol% of the NHC ligand and NaOH (50% aq.) under microwave irradiation for 1 h. In some cases, TBAB was also used as an additive to increase the reaction yield.

Table 4.15 Reactivity of **45** in the microwave-promoted Hiyama reaction.

Entry	X	Product	45 (mol%)	Additive	Yield (%)
1	Br	MeO–⬡–⬡	0.2	–	95
2	Cl	MeO–⬡–⬡	0.2	TBAB	56
3	Br	⬡–⬡	0.2	–	86
4	Br	HOOC–⬡–⬡	0.2	–	67
5	Br	N⬡–⬡	0.2	–	48
6	Br	N⬡–⬡	0.2	TBAB	79
7	Br	S⬡–⬡	0.2	TBAB	17
8	Br	S⬡–⬡	0.5	TBAB	52
9	Cl	⬡–⬡	0.5	TBAB	53
10	Cl	⬡–⬡	0.5	TBAB	66

Notably, the deactivated 4-chloroanisole was successfully coupled with phenyltrimethoxysilane in this case. Later, in 2013, the same group studied the optimal ratio between Pd and NHC to promote the reaction.[105] A ratio of 1:5 was found to be beneficial in a number of examples.

In 2012, Lu and co-workers reported the use of [Pd(IPr)(Im)Cl$_2$] (**16a**) in Hiyama coupling.[106] Reaction was promoted by TBAF · 3H$_2$O using 1 mol% of the Pd precatalyst at 120 °C in toluene. Mainly activated aryl chlorides were coupled with arylsiloxanes with good overall yields (Scheme 4.32).

Yang and Wang prepared a series of dinuclear Pd–NHC precatalysts of formula [[Pd(NHC)Cl$_2$]$_2$(μ-L)xCH$_2$Cl$_2$] for Hiyama cross-coupling.[107] In this formula, L is a linker being pyrazine or DABCO (Figure 4.21). The new systems were more active than the *in situ* systems described by Lee and Nolan[101]

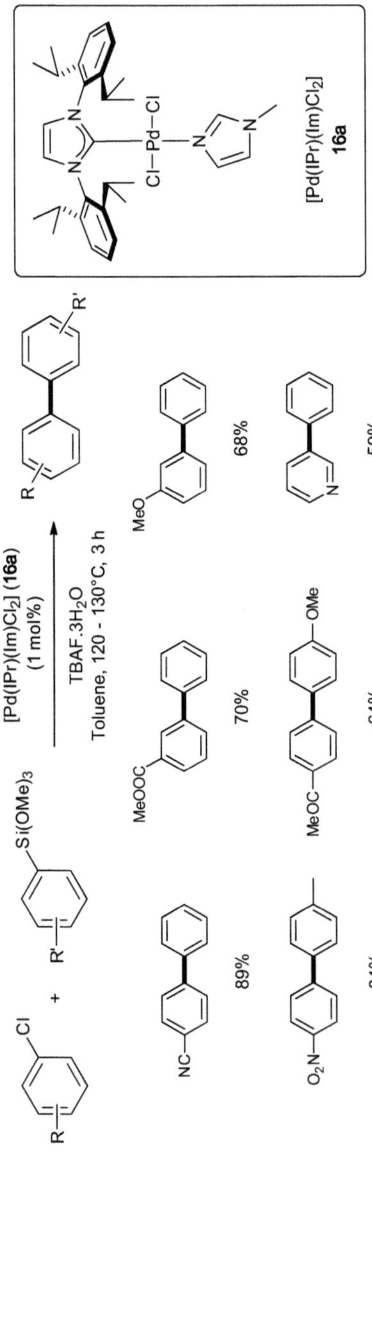

Scheme 4.32 Reactivity of [Pd(IPr)(Im)Cl₂] (**16a**) in Hiyama coupling.

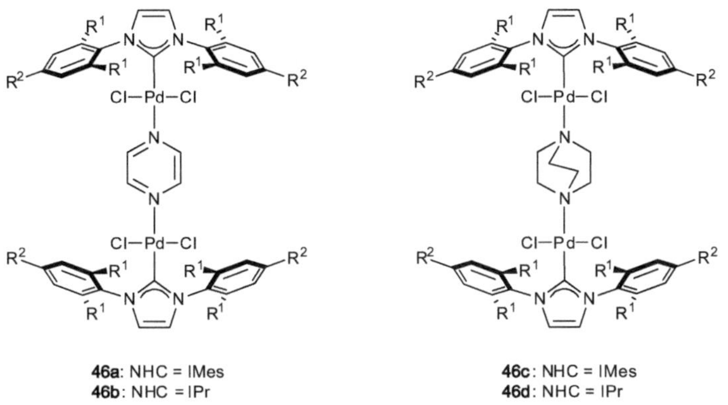

Figure 4.21 Structure of the PEPPSI-type dinuclear palladium complexes **46a–d** using pyrazine and DABCO linkers.

Table 4.16 Comparison of the activities of the new complexes with [Pd(IPr)(PEPPSI)] and some *in situ*-generated systems.

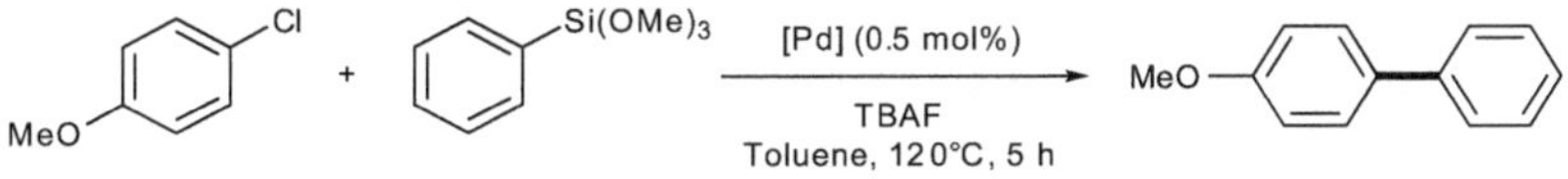

Entry	[Pd]	Yield (%)
1	**46a**	80
2	**46b**	88
3	**46c**	78
4	**46d**	84
5	[Pd(IPr)(PEPPSI)] (**15**)	86
6	Pd(OAc)$_2$ + IMes · HCl	43
7	Pd(OAc)$_2$ + IPr · HCl	51

and as good as [Pd(IPr)(PEPPSI)] for the challenging coupling between 4-chloroanisole and phenyltrimethoxysilane (Table 4.16). The scope of the reaction was studied and was found to be similar to that in previous reports in the field, affording the expected biaryl compounds in good yields with a low catalyst loading of 0.5 mol% and the use of TBAF as the promoter.

4.3.2 Heck Reaction

Heck cross-coupling is one of the most widely employed reactions to create C–C bonds. It involves the reaction between a halide or pseudohalide and an alkene.[6,108] The coupling has found many applications through the years, notably for the preparation of industrially applicable compounds[3] and natural products.[2] The mechanism of the reaction is depicted in Scheme 4.33.[9d,109] The first steps of the catalytic cycle are similar to those

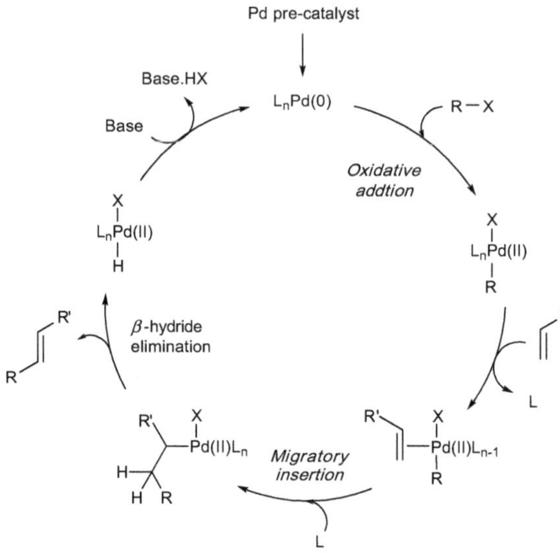

Scheme 4.33   General mechanism of the Heck reaction.

described earlier in Section 4.3.1. Indeed, the precatalyst is first activated to its active Pd(0) form, then oxidative addition of the halide or pseudohalide takes place. As has already been discussed, chlorides are both the most difficult but also the most desirable substrates to activate. The following step is fundamentally different, as it involves the coordination of the alkene to the metal centre instead of a transmetallation. Next, a migratory insertion allows the formation of the desired C–C bond. Finally, the product is re-leased after β-hydride elimination and the Pd(0) active species is regenerated by the action of a base.

4.3.2.1   The Beginning

The use of Pd–NHC systems in the Heck reaction was pioneered by Herrmann in 1995[19a] and 1998.[19b] Using complexes **47a** and **47b** and an *in situ*-generated system (**47c**), the coupling of deactivated bromides, acti-vated bromides and activated chlorides with *n*-butyl acrylate was successfully achieved at high temperatures (Figure 4.22, Table 4.17). The stability of the Pd–NHC bond was found to be essential to the success of the reaction at such high temperatures. The use of the *in situ*-generated system even per-mitted TONs up to 250 000. No induction period was observed in com-parison with the well-defined system. Notably, for the coupling of aryl chlorides, [N(*n*Bu)$_4$]Br was used as an additive, acting as a reducing agent.

   After this discovery, the Heck reaction became a proof of concept reaction for assessing the reactivity of new catalytic systems, like the Suzuki–Miyaura reaction. Many reports exist on the reactivity of Pd–NHC complexes in the Heck reaction and very interesting reviews already cover most of the work.[9c,d]

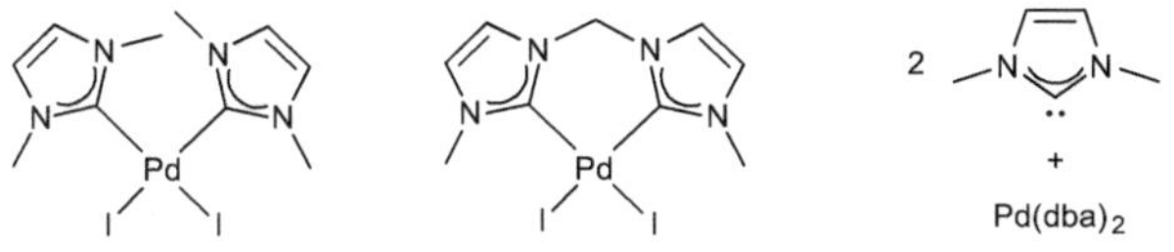

Figure 4.22 Pd–NHC catalytic systems used by Herrmann and co-workers for the Heck reaction.

Table 4.17 Heck reaction using catalysts **47a–c**.

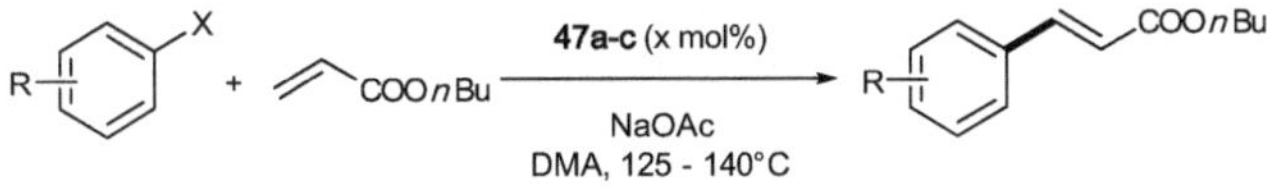

Entry	Aryl halide	Catalyst (x mol%)	Time (h)	Turnover (%)
1	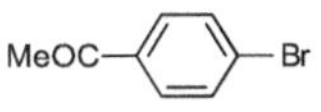 MeOC—⟨⟩—Br	**47a** (0.5)	10	>99
2	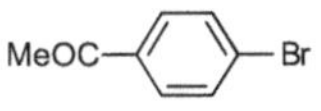 MeOC—⟨⟩—Br	**47b** (0.5)	10	>99
3	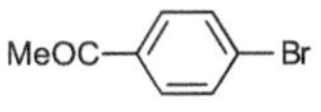 MeOC—⟨⟩—Br	**47c** (0.1)	3	>99
4	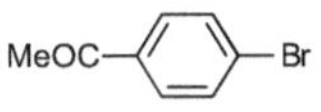 MeOC—⟨⟩—Br	**47c** (0.0004)	43	>99
5	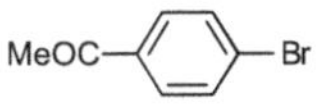 MeOC—⟨⟩—Br	**47c** (0.0002)	96	>66
6	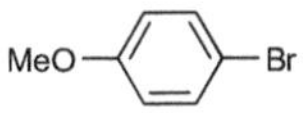 MeO—⟨⟩—Br	**47c** (0.002)	8	78
7[a]	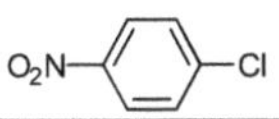 O$_2$N—⟨⟩—Cl	**47a** (0.1)	36	>99

[a][N(nBu)$_4$]Br was used as an additive.

In this section, we focus on the main developments and main challenges in the field after the award of the 2010 Nobel Prize in Chemistry.

4.3.2.2 Development of Recyclable Catalytic Systems

In 2010, one of the challenges for the reaction was to immobilize the Pd–NHC catalysts on solid supports. For example, Garcia-Verdugo and co-workers reported on the preparation of Pd catalysts immobilized on gel-supported ionic liquid-like phases (*g*-SILLPs).[110] The systems were both active in the Heck reaction and recyclable. The polymeric material was found to limit palladium leaching at high temperature, based on a *release and catch strategy*, which means that the palladium can leach but is directly recaptured by the support. Pawar and Buchmeiser described the preparation of a polymer-supported carbon dioxide-protected NHC.[111] The polymer could be loaded with a range of transition metals, including Pd. The polymer-supported catalyst was then able to promote Heck coupling with TONs up to 100 000. Bergbreiter *et al.* explored the reactivity of two polyisobutylene (PIB)-supported Pd–NHC catalysts in the Heck reaction (48, Figure 4.23).[112] The PIB-bound NHC complexes were found to be selectively soluble in heptane *versus* acetonitrile or DMF. This allowed the catalysts to be recovered by phase separation and reused in a new catalytic run. Complex 48a proved active in the coupling but could not be recycled more than three times. On the other hand, 48b was found to be less reactive but more recyclable; indeed, it could be used 10 times without any significant loss of reactivity.

In 2012, Pleixats and co-workers developed recyclable hybrid silica-based precatalysts for the Heck reaction (49, Figure 4.24).[113] The systems promoted the reaction between 4-bromoacetophenone and *n*-butyl acrylate with relatively short reaction times and high yields (Table 4.18). They were also successfully recycled up to five times.

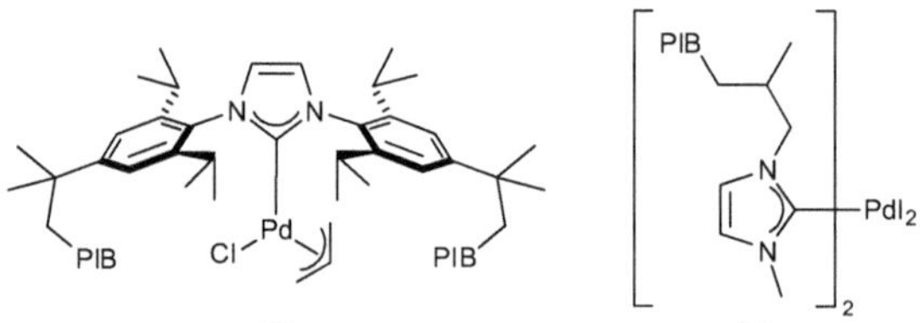

Figure 4.23 PIB-bound Pd–NHC complexes for Heck reactions.

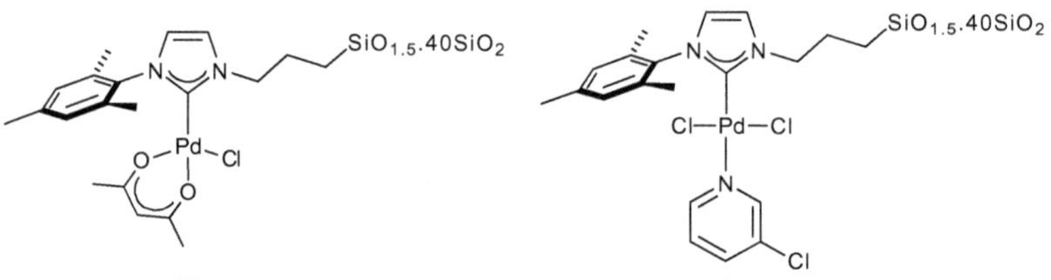

Figure 4.24 Structures of silica-based catalysts for Heck reactions.

Table 4.18 Activity and recyclability of 49a and 49b in the Heck reaction.

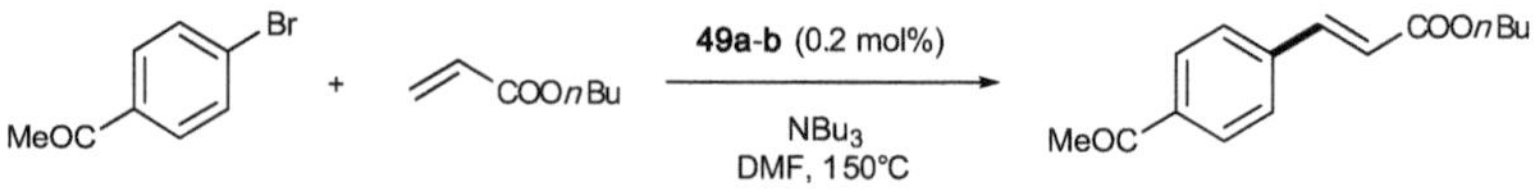

Cycle	49a		49b	
	Time (h)	Yield (%)	Time (h)	Yield (%)
1	3	98	3	98
2	2	99	2	99
3	3	98	3	99
4	3	96	3	98
5	3	97	3	95

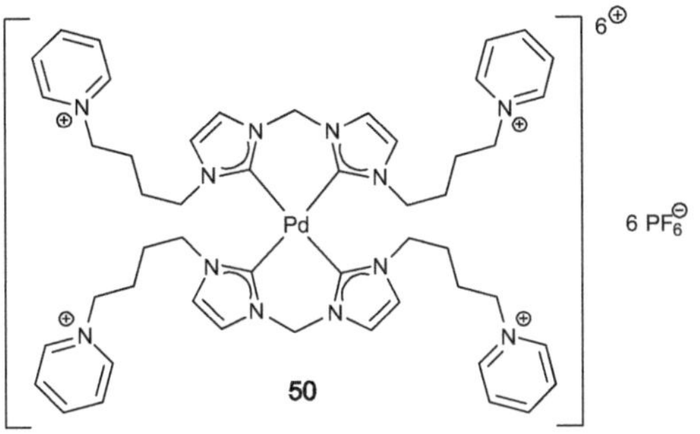

Figure 4.25 Structure of palladium catalysts bearing carbene ligands tagged with pyridinium cation.

Another significant contribution was reported by Hwang and co-workers.[114] Palladium catalysts bearing carbene ligands tagged with pyridinium cations were developed (**50**, Figure 4.25). Their use permitted the olefination of various aryl bromides and also chlorobenzene using NaOAc in DMF or Cs_2CO_3 in water. The catalytic system could be recycled up to eight times.

4.3.2.3 Development of New Well-Defined Precatalysts

Shi and co-workers prepared binuclear palladium complexes bridged by dicarbenes with linkers of different lengths (**51**, Figure 4.26).[115] The reaction between styrene and bromobenzene was studied. Complexes featuring linkers constituted of two or four carbons were the most efficient at promoting the coupling. Huynh and Jothibasu reported the preparation of three *cis*-chelating di-NHC–palladium complexes of formula $[PdX_2(di\text{-}NHC)]$,[116] where X is an anionic co-ligand (**52**, X = I, SCN or CF_3CO_2, Figure 4.26). The reactivities of the three complexes were examined and compared, notably in the Heck coupling reaction. The results revealed an increasing efficiency of the precatalysts in the order SCN $<$ I $<$ CF_3CO_2. Blakemore *et al.* developed a rare example of bidentate *trans*-chelating NHC ligands.[117] Four new complexes were prepared and were found to be active in Heck coupling (**53**, Figure 4.26). Nevertheless, their reactivity was poor for non- or deactivated bromides and inefficient for chlorides.

Lee and co-workers prepared new Pd(0) and Pd(II) complexes bearing NHCs with amide functionalities (**54**, Figure 4.27).[118] The Pd(0) complexes were found to be highly active in the coupling of activated aryl chlorides using TBAB as the ionic liquid. For the coupling of deactivated aryl chlorides and bulky aryl bromides, the *cis*-Pd(II) compounds were more effective (Table 4.19).

In 2011, Dunsford and Cavell studied the influence of the ring size of the NHCs in some Pd(0)–dvtms (dvtms = divinyltetramethyldisiloxane) complexes (**55**, Figure 4.28).[119] The reactivities of eight complexes were compared

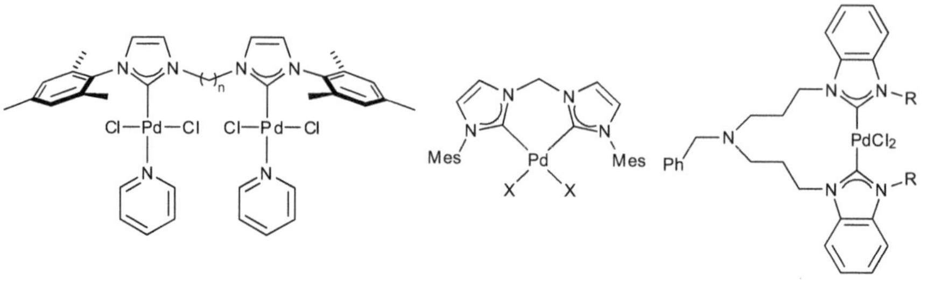

n = 2, 3, 4   X = I, NCS, CF$_3$COO   R = Me, tBu, Bn
51   **52**   **53**

Figure 4.26   Structures of the precatalysts **51**, **52** and **53**.

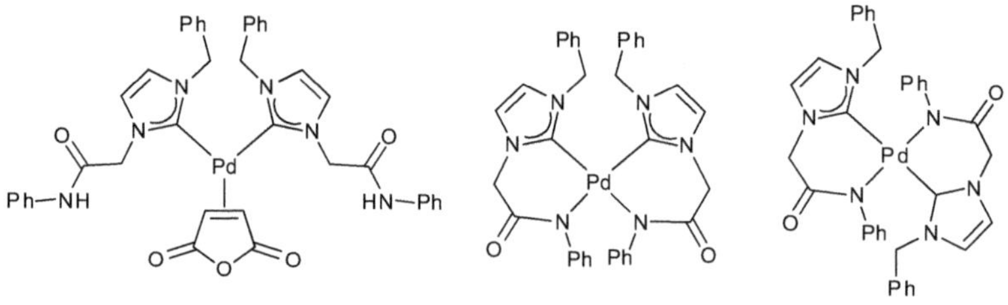

54a   *cis*-**54b**   *trans*-**54b**

Figure 4.27   Structures of some amide-based Pd–NHC complexes.

in the Heck reaction between 4-bromoacetophenone and *n*-butyl acrylate (Table 4.20). It was shown that an increase in ring size correlates with an increase in reactivity. This confirms that increased basicity and steric hindrance have a positive influence on the Heck coupling.

Shao and co-workers described the use of [Pd(IPr)(Im)Cl$_2$] (**16a**) in the Heck reaction (Scheme 4.34).[120] The catalyst was effective in the coupling of various aryl chlorides with styrenes. Both activated and deactivated chlorides were successfully used and yielded the expected alkenes in good yields using TBAB as the ionic liquid. Interestingly, the reaction could be conducted under air.

More recently, Cazin and co-workers determined the reactivity of a range of [Pd(NHC)(μ-Cl)Cl]$_2$ complexes in Heck coupling (**37**, Figure 4.17).[121] SIPr outperformed the other NHC ligands, permitting the coupling of various activated and deactivated aryl bromides at catalyst loadings as low as 0.002 mol%, with TONs up to 49 500 (Table 4.21).

4.3.2.4   Heck Coupling in Water

To avoid the use of toxic solvents, reactions in water are always desirable. In this section, some Pd–NHC systems are presented that allow Heck

Table 4.19 Activity of amide-based Pd–NHC complexes in the Heck coupling of bulky bromides and deactivated chlorides.

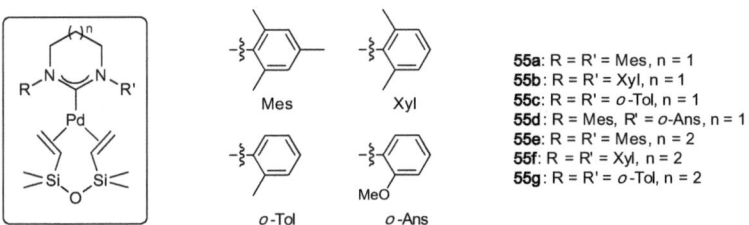

Entry	X	Product	Catalyst (*x* mol%)	Yield (%) (*trans:gem*)
1	Br	MeO / MeO / OMe — Ph	*cis*-**54b** (3)	72
2	Br	(triisopropylphenyl) — Ph	*cis*-**54b** (3)	52
3	Br	(trimethylphenyl) — COO*n*Bu	*cis*-**54b** (3)	76
4	Cl	MeO — Ph	**54a** (0.5)	9 (95:5)
5	Cl	MeO — Ph	*cis*-**54b** (0.5)	65 (91:9)
6	Cl	MeO — Ph	*trans*-**54b** (0.5)	66 (92:8)
7	Cl	MeO — COO*n*Bu	*cis*-**54b** (3)	48 (100:0)
8	Cl	OHC — COO*n*Bu	*cis*-**54b** (0.5)	98 (100:0)

55a : R = R' = Mes, n = 1
55b : R = R' = Xyl, n = 1
55c : R = R' = *o*-Tol, n = 1
55d : R = Mes, R' = *o*-Ans, n = 1
55e : R = R' = Mes, n = 2
55f : R = R' = Xyl, n = 2
55g : R = R' = *o*-Tol, n = 2

Mes   Xyl   *o*-Tol   *o*-Ans

Figure 4.28 Structures of the [Pd(NHC)(dvtms)] complexes **55**.

Table 4.20 Activity of the [Pd(NHC)(dvtms)] complexes in Heck coupling.

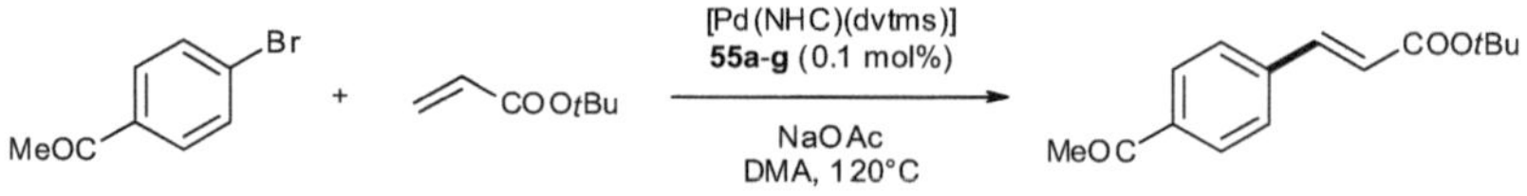

Entry	[Pd(NHC)(dvtms)]	Time (h)	Yield (%)[a]	TOF[b]
1	**55a**	3	61	217
2	**55b**	3	50	287
3	**55c**	3	59	520
4	**55d**	3	85	353
5	**55e**	3	100	1140
6	**55f**	3	100	512
7	**55g**	3	64	443
8	[Pd(IMes)dvtms)]	3	14	8

[a]By GC.
[b]Approximate turnover frequency (TOF) values calculated at 30 min.

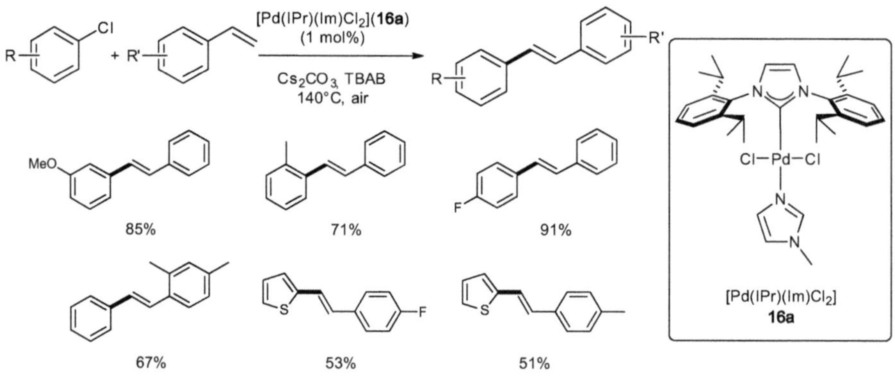

Scheme 4.34 Heck coupling of aryl chlorides using [Pd(IPr)(Im)Cl$_2$].

coupling in water. For example, Luo and Lo reported the preparation of the caffeine-based Pd–NHC complex **56** (Figure 4.29),[122] allowing the coupling of bromoacetophenone and iodobenzene at 90 °C. Shao and co-workers[123] and Ma and Lu,[124] on the other hand, prepared NHCs derived from proline to promote the coupling of acrylic acid, esters and styrenes (**57** and **58**, Figure 4.29).

Another significant contribution was reported by Tu and co-workers.[125] A robust hydrophilic pyridine-bridged bisbenzimidazolylidene–palladium pincer complex was developed (**59**), enabling Heck coupling to be performed in water at very low catalyst loadings. Aryl iodides could be coupled to *tert*-butyl acrylate, generally in good yields, using Et$_3$N as the base and TBAI as an additive at 100 °C (Scheme 4.35). Unfortunately, iodonaphthalene and 3-chloropyridine were found to be poor coupling partners.

Table 4.21 Reactivity of the $[Pd(SIPr)(\mu\text{-Cl})Cl]_2$ (**37a**) complex at very low catalyst loadings.

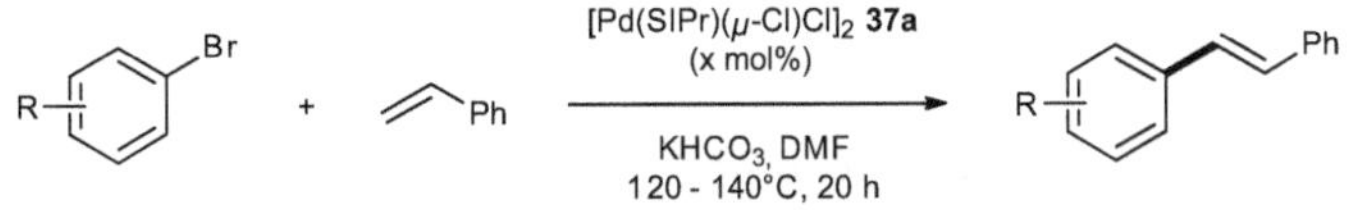

Entry	Aryl halide	[Pd] loading (mol%)	Yield (%)	TON[a]
1	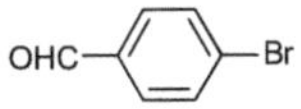 OHC—⟨⟩—Br	0.01	>99	>9900
2	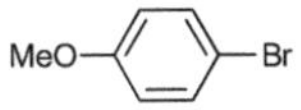 MeO—⟨⟩—Br	0.01	91	9100
3	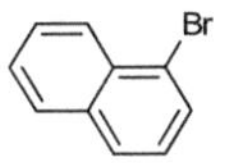 (naphthyl)—Br	0.01	96	9600
4	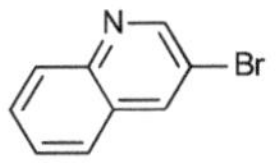 (quinolinyl)—Br	0.02	>99	>4950
5	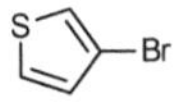 (thienyl)—Br	0.02	66	3300
6	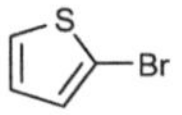 (thienyl)—Br	0.02	42	2100
7	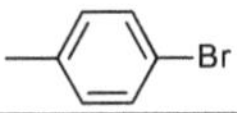 —⟨⟩—Br	0.002	99	49 500

[a]mol product/mol Pd.

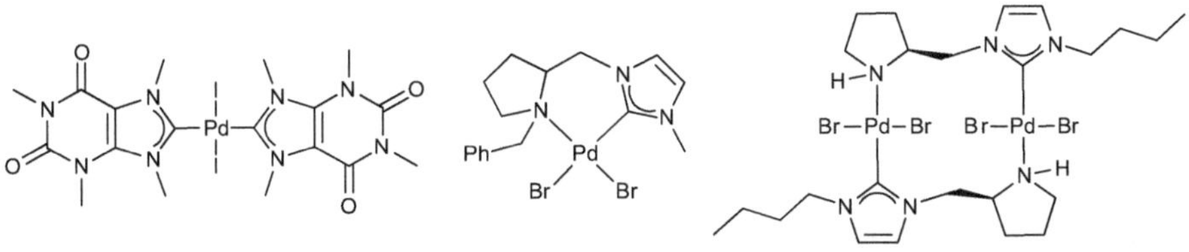

56   **57**   **58**

Figure 4.29 Structures of caffeine- and proline-derived precatalysts.

4.3.2.5 Heck Coupling Using Pseudohalides

The use of pseudohalides is rare in the Heck reaction. However, in 2012, Yus and co-workers reported the use of a hydroxyalkyl-functionalized NHC–Pd(OAc)$_2$ mixed system to promote the Heck–Matsuda coupling between arenediazonium salts and different alkenes (Scheme 4.36).[126] Interestingly in this case, the reaction does not require the use of a base and cyclohexene was found to be a suitable substrate. As an application, the preparation of the biologically active molecule U-77863 in good yield (78%) using 0.5 mol% of Pd(OAc)$_2$ and 1 mol% of the ligand was also reported. Notably, U-77863 has been reported to exhibit anti-invasive and anti-metastatic effects.

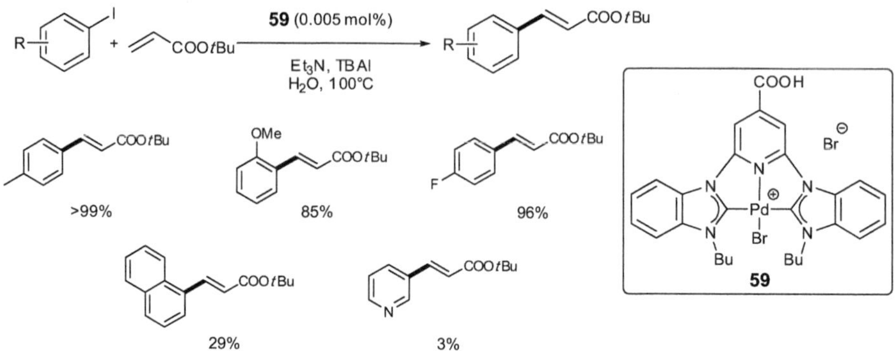

Scheme 4.35   Heck reaction at low catalyst loading in water.

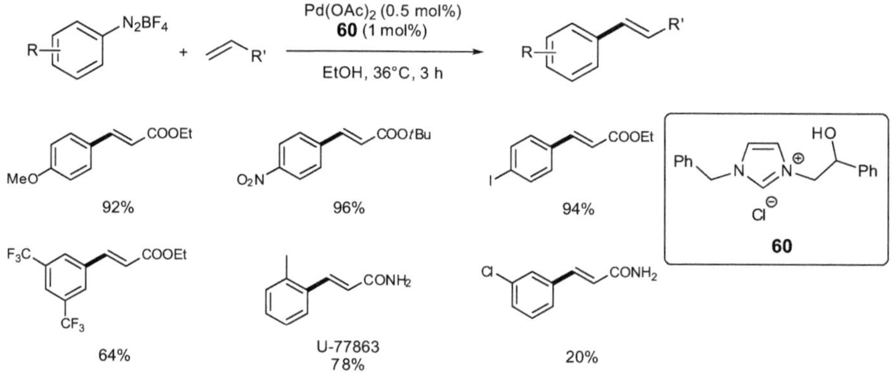

Scheme 4.36   Heck reaction using arenediazonium salts as pseudohalides.

4.3.2.6   Applications

The Heck reaction is widely used to create C–C bonds in the preparation of biologically active compounds. A recent example was reported by Duke and co-workers in 2013.[127] A library of *C*- and *O*-prenylated tetrahydroxystilbenes and *O*-prenylated cinnamates was prepared and the biological activity of the compounds towards K562 cancer cells was studied. The key step of the sequence was a decarbonylative Heck coupling using an *in situ*-formed catalytic system consisting of palladium acetate and IPr · HCl (Scheme 4.37). Compound **61** was identified as a potent growth inhibitor (IC$_{50}$ = 0.10 μM).

4.3.3   Sonogashira Coupling and Heck Alkynylation (Copper-Free Sonogashira Coupling)

The palladium-catalysed cross-coupling reaction between alkynes and aryl halides was discovered in 1975 by three independent research groups: Sonogashira,[128] Heck[129] and Cassar.[130] Heck's and Cassar's groups

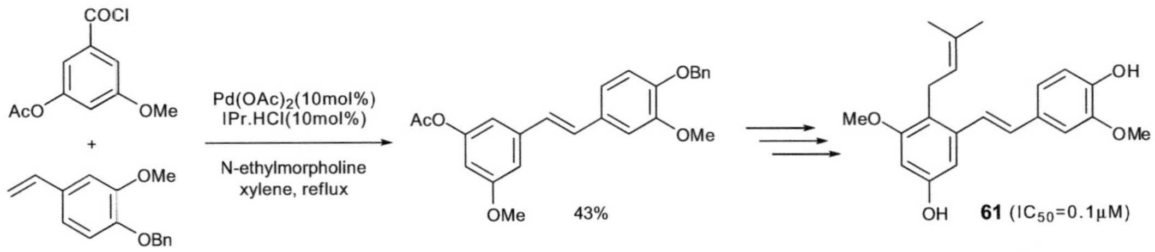

Scheme 4.37 Preparation of the potent growth inhibitor **61** using a Heck reaction catalysed by a Pd–NHC complex as the key step.

employed copper-free conditions to couple aryl bromides at high temperature. On the other hand, Sonogashira *et al.* found that the addition of copper
iodide was beneficial to promote the coupling of aryl iodides at room temperature. Since then, these reactions have been the most commonly employed methodologies to create sp^2–sp bonds.

4.3.3.1   The Beginning

The first investigations of the reaction were not highly successful using
NHCs:[19b,131] the reaction only worked using activated bromides and the
expected products were obtained in only moderate yields. One of the reasons
for this poor reactivity was attributed to the formation of a by-product resulting from the dimerization of the terminal alkyne under the reaction
conditions.[97,132]

   The first highly successful reaction was demonstrated by Caddick *et al.* in
2001 using complex **62**,[133] allowing the coupling between a vinyl iodide and
trimethylsilylacetylene in 85% yield (Scheme 4.38).

   In 2002, Yang and Nolan were also interested in investigating the utility of
Pd–NHC systems in the reaction.[134] The reactivity of *in situ*-generated catalysts constituted of Pd(OAc)$_2$ (3 mol%) and NHC · HCl (6 mol%) was investigated in the coupling between the deactivated bromoanisole and
1-phenyl-2-(trimethylsilyl)acetylene. In this case, the use of a protected
terminal alkyne was found to be necessary to avoid the dimerization of the
alkyne under the reaction conditions.[132] IMes · HCl outperformed the other
NHC ligands in the reaction. Cs$_2$CO$_3$ appeared to be the base of choice to
promote the coupling, with DMAc as the optimum solvent. Notably, Et$_3$N
was found to be completely ineffective in this case. Interestingly, the use of
copper salts could be avoided in some examples. Moreover, chlorobenzene
was also successfully used in the reaction (Scheme 4.39).

   Soon after the aforementioned protocol was published, Batey *et al.* reported on the use of a well-defined *N*-carbamoyl-substituted Pd–NHC complex in the reaction (**63**, Scheme 4.40).[135] Using a 1 : 1 ratio between the Pd–
NHC complex and PPh$_3$, the system allowed the coupling between a range of
activated and deactivated aryl iodides and bromides with terminal alkynes.
Cs$_2$CO$_3$ was used as the base and DMF as the solvent, in the presence of a

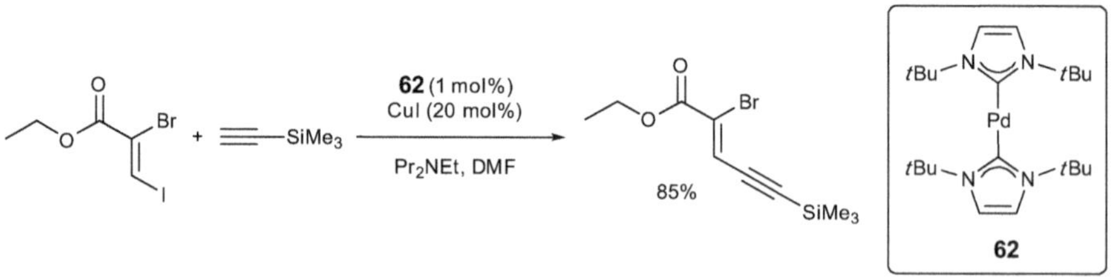

Scheme 4.38   First highly successful Sonogashira reaction using Pd–NHC
complexes.

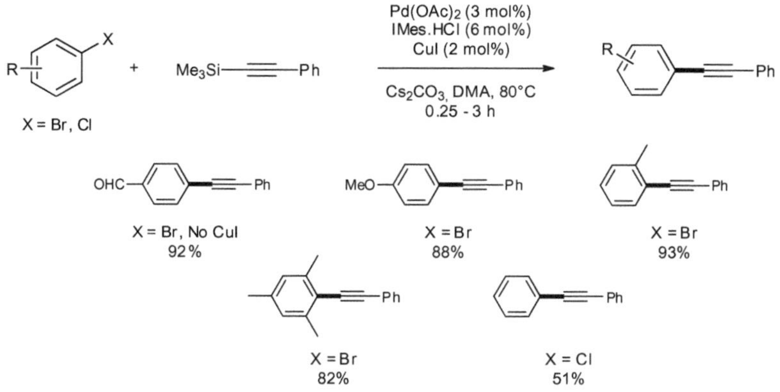

Scheme 4.39 Sila-Sonogashira reaction using an *in situ*-generated system.

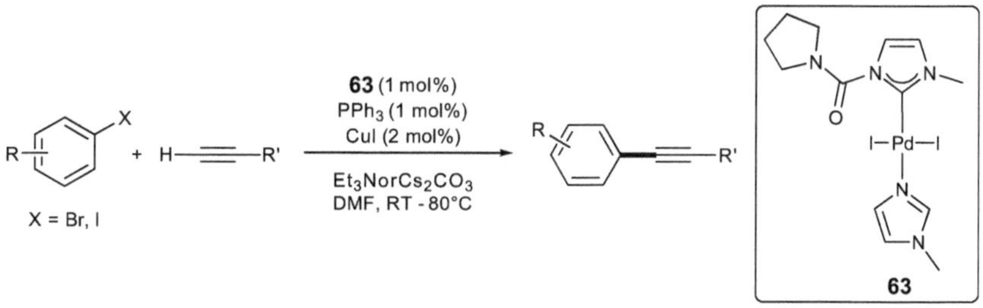

Scheme 4.40 *N*-Carbamoyl-substituted Pd–NHC complexes for the Sonogashira reaction using terminal alkynes.

catalytic amount of CuI to promote the reaction. Under these conditions, the formation of the undesired dimerization product was not mentioned by the authors.

In the following years, other protocols were developed and the main advances can be found in excellent reviews.[9d,136] In the following section, we focus on the main developments after the award of the 2010 Nobel Prize in Chemistry.

4.3.3.2 Alkynylation Reactions in Water

In 2010, Roy and Plenio reported on the preparation of sulfonated NHCs to perform the coupling in water.[60] Since many protocols were already present in the literature for the coupling of aryl bromides, the authors focused their attention on the reactivity of pyridine and thiophene halides (Scheme 4.41). The reaction involved an *in situ*-generated system consisting of Na$_2$PdCl$_4$ (0.25 mol%) and ligand **64** (0.5 mol%). The reaction occurred in a 1 : 1 water–*i*PrOH mixture at 90–95 °C using KOH as the base. Notably, the catalyst loading reported was amongst the lowest at the time for reactions in water. Interestingly, alkylacetylenes were also found to be suitable coupling partners.

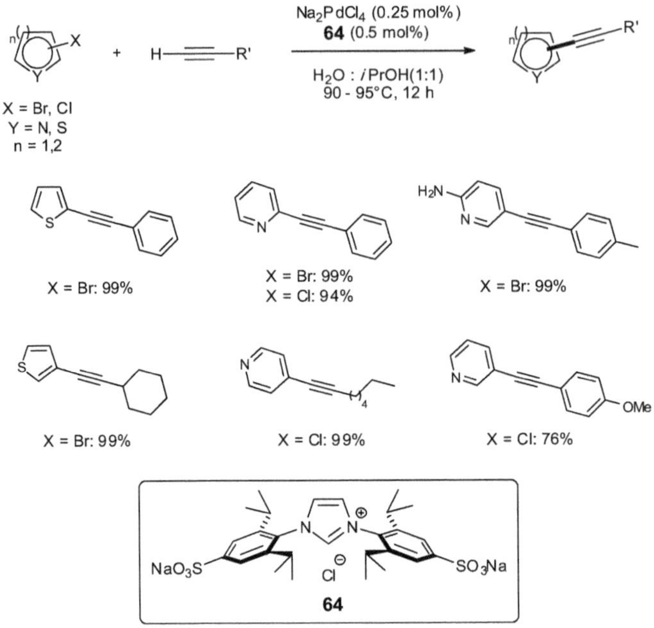

Scheme 4.41 Alkynylation of heterocycles in water using ligand **64**.

4.3.3.3   *Development of Recyclable Systems*

As with other coupling reactions, the development of recyclable systems for the Sonogashira coupling and Heck alkynylation is important. The hybrid silica-based catalysts of Pleixats and co-workers,[113] which were already active in the Heck reaction (**49**, Figure 4.24), also found application in Heck alkynylation. The systems were used to promote the coupling between 4-bromoacetophenone and phenylacetylene under copper-free conditions at 110 °C in DMF and in the presence of nBu$_4$OAc as the base. Initial experiments revealed complete reactivity within 1 h using 0.2 mol% of palladium. The reaction time had to be increased to 4 h after the first cycle, but the catalysts could be used up to five times (Table 4.22). Further experiments allowed the reduction of the catalyst loading to as low as 0.01 mol%, affording high TONs (4800-5400) and TOF (3200-3600 h^{-1}).

Huynh and Lee also developed pincer-type di(1,2,4-triazolin-5-ylidene)–palladium complexes for Heck alkynylation (Figure 4.30).[137] Precatalyst **65** was found to be active for copper- and amine-free coupling between aryl bromides and phenylacetylene. The catalyst loading could be decreased to as low as 0.01 mol% at 120 °C, affording TONs of up to 8000 for a single run. Since pincer-type complexes are known for their high thermal stability, recycling of the catalyst was investigated. More substrates were added in turn to the reaction medium. This could be done up to six times, affording an overall TON of 56 000.

Table 4.22 Activity and recyclability of **49a** and **49b** in Heck alkynylation.

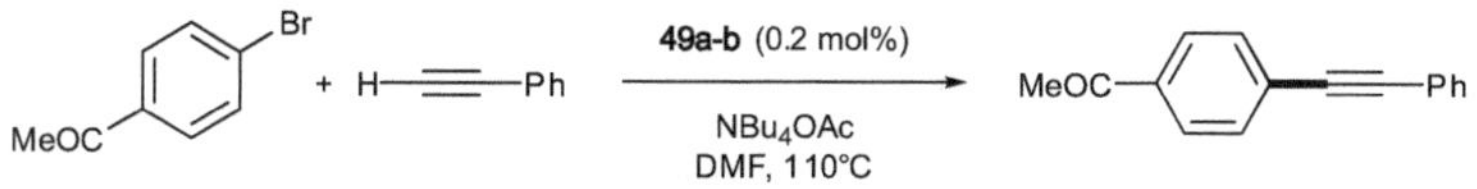

	49a		**49b**	
Cycle	Time (h)	Yield (%)^a[isolated yield]	Time (h)	Yield (%)^a[isolated yield]
1	1	90 [71]	1	87 [73]
2	4	76	4	78
3	4	79	4	82
4	4	84	4	83
5	5	62	4	86

^{a}By GC.

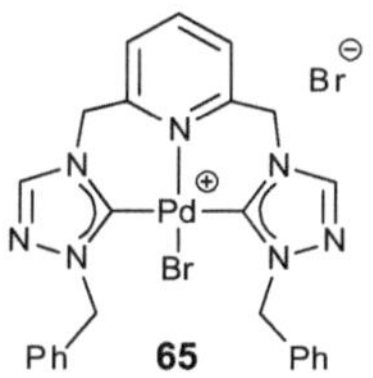

Figure 4.30 Structure of precatalyst **65**.

4.3.3.4 Development of New Pd–NHC Systems

As for other cross-coupling reactions, the modulation of the steric and electronic properties of the NHCs and also the impact of the throwaway ligand were investigated in alkylation reactions. In an attempt to improve the catalytic activity by increasing the bulk of the ligand, Straub and co-workers showed that the use of IPr** was not advantageous in copper-free cross-coupling.[55] Using previously established successful reaction conditions[102] for the coupling of phenyl iodide with phenyl acetylene, complex **20** yielded only 27% of coupling product (Scheme 4.42).

In 2012, Xu *et al.* reported the preparation of NHC-based cyclopalladated ferrocenylchloropyrimidine for Heck alkynylation.[138] Of three new complexes, the IPr-bearing complex was found to be the most active in the coupling reaction (**66**, Figure 4.31). The latter proved to be efficient under copper- and amine-free conditions, allowing the coupling of aryl bromides and activated aryl chlorides with phenyl- and *p*-tolylacetylenes. Optimized reaction conditions were the use of CsOAc as the base in DMAc at 120 °C using catalyst loadings in the range 0.1–2 mol%. Cao and co-workers[139] prepared novel chiral Pd–NHC complexes (**67**, Figure 4.31) and investigated their reactivity in the copper-free Sonogashira reaction. Complex **67a** exhibited the highest catalytic activity. However, despite success with activated bromides, the reaction did not work with chlorides. Interestingly, a similar complex (**67e**) was successfully used in a tandem Sonogashira–hydroarylation

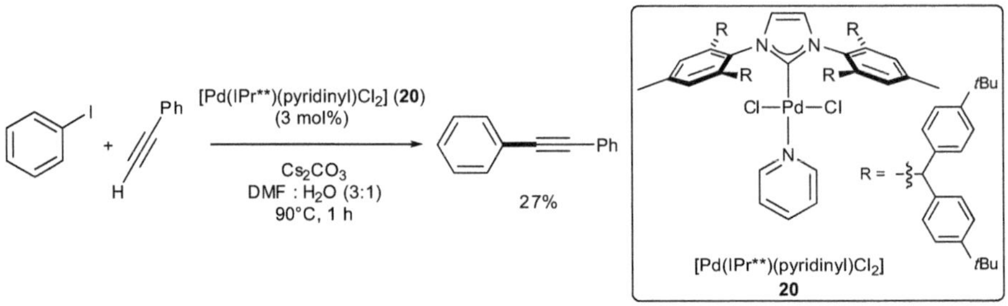

Scheme 4.42 Alkynylation reaction using the very bulky IPr** ligand.

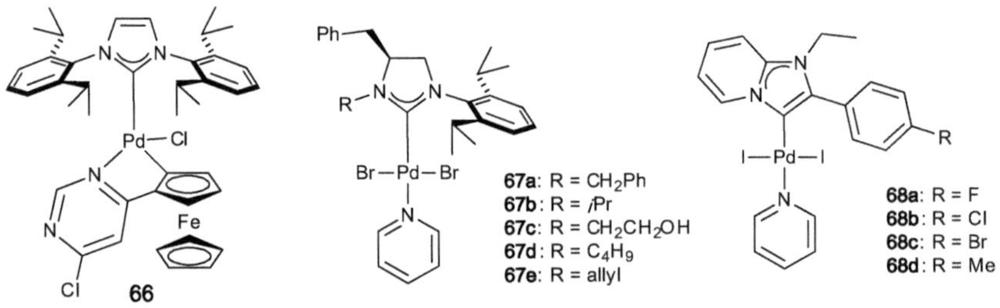

Figure 4.31 Structures of some precatalysts active in alkynylation reactions.

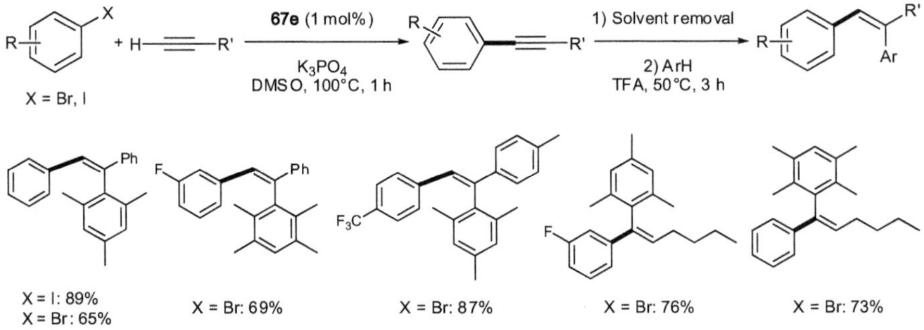

Scheme 4.43 Tandem Sonogashira–hydroarylation reaction using **67e**.

reaction (Scheme 4.43).[140] Finally, Ghosh and co-workers recently described the synthesis of a series of abnormal PEPPSI-type precatalysts for copper- and amine-free coupling under air (**68**, Figure 4.31).[141] However, only iodides and activated bromides could be used as coupling partners (Table 4.23).

4.3.4 Direct Arylation

Direct arylation has recently attracted a lot of attention for the construction of (het)aryl–(het)aryl bonds.[142] Selectivity issues can be encountered, but the

Table 4.23 Activity of complexes **68a–d** in Heck alkynylation.

			Yield (%)[a]			
Entry	X	Product	**68a**	**68b**	**68c**	**68d**
1	I	MeOC–⟨⟩–≡–Ph	99	>99	99	99
2	Br	MeOC–⟨⟩–≡–Ph	94	90	74	50
3	I	MeO–⟨⟩–≡–Ph	53	50	80	54
4	Br	OHC–⟨⟩–≡–Ph	80	81	88	81
5	Br	NC–⟨⟩–≡–Ph	90	87	91	81

[a]By GC.

direct functionalization of a (hetero)aromatic ring is now emerging as an important tool to address the recent need to develop more time-efficient and less wasteful synthetic pathways. In this section, we report the main advances in direct arylation reactions using Pd–NHC complexes, prior to and after the award of the 2010 Nobel Prize in Chemistry.

In 2005, Çetinkaya and co-workers were among the first to investigate the reactivity of Pd–NHC systems in direct arylation reactions.[143] In particular, they described the mono- and bisarylation of benzaldehyde derivatives using aryl chlorides and bromides. Pd–NHC complexes proved particularly active for the arylation of aryl chlorides, which in the past was difficult or not even possible using Pd–phosphine systems. Several *in situ*-generated catalysts were screened in the reaction, resulting from a mixture of Pd(OAc)$_2$ (1 mol%), a choice of five different imidazolium salts (2 mol%) and Cs$_2$CO$_3$ as the base in dioxane at 80 °C. The five systems had similar activities in a number of arylation reactions; however, SIMes · HCl was discovered to be slightly more active than the others and was used in some additional couplings. A large range of polyaryls was prepared in this manner (Scheme 4.44).

In the same year, Fagnou and co-workers studied the reactivity of several well-defined Pd–NHC complexes in the intramolecular formation of five- and six-membered ring biaryls.[144] Complex **69** was found to be the most active precatalyst in the reaction. The addition of IPr · HCl to the reaction medium proved to have a beneficial effect on the catalytic activity, allowing higher TONs to be reached. The scope of the reaction showed the versatility of such a method for the formation of various *N*- and *O*-containing biaryl products in excellent yields (Scheme 4.45).

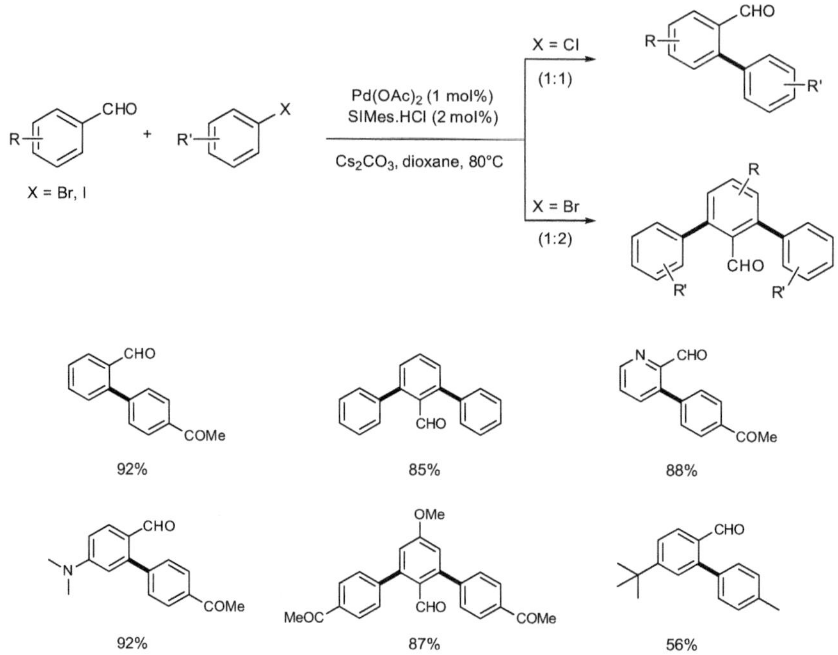

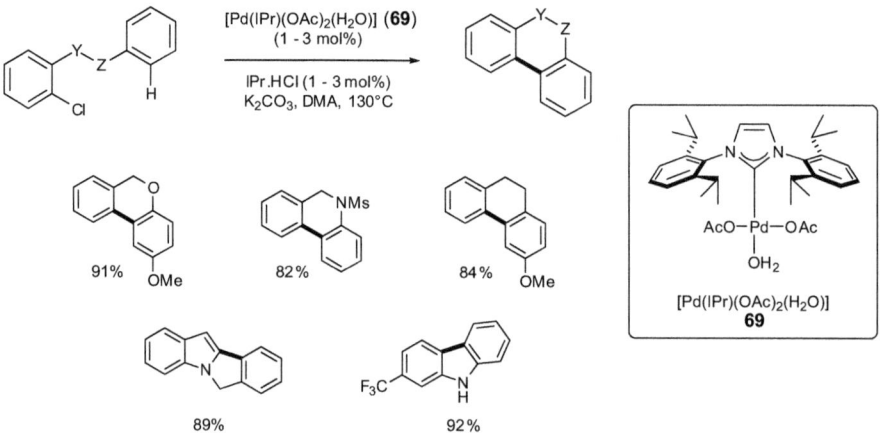

Scheme 4.44 Direct arylation of (hetero)aromatic aldehydes using an *in situ*-generated catalytic system.

Scheme 4.45 Intramolecular direct arylation using [Pd(IPr)(OAc)$_2$(H$_2$O)].

In 2005, Sames and co-workers investigated the reactivity of Pd–NHC systems in the direct arylation of *N*-containing heterocycles. First, they showed the beneficial effect of using IMes instead of PPh$_3$ for the selective C3 arylation of indole.[145] Higher yields and selectivities were observed in

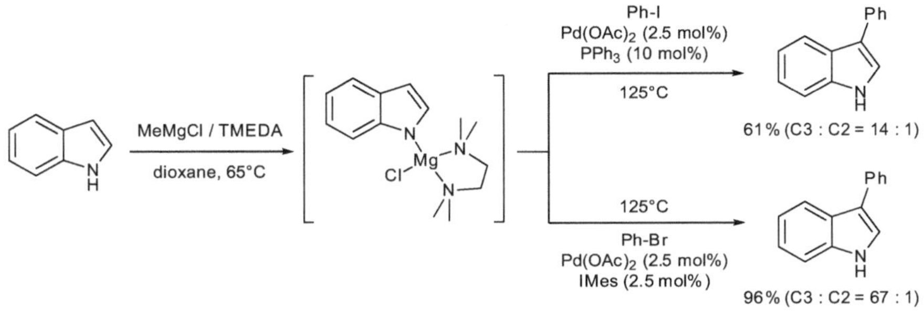

Scheme 4.46 C3 arylation of indole.

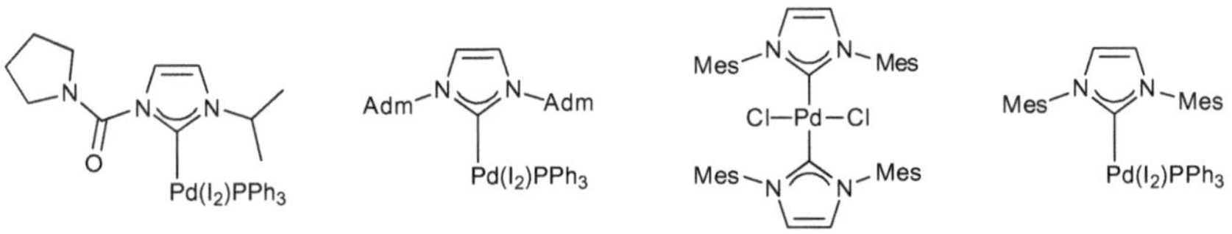

Figure 4.32 Precatalysts studied for the direct arylation of SEM-protected azoles.

comparison with the Pd–PPh$_3$ system and bromobenzene gave better results than iodobenzene (Scheme 4.46).

They then studied the arylation of SEM-protected azoles.[146] The activities of four different catalysts exhibiting different steric and electronic properties (**70**, Figure 4.32) were probed in the reaction between SEM-protected indoles and iodobenzene. Precatalyst **70d** was the most active in the coupling. However, because its synthesis was low yielding, the scope of the reaction was studied using **70a**, the second most active precatalyst. The reaction occurred at 125 °C in DMA in the presence of CsOAc as the base. Catalyst loadings ranged from 0.15 to 5 mol%. Thus, a range of *N*-containing heterocycles were selectively arylated in various positions and globally in good yields. Iodides were found to be better coupling partners than bromides (Scheme 4.47).

In 2010, Sames and co-workers also investigated the reactivity of complex **70a** for the C5 arylation of SEM-protected imidazoles.[147] The complex was active, but less efficient than the system combining Pd(OAc)$_2$ and P(nBu)Ad$_2$ (Scheme 4.48).

In 2006, Sanford and co-workers developed a system for the selective C2 arylation of indoles at room temperature.[148] The reaction took place in AcOH using 5 mol% of [Pd(IMes)(OAc)$_2$] as the precatalyst (Scheme 4.49). In this case, no aryl halide was used in the reaction. Instead, a hypervalent iodine salt was used as the arylating agent. The reaction proceeded smoothly under mild conditions (r.t.–60 °C), affording the expected 2-arylated indoles in good yields. Two examples of pyrroles were also reported.

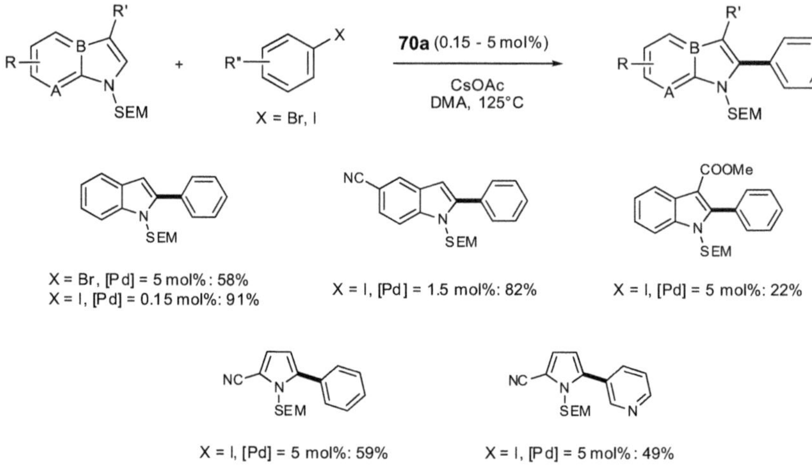

Scheme 4.47 Direct arylation of SEM-protected azoles using **70a**.

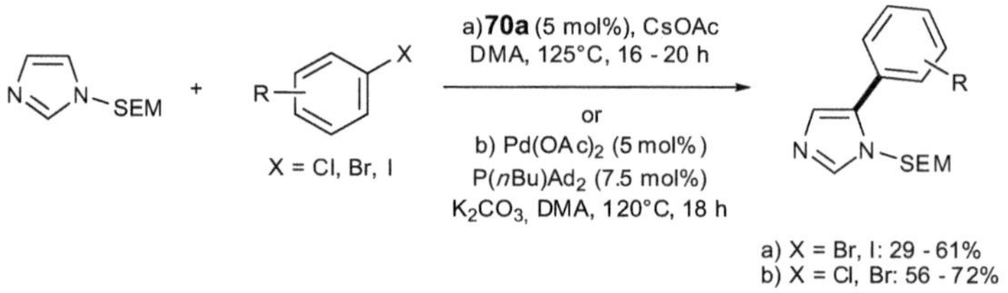

Scheme 4.48 Precatalyst **70a** *versus* Pd(OAc)$_2$/P(nBu)Ad$_2$ for the direct arylation of SEM-protected imidazoles.

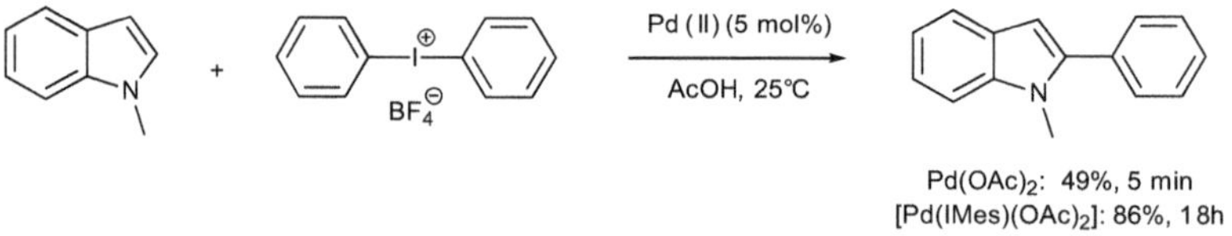

Scheme 4.49 C2-arylation of indoles using hypervalent iodine salts.

In 2010, Özdemir *et al.* started to develop a wide range of new well-defined Pd–NHC complexes for the direct arylation of heterocycles (**71–74**, Figure 4.33). First they studied the C5 direct arylation of furans, thiophenes and thiazoles using **71**, **72** and **73**.[149] Deactivated bromides were found to be suitable coupling partners. Activated aryl chlorides were also successfully used for the first time in a reaction involving a Pd–NHC complex and a heteroaromatic substrate. The corresponding heterocycles were subsequently isolated in moderate to good yields (Scheme 4.50). The various precatalysts did not show significant differences in terms of reactivity in the coupling reaction.

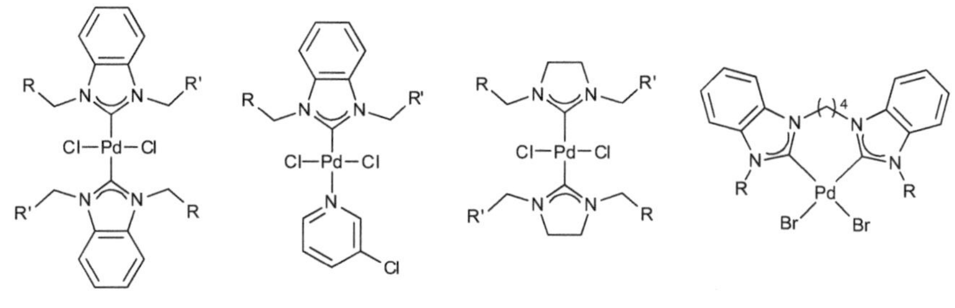

Figure 4.33 Structures of the precatalysts studied for the direct arylation of heterocycles.

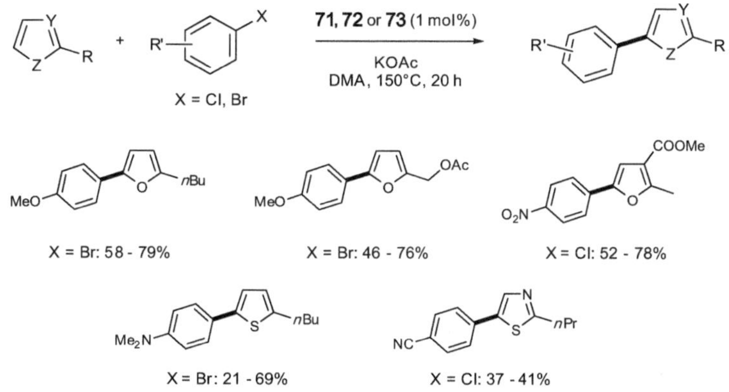

Scheme 4.50 Direct arylation of furans, thiophenes and thiazoles using precatalysts **71**, **72** and **73**.

Later, they studied the activity of complexes **74** for the C2 arylation of benzothiazole.[150] Complex **74a** was identified as an efficient precatalyst, affording the desired 2-arylated benzothiazoles in good yields (Scheme 4.51).

Finally, in a more recent contribution,[151] the arylation of pyrroles was investigated. New complexes of the type **71** were prepared (Figure 4.33), allowing the coupling of deactivated aryl chlorides using KOAc in DMA at 150 °C. Various functionalized pyrroles were prepared in good yields (Scheme 4.52). Interestingly, despite the harsh reaction conditions, aldehyde, ketone, cyano and ester functionalities were well tolerated.

Another significant contribution was made by Lee and co-workers.[152] Mixed phosphine–NHC palladium complexes were prepared for the direct C5 arylation of imidazoles (Scheme 4.53). Complex **75** bearing the electron-rich PCy$_3$ ligand exhibited the highest catalytic activity in the reaction. Notably, the coupling of deactivated 4-chloroanisole was successfully achieved in that case. Microwave irradiation was discovered to reduce the reaction times considerably, from 18 to 2 h.

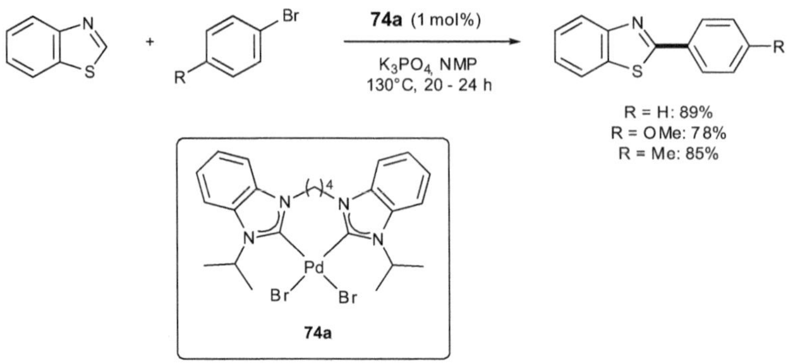

Scheme 4.51   Direct arylation of benzothiazole using **74a**.

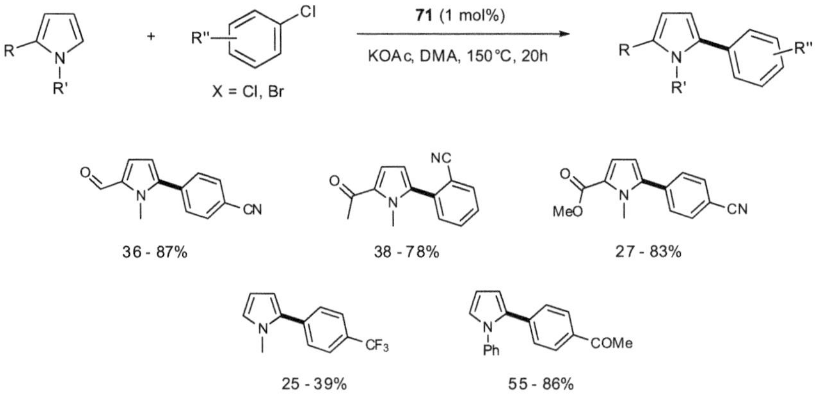

Scheme 4.52   Arylation of pyrroles using deactivated chlorides and **71**.

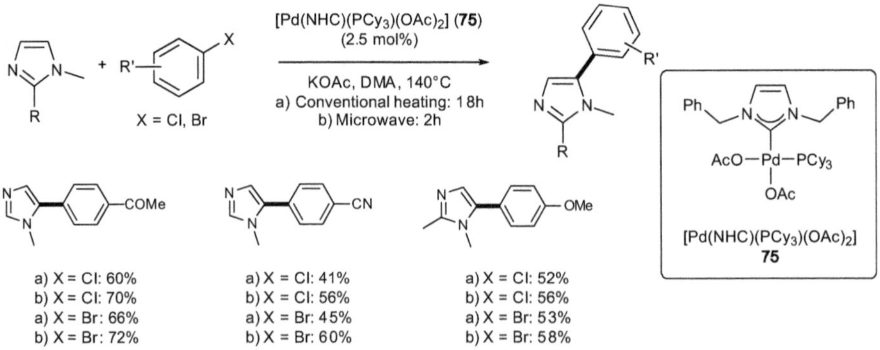

Scheme 4.53   Direct arylation of imidazoles using [Pd(NHC)(PCy$_3$)(OAc)$_2$] (**75**).

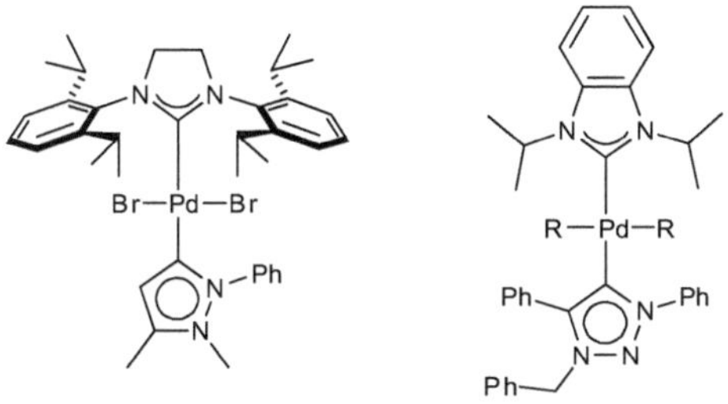

Figure 4.34 Precatalysts for the direct arylation of pentafluorobenzene.

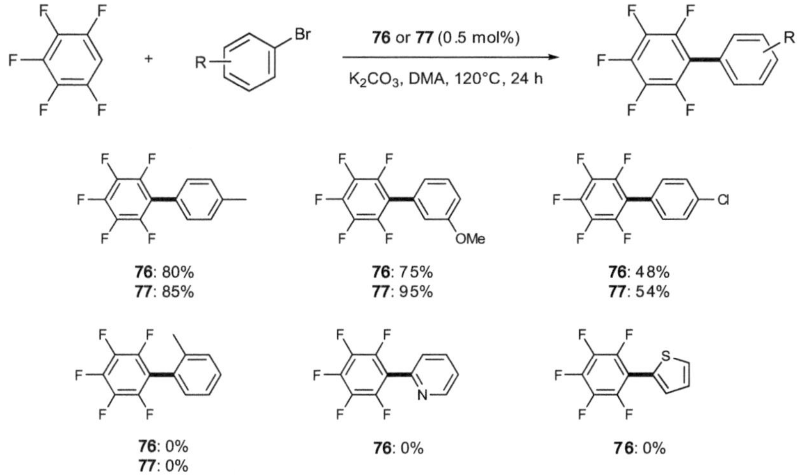

Scheme 4.54 Direct arylation of pentafluorobenzene using **76** and **77**.

Huynh's group investigated the direct arylation of pentafluorobenzene. For this purpose, two series of Pd–NHC complexes were described in two separate publications. In one contribution, the preparation of 10 new pyrazolin-5-ylidene–palladium (II) complexes was reported.[153] Complex **76** (0.5 mol%, Figure 4.34) was found to be the most active precatalyst for the coupling of pentafluorobenzene. The scope of the reaction was limited to the use of *para-* and *meta-*substituted aryl bromides. *Ortho-*substituted and heterocyclic congeners were unreactive (Scheme 4.54). In the other contribution, a series of [PdX$_2$(*i*Pr$_2$-bimy)(trz)] (X = Br or O$_2$CCF$_3$, bimy = 1,3-di-isopropylbenzimidazolin-2-ylidene, trz = 1,2,3-triazolin-5-ylidene) were synthesized.[154] The σ-donation of the trz ligands was calculated by using ^{13}C NMR spectroscopy. The complexes bearing the less-donating trz ligands were found to perform better in catalysis and the trifluoroacetato complexes were found to outperform the bromo congeners. Complex **77** (Figure 4.34)

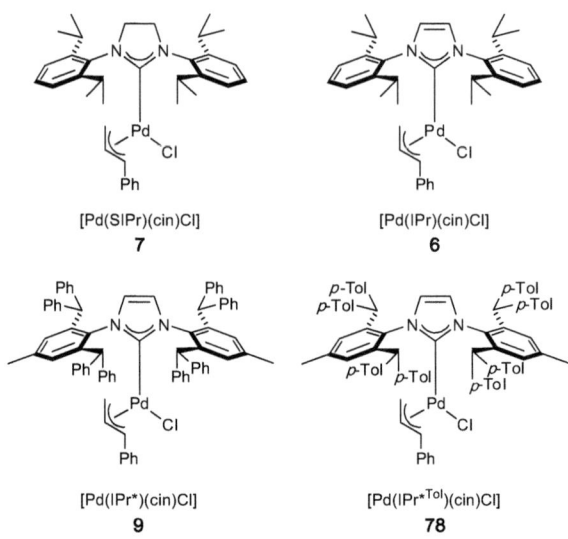

[Pd(SIPr)(cin)Cl]
7

[Pd(IPr)(cin)Cl]
6

[Pd(IPr*)(cin)Cl]
9

[Pd(IPr*^{Tol})(cin)Cl]
78

Figure 4.35 Precatalysts compared in the direct arylation of benzothiophene.

Table 4.24 C omparison of precatalysts under optimized conditions for the direct arylation of benzothiophene.

Precatalyst	Conversion (%)^a
[Pd(SIPr)(cin)Cl] (**7**)	80
[Pd(IPr)(cin)Cl] (**6**)	75
[Pd(IPr*)(cin)Cl] (**9**)	57
[Pd(IPr*^{Tol})(cin)Cl] (**78**)	58

^{a}By GC.

was used to study the scope of the reaction and similar results were obtained to those using **76** (Scheme 4.54).

In 2012, Nolan's group also investigated the use of well-defined Pd–NHC precatalysts in the arylation of heterocycles.[155] The reactivities of four different [Pd(NHC)(cin)Cl] complexes were investigated and the effect of the bulk of the NHC ligand was studied (**6, 7, 9, 78**, Figure 4.35). Bulky ligands (IPr* and IPr*^{Tol}) were not found to be advantageous and SIPr was the most efficient NHC to promote the reaction (Table 4.24). σ-Donation did not play a crucial role, as IPr and SIPr gave similar results. Benzothiophene, thiophene and imidazopyridine were successfully and efficiently arylated at low catalyst loadings (0.1–0.01 mol%) using deactivated, activated and sterically hindered aryl bromides (Scheme 4.55).

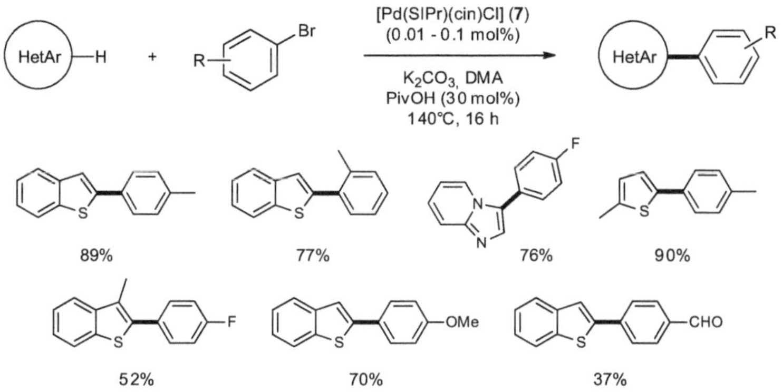

Scheme 4.55 Direct arylation of heterocycles at low catalyst loadings using [Pd(SIPr)(cin)Cl] (**7**).

4.3.5 Arylation of Ketones

The palladium-catalysed α-arylation of ketones is an elegant and effective method to form sp^2–sp^3 bonds.[156] In fact, no tin, zinc or boron derivative is required as a transmetallating reagent. Indeed, in the presence of a base, the enolate that is formed in the reaction medium can act as a suitable pseudo-transmetallating agent. The reactivity of Pd–NHC precatalysts in the reaction has hardly been investigated. In this section, we discuss the main achievements in the field prior to and after the award of the 2010 Nobel Prize in Chemistry.

In 2002, Nolan and co-workers were among the first to investigate the use of a Pd–NHC system in the α-arylation of ketones.[157] The reactivity of several [Pd(NHC)(allyl)Cl] complexes was investigated in the reaction between chlorobenzene and propiophenone. After optimization, the best conditions were found to be the use of 1 mol% of [Pd(SIPr)(allyl)Cl] as the precatalyst, NaO*t*Bu as the base, in THF (Scheme 4.56). The scope of the reaction was studied and the coupling of various deactivated and activated aryl chlorides, bromides and triflates with a wide range of aryl and alkyl ketones was successfully achieved. Notably in some cases, because the system is highly reactive, the presence of bis-arylated products was observed. The use of bulky aryl chlorides allowed this issue to be circumvented.

In 2005, Singh and Nolan investigated the use of a new well-defined complex in the reaction, [Pd(IPr)(OAc)$_2$] (**80**, Scheme 4.57).[158] The system proved highly active for the coupling of various deactivated and activated chlorides with aryl and cyclohexyl ketones. More interestingly, **80** allowed this kind of coupling to be performed at room temperature for the first time.

Nolan's group next continued to screen alternative throwaway ligands to promote the reaction and thus developed palladacycle [Pd(IPr)(dmab)Cl] (**81**, Table 4.25) (dmab = *N,N*-dimethylaminobiphenyl).[159] The α-arylation reaction was highly successful and the catalyst loading could be decreased to 0.25 mol% in refluxing THF using NaO*t*Bu as the base. Moreover, it was

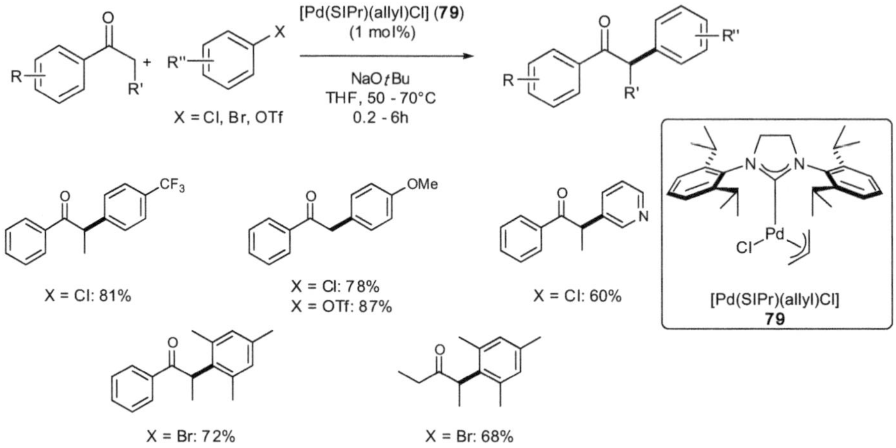

Scheme 4.56   α-Arylation of ketones catalysed by [Pd(SIPr)(allyl)Cl].

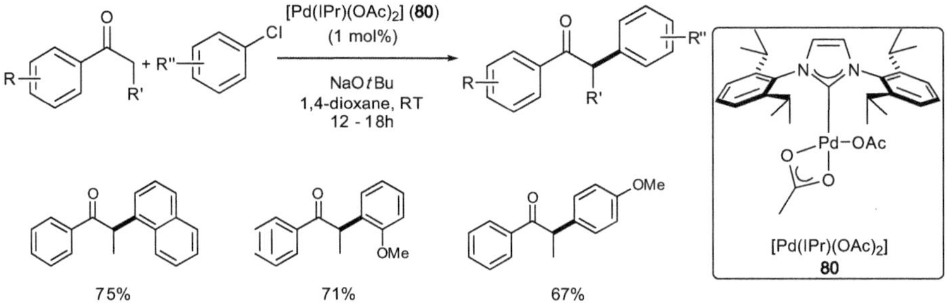

Scheme 4.57   Room temperature α-arylation using [Pd(IPr)(OAc)₂].

found that the use of microwave irradiation (130 °C) could significantly decrease the reaction time to 2 min.

Nolan and co-workers also utilized [Pd(IPr)(acac)Cl] (**82**), which showed high activity in the reaction (Figure 4.36).[160] The advantage of this precatalyst is its ease of preparation. Other groups also studied the effect of alternative throwaway ligands in this reaction. For example, Zhang *et al.* prepared the IPr-based cyclopalladated ferrocenylimine complex **83** (Figure 4.36).[161] The complex exhibited high activity for the α-arylation of various ketones using aryl chlorides and bromides, including deactivated and sterically hindered congeners. Xiao and Shao also utilized the [Pd(I-Pr)(Im)Cl₂] complex in the coupling (**16a**, Figure 4.36).[162]

Another interesting contribution was made by Navarro and co-workers,[163] who reported a domino palladium-catalysed oxidation/arylation of secondary alcohols. [Pd(IPr)(allyl)Cl] (**84**) was identified as the best precatalyst to promote the reaction. Desired coupling products were isolated in good yields (Scheme 4.58).

Table 4.25 α-Arylation at low catalyst loading using [Pd(IPr)(dmab)Cl].

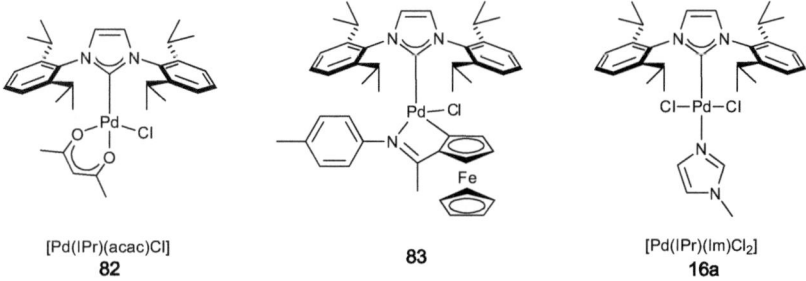

Entry	Ketone	Chloride	Product	Heating mode	T (°C)	Time	Conversion (%)[a]
1				Conventional	70	1 h	99
2				Microwave	130	2 min	98
3				Conventional	70	1 h	100
4				Microwave	130	2 min	99
5				Conventional	70	2 h	97
6				Microwave	130	2 min	100
7				Conventional	70	2 h	100
8				Microwave	130	2 min	98

[a]By GC.

Figure 4.36 Structures of some precatalysts used in α-arylation reactions.

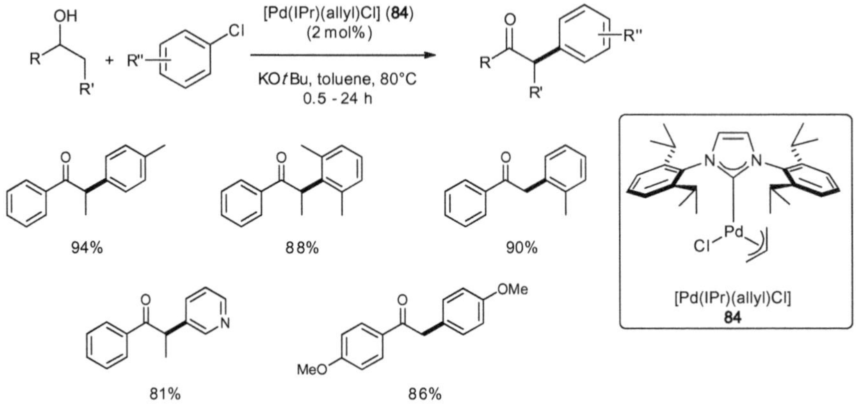

Scheme 4.58 Cascade alcohol oxidation/α-arylation using [Pd(IPr)(allyl)Cl].

4.3.6 C–Heteroatom Bond Formation

4.3.6.1 *Amination Reactions*

The palladium-catalysed amination reaction, also known as Buchwald–Hartwig cross-coupling, is one of the most efficient methods to create C–N bonds. Historically, the reaction was pioneered using aminostannanes by Migita and co-workers in 1983,[164] but the groups of Buchwald[165] and Hartwig[166] developed more convenient and efficient tin-free protocols about a decade later, in 1995. Since then, a large number of publications have appeared regarding this coupling reaction.[167] The mechanism of the reaction is described in Scheme 4.59. First the palladium source is reduced to its active Pd(0) form. The next step is the oxidative addition of the aryl halide to yield a palladium–aryl complex. The coordination of the amine to the palladium centre can occur through two main pathways, depending on the nature of the base: (a) if an alkoxide base is used, a Pd–alkoxide complex can be formed, thus deprotonating the amine, resulting in the formation of the Pd–N bond; (b) the amine can first coordinate the metal centre to afford a Pd–NHR'R'' complex, which can be deprotonated by the base to yield the Pd–NR'R'' complex. Finally, reductive elimination leads to the formation of the desired C–N bond and the active Pd(0) species is subsequently regenerated.

Since Nolan and co-workers' first report on the use of Pd–NHC systems in the amination reaction,[168] many other contributions have been presented in the literature. Some excellent reviews are available that cover most of the major advances in this field.[9c,d,167] In this section, we focus on the latest protocols reported since 2010. We pay particular attention to the new catalytic systems that permit the coupling of challenging substrates under mild conditions (*e.g.*, low catalyst loadings or low temperature).

4.3.6.1.1 Development of New Well-Defined Precatalysts Bearing Bulky NHC Ligands. Throughout this chapter, the importance of developing

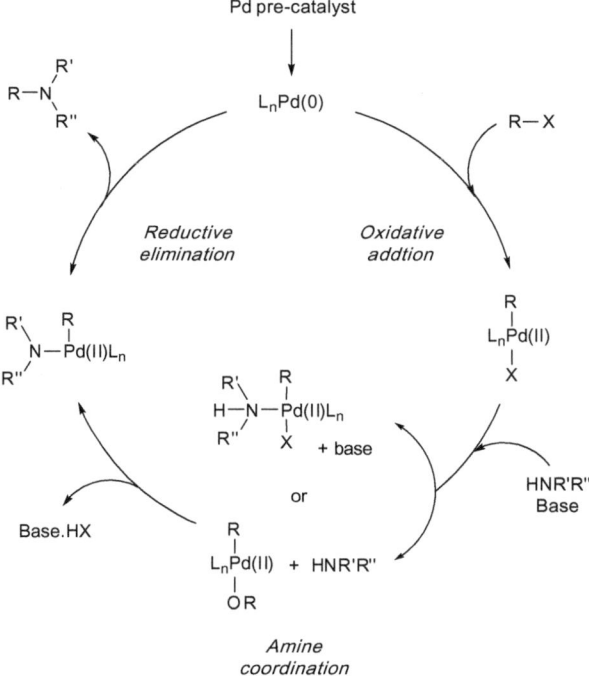

Scheme 4.59 Mechanism of the amination reaction.

new ligands with various steric and electronic properties, in order to promote cross-coupling reactions, has been repeatedly emphasized. In the palladium-catalysed amination reaction, recent efforts have been devoted to the design of new bulky ligands.[9e] For example, in 2011, Organ and co-workers reported the use of [Pd(IPent)(PEPPSI)] for amination using secondary alkyl amines.[169] The reactivity of the latter complex was directly compared with that of the [Pd(IPr)(PEPPSI)] congener. The IPent ligand had a tremendous effect on the catalysis, allowing the coupling of various aryl chlorides and amines, which was hardly possible with the IPr-based system (Scheme 4.60). The conditions described by Organ and co-workers utilized Cs_2CO_3 as the base, affording great functional group tolerance. However, high catalyst (4 mol%) and base (3 equiv.) loadings were usually required to achieve excellent results. In the same report, rate and computational calculations were also investigated to understand better the differences between the two catalytic systems. The rate studies highlighted the fact that [Pd(IPr)(PEPPSI)] is strongly dependent on the electronic nature of the chloride, whereas [Pd(IPent)(PEPPSI)] is less sensitive. The computational studies showed that amine coordination/deprotonation was the rate-limiting step. At this stage, the IPent complex was found to bind to the amine much more strongly than the IPr analogue. The combination of the two factors resulted in a much more active catalyst for amination

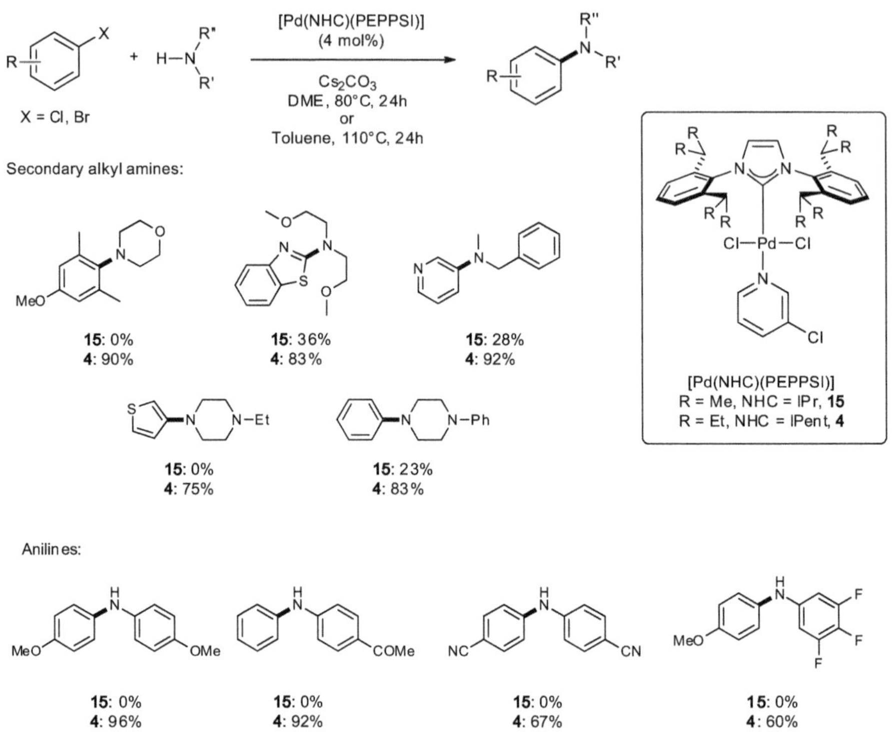

Scheme 4.60 [Pd(IPent)(PEPPSI)] *versus* [Pd(IPr)(PEPPSI)] in amination.

(Scheme 4.60). A year later, in a very similar study,[170] Organ and co-workers investigated the reactivity of [Pd(NHC)(PEPPSI)] complexes in the coupling of anilines. Rate studies showed that anilines bearing electron-withdrawing groups led to slower reactions. The authors reasoned that this was inconsistent with the deprotonation of the amine being the rate-limiting step, as was the case for the secondary alkylamines. Considering that electron-donating groups on the aniline increased the rate and that electron-withdrawing groups on the aryl chloride also increased the rate, they concluded that the reductive elimination was most probably the lim-iting step. In terms of reactivity, IPent was found to be more reactive than IPr to couple anilines, as was the case for secondary alkylamines. More interestingly, IPent was able to catalyse some reactions that were not pos-sible using IPr (Scheme 4.60).

Recently, Nolan's group has been heavily involved in the development of new well-defined Pd–NHC precatalysts to promote Buchwald–Hartwig cross-coupling. In 2012, they investigated the use of the highly bulky yet flexible IPr* ligand[171] in amination. Three new complexes were prepared: [Pd(IPr*)(cin)Cl] (9),[172] [Pd(IPr*)(acac)Cl] (85)[173] and [Pd(IPr*)(3-Cl-pyr-idinyl)Cl$_2$] (86)[56] bearing three different classical throwaway ligands de-veloped by the groups of Nolan and Organ (Figure 4.37). [Pd(IPr*)(cin)Cl] and

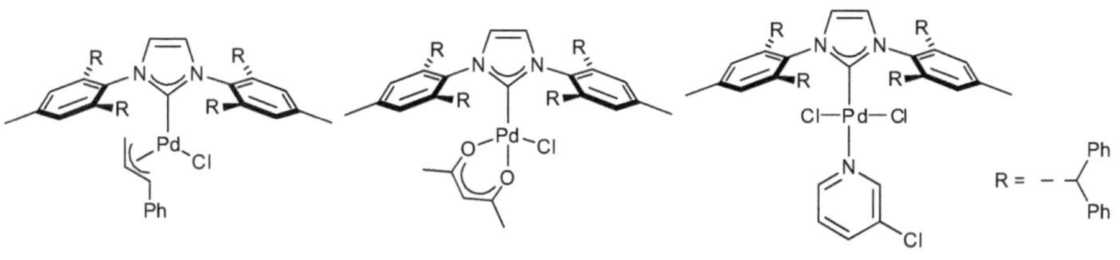

[Pd(IPr*)(cin)Cl]   [Pd(IPr*)(acac)Cl]   [Pd(IPr*)(3-Cl-pyridinyl)Cl₂]
9   85   86

Figure 4.37 [Pd(IPr*)L$_n$] complexes used in amination.

[Pd(IPr*)(3-Cl-pyridinyl)Cl$_2$] were found to be among the most active Pd–NHC precatalysts reported to date for amination reactions.

These complexes were active at room temperature using 1–2 mol% of precatalyst, allowing the coupling of deactivated and sterically hindered aryl chlorides with secondary amines. Unfortunately, at room temperature, they were not able to couple primary amines. At higher temperature (110 °C in toluene) they were capable of promoting the C–N coupling of a wide range of chlorides with a wider range of amines, including primary amines (Scheme 4.61). The catalyst loading could be decreased to as low as 0.025 mol% in these cases. Despite their structural difference, [Pd(IPr*)(cin)Cl] and [Pd(IPr*)(3-Cl-pyridinyl)Cl$_2$] exhibited similar catalytic activity in amination. This suggested that the Pd(0) active species is most likely the same in both cases. Notably, [Pd(IPr*)(cin)Cl] (**9**) was also successfully used to perform arylamination in a microflow reactor.[174]

The reactivity of [Pd(IPr*)(3-Cl-pyridinyl)Cl$_2$] was also directly compared with that of other [Pd(NHC)(3-Cl-pyridinyl)Cl$_2$] analogues.[56] It exhibited similar reactivity at room temperature but much higher catalytic activity at low catalyst loadings. The bulk of the ligand, evaluated *via* the percent buried volume (%V_{bur}),[175] was found to be an advantage in that case (Scheme 4.62).

The reactivity of [Pd(IPr*)(acac)Cl] (**85**) was also investigated in amination. The system proved active using a wide range of temperatures (r.t.–110 °C) and catalyst loadings (0.2–1 mol%) for the coupling of various aryl halides with a wide range of amines. LiHMDS was found in this case to be a more suitable base than the classical alkoxide bases (MO*t*Bu, KO*t*Am, *etc.*). Interestingly, chemoselective arylaminations of various dihalides were also performed using **85** and monoaminated products were isolated in good yields (Table 4.26).

In 2013, Nolan and co-workers developed a new ligand, IPr*^{OMe}.[176] This ligand is a simple modification of IPr*, bearing methoxy instead of methyl groups in the *para*-position of the aromatic rings. The corresponding [Pd(IPr*^{OMe})(acac)Cl] complex (**87**, Figure 4.38) was then prepared and its catalytic activity was investigated in arylamination. Optimization of the reaction conditions showed that KO*t*Am–dioxane was the best combination to

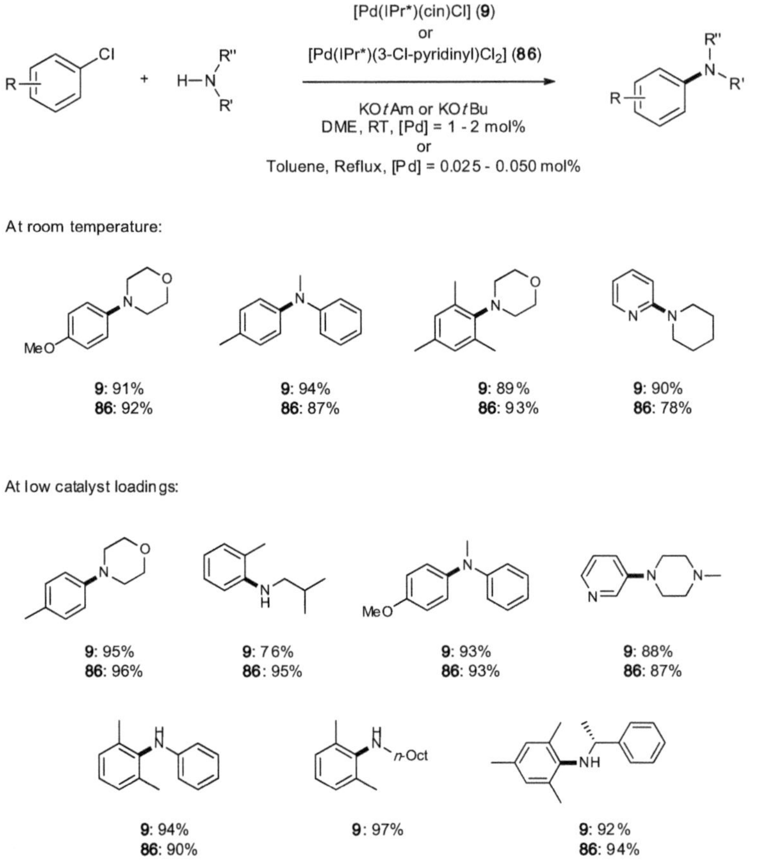

Scheme 4.61   Reactivity of [Pd(IPr*)(cin)Cl] and [Pd(IPr*)(3-Cl-pyridinyl)Cl$_2$] in amination at room temperature and low catalyst loadings.

promote the coupling. A direct comparison between IPr* and IPr*^{OMe} was then conducted. The differences in reactivity were tremendous and [Pd(IPr*^{OMe})(acac)Cl] proved to be much more active than the IPr* congener (Scheme 4.63). The scope of the reaction was studied at very low catalyst loadings (as low as 0.005 mol% for the coupling of *p*-bromotoluene and *N*-methylaniline, with a yield of 72%). The catalytic system proved highly effective, permitting the coupling of various aryl halides with a range of amines, including challenging electron-poor anilines.

Nolan and co-workers also developed a solvent-free amination protocol using the well-defined [Pd(IPr*)(cin)Cl] (**9**, Figure 4.37).[177] This complex (1 mol%) proved particularly useful for the coupling of primary amines at room temperature, which previously was not possible in the presence of solvents (Scheme 4.64). Moreover, in some cases, the reaction was complete in minutes. It is important to state that in these cases, the reaction was set

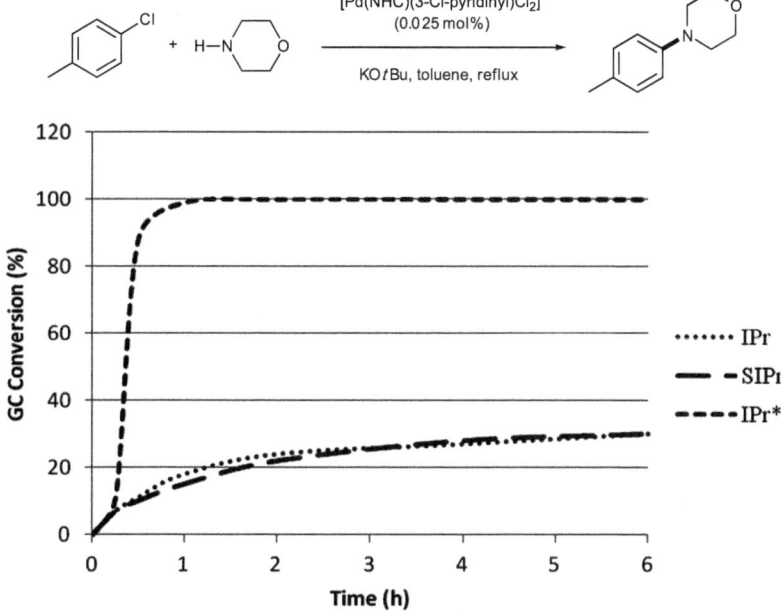

Scheme 4.62 Reactivity of [Pd(IPr*)(3-Cl-pyridinyl)Cl₂] **(86)** in comparison with other [Pd(NHC)(3-Cl-pyridinyl)Cl₂] complexes. Adapted from Ref. 56.

Table 4.26 Chemoselective amination using [Pd(IPr*)(acac)Cl].

Entry	Choride	Amine	Product	Yield (%)
1	Cl—⟨ ⟩—I	⟨ ⟩—NH₂	Cl—⟨ ⟩—N(H)—⟨ ⟩	71
2	Cl—⟨ ⟩—Br	H—N(⟨ ⟩O)	Cl—⟨ ⟩—N(⟨ ⟩O)	65
3	Cl—⟨ ⟩—Br	⟨ ⟩—NH₂	Cl—⟨ ⟩—N(H)—⟨ ⟩	76
4	⟨ ⟩(Cl)—Br	H—N(⟨ ⟩O)	⟨ ⟩(Cl)—N(⟨ ⟩O)	67
5	⟨ ⟩(Cl)—Br	⟨ ⟩—N(H)—	⟨ ⟩(Cl)—N(—)⟨ ⟩	90

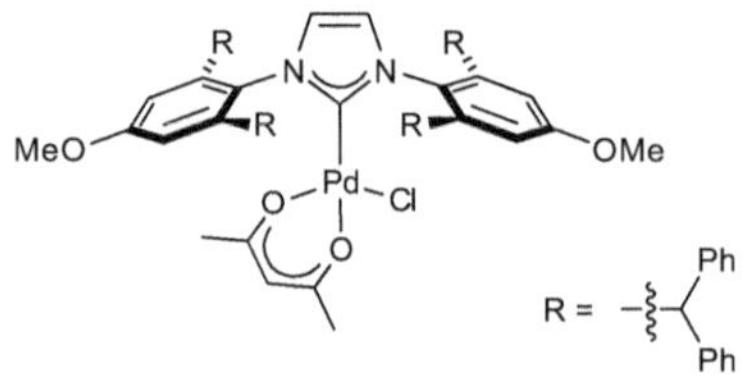

[Pd(IPr*^{OMe})(acac)Cl]
87

Figure 4.38 Structure of [Pd(IPr*^{OMe})(acac)Cl] (**87**).

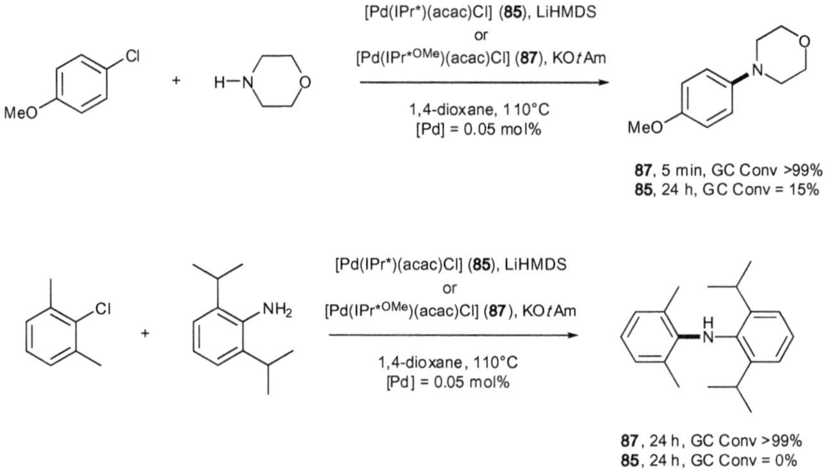

Scheme 4.63 [Pd(IPr*)(acac)Cl] *versus* [Pd(IPr*^{OMe})(acac)Cl] in amination.

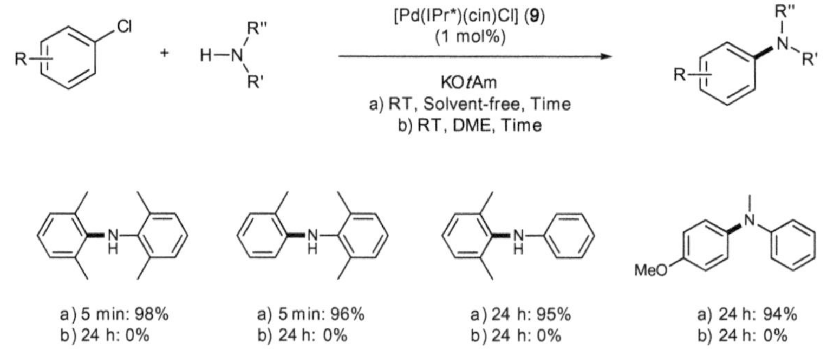

Scheme 4.64 Solvent-free amination catalysed by [Pd(IPr*)(cin)Cl].

up at room temperature. Nevertheless, during the course of the reaction, a strong exotherm was detected and promoted the coupling. There is a clear danger associated with such an exotherm and special care should be taken when attempting the reactions. The nature of the precatalysts used in the reaction was investigated. Bulky ligands were found to be essential to

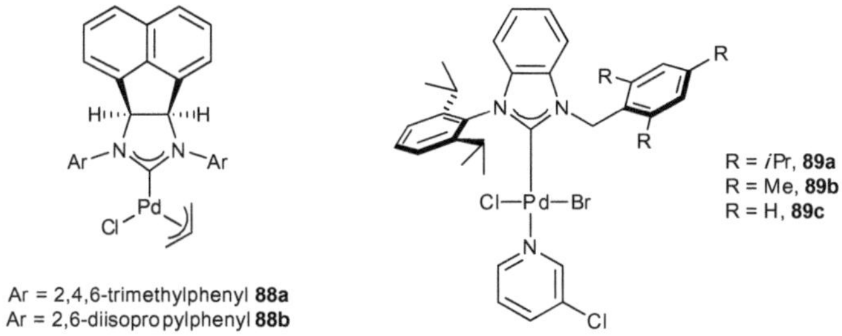

R = *i*Pr, **89a**
R = Me, **89b**
R = H, **89c**

Ar = 2,4,6-trimethylphenyl **88a**
Ar = 2,6-diisopropylphenyl **88b**

Figure 4.39 Structures of some precatalysts active in the amination reaction.

promote the coupling. The throwaway ligand also played a crucial role, as the system needs to be rapidly activated.

Other groups also developed new NHC ligands for the amination reaction. For example, Dastgir *et al.* prepared two [Pd(BIAN-SIAr)(allyl)Cl] complexes (SIAr = SIMes and SIPr, **88**, Figure 4.39).[178] Both complexes were active in the reaction using reasonably low catalyst loadings (1 mol%) at room temperature, for the coupling of various deactivated and activated aryl chlorides and bromides. Notably, sterically hindered bromomesitylene was found to be a more difficult substrate. Tu and co-workers reported the synthesis of benzimidazolylidene PEPPSI-type complexes (**89**, Figure 4.39).[179] Among three new congeners, **89a** was the most active in amination. In this structure, the metal centre is less congested and the σ-donation is modified by the benzimidazole ring. The catalytic activity of **89a** was found to be good but not exceptional.

4.3.6.1.2 Development of New Well-Defined Precatalysts Bearing New Throwaway Ligands. Jin *et al.* reported the preparation of well-defined air- and-moisture stable [Pd(NHC)(sal)Cl] (sal = salicylaldimine) complexes.[180] The influence of the steric and electronic parameters of the salicylaldimine ligands was studied in the reaction between 4-chlorotoluene and morpholine, using 1 mol% of the precatalyst and NaO*t*Bu as the base in DME at 80 °C. Sterically encumbered aryl groups on the salicylaldimine nitrogen or a methyl group in the *ortho*-position of the salicylaldimine oxygen were not found to be beneficial in the reaction. From an electronic point of view, electron-rich *N*-aryl groups were found to have a negative effect on the catalytic activity, whereas electron-deficient groups exhibited a greater effect. Consequently, **90** was identified as the most efficient precatalyst to couple a range of deactivated, activated and sterically hindered aryl chlorides with various aromatic and aliphatic primary and secondary amines (Scheme 4.65). Interestingly, the reaction could be conducted under aerobic conditions.

Nolan and co-workers investigated the modification of the structure of the acac ligand in complexes of formula [Pd(IPr)(R-acac)Cl] (**91–94**, Figure 4.40),[181] previously developed by the same group.[160a] This work was carried out in analogy with the [Pd(NHC)(R-allyl)Cl] complexes, where they

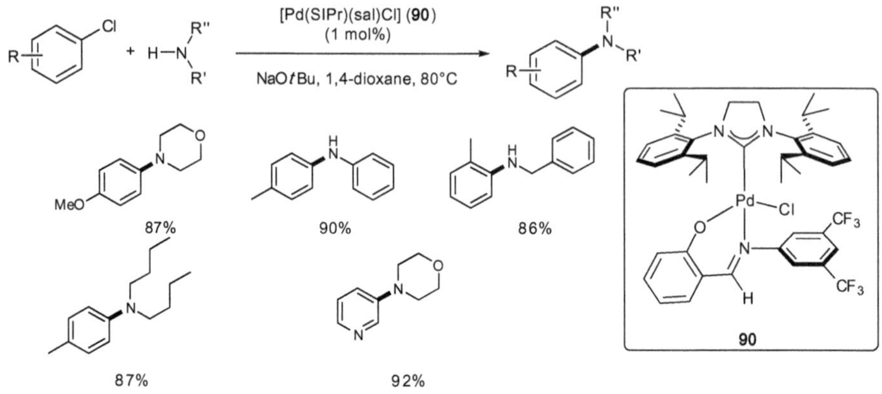

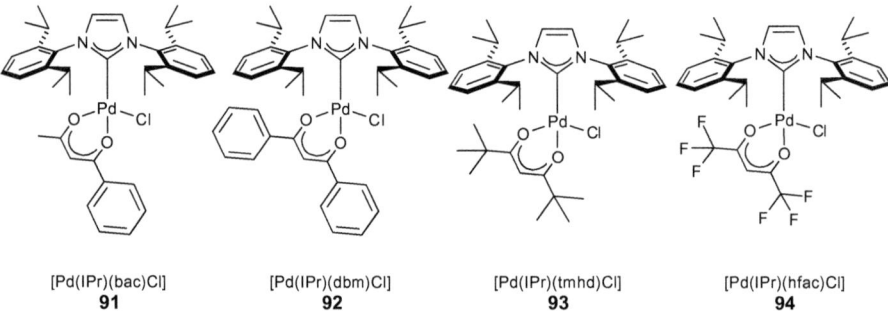

Scheme 4.65   Amination reaction using **90**.

[Pd(IPr)(bac)Cl]
91

[Pd(IPr)(dbm)Cl]
92

[Pd(IPr)(tmhd)Cl]
93

[Pd(IPr)(hfac)Cl]
94

Figure 4.40   Structures of [Pd(IPr)(R-acac)Cl] complexes used in amination.

previously showed that small modifications to the structure of the throw-away ligand could lead to tremendous improvements in catalysis.[54] Differences were found in terms of activation rates using the various precatalysts. Bulkier R-acac substituents were correlated with faster activation rates in the order acac (**82**) < bac (**91**) < dbm (**92**) < tmhd (**93**). However, electron-withdrawing substitutents such as CF_3 (hfac derivative **94**) inhibited the catalytic activity. These studies proved particularly useful for the rational design of finely tuned precatalysts.

Navarro and co-workers also used a new class of [Pd(NHC)(Et$_3$N)Cl$_2$] complexes in amination (Figure 4.41).[47] The reactivity of [Pd(SIPr)(Et$_3$N)Cl$_2$] (**95**) was directly compared with that of [Pd(SIPr)(PEPPSI)] (**96**) in the reaction between 2,6-dimethylbenzene and 2,6-diisopropylaniline; **95** exhibited higher catalytic activity at lower temperature (Table 4.27).

In 2011, Nolan and co-workers investigated the use of [Pd(NHC)(cin)OH]$_2$ complexes in cross-coupling reactions (NHC = IPr and SIPr, **97**).[182] Both complexes proved active for the coupling of 4-chloroanisole with morpholine under classical reaction conditions and good conversions were obtained in both cases (Scheme 4.66).

$[Pd(SIPr)(Et_3N)Cl_2]$
95

[Pd(SIPr)(PEPPSI)]
96

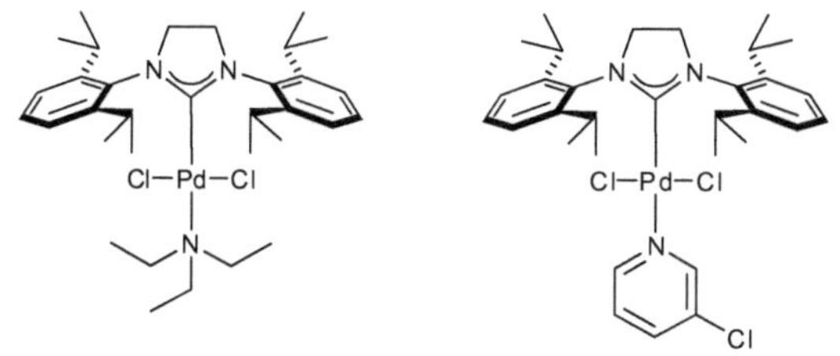

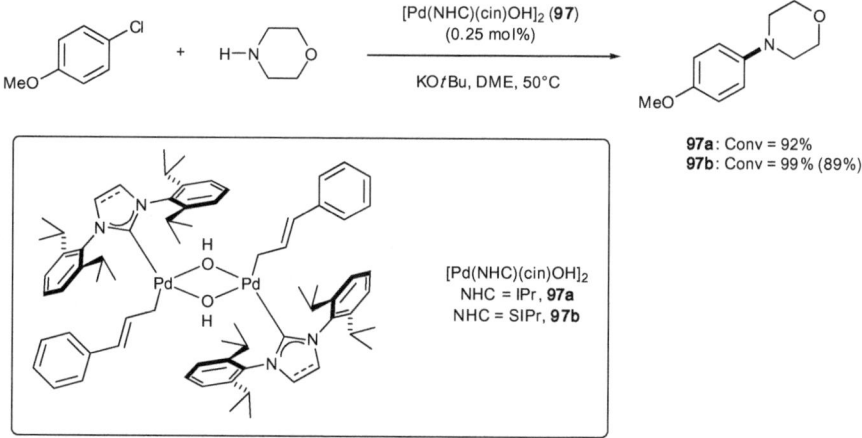

Figure 4.41 Structures of the $[Pd(SIPr)(Et_3N)Cl_2]$ and [Pd(SIPr)(PEPPSI)] complexes.

Table 4.27 Comparison of reactivities between $[Pd(SIPr)(Et_3N)Cl_2]$ and [Pd(SIPr)(PEPPSI)] for the amination of hindered substrates.

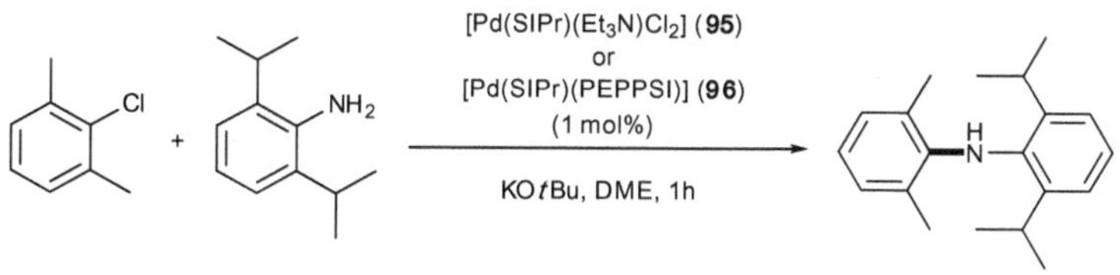

Entry	$[Pd(NHC)L_n]$	Temperature (°C)	Yield (%)
1	[Pd(SIPr)(PEPPSI)] (**96**)	25	<5
2	$[Pd(SIPr)(Et_3N)Cl_2]$ (**95**)	25	44
3	[Pd(SIPr)(PEPPSI)] (**96**)	50	83
4	$[Pd(SIPr)(Et_3N)Cl_2]$ (**95**)	50	94

Scheme 4.66 Amination using $[Pd(NHC)(cin)OH]_2$ complexes.

In another contribution dedicated to the use of the $[Pd(IPr)(Im)Cl_2]$ complex **16a**, Shao and co-workers reported its activity in amination.[183] The precatalyst was active for the coupling of various activated and deactivated

Table 4.28   C–N bond formation using amides and formamides
as substrates.

Entry	Solvent	R	Amide	Yield (%)
1	THF	4-OMe		91
2	THF	4-Me		10
3	DMF	4-Me		72
4	DMF	4-H		99
5	THF	4-CN		70
6	THF	4-OMe		92
7[a]	THF	3-OMe		92
8	THF	4-OMe		92
9[b]	THF	3-OMe		86

[a]Refluxing THF.
[b]50 °C.

chlorides with a small range of amines. The reaction conditions were typical,
using 1 mol% of precatalyst at 70–90 °C in dioxane for 3 h and KOtBu as the
base, affording the expected coupled products in excellent yields.

Interestingly, this complex was also found to create C–N bonds using
amides and formamides as substrates (Table 4.28).[184] KOtBu was found to be
important to release the amine in the reaction medium. The use of amides
or formamides instead of the classical amines allowed milder reaction
conditions to be used. The reaction thus occurred at room temperature or
with gentle heating.

4.3.6.1.3   Carbonylative Amination.   In 2012, Orellana and co-workers reported a palladium-catalysed carbonylative amination reaction of iodides
with pyrrole (Scheme 4.67).[185] Different Pd–phosphine systems were

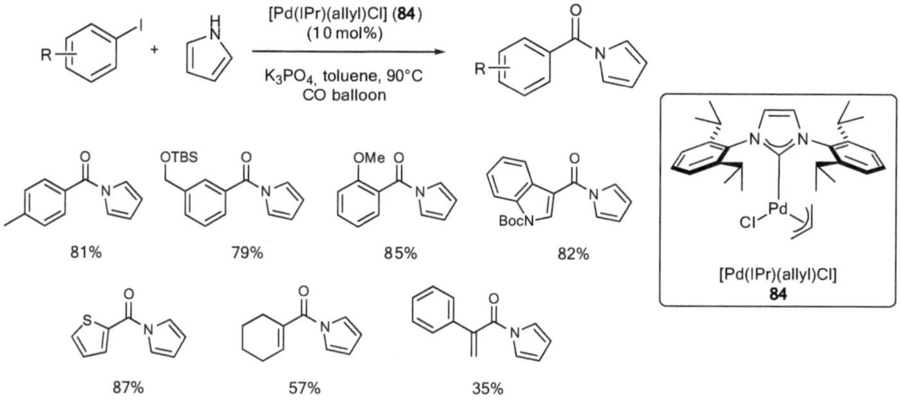

Scheme 4.67 Carbonylative amination using [Pd(IPr)(allyl)Cl].

investigated and also the [Pd(IPr)(allyl)Cl] complex (**84**). The latter (10 mol%) was found to be particularly active to promote the reaction in toluene at 90 °C in the presence of K_3PO_4 as the base and using a simple CO balloon as the carbonyl source. This permitted the synthesis of an interesting library of acylpyrrole compounds in good yields. Notably, the reaction also worked with activated aryl bromides but failed with other aryl bromide congeners.

4.3.6.1.4 Alternative Bases. The use of alternative and milder bases to increase functional group tolerance in the amination reaction is very important, because the commonly used metal alkoxide bases sometimes lead to undesired side reactions. Whereas KOtBu does not tolerate 4-chlorobenzoate, potassium 2,2,5,7,8-pentamethylchroman-6-oxide (**98**, $pK_a \approx 11.4$) was found to be tolerated by the substrate.[186] Moreover, the reactivity of the base in amination was discovered to be between those of KOtBu and Cs_2CO_3 using [Pd(IPent)(PEPPSI)] (**4**) as the precatalyst. Any attempt to modify the structure of the base resulted in lower reactivity than that of the chroman oxide derivative. A small range of compounds was prepared to show the advantage of the new base over KOtBu (Scheme 4.68).

4.3.6.2 C–S Bond Formation

The formation of C–S bonds using Pd–NHC precatalysts has hardly been reported in the literature. This is notably because sulfur species deactivate the palladium catalysts.

The first report was published in 2010 by Liu and co-workers, who prepared three complexes to investigate the reaction: the monocarbene dimer complex **99a**, the biscarbene complex **99b** and the mixed phosphine–NHC complex **99c** (Figure 4.42).[187] Reaction conditions were optimized using 4-bromoacetophenone and 4-methylbenzenethiol as substrates, highlighting the efficiency of **99c** in the reaction. KOtBu and toluene were selected as the

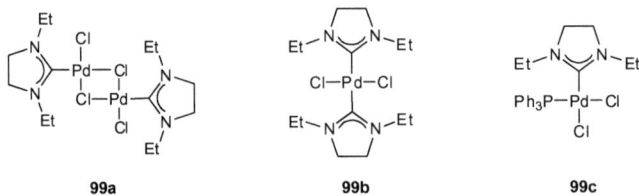

Scheme 4.68 Amination using potassium 2,2,5,7,8-pentamethylchroman-6-oxide
(**98**) as the base.

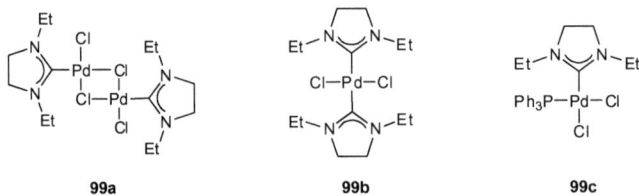

Figure 4.42 Pd-NHC complexes for C–S bond formation.

base and solvent, respectively, and the scope of the reaction was studied
using 2 mol% of complex **99c** at 110 °C. A range of bromides including activated and deactivated congeners was engaged in the coupling with various
aryl sulfides and the corresponding C–S bonds were formed in good yields
(Table 4.29). Unfortunately, aryl chlorides and aliphatic thiols were
unreactive.

A year later, Shi *et al.* reported [Pd(SIPr)(pyridinyl)Cl$_2$] (**100**) as a precatalyst
capable of promoting C–S coupling using both bromides and chlorides as
coupling partners.[188] Optimized reaction conditions utilized KOtBu as the
base in toluene at 100 °C using 2 mol% of the Pd–NHC precatalyst. Under
these conditions, a wide library of sulfide products could be prepared using
phenyl and benzyl thiols as coupling partners (Scheme 4.69).

Sayah and Organ developed a low-temperature protocol for the formation
of C–S bonds.[189] The system was composed of [Pd(IPent)(PEPPSI)] (**4**)
(2 mol%) as the precatalyst, KOtBu as the base and LiOiPr (20 mol%) as an
additive in toluene at 40 °C. In this case, LiOiPr was found to be necessary to
facilitate catalyst activation at low temperature. The system performed well

Table 4.29 C–S bond formation using [Pd(NHC)] **99c**.

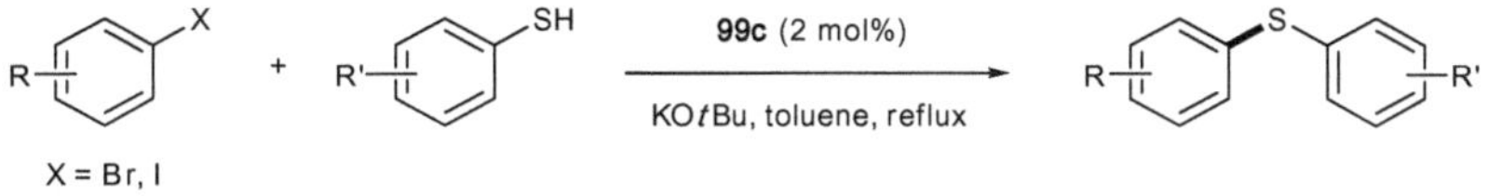

Entry	X	R	R'	Yield (%)
1	Br	4-OMe	4-Me	94
2	Br	4-Cl	4-Me	97
3	Br	4-NO$_2$	4-Me	88
4	Br	4-Me	4-MeO	75
5	Br	4-Me	4-Me	86
6	Br	H	2-Naphthalenethiol	77

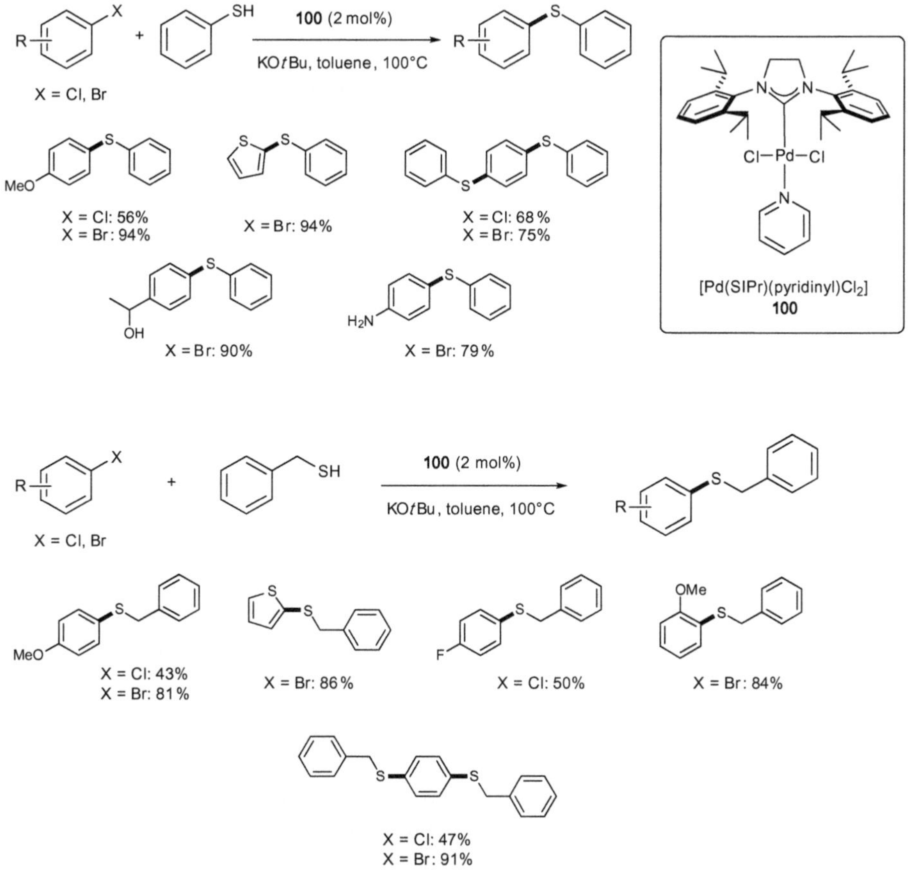

Scheme 4.69 C–S bond formation using [Pd(SIPr)(pyridinyl)Cl$_2$] (**100**).

for the coupling of aryl halides with aryl thiols (Scheme 4.70). Both bromides and chlorides were reactive, including sterically and electronically disfavoured analogues. Of note, alkanethiols were also found to be suitable

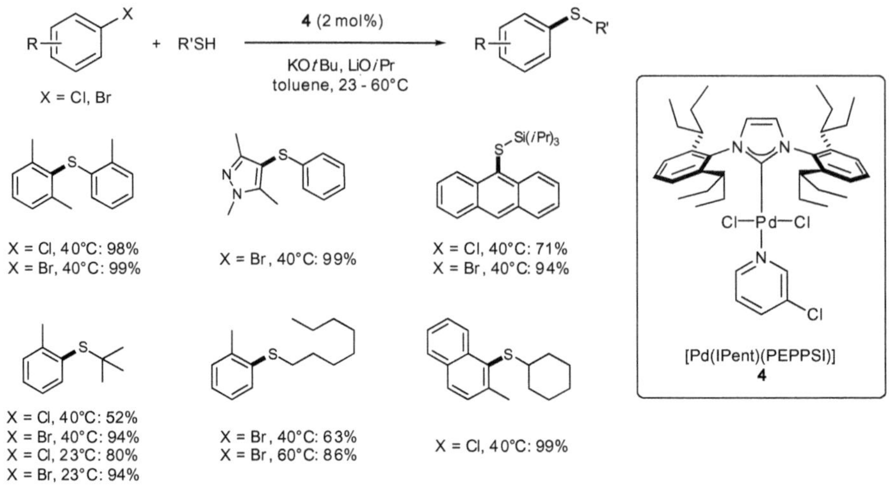

Scheme 4.70 C–S bond formation using [Pd(IPent)(PEPPSI)].

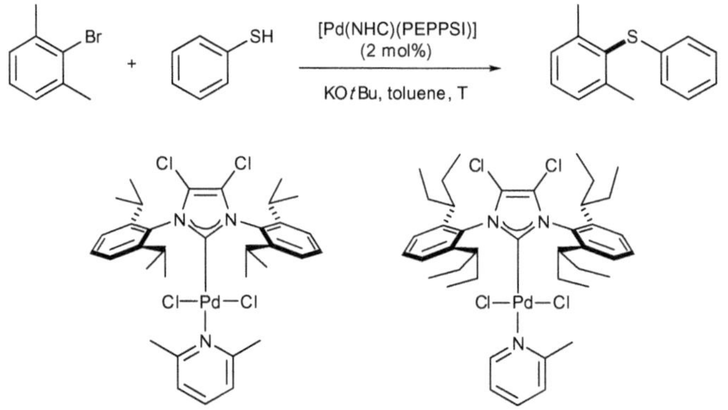

101
40°C: trace
50°C: 70%

102
23°C: -
40°C: 99%

Scheme 4.71 C–S bond formation at low temperature without additives.

coupling partners. Finally, some reactions conducted at room temperature showed similar efficiency to those at 40 °C.

Subsequently, in 2013, Organ's group investigated the mechanism of C–S bond formation using [Pd(NHC)(PEPPSI)] complexes.[190] A wide library of complexes was prepared and their reactivity towards sulfination was examined. Chlorinated backbones were notably found to be advantageous to activate the precatalysts at low temperature without the need to use additives (Scheme 4.71). The IPent ligand was also found to be more efficient than the less bulky IPr. Conducting stoichiometric reactions, they were also able to identify two Pd–NHC species involved in the catalytic cycle (Figure 4.43).

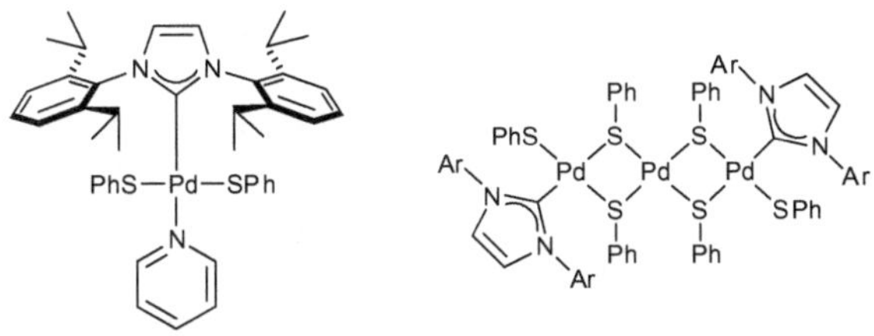

Figure 4.43 [Pd(NHC)] species involved in the catalytic cycle.

Complex **103** was found to be active under the reaction conditions whereas **104** was not, unless extra additives were added (6% pyridine or 2 equiv. of KOEt) to break the aggregate.

4.4 Conclusion

We have attempted to demonstrate that much has been done with the NHC ligand class in palladium cross-coupling catalysis. It is clear that the winners of the 2010 Nobel Prize in Chemistry laid the groundwork for these latest developments and the research community is immensely indebted to them for their still inspiring contributions. The work in this area continues to generate fascinating surprises in terms of catalyst reactivity and productivity. We expect that more is to come and look forward to more exciting developments.

References

1. (a) A. de Meijere and F. Diederich (eds), *Metal-Catalysed Cross-Coupling Reactions*, Wiley-VCH, Weinheim, 2nd edn, 2004; (b) themed collection: cross-coupling reactions in organic synthesis, *Chem. Soc. Rev.*, 2011, **40**, (10), 4877; (c) C. C. C. Johansson Seechurn, M. O. Kitching, T. J. Colacot and V. Snieckus, *Angew. Chem. Int. Ed.*, 2012, **51**, 5062; (d) H. Li, C. C. C. Johansson Seechurn and T. J. Colacot, *ACS Catal.*, 2012, **2**, 1147.
2. K. C. Nicolaou, P. G. Bulger and D. Sarlah, *Angew. Chem. Int. Ed.*, 2005, **44**, 4442.
3. J.-P. Corbet and G. Mignani, *Chem. Rev.*, 2006, **106**, 2651.
4. A. Suzuki, *Angew. Chem. Int. Ed.*, 2011, **50**, 6722.
5. E.-i. Negishi, *Angew. Chem. Int. Ed.*, 2011, **50**, 6738.
6. R. F. Heck, *Acc. Chem. Res.*, 1979, **12**, 146.
7. (a) G. C. Fu, *Acc. Chem. Res.*, 2008, **41**, 1555; (b) R. Martin and S. L. Buchwald, *Acc. Chem. Res.*, 2008, **41**, 1461; (c) D. S. Surry and S. L. Buchwald, *Angew. Chem. Int. Ed.*, 2008, **47**, 6338.
8. (a) W. A. Herrmann, *Angew. Chem. Int. Ed.*, 2002, **41**, 1290; (b) S. P. Nolan (ed.), *N-Heterocyclic Carbenes in Synthesis*, Wiley,

New York, 2006; (c) S. Diez-Gonzalez, N. Marion and S. P. Nolan, *Chem. Rev.*, 2009, **109**, 3612.

9. (a) A. C. Hillier, G. A. Grasa, M. S. Viciu, H. M. Lee, C. Yang and S. P. Nolan, *J. Organomet. Chem.*, 2002, **653**, 69; (b) E. A. B. Kantchev, C. J. O'Brien and M. G. Organ, *Angew. Chem. Int. Ed.*, 2007, **46**, 2768; (c) N. Marion and P. Nolan Steven, *Acc. Chem. Res.*, 2008, **41**, 1440; (d) G. C. Fortman and S. P. Nolan, *Chem. Soc. Rev.*, 2011, **40**, 5151; (e) C. Valente, S. Calimsiz, K. H. Hoi, D. Mallik, M. Sayah and M. G. Organ, *Angew. Chem. Int. Ed.*, 2012, **51**, 3314.

10. (a) S. Diez-Gonzalez (ed.), *N-Heterocyclic Carbenes: from Laboratory Curiosities to Efficient Synthetic Tools*, Royal Society of Chemistry, Cambridge, 2010; (b) T. Rovis and S. P. Nolan, *Synlett*, 2013, **24**, 1188.

11. N. M. Scott and S. P. Nolan, *Eur. J. Inorg. Chem.*, 2005, 1815.

12. K. Oefele, *J. Organomet. Chem.*, 1968, **12**, P42.

13. H. W. Wanzlick and H. J. Schoenherr, *Angew. Chem. Int. Ed. Engl.*, 1968, 7, 141.

14. A. J. Arduengo III, R. L. Harlow and M. Kline, *J. Am. Chem. Soc.*, 1991, **113**, 361.

15. (a) A. J. Arduengo III, J. R. Goerlich and W. J. Marshall, *J. Am. Chem. Soc.*, 1995, **117**, 11027; (b) A. J. Arduengo III, *Acc. Chem. Res.*, 1999, **32**, 913; (c) A. J. Arduengo III, R. Krafczyk, R. Schmutzler, H. A. Craig, J. R. Goerlich, W. J. Marshall and M. Unverzagt, *Tetrahedron*, 1999, **55**, 14523.

16. H. W. Wanzlick, *Angew. Chem. Int. Ed. Engl.*, 1962, **1**, 75.

17. D. Enders, K. Breuer, G. Raabe, J. Runsink, J. H. Teles, J.-P. Melder, K. Ebel and S. Brode, *Angew. Chem. Int. Ed. Engl.*, 1995, **34**, 1021.

18. A. Tudose, A. Demonceau and L. Delaude, *J. Organomet. Chem.*, 2006, **691**, 5356.

19. (a) W. A. Herrmann, M. Elison, J. Fischer, C. Koecher and G. R. J. Artus, *Angew. Chem. Int. Ed. Engl.*, 1995, **34**, 2371; (b) W. A. Herrmann, C.-P. Reisinger and M. Spiegler, *J. Organomet. Chem.*, 1998, **557**, 93.

20. H. Jacobsen, A. Correa, A. Poater, C. Costabile and L. Cavallo, *Coord. Chem. Rev.*, 2009, **253**, 687.

21. (a) D. J. Nelson and S. P. Nolan, *Chem. Soc. Rev.*, 2013, **42**, 6723; (b) S. Diez-Gonzalez and S. P. Nolan, *Coord. Chem. Rev.*, 2007, **251**, 874.

22. C. A. Tolman, *Chem. Rev.*, 1977, 77, 313.

23. G. Altenhoff, R. Goddard, C. W. Lehmann and F. Glorius, *Angew. Chem. Int. Ed.*, 2003, **42**, 3690.

24. A. Chartoire, M. Lesieur, L. Falivene, A. M. Z. Slawin, L. Cavallo, C. S. J. Cazin and S. P. Nolan, *Chem. Eur. J.*, 2012, **18**, 4517.

25. (a) H. Clavier, A. Correa, L. Cavallo, E. C. Escudero-Adan, J. Benet-Buchholz, A. M. Z. Slawin and S. P. Nolan, *Eur. J. Inorg. Chem.*, 2009, 1767; (b) A. Poater, B. Cosenza, A. Correa, S. Giudice, F. Ragone, V. Scarano and L. Cavallo, *Eur. J. Inorg. Chem.*, 2009, 1759; (c) H. Clavier and S. P. Nolan, *Chem Commun.*, 2010, **46**, 841.

26. SambVca, https://www.molnac.unisa.it/OMtools/sambvca.php (accessed 6 May 2014).

27. (a) A. Poater, F. Ragone, R. Mariz, R. Dorta and L. Cavallo, *Chem. Eur. J.*, 2010, **16**, 14348; (b) F. Ragone, A. Poater and L. Cavallo, *J. Am. Chem. Soc.*, 2010, **132**, 4249; (c) L. Wu, E. Drinkel, F. Gaggia, S. Capolicchio, A. Linden, L. Falivene, L. Cavallo and R. Dorta, *Chem. Eur. J.*, 2011, **17**, 12886.

28. Y.-R. Lou, *Comprehensive Handbook of Chemical Bond Energies*, CRC Press, Boca Raton, FL, 2007.

29. Note that the term "transmetallation" is commonly used in cases of the Suzuki–Miyaura and Hiyama reactions even though boron and silicon are not metals.

30. N. Miyaura, in *Metal-Catalysed Cross-Coupling Reactions*, Vol. 1, ed. A. de Meijere and F. Diederich, Wiley-VCH, Weinheim, 2nd edn, 2004, p. 41.

31. (a) N. Miyaura and A. Suzuki, *Chem. Rev.*, 1995, **95**, 2457; (b) N. Miyaura, *Top. Curr. Chem.*, 2002, **219**, 11.

32. J. Yin, M. P. Rainka, X.-X. Zhang and S. L. Buchwald, *J. Am. Chem. Soc.*, 2002, **124**, 1162.

33. (a) S. D. Walker, T. E. Barder, J. R. Martinelli and S. L. Buchwald, *Angew. Chem. Int. Ed.*, 2004, **43**, 1871; (b) T. E. Barder, S. D. Walker, J. R. Martinelli and S. L. Buchwald, *J. Am. Chem. Soc.*, 2005, **127**, 4685; (c) T. Hoshi, T. Nakazawa, I. Saitoh, A. Mori, T. Suzuki, J.-i. Sakai and H. Hagiwara, *Org. Lett.*, 2008, **10**, 2063; (d) L. Ackermann, H. K. Potukuchi, A. Althammer, R. Born and P. Mayer, *Org. Lett.*, 2010, **12**, 1004; (e) C. M. So, W. K. Chow, P. Y. Choy, C. P. Lau and F. Y. Kwong, *Chem. Eur. J.*, 2010, **16**, 7996; (f) W. Tang, A. G. Capacci, X. Wei, W. Li, A. White, N. D. Patel, J. Savoie, J. J. Gao, S. Rodriguez, B. Qu, N. Haddad, B. Z. Lu, D. Krishnamurthy, N. K. Yee and C. H. Senanayake, *Angew. Chem. Int. Ed.*, 2010, **49**, 5879; (g) G.-Q. Li, Y. Yamamoto and N. Miyaura, *Synlett*, 2011, 1769.

34. G. Altenhoff, R. Goddard, C. W. Lehmann and F. Glorius, *J. Am. Chem. Soc.*, 2004, **126**, 15195.

35. M. G. Organ, S. Calimsiz, M. Sayah, K. H. Hoi and A. J. Lough, *Angew. Chem. Int. Ed.*, 2009, **48**, 2383.

36. A. Schmidt and A. Rahimi, *Chem. Commun.*, 2010, **46**, 2995.

37. T. Tu, Z. Sun, W. Fang, M. Xu and Y. Zhou, *Org. Lett.*, 2012, **14**, 4250.

38. (a) N. Miyaura and A. Suzuki, *Chem. Lett.*, 1992, 691; (b) A. B. Charette and A. Giroux, *J. Org. Chem.*, 1996, **61**, 8718; (c) M. R. Netherton, C. Dai, K. Neuschütz and G. C. Fu, *J. Am. Chem. Soc.*, 2001, **123**, 10099; (d) J. H. Kirchhoff, M. R. Netherton, I. D. Hills and G. C. Fu, *J. Am. Chem. Soc.*, 2002, **124**, 13662; (e) K. Arentsen, S. Caddick, F. G. N. Cloke, A. P. Herring and P. B. Hitchcock, *Tetrahedron Lett.*, 2004, **45**, 3511.

39. D. J. Cardenas, *Angew. Chem. Int. Ed.*, 2003, **42**, 384.

40. K. Arentsen, S. Caddick, F. G. N. Cloke, A. P. Herring and P. B. Hitchcock, *Tetrahedron Lett.*, 2004, **45**, 3511.

41. C. Valente, S. Baglione, D. Candito, C. J. O'Brien and M. G. Organ, *Chem. Commun.*, 2008, 735.

42. G.-R. Peh, E. A. B. Kantchev, J.-C. Er and J. Y. Ying, *Chem. Eur. J.*, 2010, **16**, 4010.

43. J. L. Farmer, H. N. Hunter and M. G. Organ, *J. Am. Chem. Soc.*, 2012, **134**, 17470.

44. N. Debono, A. Labande, E. Manoury, J.-C. Daran and R. Poli, *Organometallics*, 2010, **29**, 1879.

45. O. Diebolt, V. Jurcik, R. Correa da Costa, P. Braunstein, L. Cavallo, S. P. Nolan, A. M. Z. Slawin and C. S. J. Cazin, *Organometallics*, 2010, **29**, 1443.

46. T. E. Schmid, D. C. Jones, O. Songis, O. Diebolt, M. R. L. Furst, A. M. Z. Slawin and C. S. J. Cazin, *Dalton Trans.*, 2013, **42**, 7345.

47. M.-T. Chen, D. A. Vicic, M. L. Turner and O. Navarro, *Organometallics*, 2011, **30**, 5052.

48. C. J. O'Brien, E. A. B. Kantchev, C. Valente, N. Hadei, G. A. Chass, A. Lough, A. C. Hopkinson and M. G. Organ, *Chem. Eur. J.*, 2006, **12**, 4743.

49. Y.-Q. Tang, J.-M. Lu and L.-X. Shao, *J. Organomet. Chem.*, 2011, **696**, 3741.

50. X.-X. Zhou and L.-X. Shao, *Synthesis*, 2011, 3138.

51. M. R. Kumar, K. Park and S. Lee, *Adv. Synth. Catal.*, 2010, **352**, 3255.

52. M. Kuriyama, S. Matsuo, M. Shinozawa and O. Onomura, *Org. Lett.*, 2013, **15**, 2716.

53. B. R. Dible, R. E. Cowley and P. L. Holland, *Organometallics*, 2011, **30**, 5123.

54. N. Marion, O. Navarro, J. Mei, E. D. Stevens, N. M. Scott and S. P. Nolan, *J. Am. Chem. Soc.*, 2006, **128**, 4101.

55. S. G. Weber, C. Loos, F. Rominger and B. F. Straub, *ARKIVOC*, 2012, 226.

56. A. Chartoire, X. Frogneux, A. Boreux, A. M. Z. Slawin and S. P. Nolan, *Organometallics*, 2012, **31**, 6947.

57. M. S. Szulmanowicz, A. Gniewek, W. Gil and A. M. Trzeciak, *ChemCatChem*, 2013, **5**, 1152.

58. D. Canseco-Gonzalez, A. Gniewek, M. Szulmanowicz, H. Mueller-Bunz, A. M. Trzeciak and M. Albrecht, *Chem. Eur. J.*, 2012, **18**, 6055.

59. T. Terashima, S. Inomata, K. Ogata and S.-i. Fukuzawa, *Eur. J. Inorg. Chem.*, 2012, **2012**, 1387.

60. S. Roy and H. Plenio, *Adv. Synth. Catal.*, 2010, **352**, 1014.

61. F. Godoy, C. Segarra, M. Poyatos and E. Peris, *Organometallics*, 2011, **30**, 684.

62. B. Karimi and P. F. Akhavan, *Chem. Commun.*, 2011, **47**, 7686.

63. E. L. Kolychev, A. F. Asachenko, P. B. Dzhevakov, A. A. Bush, V. V. Shuntikov, V. N. Khrustalev and M. S. Nechaev, *Dalton Trans.*, 2013, **42**, 6859.

64. Y. Zhang, M.-T. Feng and J.-M. Lu, *Org. Biomol. Chem.*, 2013, **11**, 2266.

65. (a) W. J. Sommer and M. Weck, *Coord. Chem. Rev.*, 2007, **251**, 860; (b) C. S. J. Cazin, *C. R. Chim.*, 2009, **12**, 1173.

66. H. Yang, X. Han, G. Li, Z. Ma and Y. Hao, *J. Phys. Chem. C*, 2010, **114**, 22221.

67. G. Li, H. Yang, W. Li and G. Zhang, *Green Chem.*, 2011, **13**, 2939.
68. H. Yang, G. Li, Z. Ma, J. Chao and Z. Guo, *J. Catal.*, 2010, **276**, 123.
69. H. Yang, Y. Wang, Y. Qin, Y. Chong, Q. Yang, G. Li, L. Zhang and W. Li, *Green Chem.*, 2011, **13**, 1352.
70. M. N. Alam and S. M. Sarkar, *React. Kinet. Mech. Catal.*, 2011, **103**, 493.
71. G. Nan, F. Ren and M. Luo, *Beilstein J. Org. Chem.*, 2010, **6**, No 70.
72. Z.-Y. Wang, G.-Q. Chen and L.-X. Shao, *J. Org. Chem.*, 2012, 77, 6608.
73. (a) A. O. King, N. Okukado and E.-i. Negishi, *J. Chem. Soc., Chem. Commun.*, 1977, 683; (b) E.-i. Negishi, A. O. King and N. Okukado, *J. Org. Chem.*, 1977, **42**, 1821; (c) E.-i. Negishi, *J. Organomet. Chem.*, 2002, **653**, 34.
74. (a) N. Hadei, E. A. B. Kantchev, C. J. O'Brien and M. G. Organ, *J. Org. Chem.*, 2005, **70**, 8503; (b) N. Hadei, E. A. B. Kantchev, C. J. O'Brien and M. G. Organ, *Org. Lett.*, 2005, 7, 3805.
75. M. G. Organ, S. Avola, I. Dubovyk, N. Hadei, E. A. B. Kantchev, C. J. O'Brien and C. Valente, *Chem. Eur. J.*, 2006, **12**, 4749.
76. G. A. Chass, C. J. O'Brien, N. Hadei, E. A. B. Kantchev, W.-H. Mu, D.-C. Fang, A. C. Hopkinson, I. G. Csizmadia and M. G. Organ, *Chem. Eur. J.*, 2009, **15**, 4281.
77. G. T. Achonduh, N. Hadei, C. Valente, S. Avola, C. J. O'Brien and M. G. Organ, *Chem. Commun.*, 2010, **46**, 4109.
78. H. N. Hunter, N. Hadei, V. Blagojevic, P. Patschinski, G. T. Achonduh, S. Avola, D. K. Bohme and M. G. Organ, *Chem. Eur. J.*, 2011, **17**, 7845.
79. L. C. McCann, H. N. Hunter, J. A. C. Clyburne and M. G. Organ, *Angew. Chem. Int. Ed.*, 2012, **51**, 7024.
80. (a) S. Calimsiz, M. Sayah, D. Mallik and M. G. Organ, *Angew. Chem. Int. Ed.*, 2010, **49**, 2014; (b) C. Valente, M. E. Belowich, N. Hadei and M. G. Organ, *Eur. J. Org. Chem.*, 2010, 4343.
81. S. Calimsiz and M. G. Organ, *Chem. Commun.*, 2011, **47**, 5181.
82. N. Hadei, G. T. Achonduh, C. Valente, C. J. O'Brien and M. G. Organ, *Angew. Chem. Int. Ed.*, 2011, **50**, 3896.
83. I. Larrosa, C. Somoza, A. Banquy and S. M. Goldup, *Org. Lett.*, 2011, **13**, 146.
84. M. Pompeo, R. D. J. Froese, N. Hadei and M. G. Organ, *Angew. Chem. Int. Ed.*, 2012, **51**, 11354.
85. M. Kitano, S.-y. Hayashi, T. Tanaka, H. Yorimitsu, N. Aratani and A. Osuka, *Angew. Chem. Int. Ed.*, 2012, **51**, 5593.
86. Y. Ookubo, A. Wakamiya, H. Yorimitsu and A. Osuka, *Chem. Eur. J.*, 2012, **18**, 12690.
87. I. N. Petersen, F. Crestey and J. L. Kristensen, *Chem. Commun.*, 2012, **48**, 9092.
88. (a) R. J. P. Corriu and J. P. Masse, *J. Chem. Soc., Chem. Commun.*, 1972, 144; (b) K. Tamao, K. Sumitani and M. Kumada, *J. Am. Chem. Soc.*, 1972, **94**, 4374; (c) M. Yamamura, I. Moritani and S.-I. Murahashi, *J. Organomet. Chem.*, 1975, **91**, C39.

89. J. Huang and S. P. Nolan, *J. Am. Chem. Soc.*, 1999, **121**, 9889.
90. A. C. Frisch, F. Rataboul, A. Zapf and M. Beller, *J. Organomet. Chem.*, 2003, **687**, 403.
91. C. E. Hartmann, S. P. Nolan and C. S. J. Cazin, *Organometallics*, 2009, **28**, 2915.
92. J. Nasielski, N. Hadei, G. Achonduh, E. A. B. Kantchev, C. J. O'Brien, A. Lough and M. G. Organ, *Chem. Eur. J.*, 2010, **16**, 10844.
93. G. Ren, X. Cui and Y. Wu, *Eur. J. Org. Chem.*, 2010, 2372.
94. Z. Jin, X.-P. Gu, L.-L. Qiu, G.-P. Wu, H.-B. Song and J.-X. Fang, *J. Organomet. Chem.*, 2011, **696**, 859.
95. J. K. Stille, *Angew. Chem. Int. Ed. Engl.*, 1986, **25**, 508.
96. T. Weskamp, V. P. W. Bohm and W. A. Herrmann, *J. Organomet. Chem.*, 1999, **585**, 348.
97. W. A. Herrmann, V. P. W. Bohm, C. W. K. Gstottmayr, M. Grosche, C. P. Reisinger and T. Weskamp, *J. Organomet. Chem.*, 2001, **617-618**, 616.
98. G. A. Grasa and S. P. Nolan, *Org. Lett.*, 2001, **3**, 119.
99. M. Dowlut, D. Mallik and M. G. Organ, *Chem. Eur. J.*, 2010, **16**, 4279.
100. (a) A. Chartoire and S. P. Nolan, in *Science of Synthesis: Cross Coupling Heck-Type Reactions*, Vol. 1, ed. C. A. Molander, Georg Thieme, Stuttgart, 2013, p. 511; (b) T. Hiyama, in *Metal-Catalysed Cross-Coupling Reactions*, ed. F. Diederich and P. J. Stang, Wiley-VCH, Weinheim, 1998, p. 421; (c) T. Hiyama and E. Shirakawa, *Top. Curr. Chem.*, 2002, **219**, 61; (d) S. E. Denmark and M. H. Ober, *Aldrichim. Acta*, 2003, **36**, 75; (e) S. E. Denmark and R. F. Sweis, in *Metal-Catalysed Cross-Coupling Reactions*, Vol. 1, ed. A. de Meijere and F. Diederich, Wiley-VCH, Weinheim, 2nd edn, 2004, p. 163; (f) Y. Nakao and T. Hiyama, *Chem. Soc. Rev.*, 2011, **40**, 4893.
101. H. M. Lee and S. P. Nolan, *Org. Lett.*, 2000, **2**, 2053.
102. C. Dash, M. M. Shaikh and P. Ghosh, *Eur. J. Inorg. Chem.*, 2009, 1608.
103. X. Zhang, Q. Xia and W. Chen, *Dalton Trans.*, 2009, 7045.
104. I. Penafiel, I. M. Pastor, M. Yus, M. A. Esteruelas, M. Olivan and E. Onate, *Eur. J. Org. Chem.*, 2011, **2011**, 7174.
105. I. Penafiel, I. M. Pastor and M. Yus, *Eur. J. Org. Chem.*, 2013, **2013**, 1479.
106. Z.-S. Gu, L.-X. Shao and J.-M. Lu, *J. Organomet. Chem.*, 2012, **700**, 132.
107. J. Yang and L. Wang, *Dalton Trans.*, 2012, **41**, 12031.
108. (a) I. P. Beletskaya and A. V. Cheprakov, *Chem. Rev.*, 2000, **100**, 3009; (b) S. Brase and A. de Meijere, in *Metal-Catalysed Cross-Coupling Reactions*, Vol. 1, ed. A. de Meijere and F. Diederich, Wiley-VCH, Weinheim, 2nd edn, 2004, p. 217.
109. K. Albert, P. Gisdakis and N. Roesch, *Organometallics*, 1998, **17**, 1608.
110. M. I. Burguete, E. Garcia-Verdugo, I. Garcia-Villar, F. Gelat, P. Licence, S. V. Luis and V. Sans, *J. Catal.*, 2010, **269**, 150.
111. G. M. Pawar and M. R. Buchmeiser, *Adv. Synth. Catal.*, 2010, **352**, 917.
112. D. E. Bergbreiter, H.-L. Su, H. Koizumi and J.-H. Tian, *J. Organomet. Chem.*, 2011, **696**, 1272.

113. G. Borja, A. Monge-Marcet, R. Pleixats, T. Parella, X. Cattoen and M. Wong Chi Man, *Eur. J. Org. Chem.*, 2012, **2012**, 3625.
114. C.-S. Lee, Y.-B. Lai, W.-J. Lin, R. R. Zhuang and W.-S. Hwang, *J. Organomet. Chem.*, 2013, **724**, 235.
115. C. Cao, Y. Zhuang, J. Zhao, Y. Peng, X. Li, Z. Shi, G. Pang and Y. Shi, *Inorg. Chim. Acta*, 2010, **363**, 3914.
116. H. V. Huynh and R. Jothibasu, *J. Organomet. Chem.*, 2011, **696**, 3369.
117. J. D. Blakemore, M. J. Chalkley, J. H. Farnaby, L. M. Guard, N. Hazari, C. D. Incarvito, E. D. Luzik and H. W. Suh, *Organometallics*, 2011, **30**, 1818.
118. J.-Y. Lee, P.-Y. Cheng, Y.-H. Tsai, G.-R. Lin, S.-P. Liu, M.-H. Sie and H. M. Lee, *Organometallics*, 2010, **29**, 3901.
119. J. J. Dunsford and K. J. Cavell, *Dalton Trans.*, 2011, **40**, 9131.
120. T.-T. Gao, A.-P. Jin and L.-X. Shao, *Beilstein J. Org. Chem.*, 2012, **8**, 1916.
121. U. I. Tessin, X. Bantreil, O. Songis and C. S. J. Cazin, *Eur. J. Inorg. Chem.*, 2013, **2013**, 2007.
122. F.-T. Luo and H.-K. Lo, *J. Organomet. Chem.*, 2011, **696**, 1262.
123. Y.-Q. Tang, C.-Y. Chu, L. Zhu, B. Qian and L.-X. Shao, *Tetrahedron*, 2011, **67**, 9479.
124. M.-T. Ma and J.-M. Lu, *Appl. Organomet. Chem.*, 2012, **26**, 175.
125. Z. Wang, X. Feng, W. Fang and T. Tu, *Synlett*, 2011, 951.
126. I. Penafiel, I. M. Pastor and M. Yus, *Eur. J. Org. Chem.*, 2012, **2012**, 3151.
127. N. Koolaji, A. Abu-Mellal, V. H. Tran, R. K. Duke and C. C. Duke, *Eur. J. Med. Chem.*, 2013, **63**, 415.
128. K. Sonogashira, Y. Tohda and N. Hagihara, *Tetrahedron Lett.*, 1975, 4467.
129. H. A. Dieck and F. R. Heck, *J. Organomet. Chem.*, 1975, **93**, 259.
130. L. Cassar, *J. Organomet. Chem.*, 1975, **93**, 253.
131. D. S. McGuinness and K. J. Cavell, *Organometallics*, 2000, **19**, 741.
132. C. Yang and S. P. Nolan, *J. Org. Chem.*, 2002, **67**, 591.
133. S. Caddick, F. G. N. Cloke, G. K. B. Clentsmith, P. B. Hitchcock, D. McKerrecher, L. R. Titcomb and M. R. V. Williams, *J. Organomet. Chem.*, 2001, **617-618**, 635.
134. C. Yang and S. P. Nolan, *Organometallics*, 2002, **21**, 1020.
135. R. A. Batey, M. Shen and A. J. Lough, *Org. Lett.*, 2002, **4**, 1411.
136. R. Chinchilla and C. Najera, *Chem. Soc. Rev.*, 2011, **40**, 5084.
137. H. V. Huynh and C.-S. Lee, *Dalton Trans.*, 2013, **42**, 6803.
138. C. Xu, X.-H. Lou, Z.-Q. Wang and W.-J. Fu, *Transition Met. Chem.*, 2012, **37**, 519.
139. L. Yang, P. Guan, P. He, Q. Chen, C. Cao, Y. Peng, Z. Shi, G. Pang and Y. Shi, *Dalton Trans.*, 2012, **41**, 5020.
140. L. Yang, Y. Li, Q. Chen, Y. Du, C. Cao, Y. Shi and G. Pang, *Tetrahedron*, 2013, **69**, 5178.
141. A. John, S. Modak, M. Madasu, M. Katari and P. Ghosh, *Polyhedron*, 2013, **64**, 20.
142. (a) D. Alberico, M. E. Scott and M. Lautens, *Chem. Rev.*, 2007, **107**, 174; (b) L. Ackermann, R. Vicente and A. R. Kapdi, *Angew. Chem. Int. Ed.*,

2009, **48**, 9792; (c) G. P. McGlacken and L. M. Bateman, *Chem. Soc. Rev.*, 2009, **38**, 2447.

143. N. Gürbüz, I. Özdemir and B. Çetinkaya, *Tetrahedron Lett.*, 2005, **46**, 2273.

144. L.-C. Campeau, P. Thansandote and K. Fagnou, *Org. Lett.*, 2005, **7**, 1857.

145. B. S. Lane, M. A. Brown and D. Sames, *J. Am. Chem. Soc.*, 2005, **127**, 8050.

146. B. B. Toure, B. S. Lane and D. Sames, *Org. Lett.*, 2006, **8**, 1979.

147. J. M. Joo, B. B. Toure and D. Sames, *J. Org. Chem.*, 2010, **75**, 4911.

148. N. R. Deprez, D. Kalyani, A. Krause and M. S. Sanford, *J. Am. Chem. Soc.*, 2006, **128**, 4972.

149. I. Özdemir, Y. Gök, O. Özeroğlu, M. Kaloğlu, H. Doucet and C. Bruneau, *Eur. J. Inorg. Chem.*, 2010, 1798.

150. (a) I. Özdemir, H. Arslan, S. Demir, D. VanDerveer and B. Çetinkaya, *Inorg. Chem. Commun.*, 2011, **14**, 672; (b) S. Demir, I. Özdemir, H. Arslan and D. VanDerveer, *J. Organomet. Chem.*, 2011, **696**, 2589.

151. I. Özdemir, N. Gürbüz, N. Kaloğlu, O. Doğan, M. Kaloğlu, C. Bruneau and H. Doucet, *Beilstein J. Org. Chem.*, 2013, **9**, 303.

152. P. V. Kumar, W.-S. Lin, J.-S. Shen, D. Nandi and H. M. Lee, *Organometallics*, 2011, **30**, 5160.

153. J. C. Bernhammer and H. V. Huynh, *Organometallics*, 2012, **31**, 5121.

154. D. Yuan and H. V. Huynh, *Organometallics*, 2012, **31**, 405.

155. A. R. Martin, A. Chartoire, A. M. Z. Slawin and S. P. Nolan, *Beilstein J. Org. Chem.*, 2012, **8**, 1637.

156. (a) M. Kawatsura and J. F. Hartwig, *J. Am. Chem. Soc.*, 1999, **121**, 1473; (b) J. M. Fox, X. Huang, A. Chieffi and S. L. Buchwald, *J. Am. Chem. Soc.*, 2000, **122**, 1360.

157. M. S. Viciu, R. F. Germaneau and S. P. Nolan, *Org. Lett.*, 2002, **4**, 4053.

158. R. Singh and S. P. Nolan, *J. Organomet. Chem.*, 2005, **690**, 5832.

159. O. Navarro, N. Marion, Y. Oonishi, R. A. Kelly III and S. P. Nolan, *J. Org. Chem.*, 2006, **71**, 685.

160. (a) N. Marion, E. C. Ecarnot, O. Navarro, D. Amoroso, A. Bell and S. P. Nolan, *J. Org. Chem.*, 2006, **71**, 3816; (b) N. Marion, P. de Fremont, I. M. Puijk, E. C. Ecarnot, D. Amoroso, A. Bell and S. P. Nolan, *Adv. Synth. Catal.*, 2007, **349**, 2380.

161. J. Zhang, X. Yang, X. Cui and Y. Wu, *Tetrahedron*, 2011, **67**, 8800.

162. Z.-K. Xiao and L.-X. Shao, *Synthesis*, 2012, **44**, 711.

163. B. Landers, C. Berini, C. Wang and O. Navarro, *J. Org. Chem.*, 2011, **76**, 1390.

164. M. Kosugi, M. Kameyama and T. Migita, *Chem. Lett.*, 1983, 927.

165. A. S. Guram, R. A. Rennels and S. L. Buchwald, *Angew. Chem. Int. Ed. Engl.*, 1995, **34**, 1348.

166. J. Louie and J. F. Hartwig, *Tetrahedron Lett.*, 1995, **36**, 3609.

167. (a) J. F. Hartwig, in *Handbook of Organopalladium Chemistry for Organic Synthesis*, Vol. 1, ed. E.-i. Negishi, Wiley, New York, 2002, p. 1051; (b) A. R. Muci and S. L. Buchwald, *Top. Curr. Chem.*, 2002, **219**, 131;

(c) R. B. Bedford, C. S. J. Cazin and D. Holder, *Coord. Chem. Rev.*, 2004, **248**, 2283; (d) L. Jiang and S. L. Buchwald, in *Metal-Catalysed Cross-Coupling Reactions*, Vol. 1, ed. A. de Meijere and F. Diederich, Wiley-VCH, Weinheim, 2nd edn, 2004, p. 699; (e) J. F. Hartwig, *Acc. Chem. Res.*, 2008, **41**, 1534.

168. J. Huang, G. Grasa and S. P. Nolan, *Org. Lett.*, 1999, **1**, 1307.
169. K. H. Hoi, S. Calimsiz, R. D. J. Froese, A. C. Hopkinson and M. G. Organ, *Chem. Eur. J.*, 2011, **17**, 3086.
170. K. H. Hoi, S. Calimsiz, R. D. J. Froese, A. C. Hopkinson and M. G. Organ, *Chem. Eur. J.*, 2012, **18**, 145.
171. G. Berthon-Gelloz, M. A. Siegler, A. L. Spek, B. Tinant, J. N. H. Reek and I. E. Marko, *Dalton Trans.*, 2010, **39**, 1444.
172. A. Chartoire, X. Frogneux and S. P. Nolan, *Adv. Synth. Catal.*, 2012, **354**, 1897.
173. S. Meiries, A. Chartoire, A. M. Z. Slawin and S. P. Nolan, *Organometallics*, 2012, **31**, 3402.
174. A. Pommella, G. Tomaiuolo, A. Chartoire, S. Caserta, G. Toscano, S. P. Nolan and S. Guido, *Chem. Eng. J.*, 2013, **223**, 578.
175. See Section 4.2.3 for more information about the percent buried volume.
176. S. Meiries, K. Speck, D. B. Cordes, A. M. Z. Slawin and S. P. Nolan, *Organometallics*, 2013, **32**, 330.
177. A. Chartoire, A. Boreux, A. R. Martin and S. P. Nolan, *RSC Adv.*, 2013, **3**, 3840.
178. S. Dastgir, K. S. Coleman, A. R. Cowley and M. L. H. Green, *Organometallics*, 2010, **29**, 4858.
179. W. Fang, J. Jiang, Y. Xu, J. Zhou and T. Tu, *Tetrahedron*, 2013, **69**, 673.
180. Z. Jin, L.-L. Qiu, Y.-Q. Li, H.-B. Song and J.-X. Fang, *Organometallics*, 2010, **29**, 6578.
181. N. Marion, O. Navarro, E. D. Stevens, E. C. Ecarnot, A. Bell, D. Amoroso and S. P. Nolan, *Chem. Asian J.*, 2010, **5**, 841.
182. J. D. Egbert, A. Chartoire, A. M. Z. Slawin and S. P. Nolan, *Organometallics*, 2011, **30**, 4494.
183. L. Zhu, T.-T. Gao and L.-X. Shao, *Tetrahedron*, 2011, **67**, 5150.
184. W.-X. Chen and L.-X. Shao, *J. Org. Chem.*, 2012, 77, 9236.
185. S. Ho, G. Bondarenko, D. Rosa, B. Dragisic and A. Orellana, *J. Org. Chem.*, 2012, 77, 2008.
186. K. H. Hoi and M. G. Organ, *Chem. Eur. J.*, 2012, **18**, 804.
187. C.-F. Fu, Y.-H. Liu, S.-M. Peng and S.-T. Liu, *Tetrahedron*, 2010, **66**, 2119.
188. Y. Shi, Z. Cai, P. Guan and G. Pang, *Synlett*, 2011, 2090.
189. M. Sayah and M. G. Organ, *Chem. Eur. J.*, 2011, **17**, 11719.
190. M. Sayah, A. J. Lough and M. G. Organ, *Chem. Eur. J.*, 2013, **19**, 2749.

CHAPTER 5

Ancillary Ligand Design in the Development of Palladium Catalysts for Challenging Selective Monoarylation Reactions

MARK STRADIOTTO

Department of Chemistry, Dalhousie University, 6274 Coburg Road, P.O. Box 15000, Halifax, Nova Scotia, Canada B3H 4R2
Email: mark.stradiotto@dal.ca

5.1 Introduction

The development and application of homogeneous transition metal catalysis has played a key role in revolutionizing modern chemical synthesis,[1] as acknowledged recently by the awarding of the Nobel Prizes for Chemistry in 2001, 2005 and 2010; the last of these was given in recognition of the pioneering work of Richard F. Heck,[2] Ei-ichi Negishi[3] and Akira Suzuki[4] in establishing effective palladium-catalyzed methodologies for the formation of C–C bonds.[5,6] In the ensuing years, a diversity of complementary palladium-catalyzed C–C and C–X (X = N, O, S, *etc.*) bond-forming protocols have been developed. Among these, the arylation of NH-containing substrates (*i.e.*, Buchwald–Hartwig amination, BHA)[7–10] and the α-arylation (AA)[11,12] of carbonyl compounds have emerged as broadly useful C_{sp^2}–N and

RSC Catalysis Series No. 21
New Trends in Cross-Coupling: Theory and Applications
Edited by Thomas J Colacot

C_{sp^2}–C_{sp^3} bond-forming methods that enjoy widespread use in both academic and industrial settings. Such synthetic protocols are of particular utility in the construction of biologically active compounds, given the ubiquitous nature of both substituted aniline and α-arylated carbonyl sub-structures within such sought-after target molecules.

Prior to the development of BHA chemistry, the construction of C_{sp^2}–N bonds was limited primarily to classical arene nitration–reduction reaction sequences, and also nucleophilic aromatic substitutions employing selected amine nucleophiles in combination with electron-poor, and thus highly activated, aryl halide electrophiles. Similarly, before the establishment of palladium-catalyzed methods, the AA of carbonyl compounds required the use of electron-poor aryl halide reaction partners and in many cases the use of preformed enolate nucleophiles. While these more conventional synthetic methods can be employed successfully in the construction of C_{sp^2}–N and C_{sp^2}–C_{sp^3} bonds, they suffer from a number of important limitations, including: low substrate scope owing to the aforementioned need for activated reactants, poor functional group tolerance given the rather harsh reaction conditions/reagents employed and the required use of air-sensitive and/or toxic reagents based on tin and other metals. The development of palladium-catalyzed BHA and AA methodologies served to circumvent these synthetic limitations, enabling reactions to be conducted efficiently and under mild, user-friendly conditions employing amines and carbonyl compounds directly in combination with structurally diverse (hetero)aryl (pseudo)halides.

5.1.1 Mechanistic Overview

The evolution of palladium-catalyzed BHA and AA chemistry has been described in a number of comprehensive reviews,[9–14] and the mechanisms[15–19] of these transformations have been elucidated. Although the precise nature of the elementary transformations, catalytic intermediates and turnover-limiting step of such catalytic cycles can vary with substrate, base and catalyst composition, the mechanistic pathways of these reactions commonly proceed as outlined in Figure 5.1. Oxidative addition of the (hetero)aryl (pseudo)halide (*i.e.*, Ar–X) to an $L_nPd(0)$ species (**A**) affords the Pd(II) intermediate **B**. In the case of BHA, amine binding to **B** to give **C** followed by base-induced hydrodehalogenation generates the amido intermediate, **D**; subsequent C–N bond reductive elimination affords the aniline derivative with concomitant regeneration of **A**. For palladium-catalyzed AA, the intermediate **B** is transformed into the enolate complex **E** (which can exist in equilibrium with the Pd–O species **F**) in the presence of an appropriate carbonyl compound and base, with C–C bond reductive elimination affording the α-arylated carbonyl compound with re-formation of **A**. Notably, the ancillary ligand(s) (*i.e.*, L_n) employed, typically phosphines or *N*-heterocyclic carbenes, have a direct influence on the course of the elementary transformations.[20–22] Electron-rich and sterically demanding ligands promote the formation of low-coordinate compounds of type **A** that are

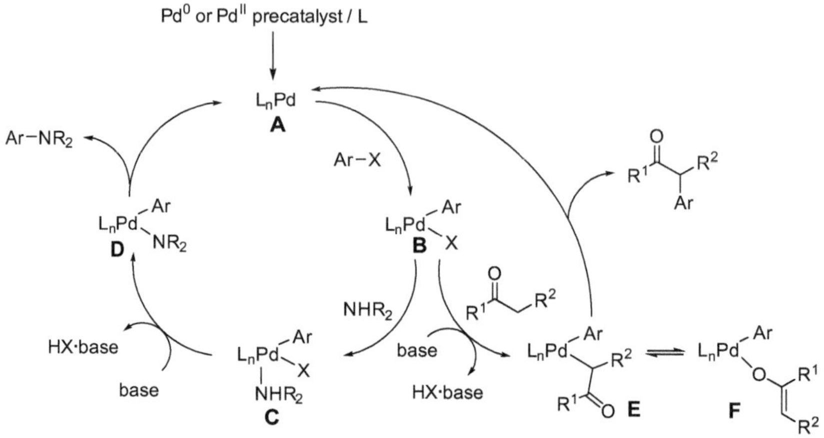

Figure 5.1   Generalized catalytic cycles for Buchwald–Hartwig amination (left) and palladium-catalyzed α-arylation reactions (right). L_n = ancillary ligand(s); Ar–X = (hetero)aryl (pseudo)halide.

predisposed to undergo Ar–X oxidative addition to give **B**. The steric and electronic properties of the ligands can serve to discourage unwanted dimerization within low-coordinate intermediates such as **B** or **D**, in addition to providing selectivity in amine binding (for BHA) or enolate complexation (for AA) proceeding from **B**. Both C–N (for BHA) and C–C (for AA) bond reductive elimination involving **D** and **E**, respectively, are facilitated by sterically demanding, yet less electron-donating, ligands; from an electronic perspective, this requirement is orthogonal to the ligand demands associated with the Ar–X oxidative addition step. Finally, an important and sometimes overlooked consideration in designing effective catalysts for BHA, AA and other palladium-catalyzed transformations relates to the efficient generation of the requisite L_nPd(0) species (**A**). Although the direct use of appropriately ligated Pd(0) species is ideal in this regard, such complexes are often cumbersome to prepare and/or exhibit air sensitivity, thereby preventing routine handling under benchtop conditions. Alternative approaches involve the formation of **A** *in situ* through the combination of an air-stable Pd(II) precursor, a chosen (ideally air-stable) ancillary ligand and a reductant. Although in some cases phosphines can be employed in excess both as the ancillary ligand and as the sacrificial reductant in this regard, the high cost of some structurally complex phosphine ligands renders such protocols unattractive. Furthermore, such *in situ* catalyst formation/reduction strategies, although operationally simple, require the efficient binding of the added ligand to palladium; when such reactions do not proceed cleanly, significant decomposition can occur, such that the actual quantity of **A** successfully formed and available for catalysis is very low. The use of preformed, well-characterized L_nPd(II) precatalyst complexes that can be reduced cleanly to **A** without consumption of the ancillary ligand offers an effective means of minimizing such problems during catalyst activation.[23] Alternatively, the direct use of a preformed catalyst intermediate (*e.g.*, **B** or structural analog), although more labor intensive than the aforementioned protocols, serves to deliver all of the added palladium to the catalytic cycle.

5.1.2 Ancillary Ligand Design Considerations

Given the profound influence that the chosen ancillary co-ligand(s) can have over the progress of these palladium-catalyzed transformations, it is not surprising that ligand design has played and continues to play a central role in advancing BHA and AA chemistry.[20,21] Early investigations made use of structurally simple triarylphosphine ancillary co-ligands; however, it was found subsequently that the use of more structurally complex ancillary ligands that serve to accelerate otherwise difficult segments of the catalytic cycle can enable more challenging substrate pairings to be accommodated, often with excellent functional group tolerance and under mild conditions (*e.g.*, low Pd/L loadings, mild reaction temperatures). In this regard, a number of effective ancillary ligation strategies have emerged, including sterically demanding monodentate ligands such as biarylmonophosphines[10,14] and *N*-heterocyclic carbenes,[21,24] and also rigid bidentate ligands including bisphosphines[9] and heterobidentate ligands[20,25] such as those featuring pairings of phosphorus and nitrogen donors. Collectively, these developments in ancillary ligand design have permitted tremendous progress to be made with regard to expanding the scope and utility of BHA and AA in terms of the (hetero)aryl (pseudo)halide and amine/carbonyl reaction partners that can be accommodated. Despite such advances, some of the most potentially useful variants of these transformations have proven to be stubbornly problematic; for example, the selective monoarylation of ammonia, hydrazine and acetone – among the simplest and most abundant of BHA/AA reaction partners – had until recently remained elusive. While each of these reactants presents a number of unique challenges in terms of their successful incorporation into productive palladium-catalyzed reaction cycles, the difficulty of obtaining monoarylation selectivity is common to all. The inherent challenge of achieving monoarylation selectivity when using these and related reaction partners in palladium-catalyzed BHA and AA chemistry arises because the initially formed monoarylation product in each case (*e.g.*, aniline, phenylhydrazine and phenylacetone for cross-coupling involving Ph–X reaction partners) competes as a preferred substrate with most known catalysts, leading to uncontrolled polyarylation even in the presence of limiting aryl electrophile.

The goal of this chapter is to highlight how the challenge of achieving monoarylation selectivity in palladium-catalyzed BHA and AA chemistry has been addressed in recent years through the design and/or implementation of sterically demanding phosphine ancillary ligands. Particular attention is given to documenting advances leading to the first reports of the selective monoarylation of ammonia, hydrazine and acetone as representative

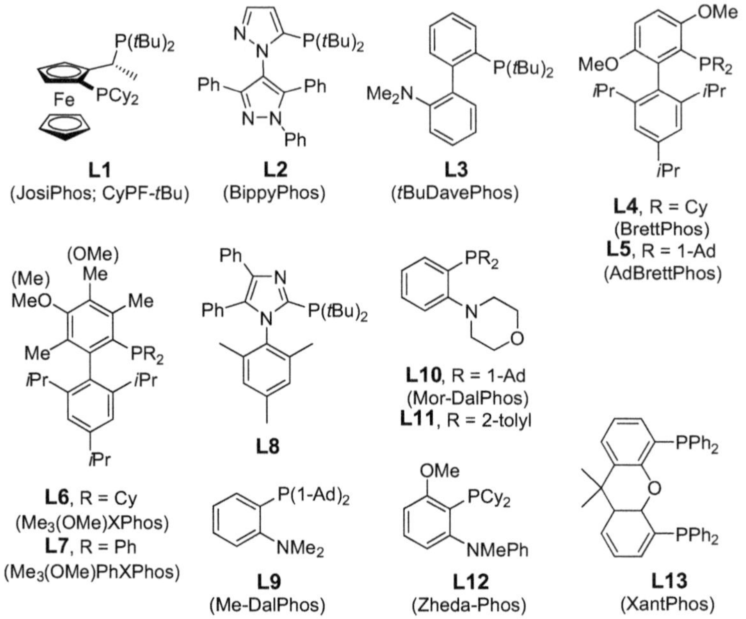

Figure 5.2   Structures of the ancillary ligands discussed in this chapter.

challenging reaction partners, given the broad synthetic potential of these novel transformations in chemical synthesis. For convenience, structural drawings of the ancillary ligands featured here, along with their common names where available, are presented in Figure 5.2.

5.2   Selective Monoarylation of Ammonia

Substituted primary anilines and related heteroaromatics represent common molecular fragments in biologically active compounds and can also serve as synthons in the rational construction of unsymmetrical secondary and tertiary arylamines. In consideration of the aforementioned synthetic limitations associated with nitration–reduction reaction sequences and nucleophilic aromatic substitutions *en route* to anilines, as outlined in Section 5.1, it is perhaps not surprising that the pursuit of BHA protocols for the construction of primary (hetero)arylamines has attracted considerable attention. Whereas the use of protected 'ammonia equivalents'[26] has been applied successfully in BHA chemistry, the direct and more atom-economical use of ammonia itself in the assembly of primary (hetero)-arylamines has proven to be particularly problematic. Indeed, many metal-catalyzed chemical transformations, including BHA, that are well established for other amine classes do not proceed with useful efficiency and selectivity when employing ammonia as a substrate using commonly employed reaction conditions.[27–29] Difficulties associated with the use of ammonia in BHA chemistry include, but are not restricted to: deactivation of the catalyst resulting from the formation of Werner-type ammine adducts; aggregation of amido intermediates of type **D** (Figure 5.1, where R = H) leading to inactive bridged polynuclear species; the slow rate of reductive elimination from sterically unencumbered intermediates of type **D**;[30] and, as alluded to in Section 5.1, uncontrolled polyarylation arising due to the competitive nature of the product (hetero)arylamines relative to ammonia when using most commonly employed BHA catalysts.[27–29] Although the use of copper-based catalysts does allow for the direct cross-coupling of (hetero)aryl bromides and iodides with ammonia,[29,31] such catalysts are not able to accommodate analogous chloride or sulfonate reagents in a useful manner, thereby restricting significantly the utility of such methods. Notwithstanding these notable challenges, the development of useful BHA protocols that allow the selective monoarylation of ammonia with a broad range of (hetero)aryl (pseudo)halides has been achieved in recent years, resulting from the design and/or application of new classes of sterically demanding ancillary ligands.

5.2.1 Development of Palladium-Catalyzed Ammonia Monoarylation

The efficient, selective monoarylation of ammonia by use of BHA protocols was first disclosed by Shen and Hartwig in 2006.[32] They successfully employed the palladium(II) precatalyst (CyPF-*t*Bu)PdCl$_2$ featuring the commercially available JosiPhos ligand, CyPF-*t*Bu (**L1**), which had been developed previously at Solvias for use in the asymmetric hydrogenation of alkenes.[33] Although the use of the chiral (non-racemic) ligand **L1** represented an unusual choice in this cross-coupling application given the lack of stereocenters in the cross-coupling products, it is worthy of note that the use of alternative monodentate phosphine [P(*t*Bu)$_3$, XPhos, QPhos] or *N*-heterocyclic carbene (IPr) ligands or bisphosphines (DPPF, BINAP) resulted in negligible reaction under the forcing reaction conditions employed (80 psi ammonia, 90 °C).[32] These observations support the idea that the rigid, sterically demanding and electron-rich nature of **L1** discourages unwanted polyarylation, in addition to possibly increasing the catalyst lifetime. In a subsequent publication 3 years later, Vo and Hartwig[34] described significant improvements to this palladium-catalyzed ammonia monoarylation chemistry through the use of Pd[P(*o*-tol)$_3$]$_2$/**L1** as the precatalyst mixture. This catalyst system allowed for the efficient monoarylation of ammonia using aryl bromides, chlorides, iodides and tosylates, including substrates featuring base-sensitive groups, without the routine need for high ammonia pressures (Figure 5.3). Notwithstanding the significance of this report[34] in terms of advancing the state-of-the-art with regard to the selective monoarylation of ammonia under BHA conditions, the need for relatively high reaction temperatures and the small demonstrated scope in the heteroaryl (pseudo)halide reaction partner left room for improvement. Klinkenberg and Hartwig[30] subsequently reported on the stoichiometric reactivity and

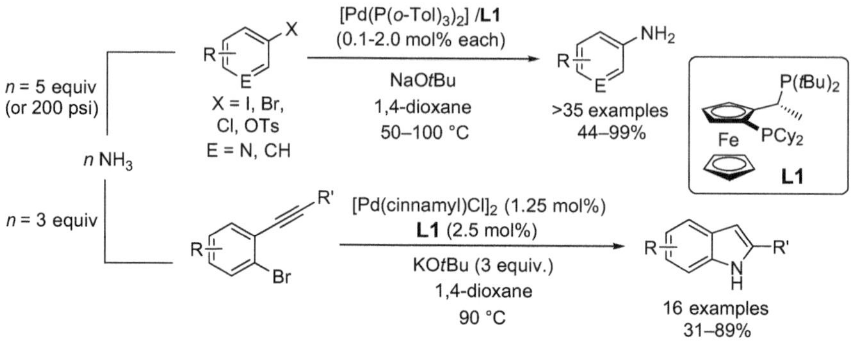

Figure 5.3 Selective palladium-catalyzed ammonia monoarylation employing the JosiPhos ligand CyPF-*t*Bu (**L1**).

kinetics of ammonia cross-coupling reactions employing Pd/**L1**, whereby the catalyst resting state was found to be an (**L1**)Pd(aryl)(NH$_2$) complex of type **D** (Figure 5.1).

5.2.2 Palladium-Catalyzed Ammonia Monoarylation in Heterocycle Synthesis

Encouraged by the seminal work of Vo and Hartwig in applying **L1** in palladium-catalyzed ammonia monoarylation, Stradiotto and co-workers demonstrated subsequently that [Pd(cinnamyl)Cl]$_2$/**L1** precatalyst mixtures could be applied successfully in analogous reactions employing functionalized 2-bromoarylacetylenes,[35] whereby KO*t*Bu-catalyzed hydroamination of the putative 2-aminoarylacetylene intermediates affords NH-indoles by way of a one-pot process (Figure 5.3). The establishment of this protocol is significant in that it represented the first reported synthesis of the biologically important indole framework directly from ammonia employing metal-catalyzed cross-coupling. Nonetheless, the lack of success in this chemistry when using 2-chloroarylacetylenes, heteroaryl halides or 2-bromoarylacetylenes featuring sp^3 substituents at the alkynyl terminus brings to light important limitations of this [Pd(cinnamyl)Cl]$_2$/**L1**-catalyzed protocol.

In 2013, Stradiotto and co-workers reported on the successful application of BippyPhos (**L2**) in a remarkably broad spectrum of BHA applications;[36] notably, [Pd(cinnamyl)Cl]$_2$/**L2** mixtures were shown to accommodate the largest diversity of NH-containing substrates reported for a single Pd–ligand catalyst system. Included in the reaction scope were a number of sterically/ electronically varied and challenging NH-containing substrates including ammonia, hydrazine and NH-indoles. The unique ability of the [Pd(cinnamyl)Cl]$_2$/**L2** catalyst system to accommodate such divergent nucleophilic partners was exploited in the development of a novel one-pot synthesis of *N*-arylindoles and related heterocyclic derivatives involving three sequential and selective C–N bond-forming steps, the first of which was the

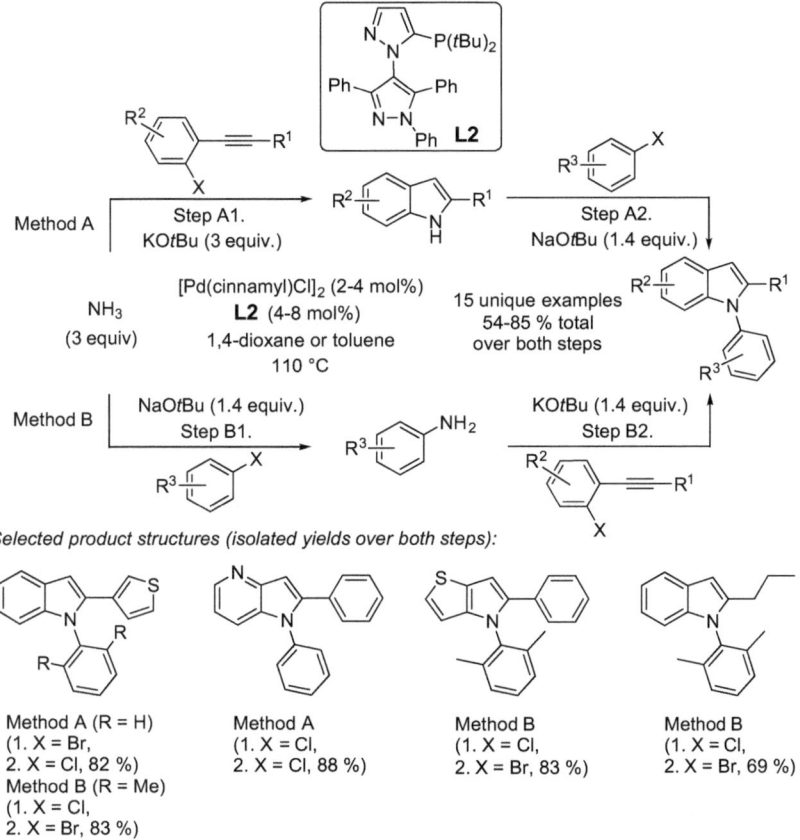

Figure 5.4 Scope of the [Pd(cinnamyl)Cl]$_2$/BippyPhos (**L2**)-catalyzed synthesis of substituted indoles and related heterocyclic derivatives involving selective ammonia monoarylation.

monoarylation of ammonia (Figure 5.4). Two complementary routes for accessing *N*-arylated indoles from ammonia in this manner were successfully established: (a) monoarylation of ammonia with a 2-halo(hetero)aryl-acetylene in the presence of excess base to form an NH-indole that was subsequently cross-coupled with an aryl halide to form the corresponding *N*-arylated indole (Method A, Figure 5.4); and (b) monoarylation of ammonia with an aryl halide to form an aniline that in turn was cross-coupled with a 2-halo(hetero)arylacetylene in the presence of excess base to form the corresponding substituted *N*-arylated indole (Method B, Figure 5.4). Collectively, the application of these selective ammonia monoarylation protocols using the [Pd(cinnamyl)Cl]$_2$/**L2** catalyst system provided access to a diversity of functionalized *N*-arylated indoles and related heterocyclic compounds in synthetically useful isolated yields [15 unique (hetero)indoles prepared; 54–85% total over both steps]. Moreover, although not exhaustively demonstrated, control experiments indicated that these one-pot indole syntheses

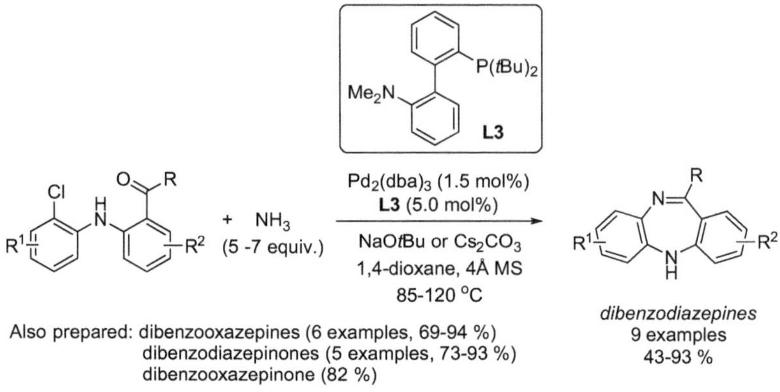

Figure 5.5   Application of ammonia monoarylation in the synthesis of dibenzodiazepines and related structural analogs.

can be conducted in air. It is worthy of mention that this ammonia monoarylation chemistry involving 2-halo(hetero)arylacetylenes does not exhibit
the substrate scope limitations encountered when using Pd(cinnamyl)Cl]₂/
L1 (Figure 5.3),[35] in that 2-chloro(hetero)arylacetylenes and substrates
featuring sp³ substituents at the alkynyl terminus were successfully transformed in the presence of [Pd(cinnamyl)Cl]₂/L2 catalyst mixtures.[36]

Biarylmonophosphine ligands developed by Buchwald and co-workers
have played a central role in the advancement of BHA chemistry,[10,14] and
their application in the selective monoarylation of ammonia has been
examined.[37–39] In preliminary experiments probing the cross-coupling of
ammonia (5 equiv.) with chlorobenzene, a selection of biarylmonophosphine ligands (5 mol%) in combination with Pd₂(dba)₃ (2 mol% Pd) as
the palladium source (dba = dibenzylideneacetone; 80 °C) were tested; the
ligand *t*BuDavePhos (L3) was found to exhibit the best performance among
the ligands examined in terms of substrate conversion and monoarylation
selectivity.[37] This Pd₂(dba)₃/L3 precatalyst system was subsequently employed with success by Tsvelikhovsky and Buchwald[38] in BHA reactions
employing ammonia, whereby the derived aniline intermediate undergoes
an intramolecular condensation reaction to afford dibenzodiazepines and
related biologically active structural analogs (Figure 5.5). The propensity of
biarylmonophosphine ligands including L3[40] to bind Pd(0) or Pd(II) *via*
phosphorus and one or more carbon atoms of the lower flanking arene ring
has been established.[10] In this context, the specific involvement of the
presumably uncoordinated dimethylamino group in L3 in promoting ammonia monoarylation selectivity remains unclear.

5.2.3   Application of Buchwald Palladacycles and Imidazolemonophosphines

The application of aminobiphenyl palladacyclic precatalysts featuring bulky
biarylmonophosphine ligands in the selective monoarylation of ammonia

was examined further by Buchwald and co-workers in 2013.[39] While Brett-Phos (**L4**) had been described in a previous publication by Fors and Buchwald as not being effective for ammonia monoarylation,[41] catalyst screening studies confirmed the superiority of AdBrettPhos (**L5**), Me$_3$(OMe)XPhos (**L6**), Me$_3$(OMe)PhXPhos (**L7**) and other structurally related and sterically demanding biarylmonophosphine ligands relative to **L3** in the selective monoarylation of ammonia with chlorobenzene using Pd$_2$(dba)$_3$ as the palladium source (3 equiv. NH$_3$; 2 mol% Pd; 5 mol% ligand).[39] Further improvements in performance were achieved through the use of the aminobiphenyl palladacyclic precatalyst complexes **P1**, **P2** and **P3** (featuring **L5**, **L6** and **L7**, respectively) (Figure 5.6). In particular, the use of **P2/L6** (2 mol% each) catalyst mixtures in place of Pd$_2$(dba)$_3$/**L6** (1 and 4 mol%, respectively) afforded faster rates of reaction and higher monoarylation product yields in the cross-coupling of ammonia with (hetero)aryl halides. The **P2/L6** catalyst system proved useful in the selective monoarylation of ammonia employing a broad selection of electron-rich, -neutral and -deficient aryl chlorides and bromides (27 examples; 53–97%; Figure 5.6). However, for more sterically hindered (hetero)aryl chlorides where the reactivity of **P2/L6** was shown to be lacking, the **P3/L7** catalyst mixture featuring Me$_3$(OMe)PhXPhos was employed as a means of obtaining higher yields of the target (hetero)aniline derivative. Reaction protocols employing **P2/L6** or **P3/L7** catalyst mixtures

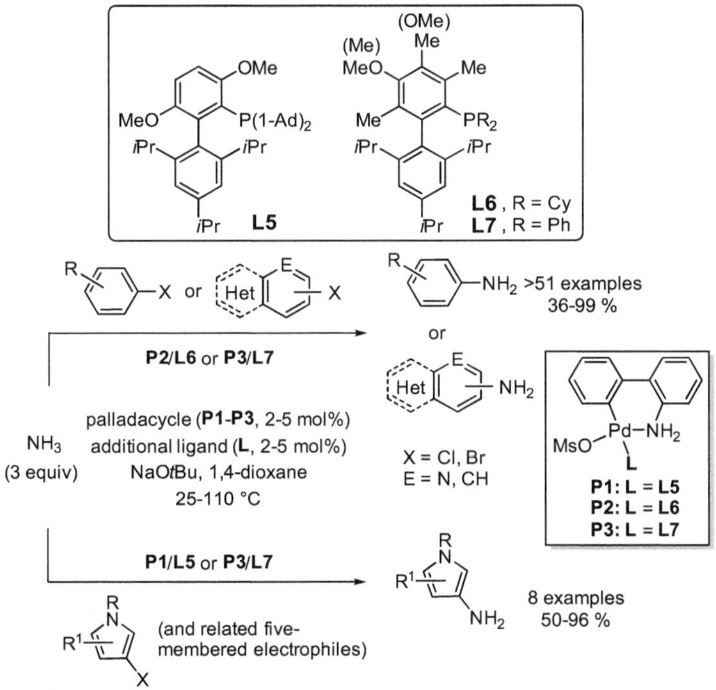

Figure 5.6   Expanding the scope of ammonia monoarylation by use of biaryl-monophosphine-ligated palladacyclic precatalysts.

also proved effective with regard to the cross-coupling of six-membered heteroaryl bromides and chlorides with ammonia, although in some particularly challenging cases diminished monoarylation selectivity was observed. A diverse array of aminopyridines, aminoquinolines and related NH$_2$-functionalized heterocycles, including benzothiophene, indole, benzothiazole, benzoxazole, pyrazine, quinoxaline, pyrimidine, pyridazine and carbazole rings, were prepared (24 examples; 36–99%; Figure 5.6). Although base-sensitive functionalities such as cyano and carbonyl groups, and also heterocyclic addenda, were reasonably well tolerated in this chemistry, room temperature reactions were limited to a relatively small number of examples featuring primarily aryl bromides and electronically activated (hetero)aryl chlorides at higher catalyst loadings (typically 5 mol%).[39]

Interest in the development of new and effective cross-coupling chemistry involving five-membered heteroaryl halide reaction partners arises from the potential utility of such reactions in the synthesis of biologically active compounds. In the pursuit of a catalyst system capable of allowing the hitherto unknown BHA of five-membered heteroaryl halides with ammonia, Buchwald and co-workers conducted further ligand screenings.[39] Whereas catalysts based on Me$_3$(OMe)XPhos (**L6**) performed poorly in the cross-coupling of ammonia with 4-bromo-1-(4-fluorophenyl)pyrazole (9% monoarylation), the di(1-adamantyl)phosphino-functionalized BrettPhos ligand variant **L5** afforded a high yield (78%) of the desired ammonia monoarylation product. The use of **P1/L5** catalyst mixtures (2 mol% each) was exploited in the monoarylation of ammonia using a range of five-membered heteroaryl bromides and chlorides including benzothiazoles, indazoles, imidazoles and pyrazoles (8 examples; 50–96%; Figure 5.6). The cross-coupling of the rather hindered 4-bromo-1,3,5-trimethylpyrazole with ammonia proved challenging when using the **P1/L5** catalyst system (40%); for this and another trisubstituted pyrazole substrate, the use of **P3/L7** afforded the desired monoarylation product in good yield (78 and 82%).[39]

Beller's group has also contributed to the development of effective palladium catalysts for the monoarylation of ammonia. In two publications,[42,43] they detailed that appropriately constructed imidazole-derived monophosphine ligands including **L8** are capable of supporting active complexes for the monoarylation of ammonia. However, high reaction temperatures (≥120 °C), pressures of inert gases (10 bar N$_2$) and catalyst loadings (1–4 mol% Pd, 4–16 mol% ligand) are commonly required when using such catalysts in order to obtain satisfactory catalytic performance, and the demonstrated scope of the reaction in the aryl electrophile is limited.

5.2.4 Heterobidentate κ^2-P,N Ligands: Chemoselectivity and Room Temperature Reactions

Given the successful application of sterically demanding monophosphine and bisphosphine ligands in both BHA and AA chemistry, it is surprising that electronically intermediate heterobidentate ligands featuring pairings of soft and hard donor atoms have received little attention in these applications; this is especially true in the light of the proven utility of κ^2-P,N ligands in alternative late metal-catalyzed chemical transformations (*e.g.*, PHOX[44] in alkene hydrogenation, allylic substitution). Sterically demanding κ^2-P,N ligands represent conceptually appealing targets of inquiry in BHA and AA chemistry, given their potential to discourage unwanted dimerization of catalytic intermediates (*e.g.*, **B**, **D** and **E** in Figure 5.1) – processes that inhibit the performance of monophosphine-based palladium catalysts. Moreover, while challenging (hetero)aryl (pseudo)halide oxidative addition reactions can be enabled *via* incorporation of an electron-rich dialkylphosphino ligand donor fragment, a κ^2-P,N ligand of this type should also render the palladium centers in key catalytic intermediates of type **D** and **E** (Figure 5.1) less electron rich than their bisphosphine-ligated counterparts, thereby providing an electronic means of promoting challenging reductive eliminations in addition to sterically promoted processes. Finally, bulky κ^2-P,N ligands may offer an effective means of achieving selectivity in challenging palladium-catalyzed BHA and AA monoarylation reactions, by favoring the binding of small, unhindered substrates, rather than their more hindered monoarylated derivatives, proceeding from catalytic intermediates of type **B**.

The successful application of sterically demanding κ^2-P,N ligands in the palladium-catalyzed selective monoarylation of ammonia was first disclosed in 2010 by Stradiotto and co-workers.[45,46] In an initial report,[45] Me-DalPhos (**L9**) was shown to be effective in BHA chemistry involving a wide range of (hetero)aryl chlorides and NH-containing substrates, including ammonia. However, whereas **L9** was observed to afford high conversions and good monoarylation selectivities in the cross-coupling of ammonia with *ortho*-substituted aryl chlorides, the use of alternative aryl chlorides lacking steric bias resulted in diminished monoarylation selectivity.[45] Subsequent ligand optimization gave rise to the Mor-DalPhos ligand (**L10**), which was shown to be highly effective for the palladium-catalyzed cross-coupling of ammonia with aryl chlorides and tosylates, including those lacking *ortho* substitution; included in this report are the first examples of room temperature BHA chemistry involving ammonia.[46] Reactions employing [Pd(cinnamyl)Cl]$_2$/**L10** catalyst mixtures proceeded with high monoarylation selectivity, generally at low catalyst loadings and under mild reaction conditions without the need for high pressures of ammonia (Figure 5.7). Aryl chlorides bearing electron-donating groups at the *para* or *meta* positions that have proven to be particularly challenging in ammonia monoarylation chemistry were effectively cross-coupled, as were substrates containing N, O, F or S heteroatoms. Sterically biased *ortho*-substituted aryl chlorides were also found to be suitable reaction partners, as were some heteroaryl chlorides. Whereas the use of [Pd(cinnamyl)Cl]$_2$/**L10** catalyst mixtures in the monoarylation of ammonia with (hetero)aryl chlorides required heating (≥ 50 °C), analogous reactions involving aryl tosylates were found to proceed at room temperature with good yields (69–83%) being obtained for both unhindered and *ortho*-substituted substrates.

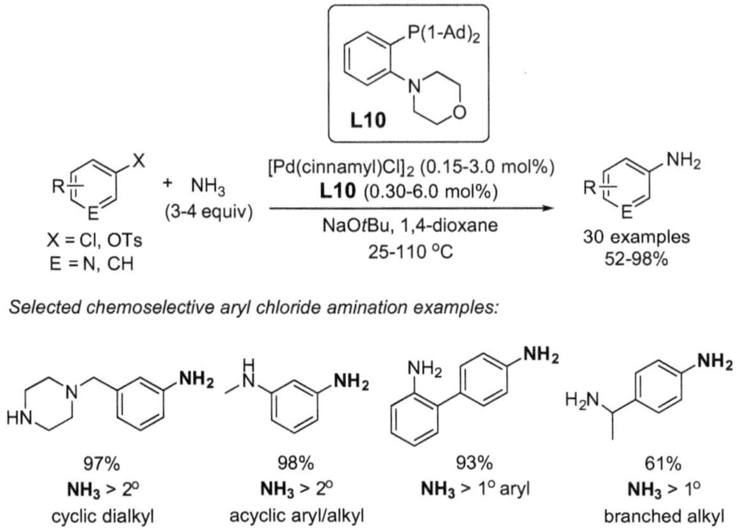

Figure 5.7 Selective monoarylation of ammonia employing the κ^2-P,N ligand Mor-DalPhos (**L10**), including chemoselective reaction variants.

The proclivity of the [Pd(cinnamyl)Cl]$_2$/**L10** catalyst system for ammonia monoarylation was exploited in unprecedented chemoselective aminations involving aryl chloride substrates featuring potentially competitive NH functionalities (Figure 5.7).[46] Reactions of aryl chlorides featuring secondary aryl/alkyl-, diaryl- or dialkylamine addenda each afforded good to excellent isolated yields (64–98%) of the ammonia-derived monoarylation product. Perhaps even more remarkable was the ability of [Pd(cinnamyl)Cl]$_2$/**L10** to promote selectively the selective monoarylation of ammonia in the presence of aryl chlorides featuring contending *primary* aryl- or alkylamino substituents.

In the same study, Stradiotto and co-workers briefly explored the coordination chemistry of **L10**, including the synthesis of putative catalytic intermediates; among the crystallographically characterized compounds prepared was the chlorobenzene oxidative addition complex (κ^2-P,N-**L10**)Pd(Ph)Cl (*cf.*, **B** in Figure 5.1).[46] Remarkably, the use of rationally prepared (κ^2-P,N-**L10**)Pd(Ph)Cl as a precatalyst enabled the first examples of room temperature ammonia monoarylation employing (hetero)aryl chlorides to be achieved. In a subsequent report, Stradiotto and co-workers established broad scope for room temperature ammonia monoarylation reactivity when using (κ^2-P,N-**L10**)Pd(Ph)Cl as a precatalyst (5 mol%), including a range of (hetero)aryl (pseudo)halides (X = Cl, Br, I, OTs) with diverse substituents (alkyl, aryl, ether, thioether, ketone, amine, fluoro, trifluoromethyl and nitrile), in addition to chemoselective monoarylations (Figure 5.8).[47] Although a complete understanding of the properties of (κ^2-P,N-**L10**)Pd(Ph)Cl that allow efficient room temperature ammonia monoarylation with (hetero)aryl (pseudo)halides is currently lacking, it is

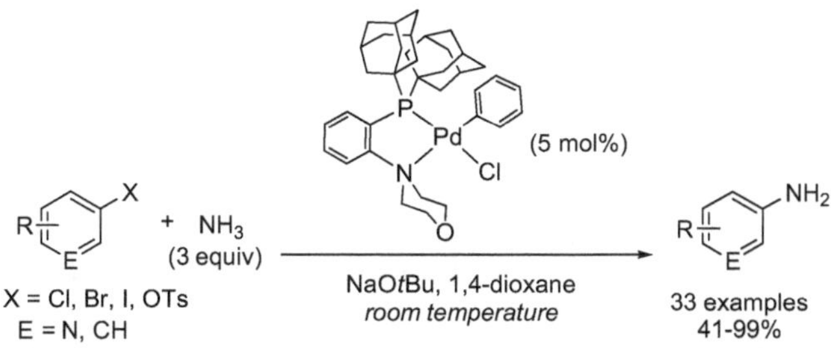

Figure 5.8 Broad scope in the room temperature selective monoarylation of ammonia with (hetero)aryl (pseudo)halides employing (κ^2-P,N-**L10**)Pd(Ph)Cl as a precatalyst.

conceivable that the direct use of this putative catalytic intermediate serves to by-pass deleterious side reactions that may otherwise occur during catalyst activation steps.

5.2.5 Summary

In less than a decade following the pioneering report of palladium-catalyzed ammonia monoarylation by Shen and Hartwig, a number of highly effective catalyst systems for such transformations have been identified, thereby permitting significant practical advances. Collectively, catalysts reported to date allow for a broad spectrum of (hetero)aryl (pseudo)halides to be accommodated, including transformations that are highly chemoselective, proceed at room temperature, and/or that can be exploited in the assembly of synthetically important heterocyclic frameworks. Although formally beyond the scope of this review, it is worthy of mention that carbonylative variants of such ammonia monoarylation reactions have also emerged that offer a novel entry point to the construction of primary aromatic amides from (hetero)aryl halides, carbon monoxide and ammonia.[48–51] The identification of useful catalysts for ammonia monoarylation laid the groundwork for the development of previously unknown BHA reactions involving the selective monoarylation of hydrazine with (hetero)aryl (pseudo)halides.

5.3 Selective Monoarylation of Hydrazine

Arylhydrazines function as important synthons in the construction of myriad heterocyclic nitrogen-containing compounds, including most notably indoles *via* the Fischer indole synthesis.[52] Given the ubiquitous nature of such heterocyclic frameworks in biologically active molecules,[53–56] it is understandable that there is considerable interest in developing efficient synthetic routes to arylhydrazines. Conventional methods for preparing arylhydrazines involve the stoichiometric oxidation of anilines to their corresponding diazonium salts, followed by reduction. Alternatively, hydrazine can in some cases react directly with electron-deficient haloarenes in nucleophilic aromatic substitution reactions as a means of producing aryl-hydrazines. However, both of these synthetic methods are limited in terms of the substrate scope and functional group tolerance that can be achieved, owing to the need for aggressive reagents and/or activated substrate molecules. In this regard, the development of BHA protocols involving the selective monoarylation of hydrazine using (hetero)aryl (pseudo)halides represents, at first glance, an obvious complementary route to substituted arylhydrazines. However, the use of hydrazine as a cross-coupling partner has proven to be among the most formidable challenges in BHA chemistry.

The utilization of hydrazine as an NH-containing reagent in BHA chemistry shares the inherent challenges associated with the use of ammonia (Section 5.2), including catalyst deactivation/aggregation, anticipated slow C–N bond reductive elimination and the potential for uncontrolled poly-arylation. Additionally, the successful use of hydrazine as a cross-coupling partner in BHA chemistry requires that unwanted hydrodehalogenation of the (hetero)aryl halide substrate, rapid reduction of the palladium pre-catalyst to give ill-defined colloidal/heterogeneous palladium materials and metal-mediated N–N bond cleavage resulting in undesired aniline by-products be circumvented. Despite these daunting challenges, reports of palladium-catalyzed hydrazine monoarylation to afford arylhydrazines have begun to appear.

5.3.1 Development of Palladium-Catalyzed Hydrazine Monoarylation

The selective monoarylation of hydrazine by use of BHA protocols was first reported by Lundgren and Stradiotto in 2010.[57] In the course of this catalytic investigation, a diversity of phosphine and *N*-heterocyclic carbene ancillary ligands were surveyed with regard to their ability to support active palladium catalysts for hydrazine monoarylation. In keeping with the challenges outlined in Section 5.3, most of the ligands examined in the test reaction involving 4-phenylchlorobenzene and hydrazine hydrate afforded either poor conversion of the chloroarene in keeping with rapid catalyst decomposition/inhibition or exclusive reduction of this substrate to biphenyl by way of palladium-catalyzed hydrodehalogenation. Conversely, **L1**, **L9** and **L10** each gave rise to the desired arylhydrazine as the major product, thereby representing the first documented examples of arylhydrazine synthesis by way of selective palladium-catalyzed hydrazine monoarylation. Further catalyst optimization revealed **L10** to be marginally better than **L1** and **L9**; in turn, [Pd(cinnamyl)Cl]$_2$/**L10** catalyst mixtures (3–10 mol% Pd; Pd:**L10** = 1:1.5) were shown to be useful in transforming various (hetero)aryl chlorides and tosy-lates, in combination with hydrazine hydrate or hydrazine hydrochloride, into the corresponding arylhydrazines with excellent monoarylation selectivity, including chemoselective examples involving substrates bearing pendant NH functionality (32 examples total; 27–97%; Figure 5.9).[57] Throughout, generally good functional group tolerance was observed, with

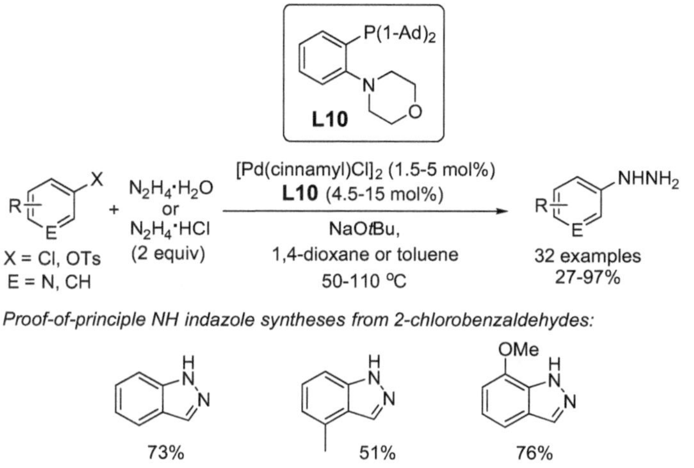

L10

$$X = Cl, OTs \qquad E = N, CH$$

$$R + N_2H_4\cdot H_2O \text{ or } N_2H_4\cdot HCl \text{ (2 equiv)} \xrightarrow[\text{NaO}t\text{Bu,} \\ \text{1,4-dioxane or toluene} \\ 50\text{-}110\ ^\circ\text{C}]{[\text{Pd(cinnamyl)Cl}]_2 \text{ (1.5-5 mol\%)} \\ \textbf{L10} \text{ (4.5-15 mol\%)}} R\text{-NHNH}_2$$

32 examples
27-97%

Proof-of-principle NH indazole syntheses from 2-chlorobenzaldehydes:

73%   51%   76%

Figure 5.9   Selective monoarylation of hydrazine with (hetero)aryl (pseudo)halides employing the [Pd(cinnamyl)Cl]$_2$/**L10** catalyst system.

oxygen-, sulfur- and nitrogen-based addenda within aryl electrophiles featuring or lacking *ortho* substitution being accommodated successfully. The successful application of 3-chloropyridine in this chemistry is noteworthy; unlike isomeric 2- or 4-halopyridine, 3-chloropyridine is not susceptible to nucleophilic aromatic substitution with hydrazine. The proof-of-principle synthesis of NH-indazoles *via* BHA/condensation reactions employing 2-chlorobenzaldehydes and hydrazine sources was also described in this report (Figure 5.9). Despite this progress, some significant substrate scope limitations were encountered when using the [Pd(cinnamyl)Cl]$_2$/**L10** catalyst system in this chemistry, including inferior yields of the target arylhydrazine when using the electron-poor substrate 4-trifluoromethylchlorobenzene (50% yield) due to competitive hydrodehalogenation, and poor conversion when employing the electron-rich substrate 4-chloroanisole (27% yield).[57] In a subsequent publication, Stradiotto and co-workers demonstrated that [Pd(cinnamyl)Cl]$_2$/**L2** catalyst mixtures are also capable of promoting the selective monoarylation of hydrazine with (hetero)aryl chlorides.[36] Although the scope in (hetero)aryl (pseudo)halide was not broadly explored, it appears that the [Pd(cinnamyl)Cl]$_2$/**L2** catalyst system may out-perform **L10**-based catalysts in the case of some electron-rich aryl chlorides (*e.g.*, 4-chloroanisole, 70%).[36]

Competition experiments were conducted in an effort to learn more about the preference of the [Pd(cinnamyl)Cl]$_2$/**L10** catalyst system for hydrazine monoarylation (Figure 5.10). The successful cross-coupling of phenylhydrazine with 2-chloro-*p*-xylene confirmed the ability of [Pd(cinnamyl)Cl]$_2$/**L10** to promote 1,2-diarylations of hydrazine. However, when analogous reactions were conducted in the presence of an equivalent of hydrazine hydrate, only the product derived from the monoarylation of hydrazine with 2-chloro-*p*-xylene was observed. Collectively, these results establish what although 1,2-diarylations of hydrazine are feasible, the [Pd(cinnamyl)Cl]$_2$/**L10**

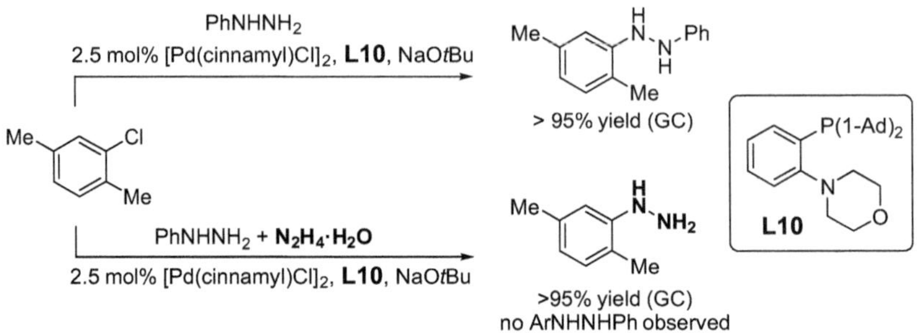

Figure 5.10 Competition experiments involving the [Pd(cinnamyl)Cl]₂/**L10** catalyst system, demonstrating high selectivity for the monoarylation of hydrazine over phenylhydrazine.

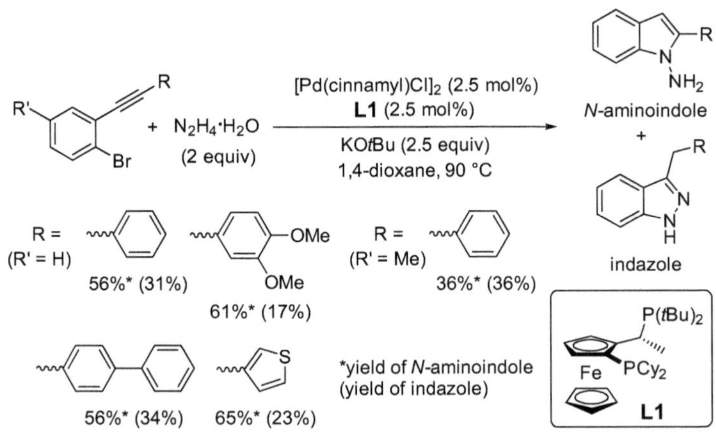

Figure 5.11 Tandem palladium-catalyzed cross-coupling–hydroamination reactions of hydrazine with 2-alkynylbromoarenes to afford *N*-aminoindole/indazole products.

catalyst system exhibits a clear preference for hydrazine over arylhydrazine as a substrate in BHA, leading to selective hydrazine monoarylation.[57]

The successful application of [Pd(cinnamyl)Cl]₂/**L1** catalyst mixtures in the BHA of hydrazine with 2-alkynylbromoarenes to afford *N*-aminoindoles and indazoles was reported by Stradiotto and co-workers (Figure 5.11).[35] Although in all cases the *N*-aminoindole product was favored, the formation of appreciable quantities of the analogous indazole serves to limit the synthetic utility of this protocol; efforts to achieve greater selectivity by altering the base or solvent or by including additives (*e.g.*, CuCl₂ or Ag₂CO₃) were unsuccessful. It is interesting that despite the efficacy of palladium catalysts featuring **L10** in the selective monoarylation of hydrazine (Figure 5.9),[57] [Pd(cinnamyl)Cl]₂/**L10** catalyst mixtures performed poorly in comparison with **L1**-based palladium catalysts in these tandem BHA–hydroamination processes.

5.3.2 Palladium-Catalyzed Hydrazine Monoarylation in Flow

Notwithstanding the synthetic potential of BHA protocols that make use of hydrazine as a direct route to synthetically useful arylhydrazines, caution must be exercised when using hydrazine in this context. Reasons for concern arise from the high energy content, flammability and explosion hazard associated with hydrazine, especially when heating in the presence of transition metals.[58] Although the use of hydrazine hydrate can provide a means of mitigating such hazards, the development of safe and potentially scalable experimental protocols that further reduce the risks associated with the use of hydrazine sources in BHA chemistry represents an important goal.

In this context, Buchwald and co-workers reported on the development of mild and effective hydrazine monoarylation protocols that make use of continuous flow reactor technology.[59] A preliminary ligand screen conducted under batch conditions using commercially available solutions of hydrazine (1.0 M in THF; 2.0 equiv.) and 4-chlorotoluene as the electrophile revealed BrettPhos (**L4**) to be particularly effective; excellent monoarylation selectivity ($\geq$95%) was achieved using only 1 mol% of the **L4**-containing precatalyst **P4** in a few minutes at room temperature. This catalyst system was then successfully adapted for use in a continuous flow process, which enabled a range of (hetero)arylhydrazines to be prepared at room temperature using short reaction times (1.5–6 min) and subsequently transformed into a range of synthetically useful target scaffolds including hydrazones, pyrazoles, pyrazolones and NH-indoles (17 unique examples; 59–95%; Figure 5.12). The substrate scope in this chemistry was found to accommodate electron-neutral and electron-poor aryl chlorides; electron-rich substrates including 4-chloroanisole (59% yield) proved more challenging, as was the case with [Pd(cinnamyl)Cl]$_2$/**LX** (X = 2 or 10), requiring 5 mol% catalyst. A diversity of functional groups were tolerated in this chemistry, including ester, fluoroether, halogen, pyrrole and alcohol groups, and also potentially competitive aniline and amide addenda. A small selection of heteroaryl chlorides were also accommodated, albeit at higher catalyst loadings (5 mol%) and reaction temperature (45 °C).[59]

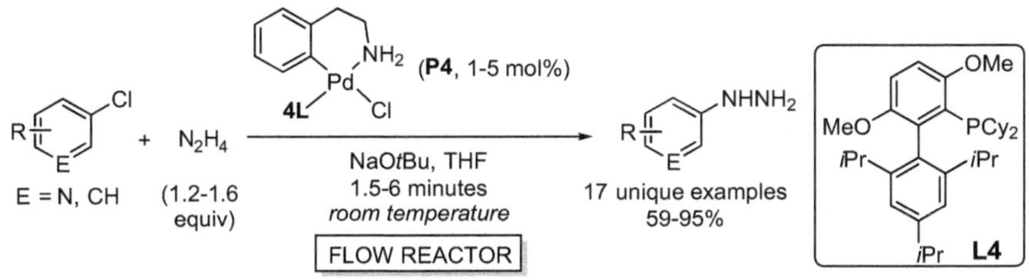

Figure 5.12 Palladium-catalyzed selective monoarylation of hydrazine with (hetero)aryl chlorides under continuous flow conditions.

5.3.3 Summary

The successful development of palladium-catalyzed protocols for the selective
monoarylation of hydrazine represents a significant advance in BHA chemistry,
providing a new route to (hetero)arylhydrazines that can be employed directly
in the construction of sought-after nitrogen heterocycles. Whereas the initial
development of this reactivity involved batch reactions, follow-on studies confirmed that palladium-catalyzed hydrazine monoarylation can be conducted
under continuous flow conditions at room temperature, thereby providing a
means of mitigating the potential hazards associated with the use of hydrazine
under BHA conditions. Although a report of a copper-based catalyst system that
is capable of promoting the selective monoarylation of hydrazine has appeared,[60] this chemistry involves the use of hydrazine hydrate as a co-solvent
and is limited to selected aryl bromides and iodides; in this regard, palladium
catalysts exhibit superior reactivity. Several of the ligands that were shown to
perform well in the palladium-catalyzed monoarylation of ammonia also
proved to be effective in hydrazine monoarylation, suggesting that these ligands may have use in the selective monoarylation of other substrates that are
inherently prone to polyarylation, including acetone in α-arylation chemistry.

5.4 Selective Monoarylation of Acetone

Despite significant progress with regard to the establishment of effective
metal-catalyzed protocols for the α-arylation of carbonyl compounds, the
development of catalysts that exhibit high levels of monoarylation selectivity
with sterically unbiased (hetero)aryl electrophiles has proven challenging.[11,12] As mentioned in Section 5.1.2, the difficulties associated with
achieving high levels of monoarylation selectivity in such transformations
can be attributed in part to the fact that the initially formed α-arylation
product possesses α-CH protons that are more acidic than those in the
starting carbonyl compound, resulting in more facile enolate formation; for
most catalysts, this scenario results in low monoarylation selectivity. Although examples of the palladium-catalyzed selective monoarylation of
methyl carbonyl compounds started to appear at the time of the initial
discovery of palladium-catalyzed α-arylation chemistry in 1997,[11,12,61–63]
such transformations involving acetone,[19] the structurally simplest ketone,
remained unknown for more than a decade thereafter. As was described in
the preceding sections regarding the successful development of palladium-
catalyzed protocols for the selective monoarylation of ammonia and hydrazine, the judicious choice of ancillary ligand proved important in achieving
monoarylation selectivity in the α-arylation of acetone.

5.4.1 Development of Palladium-Catalyzed Acetone
Monoarylation

The difficulties associated with accommodating acetone in α-arylation
chemistry have been circumvented in part through the use of stannyl or silyl

acetone enolates.[64–67] However, such strategies employing 'acetone enolate equivalents' lack atom economy and introduce added costs, given the requirement for extra synthetic steps and the use of stoichiometric additives; clearly, the direct use of acetone is preferred in the construction of phenylacetone derivatives, which are sought-after for their biological properties. In this regard, the selective monoarylation of acetone was first reported by Stradiotto and co-workers in 2011 by use of [Pd(cinnamyl)Cl]$_2$/**L10**.[68] This catalyst system was initially chosen for a test examination involving the addition of 4-chlorotoluene to acetone, on the basis of its efficacy with regard to the selective monoarylation of ammonia and hydrazine (see Sections 5.2.4 and 5.3.1).[46,57] Preliminary optimization experiments established 2 mol% Pd (Pd:**L10** = 1:2), acetone (10 equiv. or neat) and Cs$_2$CO$_3$ (2 equiv.) at 90 °C for 5 h as representing suitably effective conditions for the high-yielding formation of 4-tolylacetone (89% yield); the use of other palladium sources [Pd(OAc)$_2$ or Pd(dba)$_2$] or bases (K$_2$CO$_3$, Na$_2$CO$_3$, LiHMDS or NaOtBu) afforded comparatively poor results. In screening alternative ligands in this chemistry, it was found that replacement of the di(1-adamantyl)phosphino group in **L10** for a dicyclohexylphosphino group or the morpholino moiety for a dimethylamino donor fragment (*i.e.*, **L9**) resulted in a loss of catalyst selectivity and/or activity. A series of other sterically demanding and electron-rich monophosphines/*N*-heterocyclic carbenes were also surveyed under analogous conditions, including **L1** and **L3**, with each providing inferior results to those obtained when using **L10**.[68]

The scope of acetone monoarylation reactivity exhibited by the [Pd(cinnamyl)Cl]$_2$/**L10** catalyst system (2–5 mol% Pd; Pd:**L10** = 1:2) was found to accommodate structurally diverse aryl chlorides, bromides, iodides and tosylates (Figure 5.13, **A**) in neat acetone at 90 °C.[68] Electron-rich, -neutral and -poor aryl chlorides were employed successfully, including substrates featuring ether, alcohol, olefin, tertiary amine or *N*-heterocyclic (pyridine, pyrrole, *N*-benzylindole) addenda. Furthermore, aryl chloride electrophiles featuring potentially competitive enolizable sites, including benzyl and homobenzyl esters or an acetanilide group, were employed successfully in the chemoselective monoarylation of acetone, affording the substituted phenylacetone derivatives in synthetically useful yields. Whereas in general aryl bromides and iodides were also found to be suitable electrophiles in this chemistry, the use of electron-poor aryl bromides, such as 4-bromobenzonitrile, proved to be much more challenging (33% yield) when using the [Pd(cinnamyl)Cl]$_2$/**L10** catalyst system. Although limited to only four entries, the demonstrated scope in aryl tosylates was found to include electron-rich and sterically congested substrate variants.[68]

Scrutiny of the yield data as a function of electrophile structure revealed that electron-poor aryl halides consistently provided lower yields of the α-aryl methyl ketone product in comparison with electron-rich or -neutral aryl halides when using [Pd(cinnamyl)Cl]$_2$/**L10**; substrate competition studies indicated that reductive elimination might be rate limiting when using such aryl halides.[68] In response, a sterically hindered yet less electron-donating

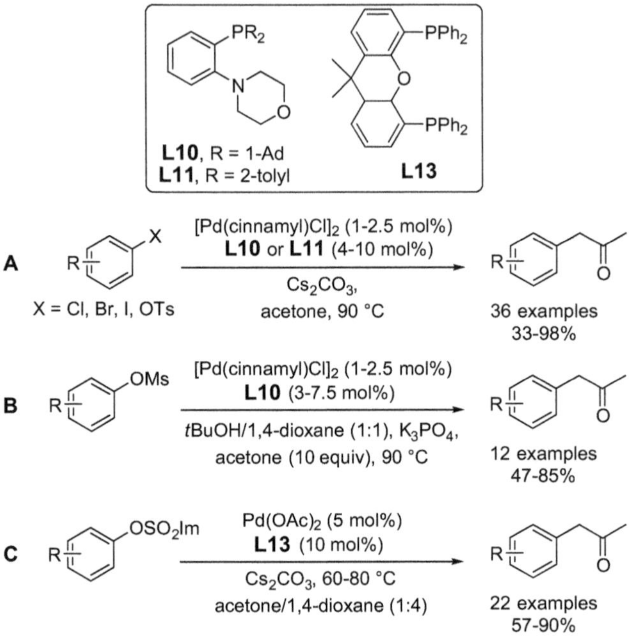

Figure 5.13 Palladium-catalyzed selective monoarylation of acetone with aryl (pseudo)halides.

DalPhos ligand variant featuring a di(2-tolyl)phosphino donor fragment (**L11**) was prepared; such structural variation is easily accommodated, given the modular nature by which DalPhos ligands are prepared using C–N and C–P cross-coupling protocols, starting from *o*-dihaloarenes. The [Pd(cinnamyl)Cl]$_2$/**L11** catalyst system was applied successfully in the selective monoarylation of acetone using aryl bromides bearing one or more electron-withdrawing group (chloro, fluoro, trifluoromethyl or acyl). The use of **L11** under similar conditions for the α-arylation of acetone with electron-deficient aryl chlorides proved unsuccessful, which is likely due to the more challenging C–Cl oxidative addition reaction when employing triarylphosphine-type ligands.[68] Collectively, this case study serves to highlight the importance of ligand modularity in the construction of ancillary ligands, as a means of rationally addressing reactivity challenges. In a subsequent study, Ma and co-workers[69] reported that the very closely related [Pd(cinnamyl)Cl]$_2$/**L12** catalyst system, under conditions identical with those described previously by Stradiotto and co-workers,[68] was also capable of promoting the selective monoarylation of acetone employing a range of (hetero)aryl chlorides.

5.4.2 Palladium-Catalyzed Acetone Monoarylation Employing Aryl Methanesulfonates

Aryl methanesulfonates (mesylates) are attractive electrophiles for palladium-catalyzed cross-coupling reactions owing to their ease of synthesis from readily available phenols and the associated increase in atom economy in comparison with the use of higher molecular weight sulfonate leaving groups, such as tosylates or triflates.[70] Furthermore, the derived cross-coupling by-product formed upon workup, methanesulfonic acid, is naturally occurring and undergoes biodegradation in wastewater processing.[71] Despite intense research efforts, the development of palladium catalysts for BHA and AA chemistry that are capable of facilitating turnover of the aryl mesylate, while circumventing phenol formation, has proven to be a considerable challenge.[70] Indeed, the successful incorporation of aryl mesylates into metal-catalyzed α-arylation was reported for the first time by Alsabeh and Stradiotto in 2013,[72] where the [Pd(cinnamyl)Cl]$_2$/**L10** catalyst system was shown to accommodate linear and cyclic dialkyl ketones, including acetone. An optimization campaign examining the palladium-catalyzed monoarylation of acetone with phenyl mesylate established the use of [Pd(cinnamyl)Cl]$_2$/**L10** (2 mol% Pd; Pd:**L10** = 2:3), acetone (10 equiv.) and K$_3$PO$_4$ (2 equiv.) in *t*BuOH–1,4-dioxane (1:1; 0.125 M) at 90 °C for 16 h as being effective conditions for the formation of phenylacetone (85% yield). Although Cs$_2$CO$_3$ was shown to be the optimal base in the [Pd(cinnamyl)Cl]$_2$/**L10**-catalyzed monoarylation of acetone using aryl halides and tosylates (see Section 5.4.1),[68] this base, and also CsF and NaO*t*Bu, afforded poor conversion to the target phenylacetone. Moreover, the use of acetone as the reaction solvent or using 1,4-dioxane or *t*BuOH individually at higher substrate concentrations (0.5 M) resulted in high conversion of the phenyl mesylate, but poor product selectivity. Alternative catalyst systems featuring **L1**, **L2**, **L4**, **L9**, **L11** or **L13** each afforded diminished yields (<40%) of phenylacetone relative to **L10** under similar conditions.[72] In focusing the discussion here on acetone monoarylation chemistry using the [Pd(cinnamyl)Cl]$_2$/**L10** catalyst system (2–5 mol% Pd; Pd:**L10** = 2:3), under the aforementioned optimized conditions a selection of electron-neutral and electron-rich aryl mesylates were accommodated, with some demonstrated tolerance for functional groups, including pyrrole, trifluoromethyl, diaryl ether and nitrile groups (12 examples; 47–85%; Figure 5.13, **B**). However, efforts to accommodate electron-poor and *ortho*-substituted aryl mesylates were unsuccessful, primarily resulting in decomposition of the starting materials to the corresponding phenol.[72]

5.4.3 Palladium-Catalyzed Acetone Monoarylation Using Aryl Imidazolylsulfonates

The application of aryl imidazolylsulfonates (ArOSO$_2$Im) as electrophiles in the monoarylation of acetone and other ketones was described by Ackermann and Mehta in 2012.[73] As with aryl mesylates, phenol-derived aryl imidazolylsulfonates are attractive in part due to the self-destructive and non-genotoxic properties of the imidazolylsulfonic acid by-product formed upon workup. Catalyst screening using Pd(OAc)$_2$–ligand mixtures (5 mol% Pd; Pd:ligand = 1:2) and Cs$_2$CO$_3$ (2 equiv.) in acetone–1,4-dioxane (1:4 by volume) at 80–110 °C for 16 h revealed Xantphos (**L13**) to be superior to a range of other commercially available monodentate and bidentate ligands. Reactions conducted in THF, toluene or MeCN were less successful. Interestingly, whereas Stradiotto and co-workers noted that Pd(OAc)$_2$ afforded vastly inferior catalytic performance relative to [Pd(cinnamyl)Cl]$_2$ in the monoarylation of acetone using aryl halides and tosylates with **L10** (Section 5.4.1),[68] Ackermann and Mehta noted the opposite trend when using **L13** in combination with aryl imidazolylsulfonates.[73] The optimized Pd(OAc)$_2$/**L13** catalyst system is attractive, given the comparatively inexpensive nature of the catalyst components. The scope of reactivity in the aryl imidazolylsulfonate was found to be very good (22 examples; 57–90%; Figure 5.13, **C**). A range of at *ortho*, *meta* and *para* groups with electron-donating and - withdrawing character were accommodated, including mono-, di- and trisubstituted substrates, and also ring-fused bicyclic systems. The successful monoarylation of acetone by use of an acetophenone imidazolylsulfonate derivative confirms the preference of the Pd(OAc)$_2$/**L13** catalyst system for acetone arylation, albeit in the presence of a significant excess of acetone. Competition experiments established that aryl imidazolylsulfonates are significantly more reactive than aryl tosylates or mesylates in this chemistry, even when they are electronically deactivated.[73]

5.4.4   Summary

The selective, palladium-catalyzed monoarylation of acetone has emerged as an efficient methodology for the construction of α-aryl methyl ketones. These newly developed cross-coupling protocols exploit acetone directly without the requirement for stoichiometric additives or preformed enolates and have been shown to accommodate a range aryl electrophiles, including chlorides, bromides, iodides, tosylates, mesylates and imidazolylsulfonates. With some notable exceptions in the case of aryl mesylates, the transformations documented thus far exhibit good functional group tolerance across several different catalyst systems, thereby suggesting that continued catalyst development will lead to further practical advances (*e.g.*, temperature, loading, scope). In this vein, the first carbonylative variants of acetone monoarylation reactions have emerged that offer synthetic inroads to 1,3-diketone derivatives from (hetero)aryl iodides, carbon monoxide and acetone.[74]

5.5   Conclusion and Outlook

Catalyst design has contributed significantly towards advancing the state-of-the-art in the selective palladium-catalyzed monoarylation of substrates that are inherently prone to polyarylation, as evidenced in the context of Buchwald–Hartwig amination and ketone α-arylation by the development of such transformations involving ammonia, hydrazine and acetone with broad substrate scope and excellent functional group tolerance. The judicious design/selection of supporting ancillary ligand has played a central role in permitting such advances, with sterically demanding mono- and bidentate phosphine-based ligands proving most effective; from a practical perspective, many of the ancillary ligands featured in this discussion are both air stable and commercially available. Although some ligand structural motifs appear to be particularly effective in promoting various palladium-catalyzed monoarylation reactions of this type, no single ligand class has emerged as being uniquely privileged. The benefits of using well-defined and rationally prepared precatalysts in which palladium is coordinated to the ancillary ligand of choice is also reaffirmed in the examples documented here; doing so has provided access to unprecedented monoarylation reaction scope under mild conditions. It is evident that further advances in ancillary ligand design and the use of rationally assembled precatalysts will figure prominently in addressing outstanding challenges in monoarylation chemistry.

Acknowledgements

The author acknowledges support from the Natural Sciences and Engineering Research Council of Canada, the Killam Trusts and Dalhousie University (Alexander McLeod Professorship).

References

1. C. A. Busacca, D. R. Fandrick, J. J. Song and C. H. Senanayake, *Adv. Synth. Catal.*, 2011, **353**, 1825–1864.
2. R. F. Heck, *Org. React.*, 1982, **27**, 345–390.
3. E. I. Negishi, *Angew. Chem. Int. Ed.*, 2011, **50**, 6738–6764.
4. A. Suzuki, *Angew. Chem. Int. Ed.*, 2011, **50**, 6722–6737.
5. X. F. Wu, P. Anbarasan, H. Neumann and M. Beller, *Angew. Chem. Int. Ed.*, 2010, **49**, 9047–9050.
6. C. C. C. Johansson Seechurn, M. O. Kitching, T. J. Colacot and V. Snieckus, *Angew. Chem. Int. Ed.*, 2012, **51**, 5062–5085.
7. A. S. Guram, R. A. Rennels and S. L. Buchwald, *Angew. Chem. Int. Ed.*, 1995, **34**, 1348–1350.
8. J. Louie and J. F. Hartwig, *Tetrahedron Lett.*, 1995, **36**, 3609–3612.
9. J. F. Hartwig, *Acc. Chem. Res.*, 2008, **41**, 1534–1544.
10. D. S. Surry and S. L. Buchwald, *Angew. Chem. Int. Ed.*, 2008, **47**, 6338–6361.
11. F. Bellina and R. Rossi, *Chem. Rev.*, 2010, **110**, 3850–3850.
12. C. C. C. Johansson and T. J. Colacot, *Angew. Chem. Int. Ed.*, 2010, **49**, 676–707.
13. D. Maiti, B. P. Fors, J. L. Henderson, Y. Nakamura and S. L. Buchwald, *Chem. Sci.*, 2011, **2**, 57–68.
14. D. S. Surry and S. L. Buchwald, *Chem. Sci.*, 2011, **2**, 27–50.
15. M. S. Driver and J. F. Hartwig, *J. Am. Chem. Soc.*, 1995, **117**, 4708–4709.
16. R. A. Widenhoefer and S. L. Buchwald, *Organometallics*, 1996, **15**, 2755–2763.
17. M. S. Driver and J. F. Hartwig, *J. Am. Chem. Soc.*, 1997, **119**, 8232–8245.

18. S. Shekhar, P. Ryberg, J. F. Hartwig, J. S. Mathew, D. G. Blackmond, E. R. Strieter and S. L. Buchwald, *J. Am. Chem. Soc.*, 2006, **128**, 3584–3591.
19. D. A. Culkin and J. F. Hartwig, *Acc. Chem. Res.*, 2003, **36**, 234–245.
20. R. J. Lundgren and M. Stradiotto, *Chem. Eur. J.*, 2012, **18**, 9758–9769.
21. C. Valente, S. Calimsiz, K. H. Hoi, D. Mallik, M. Sayah and M. G. Organ, *Angew. Chem. Int. Ed.*, 2012, **51**, 3314–3332.
22. G. C. Fortman and S. P. Nolan, *Chem. Soc. Rev.*, 2011, **40**, 5151–5169.
23. H. B. Li, C. C. C. Johansson Seechurn and T. J. Colacot, *ACS Catal.*, 2012, **2**, 1147–1164.
24. N. Marion and S. P. Nolan, *Acc. Chem. Res.*, 2008, **41**, 1440–1449.
25. R. J. Lundgren and M. Stradiotto, *Aldrichim. Acta*, 2012, **45**, 59–65.
26. X. H. Huang and S. L. Buchwald, *Org. Lett.*, 2001, **3**, 3417–3419.
27. Y. Aubin, C. Fischmeister, C. M. Thomas and J. L. Renaud, *Chem. Soc. Rev.*, 2010, **39**, 4130–4145.
28. J. I. van der Vlugt, *Chem. Soc. Rev.*, 2010, **39**, 2302–2322.
29. J. L. Klinkenberg and J. F. Hartwig, *Angew. Chem. Int. Ed.*, 2011, **50**, 86–95.
30. J. L. Klinkenberg and J. F. Hartwig, *J. Am. Chem. Soc.*, 2010, **132**, 11830–11833.
31. N. Xia and M. Taillefer, *Angew. Chem. Int. Ed.*, 2009, **48**, 337–339.
32. Q. L. Shen and J. F. Hartwig, *J. Am. Chem. Soc.*, 2006, **128**, 10028–10029.
33. H. U. Blaser, W. Brieden, B. Pugin, F. Spindler, M. Studer and A. Togni, *Top. Catal.*, 2002, **19**, 3–16.
34. G. D. Vo and J. F. Hartwig, *J. Am. Chem. Soc.*, 2009, **131**, 11049–11061.
35. P. G. Alsabeh, R. J. Lundgren, L. E. Longobardi and M. Stradiotto, *Chem. Commun.*, 2011, **47**, 6936–6938.
36. S. M. Crawford, C. B. Lavery and M. Stradiotto, *Chem. Eur. J.*, 2013, **19**, 16760–16771.
37. D. S. Surry and S. L. Buchwald, *J. Am. Chem. Soc.*, 2007, **129**, 10354–10355.
38. D. Tsvelikhovsky and S. L. Buchwald, *J. Am. Chem. Soc.*, 2011, **133**, 14228–14231.
39. C. W. Cheung, D. S. Surry and S. L. Buchwald, *Org. Lett.*, 2013, **15**, 3734–3737.
40. U. Christmann, D. A. Pantazis, J. Benet-Buchholz, J. E. McGrady, F. Maseras and R. Vilar, *J. Am. Chem. Soc.*, 2006, **128**, 6376–6390.
41. B. P. Fors and S. L. Buchwald, *J. Am. Chem. Soc.*, 2010, **132**, 15914–15917.
42. T. Schulz, C. Torborg, S. Enthaler, B. Schaffner, A. Dumrath, A. Spannenberg, H. Neumann, A. Borner and M. Beller, *Chem. Eur. J.*, 2009, **15**, 4528–4533.
43. A. Dumrath, C. Lubbe, H. Neumann, R. Jackstell and M. Beller, *Chem. Eur. J.*, 2011, **17**, 9599–9604.
44. G. Helmchen and A. Pfaltz, *Acc. Chem. Res.*, 2000, **33**, 336–345.
45. R. J. Lundgren, A. Sappong-Kumankumah and M. Stradiotto, *Chem. Eur. J.*, 2010, **16**, 1983–1991.
46. R. J. Lundgren, B. D. Peters, P. G. Alsabeh and M. Stradiotto, *Angew. Chem. Int. Ed.*, 2010, **49**, 4071–4074.
47. P. G. Alsabeh, R. J. Lundgren, R. McDonald, C. C. C. Johansson Seechurn, T. J. Colacot and M. Stradiotto, *Chem. Eur. J.*, 2013, **19**, 2131–2141.

48. P. G. Alsabeh, M. Stradiotto, H. Neumann and M. Beller, *Adv. Synth. Catal.*, 2012, **354**, 3065–3070.
49. T. Y. Xu and H. Alper, *Tetrahedron Lett.*, 2013, **54**, 5496–5499.
50. X. F. Wu, H. Neumann and M. Beller, *Chem. Asian J.*, 2010, **5**, 2168–2172.
51. X. F. Wu, H. Neumann and M. Beller, *Chem. Eur. J.*, 2010, **16**, 9750–9753.
52. B. Robinson, *Chem. Rev.*, 1963, **63**, 373–401.
53. G. R. Humphrey and J. T. Kuethe, *Chem. Rev.*, 2006, **106**, 2875–2911.
54. C. J. Ball and M. C. Willis, *Eur. J. Org. Chem.*, 2013, 425–441.
55. S. Cacchi and G. Fabrizi, *Chem. Rev.*, 2011, **111**, Pr215–Pr283.
56. G. Zeni and R. C. Larock, *Chem. Rev.*, 2004, **104**, 2285–2309.
57. R. J. Lundgren and M. Stradiotto, *Angew. Chem. Int. Ed.*, 2010, **49**, 8686–8690.
58. J. K. Niemeier and D. P. Kjell, *Org. Process Res. Dev.*, 2013, **17**, 1580–1590.
59. A. DeAngelis, D. H. Wang and S. L. Buchwald, *Angew. Chem. Int. Ed.*, 2013, **52**, 3434–3437.
60. J. M. Chen, Y. M. Zhang, W. Y. Hao, R. L. Zhang and F. Yi, *Tetrahedron*, 2013, **69**, 613–617.
61. T. Satoh, Y. Kawamura, M. Miura and M. Nomura, *Angew. Chem. Int. Ed.*, 1997, **36**, 1740–1742.
62. M. Palucki and S. L. Buchwald, *J. Am. Chem. Soc.*, 1997, **119**, 11108–11109.
63. B. C. Hamann and J. F. Hartwig, *J. Am. Chem. Soc.*, 1997, **119**, 12382–12383.
64. H. R. Chobanian, P. Liu, M. D. Chioda, Y. Guo and L. S. Lin, *Tetrahedron Lett.*, 2007, **48**, 1213–1216.
65. W. P. Su, S. Raders, J. G. Verkade, X. B. Liao and J. F. Hartwig, *Angew. Chem. Int. Ed.*, 2006, **45**, 5852–5855.
66. P. Liu, T. J. Lanza, J. P. Jewell, C. P. Jones, W. K. Hagmann and L. S. Lin, *Tetrahedron Lett.*, 2003, **44**, 8869–8871.
67. M. Kosugi, M. Suzuki, I. Hagiwara, K. Goto, K. Saitoh and T. Migita, *Chem. Lett.*, 1982, 939–940.
68. K. D. Hesp, R. J. Lundgren and M. Stradiotto, *J. Am. Chem. Soc.*, 2011, **133**, 5194–5197.
69. P. B. Li, B. Lu, C. L. Fu and S. M. Ma, *Adv. Synth. Catal.*, 2013, **355**, 1255–1259.
70. C. M. So and F. Y. Kwong, *Chem. Soc. Rev.*, 2011, **40**, 4963–4972.
71. S. C. Baker, D. P. Kelly and J. C. Murrell, *Nature*, 1991, **350**, 627–628.
72. P. G. Alsabeh and M. Stradiotto, *Angew. Chem. Int. Ed.*, 2013, **52**, 7242–7246.
73. L. Ackermann and V. P. Mehta, *Chem. Eur. J.*, 2012, **18**, 10230–10233.
74. J. Schranck, A. Tlili, H. Neumann, P. G. Alsabeh, M. Stradiotto and M. Beller, *Chem. Eur. J.*, 2012, **18**, 15592–15597.

CHAPTER 6

Transition Metal-Catalyzed Formation of C–O and C–S Bonds

JAMES P. STAMBULI[†]

Department of Chemistry and Biochemistry, The Ohio State University, 100 W. 18th Avenue, Columbus, OH 43210, USA
Email: stambuli@chemistry.ohio-state.edu

6.1 Introduction

Over the last decade, transiton metal-catalyzed cross-coupling reactions have become commonplace in small-molecule synthesis. The Nobel Committee recognized the importance of such reactions by rewarding several of the pioneering chemists who developed cross-coupling reactions with the Nobel Prize in Chemistry in 2010. The most widely used cross-coupling reactions, such as the Suzuki–Miyaura,[1,2] Heck[3,4] and Negishi[5] reactions, typically involve the formation of carbon–carbon bonds. The formation of carbon–nitrogen bonds *via* Buchwald–Hartwig aryl amination[6,7] is routinely employed in both industrial and academic laboratories, whereas cross-coupling reactions that construct carbon–oxygen and carbon–sulfur bonds are employed less frequently.

[†]Present address: AbbVie, North Chicago, IL, USA. Email: james.stambuli@abbvie.com

RSC Catalysis Series No. 21
New Trends in Cross-Coupling: Theory and Applications
Edited by Thomas J Colacot
© The Royal Society of Chemistry 2015
Published by the Royal Society of Chemistry, www.rsc.org

The development of transition metal catalysts for carbon–oxygen and carbon–sulfur bond formation has not progressed as quickly as the related cross-coupling reactions that form carbon–carbon and carbon–nitrogen bonds. A potential reason for the more limited catalyst choice for carbon–oxygen and carbon–sulfur bond formation may be attributed to the greater prevalence of C_{sp^2}–C or C_{sp^2}–N bonds in natural products and pharmaceuticals. Nonetheless, carbon–oxygen and carbon–sulfur bond formation promoted by transition metal catalysts has been used for coupling aryl halides with a wide array of oxygen and sulfur nucleophiles.

6.1.1 General Mechanism for Palladium-Catalyzed Carbon–Oxygen and Carbon–Sulfur Bond Formation

The generally proposed mechanism that describes the chemical pathway for cross-coupling reactions involving aryl halides and oxygen or sulfur nucleophiles is shown in Scheme 6.1. Some details of the proposed mechanism have varied since the first reports of these cross-couplings, but the overall catalytic process has remained largely unchanged.[8] Three fundamental organometallic reactions within the proposed mechanism are oxidative addition, transmetalation and reductive elimination. Numerous detailed mechanistic studies have been performed on the oxidative addition[9,10] and reductive elimination[11–15] reactions with late transition metals in the context of cross-coupling reactions, whereas fewer studies have examined transmetalations.[16–20]

As is the case with most cross-coupling reactions involving aryl or vinyl halides, the metal of choice for carbon–oxygen and carbon–sulfur bond forming reactions is palladium.[21,22] However, many recent investigations have involved the development of new cross-coupling catalysts based on copper. The identity and application of catalysts that promote the formation of carbon–oxygen and carbon–sulfur bonds using transition metal catalysts are presented in this chapter. The first part of the chapter involves metal-catalyzed methods to form carbon–oxygen bonds between aryl halides and

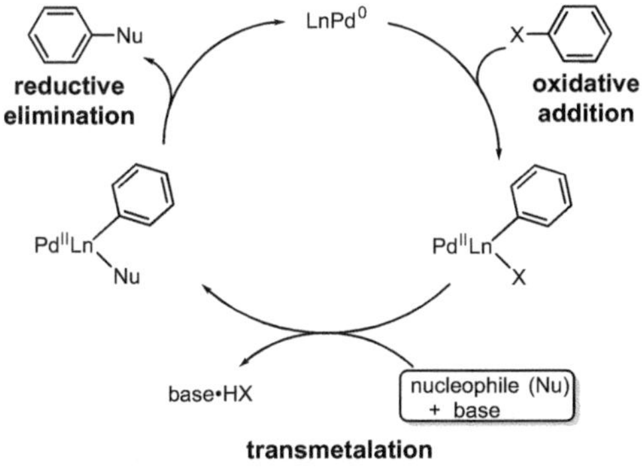

Scheme 6.1

phenols or aliphatic alcohols, and the second part describes metal-catalyzed methods to form carbon–sulfur bonds between aryl halides with aromatic and aliphatic thiols.

6.2 Palladium-Catalyzed Cross-Couplings That Form Carbon–Oxygen Bonds

Chemical methods employing transition metal catalysts are developed, among other reasons, to satisfy an unmet need in the synthetic literature that would streamline the synthesis of biologically active molecules. Many natural products contain diaryl ether linkages (Figure 6.1),[23] and exhibit interesting medicinal properties. For example, cadabicine is an anthelmintic,[24] combretastatin D_2 has exhibited cytotoxic activity,[25,26] thyroxine is one of the major hormones produced from the thyroid gland and is involved in numerous physiological processes,[27] dictyomedin A inhibits the development of cellular slime molds[28] and O-methylthalibrine has antimicrobial activity.[29]

6.2.1 Palladium-Catalyzed Formation of Diaryl Ethers

The prevalence of diaryl ether linkages found in Nature has likely contributed to the development of catalysts for palladium-catalyzed carbon–oxygen bond formation. Many active catalysts have been developed that allow the coupling of a range of electronic and sterically diverse aryl halides and phenols to form diaryl ethers. Scheme 6.2 shows some of the most reactive ligand systems for palladium-catalyzed cross-couplings of aryl halides and phenols to form diaryl ethers.[30–32] As is typical with palladium-catalyzed cross-coupling reactions, most of the indicated ligands are bulky and electron rich, allowing for stabilization of low-coordinate unsaturated palladium complexes.[33] One of the most general catalyst systems for the coupling of aryl bromides and aryl chlorides with a variety of phenols is a combination of ligand 6 and palladium acetate.[32] Representative examples

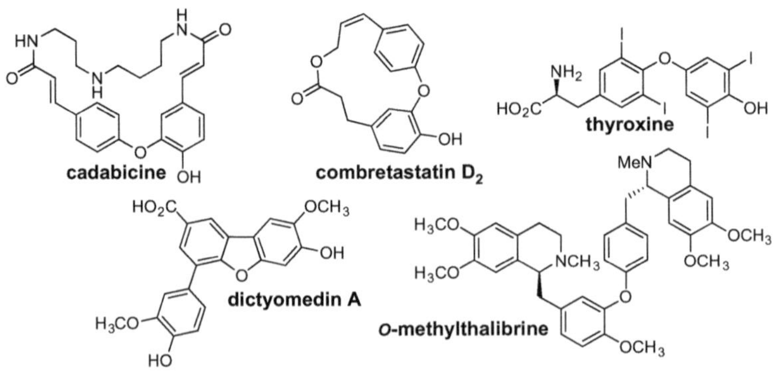

Figure 6.1   Structures of some natural products that contain diaryl ether linkages.

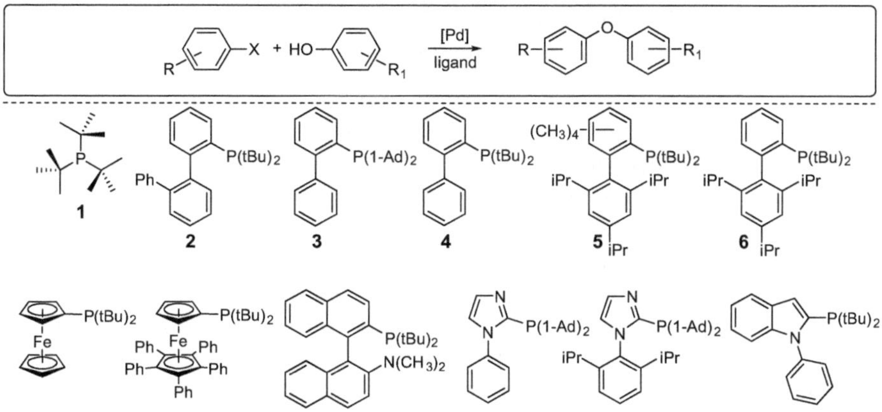

1 2 3 4 5 6

7 8 9 10 11 12

Scheme 6.2

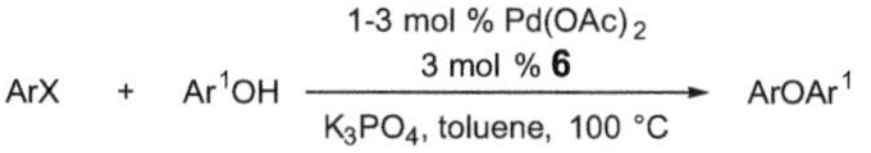

Ar	X	Ar[1]	Yield (%)
p-$CF_3C_6H_4$	Br	C_6H_5	90
4-AcC_6H_4	Br	*o*-$CH_3C_6H_4$	92
3-Pyridyl	Br	*o*-$CH_3C_6H_4$	88
p-CNC_6H_4	Br	C_6H_5	94
Pyrazin-2-yl	Cl	*p*-$CH_3C_6H_4$	87
2,5-$(CH_3)_2C_6H_3$	Cl	*p*-$CH_3OC_6H_4$	88

Scheme 6.3

of diaryl ether formations catalyzed by this catalyst system are shown in Scheme 6.3.

6.2.2 Palladium-Catalyzed Cross-Coupling Reactions of Aryl Halides With Aliphatic Alcohols

The development of catalysts for cross-coupling reactions of aryl halides with primary and secondary alcohols is less advanced than that of similar reactions of phenols. Uncovering catalysts for these reactions is important as 10 of the top 200 highest selling pharmaceuticals, such as Abilify, Cymbalta, Actos and Tricor, contain alkyl aryl ether linkages.[34] The challenges associated with suppressing potential β-hydride elimination reactions that can occur from palladium alkoxide intermediates of primary and secondary alcohols likely hindered the discovery of catalysts for these substrates. Similar issues do not exist for tertiary alcohols, so catalysts for these substrates were some of the first to be discovered for palladium carbon–oxygen bond formation reactions.[30,35–37]

6.2.2.1 Palladium-Catalyzed Cross-Coupling Reactions of Aryl Halides With Primary and Secondary Alcohols

The general structures of the ligands that have been developed for palladium-catalyzed carbon–oxygen cross-coupling between aryl halides and primary or secondary alcohols are shown in Scheme 6.4. Most of the these systems were developed in Buchwald's laboratory[38-41] whereas the interesting bispyrazole ligand **17** was developed by Beller and co-workers.[42] Currently, the most general ligand for aryl halide couplings with aliphatic alcohols is the bulky, substituted biphenyl compound **18**.[43] The obvious structural similarity amongst these ligands is the binaphthyl or biphenyl aromatic scaffolds. The aromatic moiety is thought to assist in the stabilization of highly reactive, unsaturated palladium species.[44] Catalysts employing these ligands have been developed for cross-couplings between aryl bromides and aryl chlorides and primary and secondary alcohols, producing the corresponding alkyl aryl ethers (Scheme 6.5).

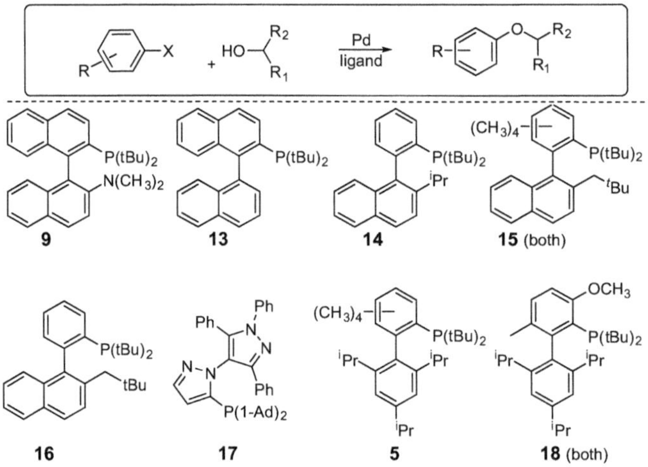

Scheme 6.4

| ArX | + | ROH | catalyst ⟶ | ArOR |

Ar	X	R	Catalyst	Yield (%)
p-CH$_3$OC$_6$H$_4$	Br	n-C$_4$H$_9$	[Pd(allyl)Cl]$_2$/**18**	80
p-PhOC$_6$H$_4$	Cl	CF$_3$CH$_2$	[Pd(allyl)Cl]$_2$/**18**	83
o-CH$_3$OC$_6$H$_4$	Br	n-C$_4$H$_9$	[Pd(allyl)Cl]$_2$/**18**	84
p-nBuC$_6$H$_4$	Br	sec-C$_4$H$_9$	Pd(OAc)$_2$/**15**	72
m-CH$_3$OC$_6$H$_4$	Br	sec-C$_4$H$_9$	Pd(OAc)$_2$/**15**	63
m-CF$_3$C$_6$H$_4$	Br	sec-C$_4$H$_9$	Pd(OAc)$_2$/**15**	80

Scheme 6.5

6.2.2.2 Palladium-Catalyzed Cross-Coupling Reactions of Aryl Halides With Tertiary Alcohols

The C–O cross-coupling reactions of aryl halides and tertiary alcohols were among the earliest reported palladium-catalyzed etherifications. The lack of β-hydrogens in tertiary alcohols is one of the main reasons why these etherification reactions were successful in the early days of C–O cross-coupling. A number of ligands that have been successfully employed in this class of coupling reaction are shown in Scheme 6.6[30,35,37,45,46] and representative products are shown in Scheme 6.7.

6.2.2.3 Intramolecular Palladium-Catalyzed Coupling Reactions of Aryl Halides With Aliphatic Alcohols

Catalysts for intramolecular etherification reactions of aryl halides with primary, secondary or tertiary alcohols have been reported. The ligands employed in these reactions are shown in Scheme 6.8 and separated into

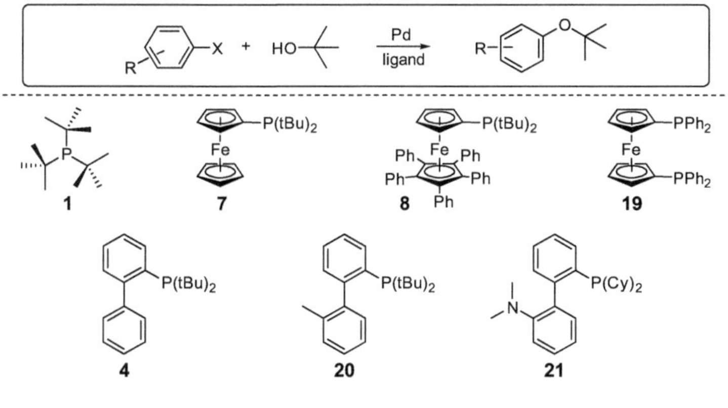

Scheme 6.6

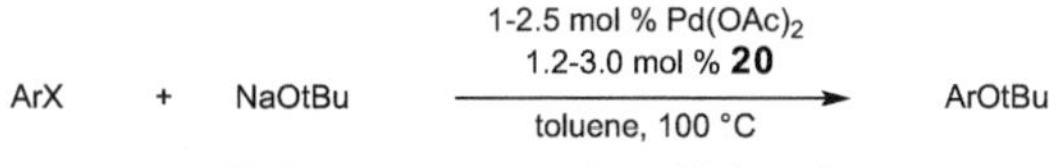

Ar	X	Yield (%)
p-tBuC$_6$H$_4$	Br	90
3,5-(CH$_3$)$_2$C$_6$H$_4$	Br	86
o-CH$_3$OC$_6$H$_4$	Cl	84
p-nBuC$_6$H$_4$	Cl	92
p-CH$_3$OC$_6$H$_4$	Cl	84
2,5-(CH$_3$)$_2$C$_6$H$_3$	Cl	87

Scheme 6.7

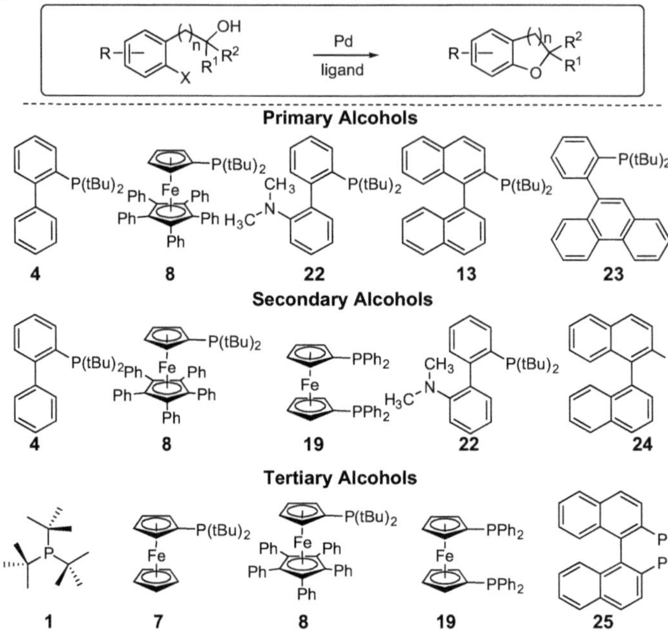

Scheme 6.8

categories of primary, secondary and tertiary alcohols. Palladium complexes ligated by **4**,[47] **8**,[36] **22**,[39] **13**[47] and **23**[39] promote intramolecular carbon–oxygen bond formation of primary alcohols with aryl halides, whereas complexes with **4**,[47] **8**,[45] **19**,[48] **22**[47] and **24**[47] and complexes with **7**,[30] **8**,[36] **19**[48] and **25**[48] are good catalysts for intramolecular reactions of aryl halides with secondary and tertiary alcohols, respectively. Once again, bulky, electron-rich phosphine ligands dominate the general ligand class. Intramolecular etherification reactions of primary and secondary alcohols typically proceed smoothly despite the presence of β-hydrogens in the intermediate alkoxide palladium complexes because this elimination from a cyclic palladium alkoxide intermediate complex is slow.

6.3 Copper-Catalyzed Cross-Coupling Reactions of Carbon–Oxygen Bonds

Although the palladium-catalyzed formation of carbon–oxygen bonds is broad in scope, the development of copper catalysts for etherification reactions is an important and growing area of research. Copper catalysts are cheaper and less toxic than palladium catalysts, hence the development of copper catalysts is of interest to synthetic chemists. One challenge in developing copper-catalyzed cross-coupling reactions is the increased complexity of the corresponding reaction mechanism. For instance, the

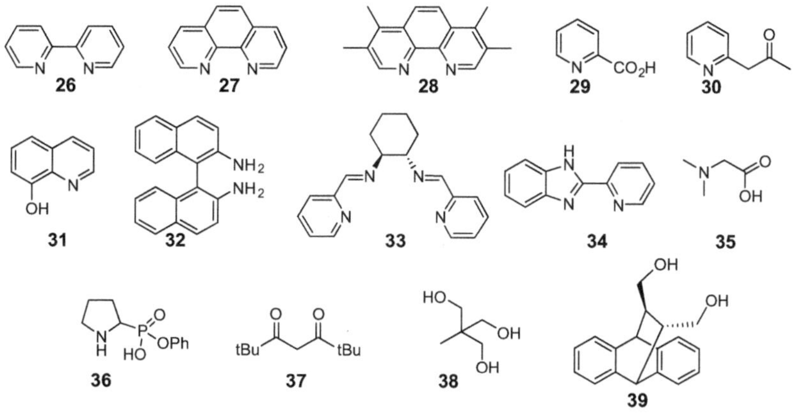

Figure 6.2 Representative ligands for copper catalysis.

proposed mechanism for copper-catalyzed etherifcations proceeds through either initial oxidative addition of the aryl halide C–X bond or through a radical anion intermediate. Many ligated copper complexes have provided carbon–oxygen bond formations with product yields similar to those with palladium catalysts; however, palladium-catalyzed etherification reactions maintain a broader scope than those using copper and therefore continue to dominate the field. Another major difference between these two catalyst systems is the identity of the ligands employed. In palladium catalysis, phosphines predominate, whereas in copper catalysis, oxygen- and nitrogen-containing ligands predominate. A number of representative ligands for copper are highlighted in Figure 6.2 (26,[49] 27,[50,51] 28,[52,53] 29,[54] 30,[55] 31,[56] 32,[57] 33,[58] 34,[59] 35,[60,61] 36,[62] 37,[63] 38,[64] 39[65]).

6.3.1 Copper-Catalyzed Formation of Diaryl Ethers

Diaryl ethers are formed from aryl bromides or aryl iodides and the corresponding phenol derivatives with copper catalysts in good yields (Scheme 6.9).[54] Many sterically and electronically diverse phenols can be employed; however, the use of aryl chlorides remains limited owing to generally poor conversion. One of the drawbacks with copper catalysis is the ineffectiveness of copper in activating aryl chlorides, hence most catalytic reactions with copper are limited to aryl bromides and aryl iodides.

6.3.2 Copper-Catalyzed Cross-Coupling Reactions of Vinyl Halides With Phenols

The use of copper catalysts to promote the coupling of vinyl halides with phenols has also been successful (Scheme 6.10).[59] The optimal catalyst system is formed *in situ* from copper iodide and an equimolar amount of benzimidazole ligand **34**. Alkene geometry may be retained or scrambled in

ArX + Ar¹OH → ArOAr¹

5-10 mol % CuI
10-20 mol % **29**
K₃PO₄, DMSO, 80-110 °C

Ar	X	Ar¹	Yield (%)
Pyrimidin-5-yl	Br	C₆H₅	70
Isoquinolin-4-yl	Br	3,4-(CH₃)₂C₆H₃	69
o-CH₃C₆H₄	I	2,6-(CH₃)₂C₆H₃	89
o-(iC₄H₉)C₆H₄	I	2,6-(CH₃)₂C₆H₃	74
o-CH₃OC₆H₄	I	C₆H₅	85
o-ClC₆H₄	I	o-CH₃C₆H₄	68

Scheme 6.9

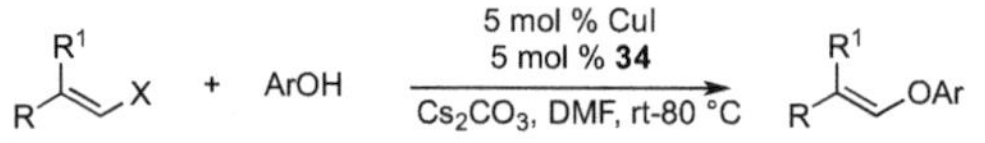

5 mol % CuI
5 mol % **34**
Cs₂CO₃, DMF, rt-80 °C

R	R¹	X	Ar	Yield (%)
Ph	H	Br	o-iPrC₆H₄	90
Ph	H	I	o-iPrC₆H₄	95
H	H	I	3,5-(OCH₃)₂C₆H₃	84
H	H	I	p-FC₆H₄	81
H	CO₂Et	I	p-tBuC₆H₄	93
H	CO₂Et	I	Benzothiazol-2-yl	92

Scheme 6.10

the vinyl ether product under the coupling conditions. Hence the proper choice of ligand in this chemistry may not only affect the product conversion, but can also affect the final geometry of the product. For example, certain copper catalysts have shown an increase in stereoselectivity of the vinyl ether with respect to the starting vinyl halide, whereas some catalysts simply retain the *E:Z* ratio of the vinyl halide starting material during the transformation to the product.

6.3.3 Copper-Catalyzed Coupling Reactions of Aryl Halides With Aliphatic Alcohols

Copper catalysts have also been shown to promote cross-coupling reactions of aryl bromides and iodides with primary and secondary alcohols (Scheme 6.11).[51,52,56] Reactions with primary alcohols are typically higher yielding and require lower catalyst loadings than related reactions of secondary alcohols. However, the requirement for neat alcohol to provide a higher yield of product is a drawback to the current methodology, as the reaction of uncommon, expensive primary alcohols would not be cost-effective.

$$\text{ArX} \quad + \quad \text{ROH} \quad \xrightarrow[\text{base, solvent, 110 °C}]{\text{cat. CuI / ligand}} \quad \text{ArOR}$$

Ar	X	R	Ligand	Base	Solvent	Yield (%)
C_6H_5	Br	$(CH_2)_2CH=CH_2$	31	K_3PO_4	Neat	92
C_6H_5	Br	n-C_4H_9	31	K_3PO_4	Neat	87
p-PhC_6H_4	Br	n-C_4H_9	31	K_3PO_4	Neat	81
p-$CH_3OC_6H_4$	Br	n-C_4H_9	31	K_3PO_4	Neat	78
p-$NO_2C_6H_4$	Br	n-C_4H_9	31	K_3PO_4	Neat	83
p-AcC_6H_4	Br	n-C_4H_9	31	K_3PO_4	Neat	92
3-Pyridyl	I	iPr	27	Cs_2CO_3	Toluene	92
p-$CH_3OC_6H_4$	I	iPr	27	Cs_2CO_3	Toluene	79
p-$CH_3OC_6H_4$	I	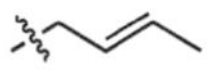 ~~\/\~~	27	Cs_2CO_3	Toluene	86
p-$CH_3OC_6H_4$	I	cyclo-C_5H_9	27	Cs_2CO_3	Toluene	75
3,5-$(CH_3)_2C_6H_4$	I	cyclo-C_5H_9	27	Cs_2CO_3	Toluene	67

Scheme 6.11

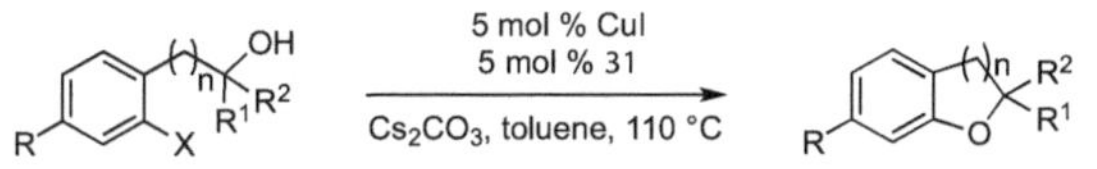

R	X	n	R^1	R^2	Yield (%)
H	Br	1	H	H	82
H	Br	1	CH_3	CH_3	83
H	Cl	1	H	H	82
H	Br	2	H	H	93
CH_3	Br	2	H	H	87

Scheme 6.12

6.3.4 Intramolecular Copper-Catalyzed Cross-Coupling Reactions of Aryl Halides With Aliphatic Alcohols

Few examples exist for the copper-catalyzed intramolecular etherification reaction of aryl halides and alcohols (Scheme 6.12).[56] Although the scope of this chemistry is limited, five- and six-membered cyclic ethers have been formed in good yields using copper iodide as catalyst in combination with the hydroxyquinoline ligand 31.

6.4 Palladium-Catalyzed Cross-Coupling Reactions of Carbon–Sulfur Bonds

The development of transition metal catalysts for carbon–sulfur bond formation between aryl halides and aryl or alkyl thiols has produced fewer catalysts than the corresponding carbon–oxygen bond formations. One of the main reasons for fewer catalysts being developed for carbon–sulfur bond

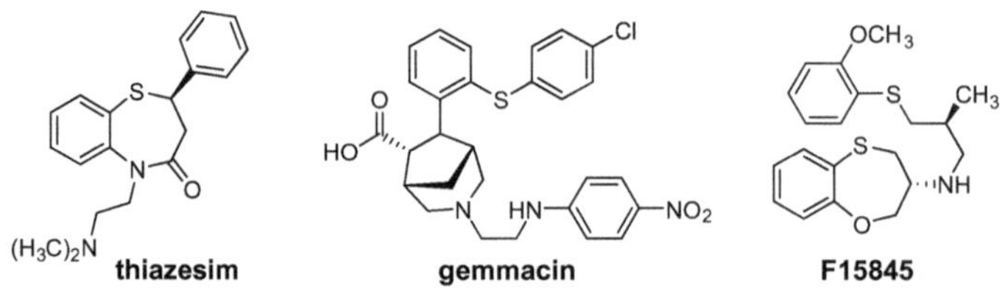

Figure 6.3 Examples of bioactive molecules containing thioether structures.

formation is likely the lesser importance of such bonds relative to carbon–oxygen bonds in natural products and pharmaceuticals. However, there are numerous reports of thioethers within the structure of bioactive molecules (Figure 6.3). For example, thiazesim acts as an antidepressant,[66] gemmacin is a highly active antibacterial,[67] and F15845 is an antianginal agent.[68]

6.4.1 Palladium-Catalyzed Cross-Coupling Reactions of Aryl Halides With Aryl or Aliphatic Thiols

The most general catalysts developed for metal-catalyzed reactions of aryl halides with alkyl or aryl thiols are shown in Scheme 6.13 (**40**,[69] **41**,[70] **42**[71,72]), reacting equally well with both aliphatic and aryl thiols and aryl bromides or chlorides. Owing to the nucleophilicity of the thiolate group, most of the general catalysts bear strongly coordinating ligands. This strong coordination prevents the thiolate anion from attacking the palladium center and displacing the ligand from the metal. If the thiolate is allowed to interact unfavorably with the palladium center, precipitation of the catalyst as palladium black can occur, which would destroy the catalytic activity of the metal center. Therefore, most catalysts for carbon–sulfur bond formation work better with aryl thiols, which are less nucleophilic than alkyl thiols. If the ligand is not strong, bulky and bidentate, zinc and/or lithium additives can be used to attenuate the reactivity of the nucleophilic thiolate and protect the palladium center.[69] This latter strategy was used in the development of the palladium(I) dimer complex **40**, one of the most general catalysts for carbon–sulfur bond formation of aryl bromides. The other two palladium catalysts shown in Scheme 6.13 have tight chelation to the metal center.

6.4.2 Palladium-Catalyzed Intramolecular Cyclization Reactions to Form Carbon–Sulfur Bonds

The intramolecular coupling reactions of aryl halides with thioketones have been reported.[73] In the presence of a palladium complex formed from Pd$_2$dba$_3$ and DPEphos, benzothiophenes are formed from the coupling of aryl bromides or aryl chlorides and thioketones in good yields.[73] Aryl bromides are more reactive than aryl chlorides in this chemistry as the formation of benzothiophene in Scheme 6.14 occurs in 30% greater yield with the aryl bromide than with the aryl chloride starting material.

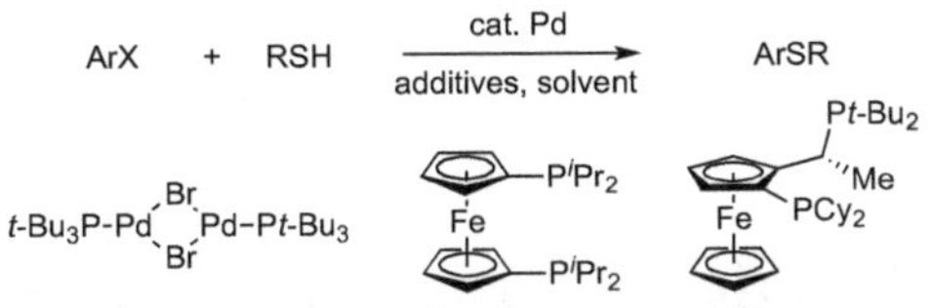

ArX	R	Pd	L	Additives	Solvent	Yield (%)
o-CF₃C₆H₄Br	C₆H₅	**40**	–	NaO*t*Bu/ ZnCl₂/LiI	THF	99
m-CO₂H C₆H₄Br	C₆H₅	**40**	–	NaO*t*Bu/ ZnCl₂/LiI	THF	77
o-CH₃OC₆H₄Br	*t*-C₄H₉	**40**	–	NaO*t*Bu/ ZnCl₂/LiI	THF	84
2,6-(CH₃)₂C₆H₃Br	Bn	**40**	–	NaO*t*Bu/ ZnCl₂/LiI	THF	99
p-CH₃OC₆H₄Br	*t*-C₄H₉	Pd(OAc)₂	**41**	NaO*t*Bu	Dioxane	99
2,5-(CH₃)₂C₆H₃Cl	*t*-C₄H₉	Pd(OAc)₂	**41**	NaO*t*Bu	Dioxane	95
p-CH₃OC₆H₄Cl	*p*-CH₃OC₆H₅	Pd(OAc)₂	**41**	NaO*t*Bu	Dioxane	93
m-CNC₆H₄Cl	C₆H₅	Pd(OAc)₂	**41**	NaO*t*Bu	Dioxane	82
m-NH₂C(O)C₆H₄Cl	C₆H₅	Pd(OAc)₂	**42**	NaO*t*Bu	DME	70
C₆H₅Cl	*t*-C₄H₉	Pd(OAc)₂	**42**	Cs₂CO₃	DME	91
m-CH₃C₆H₄Cl	*p*-CH₃C₆H₅	Pd(OAc)₂	**42**	LiHMDS	DME	93
o-CH₃OC₆H₄Cl	C₆H₅	Pd(OAc)₂	**42**	KO*t*Bu	DME	97

Scheme 6.13

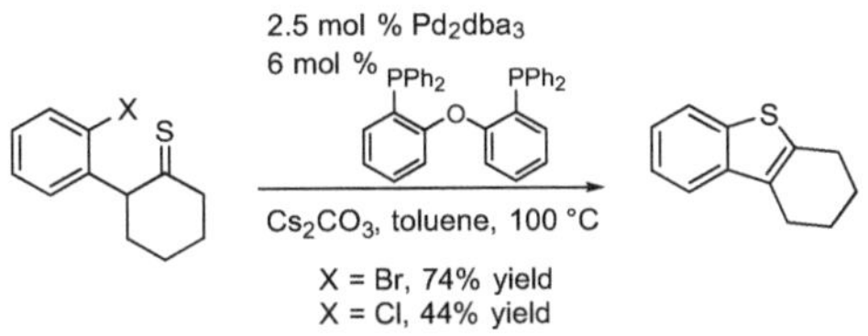

Scheme 6.14

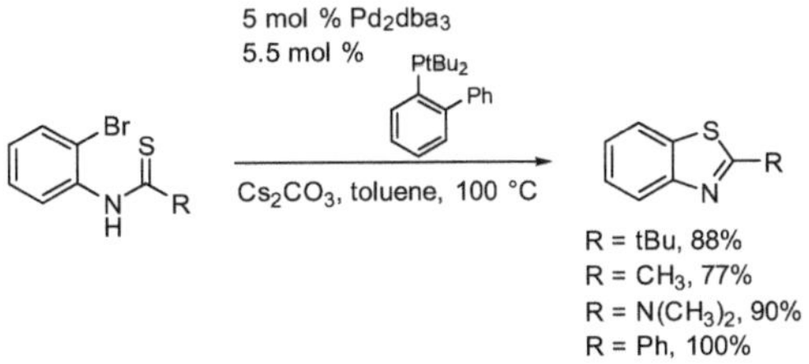

Scheme 6.15

Intramolecular reactions of aryl halides with thioamides are catalyzed by palladium and a bulky, electron-rich monodentate phosphine ligand (Scheme 6.15),[74] providing benzothiazoles in good yields. The identity of the thioamide does not greatly affect the overall yield of the reaction in most cases. Only aryl bromide coupling partners were reported to participate in the reaction.

6.5 Copper-Catalyzed Cross-Coupling Reactions of Carbon–Sulfur Bonds

Although palladium provides the most general catalyst systems for cross-coupling reactions of carbon–sulfur bonds, the development of more cost-efficient copper catalysts for this process has increased. A primary limitation with copper catalysts is their inability to effect cross-coupling reactions with aryl chlorides and the lower reaction yields obtained with aryl bromides.

Representative copper-catalyzed cross-coupling reactions to form alkyl aryl thioethers and diaryl thioethers are shown in Scheme 6.16.[75] The catalyst is composed of copper iodide and excess ethanediol. Aryl iodides are partnered with aryl thiols in high yields regardless of the functional group on the aromatic ring of the iodide. However, lower yields were seen with more hindered alkyl thiols.

Carbon disulfide has been used as a thiol equivalent in the copper-catalyzed formation of symmetrical diaryl thioethers. Several examples of this process are illustrated in Scheme 6.17.[76] The reaction proceeds with catalytic

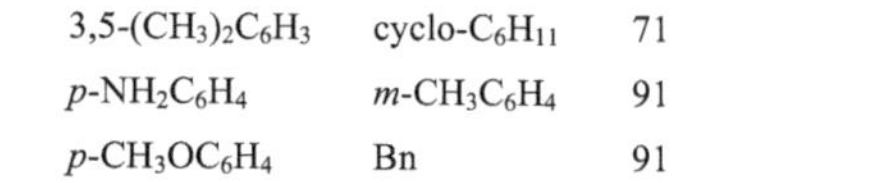

$$\text{ArI} \quad + \quad \text{RSH} \quad \xrightarrow[\text{K}_2\text{CO}_3, \text{ iPrOH}, 80\ °\text{C}]{\substack{5\ \text{mol}\ \%\ \text{CuI} \\ 2\ \text{equiv HO(CH}_2)_2\text{OH}}} \quad \text{ArSR}$$

Ar	R	Yield (%)
$m\text{-BrC}_6\text{H}_4$	C_6H_5	92
$o\text{-CO}_2\text{HC}_6\text{H}_4$	$p\text{-CH}_3\text{OC}_6\text{H}_4$	88
$o\text{-NH}_2\text{CH}_2\text{C}_6\text{H}_4$	$p\text{-CH}_3\text{OC}_6\text{H}_4$	93
$3,5\text{-(CH}_3)_2\text{C}_6\text{H}_3$	$\text{cyclo-C}_6\text{H}_{11}$	71
$p\text{-NH}_2\text{C}_6\text{H}_4$	$m\text{-CH}_3\text{C}_6\text{H}_4$	91
$p\text{-CH}_3\text{OC}_6\text{H}_4$	Bn	91

Scheme 6.16

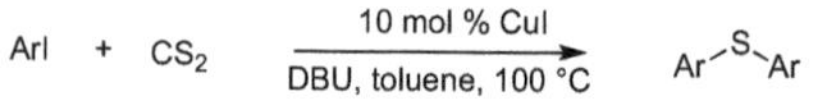

$$\text{ArI} \quad + \quad \text{CS}_2 \quad \xrightarrow[\text{DBU, toluene, 100}\ °\text{C}]{10\ \text{mol}\ \%\ \text{CuI}} \quad \text{Ar}^{\text{S}}\text{Ar}$$

Ar	Yield (%)
C_6H_5	85
$o\text{-(CH}_3)_2\text{NC}_6\text{H}_4$	78
$o\text{-CH}_3\text{OC}_6\text{H}_4$	83
$o\text{-OHC}_6\text{H}_4$	81
$o\text{-BrC}_6\text{H}_4$	67
$p\text{-ClC}_6\text{H}_4$	70

Scheme 6.17

copper iodide as catalyst, with DBU, at near reflux temperatures of toluene. The reaction is currently limited to aryl iodides.

6.6 Atypical Metal Catalysts for Cross-Coupling Reactions of Carbon–Sulfur Bonds

As is the case for most transition metal-catalyzed cross-coupling reactions, palladium is the metal of choice. This also holds true for metal-catalyzed reactions to form carbon–sulfur bonds However, several groups have replaced palladium catalysts with catalysts composed of manganese, cobalt, nickel and rhodium.

6.6.1 Manganese- and Cobalt-Catalyzed Formation of Diaryl and Alkyl Aryl Thioethers

Catalytic manganese chloride in the presence of phenanthroline ligands has been reported for the formation of carbon–sulfur bonds.[77] While only aryl iodides were successfully coupled, both aryl and alkyl thiols could be employed. Some examples of diaryl thioether syntheses are shown in Scheme 6.18. The yields of aryl thioethers were typically much higher than those for alkyl aryl thioethers. Similarly, a cobalt catalyst in the presence of stoichiometric zinc has been shown to couple carbon–sulfur bonds of aryl iodides and thiols.[78] Nearly all reported thiol substrates were aromatic, with only a few examples of reactions with alkyl thiols, and representative examples are shown in Scheme 6.19.

6.6.2 Nickel-Catalyzed Formation of Diaryl or Alkyl Aryl Thioethers

The use of nickel catalysts to promote the formation of diaryl of alkyl aryl thioethers has advanced in recent years and numerous nickel complexes

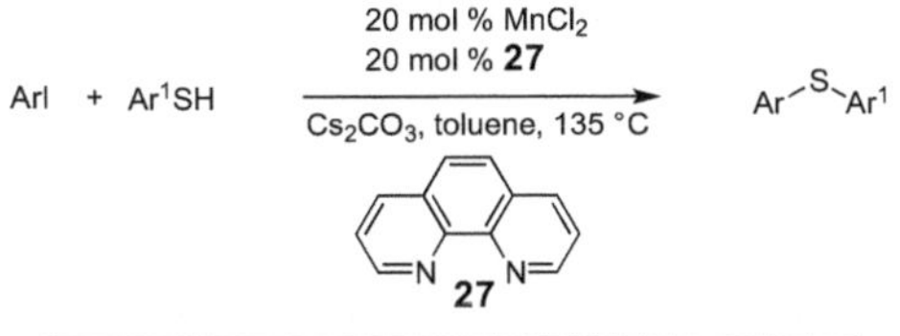

$$ArI + Ar^1SH \xrightarrow[\text{Cs}_2\text{CO}_3,\ \text{toluene, 135 °C}]{\substack{\text{20 mol \% MnCl}_2 \\ \text{20 mol \% } \mathbf{27}}} Ar^{-S}\diagdown Ar^1$$

Ar	Ar^1	Yield (%)
m-CH$_3$C$_6$H$_4$	C$_6$H$_5$	99
o-CH$_3$C$_6$H$_4$	C$_6$H$_5$	91
p-CH$_3$OC$_6$H$_4$	C$_6$H$_5$	99
o-CH$_3$OC$_6$H$_4$	C$_6$H$_5$	80
p-CH$_3$OC$_6$H$_4$	p-CH$_3$OC$_6$H$_4$	88
o-(HOCH$_2$)C$_6$H$_4$	p-ClC$_6$H$_4$	76

Scheme 6.18

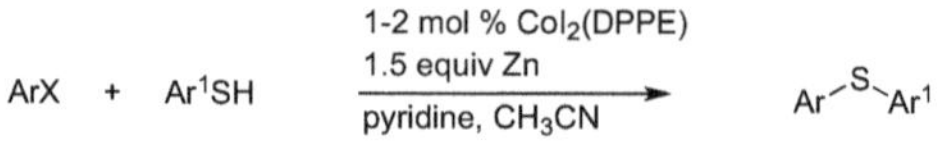

ArX + Ar¹SH $\xrightarrow[\text{pyridine, CH}_3\text{CN}]{\begin{array}{c}\text{1-2 mol \% CoI}_2\text{(DPPE)}\\ \text{1.5 equiv Zn}\end{array}}$ Ar–S–Ar¹

ArX	Ar¹	Yield (%)
C_6H_5I	C_6H_5	98
p-CNC$_6$H$_4$I	C_6H_5	84
p-FC$_6$H$_4$I	C_6H_5	94
p-CH$_3$C$_6$H$_4$Br	C_6H_5	94
2-BrC$_5$H$_4$N	C_6H_5	64
2-Br-furan	C_6H_5	82

Scheme 6.19

have been shown to facilitate C–S cross-coupling reactions. Although nickel is a more attractive cross-coupling catalyst than palladium from a cost perspective, palladium catalysts remain the system of choice for carbon–sulfur bond formation reactions. This is due in large part to generally lower yields with nickel catalysts,and also the inability to work efficiently for aryl bromides and chlorides.

A nickel phosphinite catalyst, in the presence of zinc, efficiently catalyzes the coupling of aryl iodides with both alkyl and aryl disulfides.[79] Representative examples are shown in Scheme 6.20. An earlier report by the same group described a similar nickel system, with significantly lower yields.[80]

Another report described the use of a nickel–carbene complex to form aryl thioethers from aryl bromides and thiophenol (Scheme 6.21).[81] Yields of aryl thioether products are high even at low catalyst loadings. The required temperatures are between 100 and 110 °C, which may cause compatibility issues with certain functional groups.

An recent interesting report described the cross-coupling of aryl iodides and aryl bromides with tetramethylammonium trifluoromethanethiolate to form aryl trifluoromethyl sulfides.[82] The reaction is mediated by catalytic bis-cyclooctadienylnickel and a methoxy-substituted bipyridine ligand (Scheme 6.22). Typically, yields are good for aryl iodides and moderate for aryl bromides. Aryl chlorides are not effective partners in this reaction.

In another recent report, the use of a nickel chloride catalyst at high loading and ethyl crotonate in the presence of excess zinc and pyridine, which is believed to generate a nickel catalyst containing a molecule of pyridine and two molecules of ethyl crotonate.[83] This *in situ*-generated nickel catalyst facilitates the coupling of aryl iodides with both aryl and alkyl thiols (Scheme 6.23).[84]

Similar nickel catalysts affect the intramolecular carbon–sulfur coupling reactions shown in Scheme 6.24. Interestingly, the carbon–sulfur bond formation outcompetes the carbon–carbon bond formation between the enolate and the aryl halide. The carbon–carbon bond formation is likely inhibited by the lack of a strong base in the reaction. Although the catayst

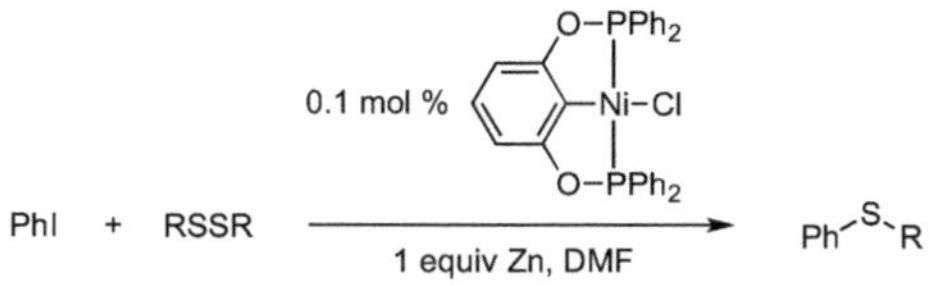

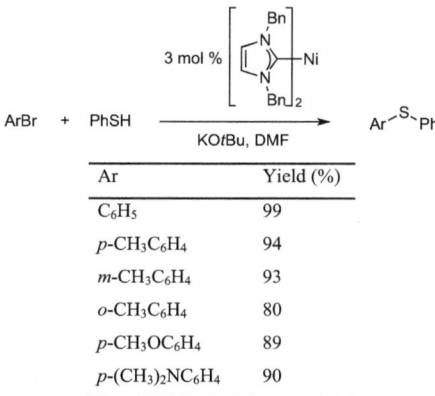

0.1 mol %
O–PPh2
Ni–Cl
O–PPh2
PhI + RSSR
1 equiv Zn, DMF
Ph–S–R

R	Yield (%)
CH_3	99
i-C_3H_7	99
n-C_4H_9	99
s-C_4H_9	99
t-C_4H_9	49
C_6H_5	86

Scheme 6.20

3 mol %
Bn
N
N
Bn
Ni
2
ArBr + PhSH
KOtBu, DMF
Ar–S–Ph

Ar	Yield (%)
C_6H_5	99
p-$CH_3C_6H_4$	94
m-$CH_3C_6H_4$	93
o-$CH_3C_6H_4$	80
p-$CH_3OC_6H_4$	89
p-$(CH_3)_2NC_6H_4$	90

Scheme 6.21

15 mol % Ni(COD)2
H3CO
OCH3
30 mol %
N
N
ArX + [N(CH3)4][SCF3]
THF
Ar–S–CF3

Ar	X	Yield (%)
p-$CH_3C_6H_4$	I	90
1,1′-Biphenyl	I	92
p-$EtO_2C_6H_4$	I	45
C_6H_5	Br	65
m-$CH_3C_6H_4$	Br	37
C_6H_5	Cl	0

Scheme 6.22

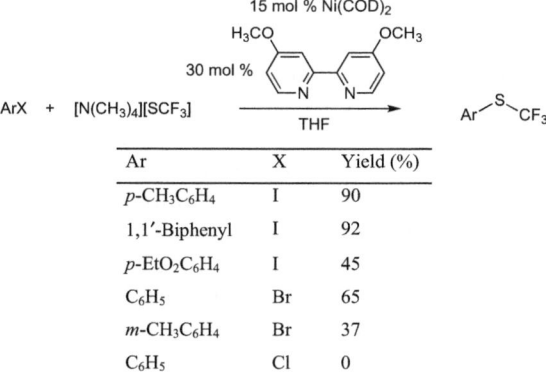

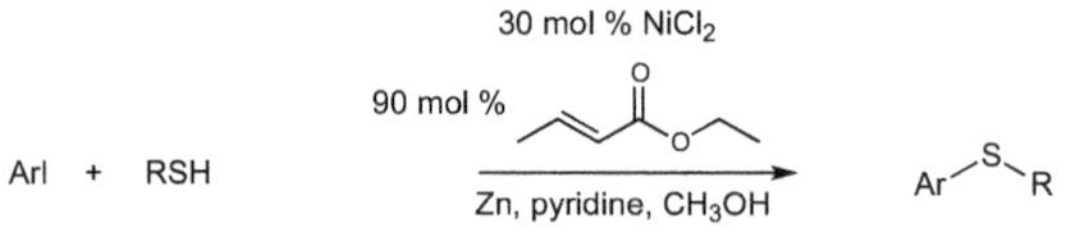

Ar	R	Yield (%)
p-$CH_3OC_6H_4$	C_6H_5	70
p-$CH_3OC_6H_4$	*p*-ClC_6H_4	67
p-$CH_3OC_6H_4$	2,4-$(CH_3)_2C_6H_3$	55
p-$CH_3OC_6H_4$	$HO(CH_2)_5$	81
p-$CH_3OC_6H_4$	$TBSO(CH_2)_5$	58
p-$CH_3OC_6H_4$	$PhthN(CH_2)_3$	55

Scheme 6.23

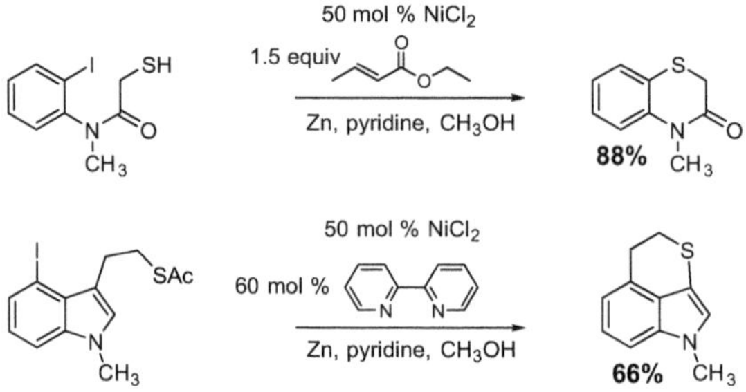

Scheme 6.24

loading is high in intramolecular reactions with β-thioketones or with thioacetates, these reactions provide structurally interesting thiazinones and thiolactones in good yields.

6.6.3 Rhodium-Catalyzed Formation of Diaryl or Alkyl Aryl Thioethers

The use of a rhodium hydride complex with dppBZ [1,2-bis(diphenylphosphino)benzene] catalyzes the cross-coupling between substituted aromatic fluorides and diaryl sulfides (Scheme 6.25).[85] Although the aromatic fluoride partners required strongly electron-withdrawing groups, the ability to employ aromatic fluorides as partners in carbon–sulfur cross-coupling reactions is interesting and expands the utility of this type of coupling.

A further use of a rhodium catalyst in carbon–sulfur bond-forming reactions was reported by Lee and co-workers (Scheme 6.26).[86] Reaction of rhodium cyclooctadiene chloride dimer, $[RhCl(COD)]_2$, and PPh_3 in the presence of sodium *tert*-butoxide at 100 °C formed an active catalyst to allow aryl iodides to couple with alkyl and aryl thiols. Whereas aryl thiols are generally good coupling partners, only a few examples of the use

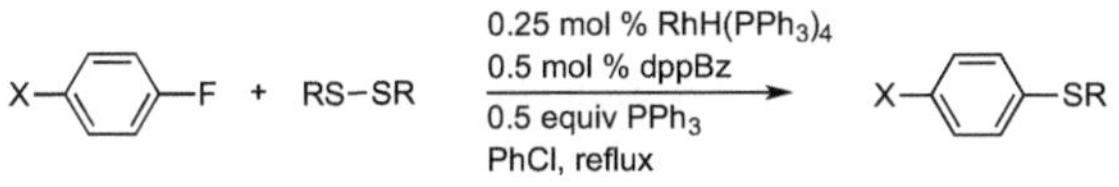

X—⟨benzene ring⟩—F + RS—SR → $\dfrac{\text{0.25 mol \% RhH(PPh}_3)_4}{\text{0.5 mol \% dppBz}}$ → X—⟨benzene ring⟩—SR
0.5 equiv PPh$_3$
PhCl, reflux

X	R	Yield (%)
C(O)Ph	*m*-CH$_3$OC$_6$H$_4$	89
C(O)Ph	*p*-ClC$_6$H$_4$	90
C(O)Ph	*p*-CH$_3$OC$_6$H$_4$	87
CN	*p*-CH$_3$OC$_6$H$_4$	91
NO$_2$	*p*-CH$_3$OC$_6$H$_4$	86
C(O)CH$_3$	*p*-CH$_3$OC$_6$H$_4$	91

Scheme 6.25

Ar-I + RSH → $\dfrac{\text{5 mol \% [RhCl(COD)]}_2}{\text{10 mol \% PPh}_3}$ → Ar-SR
toluene, NaO*t*Bu, 100 °C

Ar	R	Yield (%)
p-CH$_3$C$_6$H$_4$	*p*-ClC$_6$H$_5$	86
C$_6$H$_5$	C$_6$H$_5$	98
p-BrC$_6$H$_4$	C$_6$H$_5$	67
p-NH$_2$C$_6$H$_4$	C$_6$H$_5$	94
p-CF$_3$C$_6$H$_4$	C$_6$H$_5$	91
p-CH$_3$C$_6$H$_4$	C$_6$H$_{11}$	73

Scheme 6.26

of alkyl thiols were reported, all of which provided lower yields than aryl thiols. The use of aryl bromides in this chemistry was not reported.

6.7 Conclusion and Outlook

The transition metal-catalyzed formation of carbon–oxygen and carbon–sulfur bonds *via* cross-coupling reactions of aryl halides and alcohols or thiols has helped transform the way in which synthetic chemists design strategies to prepare small molecules. Early catalyst development for carbon–oxygen bond formation involved the use of palladium complexes containing bulky, electron rich phosphine ligands that are thought to stabilize highly reactive, coordinatively unsaturated metal complexes. New generations of catalysts led to the development of palladium complexes that can couple aryl halides with primary and secondary alcohols. For carbon–sulfur bond cross-coupling reactions, typical catalyst systems contain strongly chelating ligands, which prevent thiolate attack at the metal center and subsequent catalyst deactivation. Other methods for attenuation of thiolate nucleophilicity have also been advanced, including employing catalytic Lewis acids to chaperone the thiolate anion.

The development of copper complexes as catalysts in carbon–oxygen and carbon–sulfur bond formations has also progressed in recent years, as catalyst designers continue to focus on more cost-effective metals. Alternative, cheap metal complexes that currently have minor roles in catalyzing carbon–sulfur bond formation reactions include manganese, cobalt and nickel. Although palladium catalysts continue to provide the broadest substrate scope for both carbon–oxygen and carbon–sulfur bond-formation reactions, future catalyst design centered around non-precious metals should allow for the replacement of palladium as the catalyst of choice for these important transformations.

References

1. N. Miyaura, K. Yamada and A. Suzuki, *Tetrahedron Lett.*, 1979, **20**, 3437.
2. N. Miyaura, T. Yanagi and A. Suzuki, *Synth. Commun.*, 1981, **11**, 513.
3. R. F. Heck and J. P. Nolley, *J. Org. Chem.*, 1972, **37**, 2320.
4. T. Mizoroki, K. Mori and A. Ozaki, *Bull. Chem. Soc. Jpn.*, 1971, **44**, 581.
5. S. Baba and E. Negishi, *J. Am. Chem. Soc.*, 1976, **98**, 6729.
6. A. S. Guram, R. A. Rennels and S. L. Buchwald, *Angew. Chem. Int. Ed. Engl.*, 1995, **34**, 1348.
7. J. Louie and J. F. Hartwig, *Tetrahedron Lett.*, 1995, **36**, 3609.
8. J. F. Hartwig, *Nature*, 2008, **455**, 314.
9. F. Barrios-Landeros, B. P. Carrow and J. F. Hartwig, *J. Am. Chem. Soc.*, 2009, **131**, 8141.
10. J. P. Stambuli, C. D. Incarvito, M. Buhl and J. F. Hartwig, *J. Am. Chem. Soc.*, 2004, **126**, 1184.
11. D. A. Culkin and J. F. Hartwig, *Organometallics*, 2004, **23**, 3398.
12. A. H. Roy and J. F. Hartwig, *Organometallics*, 2004, **23**, 1533.
13. M. Yamashita and J. F. Hartwig, *J. Am. Chem. Soc.*, 2004, **126**, 5344.
14. G. Mann, D. Baranano, J. F. Hartwig, A. L. Rheingold and I. A. Guzei, *J. Am. Chem. Soc.*, 1998, **120**, 9205.
15. J. F. Hartwig, *Acc. Chem. Res.*, 1998, **31**, 852.
16. B. P. Carrow and J. F. Hartwig, *J. Am. Chem. Soc.*, 2011, **133**, 2116.
17. J. K. Huang, Y. Chen, J. Chan, M. L. Ronk, R. D. Larsen and M. M. Faul, *Synlett*, 2011, 1419.
18. J. Louie and J. F. Hartwig, *J. Am. Chem. Soc.*, 1995, **117**, 11598.
19. A. L. Casado and P. Espinet, *J. Am. Chem. Soc.*, 1998, **120**, 8978.
20. V. Farina and B. Krishnan, *J. Am. Chem. Soc.*, 1991, **113**, 9585.
21. C. C. Eichman and J. P. Stambuli, *Molecules*, 2011, **16**, 590.
22. S. Enthaler and A. Company, *Chem. Soc. Rev.*, 2011, **40**, 4912.
23. E. N. Pitsinos, V. P. Vidali and E. A. Couladouros, *Eur. J. Org. Chem.*, 2011, 1207.
24. V. U. Ahmad, A. U. R. Amber, S. Arif, M. H. M. Chen and J. Clardy, *Phytochemistry*, 1985, **24**, 2709.
25. D. Cousin, J. Mann, M. Nieuwenhuyzen and H. van den Berg, *Org. Biomol. Chem.*, 2006, **4**, 54.

26. S. B. Singh and G. R. Pettit, *J. Org. Chem.*, 1990, **55**, 2797.

27. E. C. Jorgensen, W. J. Murray and P. Block, *J. Med. Chem.*, 1974, **17**, 434.

28. Y. Takaya, H. Kikuchi, Y. Terui, J. Komiya, Y. Maeda, A. Ito and Y. Oshima, *Tetrahedron Lett.*, 2001, **42**, 61.

29. W. N. Wu, W. T. Liao, Z. F. Mahmoud, J. L. Beal and R. W. Doskotch, *J. Nat. Prod.*, 1980, **43**, 472.

30. G. Mann, C. Incarvito, A. L. Rheingold and J. F. Hartwig, *J. Am. Chem. Soc.*, 1999, **121**, 3224.

31. A. Aranyos, D. W. Old, A. Kiyomori, J. P. Wolfe, J. P. Sadighi and S. L. Buchwald, *J. Am. Chem. Soc.*, 1999, **121**, 4369.

32. C. H. Burgos, T. E. Barder, X. H. Huang and S. L. Buchwald, *Angew. Chem. Int. Ed.*, 2006, **45**, 4321.

33. J. P. Stambuli, M. Buhl and J. F. Hartwig, *J. Am. Chem. Soc.*, 2002, **124**, 9346.

34. Drugs.com. U.S. Pharmaceutical Sales-Q3 2013; c2000-13 [updated: November 2013]; available from http://www.drugs.com/stats/top100/sales.

35. C. A. Parrish and S. L. Buchwald, *J. Org. Chem.*, 2001, **66**, 2498.

36. Q. Shelby, N. Kataoka, G. Mann and J. Hartwig, *J. Am. Chem. Soc.*, 2000, **122**, 10718.

37. M. Watanabe, M. Nishiyama and Y. Koie, *Tetrahedron Lett.*, 1999, **40**, 8837.

38. K. E. Torraca, X. H. Huang, C. A. Parrish and S. L. Buchwald, *J. Am. Chem. Soc.*, 2001, **123**, 10770.

39. K. E. Torraca, S. I. Kuwabe and S. L. Buchwald, *J. Am. Chem. Soc.*, 2000, **122**, 12907.

40. A. V. Vorogushin, X. H. Huang and S. L. Buchwald, *J. Am. Chem. Soc.*, 2005, **127**, 8146.

41. X. X. Wu, B. P. Fors and S. L. Buchwald, *Angew. Chem. Int. Ed.*, 2011, **50**, 9943.

42. S. Gowrisankar, A. G. Sergeev, P. Anbarasan, A. Spannenberg, H. Neumann and M. Beller, *J. Am. Chem. Soc.*, 2010, **132**, 11592.

43. P. E. Maligres, J. Li, S. W. Krska, J. D. Schreier and I. T. Raheem, *Angew. Chem. Int. Ed.*, 2012, **51**, 9071.

44. T. E. Barder and S. L. Buchwald, *J. Am. Chem. Soc.*, 2007, **129**, 12003.

45. N. Kataoka, Q. Shelby, J. P. Stambuli and J. F. Hartwig, *J. Org. Chem.*, 2002, **67**, 5553.

46. G. Mann and J. F. Hartwig, *J. Am. Chem. Soc.*, 1996, **118**, 13109.

47. S. Kuwabe, K. E. Torraca and S. L. Buchwald, *J. Am. Chem. Soc.*, 2001, **123**, 12202.

48. M. Palucki, J. P. Wolfe and S. L. Buchwald, *J. Am. Chem. Soc.*, 1996, **118**, 10333.

49. J. J. Niu, H. Zhou, Z. G. Li, J. W. Xu and S. J. Hu, *J. Org. Chem.*, 2008, **73**, 7814.

50. R. Hosseinzadeh, M. Tajbakhsh, M. Mohadjerani and M. Alikarami, *Synlett*, 2005, 1101.

51. M. Wolter, G. Nordmann, G. E. Job and S. L. Buchwald, *Org. Lett.*, 2002, **4**, 973.
52. R. A. Altman, K. W. Anderson and S. L. Buchwald, *J. Org. Chem.*, 2008, **73**, 5167.
53. A. Shafir, P. A. Lichtor and S. L. Buchwald, *J. Am. Chem. Soc.*, 2007, **129**, 3490.
54. D. Maiti and S. L. Buchwald, *J. Org. Chem.*, 2010, **75**, 1791.
55. Q. Zhang, D. P. Wang, X. Y. Wang and K. Ding, *J. Org. Chem.*, 2009, **74**, 7187.
56. J. J. Niu, P. R. Guo, J. T. Kang, Z. G. Li, J. W. Xu and S. J. Hu, *J. Org. Chem.*, 2009, **74**, 5075.
57. A. B. Naidu and G. Sekar, *Tetrahedron Lett.*, 2008, **49**, 3147.
58. H. J. Cristau, P. P. Cellier, S. Hamada, J. F. Spindler and M. Taillefer, *Org. Lett.*, 2004, **6**, 913.
59. M. S. Kabir, M. Lorenz, O. A. Namjoshi and J. M. Cook, *Org. Lett.*, 2010, **12**, 464.
60. Q. Cai, B. L. Zou and D. W. Ma, *Angew. Chem. Int. Ed.*, 2006, **45**, 1276.
61. H. Zhang, D. W. Ma and W. G. Cao, *Synlett*, 2007, 243.
62. H. H. Rao, Y. Jin, H. Fu, Y. Y. Jiang and Y. F. Zhao, *Chem. Eur. J.*, 2006, **12**, 3636.
63. E. Buck, Z. J. Song, D. Tschaen, P. G. Dormer, R. P. Volante and P. J. Reider, *Org. Lett.*, 2002, **4**, 1623.
64. Y.-J. Chen and H.-H. Chen, *Org. Lett.*, 2006, **8**, 5609.
65. A. B. Naidu, E. A. Jaseer and G. Sekar, *J. Org. Chem.*, 2009, **74**, 3675.
66. J. Dreyfuss, A. I. Cohen and S. M. Hess, *J. Pharm. Sci.*, 1968, **57**, 1505.
67. G. L. Thomas, R. J. Spandl, F. G. Glansdorp, M. Welch, A. Bender, J. Cockfield, J. A. Lindsay, C. Bryant, D. F. J. Brown, O. Loiseleur, H. Rudyk, M. Ladlow and D. R. Spring, *Angew. Chem. Int. Ed.*, 2008, **47**, 2808.
68. B. Vie, S. Sablayrolles, R. Letienne, B. Vacher, A. Darmellah, M. Bernard, D. Feuvray and B. Le Grand, *J. Pharmacol. Exp. Ther.*, 2009, **330**, 696.
69. C. C. Eichman and J. P. Stambuli, *J. Org. Chem.*, 2009, **74**, 4005.
70. M. Murata and S. L. Buchwald, *Tetrahedron*, 2004, **60**, 7397.
71. M. A. Fernandez-Rodriguez, Q. L. Shen and J. F. Hartwig, *Chem. Eur. J.*, 2006, **12**, 7782.
72. M. A. Fernandez-Rodriguez, Q. L. Shen and J. F. Hartwig, *J. Am. Chem. Soc.*, 2006, **128**, 2180.
73. M. C. Willis, D. Taylor and A. T. Gillmore, *Tetrahedron*, 2006, **62**, 11513.
74. C. Benedi, F. Bravo, P. Uriz, E. Fernandez, C. Claver and S. Castillon, *Tetrahedron Lett.*, 2003, **44**, 6073.
75. F. Y. Kwong and S. L. Buchwald, *Org. Lett.*, 2002, **4**, 3517.
76. P. Zhao, H. Yin, H. X. Gao and C. J. Xi, *J. Org. Chem.*, 2013, **78**, 5001.
77. T. J. Liu, C. L. Yi, C. C. Chan and C. F. Lee, *Chem. Asian J.*, 2013, **8**, 1029.
78. Y. C. Wong, T. T. Jayanth and C. H. Cheng, *Org. Lett.*, 2006, **8**, 5613.
79. V. Gomez-Benitez, O. Baldovino-Pantaleon, C. Herrera-Alvarez, R. A. Toscano and D. Morales-Morales, *Tetrahedron Lett.*, 2006, **47**, 5059.

80. O. Baldovino-Pantaleon, S. Hernandez-Ortega and D. Morales-Morales, *Adv. Synth. Catal.*, 2006, **348**, 236.
81. Y. G. Zhang, K. C. Ngeow and J. Y. Ying, *Org. Lett.*, 2007, **9**, 3495.
82. C. P. Zhang and D. A. Vicic, *J. Am. Chem. Soc.*, 2012, **134**, 183.
83. C. S. Yan, Y. Peng, X. B. Xu and Y. W. Wang, *Chem. Eur. J.*, 2012, **18**, 6039.
84. X. B. Xu, J. Liu, J. J. Zhang, Y. W. Wang and Y. Peng, *Org. Lett.*, 2013, **15**, 550.
85. M. Arisawa, T. Suzuki, T. Ishikawa and M. Yamaguchi, *J. Am. Chem. Soc.*, 2008, **130**, 12214.
86. C. S. Lai, H. L. Kao, Y. J. Wang and C. F. Lee, *Tetrahedron Lett.*, 2012, **53**, 4365.

CHAPTER 7

Pd(0)-Catalyzed Carboiodination: Early Developments and Recent Advances

DAVID A. PETRONE,[†a] CHRISTINE M. LE,[†a]
STEPHEN G. NEWMAN[b] AND MARK LAUTENS[*a]

[a] Department of Chemistry, University of Toronto, 80 St George Street, Toronto, Ontario, Canada M5S 3H6; [b] Centre for Catalysis Research and Innovation, Department of Chemistry, University of Ottawa, 10 Marie Curie, Ottawa, Ontario, Canada K1N 6N5
*Email: mlautens@chem.utoronto.ca

7.1 Introduction

Palladium-catalyzed cross-coupling reactions (*e.g.*, Suzuki, Mizoroki–Heck, Negishi, Buchwald–Hartwig reactions) have become the premier approach to forge carbon–carbon and carbon–heteroatom bonds. Over the past several decades, major developments in this field have been achieved that have allowed many exceptionally challenging cross-couplings to be realized, facilitating the step-efficient synthesis of complex pharmaceuticals and biologically active natural products. Despite these enormous advances, opportunities exist to find new reactivity that can further diversify the range

---

[†] These authors contributed equally to the writing of this chapter.

RSC Catalysis Series No. 21
New Trends in Cross-Coupling: Theory and Applications
Edited by Thomas J Colacot
© The Royal Society of Chemistry 2015
Published by the Royal Society of Chemistry, www.rsc.org

of products that can be made. New reactivity is derived from combining rationally composed sets of catalytic steps and new catalysts. In 2010, Lautens and Newman reported a synthesis of 2-bromoindoles *via* C–N coupling that was only possible due to a reversible oxidative addition.[1] This subsequently led to the discovery of a Pd(0)-catalyzed carboiodination/ cycloisomerization.[2] In the reaction pathway, an ArPd(II)I species undergoes carbopalladation to an unactivated alkene. In the absence of *syn*-β-hydrogen atoms, a strictly ligand-dependent carbon–iodine reductive elimination occurs. The process retains the valuable iodine atom from the substrate, which expedites further product derivatization. This atom-economical C–C bond-forming reaction possesses a novel reaction mechanism in a very important field and we consider that a comprehensive timeline of this chemistry will help expand its utility and desirability.

The aim of this chapter is to put the origins of carbon–halogen reductive elimination from transition metal complexes into perspective, while at the same time highlighting the efforts towards developing the first Pd(0)-catalyzed carboiodination. Additionally, contributions from other researchers in this field will be highlighted and recent applications of this powerful methodology will be discussed.

7.2   Classical Reactivity of Palladium Complexes

Palladium represents a reliable and versatile transition metal, known to have great success in a wide array of catalytic reactions. Its tolerance towards diverse functional groups and harsh reaction conditions make it ideal for everyday use in organic synthesis. Yet, despite these positive characteristics, the relatively high cost of Pd compared to other transition metals, and also difficulties in separating the homogeneous catalyst from products, limit its widespread application in some instances. Nonetheless, numerous industrial processes, including the Wacker oxidation, have prevailed as some of the most important industrial reactions which provide necessary chemical feedstocks.[3–7] The intense research efforts conducted in this field have provided researchers with a strong groundwork on which modern reaction design and development can be based.

Palladium is capable of performing a number of elementary steps (*e.g.*, oxidative addition, transmetallation, carbopalladation and reductive elimination), which can be combined into useful catalytic transformations.[8] The relative rates of each elementary step can be readily influenced by modifying the steric and electronic properties of L-type phosphorus or nitrogen-based ligands, which will be a key focus in the following sections.

7.3   Oxidative Addition of Palladium to Carbon–Halogen Bonds

Oxidative addition in a cross-coupling typically involves a reaction between a Pd(0) complex and a polarized carbon–halogen bond, which undergoes

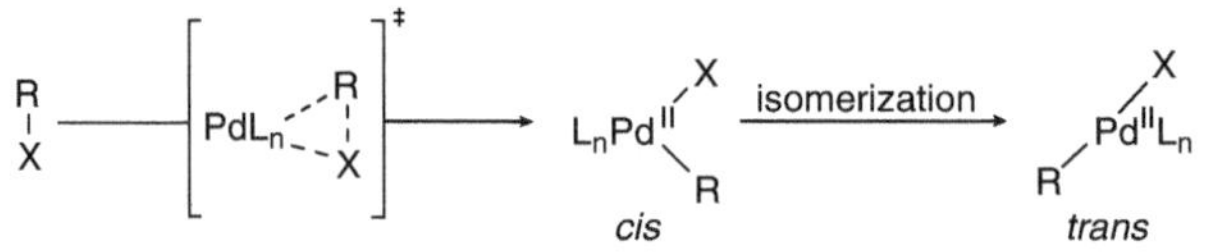

Scheme 7.1   The oxidative addition reaction.

heterolytic cleavage *via* a concerted, three-centered transition state (Scheme 7.1).[9–11] Overall, palladium donates two non-bonding electrons to form two new metal–ligand bonds with a *cis* relationship, which can undergo rearrangement to the thermodynamically favored *trans* product.[12] The classical order of reactivity with respect to aryl halides is I > Br > Cl ≫ F, where C–F bonds are essentially unreactive. However, recent efforts in the area aim to shift this longstanding paradigm.[13–15]

Oxidative addition is generally promoted by strongly σ-donating ligands that increase the electron density at palladium, thus making it more electron rich for breaking C–X bonds.[16,17] Conversely, π-accepting ligands counteract this process by rendering palladium less electron rich. Additionally, more electron-deficient carbon–halogen bonds are increasingly susceptible to undergo oxidative addition with electron-rich Pd(0) sources. Steric effects also have an important but less straightforward role in oxidative addition. While this process involves an increase in steric bulk around the metal center, large and sterically hindered ligands are typically used to favor challenging oxidative additions. This behavior has been attributed to the formation of highly active PdL_1 or PdL_2 species readily *in situ*, which have open coordination sites to allow oxidative addition to take place.[18]

7.4   Reductive Elimination from Transition Metal Complexes

Reductive elimination is the microscopic reverse of oxidative addition and can be promoted by various factors, including steric and electronic perturbations of the substrate, product, and/or ancillary ligand(s). These trends are generally opposite to those seen for oxidative addition reactions. For example, electron-poor metal centers are better suited to undergo reductive elimination than more electron-rich metal centers possessing similar steric properties, since there is an increase in electron density at the metal center over the course of the reaction. Additionally, metal complexes possessing more sterically encumbering ligands will have faster rates of reductive elimination compared to less bulky ligands with similar electronic properties. These general trends correspond well to the essence of the overall process, in which both reduction and a decrease in coordination number occur at the metal center.[19] These fundamental effects have been studied with respect to both carbon–hydrogen[20] and, in rare instances, carbon–halogen bonds.[21,22] The reductive elimination of carbon–halogen bonds from transition metal complexes represents an interesting extension, as it is typically not thermodynamically favored (compared to the case with carbon–carbon bonds). This challenging variation of reductive elimination has required the development of new catalysts and ligands.

7.4.1 Carbon–Halogen Bond Reductive Elimination

Despite the myriad of examples involving oxidative addition of transition metals to carbon–halogen bonds,[22–28] reductive elimination to form new or reform existing carbon–halogen bonds remains a rare and challenging transformation.[29] Accordingly, the oxidative addition of transition metals to carbon–halogen bonds has traditionally been considered irreversible, a supposition rooted in its high exothermicity.[30,31] As counter evidence, Ettorre reported the reductive elimination of iodobenzene from the octahedral Pt(IV) complex **7.1** to form *trans*-Pt(II) complex **7.3** [eqn (7.1)].[32] The author stated that the addition of NaI to the reaction slowed the rate of reductive elimination, suggesting that a solvolytic process involving the formation of an iodide ion occurred during the reaction coordinate to form complex **7.2**.

$$[Pt(PEt_3)_2Ph_2I_2] + solvent \xrightarrow{-I^-} [trans\text{-}[Pt(PEt_3)_2Ph_2I(solvent)]]$$

7.1   **7.2**

$$\xrightarrow{-PhI} trans\text{-}[Pt(PEt_3)_2PhI] \qquad\qquad (7.1)$$

7.3

In 1987, Echavarren and Stille reported the Pd-catalyzed coupling of aryl triflates with organostannanes. The formation of aryl chloride **7.6** was observed as a by-product during the reaction between **7.4** and sterically hindered tetrakis(trimethylsilyl)methylstannane (Scheme 7.2).[33] This result was proposed to be an artifact of the inability of the bulky stannane to transmetallate to the resulting ArPd(II)Cl species **7.5**. The exact pathway to **7.6** was not proposed; however, an S_NAr mechanism on either **7.4** or **7.5**, as well as reductive elimination from **7.5** are all possibilities.

Numerous reports of Pd(IV) intermediates in catalysis have appeared over the past few decades[34–36] and, despite resistance towards the idea of these high-valent species in certain cases, there is an active interest in both elucidating

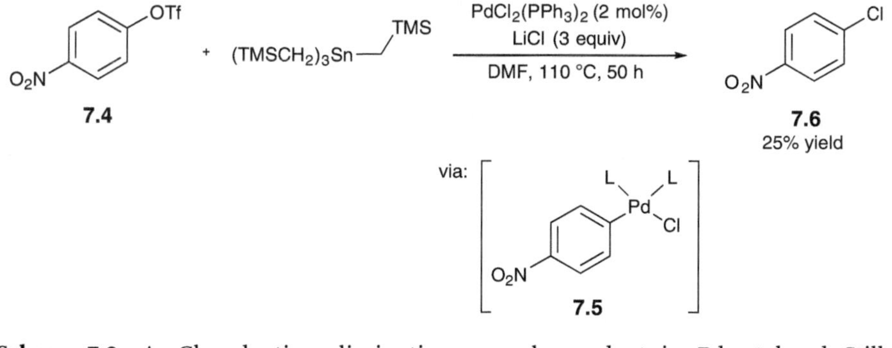

Scheme 7.2 Ar–Cl reductive elimination as a by-product in Pd-catalyzed Stille coupling of bulky stannanes.

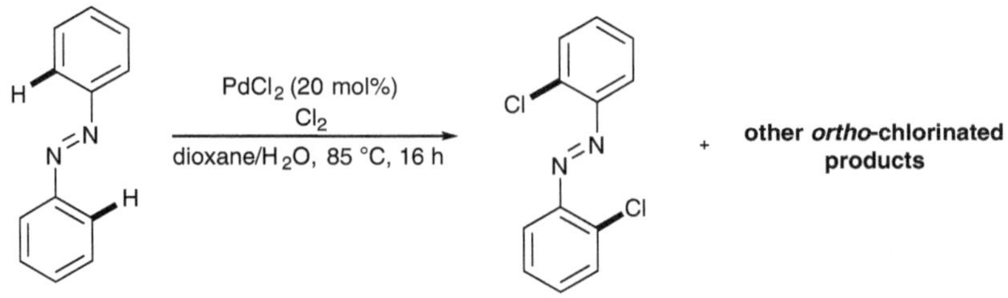

Scheme 7.3   First example of a Pd-catalyzed *ortho*-chlorination of azobenzene by Fahey.

their existence and exploiting their capacity in catalysis.[37] Carbon–halogen bond-forming reactions involving high-valent Pd(IV) species have come to light as both a useful and mechanistically interesting class of reactions. In 1970, Fahey reported the first Pd-catalyzed directed *ortho* C–H chlorination of azobenzene 7.7 to form 7.8 using Cl_2 (Scheme 7.3).[38,39] Despite the low selectivities observed for the monochlorinated product, this report established a novel and fascinating proof-of-principle, which has developed into an active field of research.

Many stoichiometric and catalytic methods have been reported since Fahey's pioneering work, including carbon–fluorine,[40] carbon–chlorine,[41] carbon–bromine[42] and carbon–iodine[43] bond formation. Many efforts in this field are directed towards gaining a better mechanistic understanding of the overall process. Nevertheless, we have elected to focus on reactions of Pd(0/II) species and interested readers are directed to relevant reviews for a discussion of higher valent palladium species.[44,45]

7.4.2   Stoichiometric Aryl Halide Reductive Elimination from Dimeric Palladium(II) Complexes

In 2001, Roy and Hartwig conducted an in-depth kinetic study regarding stoichiometric carbon–halogen bond reductive elimination from dimeric ArPd(II)X complexes 7.9a–e when reacted with excess P^tBu_3 (Table 7.1).[46] Prior to this seminal report, the oxidative addition of Pd(0) complexes to aryl–halide bonds was thought to be irreversible. As a result of this study, the authors rewrote the established rules for the kinetics and thermodynamics of oxidative addition/reductive elimination.

In addition to a 10-fold increase in K_{eq} values, dimers 7.9a–c containing *o*-Me-substituted aryl ligands provided higher yields of their respective ArX products 7.10a–c than the complexes without *ortho*-substituted aryl ligands. Notably, each successive change in halide ligand (*i.e.*, Cl to Br or Br to I) was associated with an approximate 100-fold change in K_{eq}. Excellent fit to a first-order appearance of each product was obtained, in addition to observing both a positive first-order dependence of $1/k_{obs}$ *versus* $1/[P^tBu_3]$ and an inverse dependence of k_{obs} on $[P(o\text{-tol})_3]$, both with non-zero *y*-intercepts. All data were consistent with a mechanism involving an irreversible formation of monomers from the dimeric

Table 7.1 Roy and Hartwig's stoichiometric studies on reductive elimination of aryl halides from ArPd(II)X dimers.

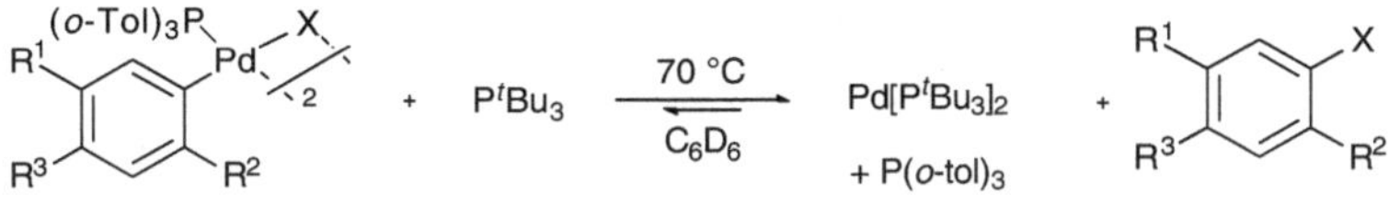

R¹ = *t*Bu, R² = Me, R³ = H
X = Cl: **7.9a**
X = Br: **7.9b**
X = I: **7.9c**

R¹ = R² = H, R³ = *t*Bu
X = Cl: **7.9d**
X = Br: **7.9e**

7.10a-e

X	Yield of **7.10** (%)	K_{eq}
7.7a (X = Cl)	70	$9(3) \times 10^{-2}$
7.7b (X = Br)	70	$2.3(3) \times 10^{-3}$
7.7c (X = I)	39	$3.7(2) \times 10^{-5}$
7.7d (X = Cl)	30	Not measured
7.7e (X = Br)	75	$3.3(6) \times 10^{-4}$

Adapted from Ref. 46. Copyright 2001 American Chemical Society.

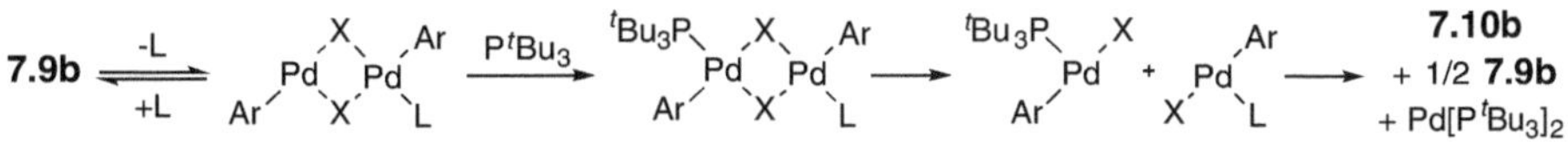

Scheme 7.4 Hartwig's proposed mechanism of reductive elimination from ArPd(II)X dimers involving ligand dissociation.
Adapted from Ref. 46. Copyright 2001 American Chemical Society.

species, which are formed from ligand substitution and subsequent reductive elimination (Scheme 7.4). The authors also reasoned that the formation of monomers and reductive elimination are faster than dissociation of P*t*Bu₃. The use of P*t*Bu₃ is crucial in this reaction and it is proposed that its steric influence outweighs its high σ-donating character, thus promoting the forward reaction. This is contrary to the information presented in previous sections discussing the factors that promote reductive elimination, which state that highly σ-donating ligands slow the rate of reductive elimination.

It was reasoned that reductive elimination from an ArPd(II)Cl species would be faster than the corresponding bromide or iodide analogues if ground state effects, such as carbon–halogen bond strengths, dominated. However, if bond strengths and electronic properties of the metal–halogen bonds controlled the rates of addition and elimination, reductive elimination from the ArPd(II)Cl would be slower than the bromide or iodide equivalents.[47]

7.4.3 Directly Observed Aryl Halide Reductive Elimination from Monomeric Palladium(II) Complexes

Most organometallic reactions proceed *via* intermediates possessing a free coordination site. Specifically, cross-coupling reactions with bulky

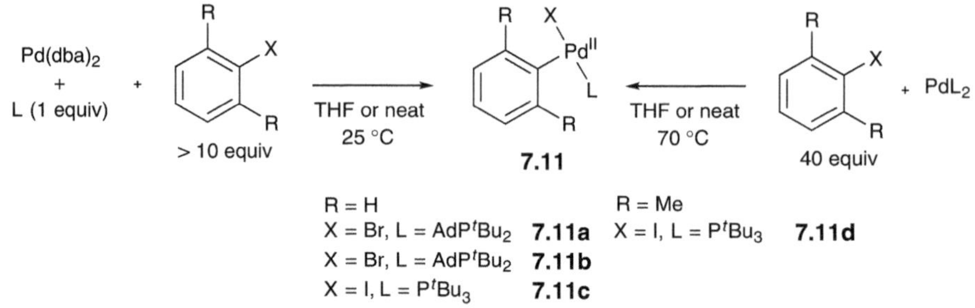
Pd(dba)$_2$
+
L (1 equiv)
> 10 equiv
THF or neat
25 °C
Pd^{II}
7.11
THF or neat
70 °C
40 equiv
+ PdL$_2$
R = H
X = Br, L = AdP^{t}Bu$_2$ 7.11a
X = Br, L = AdP^{t}Bu$_2$ 7.11b
X = I, L = P^{t}Bu$_3$ 7.11c
R = Me
X = I, L = P^{t}Bu$_3$ 7.11d

Scheme 7.5 Synthesis of the first three-coordinate 14-electron ArPd(II)X complexes by Hartwig.
Adapted from Ref. 50. Copyright 2002 American Chemical Society.

Table 7.2 Roy and Hartwig's stoichiometric studies on reductive elimination of aryl halides from monomeric ArPd(II)X complexes.

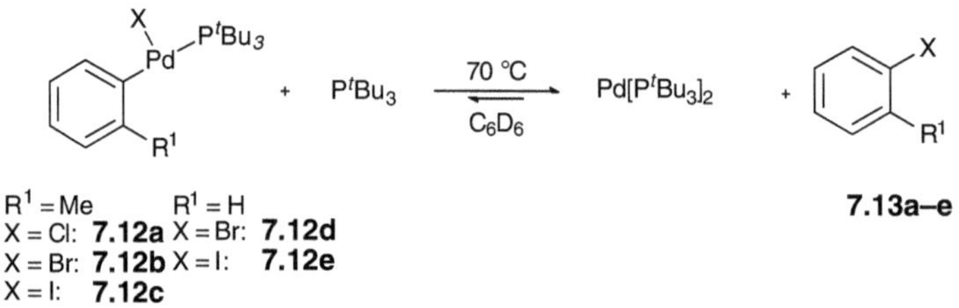
P^{t}Bu$_3$
Pd
+ P^{t}Bu$_3$
70 °C
C$_6$D$_6$
Pd[P^{t}Bu$_3$]$_2$
R^1 = Me
X = Cl: 7.12a
X = Br: 7.12b
X = I: 7.12c
R^1 = H
X = Br: 7.12d
X = I: 7.12e
7.13a–e

X	Yield of **7.13** (%)	K_{eq}
7.12a (X=Cl)	76	10.9×10^{2}
7.12b (X=Br)	98	32.7×10^{-1}
7.12c (X=I)	79	1.79×10^{-1}
7.12d (X=Br)	68	13.4×10^{-1}
7.12e (X=I)	60	0.51×10^{-1}

Adapted from Ref. 51. Copyright 2003 American Chemical Society.

phosphine ligands are commonly proposed to occur *via* three-coordinate 14-electron complexes such as **7.11**[23,48,49] and yet no such species had been isolated until a 2002 report by Hartwig and co-workers describing their synthesis, characterization and reactivity (Scheme 7.5).[50] This pioneering study led to a subsequent report in 2003 by Roy and Hartwig describing the first reductive elimination of aryl halides from monomeric ArPd(II)X complexes **7.12a-e**, which are proposed to be directly involved in palladium catalysis (Table 7.2).[51] This observation was a significant advance from their 2001 report,[46] where the species undergoing reductive elimination was not directly observed.

Yields of the respective aryl halide products were higher than those previously reported for the dimeric ArPd(II)X complex and K_{eq} values were obtained by initiating the reaction in both directions. Reductive elimination

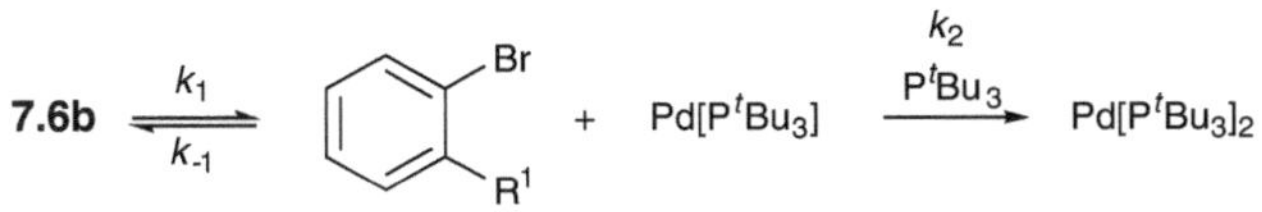

$$7.6b \underset{k_{-1}}{\overset{k_1}{\rightleftharpoons}} \quad \text{[2-R}^1\text{-C}_6\text{H}_4\text{-Br]} + Pd[P^tBu_3] \quad \xrightarrow[]{\overset{k_2}{P^tBu_3}} \quad Pd[P^tBu_3]_2$$

Scheme 7.6 Mechanism of reductive elimination involving ligand dissociation proposed by Hartwig.

from chloride **7.12a** was thermodynamically favored by a factor of 3000 over that of bromide **7.12b**, which was favored by a factor of ~ 20 over iodide **7.12c**. The authors tied this trend to the softness or nucleophilicity of the halogen in the transition state for addition and elimination.[52,53] Similarly, K_{eq} for *ortho*-substituted arenes (**7.12a–c**) was larger than that of their non-substituted analogs (**7.12d–e**). Interestingly, the kinetics were not in line with the above-mentioned thermodynamics, as reductive elimination from chloride **7.12a** was slower than that from bromide **7.12b**, despite being more energetically favorable. They also claimed that the strength of the metal–halogen bond is likely more important than the strength of the resulting carbon–halogen bond. The observed rate constant [eqn (7.4)] is inversely dependent on [ArX] and linearly dependent on [P^{t}Bu$_3$], both with non-zero y-intercepts. Based on these kinetic data, the authors proposed a mechanism that involves reversible ArBr reductive elimination followed by trapping of the of the Pd(0) intermediate by P^{t}Bu$_3$ (Scheme 7.6).

$$\text{rate} = k_{obs}[7.12b] \qquad k_{obs} = \frac{k_1 k_2 [P^tBu_3]}{k_{-1} k_2 [P^tBu_3]} \qquad (7.4)$$

These reports have shed light on the process of reductive elimination of carbon–halogen bonds from Pd(II) metal centers, in addition to laying the necessary groundwork needed for this rare elementary step to be used in new catalytic processes.

7.4.4 Alkyl Halide Reductive Elimination from Transition Metal Complexes

As highlighted in the preceding sections, the majority of carbon–halogen bond reductive eliminations from late transition metals such as Pt(IV),[54] Pd(IV) and Pd(II) involve aryl halides. A few studies of reductive elimination of alkyl halides *via* thermolysis involving Pt(IV) complexes have been reported.[55–57] The Rh(III) species [(CH$_3$CO)Rh(CO)I$_3$]$^-$ has also been shown to decompose through the loss of MeI. In the presence of a CO atmosphere, the newly formed dicarbonyl complex [(CH$_3$CO)Rh(CO)$_2$I$_3$]$^-$ reductively eliminates to form acetyl iodide.[58]

The first report of reductive elimination of alkyl halides from isolable Rh(III) alkyl complexes[59,60] was from Frech and Milstein, who described the reductive elimination of MeI from a bulky Rh(III) pincer complex in 2006 (Scheme 7.7).[61] By reacting nitrides **7.13** and **7.14** with an equimolar amount

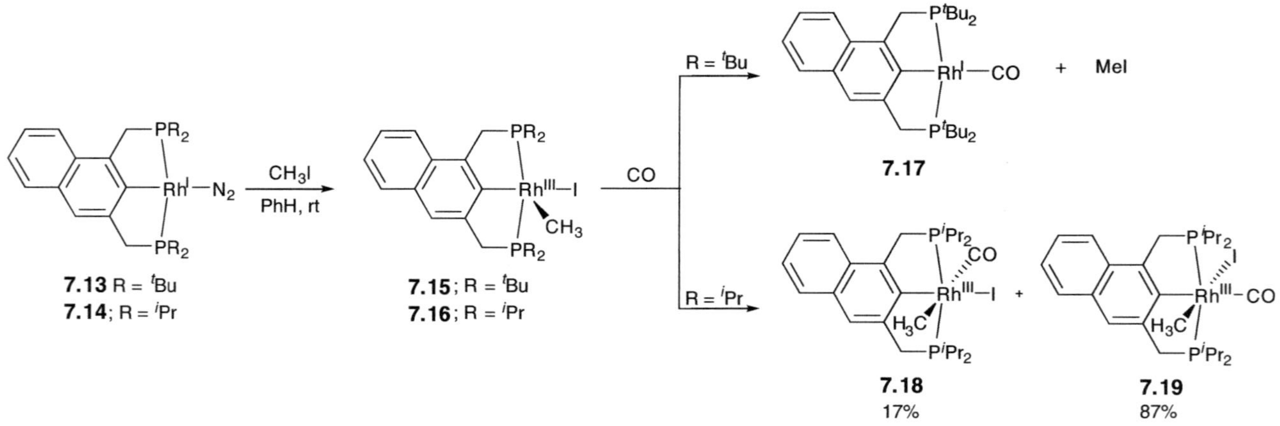

Scheme 7.7 Frech and Milstein's synthesis and reactivity of MeRh(III)I pincer complexes. Adapted from Ref. 61. Copyright 2006 American Chemical Society.

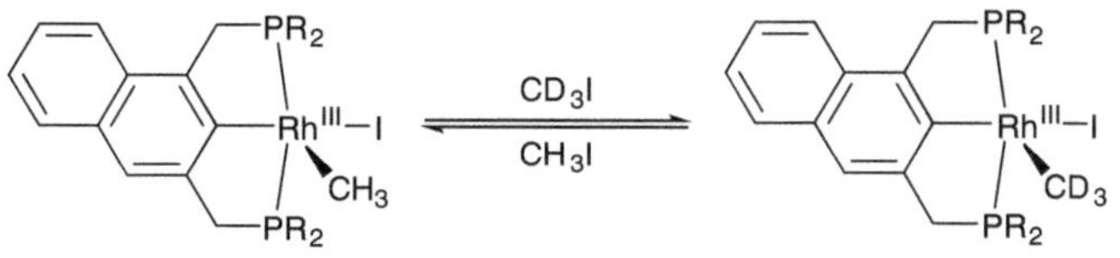

7.15; R = ^{t}Bu
7.16; R = ^{i}Pr

7.15-*d*; R = ^{t}Bu
7.16-*d*; R = ^{i}Pr

Scheme 7.8   Labeling experiments probing the reversibility of MeI reductive elimination.
Adapted from Ref. 61. Copyright 2006 American Chemical Society.

of MeI, PCP pincer complexes **7.15** and **7.16** were obtained, respectively. When complex **7.15** was treated with an excess of CO (~100 equiv), slow but almost quantitative reductive elimination to form Rh(I) carbonyl complex **7.17** occurred. Conversely, when the ^{i}Pr analog **7.16** was treated under the same reaction conditions, reductive elimination of MeI did not occur. Instead, adducts **7.18** and **7.19** were formed in 17 and 83% yield, respectively. The authors ascribed this marked switch in reactivity to the decreased sterics of the ^{i}Pr-containing P ligands. This interpretation echoes the discussion in the previous sections concerning the influence of steric congestion on carbon–halogen reductive elimination.

In order to gain insight into this rare process, the authors treated **7.15** and **7.16** with CD$_3$I. Labeled complexes **7.15**-*d* and **7.16**-*d* were detected by ^{1}H NMR spectroscopy, leading them to propose that a reversible process is operative in both instances (Scheme 7.8), despite **7.16** not undergoing thermodynamically favored reductive elimination.

The authors obtained further mechanistic insights by studying the effect of additives on the overall reductive elimination process. Adding excess ^{n}Bu$_4$NI to the reaction did not attenuate the rate of CH$_3$–CD$_3$ exchange, which contrasts with the work of Ettorre[32] by suggesting that iodide dissociation is not involved in this process. Second, no rate retardation was observed when excess MeI was added to the reaction of **7.15** to form **7.17**. The authors attributed this to the production of an Rh(I)(MeI) complex, followed by associative substitution of the MeI ligand with one CO molecule from the reaction atmosphere to yield **7.17**. Recently, the same group observed reductive elimination of MeBr and MeCl from analogous Rh(III) complexes and conducted in-depth kinetic experiments as means to gain a deeper mechanistic understanding.[62]

7.5   Polyhalogenated Substrates in Cross-Coupling

Whereas the previous sections largely focused on stoichiometric reductive elimination of carbon–halogen bonds from transition metal complexes, the following sections review catalytic reductive elimination of carbon–halogen bonds from Pd(II) centers, beginning with the seminal report on halogen exchange by Buchwald (Section 7.5.4), leading to the discovery and development of Pd(0)-catalyzed carboiodination by Lautens (Sections 7.6.2–7.8.4). Reversible oxidative addition of carbon–halogen bonds is also discussed in the context of 2-bromoindole synthesis by Lautens (Section 7.5.4)–work that ultimately led to the discovery of Pd(0)-catalyzed carboiodination. To highlight the synthetic utility and intriguing mechanistic aspects of reversible oxidative addition of carbon–halogen bonds by palladium, we believe that a brief discussion of irreversible oxidative addition in the context of polyhalogenated substrates in cross coupling is warranted.

When a starting material contains two or more of the same halogen atom, the ability to couple selectively one C–X bond over another (*i.e.*, site selectivity) has been a longstanding challenge. Nevertheless, polyhalogenated substrates have proven to be extremely useful intermediates in target-oriented synthesis and the development of increasingly selective and diverse coupling strategies remains an active area of research.[63] For example, Evans and Starr reported the total synthesis of (–)-FR182877, a cytotoxic natural product exhibiting anticancer properties, featuring an *E*-selective Suzuki cross-coupling of the *gem*-dibromoolefin **7.20** (Scheme 7.9).[64] In the later stages of the synthesis, vinyl halide **7.22** was subjected to another Suzuki cross-coupling and, following further functional group manipulations, provided the desired target. This particular synthesis highlights the power of transition metal catalysis in mediating cross-couplings of advanced intermediates with high selectivity and functional group tolerance.[65]

7.5.1 Selectivity-Governing Factors in Cross-Couplings of Polyhalogenated Substrates

In a typical palladium-catalyzed cross-coupling, oxidative addition into a C–X bond is generally considered to be irreversible and is therefore the selectivity-determining step (see Section 7.2.1). However, recent studies support the feasibility of reversible oxidative addition within certain catalytic systems – the results of which have been reviewed in Section 7.4. The majority of the methods developed for site- or chemoselective cross-couplings of polyhalogenated substrates take advantage of the subtle differences in bond strengths and steric environments between each C–X bond.

7.5.1.1 Chemoselectivity and Electronic Effects

The order of reactivity for oxidative insertion of Pd into a C–X bond has been classically described as I > Br > Cl >> F, which can be correlated with their respective C–X bond dissociation energies, with C–F bonds being the strongest. Therefore, it should follow that C–I bonds will react chemoselectively in the presence of C–Br bonds, which will react chemoselectively in preference to C–Cl bonds. Additionally, when considering two C–X bonds (where X is the same), higher rates of oxidative addition are generally observed for the more electron-deficient bond.[63] For example, Handy *et al.*

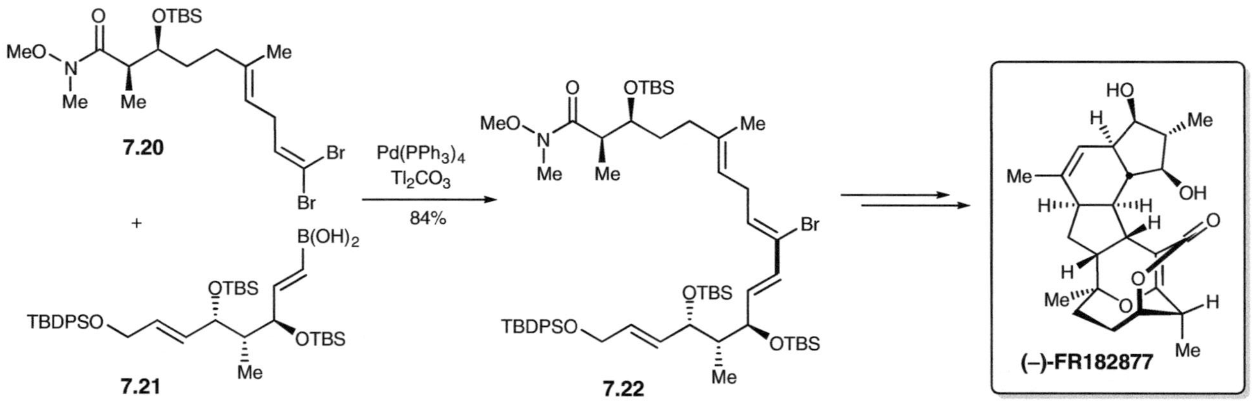

Scheme 7.9 Total synthesis of (−)-FR182877 featuring selective cross-couplings of *gem*-dibromoolefin 7.20.

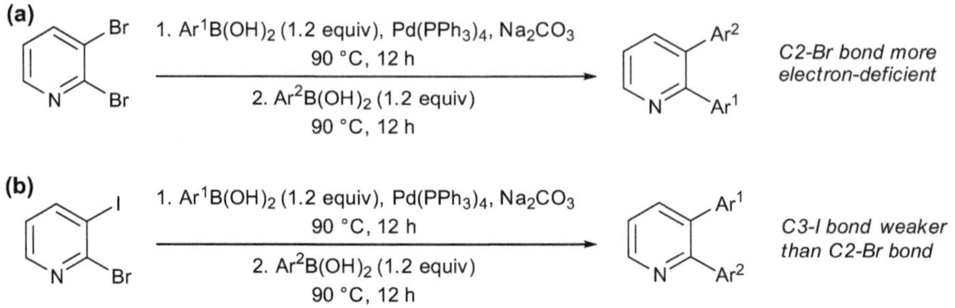

(a)

1. Ar¹B(OH)₂ (1.2 equiv), Pd(PPh₃)₄, Na₂CO₃
   90 °C, 12 h
   ————————————————————
2. Ar²B(OH)₂ (1.2 equiv)
   90 °C, 12 h

C2-Br bond more electron-deficient

(b)

1. Ar¹B(OH)₂ (1.2 equiv), Pd(PPh₃)₄, Na₂CO₃
   90 °C, 12 h
   ————————————————————
2. Ar²B(OH)₂ (1.2 equiv)
   90 °C, 12 h

C3-I bond weaker than C2-Br bond

Scheme 7.10 Switch in regioselectivity based on differences in electronic effects and bond dissociation energies.

developed a one-pot procedure towards disubstituted pyridine derivatives starting from dihalogenated pyridines *via* two sequential Suzuki cross-couplings.[66] In one example (Scheme 7.10a), the more electron-deficient C–Br bond undergoes selective monocoupling, whereas in another (Scheme 7.10b), the reverse regioselectivity is observed, as the more reactive (*i.e.*, weaker) C–I bond couples first. Selective and sequential cross-couplings of polyhalogenated aromatics are perhaps the most direct and efficient route to highly substituted aromatics. Therefore, studying the factors that govern regioselectivity in oxidative addition is a worthwhile and ongoing pursuit.

Handy and Zhang also developed a predictive guide for the order and site of coupling in polyhalogenated aromatics based on the ^{1}H NMR chemical shifts of the unsubstituted parent compounds, since chemical shifts are sensitive to electronic effects.[67] They observed that the order of reactivity can be paralleled to the chemical shift values, which are related to the intrinsic electron deficiency of each bond. In a more theoretically rigorous report, Merlic and co-workers performed extensive computational studies on the factors that govern regioselectivity in oxidative additions by Pd into poly-halogenated aromatics.[68] They proposed that distortion energies, which are related to the bond dissociation energies of the respective C–X bonds, and interaction energies, which involve the $d_{xy}(Pd) \rightarrow \pi^*(C–X)$ frontier molecular orbital interactions, are both responsible for the observed regioselectivities in cross-couplings of polyhalogenated heterocycles (Figure 7.1). These studies provide a more fundamental understanding of how electronic effects can govern site-selective oxidative additions.

7.5.1.2 Neighboring Substituents: Steric Effects and Directing Groups

Neighboring substituents or heteroatoms about the C–X bond can also affect the rate of oxidative addition due to either steric effects and/or the ability to act as directing groups. In the total synthesis of (–)-FR182877 by Evans and Starr highlighted in Section 7.5, the synthetic route took advantage of the inherent reactivity of the *gem*-dibromoolefin **7.20** (Scheme 7.9). The authors

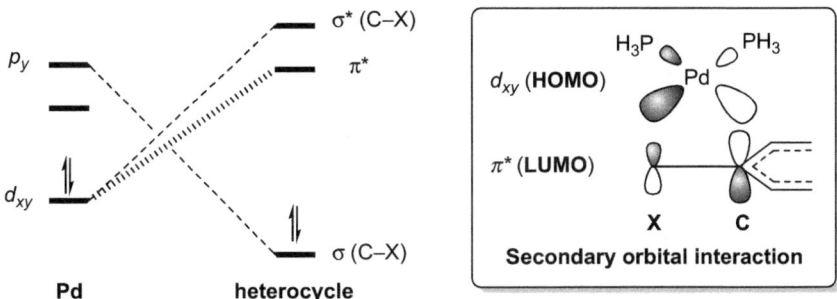

Figure 7.1 Key frontier molecular orbital interactions governing the interaction energy.
Adapted from Ref. 68*a*. Copyright 2007 American Chemical Society.

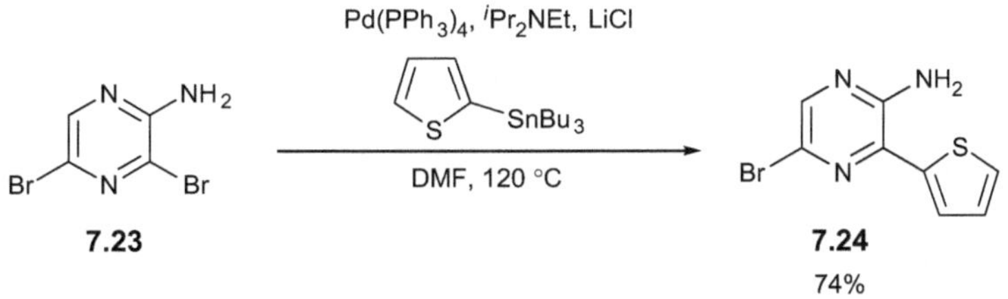

Scheme 7.11 Directing group ability overrides steric effects in Stille coupling of unsymmetrical pyrazine **7.23**.

surmised that the excellent *E*-selectivity observed for the Suzuki cross-coupling was due to the steric effect exerted by the vicinal alkyl chain, hence cross-coupling was observed at the less sterically hindered position.

In some cases, the directing ability of a neighboring functional group may override any steric effects imparted by that substituent. For example, in a study directed at synthesizing a library of chemi- or bioluminescent probes based on the 2-aminopyrazine scaffold, Nakamura *et al.* found that the amino group of **7.23** could direct oxidative addition into the *ortho* C3–Br bond in preference to the less-hindered *para* C5–Br bond to afford **7.24** in 74% yield (Scheme 7.11).[69] The use of *ortho*-directing groups in transition metal-catalyzed C–X or C–H functionalizations is common practice in synthetic chemistry.[44,70]

7.5.1.3 Tandem Intra- and Intermolecular Processes

The literature examples highlighted thus far have dealt only with systems involving stepwise cross-couplings of polyhalogenated starting materials, whereby the monofunctionalized intermediate is isolated before subjecting it to the second set of cross-coupling conditions. Examples featuring one-pot, domino cross-couplings of polyhalogenated substrates are less common, as issues of chemo- and site selectivity become more pronounced. Nevertheless, through strategic starting material design, tandem intra- and

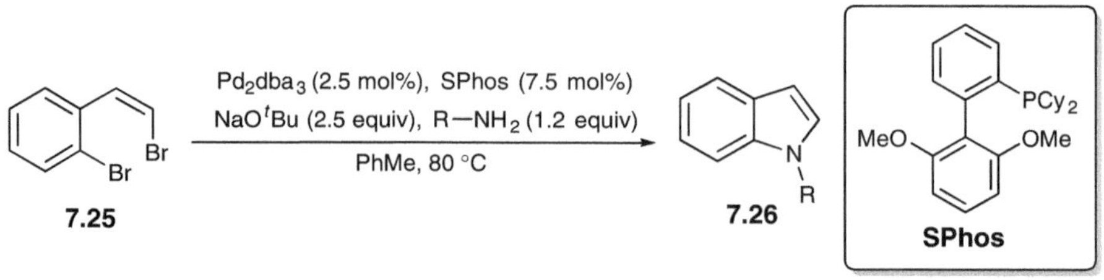

Scheme 7.12   Synthesis of indoles *via* tandem inter- and intramolecular C–N coupling.

intermolecular processes have been devised to achieve selective disubstitution in a one-pot protocol.[71] For example, Willis *et al.* developed a one-pot method for the synthesis of indoles **7.26** *via* sequential inter- and intramolecular Buchwald–Hartwig aminations starting from arylalkenyl halides **7.25** (Scheme 7.12).[72] The amine coupling partner serves as both an external and an internal nucleophile in this example. This particular strategy takes advantage of the fact that intramolecular C–N coupling should occur faster than the intermolecular process. Once the first amination has occurred at either the aryl or vinyl bromide, the second amination is more likely to occur intramolecularly rather than intermolecularly.

7.5.2   Tandem Cross-Couplings of *gem*-Dihaloolefins: Synthesis of Benzo-Fused Heterocycles

The majority of the top-selling pharmaceuticals and bioactive natural products contain at least one heterocyclic motif.[73] Fortunately, the advent of modern metal-mediated cross-coupling technologies has provided a wealth of synthetic tools to construct or derivatize heterocyclic cores. One methodology utilizes 1,1- or *gem*-dihaloolefins, which are powerful building blocks for the synthesis of highly substituted heterocycles and are readily prepared *via* Ramirez olefination from the corresponding aldehyde precursors.[74] With a long-standing interest in domino chemistry and transition metal-catalyzed heterocycle synthesis, Lautens and co-workers have utilized the inherent reactivity of *gem*-dihaloolefins in tandem intra-/intermolecular cross-couplings to access a variety of substituted indoles, benzofurans and benzothiophenes.[75–81] The seminal report utilized the *o-gem*-dihalovinylanilines **7.27** in a tandem Buchwald–Hartwig/Suzuki cross-coupling to furnish 2-substituted indoles **7.31** in a single pot (Scheme 7.13).[75,76] Mechanistic studies suggest that intramolecular carbon–heteroatom coupling precedes intermolecular cross-coupling, which implies selective oxidative addition of Pd into the (*Z*)-halide of **7.27**, as opposed to the more accessible (*E*)-halide.[76] This tandem coupling strategy was later extended to benzofurans and benzothiophenes.[77] Additionally, a variety of other nucleophilic coupling partners could be used, such as terminal alkynes,[77] alkenes,[78] heteroatoms[79] and unactivated aryl groups,[80] using either Pd or

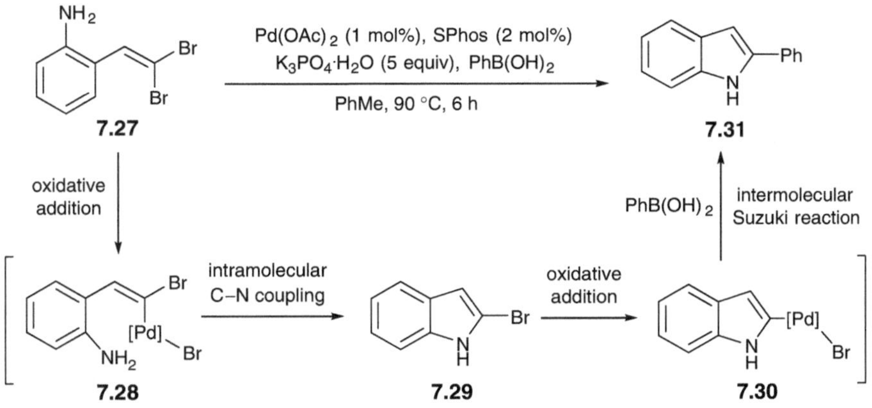

Scheme 7.13   Synthesis of 2-substituted indoles *via* Pd-catalyzed tandem C–N/C–C coupling of *gem*-dibromoolefins.

Cu catalysts. By modifying the backbone of the *gem*-dihaloolefins, azaindoles and thienopyrroles could also be prepared.[81]

7.5.3   Retention of Reactive Functionality: Synthesis of Halogenated Heterocycles *via* Intramolecular Cross-coupling of *gem*-Dihaloolefins

Halogenated heterocycles serve as particularly useful building blocks for late-stage diversification and are present in bioactive natural products themselves.[63b,82] While investigating the utility of *gem*-dihaloolefins, Lautens and co-workers noted that in the absence of an external nucleophile, entry to 2-bromoindoles 7.29 could not be achieved under the established reaction conditions.[1] It was proposed that 2-bromoindole served as a catalytic dead end *via* an irreversible oxidative addition process with Pd, forming complex 7.30. If the product is more susceptible to oxidative addition than the starting material, the Pd catalyst is rendered inactive after a single turnover. Therefore, the external nucleophile in the tandem coupling is required to regenerate the active Pd(0) catalyst and prevent product inhibition. The difficulty in decoupling the intra- and intermolecular processes warranted the development of a different catalytic system. In 2009, a mild and site-selective approach to 2-bromobenzo-fused heterocycles from *gem*-dihaloolefins *via* an intramolecular copper-catalyzed Ullmann-type coupling was developed (Scheme 7.14).[83] Whereas palladium catalysis proved to be ineffective towards this substrate class, the success of copper was thought to originative from the precoordination of the heteroatom in the substrate to the catalyst, which positions the (*Z*)-halide in close proximity for C–X coupling. Cu(0) and Cu(II) precatalysts were also effective in the reaction. Notably, this method allowed access to 2-halogenated benzo-fused heterocycles that would be difficult to obtain *via* traditional protocols (*i.e.*, aromatic

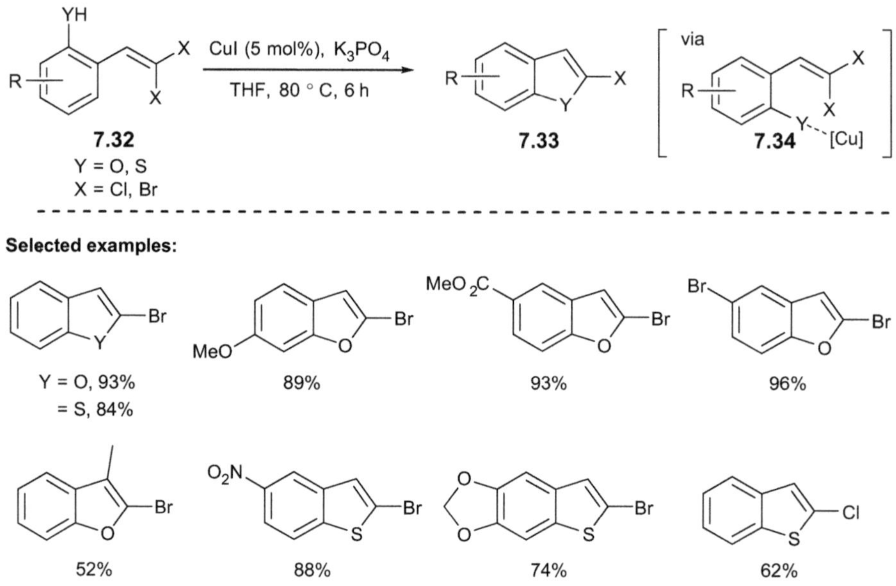

Scheme 7.14 Scope of Cu-catalyzed intramolecular C–O and C–S coupling of *gem*-dihaloolefins.

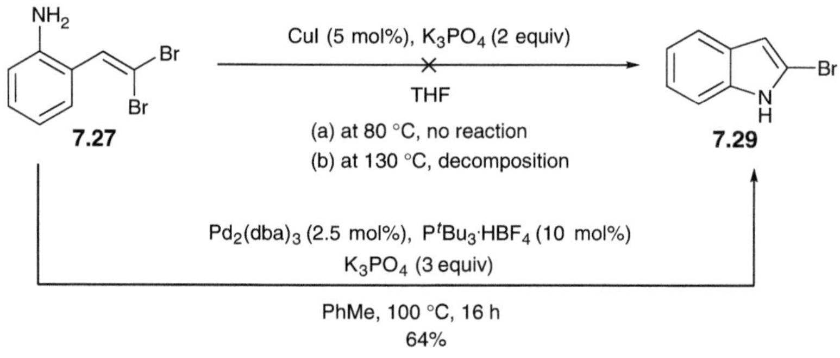

Scheme 7.15 Modified conditions for Pd-catalyzed synthesis of 2-bromoindole.

substitution with an electrophilic halogen source or selective-lithiation followed by electrophilic trapping), which often suffer from poor regioselectivity and/or functional group tolerance.

When the optimized Ullmann-type conditions were applied to the aniline **7.27**, the corresponding 2-halogenated indole **7.29** could not be obtained (Scheme 7.15).[84] Taking inspiration from the work of Hartwig and coworkers on the reductive elimination of aryl halides from Pd(II) (see Sections 7.4.2 and 7.4.3), upon switching to a Pd catalyst in combination with P^{t}Bu$_3$ as ligand, the desired 2-bromoindole was obtained. This preliminary result suggested that the steric bulk of the phosphine ligand employed may play a significant role in preventing catalyst inhibition.

7.5.4 Further Investigation into the Synthesis of Halogenated Indoles: Importance of Reversible Oxidative Addition

As mentioned previously, the selectivity-determining step in most regioselective cross-couplings of polyhalogenated substrates is believed to be oxidative addition into the C–X bond, a generally irreversible step. This is supported by computational studies, which reveal that oxidative addition of palladium complexes containing simple phosphine ligands into aryl–halide bonds is a highly exothermic process.[85] While stoichiometric studies by Hartwig and co-workers demonstrated the feasibility of C–X reductive elimination in the presence of P^tBu_3, the implications and catalytic applications of this study were only realized recently by Buchwald and co-workers, who published the first exchange reaction featuring carbon–halogen reductive elimination from Pd(II) in 2009.[86] In this study, aryl triflates could be converted into the corresponding aryl fluorides *via* a reductive elimination process using CsF and a Pd catalyst (Scheme 7.16a). The authors observed that the monoligated ArPd(II)F species could undergo reductive elimination if *t*BuBrettPhos (a bulky, electron-rich monodentate phosphine ligand) was employed. Following this seminal report, the same group reported the formation of aryl and vinyl chlorides and bromides from the corresponding triflates using similar conditions (Scheme 7.16b).[87] The proposed mechanism involves oxidative addition into the aryl–triflate bond followed by ligand exchange/transmetallation with the desired halide and, finally, reductive elimination to form the desired C–X bond. The use of *t*BuBrettPhos as ligand is crucial, as it possesses the steric and electronic properties ideal for carbon–halogen reductive elimination from the key ArPd(II)X complex.

Along the same theme of carbon–halogen reductive elimination, Lautens and co-workers reported a Pd-catalyzed intramolecular C–N coupling of *o-gem*-dibromoanilines providing access to 2-bromoindoles (Scheme 7.17).[1] While previous studies by the same group suggested the intermediacy of 2-bromoindole **7.29** in the Pd-catalyzed tandem C–N/C–C couplings of *gem*-dihaloolefins,[75] attempts to isolate this intermediate in the absence of an external nucleophile were unsuccessful (see Scheme 7.15). The catalytic dead end was believed to be irreversible oxidative addition into the product C–Br

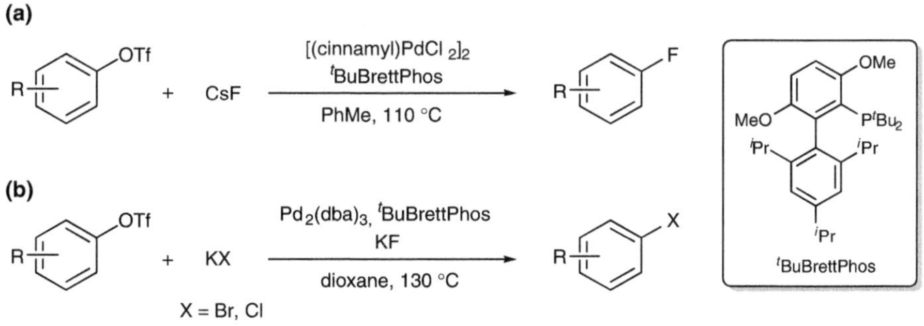

(a)

(b)

X = Br, Cl

Scheme 7.16 Pd-catalyzed conversion of aryl triflates to aryl halides.

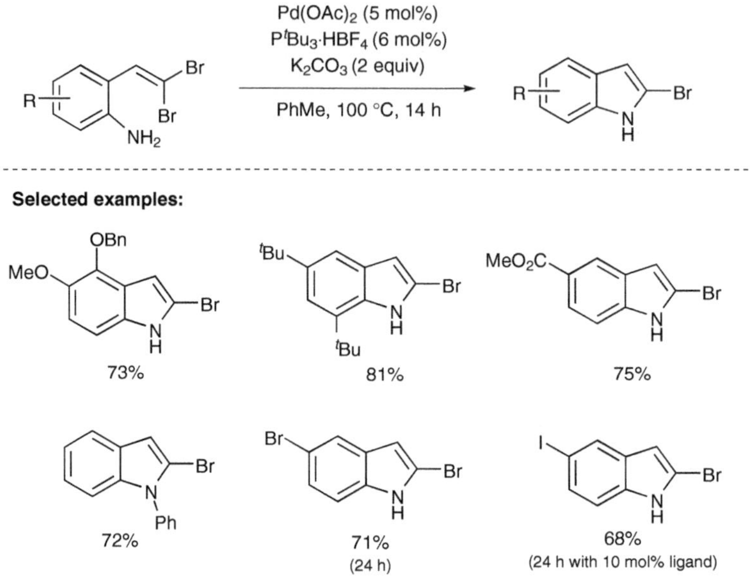

Scheme 7.17 Scope of Pd-catalyzed intramolecular C–N coupling of *o-gem*-dibromoanilines.

Table 7.3 Screen of phosphine ligands for Pd-catalyzed intramolecular C–N coupling.

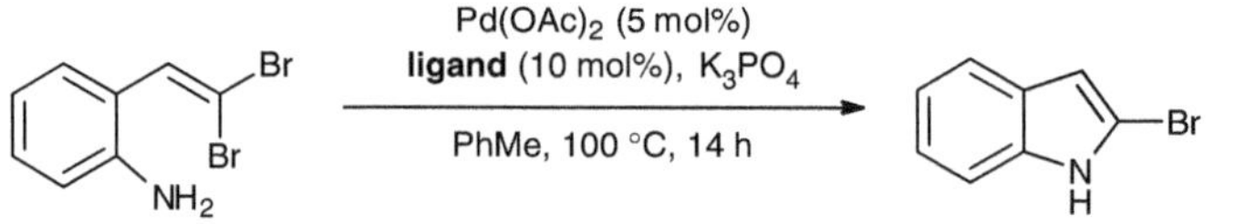

Ligand	Yield (%)
P^{t}Bu$_3$	64 (81)^a
QPhos	17
PPh$_3$, PBu$_3$, PCy$_3$, SPhos, JohnPhos, BINAP, Xantphos, P(*o*-tol)$_3$, dppf	<5

^{a}Using optimized conditions: Pd(OAc)$_2$ (5 mol%), P^{t}Bu$_3$·HBF$_4$ (6 mol%), K$_2$CO$_3$ (2 equiv), PhMe, 100 °C, 14 h.

bond. In this particular study, a screen of various phosphine ligands revealed P^{t}Bu$_3$ to be superior, furnishing **7.29** in 64% yield (Table 7.3). Fine tuning of the reaction conditions increased the yield of the desired product to 81%. The authors suggested that the steric bulk of the ligand plays a crucial role in promoting reactivity, namely regenerating the active Pd(0) catalyst by inducing carbon–halogen reductive elimination from the product.

The reaction tolerates a wide range of functional groups on the *gem*-dibromoolefin, including both electron-rich and -poor aromatics, and also sterically encumbering substituents. The reaction is also compatible with starting materials containing an additional halogen (Br or I) on the aromatic ring, which exemplifies the ability of P^tBu_3 in enabling reversible oxidative addition. This study highlights the first selective Buchwald–Hartwig cross-coupling of a vinyl bromide in the presence of an aryl iodide, which is typically difficult owing to inherent chemoselectivity issues.

To demonstrate (a) that the product **7.29** is more susceptible to oxidative addition than the starting material **7.27** and (b) that the unproductive oxidative addition complex **7.30a** can re-form the active catalyst through C–X reductive elimination (*i.e.*, oxidative addition to the product carbon–bromine bond is a reversible process), mechanistic studies were carried out. In the first study, 2-bromoindole was mixed with 1 equivalent of $Pd(P^tBu_3)_2$ in C_6D_6, stirred at room temperature and monitored by ^{31}P NMR spectroscopy over 3 days (Scheme 7.18). The disappearance of the peak corresponding to the Pd precatalyst at 84.9 ppm was accompanied by the appearance of two new peaks: one corresponding to free P^tBu_3 ligand at 62.2 ppm and the second to a new phosphorus-containing compound at 65.1 ppm in a 1:1 ratio. This finding is consistent with the formation of **7.30a** presumably arising from oxidative addition to **7.29**.

A competition experiment was carried out to test if indole **7.29** is more prone to oxidative addition than the starting *gem*-dibromoolefin **7.27**. $Pd(P^tBu_3)_2$, **7.27** and **7.29** were combined in a 1:1:1 ratio in C_6D_6 and stirred at room temperature for 3 days (Scheme 7.19). The authors observed the appearance of complex **7.30a** at 65.1 ppm in the ^{31}P NMR spectrum once again, while **7.27** was completely recovered. Under these conditions, the oxidative addition product of **7.27** was never observed by ^{31}P NMR spectroscopy, suggesting that oxidative addition into **7.29** is thermodynamically (and likely kinetically) favored.

Scheme 7.18   Stoichiometric studies on oxidative addition of **7.29**.
Adapted from Ref. 83. Copyright 2010 American Chemical Society.

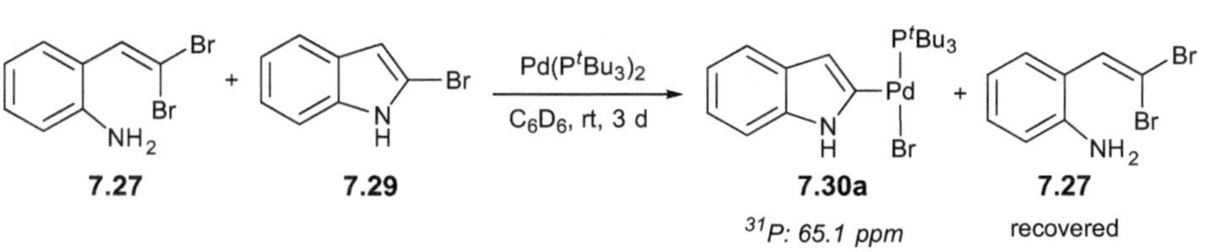

Scheme 7.19   Competition experiments between **7.27** and **7.29**.
Adapted from Ref. 83. Copyright 2010 American Chemical Society.

To determine whether or not the Pd(II) species can generate a catalytically active complex, **7.30a** (5 mol%) was reacted under the standard reaction conditions in the presence of **7.27**. The desired C–N coupling product **7.29** was afforded in a yield comparable to the case when Pd(OAc)$_2$ and P^{t}Bu$_3$ · HBF$_4$ are used as catalyst (Scheme 7.20). Therefore, **7.30a** is a competent precatalyst in the coupling reaction. Consistent with the findings of Hartwig and co-workers, oxidative addition into the product is essentially irreversible when phosphines less bulky than P^{t}Bu$_3$ are employed.

7.6 Palladium-Catalyzed Carbohalogenation

7.6.1 New Reactivity Inspired by the Mizoroki–Heck Reaction

The Mizoroki–Heck reaction is a powerful method for coupling aryl halides and alkenes. Since the coupling partner is not an organometallic reagent, as is seen in more traditional cross-coupling reactions, a unique mechanism is followed wherein the intermediate arylpalladium halide species undergoes alkene carbopalladation and β-hydride elimination to generate the new C–C bond.[88] Since its independent discovery by Heck and Mizoroki in the 1970s, the Heck reaction has bloomed into an extensive area of research, which has led to modern advances in substrate scope, enantioselective variants and catalyst and ligand design.[23,87]

The intermediate alkylpalladium halide species in a Heck reaction typically undergoes rapid β-hydride elimination. However, if no *syn*-β-hydrogen atoms are present or if an alkyne starting material is used to give a vinylpalladium intermediate, the Heck pathway may be diverted by trapping this Pd(II) species with a nucleophile (Scheme 7.21a and b). The exchange of the X-type ligand on Pd with an exogenous nucleophile (*i.e.*, transmetallation) facilitates reductive elimination from the Pd(II) complex and regenerates the active Pd(0) catalyst. Overall, the reaction constitutes a formal addition across an unsaturated functionality. This type of reactivity was first reported by Grigg and co-workers in 1988, where the vinylpalladium iodide intermediate **7.36** could be trapped by a hydride nucleophile derived from formic acid (Scheme 7.21c).[89]

This strategy was further developed by Grigg's group and others[90] to include the trapping of neopentylpalladium halide intermediates with a number of different nucleophiles such as boronic acids,[91] organotin reagents[92] and hydrides[93] (Scheme 7.22). The neopentylpalladium halide intermediates are derived from intramolecular carbopalladation of a 1,1-disubstituted olefin. Polyene cyclizations can also be combined with anionic trapping, providing access to complex hetero- and carbocyclic frameworks (Scheme 7.22c). The success of these reactions relies on the rate of intramolecular cyclization being faster than the rate of premature nucleophilic trapping.

More modern examples of carbocyclization–anion capture cascades have emerged. Zhu and co-workers combined an intramolecular carbopalladation with a direct C–H activation under Pd catalysis, which provided access to the spirodihydroquinoline **7.38** (Scheme 7.23a).[94] The first

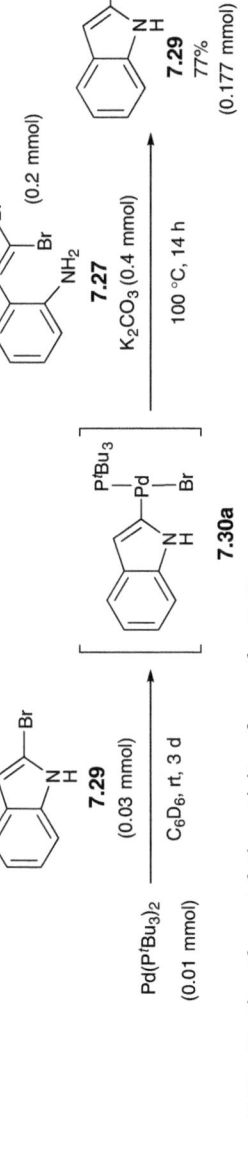

Scheme 7.20 Testing the catalytic activity of complex 7.30a.
Adapted from Ref. 83. Copyright 2010 American Chemical Society.

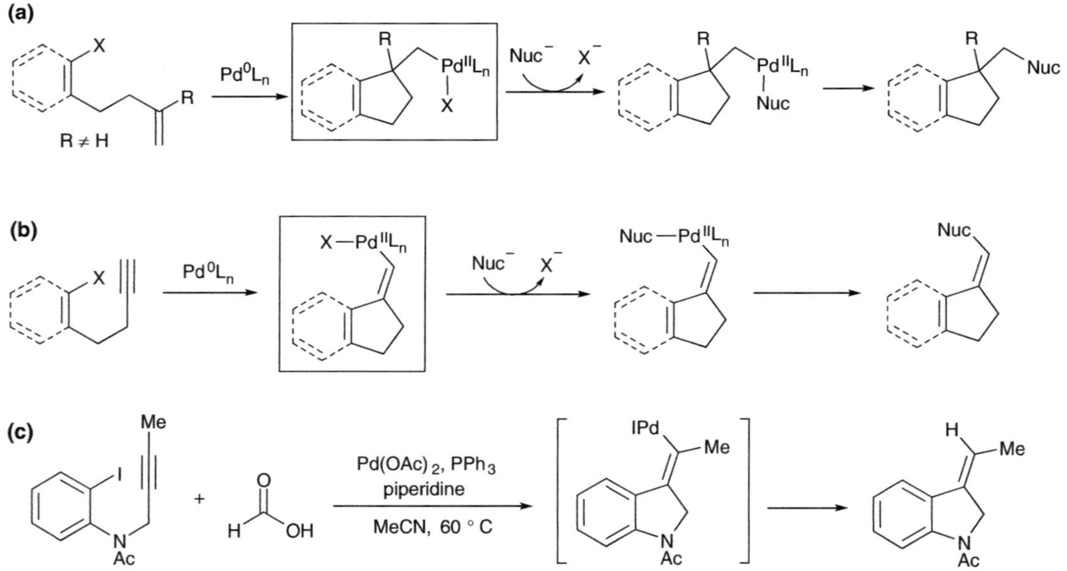

7.35 **7.36** **7.37**

Scheme 7.21 Nucleophilic trapping of (a) neopentylpalladium halide intermediates and (b) vinylpalladium halide intermediates. (c) First example of Pd-catalyzed anion capture cascade using a hydride nucleophile.

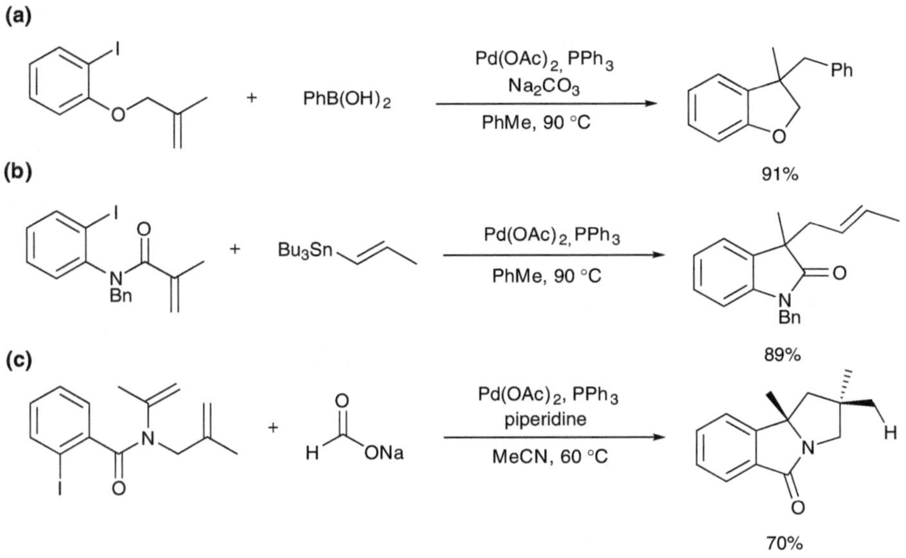

(a)

(b)

91%

(c)

89%

70%

Scheme 7.22 Extension of anion capture cascade reactions by Grigg and co-workers.

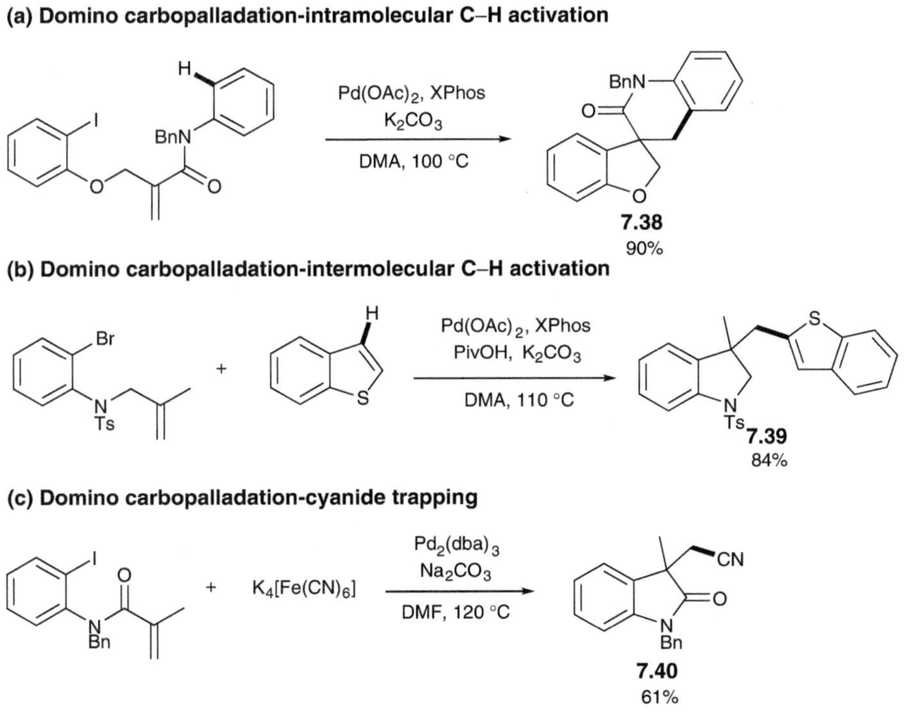

(a) Domino carbopalladation-intramolecular C–H activation

7.38
90%

(b) Domino carbopalladation-intermolecular C–H activation

7.39
84%

(c) Domino carbopalladation-cyanide trapping

7.40
61%

Scheme 7.23 Unique trapping of neopentylpalladium(II) intermediate.

carbocyclization–intermolecular direct arylation protocol was developed by Fagnou and co-workers, whereby the neopentylpalladium(II) intermediate could be trapped by a sulfur-containing heterocycle, representing an early contribution to the field of direct sp^3-sp^2 arylations (Scheme 7.23b).[95] The use of K$_4$[Fe(CN)$_6$] in tandem cyclization-cyanations was also explored by Zhu and co-workers (Scheme 7.23c).[96] This method provided access to the formal carbocyanation product 7.40. All of the examples highlighted so far involve an intramolecular carbopalladation step and thus have the advantage of high regioselectivity. Examples featuring intermolecular carbopalladation on to an olefin or alkyne followed by anionic trapping are rare. In such cases, there are several competing mechanistic pathways: (a) direct intermolecular coupling between the aryl halide and nucleophile, (b) standard Heck reaction and (c) uncontrolled oligomerization terminated with random events of nucleophilic trapping. Nevertheless, biased systems utilizing norbornene[97] or symmetrical alkynes[98] as the Heck acceptor have been reported.

7.6.2 Development of a Palladium-Catalyzed Carboiodination of Alkenes

The anion capture reactions discussed in Section 7.6.1 are a powerful method for alkene and alkyne difunctionalization. Another interesting mechanistic possibility that could arise from the RPd(II)X species would be direct carbon–halogen reductive elimination (Scheme 7.24). Overall, this *carbohalogenation* reaction would lead to the formation of two new bonds (one carbon–carbon and one carbon–halogen bond) across an olefin, leaving the reactive halogen functionality intact in the final product.

In 2011, Newman and Lautens reported the first carbohalogenation reaction invoking carbon–halogen reductive elimination from an alkyl Pd(II) halide intermediate.[2] To avoid the energetically favorable Heck reaction, their method utilized biased aryl iodides 7.41 that were bound to a pendent 1,1-disubstituted olefin [eqn (7.7)]. In this particular system, the neopentylpalladium(II) halide intermediate 7.43 that is formed after olefin insertion cannot undergo β-hydride elimination, as there are no *syn*-β-hydrogen atoms available. Notably, the use of 1,1-disubstituted olefins in domino cyclic carbopalladations is well precedented in the literature.[90] For the carbohalogenation studies, initial screening of well-defined PdL$_2$ catalysts at 90 °C revealed that bulky phosphine ligands were essential for reactivity, with QPhos giving the best results (Table 7.4, Entries 1–3). No basic additives were required since the transformation does not generate HX as a by-product, unlike the standard Heck reaction. Less bulky ligands, such a PCy$_3$ and P(o-tol)$_3$, did not provide any detectable amounts of the desired carbohalogenation product (Table 7.4, Entries 4 and 5).

At elevated temperature (100 °C), the catalyst loading could be decreased from 5.0 to 2.5 mol% (Table 7.4, entry 1). Using these optimized conditions, a range of oxygen- and nitrogen-containing five- and six-membered heterocycles were prepared in high yield (Scheme 7.25). The use of aryl chlorides

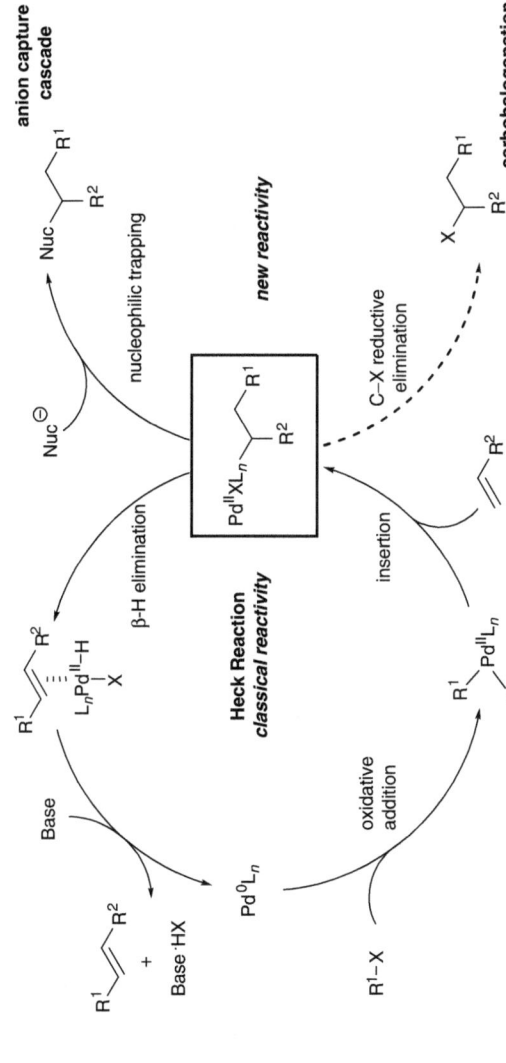

Scheme 7.24 The Heck reaction: classical *versus* new modes of reactivity.

Table 7.4 Screening of PdL$_2$ catalysts for intramolecular carboiodination.

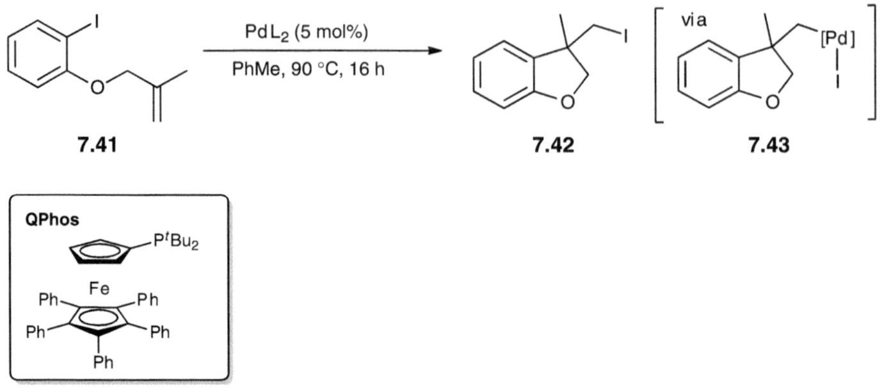

Entry	Ligand	Yield (%)
1	QPhos	82 (95)[a]
2	P^{t}Bu$_3$	61
3	PhP(P^{t}Bu$_3$)$_2$	76
4	PCy$_3$	0
5	P(o-tol)$_3$	0

[a]Reaction run with 2.5 mol% Pd(QPhos)$_2$ at 100 °C for 4 hours.

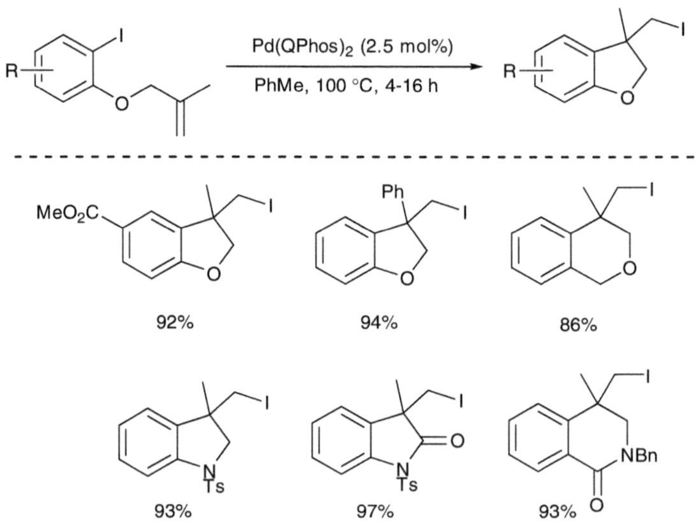

Scheme 7.25 Scope of Pd-catalyzed carboiodination.

and bromides in the reaction was unsuccessful, suggesting that C–Cl and C–Br reductive elimination may be more difficult. The proposed mechanism of the reaction involves oxidative addition to the aryl iodide, olefin insertion and unprecedented C–I reductive elimination from an alkylpalladium(II) species to regenerate the active Pd(0) catalyst.

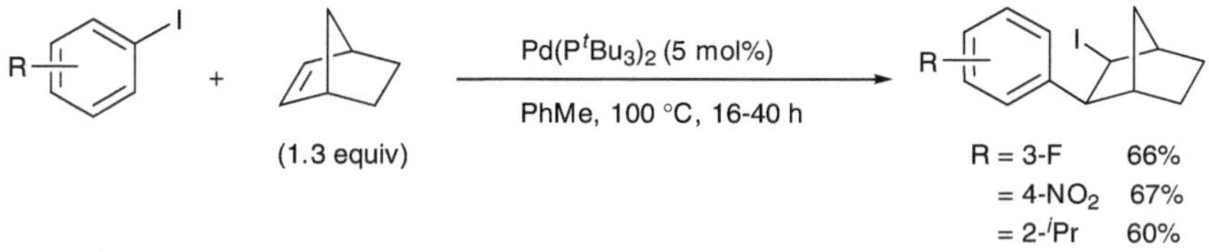

Scheme 7.26 Pd-catalyzed intermolecular carboiodination of norbornene.

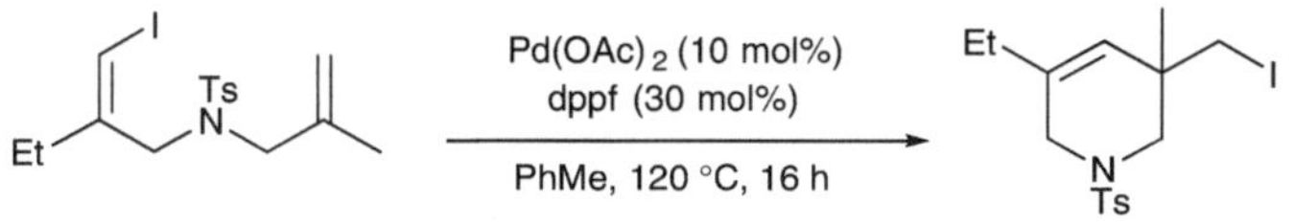

Scheme 7.27 Tong and co-workers' report of C–I reductive elimination from alkyl-palladium(II) halide complexes.

An intermolecular variant was also investigated using norbornene as the olefin coupling partner (Scheme 7.26). The strained and rigid nature of norbornene enhances the reactivity of the olefin towards addition processes and prevents unfavorable β-hydride elimination due to a lack of suitable *syn*-β-hydrogens in the key alkylpalladium(II) halide intermediate. Although still limited in scope, the use of Pd(P*t*Bu₃)₂ as catalyst provided access to the desired coupling products in modest yield. Notably, the intramolecular carboiodination was not hindered by the addition of radical scavengers such as TEMPO or Galvinoxyl, therefore discounting radical intermediates while supporting C–I reductive elimination as a key elementary step in the reaction mechanism.

Concurrent with the studies of Lautens and co-workers on Pd-catalyzed carboiodination, Tong and co-workers reported that the combination of Pd(OAc)₂ and dppf could also effect the carboiodination of alkenes starting from vinyl iodides such as **7.44** (Scheme 7.27).[99] There are notable differences between the methods developed by the groups of Lautens and Tong with respect to reaction temperature, substrate scope and catalyst composition. Most interestingly, whereas Lautens and co-workers had the greatest success with bulky monodentate phosphine ligands, Tong and co-workers' system required the use of a bidentate phosphine ligand in three-fold excess relative to the Pd precatalyst.

7.7 Computational Investigation into the Mechanism of Carbohalogenation

7.7.1 Proposed Catalytic Cycle

In agreement with the studies of Hartwig and co-workers on the reductive elimination of aryl halides from Pd(II) complexes, the carbohalogenation reaction developed by Lautens and co-workers requires the use of exceptionally bulky phosphine ligands such as QPhos and P^{t}Bu$_3$. One of the main limitations of this chemistry is the greatly attenuated reactivity of aryl chlorides and bromides under the standard conditions. The use of aryl chlorides and bromides would be beneficial, as they are typically cheaper and more readily available than the analogous aryl iodides. The carboiodination reaction also requires the use of 1,1-disubstituted olefins to prevent Heck-type processes from occurring. In substrates that possess a suitable β-hydrogen, β-hydride elimination is faster than the desired C–I reductive elimination. Towards the goal of gaining a fundamental understanding to aid in solving these problems, computational studies were undertaken by Lautens and so-workers, and the mechanism, ligand effects and the origins of reactivity and selectivity in the Pd-catalyzed carboiodination reaction were explored.[100]

In the system under study, P^{t}Bu$_3$ exhibits similar reactivity and efficiency to QPhos. Therefore, for all calculations, P^{t}Bu$_3$ was used as the ligand rather than QPhos owing to the reduced computational cost. The proposed catalytic cycle of carboiodination with aryl halide **7.46** is shown in Scheme 7.28. Starting from the active PdL$_2$ complex, substrate binding and ligand dissociation constitute the first step of the catalytic cycle, which generates the η^2 complex **7.47**. Next, oxidative addition of the aryl halide occurs, giving rise to Pd(II) complex **7.48**. Isomerization allows coordination of the pendent olefin to occur, forming **7.49**, which is followed by *syn*-carbopalladation to form a new C–C bond in the alkylpalladium(II) halide intermediate **7.50**. The C–X reductive elimination proceeds *via* a three-membered transition state, generating the new C–X bond and the product-bound Pd(0) complex **7.51**. The I→Pd coordination is weak and **7.51** readily liberates the product **7.52** to bind another molecule of ligand to regenerate the initial PdL$_2$ catalyst.

Based on the DFT calculations, the rate-limiting step of the transformation is carbon–halogen reductive elimination with a barrier of 24.9 kcal mol^{-1} for substrate **7.46a**. The high activation energy is related to the endothermicity of the transformation. Overall, the main driving force for the reaction is the formation of a C–C σ-bond from a C=C π-bond.

7.7.2 Origin of Reactivity Differences in Aryl Halides

The catalytic cycles implementing aryl bromide **7.46b** and chloride **7.46c** were also computed to understand why these halides showed poor reactivity under the reaction conditions. The carbon–halogen reductive elimination was found to be the rate-limiting step in all cases, with barriers of 27.4 and 27.9 kcal mol^{-1} for the aryl bromide and chloride, respectively. Both of these values are higher than the activation energy associated with C–I reductive elimination (24.9 kcal mol^{-1}). The reactivity of different halides can be attributed to the barrier of carbon–halogen reductive elimination, which is in turn correlated with the differences in Pd–X bond dissociation energies (Figure 7.2). Hence the breaking of a weak Pd–I bond is favored over breaking of a stronger Pd–Br or Pd–Cl bond.

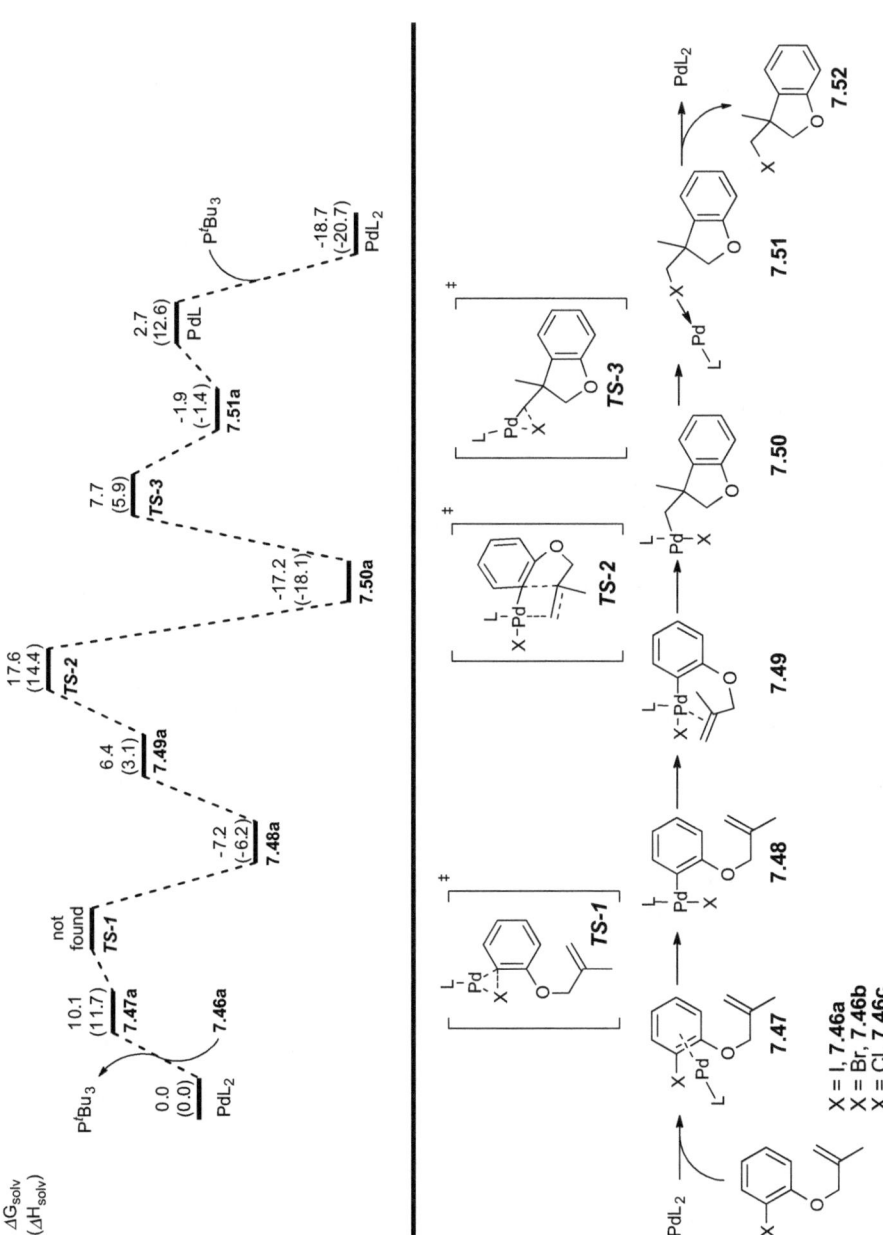

Scheme 7.28 Proposed catalytic cycle for Pd-catalyzed carbohalogenation. Adapted from Ref. 100 with permission.

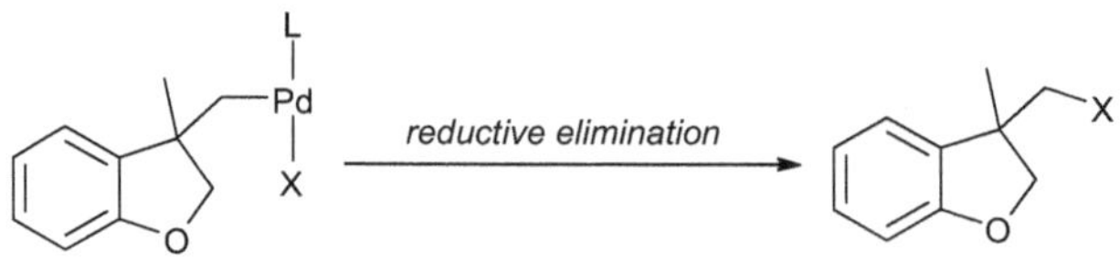

BDE (kcal/mol)		$\Delta G^{\ddagger}$ (kcal/mol)	BDE (kcal/mol)	
Pd–I:	72.8	24.9	C–I:	65.7
Pd–Br:	81.1	27.4	C–Br:	75.0
Pd–Cl:	91.9	27.9	C–Cl:	91.0

Figure 7.2 Bond dissociation energies (BDE) and activation energies ($\Delta G^{\ddagger}$) associated with carbon–halogen reductive elimination. Adapted from Ref. 100 with permission.

7.7.3 Carbon–Halogen Reductive Elimination: Role of the Ligand

To understand how the nature of the ligand affects the rate of reductive elimination, catalytic cycles were computed for a set of phosphine ligands with varying steric and electronic properties (Table 7.5). To reduce the computational cost, FcP^{t}Bu$_2$, a truncated analog of QPhos, was used for the calculations. The experimental results demonstrating that bulky phosphines are the most efficient ligands for carbohalogenation were corroborated by the computations. The lowest barriers to reductive elimination are observed for P^{t}Bu$_3$ (24.9 kcal mol^{-1}) and FcP^{t}Bu$_2$ (23.7 kcal mol^{-1}) (Table 7.5, Entries 1 and 2), whereas a higher barrier is demonstrated for the slightly less bulky P^{i}Pr$_3$ (26.3 kcal mol^{-1}) (Entry 3). Significantly smaller ligands, such as P(OMe)$_3$, PMe$_3$ and PH$_3$ (Entries 4–6), are believed to follow a different mechanism for carbohalogenation. Owing to reduced steric crowding, the alkylpalladium(II) halide intermediate can accommodate two ligands, which results in more stable and lower energy tetracoordinated intermediates of the type 7.53 and 7.54 [eqn (7.8)] Endergonic ligand dissociation to form a tricoordinated species 7.50a is required prior to reductive elimination, therefore increasing the barrier to reductive elimination with these smaller ligands.

Although P(OMe)$_3$ and PMe$_3$ possess similar degrees of steric bulk, reductive elimination is favored with P(OMe)$_3$, suggesting that, in addition to steric effects, electronic factors can also have a significant impact. It is well known that electron-withdrawing ligands facilitate reductive elimination by decreasing the electron density at palladium.

The overall results of this study suggest that steric effects in the ligand play a significant role in promoting carbohalogenation by both preventing the formation of stable tetracoordinate Pd species and causing destabilization of the neopentylpalladium halide ground state. Although electron-poor phosphine ligands also facilitate reductive elimination, ligand steric effects seem to have a more pronounced influence on the observed reactivity.

Table 7.5 Effects of ligand sterics and electronics on C–I reductive elimination.

7.53   **7.50a**   **7.54**

reductive elimination

7.52a

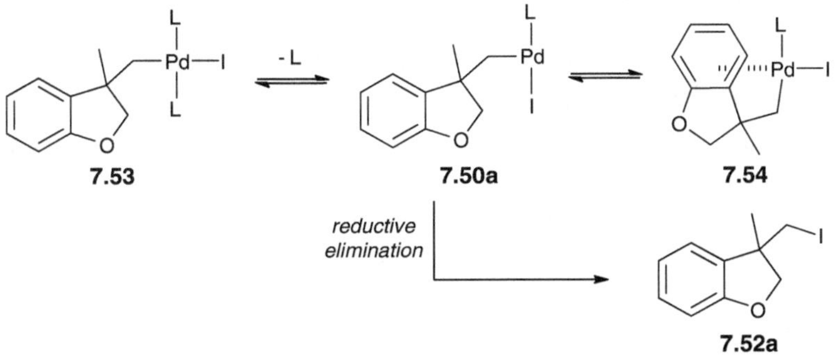

Entry	Ligand (%)	$\Delta G^{\ddagger}$ (kcal mol^{-1})
1	P^{t}Bu$_3$	24.9
2	FcP^{t}Bu$_2$	23.7
3	P^{i}Pr$_3$	26.3
4	P(OMe)$_3$^{a}	26.9
5	PMe$_3$^{a}	36.9
6	PH$_3$^{a}	30.5

^{a}Resting state is a Pd^{II}L$_2$ species.

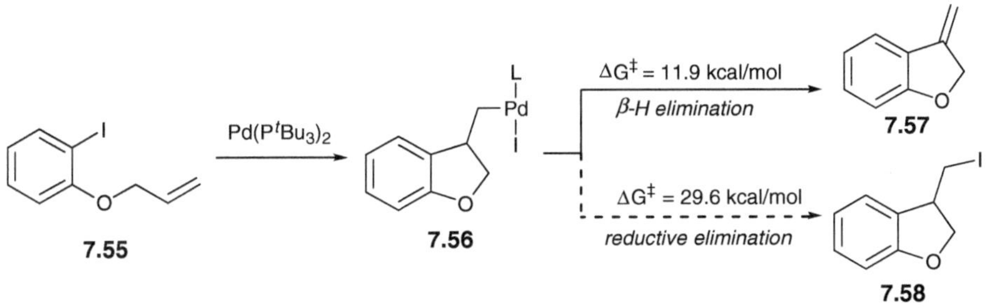

7.55   Pd(P^{t}Bu$_3$)$_2$ **7.56**

$\Delta G^{\ddagger}$ = 11.9 kcal/mol
β-H elimination   **7.57**

$\Delta G^{\ddagger}$ = 29.6 kcal/mol
reductive elimination   **7.58**

Scheme 7.29 Comparison of β-hydride elimination and reductive elimination pathways. Adapted from Ref. 100 with permission.

7.7.4 Competing β-Hydride Elimination

Experimental results demonstrate that when monosubstituted olefin **7.55** is subjected to the standard carbohalogenation conditions, only products such as **7.57**, resulting from a Heck reaction, are observed (Scheme 7.29). The competing β-hydride elimination pathway was calculated and compared with the desired carbohalogenation reaction. The computations show that β-hydride elimination is an extremely facile process with a transition-state energy of 11.9 kcal mol^{-1}. The barrier for carbon–halogen reductive elimination, on the other hand, was calculated to be 29.6 kcal mol^{-1}. Therefore, avoiding β-hydride elimination is a challenging goal that will require the development of more specialized catalyst systems.

7.8 Expanding the Scope of Pd-Catalyzed Carbohalogenation

7.8.1 Bromide to Iodide Exchange and Domino Carbohalogenation

In 2011, Lautens and co-workers reported a Pd-catalyzed carbohalogenation re-action with aryl bromides, which, in the presence of an iodide source, gave rise to formal carboiodination products (Scheme 7.30).[101] The success of this trans-formation is based on a halide exchange (bromide to iodide) at Pd, which allows carbon–iodine reductive elimination to occur from the key alkylpalladium(II) halide intermediate. This protocol is reminiscent of the work of Buchwald and co-workers on Pd-catalyzed aromatic halogen exchange reactions,[86] and permits the use of cheaper and more readily available starting materials.

In the same report,[101] the authors demonstrated that polyunsaturated aryl iodides are also suitable substrates for carbohalogenation. Upon oxidative addition of the aryl iodide, multiple olefin insertions can be terminated by C–I reductive elimination, giving rising to complex polycyclic alkyl iodide products. Using aryl iodide **7.59**, a diastereoselectivity of >20:1 could be achieved (Scheme 7.31a). Additionally, aryl bromides such as **7.61** could be used in the domino carbohalogenation by utilizing the halide exchange conditions (Scheme 7.31b).

7.8.2 Diastereoselective Pd(0)-Catalyzed Carboiodination

In 2012, Lautens and co-workers reported the application of the carboiodina-tion methodology to the diastereoselective synthesis of functionalized

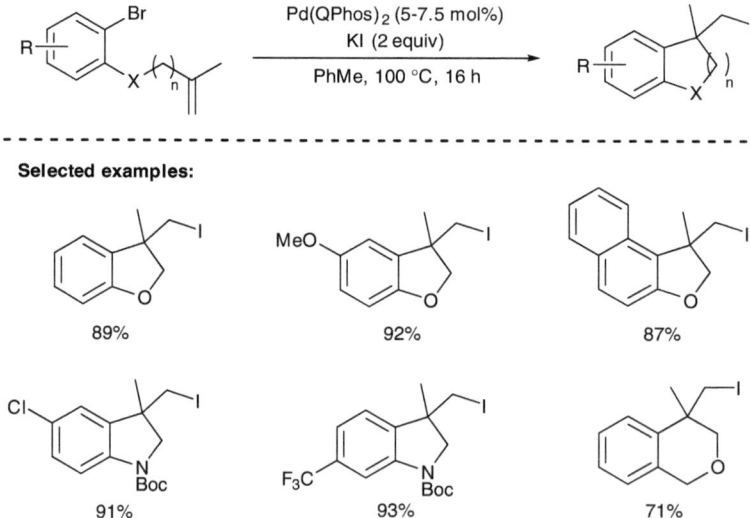

Scheme 7.30 Scope of Pd-catalyzed carboiodination *via* Br to I exchange.

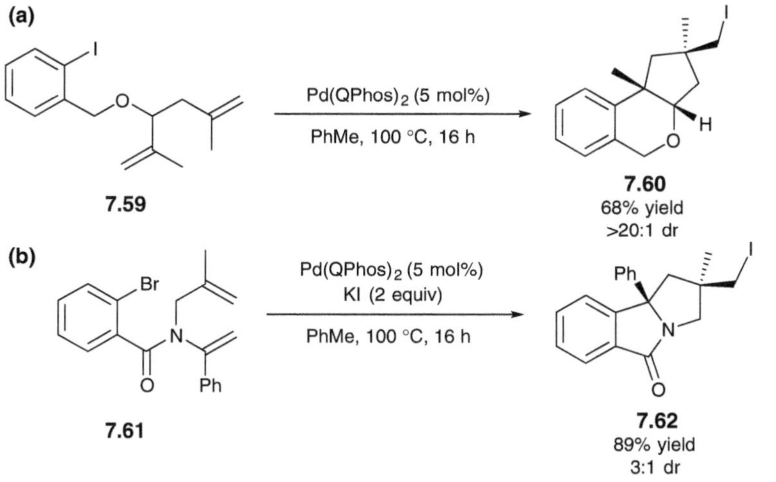

(a)

7.59

Pd(QPhos)$_2$ (5 mol%)

PhMe, 100 °C, 16 h

7.60
68% yield
>20:1 dr

(b)

7.61

Pd(QPhos)$_2$ (5 mol%)
KI (2 equiv)

PhMe, 100 °C, 16 h

7.62
89% yield
3:1 dr

Scheme 7.31   Domino carboiodination with (a) aryl iodides and (b) bromides.
Adapted from Ref. 101. Copyright 2011 American Chemical Society.

(a)

7.62

Pd(P^{t}Bu$_3$)$_2$ (5 mol%)
NEt$_3$ (1 equiv)

PhMe, 110 °C

7.63
up to 94% yield
up to >99:1 dr *cis:trans*

(b)

7.64

Pd(P^{t}Bu$_3$)$_2$ (5 mol%)
NEt$_3$ (1 equiv)

PhMe, 110 °C

7.65
up to 96% yield
down to <1:99 dr *cis:trans*

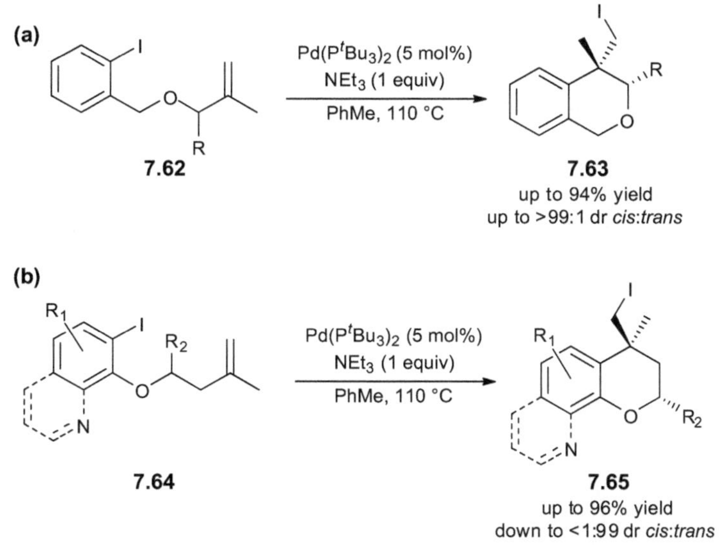

Scheme 7.32   Diastereoselective Pd(0)-catalyzed synthesis of (a) isochromans and (b)
chromans.
Adapted from Ref. 102. Copyright 2012 American Chemical Society.

isochromans **7.63** (Scheme 7.32a) and chromans **7.65** from precursors **7.62** and
7.64, respectively (Scheme 7.32b).[102] Despite many reports of their individual
syntheses,[103–106] general approaches to both frameworks are rare.[107] Hence the
development of a widely applicable, concise catalytic method was warranted.

The observed diastereoselectivities were attributed to be the result of
minimization of A^{1,2} strain (Figure 7.3a) and axial–axial interactions

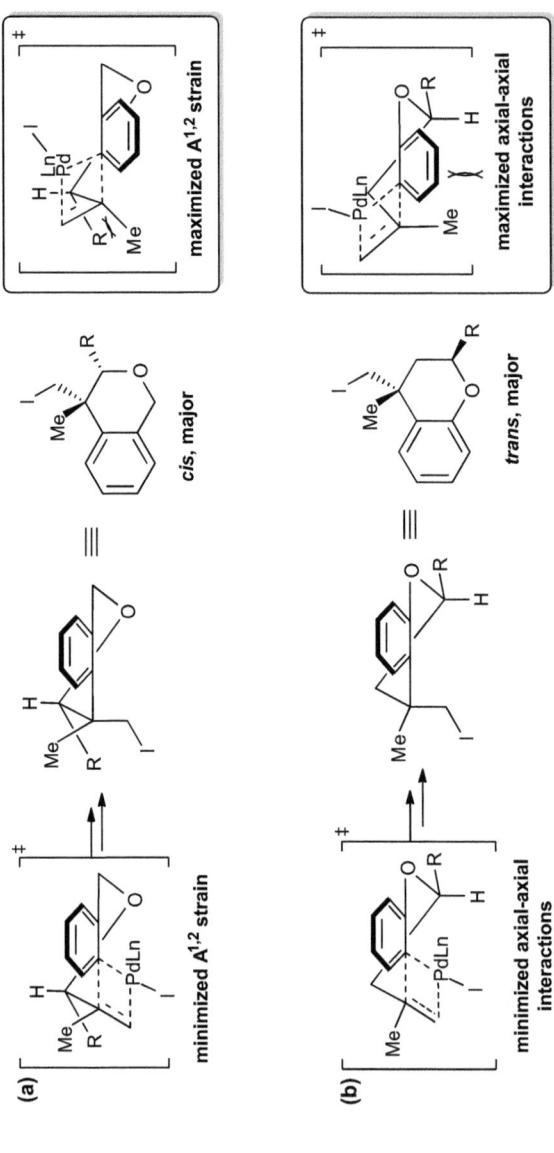

Figure 7.3 Stereochemical rationale for the synthesis of (a) isochromans and (b) chromans. Adapted from Ref. 102. Copyright 2012 American Chemical Society.

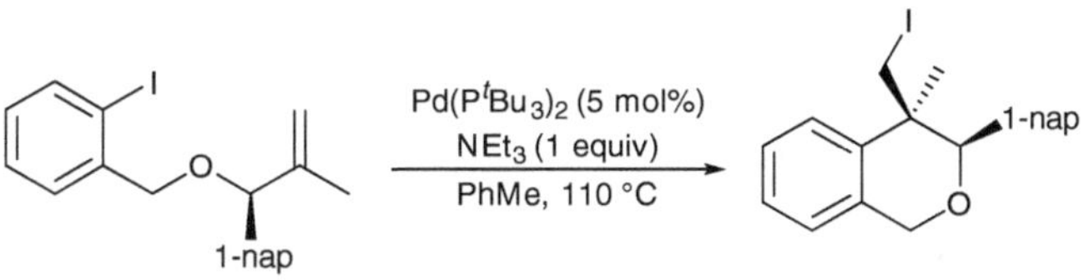

7.66
99:1 er

7.67
70% yield
99:1 er; 98:2 dr

Scheme 7.33 Synthesis of an optically enriched isochroman derivative. Adapted from Ref. 102. Copyright 2012 American Chemical Society.

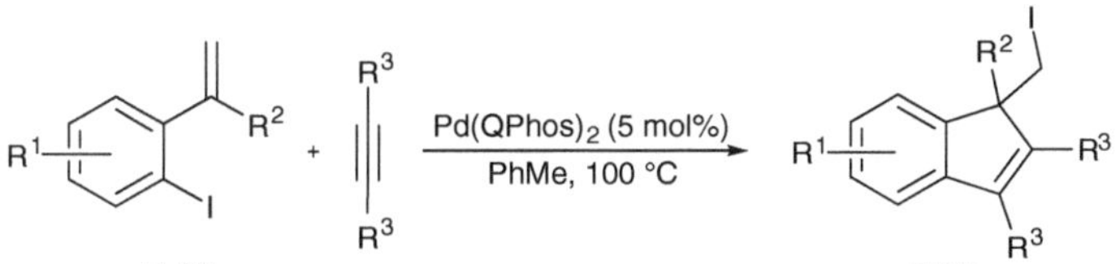

7.68

7.69
up to 96% yield

Scheme 7.34 Synthesis of indenes by Pd(0)-catalyzed carboiodination.

(Figure 7.3b) in the carbopalladation step for the isochromans and chromans, respectively. Furthermore, chiral substrate **7.66** derived from an enantioenriched allylic alcohol could be effectively cyclized to isochroman **7.67** without any loss of optical purity (Scheme 7.33).

7.8.3   Indenes *via* Pd(0)-Catalyzed Carboiodination

In 2012, Lautens and co-workers reported a conjunctive carboiodination of *o*-isopropenylaryl iodides **7.68** and alkynes that gave access to indene products **7.69** (Scheme 7.34).[108] These studies were undertaken because attempts to achieve carboiodination of an alkyne were not successful. The presence of the *o*-propenyl group allowed the vinylpalladated species, arising from intermolecular alkyne carbopalladation, to undergo cyclization to form an sp^3-Pd–I intermediate that was known to reductively eliminate.

   A possible mechanism was proposed to rationalize the product formation (Scheme 7.35). Substrate binding first occurs[100] to generate Pd(0) complex **7.70**, which then undergoes an oxidative addition to generate Pd(II) intermediate **7.71**. Intermolecular alkyne insertion occurs to product **7.72**, which can undergo cyclic alkene carbopalladation to form **7.73**. The indene product **7.74** is formed once intermediate **7.73** has undergone C–I reductive elimination. Thus far, only moderate regioselectivities and no enantioselective variant have been discussed, although these would be interesting and powerful methods.

7.8.4   Simultaneous Carboiodination/Mizoroki–Heck Reaction

Aromatics containing multiple carbon–halogen bonds are synthetically useful, as more than one bond functionalization can be executed in a predictable

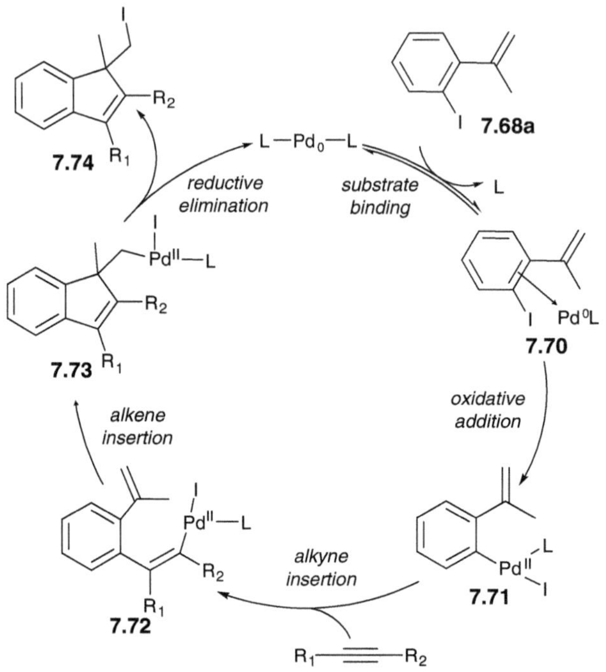

Scheme 7.35 Proposed mechanism for the Pd(0)-catalyzed conjunctive carboiodination. Adapted from Ref. 108. Copyright 2013 WILEY-VCH Verlag GmbH & Co.

fashion.[63] When two of the same halogens are present, efforts to achieve selectivity must rely on reactivity biases such as sterics, electronics or directing groups. Despite the success of these strategies, catalyst deactivation by undesired, irreversible oxidative addition to a carbon–halogen bond is an inherent setback.[109] In 2013, Lautens and co-workers hypothesized that by using Pd and QPhos, a combination well precedented to catalyze carboiodination, off-cycle ArPd(II)X species **7.76** could undergo reductive elimination to regenerate the active catalyst species (Scheme 7.36).[110]

Employing this strategy, diiodinated aromatic compounds **7.79** could undergo intramolecular carboiodination in moderate to good yields (Scheme 7.37). In the substrates described, the more sterically accessible *para* carbon–iodine bond cannot undergo intramolecular carbopalladation. Thus, if the Pd catalyst oxidatively adds at this position, reductive elimination must occur in order to regenerate the active catalytic species so the desired transformation can occur. In this reaction, the Pd/QPhos combination is required for both this reversible oxidative addition and the final sp³ carbon–iodine reductive elimination.

As a means to increase product complexity, Lautens and co-workers applied this concept to a simultaneous intermolecular/intramolecular Mizoroki–Heck/carboiodination reaction between diiodoaromatics and activated olefins (Scheme 7.38).[108] Thus, diiodoolefin **7.79** can cyclize in

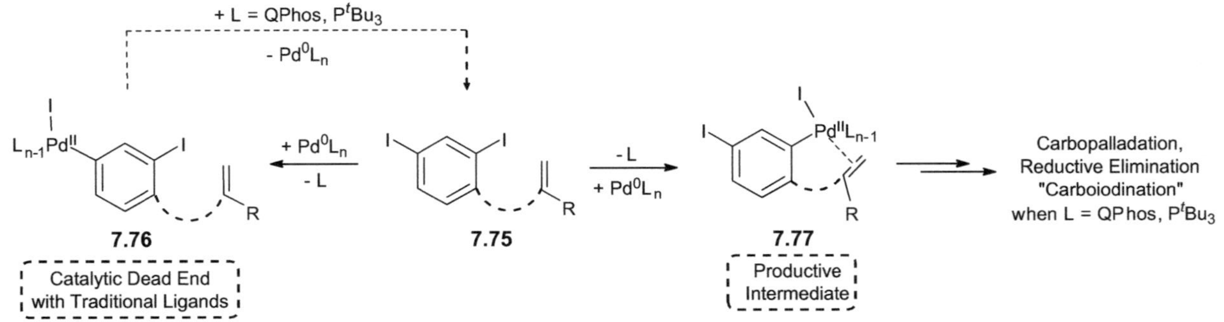

Scheme 7.36 Strategy to overcome catalyst deactivation using reversible oxidative addition. Adapted from Ref. 110. Copyright 2013 WILEY-VCH Verlag GmbH & Co.

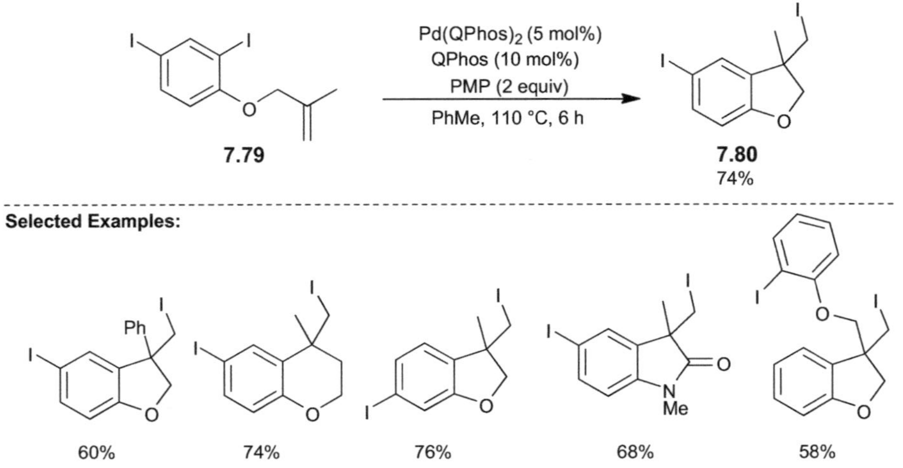

Selected Examples:

Scheme 7.37 Intramolecular carboiodination of diiodinated aromatic compounds.

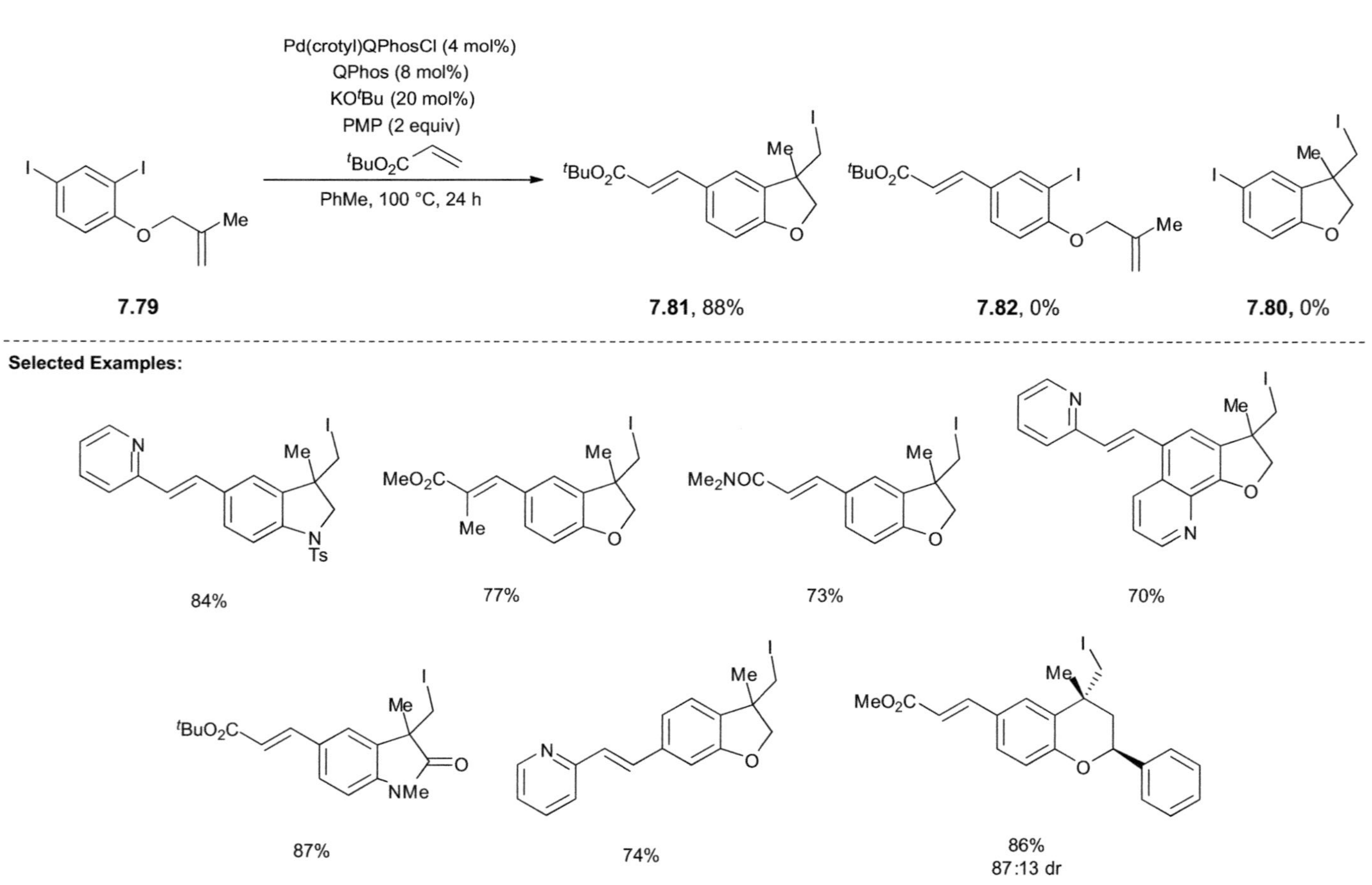

Pd(crotyl)QPhosCl (4 mol%)
QPhos (8 mol%)
KO^{t}Bu (20 mol%)
PMP (2 equiv)

^{t}BuO$_2$C

PhMe, 100 °C, 24 h

7.79

7.81, 88%   **7.82**, 0%   **7.80**, 0%

Selected Examples:

84%   77%   73%   70%

87%   74%   86%
                                      87:13 dr

Scheme 7.38   Simultaneous inter- intramolecular Mizoroki–Heck/carboiodination reaction.

the presence of *tert*-butyl acrylate to provide tandem coupling product **7.81** without any evidence of mono-Heck product **7.82** or monocyclization product **7.80**. The power of this method is highlighted by the possibility of synthesizing numerous heterocycles and using many different Mizoroki–Heck acceptors. *In situ* NMR studies revealed that both the inter- and intramolecular transformations occur simultaneously and that both resulting products converge on the desired Mizoroki–Heck/carboiodination adducts. This method can potentially facilitate the use of difficult polyhalogenated substrates, which are prone to cause catalyst deactivation.

7.9 Conclusion

Palladium-catalyzed cross-coupling has become a practical and general method in chemical synthesis. As a testament to the great impact that cross-coupling reactions have had on our society, Professor Richard F. Heck, Professor Ei-ichi Negishi and Professor Akira Suzuki were awarded the 2010 Nobel Prize in Chemistry. This intensely studied field has benefited from continuous evolution since its inception.[111] Recently, the carbon–halogen reductive elimination from transition metal complexes has come to light as a potentially useful elementary step for use in catalytic cycles, a process that was previously thought to be unachievable owing to unfavorable energetics. Early and recent examples have led us to rethink the behavior of transition metal complexes containing both metal–carbon and metal–halogen bonds. Rooted in the seminal reports by Ettorre and others, this concept has significantly matured as a result of recent studies, particularly those of Hartwig and co-workers. Similarly to other elementary transformations involving transition metals (*i.e.*, oxidative addition and transmetallation), stoichiometric experiments have been pivotal in our gaining a sound understanding of how to achieve this reactivity. These reports led to the realization of catalytic variants of carbon–halogen reductive elimination by Buchwald and co-workers, in which aromatic carbon–halogen bonds were formed during the step involving catalyst turnover. These aromatic Finkelstein-type reactions were followed up with a study by Lautens and co-workers, where cross-coupling reactions involving reversible oxidative addition to carbon–halogen bonds were conducted on polyhalogenated substrates. In 2011, extensions to Pd-catalyzed sp^3-hybridized carbon–halogen bond-forming reactions were realized by both Lautens' and Tong's groups. These reactions have emerged as an efficient method to synthesize hetero- and carbocycles from simple starting materials. With the goal of increasing the efficiency of these reactions and their corresponding catalysts, future studies will need to focus on catalyst development based on mechanistic analysis. The field awaits further contributions from the organic, inorganic and organometallic communities, who can advance our understanding of this elementary catalytic transformation and its use in synthesis.

References

1. S. G. Newman and M. Lautens, *J. Am. Chem. Soc.*, 2010, **132**, 11416.
2. S. G. Newman and M. Lautens, *J. Am. Chem. Soc.*, 2011, **133**, 1778.
3. E.-i. Negishi, in *Handbook of Organopalladium Chemistry for Organic Synthesis*, Vol. 1, ed. E.-i. Negishi and A. de Meijere, Wiley, New York, 1st edn, 2002, pp. 3–15.
4. H. Doucet and J. C. Hierso, *Curr. Opin. Drug. Discov. Devel.*, 2007, **10**, 672.
5. C. Torborg and M. Beller, *Adv. Synth. Catal.*, 2009, **351**, 3027.
6. J. Magano and J. R. Dunetz, *Chem. Rev.*, 2011, **111**, 2177.
7. J. Jiao and Y. Nishihara, in *Applied Cross-Coupling Reactions*, ed. Y. Nishihara, Lecture Notes in Chemistry, Vol. 80, Springer, Berlin, 2013, pp. 85–109.
8. E.-i. Negishi, in *Handbook of Organopalladium Chemistry for Organic Synthesis*, vol. 1, ed. E.-i. Negishi and A. de Meijere, Wiley, New York, 1st edn, 2003, pp. 127–1659.
9. J. K. Still and K. S. Y. Lau, *Acc. Chem. Res.*, 1977, **10**, 434.
10. C. Amatore and A. Jutand, *J. Organomet. Chem.*, 1999, **576**, 254.
11. J. Tsuji, *Palladium Reagents and Catalysts – New Perspectives for the 21st Century*, Wiley, Chichester, 2nd edn, 2004, pp. 1–26.
12. A. L. Casado and P. Espinet, *Organometallics*, 1998, **17**, 954.
13. E. Clot, O. Eisenstein, N. Jasim, S. A. Macgregor, J. E. McGrady and R. N. Perutz, *Acc. Chem. Res.*, 2011, **44**, 333.
14. M. Ohashi, M. Shibata, H. Saijo, T. Kambara and S. Ogoshi, *Organometallics*, 2013, **32**, 3631.
15. G. T. de Jong and F. M. Bickelhaupt, *J. Phys. Chem. A*, 2005, **109**, 9685.
16. J. P. Collman, L. S. Hegedus, J. R. Norton and R. G. Finke, *Principles and Applications of Organotransition Metal Chemistry*, 2nd edn, University Science Books, Mill Valley, CA, 1987, pp. 176–258.
17. G. O. Spessard and G. L. Meissler, *Organometallic Chemistry*, Prentice Hall, Upper Saddle River, NJ, 1996, pp. 171–175.
18. U. Christmann and R. Vilar, *Angew. Chem. Int. Ed.*, 2005, **44**, 366.
19. J. F. Hartwig, *Inorg. Chem.*, 2007, **46**, 1936.
20. W. D. Jones and V. L. Kuykendall, *Inorg. Chem.*, 1991, **30**, 2615.
21. J. F. Hartwig, S. Richards, D. Barañano and F. Paul, *J. Am. Chem. Soc.*, 1996, **118**, 3626.
22. B. C. Hamann and J. F. Hartwig, *J. Am. Chem. Soc.*, 1998, **120**, 3694.
23. J. Tsuji, *Palladium Reagents and Catalysts – New Perspectives for the 21st Century*, Wiley, Chichester, 2nd edn, 2004, pp. 1–26.
24. R. F. Heck, *Org. React.*, 1982, **27**, 345.
25. I. P. Beletskaya and A. V. Cheprakov, *Chem. Rev.*, 2000, **100**, 3009.
26. J. K. Stille, *Angew. Chem. Int. Ed. Engl.*, 1986, **25**, 508.
27. V. Farina, V. Krishnamurthy and W. J. Scott, *Org. React.*, 1997, **50**, 1.
28. N. Miyaura and A. Suzuki, *Chem. Rev.*, 1995, **95**, 2457.
29. A. Vigalok, *Chem. Eur. J.*, 2008, **14**, 5102.

30. A. L. Casado and P. Espinet, *J. Am. Chem. Soc.*, 1998, **120**, 8978.
31. L. Xue and Z. Lin, *Chem. Soc. Rev.*, 2010, **39**, 1692.
32. R. Ettorre, *Inorg. Nucl. Chem. Lett.*, 1968, **5**, 45.
33. A. M. Echavarren and J. K. Stille, *J. Am. Chem. Soc.*, 1987, **109**, 5478.
34. D. R. Fahey, *J. Organomet. Chem.*, 1971, **27**, 283.
35. R. F. Heck, *J. Am. Chem. Soc.*, 1968, **90**, 5538.
36. S. J. Tremont and H. U. Rahman, *J. Am. Chem. Soc.*, 1984, **106**, 5759.
37. A. J. Hickman and M. S. Sanford, *Nature*, 2012, **484**, 177 and references therein.
38. D. R. Fahey, *J. Chem. Soc., Chem. Commun.*, 1970, 417.
39. D. R. Fahey, *J. Organomet. Chem.*, 1971, **27**, 283.
40. For examples involving carbon–fluorine bond reductive elimination, see: (a) K. L. Hull, W. Q. Anani and M. S. Sanford, *J. Am. Chem. Soc.*, 2006, **128**, 7134; (b) S. R. Whitfield and M. S. Sanford, *J. Am. Chem. Soc.*, 2007, **129**, 15142; (c) T. Furuya and T. Ritter, *J. Am. Chem. Soc.*, 2008, **130**, 10060; (d) X. Wang, T.-S. Mei and J.-Q. Yu, *J. Am. Chem. Soc.*, 2009, **131**, 7520; (e) N. D. Ball and M. S. Sanford, *J. Am. Chem. Soc.*, 2009, **131**, 3796; (f) T. Furuya, D. Benitez, E. Tkatchouk, A. E. Strom, P. Tang, W. A. Goddard III and T. Ritter, *J. Am. Chem. Soc.*, 2010, **132**, 3793; (g) T. Furuya, D. Benitez, E. Tkatchouk, A. E. Strom, P. Tang, W. A. Goddard III and T. Ritter, *J. Am. Chem. Soc.*, 2010, **132**, 5922; (h) V. V. Grushin, *Acc. Chem. Res.*, 2010, **43**, 160; (i) Y. Ye and M. S. Sanford, *J. Am. Chem. Soc.*, 2013, **135**, 4648 and references therein.
41. For examples involving carbon–chlorine bond reductive elimination, see: (a) P. L. Alsters, P. F. Engel, M. P. Hogerheide, M. Copijn, A. L. Spek and G. van Koten, *Organometallics*, 1993, **12**, 1831; (b) J. Vicente, M.-T. Chicote, M.-C. Lagunas, P. G. Jones and E. Bembenek, *Organometallics*, 1994, **13**, 1243; (c) M.-C. Lagunas, R. A. Gossage, A. L. Spek and G. van Koten, *Organometallics*, 1998, **17**, 731; (d) D. Kalyani, A. R. Dick, W. Q. Anani and M. S. Sanford, *Tetrahedron*, 2006, **62**, 11483; (e) X. Wan, Z. Ma, B. Li, K. Zhang, S. Cao, S. Zhang and Z. Shi, *J. Am. Chem. Soc.*, 2006, **128**, 7416; (f) D. Kalyani, A. R. Dick, W. Q. Anani and M. S. Sanford, *Org. Lett.*, 2006, **8**, 2523; (g) D. Kalyani and M. S. Sanford, *J. Am. Chem. Soc.*, 2008, **130**, 2150; (h) K. J. Stowers and M. S. Sanford, *Org. Lett.*, 2009, **11**, 4584; (i) P. L. Arnold, M. S. Sanford and S. M. Pearson, *J. Am. Chem. Soc.*, 2009, **131**, 13912; (j) R. B. Bedford, C. J. Mitchell and R. L. Webster, *Chem. Commun.*, 2010, **46**, 3095; (k) D. Kalyani, A. D. Satterfield and M. S. Sanford, *J. Am. Chem. Soc.*, 2010, **132**, 8419; (l) D. C. Powers and T. Ritter, *Acc. Chem. Res.*, 2012, **45**, 840; (m) M. C. Nielsen, E. Lyngvi and F. Schoenebeck, *J. Am. Chem. Soc.*, 2013, **135**, 1978 and references therein.
42. For examples involving carbon–bromine bond reductive elimination, see: (a) R. van Belzen, H. Hoffman and C. J. Elsevier, *Angew. Chem. Int. Ed. Engl.*, 1997, **36**, 1743; (b) R. van Belzen and C. J. Elsevier, *Organometallics*, 2003, **22**, 722; (c) M. R. Manzoni, T. P. Zabawa, D. Kasi and S. R. Chemler, *Organometallics*, 2004, **23**, 5618; (d) G. Savitha, K. Felix

and P. T. Perumal, *Synlett*, 2009, **13**, 2079; (e) T.-S. Mei, R. Giri, N. Maugel and J.-Q. Yu, *Angew. Chem. Int. Ed.*, 2008, **47**, 5215 and references therein.

43. For examples involving carbon–iodine bond reductive elimination, see: (a) R. van Asselt, E. Rijnberg and C. J. Elsevier, *Organometallics*, 1994, **13**, 706; (b) A. J. Canty, J. L. Hoare, N. W. Davies and P. R. Traill, *Organometallics*, 1998, **17**, 2046; (c) H. Kodama, T. Katsuhira, T. Nishida, T. Hino and K. Tsubata, *US Pat. Appl.*, 2003181759, 2003; Chem. Abstr., 2001, **135**, 344284; (d) R. Giri, X. Chen, X.-S. Hao, J.-J. Li, J. Liang, Z.-P. Fan and J.-Q. Yu, *Tetrahedron: Asymmetry*, 2005, **16**, 3502; (e) R. Giri, X. Chen and J.-Q. Yu, *Angew. Chem. Int. Ed.*, 2005, **44**, 2112; (f) J.-J. Li, R. Giri and J.-Q. Yu, *Tetrahedron*, 2008, **64**, 6979 and references therein.

44. For a comprehensive review, see: T. W. Lyons and M. S. Sanford, *Chem. Rev.*, 2010, **110**, 1147 and references therein.

45. J. M. Racowski and M. S. Sanford, *Top. Organomet. Chem.*, 2011, **35**, 61 and references therein.

46. A. H. Roy and J. F. Hartwig, *J. Am. Chem. Soc.*, 2001, **123**, 1232.

47. A. H. Roy and J. F. Hartwig, *Organometallics*, 2004, **23**, 1533.

48. A. F. Littke, C. Dai and G. C. Fu, *J. Am. Chem. Soc.*, 2000, **122**, 4020.

49. L. M. Alcazar-Roman and J. F. Hartwig, *J. Am. Chem. Soc.*, 2000, **123**, 12905.

50. J. P. Stambuli, M. Bühl and J. F. Hartwig, *J. Am. Chem. Soc.*, 2002, **124**, 9346.

51. A. H. Roy and J. F. Hartwig, *J. Am. Chem. Soc.*, 2003, **125**, 13944.

52. V. V. Grushin and H. Alper, *Chem. Rev.*, 1994, **94**, 1047.

53. K. C. Lam, T. B. Marder and Z. Lin, *Organometallics*, 2007, **26**, 758.

54. For an example of reductive elimination of an aryl halide from a Pt(IV) complex, see: A. Yahav-Levi, I. Goldberg and A. Vigalok, *J. Am. Chem. Soc.*, 2006, **128**, 8710.

55. J. D. Ruddick and B. L. Shaw, *J. Chem. Soc. A*, 1969, 2969.

56. K. I. Goldberg, J. Y. Yan and E. L. Winter, *J. Am. Chem. Soc.*, 1994, **116**, 1573.

57. K. I. Goldberg, J. Y. Yan and E. M. Breitung, *J. Am. Chem. Soc.*, 1995, **117**, 6889.

58. P. M. Maitlis, A. Haynes, G. J. Sunley and M. J. Howard, *J. Chem. Soc., Dalton Trans.*, 1996, 2187.

59. For an example of reductive elimination of an aryl chloride from an Rh(III) complex, see: J. A. Kampmeier, R. M. Rodehorst and J. B. Phillip, *J. Am. Chem. Soc.*, 1981, **103**, 1847.

60. For an example of reversible oxidative addition of MeI to Rh(I) based on IR spectroscopy, see: L. Gonsalvi, J. A. Gaunt, H. Adams, A. Castro, G. J. Sunley and A. Haynes, *Organometallics*, 2003, **22**, 1047.

61. C. M. Frech and D. Milstein, *J. Am. Chem. Soc.*, 2006, **128**, 12434.

62. M. Feller, Y. Diskin-Posner, G. Leitus, L. J. W. Shimon and D. Milstein, *J. Am. Chem. Soc.*, 2013, **135**, 11040.

63. For reviews of transition metal-catalyzed cross-couplings with poly-halogenated substrates, see: (a) S. Schröter, C. Stock and T. Bach, *Tetrahedron*, 2005, **61**, 2245; (b) I. J. S. Fairlamb, *Chem. Soc. Rev.*, 2007, **36**, 1036; (c) J.-R. Wang and K. Manabe, *Synthesis*, 2009, **9**, 1405; (d) G. Chelucci, *Chem. Rev.*, 2012, **112**, 1344.

64. D. A. Evans and J. T. Starr, *Angew. Chem. Int. Ed.*, 2002, **41**, 1787.

65. For selected total syntheses utilizing polyhalogented substrates in cross-coupling, see: (a) T. Bach and S. Heuser, *Angew. Chem. Int. Ed.*, 2001, **40**, 3184; (b) L. S.-M. Wong, L. A. Sharp, N. M. C. Xavier, P. Turner and M. S. Sherburn, *Org. Lett.*, 2002, **4**, 1955; (c) N. F. Langille and J. S. Panek, *Org. Lett.*, 2004, **6**, 3203; (d) E. Roulland, *Angew. Chem. Int. Ed.*, 2008, **47**, 3762.

66. S. T. Handy, T. Wilson and A. Muth, *J. Org. Chem.*, 2007, **72**, 8496.

67. S. T. Handy and Y. Zhang, *Chem. Commun.*, 2006, 299.

68. (a) C. Y. Legault, Y. Garcia, C. A. Merlic and K. N. Houk, *J. Am. Chem. Soc.*, 2007, **129**, 12664; (b) Y. Garcia, F. Schoenebeck, C. Y. Legault, C. A. Merlic and K. N. Houk, *J. Am. Chem. Soc.*, 2009, **131**, 6632.

69. H. Nakamura, D. Takeuchi and A. Murai, *Synlett*, 1995, 1227.

70. (a) D. A. Colby, R. G. Bergman and J. A. Ellman, *Chem. Rev.*, 2010, **110**, 624; (b) L. Ackerman, *Chem. Commun.*, 2010, **46**, 4866; (c) P. B. Arockiam, C. Bruneau and P. H. Dixneuf, *Chem. Rev.*, 2010, **112**, 5879; (d) G. Song, F. Wang and X. Li, *Chem. Soc. Rev.*, 2012, **41**, 3651; (e) K. M. Engle, T.-S. Mei, M. Wasa and J.-Q. Yu, *Acc. Chem. Res.*, 2012, **45**, 788; (f) D. A. Colby, A. S. Tsai, R. G. Bergman and J. A. Ellman, *Acc. Chem. Res.*, 2012, **45**, 814.

71. For a recent review of cascade Pd- and Cu-catalyzed heterocycle synthesis, see: C. J. Ball and M. C. Willis, *Eur. J. Org. Chem.*, 2013, 425.

72. M. C. Willis, G. N. Brace, T. J. K. Findlay and I. P. Holmes, *Adv. Synth. Catal.*, 2006, **348**, 851.

73. J. A. Joule and K. Mills, *Heterocyclic Chemistry*, Wiley-Blackwell, Chichester, 5th edn, 2010.

74. F. Ramirez, N. B. Desai and N. McKelvie, *J. Am. Chem. Soc.*, 1962, **84**, 1745.

75. Y.-Q. Fang and M. Lautens, *Org. Lett.*, 2005, 7, 3549.

76. (a) Y.-Q. Fang and M. Lautens, *J. Org. Chem.*, 2008, **73**, 538; (b) mechanistic studies conducted for a related *gem*-dibromoolefin system suggest that *E* to *Z* isomerization of a vinylpalladium(II) halide intermediate is a plausible pathway, which would lead to a formal *Z*-selective oxidative addition product: R. Y. Huang, P. T. Franke, N. Nicolaus and M. Lautens, *Tetrahedron*, 2013, **69**, 4395

77. M. Nagamochi, Y.-Q. Fang and M. Lautens, *Org. Lett.*, 2007, **9**, 2955.

78. A. Fayol, Y.-Q. Fang and M. Lautens, *Org. Lett.*, 2006, **8**, 4203.

79. J. Yuen, Y.-Q. Fang and M. Lautens, *Org. Lett.*, 2006, **8**, 653.

80. C. S. Bryan and M. Lautens, *Org. Lett.*, 2008, **10**, 4633.

81. Y.-Q. Fang, J. Yuen and M. Lautens, *J. Org. Chem.*, 2007, **72**, 5152.

82. T. Kosjek and E. Heath, *Top. Heterocycl. Chem.*, 2012, **27**, 219.

83. S. G. Newman, V. Aureggi, C. S. Bryan and M. Lautens, *Chem. Commun.*, 2009, 5236.

84. Following the seminal report, 81 various methods for the synthesis of 2-bromoindoles from *gem*-dibromoolefins were reported. For the Cu-catalyzed method, see: (a) B. Jiang, K. Tao, W. Shen and J. Zhang, *Tetrahedron Lett.*, 2010, **51**, 6342 for metal-free conditions, see; (b) A. R. Kunzer and M. D. Wendt, *Tetrahedron Lett.*, 2011, **52**, 1815; (c) J. Liu, P. Li, W. Chen and L. Wang, *Chem. Commun.*, 2012, **48**, 10052; (d) P. Li, Y. Ji, W. Chen, X. Zhang and L. Wang, *R. Soc. Chem. Adv.*, 2013, **3**, 73.

85. L. Xue and Z. Lin, *Chem. Soc. Rev.*, 2010, **39**, 1692.

86. (a) D. A. Watson, M. Su, G. Teverovskiy, Y. Zhang, J. Garcia-Fortanet, T. Kinzel and S. L. Buchwald, *Science*, 2009, **325**, 1661; (b) T. J. Maimone, P. J. Milner, T. Kinzel, Y. Zhang, M. K. Takase and S. L. Buchwald, *J. Am. Chem. Soc.*, 2011, **133**, 18106.

87. X. Shen, A. M. Hyde and S. L. Buchwald, *J. Am. Chem. Soc.*, 2010, **132**, 14076.

88. (a) M. Oestreich (ed.), *The Mizoroki–Heck Reaction*, Wiley, Chichester, 2009; (b) M. Shibasaki, E. M. Vogl and T. Ohshima, *Adv. Synth. Catal.*, 2004, **346**, 1533; (c) A. B. Dounay and L. E. Overman, *Chem. Rev.*, 2003, **103**, 2945 and references therein.

89. B. Burns, R. Grigg, V. Sridharan and T. Worakun, *Tetrahedron Lett.*, 1988, **29**, 4325.

90. For a review of cyclic carbopalladations terminated by nucleophilic trapping, see: E.-i. Negishi, C. Copéret, S. Ma, S.-Y. Liou and F. Liu, *Chem. Rev.*, 1996, **96**, 365.

91. R. Grigg, J. M. Sansano, V. Santhakumar, V. Sridharan, R. Thangavelanthum, M. Thornton-Pett and D. Wilson, *Tetrahedron*, 1997, **53**, 11803.

92. P. Fretwell, R. Grigg, J. M. Sansano, V. Sridharan, S. Sukirthalingam, D. Wilson and J. Redpath, *Tetrahedron*, 2000, **56**, 7525.

93. B. Burns, R. Grigg, V. Santhakumar, V. Sridharan, P. Stevenson and T. Worakun, *Tetrahedron*, 1992, **48**, 7297.

94. T. Piou, L. Neuville and J. Zhu, *Org. Lett.*, 2012, **14**, 3760.

95. O. René, D. Lapointe and K. Fagnou, *Org. Lett.*, 2009, **11**, 4560.

96. A. Pinto, Y. Jia, L. Neuville and J. Zhu, *Chem. Eur. J.*, 2007, **13**, 961.

97. R. C. Larock and P. L. Johnson, *J. Chem. Soc., Chem. Commun.*, 1989, 1368.

98. (a) S. Cacchi, G. Fabrizi, A. Goggiamani and D. Periani, *Org. Lett.*, 2008, **10**, 1597; (b) C. Zhou, D. E. Emrich and R. C. Larock, *Org. Lett.*, 2003, **5**, 1579; (c) X. Zhang and R. C. Larock, *Org. Lett.*, 2003, **5**, 2993; (d) K. Shibata, T. Satoh and M. Miura, *Org. Lett.*, 2005, 7, 1781.

99. H. Liu, C. Li, D. Qiu and X. Tong, *J. Am. Chem. Soc.*, 2011, **133**, 6187.

100. Y. Lan, P. Liu, S. G. Newman, M. Lautens and K. N. Houk, *Chem. Sci.*, 2012, **3**, 1987.

101. S. G. Newman, J. K. Howell, N. Nicolaus and M. Lautens, *J. Am. Chem. Soc.*, 2011, **133**, 14916.
102. D. A. Petrone, H. A. Malik, A. Clemenceau and M. Lautens, *Org. Lett.*, 2012, **14**, 4806.
103. For use of Lewis acids, see: H. L. van Lingen, W. Zhuang, T. Hansen, F. P. J. T. Rutjes and K. A. Jørgensen, *Org. Biomol. Chem*, 2003, **1**, 1953.
104. For use of organocatalysis, see: (a) Y. Lee, S. W. Seo and S.-G. Kim, *Adv. Synth. Catal.*, 2011, **353**, 2671; (b) D.-F. Yu, Y. Wang and P. F. Xu, *Adv. Synth. Catal.*, 2011, **353**, 2960; (c) D. Enders, G. Urbanietz, R. Hahn and G. Raabe, *Synthesis*, 2012, **44**, 773; (d) F. Hu, X. Guan and M. Shi, *Tetrahedron*, 2012, **68**, 4782.
105. For use of Pd catalysis, see: (a) L. Yu, D.-H. Wang, K. M. Engle and J.-Q. Yu, *J. Am. Chem. Soc.*, 2010, **132**, 5916; (b) M. Leibeling, B. Milde, D. Kratzert, D. Stalke and D. B. Werz, *Chem. Eur. J.*, 2011, **17**, 9888; (c) A. F. Ward, Y. Xu and J. P. Wolfe, *Chem. Commun.*, 2012, **48**, 609; (d) J. S. Cannon, A. C. Olsen, L. E. Overman and N. S. Solomon, *J. Org. Chem.*, 2012, 77, 1961.
106. For other approaches, see: (a) M. R. Medeiros, R. S. Narayan, N. T. McDougal, S. E. Schaus and J. A. Porco Jr, *Org. Lett.*, 2010, **12**, 3222; (b) N. Kern, A. Blanc, J. M. Weibel and P. Pale, *Chem. Commun.*, 2011, **47**, 6665; (c) H. C. Shen, *Tetrahedron*, 2009, **65**, 3931.
107. M. Leibeling, D. C. Koester, M. Pawliczek, S. V. Schild and D. B. Werz, *Nat. Chem. Biol.*, 2010, **6**, 199.
108. X. Jia, D. A. Petrone and M. Lautens, *Angew. Chem. Int. Ed.*, 2012, **51**, 9870.
109. (a) M. Lautens and Y.-Q. Fang, *Org. Lett.*, 2003, **5**, 3679; (b) T. Saget, D. Perez and N. Cramer, *Org. Lett.*, 2013, **15**, 1354.
110. D. A. Petrone, M. Lischka and M. Lautens, *Angew. Chem. Int. Ed.*, 2013, **52**, 10635.
111. (a) C. C. C. Johansson Seechurn, M. O. Kitching, T. J. Colacot and V. Snieckus, *Angew. Chem. Int. Ed.*, 2012, **51**, 5062; (b) H. Li, C. C. C. Johansson Seechurn and T. J. Colacot, *ACS Catal.*, 2012, **2**, 1147.

CHAPTER 8

Boron Reagent Activation in Suzuki–Miyaura Coupling

ALASTAIR LENNOX AND GUY LLOYD-JONES*

School of Chemistry, West Mains Road, Edinburgh, EH9 3JJ, UK
*Email: guy.lloyd-jones@ed.ac.uk

8.1 Suzuki–Miyaura Cross-Coupling

8.1.1 Introduction

Over the last two decades, the advent of practical and scalable transition metal-catalysed cross-coupling reactions has truly revolutionized organic synthesis. Of the various metals explored, palladium has proven to be exceptional. A wide variety of Pd complexes and ligands are readily accessible or commercially available and there is usually a low thermodynamic differential between oxidation states, in this case 0 and 2, as required for efficient turnover in cross-coupling. The propensity for Pd(0) to undergo oxidative addition to a wide range of organohalides and pseudohalides, together with the development of a suite of formally nucleophilic coupling partners, has led to a very broad spectrum of applications in cross-coupling. One can subdivide the nucleophilic coupling partners into those based on organic reactants, *e.g.* alkenes (Heck)[1,2] and alkynes (Sonogashira)[3] and those based on organometallics or organometalloids, such as magnesium (Kumada),[4] zinc (Negishi)[5] tin (Stille)[6] silicon (Hiyama)[7] and boron (Suzuki–Miyaura)[8] (Scheme 8.1). Recognition of these pivotal contributions to the development of synthetic methods was recently made by the award of the 2010 Nobel Prize in Chemistry to three of the pioneers in the area.

RSC Catalysis Series No. 21
New Trends in Cross-Coupling: Theory and Applications
Edited by Thomas J Colacot
© The Royal Society of Chemistry 2015
Published by the Royal Society of Chemistry, www.rsc.org

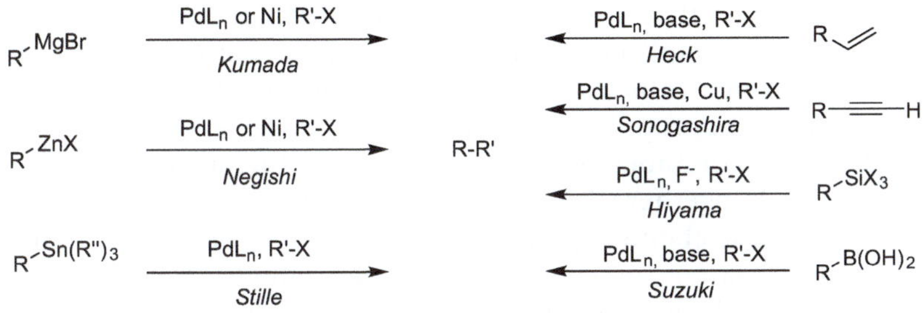

Scheme 8.1 A selection of Pd-catalysed cross-coupling reactions.

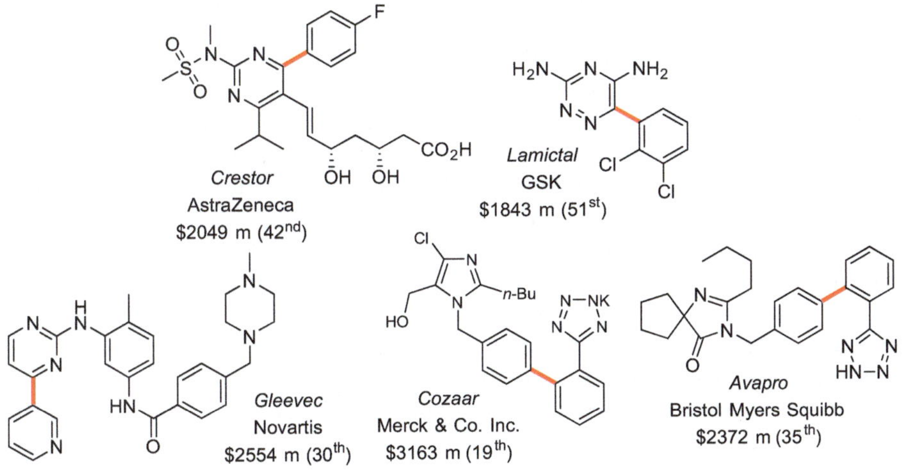

Figure 8.1 Examples of "blockbuster" drugs with biaryl moieties. Figures in parentheses are overall rank based on stated revenue in 2006.

Of the palladium-catalysed cross-coupling reactions available (Scheme 8.1), Suzuki–Miyaura (SM) coupling has emerged as the one that has enjoyed the most widespread application,[9] with numerous relatively non-toxic and stable reagents now commercially available. A vast array of developments have been reported,[9] from expansion of the substrate scope to include unactivated aryl chlorides[10] and sterically demanding substrates,[11] to reducing the catalyst loadings to very low levels,[12] *e.g.* <0.05 mol%,[13] and lower reaction temperatures.[14,15] Indeed, SM coupling has become the "gold standard" for biaryl construction, arguably resulting in the ubiquity of this moiety in modern medicinal chemistry,[16] with biaryls present in many "blockbuster" drug molecules (Figure 8.1).

8.1.2 Boron Reagents

The key to the success of the SM coupling reaction stems from the exceptional functional group tolerance of the catalyst system and the mild reaction conditions that are employed. This originates from the relatively stable, innocuous and environmentally benign nature of the boron-containing reagent. Nonetheless, transmetallation of organoboron compounds with the appropriate palladium(II) complex proceeds rapidly and efficiently. These combined features contribute to the practical upscaling of the reaction and, together with the relatively low cost of the reagent, explain its value to the fine chemical, pharmaceutical and agrochemical industries.

The hydroboration of terminal alkynes allows ready access to alkenylboranes and catecholboronic esters,[17,18] a process that provided the reagents employed by Suzuki and Miyaura for their initial discovery and investigation of the SM coupling. However, part of the power of the SM coupling lies in its diversity of reagents, as differentiated not just by the organic fragment but also by the substitution patterns at boron. In the 1980s, the borane and catecholboronic ester reagents initially explored were generally replaced by the more reactive and atom-economical boronic acids,[19] especially for aryl couplings. The scope was further expanded with the introduction of pinacolboronic esters in the 1990s, made available by the recently developed Miyaura borylation protocol. The last decade has witnessed a substantial expansion with a very wide variety of reagents for SM coupling now having been reported (Figure 8.2), with characteristics tailored for more specific tasks, such as distal manipulation or slow-release.

The outer-shell bonding electrons of neutral boron reagents ($2s^2$, $3p^1$) results in a tricoordinate sp^2 boron centre with trigonal planar geometry and a non-bonding vacant p-orbital orthogonal to this plane. The vacant p-orbital dominates the reactivity patterns and characteristics of these species as the

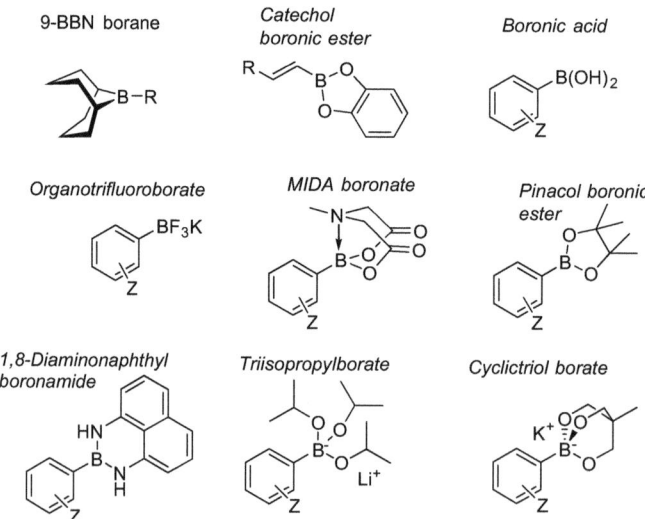

Figure 8.2   Some of the most popular boron reagents used directly or indirectly in SM coupling reactions.

electron deficiency of boron renders it susceptible to electron donation from Lewis bases. Reagents of this type are organoboranes, boronic esters and boronic acids. Upon coordination, a tetrahedral anionic "ate" complex is formed that is more stable towards side reactions, owing to "blocking" of the vacant p-orbital. Examples of reagents of this type are the organotrifluoroborates, MIDA boronate (MIDA = *N*-methyliminodiacetic acid), triisopropylborate and cyclic triolborate salts, discussed below.

8.2 Mechanism

The mechanism of the SM coupling reaction has been established as proceeding through a Pd(0)–Pd(II) manifold. The organohalide coupling partner oxidatively adds to a coordinatively unsaturated palladium(0) complex, followed by transmetallation with the organoboron reagent.[8,9,20] Reductive elimination then delivers the cross-coupled product and recycles the active palladium(0) catalyst, *e.g.* the biaryl generation exemplified in Scheme 8.2.

Palladium(0) complexes are often susceptible to decomposition, *e.g. via* ligand oxidation, meaning that their handling and storage for use as a catalyst in SM coupling are not generally convenient. Much more frequently, the active Pd complex is generated *in situ*, by reduction of a much more stable palladium(II) precatalyst. This generates a free coordination site on the more electron-rich metal centre, which can then participate in an oxidative addition event with an organohalide or pseudohalide, *e.g.* OTf. The ease of insertion into the polarized covalent bond is usually proportional to the bond dissociation enthalpy and the approximate order of reactivity is I > OTf ≈ Br >> OTs > OMs ≈ Cl, although other factors can change this dramatically.[21] Electron-withdrawing groups in a conjugated system, *e.g.* in an aryl halide, can also facilitate this process by removing electron density from the halide, thereby weakening the bond. Inductively, this effect can

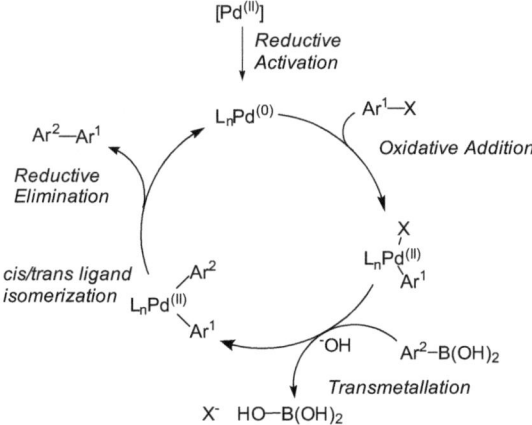

Scheme 8.2   A generic catalytic cycle for SM coupling of two aryl partners.

also aid association to the electron-rich metal centre. Oxidative addition is frequently found to be turnover limiting; therefore, the use of very electron-rich ligands, such as trialkylphosphines,[10] can be used to promote this step through their strong σ-donating ability, formally destabilizing Pd(0) and stabilizing Pd(II). In this way, organochlorides,[10] which are the cheapest but often least reactive organohalide (with the exception of organofluorides), can readily participate in cross-coupling reactions. Indeed, with the highly activating electron-rich ligands that are now widely available, the rate-limiting step can be switched from oxidative addition to the transmetallation or reductive elimination event. Moreover, the interaction between the (pseudo)-halide leaving group and the metal counter-ion to the nucleophilic boron reagent can make the transmetallation step considerably more efficient with chloride, as compared with say the bromide or iodide, thus reversing the *overall* order of "reactivity" for the aryl halides noted above.

Base is essential for efficient turnover, but mechanistically its precise role in transmetallation is not always clear; indeed, this aspect has been the subject of considerable debate.[20,22] Boronic acids containing electron-rich moieties tend to undergo the subsequent transmetallation with the oxidized palladium(II) more readily than those bearing electron-poor moieties. However, as discussed in more detail below, the multiple steps that are involved in transmetallation make a detailed analysis of such an effect non-trivial. Thus, although electron-donating substituents will make the boron centre less Lewis acidic, decreasing its propensity for association with hydroxyl, it will also accelerate the transfer of the aryl group from the resulting "ate" complex.

After transmetallation by the boron reagent, the resulting diorganopalladium complex must often undergo a series of ligand dissociation and association events to isomerize the *trans* to the *cis* isomer, which can then undergo reductive elimination. The resulting coordinatively unsaturated palladium(0) complex is then released to undergo further catalytic cycles. Typically, reductive elimination is rapid, especially with bulky ligands that sterically enhance this step. Jutand and co-workers[22] elucidated an important additional role of hydroxide as the base in SM coupling: it was found to accelerate the reductive elimination step *via* formation of a pentacoordinated intermediate (Scheme 8.3), with Berry pseudorotation then by-passing the requirement for a formal isomerization of ligands.

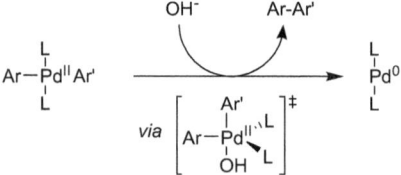

Scheme 8.3 Hydroxide-catalysed reductive elimination.

8.3 Direct Transmetallation of Boron Reagents

The mechanism of transmetallation, and the preceding activation of the boron reagent if required, has been studied in less detail than that of the oxidative addition or reductive elimination events. This is possibly due to the increased complexity involved in identifying, characterizing and measuring rates of reaction of the possible palladium(II) intermediates. Additionally, it is not a straightforward task to identify the active boron species, as this is often in equilibrium with inactive species and may thus be present in only low concentration. As a general class of reagent, organoboranes and boronic acids both undergo *direct* transmetallation, sometimes after the addition of a Lewis base. In other words, no ligands at boron need to be *substituted* prior to the delivery of the organic component to the Pd. This direct activation renders them distinct from other classes of boron reagent that can require the exchange of ligands on boron to occur in order to transmetallate effectively.

8.3.1 Organoboranes

As noted above, simple organoboranes were the first boron reagents to be employed for SM coupling, due in part to their ease in preparation by hydroboration.[23–29] The 9-BBN and disiamylboranes are the most common motifs on boron, as the secondary carbons that are appended to boron provide the differentiation required to achieve the selective delivery of the primary alkyl or alkenyl group to palladium in the transmetallation step. Dialkylboranes (HBR_2) readily add to unsaturated carbon linkages with *syn* selectivity and in an anti-Markovnikov manner to furnish the requisite organoborane for SM coupling. The reagent is often used immediately in the SM coupling as triorganoboranes can readily degrade over time.

Early mechanistic studies focused on organoboranes, as these regularly featured in SM coupling reactions. Dimeric hydroxide-bridged palladium complexes and monomeric hydroxide-ligated palladium complexes had been characterized,[30] and it was proposed that hydroxide base provided a μ-bridged hydroxide to link between the boron and palladium in a metathetical-like transition state.[31] However, there has been considerable debate regarding the sequence of events that precede the assembly of this transition state, specifically, whether the hydroxide activates the borane (*I*, Scheme 8.4) through coordination to give a more nucleophilic boronate species (*Boronate pathway*), or activates palladium (*II*, Scheme 8.4), *via* exchange of the halide on the palladium(II) centre to generate an oxo-palladium species (*oxo-palladium pathway*).[32]

In early work, Suzuki and Miyaura conducted mechanistic studies on the coupling of bromoalkenes with alkenylboranes employing alkoxide bases. Palladium tetrakis(triphenylphosphine) catalysed coupling of a tetraalkylated borane "ate" complex (Li[BR_3-alkenyl]) with a styrenyl bromide afforded the cross-coupled product in only 9% yield, and this was interpreted

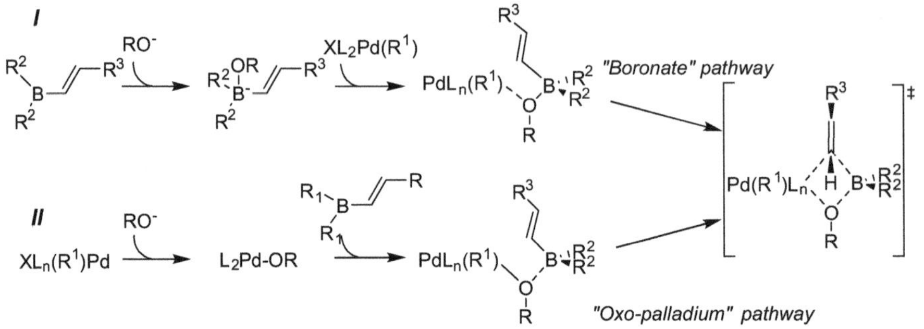

I

RO⁻ → XL₂Pd(R¹) → "Boronate" pathway

II

XLₙ(R¹)Pd → RO⁻ → L₂Pd-OR → PdLₙ(R¹) → "Oxo-palladium" pathway

Scheme 8.4 The two possible mechanisms considered for transmetallation of an alkenylborane with palladium complexes.

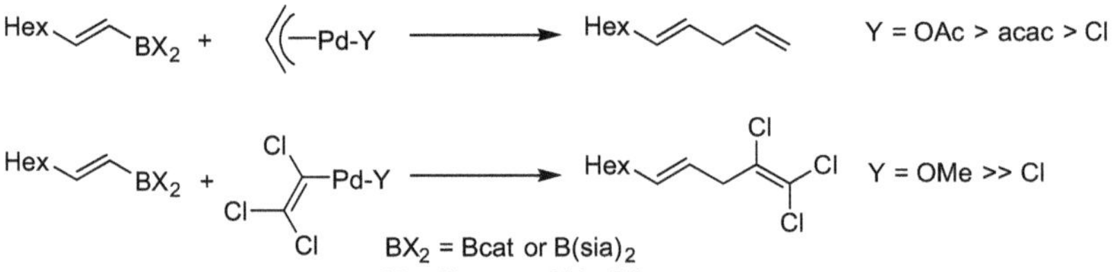

Hex—BX_2 + ⟨—Pd-Y⟩ ⟶ Hex—   Y = OAc > acac > Cl

Hex—BX_2 + Cl-vinyl-Pd-Y ⟶ Hex—   Y = OMe >> Cl

BX_2 = Bcat or B(sia)₂
Y = Cl, acac, OMe, OAc

Scheme 8.5 Studies by Suzuki and Miyaura on the reactivity of palladium complexes in stoichiometric transmetallation reactions.

as evidence against the *boronate* pathway.[33,34] However, what was not considered at the time, but is now perhaps more evident, is that the (Li[BR₃-alkenyl]) "ate" complex lacks a heteroatom at boron that can coordinate to palladium to effect the assembly of the transmetallation transition state. In contrast, further studies by Suzuki and Miyaura on *stoichiometric* couplings of palladium(II)–π-allyl and palladium(II)–trichlorovinyl complexes with 1-octenylcatecholborane or disiamylborane reagents revealed that the ligands on palladium dictated their reactivity in the order OMe >> Cl and OAc > acac > Cl, with the (alk)oxo-palladium species leading to greater yields of product. It was therefore concluded that formation of the (alk)oxo-palladium intermediate by a metathetical type displacement with, for example, sodium methoxide was important and overall interpreted as being evidence in favour of the *oxo-palladium* pathway (Scheme 8.5).[33]

Over a decade later, a very thorough mechanistic study into transmetallation of organoboranes in SM coupling was reported by Matos and Soderquist.[31] Specifically, alkylboranes were compared with alkylborinates in their cross-coupling with aryl halides in aqueous basic THF solutions; the resulting NMR, kinetic and reaction efficiency data provided compelling evidence for differing pathways. The Lewis acidic alkylboranes, *e.g.* **1**, readily formed (¹¹B NMR) boronate complexes in the presence of base and it was concluded that it was the boronate species that transmetallated with the halide complex, [PdBr(Ar)L] (Scheme 8.6). In contrast, the association of

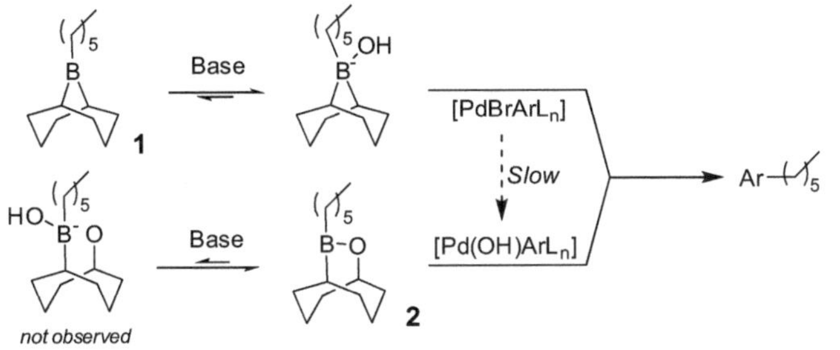

Scheme 8.6 Speciation and cross-coupling of borane **1** and borinic ester **2**.

hydroxide with the alkyl borinic esters **2** was undetectable, owing to the substantially lower Lewis acidity at boron arising from the availability of the lone-pair on oxygen to populate partially the vacant p-orbital. By competing **1** and **2** for limiting aryl bromide, it could be shown that reagent **1** was substantially more efficient than **2**, the latter reagent also undergoing slower rates of conversion when compared independently. Key to the overall analysis was a measurement of the rate of hydrolysis of [PdBr(Ar)L] by OH to give [Pd(OH)(Ar)L], which was found to be slow, thus leading to the conclusion that the neutral borinic ester **2** transmetallates with [Pd(OH)ArL$_n$], whereas the boronate complex derived from **1** transmetallates with [PdBrArL$_n$].

8.3.2 Boronic Acids

There are three general routes for the preparation of boronic acids (Scheme 8.7). The most widely used proceeds *via* organometallic inter- mediates such as organolithium[35] or organomagnesium reagents,[36,37] which are trapped with a boric acid ester [*e.g.* B(O*i*-Pr)$_3$ or B(OMe)$_3$], to provide a boronic ester that rapidly hydrolyses to a boronic acid under acidic treat- ment. The second method is through the hydrolysis of boronic esters, *e.g.* pinacolboronic ester, which are readily prepared and purified by a variety of methods. However, the liberated diol readily condenses with the boronic acid product, which can cause issues with low conversions and purification. This problem can be tackled either by intermediate formation of a distinct species that can be purified from the diol,[38] or by consuming the diol in a separate reaction that removes it from recombining with the boronic acid.[39] The third and most recently developed method is based on Pd- or Ni-cata- lysed borylation of aryl halides that employs tetrahydroxydiboron.[40,41]

   In general, boronic acids dissolve more readily in organic solvents than in neutral aqueous solutions. Under the former conditions, an equilibrium is established with the trimeric anhydride (boroxine) species, an entropically favoured transformation that liberates 3 equiv. of water (Scheme 8.8). Boroxines are stabilized through their partial aromatic character, wherein electron density is donated from oxygen to boron, formally generating a triply zwitterionic cyclic compound. Although there are reports of boroxines

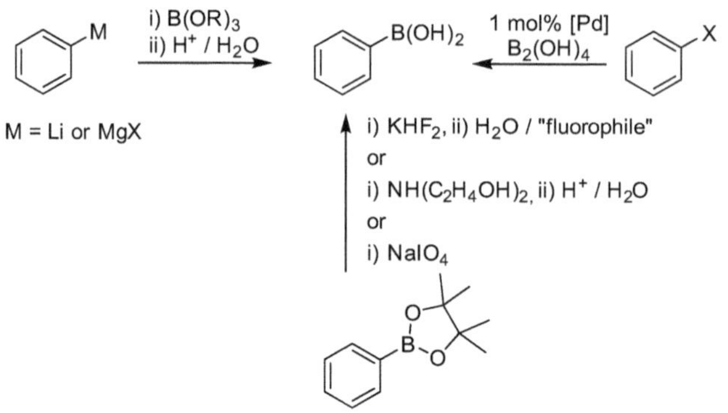

Scheme 8.7 The three general routes to prepare arylboronic acids.

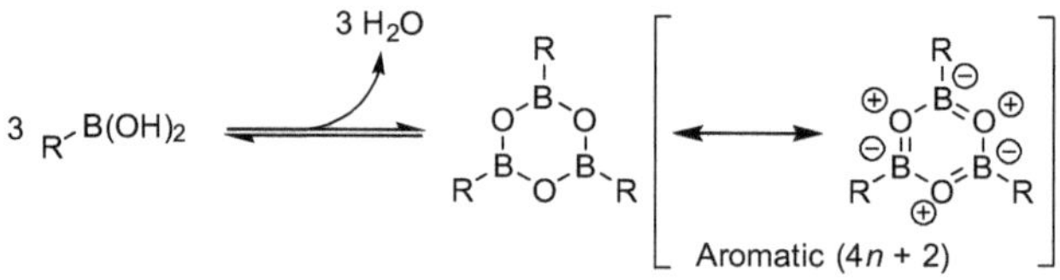

Scheme 8.8 Entropically favourable dehydration of boronic acids to form the partially aromatic boroxine species.

as active transmetallating species, *e.g.* with rhodium complexes,[42] it is not yet clear whether they are more or less reactive than boronic acids in the transmetallation of palladium(II) complexes. However, if their formation does reduce the rates of transmetallation, it could be an additional reason as to why water is a common component in SM coupling reactions. In addition, setting the correct stoichiometry of the boronic acid can sometimes be difficult, as establishing the degree of dehydration to the boroxine in the solid material is not straightforward. To account for this situation, it is common practice to add an excess of the reagent.

The mechanism of transmetallation of boronic acids to Pd(II) intermediates has been the subject of a range of computational studies.[43,44] However, owing to the ready formation of the "ate" complex ([$RB(OH)_3$]$^-$) under basic SM coupling conditions, most studies have tended to investigate only the reaction of this species with the halide complex, *i.e.* the *boronate* pathway. Maseras and co-workers,[43] however, calculated a lower energy pathway for the *oxo-palladium* pathway but could not locate any reasonable mechanism for the formation of the key oxo-palladium(II) species, despite their being precedent for such species.[45] They therefore concluded that the *boronate* pathway dominates.[43] Nevertheless, three experimental studies published in 2011 all found *kinetic* evidence in favour of the *oxo-palladium* pathway.

Jutand and co-workers[22] carefully examined the role of the base by studying voltammograms arising from generation and decay of palladium species. The concentrations of the electroactive species determined the reduction or oxidation currents and this then allowed kinetic analysis.

Four possible transmetallation scenarios were considered: (a) the base plays no role, (b) the base reacts initially with the boronic acid, (c) the base reacts initially with the palladium(II) or (d) the base reacts with both the boronic acid and palladium(II) (Figure 8.3).[22]

The electrochemical data were interpreted to reveal that the only reaction that proceeds at a sufficient rate to allow catalytic competency under SM coupling conditions is that between the oxo-palladium species and the neutral boronic acid. Moreover, in contrast to the high (undetected) barrier to palladium halide hydrolysis found computationally by Maseras and co-workers, experimentally the oxo-palladium species was found to be rapidly generated.[43] Although the trihydroxyborate species was also found to react with the halide complex, in the presence of an excess of bromide to ensure the concentration of oxo-palladium species was kept low, the rates were very low indeed. Overall, it could be concluded that the formation of the aryl trihydroxyborate is detrimental to the reaction rate. An analogous investigation was performed on the effect of fluoride in the system under anhydrous conditions. The study verified that fluoride exhibited similar activation pathways to hydroxide,[46] which is highly informative for SM coupling systems that are undertaken using KF as base under nominally dry conditions.

The rates of stoichiometric transmetallation between the halide complex $[PdXAr(PPh_3)_2]$ and aryl trihydroxyborate (*boronate* pathway), and also between the oxo-palladium and boronic acid (*oxo-palladium* pathway), were determined by Carrow and Hartwig, by way of ^{31}P NMR measurements made at low temperatures (–30 to –55 °C).[47] There was a rate differential of around four orders of magnitude between transmetallation *via* the boronate pathway (trihydroxyborate and the bromide complex) *versus* that *via* the oxo-palladium pathway (neutral boronic acid and the oxo-palladium complex), the latter being the faster. Crucially, studies of the equilibrium between the bromide complex and the oxo-palladium complex confirmed that sufficient populations of the latter are present for the dominant catalytic flux to proceed *via* the oxo-palladium pathway.

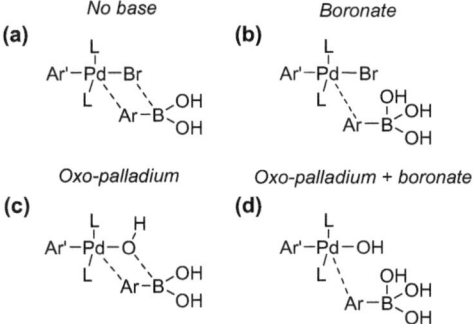

Figure 8.3 Four transmetallation scenarios considered by Amatore *et al.*[22]

In a somewhat complementary study, employing phosphine-free conditions, Schmidt *et al.* used UV spectroscopy to measure the rates of reaction between phenylboronic acid and an equilibrium mixture of $[Pd^{II}(OAc)_2]$ and base (NaOAc) *versus* the reaction of $[Pd^{II}(OAc)_2]$ with an equilibrium mixture of phenylboronic acid and base (Scheme 8.9).[48] Under the former conditions, where neutral boronic acid was added to the premixed solution of catalyst and base, thus analogous to the oxo-palladium pathway, biphenyl generation occurred around 1.3–2 times more rapidly.

Aqueous base is often employed in SM coupling; indeed, a survey of around 40 000 such couplings revealed that more than half were predicted to have an aqueous biphase present.[49] Aqueous–organic biphases can readily form upon addition of an inorganic base[47] to solvent mixtures such as THF–water.[49] Analyses of reagent speciation in such biphasic mixtures indicate that boronic acid is present in the bulk organic phase,[49] with small proportions of trihydroxyborate present in the aqueous phase. Since hydroxide can readily undergo phase transfer to the bulk medium in order to form the key catalytic intermediate, $[Pd(OH)ArL_n]$, a biphase can be particularly effective for the *oxo-palladium* pathway because it limits sequestration of boron reagent, in the form of the trihydroxyborate, in the bulk phase. Single-phase systems are likely more primed for the *boronate* pathway.

Further studies on the transmetallation event by Jutand and co-workers[50] have demonstrated that the counter-cation, which is inevitably added with an anionic base, has a decelerating effect on the overall rate. This was proposed to be due to a deactivating coordination to Ar–Pd–OH; whereby the least coordinating anion led to the greatest rate: $NBu_4^+ > K^+ > Cs^+ > Na^+$.

Additives are commonly employed to assist transmetallation of the boronic acid. This could be envisaged as proceeding *via* an assisted transmetallation or more effectively as proceeding *via* generation of a discrete reagent. Stoichiometric amounts of silver oxide have been shown to be highly effective in assisting polyfluorophenyl[51–54] and *n*-alkyl[55] couplings. Attempts to deduce the nature of the enhancing effect were made whereby oxidative addition intermediates were isolated and reacted stoichiometrically with the silver oxide and boronic acid in a toluene–water biphase.[54] It was proposed that silver oxide effects an efficient replacement of the iodo ligand on palladium with a hydroxide group, thus priming the complex

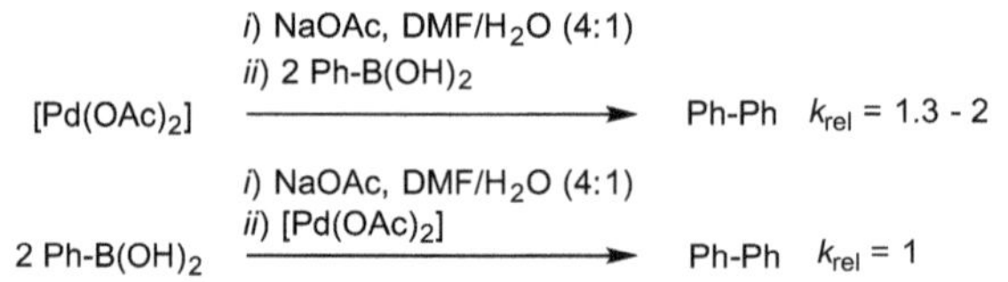

$$[Pd(OAc)_2] \xrightarrow[\text{ii) 2 Ph-B(OH)}_2]{\text{i) NaOAc, DMF/H}_2\text{O (4:1)}} \text{Ph-Ph} \quad k_{rel} = 1.3 - 2$$

$$2\ \text{Ph-B(OH)}_2 \xrightarrow[\text{ii) [Pd(OAc)}_2]]{\text{i) NaOAc, DMF/H}_2\text{O (4:1)}} \text{Ph-Ph} \quad k_{rel} = 1$$

Scheme 8.9 Relative rates of stoichiometric transmetallation in the homocoupling of phenylboronic acid, where the base is pre-equilibrated with either the boronic acid or palladium(II) catalyst.

towards transmetallation with boronic acid. A similar mechanism is proposed to be responsible for the greatly enhanced the rates of transmetallation reported when thallium bases are employed.[56] Addition of TlOH leads to three orders of magnitude faster turnover than KOH and thus TlX precipitation may encourage oxo-palladium formation. Alternatively, an efficient pre-transmetallation to form a Tl–R intermediate may be viable.[8] When Tl_2CO_3 is employed as base, it was suggested that decarboxylation generated a cationic palladium centre,[57] which would be particularly reactive towards a boronate species.

8.4 Boron Reagents Requiring Prior Activation

There are a number of boron reagents used in SM coupling that are known to require activation, usually hydrolytic, in order to undergo transmetallation. Ligand substitution has to occur as the parent reagent is stable towards palladium(II) complexes. Examples of these are organotrifluoroborate salts, MIDA boronates and 1,8-diaminonaphthylboronamides (DAN).

8.4.1 Organotrifluoroborate Salts

Potassium organotrifluoroborate (R-BF_3K) salts were discovered in 1960 by Chambers *et al.*,[58] but the following three decades saw them appearing in only a handful of publications. Since the mid-1990s, their application has grown substantially–indeed, they have rapidly become a very widely used class of organoboron reagent. They are readily prepared from the boronic acid or ester, by reaction with either KHF_2,[59] or a much less corrosive process involving sequential treatment with KF and tartaric acid[60] (Scheme 8.10).

In contrast to boronic acids and esters, organotrifluoroborates are tetrahedral in geometry and not Lewis acidic, owing to the additional ligand bound to the boron centre. This quaternization with exceptionally strong B–F bonds gives them their favourable physical characteristics of being free-flowing crystalline salts, which tend to melt and decompose only at very high temperatures. In addition to being monomeric in nature, they are stable to air and moisture in the solid state. These factors render them easy reagents to handle, unlike, for example, small alkylboronic acids that can decompose in air,[61] or pinacolboronic esters, many of which are liquids or low-melting solids. In solution, the trifluoroborate moiety is stable under anhydrous conditions, but when subjected to aqueous or protic media they hydrolyse, *via* equilibrium, to form the corresponding boronic acid or ester.[49] Upon hydrolysis, HF is formally liberated, which can cause etching of glassware

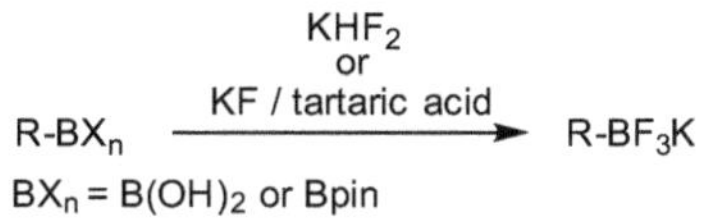

$$R\text{-}BX_n \xrightarrow{\substack{KHF_2 \\ \text{or} \\ KF \,/\, \text{tartaric acid}}} R\text{-}BF_3K$$

$$BX_n = B(OH)_2 \text{ or Bpin}$$

Scheme 8.10 Preparation of trifluoroborate salts from boronic acids or esters.

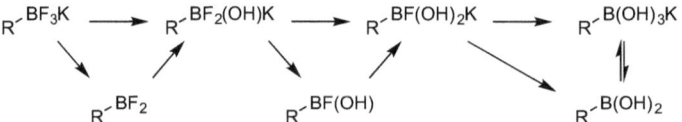

Scheme 8.11 Hydrolysis of organotrifluoroborates to boronic acids *via* mixed ligated species.

under aqueous conditions if it is not rapidly quenched by base or an alternative sacrificial fluorophile.

Initial mechanistic investigations into the origin of the high performance of the organotrifluoroborate salts in SM coupling, particularly when compared with the corresponding boronic acid,[62] indicated that the organotrifluoroborate salt itself was not the active transmetallation species.[14,63] It was shown by NMR spectroscopy that hydrolysis occurred *via* mixed fluoro–hydroxy borate intermediates (Scheme 8.11). Base titrations of the trifluoroborate and observations of the mixed ligated species by electrospray ionization mass spectrometry provided evidence in support of this proposal.[38,64]

However, it was later shown that the mixed fluoro–hydroxy borates usually have a very low population and computational analysis (DFT) of the transmetallation of the various intermediates indicated that the pathway became more favourable as all of the fluoride ligands on boron were sequentially replaced by hydroxide.[65] This is consistent with a reduction in the nucleophilicity of the organic fragment, when ligated by the highly electronegative fluoride, and also a reduction in the ability of the ligand to bridge the metal species. In other words, the mixed fluoro–hydroxy borates are solely intermediates *en route* to the boronic acid, with transmetallation primarily occurring through the latter.[63,65]

This conclusion was reinforced by a kinetic analysis of $[^2H_4]$-**3** and $[^2H_0]$-**4** under conditions where they compete for limiting **5**, confirming the boronic acid as the most reactive species.[63,65] Indeed, even when the proportion of $[^2H_4]$-**3** was only 10% in the presence of 90% of unlabelled $[^2H_0]$-**4**, after just 10% conversion the cross-coupled product **6** contained 60% of the labelled ring, *i.e.* that coming from $[^2H_4]$-**3**, before its isotopic enrichment was diluted to 10% at 100% conversion (Scheme 8.12).

Overall, it could therefore be concluded that the high performance of trifluoroborates in SM coupling was not due to the liberation of a more efficient mixed fluoro–hydroxy borate species for transmetallation, but instead due to a reduction in the amount of side products generated.[65] These side reactions stem predominantly from the boronic acid reagent (Scheme 8.13), with the use of organotrifluoroborate salts strongly suppressing them.

The palladium(II) precatalyst activation by boronic acid generates homocoupled product through a two-stage double transmetallation–reductive elimination sequence. Fluoride has been shown to effect a hydrolytic reduction of palladium(II) that yielded a monophosphine complex[66,67] and

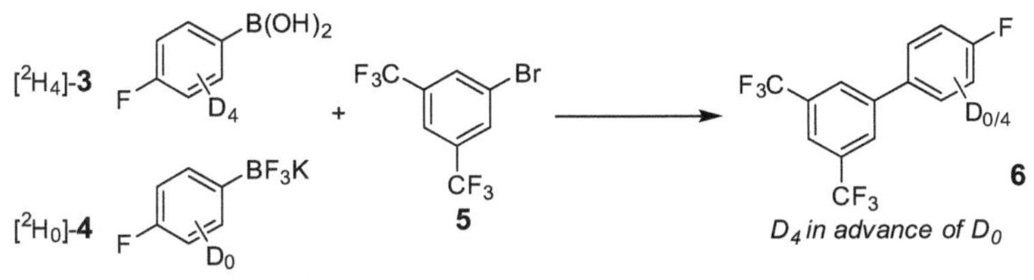

Scheme 8.12 Kinetic analysis of a competition between $[^2H_4]$-boronic acid **3** and $[^2H_0]$-trifluoroborate **4** for limiting aryl bromide **5**.

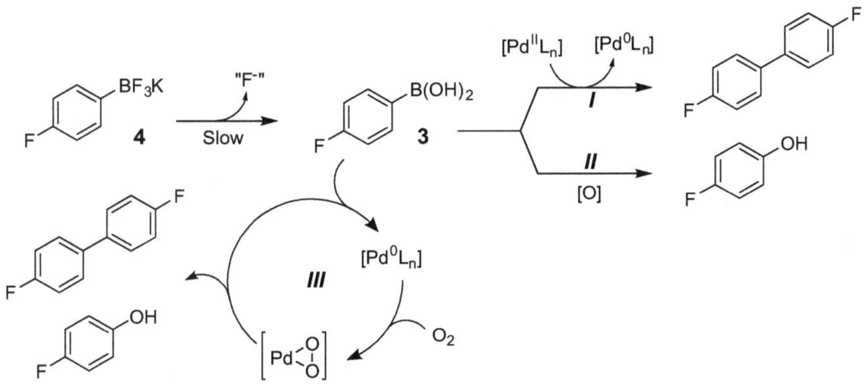

Scheme 8.13 Degradation pathways (*I*, *II* and *III*) for 4-fluorophenylboronic acid (**3**) that are all reduced through its slow formation from trifluoroborate **4** and the fluoride that is co-liberated.

^{18}O-labelling studies of organotrifluoroborate SM coupling indicated that the fluoride liberated upon hydrolysis of the organotrifluoroborate led to this palladium(II) activation pathway.[65] The same study also found that organotrifluoroborates exhibited a quenching effect of common oxidants that are found in ethereal solvents.

The final mechanism by which organotrifluoroborates were shown to reduce side reactions was in a *slow release* of the active boronic acid species.[65,68] This slow hydrolysis allows the reactive species to be maintained at a low concentration. This was shown to lead to a favourable partitioning between cross-coupling and oxidative homocoupling, whereby a lower concentration of boronic acid decreased the catalytic flux through the non-productive cycle. The absolute rate of protodeboronation is also slowed because of this effect, which is highly useful for the coupling of unstable substrates.[69] The slow-release effect was verified by a slow syringe pump addition of boronic acid to the reaction mixture conducted in air, where very low levels of homocoupling were observed.[65]

A survey of reaction conditions used to couple R-BF$_3$K salts showed that, for every class of reagent there is a unique set of optimized conditions.[49] This analysis alone suggests that there is no single mechanistic regime for hydrolysis to the transmetallating species. Optimization of the reaction temperature, solvent, precatalyst, base and time are thus a careful balance between ensuring the slowest hydrolysis in combination with the fastest turnover. Further mechanistic investigations on the hydrolysis of R-BF$_3$K salts under SM coupling conditions was undertaken to elucidate these observations so greater control and predictability can be made for future optimizations. Mechanistically, an *acid* catalysed pathway was shown to give the most rapid rates of hydrolysis to the corresponding boronic acids. The *basic* conditions of SM coupling largely suppressed this pathway. Under these regular coupling conditions, THF / water 10:1, 3 equiv. Cs$_2$CO$_3$, a biphase exists with a very basic minor aqueous phase and an organic bulk phase, Figure 8.4. Access to the acid catalysed pathway was thus found to be dependent on the efficiency of mixing of the phases. Conditions that induced good mixing, such as the vessel shape, stirring rate, ultrasound *etc.*, led to a disabling of the acid-catalysed hydrolysis and a slow background uncatalysed pathway was observed to occur. Systems that engendered poor phase mixing led to rapid rates of hydrolysis through the acid catalysed pathway. Then the release rate of boronic acid was directly correlated to the side-products formed, *i.e.,* slow hydrolysis from efficient mixing led to a low concentration of boronic acid and less side-product formation.

Rates of hydrolysis were measured for a range of R-BF$_3$K salts and were found to span five orders of magnitude. This variation was found to correlate with the DFT-derived B–F bond length of the intermediate difluoroborane [r(B--F)], as it was sensitive to the structural characteristics that dominate hydrolysis rates. Thus a reliable parameter was found to predict relative hydrolysis rates and therefore the efficiency of a particular cross-coupling reaction. In addition, the more easily sourced Swain–Lupton resonance value for R, behaving as a *para* substituent in an aromatic ring, in combination with its weighted Charton steric parameter, correlated well with rates of hydrolysis ($\log k_{rel} \propto R_{SL} - 0.09\nu$) and thus gave a rapid and simple tool for their prediction.

Organotrifluoroborates that are very electron rich, *e.g.* cyclopropyl and 4-methoxyphenyl, undergo very rapid *in situ* hydrolysis. Hydrolysis is so

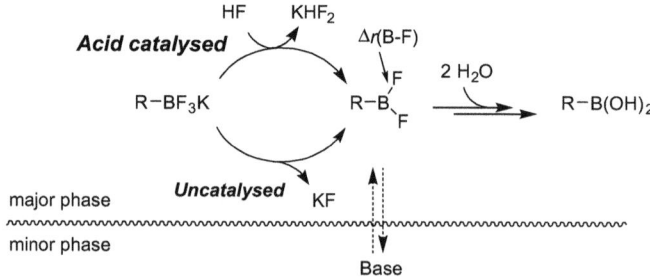

Figure 8.4 Hydrolysis of trifluoroborates to boronic acids, indicating the mechanisms of the acid catalysed and uncatalysed pathways. The acid catalysed pathway is attenuated when there is efficient mixing with the basic minor biphase

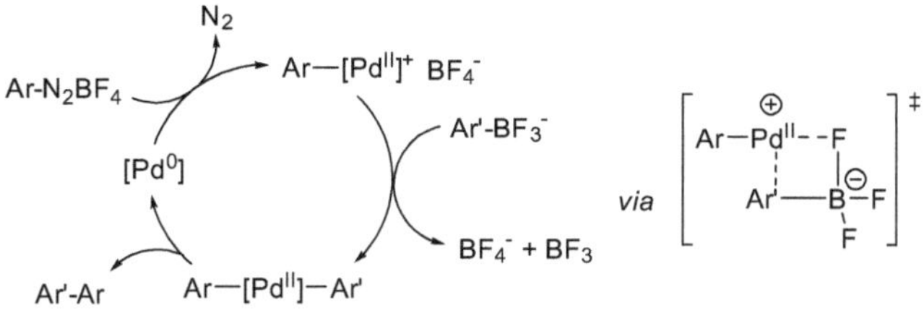

Scheme 8.14 Cationic palladium species in SM coupling with arenediazonium salts.

rapid that the benefit in reducing side reactions through "slow release" is reduced. Equally, organotrifluoroborates that are very electron poor, *e.g.* 3-nitrophenyl, 3,5-trifluoromethylphenyl or alkynyltrifluoroborates, are so stable that hydrolytic activation often will not occur sufficiently fast for the cross-coupling. These species are therefore more likely to undergo direct transmetallation with palladium(II), which requires higher temperatures and catalyst loadings and longer reaction times.[63,70] The entire set of organotrifluoroborate reagents were categorized into three classes based on these observations. Class I reagents have a half-life of less than 1 h, class II between 1 and 24 h and class II over 24 h. As Class II organotrifluoroborates hydrolyse on the same time-scale as most SM couplings, the greatest benefits of the slow-release mode of reactivity are harnessed. The other two classes do not benefit from slow release; however, there are cases where fast release has been shown to be beneficial.[71]

The use of organotrifluoroborate salts in the SM coupling with arenediazonium salts has been reported to be an efficient process under anhydrous conditions.[72] This implies that no indirect activation is required and the trifluoroborate undergoes direct transmetallation. These SM couplings gave superior yields to the corresponding coupling with boronic acids, which is intuitive considering that the cationic palladium complex formed reacts more efficiently with the anionic trifluoroborate species than the neutral boronic acid[73] (Scheme 8.14). During the transmetallation or its precursor, hydroxide has been shown to bridge B with Pd more proficiently than fluoride,[65] hence arenediazonium salts in combination with a preformed trihydroxyborate may be particularly effective partners for SM coupling.

8.4.2 MIDA Boronates

N-Methyliminodiacetic acid (MIDA) can displace water from arylboronic acids to generate aryl MIDA boronates[74] (Scheme 8.15). Preparation of heteroaryl and ethynyl MIDA boronates can be achieved with organometallic reagents,[75,76] and a transmetallation approach between vinyl TMS and BBr$_3$ with subsequent trapping by the bis-sodium salt of MIDA gives excellent yields of the vinyl MIDA boronate.[77]

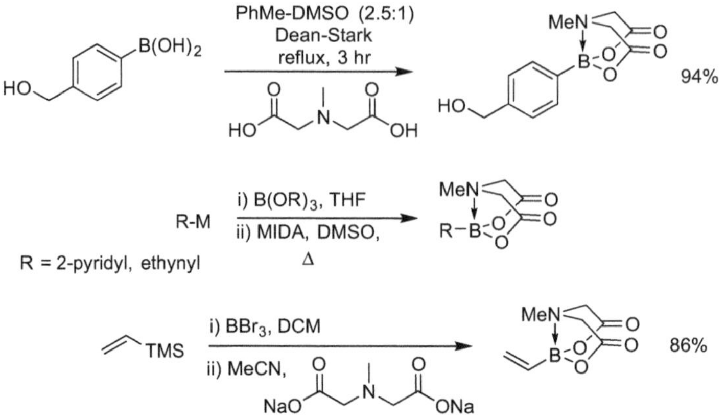

Scheme 8.15 Preparation of MIDA boronates for different substrate classes.

The quaternization at boron from nitrogen ligation usually renders the MIDA boronates free-flowing, crystalline solids that are stable to air, moisture and silica gel chromatography and can be stored without precaution "on the bench top,"[74] including the notorious 2-heterocyclic surrogates. The use of MIDA boronates in SM coupling was originally designed by Burke and co-workers for their use in iterative coupling processes, for which stability towards cross-coupling, yet ready liberation of the active species when required, is pivotal for success.[78] It has been proposed that the future of automated synthesis lies in iterative synthesis,[79] where bifunctional building blocks, with all the functionality and stereochemistry set, are coupled together using only one type of reaction. This is followed by a deprotection of latent functionality, activating the material towards further coupling and subsequent repetition. Much development is still required before iterative cross-coupling (ICC) becomes reality, but MIDA boronates have exhibited strong potential in this regard.[78,80] This is because they are stable to many reaction conditions but upon treatment with mild aqueous base the boronic acid is released for cross-coupling. The utility of MIDA boronates has been amply demonstrated in a number of natural product syntheses.[78,81,82]

Owing to the effective removal of the vacant p-orbital required for transmetallation (*via* either pathway), under anhydrous SM coupling conditions MIDA boronates are inert.[80] Dynamic ^{1}H NMR experiments have been employed to compare the stability of MIDA boronates with *N*-methyldiethanolamine boronates, which do undergo cross-coupling under anhydrous SM coupling conditions. No fluxionality was detected with the MIDA boronates, but significant broadening and shifting of the peaks was observed when the *N*-methyldiethanolamine boronates were heated, indicative that MIDA boronates are conformationally rigid whereas *N*-methyldiethanolamine boronates undergo conformational flipping (Scheme 8.16).[83]

MIDA boronates are remarkably resistant to a variety of reaction conditions and reagents, meaning that they can be carried through synthetic sequences

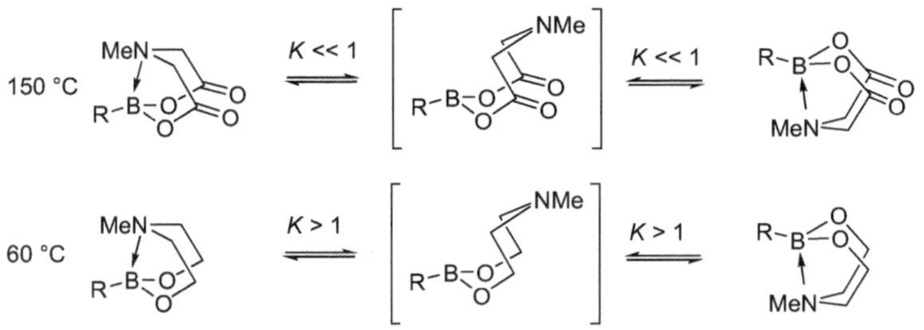

Scheme 8.16 Conformational rigidity in MIDA boronates, as determined by variable-temperature ^{1}H NMR spectroscopy.

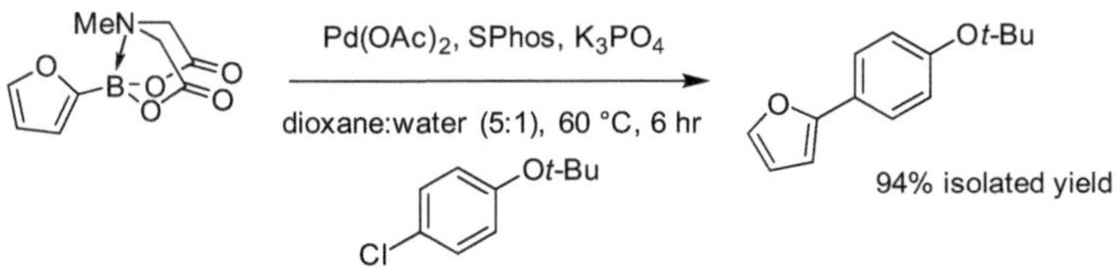

Scheme 8.17 SM coupling of a heteroaryl MIDA boronate under "slow-release" conditions.

with relative ease. They do have a few susceptibilities, however. These include protic or alcoholic solvents, in which they undergo slow solvolysis to liberate the boronic acid, although they are generally compatible with aqueous extractions employing water, brine, aqueous acid, *e.g.* NH$_4$Cl$_{(aq)}$, and even saturated NaHCO$_{3(aq)}$ in the absence of alcoholic solvents. They are not stable towards hard nucleophiles such as LiAlH$_4$, DIBAL, TBAF or metal alkoxides.[83] However, they are resistant to Swern, Dess–Martin and Jones oxidations,[83] and also iodination, Evans aldol and reductive amination reactions, Horner–Wadsworth–Evans and Takai olefinations and mild reducing agents. They are also stable towards Stille, Heck, Negishi and Sonogashira couplings, Grubbs metathesis and Miyaura borylation reactions.[78]

In a seminal publication, Burke and co-workers[84] showed that under optimized hydrolytic SM coupling conditions, MIDA boronates slowly liberate boronic acids, analogous to that occurring with organotrifluoroborates (see above), ensuring that the boronic acid concentration is maintained low enough to avoid extensive competitive side reactions, such as proto-deboronation, without this impacting on the rate of productive cross-coupling (Scheme 8.17). This slow-release strategy was demonstrated for a number of unstable (vinyl, cyclopropyl, heteroaryl) boronic acids, with excellent yields (76–99%) of the cross-coupled product, obtained solely by keeping the boronic acid at low concentration. Under "fast-release" conditions, 2-furyl MIDA boronate underwent coupling with comparable yield to the corresponding freshly prepared 2-furylboronic acid (68% *versus* 59%). Replication of the "slow-release" conditions, by syringe pump addition of 2-furylboronic acid over 3 h, restored the yield (94%) of the

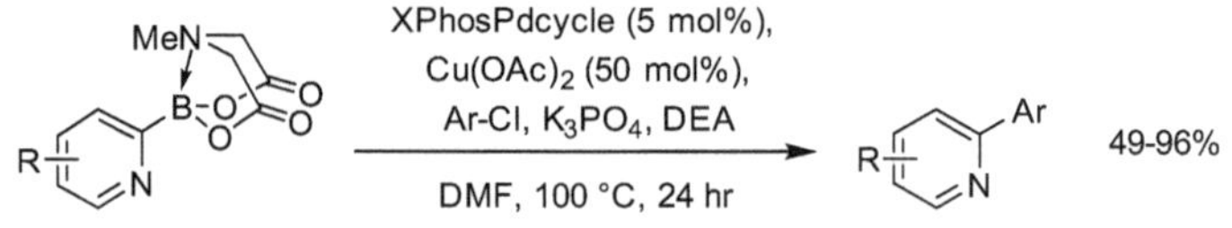

Scheme 8.18   SM coupling of 2-pyridyl MIDA boronate with aryl chlorides.

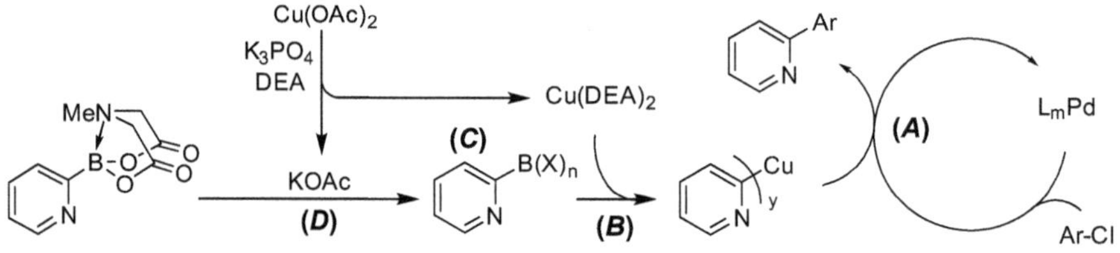

Scheme 8.19   Coupling of the 2-pyridyl moiety employing four strategies in concert to minimize degradation of the intermediate boronic acid and derivatives.

cross-coupled product to that when the MIDA boronate was employed, Scheme 8.17. The total synthesis of (+)-dictyosphaeric acid A provided excellent validation of the approach, in which a vinylic MIDA boronate, whose boronic acid can be unstable towards polymerization at high concentrations, was cross-coupled with an alkenyl iodide with an isolated yield of 82%.[85]

This work has also been extended to include the 2-pyridyl moiety, which was successfully (49–96%) cross-coupled with unactivated aryl chlorides (Scheme 8.18), arguably one of the most difficult aryl cross-couplings.[86]

Side reactions in SM coupling are not exclusively, but predominantly, centred around the boron reagent and a number of strategies for their attenuation have been developed.[68] The 2-pyridyl moiety is notoriously difficult to cross-couple (as the nucleophilic component) and the MIDA system applies four strategies in concert for effective coupling (Scheme 8.19):

A. *Catalyst activation* – Employing a rapidly activated precatalyst liberates an XPhos-ligated palladium(0) complex under the reaction conditions.[87] These electron-rich Buchwald catalyst systems shift the turnover-limiting step from oxidative addition towards transmetallation, thereby making the resting state Pd(II) rather than Pd(0), and accelerating transmetallation. Faster turnover then decreases the net reaction time, thus minimizing the extent of exposure of the boronic acid to the reaction medium.

B. *Boron reagent activation* – Silver[51–53] and copper[88] salts have been shown to increase the rate of transmetallation to palladium. Silver may effect more rapid halogen–hydroxide exchange on palladium to generate the key oxo-palladium intermediate, whereas copper appears to effect a more efficient transmetallation with boron. In the 2-pyridyl

MIDA boronate system, $Cu(OAc)_2$ in combination with diethanolamine was found to increase yields substantially and mechanistic studies indicated that a $Cu(DEA)_2$species is probably formed.

C. *Boron reagent masking* – The "masking" of the Lewis acidic boronic acids in unstable substrates *via* addition of more Lewis-basic ligands, *e.g.* alkoxides, has been shown to improve cross-coupling.[89] Diethanolamine can potentially coordinate to the 2-pyridylboronic acid following hydrolytic liberation from the MIDA boronate. This intermediate, or one involving acetate, can then undergo transmetallation with copper, prior to reaction with palladium(II).

D. *Slow release* – The release of the 2-pyridylboronic acid from the stable 2-pyridyl MIDA boronate can be tuned to be sufficiently slow to stop the boronic acid accumulating, thus reducing the extent of protodeboronation.

8.4.3   DAN Boronamides

1,8-Diaminonaphthalene (DAN) ligands and related systems have been developed primarily by Suginome's group[90–92] for iterative cross-coupling (ICC) and are readily prepared by condensing the free ligand with a boronic acid. Water is removed azeotropically in refluxing toluene (Scheme 8.20).[93]

The DAN-protected boronic acids exhibit particular stability due to electron donation from the very Lewis-basic nitrogen atoms. This strongly stabilizes the boron centre and the reagent is stable towards aqueous work-up and column chromatography. In contrast to organotrifluoroborate salts and MIDA boronates, DAN boronamides are stable towards basic conditions, which makes them unreactive under all SM coupling conditions. Deprotection is readily effected under mild acidic conditions (Scheme 8.21). Presumably protonation of nitrogen is necessary to weaken the B–N bond and liberate the p-orbital on boron for hydrolytic attack; equilibrium is then driven to the boronic acid *via* protonation of the liberated DAN ligand.

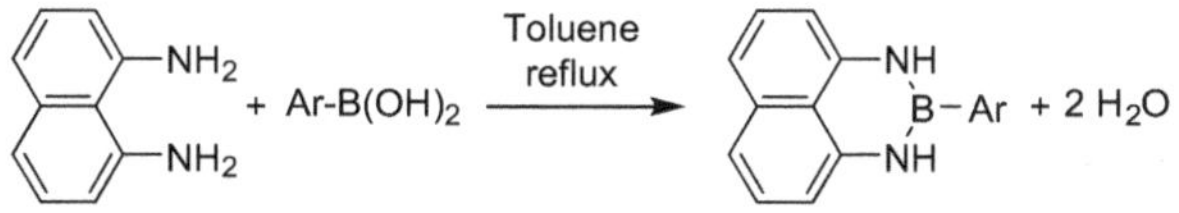

Scheme 8.20   Preparation of DAN boronamides from boronic acids.

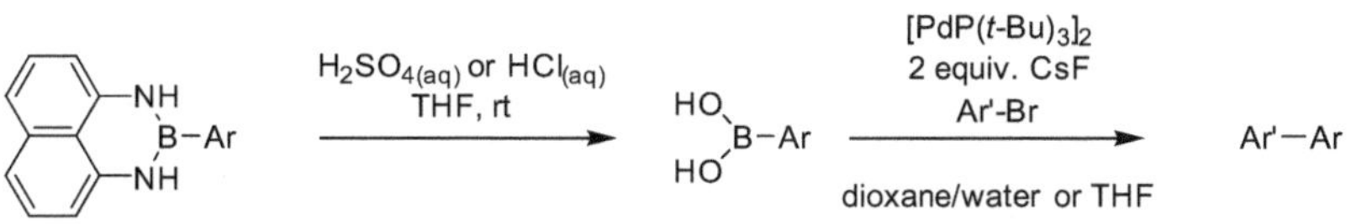

Scheme 8.21   Deprotection and cross-coupling of aryl DAN boronamides.

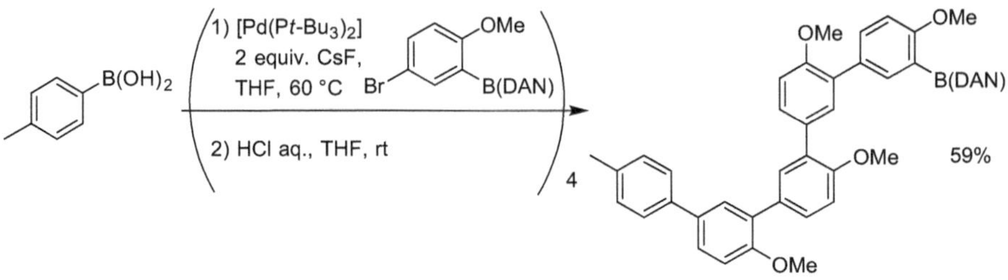

Scheme 8.22   Iterative cross-coupling to give a polyaromatic compound.

Bifunctional building blocks, composed of halide and DAN boronamide functional groups, are used for iterative cross-coupling (ICC). The halide motif undergoes the first coupling with a boronic acid; this can be conducted in aqueous basic conditions as the DAN boronamide present is stable towards these conditions. In contrast, the MIDA boronates require anhydrous coupling conditions to retain its functionality.[94] This step is then followed by an acidic deprotection of the DAN protecting group, which reveals the boronic acid that can then undergo further coupling. This was elegantly demonstrated through the synthesis of polyaromatic conjugated systems from simple bifunctional benzene-based building blocks (Scheme 8.22).[93] Up to four cycles of deprotection–cross-coupling were demonstrated.

8.5   Boron Reagents for Which the Activation Mode Is Yet To Be Established

Preformed tetrahedral borate salts and trigonal boronic esters are two classes of SM coupling reagents that undergo cross-coupling under regular SM coupling conditions, *i.e.* with no individual activating procedures, but for which there has not been sufficient mechanistic investigation to establish their specific activation mechanisms. In essence, it is not yet clear whether ligand substitution is required prior to transmetallation or whether the reagents can transmetallate directly.

8.5.1   Boronic Esters

Boronic esters tend to exhibit greater chemical stability than their corresponding boronic acids due to the lower Lewis acidity and greater steric shielding at boron. There are a number of methods to prepare boronic esters (Scheme 8.23) and the products are generally amenable to column chromatography. Hydroboration of unsaturated carbon bonds by pinacolborane (HBPin) or catecholborane (HBcat) is difficult, but with the addition of a suitable metal catalyst they readily add to alkenes[95,96] and alkynes.[97] Systems have been reported that employ non-precious metal catalysts[98–100] and even metal-free catalysts.[101,102] The method can also be conducted

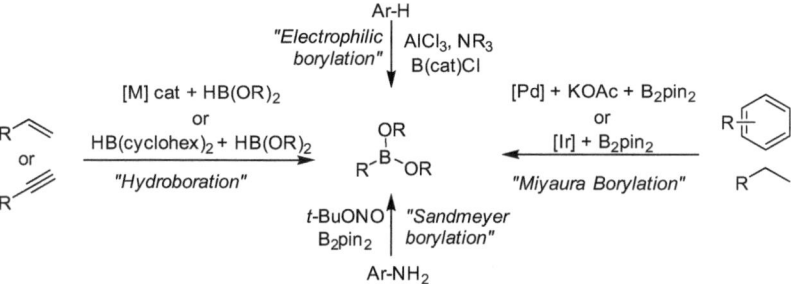

Scheme 8.23   Selected methods for the preparation of boronic esters.

enantioselectively,[103–105] with high regio-control for the unusual Markovnikov addition[106] and to give products from *trans*-hydroboration rather than the usual *cis* route.[107,108] The second major method used is a palladium-catalysed borylation of aryl halides[109,110] or alkenyl halides,[111] *i.e.* the Miyaura borylation. Iridium complexes can catalyse the borylation of arenes[112] and alkanes[113,114] and rhodium can also catalyse the borylation of arenes.[115] These reactions are conducted under remarkably mild conditions that made it very functional group tolerant. Boronic esters can also be prepared from arylamines using a recently developed method based on the Sandmeyer reaction[116,117] or *via* an electrophilic arene borylation protocol that employs a very strong Lewis base (Scheme 8.23).[118,119]

It is unclear at present what the precise active transmetallating species is during the SM coupling of boronic esters. Boronic esters have the capacity to behave analogously to boronic acids, whereby they transmetallate directly with an oxo-palladium species or become activated by hydroxide to undergo transmetallation *via* the *boronate* pathway. Alternatively, boronic esters could undergo complete or partial hydrolysis to form a more reactive species that can also transmetallate *via* the *oxo-palladium* or *boronate* pathways (Scheme 8.24). A complete mechanistic investigation to elucidate this aspect has not yet been reported. Carrow and Hartwig, however, demonstrated that catechol-, neopentyl- and pinacolboronic esters react rapidly, even at –55 °C, with a stoichiometric quantity of oxo-palladium species, generated *in situ* from the bridged hydroxy-palladium dimer and PPh$_3$.[47] Kinetic data showed that the catechol- and neopentylboronic esters reacted at a similar rate to the boronic acid but the bulky pinacolboronic ester was the slowest. However, all three reacted faster than the trihydroxyborate and a palladium halide complex, for which no reaction was detected over the same period even at –30 °C. It was therefore concluded that these boronic esters transmetallate *via* the *oxo-palladium* pathway. Measuring rates of hydrolysis and rates of transmetallation for the boronate species of a wide range of boronic esters would be a valuable addition in this area.

Pinacolboronic esters can be coupled under nominally anhydrous conditions.[120] However, the presence of trace amounts of adventitious water, or indeed small proportions of deliberately added water,[89,121] a common

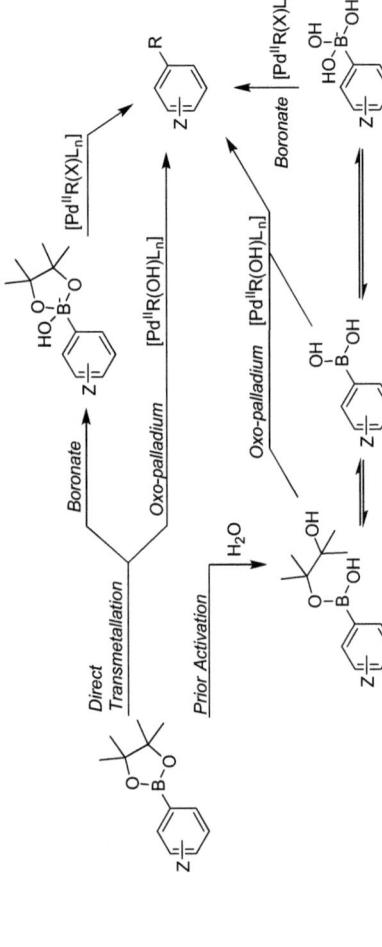

Scheme 8.24 Possible transmetallation mechanisms for the coupling of an arylpinacolboronic ester.

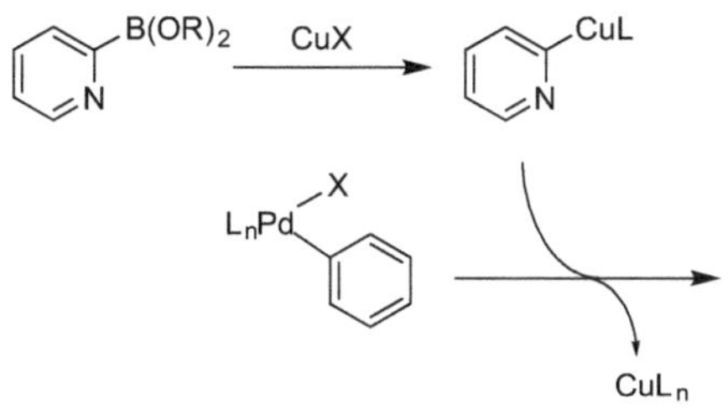

Scheme 8.25 Copper-assisted transmetallation.

feature in SM couplings, means that conclusions regarding the mode of transmetallation must be drawn with caution: water may effect a prior hydrolysis of the boronic ester or facilitate the generation of the oxo-palladium(II) intermediate.

Copper salts have been reported to increase the rate of transmetallation of SM coupling reactions.[88,122] In these cases, it is proposed that the boronic acid initially reacts with the copper salt, which then delivers an organo-copper reagent to palladium more efficiently than the boronic ester. The addition of copper salts was shown to be an effective way to reduce proto-deboronation in the coupling of pinacolheteroarylboronic esters[88] (Scheme 8.25).

8.5.2 Trihydroxyborate Salts

Sodium aryl trihydroxyborate salts can be efficiently prepared by dropwise addition of a saturated aqueous sodium hydroxide solution to the parent boronic acid in toluene; the product is isolated simply by filtration. These trihydroxyborate salts have been reported to undergo clean SM coupling under nominally base-free conditions (Scheme 8.26).[123] This may suggest that the reaction proceeds by direct transmetallation, *i.e.* through the *boronate* pathway. However, it should be noted that the solubility of the trihydroxyborate salts in dry toluene is low. Indeed, this is a key aspect of their preparation. Equilibrium between the trihydroxyborate salt and the boronic acid and sodium hydroxide must also be considered and, moreover, under nominally anhydrous conditions, dehydration of boronic acid to the corresponding boroxine liberates water (one per boron), thus potentially solubilizing the sodium hydroxide and facilitating the formation of the oxo-palladium intermediate. In applications of trihydroxyborate salts to SM couplings under aqueous conditions,[124] the liberation of an equivalent of base plus the boronic acid is even more likely.

8.5.3 Triisopropylborate Salts

Lithium aryl triisopropylborate reagents have been successfully applied to the SM coupling of unstable heteroaryl systems. The triisopropylborates are

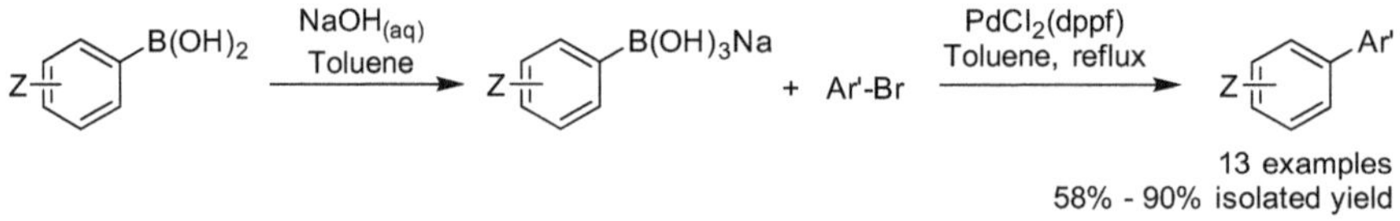

Scheme 8.26 SM coupling of aryl trihydroxyborate salts with aryl bromides.

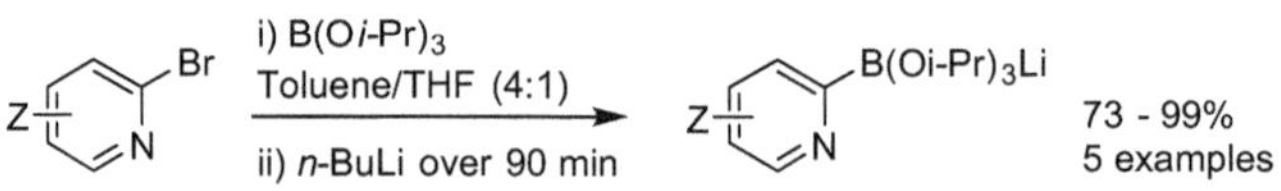

Scheme 8.27 Preparation of lithium triisopropylborate salts.

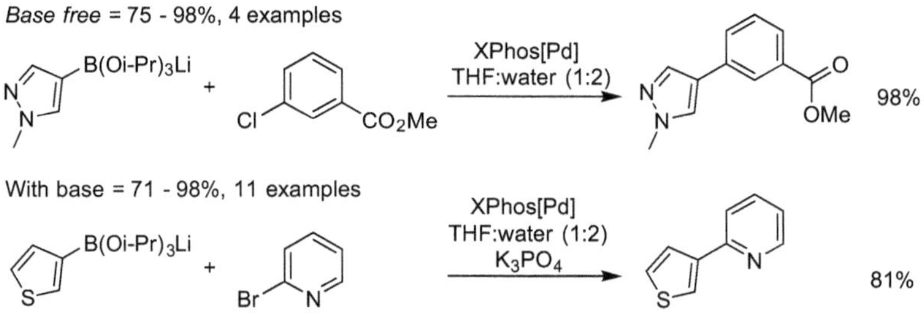

Scheme 8.28 SM coupling of lithium heteroaryl triisopropylborate salts.

prepared from the borylation of organohalides, following lithium–halogen exchange by treatment with BuLi. The product is isolated simply by evaporating the solvent and bromobutane co-product (Scheme 8.27).

Phosphine oxide ligands were originally employed[125] for the SM coupling of the notoriously difficult 2-pyridyl systems, but the introduction of the XPhos precatalyst (generation II) has substantially expanded the substrate scope and afforded conditions that are general for the coupling of heteroarylborates (Scheme 8.28).[126] A key observation is that under anhydrous conditions, no coupling of the triisopropylborate reagent was observed, suggesting that hydrolytic reagent activation is required for transmetallation. Although SM coupling of triisopropylborate reagents does proceed without any added base and was shown to be effective for base-sensitive organohalide coupling partners, such as methyl esters or oxazoles,[126] the pH of these couplings (THF–water, no added base) (Scheme 8.28) is between 12 and 13 and strongly suggests liberation of 2-propanol and hydroxide. The addition of potassium phosphate and CuCl in combination with $ZnCl_2$ led to improved yields, although the origins of these effects have not been investigated.[127]

8.5.4 Cyclic Triol Borate Salts

Cyclic triol borates are tetrahedral at boron as they are ligated by three tethered alkoxides.[128] They are prepared by condensing the triol with the

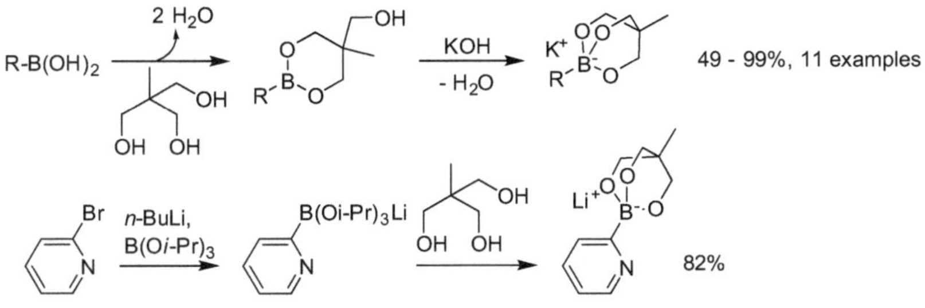

Scheme 8.29 Methods for preparing the potassium and lithium salts of cyclic triol borates.

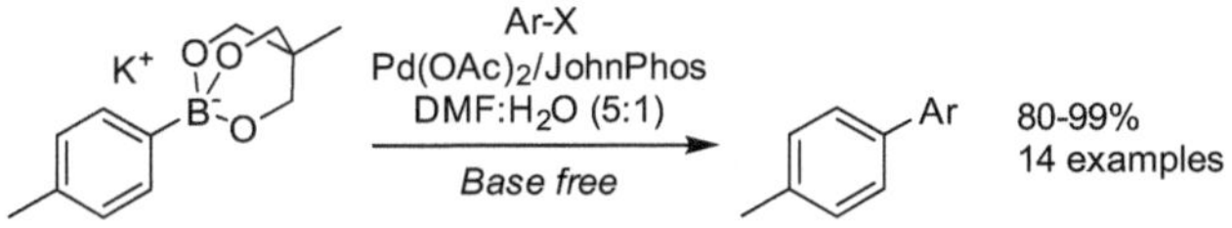

Scheme 8.30 SM coupling of cyclic triol borates with aryl halides.

parent boronic acid, with 2 equiv. of liberated water being removed azeo-tropically. Pyramidalization is then achieved by treatment with KOH to furnish the potassium salt of the borate. The lithium salt is readily accessed by the organometallic pathway (Scheme 8.29).

Despite the stability gained from blocking the vacant p-orbital on boron, the nucleophilicity of the organic moiety is the highest of all the SM coupling boron reagents. This was illustrated by comparing the rates of their metal-free reaction with the benzhydrylium ion electrophile.[129] The cyclic triol-borate salt reacted 11 orders of magnitude faster than MIDA boronates, which were the least nucleophilic. This analysis may indicate that hydrolysis prior to transmetallation is not necessary, as the boron reagent is reactive enough to proceed through the *boronate* pathway. However, although cross-coupling was achieved under base-free conditions (Scheme 8.30), the aqueous DMF solvent mixture employed still has the potential for prior hydrolysis to liberate potassium hydroxide and thus reaction could proceed through an oxo-palladium intermediate or through a tethered palladium alkoxy species.

Other sterically congested systems were prepared using these coupling partners, also under base-free conditions.[130,131] Optimized conditions used for the tetra-*ortho*-substituted biaryl preparation involved an anhydrous/base-free DMF system with a copper(I) co-catalyst. Under these conditions, it is difficult to envisage significant prior hydrolysis of the cyclic triol borate salt. However, formation of an arylcopper species with the nucleophilic boronate may be facile, which then undergoes reaction with palladium(II). Cyclic triol borates have been compared with boronic acids in the SM coupling of dibromoarenes and found to provide superior yields of the double-arylation product (Scheme 8.31).[131] The presence of a copper salt (CuCl) was again found to be key for the generation of the sterically

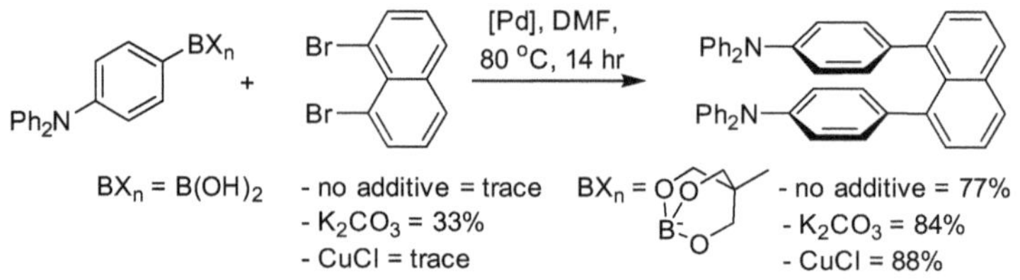

BXn = B(OH)2   - no additive = trace   BXn = ⟨boron triolate structure⟩   - no additive = 77%
                - K2CO3 = 33%   - K2CO3 = 84%
                - CuCl = trace   - CuCl = 88%

Scheme 8.31   SM coupling of both cyclic triolates and boronic acids with dibromonaphthalene.

congested aryl systems. In contrast to the couplings outlined above (Scheme 8.30), the addition of base (K_2CO_3, 2 equiv.) was found to be beneficial, suggesting that prior hydrolytic activation may be necessary.

Interestingly, the triol unit can be tailored to provide the requisite physical or chemical properties. For example, when the bridging carbon bears a methyl group, the reagents are more soluble in organic solvents than organotrifluoroborate salts, and when it bears a polar sulfonate group, the solubility in aqueous solutions becomes appreciable.[132]

8.6   Conclusion

Overall, it is concluded that the boron reagents used in SM coupling can be categorized into three distinct groups (Figure 8.5). It must be emphasized that such a categorization is subject to future mechanistic studies that will undoubtedly reveal further information necessary to refine it.

For organoboranes, such as 9-BBN-borane and boronic acids, it is clear that direct transmetallation can occur, either through reaction of the borane or boronic acid with an oxo-palladium intermediate or by simple association of hydroxide to form a boronate species that reacts with a palladium(II) halide complex or a cationic palladium(II) complex. With their high Lewis acidity and low steric hindrance around the boron, boronic acids are the most reactive boron reagent for transmetallation in SM coupling.

The next class of reagents includes organotrifluoroborate salts, MIDA boronates and DAN boronamides, where it is now established that in the majority of cases prior activation must occur for transmetallation to proceed. The reagents themselves are generally very unreactive towards palladium(II), hence ligand exchange has to occur to prime them for efficient transmetallation by either the *oxo-palladium* or the *boronate* pathway. For the particular reagents detailed in this chapter, the activation is hydrolytic. Organotrifluoroborate hydrolysis is catalysed by acid, but also proceeds to completion under basic aqueous conditions due to the equilibrium being driven toward the boronic acid by consumption of the liberated HF. Electron-poor aryltrifluoroborates and alkynyltrifluoroborates hydrolyse more slowly than electron-rich aryltrifluoroborates and alkyltrifluoroborates. In fact, there are instances where their hydrolysis is so slow that direct transmetallation occurs and therefore should be placed in the direct

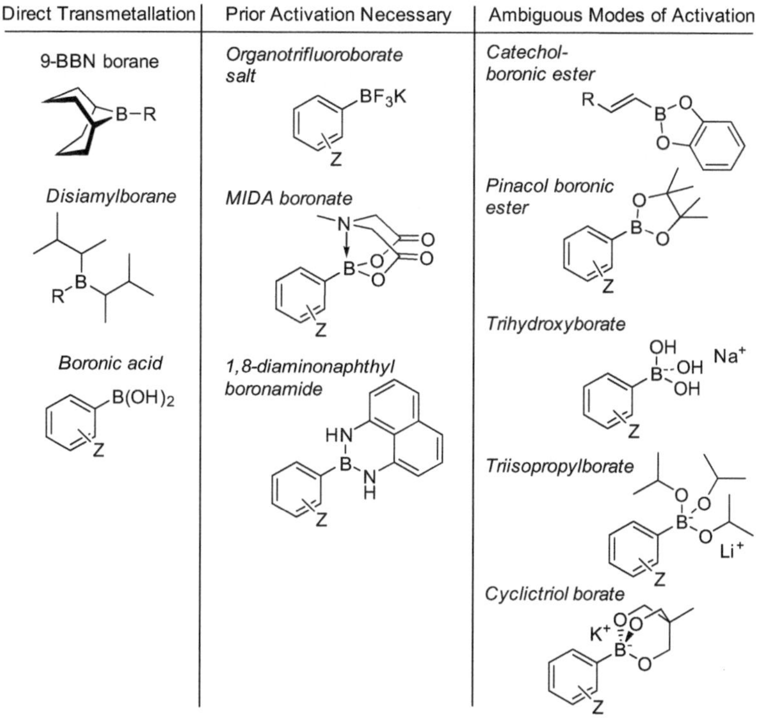

Direct Transmetallation	Prior Activation Necessary	Ambiguous Modes of Activation
9-BBN borane	*Organotrifluoroborate salt*	*Catechol-boronic ester*
Disiamylborane	*MIDA boronate*	*Pinacol boronic ester*
Boronic acid	*1,8-diaminonaphthyl boronamide*	*Trihydroxyborate*
		Triisopropylborate
		Cyclictriol borate

Figure 8.5 Categorization of boron reagents in SM coupling into three groups: those that transmetallate directly, those that require prior ligand substitution and those whose activation mechanisms are ambiguous.

transmetallation category. MIDA boronates undergo activation under mild aqueous basic conditions [NaOH$_{(aq)}$, THF]. The reaction rate is rapid but can be slowed by changing the base and solvent system [K$_3$PO$_4$, dioxane–water (5 : 1)]. This rate variation between R groups is very much less sensitive than with organotrifluoroborate salts and these reagents are therefore amenable to the cross-coupling of a wide range of unstable boronic acids using a simple and reliable method. Complementary to MIDA boronates are DAN boronamides, which are stable to basic conditions but rapidly liberate the boronic acid under acidic conditions.

The final category of reagents includes boronic esters and borate salts and here there is currently ambiguity surrounding whether activation occurs prior to transmetallation. Despite numerous clues throughout the chemical literature, insufficient mechanistic studies have been conducted to conclude accurately which category they belong to. Borate salts have the capacity to be included in either category and it may depend on the nature of the particular ligand on boron. Indeed, there may also be variation within a class of boron reagent, *e.g.* some especially electron-poor aromatic pinacolboronic esters may undergo direct transmetallation with an oxo-palladium(II) species due to the increased Lewis acidity at boron, whereas other electron-rich aromatic pinacolboronic esters may require prior hydrolysis to the more reactive boronic acid. The results of such a mechanistic study are expected to be highly interesting and useful for the future development of their chemistry.

References

1. R. F. Heck and J. P. Nolley, *J. Org. Chem.*, 1972, **37**, 2320.
2. I. P. Beletskaya and A. V. Cheprakov, *Chem. Rev.*, 2000, **100**, 3009.
3. R. Chinchilla and C. Najera, *Chem. Rev.*, 2007, **107**, 874.
4. M. Kumada, K. Tamao and K. Sumitani, *J. Am. Chem. Soc.*, 1972, **94**, 4374.
5. E. Negishi, A. O. King and N. Okukado, *J. Org. Chem.*, 1977, **42**, 1821.
6. J. K. Stille, *Angew. Chem. Int. Ed. Engl.*, 1986, **25**, 508.
7. Y. Hatanaka and T. Hiyama, *J. Org. Chem.*, 1988, **53**, 918.
8. N. Miyaura and A. Suzuki, *Chem. Rev.*, 1995, **95**, 2457.
9. (a) C. C. C. Johansson Seechurn, M. O. Kitching, T. J. Colacot and V. Snieckus, *Angew. Chem. Int. Ed.*, 2012, **51**, 5062; (b) H. Li, C. C. C. Johansson Seechurn and T. J. Colacot, *ACS Catal.*, 2012, **2**, 1147.
10. A. F. Littke and G. C. Fu, *Angew. Chem. Int. Ed.*, 2002, **41**, 4176.
11. G. Altenhoff, R. Goddard, C. W. Lehmann and F. Glorius, *J. Am. Chem. Soc.*, 2004, **126**, 15195.
12. J. P. Wolfe, R. A. Singer, B. H. Yang and S. L. Buchwald, *J. Am. Chem. Soc.*, 1999, **121**, 9550.
13. A. Alimardanov, L. Schmieder-van de Vondervoort, A. H. M. De Vries and J. G. De Vries, *Adv. Synth. Catal.*, 2004, **346**, 1812.
14. G. C. Fu, A. F. Littke and C. Dai, *J. Am. Chem. Soc.*, 2000, **122**, 4020.
15. D. Zim, A. S. Gruber, G. Ebeling, J. Dupont and A. L. Monteiro, *Org. Lett.*, 2000, **2**, 2881.
16. (a) T. W. J. Cooper, I. B. Campbell and S. J. F. Macdonald, *Angew. Chem. Int. Ed.*, 2010, **49**, 8082; (b) see also *e.g.* the proportion of biaryls present in the best-selling five-membered ring heterocyclic pharmaceuticals: M. Baumann, I. R. Baxendale, S. V. Ley and N. Nikbin, *Beilstein J. Org. Chem.*, 2011, 7, 442.
17. N. Miyaura and A. Suzuki, *J. Chem. Soc., Chem. Commun.*, 1979, 866.
18. N. Miyaura, K. Yamada and A. Suzuki, *Tetrahedron Lett.*, 1979, **20**, 3437.
19. N. Miyaura, T. Yanagi and A. Suzuki, *Synth. Commun.*, 1981, **11**, 513.
20. (a) A. Suzuki, *J. Organomet. Chem.*, 1999, **576**, 147; (b) N. Miyaura, *J. Organomet. Chem.*, 2002, **653**, 54; (c) A. J. J. Lennox and G. C. Lloyd-Jones, *Angew. Chem. Int Ed.*, 2013, **52**, 7362.
21. See *e.g.* (a) T. Kamikawa and T. Hayashi, *Tetrahedron Lett.*, 1997, **38**, 7087; (b) G. Espino, A. Kurbangalieva and J. M. Brown, *Chem. Commun.*, 2007, 1742.
22. (a) C. Amatore, A. Jutand and G. Le Duc, *Chem. Eur. J.*, 2011, **17**, 2492; (b) C. Amatore, G. Le Duc and A. Jutand, *Chem. Eur. J.*, 2013, **19**, 10082.
23. R. S. Dhillon, *Hydroboration and Organic Synthesis*, Springer, Berlin, 2007.

24. H. C. Brown, *Hydroboration*, Wiley-Interscience, New York, 1962.
25. I. Beletskaya and A. Pelter, *Tetrahedron*, 1997, **53**, 4957.
26. P. C. Keller, *Synth. React. Inorg. Met.-Org. Chem.*, 1976, **6**, 77.
27. A. Pelter, K. Smith and H. C. Brown, *Borane Reagents*, Academic Press, New York, 1988.
28. K. Burgess and M. J. Ohlmeyer, *Chem. Rev.*, 1991, **91**, 1179.
29. E. R. Burkhardt and K. Matos, *Chem. Rev.*, 2006, **106**, 2617.
30. V. V. Grushin and H. Alper, *Organometallics*, 1996, **15**, 5242.
31. K. Matos and J. A. Soderquist, *J. Org. Chem.*, 1998, **63**, 461.
32. A. J. J. Lennox and G. C. Lloyd-Jones, *Angew. Chem. Int. Ed.*, 2013, **52**, 7362.
33. N. Miyaura, K. Yamada, H. Suginome and A. Suzuki, *J. Am. Chem. Soc.*, 1985, **107**, 972.
34. A. Suzuki, *J. Organomet. Chem.*, 1999, **576**, 147.
35. S. O. Lawesson, *Acta Chem. Scand.*, 1957, **11**, 1075.
36. E. Khotinsky and M. Melamed, *Ber. Dtsch. Chem. Ges.*, 1909, **42**, 3090.
37. D. C. Gerbino, S. D. Mandolesi, H.-G. Schmalz and J. C. Podestá, *Eur. J. Org. Chem.*, 2009, 3964.
38. C. A. Hutton and A. K. L. Yuen, *Tetrahedron Lett.*, 2005, **46**, 7899.
39. H. Nakamura, M. Fujiwara and Y. Yamamoto, *J. Org. Chem*, 1998, **63**, 7529.
40. G. A. Molander, S. L. J. Trice and S. D. Dreher, *J. Am. Chem. Soc.*, 2010, **132**, 17701.
41. (a) G. A. Molander, S. L. J. Trice, S. M. Kennedy, S. D. Dreher and M. T. Tudge, *J. Am. Chem. Soc.*, 2012, **134**, 11667; (b) G. A. Molander, L. N. Cavalcanti and C. García-García, *J. Org. Chem.*, 2013, **78**, 6427.
42. R. Shintani, M. Takeda, T. Nishimura and T. Hayashi, *Angew. Chem. Int. Ed.*, 2010, **49**, 3969.
43. (a) A. A. C. Braga, N. H. Morgon, G. Ujaque and F. Maseras, *J. Am. Chem. Soc.*, 2005, **127**, 9298; (b) A. A. C. Braga, N. H. Morgon, G. Ujaque, A. Lledós and F. Maseras, *J. Organomet. Chem.*, 2006, **691**, 4459.
44. R. Glaser and N. Knotts, *J. Phys. Chem. A*, 2006, **110**, 1295.
45. V. V. Grushin and H. Alper, *Organometallics*, 1993, **12**, 1890.
46. C. Amatore, A. Jutand and G. Le Duc, *Angew. Chem. Int. Ed.*, 2012, **51**, 1379.
47. B. P. Carrow and J. F. Hartwig, *J. Am. Chem. Soc.*, 2011, **133**, 2116.
48. A. F. Schmidt, A. A. Kurokhtina and E. V. Larina, *Russ. J. Gen. Chem*, 2011, **81**, 1573.
49. A. J. J. Lennox and G. C. Lloyd-Jones, *J. Am. Chem. Soc.*, 2012, **134**, 7431.
50. C. Amatore, A. Jutand and G. Le Duc, *Chem. Eur. J.*, 2012, **18**, 6616.
51. N. Y. Adonin, D. E. Babushkin, V. N. Parmon, V. V. Bardin, G. A. Kostin, V. I. Mashukov and H.-J. Frohn, *Tetrahedron*, 2008, **64**, 5920.
52. J. Chen and A. Cammers-Goodwin, *Tetrahedron Lett.*, 2003, **44**, 1503.
53. T. Korenaga, T. Kosaki, R. Fukumura, T. Ema and T. Sakai, *Org. Lett.*, 2005, 7, 4915.
54. Y. Nishihara, H. Onodera and K. Osakada, *Chem. Commun.*, 2004, 192.

55. G. Zou, Y. K. Reddy and J. R. Falck, *Tetrahedron Lett.*, 2001, **42**, 7213.

56. J. Uenishi, J. M. Beau, R. W. Armstrong and Y. Kishi, *J. Am. Chem. Soc.*, 1987, **109**, 4756.

57. I. E. Markó, F. Murphy and S. Dolan, *Tetrahedron Lett.*, 1996, **37**, 2507.

58. R. D. Chambers, H. C. Clark and C. J. Willis, *J. Am. Chem. Soc.*, 1960, **82**, 5298.

59. E. Vedejs, R. W. Chapman, S. C. Fields, S. Lin and M. R. Schrimpf, *J. Org. Chem.*, 1995, **60**, 3020.

60. A. J. J. Lennox and G. C. Lloyd-Jones, *Angew. Chem. Int. Ed.*, 2012, **51**, 9385.

61. D. G. Hall, *Structure, Properties and Preparation of Boronic Acid Derivatives. Overview of Their Reactions and Applications*, Wiley-VCH, Weinheim, 2005.

62. G. A. Molander and B. Canturk, *Angew. Chem. Int. Ed.*, 2009, **48**, 9240.

63. (a) G. A. Molander and T. Ito, *Org. Lett.*, 2001, **3**, 393; (b) G. A. Molander and B. Biolatto, *J. Org. Chem.*, 2003, **68**, 4302.

64. R. A. Batey and T. D. Quach, *Tetrahedron Lett.*, 2001, **42**, 9099.

65. M. Butters, J. N. Harvey, J. Jover, A. J. J. Lennox, G. C. Lloyd-Jones and P. M. Murray, *Angew. Chem. Int. Ed.*, 2010, **49**, 5156.

66. J. G. Verkade and M. R. Mason, *Organometallics*, 1990, **9**, 864.

67. J. G. Verkade and P. A. McLaughlin, *Organometallics*, 1998, **17**, 5937.

68. A. J. J. Lennox and G. C. Lloyd-Jones, *Isr. J. Chem.*, 2010, **50**, 664.

69. W. Ren, J. Li, D. Zou, Y. Wu and Y. Wu, *Tetrahedron*, 2012, **68**, 1351.

70. G. A. Molander, B. W. Katona and F. Machrouhi, *J. Org. Chem.*, 2002, **67**, 8416.

71. J.-K. Lee, M. C. Gwinner, R. Berger, C. Newby, R. Zentel, R. H. Friend, H. Sirringhaus and C. K. Ober, *J. Am. Chem. Soc.*, 2011, **133**, 9949.

72. S. Darses, J.-P. Genêt, J.-L. Brayer and J.-P. Demoute, *Tetrahedron Lett.*, 1997, **38**, 4393.

73. P. Mastrorilli, N. Taccardi, R. Paolillo, V. Gallo, C. F. Nobile, M. Räisänen and T. Repo, *Eur. J. Inorg. Chem.*, 2007, 4645.

74. E. P. Gillis and M. D. Burke, *Aldrichim. Acta*, 2009, **42**, 17.

75. G. R. Dick, D. M. Knapp, E. P. Gillis and M. D. Burke, *Org. Lett.*, 2010, **12**, 2314.

76. J. R. Struble, S. J. Lee and M. D. Burke, *Tetrahedron*, 2010, **66**, 4710.

77. B. E. Uno, E. P. Gillis and M. D. Burke, *Tetrahedron*, 2009, **65**, 3130.

78. S. J. Lee, K. C. Gray, J. S. Paek and M. D. Burke, *J. Am. Chem. Soc.*, 2008, **130**, 466.

79. C. Wang and F. Glorius, *Angew. Chem. Int. Ed.*, 2009, **48**, 5240.

80. E. P. Gillis and M. D. Burke, *J. Am. Chem. Soc.*, 2007, **129**, 6716.

81. E. M. Woerly, A. H. Cherney, E. K. Davis and M. D. Burke, *J. Am. Chem. Soc.*, 2010, **132**, 6941.

82. S. J. Lee, T. M. Anderson and M. D. Burke, *Angew. Chem. Int. Ed.*, 2010, **49**, 8860.

83. E. P. Gillis and M. D. Burke, *J. Am. Chem. Soc.*, 2008, **130**, 14084.

84. D. M. Knapp, E. P. Gillis and M. D. Burke, *J. Am. Chem. Soc.*, 2009, **131**, 6961.

85. A. R. Burns, G. D. McAllister, S. E. Shanahan and R. J. K. Taylor, *Angew. Chem. Int. Ed.*, 2010, **49**, 5574.

86. G. R. Dick, E. M. Woerly and M. D. Burke, *Angew. Chem. Int. Ed.*, 2012, **51**, 2667.

87. T. Kinzel, Y. Zhang and S. L. Buchwald, *J. Am. Chem. Soc.*, 2010, **132**, 14073.

88. J. Z. Deng, D. V. Paone, A. T. Ginnetti, H. Kurihara, S. D. Dreher, S. A. Weissman, S. R. Stauffer and C. S. Burgey, *Org. Lett.*, 2009, **11**, 345.

89. D. W. Robbins and J. F. Hartwig, *Org. Lett.*, 2012, **14**, 4266.

90. H. Ihara, M. Koyanagi and M. Suginome, *Org. Lett.*, 2011, **13**, 2662.

91. H. Ihara and M. Suginome, *J. Am. Chem. Soc.*, 2009, **131**, 7502.

92. M. Koyanagi, N. Eichenauer, H. Ihara, T. Yamamoto and M. Suginome, *Chem. Lett.*, 2013, **42**, 541.

93. H. Noguchi, K. Hojo and M. Suginome, *J. Am. Chem. Soc.*, 2007, **129**, 758.

94. N. Iwadate and M. Suginome, *Org. Lett.*, 2009, **11**, 1899.

95. D. Mannig and H. Noth, *Angew. Chem. Int. Ed. Engl.*, 1985, **24**, 878.

96. X. He and J. F. Hartwig, *J. Am. Chem. Soc.*, 1996, **118**, 1696.

97. S. Pereira and M. Srebnik, *Organometallics*, 1995, **14**, 3127.

98. L. Zhang, D. Peng, X. Leng and Z. Huang, *Angew. Chem. Int. Ed.*, 2013, **52**, 3676.

99. H. R. Kim, I. G. Jung, K. Yoo, K. Jang, E. S. Lee, J. Yun and S. U. Son, *Chem. Commun.*, 2010, **46**, 758.

100. X. Feng, H. Jeon and J. Yun, *Angew. Chem. Int. Ed.*, 2013, **52**, 3989.

101. A. Arase, M. Hoshi, A. Mijin and K. Nishi, *Synth. Commun.*, 1995, **25**, 1957.

102. K. Shirakawa, A. Arase and M. Hoshi, *Synthesis*, 2004, 1814.

103. C. M. Crudden and D. Edwards, *Eur. J. Org. Chem.*, 2003, 4695.

104. A.-M. Carroll, T. O'sullivan and P. Guiry, *Adv. Synth. Catal.*, 2005, **347**, 609.

105. R. Corberán, N. W. Mszar and A. H. Hoveyda, *Angew. Chem. Int. Ed.*, 2011, **50**, 7079.

106. H. Jang, A. R. Zhugralin, Y. Lee and A. H. Hoveyda, *J. Am. Chem. Soc.*, 2011, **133**, 7859.

107. T. Ohmura, Y. Yamamoto and N. Miyaura, *J. Am. Chem. Soc.*, 2000, **122**, 4990.

108. C. Gunanathan, M. Hölscher, F. Pan and W. Leitner, *J. Am. Chem. Soc.*, 2012, **134**, 14349.

109. T. Ishiyama, M. Murata and N. Miyaura, *J. Org. Chem.*, 1995, **60**, 7508.

110. T. Ishiyama, K. Ishida and N. Miyaura, *Tetrahedron*, 2001, **57**, 9813.

111. J. Takagi, K. Takahashi, T. Ishiyama and N. Miyaura, *J. Am. Chem. Soc.*, 2002, **124**, 8001.

112. T. Ishiyama, J. Takagi, K. Ishida, N. Miyaura, N. R. Anastasi and J. F. Hartwig, *J. Am. Chem. Soc.*, 2002, **124**, 390.

113. H. Chen and J. Hartwig, *Angew. Chem. Int. Ed.*, 1999, **38**, 3391.
114. C. N. Iverson and M. R. Smith, *J. Am. Chem. Soc.*, 1999, **121**, 7696.
115. S. Shimada, A. S. Batsanov, J. A. K. Howard and T. B. Marder, *Angew. Chem. Int. Ed.*, 2001, **40**, 2168.
116. F. Mo, Y. Jiang, D. Qiu, Y. Zhang and J. Wang, *Angew. Chem. Int. Ed.*, 2010, **49**, 1846.
117. D. Qiu, L. Jin, Z. Zheng, H. Meng, F. Mo, X. Wang, Y. Zhang and J. Wang, *J. Org. Chem.*, 2013, **78**, 1923.
118. M. J. Ingleson, *Synlett*, 2012, **23**, 1411.
119. A. Del Grosso, M. D. Helm, S. A. Solomon, D. Caras-Quintero and M. J. Ingleson, *Chem. Commun.*, 2011, **47**, 12459.
120. S. Fujii, S. Y. Chang and M. D. Burke, *Angew. Chem. Int. Ed.*, 2011, **50**, 7862.
121. T. E. Barder, S. D. Walker, J. R. Martinelli and S. L. Buchwald, *J. Am. Chem. Soc.*, 2005, **127**, 4685.
122. N. Leconte, A. Keromnes-Wuillaume, F. Suzenet and G. Guillaumet, *Synlett*, 2007, 204.
123. A. N. Cammidge, V. H. M. Goddard, H. Gopee, N. L. Harrison, D. L. Hughes, C. J. Schubert, B. M. Sutton, G. L. Watts and A. J. Whitehead, *Org. Lett.*, 2006, **8**, 4071.
124. B. Basu, K. Biswas, S. Kundu and S. Ghosh, *Green Chem.*, 2010, **12**, 1734.
125. K. L. Billingsley and S. L. Buchwald, *Angew. Chem. Int. Ed.*, 2008, **47**, 4695.
126. M. A. Oberli and S. L. Buchwald, *Org. Lett.*, 2012, **14**, 4606.
127. K. Chen, R. Peterson, S. K. Math, J. B. LaMunyon, C. A. Testa and D. R. Cefalo, *Tetrahedron Lett.*, 2012, **53**, 4873.
128. Y. Yamamoto, M. Takizawa, X.-Q. Yu and N. Miyaura, *Angew. Chem. Int. Ed.*, 2008, **47**, 928.
129. G. Berionni, B. Maji, P. Knochel and H. Mayr, *Chem. Sci.*, 2012, **3**, 878.
130. G.-Q. Li, Y. Yamamoto and N. Miyaura, *Synlett*, 2011, 1769.
131. G.-Q. Li, Y. Yamamoto and N. Miyaura, *Tetrahedron*, 2011, **67**, 6804.
132. D. S. Matteson and H.-W. Man, *J. Org. Chem.*, 1996, **61**, 6047.

CHAPTER 9

Modern Heck Reactions

IRINA P. BELETSKAYA* AND ANDREI V. CHEPRAKOV

Department of Chemistry, Moscow State University, Moscow, 119991,
Russia
*Email: beletska@org.chem.msu.ru

9.1   Introduction

In recent years, the chemistry owing its birth to Richard Heck, joint winner
of the Nobel Prize in Chemistry in 2010, has shown vigorous advances. In
fact, the name of Heck is linked primarily with the now classical Mizoroki–
Heck reaction, and the book[1] that appeared just before the Nobel award is
entitled, rightfully so, *The Mizoroki–Heck Reaction*. However, Heck himself
developed not one, but rather a whole family of transition metal-catalysed
transformations, only one of which is the Mizoroki–Heck reaction. The development of ideas contained in the papers by Heck and his contemporaries
gave rise to a whole new field of chemistry, mediated by transition metal
intermediates, involving both truly catalytic and stoichiometric processes,
and extremely flexible with respect to substrates, precatalysts, reaction
media, synthetic tools and goals achieved, but based on the same fundamental elementary process, the carbometallation of multiple bonds. Such
processes can be referred to as the generic Heck reaction.

   The history of Heck chemistry begins not with the development of catalytic arylation driven by the Pd(0)/Pd(II) catalytic cycle, but instead with
stoichiometric reactions with a Pd(II) reagent, initially understood as an
electrophile, similar to such well-known metallic electrophiles as Hg(II)
salts. The chemistry was initiated and developed simultaneously by Heck
working for Hercules Inc. and Fujiwara, Moritani and co-workers in Japan.

RSC Catalysis Series No. 21
New Trends in Cross-Coupling: Theory and Applications
Edited by Thomas J Colacot
© The Royal Society of Chemistry 2015
Published by the Royal Society of Chemistry, www.rsc.org

Arylpalladium complexes, generated from organomercuric compounds (organotin and organolead compounds were also tried[2]) were found to arylate olefins.[3,4] These studies set the range of olefins that later became the main targets of research for over 40 years – styrenes, acrylates, allylic alcohols and enol ethers. Reoxidation of Pd(0) formed by cupric salts was introduced to make the oxidative arylation catalytic in palladium.

The classical mechanism of arylation involving three major steps – oxidative addition (transmetallation in oxidative cases), migratory insertion and reductive elimination to generate an XPdH intermediate – was postulated and experimentally substantiated to become the main mechanistic paradigm, which has remained practically unchanged except in minor details.[5,6] Heck discovered and interpreted double-bond shifts, intramolecular chelation in heteroatom-substituted olefins and other detailed effects, which later became subjects of numerous studies.[6]

Simultaneously and independently, Fujiwara, Moritani and co-workers started to exploit the well-known ability of simple Pd(II) salts to metallate benzene and other arenes, with the thus generated arylpalladium intermediates engaged in reactions with olefins. Reactions were conducted only with stoichiometric amounts of palladium salts, without attempts to recycle Pd(II) with cheaper reoxidants and, except in a few special cases, were fairly unselective and gave poor to modest yields.[7–20]

Thus, up to the end of the 1960s, a thorough understanding of the chemistry involved in the reaction of arylpalladium intermediates, generated *via* stoichiometric transmetallation, with olefins was accumulated, primarily due to the masterful in-depth studies of Heck. Simultaneously, the generation of arylpalladium intermediates by oxidative addition of Pd(0) to aryl iodides using stable phosphine complexes of Pd(0) was described by Fitton and co-workers.[21–23] The stage was thus set for a major breakthrough.

The story of a true catalytic process began in 1971 when Mizoroki and co-workers published a one-page preliminary communication (the full paper appeared in 1973)[24,25] on the arylation of olefins (ethylene, propylene, styrene, methyl acrylate) by iodobenzene in the presence of 1 mol% of PdCl$_2$ with KOAc as base in methanol at 120 °C in an autoclave. Palladium black, formed by prereduction of PdCl$_2$, was shown to deliver the same level of catalytic activity as the soluble precatalyst.

In 1972, a new series of papers by Heck and co-workers describing the palladium-catalysed reaction of olefins with aryl and vinyl halides began to appear. In the first paper, a system using Pd(OAc)$_2$ in the presence of tertiary amine base in a polar aprotic solvent was elaborated by Heck and Nolley to become, with minor variations, the classical entry-level catalytic system for generations of researchers to follow.[26] In subsequent studies, the effect of phosphine ligands was discovered by Dieck and Heck, which permitted the reaction with less reactive and readily available aryl bromides[27] and vinyl bromides.[28,29] Further papers described the arylation and vinylation of allylic alcohols,[30] acetals of unsaturated aldehydes and ketones,[31] ethylene,[32] heteroaryl-substituted alkenes,[33] *N*-vinylpyrrolidone[34] and dienes,[35,36]

The possibility of using the reaction for heterocyclization to afford a flexible route to various heterocyclic systems was realized.[37,38] Basic principles of what were later to become known as *domino* reactions, in which organo-palladium intermediates formed *via* carbopalladation are captured by various follow-up reactions, *e.g.* intramolecular carbopalladation and allylic substitution, were discovered.[39-43] In these studies, Heck and co-workers revealed that the modification of basic triphenylphosphine ligand gave new ligands with improved performance and greater effects on catalytic activity.

Tris(*o*-tolylphosphine) was identified as a ligand useful in performing reactions with the least reactive, electron-rich aryl bromides, such as *p*-bromo-*N,N*-diethylaniline or *p*-bromophenol, and the effect of the ligand was traced to catalyst inactivation mechanisms.[44] This ligand, reintroduced later by Spencer,[45,46] became one of the key indispensable ancillaries for the Mizoroki–Heck reaction. The formation of a palladacycle from tris(*o*-tolyl-phosphine) and Pd(OAc)$_2$ was observed and its structure established.[47] This palladacycle would be rediscovered a decade later by Herrmann and co-workers and serve as a model precatalyst in further intense efforts to develop superior catalyst precursors.[48,49]

Thus, the chemistry rapidly grew into a fully fledged synthetic strategy and in 1979 an impressive account of the newly born chemistry, explicitly showing its enormous preparative potential, versatility and promise for further development, was written by Heck,[50] followed by a chapter in *Organic Reactions*.[51]

Heck and co-workers in fact discovered (alone or simultaneously with the Mizoroki and Fujiwara–Moritani teams in Japan) not a single reaction, but a general type of transition metal-mediated arylation/alkenylation of unsaturated bonds, and its realization in no less than three different catalytic processes: (a) the oxidative process, stoichiometric in palladium in the absence of an external oxidant, which should be given the name Heck–Fujiwara–Moritani reaction; (b) the catalytic process taking place in the presence of Pd complexes not requiring specific stable ligands (often referred to as phosphine-free or even ligand-free, both being misnomers); and (c) the catalytic reaction taking place in the presence of Pd complexes involving stable ligands (almost always phosphines), which exert a critically important influence on the performance (the activity or selectivity) of the catalysts.

In a general sense, the Heck reaction can be defined as a process involving the addition of a σ-organylpalladium complex (carbopalladation) of unsaturated (usually double bonds, but triple bonds cannot be discarded from consideration). In fact, this chemistry is not necessarily restricted to palladium, as was shown by Heck himself – it can be performed with other transition metals and this story is under investigation, albeit much more slowly than Heck's chemistry of palladium.

Heck's chemistry of today includes catalytic and stoichiometric reactions, and about a dozen named processes (Mizoroki–Heck, Fujiwara–Moritani, Blaser, Myers, Matsuda–Heck, *etc.*) are included in this context. Further generalizations can involve the addition of other σ-bonded complexes of palladium, formally similar to what happens in carbopalladation, an already well-established technique, *e.g.* in the aza-Heck (Narasaka–Heck cyclization) reaction based on intramolecular aza-palladation of the double bond.[52–56]

In Scheme 9.1, a minimalistic overview of the Heck chemistry manifold is shown to demonstrate the main stages involved in the majority of Heck reactions. The process should involve the generation of a reactive but relatively stable (isolable or detectable) σ-organylpalladium complex by (i) oxidative addition from organic electrophile with an appropriate nucleofugic group or (ii) transmetallation or electrophilic palladation of C–H or C–metal bonds or (iii) various addition reactions to triple bonds. This complex next adds to a C=C double bond either inter- or intramolecularly. The resulting adduct is either labile, if there is a vicinal H-atom, which can rotate into the *syn*-position relative to the Pd centre, to undergo *syn* elimination of Pd hydride giving the product and the palladium hydride species is recycled *via* deprotonation or oxidation. Such a simple scheme describes both regular and oxidative Heck reactions. Each of the stages allows for multiple alternatives, which accounts for the amazing diversity of Heck reactions.

9.1.1   The Diversity of Heck Reactions

Regarding the alternatives to *syn*-PdH elimination, the most common product-forming step, we can obtain an interesting classification based on net stoichiometry. In the cases when there is a nucleofugic group at the adjacent atom, the alternative termination may involve the elimination of Pd^{2+} species and this nucleofuge. If we take into account this possibility, four distinct stoichiometries can be considered, two of which are redox processes involving Pd(0)/Pd(II) and Pd(II)/Pd(0) pairs and the other two are isohypsic involving the same oxidation states at the termini (Scheme 9.2). All four are known and are pairwise complementary.

The elimination of nucleofuge was initially regarded as a minor pathway, occurring only in specific cases. Such an elementary step is actually equivalent to a retro-Wacker reaction (Scheme 9.3).

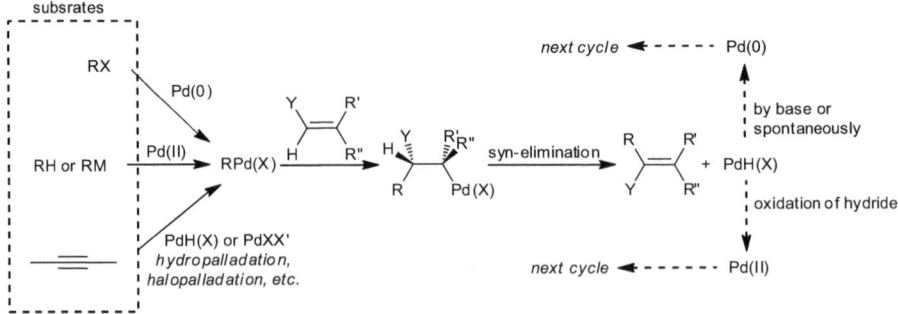

Scheme 9.1   The core of the Heck reaction paradigm.

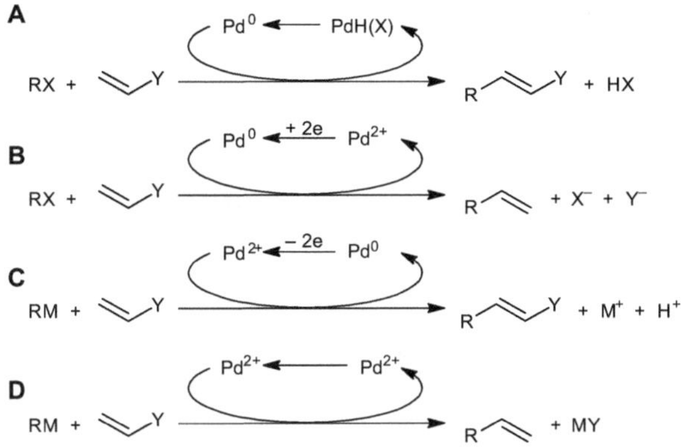

Scheme 9.2 Basic stoichiometric types of Heck reactions (X, Y = nucleofugic groups, M = electrofugic group). (A) Common Mizoroki–Heck reaction, an isohypsic process requiring base for regeneration of catalyst; a spontaneous base-free version is also possible. (B) Reductive Heck reaction, requires reductant for regeneration of Pd(0) catalyst. (C) Oxidative Heck reaction (Fujiwara–Moritani reaction if M=H), requires oxidant for regeneration of catalyst. D A rare type of Heck reaction, isohypsic and not requiring any auxiliary reagents for regeneration of catalyst, preparatively identical with cross-coupling but catalysed by Pd(II) and not by Pd(0).

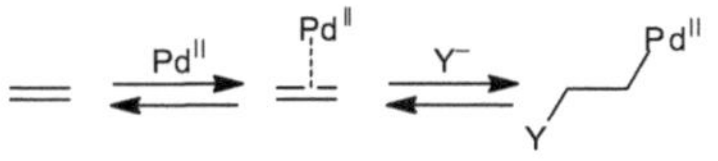

Scheme 9.3 The β-eliminaition of Pd(II) and nucleofuge viewed as the retro-Wacker reaction.

Historically, the process was revealed as early as 1977 by Kasahara *et al.*[57] The importance of this chemistry in the Heck reaction was demonstrated by Cabri *et al.*,[58] who actually showed that this can be realized as a special version of the Heck reaction, which corresponds to the stoichiometric type B in Scheme 9.2. In recent studies, this termination was revealed as a common phenomenon occurring in many reactions involving carbopalladation. Sufficient cases were accumulated to make possible a thorough review of factors governing the selectivity of elimination and the competition between the normal PdH elimination and the Cabri–Kasahara termination.[59]

Such a process occurs, for example, in the arylation of vinyl acetate leading to styrenes, discovered by Cabri *et al.*[58] As the process ends with Pd(II), an external reducing agent is required to make the process catalytic. In reality this is not a problem, as both olefin and Et$_3$N commonly used in such protocols do reduce Pd(II), but nevertheless the participation of a reducing agent should be explicitly taken into consideration in the stoichiometry and it should not be forgotten that the reductant is actually consumed in the reaction in equivalent quantities (1 equiv. of two-electron reductant per equivalent of ArX), which may become a serious cause of the formation of unexpected side products.

The reactions obeying this paradigm should be called *reductive Heck reactions*, which can be formally considered as the opposite case to *oxidative Heck reactions.*

Examples of reductive Heck reactions can be found in intramolecular transformations of formidable utility and complexity. Thus, Lautens *et al.* observed a stereoselective ring closure (Scheme 9.4), corresponding to pathway B in Scheme 9.2.[60]

The protocol was extended to intermolecular reactions.[61,62] This version is interesting because as Pd(0) cannot be regenerated *via* the regular pathway by PdH deprotonation, the reaction could be considered as a base-free Heck reaction and the amine used in the catalytic system probably served as a reducing agent for Pd(II).

Another example of pathway B shows that the methodology can come close to performing miracles! The C–F bond is among the least reactive bonds. However, the case of termination by Pd–F elimination was reported by Ichikawa *et al.* in an intramolecular reaction (Scheme 9.5).[63] In the absence of reducing agents the reaction is apparently stoichiometric in Pd(0) complex.

The last stoichiometric type corresponding to pathway D in Scheme 9.2 is so far very rare. Zaitsev and Daugulis showed in 2005 that acetanilides can be made to react with 2-bromoacrylates in the presence of a Pd catalyst and silver salt to afford the same cinnamates as in the directed oxidative Heck reaction terminated not by PdH elimination but by Pd–Br elimination, with silver triflate apparently serving the role of removal of the strongly bonded Br ligand, which would otherwise suppress the reaction[64] (*e.g.* Scheme 9.6).

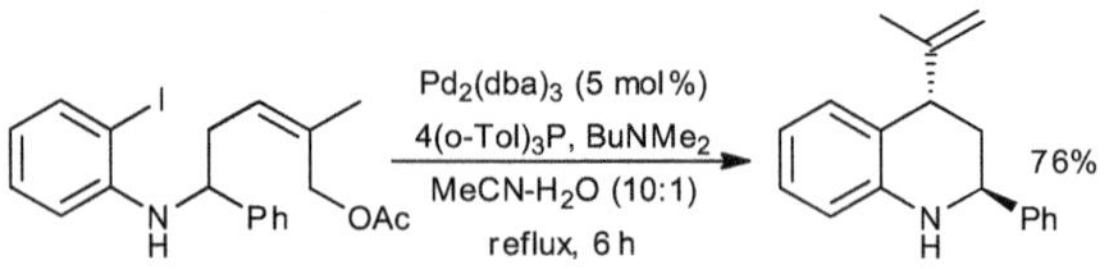

Scheme 9.4   Intramolecular arylation terminated by β-elimination of acetate.

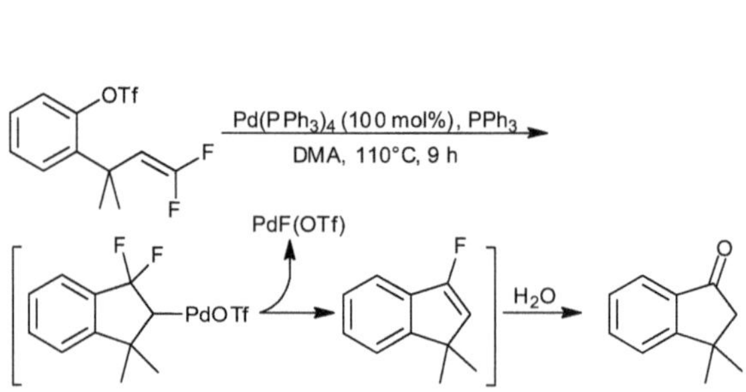

Scheme 9.5   Intramolecular arylation terminated by fluoride elimination. A stoichiometric case.

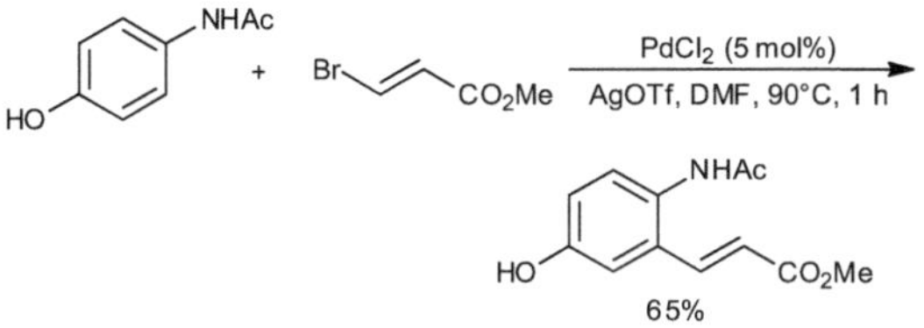

Scheme 9.6 Isohypsic alkenylation of an electron-rich arene.

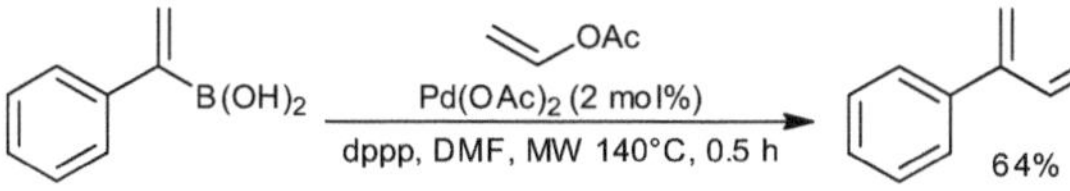

Scheme 9.7 Example of deborylative isohypsic alkenylation.

Scheme 9.8 Deborylative isohypsic alkenylation in a continuous flow reactor.

This case is interesting because in spite of a close relation to the Fujiwara–Moritani reaction, in fact it is a catalytic process without the need to re-generate active palladium catalyst either by oxidation or by base.

Another case of the same methodology involves the reaction of aryl- and alkenylboronic acids or aryltrifluoroborates with vinyl acetate, developed by Larhed and co-workers,[65] which is an instance of a catalytic process that is a complete inversion of the regular Mizoroki–Heck scheme (Scheme 9.7). Unlike the synthesis of styrenes by arylation of ethylene or vinyl acetate, which is plagued by diarylation, this reaction cannot lead to stilbenes, thus being more easily controllable. Further, it is fully catalytic, not requiring any auxiliary reagents (bases, oxidants, reductants).

The reaction was realized in a continuous flow reactor (Scheme 9.8), which allows for a dramatic decrease in the reaction time and easy scale-up, which, in our opinion, confirms that the mechanism is closer to that expected than to indirect arylation of ethylene formed *in situ*.[66]

The other potent source of diversity comes from consideration of the initial step, the generation of σ-organylpalladium species (organyl is nor-mally aryl or alkenyl, but can also be some other types involving sp³, sp and other valence states of carbon, although so far examples are few and am-biguous). In addition to oxidative addition, transmetallation and electro-philic palladation involved in the Mizoroki–Heck and oxidative Heck reactions, the RPdX intermediates can be generated by other means, among which the addition of Pd(II) species to a triple bond is a now well-recognized family of processes. These processes are more often used in cascade

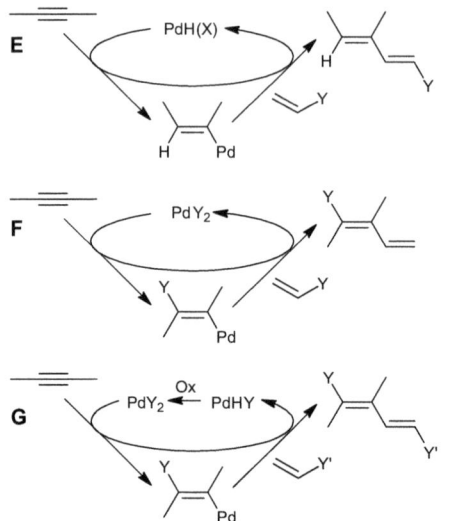

Scheme 9.9   Heck reactions involving acetylenes. Reactions E and F are two truly
catalytic pathways not requiring auxiliary reagents to regenerate the
catalyst. Pathway G is a combination of E and F requiring an auxiliary
oxidant. The inverse case (not shown) can also be formulated, requiring
auxiliary reductant for catalyst regeneration.

processes as intermediate steps. A formally electrophilic *trans* addition of
X′Pd–X reagent to a triple bond (halopalladation, acetoxypalladation, *etc.*), so
far less exploited reactions, can be used to initiate a Heck-like sequence of
transformations, interestingly often terminated by a reverse reaction – the
elimination of X′Pd–X species (Scheme 9.9).

Thus, an interesting example corresponding to pathway G in Scheme 9
was reported by Jiang *et al.*[67] In spite of the complex multistep mechanism of
the transformation of carbopalladated intermediate into final product,
complicated by double-bond shifts and terminated by β-elimination of XPdZ,
where Z is a common nucleofugic group, such as OH or OAc, the process
overall is highly regio- and stereoselective with a wide range of starting
alkynols and alkenes (Scheme 9.10). In the absence of alkene, alkynols are
regio- and stereoselectively dimerized into 2,4-dihalo-1,3,5-trienes.[68] A similar intramolecular case was described by Alvarez and co-workers.[69]

Another recent example (pathway F, Scheme 9.9) showing a similar
transformation initiated by chloropalladation of alkyne, followed by Heck-
type alkenylation of allyl halide and terminated by Pd(II) halide elimination,
was reported by Zhu and co-workers[70] (Scheme 9.11).

An example using acetoxypalladation in the first stage and termination
through Pd–OAc elimination was reported by Lu and co-workers[71]
(Scheme 9.12).

Skrydstrup and co-workers developed a methodology in which an alkenylpalladium intermediate is generated by hydropalladation of an internal

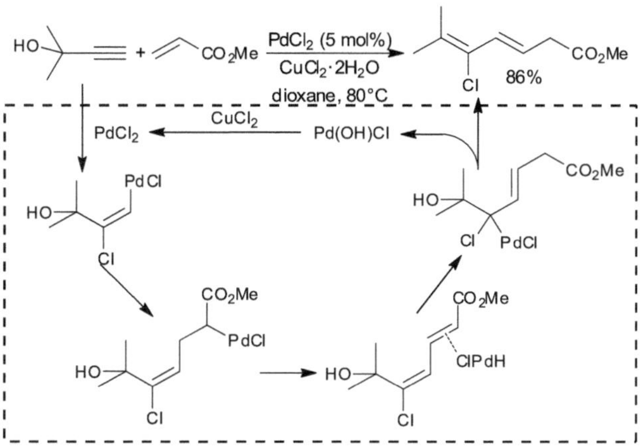

Scheme 9.10 Example of alkenylation initiated by chloropalladation of an alkyne, and mechanism (in dashed box).

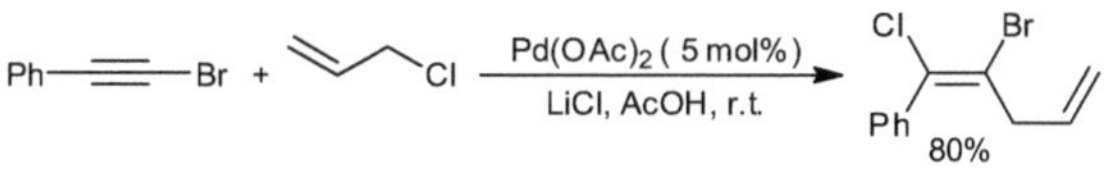

Scheme 9.11 Alkenylation initiated by chloropalladation: an example of the protocol developed by Zhu and co-workers.[70]

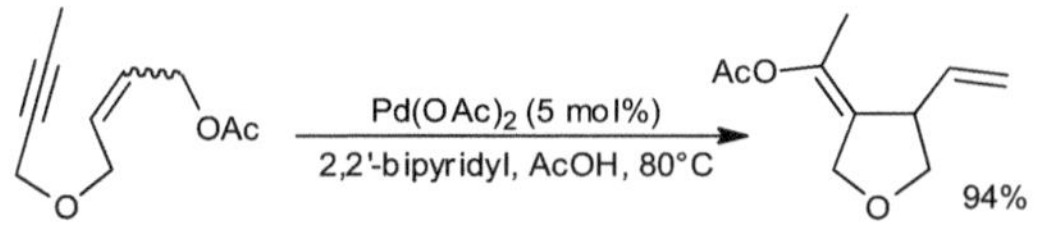

Scheme 9.12 Example of intramolecular alkenylation initiated by acetoxypalladation.

triple bond, corresponding to pathway E in Scheme 9.9.[72] Thus, the catalytic cycle has a PdH complex at the beginning, being formed either spontaneously or by decarbonylation–elimination from isobutyryl chloride. Regeneration of PdH at the end of each cycle makes this catalytic process self-sustaining without bases, oxidants or any other reagents (Schemes 9.13 and 9.14).

Similar couplings were reported to be catalysed by Co,[73] Ni,[74] Ru[75] and Rh,[76,77] although an alternative mechanism involving metallacycles is often considered for such transformations.

9.1.2 Heck Reaction *Versus* Cross-Coupling

In palladium- and, more generally, transition metal-catalysed C–C bond formation methodology, the Heck reaction is a major type of process, compared with cross-coupling, in which a new C–C bond is formed by

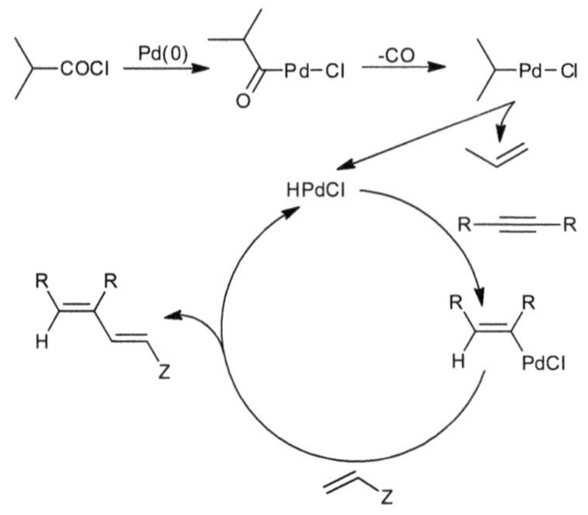

Scheme 9.13 Alkenylation initiated and terminated by PdH species: outline of the mechanism.

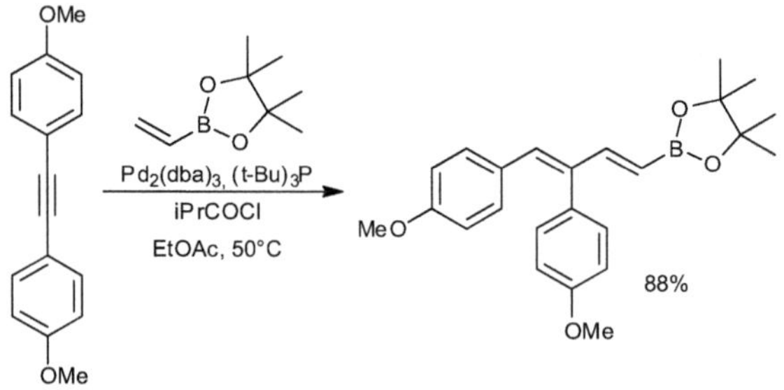

Scheme 9.14 Example of alkenylation according to Skrydstrup and co-workers.[72]

reductive elimination from R–Pd–R′ complexes. There are cases, however, where the difference is not definite and, at least until the mechanisms of each particular case become established unambiguously, the margin for interpretations will remain. Such borderline cases are excellent manifestations of the diversity of Heck reactions, as in these cases the specific behaviour of common intermediates is of critical importance for differentiation between the mechanistic paradigms.

Thus, the reactions of alkenyl- and arylpalladium intermediates with unsaturated aromatic substrates leading to arylation and alkenylation can formally be viewed as carbopalladation of double bonds involved in aromatic systems. Although the detailed mechanisms of such reactions are unknown, it seems unlikely that such transformations can really involve carbopalladation. More likely is an electrophilic pathway, which is closer to the cross-coupling paradigm. Nevertheless, true carbopalladation can take place in annelated, particularly five-membered systems, where the double bonds are substantially localized and indeed can behave as regular double bonds.

Particularly relevant, this formal analogy appears in alkenylation reactions, well known in heterocyclic chemistry (reviewed by Rossi *et al.*[78]), and

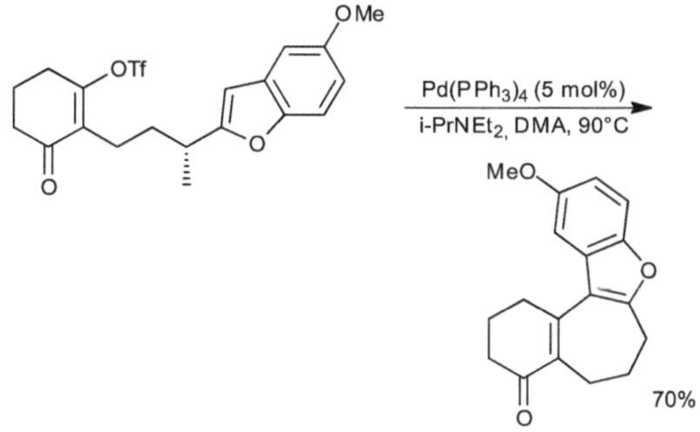

Scheme 9.15 Heck reaction or cross-coupling? The intramolecular alkenylation of benzofuran.

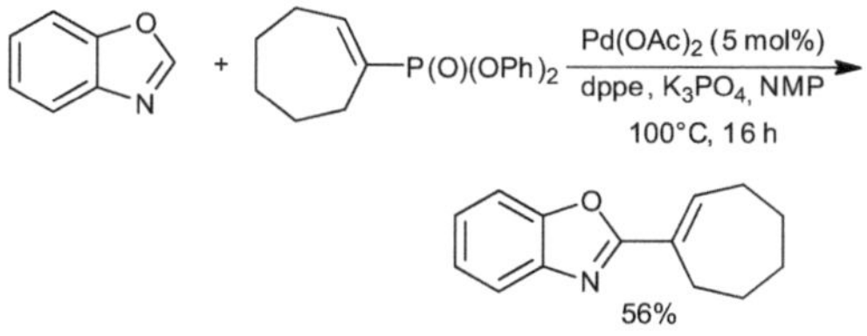

Scheme 9.16 Example of alkenylation of benzoxazole.

first described in an intramolecular version by Hughes and Trauner[79] (Scheme 9.15).

An interesting intermolecular realization of this methodology was described recently by Ackermann and co-workers using alkenyl phosphates (Scheme 9.16)[80] or alkenyl tosylates.[81]

This approach belongs to the more general methodology of arylation (vinylation) by Pd-catalysed reaction of aryl (vinyl) halides with arenes or hetarenes, in which a new C–C bond is formed not *via* carbopalladation, but by the attack of an organopalladium intermediate at a C–H bond, similarly to what happens in the Fujiwara–Moritani reaction (see below). Such reactions bear considerable potential for organic synthesis (selected reviews[82–84]), but fall outside the scope of Heck chemistry, rather belonging to the cross-coupling paradigm, both formally because a new C–C bond is formed by reductive elimination and not in the carbopalladation stage, and experimentally, as the reactions benefit from special ligands developed for cross-coupling reactions.[85]

The borderline between the two mechanistic domains is vague, however; both employ the same Pd(0)/Pd(II) chemistry, dependent on bases to become catalytic. This particularly applies with weakly aromatic heterocycles with largely localized double bonds (*e.g.* furans, pyrones, benzannelated heterocycles), when carbopalladation can be involved in a typical intramolecular Heck reaction, as was suggested by *e.g.* Taylor and co-workers for the intramolecular arylation of pyrones[86] (Scheme 9.17). The stereochemical

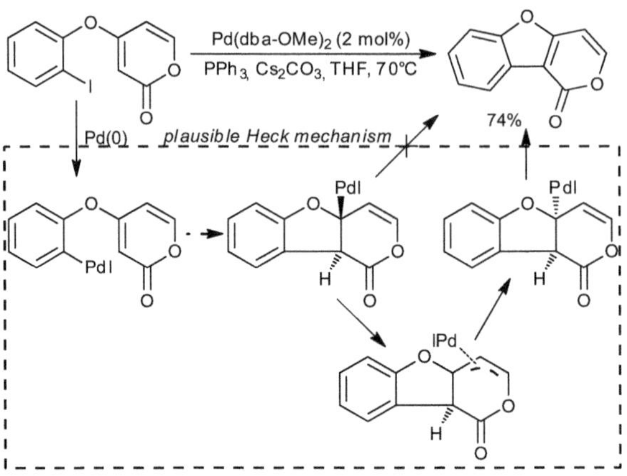

Scheme 9.17 Heck reaction or cross-coupling? The intramolecular arylation of α-pyrone.

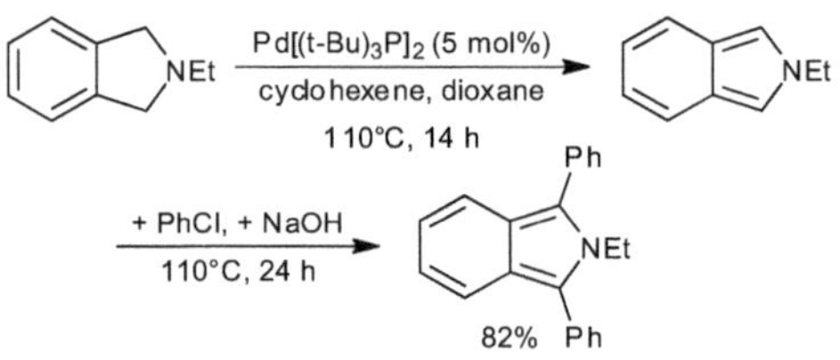

Scheme 9.18 Arylation of *N*-alkyl-*2H*-isoindole.

restriction of the unavailability of a proton for *syn* elimination of PdH in the hypothetical primary carbopalladation adduct can be lifted *via* rearrangement involving a configurationally fluxional η³-complex (the possibility of such transformations *via* similar η³-benzylic complexes in Heck reactions with conformationally rigid systems was noted earlier[87]).

Another interesting example of such chemistry showing how subtle the borderline between Heck reaction and arylation *via* C–H activation sometimes is involves the arylation of *N*-protected isoindoles, highly reactive heterocyclic compounds, the structure of which is closer to that of non-aromatic *o*-quinodimethanes with localized double bonds than to regular aromatic heterocycles. Ohmura *et al.* showed that isoindoles are generated *in situ* by dehydrogenation of readily available isoindolines and are then arylated by aryl chlorides (Scheme 9.18).[88] Owing to the high reactivity of the double bonds in this system, bis-arylation predominates.

There is at least one another major type of unsaturated substrates where the Heck reaction/cross-coupling dichotomy is not readily resolved, leading to interesting questions. The carbonyl group and its equivalents in aldehydes, ketones, imines, nitriles, *etc.*, can participate in both carbopalladation and cross-coupling reactions. These substrates can take part either in direct carbopalladation of a carbon–heteroatom multiple bond, followed by protolysis of the heteroatom–Pd bond (the addition pathway, left side in Scheme 9.19), or HPd(X) elimination to regenerate a C=Y bond. Enolizable compounds and preformed enol derivatives (ethers, esters, organometallic enolates, enamines, *etc.*) can take part either in the Heck reaction as typical electron-rich olefins (centre pathway in Scheme 9.19) or in cross-coupling *via* palladation of nucleophilic α-carbon (α-arylation pathway, right side in Scheme 9.19).

All of these pathways are known, but their relative contributions are very different. The addition (carbopalladation followed by protolysis) is either a realization of a reductive Heck reaction, equivalent to pathway B in Scheme 9.2, initiated by Pd(0) and terminated by Pd(II), which would require a two-electron reductant to make the process catalytic, or an isohypsic Heck reaction, equivalent to pathway D in Scheme 9.2 and catalytic in Pd(II). A few examples of the latter have been reported so far, using arylboronic acids as sources of ArPdX species, including an enantioselective one,[89] based on the work of Wu, Cheng and co-workers.[90–93] These reactions require special phosphine ligands such as trinaphthylphosphine, which have found little application in more conventional Heck reactions, and base (Scheme 9.20). Faster reactions and higher yields were consistently obtained with electron-deficient aldehydes. On the other hand, enolizable aldehydes give the same

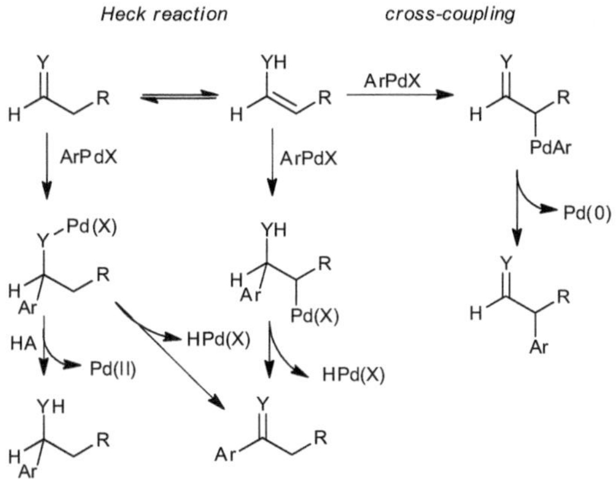

Scheme 9.19 Probable transformations of aldehydes and their analogues effected by ArPd(X) species. For ketones and their analogues the same pathways can be operative except that in which HPd(X) elimination needs the aldehyde proton.

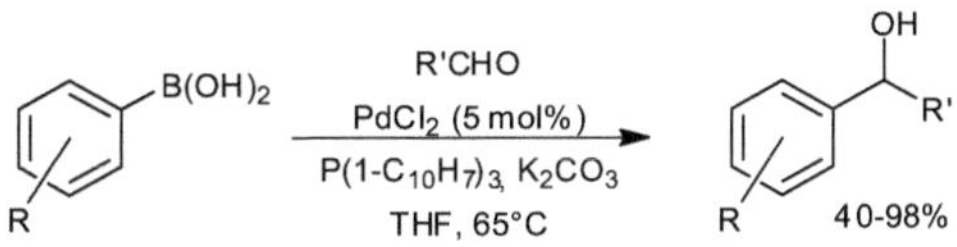

Scheme 9.20 Arylation of C=O bonds of aldehydes [isohypsic Heck reaction catalysed by Pd(II)].

type of products without side reactions associated with arylation of enolic forms.

Concerning the enolic forms of enolizable carbonyl compounds, both cross-coupling and Heck reactions can be performed depending on the particular type of derivative and conditions used. These two directions differ in the products formed – the cross-coupling affords α-arylation. Ionic enolates, silyl ethers and similar compounds have been developed in the last decade as versatile nucleophiles suitable for the cross-coupling methodology, depending mainly on the design of special ligands (reviewed in 2009[94]).

On the other hand, enol ethers and similar compounds are well-known targets for Heck reactions, both regular and oxidative, being considered as electron-rich olefins subject to special Cabri–Larhed methodology using bidentate phosphines for enforcing high arylation regioselectivity. This method actually transforms aldehydes into ketones, thus effectively extending the classical *Umpolung* strategy of organic synthesis (acyl transfer from aldehydes) to aromatic and unsaturated compounds.

Whereas the Cabri–Larhed approach uses preformed enol derivatives, new protocols successfully employ raw aldehydes for acylation. Such a breakthrough was reported by Xiao and co-workers[95,96] using organocatalysis coupled with Heck arylation of enamines generated *in situ* from enolizable aldehydes (Scheme 9.21). High α-regioselectivity of enamine arylation was confirmed by reaction with a preformed enamine.

Aryl bromides are used with a typical catalytic system [Pd(dba)$_2$, dppp, pyrrolidine, 4A molecular sieves, DMF, 115 °C] for arylation of electron-rich olefins.[95] Ruan and co-workers showed that the same goal can be achieved in a simple "phosphine-free" system using palladium acetate as precatalyst and Bu$_4$NBr, the standard agent for stabilization of Pd(0) species (Scheme 9.22). Palladium sol was detected in the reaction medium, which apparently served as the actual precatalyst (see below). The system is remarkable from many points of view, not only because of the apparent preparative value, but also because it serves well for a wide range of aryl bromides, including many electron-rich compounds, thus being one of the

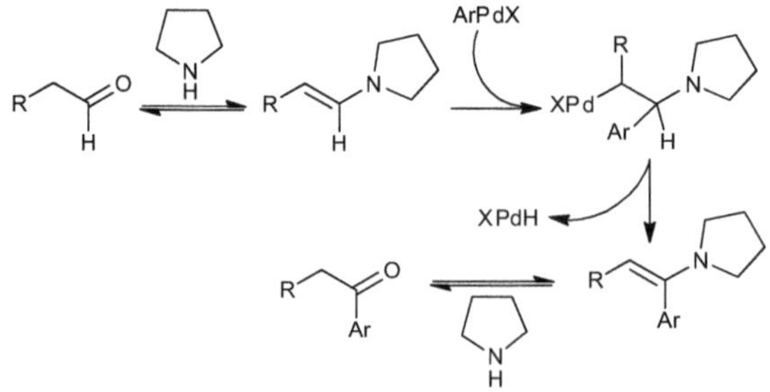

Scheme 9.21 Mechanism of arylation of enolizable aldehydes using organocatalysis.

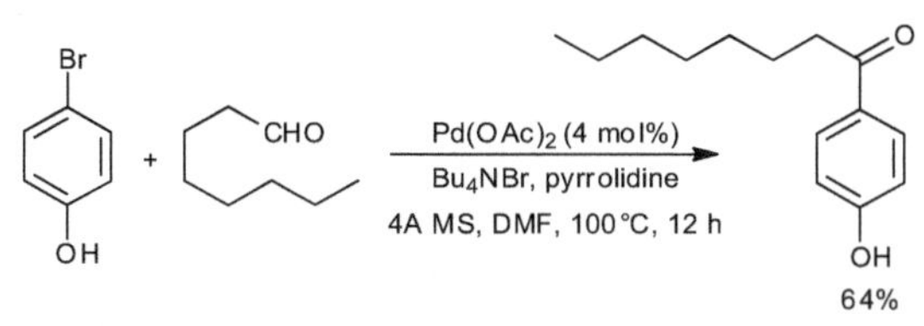

Scheme 9.22 Example of acylation *via* arylation of enamine formed *in situ* in a "phosphine-free" system.

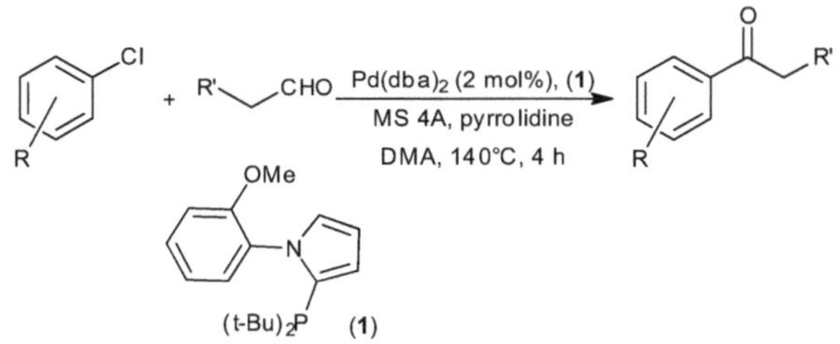

Scheme 9.23 Arylation of enolizable aldehydes by aryl chlorides using Beller and co-workers' phosphine ligand (**1**).

most reliable and wide-scope "phosphine-free" systems described so far for deactivated aryl bromides.

With aryl chlorides the protocol is highly ligand dependent,[96] requiring particular ligands of Beller and co-workers' phosphine series,[97] thus providing a useful versatile acyldechlorination method (Scheme 9.23).

9.1.3 Carbonylative and Decarbonylative Reactions

Carbonylative cross-coupling reactions are well known,[98] whereas their counterpart in Heck chemistry has been elusive except for sporadic reports of intramolecular examples.[99–101] On the other hand, a closely related decarbonylative Heck reaction (Blaser reaction) is well known although still rather unpopular in the synthesis and can involve aroyl chlorides,[102] but also the respective anhydrides and *p*-nitrophenyl or enol esters,[103,104] the leaving groups of which perform the role of base in the regeneration of Pd(0) species, permitting rare cases of base-free (and potentially waste-minimized[105]) Heck reactions.

The intermolecular carbonylative Heck reaction was reported by Miura and co-workers,[106] but no protocol suitable for a wider range of substrates was then elaborated. To discover appropriate conditions and catalytic systems in which sequential binding of carbon monoxide and olefin to an arylpalladium intermediate can take place in the correct order is a challenging task. At least three competitive pathways can be identified in the system involving CO (Scheme 9.24). Carbonylative and decarbonylative pathways are intersected in the steps involving reversible binding/extrusion

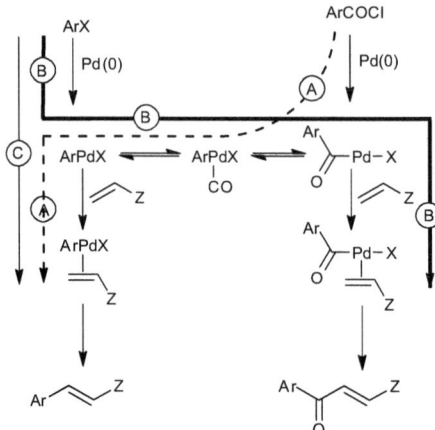

Scheme 9.24 Three pathways associated with carbonylative Heck reaction: (A) decarbonylative Heck reaction (Blaser reaction); (B) carbonylative Heck reaction; (C) regular Mizoroki–Heck reaction.

of CO. These equilibria are highly dynamic and their shift in one or the other direction is controlled by concentration of CO in the reaction medium and ancillary ligands. The competitive regular Mizoroki–Heck pathway is irreversible and therefore usually wins the competition, making the realization of the carbonylative Heck reaction a challenge with regard to the formulation of the catalytic system and choice of proper stable ancillary ligands.

The problem was recently solved by Beller and co-workers, who developed the first general protocols for the carbonylative Heck reaction. The synthesis of chalcones from aryl triflates and styrenes[107] is critically dependent on the choice of ligand and conditions. The polar pathway suits this specific task best – aryl triflates or nonaflates as substrates and the bidentate phosphine ligand dppp secure a free coordination place needed for consecutive binding of CO and olefin (Scheme 9.25). A slightly increased pressure of CO (2–10 bar) is used to promote the carbonylative process.

The protocol is applicable without modifications to a wide range of triflates, including electron-rich and electron-deficient aryl derivatives and enol esters (Scheme 9.26).

The method was subsequently extended to aryl iodides and bromides through the use of Beller and co-workers' ligands, bulky and electron-rich monophosphines (Scheme 9.27).[108–110] Bulky monophosphines of such a type are believed to ensure that the coordination shell contains only a single stable ancillary ligand (phosphine itself), thus effectively pursuing in the non-polar pathway the same goal as the diphosphine dppp does in the polar pathway – keeping two coordination sites available for catalytic transformation.

For the reactions involving bromoarenes and styrenes, Beller and co-workers showed that the classical Heck catalytic system using PPh$_3$ and NEt$_3$ unexpectedly outperformed advanced phosphines.[111]

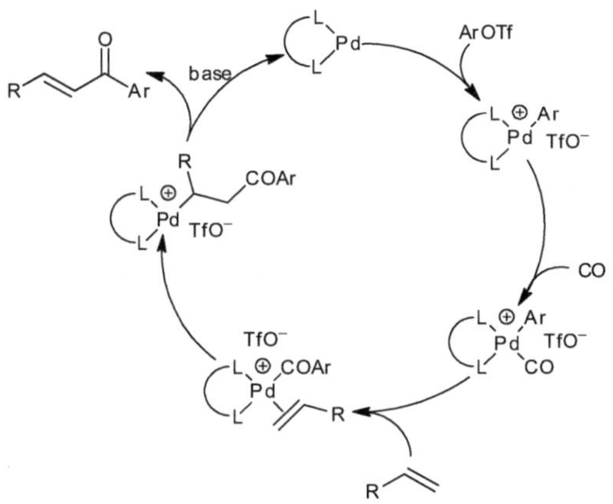

Scheme 9.25 The hypothetical catalytic cycle for carbonylative Heck reactions of aryl triflates.

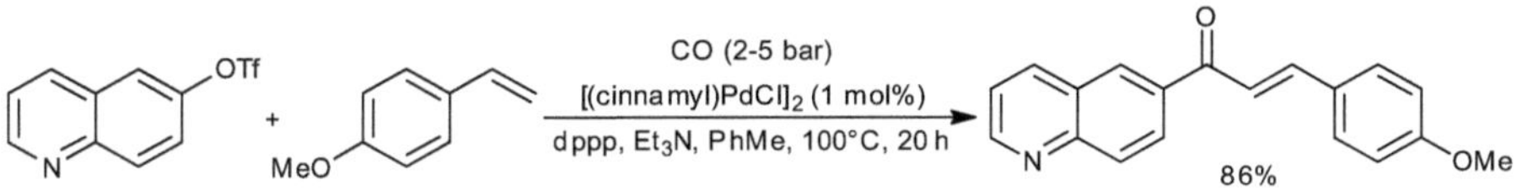

Scheme 9.26 Carbonylative Heck reaction of aryl triflates.

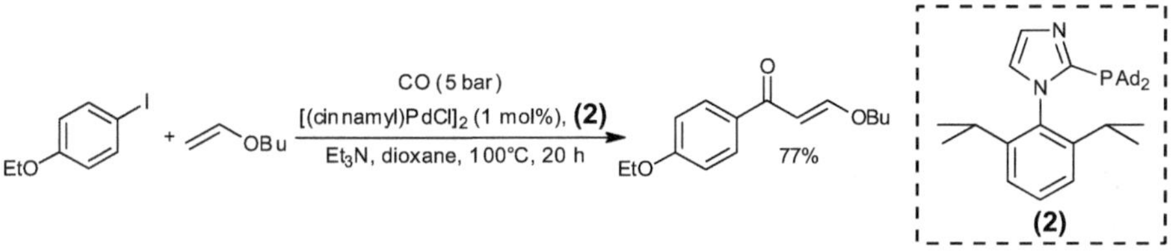

Scheme 9.27 Carbonylative Heck reaction of aryl halides using a bulky electron-rich phosphine (2).

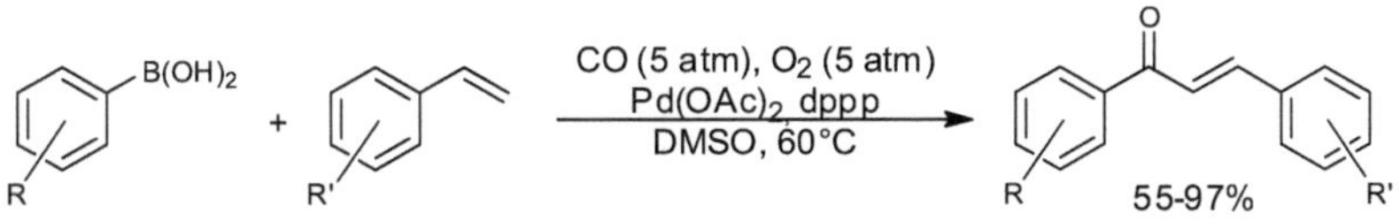

Scheme 9.28 Oxidative deborylative carbonylative Heck reaction.

Beller and co-workers found a further extension of the carbonylative Heck methodology and realized oxidative carbonylative Heck reactions using arylboronic acids as nucleophilic coupling partners and oxygen as stoichiometric oxidant (Scheme 9.28).[112]

Intramolecular carbonylative cyclization involving an iodoalkane residue is one of the rarest examples of the Heck reaction involving an sp³ carbon,[113] although the Heck process itself is the attack of acylpalladium at the double bond and is therefore not much different from other carbonylative Heck reactions.

9.2   The Mizoroki–Heck Reaction

9.2.1   Ancillary Ligands in Heck Chemistry

From the very beginning of Heck chemistry, from the work of the Heck, Mizoroki, Fujiwara and Moritani teams it became evident that this chemistry in general is not dependent on special ancillary ligands – all starting catalytic systems relied on a random choice of catalyst precursors, the role of which was excellently performed by the first available Pd(II) salts. Meanwhile, the mechanism of these processes contains many distinct steps involving the preactivation of the precatalyst [reduction to Pd(0) in the Mizoroki–Heck reaction, formation of monomeric Pd complexes from normally oligomeric $PdCl_2$, $Pd(OAc)_2$, *etc.*, in oxidative arylation], oxidative addition or (trans)-metallation, binding of olefin, migratory insertion, reductive elimination, regeneration of active Pd species – and none of these steps depends on specific stable ancillaries (detailed discussions of the mechanism can be found elsewhere[87,114–116]), in cases when specific tasks (reactions of less reactive substrates, controlling regio- or stereoselectivity, *etc.*) are not pursued. It should be stated that from this viewpoint, Heck chemistry is different from the majority of other important transition-metal-catalysed reactions critically dependent on the correct choice of ancillaries – an archetypical example is the Buchwald–Hartwig C–N and C–O, C–C (carbanion arylation) cross-coupling reactions in which the effect of specific stable ancillaries is critical in the nucleophile binding and reductive elimination steps, and also for suppression of side reactions.

Although it is not commonly realized, the opposite seems to be true in Heck chemistry. Most Heck reactions – both oxidative and regular types – are performed without specific ancillary ligands. It should be stressed that this concerns only those ligands for which the binding mode is considered to be *stable* – as opposed to *labile* and *hemi-labile* ligands. Stable ligands survive in the coordination shell of the metal centre throughout all the reactions involved in the transformation and in the catalytic process they survive for many turns of the cycle. The processes performed in the absence of stable ancillary ligands are commonly called *phosphine-free* or even *ligand-free*. The latter term is an inappropriate misnomer, and the former is also a misnomer (Heck chemistry can involve non-phosphine stable ancillaries, such as arsines, carbenes, *N,N*-chelators and others) but it tacitly reflects the common impression that phosphines are by far the most important archetypical ligands for palladium catalysis as a whole and Heck chemistry in particular. Therefore, in further discussion we reserve the use of the term "phosphine-free" to designate the reactions served by palladium complexes lacking definite stable ancillary ligands.

The possibilities for control *via* stable ancillaries in Heck processes are limited, and this is well seen particularly in comparison of the Heck process with the cross-coupling process; the latter is chosen not only because of its fundamental role in Pd catalysis, but also because it is an example of a stable ancillary ligand-controlled pathway. The *cross-coupling processes* involve ligand exchange of halide or other leaving group for strong nucleophiles or reactive organometallic compound. The chemistry of cross-coupling is highly determinate as the reagents – the electrophile and the nucleophile – can be bonded to the Pd centre only in the right order and the processes involved are commonly irreversible. In this process, a single coordination site of a formally neutral Pd complex is involved without the need for pre-dissociation of existing ligands, which opens up ample possibilities *via* controlling the reaction by stable mono- or bidentate ancillaries (Scheme 9.29). It is not surprising that the ligand design became a highly successful strategy in cross-coupling and the implementation of this strategy led to exceptional progress in the 2000s. On the other hand, the Heck processes are much more challenging in implementing a similar level of control by the ancillaries. One of the reagents, the olefin, can be and is bonded prematurely and this extra equilibrium is an unavoidable factor, decreasing the concentration of active Pd species and thus the activity of the catalytic systems.[117] Further, the Heck pathway involves the ligand exchange of an anionic ligand (actually either σ-bonded halides or labile solvent ligand in the case of an ionic bond with nucleofugic anions such as OTf⁻ and BF_4^-) for neutral olefin. This exchange can be realized only *via* predissociation with two possibilities existing – through the neutral pathway *via* exchange with a neutral ancillary or through the polar pathway *via* exchange with an anionic ligand or labile ancillary in a cationic complex.

Three distinct scenarios for the roles of stable ancillaries in Heck processes are outlined in Scheme 9.30. With stable ancillaries, the neutral

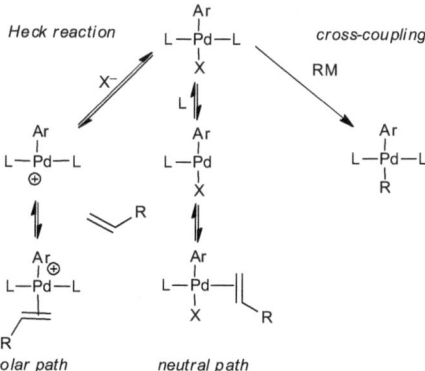

Scheme 9.29 Comparison of Heck and cross-coupling processes with respect to the ways of binding of the second reagent (olefin in the Heck reaction, organometallic compound or other nucleophiles in cross-coupling).

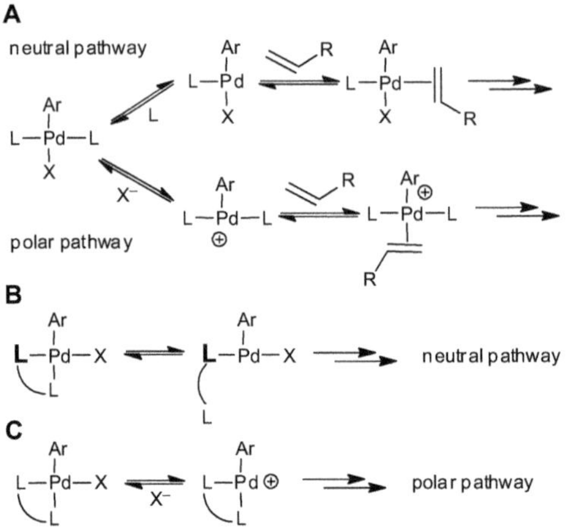

A

Scheme 9.30 Dominant scenarios of dependence of pathway type on stable mono- and bidentate ancillary ligands.

pathway requires the dissociation of a ligand that is favoured by the steric bulk. Such a pathway can occur either with monodentate ligands or un-symmetrical bidentate ligands involving two coordination sites – a stronger one, usually phosphine, and a weaker one, such as N or O atoms, electron-rich aryls furnishing the η^2-binding mode, and even the second but more bulky phosphine centre. Such unsymmetrical bidentate ligands, developed particularly by the Buchwald, Beller and Hartwig groups, achieved great success in the cross-coupling chemistry (reviewed recently[118]), but their impact on Heck chemistry has so far been modest.

True bidentate ligands have a separate and very profound impact on Heck chemistry. Whereas all other ligands can be involved in both polar and neutral pathways, true bidentate ligands form complexes which can operate only *via* the polar pathway, so their effect is to direct the process into a definite route. This property is used in two major branches of Heck chemistry – the regioselective arylation of electron-rich olefins[119] and enantiose-lective Heck reaction.[120–123]

The detailed analysis and the classification of the roles of ancillary ligands in the Mizoroki–Heck reaction[124] resulted in the conclusion that practically all the data accumulated on this reaction can be fitted into four major types of catalytic systems according to the role of the ancillary ligands. The de-tailed analysis and arguments can be found in Ref. 124 and here only a brief description of each is given. It should be noted that the Mizoroki–Heck re-action, particularly the reactions of iodoarenes and selected activated bro-moarenes with standard olefins (acrylates and styrene), was for a decade and a half used as a testing ground for developing new catalytic systems, pre-catalysts and ligands and new ways of performing reactions,[125] and many of the thus developed systems were tried with variable success and with un-certain trends for less reactive bromo- and even chloroarenes. Such a huge collection of randomly accumulated data provides a unique, unbiased basis for conclusions about real activity and catalytic efficiency of palladium systems towards haloarenes of variable reactivity.

The *first* type of catalytic system is used with highly reactive substrates, iodoarenes and, partially, the electron-deficient bromoarenes, although the behaviour of the latter is on the borderline. With such substrates, the oxidative addition is very fast and not rate limiting – Pd(0) species are rapidly consumed, keeping the concentration of such species very low. Most probably none of the stages of the catalytic cycle is rate limiting and this role is delegated to catalyst preactivation, delivering Pd in a zerovalent oxidation state with a sufficiently labile coordination shell. The reactions run on ill-defined (as opposed to well-defined) Pd species. Virtually any Pd compound or Pd-containing material (including traces of Pd on laboratory glassware and in the reagents used) can support such reactions, the composition of some of such mixtures being so complex and obscure in functionality that they deserve to be referred to as catalytic "cocktails."[126,127]

The *second* type is also phosphine free and probably is the most important and abundant in the Mizoroki–Heck reactions. Unactivated bromoarenes and even chloroarenes were from the very beginning established to react under phosphine-free conditions, although usually giving incomplete conversions and modest to poor yields. The mainstream research placed such substrates in the area requiring phosphine ligands and research to design such ligands was launched. Meanwhile, more and more data showing that unreactive substrates can be processed using "phosphine-free" systems accumulated. Indeed, if some reagent can take part in a few catalytic cycles, why cannot it be made to react further? The answer is absolutely evident – the problem is not the reactivity, but rather the stability of the catalyst. If the stability of catalytic systems can be controlled, then bromo- and even chloroarenes could be processed without special ligands.

The main factor governing the stability is also clear – this is the predisposition of Pd(0) species lacking stable ancillaries to aggregate into clusters, nanoparticles and then macroparticles. Therefore, a solid understanding of a tight association between the phosphine-free reactions and the occurrence and metamorphoses of the nanoparticles in such systems has been developed simultaneously by many researchers in the field.

Under the conditions of real Mizoroki–Heck reactions, such metallic particles can partially redissolve, but the rate of such catalyst reactivation falls rapidly with increasing size of the particles as an apparent consequence of the decrease in available surface area. Thus, if the aggregation is left uncontrolled, the result would be a decrease in net reaction rate and incomplete conversions. Owing to the well-known effect, Ostwald ripening, the size of metal particles under the conditions of reversible dissolution–resedimentation tends to increase at the expense of actively dissolving small particles.

To avoid misunderstanding, the critical importance of nanoparticles in reactions with less-reactive substrates does not mean that Pd sols cannot occur in the reactions of aryl iodides and other reactive substrates. The formation of such particles is often observed in these reactions, but in these cases this only means that too much simple and readily reducible Pd pre-catalyst is loaded or that the reactions are near completion. In such cases, the formation of metal particles is not as important. This can only be re-garded as a signal that the composition of the catalytic system is far from optimal.

As soon as the rate of oxidative addition is low to very low for the typical substrates in this category, the precatalyst loading and the rate of pre-activation and generation of Pd(0) species should not be higher than the rate of the first step of the catalytic cycle, as in this case Pd(0) would be accu-mulated, the aggregation begins and Ostwald ripening contributes to with-drawing Pd from the catalytic cycle until too little is left to support the cycle running at an appreciable rate. On the other hand, if the loading or pre-activation rate is too low, the rate of the catalytic process would be limited by it and the overall rate again would be inappropriate.

Even with the most careful tuning of such systems and achieving a perfect match between release and consumption rates, the metal particles would appear near the reaction completion when the Pd(0) consumption rate naturally decreases and aggregation begins. The freshly formed nano-particles would serve as alternative precatalysts and maintain the pool of accessible palladium until their quality worsens irreversibly due to Ostwald ripening. Two things should be borne in mind:

1. palladium nanoparticles are never true catalysts, but rather function as secondary precatalysts (*cf.* the detailed discussion in Ref. 124);
2. if they work (as the precatalysts) they are subject to Ostwald ripening and therefore to gradual deterioration of performance because of un-favourable changes in morphology (a similar effect was diagnosed in other Pd-catalysed reactions accompanied by the formation of nanoparticles[128]).

The recyclable systems for the Mizoroki–Heck reaction (and the very closely related, with respect to the design and operation of phosphine-free catalytic systems, Suzuki–Miyaura reaction) can be built around Pd nano-particles only if some mechanism of continuous counter-Ostwald redisper-sion is implemented and, as far as we know, this trick remains unaccomplished, although efforts have been made, to realize such systems, *e.g. via* the use of nanoparticles supported on various sophisticated supports including biopolymers, nanocomposites and structured materials,[129–141] in which case the surface or the pores of the support to a certain extent regulate the size of the dispersions. The success of such systems has so far been disappointing, as the Ostwald ripening cannot be switched off when needed (a discussion of earlier results on supported nanoparticles and similar tricks in Mizoroki–Heck reactions can be found elsewhere[124,125]). Among other attempts to control dispersion, the use of ultrasonic stimulation in aqueous media in the presence of the phase-transfer and colloid protector agent Aliquat-336 was reported.[142]

The catalytic systems of the second type are serviced, similarly to the first type, by Pd species with ill-defined coordination shells, without distinct stable ancillaries, usually under rather harsh conditions, and are often helped by additives that prolong the lifetime of Pd(0) species by labile ligation. The best known of such additives are quaternary ammonium salts, particularly Bu_4NBr. The role of such additives is likely to be simple – by forming labile anionic Pd(0) complexes the aggregation is delayed and thus the margin for entering the narrow entry gate of the catalytic cycle is widened. The other known role is the well-established Amatore–Jutand effect, the increase in reactivity towards oxidative addition,[143,144] equivalent to widening the entry gate. Both effects thus act in the same direction.

Although the sources of palladium can be as diverse as in the type 1 catalytic systems, here the role of the precatalyst is more critical than just to supply some Pd(0) species. Indeed, the need to match the rate of pre-activation with the rate of Pd(0) consumption would be best met by pre-catalysts able to release Pd(0) species at a controllable rate, so that by optimization of the conditions, adjusting the temperature and supplying additives this rate can be tuned to the requirements of a particular catalytic cycle. This need is best suited by stable palladium complexes of various sorts capable of slow decomposition under the conditions of the Heck reaction. Such palladium-containing materials can be regarded as *slow-release precatalysts*.[124]

The *third* type of catalytic system is processes that rely on specific stable ancillary ligands to increase the reactivity of Pd(0) species towards oxidative addition and simultaneously to suppress altogether the aggregation of Pd(0) species and deactivation through the formation of metallic particles. This type of Heck reaction was discovered by Heck himself, who noted that the addition of Ph_3P substantially increased the yields of reactions with less reactive aryl bromides. Soon afterwards, he discovered the second classical ligand – tris(*o*-tolyl)phosphine, revealing that increasing the bulk of the ligand is a favourable factor, probably because the coordination equilibria shift to species bearing less stable ligands per metal atom and we need no more than two of such species for active Pd particles. Hence the classical Heck protocols – Pd acetate or dba complex plus Ph_3P or $(o\text{-tol})_3P$ plus Et_3N or Bu_3N base in a polar solvent (DMF, DMA, NMP, MeCN) – were established and these protocols with minor changes still dominate Heck chemistry with aryl bromides and also aryl iodides, when reliable operation and yields are more important than economical value or very high turnover number (TON).

The systems of this type can be called *ligand-accelerated systems*, bearing in mind that the effects of the stable ancillary are not only an increase in rate and the use of less reactive substrates, but also the improvement of stability and convenience, most important in product-oriented studies when the target compound of a complex synthesis is what matters and optimization of the catalytic system is a waste of time and labour.

The *fourth* type of catalytic systems is those which also rely on stable ancillaries, but not to increase the reactivity towards oxidative addition. The role of ligands in this approach is to direct the reaction on to a definite pathway with high regio- or stereoselectivity and to suppress all alternative pathways of lower selectivity. Such systems can be defined as *ligand-restricted* and are best served by bidentate *P,P*-, *P,N*- and *N,N*-chelators. Such pathways are by default polar (pathway C in Scheme 9.30), but it is the nature of the ligands and their effect on reactivity and selectivity that matter, not the polarity of the pathway. The other important feature of ligand-restricted Heck reactions is their validity not only for Mizoroki–Heck reactions but also for oxidative Heck reactions and other extensions of Heck chemistry. The widest known implementations of ligand-restricted Heck chemistry are the regioselective reactions of electron-rich olefins and enantioselective Heck reactions.

9.2.2   Ligand-Accelerated Reactions

As stated above, such reactions rely on stable ancillary ligands, usually phosphines. However, even the presence of such ligands is not a guarantee of a ligand-acceleration effect and many such reactions actually belong to the first and second types – indeed, there are many "phosphine-free" reactions using precatalysts bearing phosphine ligands or their peer analogues (see below). In order to be treated as ligand-accelerated, the ancillary ligand effects should be explicitly manifested. The most important manifestations are as follows:

1. the ability of a given catalytic system to process less reactive substrates under mild conditions, near room temperatures (20–50 °C) for reactions of deactivated aryl bromides and modest temperatures (100–110 °C or below) for reactions of deactivated aryl chlorides;
2. the ability to process triflates and nonaflates;
3. a wide substrate scope allowing both more reactive and less reactive substrates to be processed;
4. *robustness* of the protocols, predictable trends in the dependence of yields on the reactivity of substrates. This is the robustness which makes ligand-accelerated reactions the tools of choice for product-oriented synthetic studies, which indeed often prefer reliable "load-and-forget" protocols, rather than the prospect of tedious tailoring of some exceptionally efficient and economical procedure developed and polished for a handful of model substrates.

As already stated, the classical ligand-accelerated protocol developed by Heck uses Ph$_3$P or (*o*-tol)$_3$P to ensure stable reproducible procedures involving iodo- and bromoarenes. Stability and reproducibility are the key notions here – it is well known that Mizoroki–Heck reactions of iodoarenes catalysed by phosphine complexes are many orders of magnitude slower than "phosphine-free" reactions, which makes the term "ligand-accelerated" somewhat ambiguous in this context. TON and turnover frequency (TOF) records are always never obtained in reactions involving phosphine complexes of palladium. Nevertheless, synthetic chemists often choose phosphine-assisted catalysis for their syntheses. For example, in the review by Torborg and Beller of industrial applications of Pd-catalysed reactions,[145] most of the cited processes use (*o*-tol)$_3$P-based catalytic systems for multi-kilogram syntheses, the few remaining being "phosphine-free" reactions. Apparently, from the viewpoint of technology, the reliability and reproducibility of the process outweigh the expense of the ligand and generous precatalyst loading and such merits are inherent in ligand-accelerated Heck chemistry. The same is true in complex organic syntheses. A recent example of the use of the classical Heck protocol for selectively linking two complex natural compounds, each abundant in functionality and with chiral centres and a second double bond, is illustrated in Scheme 9.31.[146]

Simple phosphine complexes can sometimes solve challenging tasks. For example, although sp^3 substrates remain in general an unresolved target, in specific cases, *e.g.* involving electron-deficient arylsulfonyldifluoromethyl bromide, a Mizoroki–Heck reaction at the sp^3 centre was realized by Reutrakul and co-workers through the use of the Pd(PPh$_3$)$_4$ complex at a high initial loading.[147]

The first classical ligand-accelerated protocols using basic triarylphosphine ligands exhibited low reactivity with deactivated aryl bromides and zero reactivity with aryl chlorides. The search for more reactive catalytic systems was intensive and advanced in parallel with similar studies in the cross-coupling chemistry. The latter brought forward multiple effective protocols serviced by many advanced ligands of the phosphine and carbene families optimized for solving specific tasks for different combinations of substrates, and gave exceptionally mild room-temperature methods for processing the least reactive electron-rich and sterically hindered chloro

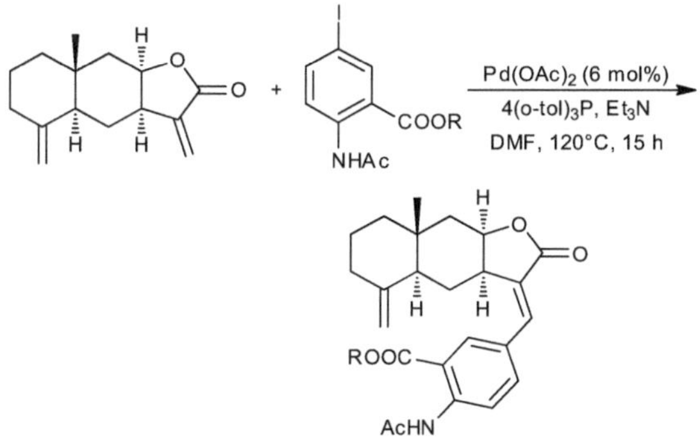

Scheme 9.31 Classical Heck protocol for the synthesis of a hybrid of two natural molecules, isoalantolactone and lappaconitine (R is the diterpene part of this alkaloid).

derivatives. Comparison with the current status of cross-coupling excellently highlights the difference between it and Heck chemistry, in which the course of development was closely similar, but the results were much more modest. In fact, there is just a single well-established new-generation protocol, discovered by Hartwig and co-workers by high-throughput ligand screening,[148] and thoroughly investigated by Littke and Fu[126] to rely on a single ligand – tris(*tert*-butyl)phosphine. None of the other state-of-the-art ligands that gave a powerful push to C–C and C–heteroatom cross-coupling chemistry turned out to be effective and generally useful in Heck chemistry.

However, even with the advent of the new-generation phosphine-accelerated Heck chemistry there are still no peer protocols for

(a) room-temperature reactions of chloro derivatives;
(b) processing of sterically hindered substrates, *e.g.* aryls with a 2,6-disubstitution pattern;
(c) protocols for less reactive aryl tosylates, mesylates and similar substrates (alkenyl tosylates are reactive; see below). Chloroarenes have so far failed to achieve a decent place in Heck chemistry, whereas in cross-coupling reactions the availability of good mild protocols has made chloroarenes common substrates, no less useful than bromides and triflates.

Ligand-accelerated protocols use rather high loadings of Pd precatalysts, rarely less than 1 mol% and often higher. This could be regarded as a paradox – indeed, if such ancillaries accelerate the rate-limiting oxidative addition stage, then why are such systems capable of running no more than 100 catalytic cycles? As a matter of common sense, the reason for such a low efficiency of "well-defined" Pd catalysts should be associated with the coordination chemistry of palladium. Active species most likely contain a single phosphine ligand. Hartwig and co-workers[149] established that monophosphine species easily lose the phosphine ligand after oxidative addition to afford an anionic phosphine-free species, existing in solution as a dimer. The anionic complex is more reactive towards olefins than phosphine complex (Scheme 9.32).

Thus, in the ligand-accelerated catalytic cycles monophosphine complexes are unstable due to displacement of phosphine from the shell, and also

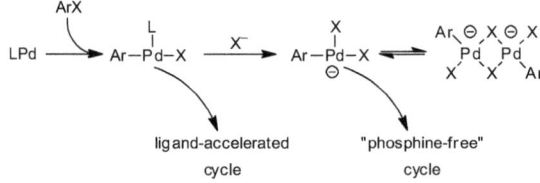

Scheme 9.32 Switching from a ligand-accelerated to a phosphine-free catalytic cycle.

other well-known reactions leading to phosphine depletion (catalysed oxidation, aryl scrambling, *etc.*). The cycle is supported for some time because 2 mol of phosphine are usually loaded, so the monophosphine complex would be partially recovered, but the concentration of such species would be low.

Hartwig and co-workers experimentally demonstrated a very important feature of ligand-accelerated processes.[149] Such processes readily and spontaneously switch to "phosphine-free" processes if (a) more reactive aryl halides are used and (b) the reactions are run at higher temperatures, permitting "phosphine-free" oxidative addition even of less reactive substrates. Phosphine-free and ligand-accelerated processes can run competitively. Therefore, even the phosphine complexes of palladium can serve as precatalysts for "phosphine-free" catalytic processes.

Murray *et al.*, by screening a large set of Mizoroki–Heck reactions of aryl bromides with a wide range of electronic and steric demands with terminal and internal olefins under widely varied conditions and a large set of both phosphine-free precatalysts and palladium–phosphine complexes, identified a general protocol suitable for the majority of substrate combinations (Scheme 9.33).[150] Not surprisingly, the protocol thus identified is very close to the Littke–Fu protocol with some variations of solvent/additive. In addition to tris(*tert*-butyl)phosphine, taken as preformed Pd^0 (PdL$_2$) or Pd^I ([Pd(L)Br]$_2$) complexes, a bulky electron-rich analogue of dppf was found to be an equally effective ligand delivering somewhat more uniform results across the series.

A few recent examples are presented to show the possibilities of the Littke–Fu protocol, with particular attention to a few instances of the use of other ligands. Thus, the arylation of amidoacrylates by highly electron-rich substituted *p*-bromoanilines using the Littke–Fu protocol at room temperature gives amidocinnamates, further used in enantioselective hydrogenation to give optically active amino acids with high yields and enantioselectivities[151] (Scheme 9.34).

Very interesting phenomena throwing light on the detailed operation of the ligand-accelerated Heck reaction and ligand control over selectivity were established by Skrydstrup and co-workers in a series of papers devoted to the use of alkenyl tosylates and phosphonates.[152–154] It should be noted that the application of such leaving groups in Heck chemistry is practically unknown, although due to the development of new-generation phosphine ligands they are fairly well accommodated in cross-coupling chemistry. Skrydstrup and

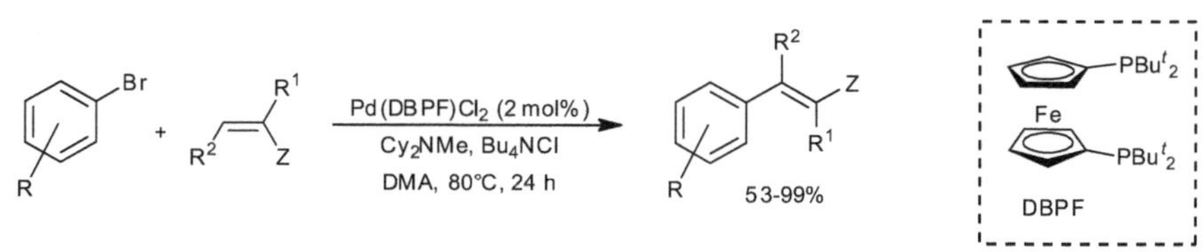

Scheme 9.33 Generic ligand-accelerated protocol suitable for a broad range of aryl bromides and olefins.

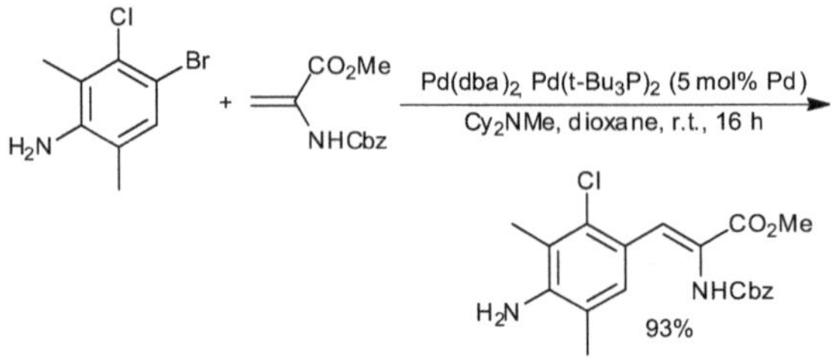

Scheme 9.34 Room-temperature arylation of an amidoacrylate by an electron-rich aryl bromide using the Littke–Fu method.

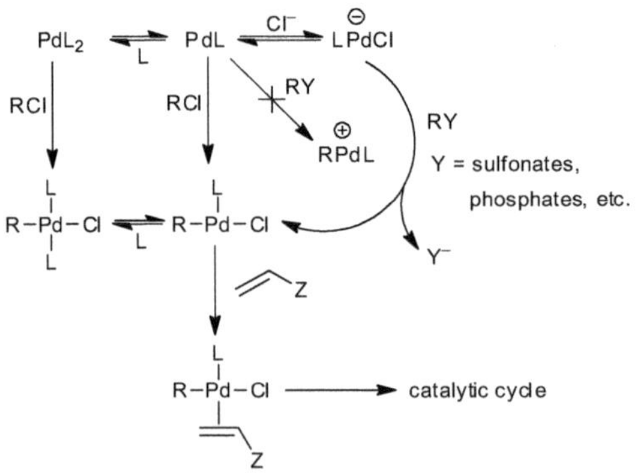

Scheme 9.35 Ligand-accelerated catalytic cycle initiation in the case of a halide substrate (usual neutral pathway, RCl shown for uniformity) *versus* a substrate with a highly nucleofugic leaving group.

co-workers discovered that with a small but very essential modification, the Littke–Fu protocol can be applied to alkenyl tosylates and phosphonates. The original Littke–Fu method was applied to halide substrates and is believed to follow the neutral pathway. With tosylates and similar substrates bearing highly nucleofugic leaving groups, a switch-over to the polar pathway would seem to be inevitable, but such a pathway is probably energetically disfavoured. The addition of a soluble chloride salt is likely to lead to a switch-back to the neutral pathway, most likely *via* prebinding of chloride and formation of an anionic complex, as the alternative, post-binding after oxidative addition, is less probable, as it does not explain the facilitation of the process in the presence of chloride. In any case, a polar solvent is required for ionic ligand binding to take place (Scheme 9.35).

With bulky alkenyltosylates (phosphonates), another intriguing feature was disclosed, namely rearrangement of the alkenyl residue taking place *via* a very rare PdH elimination from sp^2 centres (unambiguously proven by experiments with deuterated substrates). In this case, the direct carbo-palladation of the initially formed alkenylpalladium complex is probably sterically retarded to make possible the establishment of an elimination–addition equilibrium to bring into play the other channel with rearranged

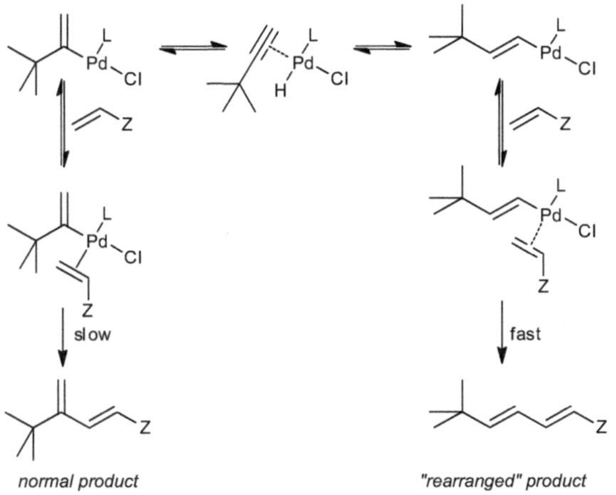

Scheme 9.36 Reversible addition–elimination of PdH in the rearrangement of an alkenyl residue, the Curtin–Hammett regime.

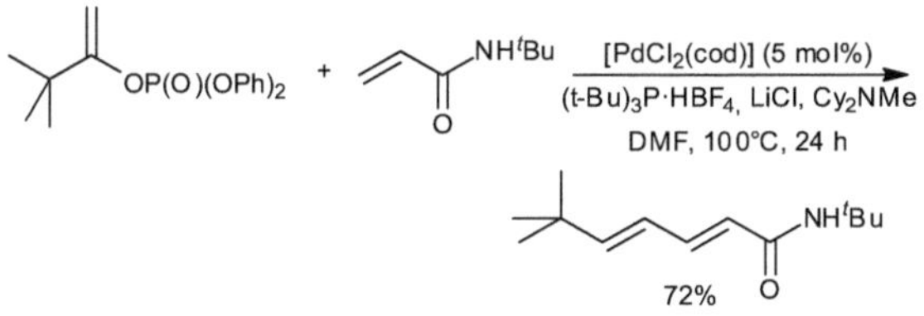

Scheme 9.37 Example of alkenylation with rearrangement.

alkenylpalladium intermediate (Scheme 9.36). The selectivity in such cases is known to obey Curtin–Hammett control (the other case of Curtin–Hammett control in Mizoroki–Heck reactions is discussed in Ref. 87).

The result is a spectacular rearrangement applicable to a wide range of bulky alkenyl tosylates and phosphonates and various olefins (Scheme 9.37).

Moreover, Skrydstrup and co-workers[152] showed that the rearrangement can be controlled by a judicious choice of ligand. This is one of the rarest cases in Heck reaction chemistry where such tight control over the course of the reaction is established through a monodentate ligand. The use of the bulky phosphine XPhos allowed the rearrangement to be blocked. In this case, XPhos most likely interferes with the alkenyl ligand rotating and occupying the in-plane conformation in which Pd and H atoms are close to each other to allow *syn* elimination (Scheme 9.38).

The result is a complete absence of rearrangement in practically the same system as shown in Scheme 9.37 with only the ligand being changed to XPhos (Scheme 9.39).

The application of the ligand-accelerated Heck reaction in enantioselective arylation/alkenylation was demonstrated by Datta and Larhed[155] using

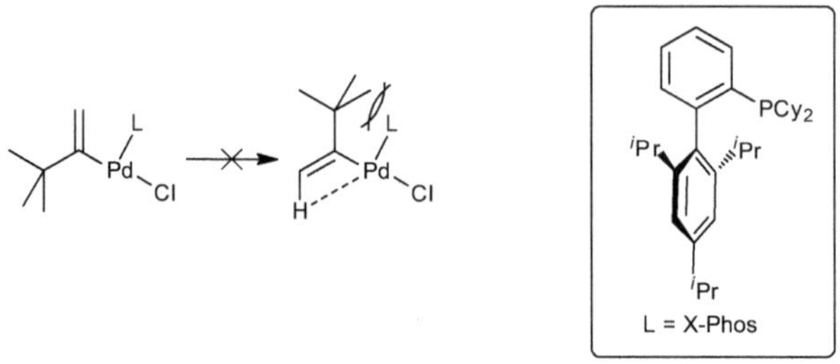

Scheme 9.38 Ligand control over *syn* elimination.

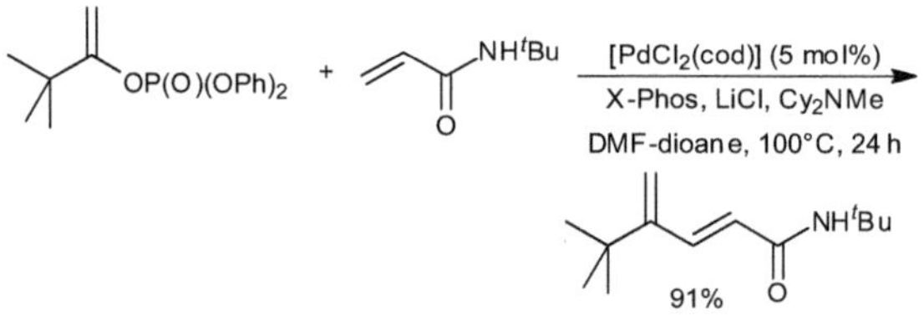

Scheme 9.39 Alkenylation without rearrangement, blocked by the XPhos ligand.

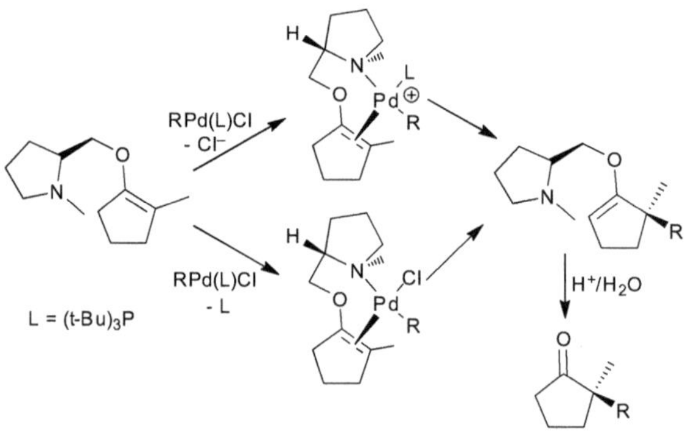

Scheme 9.40 Asymmetric induction by chiral pendant serving as ancillary ligand for Pd.

olefins modified by chiral pendants capable of coordination to the Pd centre (Scheme 9.40). The protocol developed was highly successful, affording high yields of arylated and alkenylated ketones with high optical purity (90% and more, Scheme 9.41). A broad preparative scope with consistently high yields obtained with a wide range of aryl and vinyl chlorides and vinyl triflates unambiguously reveals the key role of the (*t*-Bu)$_3$P ancillary. The same protocol without modification applies to aryl bromides.[156]

A similarly high enantioselectivity with the same olefin was obtained with aryl triflates in a system based on Ph$_3$P ligand, with aryl bromides in a system based on JohnPhos ligand and palladacycle precatalyst for aryl bromides and with a "phosphine-free" system for aryl iodides, with media and additives kept the same (Scheme 9.42).[157]

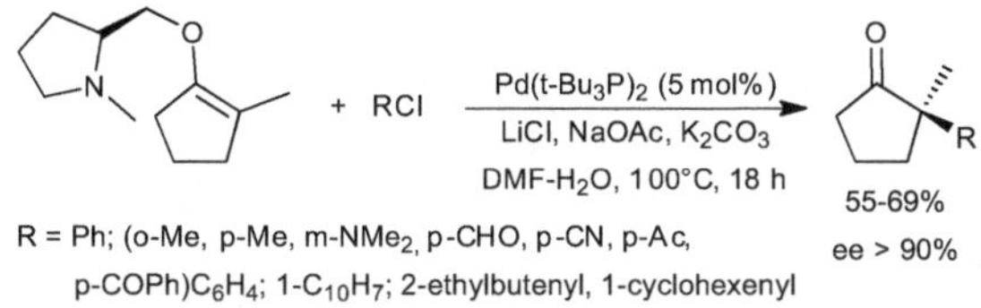

Scheme 9.41 Enantioselective arylation and alkenylation by the ligand-accelerated Heck reaction.

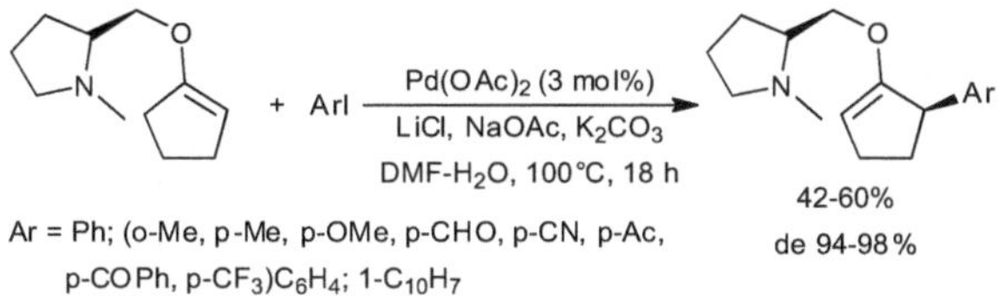

Scheme 9.42 Enantioselective arylation by aryl iodides in a "phosphine-free" catalytic system.

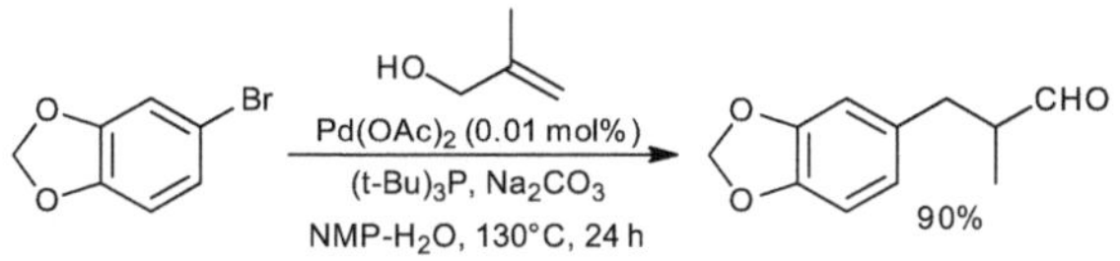

Scheme 9.43 Example of a high TON reaction in the presence of $(t\text{-Bu})_3\text{P}$ ligand.

The use of tris(*tert*-butyl)phosphine, however, does not guarantee that a given reaction with some new olefin will run in the ligand-accelerated mode (*cf.* Scheme 9.32). An interesting example of such behaviour can be seen in a study of the arylation of methallyl alcohol by aryl bromides by Scrivanti *et al.*[158] The initial attempt to use the standard Littke–Fu protocol using $(t\text{-Bu})_3\text{P}$ ligand and Cy_2NEt base at temperatures up to 120 °C gave modest yields only marginally better than those obtained with Ph_3P ligand. A sharp increase in yield was observed on increasing the temperature to 130–140 °C, changing the base to Na_2CO_3 and using highly polar aqueous *N*-methylpyrrolidone (NMP) solvent. Such an effect can probably be interpreted as that in this case the Pd complex with $(t\text{-Bu})_3\text{P}$ ligands behaves as a typical slow-release precatalyst, with phosphine ligands completely shed at the temperature of reaction. This is additionally manifested by the observation of high TON and low Pd loading (TON up to 20 000, TOF exceeding 1000 h^{-1}) (Scheme 9.43), which is a behaviour not so far observed for ligand-accelerated Mizoroki–Heck reactions.

Among phosphine ligands that have been used for performing Mizoroki–Heck reactions, one ligand – the tetrapodal phosphine Tedicyp,[159] developed by Doucet, Santelli and co-workers, occupies a special place. Probably no other ligand, including the classical Ph_3P and $(o\text{-tol})_3\text{P}$, has been tested so widely towards aryl and alkenyl bromides[160] and an overwhelming variety of alkenes, including not only the standard olefins,[160–162] but also terminal alkenes,[163] alkenoic acids,[164] vinyl sulfides, sulfoxides, sulfones, sulfonates,[165] allylic alcohols,[166–168] enol ethers,[169–171] vinyl ketones,[172,173] allenes[174] and even such exotic species as alkenylidenecyclopropanes (Scheme 9.44).[175]

The reactions with practically all the substrates tested were run under similar conditions using common bases (K_2CO_3, $NaHCO_3$ or NaOAc) in polar aprotic solvents (DMF, DMA) at temperatures above 100 °C. The substrate scope included electron-rich and sterically hindered aryl bromides including some really challenging 2,6-disubstituted molecules (Scheme 9.45).

Reactions were typically run with moderate palladium loadings, but some reactions were effectively performed with as low as 0.001 mol% Pd and lower, giving very high TON and TOF values, reaching 10^8 and 5×10^6 h^{-1}, respectively, for selected activated bromoarenes and iodoarenes.[176] Thus, the behaviour of Tedicyp ligand in catalytic systems, particularly the ability to process a broad range of different substrates under similar conditions, agrees with the ligand-acceleration paradigm. On the other hand, the active form of the catalyst is unknown. Phosphine centres are too remote from each other to allow tetra- or tridentate chelation, hence the ligand is most likely engaged in bidentate chelation with either of the phosphine centre pairs to form seven- or eight-membered rings.[177] Such chelates are likely to serve as precatalysts. It is noteworthy that even the reactions with highly reactive aryl iodides require temperatures above 100 °C and such switching on of the activity implies that the original chelates are inactive and that heating leads to chelate cleavage. The systems fail with electron-rich aryl

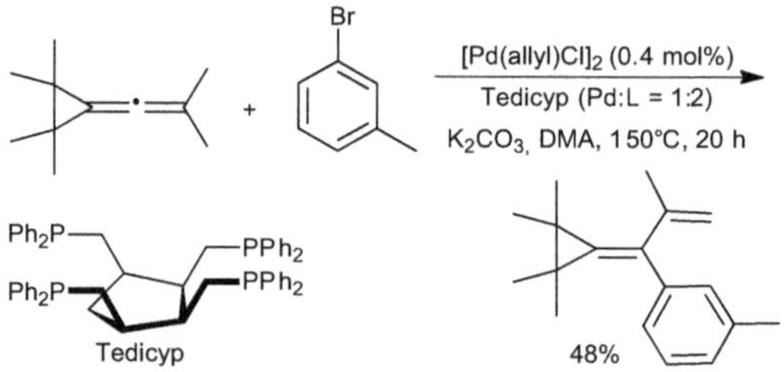

Scheme 9.44 Arylation of alkenylidenecyclopropanes using the tetraphosphine Tedicyp.

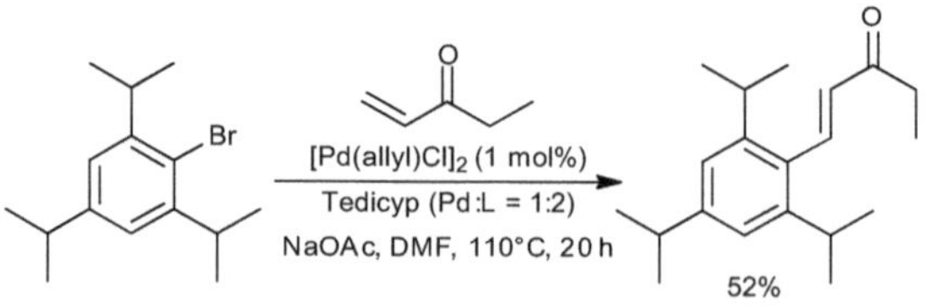

Scheme 9.45 Arylation of an enone by a sterically hindered aryl bromide in the presence of the Pd complex of Tedicyp.

chlorides and actually the results with typical aryl bromides are not much different from those obtained in "phosphine-free" systems, thus obviously falling behind in activity (see below), although undoubtedly being superior in scope. Thus, on the one hand the system works very similarly to slow-release precatalyst (SRPC)-based systems (activation at high temperature, peak performance with reactive substrates), but on the other hand is un-precedentedly versatile towards substrates of different reactivity. Therefore, it could be hypothesized that the resting stable chelate, which maintains a pool of palladium protected against decomposition and deactivation, is thermally activated to release the monophosphine complex (Scheme 9.46), very stable and active to support the reaction of deactivated aryl bromides. In this respect, Tedicyp-based systems provide an interesting and special case of phosphine-assisted catalysis, well supported by ample evidence.

Heterocyclic carbene ligands are apparently the most important major group of non-phosphine ligands introduced so far into transition metal catalysis. A large series of carbene ligands including mono- and bidentate, normal and sterically hindered ligands giving homo- and heteroleptic complexes, and also pincer-type structures, were obtained and tested in palladium-catalysed reactions. Bulky electron-rich carbenes effectively mimic the new-generation phosphine ligands, but also have their own unique features that find them a special and important place in transition metal catalysis.[178] Indeed, in C–C and C–heteroatom cross-coupling reactions, and also in some other important Pd-catalysed reactions, many new, effective protocols involving carbene ligands and preformed carbene complexes have been introduced to demonstrate that the expectations are in general satisfied and a serious competitor to the phosphine family does exist.[179–181] The ability of heterocyclic carbene ligands, particularly the most often used ones, IMes, IPr, SIMes and SIPr (Schemes 9.47 and 9.48), to catalyse cross-coupling reactions with the least reactive electron-rich and bulky aryl chlorides and bromides under mild conditions at room temperature or gentle heating below 60 °C is well documented.

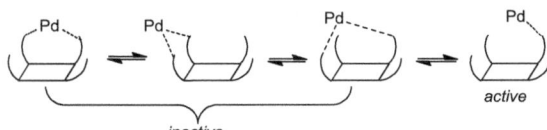

Scheme 9.46 Probable coordination modes of the tetraphosphine Tedicyp.

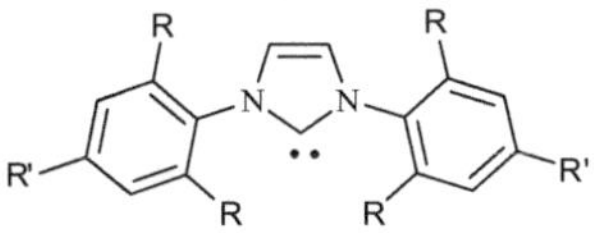

Scheme 9.47 Heterocyclic carbene ligands IPr (**3**) (R = ^{i}Pr, R′ = H) and IMes (**4**) (R, R′ = Me).

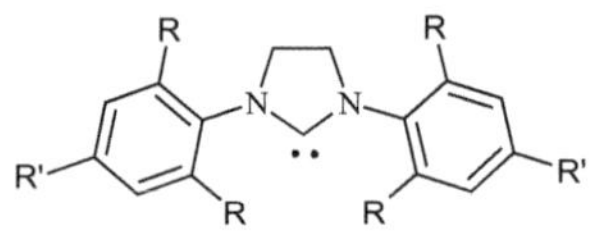

Scheme 9.48   Saturated heterocyclic carbene ligands SIPr (**5**) (R = ^{i}Pr, R′ = H) and SIMes (**6**) (R, R′ = Me).

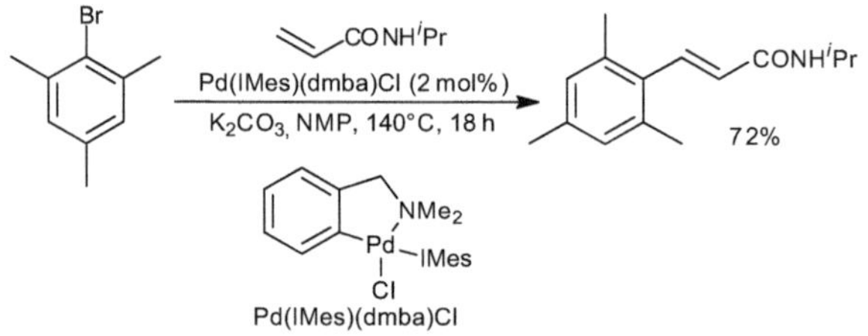

Pd(IMes)(dmba)Cl

Scheme 9.49   Example of probable ligand-accelerated Mizoroki–Heck reaction using the heterocyclic carbene ligand IMes delivered in a hybrid palladacyclic precatalyst.

Strange as it might seem at first glance, there is an exclusion from the list of victories held by carbene ligands in catalysis – namely the Mizoroki–Heck reaction, in which heterocyclic carbene ligands and their preformed complexes showed a rather typical performance, which is roughly similar to the behaviour of various SRPCs based on other ligands, and no clear indications of ligand acceleration was demonstrated (a detailed analysis can be found elsewhere[124]). The carbene complexes served as precatalysts for reactions of aryl bromides and activated aryl chlorides,[179,182–191] but no reports involving electron-rich bulky aryl bromides or chlorides or other types of substrates in wide-scope protocols under mild conditions, requiring unambiguous ligand acceleration, have been reported. Given that carbene-based catalysts readily process unreactive chloroarenes in cross-coupling reactions, including the Sonogashira–Hagihara reaction, which is sometimes perceived as a close relative of the Mizoroki–Heck reaction, such a modest performance in the latter is unexpected. Therefore, most probably in the Mizoroki–Heck reaction the heterocyclic carbene complexes commonly serve as slow-release palladium precatalysts.

A few recent reports have shown that carbene ligands can sometimes deliver ligand acceleration in Mizoroki–Heck reactions. For example, Kantchev and co-workers[192] showed that a hybrid palladacycle–carbene complex can be used as precatalyst for reactions of a wide range of aryl iodides and bromides, including hindered and electron-rich compounds, with acrylates, acrylamides and styrenes, giving consistently good to high yields (Scheme 9.49), although the reaction conditions were rather harsh, fitting the SRPC better than the ligand-acceleration paradigm.

The system was useful for triflates, the substrates for which "phosphine-free" systems are generally not effective, thus providing solid evidence in favour of the stable ancillary effect. The reaction, however, was highly dependent of additives, typical for "phosphine-free" reactions (Scheme 9.50).

With aryl bromides, diarylation was also developed into a useful protocol applicable to various aryl bromides. All these data suggest that the palladacycle–carbene complex on heating releases active Pd(IMes) complex, serving as catalyst. The high temperature required for aryl iodides, bromides and triflates, that is disregarding the reactivity of the substrate, is thus accounted for by precatalyst thermal activation *via* stripping off the palladacycle ligand shell. In a more recent study, a similar complex of IPr carbene with ferrocenylimine palladacycle showed a more typical performance limited to aryl and alkenyl bromides and iodides and standard olefins, but also showing consistently high yields for about two dozen different substrates, many of them electron-rich (system 0.2 mol% Pd, K_3PO_4, Bu_4NBr, DMF, 140 °C).[193]

Whereas in the conventional Mizoroki–Heck reactions the heterocyclic carbene ligands have so far failed to find a reasonable role, in the very exotic Heck alkylation one of the few existing protocols does rely on a carbene ligand. Intramolecular Heck alkylation was demonstrated by Firmansjah and Fu to afford exomethylenecyclopentenes *via* ligand-controlled selective carbopalladation using an alkyl bromide or chloride reactive centre (an example is shown in Scheme 9.51).[194] Only the heterocyclic carbene SIMes was found to be able to control the process by suppressing premature β-hydride elimination.

Among new additions to the arsenal of ligands, an unprecedented phosphazane system was reported by Iranpoor *et al.*,[195] which showed very unusual performance. The phosphazanes are structurally different from common phosphorus-based ligands used in Pd catalysis, bearing a certain similarity only to Verkade's triaminophosphine and proazaphosphatrane ligands, which, however, have not been found particularly useful in Heck chemistry. The new ligand can be used not only in common polar solvents (DMF, DMSO) but also in the non-polar solvent toluene and even in neat water in heterogeneous systems not using any solubilizing or phase-transfer

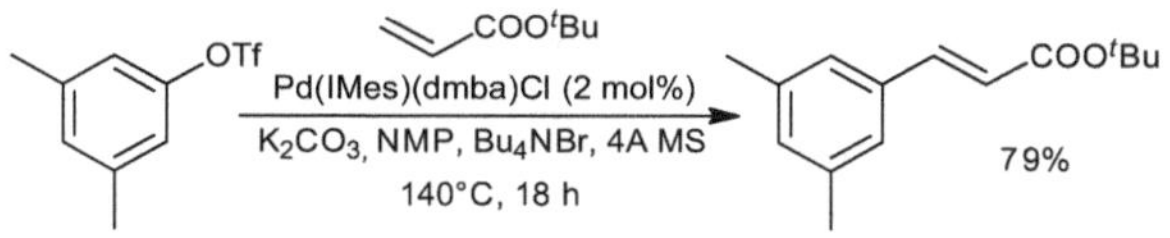

Scheme 9.50 Example of arylation by triflates in the presence of IMes complex.

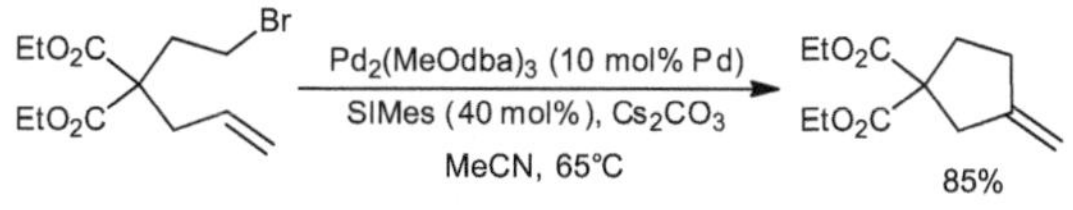

Scheme 9.51 Example of an intramolecular Heck reaction involving an sp^3-carbon centre.

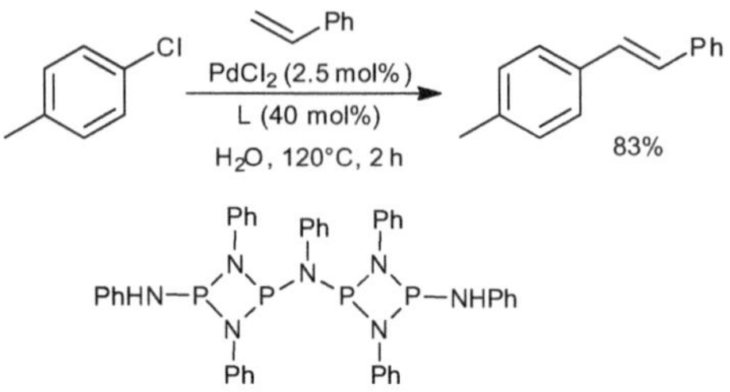

Scheme 9.52 Example of a phosphazane ligand in the "base-free" Mizoroki–Heck reaction of electron-rich chloroarenes.

agents. Moreover, even the addition of base was not needed because the ligand was observed to serve as such. Such a dual function makes it necessary to use a huge excess of such a ligand (2.5 mol% Pd per 40 mol% ligand) comparable to the substrate loading, assuming that such compounds can take several protons (Scheme 9.52). Such systems were used with a wide range of aryl iodides and bromides at 100 °C and chlorides at 120 °C. Hence it can be safely concluded that this unusual ligand indeed shows unambiguous evidence of ligand acceleration combined with excellent performance in neat aqueous media. The catalyst was even found to be recyclable up to 10 times.[195]

9.2.3 Systems with Ill-Defined Catalytic Species

Mizoroki–Heck reactions involving Pd species without definite stable ancillary ligands, as already introduced, fall into two distinct categories. Those using iodoarenes and selected reactive electron-deficient bromoarenes can take place in the presence of any palladium source. On the other hand, the reactions of less reactive bromo- and chloroarenes require careful optimization of simple catalytic systems for a particular application and a proper choice of palladium precatalyst.

There were two seminal works that set the milestones in this chemistry, both involving precatalysts containing the well-known tris-*o*-tolylphosphine ligand, but delivering activity far beyond the expectations based on the knowledge gained previously on systems with this and other phosphine ligands. Spencer was the first to discover that even less reactive substrates such as reactive electron-deficient aryl chlorides or bromides can be made to react in a system that utilizes a simple palladium precatalyst, if conditions such as base, solvent, temperature and palladium loading are thoroughly optimized.[45,46,196,197] The idea was not immediately appreciated, however, because such catalytic systems were unstable and inconvenient for practical use. The reactions in such systems often end up swiftly in the formation of palladium black and sudden termination, hence many researchers considered them to be poorly reproducible and give them up for more reliable ligand-accelerated systems.

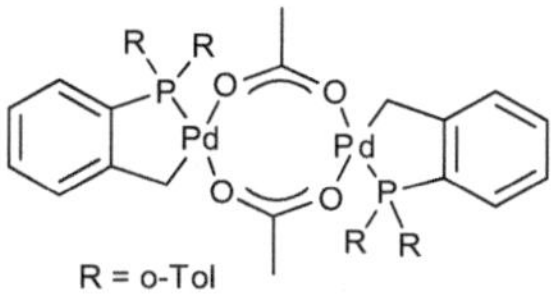

R = o-Tol

Scheme 9.53 The Herrmann–Beller palladacycle (trade name CataCXium C) (7).

Strong interest was drawn to these phenomena with the publication by Herrmann, Beller and co-workers on the catalytic properties of the palla- dacycle readily formed in the reaction of palladium acetate with (*o*-tol)$_3$P (Scheme 9.53).[198] This material had been known before, being discovered first by Heck himself and then used by Spencer, but only in this last study[198] was it noted that the material is not just some interesting Pd complex, but a potent source of catalytic activity.

The compound attracted intense interest and brought forward several new concepts, including the following:

(a) The idea that at least some well-defined catalysts can operate not *via* the standard Pd(0)/Pd(II) catalytic cycle, but *via* the Pd(II)/Pd(IV) route, if the structure of the precatalyst (palladacycle or pincer palladabi- cycle) is presumed to be conserved throughout the cycle.

(b) Monophosphine catalysis, if the palladacyclic structure is presumed to be partially cleaved to release the very reactive PdL complex.

(c) Catalysis by nanoparticles, if the precatalyst is presumed to be cleaved completely, giving rise to unstable Pd(0) species that rapidly form fine and highly reactive Pd sols.

The study of each of these concepts, particularly that of monophosphine catalysis, gave many interesting results in other areas. In reality, and this was eventually directly demonstrated by de Vries and co-workers,[199–201] various palladium precatalysts from simple salts to sophisticated palladacycles give essentially similar results when reactions are carried out under comparable conditions, which means that the reactions are actually catalysed by Pd species bearing no specific ancillary ligands, but instead containing arbi- trary ligands from the reaction media, of which the anions coming from bases and salt additives used perform an important role in inhibiting early aggregation and deactivation of the catalyst. Palladium sols (nanoparticles) play only a subsidiary role, serving as pools of palladium species *via* accretion–redissolution processes. Simple palladium precatalysts such as Pd(OAc)$_2$ alone can serve as well as any sophisticated precatalyst if a suf- ficiently high temperature is applied and the reaction conditions are care- fully optimized. Sol-stabilizing additives, such as tetraalkylammonium salts, are known to facilitate such reactions, making them less critically dependent on specific conditions. This approach is well known as the Jeffery or Jeffery– Larock method.[202–205] Various technical improvements allowing fast heating and mixing of reagents, *e.g.* microwave and microflow techniques, further facilitate such reactions.[206,207]

In some cases, even with aryl iodides, the net rate of the catalytic cycle can be slow owing to other factors,[208] *e.g.* weak binding of electron-deficient olefins and/or low solubility of aryl iodides in the reaction medium. In this case, the TOF would be low and withdrawal of palladium from the catalytic process occurs. The addition of an extra amount of precatalyst was required to ensure high yields, probably because the generation of a reactive olefin by elimination lags behind the rate of the Mizoroki–Heck catalytic process, hence the Pd(0) concentration increases up to levels where the nucleation and growth of inactive metal withdraws the catalyst. Such systems nevertheless can be useful even in challenging cases.[209]

Studies on new catalysts meanwhile continued and hundreds of different complexes, including simple and pincer palladacycles, carbene complexes, various chelates and hybrid complexes involving palladacyclic, carbene and phosphine ligands, were obtained and assayed for catalytic activity in model Mizoroki–Heck reactions (for a comparative compilation of results, see Ref. 124). Almost all of the results, except for a very few,[124] could be fitted within the concept that disregarding the initial form used, during the pre-activation stage (marked by induction periods of variable length) all specific ligands were stripped and the catalytic cycles were led by ill-defined Pd species with coordination shells filled by solvent molecules and ions from the respective reaction media. Hence the sophisticated ligands behaved like "disposable wrappers" for delivery of Pd, and therefore the simplest possible structures based on very cheap ligands ("junk ligands") can behave as well as new, sophisticated, specially synthesized ligands.[210] The role of the precatalyst is just to supply Pd species at a rate compatible with the consumption rate (see above), thus avoiding deactivation through aggregation to clusters, then nanoparticles and their death *via* Ostwald ripening.

To describe this behaviour, it is convenient to use the concept of slow-release precatalysts (SRPCs),[124] a notion actually inspired by agriculture, where slow-release fertilizers are becoming popular, as they continuously and slowly feed the crops when administered just once at the start of the season. The key idea is to make the rate of release compatible with the rate of consumption. In contrast, common (fast-release) fertilizers give a sudden large overdosage of nutrients, which cannot be consumed and the unconsumed excess fails to feed the roots but instead poisons the environment. A similar thing happens in catalysis by palladium in the catalytic cycles involving less reactive substrates, as simple Pd precatalysts (salts, dba complex, *etc.*) supply Pd(0) at such a rate that it cannot be consumed owing to limitations of the net rate of the catalytic reaction, whereas the rate of release from an SRPC can be compatible with the rate of consumption by the catalytic cycle.

The mechanisms of release involve common reactions at the Pd centre invoking reductive elimination and sequential rupture of chelate bonds. The mechanisms of release have nothing to do with the inherent thermal stability of the complexes, which is often very high, particularly for palladacycles and pincer-type chelates, which in individual form withstand heating above 200 °C. In the reaction medium, the disassembly of such "stable" complexes takes place at modest temperatures, as was recently shown by Campora and co-workers[211] for the well-known Milstein's pincer complexes,[212] which made a great impact on Heck chemistry, initially presumed to serve as the long-awaited well-defined catalysts operating *via* the Pd(IV)/Pd(II) catalytic cycle (*cf.* Scheme 58 in Ref. 87). The disassembly occurs under mild conditions at 50 °C to involve the PdH intermediate, readily formed by β-hydride elimination, *e.g.* from an alcohol (or similarly from an amine or a Wacker-type reaction with an olefin) (Scheme 9.54).

Until recently, the scope of "phosphine-free" procedures was commonly believed to be limited to reactive substrates, mainly aryl iodides and reactive bromides. However, low to modest yields were occasionally recorded for electron-rich aryl bromides and even chlorides, so it was only a matter of time before the stability of catalytic systems would be tuned to afford preparatively meaningful processing of less reactive substrates. Indeed, a number of impressive protocols have appeared in recent years. Such protocols are an excellent demonstration of the capabilities of ill-defined catalysts, but the enthusiasm in their favour should be moderated. Systems of this sort cannot get rid of inherent deficiencies of the approach, the main one of which is poor robustness of the protocols – any variation from the published procedures, such as the choice of substrates, olefins, additives, concentrations, temperatures, *etc.*, would probably be deleterious or even lead to complete failure. All users of the "phosphine-free" approach should take into consideration that it is an art, and therefore before complaining about the irreproducibility, the original protocol should first be tried exactly as published. Due care should be exercised with new precatalysts and their coordination chemistry should be verified in sources not associated with catalysis. The role of impurities should also be taken into consideration, as

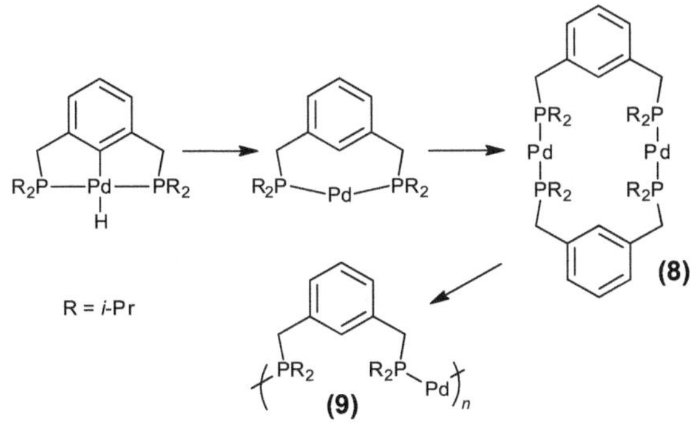

Scheme 9.54 Mechanism of release of Pd(0) from Milstein's pincer complex. Dimeric (**8**) and polymeric (**9**) Pd(0) complexes were isolated and characterized.[211]

"phosphine-free" Mizoroki–Heck reactions can be catalysed by very small amounts of active palladium (the so-called "homeopathic" catalysis[213]).

Palladium complexes of tetraphenylporphyrin and phthalocyanine were shown to sustain the Mizoroki–Heck reaction of aryl bromides and even chlorides with standard olefins, giving incomplete but in some cases fairly high conversions.[214] However, it is known that the stability of Pd complexes of porphyrins and phthalocyanines is enormous; the former can slowly release Pd when heated with concentrated sulfuric acid[215] and the latter is even more robust. Both complexes, however, because of the very harsh conditions of synthesis and if improperly purified, may contain Pd-containing impurities, which can release active species. The results obtained with highly stable Pd complexes should be treated with due caution.

Several phosphine-free systems have been reported that gave very good results not only for electron-rich bromoarenes, but also chloroarenes. Electron-rich chloroarenes were reported earlier from time to time to be reactive under the conditions of "phosphine-free" catalysis, but with low TON values and incomplete conversions (for the analysis of data obtained before 2008 and predictions on the perspectives of SRPCs for less reactive chloroarenes, see Ref. 124). Several recent studies showed that by further optimization and development of new precatalysts, this task can be solved and electron-rich chloroarenes can be made to react at least with standard olefins with high conversions and yields. The conditions, however, are much more important than the nature of the precatalyst and spectacular results can be obtained using common Pd sources.

Thus, molten mixtures of Bu_4NBr (TBAB) and Bu_4NAc (TBAA) were shown by Nacci and co-workers to serve as excellent ionic liquid media for phosphine-free reactions.[216,217] In such media, the halide effect and support of Pd(0) species (see above) should be manifested most strongly due to the very high concentration of ions. These media were used in the protocol, allowing the reaction of chloroarenes, including electron-rich compounds, with a wide range of olefins, including not only standard styrenes and acrylates, but also terminal alkenes, vinyl ethers, *vic-* and *gem-*disubstituted olefins such as methacrylate, crotonate, benzalacetone, cinnamate and stilbene.[218] Pd acetate (1.5 mol%) was used as a precatalyst, which under the conditions applied was rapidly reduced to a fine Pd sol, maintaining an active pool of SRPC for effective processing even of the most stubborn and demanding substrate pairs (Scheme 9.55).

The performance of this system is truly impressive in clearly showing the huge reserves of phosphine-free catalyst systems. Furthermore, a sharp discrimination of reactivity between chloro and bromo substituents can be achieved by simply changing the temperature and reaction time, hence a selective stepwise one-pot transformation becomes possible[218] (Scheme 9.56).

Another system, reported by Röhlich and Köhler,[214,219] takes advantage of a highly stable Pd chelate apparently used as an SRPC. Optimization of the conditions gave a system in which a typical electron-rich chloroarene reacts

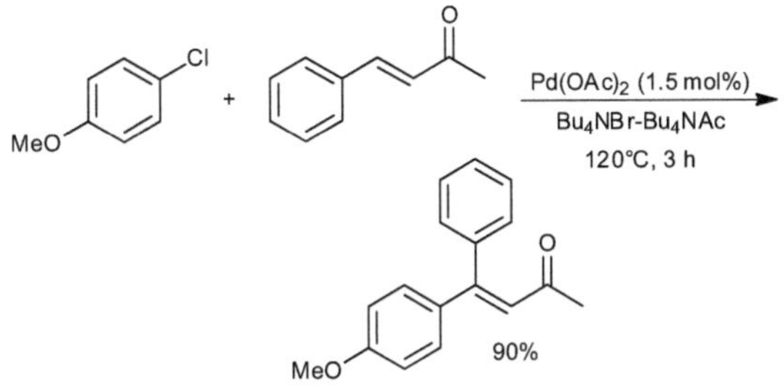

Scheme 9.55 Example of arylation by an electron-rich aryl chloride in a "phosphine-free" system in an ionic liquid medium.

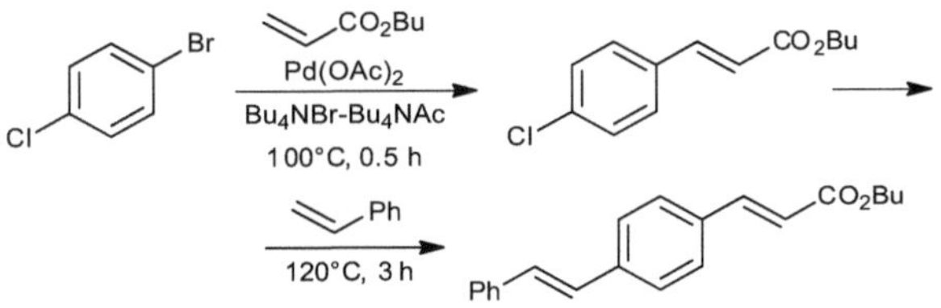

Scheme 9.56 Example of sequential alkenylation in an ionic liquid.

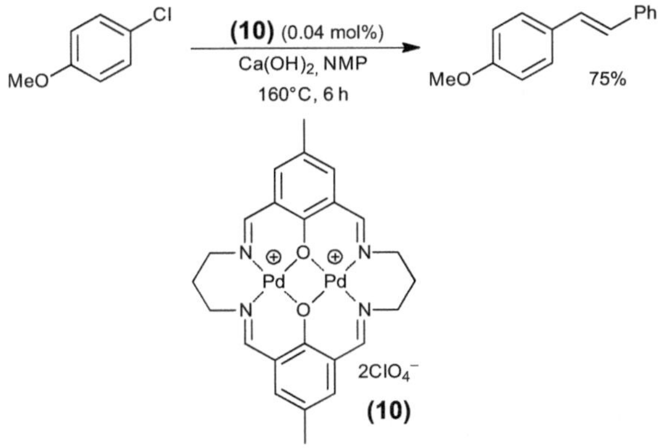

Scheme 9.57 Stable Pd chelate with macrocyclic ligand (**10**) as precatalyst in the Mizoroki–Heck reaction of electron-rich chloroarenes.

with styrene to give the respective stilbene in high yield without the need to use the standard stabilizing agent Bu$_4$NBr (Scheme 9.57). A very low loading of precatalyst (0.0375 mol% Pd) is used. Given that apparently active Pd species are slowly and partially released from the precatalyst, it is evident that Pd species can be fairly reactive in the oxidative addition to an unactivated C–Cl bond without special stable accelerating ancillary ligands.

Another recent example of fitting the precatalyst to the reaction of unactivated aryl chlorides and bromides was reported by Pons and co-workers, who used macromolecular chelators (**11–13**) (Scheme 9.58).[220] Chloro- and

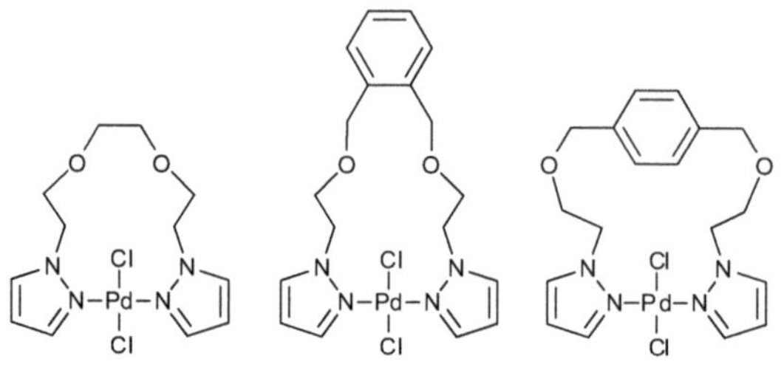

(11)   (12)   (13)

Scheme 9.58 Palladium complexes with macromolecular chelators (**11–13**) used as precatalysts for Mizoroki–Heck reactions of chloro- and bromoarenes.

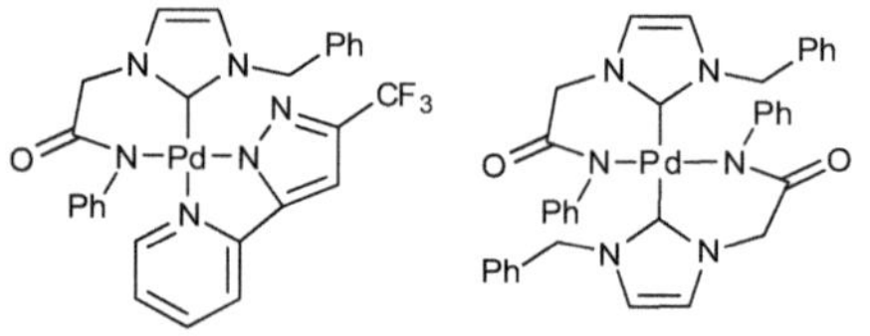

(14)   (15)

Scheme 9.59 Heteroleptic complex (**14**) and homoleptic complex (**15**) used as precatalysts by Lee and co-workers.[221]

bromobenzene reacted with standard olefins in the presence of such complexes (0.1–0.01 mol% Pd, Et$_3$N or K$_2$CO$_3$, Bu$_4$NBr, DMF, 140 °C, 2-72 h) to give good to high yields (56–93% for PhBr) or modest to high yields (20–89%, for PhCl) of the respective products. TON and TOF values for PhBr were up to 7300 and 600 h^{-1} and for PhCl up to 6800 and 120 h^{-1}, respectively, revealing the good activity and stability of systems based on such precatalysts.

A further important achievement in the design of precatalysts was reported by Lee and co-workers.[221] They created a series of sophisticated heteroleptic complexes containing two *N,N*-bidentate ligands – one based on a heterocyclic carbene with attached flexible amide arm being strongly bonded, and the other being more weakly bonded rigid 2-(3-pyrazolyl)pyridine (Scheme 9.59).

This complex, if used in a system with an ionic liquid medium (0.1–0.5 mol% Pd, NaOAc, Bu$_4$NBr, 140 °C), provides an efficient precatalyst for the arylation of standard olefins by aryl bromides and chlorides, including electron-rich and hindered 2,6-disubstituted compounds, with good to high yields. Similar results (a slightly lower yield in half the time) were obtained with homoleptic complexes (the *trans*-complex **15** is shown; the *cis*-isomer exhibited the same performance) of the same ligand[222] (Scheme 9.60).

This system could even be considered as a true ligand-accelerated case with heterocyclic carbene as a stable ancillary, but the behaviour – high reaction temperature and induction period revealing a characteristic preactivation phenomenon – indicates that the complex works as a typical SRPC.

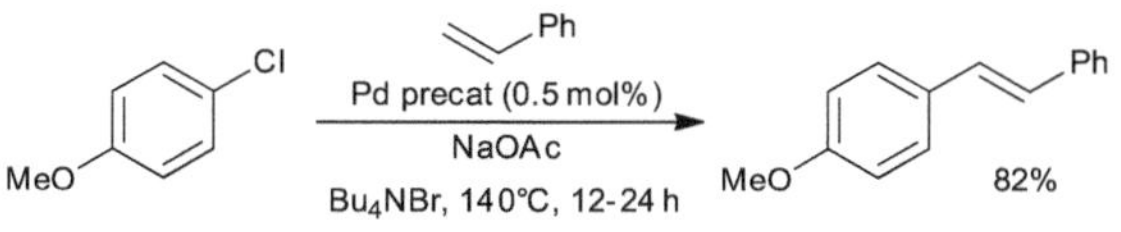

Scheme 9.60 The use of precatalysts **14** and **15** in the arylation of styrene by an electron-rich chloroarene.

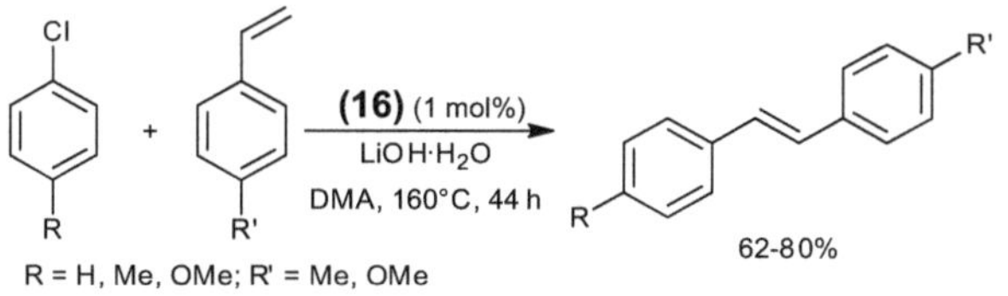

Scheme 9.61 Arylation of styrenes by electron-rich chloroarenes using precatalyst **16**.

Exceptionally thermally stable complexes, such as **16**, were shown by Kantam and co-workers to be among the best so far known precatalysts to afford the Mizoroki–Heck reaction of electron-rich chloroarenes with styrenes[223,224] (Scheme 9.61). These precatalysts work at 160 °C provided that LiOH·H$_2$O is used as base in a polar solvent (DMA or NMP). These precatalysts under such conditions apparently provide optimal levels of active Pd release, so no common additives are required for extra stabilization. The reactions were very slow, however, which reflects both the rate of release and the rate of oxidative addition to C–Cl bonds, which are presumably matched to each other. Reactions were carried out with a 1 mol% Pd loading, but 0.1 mol% could also be used with an insignificant deterioration of the yield to achieve a TON of 620.

These new systems, capable of involving electron-rich chloroarenes, are evidently too different to allow for identification of general trends to account for a high reactivity and performance. Each individual parameter (nature of precatalyst, loading, media, additives, base, temperature, concentrations, *etc.*) can be varied, with the proper combination and conditions to be arrived at only by optimization of a chosen system, and remain valid only for a relatively narrow set of substrates.

For more reactive bromo- and iodoarenes, the offer of new systems is naturally broader. Such systems may not seem very practical, but each shows something important about the operation of "phosphine-free" catalytic cycles based on both regular and slow-release precatalysts. Moreover, it could be argued that the difference between the chloroarene-enabled systems and those developed for more reactive substrates is not dramatic. In fact, further optimization of any less active system can result in a substantial increase in catalytic activity. A cross-comparison of parameters may give such clues, if desired.

Salicylaldehyde thiosemicarbazones as ONS-tridentate ligands, such as **17**, were used for the preparation of a series of heteroleptic Pd complexes. These complexes were used as precatalysts for the Mizoroki–Heck reaction of aryl iodides and bromides with acrylates showing a modest level of activity (system: 0.1–1 mol% Pd, Na_2CO_3, DMF, 130–145 °C, 24–36 h). The least reactive substrate studied was *p*-bromotoluene, giving a 78% yield of methyl *p*-methylcinnamate with a 0.1 mol% loading after 36 h at 145 °C.[225]

In addition to preformed complexes, new ligands are also appearing. The conditions of the catalytic processes, however, show that no apparent ligand-acceleration effect is exhibited and the ligands just bind Pd, forming an SRPC *in situ*. Thus, a bisphosphite derivative of pentaerythritol (**18**) (Scheme 9.62) was tested as a ligand for the Mizoroki–Heck reaction of aryl iodides and bromides with standard olefins. The least reactive substrates used were *o*- and *p*-bromoanisole and *p*-bromo-*N*,*N*-dimethylaniline, giving only modest yields (35–65%) of the respective cinnamates [system: $Pd(OAc)_2$ (1 mol%), L, Na_2CO_3, DMF, 140 °C, 12 h].[226]

Another interesting example was reported by Buckley and Neary,[227] namely the sterically hindered thiadiazolidine *S*-oxide **19**, a structural analogue of the most useful carbene ligands. This ligand was used in a typical catalytic system containing NaOAc and Bu_4NBr for reactions of aryl iodides and bromides with standard olefins to give high yields of the respective products with high TON values. Even the least reactive substrate used, *p*-bromoanisole, was processed with an impressive TON of 740 000 (Scheme 9.63). The reactions were run at high temperatures, achieved by microwave heating.

A tetraselenide forms two different complexes – a bis-pincer (**20**) and a regular chelate (**21**) involving seven-membered rings (Scheme 9.64). Both served as precatalysts for the Mizoroki–Heck reaction of a series of aryl bromides with acrylate and styrene (0.001 mol% Pd as **20** or **21**, *n*-BuNH$_2$, DMA, 100 °C).[228] In spite of the great difference in structure – the pincer Se–C–Se complex should be much more stable than the seven-membered simple

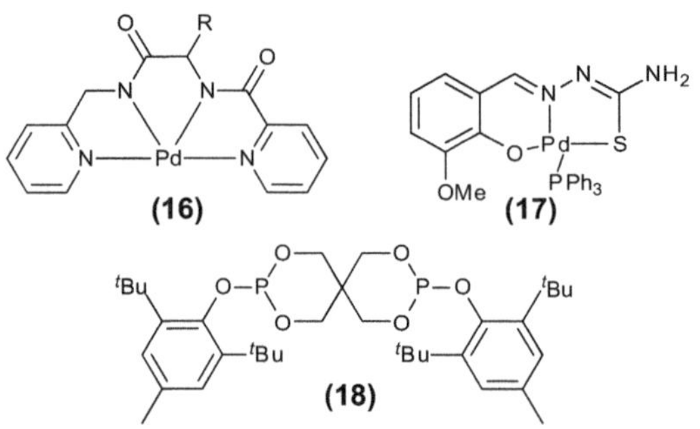

Scheme 9.62   Precatalysts 16–18.

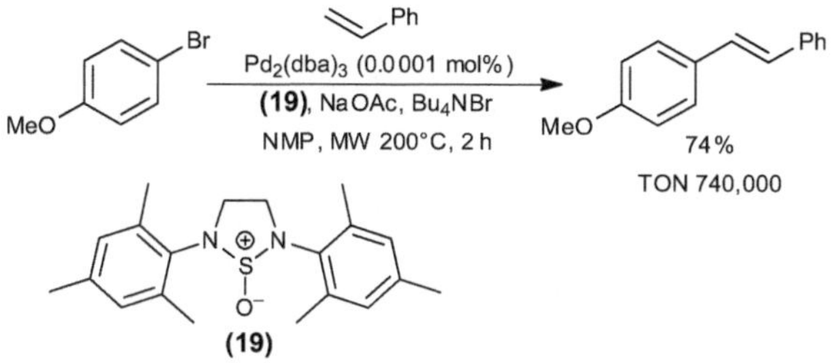
Ph
Br
Pd2(dba)3 (0.0001 mol%)
(19), NaOAc, Bu4NBr
NMP, MW 200°C, 2 h
MeO
MeO
Ph
74%
TON 740,000
(19)

Scheme 9.63 High-TON arylation of styrene by an electron-rich aryl bromide in the presence of ligand **19**.

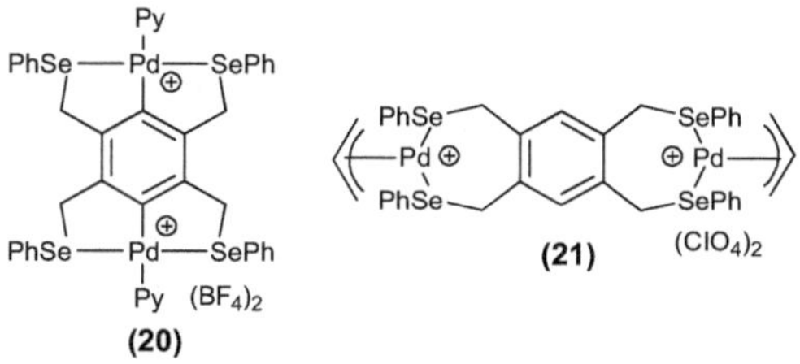
(20)
(21)
(ClO4)2
(BF4)2

Scheme 9.64 Two precatalysts of similar catalytic activity formed from the same tetraselenide chelator.

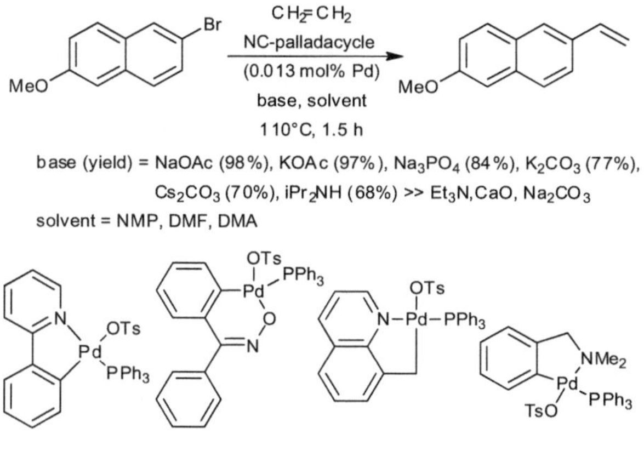
CH2=CH2
NC-palladacycle
(0.013 mol% Pd)
base, solvent
110°C, 1.5 h
base (yield) = NaOAc (98%), KOAc (97%), Na3PO4 (84%), K2CO3 (77%),
Cs2CO3 (70%), iPr2NH (68%) >> Et3N, CaO, Na2CO3
solvent = NMP, DMF, DMA

Scheme 9.65 Monoarylation of ethylene giving naproxen precursor: a comparative study using *N,C*-palladacycles.

Se–Se chelate – the catalytic activity (yields, TON and TOF values) were closely similar, thus showing once again that catalytic activity is associated not with the structure of the precatalyst, but rather with the composition of the catalytic system.

A study giving an unambiguous example of the typical operation of a catalytic system based on SRPC was reported by Kelkari and co-workers[229a] (Scheme 9.65). Working with carefully standardized reaction conditions (reagent concentrations, palladium loading, time, temperature), they showed that the performance of the catalytic system, including palladacyclic precursors based on different scaffolds, is practically insensitive to the nature of the precatalyst, and also base and solvent taken from a series commonly used for such reactions. The structural type of palladacycles taken for the study (hybrid monomeric complexes with a highly labile tosylate ligand to facilitate ligand exchange) permitted effective preactivation under relatively mild conditions. This study was performed not with a reactive aryl halide and standard olefin, but with a more challenging case using an electron-rich aryl bromide and ethylene, which even better reflects the relevance of the results for adequate positioning of catalytic systems based on such precatalysts.

Among the most unusual SRPCs is palladium chloride deposited on wool, reported by Lei and co-workers.[229b] However exotic and unusual this material might seem, its reported performance is very high. The precatalyst was able to service the reactions of unactivated bromoarenes not only with standard olefins, but also with some disubstituted alkenes (methacrylates, 1,1-diphenylethylene, α-methylstyrene) in high yields under remarkably mild conditions at 80 °C in neat water in the presence of a very small amount of PEG-400 (0.2% solution, which is much weaker than the usually used concentrations of such phase-transfer agents) (Scheme 9.66). Under such conditions, the reaction systems are heterogeneous and probably the only way for such a system to work is *via* absorption of organic reagents on the surface of the wool.

Among other systems that have been reported since 2008 is 2,2'-diamino-6,6'-dimethyldiphenyl, used as a ligand in a typical system of Pd(OAc)$_2$ (1 mol%), L, K$_2$CO$_3$, DMF, 140 °C, exhibiting modest activity for Mizoroki–Heck reactions of aryl iodides and bromides with standard olefins and just a single example of an unactivated aryl bromide.[230] Similar activity was reported for the Pd complex of the tetraphosphine *N,N,N',N'*-tetra(diphenylphosphinomethyl)-1,2-ethylenediamine.[231] N–C–N pincers, derivatives of isoxazole, such as **22** (Scheme 9.67) and isothiazole showed typical performances

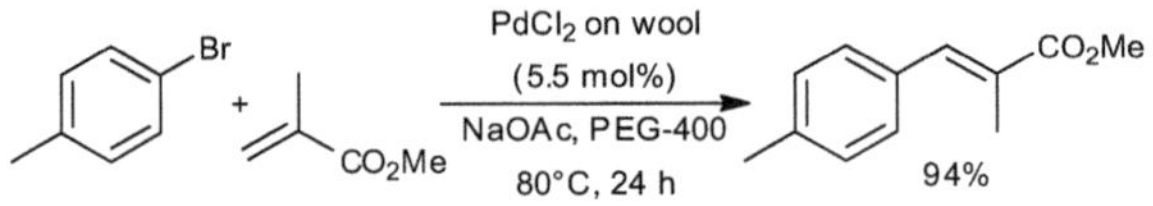

Scheme 9.66 Palladium on wool as an effective precatalyst.

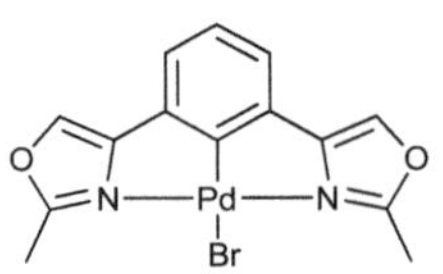

Scheme 9.67 New precatalyst (**22**) for the Mizoroki–Heck reaction of reactive aryl bromides and iodides with standard olefins.

in reactions of reactive aryl bromides and iodides with standard olefins (0.0001–1 mol% Pd, KOAc or DIPEA, NMP, 120–160 °C, 5–40 h).[232]

A pincer Se–N–Se complex behaved as an SRPC with modest activity, showing high TON values for aryl iodides and bromides, but only incomplete conversions and low yields with electron-rich aryl bromides.[233] Several other systems, including phenethylamine palladacycle[234] and amidoamine dendrimer–Pd complex,[235] were tested as typical SRPCs for standard Mizoroki–Heck reactions.

9.2.4 Pd(II)/Pd(IV) Mechanism

The Pd(II)/Pd(IV) mechanism for Mizoroki–Heck reactions has been intensely discussed in connection with palladacyclic precatalysts, but in the absence of decisive evidence the hypothesis was dropped, but recently revived again. Recently, the arguments were partially revived because of increasing interest in Pd(IV) chemistry (see recent detailed reviews[236–238]). The existing manifestations of Pd(IV) complexes in organic reactions explicitly state that so far no evidence on β-elimination from Pd(IV) complexes has been documented, which alone makes the interpretation of Heck reactions with Pd(IV)/Pd(II) catalytic cycles highly speculative.

Concerning the other key step in Heck chemistry, oxidative addition, some Pd(II) chelates,[239] and also P–C–P and N–C–N pincer complexes, were shown to undergo oxidative addition of hypervalent iodine compounds, leading to Pd(IV) intermediates bearing aryl or alkynyl ligands.[240] An aryl–Pd(IV) intermediate thus indeed can form in the interaction of diaryliodonium with Pd(II) pincers, while further reaction of this intermediate with olefins leading to Mizoroki–Heck products has not so far been documented experimentally, although the potential feasibility of such process was confirmed by computational modelling of the Pd(II)/Pd(IV) Heck pathway[241,242]

Diaryliodonium salts are among the most reactive compounds in Pd-catalysed reactions,[243] similarly to arenediazonium salts,[244] but more convenient in handling as they are not so prone to uncontrolled decomposition and side reactions. In simple cases, extremely fast reactions can be performed using neat water as solvent.[245] However, as these substrates are the least atom economical, expensive and severely limited in scope, their application lags behind the chemistry of arenediazonium compounds and the gap is swiftly widening. Both types of substrates are used also because both enforce a polar pathway. The mechanisms of such Mizoroki–Heck reactions nevertheless almost certainly conform to the common Pd(0)/Pd(II) pathways.

Pincer complexes are able to catalyse Mizoroki–Heck reactions of various olefins with diaryliodonium salts under mild conditions. Although the catalytic activity (both TON and TOF) is rather modest, a broad scope of olefins forming normal Mizoroki–Heck products without the usual double bond shifts and other complications makes the protocol useful for complex synthetic tasks (Scheme 9.68). Such behaviour may be a starting point for the

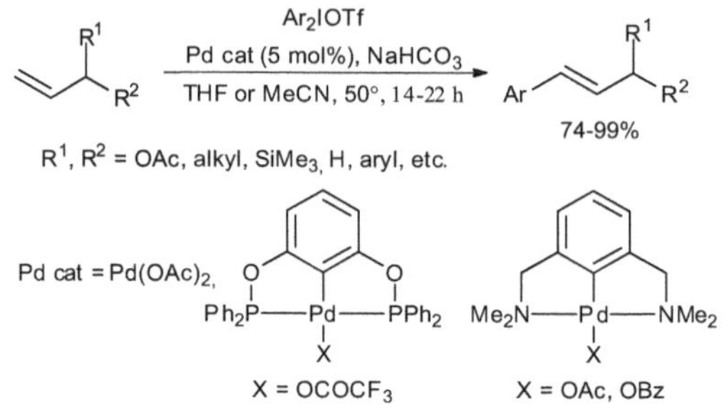

Scheme 9.68 Arylation of allylic substrates by diaryliodonium salts under mild conditions and in the presence of pincer complexes.

probable realization of the long-awaited Pd(IV)/Pd(II) pathway in Mizoroki–Heck reactions.[246]

Important evidence in favour of a non-Pd(0) pathway was declared to be the formation of regular Mizoroki–Heck products with olefins such as allyl acetates, which otherwise could be considered to undergo oxidative addition to form η^3-allylpalladium complexes. This evidence is not decisive, however, as the formation of regular arylation products from allyl acetates has actually been documented in the presence of Pd(0) catalysts.[247]

On the other hand, it should be noted that the same results were obtained with a simple salt, Pd(OAc)$_2$, which can hardly be expected to operate *via* Pd(IV) intermediates due to poor stabilization of this oxidation state by electronegative ligands. Therefore, we have to conclude that in spite of these new data there is still too little conclusive evidence in favour of Pd(II)/Pd(IV) Heck chemistry. Moreover, it becomes clear that the real coordination and organometallic chemistry of Pd(IV) state is not suitable for involvement in Mizoroki–Heck reactions.

The recent work of Vicente and co-workers shows an unambiguous case of both oxidative addition of an aryl iodide to a Pd(II) complex and the reaction of the arylpalladium(IV) intermediate with a typical Mizoroki–Heck olefin.[248,249] Both processes were subject to critically important restrictions: the oxidative addition was possible only in an intramolecular (*ortho*-directed) fashion and the reaction with olefin was enabled by soft Lewis acid-assisted ligand exchange, which is equivalent to switching the process to the polar mode. Moreover, the long-awaited catalytic olefin arylation process catalysed by a Pd(IV) complex was unambiguously observed. The process took place under mild conditions at room temperature and base free. Although only a few catalytic cycles were observed (Scheme 9.69), this important result shows that such chemistry is indeed possible, but only with a very specific combination of substrate and conditions.

9.2.5 Other Metals in the Heck Reaction

The first evidence that other metals can catalyse the arylation of olefins was provided in 1976 by Mizoroki and co-workers, who showed that rhodium

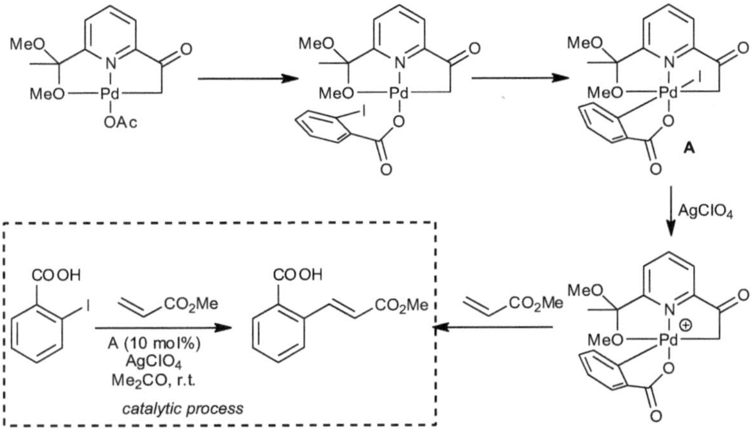

Scheme 9.69 The first proven case of a Pd(II)/Pd(IV) Heck reaction.

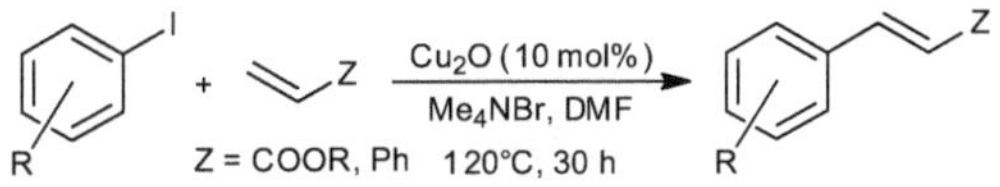

Z = COOR, Ph   120°C, 30 h

Scheme 9.70 Copper-catalysed arylation of olefins.

salts catalyse the arylation of olefins by aryl iodides.[250] Subsequently, evidence on Heck-like reactions using derivatives of Ru, Rh, Cu, Ni, Fe and some other metals occasionally appeared. The chemistry was comprehensively reviewed fairly recently,[251] so here we cite only a few more recent interesting additions.

Probably the most interesting other metal is copper, because copper-catalysed reactions are steadily advancing as a strong competitor for palladium in C–C and C–heteroatom cross-coupling reactions,[252] the development of which, as was stated in the Introduction, is in parallel with the development of Heck chemistry.

Heck reactions using copper precatalysts are indeed known, although the data published are sparse and the results are so irregular that no trends can be identified. In most cases, copper catalysts are effective only in simple reactions between iodoarenes and standard olefins. The systems lack specific ancillary ligands and often use heterogeneous supported precatalysts.[253] Arylation of acrylates by aryl iodides can be performed in the presence of CuI (10 mol%, K_2CO_3) in molten PEG-3400 with microwave heating at 180 °C.[254] Copper(I) oxide was recently shown to effect the Mizoroki–Heck reaction of aryl iodides with standard olefins.[255] The reactions were run under conditions closely similar to those used in regular Pd-catalysed Mizoroki–Heck reactions, but with a serious and highly intriguing difference – copper-catalysed reactions run well in the absence of base, which was successfully replaced with inert Me_4NBr salt (Scheme 9.70).

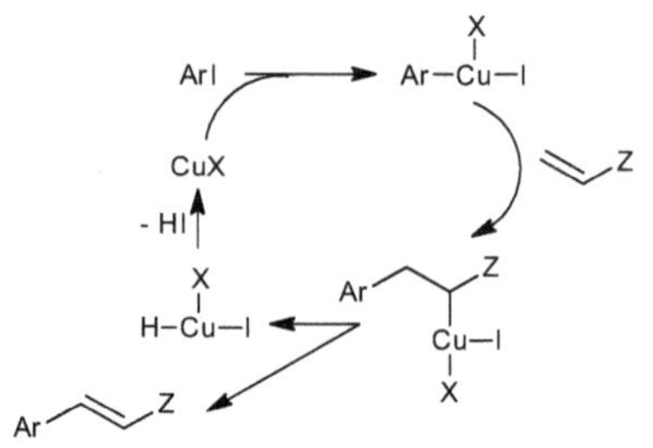

X
|
ArI ———→ Ar—Cu—I

CuX   Z

- HI↑

X   Z
| |
H—Cu—I ←——   Ar   Cu—I
                            |
                            X

Z
Ar

Scheme 9.71   Hypothetical mechanism of base-free copper-catalysed Heck reaction.

This interesting transformation apparently ignores the build-up of acidity, which is not tolerated in Pd-catalysed regular Mizoroki–Heck reactions using aryl halide substrates. This is most probably somehow related to the mechanism of regeneration of the catalytically active low-valent state of Cu. If we presume that the mechanisms overall are similar, the difference in the Cu version is probably due to spontaneous decomposition of a copper(III) hydride intermediate. PdH intermediates in the Mizoroki–Heck reaction also decompose spontaneously, which allows a few turns of the catalytic cycle, but this reaction is choked by increasing acidity, whereas the Cu-catalysed counterpart apparently is not (Scheme 9.71). In this respect, copper-catalysed Heck reaction appears similar to palladium-catalysed arylation by arenediazonium ions (see below).

Diaryliodonium salts except iodoarenes have also been used in copper-catalysed Heck reactions.[256–258]

Hence these new achievements with copper promise important new findings and as such should be thoroughly investigated further.

9.2.6   Arylation by Arenediazonium Salts, the Matsuda–Heck Reaction

Among the less usual substrates for the Mizoroki–Heck reaction, arenediazonium salts[244] occupy a special place. These compounds are by default the most reactive aromatic electrophiles, with reactivity far exceeding that of all common substrates. The other important feature of diazonium salts is the absolutely non-nucleophilic leaving group, hence the oxidative addition of Pd(0) to these compounds always gives cationic Pd complexes, so the polar pathway would always be involved in the reactions of arenediazonium salts, unless strong anionic ligands (such as halides) are deliberately added.

Arenediazonium salts came rather late in palladium catalysis, and at first were viewed with some scepticism. The development was very slow (a few initial steps were treated in Ref. 87) until recently. However, in the last 3–4 years, studies have begun to multiply and we can now see that arylation by arenediazonium salts is among the most fruitful source of new useful protocols (for a recent review, see Ref. 259), an intensely growing branch of

Heck chemistry. Moreover, diazonium salts are notably different from all other types of organic electrophiles used for Mizoroki–Heck reactions with respect to both reactivity and tolerance to base-free conditions, allowing for other means of regeneration of the active catalytic species. In this respect, the reactions of diazonium salts bear parallels with oxidative Heck reactions. Ancillary ligands in these reactions play different roles than in more common Mizoroki–Heck reactions.

The main problem with the reactions of arenediazonium salts is their low stability in complex reaction multicomponent systems used for catalytic processes. Uncontrolled decomposition of arenediazonium salts leading to various by-products and coloured tars is known to be triggered by heating, even gently, in the presence of bases, nucleophiles and particularly transition metals. Therefore, the choice of composition of a catalytic system suitable for carrying out useful and high-yield catalytic transformations is a challenging task in itself, which explains why so useful and reactive substrates were only introduced into Heck chemistry very recently. However, the arenediazonium salts are attracting more and more interest as highly reactive and readily available electrophiles for transition metal-catalysed reactions, and in the course of such studies many prejudices against these compounds, routinely believed to be highly unstable, dangerous, prone to undergoing undesirable side-reactions, *etc.*, are being revoked one by one. New methods of arenediazonium salt generation are being discovered to contribute to the realization by synthetic chemists that these reagents are readily available, safe and definitely worth trying in the pursuit of sophisticated synthetic goals. Most remarkably, Filimonov *et al.* discovered that pure tosylates of arenediazonium salts, prepared in pure form by diazotization with nitrite-exchanged anion-exchange resin, are astonishingly thermally stable with distinct melting points above 100 °C without decomposition.[260] Such salts are very reactive in nucleophilic substitution reactions, *e.g.* to give the respective substitution products in high yields not only with iodide, but also with bromide and nitrite without copper catalysis. This study clearly shows that the resources of arenediazonium salts are impressive and much yet remains to be discovered.

The stability of arenediazonium salts depends on the substituents on the benzene ring, with both electron-donating and electron-withdrawing substituents stabilizing the salt against spontaneous thermal decomposition (but not necessarily towards side-reactions induced by bases, nucleophiles and metals). As a result, the parent benzenediazonium salt often fails to behave and give good yields of target products of arylation, thus requiring special tricks such as addition in small portions (rarely used in catalytic reactions overall, which are customarily performed by charging all components at once into a reaction vessel). This is likely to become most important when scaling-up and ensuring reproducibility of procedures initially developed for micro- or millimolar scales.

In other cases, amazing effects are observed. For instance, phenoldiazonium salts are somewhat exotic as they are not easily prepared by common diazotization methods. Nevertheless, once formed the salts show enhanced thermal stability because a strong donor effect of the hydroxyl group increases the strength of the C–N bond, so that the respective phenolates behave more like a quinoid diazo derivative. Such diazonium salts were shown by Schmidt *et al.* to give dramatically better yields of Heck arylation products in a simple catalytic system – both base-assisted and base-free conditions work comparably well, as compared with simpler alkoxy derivatives.[261] With standard olefins and acrylates, the yields are close to quantitative, with only 2.5 mol% Pd being used either as $Pd(OAc)_2$ or $Pd_2(dba)_3$ precatalysts. Arylation of more challenging olefins such as acetamidoacrylates also gives good yields (Scheme 9.72). However, with *o*- or *m*-aminophenols this trend failed.[262]

Successful arylation of amidoacrylates opens up a new route to amino acids and an effective protocol not limited to some specific diazonium salt was developed by de Azambuja and Correia.[263] The reaction required the non-nucleophilic sterically hindered base di-*tert*-butylmethylpyridine (DTBMP) and methanol as solvent. Interestingly, in this case the products were not the expected amidocinnamates, but their tautomers instead, as if the elimination of PdH takes place not in a regular way to give a C=C bond, but rather to give a C=N bond, and the intermediate imine binds methanol (Scheme 9.73). The formation of these and similar adducts often takes place in base-free reactions of arenediazonium salts and in this case the addition is likely to be catalysed by the conjugate acid of weak base DTBMP.

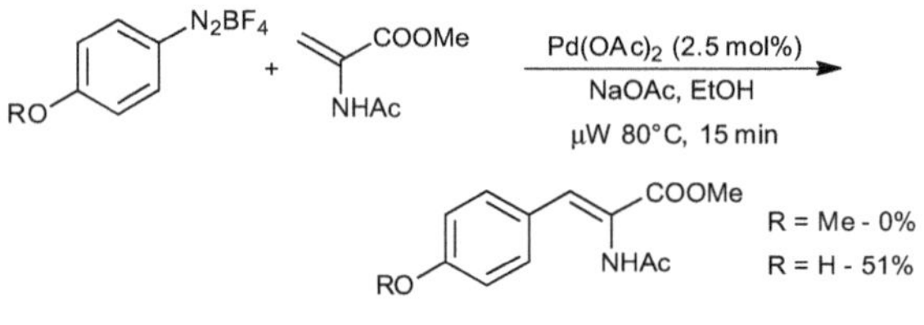

Scheme 9.72 Example of arylation by phenoldiazonium salts.

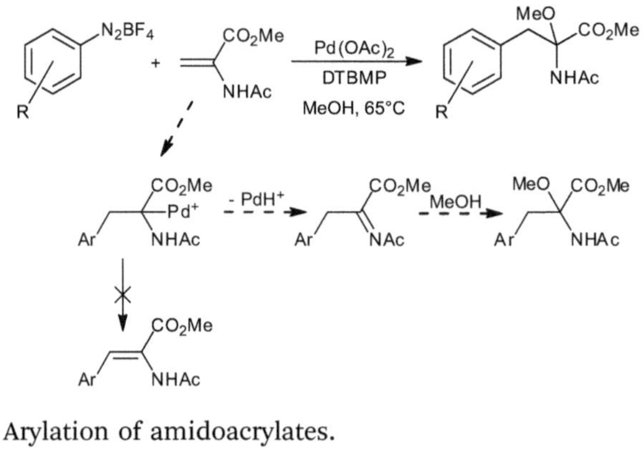

Scheme 9.73 Arylation of amidoacrylates.

As the use of arenediazonium salts forces the reactions to follow a polar pathway, it is particularly justified when the polar mechanism is known to confer beneficial effects. Arylation of allylic acetates and related vinyl derivatives of pentenolides with retention of the double-bond position was realized by Correia and co-workers using diazonium salts in a simple catalytic system taking advantage of PhCN as solvent, which is very likely to provide labile ligands for stabilization of Pd species in the absence of phosphines or other stable ancillaries. This useful protocol was successfully used for the syntheses of a number of interesting plant metabolites (Scheme 9.74).[264,265] Arylation by arenediazonium salts can be applied for challenging multisubstituted olefins.[266]

The same protocol using PhCN as solvent, providing a labile ligand support for transient Pd intermediates, works well for the arylation of styrenes. These reactions give better yields if carried out at room temperature to reduce side reactions, and this simple trick enables the precatalyst loading to be decreased to 1 mol% and even lower. Favourable results were also achieved when the reactions were carried out under a CO atmosphere, which is likely to assist the reduction of Pd(II) in the system where the generation of Pd(0) species may be a limiting factor. The protocol was successfully used in the synthesis of resveratrol precursor (Scheme 9.75).[267]

The failure to produce amidoacrylates, the classical precursors of chiral amino acids, by enantioselective hydrogenation is a drawback of this

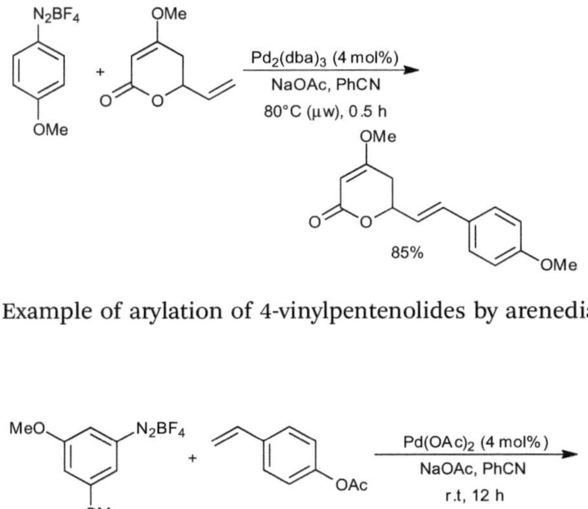

Scheme 9.74 Example of arylation of 4-vinylpentenolides by arenediazonium salts.

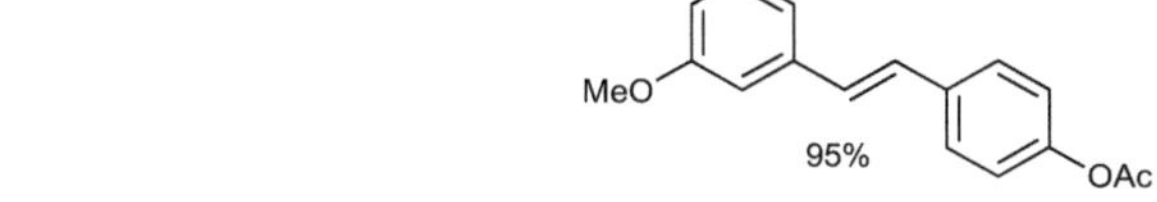

Scheme 9.75 Example of arylation of styrenes by arenediazonium salts.

approach. However, racemic amino acids can be obtained by one-pot procedure of arylation in the presence of a silane reductant under very mild conditions, the latter being compatible with all components of the catalytic system, although the yields were only modest (Scheme 9.76).[263]

Compatibility of arenediazonium salts with silanes was demonstrated in yet another protocol. A special case of base-free Heck arylation is the arylation of olefins, *e.g.* norbornene, in which PdH elimination is blocked for steric reasons and therefore base is not required as no acid is liberated, but some hydride-transfer reducing agent is needed to terminate the cascade. Arenediazonium salts were shown by Cacchi and co-workers to be compatible with silanes and hydroarylation was realized under very mild conditions.[268] The arylation is *exo*-stereospecific (Scheme 9.77).

High diastereoselectivity of carbopalladation and full retention of configuration of the existing chiral centre, manifesting a short lifetime of the PdH intermediate, was observed by Correia and co-workers in the arylation of a chiral dihydrofuran precursor derived from L-glutamic acid and developed as a flexible approach to styryllactone metabolites (Scheme 9.78).[269]

Arylation of methyl vinyl ketone is a special task, as this olefin is very reactive towards common bases and nucleophiles, hence common Mizoroki–Heck protocols often give poor results. The use of aryldiazonium salts, described by Brunner and co-workers, allows the development of a very

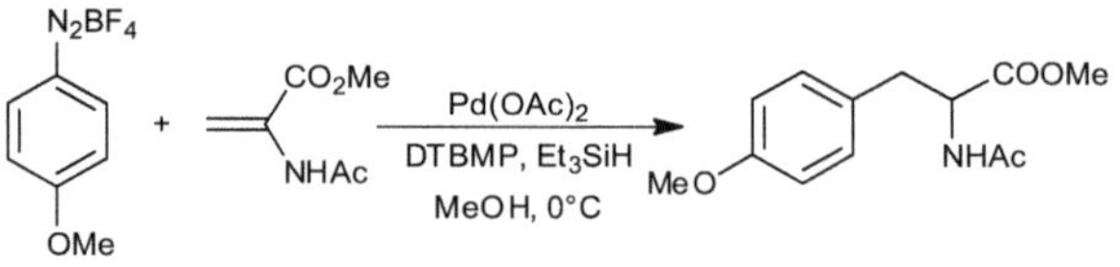

Scheme 9.76   Example of hydroarylation of amidoacrylates.

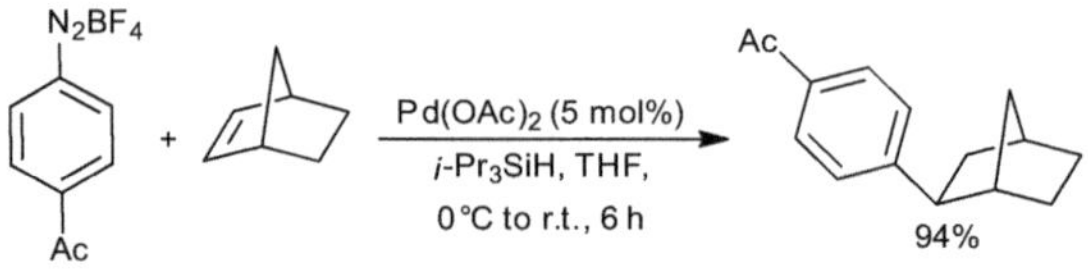

Scheme 9.77   Hydroarylation of norbornene by arenediazonium salts.

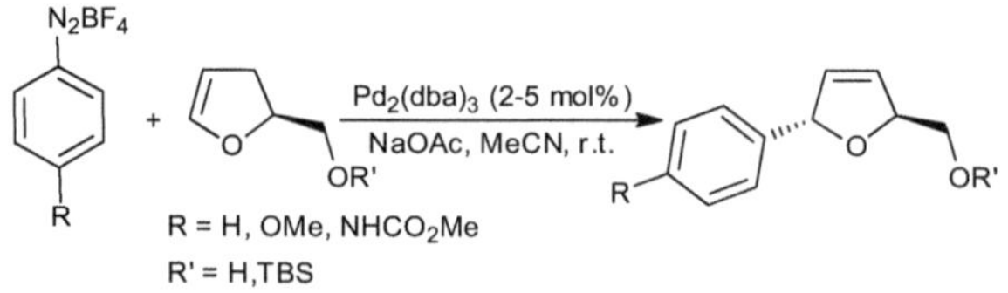

Scheme 9.78   Example of diastereospecific arylation of a chiral dihydrofuran derivative.

mild protocol using $CaCO_3$ as base and $Pd(OAc)_2$ or Pd/C in MeOH.[270] As with other reactions involving diazonium salts, I and Br derivatives are not reactive under the mild conditions employed, thus allowing the retention of halogen atoms in the products (Scheme 9.79).

Arylation of bis-Boc-protected allylamine requires the use of Pd(0) pre-catalyst and different conditions for electron-rich and electron-deficient aryls using NaOAc or $CaCO_3$ bases (Scheme 9.80).[271]

This protocol was used for the synthesis of indoles by consecutive Heck reactions – intermolecular arylation of N,N-disubstituted allylamine followed by intramolecular *exo-trig* cyclization. Worth of note is that the second Heck reaction requires different catalytic system and re-charging of Pd precatalyst (Scheme 9.81).[271]

Whereas intramolecular Mizoroki–Heck reactions have been known for decades and provide a well-established tool for the synthesis of various heterocycles, such an approach with arenediazonium salts was unknown

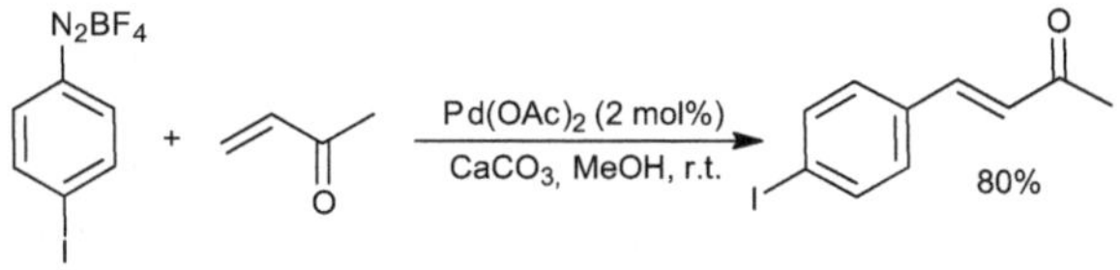

Scheme 9.79 Example of mild arylation of methyl vinyl ketone by arenediazonium salts, showing retention of iodine substituent.

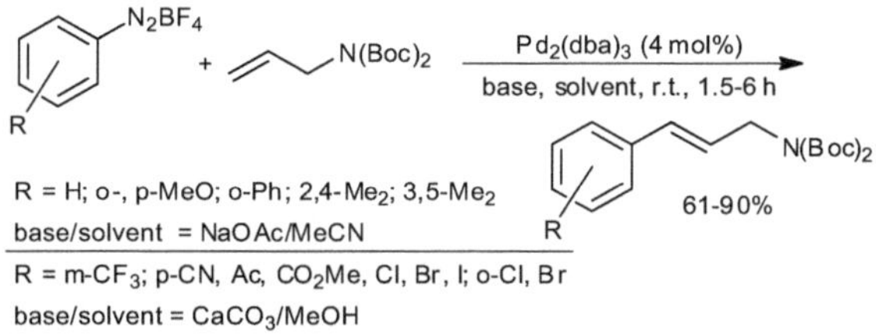

Scheme 9.80 Arylation of protected allylamine.

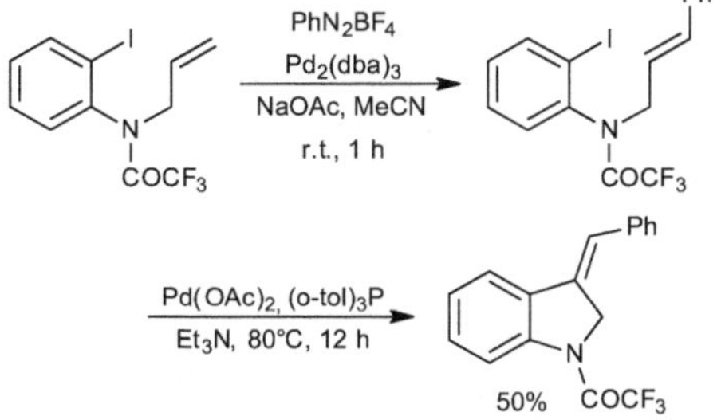

Scheme 9.81 Intramolecular arylation of allylamine, an example.

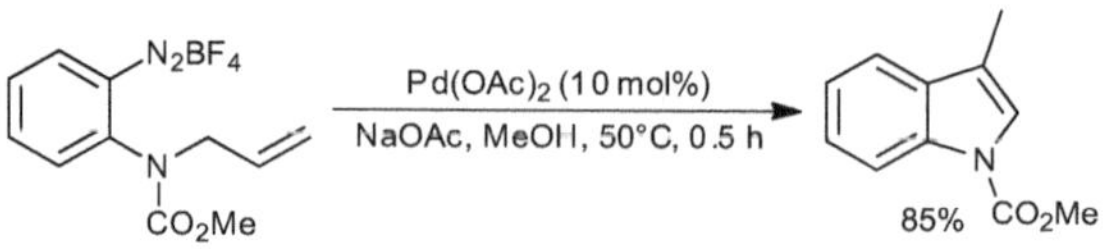

Scheme 9.82   The first intramolecular arylation using a diazonium substituent formed *in situ*.

until recently. The intramolecular Heck reaction with arenediazonium salts was discovered by Correia and co-workers.[272] Diazotization is compatible with various alkenyl pendants, thus opening up the possibility of performing ring closures. In this way, a number of indoles, benzofurans, chromenes and similar heterocycles were prepared. In this case, the double bond migrates into the new ring (Scheme 9.82).

The catalytic systems used for different diazonium salts had to be optimized separately in each case, with either Pd(OAc)$_2$ or Pd$_2$(dba)$_3$, NaOAc as base or base free, and the presence or absence of additives securing the regeneration of Pd(0) being required to achieve good yields. Very high sensitivity towards remote substituents in the benzene ring was also noted, with yields varying dramatically along series of similar substrates. Such a thorough study is very important for understanding not only the clear and spectacular benefits of arenediazonium salts, but also the need for further investigation and understanding of the factors that control the performance of catalytic systems involving such reactive substrates and the nature of side reactions decreasing the output of the desired catalytic process.

9.2.6.1   Base-Free Arylation by Arenediazonium Salts

A special feature of arylation by arenediazonium salts is the possibility of running the reactions in the absence of base. This possibility was observed not simply as a strange irregularity, but as a valuable feature opening up new possibilities only recently, and such base-free Heck reactions have been swiftly developing since 2008, showing many benefits and special features. The first examples of base-free reactions were attributed to the use of special donor ligands, which were considered to increase the acidity of LPdH intermediates to the extent of permitting spontaneous dissociation and release of Pd(0) species entering a new turn of the catalytic cycle (for a discussion, see Ref. 124). Now it is evident that the interpretation was most likely wrong, and a number of base-free procedures not involving any special ancillary ligands have been discovered since then. It is therefore more probable that all PdH intermediates can behave as Brønsted protic acids to be in equilibrium with Pd(0) and the conjugate acid of whatever is capable of protonation among the system components. The high reactivity of diazonium salts enables them to capture low concentrations of Pd(0), even when the increasing acidity of the reaction medium in base-free processes shifts the PdH deprotonation equilibrium to the left and therefore the build-up of acidity cannot suppress the reaction before it achieves high conversions. This works reasonably well also because all the published base-free protocols involving diazonium salts employ rather high loadings of Pd precatalysts. In connection with the rapid development of base-free protocols involving diazonium acids, it would be interesting to consider the question of why there is no working base-free protocol for aryl iodides, the reactivity of which towards Pd(0) is also known to be very high (indeed, what else can account for incredible TONs of up to 10^7 and high-yield reactions performed with "homeopathic" loadings of palladium?). The answers to this strange question might be as simple as (a) nobody tried (although indirect evidence on this does exist because iodine is conserved when iodobenzenediazonium salts are used in the Matsuda–Heck reactions – therefore it is clear that iodine is much less reactive than the diazonium group); (b) IPdH complexes are probably less acidic then PdH$^+$ complexes; (c) in reactions with aryl iodides, the palladium loadings are usually much lower than those used for diazonium salts, thus the relative concentrations of spontaneous Pd(0) are too low.

Arenediazonium tosylates give stilbenes in high yields using a very simple base-free protocol, although the utility of this form of diazonium salt with a broader range of olefins was not explored (Scheme 9.83).[260]

High loadings of palladium precatalysts (5–10% are common) have so far been a common feature of various protocols of the Matsuda–Heck approach. If arenediazonium salts are so reactive, why are so few catalytic cycles observed? Although this issue has never been specially investigated, it is clear that the conditions used fail to support catalytically active Pd(0) species effectively and most of the precatalyst added is just wasted. A good solution to this problem is yet to be found, but some interesting ideas are already available. Yus and co-workers optimized the system using a derivative of a heterocyclic carbene ligand and base-free conditions and arrived to 0.1–0.5 mol% loadings without a decrease in activity, reliably working for a very broad selection of arenediazonium salts.[273]

A palladium salt deposited on charcoal [Pd(II)/C] was shown by Felpin *et al.* to serve as a very efficient precatalyst for the arylation of acrylates[274] and α-aryl acrylates.[275] Non-reduced supported precatalyst was much more active than common reduced Pd/C materials. Supported precatalyst work *via* dissolution–resedimentation, which is manifested by a change in material morphology during the reaction.[274] An interesting variation of this procedure is to use the soluble precatalyst Pd(OAc)$_2$ in the presence of charcoal, in which case two goals are pursued – to achieve optimal activity of a homogeneous system and to withdraw reduced palladium from the

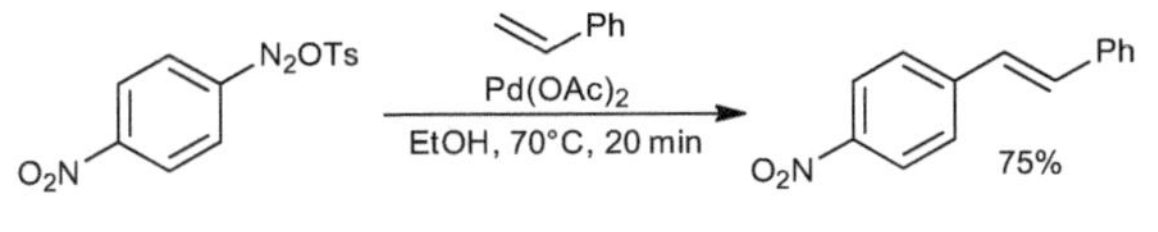

Scheme 9.83 Example of base-free arylation using arenediazonium tosylates.

products.[276] As the reactions are run under very mild conditions at room temperature, Pd(0) apparently cannot be dissolved. The very low loading of precatalyst, contrasting with the 5–10% usually needed for other protocols using arenediazonium salts, draws attention to this interesting protocol (Scheme 9.84).

In a separate study, Felpin *et al.* investigated the possibilities of lowering the precatalyst loading using soluble palladium precatalysts and found that methanol is by far the best solvent for such reactions, if the temperature is brought down to ambient.[277] Apparently, at higher temperatures the rate of decomposition of diazonium salt is too high and simultaneously Pd(II) is reduced to Pd(0) and aggregates of inactive black particles. At lower temperatures, the diazonium salt is stable enough to enter the catalytic cycle almost quantitatively. However, at low temperatures the choice of solvent becomes the other critical factor, as too strongly coordinating solvents such as nitriles apparently block the coordination shell – the lability of the respective ligands is likely to be insufficient for ligand exchange with olefin to take place at a reasonable rate. As little as 0.05–0.1 mol% of Pd(OAc)$_2$ is sufficient for arylation not only of standard olefins, but also of such challenging olefins as various α-aryl acrylates. With the latter, very high diastereoselectivity is observed to result in the formation of the *E*-isomer (Scheme 9.85).[277]

Najera and co-workers recently discovered that if neat water is used as solvent, the protocol can be substantially simplified by exclusion of everything except arenediazonium salt, olefin and 1 mol% Pd(OAc)$_2$ and the reaction is performed at room temperature.[278] Interestingly, such a protocol works not only with acrylates and similar polar olefins that are fairly soluble in water, but also with styrenes and other hydrophobic alkenes, the solubility

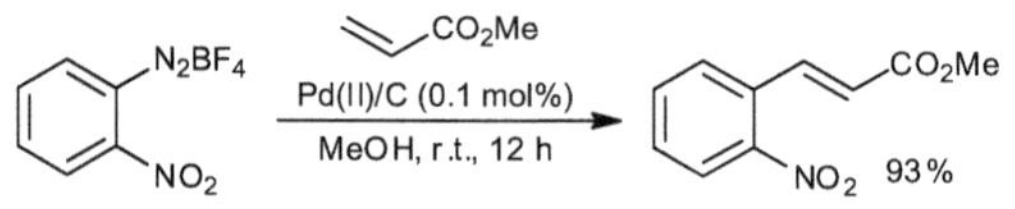

Scheme 9.84 Base-free arylation of acrylates using a low loading of supported palladium precatalyst.

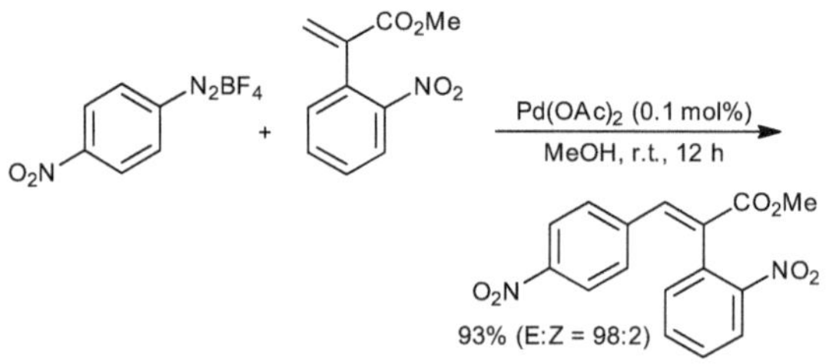

Scheme 9.85 Base-free low-loading arylation using a soluble palladium precatalyst.

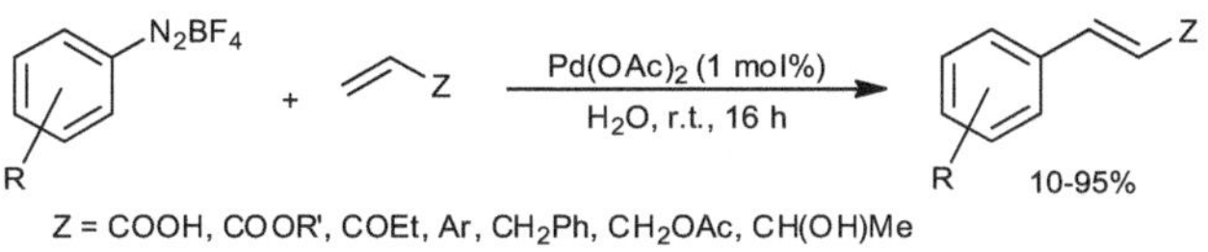

Scheme 9.86 Aqueous protocol for Matsuda–Heck reaction.

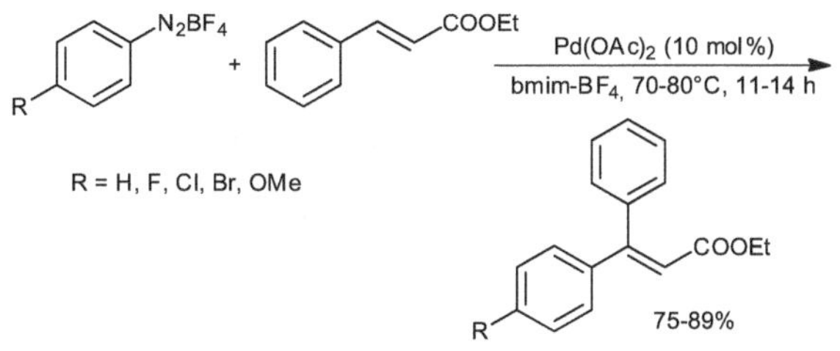

Scheme 9.87 Base-free arylation of cinnamates by arenediazonium salts in an ionic liquid.

of which in water is apparently negligible (Scheme 9.86). Gaikwad and Pore addressed the solubility problem by the addition of a solubilizing non-ionic surfactant to otherwise the same system.[279] Simultaneously, Gholinejad described a similar protocol, the only difference being that palladium nanoparticles supported on agarose were used as precatalyst.[129]

Base-free arylation of standard olefins, and also disubstituted alkenes such as cinnamates and methacrylates, was realized using an ionic liquid. This protocol requires a high loading of Pd precatalyst (up to 20 mol%), hence only a few catalytic cycles take place, which may be the outcome of acidity build-up in the solvent not capable of alleviating its effect. Nevertheless, the reaction is highly selective, *e.g.* giving *E*-isomers in the arylation of cinnamates, with the exception of the *p*-methoxy derivative, which gives a mixture of both diastereomers (Scheme 9.87).[280]

The stereoselectivity of the arylation of cinnamates by arenediazonium salts was investigated by Pastre and Correia using a more conventional catalytic system with alcohols as solvent run in either the presence or absence of bases.[281] If both carbopalladation and hydropalladation/PdH elimination strictly obey *syn* stereochemistry (this statement is not guaranteed always to be true because of *e.g.* the probable involvement of η^3-benzyl complexes; see below) the normal (kinetically controlled) product is *E*, whereas reversible hydropalladation leads to equilibration of both diastereomers. Purportedly, either electron-withdrawing substituents in the arenediazonium salt or the presence of base-scavenging PdH are the factors disfavouring equilibration and loss of stereospecificity (Scheme 9.88).[281]

The approach was developed by Taylor and Correia into a flexible protocol for the synthesis of β,β-diaryl acrylates (Scheme 9.89).[282]

Ionic liquids were used for the arylation of cyclic allylic amines, such as *N*-protected tetrahydropyridines. The reactions were successful only in

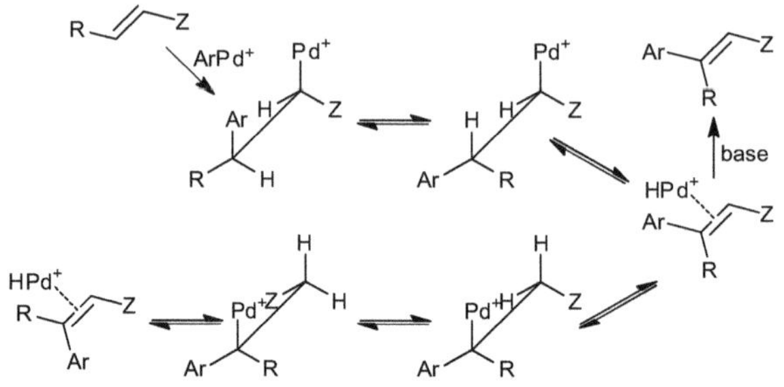

Scheme 9.88 Mechanistic rationale for stereoselectivity.

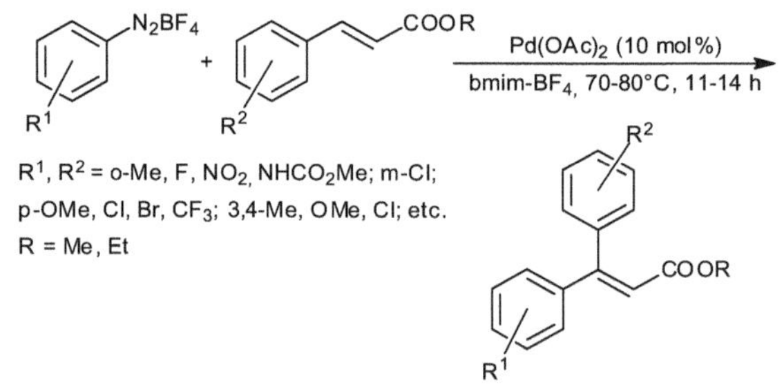

R^1, R^2 = o-Me, F, NO$_2$, NHCO$_2$Me; m-Cl;
p-OMe, Cl, Br, CF$_3$; 3,4-Me, OMe, Cl; etc.
R = Me, Et

Scheme 9.89 Stereoselective synthesis of β,β-diaryl acrylates.

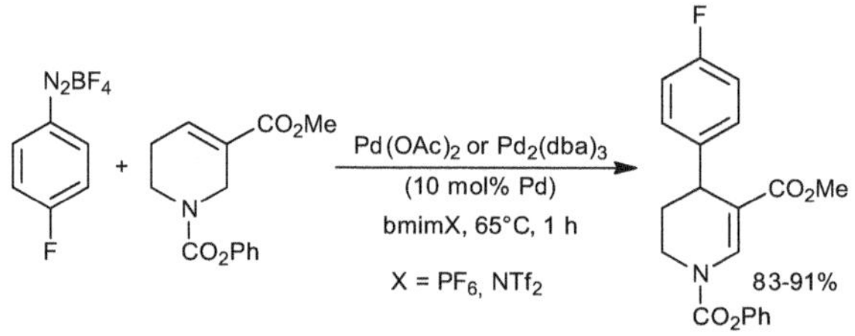

Scheme 9.90 Base-free arylation of cyclic allylic amides.

the base-free mode; the addition of base, even NaOAc commonly used in other protocols involving diazonium salts, suppressed the arylation (Scheme 9.90).[283]

The base-free arylation of 1-acyl-2-trifluoromethylethylenes was shown by Konno *et al.* to give the products in which the incoming aryl selectively went to the β-position relative to the CF$_3$ group and not to acyl, as might be expected.[284] The reactions were performed either in the presence of phosphine ligand or in a phosphine-free mode (in THF in place of ethanol). With some substrate pairs the former gave better results and with some other substrate pairs the latter, with a common trend that phosphine-free reactions gave lower stereoselectivity (Scheme 9.91).

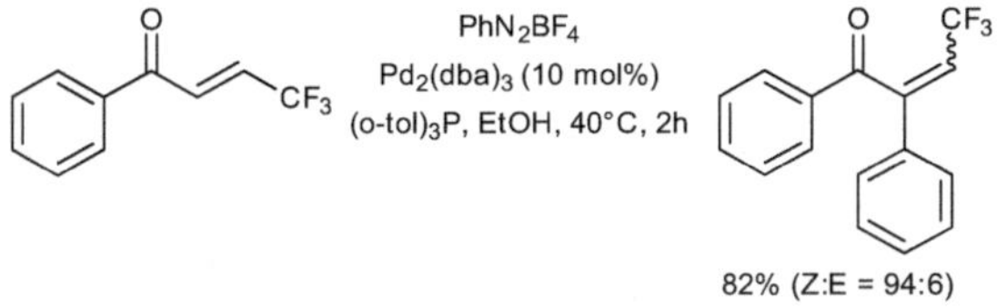

Scheme 9.91 Base-free arylation of 1-acyl-2-trifluoromethylethylenes.

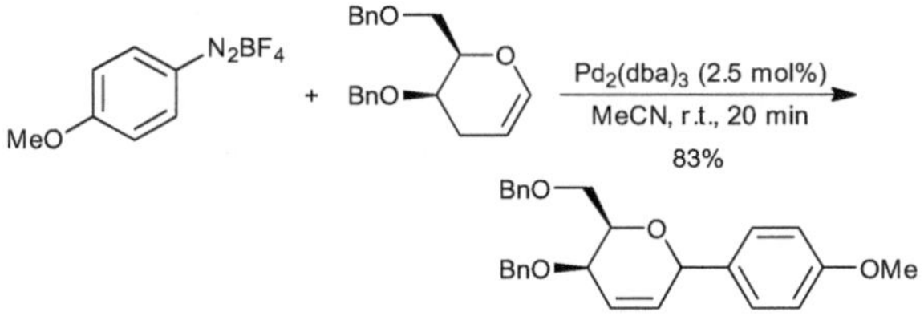

Scheme 9.92 Example of arylation of a protected glycal.

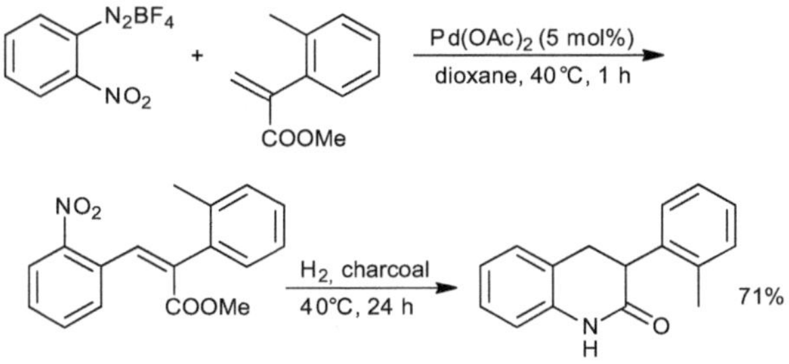

Scheme 9.93 Acid-catalysed cyclization cascaded with the base-free arylation of α-aryl acrylates by arenediazonium salts.

Arylation of enol ethers, including cyclic compounds, nicely follows the polar pathway paradigm.[285] Arylation of glycals is a unique route to *C*-aryl-glycosides. Glycals are electron-rich internal olefins, hardly compatible with common Mizoroki–Heck procedures. Aryldiazonium salts allowing for very mild reaction conditions were shown by Schmidt and Biernat[286] to arylate effectively protected glycal in a *base-free reaction* using only 2.5 mol% Pd$_2$(dba)$_3$; more than 30 catalytic cycles run within 20 min in the increasingly acidified environment (Scheme 9.92).

The base-free reaction using diazonium tetrafluoroborates generates a strong acid, HBF$_4$, which may participate as an acid catalyst in follow-up or side reactions involving the components of the reaction system. An interesting example of such behaviour was discovered by Felpin *et al.*, who investigated a one-pot Heck arylation–hydrogenation–cyclization sequence (Scheme 9.93).[287]

In this method, *o*-nitrobenzenediazonium salt arylated acrylates or α-substituted acrylates in a base-free system and the product formed was subject to hydrogenation in the presence of the Pd black formed, which resulted in spontaneous cyclization of the intermediate. Hydrogenation and cyclization reactions were found to be catalysed by the liberated HBF_4. If the Heck product was isolated and subjected to further transformations in a separate vessel with all the same reagents but lacking an equivalent amount of acid, the yields were noticeably lower.[287]

Another example of interesting reactivity in base-free reaction is the arylation of L-dehydroproline by diazonium salts described by Correia and co-workers.[288] In this case, the addition of base was explicitly shown to have an adverse effect on the reaction outcome. A reproducible and wide-scope (with respect to substituents in the diazonium salts) process was realized in methanol or aqueous solvents containing acetic acid. Under such conditions, the Heck reaction is terminated by acid-catalysed addition of water or alcohol to the 2-dehydropyrrole system.[289] The arylation is diastereospecific and pure enantiomers are obtained in high material and optical yields (Scheme 9.94).

The same method was used for the synthesis of 3-arylpyrrolidine *via* the respective 2-hydroxypyrrolidine derivative.[290]

Methanolysis, catalysed by liberated acid, accounts for the formation of cyclic acetals in the arylation of protected 2-butene-1,4-diol.[291] This approach was applied by Correia and co-workers for the enantioselective synthesis of 4-aryl-γ-lactones by arylation of (Z)-2-butene-1,4-diol (Scheme 9.95).[292]

Similar acid-catalysed cyclization takes place in the arylation of 4-hydroxy-2-butenoates (Scheme 9.96).[293]

Sigman and co-workers discovered an enantioselective arylation of unsaturated alcohols leading to the creation of a tertiary chirogenic atom, with the double bond shifted from the initial position into enol rearranged into ketone (the redox-relay strategy, Scheme 9.97).[294]

From a technical viewpoint, the base-free Heck reaction of arenediazonium salts enables one-pot diazotization–Heck protocols to be realized using an acidic system for both steps without the need to change the medium or add base. Simplification of the protocol is vital for realization in advanced technical setups, such as microflow reactors (Scheme 9.98).[295]

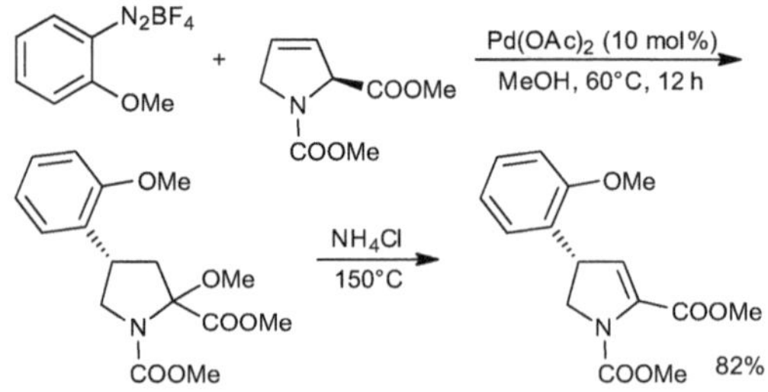

Scheme 9.94 Diastereospecific arylation of 3-dehydropyrrole derivative in a base-free system.

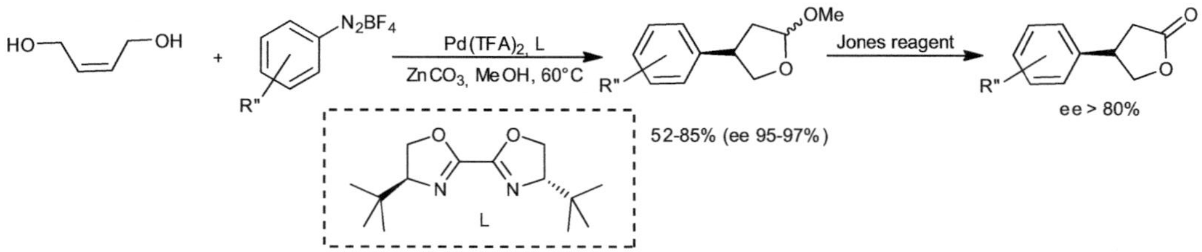

Scheme 9.95 Acid-catalysed methanolysis of the primary product of base-free arylation.

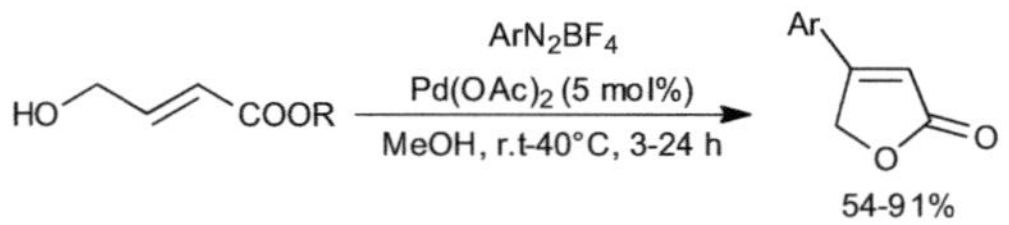

Scheme 9.96 Acid-catalysed cyclization in the base-free arylation of 4-hydroxy-2-butenoates by arenediazonium salts.

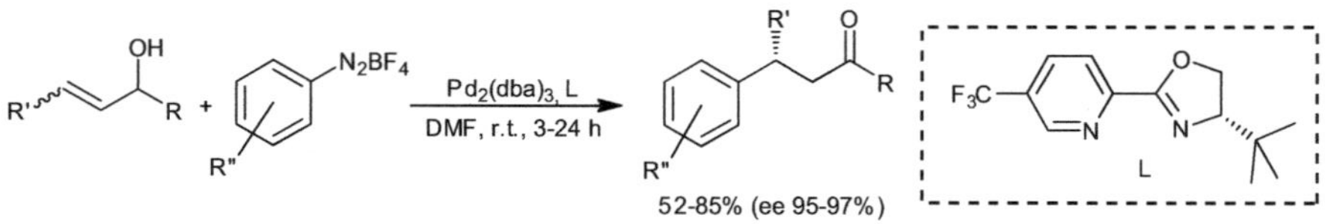

Scheme 9.97 Highly enantioselective arylation of allylic alcohols under mild base-free conditions.

Scheme 9.98 One-pot diazotization–arylation protocol realized using a microflow reactor, an example of chemoselective substitution of a diazonium group in the presence of iodine.

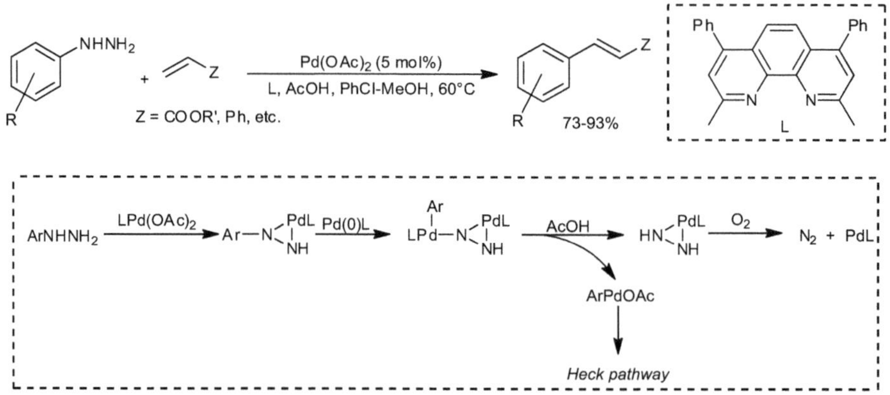

Scheme 9.99 Arylhydrazines as substrates in the Mizoroki–Heck reaction. A tentative mechanism involving the activation of the C–N bond through the formation and cleavage of a palladadiazirine complex is shown in the dashed box.

Moreover, as acidity is generated in the base-free Heck reaction, a catalytic amount of strong acid can be used for *in situ* generation of arenediazonium salts. Felpin and co-workers developed an efficient protocol using a sub-stoichiometric amount of methanesulfonic acid, which turned out to be the most effective acidic additive.[296] The protocol, unexpectedly, works the best for *ortho*-substituted anilines.[297] Schmidt and Berger introduced acet-anilides as substrates, which are deacetylated *in situ*, diazotized and used in the Matsuda–Heck reaction in a one-pot manner.[298]

Another approach was demonstrated by Loh and co-workers.[299] Arylhy-drazines were shown to react with typical Heck olefins in the presence of Pd(OAc)$_2$ to afford arylation products in good to high yields in the presence of acetic acid (Scheme 9.99). Under aerobic conditions, the reaction is made catalytic in Pd(II). In this process, palladium plays a dual role – in addition to catalysing the normal Mizoroki–Heck catalytic cycle, it serves for oxidative generation of an arylpalladium intermediate *via* cleavage of the C–N bond. Many suggestions can be made to explain the activation of the C–N bond towards the oxidative addition by oxidation of a hydrazine residue, from any form of diazo group to palladadiaziridine species, the latter being the choice of the authors. The mild conditions of this new method allowed its appli-cation in the selective arylation of acetyl-protected glycals, in which case, as was shown by Liu and co-workers, the protocol can even be further simplified to rely on an unsubstituted phenanthroline ligand.[300]

This encouraging metamorphosis, however, should not conceal the problems with this chemistry. Moreover, these problems should in fact be welcomed because any progress in solving them would inevitably reciprocate in Heck chemistry as a whole. One of the most intriguing problems, in our opinion, is the low efficiency of the existing protocols, the need for high loadings of Pd, commonly in the 5–10% range, but very rarely decreasing even to 1 mol%. But why is so much palladium used in these reactions? Indeed, arenediazonium salts can hardly be less reactive than aryl iodides, the reactions of which are reliably conducted with 0.01–0.1 mol% Pd under mild conditions (record-making TON/TOF runs are not good reference points for comparison, as they usually require harsh conditions that cannot be reproduced in the reactions with arenediazonium salts). In the case of arenediazonium salts, neither oxidative addition, carbopalladation, nor reductive elimination can be rate-limiting steps. The only step that is substantially different is the regeneration of catalytically active Pd(0) species. In the base-free protocols, it relies on spontaneous deprotonation and thus on *a priori* unknown Brønsted acidity of PdH depending on a dozen different factors (*e.g.* the nature of ancillary ligands, media effects). This process is indeed very likely to be ineffective, which may prompt a further search for reactivation methods in addition to what is occasionally applied now (the addition of various reducing agents, such as CO, carbonyls, *etc.*). In the base-assisted reactions, the properties of arenediazonium salts dictate the choice of weak and often heterogeneous bases, such as $CaCO_3$. Needless to say, such bases, insoluble in the reaction medium, cannot take part in the reaction directly, rather serving as scavengers of the liberated strong acid. Therefore, the means of regeneration of Pd(0) is actually the same as in the base-free reactions – the spontaneous deprotonation of PdH, and the same restrictions apply.

To conclude, we believe that further progress in Heck reactions with arenediazonium salts has to rely on further judicious refinement of the methodology, probably through the search for stable ancillary ligands that would allow for tuning of the PdH acidity and more effective regeneration of Pd(0). There is also an encouraging parallel between Heck chemistry involving arenediazonium salts and oxidative Heck reactions – the two types actually differ only in the way in which the catalytic cycle is initiated and terminated, while everything in between is very similar, including mild conditions, low catalytic efficiency, a clear preference for a polar pathway, regioselectivity, the role of stable ancillary ligands, if any are used, and the fate of PdH species in base-free environments, where they are not trapped but rather left to choose their fate themselves. The parallel development and analysis of regular Heck reactions of arenediazonium salts and oxidative Heck reactions (particularly the demetallative reactions) would definitely bring forward new ideas and trends in Heck chemistry overall.

9.3   Oxidative Heck Reactions

In the oxidative Heck reactions,[301] arylpalladium intermediates are formed
not by oxidative addition in which Pd changes oxidation state, but by isohypsic transformation, which can be formally viewed as electrophilic palladation of an aromatic substrate or *ipso*-palladodemetallation of an
organometallic substrate by Pd(II) species. This process was actually discovered before the oxidative addition involved in the classical catalytic Heck
or Mizoroki–Heck reactions, but until recently remained in the shadow of
the former, being tacitly considered as an exotic and impractical version. The
common name oxidative Heck reactions can be ascribed to such processes,
which would thus encompass both reactions involving palladation of C–H
bonds bearing the name of the Fujiwara–Moritani reaction and reactions
involving transmetallation, discovered by Heck. Both types of processes
represent the same general methodology and are likely to borrow their ideas
and catalyst design principles from one another. It is important to realize,
however, that oxidative Heck reactions are not a branch of the Mizoroki–
Heck reaction, but rather a parallel phenomenon within the context of
general Heck chemistry (a similar situation exists in Cu catalysis, Ref. 302).

Nevertheless, a steady growth of interest in the oxidative Heck reactions
has been evident in the last few years. The potential of this branch is being re-
evaluated with excellent perspectives for fast growth. Oxidative Heck reactions cannot compete with catalytic Mizoroki–Heck reactions in generality
and versatility, but they help to broaden the scope of substrates. In recent
research, the highly toxic organomercuric and organotin compounds used in
the original research have given way to organoboron compounds, the
chemistry of which has advanced enormously, and, unexpectedly, to carboxylic acids, considered as crypto-carbanion equivalents *via* a decarboxylative
palladation pathway. The Fujiwara–Moritani reaction, on the other hand,
instead of pursuing the innately unselective palladation of common aromatic
substrates, switched to a search for specific molecules the palladation of
which is governed by directive effects to ensure high regioselectivity.

From a mechanistic viewpoint, the regular Mizoroki–Heck reaction and
oxidative Heck reaction represent the same transformation, in which only
the origin of ArPdX species differs. The Mizoroki–Heck reaction is begun by
Pd(0) species, which reappear on the path at the end of the process to form
from PdH species, routinely *via* the action of base but also, at least in some
cases, also spontaneously. Therefore, the Mizoroki–Heck reaction is by default a catalytic process. In oxidative Heck reactions, PdH species must be
oxidized into reactive PdX_2 and this process cannot take place spontaneously, but necessarily requires an oxidant to initiate the next cycle
(Scheme 9.100) [for a discussion of Pd(0)–Pd(II) reoxidation by various oxidants, see Ref. 303]. In the absence of oxidant, the reactions are stoichiometric with respect to Pd(II).

The need for an oxidant is a substantial limiting factor for oxidative Heck
reactions, because phosphine ligands – the useful means of control in

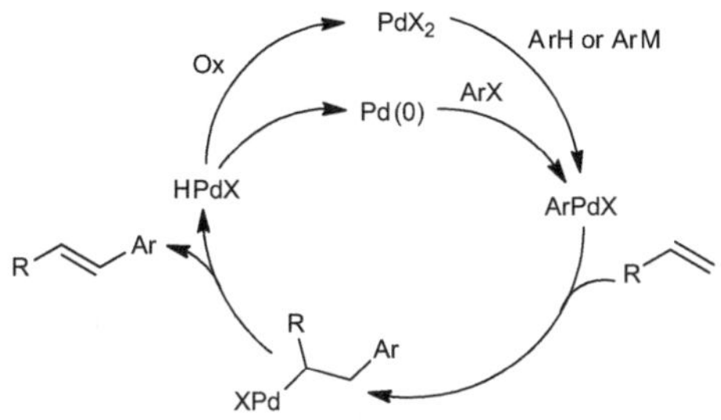

Scheme 9.100 Comparison of regular and oxidative Heck catalytic cycles.

Mizoroki–Heck reactions – are unlikely to be compatible owing to sensitivity towards oxidation. In oxidative Heck reactions, *P,P*-chelators can be replaced by *N,N*-chelators, but it remains a question whether the stability of *N,N*-chelated Pd complexes is sufficient for reliably restricting parallel reactions of unchelated, and thus uncontrolled, Pd species. As a result, the development of an enantioselective version of oxidative Heck reactions is a challenge that remains still practically unaddressed with the exception of a few reports on directed alkenylation.

As this reaction is base free, re-entry of Pd(II) in the actual catalytic cycle should be serviced by oxidation. It is well known, however, that air and oxygen are ineffective oxidants for Pd(0) species, which usually require an intermediate oxidant such as copper(II) salts. Many of the reported protocols, however, are effectively serviced by oxygen (or even just air at the expense of lower yields and prolonged reaction times). Therefore, probably in the protocol reported reoxidation intercepts the PdH intermediate before it decomposes to release Pd(0) species, which almost inevitably in the absence of a more potent oxidant would begin to form metallic Pd particles and leave the catalysis. Reoxidation of metallic Pd by air under the mild conditions reported is hardly imaginable.

Scavenging of the PdH intermediate by oxygen would have another important consequence, as in this case no acid is liberated and the build-up of free acidity does not take place, so base-free conditions are justified.

Reactions in the presence of Pd catalysts and using dioxygen as terminal reoxidant are numerous and extremely interesting, as shown in a highly relevant review by Stahl a decade ago,[304] although so far they have apparently been underinvestigated and promise a rich further development. Concerning Mizoroki–Heck and related chemistry, we need to understand why regular Mizoroki–Heck reactions, particularly those using phosphine-free systems, are not particularly sensitive to air, whereas in oxidative Heck reactions air and oxygen were shown to be able to serve as a Pd reoxidant without the need to employ intermediate reoxidizers such as Cu(II) salts or quinones. Closely related is the fate of Pd hydride intermediates in the presence of oxygen, as this may explain such intricacies as the functioning of base-free catalytic systems – whether acidity build-up can be avoided, and also the lifetime of such hydrides, which often underlies the regioselectivity issues – whether the double bond shifts take place or not.

In fact, substantial insight into the functioning of such systems can be gained from the catalytic oxidation of alcohols by Pd(II) complexes, actively investigated because of its importance for the development of fuel cells and in relation to biocatalysis.

PdH can directly bind O_2 to give hydroperoxide complexes Pd(OOH), thus completely avoiding the Pd(0) state and the risk of deactivation of catalyst due to the formation of Pd black.. Such complexes can take part in ligand exchange to liberate hydrogen peroxide, which is rapidly decomposed.[305] The process was explicitly observed by Goldberg and co-workers using a pincer PCP–Pd complex.[306,307] Apart from hydroperoxide, a hydroxyl complex was observed, which is likely to be formed from facile decomposition of the former (Scheme 9.101).

Similar observations were made by Stahl and co-workers using Pd complex with electron-rich bulky monodentate IMes ligands.[308–310] In this case, richer chemistry is involved and two pathways are competing: the Pd(0) complex can either bind dioxygen to form an η^2-complex split by acid to form hydroperoxide or it can undergo oxidative addition to acid to form hydride, which reacts with dioxygen to form hydroperoxide (Scheme 9.102).

A few conclusions relevant in the context of oxidative regeneration of Pd(II) species for oxidative Heck reactions can be drawn:

1. The PdH lifetime depends on ancillary ligands – electron-rich ligands enhance the stability and such complexes may even reverse the reductive deprotonation of PdH. Therefore, the reactions in the presence of Pd complexes with such ligands should be and usually are air

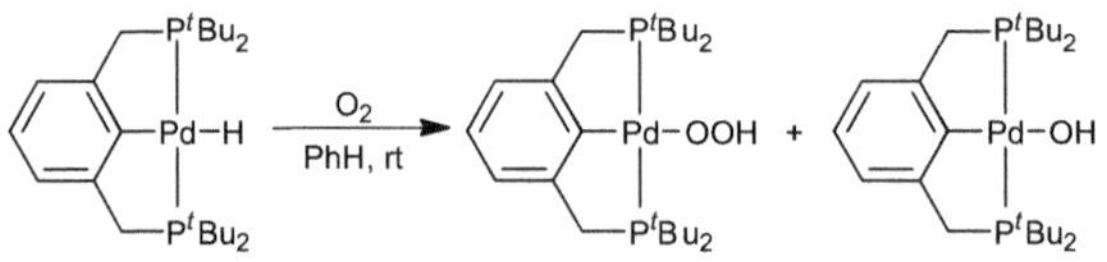

Scheme 9.101 Binding of dioxygen by Pd hydride complex.

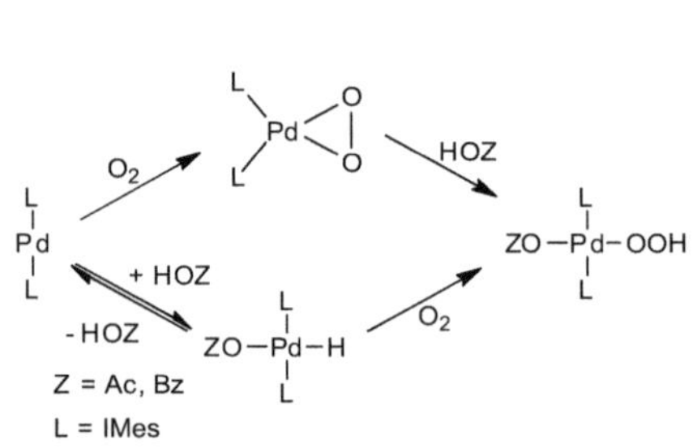

Scheme 9.102 Convergent pathways for scavenging of Pd(0) and PdH intermediates by dioxygen.

sensitive. Particularly sensitive are phosphines, which are known to be catalytically oxidized by oxygen.[311]

2. In the absence of electron-rich ancillaries, PdH intermediates are likely to be highly acidic to release proton spontaneously to weakly basic components of the reaction system (*e.g.* solvent), thus permitting base-free systems. There is some controversy regarding this subject, however, as strong donor ligands were also observed to enhance the acidity of PdH (for a discussion, see Ref. 124).

3. Direct recycling of catalytically active Pd species by oxygen without the need for bases and other oxidants is possible, although the effectiveness of such pathways critically depends on other ligands at the Pd centre. Underligated complexes bearing highly labile ligands and cationic palladium centres are good candidates for such systems. A single stable *N,N*-bidentate ligand can be present to ensure overall stability of the catalytic species.

The rest of both processes involve exactly the same chemistry, hence common trends in reactivity and selectivity should also be the same, which makes the comparisons between reactions of the two types a promising source of helpful conclusions.

9.3.1 The Fujiwara–Moritani Reaction

Direct C–H bond activation without the need to use various leaving groups is the ultimate goal of any synthetic strategy based on substitution reactions. Only such an approach would meet the atom economy principle,[312] as otherwise the material of the leaving group is waste. Therefore, the possibility of using benzene and other aromatic substrates for the arylation of olefins, discovered in 1967 by Fujiwara, Moritani and co-workers,[7] was highly motivating. Nevertheless, in reality, the reactions discovered simultaneously by the Heck and Mizoroki teams and using the worst possible (from the economy principles viewpoint) leaving groups, heavy and expensive iodine, mercury, *etc.*, progressed rapidly and captured the area, whereas the original Fujiwara–Moritani reaction had been practically abandoned for quite a period.

The original method suffered from at least two serious drawbacks – the need to use a stoichiometric amount of Pd(II) salt and low yields and selectivity. Substituted benzenes gave mixtures of isomers (Scheme 9.103).[13]

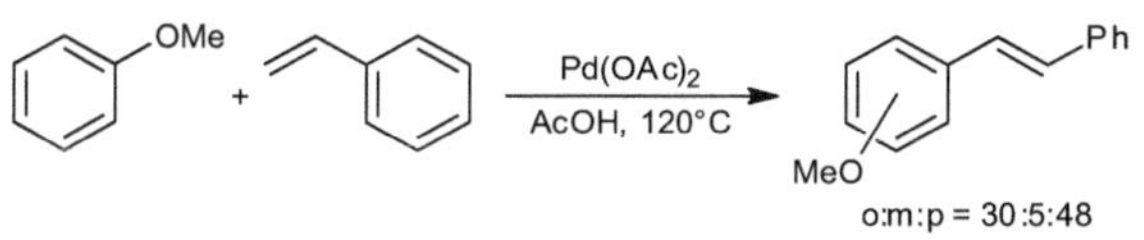

o:m:p = 30:5:48

Scheme 9.103 Low regioselectivity of the Fujiwara–Moritani reaction of simple benzene derivatives.

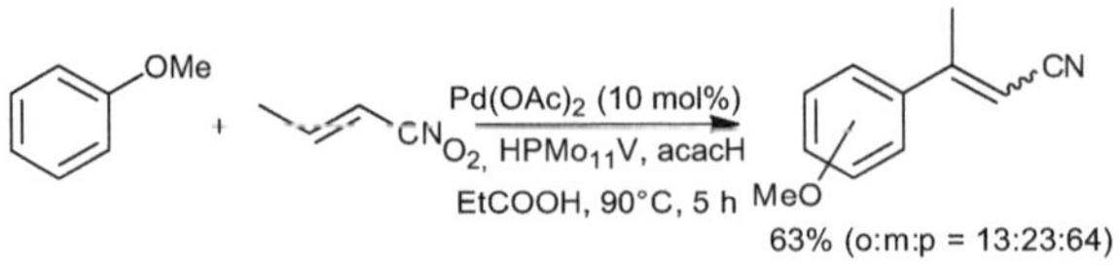

63% (o:m:p = 13:23:64)

Scheme 9.104   Oxidative arylation by simple arenes in the presence of substoichio-metric amounts of Pd salts.

Further development of this reaction applied to simple arenes has been rather slow, as the reaction fails to compete with the regular Mizoroki–Heck reaction. A few improvements have been targeted on making the reaction more economical through the use of catalytic amounts of Pd salt. Thus, Obora *et al.* showed that oxygen can be used as a terminal oxidant if the reoxidation is promoted by molybdovanadophosphoric heteropolyacid (Scheme 9.104).[313,314]

The original Fujiwara–Moritani reaction is applied to electron-rich arenes. This limitation seems natural as the palladation seemed to be a regular electrophilic substitution reaction obeying typical trends common for text-book reactions of this type, the closest analogues being mercuration, thal-lation, *etc.* However, organometallic reactions involving transition metals such as Pd often follow concerted pathways not involving carbenium ions, carbanions and other discreet ionic species common in non-transition metal chemistry. The palladation of arenes most probably is no different – the rupture of a C–H bond and the formation of a Pd–C bond are most probably concerted, taking place within the coordination shell of palladium with one of the ligands probably lending assistance for deprotonation,[315–318] hence the process is an inner-sphere reorganization of bonds. As a result, the effect of substituents is mitigated (in common electrophilic substitution the dif-ference in reactivity between PhH, PhMe, PhOMe, PhNMe$_2$ is extremely large, exceeding 10–15 orders of magnitude along the series, whereas in palladation chemistry the difference is only modest). On the other hand, by varying the transition state it is possible to reach the "carbanion-like" transition state (rupture of the C–H bond is slightly ahead of Pd–C bond formation) and to observe the reaction with electron-deficient substrates (this can be considered as a switch-over of distinct mechanisms,[319] but in common physical organic chemistry the response of the transition state intimate structure to variations of substituent effects, substrate structure and environment effects which are ubiquitous in classical mechanistic paradigms, such as S_N2, $E2$, *etc.*, are conveniently discussed as a mechanistic continuum of a variable transition state within a single mechanism).

Such abuse of the electrophilic palladation paradigm often happens in directed palladation (see below), but in non-assisted palladation it seems not to have been observed until recently. The reason why non-directed substi-tution with electron-deficient substrates is rare, whereas in directed sub-stitution the difference between electron-rich and electron-deficient substrates is less manifested, is probably very simple – the attack of the Pd centre at a C–H bond is initiated by coordination involving fairly weak, probably agostic interactions, hence in the non-directed intermolecular case the probability of bonding two reagents in the reacting complex is too small. In directed palladation, the attack of the bonded Pd centre at the proximal C–H bond is more facile, hence no restrictions due to substituent effects apply.

Recently, it was discovered that electron-deficient polyfluorinated molecules are activated towards reactions involving direct or *ipso*-palladation. The use of polyfluorobenzenes for palladium-assisted arylation (oxidative cross-coupling) was described by Fagnou and co-workers.[319–321] This reaction involves the attack of an arylpalladium intermediate formed by oxidative addition of a Pd–phosphine complex on an electron-deficient polyfluoroarene (theoretical analysis of the reaction mechanism by high-level DFT calculations was performed[322]). The palladium centre bearing σ-aryl and electron-rich phosphine ligands apparently prefers the attack of an electron-deficient substrate.

The use of electron-deficient polyfluoroarenes in the Fujiwara–Moritani reaction was observed only fairly recently. Zhang and co-workers demonstrated that arylation of terminal alkenes takes place in the presence of a palladium catalyst and 2 equiv. of Ag_2CO_3, which apparently has a dual role, being both a terminal oxidant and very probably the base required for deprotonation (Scheme 9.105).[323]

This study explored how many fluorine atoms are required for reaction to take place. Just two fluorine atoms are sufficient to activate the molecule towards palladation, in which case the major product is the expected "between-fluorines" isomer, whereas the isomer in which palladation takes place is the "*ortho*-to-a-single-fluorine" site (Scheme 9.106).

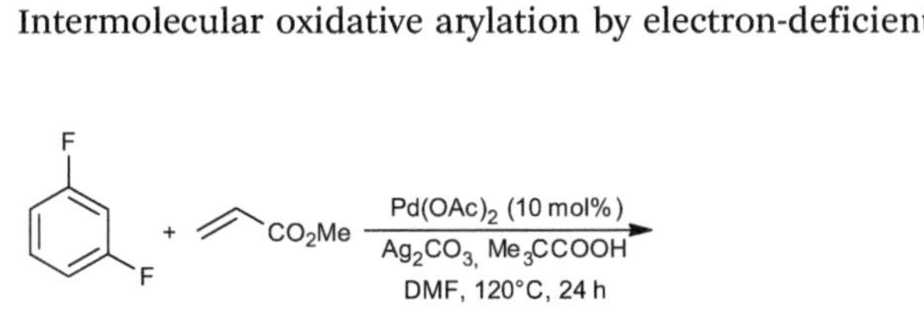
F
F
F
F
F
R'
R
+
Pd(OAc)2 (10 mol%)
Ag2CO3
DMSO-DMF, 120°C, 12 h
R'
R
R' = H, R = CO2Me, CO2nBu, CO2tBu, CONMe2, CN, COEt, COPh,
Ph, tBu, nBu; R' = Me, R = CO2Me, Ph

Scheme 9.105 Intermolecular oxidative arylation by electron-deficient arenes.

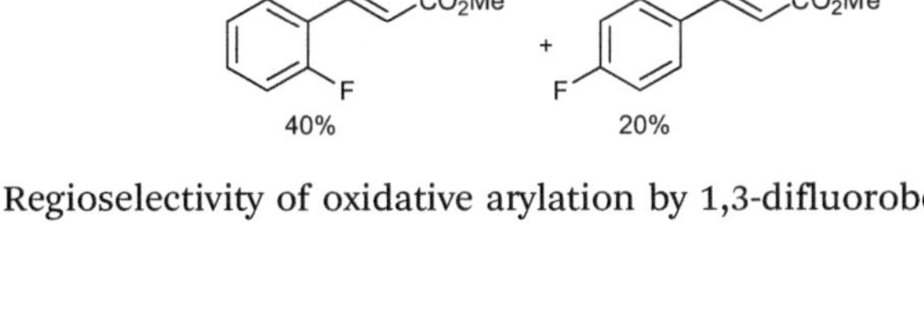
F
F
+
CO2Me
Pd(OAc)2 (10 mol%)
Ag2CO3, Me3CCOOH
DMF, 120°C, 24 h
CO2Me
40%
CO2Me
20%

Scheme 9.106 Regioselectivity of oxidative arylation by 1,3-difluorobenzene.

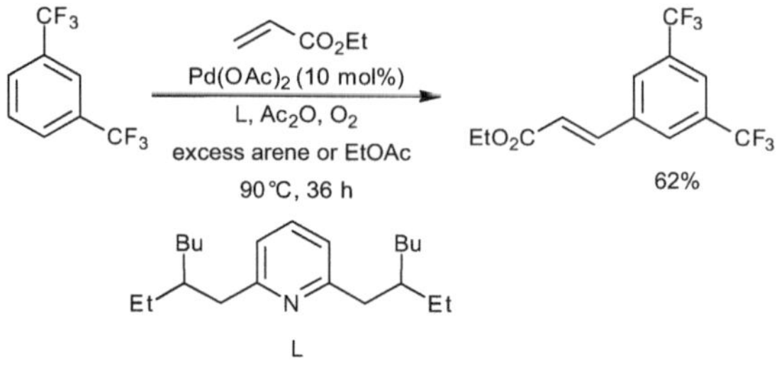

Scheme 9.107   Oxidative arylation by electron-deficient arenes in the presence of 2,6-dialkylpyridine.

Pyridine derivatives used as ligands allow the alkenylation of common electron-deficient substrates containing one or two electron-withdrawing groups, such as CF_3, NO_2, CO_2R and COMe, as in the system described by Yu and co-workers (Scheme 9.107).[324] Such substituents direct the reaction mainly at *meta*-positions, thus supporting the electrophilic nature of the mechanism of Pd(II) attack.

A very fine balance of steric bulk is required for the system to work well. Less bulky pyridines (pyridine itself, 2,6-lutidine, ethyl nicotinate) were unreactive, similarly to the much bulkier 2,6-di-*tert*-butyl- and 2,6-dineopentylpyridines. Good yields were obtained with moderately bulky primary *n*-pentyl and 2-ethylhexyl 2,6-disubstituted pyridines. Such a requirement for a fine balance of steric bulk and, probably, ligand nucleophilicity indicates that stable monoligation of the Pd centre by a pyridine ligand is what is needed. Probably, indeed, such ligands favour the formation of $LPd(OAc)_2$ complexes, but why such complexes may exhibit high electrophilicity sufficient for attacking strongly deactivated arenes remains a question. Sanford and co-workers further showed that pyridine ligands, such as that shown in Scheme 9.107 taken in a 1 : 1 Pd:L ratio are highly effective in increasing the activity and selectivity of catalytic systems, thus making the intermolecular Fujiwara–Moritani reaction much closer to practical application.[325]

On the other hand, although competitive experiments were not performed, from the data on yields and reaction times it is evident that the differences between electron-rich benzenes, benzene itself and electron-deficient arenes were minimal, which is hardly compatible with true electrophilic metallation. The ability to attack nitrobenzene and 1,3-bis-(trifluoromethyl)benzene at the *meta* position indicates high electrophilicity, whereas the substrate selectivity does not, being more common for directed palladation (see below). Moreover, if we take electrophilic mercuration as an archetypal metallation reaction with a highly reactive metal electrophile, the positional selectivity of mercuration is different to be highly typical for common electrophilic reactions.[326] The palladation involved in the reaction under consideration is less positionally selective than mercuration. The other analogy to be drawn is with palladium-catalysed oxidative homo-coupling of arenes, described by Labinger and co-workers to involve the palladation of an electron-deficient arene ($C_6H_5CF_3$) by a cationic palladium complex, which shows the formation of *meta* + *para* isomers without *ortho* substitution, thus being a clear prototype of this alkenylation method. However, even in this case, undoubtedly showing that electrophilic palladation of electron-deficient arenes is indeed possible, the distribution of isomers was different from what could be reasonably explained by the interplay of a static inductive effect and steric bulk.[327] We have paid considerable attention to this case because it explicitly reveals that the trends in reactivity of palladium complexes in palladation reactions are not properly understood and interesting chemistry promising substantial widening of the reaction scope is most likely still hidden. More research is required to understand better the reactivity of various palladium complexes in reactions involving undirected palladation.

However, although electron-deficient arenes are reactive, the majority of arenes used in the intermolecular Fujiwara–Moritani reaction are electron rich and the positions attacked are the same as in common electrophilic substitution reactions The problem of low regioselectivity is commonly overcome by using multisubstituted compounds in which only a single option for the substitution is possible, as in, for example, the arylation of substituted fluorofurans developed by Zhao and co-workers (Scheme 9.108),[328] reaction of 2-substituted thiazoles and oxazoles with standard olefins in the presence of 10–20 mol% Pd(OAc)$_2$ and silver acetate in the presence of carboxylic acids,[329] reaction of 1,2,3-triazoles with standard olefins realized in a Pd/Cu/O$_2$ system,[330] the alkenylation of imidazo[1,2-*a*]pyridines by standard olefins taking place in the presence of a Pd/Cu system in a non-polar solvent,[331] alkenylation of 5-pyrazolones under mild conditions using $K_2S_2O_8$ as reoxidant[332] and the arylation of quinones.[333]

In other cases, the reactions of less substituted heterocyclic compounds are selective enough to afford definite products, as *e.g.* in the alkenylation of indolizines taking place in a complex system using a silver salt, a base and the bidentate ligand 2,2′-bipyridyl (Scheme 9.109),[334] the arylation of

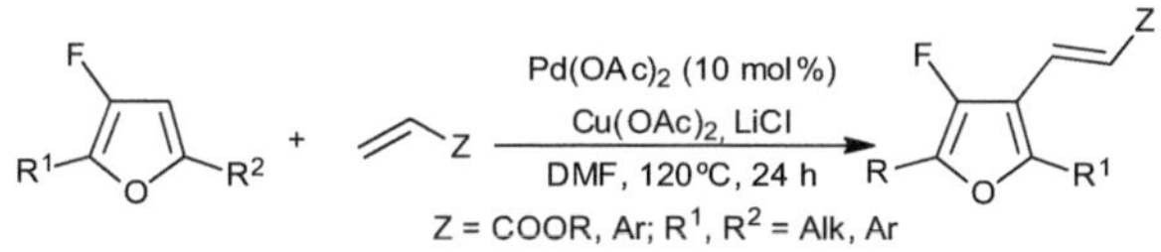

Scheme 9.108   Oxidative arylation by trisubstituted furans.

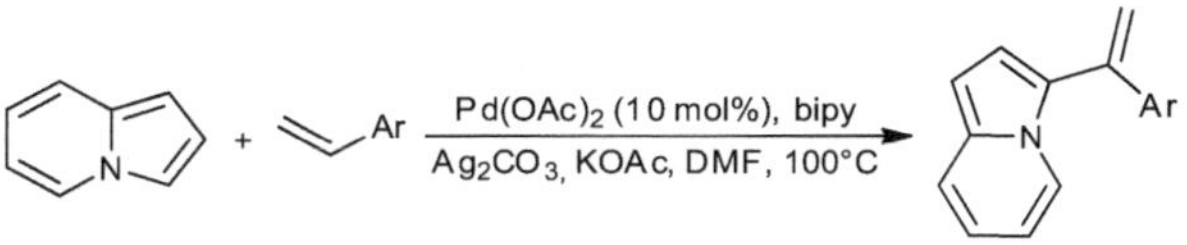

Scheme 9.109   Alkenylation of indolizines in the presence of a Pd complex of 2,2′-bipyridyl ligand.

partially substituted furans in a simple system using Pd(OAc)₂ in DMSO–AcOH mixture in the absence of metal salt additives[335] or the alkenylation of coumarins taking place in pivalic acid solution under an oxygen atmosphere.[336] The key role of DMSO in the activation of palladium acetate through cleavage of a trimeric structure and formation of more reactive monomeric and dimeric species was elucidated by Le Bras and co-workers.[337] Alkenylation of indoles at the 3-position was achieved with high yields in a Pd(OAc)₂ (10 mol%)–H₃PMo₁₂O₄₀–DMAP system under an oxygen atmosphere in DMF–DMSO solvent mixture.[338]

Although still very rare, a few examples of oxidative Heck reactions not with aromatic substrates but with alkenes have appeared, as in the report by Gillaizeau and co-workers on the alkenylation of protected enamides (Scheme 9.110).[339]

The other important direction of development is the search for more economical and practical systems for Pd(II) recycling. In many cases this chemistry can employ oxygen gas, but often an elevated pressure is required, which is both hazardous and prone to cause oxidative side reactions. Bäckvall and co-workers described an approach to solve these problems by employing electron-transfer mediators, which dramatically increase the effectiveness of reoxidation, thus reducing the required pressure of O₂ to ambient.[340] The system used relied upon iron phthalocyanine as the primary oxidant passing electrons to dioxygen and benzoquinone as the secondary oxidant taking electrons from Pd(0) and passing them to iron phthalocyanine. The system can affect both mono- and diarylation (Scheme 9.111).

Concerning olefins, the Fujiwara–Moritani reaction is usually applied to the same manifold of alkenes as the Mizoroki–Heck reaction – acrylic acid derivatives being by far the most popular, which is not surprising as the mechanism of olefin binding is the same in both reactions. However, a certain broadening of scope is evident, as in many of the published oxidative olefination protocols the use of trisubstituted olefins is fairly widespread,

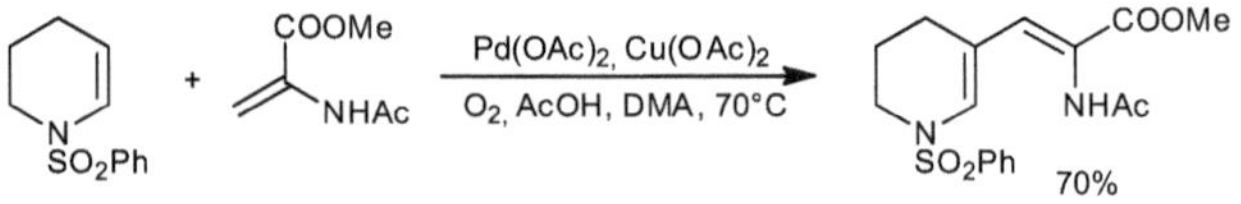

Scheme 9.110 Example of oxidative alkenylation of an enamide.

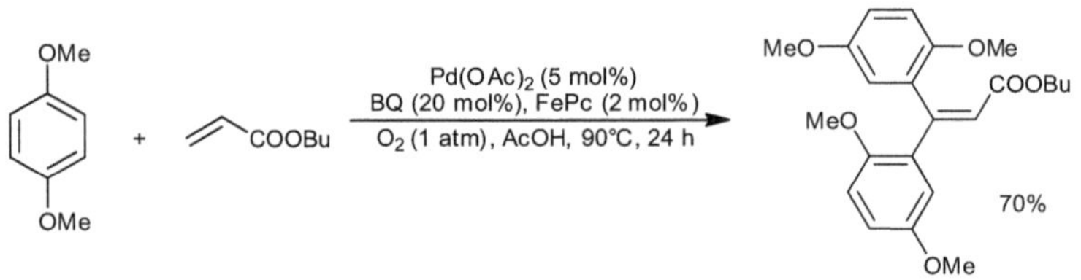

Scheme 9.111 Diarylation in a biomimetic system.

whereas in the Mizoroki–Heck reaction such olefins are rare and often require special protocols and ancillary ligands. The Fujiwara–Moritani reaction has been applied to rather complex or challenging olefins, such as α-methylenebutyrolactone,[341] cinnamylphosphonates,[342] β-substituted α,β-unsaturated Weinreb amides,[343] Boc-protected allylamines[344] and others.[345] The reaction with allyl esters was reported by Liu and co-workers to afford selectively the normal products formed by [PdH] elimination.[346] Kim and co-workers described the stereoselective arylation of the trisubstituted alkene bond in oxindole derivatives, which gave much better results than the regular Mizoroki–Heck arylation of the same substrates by aryl iodides (Scheme 9.112).[347]

Intramolecular reactions can also be unambiguous, if alternative positions are properly protected. Gaunt and co-workers developed a highly regioselective procedure for the alkenylation of *N*-protected pyrroles using a peroxy ester as a terminal oxidant – *N*-acylpyrroles underwent 2-alkenylation, whereas *N*-trialkylsilyl-protected pyrroles gave 3-isomers.[348] This protocol in an intramolecular version was used by the same group in the elegant synthesis of important natural metabolite rhazinicine (Scheme 9.113).[349]

Intramolecular alkenylation leading to the formation of quaternary carbon atom was investigated in detail by Schiffner and Oestreich using anilides of α,β-unsaturated acids.[350] A useful protocol remarkable for an unusually low (for Fujiwara–Moritani reactions) palladium loading, using pyridine ligands (pyridine itself or methyl nicotinate) was developed to afford the respective indolin-2-ones, although substantial limitations are imposed on the anilides used, which should be electron rich in both aromatic and alkene residues (Scheme 9.114).

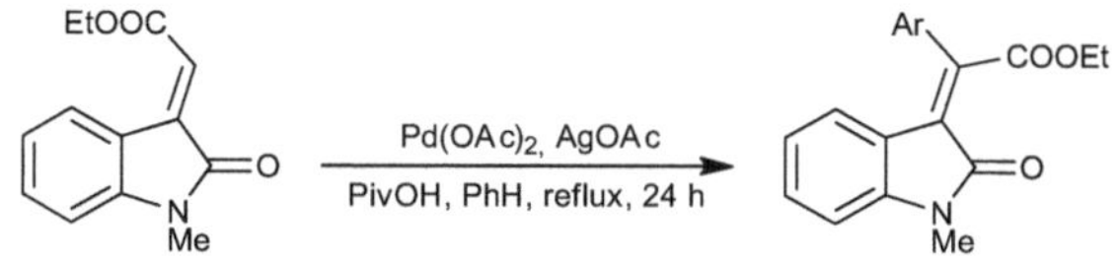

Scheme 9.112 Arylation of a trisubstituted alkene residue (only the major stereoisomer is shown).

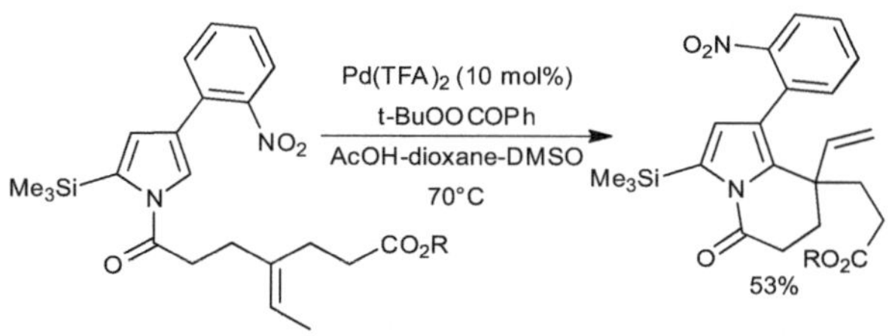

Scheme 9.113 Intramolecular Fujiwara–Moritani reaction in the synthesis of important metabolites.

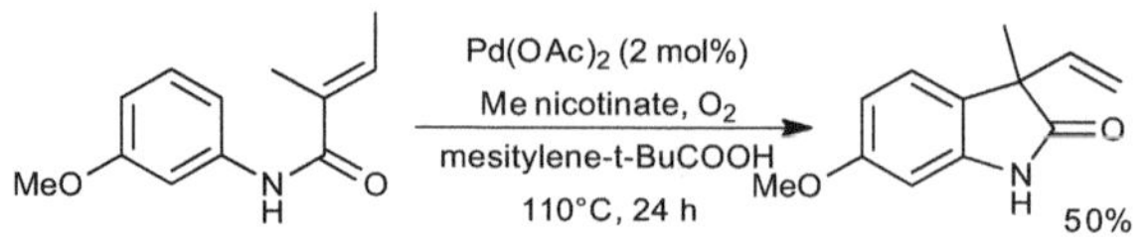

Scheme 9.114 Low Pd loading intramolecular Fujiwara–Moritani reaction.

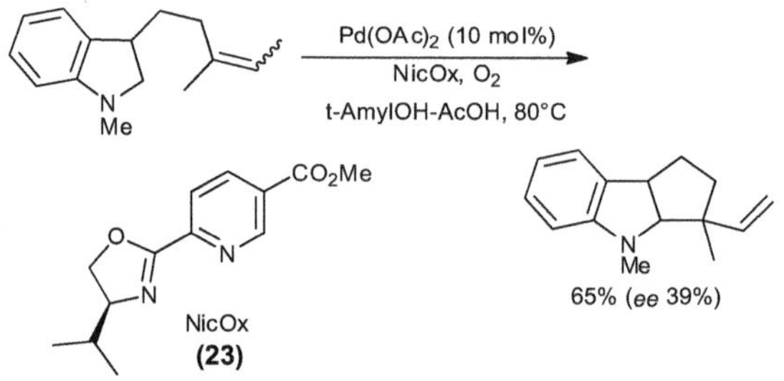

Scheme 9.115 Enantioselective intramolecular Fujiwara–Moritani reaction and chiral oxazoline ligand used.

The application of this approach to the development of enantioselective oxidative Heck reactions was demonstrated. A chiral bidentate ligand, oxazoline-appended nicotinate (**23**), was explored by Oestreich and co-workers for the development of enantioselective cyclization.[351,352] Optical yields were only low to modest (Scheme 9.115), which is understandable, as in the prototypical protocol (Scheme 9.114) monodentate pyridines were used as ligands, which probably indicates that the mechanism of these two processes depends on monodentate-bonded stable ancillaries and thus NicOx can also be bonded in such a mode (which can be additionally enforced by using an acidic medium and pyridyloxazoline being monoprotonated). Therefore, asymmetric induction from a monodentate ligand is imperfect.

Moreover, in fact, the role of ligands in oxidative Heck reactions is uncertain. Most probably, even in the presence of ligands, the reaction can proceed by multiple pathways, with or without chelating ancillaries in the coordination shell; as a result, the outcome of the process is an arbitrary sum of various processes, with individual contributions being dependent on subtle variations of the conditions and components. Thus, unlike what occurs in the enantioselective Mizoroki–Heck reaction, in the Fujiwara–Moritani reaction chelating ligands cannot play a restrictive role and decouple all "ligandless" pathways from the main reaction course. More work is needed to understand the role of stable ancillaries in oxidative Heck reactions and arriving at really enantioselective protocols, if any are even possible. Earlier attempts[353] to design an enantioselective intermolecular Fujiwara–Moritani reaction gave similar results.

Further application of this approach was the development of alkenylation *via* intramolecular Fujiwara–Moritani reaction followed by cleavage of the intermediate cyclic product (Scheme 9.116).[354]

Oxidative Heck reactions can be involved in tandem reactions in which the substrate for oxidative olefination is formed through catalysed isomerization. A very interesting example of this approach was discovered by Wang and co-workers involving prior copper(II) catalysed isomerization of cyclopropene derivatives to give furans (or rather their organocopper derivatives), which further react with olefins (Scheme 9.117).[355]

Another interesting example of the tandem isomerization–oxidative alkenylation strategy exploits the dehydrogenative aromatization of cyclohexanones. This tandem process, described by Hong and co-workers, is versatile and regioselective (Scheme 9.118).[356]

So far, aromatic substrates constitute the bulk of published examples of oxidative Heck reaction *via* CH activation. Recent studies showed that this restriction is not absolute. Several examples of the use of olefins, although not common but rather specific ones, have been reported. Thus, Kapur and co-workers succeeded in observing reactions between dihydropyrans and a broad range of olefins[357] and Lei and co-workers demonstrated that specific sp[3] CH bonds can be involved (Scheme 9.119).[358]

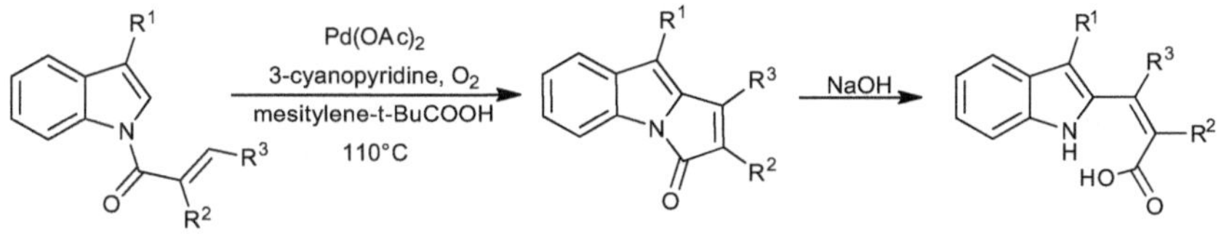

Scheme 9.116 Emulation of intermolecular Fujiwara–Moritani reaction *via* intramolecular reaction.

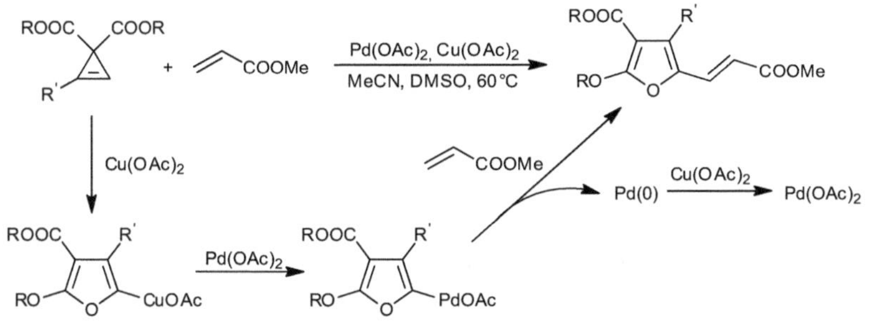

Scheme 9.117 Tandem isomerization–oxidative Heck reaction.

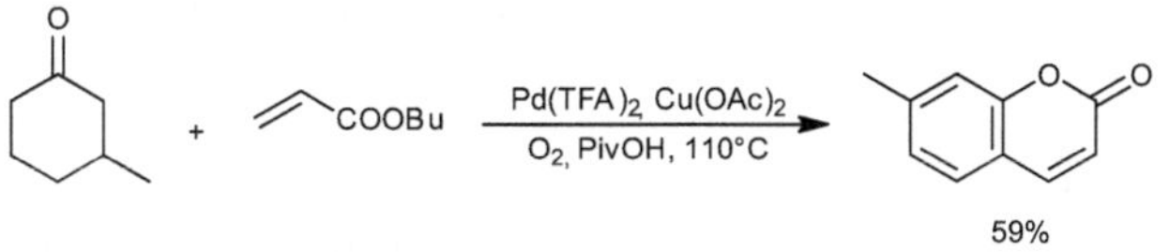

Scheme 9.118 Tandem isomerization–dehydrogenative aromatization–oxidative alkenylation–lactonization process.

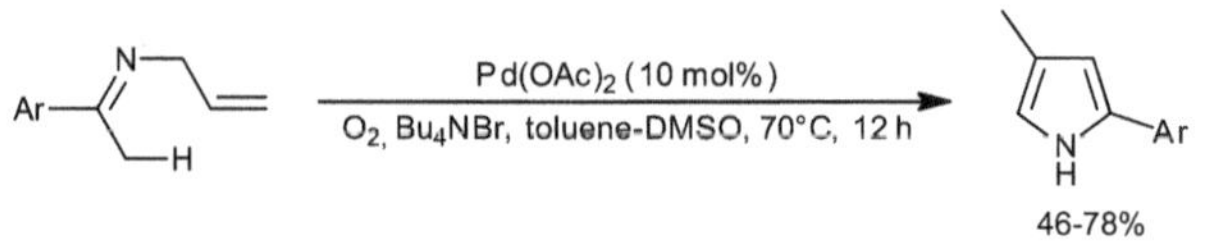

$$\text{Scheme 9.119} \quad sp^3 \text{ CH activation in an intramolecular oxidative Heck reaction.}$$

In addition to palladium, the oxidative alkenylation of arenes is also effectively catalysed by ruthenium complexes, reported first by Milstein and co-workers[359] and comprehensively reviewed recently by Kozhushkov and Ackermann.[360] Rhodium complexes are also known as catalysts for this reaction, although in most cases only directed processes take place; however, Glorius and co-workers recently reported a case of non-directed oxidative alkenylation of bromobenzenes in the presence of Rh(III) catalyst.[361] The bromine atoms in the substrates in this case take part in specific activation of the rhodium catalyst, but play no role in directing the reaction.

9.3.1.1 Directed Fujiwara–Moritani Reaction

Palladation is well known to be directed by various proximal donor groups – this phenomenon is the basis of popular and well-developed palladacycle chemistry.[362] Palladacycles are well known to be reactive in reactions with olefins under the conditions of the Mizoroki–Heck reaction – which is most probably the process to account for palladacycle disassembly and release of free catalytically active Pd species, and the operation of palladacycles as precatalysts in Mizoroki–Heck reactions.[363,364] Seen from the other side, these transformations account for the directive effect of various groups to make the Fujiwara–Moritani reaction in such cases highly regioselective. Moreover, directed palladation takes place at both electron-rich and electron-deficient aromatic rings,[365] hence the limitations of undirected palladation are lifted.

It should be noted that the mechanism of directed palladation is likely to involve a concerted attack of the Pd centre at an aromatic carbon with simultaneous (but not necessarily synchronized) abstraction of proton either by external base or a ligand, similar to what takes place in non-directed palladation (see below). As in any other concerted reaction, electronic preferences (nucleophile–electrophile dichotomy) of reactants are not strictly set, but adapt themselves to a given situation, the nature of substituents, ligands, etc.

An exemplary case of such substituent indiscriminacy can be found in e.g. a detailed study devoted to the Fujiwara–Moritani reaction of phenylacetic acids with terminal olefins reported by Yu and co-workers, which takes place in a system containing N-acylamino acids as ligands, weak base $KHCO_3$ and O_2 as terminal oxidant.[366] Both electron-donating and electron-withdrawing substituents can be present in phenylacetic acid to afford the products in high yields under the same conditions(Scheme 9.120).

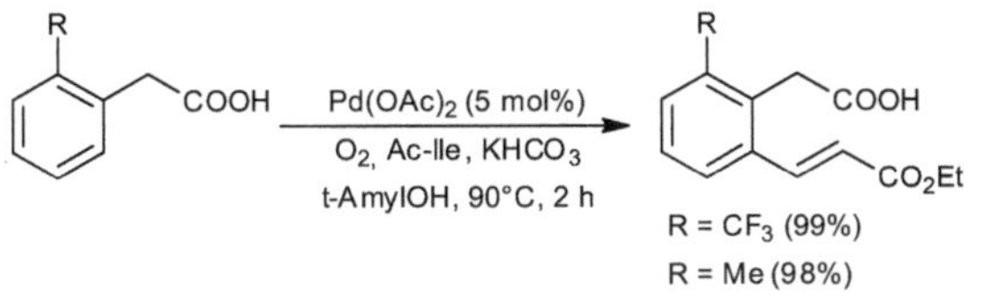

Scheme 9.120 Alkenylation directed by a carboxylic group. Ac-Ile = *N*-acetyl-L-isoleucine.

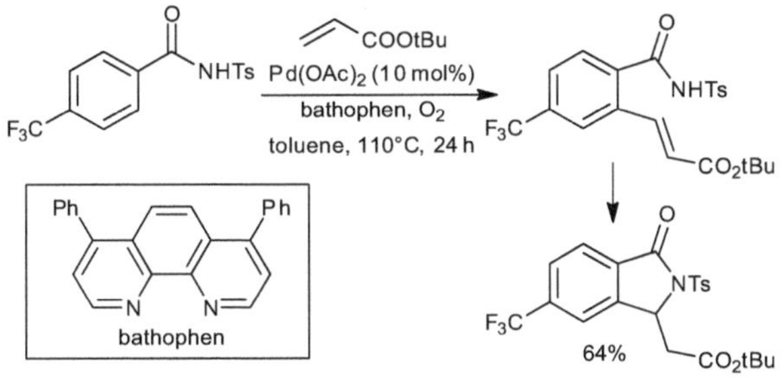

Scheme 9.121 *N*-Tosylamide-directed alkenylation.

Subtle effects were revealed on comparing kinetic parameters – relative rate constants and H/D isotope effects, which were established to be dependent on the substituents in phenylacetic acid and the ligand used, *e.g.* with *N*-Boc-L-valine ligand *o*-methylphenylacetic acid is slightly more reactive than the *o*-trifluoromethyl derivative ($k_{CF_3}/k_{Me} = 0.9$), whereas with *N*-acetyl-L-isoleucine the opposite was observed ($k_{CF_3}/k_{Me} = 1.38$). Given that the variations observed are rather small and dependent on a slight variation in reactivity, this case nicely illustrates the variable nature of the reaction pathway of directed palladation.

N-Tosylbenzamides were demonstrated by Zhu and Falck to be excellent substrates for directed alkenylation, allowing for both electron-rich and electron-withdrawing aryl rings and both electron-rich and electron-withdrawing alkenes.[367] Oxygen is used as terminal oxidant, being superior to benzoquinone, and 4,7-diphenylphenanthroline (bathophenanthroline) derivative as the ligand of choice. The products of the reaction are not the initially formed alkenes, but the products of cyclization. Electron-deficient alkenes give the respective Michael adducts (Scheme 9.121).

Electron-rich alkenes give the products of Wacker-type palladium-assisted conjugate addition as a mixture of *exo* and *endo* cyclization products, with the *exo* adduct being the major one (Scheme 9.122).

The involvement of a palladacycle was explicitly established and studied by Brown and co-workers.[368] The reaction of acetanilides with acrylate in the presence of strong acid and benzoquinone as terminal oxidant involves the respective palladacycles, which can be isolated and reintroduced into the catalytic cycle (Scheme 9.123).

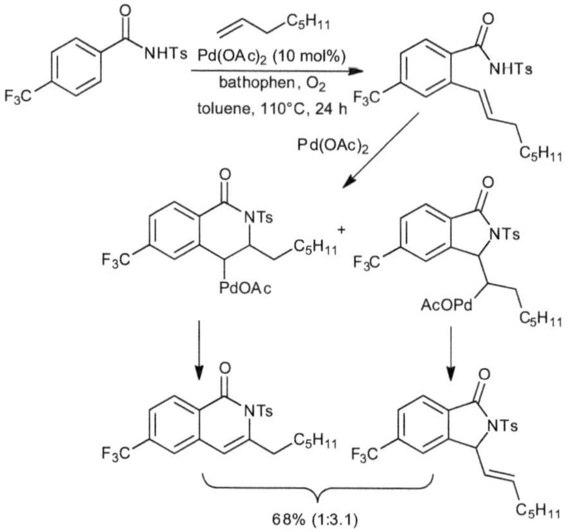

Scheme 9.122 *N*-Tosylamide-directed alkenylation by electron-rich olefins.

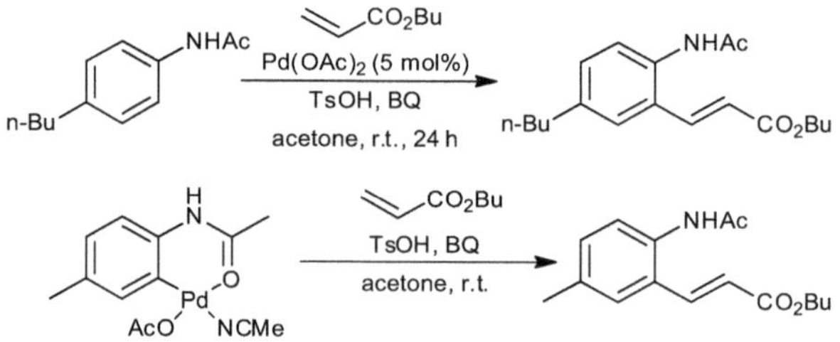

Scheme 9.123 The involvement of a palladacycle in *N*-acetylamine-directed alkenylation.

Directed olefination is often accompanied by various ring closure re-actions, such as Wacker-type palladium-assisted nucleophilic attack at the double bond of the new fragment. Thus, for example, amino acid derivatives were found by Yu and co-workers to be effective ligands for directed olefi-nation of arylethyl alcohols using a very complex catalytic system.[369] The process is highly sensitive to the nature of the ligand, by far the best being the *O*-menthyl ester of L-leucine, all tested *N*-protected amino acids being much less effective. The systems used a silver salt as terminal reoxidant, a weak heterogeneous base and the exotic solvent C_6F_6 to afford the products of Wacker-type conjugate addition to the initially formed alkene (Scheme 9.124).

Tandem directed alkenylation–aminopalladation to afford a variety of heterocylic structures was reported by Zhang and co-workers (Scheme 9.125).[370]

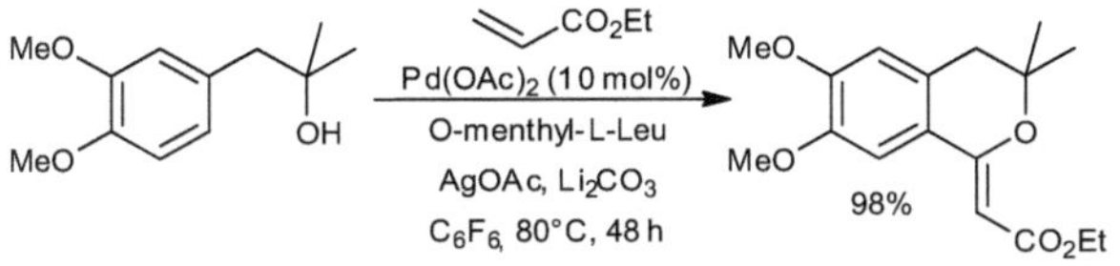

Scheme 9.124 Directed alkenylation in tandem with oxypalladation.

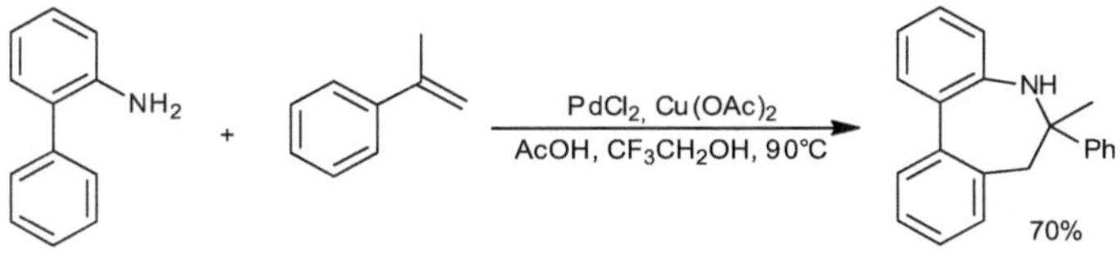

Scheme 9.125 Directed alkenylation in tandem with aminopalladation.

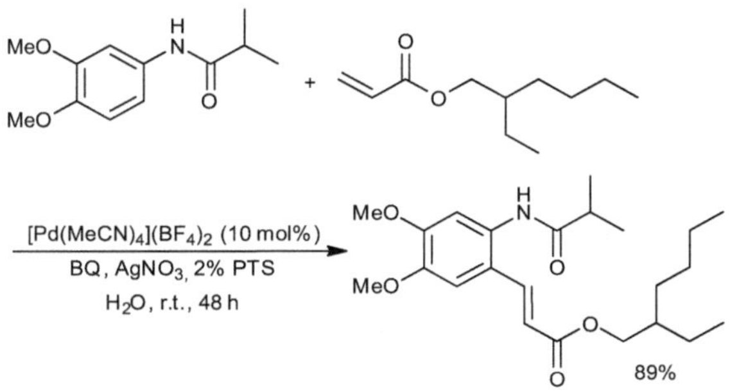

Scheme 9.126 *N*-Acylamine-directed alkenylation in aqueous micellar media.

The use of water and aqueous media is well known in regular Mizoroki–Heck reactions. Nishikata and Lipshutz showed that water can serve as a good solvent also for directed Fujiwara–Moritani reactions.[371] Moreover, in the presence of the non-ionic surfactant PTS (polyoxyethylated tocopheryl sebacinate), directed alkenylation of anilides with hydrophobic acrylates bearing C_4–C_{12} alkyl tails takes place in the presence of a water-soluble silver salt and benzoquinone as terminal oxidant. High yields of alkenylation products are obtained either using $Pd(OAc)_2$ and strong acid HBF_4 or the cationic palladium complex without acid (Scheme 9.126).

Under the conditions used, the substrates are partly solubilized in PTS micelles and partly form a hydrophobic phase, hence vigorous stirring is required to ensure effective mass transfer. High yields and monoalkenylation selectivity are obtained in this system, which demonstrates the good prospects of aqueous systems for oxidative Heck reactions. A highly polar aqueous system is likely to enhance the reactivity of cationic palladium species.

A directed Fujiwara–Moritani reaction is more suitable for the development of an enantioselective version, as the configuration of intermediate

complexes is more ordered and chirality transfer from an appropriate ancillary can be expected. Indeed, a series of papers by Yu and co-workers
described this approach, using protected amino acids as chiral ancillaries
and rather sophisticated substrates. The most interesting feature of this
process was an extremely strong influence of alkali metal counterions on
both material and optical yields,[372] showing a strong dependence of the
outcome on the structure of intermediate complexes in which the substrate
and chiral ancillary are bonded in a proper configuration through chelation
of counterions (Scheme 9.127).[373]

The same protected amino acids were shown by Cui and co-workers to be
fairly effective for induction of planar chirality in the oxidative alkenylation
of dimethylaminomethylferrocene (Scheme 9.128).[374]

A promising development of directed palladation methodology uses
temporarily installed directing groups that are subsequently removed from
the products. Thus, a 2-pyridylsulfoxide was reported by Carretero and co-
workers as a general and removable directing group in alkenylations performed in the presence of $K_2S_2O_8$ or $PhI(OAc)_2$ as terminal oxidants.[375,376]
The nitrogen atom of the pyridine ring and not the sulfoxide oxygen lend the
directing assistance, as *e.g.* phenylsulfoxide residues failed to show the effect. This methodology was successfully used in the synthesis of resveratrol
(Scheme 9.129).

A similar 2-pyridylsulfonyl substituent at the pyrrole or indole nitrogen
was established by Carretero and co-workers to serve as an excellent directing substituent. Although other acyls and sulfonyls are known also to
direct alkenylation at a 2-site, the use of the 2-pyridylsulfonyl group was
reported to afford a more reliable alkenylation protocol, useful for a broad
range of indoles, pyrroles and olefins. The method uses Cu(II) as terminal
oxidant and in addition to monoalkenylation also affords bisalkenylation
(Scheme 9.130).[377]

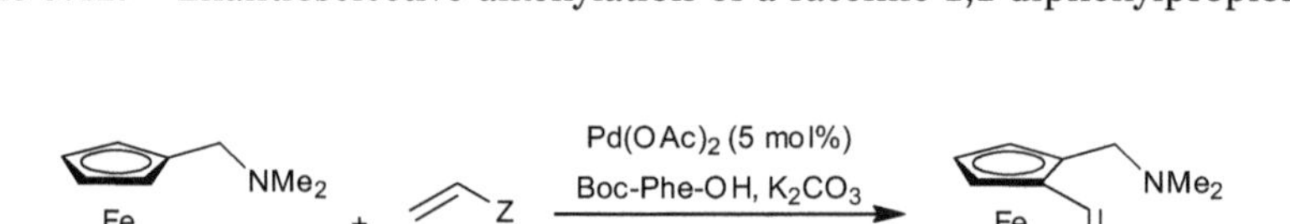

73% (ee 97%)

Scheme 9.127   Enantioselective alkenylation of a racemic 1,1-diphenylpropionate.

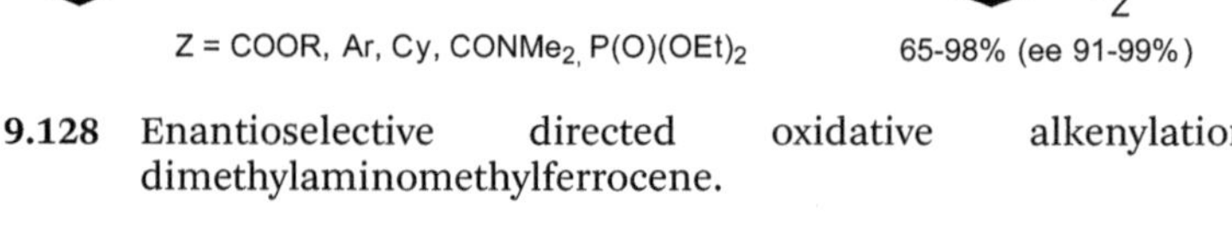

Z = COOR, Ar, Cy, CONMe$_2$, P(O)(OEt)$_2$ 65-98% (ee 91-99%)

Scheme 9.128   Enantioselective   directed   oxidative   alkenylation   of
dimethylaminomethylferrocene.

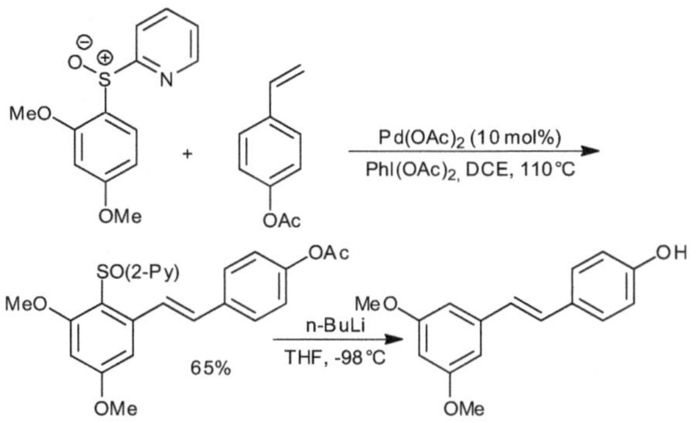

Scheme 9.129 Alkenylation directed by a temporary 2-pyridylsulfoxide residue.

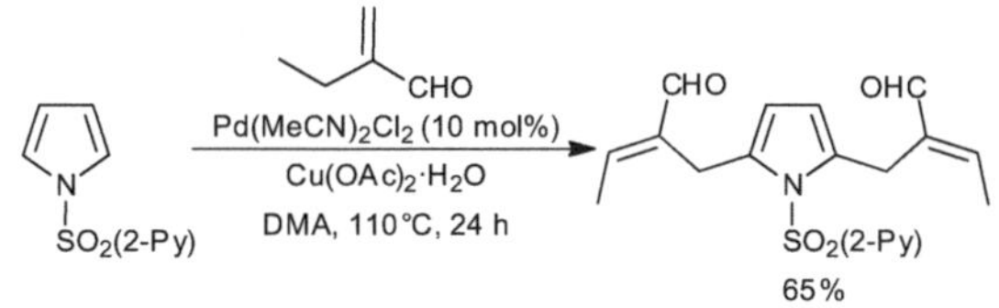

Scheme 9.130 2-Pyridylsulfonyl residue as a directing group in oxidative alkenylation.

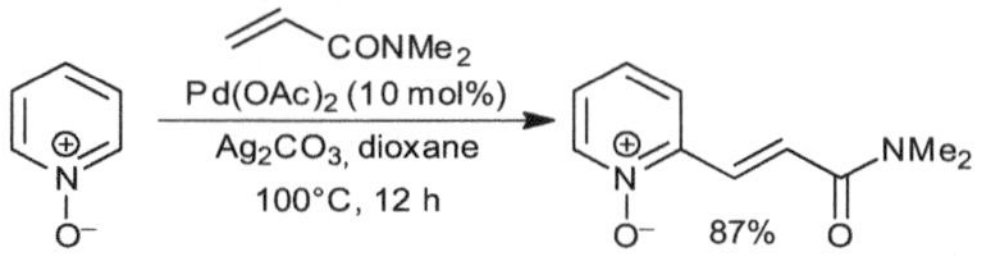

Scheme 9.131 *N*-Oxide as a directing group in oxidative alkenylation.

The *N*-oxide functionality can also be regarded as a temporarily installed directing group. Alkenylation at the α-position takes place in a simple system using a silver salt as terminal oxidant (Scheme 9.131).[378] Recently, Lee and co-workers reported a mono-*O*-methylated phosphate group installed on phenolic hydroxyl as an effective directing group for *ortho*-directed oxidative alkenylation.[379]

Sequential dialkenylation *via* a directed Fujiwara–Moritani reaction, followed by a Matsuda–Heck reaction, was described by Schmidt and Elizarov.[380] In addition to palladium, the directed Fujiwara–Moritani reactions are effectively catalysed also by ruthenium[360,381–383] and rhodium[382,384–387] complexes.

9.3.2 Deborylative Heck Reaction

Transmetallation of organoboron compounds by Pd(II) derivatives makes oxidative Heck reactions involving aryl- and alkenylboronic acids probably the most interesting choice for developing new Heck-type olefin arylation methods. These reactions are generally run under very mild conditions in simple catalytic systems, thus allowing for a better level of control over reactivity and selectivity by specific ancillary ligands.

The deborylative Heck reaction was discovered by Dieck and Heck in 1975 in a stoichiometric mode.[28] The reaction was run at room temperature in the presence of base in excess of liquid reagents without solvent and phosphine ligands (Scheme 9.132).

In spite of mild conditions, high yields and stereoselectivity, the reaction then remained practically unnoticed and had to wait almost 20 years for Uemura and co-workers to reinvestigate its potential and try to realize the process in the presence of substoichiometric amounts of Pd(II) using acetic acid as solvent and silver carbonate as terminal oxidant.[388,389]

Major progress with this method had to wait until the 2000s and in recent years the popularity of this chemistry has been rapidly growing, as it turned out that it conceals great possibilitites for tuning up and control.[390]

First, Mori and co-workers reported a general protocol using $Cu(OAc)_2$ as terminal oxidant.[391] Jung and co-workers suggested the use of oxygen as terminal oxidant and proposed a versatile wide-scope protocol applicable for various boronic acids, boronates and olefins (Scheme 9.133).[392,393]

Simultaneously this chemistry was developed by Larhed and co-workers.[394] An effective protocol for the arylation of electron-rich olefins (enamides) employed oxygen as terminal oxidant, N-methylmorpholine base and the bidentate N,N-chelator 2,9-dimethyl-1,10-phenanthroline (dmphen).[395,396] The system was further developed into a general, mild and wide-scope protocol. These major advances clearly showed the potential and versatility of the deborylative Heck reaction.[397]

Further development of the deborylative Heck reaction showed that of all known oxidative Heck reactions it is the best suited for the development of new methodology and the elaboration of synthetically appealing procedures.

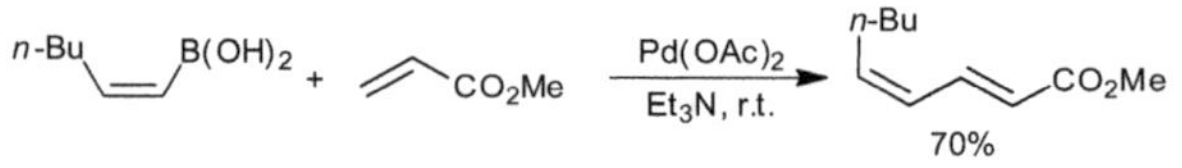

Scheme 9.132   The first deborylative alkenylation discovered by Dieck and Heck in 1975.

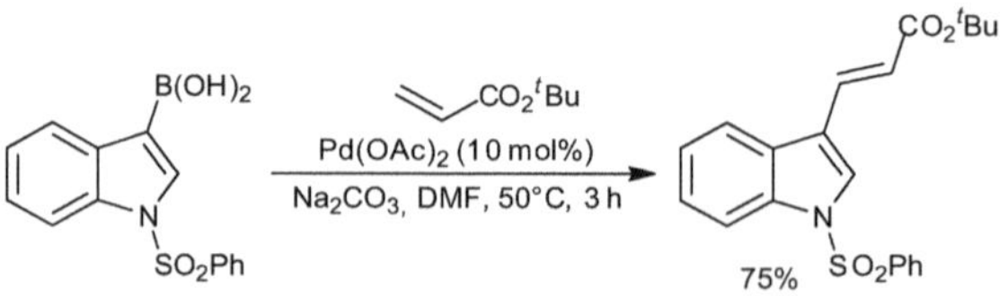

Scheme 9.133   Deborylative Heck reaction using oxygen as terminal oxidant.

An interesting protocol using a ferrocenylimine *C,N*-palladacycle as pre-catalyst for the oxidative arylation of standard olefins (styrenes, acrylates, but also methacrylate, a single example) by arylboronic acids was reported by Wu and co-workers.[398] The protocol is remarkable for a combination of unusual features: (i) the extreme simplicity of the catalytic system, involving only Pd precatalyst with a low loading (0.5 mol%); (ii) the use of oxygen as the only oxidant; and (iii) the mildness of the conditions (50 °C, DMF) (Scheme 9.134). The use of a palladacycle for the oxidative Heck reaction is particularly notable.

The reaction was realized by Larhed and co-workers in a continuous-flow reactor, which allows for very fast mixing and heating of the reagents, resulting in a dramatic shortening of the reaction time, and provides protocols that are easily scalable and reproducible, with both standard (Scheme 9.135) and electron-rich olefins (Scheme 9.136).[66]

Depending on the nature of the palladium precatalyst, deborylative Heck reactions can follow both polar and neutral pathways. In many cases, deborylative reactions are subject to the same modes of regioselectivity

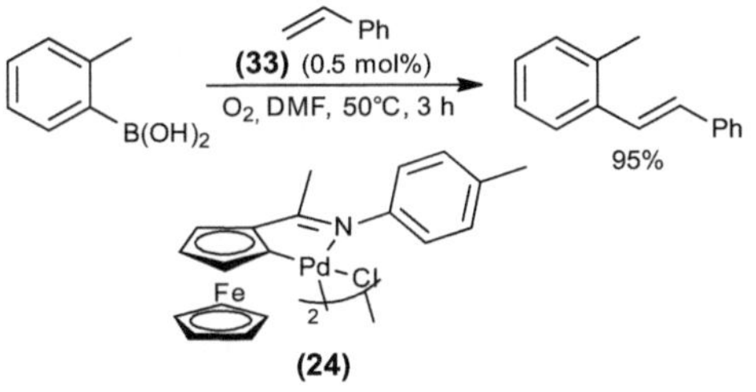

Scheme 9.134 Deborylative Heck reaction in the presence of a palladacyclic pre-catalyst (**24**).

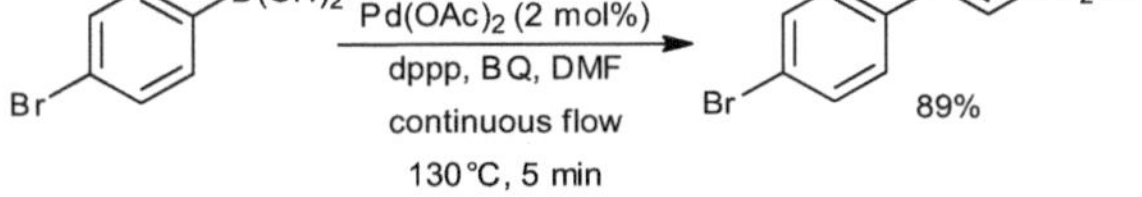

Scheme 9.135 Example of deborylative arylation of acrylate in a continuous flow reactor.

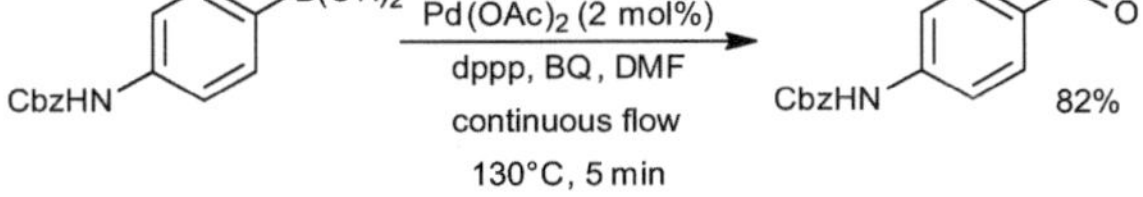

Scheme 9.136 Example of deborylative arylation of electron-rich olefins in a continuous flow reactor.

control by intramolecular chelation as are known for other versions of regular and oxidative Heck reactions. In deborylative reactions, palladium acetate is often used, which, at the first glance, would not favour the polar pathway, particularly when copper acetate is used as terminal oxidant to supply a good excess of acetate ligands for Pd species. However, it is very likely that the polar pathway is nevertheless enforced by using such neutral labile ligands as sulfoxides (the addition of DMSO to the solvent or special sulfoxide ligands), which may be expected to effect acetate–sulfoxide ligand exchange, which, given the coordination lability of sulfoxide ligands, can be considered as an effective means of triggering the polar pathway (Scheme 9.137).

Bidentate disulfoxide can perform this task even without being used in huge excess because of the chelating effect. Intramolecular chelation of cationic palladium intermediates is likely to exert an excellent degree of control over regioselectivity in the arylation of various functionalized olefins by arylboronic acids[399] (Scheme 9.138).

This interesting and highly fruitful approach implicitly exploits directed carbopalladation (various O, N and S centres do the task; for a review on chelation-controlled Heck reactions, see Ref. 400), in which case steric reasons dictate (a) the attack of the Pd centre at the proximal atom of the double bond and (b) fixation of the chelate ring restricting PdH elimination from an inner atom, thus suppressing the double-bond shift. Regular Mizoroki–Heck reactions involving neutral arylpalladium intermediates in such cases are non-selective, giving variable regioselectivity and double-bond shifts (as shown by pathways marked with dashed arrows in Scheme 9.139).

The retention of the double-bond position in the arylation of allylic acetates was reported by Su and Jiao to require a sophisticated catalytic system including both silver and copper salts and also KHF_2 additive.[401] The effect was attributed to chelation control (Scheme 9.140).

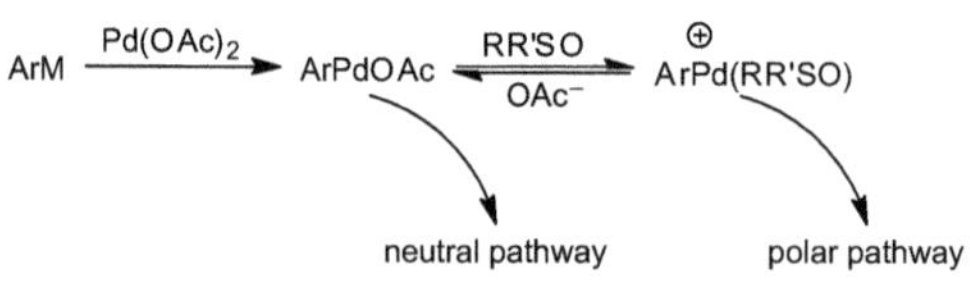

Scheme 9.137 Multiplicity of pathways in deborylative (and other oxidative) Heck reactions in the presence of (hemi)labile sulfoxide ligands.

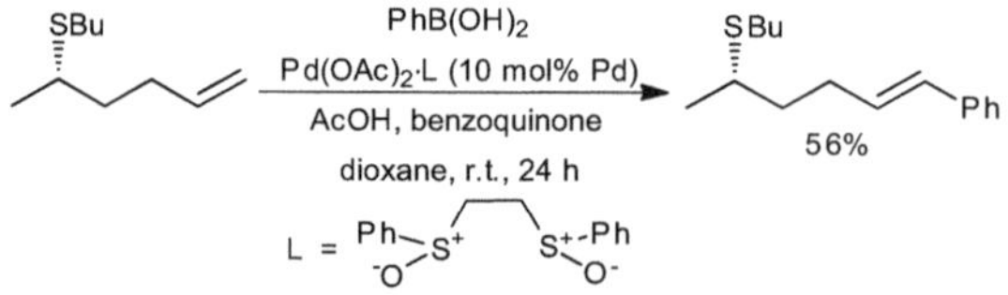

Scheme 9.138 Regioselective deborylative arylation of remote double bonds in chelation-enabled alkenes.

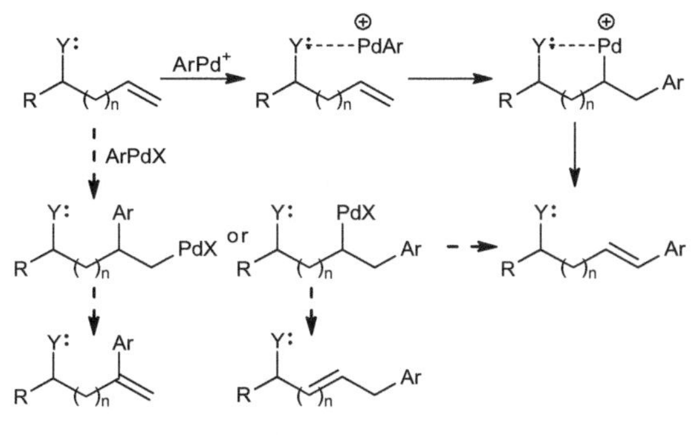

Scheme 9.139 Chelation control over regioselectivity.

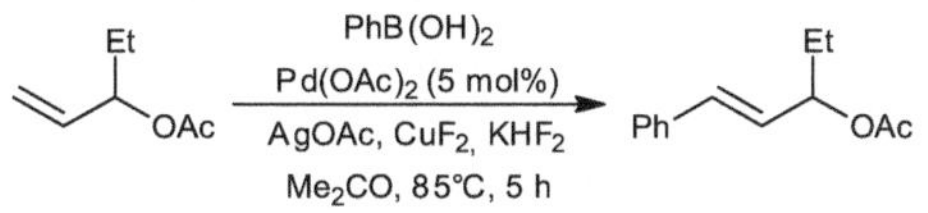

Scheme 9.140 Retention of double-bond position in the deborylative arylation of allylic acetates.

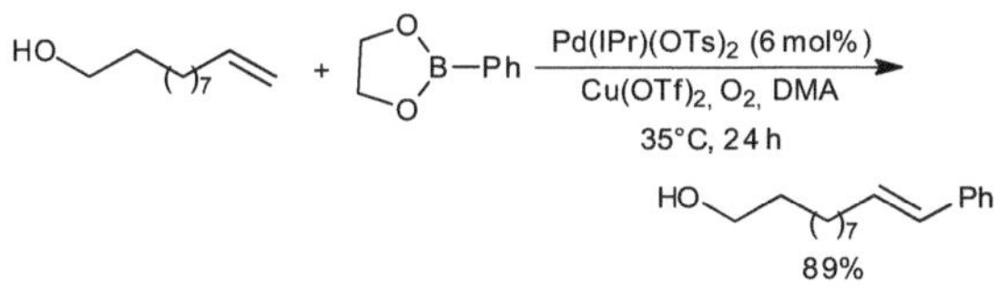

Scheme 9.141 Retention of double-bond position in alkenes not capable of intra-molecular chelation.

Probably, however, the regioselectivity is governed by other factors, besides involving intramolecular chelation in intermediates. A very interesting system was reported by Werner and Sigman.[402] The complex Pd(I^{i}Pr)(OTs)$_2$[403] was shown to serve in the system with Cu(OTf)$_2$ (20 mol%) as an intermediate reoxidizer and O$_2$ as terminal reoxidizer in DMA at 40 °C for the arylation of a wide range of functionally ω-substituted terminal olefins by ethylene glycol esters of arylboronic acids to yield the arylation products with high regioselectivity (conservation of double-bond position) and stereoselectivity (*E*-isomers were the major products), disregarding the nature of functionality, which excludes intramolecular chelation as the main factor to account for the high regioselectivity of PdH elimination (Scheme 9.141).

Even more interesting is the complete absence of racemization in a proximal chiral centre, which would necessarily have occurred if double-bond migration took place. This finding permits thermodynamic control with equilibrium bond shifts favouring the most stable conjugated position to be completely ruled out (Scheme 9.142).

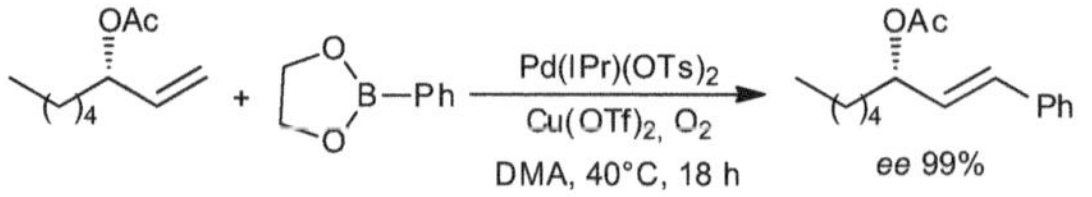

Scheme 9.142 Retention of configuration of a chiral centre sitting in the path of probable a double-bond shift.

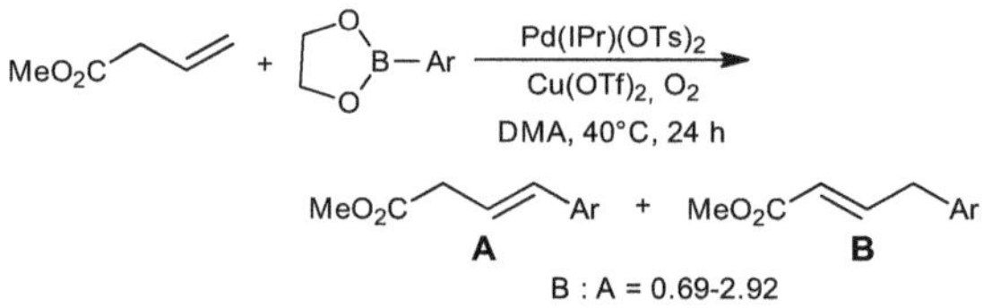

Scheme 9.143 Regioselectivity of arylation of methyl 3-butenoate by arylboronates in the presence of a cationic Pd complex with heterocyclic carbene ligand.

This work clearly showed the importance of a stable ancillary ligand in Pd complexes with respect to both catalyst stability and regioselectivity. The complex $Pd(MeCN)_2(OTs)_2$ showed comparable activity, but much lower regioselectivity. Apparently, labile ligands in the coordination shell give rise to multiple variably ligated Pd species and thus multiple pathways can be taken.

The suppression of a bond shift, however, is not a general phenomenon – it is valid only if travel of the double bond along the chain must pass through unsubstituted CH_2 links or a link bearing donor O or N atoms destabilizing the emerging double bond. In these cases, the η^3-benzylic complex serves as a thermodynamically favoured anchor for the double-bond migrations. On the other hand, the arylation of methyl 3-butenoate gave mixtures of two products in which the product with conserved double-bond position was the minor isomer formed. The ratio was found to depend on the substituent in the boronic acid aryl group – the amount of rearranged product was the highest for phenyl boronate (B : A = 2.92), with both electron-donating and electron-withdrawing substituents resulting in a loss of selectivity (Scheme 9.143). In this case, both double-bond positions are thermodynamically favoured and the product ratio obtained would depend on the interplay of equilibria.

On the other hand, in other catalytic systems double-bond shifts take place which can be employed for the introduction of remote aryl groups into carbonyl compounds: in the arylation of unsaturated alcohols, the double bond readily travels from its initial position to a hydroxyl group and converts it into carbonyl, skipping 1–4 carbon atoms. Sigman and co-workers developed a regio- and enantioselective version of this process.[404] In the presence of a chiral oxazoline ligand, the emerging tertiary carbon accepts the chirality. High optical yields are generally obtained despite wide variations of both arylboronic acids and unsaturated alcohols (Scheme 9.144).

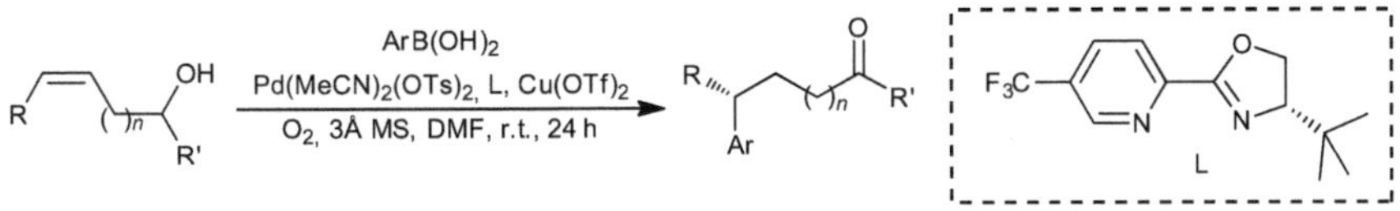

Scheme 9.144 Enantioselective oxidative arylation of unsaturated alcohols (only the major product is shown).

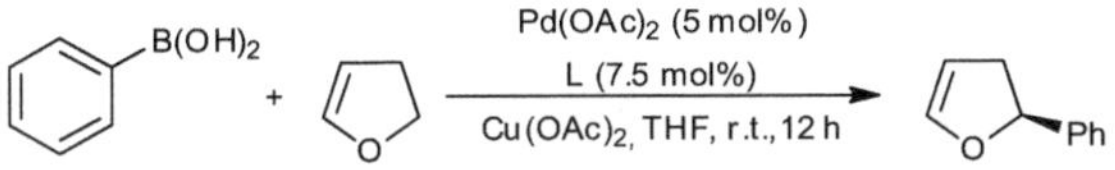

L	Yield, % (*ee*)
(R)-BINAP	71 (57)
(R)-MeOBiphep	67 (82)

Scheme 9.145 Enantioselective deborylative Heck reaction.

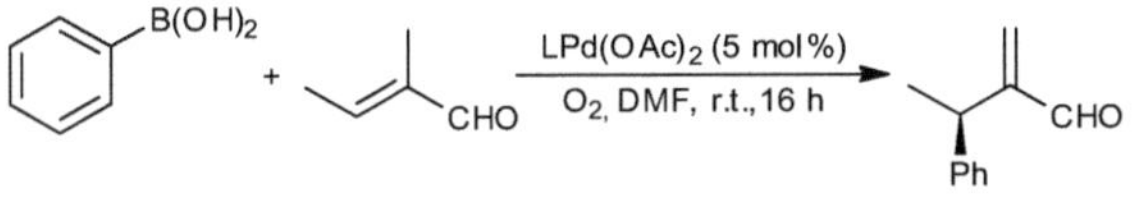

74% (*ee* 75%)

Scheme 9.146 Enantioselective deborylative arylation of an enal.

The hypothesis that oxidative Heck reactions involving deborylation must proceed *via* the polar pathway prompted Gelman and co-workers[405] to investigate the application of this reaction in enantioselective mode using chiral ligands normally used with success in the enantioselective Mizoroki–Heck reaction, including BINAP and oxazoline chelators.[120] The results were, however, inferior with respect to both material and optical yields and some ligands highly effective in enantioselective Mizoroki–Heck reactions failed in the deborylative version (Scheme 9.145). However, the new reaction took place under substantially milder conditions and was significantly faster, which promises further improvements.

Similar results were obtained by Jung and co-workers using a different system with *N,N*-chelators, O_2 as terminal oxidant and an open-chain alkene.[406] The optical yields obtained were fairly high (around 70%) provided that a preformed complex was used as precatalyst to minimize the ligand-unrestricted pathways (Scheme 9.146).

In this respect, a few comments are should be made. The use of bidentate ligands is not a guarantee of the realization of the polar pathway. When using Pd(OAc)$_2$ as a Pd source, the catalytic species would contain an OAc ligand, which is a good ligand for palladium with a stable or hemi-labile mode of binding, not easily displaced by olefin. Therefore, the reaction may (or even must) take place *via* a neutral pathway with dechelation, leaving the chiral ligand in the transition state and intermediates bonded in a

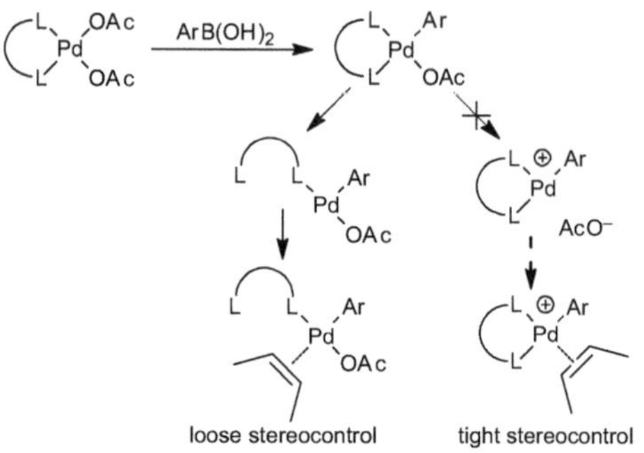

Scheme 9.147   Bidentate ligands do not guarantee bidentate chelation.

monodentate fashion, which is a likely source of less effective stereo-differentiation (Scheme 9.147).

The other probable source is random wandering of the double bond, leading to partial racemization, which may take place because in oxidative Heck reactions PdH intermediates are not rapidly scavenged by base and are more long-lived, bringing havoc into the orderly course of the enantiose-lective pathway.

Therefore, more efforts are required to develop enantioselective oxidative Heck reactions, both by ligand design to seek ligands that would not be dechelated during the reaction and the use of nucleofugic counterions to permit the polar pathway with a stable bidentate chiral ancillary and a free coordination location for catalytic action. It should be taken into consideration that there are more problems in oxidative pathways to be dealt with: (a) in the case of phosphine ancillaries, oxidative conditions lead to oxidation of phosphine sites and thus to disruption of bidentate binding and deterioration of stereocontrol; and (b) the oxidant itself often provides non-labile ligands for palladium (e.g. Cu acetate gives acetate, O_2 gives OH, OOH or –O– ligands, quinones may give semiquinone or phenolate ligands, etc.), which would block the polar pathway, hence a judicious choice of terminal oxidant is essential if tighter stereocontrol is desired.

In a later study, Jung and co-workers obtained very important clues on which direction the ligand design for enantioselective deborylative reactions can take.[407] A tridentate CNO ligand based on a heterocyclic carbene modified with a chiral amino alcohol was used as a preformed complex to deliver high optical yields although at the expense of reaction rate and material yields. The idea is indeed highly motivating – if one cannot reliably control the anionic ligand in the coordination shell, such can be provided in advance as the third arm of a tridentate ligand. The third site should be chosen to be able to behave only as a hemi-labile ligand capable of binding the olefin. Such hemi-lability is likely to be enhanced by binding borate during transmetallation (Scheme 9.148).

This work also provides an instructive example demonstrating that it is the bidentate ligand and not the polar (cationic) state of the Pd centre which

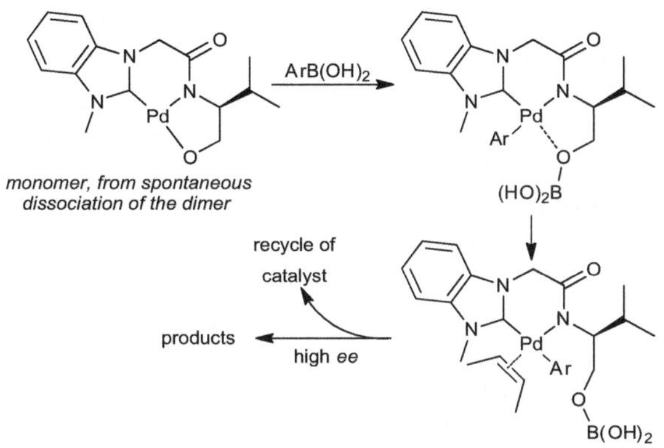

Scheme 9.148 Ligand design for enantioselective deborylative arylation – improving control over the coordination shell.

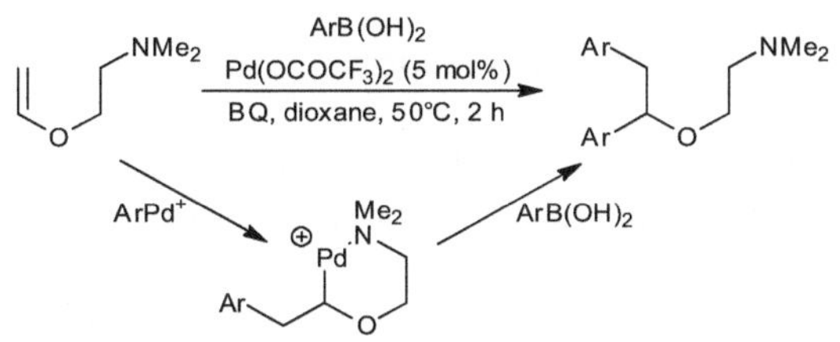

Scheme 9.149 Diarylation *via* tandem carbopalladation–cross-coupling.

matters for the ligand-restricted (here, enantioselective) pathway to operate. In this case, both the original ligand and all intermediates are neutral, bearing two σ-bonded anionic ligands. The most serious restriction of this outstanding system is the low material yield of the arylation product, in other words, a low TON value, which is likely to be associated with problems in recycling the reactive form of the catalyst, as the O-sidearm may not get rid of borate and thus the third coordination location loses the ability to recoil.

The carbopalladation intermediate can undergo cross-coupling with excess of organometallic reagent. Such a tandem process takes place readily with more reactive organotin compounds (see below), but with organoboron compounds no instances were known until recently. Cross-coupling with boronic acids usually requires the addition of a hard Lewis base to activate the C–B bond towards transmetallation and PdH elimination is therefore faster than transmetallation under the usual conditions of oxidative Heck reactions. Larhed and co-workers found an interesting class of olefins bearing a pendant moiety with a basic NMe$_2$ group, which undergo diarylation (Scheme 9.149).[408] The role of the pendant moiety is probably to stabilize the carbopalladation intermediate by chelation, increasing its lifetime to match it with the rate of transmetallation. The other probable explanation may involve the activation of boronic acid by the basic $RNMe_2$ group, although in this hypothesis no six-membered rings can be drawn in the intermediates. An increase in the length of the pendant moiety or a change to a bulkier NEt_2 group or to secondary $NHMe$ ruins the diarylation.

Another interesting feature of these reactions was manifested in the case of Michael acceptors used as olefins is the competition between the normal Heck pathway and the conjugate addition pathway. This is particularly important for cyclic enones, the primary products of the carbopalladation of which cannot eliminate PdH directly, but are engaged in metallotropic rearrangements leading to either of two products. This behaviour is in fact well known in regular Mizoroki–Heck chemistry (for a discussion, see Ref. 87). Lee and co-workers showed that in the arylation of cyclohexenones and similar enones by arylboroxines, the outcome depends dramatically on the conditions used, and pathway can be induced as needed by changing the conditions (Scheme 9.150).[409]

Stahl and co-workers developed this approach further by performing a one-pot arylation of cyclohexenones with subsequent isomerization induced by the addition of DMSO to the reaction mixtures to afford 3-arylated phenols (Scheme 9.151).[410]

This mechanistic dichotomy in the reactions of organoboron compounds with enones and other Michael acceptors was observed also with catalysis by Ni, Co[411] and Rh[412,413] complexes.

An interesting reaction formally looking like an oxidative deborylative Heck reaction was described by Liwosz and Chemler.[414] Alkylation of alkenes by alkyltrifluoroborates was achieved in the presence of copper(II) triflate (Scheme 9.152). Both the unusual substrate profile and mechanistic studies, however, revealed a free-radical pathway without the involvement of organometallic species.

Deborylative Heck reactions have been applied to a variety of interesting methodologies, *e.g.* for the synthesis of 4-arylated coumarins,[415] selective arylation of allylamines[416] and allylic ethers[417] and cascaded arylation–oxypalladation to afford 2-aryltetrahydrofurans from homoallylic alcohols.[418]

9.3.3 Desilylative Heck Reaction

In cross-coupling chemistry, organosilicon compounds belong to the "Big Five" major organometallics used as transmetallation agents (B, Sn, Zn, Mg, Si). In oxidative Heck reactions, so far organoboron compounds are beyond competition, but there are some data on Sn and Si compounds showing that the potential exists, but is largely unexploited.

Organosilicon compounds are generally less reactive than organoboron compounds and require more energetic activation, usually by fluoride-ions (the Hiyama cross-coupling reaction), as compared with common bases used in the Suzuki–Miyaura reaction. Concerning the oxidative Heck reaction, where the transmetallation is believed to be effected by more reactive

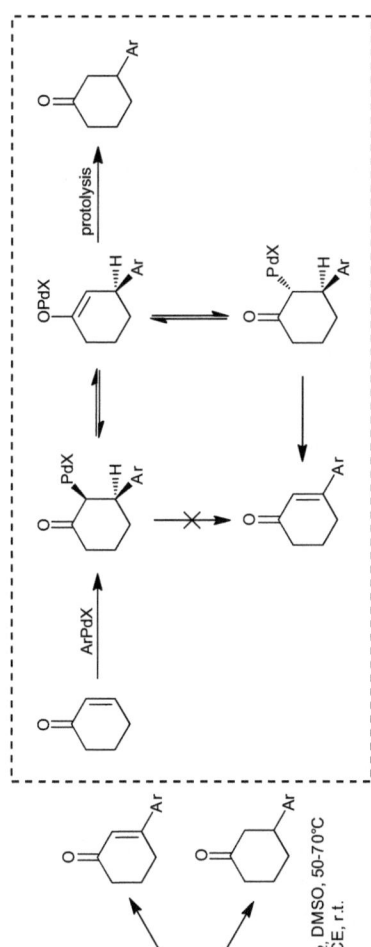

Scheme 9.150 Formation of regular Heck products *versus* conjugate addition.

A. (MeCN)₄Pd(OTf)₂, O₂, DMSO, 50-70°C
B. (MeCN)₄Pd(OTf)₂, DCE, r.t.

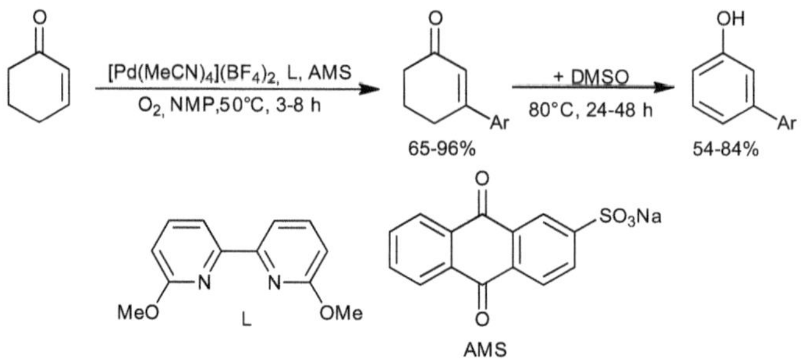

Scheme 9.151   Synthesis of 3-arylphenols *via* deborylative Heck reaction.

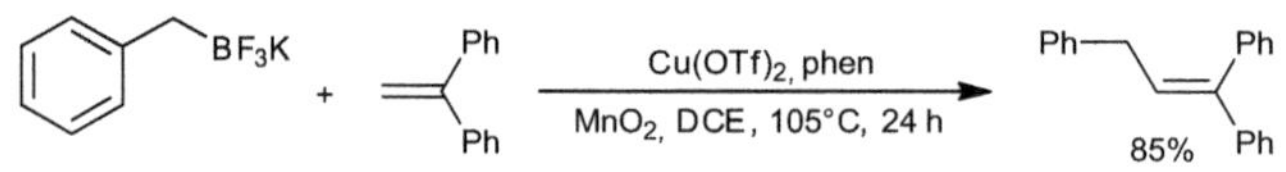

Scheme 9.152   Copper(II)-assisted Heck-like alkylation of alkenes'

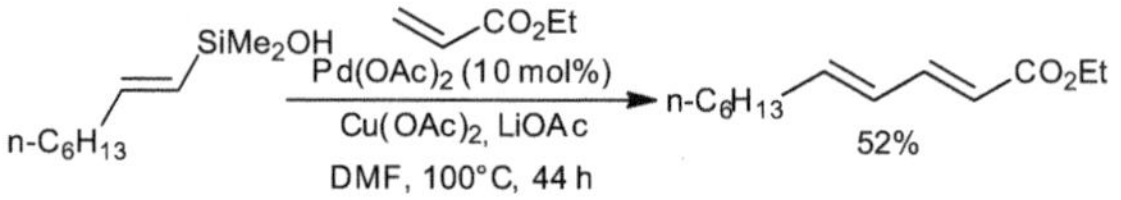

Scheme 9.153   Example of a desilylative alkenylation reaction.

non-organometallic Pd(II) complexes, organoboron compounds do not need activation by base. What about organosilicon compounds?

Unfortunately, so far too few data exist. Organosilicon compounds are more often used in the regular Mizoroki–Heck reaction, where organosilicon residues compete with hydrogen in the elimination, bringing forward the problem of selectivity.[419]

Organosilicon compounds were shown to take part in oxidative Heck reactions by Mori and co-workers,[420] who used silanols, compounds which are known not to require activation by fluoride to take part in transmetallation with organopalladium species, which is taken advantage of in a special cross-coupling method (the Hiyama–Denmark reaction).[421] The transmetallation with more reactive Pd(II) complexes takes place readily to allow the oxidative Heck procedure, which was performed either stoichiometrically in a very simple system containing only Pd(OAc)₂ or using a cupric salt as terminal oxidant (Scheme 9.153).

The reactions were rather slow and gave modest yields. A rhodium-catalysed version of this reaction was discovered,[421] but no more interest in the method was evident for many years.

This underexplored method was recently revisited by Cheng and co-workers, who used trimethoxysilanes, which are easily activated by fluoride anion to enter the Heck process using 1,10-phenanthroline as ligand and AgF as both reoxidant and fluoride source.[422] The method was applicable not

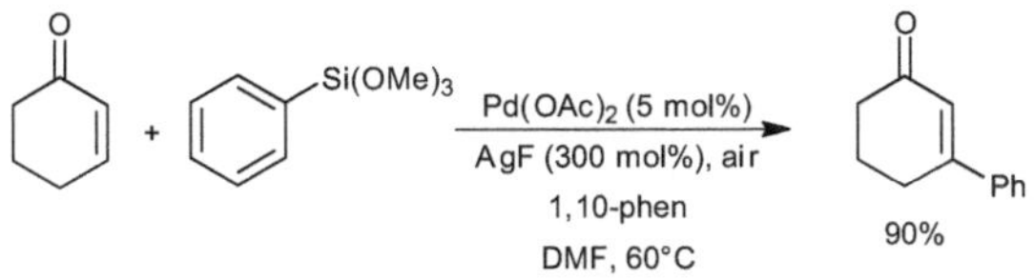

Scheme 9.154 Desilylative arylation reaction using fluoride activation.

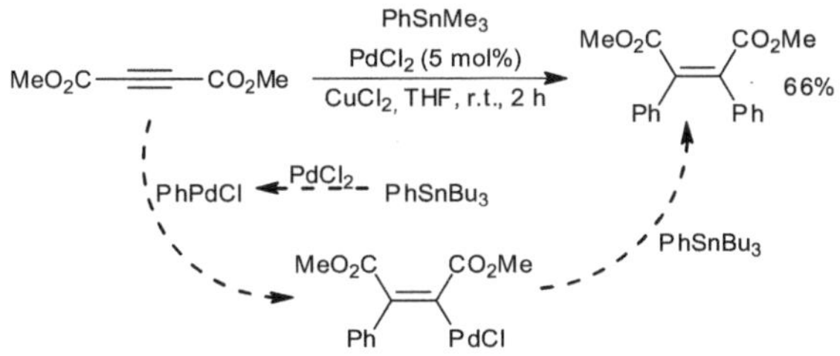

Scheme 9.155 Alkyne diarylation *via* destannylative Heck reaction cascaded with Stille cross-coupling.

only to standard olefins, but also to cyclohexenone (Scheme 9.154), acrylamide, vinyl acetate and 2-vinylpyridine. A similar system employing a less common ligand, the bis(2-pyridylhydrazone) of glyoxal, was introduced by Mino *et al.*[423]

9.3.4 Destannylative Heck Reaction

The oxidative Heck reaction with organotin compounds was first performed by Heck, who carried out a few runs with tetraphenyltin within studies on organomercuric compounds. As the results were disappointing, tin compounds were practically abandoned. Closely related is an alkyne diarylation procedure developed by Kosugi and co-workers, who showed that the reactivity of organotin compounds towards transmetallation allows trapping of the carbopalladation intermediate by the cascade cross-coupling event (Scheme 9.155).[424] A similar diarylation procedure was used for norbornene.[425]

The first protocol for destannylative oxidative arylation was reported by Mori and co-workers using Cu(II) as terminal oxidant to afford the expected products in good yields, although the reaction required prolonged heating (Scheme 9.156).[426]

A major improvement was subsequently reported by Jung and co-workers, who employed roughly the same system, but managed to perform the reactions at room temperature.[427] However, an even a more important advance was the use of oxygen as the sole terminal reoxidant, which was shown to be as effective as cupric salts and in some cases significantly more effective.

More recent reinvestigation of the behaviour of organotin compounds under conditions of the oxidative Heck reaction was initiated by Sigman and

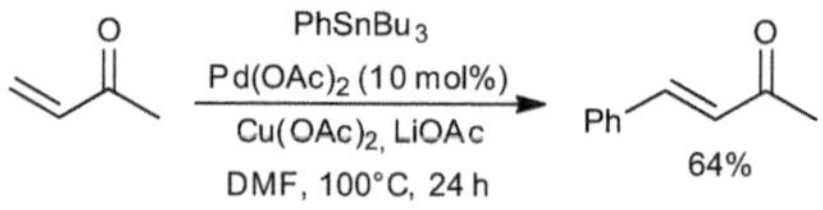

Scheme 9.156   Example of destannylative arylation.

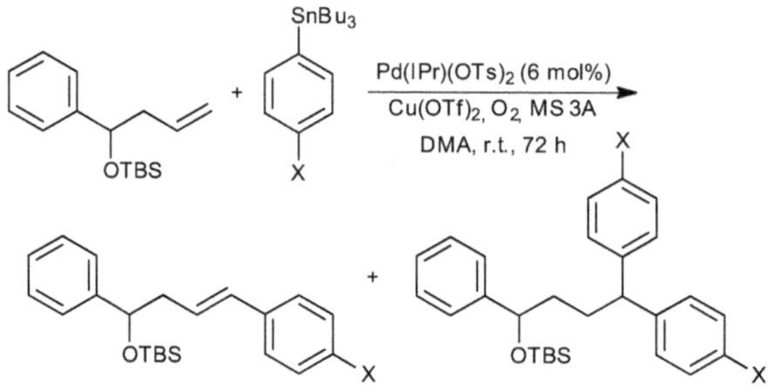

X			
OMe	13%		82%
Cl	36%		46%

Scheme 9.157   Destannylative arylation *versus* 1,1-diarylation.

co-workers,[428] which led to discovery of interesting trends seeming highly relevant in the more common context of Heck chemistry.

The reaction, performed with a cationic palladium complex with a heterocyclic carbene ligand, led, in addition to the expected Heck product, to 1,1-diarylation (Scheme 9.157).[428] The distribution of products depended on the substituent in the organotin compound – electronegative substituents favoured the Heck pathway, whereas electron-rich organotin compounds gave mostly 1,1-diarylation, which in these cases can be regarded as a selective reaction.

The results are extremely intriguing and reveal important aspects of the internal functioning of Heck pathways. First, the formation of diarylation products is an evident outcome of the Heck–Stille cascade, with reactive organotins being able to trap carbopalladation intermediates. In deborylative reactions, such cascades do not occur because the activation of organoboron compounds to transmetallation requires the presence of base, which usually is not added to reaction mixtures in oxidative deborylative reactions.

In Sigman and co-workers' process, the primary carbopalladation adduct is not trapped to a significant extent, but instead undergoes PdH elimination giving either the monoarylation product or the isomeric adduct, which is what is captured by the second molecule of the organotin compound (Scheme 9.158).

Such "strange" behaviour is likely to be accounted for by the kinetics of the stages. Transmetallation with organotins should be slow, particularly with electron-deficient or sterically hindered compounds. This gives a

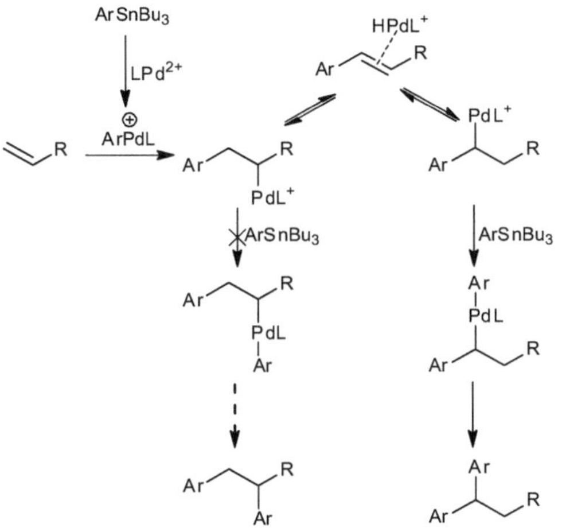

Scheme 9.158 Transformations of carbopalladation adduct in the destannylative Heck reaction.

chance for the formation of Heck product in the base-free system, where the PdH intermediate is likely to be a long-lived species, waiting for spontaneous deprotonation or oxidation by O_2. Under such conditions, the isomeric adduct is given the chance to define the predominant pathway. In order to do so, this adduct should be relatively the most stable of all the intermediates involved, hence it can be trapped in the cascaded Stille cross-coupling.

Moreover, such a mechanism involves the reversible elimination–addition of PdH. What is unusual (although not without precedent – *cf.* the similar chemistry in the deborylative reactions, *e.g.* Scheme 9.142) is that these elimination and addition events seem to stay bound to the initial reaction site and do not lead to shifts of the double bond. The experiment with a substrate containing a chiral centre next to the reaction centre shows that elimination in this direction does not take place.

This result certainly can be explained by the suppression of PdH *syn* elimination by the coordinating group, which is a common feature of Heck reactions. However, in other substrates without such groups, the double-bond shift would be manifested by the formation of 1,3-, 1,4-, *etc.* arylation products, which were not observed. Thus, all the evidence indeed demonstrates that the isomeric adduct is somehow stabilized. This feature was explained by appending an η^3-benzylpalladium complex to the pathway, being in equilibrium with the normal η^1-benzylpalladium complex (Scheme 9.159).

The idea is probably relevant for other protocols in Heck chemistry, where the double-bond migrations are suppressed. The formation of relatively stable η^3-benzyl complexes is probably favoured by cationic palladium centres possessing "free" coordination sites, but more research efforts are required to clarify the factors that favour such a coordination mode.

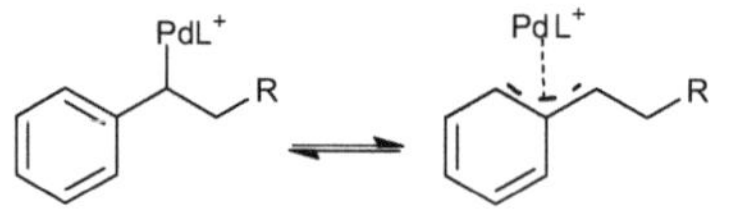

Scheme 9.159   Equilibrium between η^1- and η^3-benzylpalladium complexes.

9.3.5   Decarboxylative Heck Reaction

The idea that carboxylate ions can be regarded as masked carbanions is a matter of common knowledge going back to the distant past. The use of carboxylates as equivalents of organometallic compounds is, however, a rather late addition to the domain of palladium catalysis, but rapidly gaining in popularity. An arylpalladium intermediate is formed by decarboxylation presumably of a Pd carboxylate complex[429] (for recent reviews on transition metal decarboxylative reactions, see Refs 430 and 431). The decarboxylative Heck reaction was first described by Myers and co-workers, who used an excess of silver carbonate both as base and as terminal oxidant.[429,432,433]

Experimental mechanistic studies[429] revealed distinct steps involved in the catalytic cycle: (a) ligand exchange in which arenecarboxylate enters the coordination sphere; (b) extrusion of carbon dioxide from the thus formed Pd carboxylate; and (c) the rest faithfully follows the standard Heck catalytic cycle with oxidative regeneration of Pd(II) catalyst (Scheme 9.160).

It should be noted that from the beginning, palladodecarboxylation was regarded as an electrophilic *ipso*-substitution reaction. The course of such a reaction would critically depend on maintaining an adequate level of electrophilicity of the Pd centre, which requires that the system does not contain strongly bonding ligands, such as donor solvents, halide ions, amines, phosphines, *etc.* These limitations made the formulation of the reaction systems (choice of solvents and base) rather tricky. The presence of DMSO in the reaction medium is dictated by the finding of Myers and co-workers[429] that this solvent serves as a labile ligand adequately protecting the Pd centre from deactivation. Besides, DMSO can serve as a labile ligand ensuring the polar pathway mechanism of subsequent Heck transformations by keeping anionic ligands (acetate, hydrocarbonate, *etc.*, often being major constituents of the catalytic systems through Pd and Cu acetates, silver salts, *etc.*), similarly to what we commented upon in the discussion of the deborylative Heck reaction (see above).

The system can be as simple as just some $Pd(OAc)_2$ under an oxygen atmosphere without any extra additives. Such a system works well both for standard Heck olefins (acrylates and styrenes) and for more challenging olefins with internal double bonds, in which case complete inversion of configuration takes place as in other Heck reactions with such olefins (Scheme 9.161).[434]

Su and co-workers developed an alternative protocol involving Pd trifluoroacetate in the presence of benzoquinone and 1-adamantanecarboxylic acid in a complex mixture of solvents, applicable to electron-rich

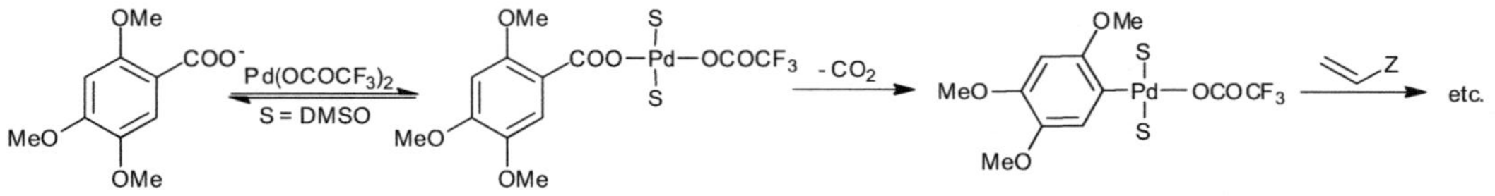

Scheme 9.160 Decarboxylative palladation involves the formation of a palladium carboxylate complex.

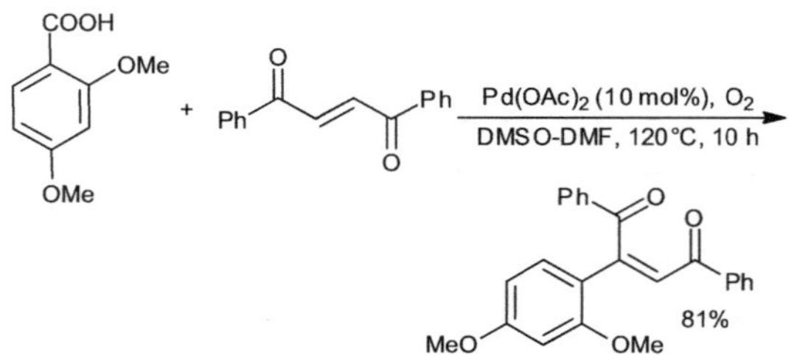

Scheme 9.161 A simple decarboxylative arylation protocol.

arenecarboxylic acids preferably containing an *o*-MeO group, which suggests directed *ipso*-palladodecarboxylation.[435] The thus optimized protocol is highly regioselective, giving only regular Mizoroki–Heck products with negligible double-bond shifts, and also is highly tolerant of varying functionality in both reagents (Scheme 9.162).

The method is suitable not only for standard olefins, but also for more challenging cases such as the arylation of allylic alcohols giving β-aryl ketones[436] and of enones by electron-rich aryls or heteroaryls (Scheme 9.163).[433]

As *ipso*-palladodecarboxylation in the Myers system is a reaction obeying the electrophilic substitution reactivity trends and it is facilitated by electron-rich substituents in the benzene ring,[437,438] which is the major scope-limiting factor. The extension of the decarboxylative Heck reaction to electron-deficient substrates followed a similar path to the same steps undertaken in the other oxidative Heck reactions. There are, however, interesting particular cases when the reaction takes place with substrates for which electrophilic attack should not be possible (Scheme 9.164)[439] Although the yield is only modest, the very possibility of carboxyl cleavage from pentafluorobenzoic acid shows that an alternative pathway of uncertain nature does exist.

The reaction with electron-deficient benzoic acid is promoted by silver salts, suggesting that a different pathway involving thermal decomposition of silver carboxylate in the presence of Pd species may lead to the generation

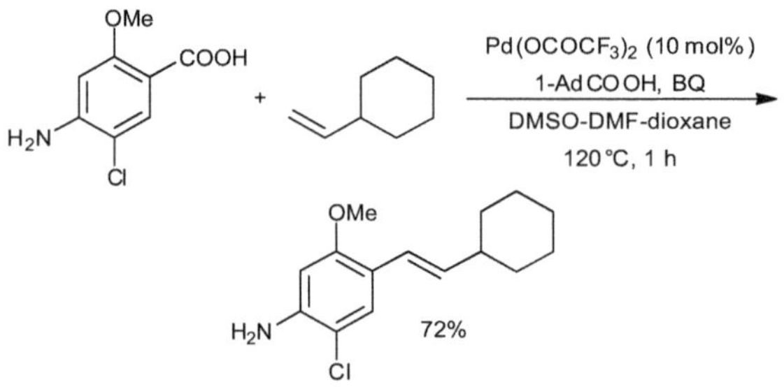

Scheme 9.162   Example of decarboxylative arylation using *p*-benzoquinone (BQ) as terminal oxidant.

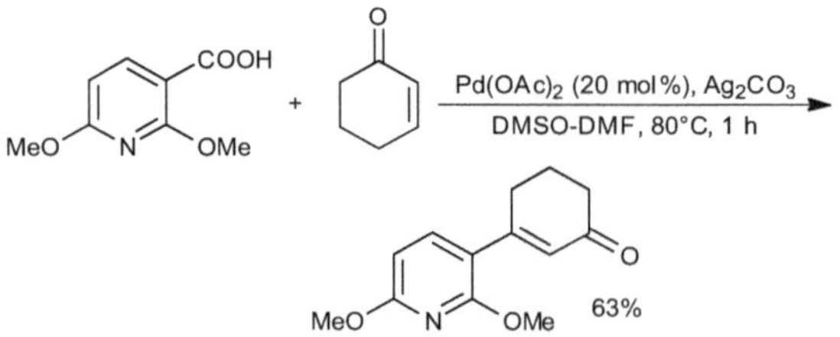

Scheme 9.163   Example of decarboxylative arylation of enones.

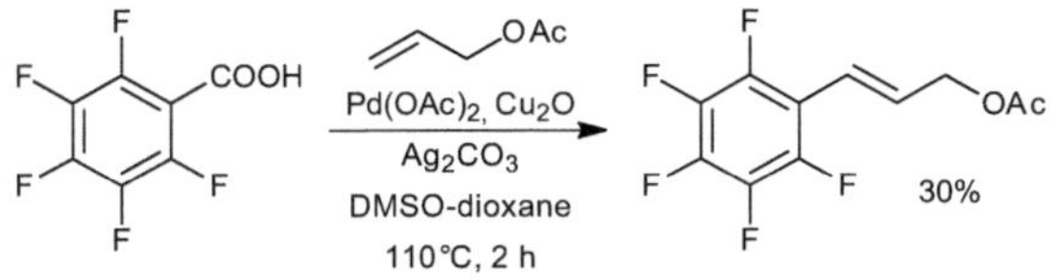

Scheme 9.164 Example of decarboxylative arylation by electron-deficient carboxylic acids.

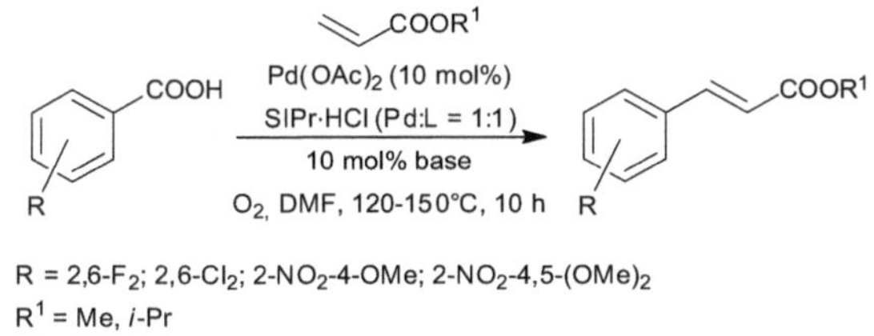

R = 2,6-F₂; 2,6-Cl₂; 2-NO₂-4-OMe; 2-NO₂-4,5-(OMe)₂
R¹ = Me, i-Pr

Scheme 9.165 Decarboxylative arylation using electron-deficient benzoic acids.

Scheme 9.166 Decarboxylative arylation using *o*-nitrobenzoate.

of arylpalladium intermediates. Further interesting achievements in this area revealed that reaction can take place in the presence of a Pd complex with an electron-rich heterocyclic carbene ligand under an oxygen atmosphere and rather harsh heating (Scheme 9.165). Intriguingly, the reaction takes place in the absence of a stoichiometric amount of base, thus actually under conditions of increasing acidity.[434]

o-Nitrobenzoates were shown by Goossen and co-workers to arylate standard olefins (styrenes, acrylates, also *N*-isopropylacrylamide) in a multicomponent system including cupric fluoride, benzoquinone, ligand [1,4,5-triazanaphthalene, TAN (**25**)] and molecular sieves, in which all components are important to achieve good yields (Scheme 9.166).[440]

A suggestion was made that the role of CuF₂ in this process is different from just being a terminal stoichiometric oxidant, rather it is involved in oxidative decarboxylation to give an ArCuX intermediate (an anionic complex is shown in the scheme as the release of fluoride anion in an anhydrous aprotic medium is extremely unlikely), which is later transmetallated by Pd(II) and enters the catalytic cycle (Scheme 9.167).

The experimental data and theoretical modelling of reaction pathways[437,441] revealed that the transformation of a carboxylate complex into an arylpalladium intermediate is a concerted process developing a typical migratory rearrangement (similar to what happens, for example, in

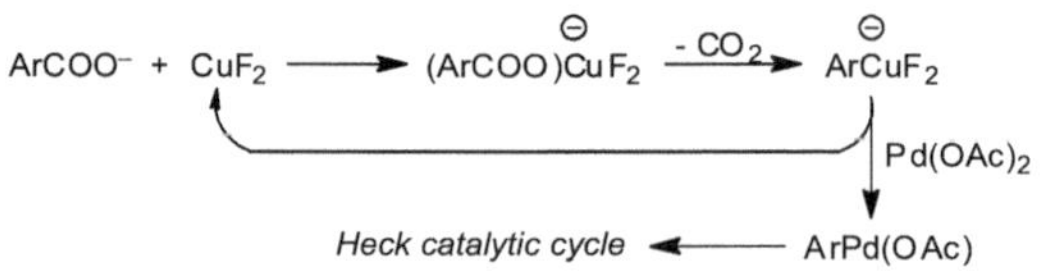

Scheme 9.167   Hypothetical mechanism explaining the particular role of CuF_2 in the decarboxylative arylation by *o*-nitrobenzoates.

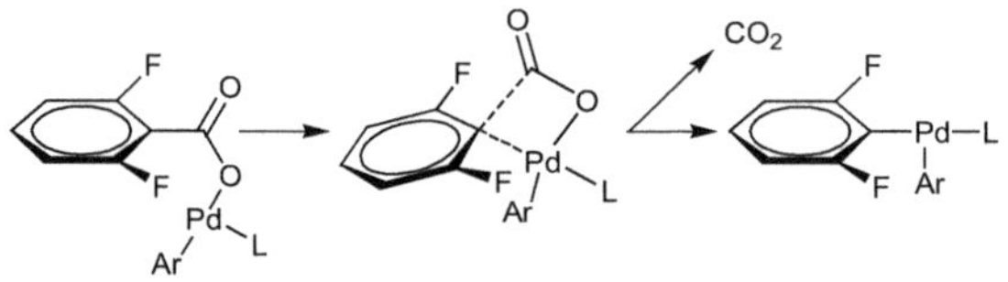

Scheme 9.168   The palladodecarboxylation transition state.

the well-known decarbonylation pathway, with the only essential difference being that CO_2 unlike CO is a very poor ligand immediately leaving the co-ordination shell) (Scheme 9.168).

This reasoning shows that in general, the transformation of an arene-carboxylate palladium complex into an arylpalladium complex with extrusion of carbon dioxide is a concerted process – a continuous rearrangement taking place *via* a four-centred cyclic transition state. As was noted above for other oxidative palladation processes, the concerted pathways often involve what is known as a variable transition state in which the degree of bond rupture/bond formation may take place in a non-synchronous fashion, thus adjusting in each particular case to the electronic requirements of the particular reagents. Concerning decarboxylative palladation, two boundary cases should be considered. With electron-rich benzoic acids, the process resembles electrophilic substitution with Pd–C bond formation taking the lead over rupture of the C–carboxylate bond. This case is favoured by the enhanced electrophilicity of the Pd centre in the absence of strongly bonded donor ligands and additionally is apparently guided by an *ortho*-directing effect from weakly coordinating *ortho* substituents, such as MeO.

In the other boundary case, cleavage of carboxylate leads in the process, with C–Pd bond formation lagging behind, hence in the transition state a carbanion-like configuration emerges. This case is favoured by a Pd centre the electrophilicity of which is suppressed by donor ligands, such as phosphines and electron-deficient arenecarboxylic acids.

Theoretical modelling of reaction pathways give clues as to why *ortho* substituents, even such simple ones as methyl, apparently favour the reaction. In addition to an obvious suggestion that *o*-methoxy and similar groups simply direct *ipso-ortho* metallation, *ortho* substituents for steric reasons favour intramolecular concerted attack with the four-centred transition state cycle being perpendicular to the aryl ring – thus *ortho* substituents push the Pd centre towards the correct position for the attack and accordingly sterically destabilize all other configurations along the pathway (Scheme 9.169).[441] Moreover, this hypothesis predicts that two *ortho* substituents are required, but their nature is not important (indeed, 2,6-dimethoxy, difluoro, dimethyl, 2-fluoro-6-trifluoromethyl, 2-chloro-6-fluoro, *etc.* patterns can be found in papers dealing with decarboxylative reactions giving good results, whereas only one *ortho* substituent rarely suffices, except, probably for a nitro group (see below).

The protocols described so far for decarboxylative reactions can be grouped into two types – those useful for electron-rich acids and those useful for electron-deficient acids. Additional clues as to the difference between decarboxylative reactions of electron-rich and electron-deficient benzoic acids can be drawn from decarboxylative cross-coupling reactions. In such reactions, by analogy with regular cross-coupling, benzoate is a nucleophilic coupling partner, serving the role of organometallic reagent, a carbanion equivalent. If this is true, the decarboxylative step should come after oxidative addition to aryl halide or triflate. The latter requires phosphine ligands and therefore it can be expected that electron-deficient benzoic acids would be preferred over electron-rich acids – and this is indeed the case, as was shown by Liu and co-workers (Scheme 9.170).[442]

Heck-like carbopalladation of C=O or C=N bonds was also recently described to involve a typical Myers-style Pd–Ag system, although in the absence of base and with catalytic amounts of both $PdCl_2$ (10 mol%) and AgOTf (20 mol%) (Scheme 9.171).[443] This process does not need a terminal oxidant, because it is fully serviced by Pd(II), as unlike the carbopalladation of a C=C bond here PdH is not generated and the product is released from the carbopalladation adduct by protolysis (Scheme 9.172) and not by reductive elimination.

Similar deborylative reactions were observed by Yamamoto and coworkers to take place in the presence of ruthenium and rhodium catalysts,[444,445] and palladium- and copper-catalysed versions by Wu and co-workers[90,446] and platinum- and palladium-catalysed versions by Hu and

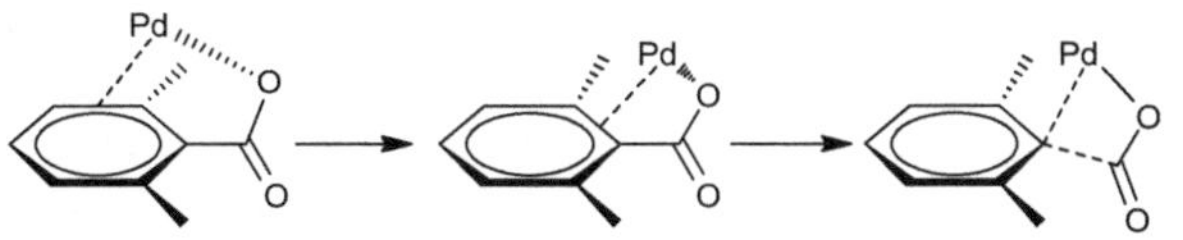

Scheme 9.169 Effect of *ortho* substituents in decarboxylative palladation – steering the Pd centre away from alternative coordination modes.

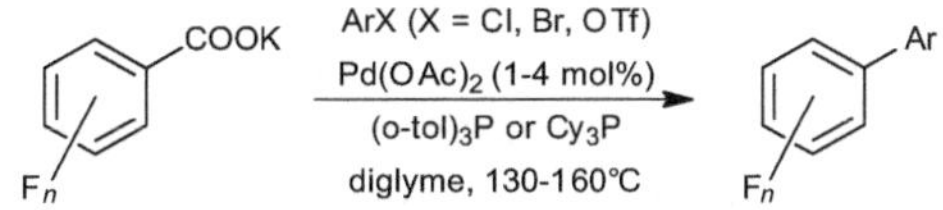

Scheme 9.170 Decarboxylative cross-coupling using electron-deficient polyfluorobenzoates.

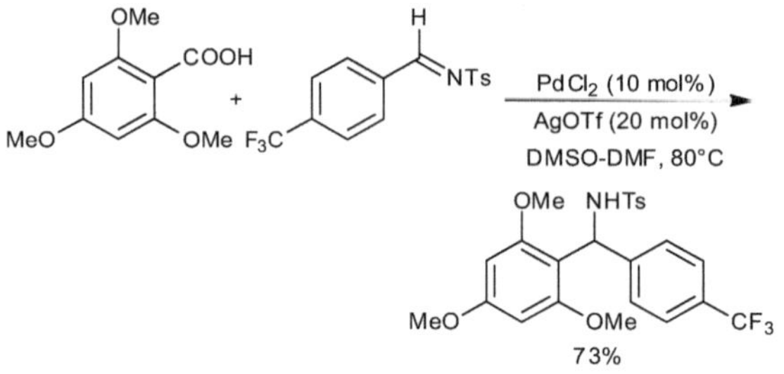

Scheme 9.171   Example of decarboxylative hydroarylation of imines.

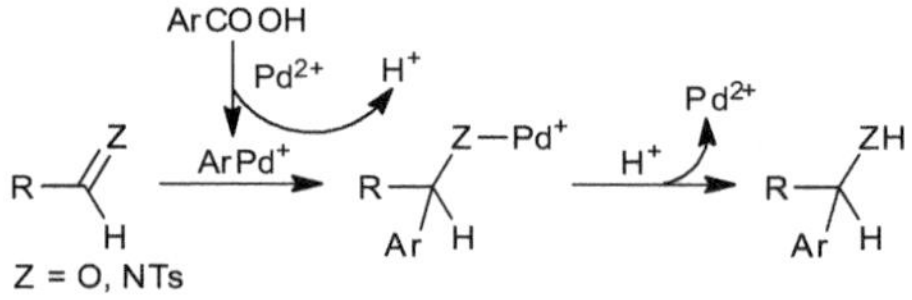

Scheme 9.172   Carbopalladation of C=heteroatom bonds terminated by protolysis.

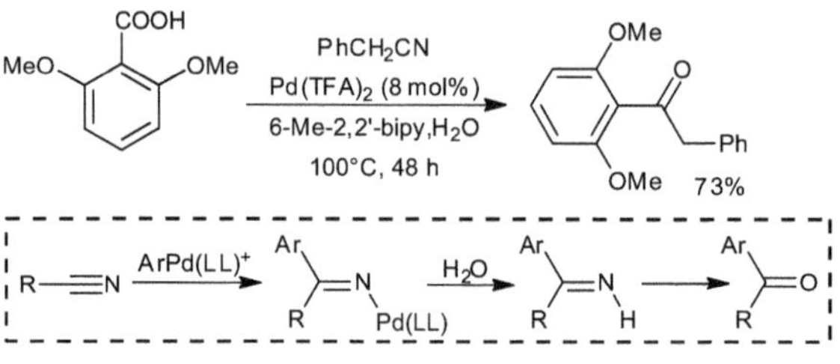

Scheme 9.173   Decarboxylative arylation of C≡N triple bonds.

co-workers,[447] to mention only a few examples to show that this extension of the Heck pathway bears a huge potential for development.

Larhed and co-workers extended the method to C≡N triple bonds and used this methodology to develop a new method of ketone synthesis by decarboxylative carbopalladation of nitriles by electron-rich acids using 2,2′-bipyridyls as ligands and excess nitrile as solvent (Scheme 9.173).[448] Primary adducts, imines and their palladium derivatives were observed by ESI mass spectrometry of the reaction mixtures.

An interesting reaction indirectly related to the topic under consideration is the decarboxylative cross-coupling of cinnamic acid with aryl iodides reported by Wu and co-workers.[449] The reaction is synthetically equivalent to the archetypal Mizoroki–Heck reaction of aryl iodides with styrene. The protocol shows that decarboxylative processes are not limited to particularly substituted benzoic acids, although the reaction with cinnamic acids requires harsh conditions and a special phosphine ligand. Ligands of this sort were designed by Buchwald and co-workers for carrying out C–C, C–N, C–O and other cross-coupling reactions with less reactive electrophiles (aryl and vinyl chlorides, tosylates, *etc.*) under mild conditions (see above). Here, its role is apparently associated with decarboxylative palladation. The other interesting feature of this reaction is that under the conditions developed, no regular arylation of cinnamic acid by aryl iodide takes place (Scheme 9.174).[449]

Of further interest with respect to the use of various carboxylic acids in palladium-catalysed decarboxylative processes is the cross-coupling of 2-pyridylacetates with aryl bromides, in which effective decarboxylative palladation is ensured by the use of Xantphos or similar wide bite angle bidentate ligands (DPEPhos, BINAP, TolBINAP) and similarly harsh conditions [up to 2 mol% $Pd_2(dba)_3$, 3L (L = Xantphos), diglyme, 150 °C].[450] The process can be viewed as a Heck-type arylation of enamine equivalent.

9.3.6 Arylation by Arenesulfinates

The use of chlorosulfonylarenes in the regular Mizoroki–Heck reaction was described as early as in 1989 by Miura and co-workers.[451,452] The reaction involves extrusion of SO_2 from an arenesulfonylpalladium complex. The evident alternative approach to such complexes – by ligand exchange from Pd(II) species, which allows an alternative approach to the same goal (Scheme 167) – had to wait until 2011 to be added to the Heck family of reactions. Deng and co-workers discovered a catalytic system that permits the use of arenesulfinates in the oxidative Heck reaction.[453] Formal comparison of two pathways – regular Mizoroki–Heck reaction initiated by oxidative addition of Pd(0) and oxidative Heck reaction initiated by nucleophilic substitution (ligand exchange) at Pd(II) – shows that these two pathways are perfectly complementary (Scheme 9.175).

Nevertheless, unlike the reaction of chlorosulfonylarenes which is performed in a "phosphine-free" reaction, its oxidative counterpart gives the

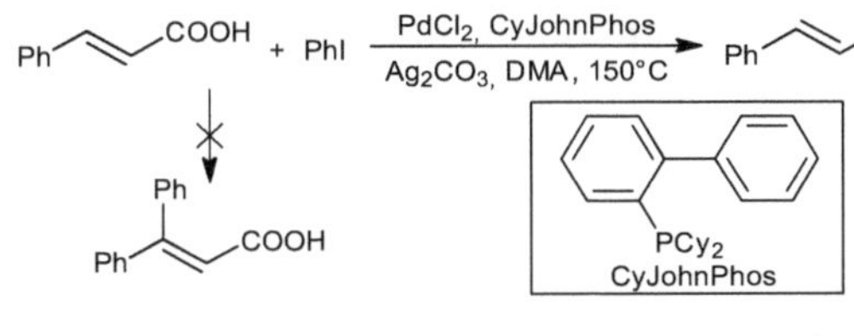

Scheme 9.174 Decarboxylative cross-coupling preparatively equivalent to the Mizoroki–Heck reaction.

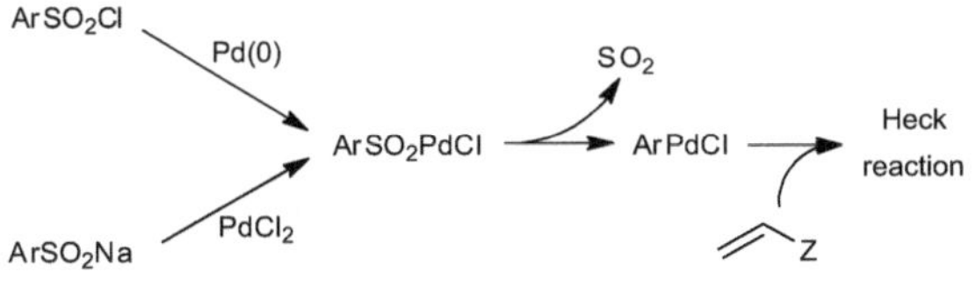

Scheme 9.175 Complementary pathways in Heck reactions *via* SO_2 extrusion.

best results when the bidentate phosphine dppe is used, which under an oxygen atmosphere is probably rapidly oxidized to monodentate dppe-monooxide ligand. The main problem in this reaction is the high nucleophilicity of sulfinates, which competitively form Michael adducts with acrylates and similar electronegative olefins. Further, in the absence of a phosphine ancillary, the second sulfinate ligand may enter the coordination sphere of Pd(II) to inhibit the Heck reaction at the expense of side reactions. However, overall, the new process is impressive in scope, being applicable to various olefins and sulfinates bearing both electron-donating and electron-withdrawing substituents in the aryl ring (Scheme 9.176). As the ligand exchange with nucleophilic sulfinate ion is indeed likely to be weakly sensitive to the nature of the arene substituents and the extrusion of SO_2 is more facile than that of CO_2 from formally analogous carboxylates, the reaction shows relatively lower sensitivity to electronic effects among the oxidative Heck reactions. The process definitely invites further research and refinement of the protocol, *e.g.* to find a more convenient solvent and to establish whether the phosphine ancillaries are indeed needed, as the beneficial effect of addition of the ligand reported[453] is actually rather marginal.

Arylation of enones by sodium arenesulfinates gives the products of conjugate addition, similarly to what is observed in other oxidative Heck reactions (*cf.* Scheme 9.150). The reactions take place in an unusual reaction system using Pd(OH)$_2$ precatalyst, DABCO and an excess of trifluoroacetic acid, which probably facilitates the protolysis of the palladium enolate intermediate, in aqueous dioxane. Being an isohypsic reaction, this process avoids the use of reoxidants.[454] Further, Tian and co-workers introduced an interesting modification of this reaction by applying readily available arenesulfonylhydrazides in the reaction with Pd(II) catalyst.[455] Deng and co-workers further showed that in the reaction with enones these substrates gave conjugate addition products in high yields and in a very simple catalytic system (Scheme 9.177).[456] Apparently, the hydrazide is first oxidized to some

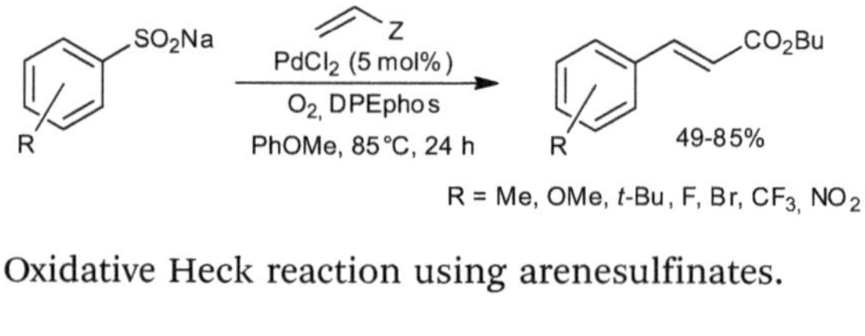

Scheme 9.176 Oxidative Heck reaction using arenesulfinates.

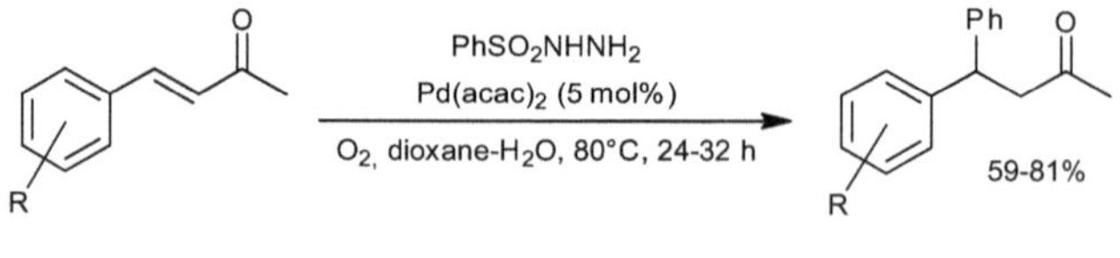

Scheme 9.177 Conjugate arylation of enones by arenesulfonylhydrazides.

intermediate, further undergoing extrusion of N_2 and SO_2 to generate an arylpalladium species. Oxygen is thus used for palladium-catalysed pre-oxidation of hydrazide, whereas the arylation itself is more likely to be realized *via* an isohypsic Heck pathway.

9.4 Conclusion

The Nobel Prize in Chemistry is awarded for outstanding discoveries, which change the way in which researchers think and solve problems. The discovery of Richard Heck of the whole family of transformations taking place when an organopalladium complex formed from various precursors adds across a double bond is undoubtedly one of such major break-throughs. At first looking like yet another example of the emerging chemistry of transition metal-catalysed reactions, the Heck reaction soon became a tool (a sharpening stone) for exploring the intricacies and challenges of transition metal catalysis. Hypotheses on the mechanistic pathways, principles of catalyst design and operation of catalytic mech-anisms were put forward, checked, rejected or enforced in the frames of Heck chemistry and closely related areas and reciprocated in the other fundamental branches of this science, such as cross-coupling and carbo-nylation. Heck chemistry has grown enormously to include dozens of derivative reactions, in addition to the archetypal Mizoroki–Heck reaction. The diversity, innately present in this basic organometallic transformation and stemming from various complementary and competing ways of gen-eration of oranopalladium (and even broader – organometallic, as in addition to palladium other metals including Rh, Ru, Ni, Co, Fe and Cuwere noted to be able to take part in Heck-like processes) intermedi-ates, the involvement of other types of multiple bonds and variable fates of the carbopalladation adducts account for an overwhelming variety of derivative processes, and the availability of alternative approaches to the same synthetic goals (*e.g.* through the choice of leaving groups, oxidative or isohypsic reactions). Interestingly and stimulatingly, Heck chemistry is still full of unanswered questions and unsolved challenges, probably even of misconceptions awaiting to be resolved. This chapter has attempted to show the diversity and synthetic might of this brilliant chemistry, al-though of necessity only a rough cross-section, very far from being comprehensive.

References

1. M. Oestreich (ed.), *The Mizoroki–Heck Reaction*, Wiley, Chichester, 2009.
2. R. F. Heck, *J. Am. Chem. Soc.*, 1968, **90**, 5538–5542.
3. R. F. Heck, *J. Am. Chem. Soc.*, 1968, **90**, 5526–5531.
4. R. F. Heck, *J. Am. Chem. Soc.*, 1968, **90**, 5535–5538.
5. R. F. Heck, *J. Am. Chem. Soc.*, 1969, **91**, 6707–6714.
6. R. F. Heck, *J. Am. Chem. Soc.*, 1971, **93**, 6896–6901.

7. I. Moritani and Y. Fujiwara, *Tetrahedron Lett.*, 1967, 1119.
8. Y. Fujiwara, I. Moritani, R. Asano and S. Teranishi, *Tetrahedron Lett.*, 1968, 6015.
9. Y. Fujiwara, I. Moritani, M. Matsuda and S. Teranishi, *Tetrahedron Lett.*, 1968, 3863.
10. Y. Fujiwara, I. Moritani and M. Matsuda, *Tetrahedron*, 1968, **24**, 4819.
11. Y. Fujiwara, I. Moritani, M. Matsuda and S. Teranishi, *Tetrahedron Lett.*, 1968, 633.
12. S. Danno, I. Moritani and Y. Fujiwara, *Tetrahedron*, 1969, **25**, 4809.
13. Y. Fujiwara, I. Moritani, R. Asano, H. Tanaka and S. Teranishi, *Tetrahedron*, 1969, **25**, 4815.
14. S. Danno, I. Moritani and Y. Fujiwara, *Tetrahedron*, 1969, **25**, 4819.
15. R. Asano, I. Moritani, A. Sonoda, Y. Fujiwara and S. Teranishi, *J. Chem. Soc. C*, 1971, 3691.
16. I. Moritani, Y. Fujiwara and S. Danno, *J. Organomet. Chem.*, 1971, **27**, 279.
17. S. Danno, I. Moritani, Y. Fujiwara and S. Teranishi, *J. Chem. Soc. B*, 1971, 196.
18. I. Moritani and Y. Fujiwara, *Synthesis*, 1973, 524–533.
19. M. Yamamura, I. Moritani, A. Sonoda, S. Teranishi and Y. Fujiwara, *J. Chem. Soc., Perkin Trans. 1*, 1973, 203–205.
20. Y. Fujiwara, R. Asano, I. Moritani and S. Teranishi, *J. Org. Chem.*, 1976, **41**, 1680–1683.
21. P. Fitton, M. P. Johnson and J. E. McKeon, *Chem. Commun.*, 1968, 6–7.
22. P. Fitton and E. A. Rick, *J. Organomet. Chem.*, 1971, **28**, 287–291.
23. D. R. Coulson, *Chem. Commun.*, 1968, 1530–1531.
24. T. Mizoroki, K. Mori and A. Ozaki, *Bull. Chem. Soc. Jpn.*, 1971, **44**, 581.
25. K. Mori, T. Mizoroki and A. Ozaki, *Bull. Chem. Soc. Jpn.*, 1973, **46**, 1505–1508.
26. R. F. Heck and J. P. Nolley, *J. Org. Chem.*, 1972, **37**, 2320–2322.
27. H. A. Dieck and R. F. Heck, *J. Am. Chem. Soc.*, 1974, **96**, 1133–1136.
28. H. A. Dieck and R. F. Heck, *J. Org. Chem.*, 1975, **40**, 1083–1090.
29. J.-I. I. Kim, B. A. Patel and R. F. Heck, *J. Org. Chem.*, 1981, **46**, 1067–1073.
30. J. B. Melpolder and R. F. Heck, *J. Org. Chem.*, 1976, **41**, 265–272.
31. T. C. Zebovitz and R. F. Heck, *J. Org. Chem.*, 1977, **42**, 3907–3909.
32. J. E. Plevyak and R. F. Heck, *J. Org. Chem.*, 1978, **43**, 2454–2456.
33. W. C. Frank, Y. C. Kim and R. F. Heck, *J. Org. Chem.*, 1978, **43**, 2947–2949.
34. C. B. Ziegler and R. F. Heck, *J. Org. Chem.*, 1978, **43**, 2949–2952.
35. B. A. Patel, J. E. Dickerson and R. F. Heck, *J. Org. Chem.*, 1978, **43**, 5018–5020.
36. B. A. Patel, L.-C. Kao, N. A. Cortese, J. V. Minkiewicz and R. F. Heck, *J. Org. Chem.*, 1979, **44**, 918–921.

37. N. A. Cortese, C. B. Ziegler, B. J. Hrnjez and R. F. Heck, *J. Org. Chem.*, 1978, **43**, 2952–2958.
38. M. O. Terpko and R. F. Heck, *J. Am. Chem. Soc.*, 1979, **101**, 5281–5283.
39. B. A. Patel, J.-I. I. Kim, D. D. Bender, L.-C. Kao and R. F. Heck, *J. Org. Chem.*, 1981, **46**, 1061–1067.
40. L. C. Kao, F. G. Stakem, B. A. Patel and R. F. Heck, *J. Org. Chem.*, 1982, **47**, 1267–1277.
41. D. D. Bender, F. G. Stakem and R. F. Heck, *J. Org. Chem.*, 1982, **47**, 1278–1284.
42. C. K. Narula, K. T. Mak and R. F. Heck, *J. Org. Chem.*, 1983, **48**, 2792–2796.
43. L. Shi, C. K. Narula, K. T. Mak, L. Kao, Y. Xu and R. F. Heck, *J. Org. Chem.*, 1983, **48**, 3894–3900.
44. C. B. Ziegler and R. F. Heck, *J. Org. Chem.*, 1978, **43**, 2941–2946.
45. A. Spencer, *J. Organomet.Chem.*, 1983, **258**, 101–108.
46. A. Spencer, *J. Organomet.Chem.*, 1984, **270**, 115–120.
47. T. Mitsudo, W. Fischetti and R. F. Heck, *J. Org. Chem.*, 1984, **49**, 1640–1646.
48. J. E. Plevyak, J. E. Dickerson and R. F. Heck, *J. Org. Chem.*, 1979, **44**, 4078–4080.
49. W. Tao, S. Nesbitt and R. F. Heck, *J. Org. Chem.*, 1990, **55**, 63–69.
50. R. F. Heck, *Acc. Chem. Res.*, 1979, **12**, 146–151.
51. R. F. Heck, *Org. React.*, 1982, **27**, 345–390.
52. H. Tsutsui and K. Narasaka, *Chem. Lett.*, 1999, 45–46.
53. H. Tsutsui, M. Kitamura and K. Narasaka, *Bull. Chem. Soc. Jpn.*, 2002, **75**, 1451–1460.
54. K. Narasaka and M. Kitamura, *Eur. J. Org. Chem.*, 2005, 4505.
55. J. Ichikawa, R. Nadano and N. Ito, *Chem. Commun.*, 2006, 4425–4427.
56. A. Faulkner, J. S. Scott and J. F. Bower, *Chem. Commun.*, 2013, **49**, 1521–1523.
57. A. Kasahara, T. Izumi and N. Fukuda, *Bull. Chem. Soc. Jpn.*, 1977, **50**, 551–552.
58. W. Cabri, I. Candiani, A. Bedeschi and R. Santi, *J. Org. Chem.*, 1992, **57**, 3558–3563.
59. J. Le Bras and J. Muzart, *Tetrahedron*, 2012, **68**, 10065–10113.
60. M. Lautens, E. Tayama and C. Herse, *J. Am. Chem. Soc.*, 2004, **127**, 72–73.
61. B. Mariampillai, C. Herse and M. Lautens, *Org. Lett.*, 2005, 7, 4745–4747.
62. Y. Liu, B. Yao, C.-L. Deng, R.-Y. Tang, X.-G. Zhang and J.-H. Li, *Org. Lett.*, 2011, **13**, 1126–1129.
63. J. Ichikawa, K. Sakoda, J. Mihara and N. Ito, *J. Fluorine Chem.*, 2006, **127**, 489–504.
64. V. G. Zaitsev and O. Daugulis, *J. Am. Chem. Soc.*, 2005, **127**, 4156–4157.
65. J. Lindh, J. Savmarker, P. Nilsson, P. J. R. Sjoberg and M. Larhed, *Chem. Eur. J.*, 2009, **15**, 4630–4636.

66. L. R. Odell, J. Lindh, T. Gustafsson and M. Larhed, *Eur. J. Org. Chem.*, 2010, 2270–2274.
67. H. F. Jiang, C. L. Qiao and W. B. Liu, *Chem. Eur. J.*, 2010, **16**, 10968–10970.
68. H. F. Jiang, Y. Gao, W. Q. Wu and Y. B. Huang, *Org. Lett.*, 2013, **15**, 238–241.
69. C. Martinez, J. M. Aurrecoechea, Y. Madich, J. G. Denis, A. R. de Lera and R. Alvarez, *Eur. J. Org. Chem.*, 2012, 99–106.
70. X. Y. Chen, W. Kong, H. T. Cai, L. C. Kong and G. G. Zhu, *Chem. Commun.*, 2011, **47**, 2164–2166.
71. Q. G. Zhang, W. Xu and X. Y. Lu, *J. Org. Chem.*, 2005, **70**, 1505–1507.
72. A. T. Lindhardt, M. L. H. Mantel and T. Skrydstrup, *Angew. Chem. Int. Ed.*, 2008, **47**, 2668–2672.
73. S. Mannathan and C. H. Cheng, *Chem. Commun.*, 2010, **46**, 1923–1925.
74. H. Horie, I. Koyama, T. Kurahashi and S. Matsubara, *Chem. Commun.*, 2011, **47**, 2658–2660.
75. H. Miura, S. Shimura, S. Hosokawa, S. Yamazoe, K. Wada and M. Inoue, *Adv. Synth. Catal.*, 2011, **353**, 2837–2843.
76. Y. Shibata, M. Hirano and K. Tanaka, *Org. Lett.*, 2008, **10**, 2829–2831.
77. M. Kobayashi and K. Tanaka, *Chem. Eur. J.*, 2012, **18**, 9225–9229.
78. R. Rossi, F. Bellina and M. Lessi, *Synthesis*, 2010, 4131–4153.
79. C. C. Hughes and D. Trauner, *Angew. Chem. Int. Ed.*, 2002, **41**, 1569–1572.
80. L. Ackermann, S. Barfusser and J. Pospech, *Org. Lett.*, 2010, **12**, 724–726.
81. L. Ackermann and S. Fenner, *Chem. Commun.*, 2011, **47**, 430–432.
82. L. Ackermann, R. Vicente and A. R. Kapdi, *Angew. Chem. Int. Ed.*, 2009, **48**, 9792–9826.
83. G. P. McGlacken and L. M. Bateman, *Chem. Soc. Rev.*, 2009, **38**, 2447–2464.
84. S. Pascual, P. de Mendoza and A. M. Echavarren, *Org. Biomol. Chem.*, 2007, **5**, 2727–2734.
85. M. Yagoubi, A. C. F. Cruz, P. L. Nichols, R. L. Elliott and M. C. Willis, *Angew. Chem. Int. Ed.*, 2010, **49**, 7958–7962.
86. M. J. Burns, R. J. Thatcher, R. J. K. Taylor and I. J. S. Fairlamb, *Dalton Trans.*, 2010, **39**, 10391–10400.
87. I. P. Beletskaya and A. V. Cheprakov, *Chem. Rev.*, 2000, **100**, 3009–3066.
88. T. Ohmura, A. Kijima and M. Suginome, *Org. Lett.*, 2011, **13**, 1238–1241.
89. R. Zhang, Q. Xu, X. C. Zhang, T. Zhang and M. Shi, *Tetrahedron: Asymmetry*, 2010, **21**, 1928–1935.
90. C. Qin, H. Wu, J. Cheng, X. A. Chen, M. Liu, W. Zhang, W. Su and J. Ding, *J. Org. Chem.*, 2007, **72**, 4102–4107.
91. C. M. Qin, J. X. Chen, H. Y. Wu, J. Cheng, Q. Zhang, B. Zuo, W. K. Su and J. C. Ding, *Tetrahedron Lett.*, 2008, **49**, 1884–1888.
92. C. M. Qin, H. Y. Wu, J. X. Chen, M. C. Liu, J. Cheng, W. K. Su and J. C. Ding, *Org. Lett.*, 2008, **10**, 1537–1540.

93. C. M. Qin, H. Y. Wu, J. Cheng, X. A. Chen, M. C. Liu, W. W. Zhang, W. K. Su and J. C. Ding, *J. Org. Chem.*, 2007, **72**, 4102–4107.
94. F. Bellina and R. Rossi, *Chem. Rev.*, 2009, **110**, 1082–1146.
95. J. Ruan, O. Saidi, J. A. Iggo and J. Xiao, *J. Am. Chem. Soc.*, 2008, **130**, 10510–10511.
96. P. Colbon, J. Ruan, M. Purdie and J. Xiao, *Org. Lett.*, 2010, **12**, 3670–3673.
97. T. Schulz, C. Torborg, S. Enthaler, B. Schäffner, A. Dumrath, A. Spannenberg, H. Neumann, A. Börner and M. Beller, *Chem. Eur. J.*, 2009, **15**, 4528–4533.
98. I. P. Beletskaya and A. V. Cheprakov, in *Comprehensive Organometallic Chemistry III*, ed. H. C. Robert and D. M. P. Mingos, Elsevier, Oxford, 2007, pp. 411–433.
99. E.-i. Negishi, S. Ma, J. Amanfu, C. Copéret, J. A. Miller and J. M. Tour, *J. Am. Chem. Soc.*, 1996, **118**, 5919–5931.
100. T. Hayashi, J. Tang and K. Kato, *Org. Lett.*, 1999, **1**, 1487–1489.
101. X. Wu, P. Nilsson and M. Larhed, *J. Org. Chem.*, 2004, **70**, 346–349.
102. H.-U. Blaser and A. Spencer, *J. Organomet.Chem.*, 1982, **233**, 267–274.
103. L. J. Goossen and J. Paetzold, *Angew. Chem. Int. Ed.*, 2002, **41**, 1237–1241.
104. L. J. Goossen and J. Paetzold, *Angew. Chem. Int. Ed.*, 2004, **43**, 1095–1098.
105. L. Goossen and K. Goossen, in *The Mizoroki–Heck Reaction*, ed. M. Oestreich, Wiley, Chichester, 2009, pp. 163–178.
106. T. Satoh, T. Itaya, K. Okuro, M. Miura and M. Nomura, *J. Org. Chem.*, 1995, **60**, 7267–7271.
107. X. F. Wu, H. Neumann and M. Beller, *Angew. Chem. Int. Ed.*, 2010, **49**, 5284–5288.
108. X. F. Wu, H. Neumann, A. Spannenberg, T. Schulz, H. J. Jiao and M. Beller, *J. Am. Chem. Soc.*, 2010, **132**, 14596–14602.
109. T. M. Gogsig, D. U. Nielsen, A. T. Lindhardt and T. Skrydstrup, *Org. Lett.*, 2012, **14**, 2536–2539.
110. P. Hermange, T. M. Gogsig, A. T. Lindhardt, R. H. Taaning and T. Skrydstrup, *Org. Lett.*, 2011, **13**, 2444–2447.
111. X. F. Wu, H. J. Jiao, H. Neumann and M. Beller, *ChemCatChem*, 2011, **3**, 726–733.
112. X. F. Wu, H. Neumann and M. Beller, *Chem. Asian J.*, 2012, **7**, 282–285.
113. K. S. Bloome and E. J. Alexanian, *J. Am. Chem. Soc.*, 2010, **132**, 12823–12825.
114. A. Jutand, in *The Mizoroki–Heck Reaction*, ed. M. Oestreich, Wiley, Chichester, 2009, pp. 1–50.
115. J. P. Knowles and A. Whiting, *Org. Biomol. Chem.*, 2007, **5**, 31–44.
116. W. Cabri and I. Candiani, *Acc. Chem. Res.*, 1995, **28**, 2–7.
117. C. Amatore, E. Carre, A. Jutand and Y. Medjour, *Organometallics*, 2002, **21**, 4540–4545.
118. C. A. Fleckenstein and H. Plenio, *Chem. Soc. Rev.*, 2010, **39**, 694–711.

119. P. Nilsson, K. Olofsson and M. Larhed, in *The Mizoroki–Heck Reaction*, ed. M. Oestreich, Wiley, Chichester, 2009, pp. 133–162.

120. A. G. Coyne, M. O. Fitzpatrick and P. J. Guiry, in *The Mizoroki–Heck Reaction*, ed. M. Oestreich, Wiley, Chichester, 2009, pp. 405–432.

121. M. Shibasaki and T. Ohshima, in *The Mizoroki–Heck Reaction*, ed. M. Oestreich, Wiley, Chichester, 2009, pp. 463–483.

122. M. Shibasaki, C. D. J. Boden and A. Kojima, *Tetrahedron*, 1997, **53**, 7371–7395.

123. M. Shibasaki, E. M. Vogl and T. Ohshima, *Adv. Synth. Catal.*, 2004, **346**, 1533–1552.

124. I. P. Beletskaya and A. V. Cheprakov, in *The Mizoroki–Heck Reaction*, ed. M. Oestreich, Wiley, Chichester, 2009, pp. 51–132.

125. F. Alonso, I. P. Beletskaya and M. Yus, *Tetrahedron*, 2005, **61**, 11771–11835.

126. A. F. Littke and G. C. Fu, *J. Am. Chem. Soc.*, 2001, **123**, 6989–7000.

127. V. P. Ananikov and I. P. Beletskaya, *Organometallics*, 2012, **31**, 1595–1604.

128. D. Pun, T. N. Diao and S. S. Stahl, *J. Am. Chem. Soc.*, 2013, **135**, 8213–8221.

129. M. Gholinejad, *Appl. Organomet. Chem.*, 2013, **27**, 19–22.

130. W. Zhu, Y. Yang, S. Hu, G. Xiang, B. Xu, J. Zhuang and X. Wang, *Inorg. Chem.*, 2012, **51**, 6020–6031.

131. F. Zhang, J. Niu, H. Wang, H. Yang, J. Jin, N. Liu, Y. Zhang, R. Li and J. Ma, *Mater. Res. Bull.*, 2012, **47**, 504–507.

132. Y. Zeng, Y. Wang, Y. Xu, Y. Song, J. Jiang and Z. Jin, *Catal. Lett.*, 2013, **143**, 200–205.

133. D. Wang, D. Denux, J. Ruiz and D. Astruc, *Adv. Synth. Catal.*, 2013, **355**, 129–142.

134. P. M. Uberman, L. A. Perez, G. I. Lacconi and S. E. Martin, *J. Mol. Catal. A: Chem.*, 2012, **363**, 245–253.

135. B. Tamami and F. N. Dodeji, *J. Iranian Chem. Soc.*, 2012, **9**, 841–850.

136. D. Sharma, S. Kumar, A. K. Shil, N. R. Guha, Bandna and P. Das, *Tetrahedron Lett.*, 2012, **53**, 7044–7051.

137. G. Nie, L. Zhang and Y. Cui, *React. Kinet. Mech. Catal.*, 2013, **108**, 193–204.

138. A. P. Kumar, B. P. Kumar, A. B. V. K. Kumar, B. T. Huy and Y.-I. Lee, *Appl. Surf. Sci.*, 2013, **265**, 500–509.

139. A. Kamal, V. Srinivasulu, B. N. Seshadri, N. Markandeya, A. Alarifi and N. Shankaraiah, *Green Chem.*, 2012, **14**, 2513–2522.

140. R. J. Kalbasi and N. Mosaddegh, *C. R. Acad. Sci.*, 2012, **15**, 988–995.

141. X. Jin, K. Zhang, J. Sun, J. Wang, Z. Dong and R. Li, *Catal. Commun.*, 2012, **26**, 199–203.

142. K. Saïd, Y. Moussaoui, M. Kammoun and R. B. Salem, *Ultrason. Sonochem.*, 2011, **18**, 23–27.

143. C. Amatore, E. Carre, A. Jutand, M. A. Mbarki and G. Meyer, *Organometallics*, 1995, **14**, 5605–5614.

144. C. Amatore and A. Jutand, *Acc. Chem. Res.*, 2000, **33**, 314–321.
145. C. Torborg and M. Beller, *Adv. Synth. Catal.*, 2009, **351**, 3027–3043.
146. A. Belovodskii, E. Shul'ts, M. Shakirov, V. Romanov, B. Elmuradov, K. Shakhidoyatov and G. Tolstikov, *Chem. Nat. Compd.*, 2011, **46**, 880–885.
147. N. Surapanich, C. Kuhakarn, M. Pohmakotr and V. Reutrakul, *Eur. J. Org. Chem.*, 2012, 5943–5952.
148. K. H. Shaughnessy, P. Kim and J. F. Hartwig, *J. Am. Chem. Soc.*, 1999, **121**, 2123–2132.
149. F. Barrios-Landeros, B. P. Carrow and J. F. Hartwig, *J. Am. Chem. Soc.*, 2009, **131**, 8141–8154.
150. P. M. Murray, J. F. Bower, D. K. Cox, E. K. Galbraith, J. S. Parker and J. B. Sweeney, *Org. Process Res. Dev.*, 2013, **17**, 397–405.
151. X. Han, X.-J. Jiang, R. L. Civiello, A. P. Degnan, P. V. Chaturvedula, J. E. Macor and G. M. Dubowchik, *J. Org. Chem.*, 2009, **74**, 3993–3996.
152. A. L. Hansen, J.-P. Ebran, M. Ahlquist, P.-O. Norrby and T. Skrydstrup, *Angew. Chem. Int. Ed.*, 2006, **45**, 3349–3353.
153. J. P. Ebran, A. L. Hansen, T. M. Gogsig and T. Skrydstrup, *J. Am. Chem. Soc.*, 2007, **129**, 6931–6942.
154. A. T. Lindhardt and T. Skrydstrup, *Chem. Eur. J.*, 2008, **14**, 8756–8766.
155. G. K. Datta and M. Larhed, *Org. Biomol. Chem.*, 2008, **6**, 674–676.
156. G. K. Datta, P. Nordeman, J. Dackenberg, P. Nilsson, A. Hallberg and M. Larhed, *Tetrahedron: Asymmetry*, 2008, **19**, 1120–1126.
157. A. Trejos, J. Savmarker, S. Schlummer, G. K. Datta, P. Nilsson and M. Larhed, *Tetrahedron*, 2008, **64**, 8746–8751.
158. A. Scrivanti, M. Bertoldini, V. Beghetto and U. Matteoli, *Tetrahedron*, 2008, **64**, 543–548.
159. H. Doucet and M. Santelli, *Synlett*, 2006, 2001–2015.
160. M. Lemhadri, A. Battace, F. Berthiol, T. Zair, H. Doucet and M. Santelli, *Synthesis*, 2008, 1142–1152.
161. M. Feuerstein, H. Doucet and M. Santelli, *Tetrahedron Lett.*, 2002, **43**, 2191–2194.
162. I. Kondolff, M. Feuerstein, H. Doucet and M. Santelli, *Tetrahedron*, 2007, **63**, 9514–9521.
163. Y. Fall, F. Berthiol, H. Doucet and M. Santelli, *Synthesis*, 2007, 1683–1696.
164. M. Lemhadri, A. Battace, T. Zair, H. Doucet and M. Santelli, *J. Organomet. Chem.*, 2007, **692**, 2270–2281.
165. A. Battace, T. Zair, H. Doucet and M. Santelli, *Synthesis*, 2006, 3495–3505.
166. F. Berthiol, H. Doucet and M. Santelli, *Appl. Organomet. Chem.*, 2006, **20**, 855–868.
167. F. Berthiol, H. Doucet and M. Santelli, *Tetrahedron*, 2006, **62**, 4372–4383.
168. F. Berthiol, H. Doucet and M. Santelli, *Tetrahedron*, 2006, **62**, 4372–4383.

169. A. Battace, M. Feuerstein, M. Lemhadri, T. Zair, H. Doucet and M. Santelli, *Eur. J. Org. Chem.*, 2007, 3122–3132.
170. I. Kondolff, H. Doucet and M. Santelli, *Eur. J. Org. Chem.*, 2006, 765–774.
171. A. Battace, T. Zair, H. Doucet and M. Santelli, *Tetrahedron Lett.*, 2006, **47**, 459–462.
172. M. Lemhadri, H. Doucet and M. Santelli, *Synlett*, 2006, 2935–2940.
173. M. Lemhadri, Y. Fall, H. Doucet and M. Santelli, *Synthesis*, 2009, 1021–1035.
174. Y. Fall, H. Doucet and M. Santelli, *Tetrahedron Lett.*, 2007, **48**, 3579–3581.
175. Y. Fall, H. Doucet and M. Santelli, *Tetrahedron*, 2010, **66**, 2181–2188.
176. M. Feuerstein, H. Doucet and M. Santelli, *J. Org. Chem.*, 2001, **66**, 5923–5925.
177. C. Reynaud, Y. Fall, M. Feuerstein, H. Doucet and M. Santelli, *Tetrahedron*, 2009, **65**, 7440–7448.
178. N. Marion and S. P. Nolan, *Acc. Chem. Res.*, 2008, **41**, 1440–1449.
179. W. A. Herrmann, *Angew. Chem. Int. Ed.*, 2002, **41**, 1290–1309.
180. J. A. Mata, M. Poyatos and E. Peris, *Coord. Chem. Rev.*, 2007, **251**, 841–859.
181. E. A. B. Kantchev, C. J. O'Brien and M. G. Organ, *Angew. Chem. Int. Ed.*, 2007, **46**, 2768–2813.
182. W. A. Herrmann, M. Elison, J. Fischer, C. Koecher and G. R. J. Artus, *Angew. Chem. Int. Ed. Engl.*, 1995, **34**, 2371–2374.
183. V. Calo, A. Nacci, L. Lopez and N. Mannarini, *Tetrahedron Lett.*, 2000, **41**, 8973–8976.
184. D. S. McGuinness and K. J. Cavell, *Organometallics*, 2000, **19**, 741–748.
185. V. Calo, A. Nacci, A. Monopoli, L. Lopez and A. di Cosmo, *Tetrahedron*, 2001, **57**, 6071–6077.
186. S. Grundemann, M. Albrecht, J. A. Loch, J. W. Faller and R. H. Crabtree, *Organometallics*, 2001, **20**, 5485–5488.
187. E. Peris, J. A. Loch, J. Mata and R. H. Crabtree, *Chem. Commun.*, 2001, 201–202.
188. J. A. Loch, M. Albrecht, E. Peris, J. Mata, J. W. Faller and R. H. Crabtree, *Organometallics*, 2002, **21**, 700–706.
189. K. Selvakumar, A. Zapf, A. Spannenberg and M. Beller, *Chem. Eur. J.*, 2002, **8**, 3901–3906.
190. H. Lebel, M. K. Janes, A. B. Charette and S. P. Nolan, *J. Am. Chem. Soc.*, 2004, **126**, 5046–5047.
191. V. Polshettiwar, P. Hesemann and J. J. E. Moreau, *Tetrahedron Lett.*, 2007, **48**, 5363–5366.
192. G.-R. Peh, E. A. B. Kantchev, C. Zhang and J. Y. Ying, *Org. Biomol. Chem.*, 2009, 7, 2110–2119.
193. G. Ren, X. Cui, E. Yang, F. Yang and Y. Wu, *Tetrahedron*, 2010, **66**, 4022–4028.
194. L. Firmansjah and G. C. Fu, *J. Am. Chem. Soc.*, 2007, **129**, 11340–11341.

195. N. Iranpoor, H. Firouzabadi, A. Tarassoli and M. Fereidoonnezhad, *Tetrahedron*, 2010, **66**, 2415–2421.
196. A. Spencer, *J. Organomet.Chem.*, 1983, **247**, 117–122.
197. A. Spencer, *J. Organomet.Chem.*, 1984, **265**, 323–332.
198. M. Beller, H. Fischer, W. A. Herrmann, K. Öfele and C. Brossmer, *Angew. Chem. Int. Ed. Engl.*, 1995, **34**, 1848–1849.
199. M. T. Reetz and J. G. de Vries, *Chem. Commun.*, 2004, 1559–1563.
200. A. H. M. de Vries, J. Mulders, J. H. M. Mommers, H. J. W. Henderickx and J. G. de Vries, *Org. Lett.*, 2003, **5**, 3285–3288.
201. J. G. de Vries, *Dalton Trans.*, 2006, 421–429.
202. T. Jeffery, *Tetrahedron*, 1996, **52**, 10113–10130.
203. T. Jeffery, in *Advances in Metal-Organic Chemistry*, Vol. 5, ed. L. S. Liebeskind, Jai Press, Greenwich, CT, 1996, pp. 153–260.
204. T. Jeffery, *Tetrahedron Lett.*, 1999, **40**, 1673–1676.
205. T. Jeffery and M. David, *Tetrahedron Lett.*, 1998, **39**, 5751–5754.
206. T. N. Glasnov, S. Findenig and C. O. Kappe, *Chem. Eur. J.*, 2009, **15**, 1001–1010.
207. P. Cyr, S. T. Deng, J. M. Hawkins and K. E. Price, *Org. Lett.*, 2013, **15**, 4342–4345.
208. W. Rauf and J. M. Brown, *Chem. Commun.*, 2013, **49**, 8430–8440.
209. K. Hirotaki and T. Hanamoto, *J. Org. Chem.*, 2011, **76**, 8564–8568.
210. I. P. Beletskaya, A. N. Kashin, N. B. Karlstedt, A. V. Mitin, A. V. Cheprakov and G. M. Kazankov, *J. Organomet. Chem.*, 2001, **622**, 89–96.
211. C. Melero, L. M. Martinez-Prieto, P. Palma, D. del Rio, E. Alvarez and J. Campora, *Chem. Commun.*, 2010, **46**, 8851–8853.
212. M. Ohff, A. Ohff, M. E. van der Boom and D. Milstein, *J. Am. Chem. Soc.*, 1997, **119**, 11687–11688.
213. A. Alimardanov, L. S. V. de Vondervoort, A. H. M. de Vries and J. G. de Vries, *Adv. Synth. Catal.*, 2004, **346**, 1812–1817.
214. C. Röhlich and K. Köhler, *Adv. Synth. Catal.*, 2010, **352**, 2263–2274.
215. J. W. Buchler, in *Porphyrins and Metalloporphyrins*, ed. K. M. Smith, Elsevier, Amsterdam, 1975, pp. 157–231.
216. V. Calo, A. Nacci and A. Monopoli, *Eur. J. Org. Chem.*, 2006, 3791–3802.
217. V. Calo, A. Nacci, A. Monopoli and V. Ferola, *J. Org. Chem.*, 2007, **72**, 2596–2601.
218. V. Calo, A. Nacci, A. Monopoli and P. Cotugno, *Angew. Chem. Int. Ed.*, 2009, **48**, 6101–6103.
219. C. Röhlich and K. Köhler, *Chem. Eur. J.*, 2010, **16**, 2363–2365.
220. M. Guerrero, J. Pons and J. Ros, *J. Organomet. Chem.*, 2010, **695**, 1957–1960.
221. M. H. Sie, Y. H. Hsieh, Y. H. Tsai, J. R. Wu, S. J. Chen, P. V. Kumar, J. H. Lii and H. M. Lee, *Organometallics*, 2010, **29**, 6473–6481.
222. J. Y. Lee, P. Y. Cheng, Y. H. Tsai, G. R. Lin, S. P. Liu, M. H. Sie and H. M. Lee, *Organometallics*, 2010, **29**, 3901–3911.
223. P. Srinivas, P. R. Likhar, H. Maheswaran, B. Sridhar, K. Ravikumar and M. L. Kantam, *Chem. Eur. J.*, 2009, **15**, 1578–1581.

224. P. Srinivas, K. Srinivas, P. R. Likhar, B. Sridhar, K. V. Mohan, S. Bhargava and M. L. Kantam, *J. Organomet. Chem.*, 2011, **696**, 795–801.

225. G. Xie, P. Chellan, J. Mao, K. Chibale and G. S. Smith, *Adv. Synth. Catal.*, 2010, **352**, 1641–1647.

226. E. Jung, K. Park, J. Kim, H.-T. Jung, I.-K. Oh and S. Lee, *Inorg. Chem. Commun.*, 2010, **13**, 1329–1331.

227. B. R. Buckley and S. P. Neary, *Tetrahedron*, 2010, **66**, 7988–7994.

228. D. Das, P. Singh, M. Singh and A. K. Singh, *Dalton Trans.*, 2010, **39**, 10876–10882.

229. (a) S. B. Atla, A. A. Kelkar, V. G. Puranik, W. Bensch and R. V. Chaudhari, *J. Organomet. Chem.*, 2009, **694**, 683–690; (b) S. Wu, H. Ma, X. Jia, Y. Zhong and Z. Lei, *Tetrahedron*, 2011, **67**, 250–256.

230. J. M. Lu, H. Ma, S. S. Li, D. Ma and L. X. Shao, *Tetrahedron*, 2010, **66**, 5185–5189.

231. X.-J. Yu, R. Zhou, Y. Zhang, H.-Y. Fu, R.-X. Li, H. Chen and X.-J. Li, *Catal. Commun.*, 2010, **12**, 222–225.

232. Q.-L. Luo, J.-P. Tan, Z.-F. Li, Y. Qin, L. Ma and D.-R. Xiao, *Dalton Trans.*, 2011, **40**, 3601–3609.

233. D. Das, G. K. Rao and A. K. Singh, *Organometallics*, 2009, **28**, 6054–6058.

234. A. R. Hajipour, K. Karami, A. Pirisedigh and A. E. Ruoho, *J. Organomet. Chem.*, 2009, **694**, 2548–2554.

235. S. Dietrich, A. Nicolai and H. Lang, *J. Organomet. Chem.*, 2011, **696**, 739–747.

236. L.-M. Xu, B.-J. Li, Z. Yang and Z.-J. Shi, *Chem. Soc. Rev.*, 2010, **39**, 712–733.

237. P. Sehnal, R. J. K. Taylor and I. J. S. Fairlamb, *Chem. Rev.*, 2010, **110**, 824–889.

238. A. J. Canty, *Dalton Trans.*, 2009, 10409–10417.

239. A. J. Canty, J. Patel, T. Rodemann, J. H. Ryan, B. W. Skelton and A. H. White, *Organometallics*, 2004, **23**, 3466–3473.

240. A. J. Canty, T. Rodemann, B. W. Skelton and A. H. White, *Organometallics*, 2006, **25**, 3996–4001.

241. O. Blacque and C. M. Frech, *Chem. Eur. J.*, 2010, **16**, 1521–1531.

242. K. J. Szabo, *J. Mol. Catal. A: Chem.*, 2010, **324**, 56–63.

243. N. R. Deprez and M. S. Sanford, *Inorg. Chem.*, 2007, **46**, 1924–1935.

244. A. Roglans, A. Pla-Quintana and M. Moreno-Manas, *Chem. Rev.*, 2006, **106**, 4622–4643.

245. M. Zhu, C. Shentu and Z. S. Zhou, *Chin. Chem. Lett.*, 2007, **18**, 272–274.

246. J. Aydin, J. M. Larsson, N. Selander and K. J. Szabo, *Org. Lett.*, 2009, **11**, 2852–2854.

247. D. L. Pan, A. J. Chen, Y. J. Su, W. Zhou, S. Li, W. Jia, J. Xiao, Q. J. Liu, L. R. Zhang and N. Jiao, *Angew. Chem. Int. Ed.*, 2008, **47**, 4729–4732.

248. J. Vicente, A. Arcas, F. Julia-Hernandez and D. Bautista, *Angew. Chem. Int. Ed.*, 2011, **50**, 6896–6899.

249. F. Julia-Hernandez, A. Arcas and J. Vicente, *Chem. Eur. J.*, 2012, **18**, 7780–7786.

250. K. Mori, T. Mizoroki and A. Ozaki, *Bull. Chem. Soc. Jpn.*, 1976, **49**, 758–761.
251. L. Ackermann and R. Born, in *The Mizoroki–Heck Reaction*, ed. M. Oestreich, Wiley, Chichester, 2009, pp. 383–404.
252. I. P. Beletskaya and A. V. Cheprakov, *Organometallics*, 2012, **31**, 7753–7808.
253. S. G. Babu, N. Neelakandeswari, N. Dharmaraj, S. D. Jackson and R. Karvembu, *RSC Adv.*, 2013, **3**, 7774–7781.
254. V. Declerck, J. Martinez and F. Lamaty, *Synlett*, 2006, 3029–3032.
255. Y. Peng, J. X. Chen, J. C. Ding, M. C. Liu, W. X. Gao and H. Y. Wu, *Synthesis*, 2011, 213–216.
256. R. J. Phipps, L. McMurray, S. Ritter, H. A. Duong and M. J. Gaunt, *J. Am. Chem. Soc.*, 2012, **134**, 10773–10776.
257. N. Gigant, L. Chausset-Boissariei, M. C. Belhomme, T. Poisson, X. Pannecoucke and I. Gillaizeu, *Org. Lett.*, 2013, **15**, 278–281.
258. Z. F. Xu, C. X. Cai and J. T. Liu, *Org. Lett.*, 2013, **15**, 2096–2099.
259. J. G. Taylor, A. V. Moro and C. R. D. Correia, *Eur. J. Org. Chem.*, 2011, 1403–1428.
260. V. D. Filimonov, M. Trusova, P. Postnikov, E. A. Krasnokutskaya, Y. M. Lee, H. Y. Hwang, H. Kim and K. W. Chi, *Org. Lett.*, 2008, **10**, 3961–3964.
261. B. Schmidt, F. Holter, R. Berger and S. Jessel, *Adv. Synth. Catal.*, 2010, **352**, 2463–2473.
262. B. Schmidt, N. Elizarov, R. Berger and F. Holter, *Org. Biomol. Chem.*, 2013, **11**, 3674–3691.
263. F. de Azambuja and C. R. D. Correia, *Tetrahedron Lett.*, 2011, **52**, 42–45.
264. A. V. Moro, F. S. P. Cardoso and C. R. D. Correia, *Org. Lett.*, 2009, **11**, 3642–3645.
265. C. Soldi, A. V. Moro, M. G. Pizzolatti and C. R. D. Correia, *Eur. J. Org. Chem.*, 2012, 3607–3616.
266. T. Nakano, M. Miyahara, T. Itoh and A. Kamimura, *Eur. J. Org. Chem.*, 2012, 2161–2166.
267. A. V. Moro, F. S. P. Cardoso and C. R. D. Correia, *Tetrahedron Lett.*, 2008, **49**, 5668–5671.
268. G. Bartoli, S. Cacchi, G. Fabrizi and A. Goggiamani, *Synlett*, 2008, 2508–2512.
269. P. R. R. Meira, A. V. Moro and C. R. D. Correia, *Synthesis*, 2007, 2279–2286.
270. T. Stern, S. Ruckbrod, C. Czekelius, C. Donner and H. Brunner, *Adv. Synth. Catal.*, 2010, **352**, 1983–1992.
271. S. Cacchi, G. Fabrizi, A. Goggiamani and A. Sferrazza, *Org. Biomol. Chem.*, 2011, **9**, 1727–1730.
272. F. A. Siqueira, J. G. Taylor and C. R. D. Correia, *Tetrahedron Lett.*, 2010, **51**, 2102–2105.
273. I. Penafiel, I. M. Pastor and M. Yus, *Eur. J. Org. Chem.*, 2012, 3151–3156.

274. F. X. Felpin, E. Fouquet and C. Zakri, *Adv. Synth. Catal.*, 2008, **350**, 2559–2565.
275. F. X. Felpin, O. Ibarguren, L. Nassar-Hardy and E. Fouquet, *J. Org. Chem.*, 2009, **74**, 1349–1352.
276. C. Rossy, E. Fouquet and F. X. Felpin, *Synthesis*, 2012, **44**, 37–41.
277. F. X. Felpin, K. Miqueu, J. M. Sotiropoulos, E. Fouquet, O. Ibarguren and J. Laudien, *Chem. Eur. J.*, 2010, **16**, 5191–5204.
278. J. Salabert, R. M. Sebastian, A. Vallribera, J. F. Civicos and C. Najera, *Tetrahedron*, 2013, **69**, 2655–2659.
279. D. S. Gaikwad and D. M. Pore, *Synlett*, 2012, 2631–2634.
280. R. G. Kalkhambkar and K. K. Laali, *Tetrahedron Lett.*, 2011, **52**, 1733–1737.
281. J. C. Pastre and C. R. D. Correia, *Adv. Synth. Catal.*, 2009, **351**, 1217–1223.
282. J. G. Taylor and C. R. D. Correia, *J. Org. Chem.*, 2011, **76**, 857–869.
283. J. C. Pastre, Y. Genisson, N. Saffon, J. Dandurand and C. R. D. Correia, *J. Braz. Chem. Soc.*, 2010, **21**, 821–836.
284. T. Konno, S. Yamada, A. Tani, M. Nishida, T. Miyabe and T. Ishihara, *J. Fluorine Chem.*, 2009, **130**, 913–921.
285. A. H. L. Machado, M. A. de Sousa, D. C. S. Patto, L. F. S. Azevedo, F. I. Bombonato and C. R. D. Correia, *Tetrahedron Lett.*, 2009, **50**, 1222–1225.
286. B. Schmidt and A. Biernat, *Eur. J. Org. Chem.*, 2008, 5764–5769.
287. F. X. Felpin, J. Coste, C. Zakri and E. Fouquet, *Chem. Eur. J.*, 2009, **15**, 7238–7245.
288. K. P. da Silva, M. N. Godoi and C. R. D. Correia, *Org. Lett.*, 2007, **9**, 2815–2818.
289. M. J. S. Carpes and C. R. D. Correia, *Synlett*, 2000, 1037–1039.
290. R. D. Barreto, L. Nascimbem and C. R. D. Correia, *Synth. Commun.*, 2007, **37**, 2011–2018.
291. S. Cacchi, G. Fabrizi, A. Goggiamani and A. Sferrazza, *Synlett*, 2009, 973–977.
292. C. C. Oliveira, R. A. Angnes and C. R. D. Correia, *J. Org. Chem.*, 2013, **78**, 4373–4385.
293. S. Cacchi, G. Fabrizi, A. Goggiamani and A. Sferrazza, *Synlett*, 2009, 1277–1280.
294. E. W. Werner, T. S. Mei, A. J. Burckle and M. S. Sigman, *Science*, 2012, **338**, 1455–1458.
295. B. Ahmed-Omer, D. A. Barrow and T. Wirth, *Tetrahedron Lett.*, 2009, **50**, 3352–3355.
296. F. Le Callonnec, E. Fouquet and F. X. Felpin, *Org. Lett.*, 2011, **13**, 2646–2649.
297. N. Susperregui, K. Miqueu, J. M. Sotiropoulos, F. Le Callonnec, E. Fouquet and F. X. Felpin, *Chem. Eur. J.*, 2012, **18**, 7210–7218.
298. B. Schmidt and R. Berger, *Adv. Synth. Catal.*, 2013, **355**, 463–476.
299. M. K. Zhu, J. F. Zhao and T. P. Loh, *Org. Lett.*, 2011, **13**, 6308–6311.

300. Y. Bai, L. M. H. Kim, H. Liao and X.-W. Liu, *J. Org. Chem.*, 2013, **78**, 8821–8825.
301. E. M. Ferreira, H. Zhang and B. M. Stoltz, in *The Mizoroki–Heck Reaction*, ed. M. Oestreich, Wiley, Chichester, 2009, pp. 345–382.
302. I. P. Beletskaya and A. V. Cheprakov, *Coord. Chem. Rev.*, 2004, **248**, 2337–2364.
303. J. Piera and J.-E. Bäckvall, *Angew. Chem. Int. Ed.*, 2008, **47**, 3506–3523.
304. S. S. Stahl, *Angew. Chem. Int. Ed.*, 2004, **43**, 3400–3420.
305. T. Nishimura, T. Onoue, K. Ohe and S. Uemura, *J. Org. Chem.*, 1999, **64**, 6750–6755.
306. M. C. Denney, N. A. Smythe, K. L. Cetto, R. A. Kemp and K. I. Goldberg, *J. Am. Chem. Soc.*, 2006, **128**, 2508–2509.
307. J. M. Keith, R. P. Muller, R. A. Kemp, K. I. Goldberg, W. A. Goddard and J. Oxgaard, *Inorg. Chem.*, 2006, **45**, 9631–9633.
308. M. M. Konnick, B. A. Gandhi, I. A. Guzei and S. S. Stahl, *Angew. Chem. Int. Ed.*, 2006, **45**, 2904–2907.
309. B. V. Popp, J. E. Wendlandt, C. R. Landis and S. S. Stahl, *Angew. Chem. Int. Ed.*, 2007, **46**, 601–604.
310. M. M. Konnick and S. S. Stahl, *J. Am. Chem. Soc.*, 2008, **130**, 5753–5762.
311. V. V. Grushin, *J. Am. Chem. Soc.*, 1999, **121**, 5831–5832.
312. B. M. Trost, *Acc. Chem. Res.*, 2002, **35**, 695–705.
313. Y. Obora, Y. Okabe and Y. Ishii, *Org. Biomol. Chem.*, 2010, **8**, 4071–4073.
314. Y. Obora and Y. Ishii, *Molecules*, 2010, **15**, 1487–1500.
315. M. Lafrance and K. Fagnou, *J. Am. Chem. Soc.*, 2006, **128**, 16496–16497.
316. D. García-Cuadrado, A. A. C. Braga, F. Maseras and A. M. Echavarren, *J. Am. Chem. Soc.*, 2006, **128**, 1066–1067.
317. D. Lapointe and K. Fagnou, *Chem. Lett.*, 2010, **39**, 1119–1126.
318. M. Livendahl and A. M. Echavarren, *Isr. J. Chem.*, 2010, **50**, 630–651.
319. M. Lafrance, C. N. Rowley, T. K. Woo and K. Fagnou, *J. Am. Chem. Soc.*, 2006, **128**, 8754–8756.
320. M. Lafrance, D. Shore and K. Fagnou, *Org. Lett.*, 2006, **8**, 5097–5100.
321. O. René and K. Fagnou, *Org. Lett.*, 2010, **12**, 2116–2119.
322. J. Guihaume, E. Clot, O. Eisenstein and R. N. Perutz, *Dalton Trans.*, 2010, **39**, 10510–10519.
323. X. G. Zhang, S. L. Fan, C. Y. He, X. L. Wan, Q. Q. Min, J. Yang and Z. X. Jiang, *J. Am. Chem. Soc.*, 2010, **132**, 4506–4507.
324. Y.-H. Zhang, B.-F. Shi and J.-Q. Yu, *J. Am. Chem. Soc.*, 2009, **131**, 5072–5074.
325. A. Kubota, M. H. Emmert and M. S. Sanford, *Org. Lett.*, 2012, **14**, 1760–1763.
326. Y. Ogata and M. Tsuchida, *J. Org. Chem.*, 1955, **20**, 1637–1643.
327. L. J. Ackerman, J. P. Sadighi, D. M. Kurtz, J. A. Labinger and J. E. Bercaw, *Organometallics*, 2003, **22**, 3884–3890.
328. P. Li, J.-W. Gu, Y. Ying, Y.-M. He, H.-F. Zhang, G. Zhao and S.-Z. Zhu, *Tetrahedron*, 2010, **66**, 8387–8391.

329. M. Miyasaka, K. Hirano, T. Satoh and M. Miura, *J. Org. Chem.*, 2010, **75**, 5421–5424.

330. H. Jiang, Z. Feng, A. Wang, X. Liu and Z. Chen, *Eur. J. Org. Chem.*, 2010, 1227–1230.

331. J. Koubachi, S. Berteina-Raboin, A. Mouaddib and G. Guillaumet, *Synthesis*, 2009, 271–276.

332. Y. Yang, H. Gong and C. Kuang, *Eur. J. Org. Chem.*, 2013, 5276–5281.

333. S. Zhang, F. Song, D. Zhao and J. You, *Chem. Commun.*, 2013, **49**, 4558–4560.

334. Y. Z. Yang, K. Cheng and Y. H. Zhang, *Org. Lett.*, 2009, **11**, 5606–5609.

335. A. Vasseur, D. Harakat, J. Muzart and J. Le Bras, *Adv. Synth. Catal.*, 2013, **355**, 59–67.

336. M. Min, Y. Kim and S. Hong, *Chem. Commun.*, 2013, **49**, 196–198.

337. A. Vasseur, D. Harakat, J. Muzart and J. Le Bras, *J. Org. Chem.*, 2012, **77**, 5751–5758.

338. Q. F. Huang, Q. Q. Song, J. Cai, X. F. Zhang and S. Lin, *Adv. Synth. Catal.*, 2013, **355**, 1512–1516.

339. N. Gigant, L. Chausset-Boissarie, R. Rey-Rodriguez and I. Gillaizeau, *C. R. Chim.*, 2013, **16**, 358–362.

340. B. P. Babu, X. Meng and J. E. Backvall, *Chem. Eur. J.*, 2013, **19**, 4140–4145.

341. S. H. Kim, K. H. Kim, H. J. Lee and J. N. Kim, *Tetrahedron Lett.*, 2013, **54**, 329–334.

342. H. S. Lee, C. H. Lim, H. J. Lee and J. N. Kim, *Bull. Korean Chem. Soc.*, 2012, **33**, 3817–3822.

343. K. H. Kim, S. Lee, S. H. Kim, C. H. Lim and J. N. Kim, *Tetrahedron Lett.*, 2012, **53**, 5088–5093.

344. Z. Jiang, L. J. Zhang, C. N. Dong, Z. Z. Cai, W. J. Tang, H. R. Li, L. J. Xu and J. L. Xiao, *Adv. Synth. Catal.*, 2012, **354**, 3225–3230.

345. K. H. Kim, J. W. Lim, H. J. Lee and J. N. Kim, *Tetrahedron Lett.*, 2013, **54**, 419–423.

346. Y. X. Zhang, Z. J. Li and Z. Q. Liu, *Org. Lett.*, 2012, **14**, 226–229.

347. H. J. Lee, K. H. Kim, S. H. Kim and J. N. Kim, *Tetrahedron Lett.*, 2013, **54**, 170–175.

348. E. M. Beck, N. P. Grimster, R. Hatley and M. J. Gaunt, *J. Am. Chem. Soc.*, 2006, **128**, 2528–2529.

349. E. M. Beck, R. Hatley and M. J. Gaunt, *Angew. Chem. Int. Ed.*, 2008, **47**, 3004–3007.

350. J. A. Schiffner and M. Oestreich, *Eur. J. Org. Chem.*, 2011, 1148–1154.

351. J. A. Schiffner, T. H. Wöste and M. Oestreich, *Eur. J. Org. Chem.*, 2010, 174–182.

352. J. A. Schiffner, A. B. Machotta and M. Oestreich, *Synlett*, 2008, 2271–2274.

353. K. Mikami, M. Hatano and M. Terada, *Chem. Lett.*, 1999, 55–56.

354. S. R. Kandukuri, J. A. Schiffner and M. Oestreich, *Angew. Chem. Int. Ed.*, 2012, **51**, 1265–1269.

355. C. L. Song, L. Ju, M. C. Wang, P. C. Liu, Y. Z. Zhang, J. W. Wang and Z. H. Xu, *Chem. Eur. J.*, 2013, **19**, 3584–3589.
356. D. Kim, M. Min and S. Hong, *Chem. Commun.*, 2013, **49**, 4021–4023.
357. G. G. Pawar, G. Singh, V. K. Tiwari and M. Kapur, *Adv. Synth. Catal.*, 2013, **355**, 2185–2190.
358. L. K. Meng, K. Wu, C. Liu and A. W. Lei, *Chem. Commun.*, 2013, **49**, 5853–5855.
359. H. Weissman, X. P. Song and D. Milstein, *J. Am. Chem. Soc.*, 2001, **123**, 337–338.
360. S. I. Kozhushkov and L. Ackermann, *Chem. Sci.*, 2013, **4**, 886–896.
361. F. W. Patureau, C. Nimphius and F. Glorius, *Org. Lett.*, 2011, **13**, 6346–6349.
362. A. D. Ryabov, *Synthesis*, 1985, 233–252.
363. I. P. Beletskaya and A. V. Cheprakov, *J. Organomet. Chem.*, 2004, **689**, 4055–4082.
364. J. Dupont, C. S. Consorti and J. Spencer, *Chem. Rev.*, 2005, **105**, 2527–2571.
365. M. J. Tredwell, M. Gulias, N. Gaunt Bremeyer, C. C. C. Johansson, B. S. L. Collins and M. J. Gaunt, *Angew. Chem. Int. Ed.*, 2011, **50**, 1076–1079.
366. K. M. Engle, D.-H. Wang and J.-Q. Yu, *J. Am. Chem. Soc.*, 2010, **132**, 14137–14151.
367. C. Zhu and J. R. Falck, *Org. Lett.*, 2011, **13**, 1214–1217.
368. W. Rauf, A. L. Thompson and J. M. Brown, *Dalton Trans.*, 2010, **39**, 10414–10421.
369. Y. Lu, D.-H. Wang, K. M. Engle and J.-Q. Yu, *J. Am. Chem. Soc.*, 2010, **132**, 5916–5921.
370. Z. J. Liang, L. Ju, Y. J. Xie, L. H. Huang and Y. H. Zhang, *Chem. Eur. J.*, 2012, **18**, 15816–15821.
371. T. Nishikata and B. H. Lipshutz, *Org. Lett.*, 2010, **12**, 1972–1975.
372. B.-F. Shi, Y.-H. Zhang, J. K. Lam, D.-H. Wang and J.-Q. Yu, *J. Am. Chem. Soc.*, 2009, **132**, 460–461.
373. D. G. Musaev, A. Kaledin, B.-F. Shi and J.-Q. Yu, *J. Am. Chem. Soc.*, 2012, **134**, 1690–1698.
374. C. Pi, Y. Li, X. Cui, H. Zhang, Y. Han and Y. Wu, *Chem. Sci.*, 2013, **4**, 2675–2679.
375. A. García-Rubia, M. Á. Fernández-Ibáñez, R. Gómez Arrayás and J. C. Carretero, *Chem. Eur. J.*, 2011, **17**, 3567–3570.
376. B. Urones, R. G. Arrayas and J. C. Carretero, *Org. Lett.*, 2013, **15**, 1120–1123.
377. A. García-Rubia, B. Urones, R. Gómez Arrayás and J. C. Carretero, *Chem. Eur. J.*, 2010, **16**, 9676–9685.
378. S. H. Cho, S. J. Hwang and S. Chang, *J. Am. Chem. Soc.*, 2008, **130**, 9254–9256.
379. L. Y. Chan, S. Kim, T. Ryu and P. H. Lee, *Chem. Commun.*, 2013, **49**, 4682–4684.
380. B. Schmidt and N. Elizarov, *Chem. Commun.*, 2012, **48**, 4350–4352.

381. M. C. Reddy and M. Jeganmohan, *Eur. J. Org. Chem.*, 2013, 1150–1157.
382. Q. Zhang, H. Z. Yu, Y. T. Li, L. Liu, Y. Huang and Y. Fu, *Dalton Trans.*, 2013, **42**, 4175–4184.
383. S. R. Chidipudi, M. D. Wieczysty, I. Khan and H. W. Lam, *Org. Lett.*, 2013, **15**, 570–573.
384. A. S. Tsai, M. l. Brasse, R. G. Bergman and J. A. Ellman, *Org. Lett.*, 2010, **13**, 540–542.
385. F. W. Patureau and F. Glorius, *J. Am. Chem. Soc.*, 2010, **132**, 9982–9983.
386. S. Rakshit, C. Grohmann, T. Besset and F. Glorius, *J. Am. Chem. Soc.*, 2011, **133**, 2350–2353.
387. B. Li, J. Ma, Y. Liang, N. Wang, S. Xu, H. Song and B. Wang, *Eur. J. Org. Chem.*, 2013, 1950–1962.
388. C. S. Cho, K. Itotani and S. Uemura, *J. Organomet. Chem.*, 1993, **443**, 253–259.
389. C. S. Cho and S. Uemura, *J. Organomet. Chem.*, 1994, **465**, 85–92.
390. M. S. Sigman and E. W. Werner, *Acc. Chem. Res.*, 2011, **45**, 874–884.
391. X. L. Du, M. Suguro, K. Hirabayashi, A. Mori, T. Nishikata, N. Hagiwara, K. Kawata, T. Okeda, H. F. Wang, K. Fugami and M. Kosugi, *Org. Lett.*, 2001, **3**, 3313–3316.
392. Y. C. Jung, R. K. Mishra, C. H. Yoon and K. W. Jung, *Org. Lett.*, 2003, **5**, 2231–2234.
393. C. H. Yoon, K. S. Yoo, S. W. Yi, R. K. Mishra and K. W. Jung, *Org. Lett.*, 2004, **6**, 4037–4039.
394. M. M. S. Andappan, P. Nilsson and M. Larhed, *Chem. Commun.*, 2004, 218–219.
395. M. M. S. Andappan, P. Nilsson, H. von Schenck and M. Larhed, *J. Org. Chem.*, 2004, **69**, 5212–5218.
396. P. A. Enquist, P. Nilsson, P. Sjoberg and M. Larhed, *J. Org. Chem.*, 2006, **71**, 8779–8786.
397. J. Lindh, P.-A. Enquist, Å. Pilotti, P. Nilsson and M. Larhed, *J. Org. Chem.*, 2007, **72**, 7957–7962.
398. Y. T. Leng, F. Yang, K. Wei and Y. J. Wu, *Tetrahedron*, 2010, **66**, 1244–1248.
399. J. H. Delcamp, A. P. Brucks and M. C. White, *J. Am. Chem. Soc.*, 2008, **130**, 11270–11271.
400. K. Itami and J.-i. Yoshida, in *The Mizoroki–Heck Reaction*, ed. M. Oestreich, Wiley, Chichester, 2009, pp. 259–280.
401. Y. J. Su and N. Jiao, *Org. Lett.*, 2009, **11**, 2980–2983.
402. E. W. Werner and M. S. Sigman, *J. Am. Chem. Soc.*, 2010, **132**, 13981–13983.
403. K. B. Urkalan and M. S. Sigman, *Angew. Chem. Int. Ed.*, 2009, **48**, 3146–3149.
404. T. S. Mei, E. W. Werner, A. J. Burckle and M. S. Sigman, *J. Am. Chem. Soc.*, 2013, **135**, 6830–6833.
405. L. Penn, A. Shpruhman and D. Gelman, *J. Org. Chem.*, 2007, 72, 3875–3879.

406. K. S. Yoo, C. P. Park, C. H. Yoon, S. Sakaguchi, J. O'Neill and K. W. Jung, *Org. Lett.*, 2007, **9**, 3933–3935.
407. K. S. Yoo, J. O'Neill, S. Sakaguchi, R. Giles, J. H. Lee and K. W. Jung, *J. Org. Chem.*, 2009, **75**, 95–101.
408. S. Yahiaoui, A. Fardost, A. Trejos and M. Larhed, *J. Org. Chem.*, 2011, **76**, 2433–2438.
409. S. E. Walker, J. Boehnke, P. E. Glen, S. Levey, L. Patrick, J. A. Jordan-Hore and A. L. Lee, *Org. Lett.*, 2013, **15**, 1886–1889.
410. Y. Izawa, C. W. Zheng and S. S. Stahl, *Angew. Chem. Int. Ed.*, 2013, **52**, 3672–3675.
411. M. H. Chen, S. Mannathan, P. S. Lin and C. H. Cheng, *Chem. Eur. J.*, 2012, **18**, 14918–14922.
412. R. Jana and J. A. Tunge, *J. Org. Chem.*, 2011, **76**, 8376–8385.
413. N. Kuuloja, M. Vaismaa and R. Franzen, *Tetrahedron*, 2012, **68**, 2313–2318.
414. T. W. Liwosz and S. R. Chemler, *Org. Lett.*, 2013, **15**, 3034–3037.
415. Y. M. Li, Z. S. Qi, H. F. Wang, X. M. Fu and C. Y. Duan, *J. Org. Chem.*, 2012, **77**, 2053–2057.
416. L. J. Zhang, C. N. Dong, C. J. Ding, J. Chen, W. J. Tang, H. R. Li, L. J. Xu and J. L. Xiao, *Adv. Synth. Catal.*, 2013, **355**, 1570–1578.
417. T. Mino, T. Kogure, T. Abe, T. Koizumi, T. Fujita and M. Sakamoto, *Eur. J. Org. Chem.*, 2013, 1501–1505.
418. C. Zhu and J. R. Falck, *Angew. Chem. Int. Ed.*, 2011, **50**, 6626–6629.
419. K. Olofsson, M. Larhed and A. Hallberg, *J. Org. Chem.*, 1998, **63**, 5076–5079.
420. A. Mori, Y. Danda, T. Fujii, K. Hirabayashi and K. Osakada, *J. Am. Chem. Soc.*, 2001, **123**, 10774–10775.
421. S. E. Denmark and C. S. Regens, *Acc. Chem. Res.*, 2008, **41**, 1486–1499.
422. Z. S. Ye, F. Chen, F. Luo, W. H. Wang, B. D. Lin, X. F. Jia and J. Cheng, *Synlett*, 2009, 2198–2200.
423. T. Mino, M. Shibuya, S. Suzuki, K. Hirai, M. Sakamoto and T. Fujita, *Tetrahedron*, 2012, **68**, 429–432.
424. H. Oda, M. Morishita, K. Fugami, H. Sano and M. Kosugi, *Chem. Lett.*, 1996, 811–812.
425. K. Fugami, S. Hagiwara, H. Oda and M. Kosugi, *Synlett*, 1998, 477–478.
426. K. Hirabayashi, J.-i. Ando, Y. Nishihara, A. Mori and T. Hiyama, *Synlett*, 1999, 99–101.
427. J. P. Parrish, Y. C. Jung, S. I. Shin and K. W. Jung, *J. Org. Chem.*, 2002, **67**, 7127–7130.
428. E. W. Werner, K. B. Urkalan and M. S. Sigman, *Org. Lett.*, 2010, **12**, 2848–2851.
429. D. Tanaka, S. P. Romeril and A. G. Myers, *J. Am. Chem. Soc.*, 2005, **127**, 10323–10333.
430. L. J. Goossen, F. Collet and K. Goossen, *Isr. J. Chem.*, 2010, **50**, 617–629.
431. T. Satoh and M. Miura, *Synthesis*, 2010, 3395–3409.

432. A. G. Myers, D. Tanaka and M. R. Mannion, *J. Am. Chem. Soc.*, 2002, **124**, 11250–11251.
433. D. Tanaka and A. G. Myers, *Org. Lett.*, 2004, **6**, 433–436.
434. Z. J. Fu, S. J. Huang, W. P. Su and M. C. Hong, *Org. Lett.*, 2010, **12**, 4992–4995.
435. P. Hu, J. Kan, W. P. Su and M. C. Hong, *Org. Lett.*, 2009, **11**, 2341–2344.
436. L. B. Huang, J. Qi, X. Wu, K. F. Huang and H. F. Jiang, *Org. Lett.*, 2013, **15**, 2330–2333.
437. S. L. Zhang, Y. Fu, R. Shang, Q. X. Guo and L. Liu, *J. Am. Chem. Soc.*, 2010, **132**, 638–646.
438. J. S. Dickstein, C. A. Mulrooney, E. M. O'Brien, B. J. Morgan and M. C. Kozlowski, *Org. Lett.*, 2007, **9**, 2441–2444.
439. J. T. Wang, Z. L. Cui, Y. X. Zhang, H. J. Li, L. M. Wu and Z. Q. Liu, *Org. Biomol. Chem.*, 2010, **9**, 663–666.
440. L. J. Goossen, B. Zimmermann and T. Knauber, *Beilstein J. Org. Chem.*, 2010, **6**, 43.
441. L. Q. Xue, W. P. Su and Z. Y. Lin, *Dalton Trans.*, 2010, **39**, 9815–9822.
442. R. Shang, Q. Xu, Y. Y. Jiang, Y. Wang and L. Liu, *Org. Lett.*, 2010, **12**, 1000–1003.
443. Y. Luo and J. Wu, *Chem. Commun.*, 2010, **46**, 3785–3787.
444. Y. Yamamoto, K. Kurihara and N. Miyaura, *Angew. Chem. Int. Ed.*, 2009, **48**, 4414–4416.
445. K. Kurihara, Y. Yamamoto and N. Miyaura, *Adv. Synth. Catal.*, 2009, **351**, 260–270.
446. H. Zheng, Q. Zhang, J. Chen, M. Liu, S. Cheng, J. Ding, H. Wu and W. Su, *J. Org. Chem.*, 2008, **74**, 943–945.
447. Y.-X. Liao, C.-H. Xing, P. He and Q.-S. Hu, *Org. Lett.*, 2008, **10**, 2509–2512.
448. J. Lindh, P. J. R. Sjöberg and M. Larhed, *Angew. Chem. Int. Ed.*, 2010, **49**, 7733–7737.
449. Z. Y. Wang, Q. P. Ding, X. D. He and J. Wu, *Org. Biomol. Chem.*, 2009, **7**, 863–865.
450. R. Shang, Z. W. Yang, Y. Wang, S. L. Zhang and L. Liu, *J. Am. Chem. Soc.*, 2010, **132**, 14391–14393.
451. M. Miura, H. Hashimoto, K. Itoh and M. Nomura, *J. Chem. Soc., Perkin Trans. 1*, 1990, 2207–2211.
452. M. Miura, H. Hashimoto, K. Itoh and M. Nomura, *Tetrahedron Lett.*, 1989, **30**, 975–976.
453. X. Zhou, J. Luo, J. Liu, S. Peng and G.-J. Deng, *Org. Lett.*, 2011, **13**, 1432–1435.
454. W. Chen, X. Y. Zhou, F. H. Xiao, J. Y. Luo and G. J. Deng, *Tetrahedron Lett.*, 2012, **53**, 4347–4350.
455. F. L. Yang, X. T. Ma and S. K. Tian, *Chem. Eur. J.*, 2012, **18**, 1582–1585.
456. W. Chen, H. Chen, F. H. Xiao and G. J. Deng, *Org. Biomol. Chem.*, 2013, **11**, 4295–4298.

CHAPTER 10

Palladium-Catalyzed Carbonylative Coupling and C–H Activation

XIAO-FENG WU*[a] AND CHRISTOPHER F. J. BARNARD*[b]

[a] Leibniz-Institut für Katalyse eV, Albert-Einstein-Strasse 29a, Rostock, D-18059, Germany; [b] Johnson Matthey Technology Centre, Blount's Court, Sonning Common, Reading, RG4 9NH, UK
*Email: xiao-feng.wu@catalysis.de; barnacfj@matthey.com

10.1 Introduction

Early studies by Henry, Heck and others developed the carbonylation of aryl halides to give a wide variety of functionalized derivatives, providing useful compounds and intermediates for further reactions in chemical synthesis.[1] As palladium-catalysed coupling chemistry was developed to employ a wide range of reactants (e.g. alkenes – Heck, arenes – Suzuki, alkynes – Sonogashira), so the mechanisms of these reactions became better understood and the potential for extending these reactions was appreciated. Thus, the development of carbonylative versions of these classic coupling reactions was achieved (Figure 10.1). Here, examples are given of carbonylative Heck, Suzuki and Sonogashira reactions, where control of the catalyst and the reaction conditions can lead to high selectivity favouring the carbonyl insertion product over the direct coupling product.

The above-described reactions still require the preparation of aryl halides and potentially other reagents as their starting materials, and such preparations can, on occasion, be difficult owing to the poor stability of some of the

RSC Catalysis Series No. 21
New Trends in Cross-Coupling: Theory and Applications
Edited by Thomas J Colacot
© The Royal Society of Chemistry 2015
Published by the Royal Society of Chemistry, www.rsc.org

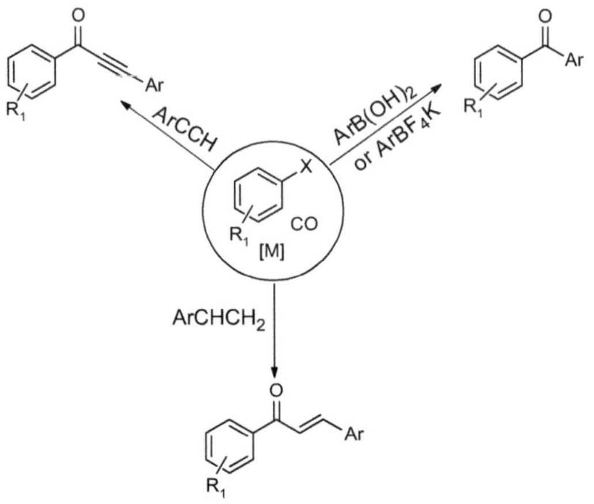

Figure 10.1 Carbonylative coupling reactions.

materials. In addition, they also involve costs and waste generation that could be avoided if the reactions could be conducted on the original arene or heteroarene without exchange of H for halide. This goal of reactions using C–H bonds has been investigated in many areas of organic chemistry in recent years and much progress has been made.[2] The adaptation of such reactions to carbonylation processes in conjunction with CO was perceived as difficult owing to the potential of CO to reduce Pd(II) to Pd(0), but steady progress has been made by many contributors over a number of years such that suitable protocols now exist for these oxidative carbonylation reactions for a variety of substrates. A selection of additives have been used to reoxidize Pd(0) to Pd(II), a step that is normally achieved in conventional coupling reactions by oxidative addition with aryl halide. These reactions can be classified by the hybridization of the activated bond (sp, sp^2 and sp^3) and the supporting functionality of the substrate that contributes to directing the reaction and thus providing specificity. In the following, progress in this area is reviewed on this basis, illustrating the potential for further developments of this chemistry.[3]

Extending carbonylation from the aryl halide substrates described above to less reactive C–H compounds represents a significant challenge owing to binding of CO to the Pd(II) centre in competition with the weak interaction with C–H. Also, π-backbonding to a carbonyl ligand may reduce the nucleophilicity of the Pd centre for interaction with C–H, in addition to the aforementioned potential of CO to promote the reduction of Pd(II) to Pd(0) leading to catalyst decomposition. In order to maximize the efficiency of the transformation, many methods have been developed using a variety of additives. Also, the choice of oxidant to return Pd(0) to Pd(II) is complex, resulting in a number of different choices from different research groups. Ultimately, it would be advantageous to use air (oxygen) as the terminal oxidant, and attempts to achieve this continue (Figure 10.2).

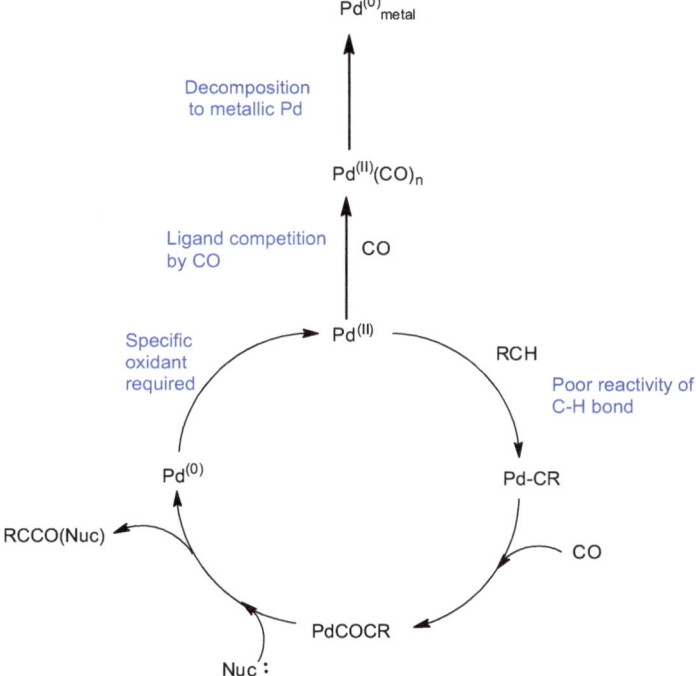

Figure 10.2 Palladium-catalysed oxidative carbonylations.

10.2   Carbonylative Heck Reactions

In the late 1960s, Heck and co-workers developed several coupling reactions of arylmercury compounds in the presence of either stoichiometric or catalytic amounts of palladium salts.[4a–g] Based on this work, in 1972 Heck and Nolley described a protocol for the coupling of iodobenzene with styrene, which today is known as the Heck–Mizoroki reaction.[4h] Two years later, Heck and co-workers described their pioneering work on palladium-catalysed alkoxy-, hydroxy- and aminocarbonylation of aryl iodides and bromides, which is nowadays called the Heck carbonylation reaction.[1] The first palladium-catalysed copolymerization of CO with olefins was described in 1982,[4i] and consequently carbonylative coupling reactions with alkenes were reported soon after. Notably, it was Negishi and Miller who discovered the first two examples of the intramolecular carbonylative Heck reaction of 1-iodopenta-1,4-dienes by applying stoichiometric amounts of palladium.[5] The product 5-methylenecyclopent-2-enones were produced in moderate yields (Scheme 10.1). However, using **1a–e** (Scheme 10.2) as substrates, no desired carbonylation products were detected, although complete conversion of starting material occurred. Presumably, polymerization of **1a–e** (Scheme 10.2) took place.[6]

The same group improved the methodology 2 years later,[6] where they applied **1a–e** as starting materials using catalytic amounts of palladium salts in

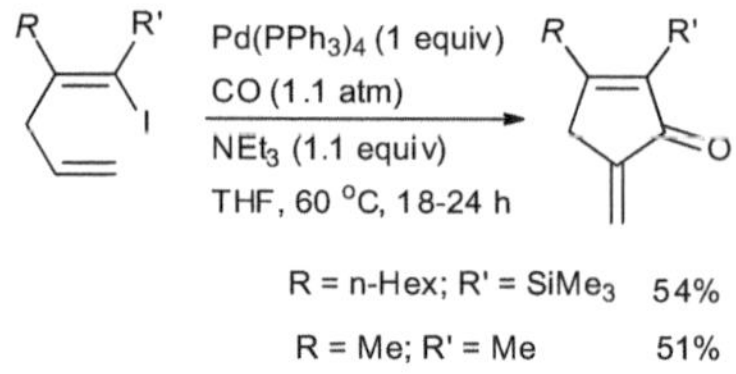
Pd(PPh3)4 (1 equiv)
CO (1.1 atm)
NEt3 (1.1 equiv)
THF, 60 °C, 18-24 h
R = n-Hex; R' = SiMe3 54%
R = Me; R' = Me 51%

Scheme 10.1 The first examples of palladium-mediated intramolecular carbonylative Heck reactions.

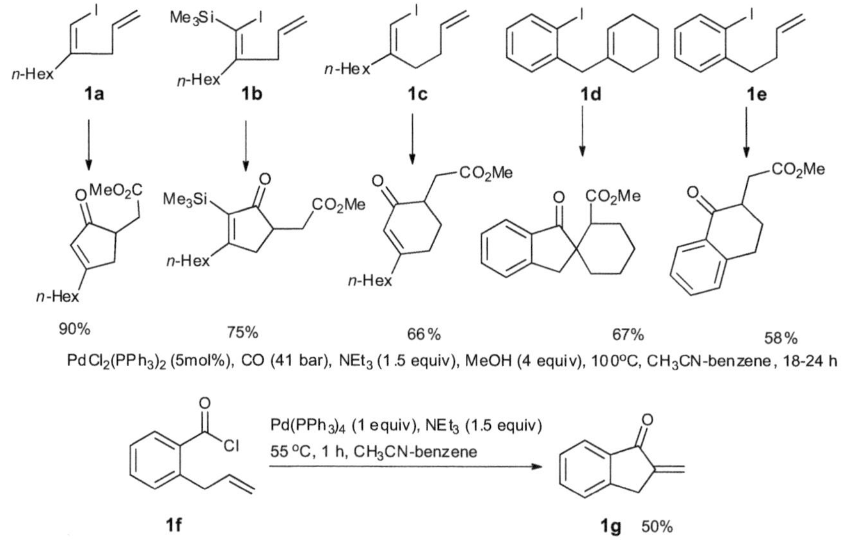
1a
1b
1c
1d
1e
90%
75%
66%
67%
58%
PdCl2(PPh3)2 (5mol%), CO (41 bar), NEt3 (1.5 equiv), MeOH (4 equiv), 100°C, CH3CN-benzene, 18-24 h
Pd(PPh3)4 (1 equiv), NEt3 (1.5 equiv)
55 °C, 1 h, CH3CN-benzene
1f
1g 50%

Scheme 10.2 Pd-catalysed intramolecular carbonylative Heck reactions.

the presence of MeOH. The corresponding products were obtained in moderate to excellent yields (Scheme 10.2). A possible reaction mechanism was proposed and "CO-free" carbonylative Heck reactions were realized. 2-Methylene-2,3-dihydroinden-1-one (**1g**) was produced from the corresponding acid chloride **1f** in 50% yield under "CO-free" conditions (Scheme 10.2).

Negishi's group continued their interest in this topic by synthesizing various quinones from *o*-iodoaryl cyclohexyl ketones as starting materials.[7a] In the presence of $Pd(dba)_2$ (5 mol%) and CO (41 bar), quinones were produced in good yields with 100% regioselectivity (Scheme 10.3). Here, the selection of the catalyst system was important, because 58% of **3a** and **3b** were formed instead of quinones if a $Pd(OAc)_2/PPh_3$ system was used. This latter work can be considered as the first real Pd-catalysed intramolecular carbonylative Heck reaction. In 1996, a full account using different vinyl iodides was published by the same group.[7b]

In 2002, Ryu *et al.* reported similar reactions which were catalysed by Pd in the presence of light.[7e] Similar products are obtained by the palladium-catalysed carbonylative coupling of allyl acetate with benzynes.[7f]

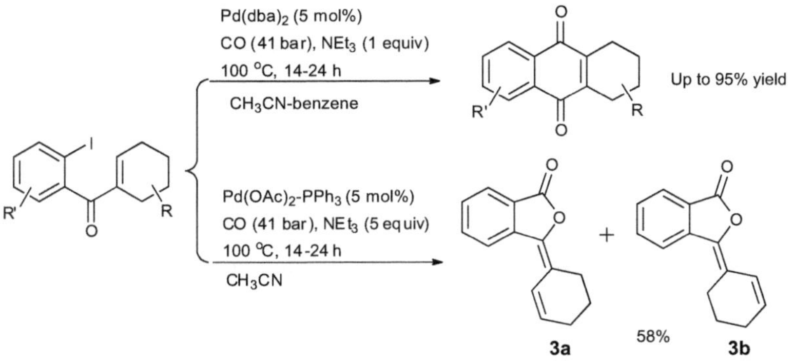

Scheme 10.3 First Pd-catalysed intramolecular carbonylative Heck reaction.

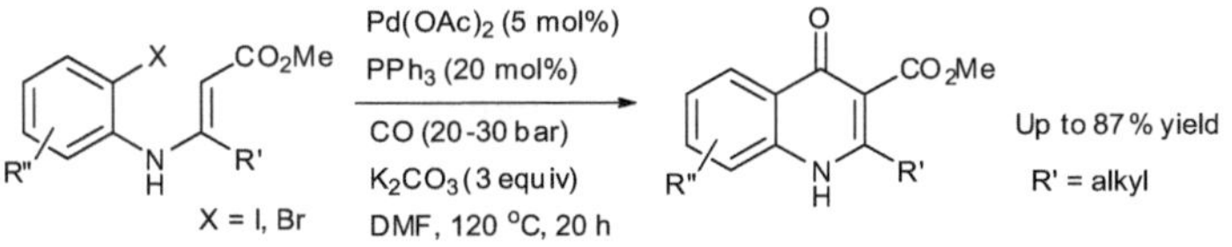

Scheme 10.4 Pd-catalysed carbonylative Heck reaction to give quinolinones.

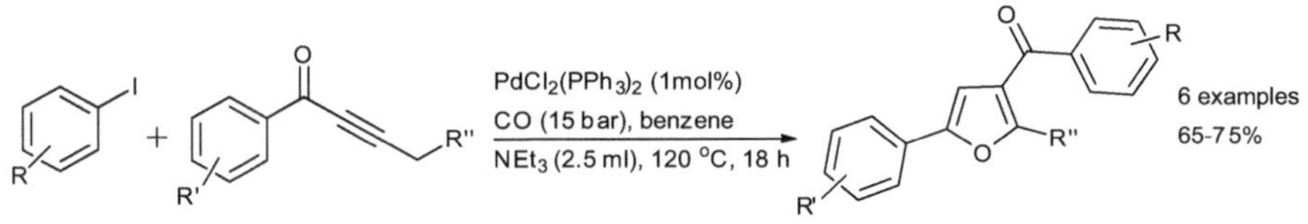

Scheme 10.5 Pd-catalysed carbonylative cross-coupling reactions of aryl iodides with alkynones to give furans.

Notably, Torii *et al.* reported the intramolecular carbonylative Heck coupling of 3-(2-haloarylamino)prop-2-enoates to the corresponding quinolinone derivatives.[8] In the presence of Pd(OAc)$_2$ under a CO atmosphere of 20 bar at 120 °C, quinolinones were synthesized in good yields (Scheme 10.4). A related carbonylative cross-coupling of aryl iodides with alkynones was reported by Miura and co-workers.[9] They showed that in the presence of CO and a palladium catalyst, various furans were produced in good yields (Scheme 10.5).

In 1995, Miura and co-workers reported a palladium-catalysed carbonylative cross-coupling of aryl iodides with five-membered cyclic olefins.[10] This represents the first palladium-catalysed *intermolecular* carbonylative cross-coupling of aryl iodides with olefins. Various benzoylated cyclic olefins were isolated in good yields (Scheme 10.6). Unfortunately, the reaction with cyclopentene led to a mixture of three regioisomers of benzoylcyclopentene. During their investigations, Miura's group found that the amount of PPh$_3$ added to the reaction mixture markedly influenced the product yields.

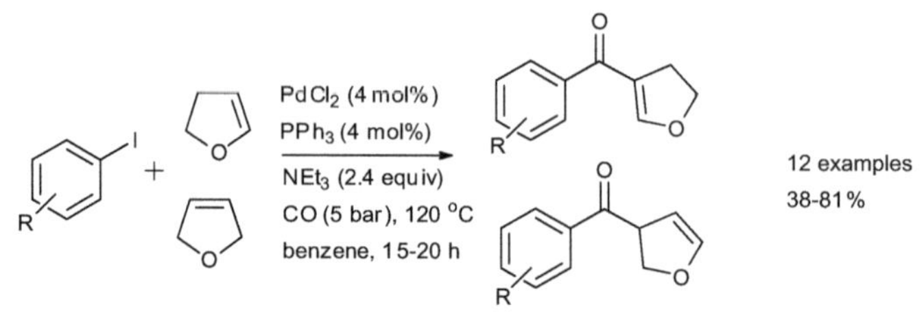

PdCl$_2$ (4 mol%)
PPh$_3$ (4 mol%)
NEt$_3$ (2.4 equiv)
CO (5 bar), 120 °C
benzene, 15-20 h

12 examples
38-81%

Scheme 10.6 Pd-catalysed carbonylative cross-coupling of aryl iodides with cyclic olefins.

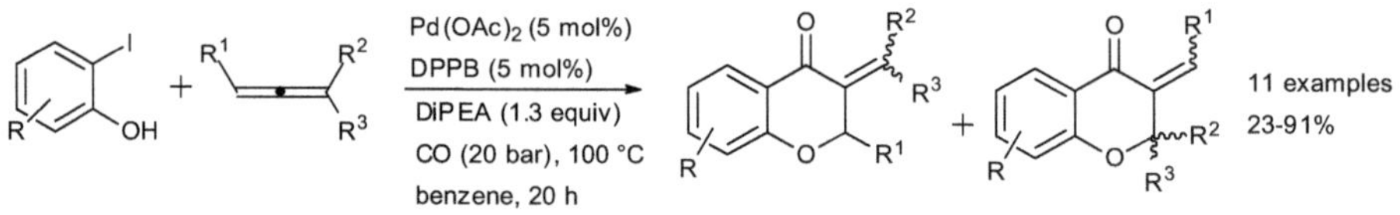

Pd(OAc)$_2$ (5 mol%)
DPPB (5 mol%)
DiPEA (1.3 equiv)
CO (20 bar), 100 °C
benzene, 20 h

11 examples
23-91%

Scheme 10.7 Pd-catalysed carbonylative coupling of *o*-iodophenols with allenes.

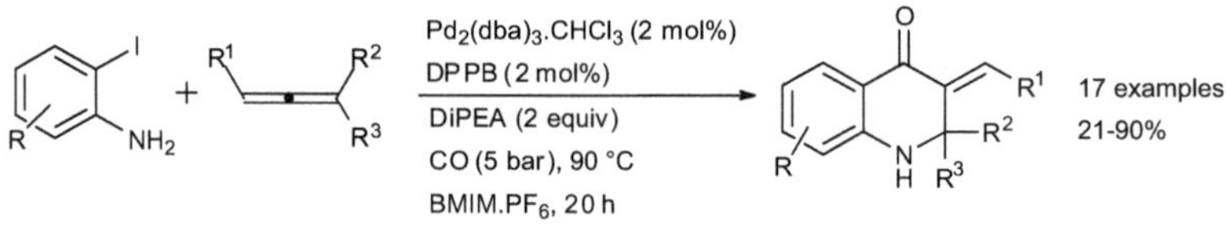

Pd$_2$(dba)$_3$.CHCl$_3$ (2 mol%)
DPPB (2 mol%)
DiPEA (2 equiv)
CO (5 bar), 90 °C
BMIM.PF$_6$, 20 h

17 examples
21-90%

Scheme 10.8 Pd-catalysed carbonylative coupling of *o*-iodoanilines with allenes.

In 1997, allenes, as a special family of alkenes, were employed in palladium-catalysed carbonylative reactions with *o*-iodophenols by Okuro and Alper.[11] Since the reaction was highly regioselective, only a single benzopyranone isomer was obtained in good yield (Scheme 10.7). In the same year, Grigg and Pratt reported another carbonylative cascade reaction. Starting from 2-methallyliodobenzene, allenes were incorporated during the cascade.[12]

Similarly to their previous work, Alper's group also described a palladium-catalysed carbonylative coupling of *o*-iodoanilines with allenes.[13] Consequently, they synthesized quinolinones in moderate to good yields under a low pressure of CO (5 bar) (Scheme 10.8). Here, an ionic liquid was used as the solvent and promoter to enhance the efficiency of the cyclization protocol. Interestingly, the recyclability of the system was also demonstrated.

In 1999, Iwasawa and Satoh reported a new Heck-type coupling for the synthesis of 4,5-didehydrotropone–Co$_2$(CO)$_4$·dppm complexes.[14] The palladium-catalysed carbonylation was promoted by a diphenylacetylene–cobalt complex (**A**), furnishing the desired complexes in high yields (Scheme 10.9).

In the same year, an enantioselective palladium-catalysed intramolecular carbonylative coupling of aryl and alkenyl triflates was reported by Hayashi *et al.*[15] Based on a Pd(TFA)$_2$/(*S*)-BINAP catalyst system, asymmetric carbonylative cyclization of prochiral *o*-allylaryl triflates and 2-allylalkenyl triflates

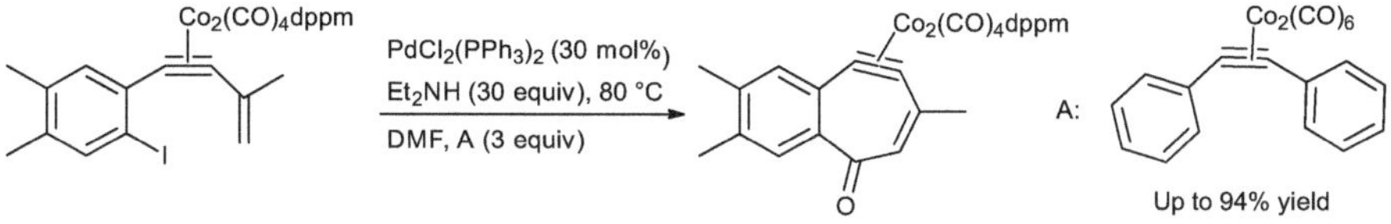

Scheme 10.9 Palladium-catalysed cyclization of 4,5-didehydrotropone–$Co_2(CO)_4 \cdot$ dppm complexes *via* carbonylative Heck reaction.

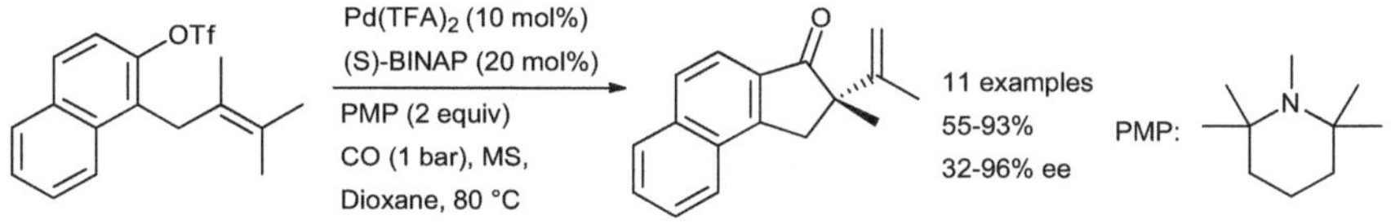

Scheme 10.10 Palladium-catalysed intramolecular carbonylative coupling of aryl and alkenyl triflates to give cyclopentenones.

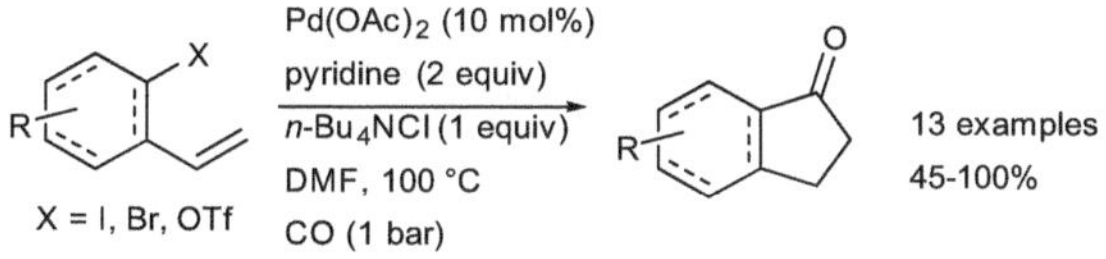

Scheme 10.11 Palladium-catalysed carbonylative indanone synthesis.

proceeded to give enantiomerically enriched cyclopentenones in high yields (Scheme 10.10).

In 2003, Gagnier and Larock discovered a general and efficient methodology for the synthesis of indanones and 2-cyclopentenones using *o*-iodostyrene as the model system (Scheme 10.11).[16] In the same way, Negishi *et al.* reported the synthesis of indenones from *o*-iodostyrene.[7c] Subsequently, they extended this methodology to dienyl triflates, iodides and bromides. All products were isolated in good yields. Regarding the mechanism, traces of water in the solvent may play an important role as a proton source to favour the formation of indanones instead of indenones.

More recently, Larhed and co-workers developed a metal carbonyl-mediated and microwave (MW)-supported intramolecular carbonylative coupling reaction of *o*-bromostyrenes.[17] Indanones and 3-acylaminoindanones were produced in good yields. In the presence of $Pd(OAc)_2/(t\text{-Bu})_3P \cdot HBF_4$ with $Mo(CO)_6$ as the CO source and under MW irradation, indanones were obtained in 20 min from the corresponding aryl bromides and chlorides. Both electron-withdrawing and electron-donating groups on the arene were tolerated, but electron-poor *o*-bromocinnamic acid derivatives furnished only the corresponding lactones *via* a hydroxycarbonylation–Michael addition sequence (Scheme 10.12).

In 2010, Bloome and Alexanian described a palladium-catalysed carbonylative Heck-type cyclization of alkyl halides.[18] Treatment of a range of

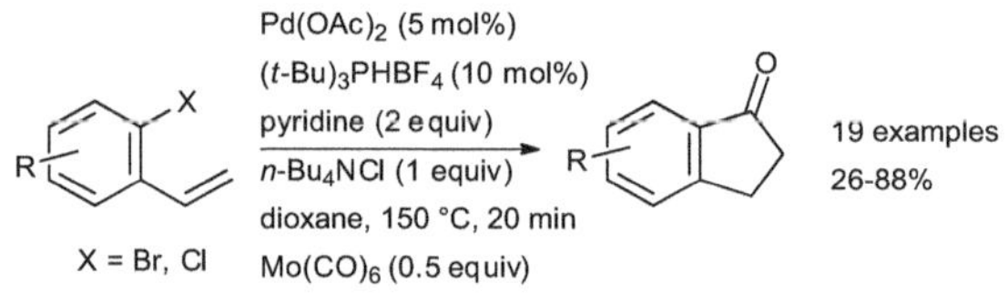

Scheme 10.12   Pd-catalysed carbonylative synthesis of indanones using microwave irradiation.

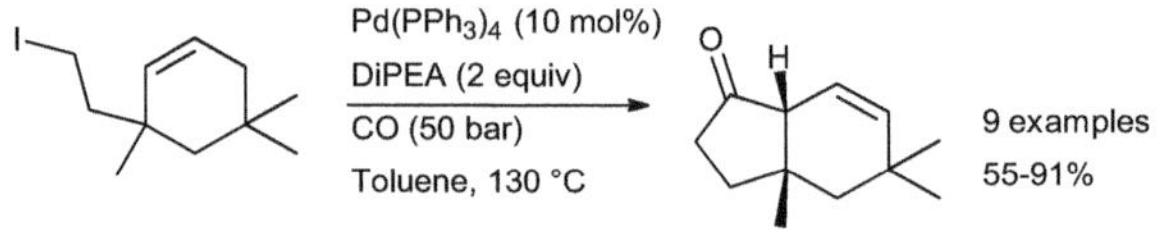

Scheme 10.13   Palladium-catalysed carbonylative Heck-type cyclization of alkyl halides.

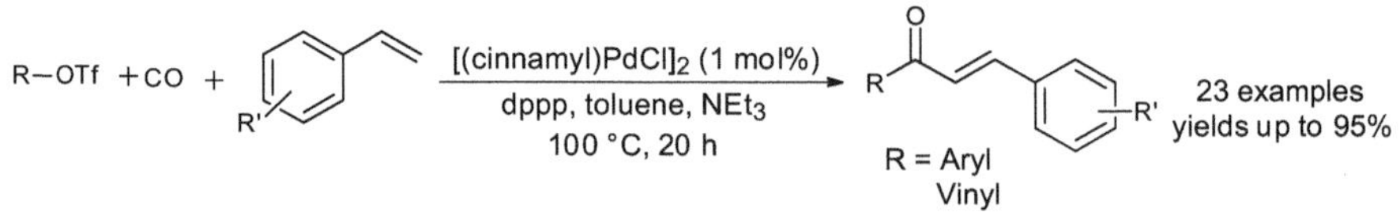

Scheme 10.14   Palladium-catalysed carbonylative Heck reaction of aryl triflates with styrenes.

primary and secondary alkyl iodides in the presence of a palladium catalyst under CO pressure yielded a variety of synthetically versatile enone products. This novel palladium-catalysed Heck-type cyclization is a rare example where unactivated alkyl halides with β-hydrogens were involved. Different substituted alkenes were well tolerated and both monocyclic and bicyclic carbocycles were easily accessed (Scheme 10.13).

Clearly, so far efforts in this area have been focused on intramolecular carbonylative Heck reactions. A general intermolecular carbonylative coupling of aryl halides or triflates with terminal olefins was not known until the recent work of Beller's group. Initially, Beller and co-workers succeeded in carbonylative Heck couplings of aryl triflates with styrenes.[19] Starting from readily available aryl and alkenyl triflates, the corresponding unsaturated ketones are obtained in good yields (Scheme 10.14). The products represent useful building blocks for the synthesis of a variety of biologically active compounds.

Shortly afterwards, a more general palladium-catalysed carbonylative Heck reaction of aryl halides was developed by the same group.[20] For the first time, various aromatic and aliphatic alkenes were employed successfully in this system and good yields of the corresponding α,β-unsaturated ketones were obtained (41–90%). Starting from readily available aryl iodides and bromides, interesting building blocks were obtained under mild conditions (Scheme 10.15). With respect to the reaction mechanism, the aryl–palladium and acyl–palladium complexes were characterized by X-ray diffraction and the

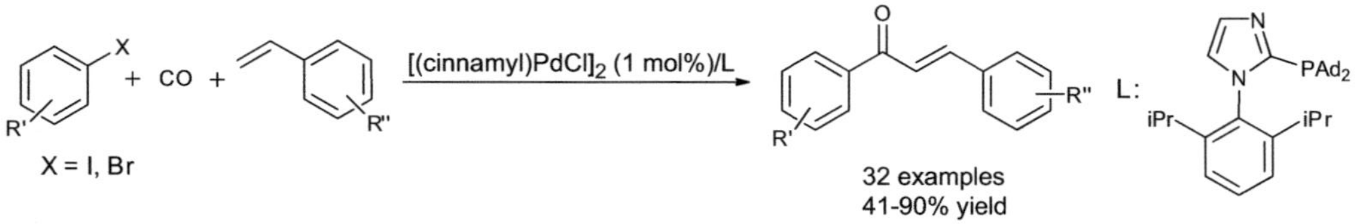

Scheme 10.15 Palladium-catalysed carbonylative Heck reaction of aryl halides with alkenes.

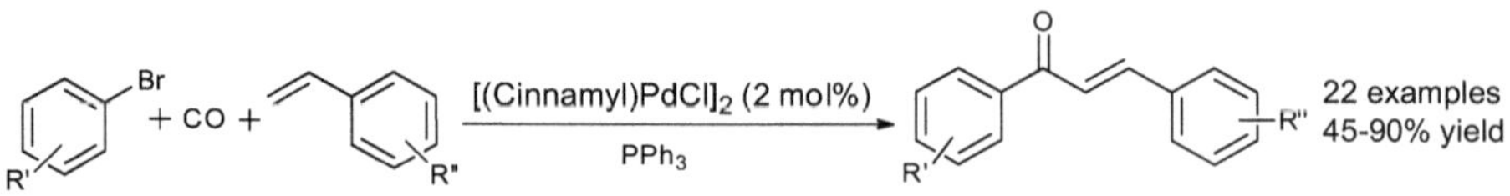

Scheme 10.16 Palladium-catalysed carbonylative Heck reaction of aryl bromides.

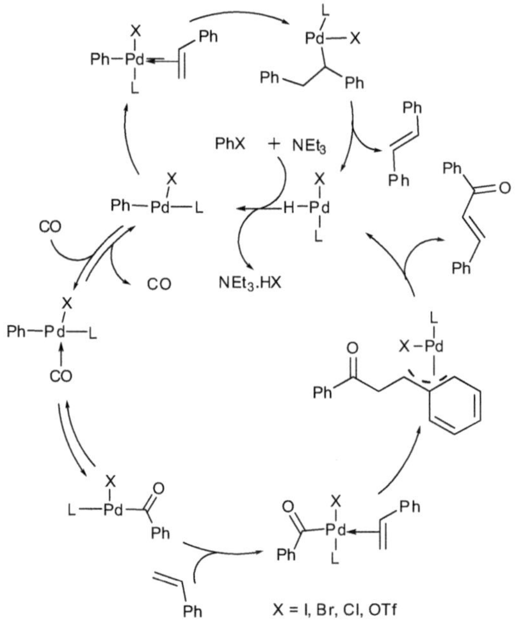

Scheme 10.17 General reaction mechanism for the carbonylative Heck reaction and non-carbonylative Heck reaction.

mechanism was studied step by step. The results fitted well with DFT calculations.

Most recently, the synthesis of chalcones from aryl bromides in the presence of PPh$_3$ as ligand was achieved (Scheme 10.16).[21] Based on both experimental results and DFT calculations, the proposed mechanism for this reaction is shown in Scheme 10.17. It starts with oxidative addition of ArX to the Pd^0 center to form the corresponding aryl–palladium complex, followed by the coordination and insertion of CO to produce the respective acyl–palladium complex. After coordination, addition and elimination processes, the desired chalcone is obtained. With the help of base, Pd^0 is formed again and the next reaction cycle can start.

10.3 Carbonylative Sonogashira Reactions

The Sonogashira reaction is generally known as a coupling reaction of terminal alkynes with aryl or vinyl halides. This reaction was first reported by

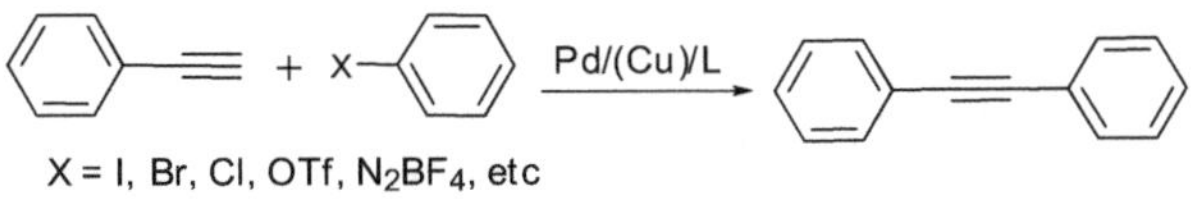

X = I, Br, Cl, OTf, N₂BF₄, etc

Scheme 10.18 Palladium-catalysed Sonogashira coupling reaction.

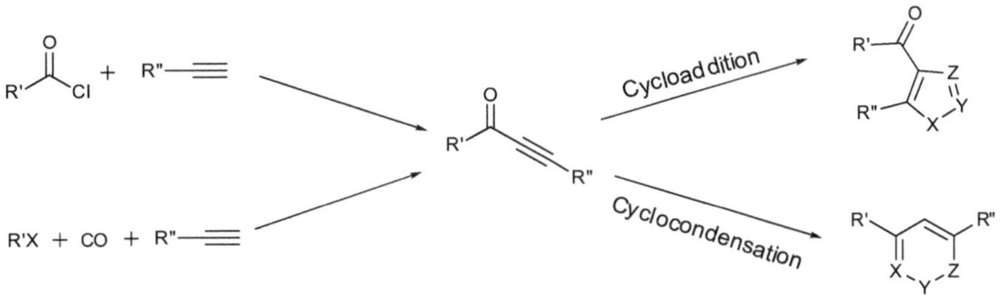

Scheme 10.19 Selected synthesis and applications of alkynones.

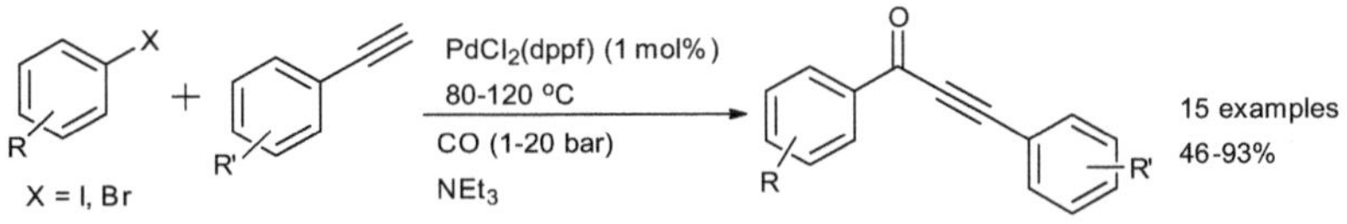

Scheme 10.20 First Pd-catalysed carbonylative Sonogashira coupling of organic halides.

Sonogashira *et al.* in 1975 (Scheme 10.18).[22] Today, the Sonogashira coupling reaction is one of most powerful processes for C–C bond formation, especially for the synthesis of substituted alkynes.[23]

Carbonylative Sonogashira reactions lead to alkynones, which represent an interesting structural motif found in numerous biologically active molecules.[24] Notably, such compounds play a crucial role as key intermediates in the synthesis of natural products[25] and for the efficient formation of several heterocycles.[26] Traditionally, alkynones have been synthesized by transition metal-catalysed cross-coupling reactions of acid chlorides and terminal alkynes (Scheme 10.19).[27] However, the stability of the respective acid chlorides is limited and a lack of functional tolerance is another problem with this methodology. Without doubt, the carbonylative Sonogashira coupling of corresponding terminal alkynes and aryl halides represents the most straightforward way to construct alkynones.

The first palladium-catalysed carbonylative Sonogashira coupling was reported in 1981 by Kobayashi and Tanaka.[28] Aryl, heterocyclic and vinylic halides reacted with CO (80 bar) and terminal acetylenes at 120 °C in the presence of NEt₃ and a catalytic amount of a palladium(II) complex to form alkynones in 46–93% yield (Scheme 10.20). Remarkably, aryl bromides and aliphatic alkynes were also included in the range of substrates.

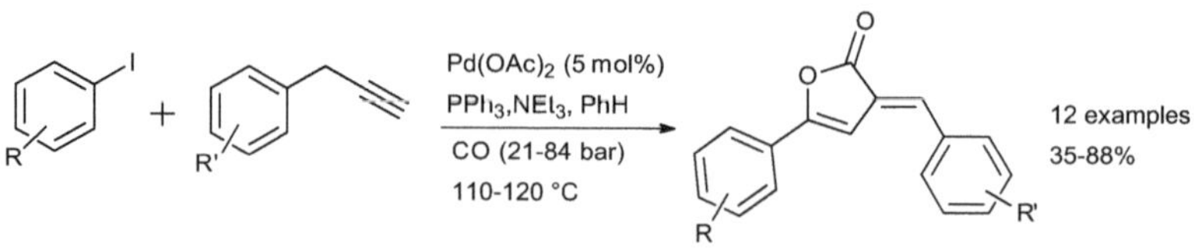

Scheme 10.21   Pd-catalysed carbonylation of benzylacetylenes to give furanones.

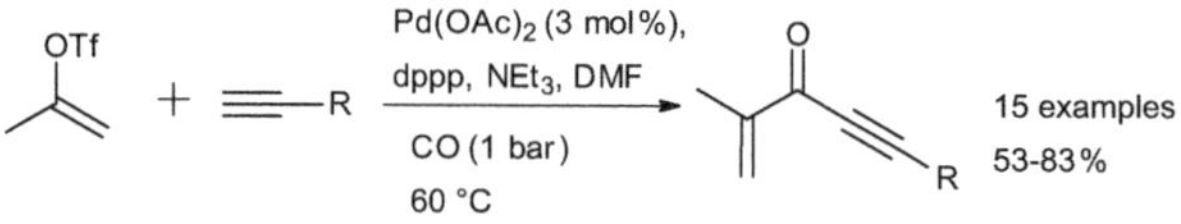

Scheme 10.22   Pd-catalysed carbonylative Sonogashira coupling of vinyl triflates with 1-alkynes.

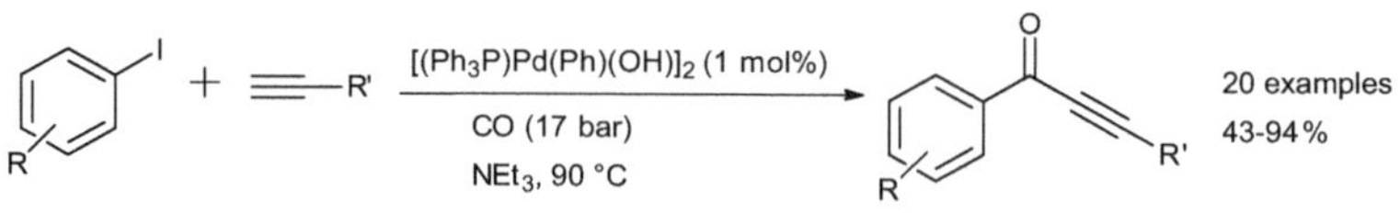

Scheme 10.23   Dimeric palladium hydroxide-catalysed carbonylative Sonogashira coupling.

Interestingly, in 1991, Huang and Alper reported another type of palladium-catalysed carbonylative Sonogashira coupling of aryl iodides with benzyl acetylenes, where furanones were isolated as terminal products and not the predicted alkynones.[29a] In the presence of Pd(OAc)$_2$/PPh$_3$, aryl iodides and benzylacetylenes were transformed into furanones in 33–88% yields (Scheme 10.21). Later, Beller's group proved that this transformation can be done at room temperature under 1 bar of CO.[29b] They also extended their methodology to aryl bromides and benzyl chlorides.[29c,d] Palladium-catalysed carbonylative Sonogashira coupling reactions of iodobenzene and 2-methyl-3-butyn-2-ol under biphasic conditions yielding furanones were also reported by Kiji *et al.*[29e]

In 1991, Ortar and co-workers published a general procedure for carbonylative Sonogashira couplings of vinyl triflates with terminal acetylenes.[30] Various alkynyl ketones were produced in moderate to good yields (Scheme 10.22). However, this methodology failed in the case of activated alkynes or aryl triflates.

The catalytic ability of dimeric palladium hydroxide in carbonylative Sonogashira coupling was demonstrated by Alper and co-workers in 1994.[31] In this report, terminal alkynes and alkynols were coupled with aryl iodides in the presence of carbon monoxide with moderate to good yields (Scheme 10.23).

In 1995, Cacchi and co-workers presented a general methodology for the synthesis of 5-(2-acylethynyl)-3′,5′-di-*O*-acetyl-2′-deoxyuridines.[32] In the

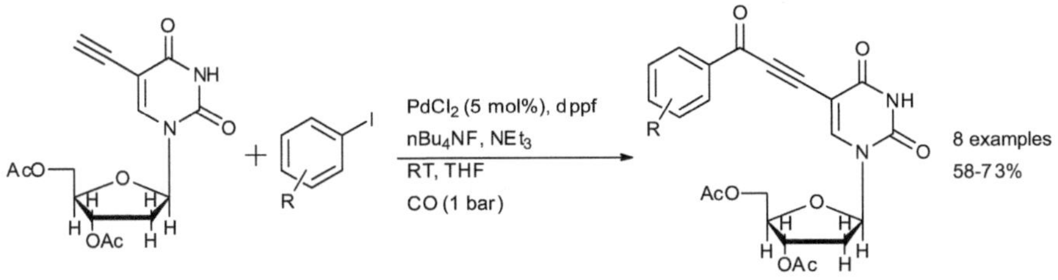

PdCl2 (5 mol%), dppf
nBu4NF, NEt3
RT, THF
CO (1 bar)
8 examples
58-73%

Scheme 10.24 Pd-catalysed carbonylative Sonogashira coupling of aryl iodides.

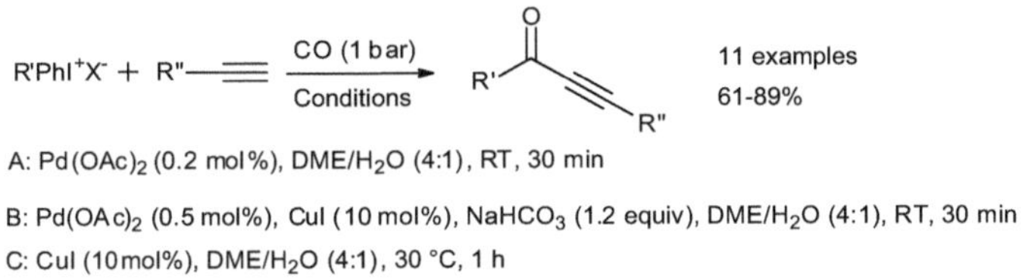

CO (1 bar)
Conditions
11 examples
61-89%
A: Pd(OAc)2 (0.2 mol%), DME/H2O (4:1), RT, 30 min
B: Pd(OAc)2 (0.5 mol%), CuI (10 mol%), NaHCO3 (1.2 equiv), DME/H2O (4:1), RT, 30 min
C: CuI (10mol%), DME/H2O (4:1), 30 °C, 1 h

Scheme 10.25 Pd-catalysed carbonylative Sonogashira coupling of iodonium salts with and without Cu co-catalysts.

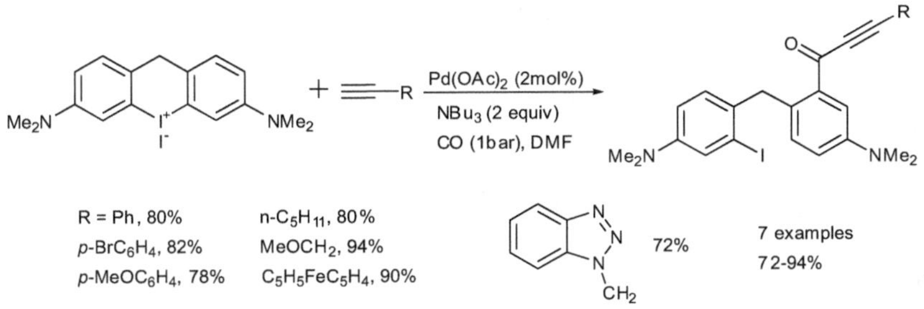

Pd(OAc)2 (2mol%)
NBu3 (2 equiv)
CO (1bar), DMF
R = Ph, 80%
p-BrC6H4, 82%
p-MeOC6H4, 78%
n-C5H11, 80%
MeOCH2, 94%
C5H5FeC5H4, 90%
72%
7 examples
72-94%

Scheme 10.26 Pd-catalysed carbonylative Sonogashira coupling of iodinium iodide.

presence of a palladium catalyst, the corresponding alkynones were synthesized from aryl iodides and alkynes (Scheme 10.24).

The carbonylative Sonogashira reaction of iodonium salts with terminal alkynes was described by Kang *et al.*[33] Both palladium/copper and palladium catalyst systems alone could be used and various alkynones were synthesized in moderate to good yields in aqueous media (Scheme 10.25).

Another example of carbonylative Sonogashira coupling reactions with iodinium iodide and 1-alkynes was published by Ma and co-workers in 2001.[34] Under mild conditions, iodine-substituted alkynones were produced in good yields (Scheme 10.26). Both aromatic, aliphatic and heterocylic terminal acetylenes could be applied as substrates.

An interesting room temperature carbonylation using a palladium/copper catalyst system was published by Ahmed and Mori in 2003.[35a,b] As shown in Scheme 10.27, various aromatic alkynones were produced in moderate to

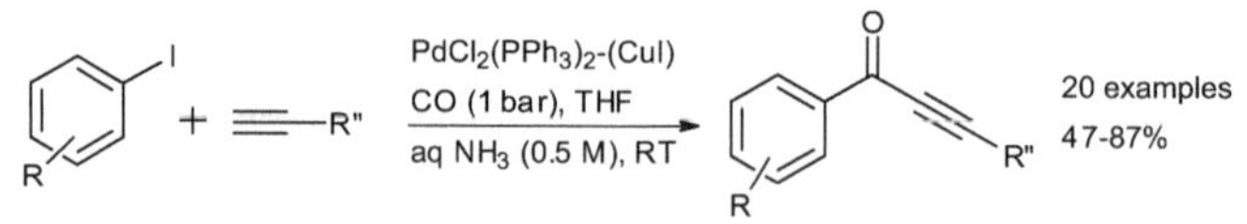

Scheme 10.27   Carbonylative room temperature Sonogashira reaction in aqueous ammonia.

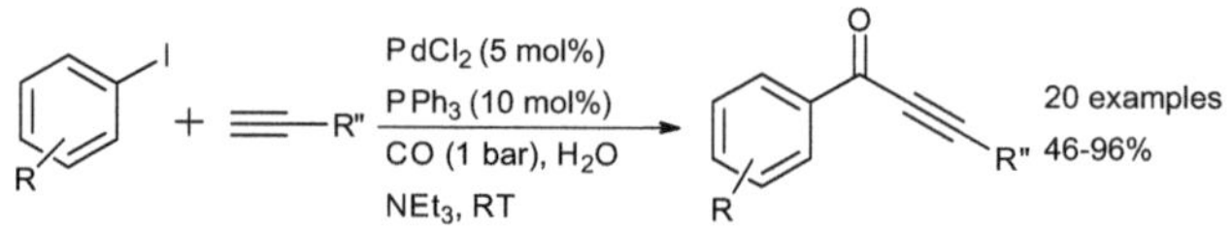

Scheme 10.28   Pd-catalysed carbonylative Sonogashira coupling in water.

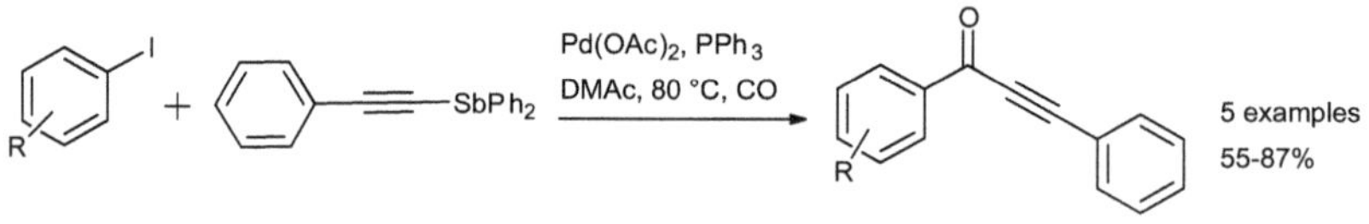

Scheme 10.29   Pd-catalysed carbonylative Sonogashira coupling of an ethynylstibane with aryl iodides.

good yields using aqueous ammonia as base. Surprisingly, no competitive amination reaction occurred. This methodology was later exploited by Bishop *et al.* to generate pyrazoles.[35c]

Water as a green solvent has been applied successfully as the reaction medium for palladium-catalysed carbonylative Sonogashira reactions (Scheme 10.28).[36] In addition to alkynes, activated acetylenic stibanes can be applied as coupling partners in carbonylations. An example is the palladium-catalysed carbonylative Sonogashira coupling of alkynylstibanes with aryl iodides published by Kakusawa and Kurita in 2006.[37] The reaction was carried out under 1 bar of CO in DMAc using 5 mol% of Pd(OAc)$_2$ and 20 mol% of PPh$_3$. Alkynones were obtained in good yields along with a small amount of non-carbonylative coupling products (Scheme 10.29). However, this side reaction could be completely suppressed by increasing the CO pressure to 20 bar.

The use of ionic liquids and flow chemistry technologies are attracting increasing attention. Consequently, these novel tools were also successfully applied in carbonylative Sonogashira reactions by Ryu and co-workers.[38] Various alkynones were synthesized in moderate to good yields at low pressure of CO in *n*-butylmethylimidazolium hexafluorophosphate. The microreactor-based flow system was compared with typical batch conditions and higher yields could be achieved in the former case (Scheme 10.30).

In 2006, Chen and co-workers reported a convenient, effective method for the carbonylative Sonogashira coupling of aryl iodides with ethynylferrocene under 1 atm of CO.[39] Various aryl ferrocenylethynyl ketones were

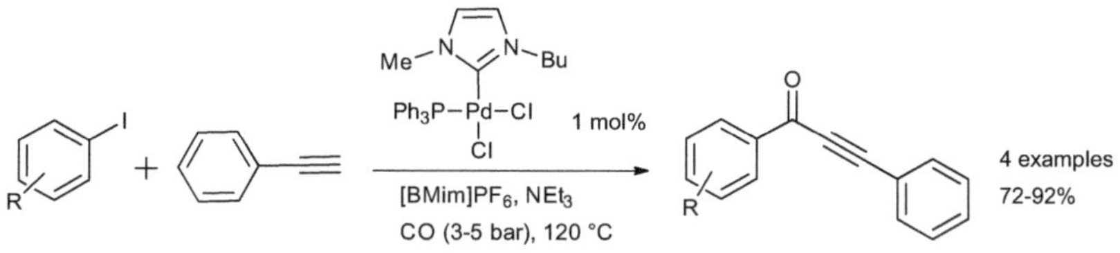

Scheme 10.30 Pd-catalysed carbonylative Sonogashira reaction in ionic liquids.

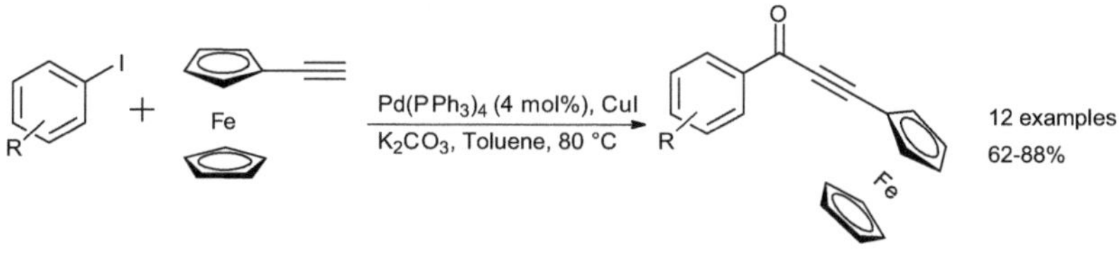

Scheme 10.31 Pd-catalysed carbonylative Sonogashira coupling of ethynylferrocene with aryl iodides.

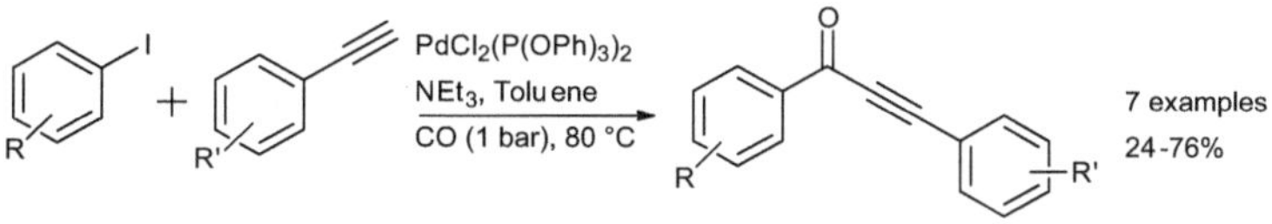

Scheme 10.32 PdCl$_2$[P(OPh)$_3$]$_2$-catalysed carbonylative Sonogashira coupling of aryl iodides with alkynes.

synthesized in 62–88% yield (Scheme 10.31). Unexpectedly, strongly activated aryl iodides (4-Ac, 4-NO$_2$) and iodopyridine gave no desired carbonylation product. However, this methodology was also applied for a two-step synthesis of ferrocenylpyrazole and pyrimidine derivatives by Skoda-Foldes and co-workers.[40] In 2009, Chen's group also reported another protocol for the synthesis of ferrocenylethynyl ketones in water.[41]

The use of phosphites, *e.g.* P(OPh)$_3$, as ligands in palladium-catalysed carbonylative Sonogashira coupling reactions was first reported by Trzeciak and co-workers.[42] Using the defined complex PdCl$_2$[P(OPh)$_3$]$_2$ as catalyst, alkynones were produced in low to moderate yields at 1 bar of CO (Scheme 10.32). When the reaction was conducted in an ionic liquid, the catalyst could be reused in four consecutive catalytic runs with high activity. Notably, benzyl bromide was reported as substrate for the first time, but 2 equiv. of the acetylene were required in this system.

Iizuka and Kondo presented a palladium-catalysed "CO-free" method for the synthesis of alkynones, which applied stoichiometric amounts of Mo(CO)$_6$ as CO source.[43] The reaction was carried out at room temperature and P*t*Bu$_3$ was found to be an essential ligand under these conditions. When strongly electron-withdrawing substituted aryl iodides were employed as substrates in this protocol, the corresponding alkynones were produced in

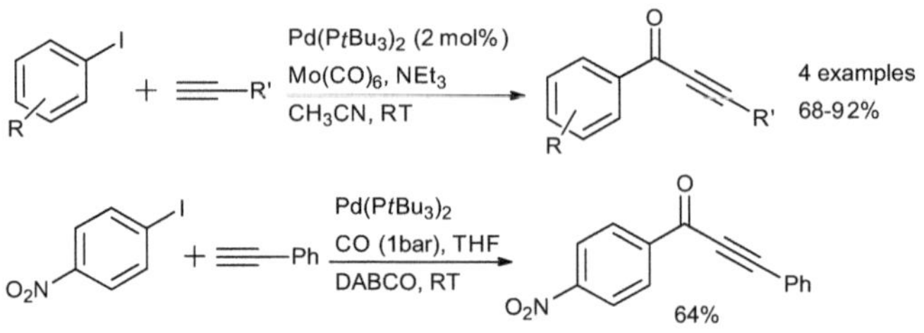

Scheme 10.33 Pd-catalysed carbonylative Sonogashira coupling reaction with Mo(CO)$_6$ as CO source.

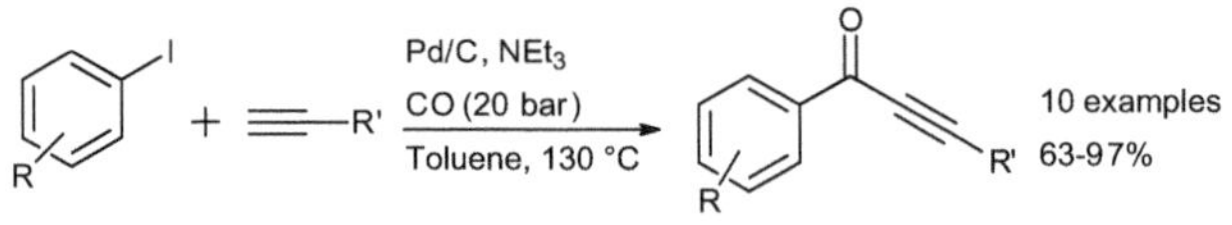

Scheme 10.34 Pd/C-catalysed carbonylative Sonogashira reaction of aryl iodides.

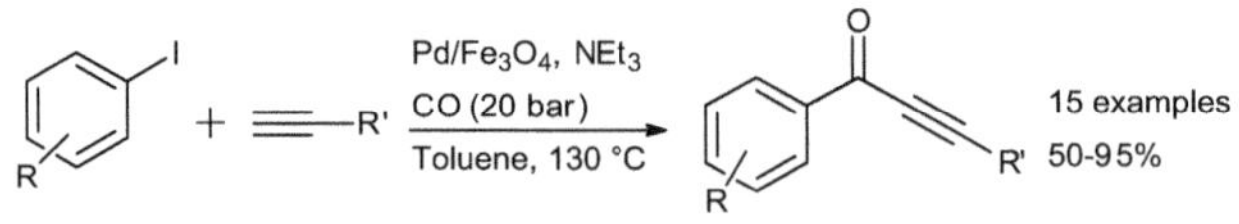

Scheme 10.35 Pd/Fe$_3$O$_4$-catalysed carbonylative Sonogashira coupling of aryl iodides.

good to excellent yields (Scheme 10.33). Again, a one-pot synthesis of pyrazoles *via* condensation of corresponding alkynones with hydrazine was also conducted and the corresponding products were obtained in good yields at room temperature.

In 2008, Xia and co-workers reported a recyclable phosphine-free catalyst system for alkynone synthesis.[44] Using palladium on charcoal (Pd/C) and NEt$_3$, the carbonylative Sonogashira coupling of aryl iodides with alkynes was carried out smoothly and the desired products were isolated in moderate to excellent yields (Scheme 10.34).

Later on, the same group presented an unusual variation of the palladium-catalysed carbonylative Sonogashira coupling reaction.[45] Here, a magnetically separable palladium catalyst was synthesized by combining palladium nanoparticles and superparamagnetic Fe$_3$O$_4$ nanoparticles in KBH$_4$ solution. This catalyst proved to be efficient for the carbonylation reaction of aryl iodides with alkynes under phosphine-free conditions. Because of the magnetic behaviour of Fe$_3$O$_4$, the catalyst could be reused with sustained selectivity and activity. Various alkynones were synthesized in good to excellent yields (Scheme 10.35).

Another approach applying a heterogeneous palladium catalyst was reported by Cai and co-workers, who described an MCM-41-supported bidentate phosphine–palladium complex [MCM-41-2p-Pd(0)] as a

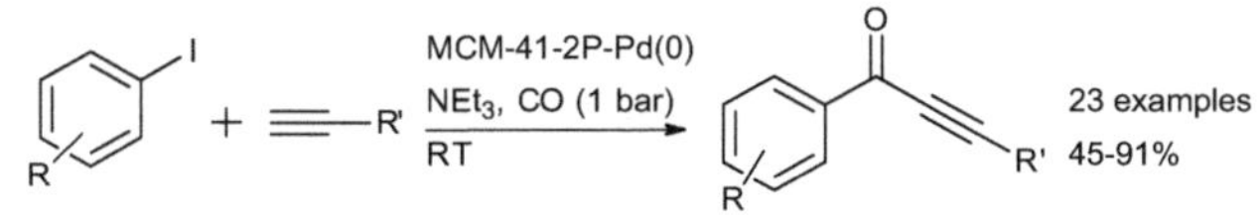

Scheme 10.36 [MCM-41-2p-Pd(0)]-catalysed carbonylative Sonogashira coupling reaction of aryl iodides.

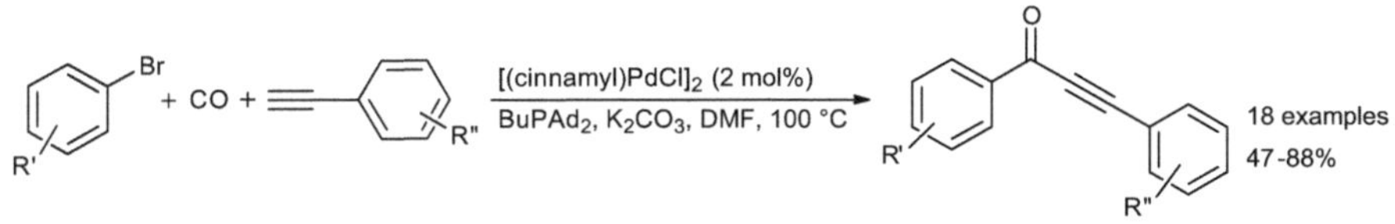

Scheme 10.37 Pd-catalysed carbonylative Sonogashira coupling of aryl bromides.

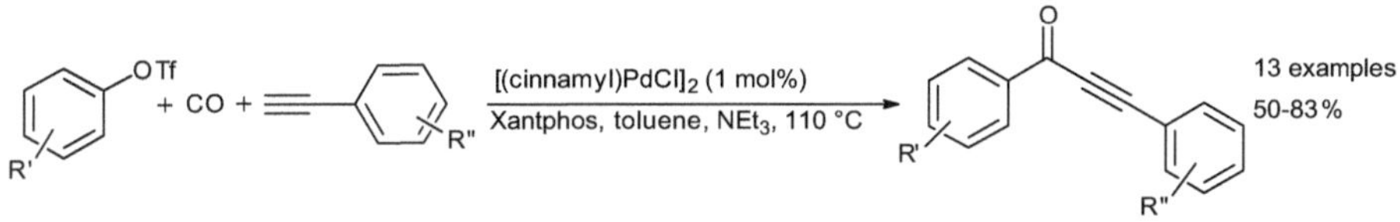

Scheme 10.38 Pd-catalysed carbonylative Sonogashira coupling of aryl triflates.

polymer-supported palladium catalyst.[46] Terminal alkynes were converted with aryl iodides under 1 bar of CO to give alkynones in good to high yields (Scheme 10.36). It is noteworthy that the use of a polymer as support in a Sonogashira coupling reaction was earlier reported by Takahashi and co-workers in 2008, where the products could be released from the polymer by adding acid.[47]

So far, basically all methodological developments in this area were focused on the use of expensive and easy-to-activate aryl iodides. It was therefore interesting that in 2010 Beller and co-workers reported a general and convenient palladium-catalysed carbonylative Sonogashira coupling of aryl bromides.[48] Key to success was the application of BuPAd$_2$ as ligand in the presence of K$_2$CO$_3$. Alkynones were generated in moderate to good yields from the corresponding aryl bromides and terminal alkynes (Scheme 10.37). The one-pot synthesis of isoxazolines and pyrazoles was also successful.

Since aryl triflates can be easily generated from the corresponding phenols, in 2011 Beller's group developed a palladium-catalysed carbonylative Sonogashira coupling of aryl triflates.[49a] This is the first carbonylative Sonogashira protocol to apply aryl triflates as substrates. Various alkynones were produced in moderate to good yields under a low pressure of CO (Scheme 10.38). A one-pot synthesis of enaminones was also achieved by running the reaction in the presence of primary amines. As arylamines are abundantly available and relatively inexpensive, the authors also succeeded in using amine as a precursor of electrophiles.[49b] In the presence of *t*BuONO

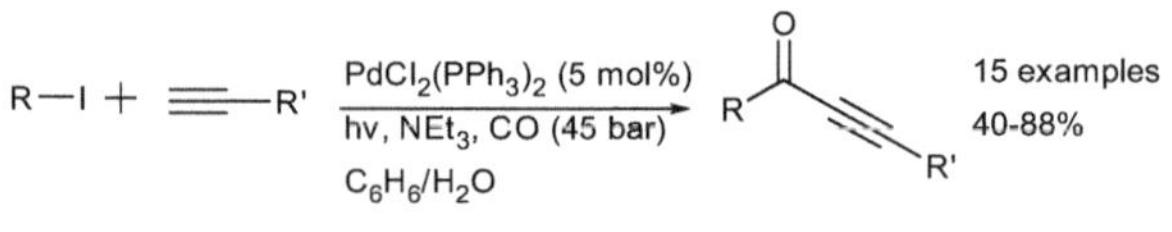

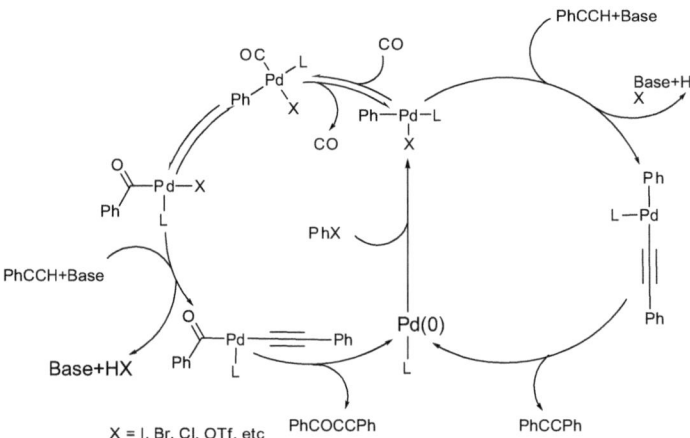

Scheme 10.39 Pd-catalysed carbonylative Sonogashira coupling of iodoalkanes using xenon light.

Scheme 10.40 Proposed reaction mechanism for the carbonylative Sonogashira reaction (left) *versus* non-carbonylative Sonogashira reaction (right).

as oxidant and tri(2-furyl)phosphine as ligand, various arylamines were coupled with terminal alkynes and gave the corresponding alkynones in moderate to good yields.

Ryu and co-workers reported the synthesis of alkyl alkynyl ketones *via* Pd/ light-induced carbonylative Sonogashira coupling of iodoalkanes with terminal alkynes.[50] Using xenon light, in the presence of a catalytic amount of PdCl$_2$(PPh$_3$)$_2$ and NEt$_3$, alkynones were produced in good yields (Scheme 10.39). This represents the first example of Sonogashira carbonylations of iodoalkanes (alkyl iodides).[51]

Despite all the synthetic developments to date, relatively little detailed mechanistic work has been performed on Sonogashira carbonylations. The generally accepted mechanism is shown in Scheme 10.40. The typical reaction starts with the oxidative addition of ArX to a palladium(0) complex to form an arylpalladium(II) intermediate. Subsequent insertion of CO leads to the respective acyl–palladium complex. Transmetallation and finally reductive elimination release the product and a new catalytic cycle can be started. Notably, all species passing through the cycle are believed to be in a reversible equilibrium.

Intramolecular Sonogashira carbonylations offer various possibilities for the preparation of interesting heterocycles. Typically, in these reactions 2-halophenols and 2-haloanilines or their derivatives are employed with

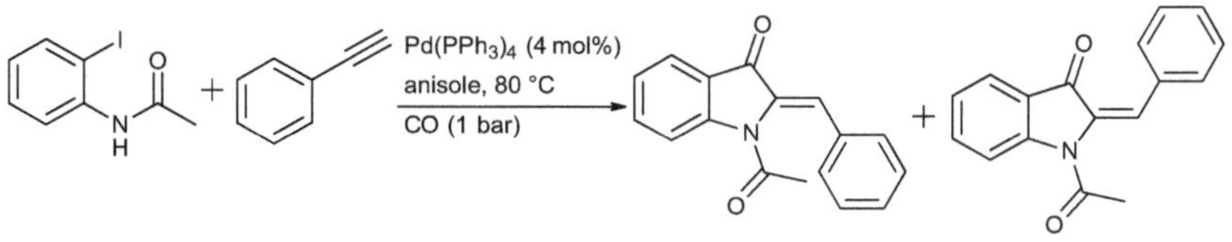

Scheme 10.41 Pd-catalysed carbonylative synthesis of indoxyls.

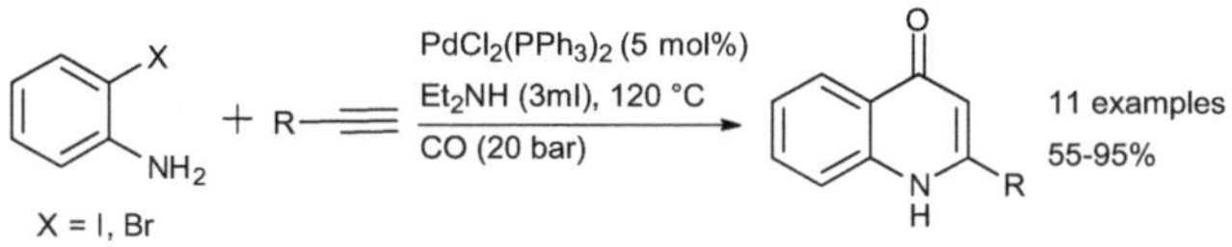

Scheme 10.42 Pd-catalysed carbonylative synthesis of quinolines.

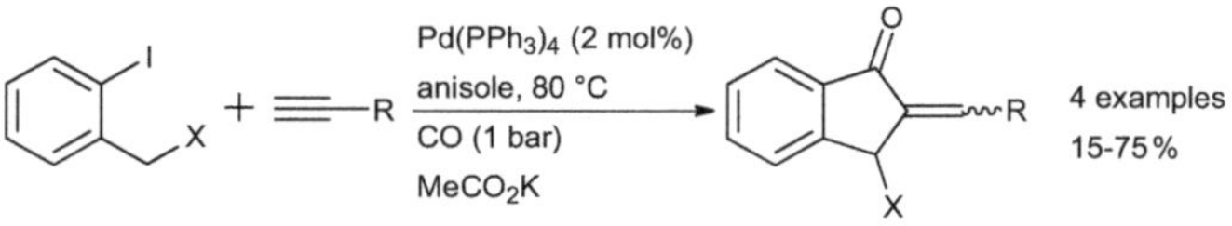

Scheme 10.43 Pd-catalysed carbonylative synthesis of indenones.

terminal alkynes. As early as 1990, Chiusoli and co-workers reported the palladium-catalysed synthesis of indoxyl derivatives (Scheme 10.41).[52]

Shortly afterwards, Torii and co-workers reported a novel methodology for the synthesis of quinolines.[53a,b] Quinolines were produced in good yields *via* palladium-catalysed carbonylation of 2-haloaniline with terminal alkynes in the presence of CO (Scheme 10.42).

When the amino group of the 2-haloaniline substrate is primary, the cyclization proceeds without problems, but when using alkylated anilines under the same conditions, the yield of the corresponding cyclization product decreases dramatically. A similar methodology was reported in 1992 by Kalinin and co-workers using PdCl$_2$(dppf) as palladium precursor.[53c] This cyclization was applied to synthesize the quinolone substructure of BILN 2061, a serine protease inhibitor.[53d] In the same year, Chiusoli and co-workers published an interesting methodology for indenone synthesis.[54] Sequential oxidative addition of *o*-alkoxycarbonylmethylene or alkylamido-methylene-substituted aryl iodides, CO insertion, reductive coupling with terminal alkynes, nucleophilic attack by the activated methylene group and protonation with metal elimination afforded the indenones in high yields in a one-pot process (Scheme 10.43).

In 2000, Miao and Yang reported a novel method for the preparation of flavones.[55] Various flavones are easily synthesized *via* palladium-catalysed carbonylative annulation of iodophenol acetates with terminal acetylenes in high yields (Scheme 10.44). This novel reaction provides the possibility for a combinatorial synthesis of flavones on solid supports.

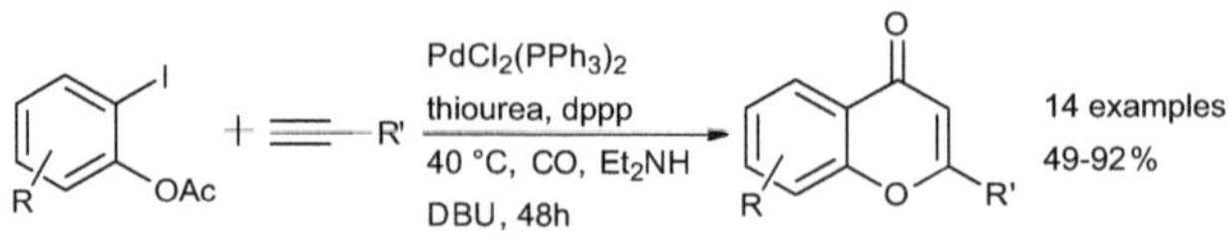

Scheme 10.44   Pd-catalysed carbonylative synthesis of flavones.

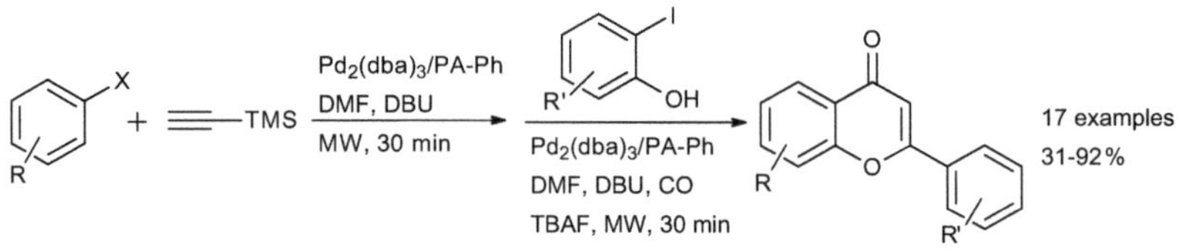

Scheme 10.45   Pd-catalysed carbonylative synthesis of flavones.

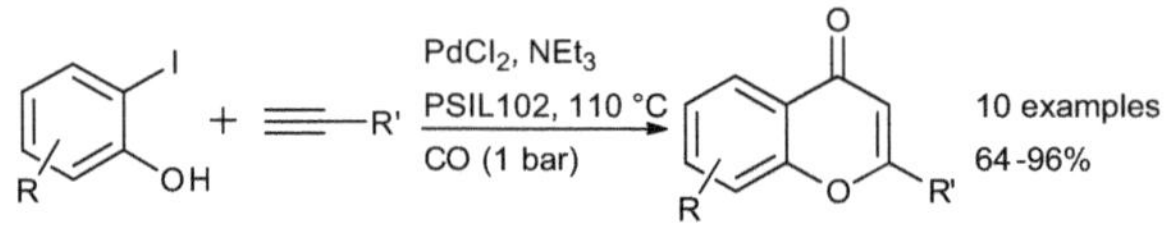

Scheme 10.46   Pd-catalysed carbonylative synthesis of flavones in PSIL102.

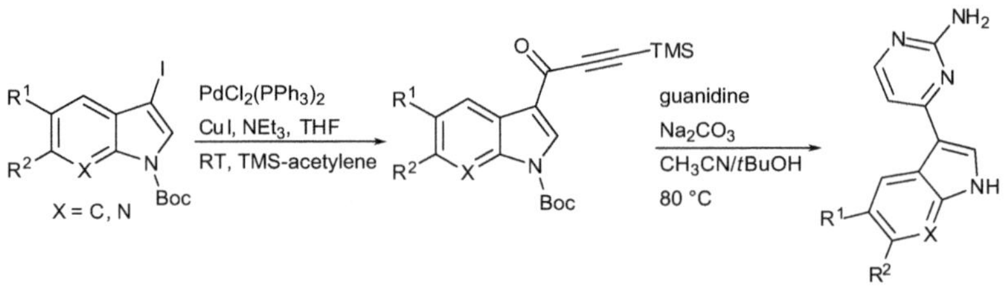

Scheme 10.47   Pd-catalysed carbonylative synthesis of meridianins.

More recently, Awuah and Capretta reported a microwave-assisted, one-pot palladium-catalysed carbonylative Sonogashira annulation reaction.[56] Various flavones were produced in moderate to good yields (Scheme 10.45). Yang and Alper reported another example of carbonylations of *o*-iodophenols with terminal acetylenes to obtain flavones.[57] Their reaction proceeded under 1 bar of CO in ionic liquids based on a phosphonium salt [PSIL102, $C_{14}H_{29}(C_6H_{13})_3)P^+Br^-$] (Scheme 10.46). It should be noted that by using PSIL102 as an ionic liquid no phosphine ligand was required.

Elegant synthetic applications of carbonylative Sonogashira reactions were reported by Müller's group. For example, in 2005 they succeeded in palladium-catalysed one-pot four-component carbonylations for the synthesis of meridianins,[58] which are natural and bioactive compounds (Scheme 10.47).

10.4   Carbonylative Suzuki Reactions

The palladium-catalysed cross-coupling of organohalides with organoboranes is known as the Suzuki reaction and is one of the most popular coupling reactions. When a Suzuki reaction is performed under an atmosphere of CO, one carbonyl group will be inserted into the two coupling partners (carbonylative Suzuki reaction).

As early as 1986, Kojima and co-workers reported the carbonylative coupling of aryl iodides or benzyl halides with organoboranes in the presence of a catalytic amount of palladium catalyst.[59] This was the first application of organoboranes in carbonylative coupling reactions mediated by 1.1 equiv. of Zn(acac)$_2$ to favour the transmetallation. Various ketones were produced in good yields starting from aryl iodides and benzyl chloride (Scheme 10.48).

Later, in 1991, Suzuki and co-workers developed another methodology for the carbonylative coupling of vinyl halides with organoboranes using Pd(PPh$_3$)$_4$ and K$_3$PO$_4$ as a base to synthesize vinyl ketones in moderate to excellent yields[60] (Scheme 10.49). However, for iodoalkenes bearing electron-withdrawing substituents, the chemoselectivity decreased and a mixture of carbonylated and non-carbonylated products was observed. In the same year, this methodology was extended to iodoalkanes with the assistance of light.[61]

In 1993, Suzuki and co-workers reported the palladium-catalysed carbonylative coupling of aryl iodides with aryl boronic acids.[62a] Various diaryl ketones were produced in high yields (Scheme 10.50). The correct choice of base and solvent was essential to obtain the desired ketones without biaryl by-products. The coupling of benzyl bromide was also described. In 1998, the same group extended this methodology to aryl bromides and triflates.[62b] In the case of aryl bromides, NaI or KI was required as an additive.

Ishikura and Terashima described an application of the carbonylative coupling of organohalides or vinyl triflates in a one-pot procedure using indolyl borates as coupling partner to achieve indol-2-yl ketones in good

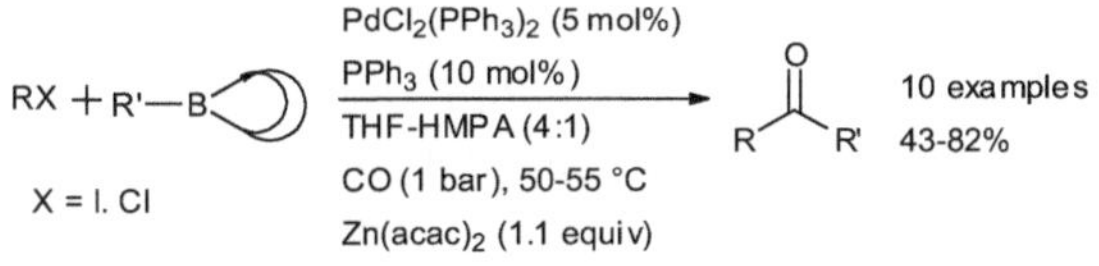

Scheme 10.48   First Pd-catalysed carbonylative coupling of organoboranes.

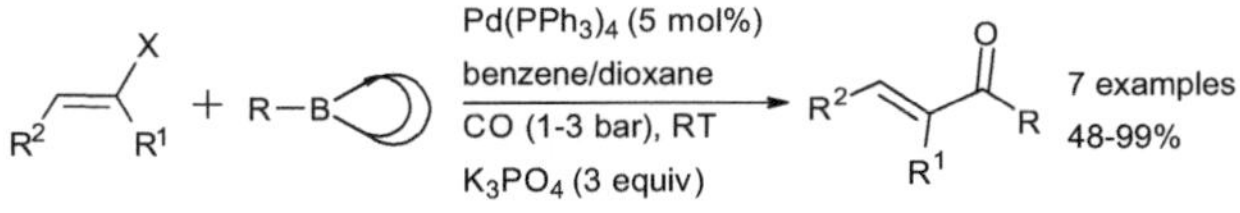

Scheme 10.49   Pd-catalysed carbonylative Suzuki coupling of organoboranes with vinyl halides.

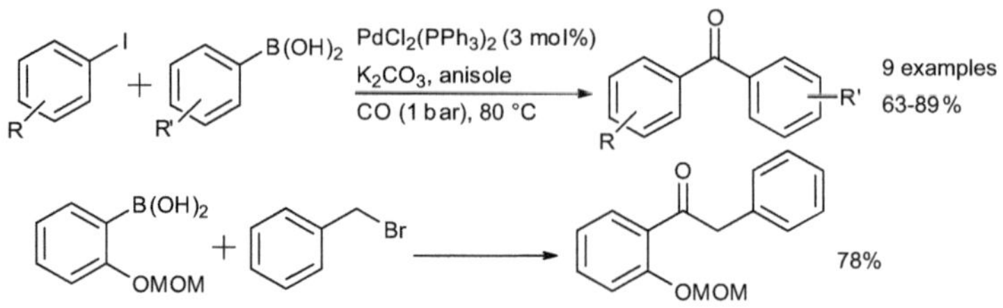

Scheme 10.50   Pd-catalysed carbonylative Suzuki coupling of aryl iodides with arylboronic acids.

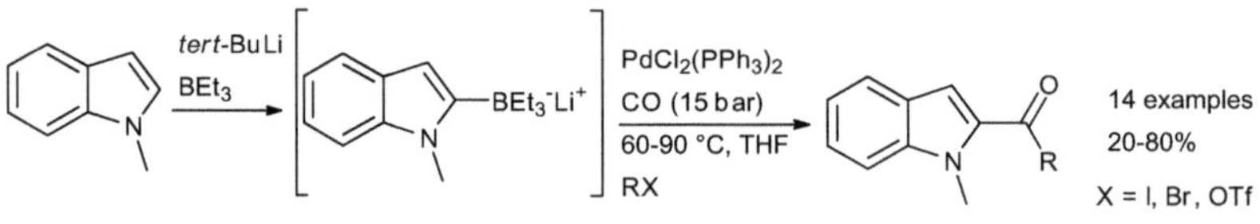

Scheme 10.51   Pd-catalysed synthesis of indol-2-yl ketones.

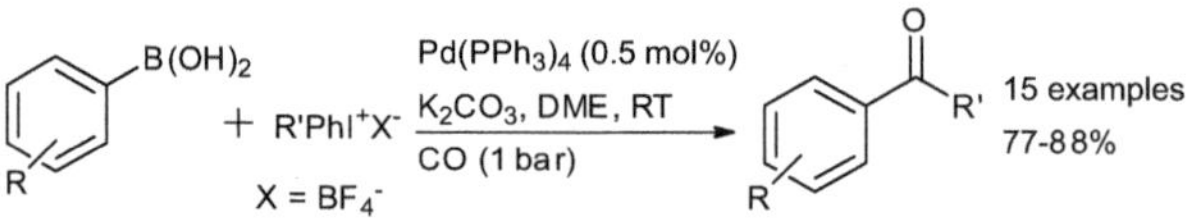

Scheme 10.52   Pd-catalysed carbonylative Suzuki coupling of arylboronic acids with hypervalent iodonium salts.

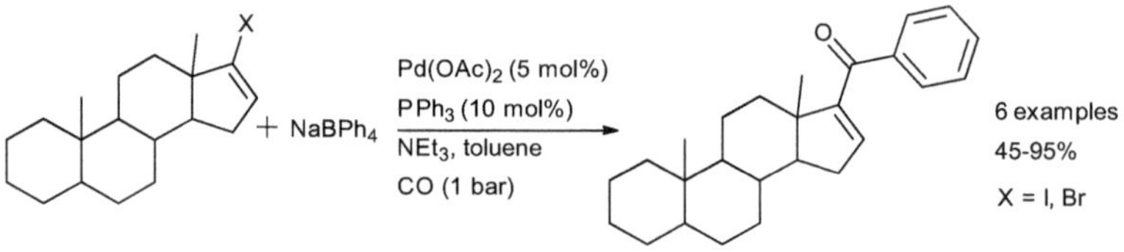

Scheme 10.53   Pd-catalysed carbonylative synthesis of steroidal ketones.

yield (Scheme 10.51).[63] In 1998, Kang *et al.* used hypervalent iodonium salts instead of aryl halides to couple organoboronic acids in the presence of CO to obtain unsymmetric aromatic ketones.[64] Moderate yields were achieved in the carbonylation of arylboronic acids with aryl-, alkenyl- and alkynyliodonium salts at room temperature under 1 bar of CO (Scheme 10.52). Starting from aryl trifluoroborates Xia and Chen were able to extend this methodology using Pd(OAc)$_2$ as precursor.[65]

Steroidal phenyl ketones were synthesized by Skoda-Foldes *et al. via* a related carbonylation pathway.[66] The ketones were produced in high yields by carbonylation of 17-iodo-androst-16-ene derivatives in the presence of NaBPh$_4$ (Scheme 10.53). Alkenyl bromides or enol triflates gave lower yields under the same reaction conditions.

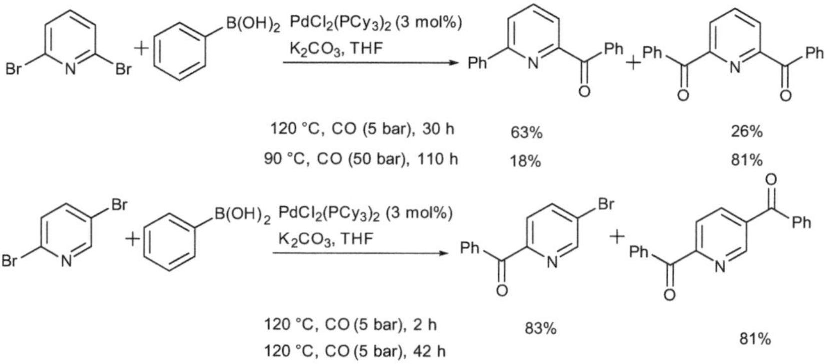

Scheme 10.54   Pd-catalysed Suzuki carbonylation of halopyridines.

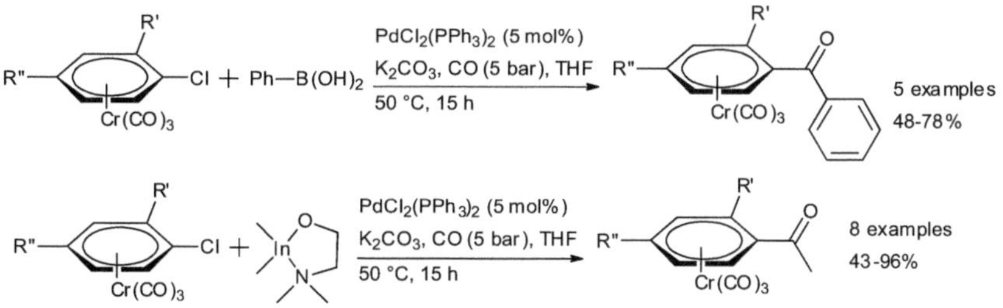

Scheme 10.55   Pd-catalysed   carbonylative   coupling   of   chloroarene–Cr(CO)₃ complexes.

In 2001, Castanet and co-workers demonstrated that in a palladium-catalysed carbonylative Suzuki reaction, pyridine halides react with aryl-boronic acids to give 2-pyridyl ketones in good yields (81–95%) (Scheme 10.54).[67a] The proper choice of solvent, catalyst precursor and CO pressure permitted the selective transformation of mono- and dihalopyridines. Later, they extended this methodology to pyridine chlorides by applying an NHC ligand and Cs₂CO₃ as base.[67b]

Schmalz and co-workers investigated the carbonylative Suzuki reaction of chloroarene–Cr(CO)₃ complexes with phenylboronic acid.[68] Using PdCl₂(PPh₃)₂ as catalyst precursor, benzophenone derivatives were obtained in good yields (Scheme 10.55). Palladium-catalysed carbonylative cross-methylation reactions of different chloroarene–Cr(CO)₃ complexes mediated by a stabilized dimethylindium(III) reagent were also described.

The palladium-catalysed carbonylative coupling of aryldiazonium ions with arylboronic acids was published in 2002.[69] Various aryl ketones were produced in moderate to high yields under mild conditions (Scheme 10.56). One benefit of applying aryldiazonium ions as electrophiles is the potential to perform coupling reactions under base-free conditions. Both electron-donating and electron-withdrawing substituents on the aryldiazonium did

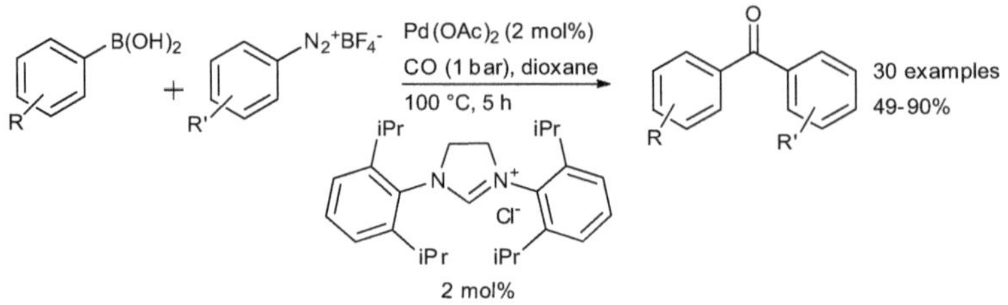

Scheme 10.56 Pd-catalysed carbonylative coupling of aryldiazonium salts.

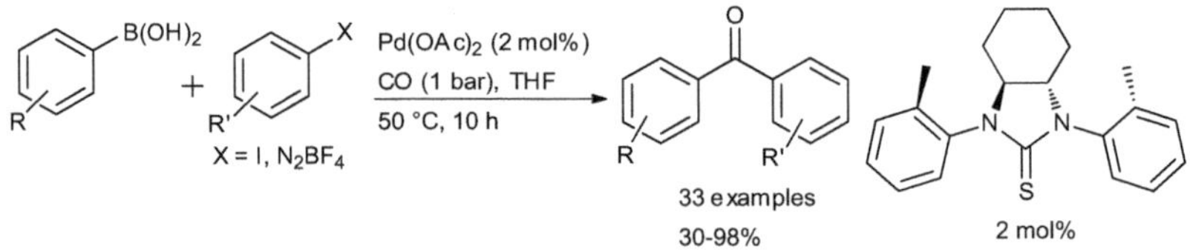

Scheme 10.57 A thiourea-based ligand in carbonylative Suzuki reactions.

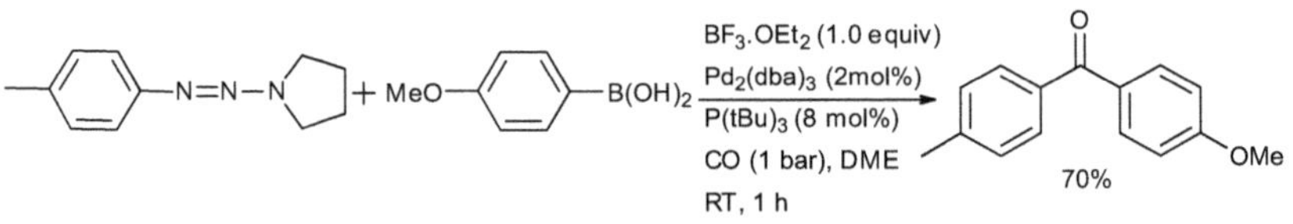

Scheme 10.58 Pd-catalysed carbonylative coupling of a 1-aryltriazene.

not reduce the product yields and also *ortho*-substituted arylboronic acids resulted in good yields. In 2004, Yang and co-workers described a thiourea-based ligand for the palladium-catalysed carbonylative coupling of aryl iodides or aryldiazonium salts with arylboronic acids.[70] This sterically bulky thiourea ligand was successfully applied under aerobic conditions and ketones were synthesized in good to excellent yields (Scheme 10.57).

Tamao and co-workers reported an example of the carbonylative coupling of a 1-aryltriazene with a boronic acid in 2004.[71] The corresponding diaryl ketone was produced in 70% yield under 1 bar of CO, in the presence of a catalytic amount of Pd$_2$(dba)$_3$ and P(*t*Bu)$_3$ (Scheme 10.58).

Långström and co-workers reported the synthesis of [11]C- and ^{13}C-labelled ketones by palladium-catalysed carbonylative Suzuki coupling.[72] Aryl triflates with methyl- or arylboronic acids and a low concentration of [11]CO were employed in the synthesis of ^{11}C-labelled ketones using Pd(PPh$_3$)$_4$ as catalyst. The ^{11}C-labelled products were obtained with decay-corrected radiochemical yields in the range 10–70% after a reaction time of 5 min at 150 °C.

The first report of the synthesis of β-keto sulfoxides using a palladium-catalysed carbonylative Suzuki reaction of an α-bromosulfoxide with arylboronic acids appeared in 2005.[73] Asensio and co-workers reported the synthesis of 12 β-keto sulfoxides in moderate to excellent yields using

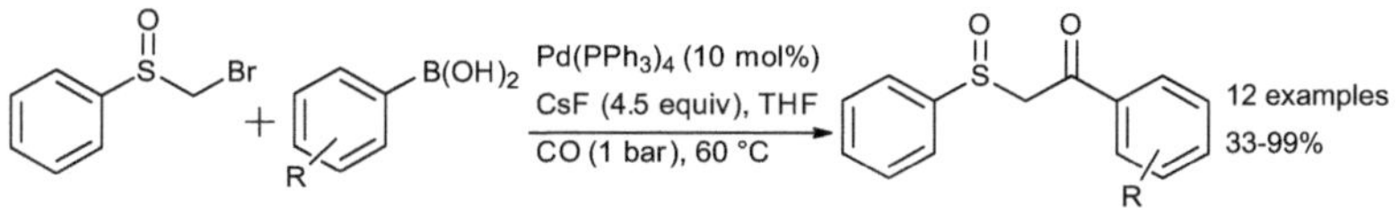

Scheme 10.59 Pd-catalysed synthesis of β-keto sulfoxides.

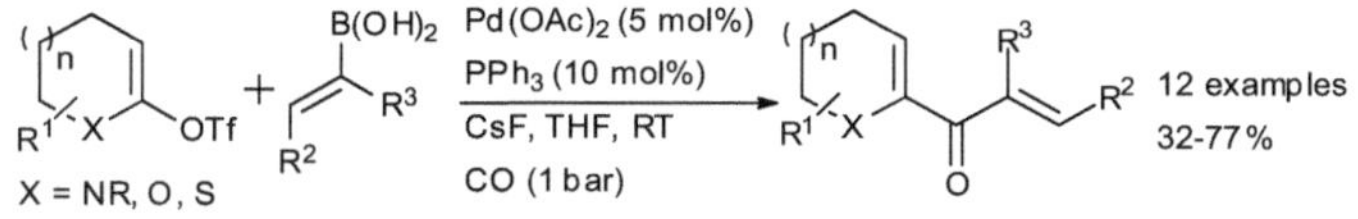

Scheme 10.60 Pd-catalysed carbonylative coupling of enol triflates.

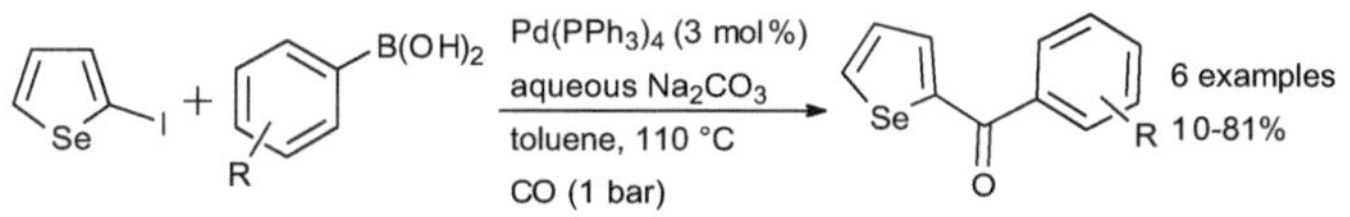

Scheme 10.61 Pd-catalysed carbonylative coupling of 2-iodoselenophenes.

Pd(PPh$_3$)$_4$ and CsF as base in the presence of 1 bar of CO (Scheme 10.59). The carbonylative coupling reaction is strongly favoured over the competing non-carbonylation and homocoupling processes. However, in case of boronic acids carrying strongly electron-withdrawing substituents, the chemoselectivity decreased and by-products were observed.

Occhiato and co-workers described the carbonylative coupling of lactam-, lactone- and thiolactone-derived enol triflates with boronic acids.[74] Several unsymmetrical dienones were formed in moderate to good yields at room temperature under 1 bar of CO with 1–5 mol% of palladium catalyst (Scheme 10.60). This methodology allows for a convergent and rapid preparation of substrates, which are useful in conjugate additions and Nazarov reactions.

Kollar and co-workers discussed the carbonylative coupling of 1-iodocyclohexene with arylboronic acids.[75] Among all the possible products, alkenyl ketones were synthesized selectively under appropriate conditions. The yields were strongly dependent on both the reaction conditions and the type of arylboronic acid. Zeni and co-workers reported the coupling of 2-iodoselenophenes with arylboronic acids and CO in 2006 (Scheme 10.61).[76] Interestingly, the reaction proceeded with aqueous Na$_2$CO$_3$ as base under 1 bar of CO. Concerning the type of arylboronic acids, those substituted with strong electron-withdrawing groups and *ortho*-substituted arylboronic acids gave low or no yields.

Palladium-catalysed carbonylative couplings of iodoferrocene with arylboronic acids were reported by Yang and co-workers.[77] A series of aryl ferrocenyl ketones were prepared in good yields under mild conditions (Scheme 10.62). Both strongly activated and *ortho*-substituted arylboronic acids gave the corresponding ketones in moderate yields.

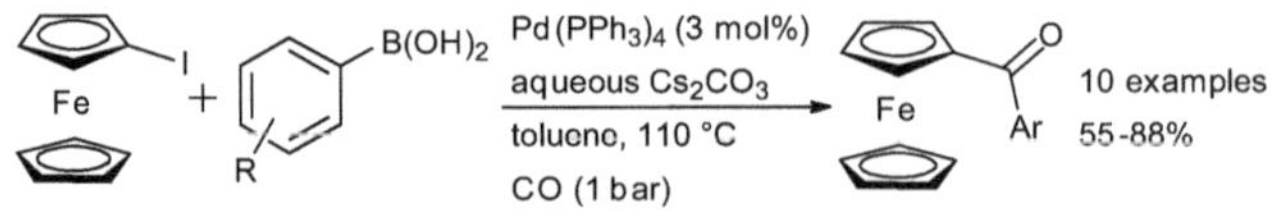

Scheme 10.62 Pd-catalysed carbonylative coupling of iodoferrocene.

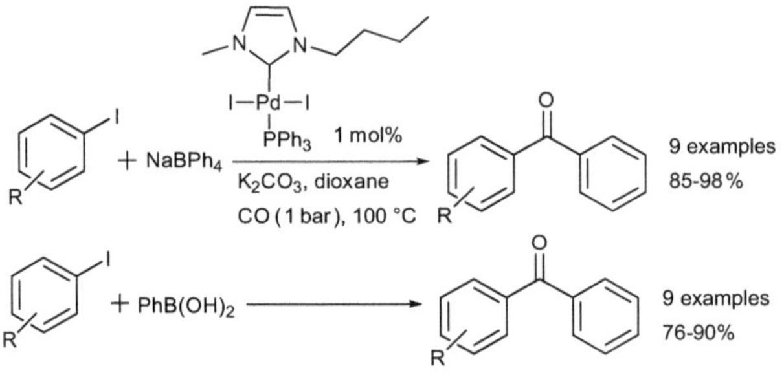

Scheme 10.63 NHC–Pd-catalysed carbonylative coupling of aryl iodides.

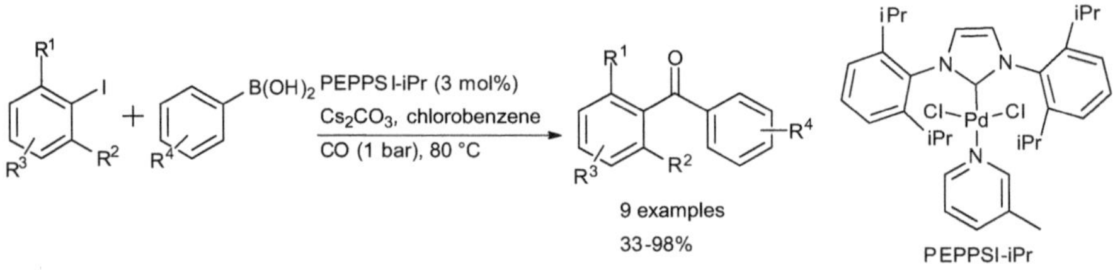

Scheme 10.64 Pd-catalysed carbonylative coupling of *ortho*-disubstituted aryl iodides.

NHC–palladium complexes as efficient catalysts for carbonylative Suzuki coupling of aryl iodides with organoboranes were reported by Xia and co-workers.[78] Aryl ketones were produced in high yields under mild conditions (Scheme 10.63). Interestingly, both $NaBPh_4$ and $ArB(OH)_2$ could be used as phenylating reagent.

Ortho-disubstituted aryl iodides, as representative examples of sterically hindered compounds, are more challenging substrates in palladium-cata-lysed coupling reactions. In this respect, it is interesting that Martin and co-workers reported a synthesis of sterically hindered aryl ketones by using the NHC–palladium complex PEPPSI-*i*Pr.[79] Several diaryl ketones were produced in moderate to good yields by this method (Scheme 10.64).

In 2008, Beller and co-workers developed a general method for diaryl ke-tone synthesis by palladium-catalysed carbonylative coupling of aryl brom-ides with arylboronic acids.[80] The combination of $Pd(OAc)_2$ and $BuPAd_2$ allowed the coupling of aryl/heteroaryl bromides with arylboronic acids to give a broad range of ketones in good yields (Scheme 10.65). With this catalyst in hand, they were able to synthesize suprofen, a non-steroidal anti-inflammatory drug.

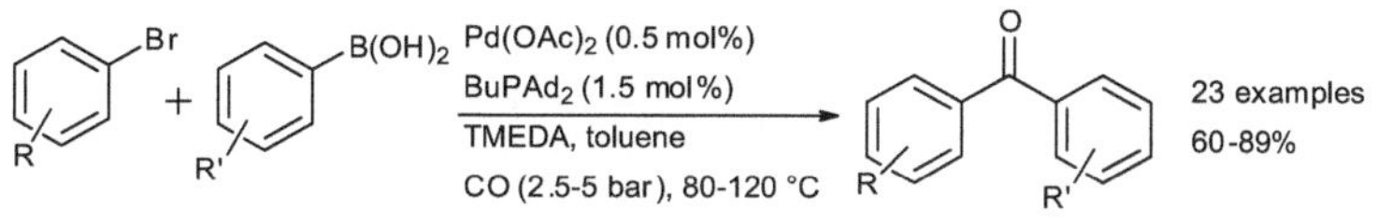

Scheme 10.65 Pd-catalysed carbonylative Suzuki reaction of aryl bromides.

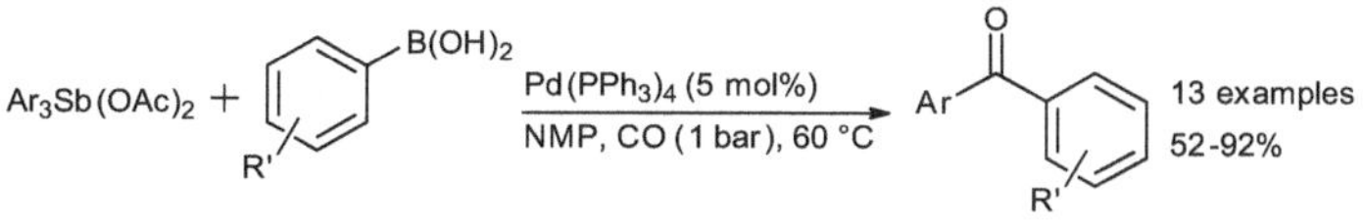

Scheme 10.66 Pd-catalysed carbonylative coupling of triarylantimony dicarboxylates.

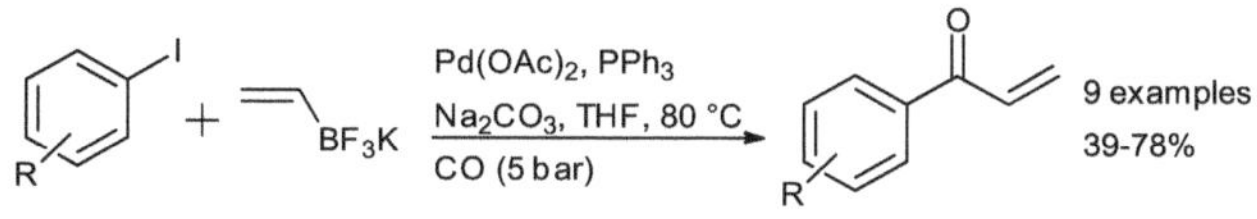

Scheme 10.67 Pd-catalysed carbonylative synthesis of aryl vinyl ketones.

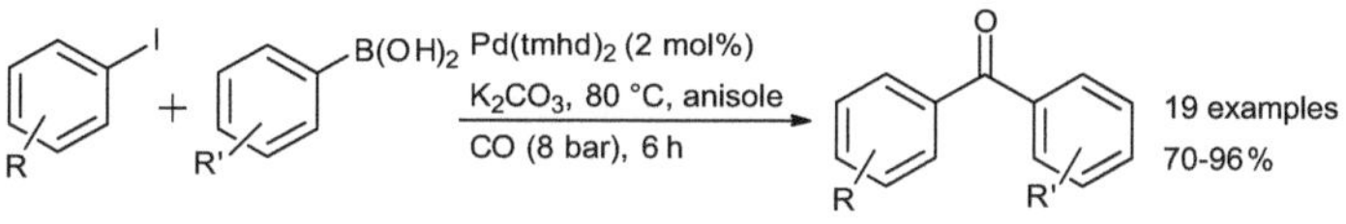

Scheme 10.68 Pd(tmhd)$_2$-catalysed carbonylative coupling of aryl iodides.

Kurita and co-workers described a carbonylative coupling of triarylantimony dicarboxylates with arylboronic acids (Scheme 10.66).[81] Remarkably, no base was needed in this system.

In 2009, Castanet and co-workers reported a useful protocol for aryl vinyl ketone synthesis.[82] The carbonylative cross-coupling of potassium vinyltrifluoroborate or 2,3,6-trivinylcycloboroxane with aryl iodides affords vinyl ketones in moderate yields (Scheme 10.67). This reaction is strongly influenced by the substituents: whereas activated and *ortho*-substituted aryl iodides gave low yields, good yields were achieved in the case of electron-donating decorated aryl iodides.

In 2009, Cai and co-workers reported the use of an MCM-41-supported bidentate phosphine–palladium(0) complex as catalyst for the carbonylative Suzuki coupling of aryl iodides with arylboronic acids.[83]

Pd(tmhd)$_2$ (thmd = 2,2,6,6-tetramethyl-3,5-heptanedionate) as another catalyst for carbonylative coupling of aryl iodides with arylboronic acids was reported by Bhanage and co-workers (Scheme 10.68).[84] The diketone (tmhd) can act as ligand in this methodology and no additional ligand is required.

Pontikis and co-workers reported a convenient one-pot procedure for the synthesis of 2-aroylindoles.[85] Using a domino palladium-catalysed CN coupling–carbonylation–CC coupling sequence, 2-aroylindoles were

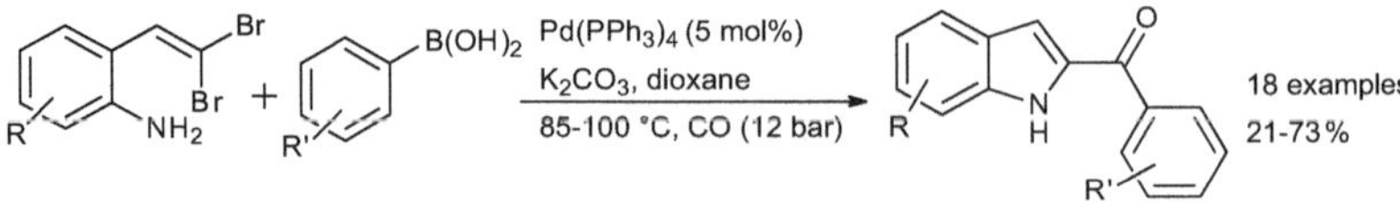

Scheme 10.69   Pd-catalysed synthesis of 2-aroylindoles.

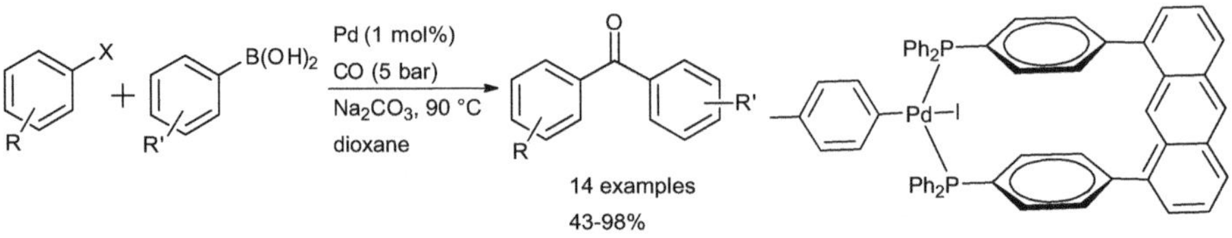

Scheme 10.70   Pd-catalysed carbonylative Suzuki coupling of aryl halides with arylboronic acids.

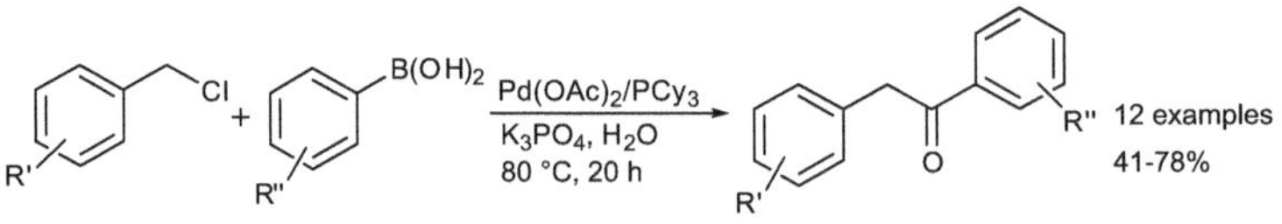

Scheme 10.71   Pd-catalysed carbonylative coupling of benzyl chlorides.

produced in good yields (Scheme 10.69). Since this reaction tolerates various functional groups, a practical route to a wide range of 2-aroylindoles is possible starting from 2-*gem*-dibromovinylanilines.

In 2010, Gelman and co-workers reported a new bidentate phosphine ligand for palladium-catalysed carbonylative Suzuki coupling reactions.[86] Aryl iodides and bromides were coupled with arylboronic acids in the presence of 0.01–1 mol% of catalyst. Ketones were produced with high selectivity in good yields (Scheme 10.70).

Recently, Beller's group reported a novel carbonylative coupling of benzyl chlorides with arylboronic acids.[87a] This was the first report on carbonylative Suzuki couplings of benzyl chlorides with arylboronic acids (Scheme 10.71). The reaction is carried out using a commerically available Pd(OAc)$_2$/PCy$_3$ catalyst in the presence of K$_2$CO$_3$ and water as solvent. Twelve ketones were synthesized in good yields. Later, this was extended to reactions with ArBF$_3$K.[87b]

10.5   Carbonylative C–H Activation Reactions

As noted in the Introduction, it would be advantageous to use C–H derivatives rather than the C–halogen substrates described thus far. The activation of 1-alkynes by Pd/Cu catalysts is an established protocol following the work of Sonogashira, as noted above. If the acyl palladium intermediate is trapped by a nucleophile other than an aryl group derived from an aryl halide, then a further range of derivatives can be obtained. Thus, for example, alcohols and amines can be used for the preparation of esters and amides. Tsuji *et al.* reported the preparation of esters as early as 1980.[88] The reaction proceeded readily at ambient temperature under balloon pressure of CO.

More recently, Gabriele *et al.* reported the synthesis of 2-ynamides.[89] Initial attempts to use the conditions described by Tsuji *et al.* were ineffective with amines, but under more forcing conditions reasonable product yields (50–75%) were obtained for a variety of alkyl- and arylacetylenes (Scheme 10.72). Effective reaction was limited to nucleophilic secondary amines, hindered amines being unreactive and primary amines yielding a complex mixture of products.

The Tsuji–Trost reaction, *i.e.* palladium-catalysed substitution at allylic carbon, has been widely applied in synthetic chemistry owing to the high level of control over chemo-, regio- and steroeselectivity.[90] Normally, some functionality at the allylic reaction site is required, so the extension of this methodology to C–H compounds is particularly useful. Progress in this direction was reported by Jiang and co-workers, where alkenyl carboxylate esters were obtained by oxidative alkoxycarbonylation (Scheme 10.73).[91] Benzoquinone was the preferred oxidant with an additive, 2,3-dichloro-5,6-dicyanobenzoquinone (DDQ), required to limit the second carboxylation to the diester.

The availability of a carboxylate directing group in α,β-unsaturated carboxylic acids also allowed the formation of *cis*-1,2-dicarboxylic acids (Scheme 10.74).[92]

The assistance of coordinating functions in stabilizing the interaction of the C–H bond with the palladium centre is often crucial to achieving effective reaction. Many examples of this approach have been reported in the

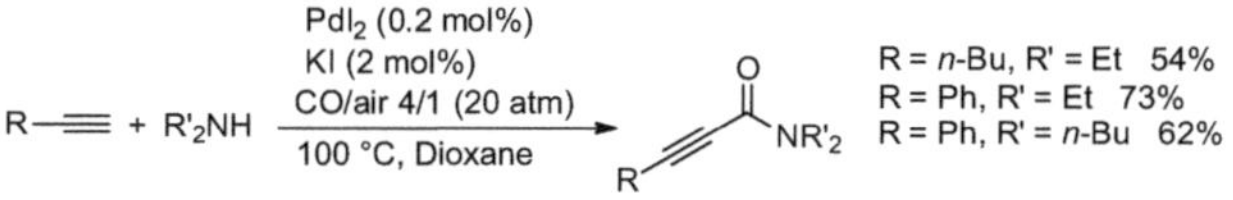

Scheme 10.72 Oxidative aminocarbonylation of 1-alkynes.

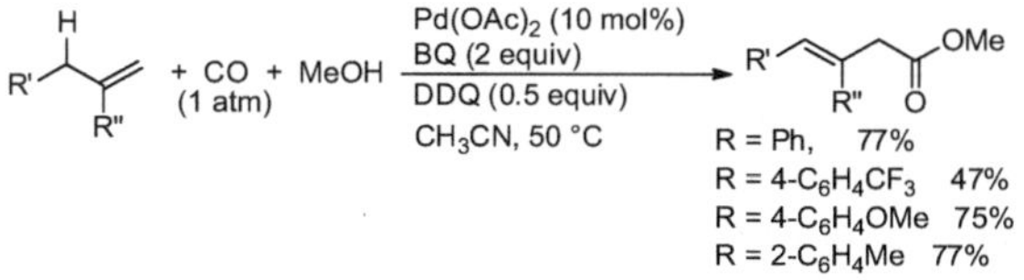

Scheme 10.73 Oxidative allylic carbonylation of alkenes.

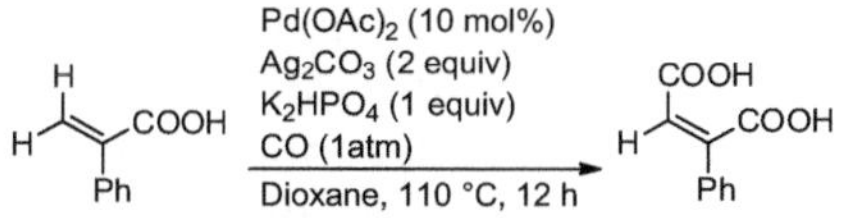

Scheme 10.74 Oxidative carboxylation of an α,β-unsaturated carboxylic acid.

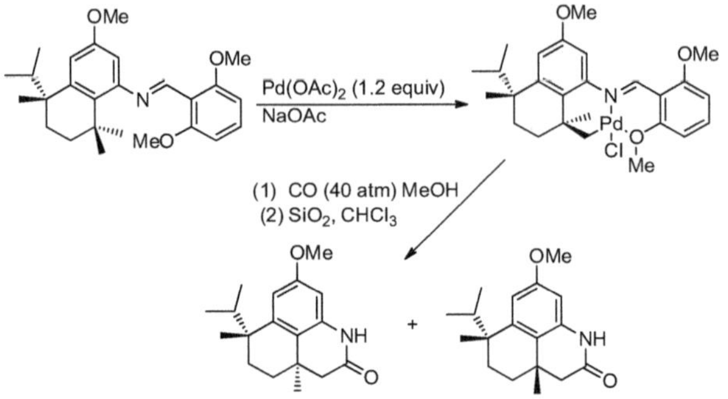

Scheme 10.75   Oxidative carbonylation for lactam synthesis.

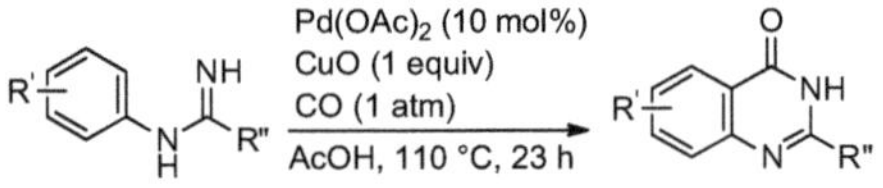

Scheme 10.76   Intramolecular carboxamidation.

recent literature. A wide variety of functional groups, such as amide, pyridine, imine, ketone and carboxylate, can function in this way. In addition, this coordination also serves to direct reaction to a specific C–H site and thus deliver the regioselectivity that is so important to the transformation of the complex molecules that are of interest to the pharmaceutical industry. An early application of this approach in natural product synthesis was provided by Sames and co-workers (Scheme 10.75).[93] They used the stoichiometric reaction of a Schiff base with palladium chloride to form the palladacycle, with subsequent reaction with CO (40 atm) and then acid hydrolysis to give the lactam in a one-pot reaction.

The ready formation of palladacycles from *N*-alkyl-ω-arylalkylamines is another example of C_{sp^2}–H activation. Combining this with carbon monoxide (balloon pressure), oxygen and $Cu(OAc)_2$ while heating in toluene proved effective for the formation of benzolactams.[94] The reaction proceeds much more rapidly when forming the five- rather than the six-membered ring and is also accelerated by the presence of other potentially chelating donor atoms such as oxygen in a 3,4-methylenedioxy group. This study was later extended as a method to provide ureas, carbamates and 1,3-oxazolidinones and also benzolactams.[95] With ethanol as the nucleophile, *N*,*N*-dimethylbenzylamines are carbonylated to yield *ortho*-substituted carboxylate esters with lithium chloride being added as a Lewis acid to promote the insertion of CO into the Pd–C bond.[96] In a similar reaction, Zhu and co-workers reported the intramolecular carboxamidation of *N*-arylamidines to yield quinazolin-4(3*H*)-one derivatives (Scheme 10.76).[97] Enabling these reactions to be carried out under milder conditions is advantageous for maximizing the selectivity of the reaction, in addition to avoiding the degradation of sensitive substrates.

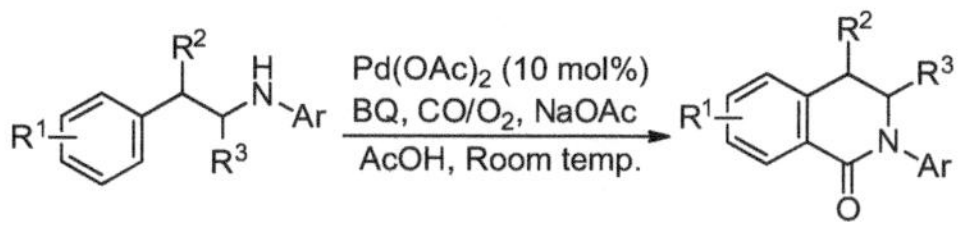

Scheme 10.77 Amine-directed C–H carbonylation.

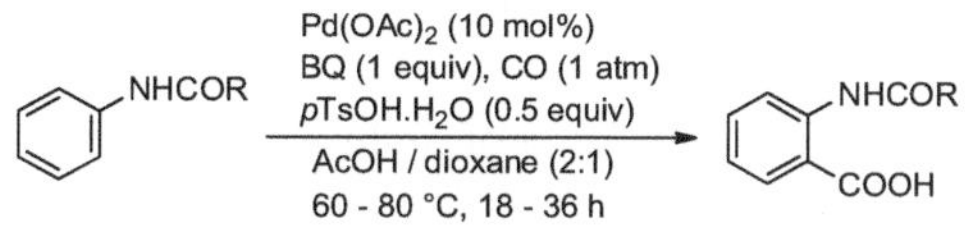

Scheme 10.78 Oxidative carbonylation for the preparationof anthranilic acids.

Recently, Gaunt and co-workers reported the room temperature carbonylation of *N*-aryl-β-arylethylamines.[98] *N*-Aryl substitution using a 4-methoxyphenyl group was anticipated to reduce the nucleophilicity of the nitrogen atom and reduce the likelihood of bis-amino–Pd(II) complex formation that could prevent carbopalladation and at the same time increase the acidity of the aryl N–H bond. This approach proved successful, with the reaction being effective at room temperature under 1 atm of CO and O_2 (Scheme 10.77). Studies of the stoichiometric reaction of cyclopalladated complexes containing primary arylalkylamines to form lactams were reported by Vicente and co-workers.[99]

As noted above for allylic substrates, the use of carboxylate as a directing group was explored by Yu's group.[92] Optimized conditions for converting benzoic to phthalic acids were found to be Pd(OAc)$_2$ as catalyst with Ag$_2$CO$_3$ as oxidant with 2 equiv. of NaOAc in 1,4-dioxane at 130 °C under 1 atm of CO. Significant steric and electronic effects were noted, suggestive of an electrophilic palladation pathway. Different conditions were required to carbonylate   anilides   for   the   formation   of   anthranilic   acids (Scheme 10.78).[100] Without the addition of acid (*e.g.* acetic, trifluoroacetic or *p*-toluenesulfonic acid), the presence of CO inhibited the C–H activation process and promoted the reduction of Pd(II) to Pd(0). A number of minor changes to the synthetic protocol in terms of solvent and additives were required to optimize the reaction for different substrates. The reaction could also be carried out using benzanilides with reaction exclusively on the aniline fragment giving *N*-benzoylanthranilic acids.

Similarly, Lloyd-Jones and co-workers identified conditions for the reaction of arylureas to form anthranilates by methoxycarbonylation.[101] Good yields were obtained at room temperature when using Pd(OTs)$_2$(MeCN)$_2$as catalyst with benzoquinone as oxidant and *p*-toluenesulfonic acid (0.5 equiv.) as additive in 1 : 1 THF–MeOH under CO (1 atm). If the reaction was carried out in CH$_2$Cl$_2$instead of THF–MeOH, cyclic imidates were obtained (Scheme 10.79). Control experiments suggested that anthranilate formation occurred *via* methanolysis of the imidate.

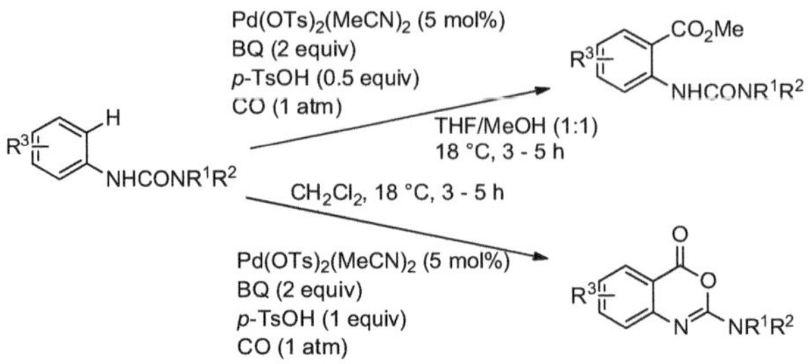

Scheme 10.79 Oxidative carbonylation of arylureas.

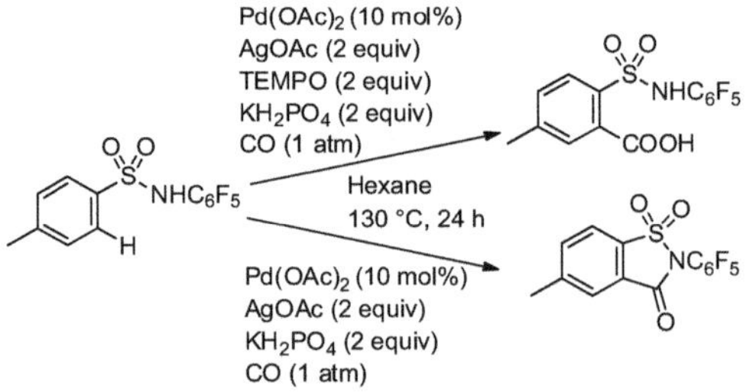

Scheme 10.80 Sulfonamide-drected oxidative carbonylation.

There is an evident need to maximize the selectivity of C–H activation to ensure regioselectivity. Recently, Yu and co-workers studied the benefits of sulfonamide as a directing group for a number of reactions, including carbonylation.[102] Reactions occurred exclusively at sites *ortho* to the sulfonamide group. Using Pd(OAc)$_2$/AgOAc/KH$_2$PO$_4$/CO (1 atm) in *n*-hexane at 130 °C for 24 h, intramolecular ring closure occurred, and in the presence of TEMPO (2,2,6,6-tetramethylpiperidine-1-oxyl) the carboxylate product was isolated (Scheme 10.80).

Examples of the use of hydroxyl functions as directing groups in C–H carbonylation are less common, but Yu and co-workers illustrated that it is possible to convert α-disubstituted phenethyl alcohols to 1-isochromanones.[103] They found that it was necessary to add a protecting ligand [the preferred example was (+)-menthyl(O$_2$C)-Leu-OH] to prevent palladium black formation under the reductive CO atmosphere (Scheme 10.81). Primary and secondary phenethyl alcohols were unstable under the reaction conditions, leading to reduced yields and oxidative decomposition products.

The carboxylation of anisoles was reported by Ishii and co-workers.[104] In this case, molybdovanadophosphate was employed as the acid. With Pd(OAc)$_2$ as catalyst, under 0.5 atm of CO and 0.5 atm of O$_2$ in acetic acid at 70 °C for 15 h, the major product was the *para* derivative with *ortho* substitution forming the minor product (∼3:1 ratio).

The oxidative carbonylation of heteroarenes, and in particular *N*-substituted indoles, to yield 3-substituted indole acids and esters was first reported in

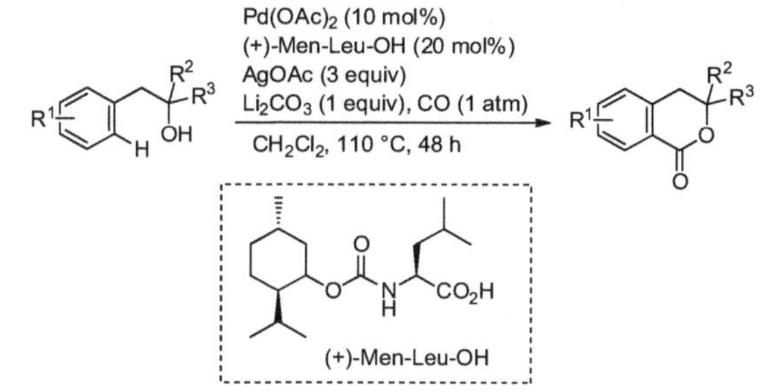

Scheme 10.81 Oxidative carbonylation of phenethyl alcohols.

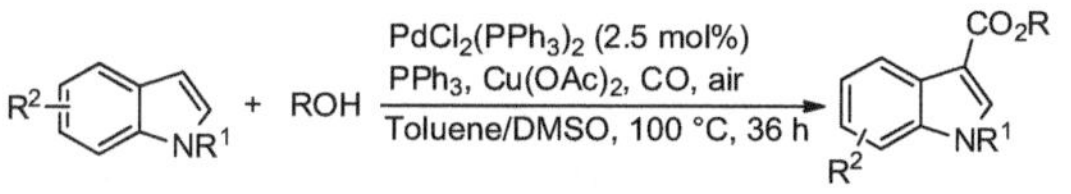

Scheme 10.82 Oxidative carbonylation of indoles.

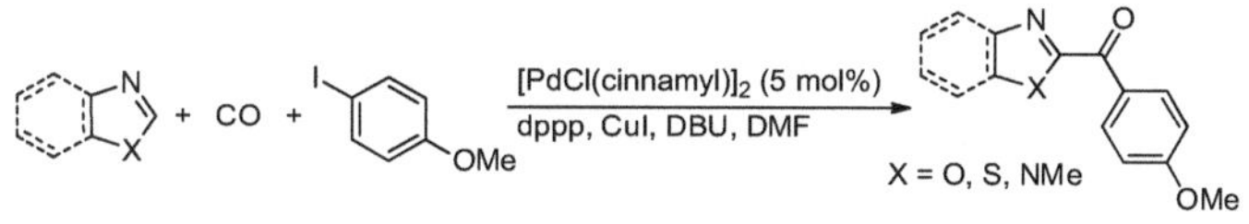

Scheme 10.83 Carbonylative coupling of heteroarenes.

1982 by Itahara[105] and recently studied by Lei and co-workers.[106] Itahara used a simple treatment involving Pd(OAc)$_2$ and Na$_2$S$_2$O$_8$ and 1 atm of CO in acetic acid at reflux to convert 1-acylindoles to 1-acylindole-3-carboxylic acids. A more complex reaction system using PdCl$_2$(PPh$_3$)$_2$/2PPh$_3$/Cu(OAc)$_2$ under 1 atm CO–air (7:1) in toluene–DMSO (15:1 v/v) at 100 °C for 36 h was found by Lei and co-workers to give moderate to good isolated yields of indole-3-carboxylate esters (Scheme 10.82). For unsubstituted indoles, reaction occurs at the N atom to give the indole carbamates. The substrate scope was extended to thiophenes and benzothiophenes with reaction occurring mainly at C2.[105,106]

The formation of ketones by carbonylative coupling processes, as outlined in the sections above, can be enhanced by C–H activation of the coupling partner, rather than using organometallic derivatives such as ArB(OH)$_2$. Beller and co-workers studied reactions of this type using the carbonylative coupling of iodobenzene and benzoxazole to give 2-(benzoyl)benzoxazole as their model.[107] The catalytic system was [PdCl(cinnamyl)]$_2$/phosphine with CuI. It was found that chelating bidentate phosphines performed better than monophosphines and a high CO pressure was necessary to minimize the formation of the non-carbonylative coupled product. Optimized conditions were [PdCl(cinnamyl)]$_2$(5 mol%)/1,3-bis(diphenylphosphino)propane (10 mol%) with CuI (1.5 equiv.) with DBU (1 equiv.) under CO (40 bar) at 120 °C in DMF for 30 h (Scheme 10.83).

Without the effect of a directing group to activate the C–H bond and promote regioselectivity, reactions of simple arenes often require harsher conditions and give lower yields and poorer selectivity than their directed counterparts. Early studies of the carbonylation of arenes by Fujiwara *et al.* using just Pd(OAc)$_2$ under CO (15 atm) indicated the potential for generating carboxylic acids but without achieving catalysis.[108] With the addition of 1,2-dibromomethane, acid anhydrides were formed.[109] Attempts to develop catalytic processes were reported later.[110,111] The catalytic systems developed included Pd(OAc)$_2$ with 1,10-phenanthroline in acetic acid/O$_2$ and Pd(OAc)$_2$/tBuOOH/CH$_2$=CHCH$_2$Cl with acetic acid. Improved results, but still with moderate turnover numbers (<20), for the conversion of benzene to benzoic acid were reported in 1999.[112] Under a low CO pressure (1 atm) with Pd(OAc)$_2$/TFA solvent/K$_2$S$_2$O$_8$, reactions could be conducted at room temperature in ~20 h.

Continuing their studies of palladium-catalysed oxidative carbonylation using molybdovanadophosphate as acid and oxygen as the terminal oxidant, Ishii and co-workers were able to show that unsubstituted benzene could be reacted to form benzoic acid.[113] They noted that the method of preparation of H$_5$PMo$_{10}$V$_2$O$_{40}$·nH$_2$O was crucial to obtaining the best yield, with contamination by sodium ions being a possible explanation.

A study of the oxidative carbonylation of benzene trifluoride with ammonium vanadate as an additive was reported by Zakzeski and Bell using either rhodium or palladium catalysis.[114] Tris(acetylacetonato)rhodium(III) proved to be the best catalyst.

A double C–H activation with carbonylation was described by Lei and co-workers for the preparation of xanthones (Scheme 10.84).[115] Preferred conditions were Pd(OAc)$_2$ with K$_2$S$_2$O$_8$ as the oxidant in trifluoroacetic acid at 50 °C under 1 atm of CO. Kinetic studies using IR spectroscopy indicated that neither the initial C–H activation nor the carbonyl insertion were rate limiting, suggesting that the slow step was the C–H functionalization of the second ring.

As with the studies on simple arenes, early investigations of the conversion of simple alkane derivatives such as methane and ethane to carboxylic acids were carried out by Fujiwara and co-workers.[116] Their conditions required the use of Pd(OAc)$_2$/Cu(OAc)$_2$/K$_2$S$_2$O$_8$/TFA to obtain low levels of catalysis (<10 turnovers in 20 h) and generally low conversion of alkane (<5%). More recently, Murai and co-workers reported carbonylation at C$_{sp^3}$–H adjacent to the nitrogen atom of alkylamines using [RhCl(COD)]$_2$ as the catalyst[117] and with the assistance of chelation through amide and pyridine nitrogens when using Ru$_3$(CO)$_{12}$ as the catalyst.[118] In the latter

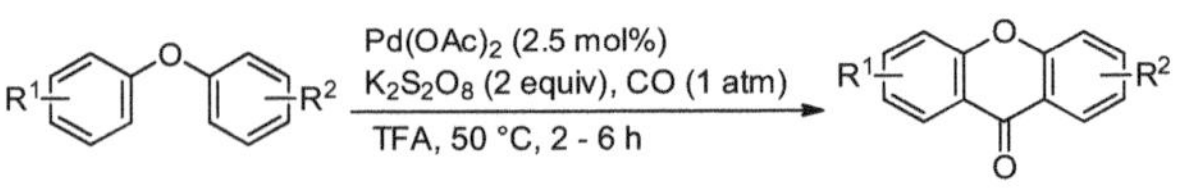

Scheme 10.84 Oxidative carbonylation with double C–H activation for diaryl ethers.

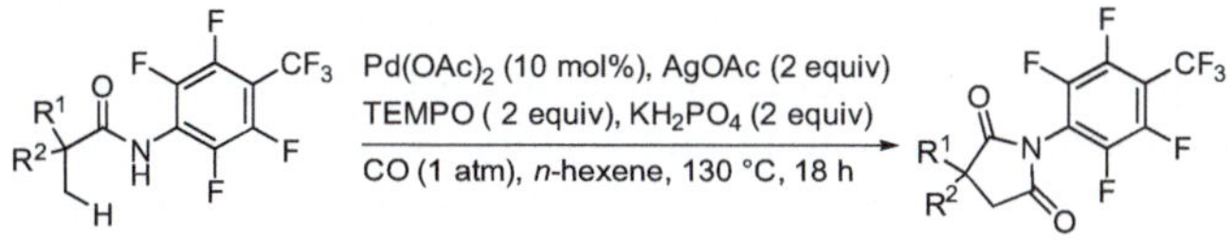

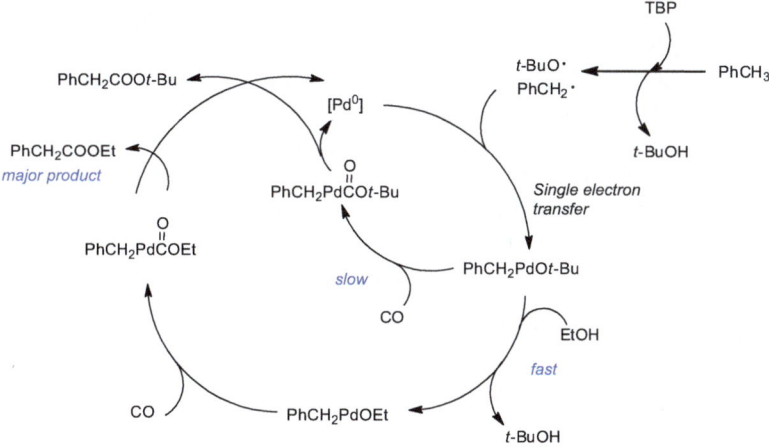

Scheme 10.85   Oxidative carbonylation of C_{sp^3}–H in amides.

Scheme 10.86   Reaction mechanism for oxidative carbonylation of toluene.

work, reaction required the presence of ethene, presumed to act as a hydrogen acceptor, and water, to activate the catalyst resting state. In a substrate that offers the potential for either C_{sp^2}–H or C_{sp^3}–H conversion, the former shows preferential reactivity.[118,119] Yu and co-workers also reported on the carbonylation of C_{sp^3}–H of amides leading to a range of succinimides that could be further hydrolysed to provide 1,4-dicarbonyl compounds (Scheme 10.85).[120] They found that the choice of TEMPO as co-oxidant (with AgOAc) was crucial.

The reaction of C_{sp^3}–H bonds in toluene derivatives was reported by Huang and co-workers (Scheme 10.86).[121] In this case, the preferred oxidant was di-*tert*-butyl peroxide, leading to a radical-involved benzylic C–H functionalization. Carbonylation in the presence of ethanol as the preferred nucleophile led to the synthesis of a range of ethyl 2-phenylacetate derivatives. Under optimized conditions, turnover numbers >100 were achieved. The absence of product when the reaction is carried out in the presence of radical scavengers such as TEMPO and 1,1-diphenylethylene suggests that activation of the benzylic substrate by removal of a hydrogen atom to form a benzyl radical precedes two-step oxidation of Pd(0) in a single electron transfer process.

10.6   Conclusion

Carbon–carbon and carbon–heteroatom bond formation using carbon monoxide has been widely exploited in organic synthesis during the last

30 years. However, this is still a rapidly developing area with much to be done to improve the efficiency and economics of the reactions for commercial purposes. In particular, the results of ligand development and improved mechanistic understanding for non-carbonylative coupling can be expected to result in improved processes involving carbonylation. Also, the rapid developments in direct arylation and C–H activation can be expected to be fruitfully applied to oxidative carbonylation. In all these reactions, carbonylation products are formed in competition with by-products arising from non-carbonylative processes, so an improved understanding of the mechanism of these reactions will aid in maximizing specificity. In addition, a greater understanding of the role of additives in stabilizing Pd with respect to reduction and precipitation will be helpful. Clearly, from the results discussed above, this includes details such as choice of solvent and the nature of acids and bases and of oxidants. Regarding the substrates applied, iodide and bromide compounds and triflates have been extensively studied, but the cheaper but less activated chlorides, tosylates and mesylates are less commonly reported, and improved conditions for these are needed. In the case of alkyl halides, radical reactions are the main pathway in carbonylation reactions, with more effort still needed to improve the efficiency.

References

1. (a) P. M. Henry, *Tetrahedron Lett.*, 1968, 2285; (b) A. Schoenberg, I. Bartoletti and R. F. Heck, *J. Org. Chem.*, 1974, **39**, 3318; (c) A. Schoenberg and R. F. Heck, *J. Org. Chem.*, 1974, **39**, 3327; (d) A. Schoenberg and R. F. Heck, *J. Am. Chem. Soc.*, 1974, **96**, 7761; For further details on carbonylation processes, see L. Kollar (ed.), *Modern Carbonylation Methods*, Wiley-VCH, Weinheim, 2008.
2. D. Alberico, M. E. Scott and M. Lautens, *Chem. Rev.*, 2007, **107**, 174.
3. (a) Q. Liu, H. Zhang and A. Lei, *Angew. Chem. Int. Ed.*, 2011, **50**, 10788; (b) Y. Fujiwara, K. Takaki and Y. Taniguchi, *Synlett.*, 1996, 591; (c) C. Jia, T. Kitamura and Y. Fujiwara, *Acc. Chem. Res.*, 2001, **34**, 633; (d) K. M. Engle, T.-S. Mei, M. Wasa and J.-Q. Yu, *Acc. Chem. Res.*, 2012, **45**, 788.
4. (a) R. F. Heck, *J. Am. Chem. Soc.*, 1968, **90**, 5518; (b) R. F. Heck, *J. Am. Chem. Soc.*, 1968, **90**, 5526; (c) R. F. Heck, *J. Am. Chem. Soc.*, 1968, **90**, 5531; (d) R. F. Heck, *J. Am. Chem. Soc.*, 1968, **90**, 5535; (e) R. F. Heck, *J. Am. Chem. Soc.*, 1968, **90**, 5538; (f) R. F. Heck, *J. Am. Chem. Soc.*, 1968, **90**, 5542; (g) R. F. Heck, *J. Am. Chem. Soc.*, 1968, **90**, 5546; (h) R. F. Heck and J. P. Nolley, *J. Org. Chem.*, 1972, **37**, 2320; (i) A. Sen and T.-W. Lai, *J. Am. Chem. Soc.*, 1982, **104**, 3520.
5. E.-i. Negishi and J. A. Miller, *J. Am. Chem. Soc.*, 1983, **105**, 6761.
6. (a) J. M. Tour and E.-i. Negishi, *J. Am. Chem. Soc.*, 1985, **107**, 8289; (b) E.-i. Negishi, G. Wu and J. M. Tour, *Tetrahedron Lett.*, 1988, **29**, 6745.

7. (a) E.-i. Negishi and J. M. Tour, *Tetrahedron Lett.*, 1986, **27**, 4869; (b) E.-i. Negishi, S. Ma, J. Amanfu, C. Copéret, J. A. Miller and J. M. Tour, *J. Am. Chem. Soc.*, 1996, **118**, 5919; (c) E.-i. Negishi, C. Copéret, S. Ma, T. Mita, T. Sugihara and J. M. Tour, *J. Am. Chem. Soc.*, 1996, **118**, 5904; (d) C. Copéret, S. Ma and E.-i. Negishi, *Angew. Chem. Int. Ed. Engl.*, 1996, **35**, 2125; (e) I. Ryu, S. Kreimerman, F. Araki, S. Nishitani, Y. Oderaotoshi, S. Minakata and M. Komatsu, *J. Am. Chem. Soc.*, 2002, **124**, 3812; (f) N. Chatani, A. Kamitani, M. Oshita, Y. Fukumoto and S. Murai, *J. Am. Chem. Soc.*, 2001, **123**, 12686.

8. S. Torii, H. Okumoto and L. H. Xu, *Tetrahedron Lett.*, 1990, **31**, 7175.

9. K. Okuro, M. Furuune, M. Miura and M. Nomura, *J. Org. Chem.*, 1992, **57**, 4754.

10. T. Satoh, T. Itaya, K. Okuro, M. Miura and M. Nomura, *J. Org. Chem.*, 1995, **60**, 7267.

11. K. Okuro and H. Alper, *J. Org. Chem.*, 1997, **62**, 1566.

12. R. Grigg and R. Pratt, *Tetrahedron Lett.*, 1997, **38**, 4489.

13. F. Ye and H. Alper, *J. Org. Chem.*, 2007, **72**, 3218.

14. N. Iwasawa and H. Satoh, *J. Am. Chem. Soc.*, 1999, **121**, 7951.

15. T. Hayashi, J. Tang and K. Kato, *Org. Lett.*, 1999, **1**, 1487.

16. S. V. Gagnier and R. C. Larock, *J. Am. Chem. Soc.*, 2003, **125**, 4804.

17. X. Wu, P. Nilsson and M. Larhed, *J. Org. Chem.*, 2005, **70**, 346.

18. K. S. Bloome and E. J. Alexanian, *J. Am. Chem. Soc.*, 2010, **132**, 12823.

19. X.-F. Wu, H. Neumann and M. Beller, *Angew. Chem. Int. Ed.*, 2010, **49**, 5284.

20. X.-F. Wu, H. Neumann, A. Spannenberg, T. Schulz, H. Jiao and M. Beller, *J. Am. Chem. Soc.*, 2010, **132**, 14596.

21. X.-F. Wu, H. Jiao, H. Neumann and M. Beller, *ChemCatChem*, 2011, **3**, 726.

22. (a) K. Sonogashira, Y. Tohda and N. Hagihara, *Tetrahedron Lett.*, 1975, **16**, 4467; (b) X. Pu, H. Li and T. J. Colacot, *J. Org. Chem.*, 2013, **78**, 568.

23. (a) H. Doucet and J.-C. Hierso, *Angew. Chem. Int. Ed.*, 2007, **46**, 834; (b) K. C. Nicolaou, P. G. Bulger and D. Sarlah, *Angew. Chem. Int. Ed.*, 2005, **44**, 4442; (c) E.-i. Negishi and L. Anastasia, *Chem. Rev.*, 2003, **103**, 1979.

24. (a) C. H. Faweett, R. D. Firu and D. M. Spencer, *Physiol. Plant Pathol.*, 1971, **1**, 163; (b) K. Imai, *J. Pharm. Soc. Jpn.*, 1956, **76**, 405; (c) C. A. Quesnelle, P. Gill, M. Dodier, D. St Laurent, M. Serrano-Wu, A. Marinier, A. Martel, C. E. Mazzucco, T. M. Stickle, J. F. Barrett, D. M. Vyas and B. N. Balasubramanian, *Bioorg. Med. Chem. Lett.*, 2003, **13**, 519.

25. (a) A. S. Karpov, E. Merkul, F. Rominger and T. J. J. Müller, *Angew. Chem. Int. Ed.*, 2005, **44**, 6951; (b) D. M. D'souza and T. J. J. Müller, *Nat. Protocols*, 2008, **3**, 1660; (c) J. Marco-Contelles and E. de Opazo, *J. Org. Chem.*, 2002, **67**, 3705; (d) C. J. Forsyth, J. Xu, S. T. Nguyen, I. A. Samdai, L. R. Briggs, T. Rundberget, M. Sandvik and C. O. Miles, *J. Am. Chem.*

Soc., 2006, **128**, 15114; (e) L. F. Tietze, R. R. Singidi, K. M. Gericke, H. Bockemeier and H. Laatsch, *Eur. J. Org. Chem.*, 2007, 5875.

26. (a) B. Willy and T. J. J. Müller, *Arkivoc*, 2008, 195; (b) A. Aradi, M. Aschi, F. Marinelli and M. Verdecchia, *Tetrahedron*, 2008, **64**, 5354; (c) P. Bannwarth, A. Valleix, D. Gree and R. Gree, *J. Org. Chem.*, 2009, **74**, 4646.

27. (a) K. Y. Lee, M. J. Lee and J. N. Kim, *Tetrahedron*, 2005, **61**, 8705; (b) H. A. Stefani, R. Cella, F. A. Dorr, C. M. P. de Pereira, F. P. Gomes and G. Zeni, *Tetrahedron Lett.*, 2005, **46**, 2001; (c) S. S. Palimkar, P. H. Kumar, N. R. Jogdand, T. Daniel, R. J. Lahoti and K. V. Srinivasan, *Tetrahedron Lett.*, 2006, **47**, 5527; (d) S. J. Yim, C. H. Kwon and D. K. An, *Tetrahedron Lett.*, 2007, **48**, 5393; (e) M. M. Jackson, C. Leverett, J. F. Toczko and J. C. Roberts, *J. Org. Chem.*, 2002, **67**, 5032; (f) D. A. Alonso, C. Nájera and M. C. Pacheco, *J. Org. Chem.*, 2004, **69**, 1615; (g) B. Wang, M. Bonin and L. Micouin, *J. Org. Chem.*, 2005, **70**, 6126; (h) L. Chen and C. Li, *Org. Lett.*, 2004, **6**, 3151; (i) N. Kakusawa, K. Yamaguchi, J. Kurita and T. Tsuchiya, *Tetrahedron Lett.*, 2000, **41**, 4143.

28. T. Kobayashi and M. Tanaka, *J. Chem. Soc., Chem. Commun.*, 1981, 333.

29. (a) Y. Huang and H. Alper, *J. Org. Chem.*, 1991, **56**, 4534; (b) X.-F. Wu, H. Jiao, H. Neumann and M. Beller, *Chem. Eur. J.*, 2012, **18**, 16177; (c) X.-F. Wu, B. Sundararaju, P. Anbarasan, H. Neumann, P. H. Dixneuf and M. Beller, *Chem. Eur. J.*, 2011, **17**, 8014; (d) X.-F. Wu, H. Neumann and M. Beller, *Org. Biomol. Chem.*, 2011, **9**, 8003; (e) J. Kiji, T. Okano, H. Kimura and K. Saiki, *J. Mol. Catal. A: Chem.*, 1998, **130**, 95.

30. P. G. Ciattini, E. Morera and G. Ortar, *Tetrahedron Lett.*, 1991, **32**, 6449.

31. L. Delaude, A. M. Masdeu and H. Alper, *Synthesis*, 1994, 1149.

32. A. Areadi, S. Cacchi, F. Marinelli, P. Pace and G. Sanzi, *Synlett*, 1995, 823.

33. S.-K. Kang, K.-H. Lim, P.-S. Ho and W.-Y. Kim, *Synthesis*, 1997, 874.

34. S.-L. Luo, Y.-M. Liang, C.-M. Liu and Y.-X. Ma, *Synth. Commun.*, 2001, **31**, 343.

35. (a) M. S. M. Ahmed and A. Mori, *Org. Lett.*, 2003, **5**, 3057; (b) M. S. M. Ahmed, A. Sekiguchi, K. Masui and A. Mori, *Bull. Chem. Soc. Jpn.*, 2005, **78**, 160; (c) B. C. Bishop, K. M. J. Brands, A. D. Gibb and D. J. Kennedy, *Synthesis*, 2004, 43.

36. B. Liang, M. Huang, Z. You, Z. Xiong, K. Lu, R. Fathi, J. Chen and Z. Yang, *J. Org. Chem.*, 2005, **70**, 6097.

37. N. Kakusawa and J. Kurita, *Chem. Pharm. Bull.*, 2006, **54**, 699.

38. (a) Md. T. Rahman, T. Fukuyama, N. Kamata, M. Sato and I. Ryu, *Chem. Commun.*, 2006, 2236; (b) T. Fukuyama, R. Yamaura and I. Ryu, *Can. J. Chem.*, 2005, **83**, 711.

39. W. Ma, X. Li, J. Yang, Z. Liu, B. Chen and X. Pan, *Synthesis*, 2006, 2489.

40. C. Feher, A. Kuik, L. Mark, L. Kollar and R. Skoda-Foldes, *J. Organomet. Chem.*, 2009, **694**, 4036.

41. C. Li, X. Li, Q. Zhu, H. Cheng, Q. Lv and B. Chen, *Catal. Lett.*, 2009, **127**, 152.

42. V. Sans, A. M. Trzeciak, S. Luis and J. J. Ziolkowski, *Catal. Lett.*, 2006, **109**, 37.

43. M. Iizuka and Y. Kondo, *Eur. J. Org. Chem.*, 2007, 5180.

44. J. Liu, J. Chen and C. Xia, *J. Catal.*, 2008, **253**, 50.

45. J. Liu, X. Peng, W. Sun, Y. Zhao and C. Xia, *Org. Lett.*, 2008, **10**, 3933.

46. W. Hao, J. Sha, S. Sheng and M. Cai, *J. Mol. Catal. A: Chem.*, 2009, **298**, 94.

47. T. Doi, H. Inous, M. Tokita, J. Watanabe and T. Takahashi, *J. Comb. Chem.*, 2008, **10**, 135.

48. X.-F. Wu, H. Neumann and M. Beller, *Chem. Eur. J.*, 2010, **16**, 12104.

49. (a) X.-F. Wu, B. Sundararaju, H. Neumann, P. H. Dixneuf and M. Beller, *Chem. Eur. J.*, 2011, **17**, 106; (b) X.-F. Wu, H. Neumann and M. Beller, *Angew. Chem. Int. Ed.*, 2011, **50**, 11142.

50. A. Fusano, T. Fukuyama, S. Nishitani, T. Inouye and I. Ryu, *Org. Lett.*, 2010, **12**, 2410.

51. A. C. Frisch and M. Beller, *Angew. Chem. Int. Ed.*, 2005, **44**, 674.

52. Z.-W. An, M. Catellani and G. P. Chiusoli, *J. Organomet. Chem.*, 1990, **397**, C31.

53. (a) S. Torii, H. Okumoto and L. H. Xu, *Tetrahedron Lett.*, 1991, **32**, 237; (b) S. Torii, H. Okumoto, L. H. Xu, M. Sadkane, M. V. Shostakovsky, A. B. Ponomaryov and V. N. Kalinin, *Tetrahedron*, 1993, **49**, 6773; (c) V. N. Kalinin, M. V. Shostakovsky and A. B. Ponomaryov, *Tetrahedron Lett.*, 1992, **33**, 373; (d) N. Haddad, J. Tan and V. Farina, *J. Org. Chem.*, 2006, **71**, 5031; (e) M. Genelot, A. Bendjeriou, V. Dufaud and L. Djakovitch, *Appl. Catal. A: Gen.*, 2009, **369**, 125.

54. E. Brocato, C. Cstagnoli, M. Catellani and G. P. Chiusoli, *Tetrahedron Lett.*, 1992, **33**, 7433.

55. H. Miao and Z. Yang, *Org. Lett.*, 2000, **2**, 1765.

56. E. Awuah and A. Capretta, *Org. Lett.*, 2009, **11**, 3210.

57. Q. Yang and H. Alper, *J. Org. Chem.*, 2010, 75, 948.

58. A. S. Karpov, E. Merkul, F. Rominger and T. J. J. Müller, *Angew. Chem. Int. Ed.*, 2005, **44**, 6951.

59. Y. Wakita, T. Yasunaga, M. Akita and M. Kojima, *J. Organomet. Chem.*, 1986, **301**, C17.

60. T. Ishiyama, N. Miyaura and A. Suzuki, *Bull. Chem. Soc. Jpn.*, 1991, **64**, 1999.

61. (a) T. Ishiyama, N. Miyaura and A. Suzuki, *Tetrahdron Lett.*, 1991, **32**, 6923; (b) T. Ishiyama, M. Murata, A. Suzuki and N. Miyaura, *J. Chem. Soc., Chem. Commun.*, 1995, 295.

62. (a) T. Ishiyama, H. Kizaki, N. Miyaura and A. Suzuki, *Tetrahedron Lett.*, 1993, **34**, 7595; (b) T. Ishiyama, H. Kizaki, T. Hayashi, A. Suzuki and N. Miyaura, *J. Org. Chem.*, 1998, **63**, 4726.

63. M. Ishikura and M. Terashima, *J. Org. Chem.*, 1994, **59**, 2634.

64. S.-K. Kang, K.-H. Lim, P.-S. Ho, S.-K. Yoon and H.-J. Son, *Synth. Commun.*, 1998, **28**, 1481.

65. M. Xia and Z. Chen, *J. Chem. Res.*, 1999, 400.

66. R. Skoda-Foldes, Z. Szekvoelgyi, L. Kollar, Z. Berente, J. Horvath and Z. Tuba, *Tetrahedron*, 2000, **56**, 3415.

67. (a) S. Couve-Bonnaire, J.-F. Carpentier, A. Mortreux and Y. Castanet, *Tetrahedron Lett.*, 2001, **42**, 3689; (b) E. Maerten, F. Hassouna, S. Couve-Bonnaire, A. Mortreux and J.-F. Carpentier, *Synlett*, 2003, 1874; (c) S. Couve-Bonnaire, J.-F. Carpentier, A. Mortreux and Y. Castanet, *Tetrahedron*, 2003, **59**, 2793; (d) E. Maerten, M. Sauthier, A. Mortreux and Y. Castanet, *Tetrahedron*, 2007, **63**, 682.

68. B. Gotov, J. Kaufmann, H. Schumann and H.-G. Schmalz, *Synlett*, 2002, 1161.

69. M. B. Andrus, Y. Ma, Y. Zang and C. Song, *Tetrahedron Lett.*, 2002, **43**, 9137.

70. D. Mingli, B. Liang, C. Wang, Z. You, J. Xiang, G. Dong, J. Chen and Z. Yang, *Adv. Synth. Catal.*, 2004, **346**, 1669.

71. T. Saeki, E.-C. Son and K. Tamao, *Org. Lett.*, 2004, **6**, 617.

72. O. Rahman, T. Kihlberg and B. Långström, *Eur. J. Org. Chem.*, 2004, 474.

73. M. Medio-Simon, C. Mollar, N. Rodriguez and G. Asensio, *Org. Lett.*, 2005, 7, 4669.

74. L. Bartali, A. Guarna, P. Larini and E. G. Occhiato, *Eur. J. Org. Chem.*, 2007, 2152.

75. A. Petz, G. Peczely, Z. Pinter and L. Kollar, *J. Mol. Catal. A: Chem.*, 2006, **255**, 97.

76. P. Prediger, A. V. Moro, C. W. Nogueira, L. Saegnago, P. H. Menezes, J. B. T. Rocha and G. Zeni, *J. Org. Chem.*, 2006, **71**, 3786.

77. D. Yang, Z. Liu, Y. Li and B. Chen, *Synth. Commun.*, 2007, **37**, 3759.

78. S. Zheng, L. Xu and C. Xia, *Appl. Organomet. Chem.*, 2007, **21**, 772.

79. B. M. O'Keefe, N. Simmons and S. F. Martin, *Org. Lett.*, 2008, **10**, 5301.

80. H. Neumann, A. Brennführer and M. Beller, *Chem. Eur. J.*, 2008, **14**, 3645.

81. W. Qin, S. Yasuike, N. Kakusawa and J. Kurita, *J. Organomet. Chem.*, 2008, **693**, 2949.

82. C. Pirez, J. Dheur, M. Sauthier, Y. Castanet and A. Mortreux, *Synlett*, 2009, 1745.

83. (a) M. Cai, G. Zheng, L. Zha and J. Peng, *Eur. J. Org. Chem.*, 2009, 1585; (b) G. Zheng, P. Wang and M. Cai, *Chin. J. Chem.*, 2009, **27**, 1420.

84. P. J. Tambade, Y. P. Patil, A. G. Panda and B. M. Bhanage, *Eur. J. Org. Chem.*, 2009, 3022.

85. M. Arthuis, R. Pontikis and J.-C. Florent, *Org. Lett.*, 2009, **11**, 4608.

86. L. Kaganovsky, D. Gelman and K. Rueck-Braun, *J. Organomet. Chem.*, 2010, **695**, 260.

87. (a) X.-F. Wu, H. Neumann and M. Beller, *Tetrahedron Lett.*, 2010, **51**, 6146; (b) X.-F. Wu, H. Neumann and M. Beller, *Adv. Synth. Catal.*, 2011, **353**, 788.

88. J. Tsuji, M. Takahashi and T. Takahashi, *Tetrahedron Lett.*, 1980, **21**, 849.
89. B. Gabriele, G. Salerno, L. Veltri and M. Costa, *J. Organomet. Chem.*, 2001, **622**, 84.
90. (a) J. Tsuji, *Palladium Reagents and Catalysts: New Perspectives for the 21st Century*, Wiley, Chichester, 2004, Ch. 4; (b) B. M. Trost, *J. Org. Chem.*, 2004, **69**, 5813.
91. H. Chen, C. Cai, X. Liu, X. Li and H. Jiang, *Chem. Commun.*, 2011, **47**, 12224.
92. R. Giri and J.-Q. Yu, *J. Am. Chem. Soc.*, 2008, **130**, 14082.
93. B. D. Dangel, K. Godula, S. W. Youn, B. Sezen and D. Sames, *J. Am. Chem. Soc.*, 2002, **124**, 11856.
94. K. Orito, A. Horibata, T. Nakamura, H. Ushito, H. Nagasaki, M. Yuguchi, S. Yamashita and M. Tokuda, *J. Am. Chem. Soc.*, 2004, **126**, 14342.
95. K. Orito, M. Miyazawa, T. Nakamura, A. Horibata, H. Ushito, H. Nagasaki, M. Yuguchi, S. Yamashita, T. Yamazaki and M. Tokuda, *J. Org. Chem.*, 2006, **71**, 5951.
96. H. Li, G.-X. Cai and Z.-J. Shi, *Dalton Trans.*, 2010, **39**, 10442.
97. B. Ma, Y. Wang, J. Peng and Q. Zhu, *J. Org. Chem.*, 2011, **76**, 6362.
98. B. Haffemayer, M. Gulias and M. J. Gaunt, *Chem. Sci.*, 2011, **2**, 312.
99. J. Vicente, I. Saura-Llamas and J.-A. Garcia-Lopez, *Organometallics*, 2009, **28**, 448.
100. R. Giri, J. K. Lam and J.-Q. Yu, *J. Am. Chem. Soc.*, 2010, **132**, 686.
101. C. Houlden, M. Hutchby, C. D. Bailey, J. G. Ford, S. N. G. Tyler, M. R. Gagné, G. C. Lloyd-Jones and K. I. Booker-Milburn, *Angew. Chem. Int. Ed.*, 2009, **48**, 1830.
102. H.-X. Dai, A. F. Stepan, M. S. Plummer, Y.-H. Zhang and J.-Q. Yu, *J. Am. Chem. Soc.*, 2011, **133**, 7222.
103. Y. Lu, D. Leow, X. Wang, K. M. Engle and J.-Q. Yu, *Chem. Sci.*, 2011, **2**, 967.
104. S. Ohashi, S. Sakaguchi and Y. Ishii, *Chem. Commun.*, 2005, 486.
105. T. Itahara, *Chem. Lett.*, 1982, 1151.
106. H. Zhang, D. Liu, C. Chen, C. Liu and A. Lei, *Chem. Eur. J.*, 2011, **17**, 9581.
107. X.-F. Wu, P. Anbarasan, H. Neumann and M. Beller, *Angew. Chem. Int. Ed.*, 2010, **49**, 7316.
108. Y. Fujiwara, T. Kawauchi and H. Taniguchi, *J. Chem. Soc., Chem. Commun.*, 1980, 220.
109. Y. Fujiwara, I. Kawata, T. Kawauchi and H. Taniguchi, *J. Chem. Soc., Chem. Commun.*, 1982, 132.
110. Y. Fujiwara, I. Kawata, H. Sugimoto and H. Taniguchi, *J. Organomet. Chem.*, 1983, **256**, C35.
111. T. Jintoku, H. Taniguchi and Y. Fujiwara, *Chem. Lett.*, 1987, 1159.
112. W. Lu, Y. Yamaoka, Y. Taniguchi, T. Kitamura, K. Takaki and Y. Fujiwara, *J. Organomet. Chem.*, 1999, **580**, 290.

113. S. Yamada, S. Ohashi, Y. Obora, S. Sakaguchi and Y. Ishii, *J. Mol. Catal. A: Chem.*, 2008, **282**, 22.
114. J. Zakzeski and A. T. Bell, *J. Mol. Catal. A: Chem.*, 2009, **302**, 59.
115. H. Zhang, R. Shi, P. Gan, C. Liu, A. Ding, Q. Wang and A. Lei, *Angew. Chem. Int. Ed.*, 2012, **51**, 5204.
116. K. Nakata, Y. Yamaoka, T. Miyata, Y. Taniguchi, K. Takaki and Y. Fujiwara, *J. Organomet. Chem.*, 1994, **473**, 329.
117. N. Chatani, T. Asaumi, T. Ikeda, S. Yorimitsu, Y. Ishii, F. Kakiuchi and S. Murai, *J. Am. Chem. Soc.*, 2000, **122**, 12882.
118. N. Hasegawa, V. Charra, S. Inoue, Y. Fukumoto and N. Chatani, *J. Am. Chem. Soc.*, 2011, **133**, 8070.
119. S. Inoue, H. Shiota, Y. Fukumoto and N. Chatani, *J. Am. Chem. Soc.*, 2009, **131**, 6898.
120. E. J. Yoo, M. Wasa and J.-Q. Yu, *J. Am. Chem. Soc.*, 2010, **132**, 17378.
121. P. Xie, Y. Xie, B. Qian, H. Zhou, C. Xia and H. Huang, *J. Am. Chem. Soc.*, 2012, **134**, 9902.

CHAPTER 11

Stereospecific and Stereoselective Suzuki–Miyaura Cross-Coupling Reactions

BEN W. GLASSPOOLE, ERIC C. KESKE AND
CATHLEEN M. CRUDDEN*

Department of Chemistry, Queen's University, Kingston, Ontario, Canada
K7L 3N6
*Email: cathleen.crudden@chem.queensu.ca

11.1 Introduction

The Suzuki–Miyaura reaction[1–3] has changed the way in which certain classes of molecules are made, and in the process it has become one of the most often employed reactions for carbon–carbon bond formation in the pharmaceutical industry.[4] Among all other cross-coupling reactions, the Suzuki–Miyaura reaction stands out for its mildness and low environmental impact since it does not employ any toxic main group species, can even be performed in water, open to air and with parts per billion loadings of Pd.[3,5] Despite its obvious broad utility, until recently, the Suzuki–Miyaura reaction was virtually unheard of for the coupling of stereochemistry-bearing C–C bonds with the exception of cyclopropyl systems (Scheme 11.1).[6,7] Considering the ubiquity of the transformation and the importance of chiral molecules to the pharmaceutical industry specifically and organic chemistry in general, this is a remarkable statement.

RSC Catalysis Series No. 21
New Trends in Cross-Coupling: Theory and Applications
Edited by Thomas J Colacot
© The Royal Society of Chemistry 2015
Published by the Royal Society of Chemistry, www.rsc.org

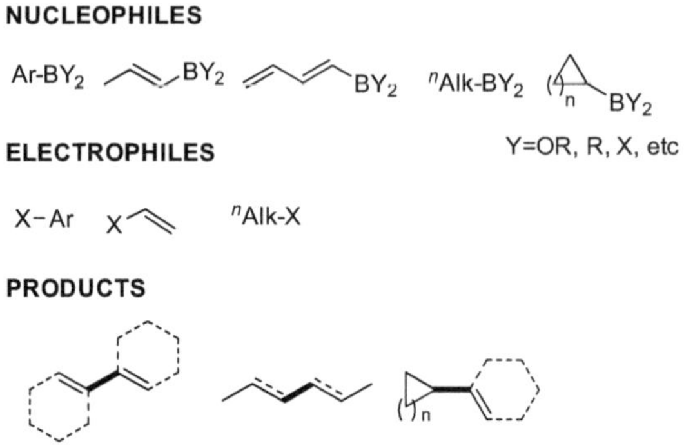

NUCLEOPHILES

Ar-BY$_2$ BY$_2$ BY$_2$ ^{n}Alk-BY$_2$ BY$_2$

ELECTROPHILES   Y=OR, R, X, etc

X–Ar   X ^{n}Alk-X

PRODUCTS

Scheme 11.1   Typical nucleophiles, electrophiles and products in Suzuki–Miyaura coupling chemistry.

Over the last few years, this deficiency has begun to be addressed.[8,9] Methods to affect the stereospecific coupling of certain classes of chiral organoboron nucleophiles have been reported over the last 5 years and development continues in this area. In addition, chiral electrophiles have also shown promise as partners in stereospecific coupling reactions. Finally, racemic electrophiles have been extensively examined in dynamic kinetic resolution strategies. This chapter describes recent advances in these three areas, starting first with stereospecific couplings of chiral, enantioenriched organoboron species. Advances in cross-coupling reactions that are not of the Suzuki–Miyaura variant will not be described in this chapter unless they are critical to the discussion. For these reactions, the reader is referred to several recent reviews.[8–11]

11.2   Chiral Nucleophiles: Stereospecific Suzuki–Miyaura Cross-Couplings

As early as 1986,[2] primary sp^3-hybridized boron nucleophiles (**1**) were shown to be effective partners in the Suzuki–Miyaura cross-coupling reaction [eqn (11.1)]. Since that time, this reaction has been employed extensively in the synthesis of complex organic molecules.[12] The transformation is made even more useful by the ease of synthesis of the primary borane by simple hydroboration of the corresponding alkene, usually with 9-BBNH, a reaction often performed back-to-back with the coupling itself, providing an interesting overall transformation [eqn (11.2)].[13] Other iterations of this reaction employ primary boronic esters or alternative trialkylboranes.[12–16]

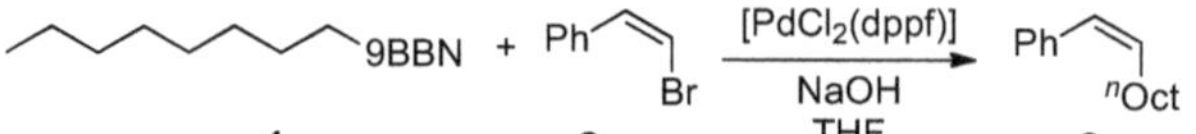

$$\text{9BBN} + \text{Ph} \xrightarrow[\text{Br}]{} \xrightarrow[\substack{\text{NaOH} \\ \text{THF}}]{[\text{PdCl}_2(\text{dppf})]} \text{Ph}\,{^n}\text{Oct} \qquad (11.1)$$

1   **2**   **3**

$$\text{R} \xrightarrow{\text{9BBN-H}} \left[\text{R}\,\text{9BBN} \right] \xrightarrow[\text{Ar-X}]{\text{Pd/Base}} \text{R}\,\text{Ar} \qquad (11.2)$$

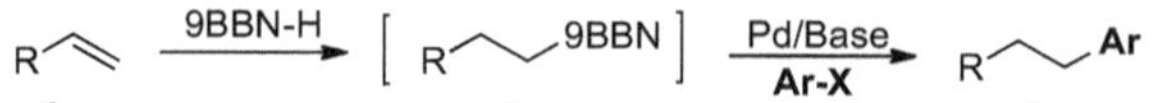

4   **5**   **6**

Secondary aliphatic boranes or boronic acids and esters are exceptionally interesting substrates since they provide the opportunity to prepare enantiomerically enriched products. However, these compounds have been extremely difficult compounds to engage in the Suzuki–Miyaura reaction. Transmetallation to Pd is problematic for branched substrates, likely because of the increased steric hindrance of the secondary site. In an elegant demonstration of this, Zou and Falck prepared the dialkyl pinacol boronate 7, in which a primary and a secondary butyl group both compete for transmetallation to Pd.[17] In all cases, the only product observed was that resulting from transmetallation and then coupling of the *primary* alkyl substituent [eqn (11.3)]. This follows from early work of Miyaura *et al.*, who similarly demonstrated that in triorganoboranes containing both primary and secondary alkyl groups, only the primary group participates in coupling [eqn (11.4)].[13] In the absence of *n*-alkyl substituents, secondary alkyl groups do react,[13] but the involvement of β-hydride elimination/addition sequences is not possible to assess, as will be elaborated on later in this chapter [eqn (11.5)]. Interestingly, this early work also demonstrated the relative resistance of the corresponding boronic ester derivative to coupling in comparison with trialkylboranes.

$$(11.3)$$

$$(11.4)$$

$$(11.5)$$

The second challenge for the successful coupling of secondary organoboranes is the propensity of secondary alkyl transition metal complexes to undergo β-hydride elimination.[8,18–20] The overall consequences of this side-reaction are the loss of organic groups upon reductive elimination with the hydride generated or conversely the readdition of the hydride to give a linear alkyl substituent and subsequent coupling to provide the undesired, linear product, 17 [eqn (11.6)].[21] In cyclic symmetrical systems, such addition/elimination sequences are invisible and therefore difficulties only appear when chiral or unsymmetrical substrates are employed.

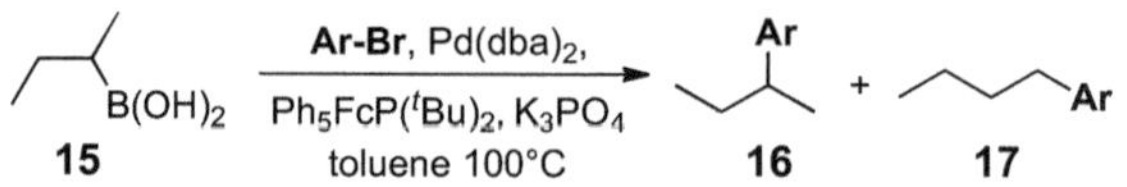

$$\text{(11.6)}$$

However, despite the associated challenges with the coupling of chiral secondary organoboron species, the variety of methods available for the preparation of enantioenriched secondary boronic acid and borane derivatives[22–32] and the obvious utility of the products make this a transformation of potentially high utility.

Initial reports of successful Suzuki–Miyaura cross-coupling reactions of secondary organoboron nucleophiles were confined to cyclopropyl derivatives.[6] These privileged structures have increased sp^2 character in cyclopropyl bonds, which likely facilitates transmetallation, and the high strain energy in cyclopropenes provides an impediment to post–transmetallation β-hydride elimination. Cyclopropylboronic acids were the first to be successfully coupled to aryl halides under Suzuki–Miyaura conditions [eqn (11.7) and (11.8)],[6] followed thereafter by potassium trifluoroborate[33,34] and boronic ester[35,36] analogs. When combined with the highly enantioselective hydroboration of cyclopropene derivatives reported by Gevorgyan and co-workers[37] (Scheme 11.2), this method represents an important route to chiral cyclopropanes. This method has proved invaluable for the incorporation of cyclopropyl groups.

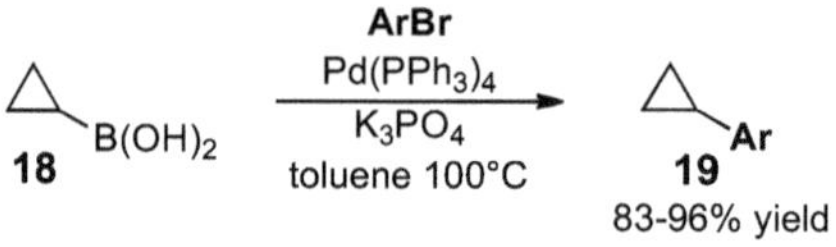

$$\text{(11.7)}$$

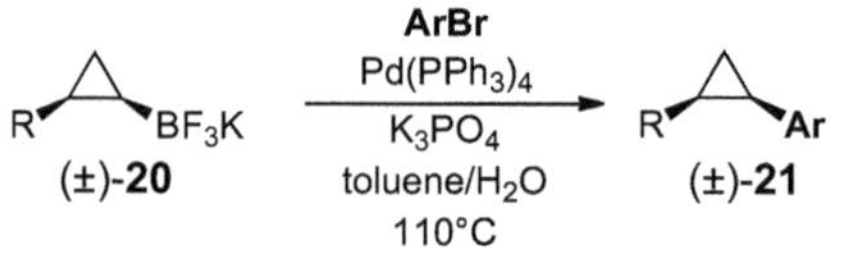

$$\text{(11.8)}$$

Cyclobutyl trifluoroborates[33] have also been employed effectively in Suzuki–Miyaura couplings [eqn (11.9)], but the more general case of cyclopentyl systems was first reported by Fu's group.[38] In a paper otherwise dedicated to the coupling of arylboronic acids, one example of the coupling of cyclopentylboronic acid (28) was included [eqn (11.10)].[39] This area lay

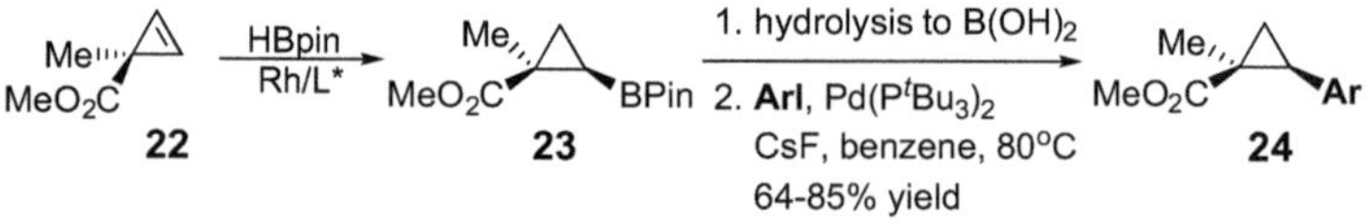

Scheme 11.2 Hydroboration/Suzuki–Miyaura coupling conditions for the synthesis of stereodefined cyclopropanes.

dormant for a remarkable 8 years until van den Hoogenband *et al.* described the cross-coupling of cyclopentyl-BF$_3$K derivatives using RuPhos as the ligand in combination with Pd(OAc)$_2$ to yield the corresponding coupling products in moderate to good yields, [eqn (11.11)].[40]

$$\text{25} \quad \text{BF}_3\text{K} + \text{Ar-Cl} \quad \xrightarrow[\substack{\text{Cs}_2\text{CO}_3 \\ \text{toluene/H}_2\text{O} \\ 100°\text{C}}]{\text{Pd(OAc)}_2, \, ^n\text{BuPAd}_2} \quad \text{27} \quad \text{Ar}$$

$$\tag{11.9}$$

26

$$\text{28} \quad \text{B(OH)}_2 + \text{Tolyl-Cl} \quad \xrightarrow[\substack{\text{KF, THF} \\ 100°\text{C}}]{\text{Pd(OAc)}_2, \, ^t\text{Bu}_3\text{P}} \quad \text{30} \quad \text{CH}_3$$

29

75% yield

$$\tag{11.10}$$

$$\text{31} \quad \text{BF}_3\text{K} + \text{Ar-Br} \quad \xrightarrow[\substack{\text{K}_3\text{PO}_4, \, 115 \, °\text{C} \\ \text{toluene/water}}]{\text{Pd(OAc)}_2, \, \text{RuPhos}} \quad \text{33} \quad \text{Ar}$$

32

$$\tag{11.11}$$

Note that in both of these cases, symmetrical nucleophiles are employed and therefore β-hydride elimination/reinsertion events will not be visible. Closely following after the van den Hoogenband report,[40] Molander and co-workers showed clearly that these events do occur [eqn (11.12)],[41] and they result in chain walking of the Pd and subsequently aryl incorporation throughout the entire structure of the organoborane. This sequence of events can be minimized by the use of bulky electron-rich phosphine ligands such as PhP^{t}Bu$_2$ and, for symmetrical organoborates such as **39**, high yields can be obtained [eqn (11.13)].[41]

$$\tag{11.12}$$

ratio = 72 : 4 : 3 : 21

$$\text{39} \quad \text{BF}_3\text{K} \quad \xrightarrow[\substack{\text{Cs}_2\text{CO}_3 \\ \text{toluene/water} \\ 100°\text{C}}]{\text{Pd(OAc)}_2, \, \text{PhP}^t\text{Bu}_2} \quad \text{40} \quad \text{Ar}$$

up to 94% yield

$$\tag{11.13}$$

Interestingly, for the simple isopropyl trifluoroborates (**41**), the amount of direct coupling *versus* β-hydride elimination followed by coupling is dependent on the structure of the aryl halide, illustrating the sensitivity of this reaction to sterics (Scheme 11.3). Note that these elimination/addition

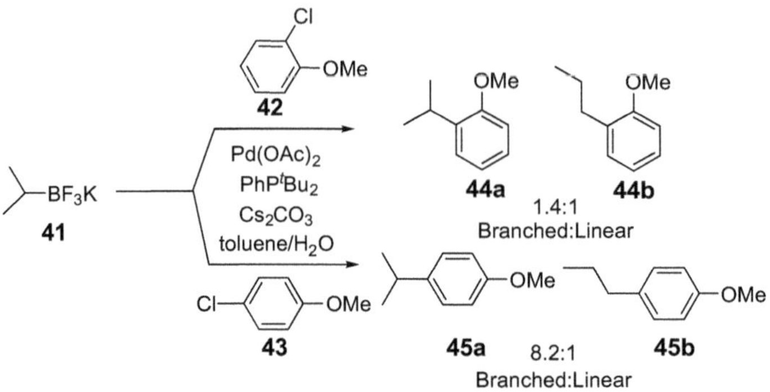

Scheme 11.3 Effect of aryl halide on β-hydride elimination and subsequent product distribution in coupling of ^{i}PrBF$_3$K (**41**).

pathways were first reported by Hartwig and co-workers in the coupling of ^{s}BuB(OH)$_2$ to aryl iodides as previously noted [eqn (11.6)].[21]

In 2009, the first example of a cross-coupling of chiral boronic esters that did not undergo detectable β-hydride elimination/addition was reported by our group.[42] Using boronic ester **46**, prepared by the regio- and enantio-selective hydroboration of styrene, we effected Suzuki–Miyaura cross-coupling with aryl iodides leading to diarylethanes (**47**) with high enantiospecificity [eqn (11.14)].[42] Critically, it was determined that the main difficulty to be overcome for this class of substrate was not β-hydride elimination, but a sluggish transmetallation. Thus, in the presence of Pd catalysts and typical inorganic and organic bases, boronic ester **46** remained untouched. This problem was solved by employing Ag$_2$O as the base, which was reported by Kishi and co-workers to promote transmetallation in difficult achiral systems.[43]

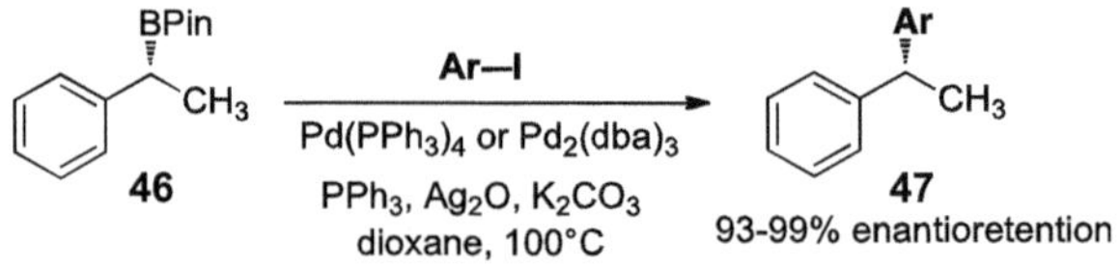

(11.14)

Simple adjustments to the reaction conditions, including the addition of triphenylphosphine in mild excess relative to Pd, resulted in a high-yielding, highly enantiospecific reaction producing 1,1-diarylethanes that are difficult to prepare by other methods.[44–48] Interestingly, the related BF$_3$K salts are completely unreactive under these conditions and under other conditions typically employed for the coupling of this type of nucleophile.[49–52] Comparisons with known compounds from the literature were performed to demonstrate that this coupling proceeds with retention of stereochemistry. This reaction was subsequently employed by Li and Burke in the synthesis of **48**, which is a glucagon receptor antagonist being examined for the treatment of type II diabetes [eqn (11.15)].[53]

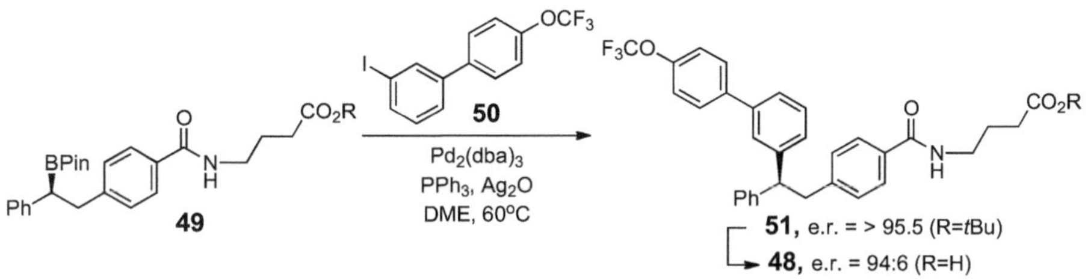

$$(11.15)$$

Remarkably, the Ag$_2$O-mediated protocol, so effective at coupling benzylic boronic esters, was found to be completely ineffective with purely aliphatic primary or secondary nucleophiles in the absence of other bases [eqn (11.16)]. Thus the linear hydroboration product (52) is unreactive whereas the branched isomer reacts in good yield.[42]

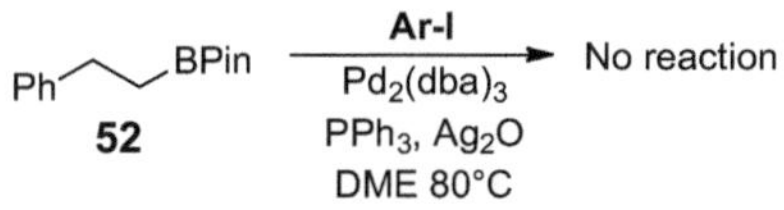

$$(11.16)$$

As a clue to the origin of this effect, although simple *n*-alkyl- or even *sec*-alkylboronic esters do not couple under the silver(I) oxide conditions, installing a single site of π-unsaturation proximal to the secondary boronic ester renders the resulting allylic boronic esters reactive under optimized conditions (Scheme 11.4).[54] This discovery led to the first general protocol for the cross-coupling of secondary allylic boronic esters. The position of arylation is dictated by the nature of the substituents on the allylic boronic ester, with coupling taking place typically at the γ-position, except in the case of styrenyl systems such as 58, in which α-selective coupling is observed (Scheme 11.5).

In a collaboration between the Aggarwal and Crudden groups,[55] this reaction was expanded to include enantiomerically enriched boronic esters and mechanistic details elucidated by the use of deuterated substrates. Interestingly, unlike the case of benzylic boronic esters (17), optimal reaction

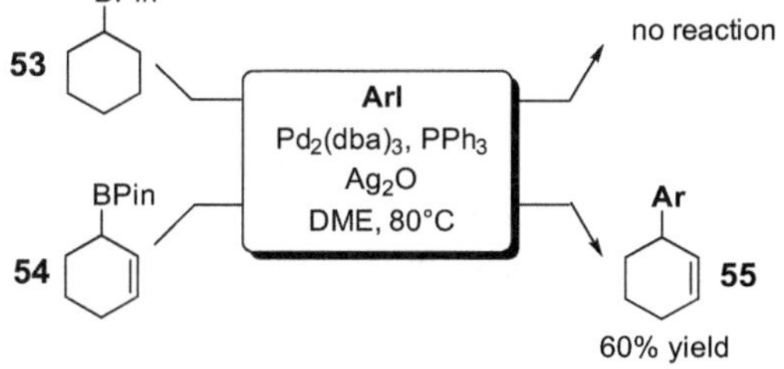

Scheme 11.4 Effect of allylic unsaturation on Suzuki–Miyaura cross-coupling of secondary boronic esters.

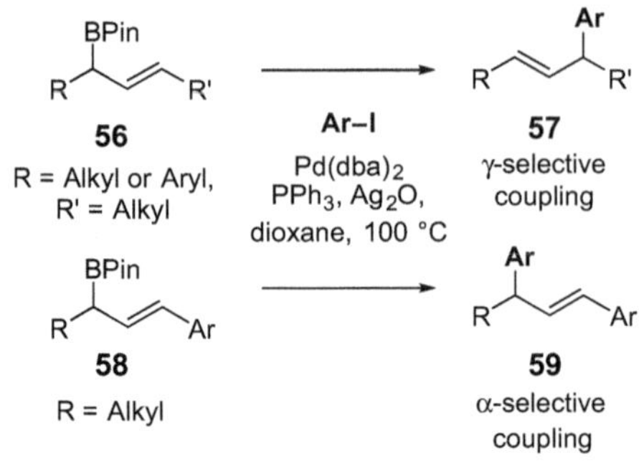

Scheme 11.5   Suzuki–Miyaura coupling of unsymmetrical allylboronic esters.

conditions required the use of significantly lower amounts of PPh$_3$, although this still proved to be the ligand of choice. Thus, using only 2 equiv. of phosphine to Pd, moderate to good yields were obtained for the desired coupling products with high to perfect enantioselectivities (Table 11.1).

For simple aliphatic substrates, $\gamma:\alpha$ ratios from 78:22 to 95:5 were obtained, but consistently higher regioselectivities were observed for systems containing an aryl group in the allylic position (Table 11.1, Entries 1–3; Table 11.2, Entry 1). Interestingly, if this aryl substituent is on the olefin, the reactions are uniformly α-selective (Table 11.1, Entries 4–6). In order to examine this reaction in more detail and to assess the potential involvement of π-allyl-Pd intermediates in the reaction, deuterated substrate **71** was prepared, in which 90% of the deuterium label is at the vinylic position and the allyl unit is substituted with two identical methyl groups [eqn (11.17)].[55] If a fully equilibrated π-allyl-Pd species is implicated, a 50:50 mixture of **72** and **73** would be expected. In the event, an 85:15 ratio of **72**:**73** was observed, with the major product resulting from γ-selective coupling.[55]

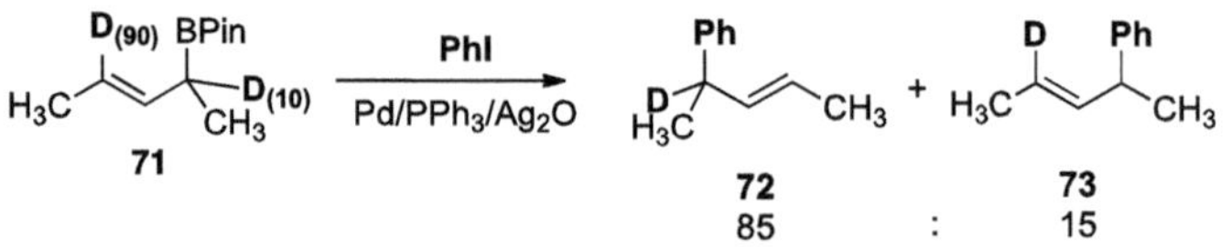

$$(11.17)$$

This implies that the reaction proceeds *via* a facile reductive elimination following γ-transmetallated intermediate **II**. Based on the deuterium label distribution in the product, less than 5% of intermediate **II** is isomerized to **IV** *via* π-allyl intermediate **III** (Scheme 11.6).[55]

Although the related deuterated substrate flanked by two phenyl groups could be prepared, it underwent facile borotropic shifts that resulted in scrambling of the deuterium and rendered this same analysis impossible. However, analyzing the product distribution from substrates **74** and **75** (Scheme 11.7) provides some interesting insight. As expected, (*E*)-**74** undergoes a highly γ-selective coupling giving the expected product (**76**) in a

Table 11.1 Cross-coupling of racemic allylic boranes.[55,a]

Entry	Compound	$\gamma : \alpha^a$	$E : Z^b$	Yield $(\%)^c$
1	(E)-60 (H₃C···BPin···Ph)	97 : 3	99 : 1	65
2	(Z)-61 (CH₃ BPin / Ph)	92 : 8	99 : 1	70
3	62 (H₃C···CH₃ BPin···Ph)	92 : 8	99 : 1	40
4	(E)-63 (Ph···BPin···Ph)	39 : 61	86 : 14	28^d
5	(E)-64 (Ph···BPin···C₄H₉)	8 : 92	n.d.^{e}	58
6	(E)-65 (Ph···BPin···C₆H₁₃)	19 : 81	n.d.^{e}	53

^{a}Conditions: PhI (1 equiv.), Pd(dba)$_2$ (5%), PPh$_3$ (10%), Ag$_2$O (1.5 equiv.), DME (0.1 M), 90 °C, 16 h.
$^b\gamma : \alpha$ ratio determined by GC and/or ^{1}H NMR spectroscopy and $E : Z$ ratio given for the major isomer.
^{c}Isolated yield of major isomer unless noted otherwise.
^{d}Yield of a mixture of regioisomers.
^{e}Not determined.

98:2 ratio with the corresponding α-product (Scheme 11.7). When the isomeric substrate **75** was employed, the same major product was obtained, but the selectivity was significantly lower. This implies that intermediate **II** formed by transmetallation *via* allylic transposition is more prone to equilibration *via* π-allyl-Pd intermediates in cases where this results in re-establishment of stable styrene-type units.[55]

Aggarwal's group has also demonstrated that propargylic boronic esters also react with high yield and perfect enantiospecificity under the conditions initially developed for benzylic systems, even in the case when the boron substituent is at a fully substituted tertiary carbon atom [eqn (11.18)].[56] This method leads to a very valuable synthetic route for the preparation of chiral tetrasubstituted allenes.

Table 11.2 Cross-coupling of enantioenriched allylic boranes.[55,a]

Entry	Compound	γ:α[b]	E:Z[b]	Yield (%)[c]	Enantioenrichment (S.M.)[d]	Enantiospecificity (%)[e]
1	(R,E)-66	98:2	99:1	78	98:2	100
2	(R,Z)-67	83:17	94:6	75[f]	98:2	96
3	(R,E)-68	94:6	78:22	81	98:2	100
4	(R,Z)-69	90:10	99:1	77[f]	96:4	96
5	(R,E)-70	92:8	99:1	71	96:4	100

[a]Conditions as in Table 11.1.
[b]γ:α ratio determined by ^{1}H NMR spectroscopy and E:Z selectivity is given for the major regioisomer.
[c]Isolated yield of the major isomer unless stated otherwise.
[d]Enantioenrichment of the starting material.
[e]Enantiospecificity = ee product/ee starting material.
[f]Isolated as mixtures of γ and α products.

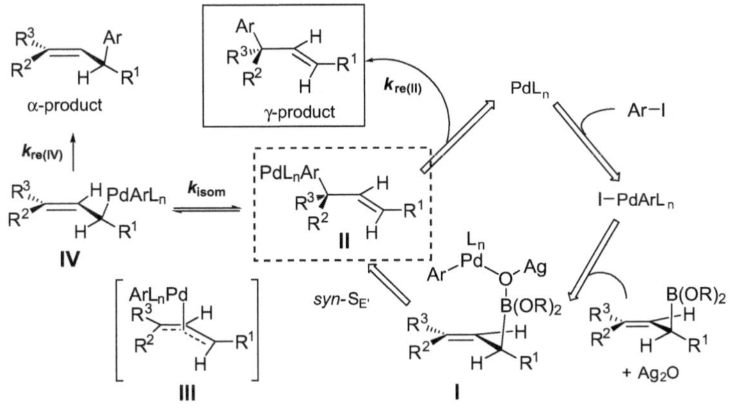

Scheme 11.6 Plausible mechanism for the Suzuki–Miyaura coupling of unsymmetrical allylic boronic esters.

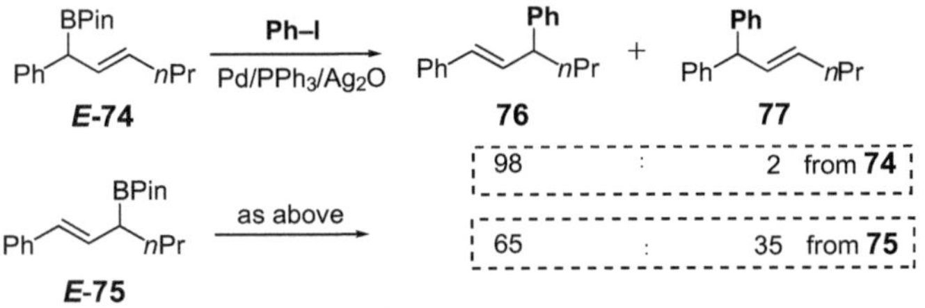

Scheme 11.7 Effect of styrene units on positional selectivity in allylic coupling.

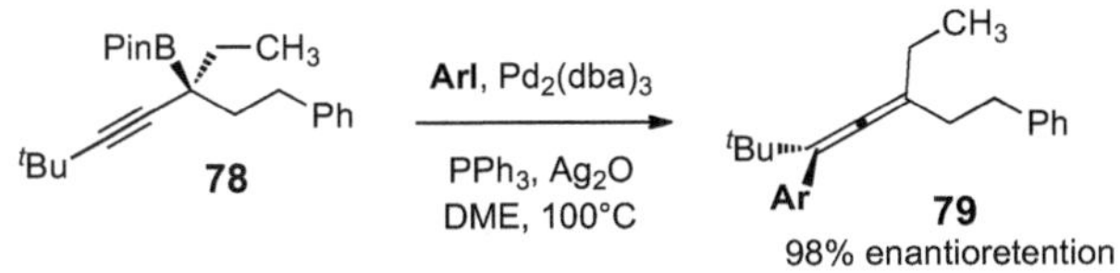

$$(11.18)$$

In addition to the coupling of allylic, benzylic and propargylic boronic esters, dibenzylic boronic esters are also effective partners.[57] These substrates are interesting since they are immune from any β-hydride elimination pathways, but the challenge in this case is their enantioselective synthesis. In an attempt to design enantioselective triarylmethane synthesis, Crudden's group examined a variety of methods that would yield enantiomerically enriched dibenzylic boronic esters, ultimately settling on a variant of Aggarwal's chemistry (Scheme 11.8).[57]

The challenge in this case was the unprecedented use of benzylic carbamates in the 1,2-metallate rearrangement (Scheme 11.8, **80b**, R = aryl). Once deprotonated, the benzylic organolithium species derived from **80b** is more stable than typical alkyl organometallics and, when combined with the increased steric hindrance of the borate resulting from the reaction of

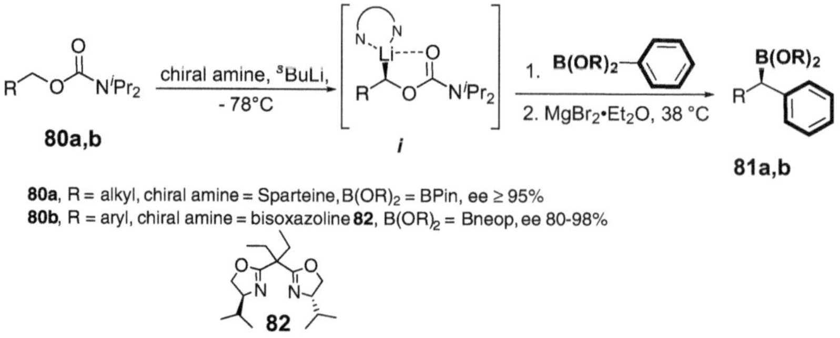

Scheme 11.8 Synthesis of optically enriched boronic esters **81a** and **81b**.

intermediate *i* with the boronic ester, this leads to re-equilibration back to *i* at higher temperature and resulting racemization. However, consistent with reports from Aggarwal *et al.* in other systems sensitive to racemization,[58] it was found that employing less hindered neopentyl boronic esters along with a bis-oxazoline ligand[59] in place of the typical sparteine ligand in addition to the smaller neopentyl diol-derived boronate ester led to high enantioselectivities (Scheme 11.8).[57]

Once the synthesis of the starting boronic esters was developed, the coupling reaction proceeded with good to excellent enantiospecificities [eqn (11.19)], typically on the order of 90–100%. Consistent with related chemistry, the Suzuki–Miyaura cross-coupling proceeded with retention of configuration.[57]

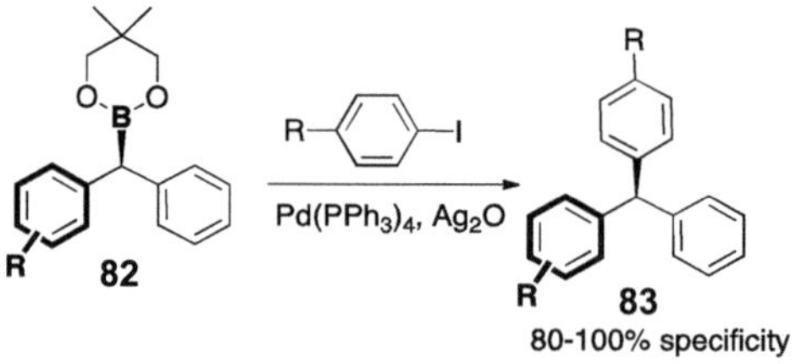

(11.19)

An outlier in terms of the requirement for silver(I) oxide in the coupling of secondary boronic ester derivatives, homoallylic boronic esters also undergo coupling in the specific case where there is activation by an internal base.[60] Thus, Masaki and Suginome demonstrated that **84a** and **84b** undergo coupling with aryl iodides in the presence of typical Pd catalysts and bases with retention of configuration (Scheme 11.9). Interestingly, the corresponding *O*-acetylated BPin derivatives (**86a** and **86b**) are inert under identical conditions, demonstrating the importance of the specific substituents on boron in this case.[60]

The final class of substrate that has been demonstrated to undergo stereoretentive coupling is benzylic ethers. In 2012, Molander's group reported that enantioenriched benzyl ethers **87** substituted at the α-position with a BF$_3$K group undergo stereoretentive Suzuki–Miyaura coupling [eqn (11.20)].[61]

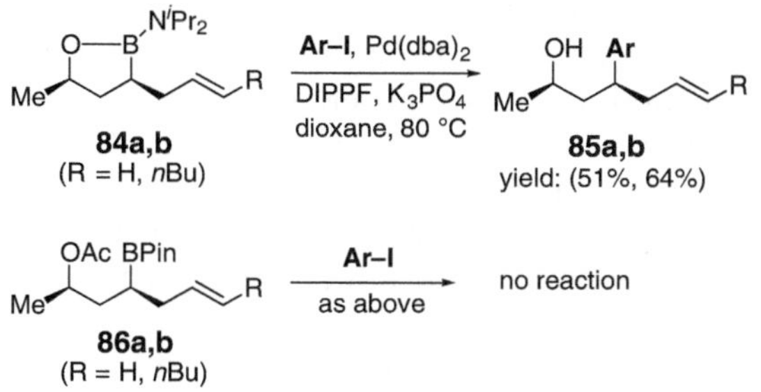

Scheme 11.9 Effect of internal chelating group on reactivity of homoallylic boronic ester derivatives **84** and **86**.

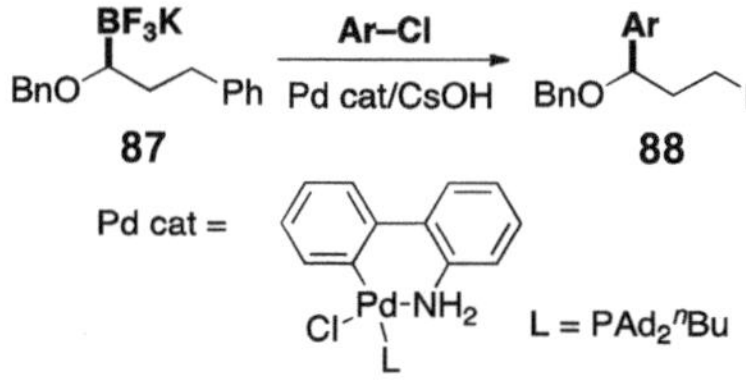

$$(11.20)$$

Remarkably, the reaction showed considerable sensitivity to the nature of the protecting group. Of those tried, only the benzyl group showed any reactivity. This is interesting in the light of the fact that the reaction occurs with retention of configuration, since it implies that some type of π-coordination is occurring in this system, in analogy with the other examples above. Unlike the bulk of examples of π-activated systems, the most effective base was found to be CsOH, although the reactivity of Ag$_2$O was not reported.

In addition to those substrates described above, amide-containing substrates are another important class of chiral compounds that can be made to react under Suzuki–Miyaura cross-coupling conditions. Suginome's group[62] was the first in this field, demonstrating that α-borylamides undergo coupling with aryl bromides leading to diarylamides with high levels of stereochemical *inversion* [eqn (11.21)]. They proposed that the carbonyl of the amide plays a critical role in activating the boron center by binding to it and then promoting an S_N2-like transmetallation. Interestingly, and in support of this concept, the addition of a Lewis acid that can bind to the amide carbonyl results in a process that occurs with high *retention* of configuration [eqn (11.21)].[63]

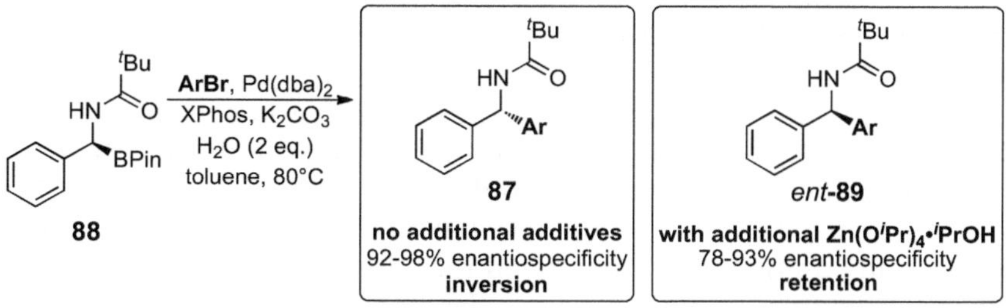

$$(11.21)$$

Molander and co-workers[64] demonstrated that systems in which the amide is further removed from the boryl substituent also undergo Suzuki–Miyaura couplings with high enantiospecficity [eqn (11.22)]. Using chiral, enantiomerically enriched β-boryl amides (**90**) under conditions similar to those used by Suginome's group, Molander and co-workers showed that this reaction also proceeds with inversion of configuration. Highlighting the importance of the amide group, when this substituent was replaced with an ester, no reaction was observed. Importantly, the starting materials are readily available in enantiomerically pure form by hydroboration of the corresponding alkene using conditions developed by Takacs and co-workers[28,65] or alternatively NHC[66] or Cu–cat borylations[67] (Scheme 11.10).

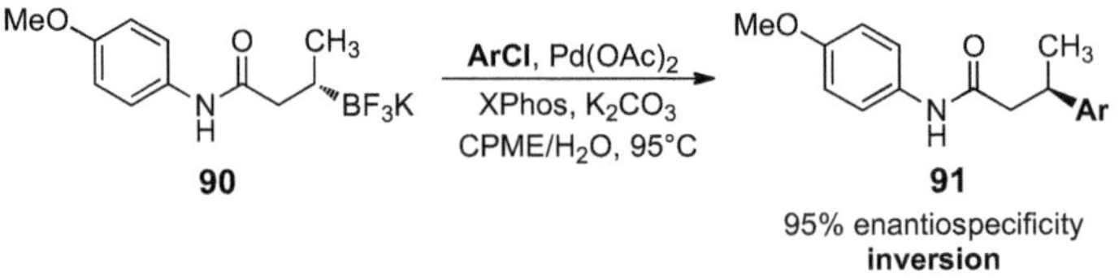

(11.22)

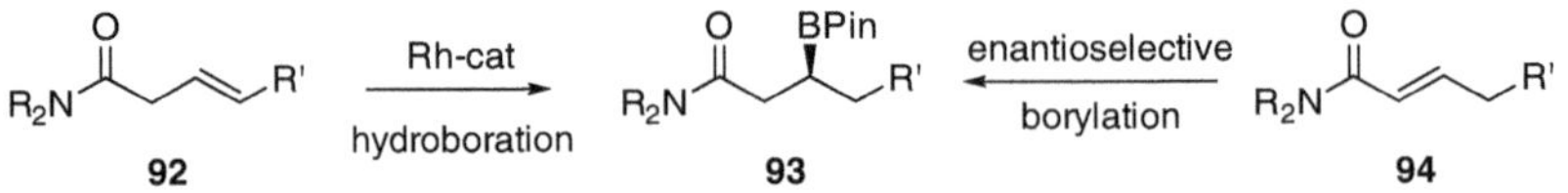

Scheme 11.10   Synthetic methods for the preparation of β-borylated amides.

Taking advantage of the protecting features of Suginome and co-workers' dansyl group (1,8-diaminonaphthalene, dan),[68] Hall and co-workers synthesized chiral bis-boronate **95** and demonstrated the enantiospecific coupling of one geminal boronic ester chemoselectively in the presence of the other (Scheme 11.11).[69] As in other amide-directed methods, the reaction proceeds with inversion of stereochemistry and virtually perfect enantiospecificity was observed. After removal of the dan group and conversion to an amide, the remaining boron group was coupled with a second aryl halide, also stereospecifically (Scheme 11.11).[69] Interestingly, an ester is sufficiently activating in the case of the geminal diboron compound **95**, but once the first Suzuki–Miyaura coupling has been performed, this group must be converted to amide **99** prior to the final coupling. This higher activity of geminal diboron species **95** is consistent with the work of Shibata and co-workers, which showed that racemic geminal bis-boron compounds such as **101** react without any directing or activating groups [eqn (11.23)].[70]

Consistent with the importance of the second boron substituent, Shibata and co-workers showed that compounds with simple aliphatic C–B bonds do not couple under these conditions and furthermore that a silicon substituent is not able to activate a geminal C–B bond for coupling.[70] Subsequently, Morken and co-workers demonstrated that this reaction can be carried out with enantiotopic group selection, leading to chiral enantiomerically

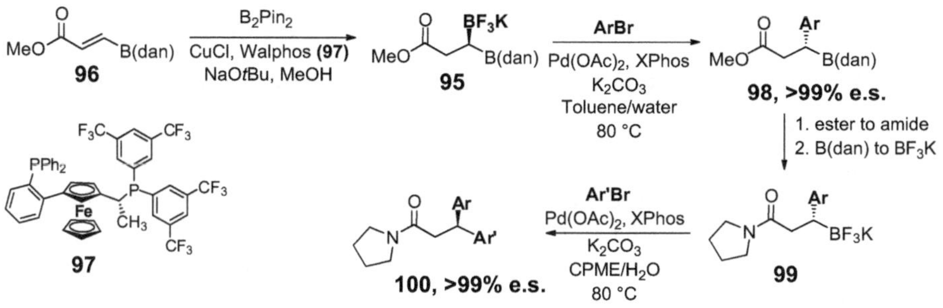

Scheme 11.11 Hall method for the sequential coupling of β-diboryl carbonyl compounds.

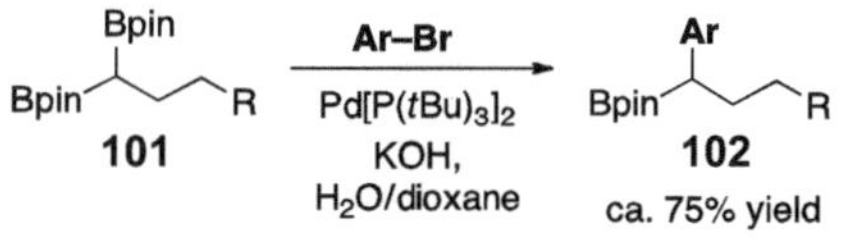

$$(11.23)$$

enriched boronic esters **104** [eqn (11.24)].[32] This provides a very significant advance in the synthesis of chiral organoboron compounds.

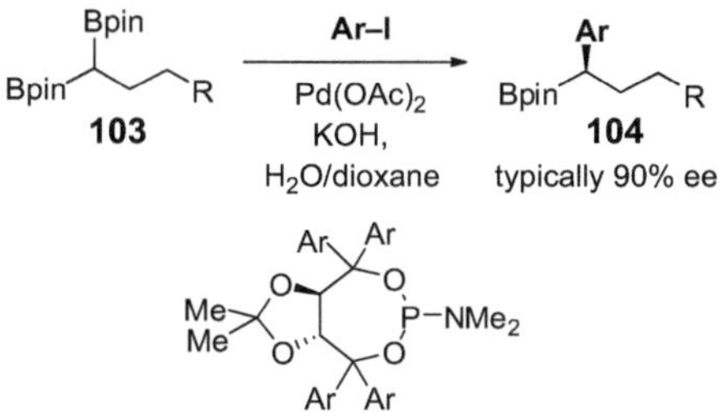

$$(11.24)$$

Hence although the enantiospecific cross-coupling of secondary boronic esters is still limited in scope, there have been significant advances in the past few years such that this reaction is now feasible with a variety of substrates, expanding the utility of this reaction past the initially discovered cyclopropyl systems. Importantly, these advances are starting to find practical applications in complex synthesis. For example, the silver-mediated cross-coupling of benzylic boronic esters under Crudden conditions has been used by Li and Burke to effect the critical, stereoretentive coupling step in the total synthesis of the 1,1-diarylethane **48**, which is a glucagon receptor antagonist being examined for treatment of type II diabetes [eqn 11.15)]. [53] Clearly, these new advances in the Suzuki–Miyaura reaction are already providing facile access to compounds that are difficult to prepare by other routes. Further developments in this area will only increase the scope and potential applications of this already powerful reaction.

11.3 Chiral Electrophiles

11.3.1 Stereospecific Cross-Coupling of Alkyl Electrophiles

As noted in the Introduction, primary alkyl electrophiles have been successfully used as coupling partners in the Suzuki–Miyaura reaction for decades.[71–73] Thus, in addition to coupling reactions of secondary boronic esters and their derivatives as a route to the creation of stereocenters *via* Suzuki–Miyaura chemistry, the use of secondary electrophiles is another possibility.[10] Since the oxidative addition of alkyl bromides with Pd typically proceeds *via* an S_N2 manifold,[74–77] this provides a potentially valuable method for the use of these and related electrophiles in cross-coupling chemistry.

The first examples of secondary alkyl electrophiles used competently in C–C bond-forming events were in the Negishi manifold of coupling reactions. In 2003, Rovis and co-workers reported a Ni-catalyzed decarbonylative C–C coupling of *meso*-anhydrides with diphenylzinc.[78] The decarbonylation of the initially formed RCONi species was driven by the *in situ* formation of an Ni–CO complex (Scheme 11.12), which, although effective, requires a super-stoichiometric amount of Ni as the CO trap. The observation of only one diastereomer and the lack of *exo–endo* inversions demonstrate the configurational stability of the C–Ni bond under the specific reaction conditions (Scheme 11.12).

In the realm of alkyl halides themselves, building on the work of Stille and co-workers, who showed that enantiomerically enriched 2-phenethyl bromide (**111**) undergoes stereospecific Pd-catalyzed carbonylation,[74] Carretero and co-workers demonstrated that the same enantiomerically enriched bromide could undergo Kumada–Corriu coupling with almost perfect stereoinversion.[79] The use of Xantphos as a ligand was shown to be important to prevent β-hydride elimination and, under optimized conditions, the reaction proceeded with very high yield, providing another interesting method for the synthesis of 1,1-diarylethanes [eqn (11.25)].

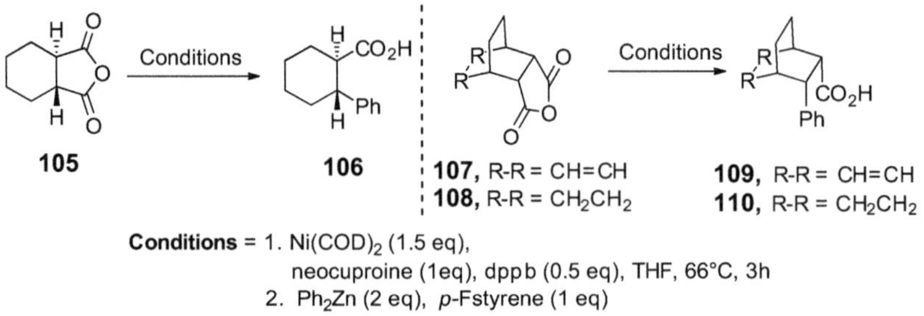
Conditions
Conditions
CO2H
Ph
CO2H
Ph
R
R'
105
106
107, R-R = CH=CH
108, R-R = CH2CH2
109, R-R = CH=CH
110, R-R = CH2CH2
Conditions = 1. Ni(COD)2 (1.5 eq),
neocuproine (1eq), dppb (0.5 eq), THF, 66°C, 3h
2. Ph2Zn (2 eq), p-Fstyrene (1 eq)

Scheme 11.12 Observation of net retention of configuration in Ni-mediated Negishi coupling of anhydrides.

$$\text{(11.25)}$$

111
85% ee

→ ArMgBr, Pd(CH₃CN)₂Cl₂, Xantphos, THF, rt, 14 h →

112
97% yield
>98%% e.s.

However, there are three main difficulties with this as a general approach: (1) the dearth of methods available for the synthesis of enantiomerically pure alkyl halides; (2) the possibility of competing radical mechanisms for oxidative addition that would be expected to occur with racemization; and (3) the increased ease of β-hydride eliminations in secondary alkyl–metal complexes relative to their primary congeners, which may lead to scrambling of positional selectivity even after a successful oxidative addition event. As shown in the examples below, the use of specific ligands can surmount difficulties with the β-hydride elimination reaction, but lack of S_N2-type re-activity and substrate limitations may be a more significant problem.

In fact, Fu and co-workers demonstrated that sterics have a strong effect on the rate of oxidative addition of even primary alkyl halides to Pd catalysts.[76] Systematically adding steric bulk to the γ- and β-positions of a primary alkyl bromide slows the rate of oxidative addition by factors of 5 and 20. respectively (Scheme 11.13), and the use of a secondary alkyl bromide was found to completely shut down reactivity with Pd catalysts. This latter fact seems to suggest that only activated secondary bromides are reasonable partners, although it should be noted that a relatively bulky phosphine ligand was employed in this study.

The first report of a stereospecific Suzuki–Miyaura reaction with enantiomerically pure electrophiles involved the use of α-bromosulfoxides,[80] in which the bromide is activated by virtue of the adjacent sulfoxide in the same manner as the benzyl substituent increases the reactivity of the bromide in **111**. As shown in eqn (11.26), the reaction of enantio- and dia-stereo-defined bromide **113** with phenylboronic acid leads to the desired product **114** with perfect inversion of stereochemistry. The authors proposed that the inversion is related to the S_N2 nature of the oxidative addition step and that the remainder of the catalytic cycle proceeds with retention of configuration.[80] The presence of *tert*-amyl alcohol was shown to be im-portant to suppress β-hydride elimination as a non-oxidizable ligand for Pd.

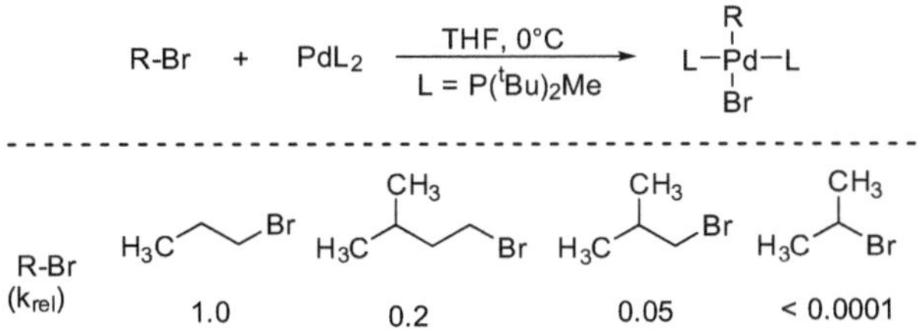

R-Br + PdL₂ → THF, 0°C, L = P(^{t}Bu)₂Me →

R
|
L–Pd–L
|
Br

R-Br
(k$_{rel}$)

1.0 0.2 0.05 < 0.0001

Scheme 11.13 Relative rates for oxidative addition of various alkyl halides to Pd⁰.

Interestingly, the reaction was highly sensitive to chirality at sulfur, such that the diastereomer of **113** was unreactive.

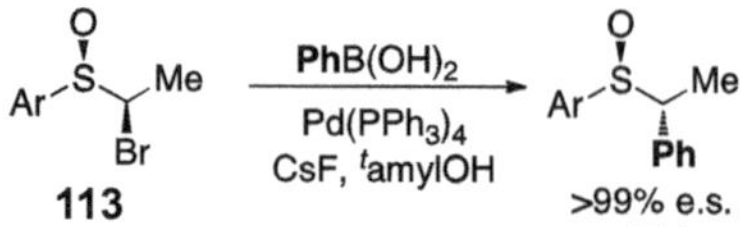

$$\underset{\textbf{113}}{\underset{\overset{|}{\underset{Br}{\overset{O}{\overset{\|}{Ar\overset{}{-}S}}}}}{}}\overset{}{\underset{}{Me}} \quad\xrightarrow[\substack{Pd(PPh_3)_4 \\ CsF, \,^tamylOH}]{PhB(OH)_2}\quad \underset{\substack{>99\% \text{ e.s.} \\ \textbf{114}}}{\underset{\overset{|}{\underset{Ph}{\overset{O}{\overset{\|}{Ar\overset{}{-}S}}}}}{}}\overset{}{\underset{}{Me}} \qquad (11.26)$$

Following on this work, He and Falck described the stereospecific Suzuki–Miyaura coupling of α-cyanotriflates **115**.[81] As shown in eqn (11.27), boronic acids couple readily with these substrates under mild conditions with high levels of enantiospecificity. Again, the reaction proceeds with inversion of configuration.

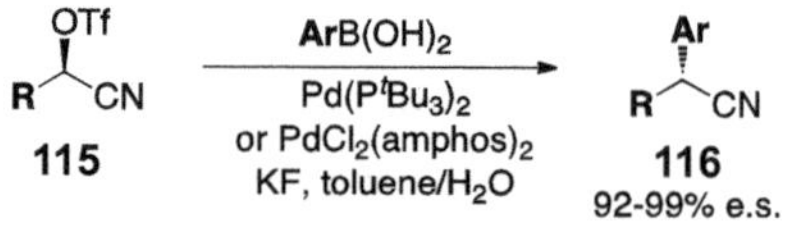

$$\underset{\textbf{115}}{\overset{OTf}{\underset{|}{R\overset{}{\frown}CN}}} \quad\xrightarrow[\substack{Pd(P^tBu_3)_2 \\ \text{or } PdCl_2(amphos)_2 \\ KF, \text{ toluene}/H_2O}]{ArB(OH)_2}\quad \underset{\substack{\textbf{116} \\ 92\text{-}99\% \text{ e.s.}}}{\overset{Ar}{\underset{|}{R\overset{}{\frown}CN}}} \qquad (11.27)$$

Building on these precedents and those of racemic couplings of benzylic carbonates,[82] Jarvo and co-workers demonstrated that stereodefined alkyl ethers could serve as useful substrates in Kumada–Corriu coupling [eqn (11.28)].[83] As shown below, the reaction proceeds *via* Ni catalysis and the presence of a π-extended aromatic in the electrophile is critical. However, considering the wide variety of enantiomerically enriched alcohols that can be easily prepared, this method is an extremely important advance in the area of stereospecific coupling of chiral nucleophiles. In subsequent work, the nature of the ether substituent was shown to play a significant role, such that by employing chelating groups such as the ethylene glycol derivative in **117b**, the scope of the reaction was widened to include aryl Grignard reagents.[84,85] This is important since it permits the synthesis of enantiomerically enriched triarylmethanes, which, as previously noted, are difficult to prepare by other methods. As in the other examples described to date, the reaction proceeded with inversion of configuration.

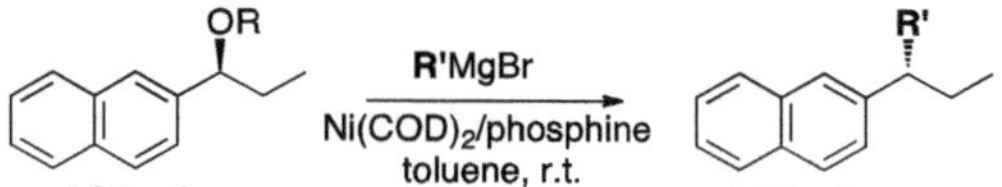

$$\underset{\substack{\textbf{117a, b}}}{}\qquad\xrightarrow[\substack{Ni(COD)_2/phosphine \\ toluene, \text{ r.t.}}]{R'MgBr}\qquad \underset{\substack{\textbf{118a, b}}}{} \qquad (11.28)$$

a, R = Me, R' = alkyl only
b, R = CH₂CH₂OMe, R' = alkyl or aryl

This reaction was later expanded to include organozinc nucleophiles,[86] but more relevant to the topic of this chapter, this same concept has also been applied to the Suzuki–Miyaura reaction of chiral alcohol-derived electrophiles. In 2013, Jarvo's group demonstrated that a variety of enantioenriched electrophiles could participate in the Ni-catalyzed Suzuki-Miyaura reaction with neopentylboronic esters.[87] Remarkably, by switching the nature of the ligand on Ni, the reaction could be made to proceed with high levels of inversion or retention of configuration. Thus, as shown in eqn (11.29), when PCy$_3$ is used as a ligand for Ni, the reaction proceeds with retention, whereas with SIMes it proceeds with inversion of configuration. *n*-Butanol proved to be a key additive in the reaction in order to achieve high levels of stereospecificity. Of the derivatives employed, the carbamate shown in **119** gives the best combination of high yield and enantiospecificity, with other protecting groups for the alcohol giving lower yields or selectivities. Interestingly, however, when the pivalate ester **121** is employed instead, the requirement for an extended aromatic ring on the starting electrophile is lifted, dramatically expanding the scope of this important transformation [eqn (11.30)].

$$(11.29)$$

$$(11.30)$$

In related work appearing soon after the Jarvo study, Watson and co-workers also demonstrated that the pivalate group could be effectively employed in the stereospecific Suzuki–Miyaura coupling reaction shown in eqn (11.31).[88] Again, nickel catalysis was employed but in this case an added ligand was not required. Optimal conditions included the use of arylboroxines, NaOMe as a base at 70 °C in toluene with 5% Ni(COD)$_2$ as catalyst. Under these conditions, a variety of diarylethanes were produced in 86–99% *ee*. Interestingly, as under the conditions used by Jarvo and co-workers, the presence of a π-system with extended conjugation was typically required, but one example was given a in which a biphenyl substituent was employed. Again, the reaction was found to proceed with inversion of configuration.

$$(11.31)$$

In addition to the use of C–O bonds as nascent electrophiles, Watson and co-workers also described the cleavage of C–N bonds, activated as ammonium groups.[89] As shown in eqn (11.32), the Suzuki–Miyaura coupling of benzylic ammonium compounds, readily prepared by alkylation of the corresponding amine, provides another very valuable route for the synthesis of chiral hydrocarbons. Simple monocyclic aromatic derivatives also work well in this reaction, although lower yields are typically observed. Both of these methods are important in that they permit the use of compounds that are readily available in enantiomerically pure form as electrophiles for the stereospecific Suzuki–Miyaura reaction.

$$
\begin{array}{ccc}
\underset{\substack{\mathbf{125}\\ >98\%\ ee}}{\overset{\mathrm{NMe_3^+}}{\underset{\mathrm{Ar^1}}{\big|}}\!\!\!\!\!\!\!\!\!\!\!\!\underset{}{\diagup\!\!\!\searrow}\mathrm{CH_3}} & \xrightarrow[\substack{\mathrm{Ni(COD)_2}\\ \mathrm{P(oTol)_3\ or\ ^tBuXantphos}\\ \mathrm{K_3PO_4,\ 70\ ^\circ C,\ 6\ h}}]{\mathrm{Ar^2B(OH)_2}} & \underset{\substack{\mathbf{126}\\ \text{e.s. } 88\text{-}99\%\\ \text{yield } 54\text{-}85\%}}{\overset{\mathrm{Ar^2}}{\underset{\mathrm{Ar^1}}{\big|}}\!\!\!\!\!\!\!\!\!\!\!\!\underset{}{\diagup\!\!\!\searrow}\mathrm{CH_3}}
\end{array}
\qquad (11.32)
$$

11.3.2 Stereoconvergent Cross-Coupling of Secondary Alkyl Electrophiles

An alternative to attempting to restrict the oxidative addition of secondary alkyl electrophiles to stereoinvertive S_N2-like mechanisms would be to harness the power of radical intermediates in the oxidative addition, permit racemization and employ this in a dynamic kinetic resolution sense. This concept would obviate the need for enantioselective synthesis of the electrophile, which is certainly problematic in the case of halides and also might widen the scope of the electrophiles that could be employed.

The greater activity of nickel catalysts towards oxidative addition of a wide variety of halides and pseudohalides has long been exploited in the coupling of sp^2 carbon centers.[90] In addition, nickel is more prone to radical mechanisms, which should not be as sensitive to steric bulk in the alkyl halide as Pd and, most importantly, should permit the development of an enantioconvergent mechanism. It is therefore no surprise that in the cross-coupling of secondary alkyl halides, nickel catalysts have dominated the literature.[91]

Building on the work of the groups of Rovis[78] and Knochel,[92,93] Zhou and Fu were able to employ a catalytic amount of an Ni-PyBOX complex to effect the Negishi cross-coupling of racemic secondary alkyl halides [eqn (11.33)].[94]

$$
\begin{array}{c}
\underset{(\pm)\text{-}\mathbf{127}}{\overset{\mathrm{Alk}}{\underset{\mathrm{Alk}}{\big|}}\!\!\!\!\!\!\!\!\!\!\!\!\overset{}{\diagdown}\!\!\!\!\mathrm{Br/I}}\ +\ 1.6\ \mathrm{XZn}\!\frown\!\mathrm{R}
\end{array}
\xrightarrow[\substack{\mathrm{DMA,\ RT,\ 20\ h}}]{\substack{4\ \mathrm{mol\%\ Ni(COD)_2}\\ 8\ \mathrm{mol\%\ ^sBu\text{-}PyBox}}}
\underset{\substack{(\pm)\text{-}\mathbf{128}\\ 62\text{-}91\%\ \mathrm{yield}}}{\overset{\mathrm{Alk}}{\underset{\mathrm{Alk}}{\big|}}\!\!\!\!\!\!\!\!\!\!\!\!\overset{}{\diagup\!\!\!\searrow}\mathrm{R}}
\qquad (11.33)
$$

$$
\underset{\substack{\mathrm{Bu}\qquad\qquad\mathrm{Bu}\\ \text{\textit{s}Bu-PyBox}}}{\text{(oxazoline-pyridine-oxazoline structure)}}
$$

Applications of these findings to the Suzuki–Miyaura cross-coupling system were not long in coming. However, the catalyst system described by Zhou and Fu to couple secondary alkyl bromides to organozincs was not effective for the analogous organoboron nucleophiles. They subsequently found that by changing the ligand to bathophenanthroline (**129**), the Ni-catalyzed cross-coupling of a variety of secondary alkyl bromides to phenylboronic acid could be achieved in good yield [eqn (11.34)].[95] Notable drawbacks to the initially disclosed catalytic system were the inability to couple alkyl chlorides (primary or secondary), primary alkyl bromides or sterically hindered aryl boronic acids.

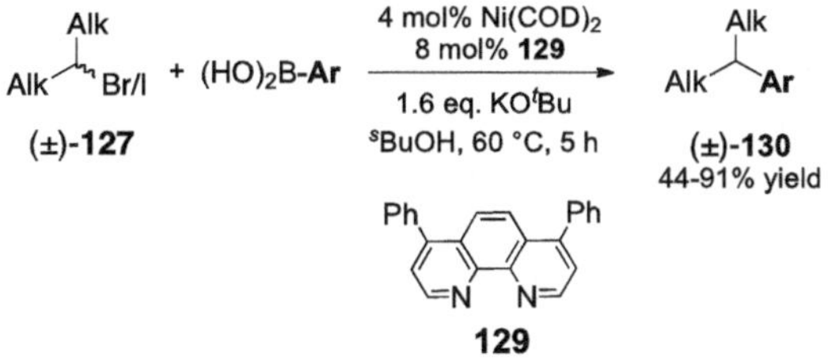

$$ (11.34) $$

These limitations were addressed in subsequent work by Fu's group by the introduction of amino alcohols as ligands for the Ni-catalyzed system.[96] Remarkably, in the presence of an air-stable $NiCl_2 \cdot$ glyme precatalyst and the amino alcohol prolinol, secondary alkyl chlorides were successfully cross-coupled with phenylboronic acid. Interestingly, the application of a similar amino alcohol-ligated Ni-catalyzed system for the coupling of *endo*-2-bromonorbornane (**131**) to phenylboronic acid led exclusively to the thermodynamically favored *exo*-cross-coupled product **132** (Scheme 11.14),[97] corroborating evidence[98] that oxidative addition to Ni proceeds *via* a radical pathway under these conditions.[99]

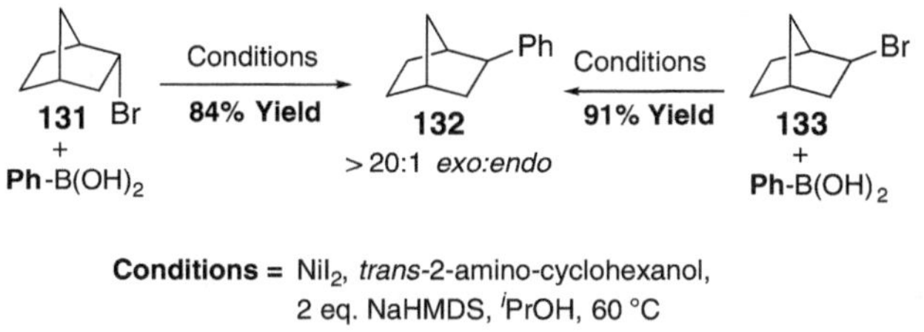

Conditions = NiI_2, *trans*-2-amino-cyclohexanol,
2 eq. NaHMDS, ^iPrOH, 60 °C

Scheme 11.14 Ni-catalyzed Suzuki–Miyaura coupling demonstrating configurational instability of the alkyl halide during the overall reaction sequence.

Alkyl–alkyl Suzuki–Miyaura couplings were made possible with only minor modifications to the reaction conditions. Thus alkyl-9-BBN boranes such as **134** were successfully coupled with secondary alkyl iodides, bromides and even chlorides at room temperature [eqn (11.35)].[100,101] Key to the successful coupling of the alkyl-9-BBN reagents was the use of KO^tBu as a base and ^tBuOH as an additive, which the authors speculated may be involved in the formation of reactive four-coordinate borate species that were detectable by ^{11}B NMR spectroscopy. Interestingly, under previously optimized conditions,[95,102] less than 5% yield was observed.

$$\text{(11.35)}$$

Taking advantage of the radical pathway observed for the aforementioned Ni-catalyzed processes, Fu and co-workers were able to develop methods for the stereoconvergent cross-coupling of *racemic* secondary alkyl halides to organoboron nucleophiles using a chiral ligand. As shown in eqn (11.36), primary alkylboranes were coupled to a series of racemic secondary alkyl halides in the presence of $[Ni(COD)_2]$ and chiral diamine ligands. Initially, homobenzylic secondary bromides were coupled to alkylboranes in high yield and enantioselectivity with a Ni–chiral diamine ligand catalyst system.[103] This advance was followed shortly thereafter by the asymmetric arylation of racemic β-chloroamides.[104] Notably, the precise conditions (solvent, additive) vary from substrate to substrate and can have a significant impact on the reaction outcome.

$$\text{(11.36)}$$

$$\text{(11.37)}$$

A possible mechanism for the Ni-catalyzed alkyl Suzuki reaction is provided in Scheme 11.15. The reaction is believed to occur *via* a Ni^I–Ni^{III} manifold, with the oxidative addition occurring by a radical mechanism.[105–108] Competition studies have demonstrated that oxidative addition is likely not rate limiting in the case of cyclohexyl iodides or bromides, whereas it is likely rate limiting for cyclohexyl chlorides.[101] Interestingly, in the case of α-chloro-amides as coupling partners, modest enantioenrichment was observed in unreacted starting material, leading the authors to suggest that the

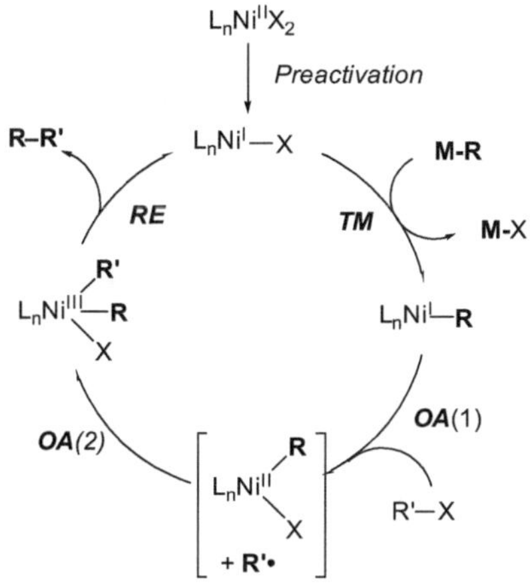

Scheme 11.15 Proposed mechanism for the Ni-catalyzed Suzuki–Miyaura coupling of alkyl halides.

enantiopure catalyst may be capable of differentiating between the two enantiomers of the starting material.[104] Reactions with enantiopure α-chloroamides display little or no erosion of enantiopurity of the starting alkyl halide (see below), implying that there is no apparent reversibility in the oxidative addition with these substrates.[104,109] However, in the case of α-bromoamides, the unreacted starting material remains racemic throughout the reaction.

Elegant studies by Taylor and Jarvo demonstrated that at room temperature, non-benzylic Ni–C bonds are configurationally stable under typical Suzuki–Miyaura reaction conditions [eqn (11.38)],[110] providing further evidence for the proposal that the scrambling of stereochemistry observed in Fu-type couplings of secondary alkyl halides is a feature of the oxidative addition step, not instability of the resulting alkyl–metal intermediate.

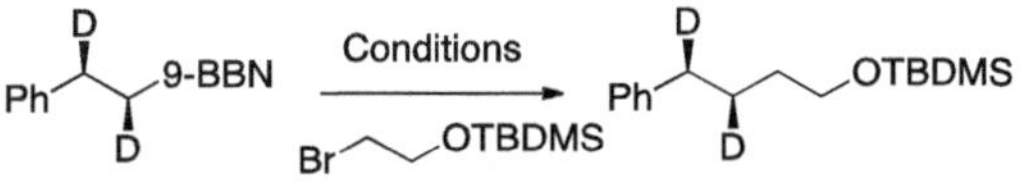

(11.38)

Conditions = NiCl$_2$·DME or Ni(COD)$_2$/diamine ligands
KO^{t}Bu, ^{i}BuOH, dioxane, rt 24 h

Although activated alkyl bromides are not needed in the chemistry reported thus far, the observation of some levels of substrate specificity led the authors to propose that some interaction between functionality on the alkyl halide and the catalyst was important.[103] Namely, it was observed that homobenzylic halides undergo Suzuki–Miyaura cross-coupling with high enantioselectivity, but dramatically reduced enantiomeric excess is observed in the one-carbon homologs (Scheme 11.16).[103] These results suggest the presence of an interaction between the aryl group and the chiral catalyst, enabling the Ni complex

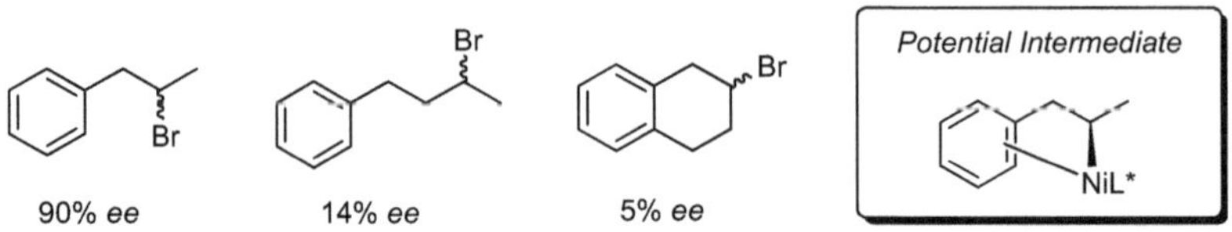

90% *ee*   14% *ee*   5% *ee*

Scheme 11.16 Relationship between position of aromatic group and enantioselectivity of the cross-coupling reaction.

to differentiate between the two alkyl groups attached to the secondary halide, thus ensuring the enantioselectivity of the C–C bond formation event. Understanding these substrate–catalyst interactions allowed the authors to expand the scope of highly selective halides from the original homobenzylic systems to include acylated halohydrins[111] and arylated amines[112] for this substrate-directed process. It should be noted, however, that the halides employed are still unactivated by comparison with previous work.

Pushing the scope further, the enantioselective Suzuki–Miyaura reaction of secondary halides has recently been applied to the γ- and δ-alkylation of amides [eqn (11.39)].[109] Since there are a variety of methods for α and β-functionalization of carbonyl compounds, this method provides a highly useful new retrosynthetic approach to carbonyl synthesis. Remarkably, γ-chloroamides were shown to react not only with alkyl boranes, but also with arylboranes and -boronic esters, giving modest enantioselectivity. Most notably, a secondary organoborane (cyclopropyl-9-BBN) was also shown to cross-couple successfully in high yield and enantioselectivity. Although the cyclopropyl group is a privileged substrate, the enantioselective cross-coupling of two secondary coupling partners is truly groundbreaking.[113]

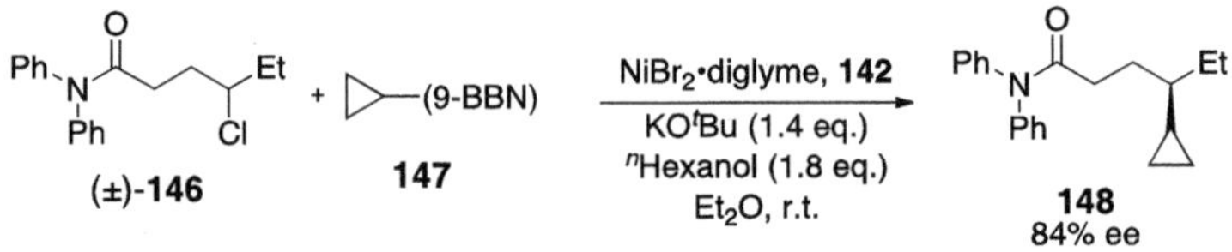

$$(11.39)$$

Recently, the scope of the effective directing groups has been expanded even further to include carbamates and sulfamides.[114] This provides a syn-thetically useful strategy for organic synthesis, as the carbamate acts as both a protecting group and a directing group for the coupling reaction and can be removed in high yield by hydrolysis. As observed previously, the distance of the directing group to the halide has a significant effect on the observed enantioselectivity. The replacement of the alkylborane with an aryl sub-stituent resulted in the first example of arylation of an alkyl electrophile in high yield and high enantioselectivity.

In another significant development, tertiary halides were also shown to be competent partners for the Ni-catalyzed Suzuki–Miyaura coupling [eqn (11.40)]. This concept was first demonstrated in the Miyaura borylation of unactivated tertiary alkyl halides,[115] then subsequently in the Suzuki–Miyaura reaction.[116] Specifically, tertiary alkyl bromides were successfully reacted with Ar-9-BBN coupling partners using a catalyst system consisting of NiBr$_2$–diglyme and 4,4′-di-*tert*-butyl-2,2′-bipyridine (**150**) in the presence of alkoxide activators. Notably, synthetically useful yields were obtained in the products of *meta*-substituted phenyl-9-BBN reagents that are not otherwise accessible through Friedel–Crafts alkylation procedures. Although no asymmetric Suzuki–Miyaura couplings of these substrates have yet been reported, this is an exciting contribution to the field that bears considerable promise.

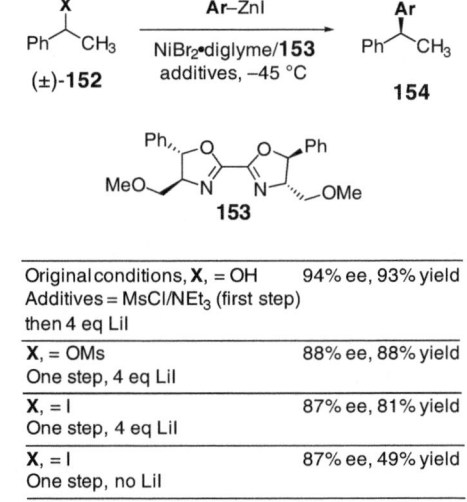

$$\text{(11.40)}$$

Finally, in a recent development in the Negishi manifold, Fu and co-workers have shown that racemic *alcohols* are also viable partners in the stereoconvergent Suzuki–Miyaura coupling reaction (Scheme 11.17).[117] Three key features of this approach are (1) the *in situ* generation of reactive mesylates, (2) the fact that successful substrates are benzylic and (3) the use of LiI as an essential additive. Putting these three facts together along with comparative reactions in which the corresponding benzylic *iodide* was employed led the authors to conclude that the reaction actually

Original conditions, **X**, = OH Additives = MsCl/NEt$_3$ (first step) then 4 eq LiI		94% ee, 93% yield
X, = OMs One step, 4 eq LiI		88% ee, 88% yield
X, = I One step, 4 eq LiI		87% ee, 81% yield
X, = I One step, no LiI		87% ee, 49% yield

Scheme 11.17   Stereoconvergent Negishi coupling of *in situ*-generated iodides.

proceeds *via* the *in situ* generation of a reactive benzylic iodide (Scheme 11.17). These reaction conditions are exceptionally interesting for the development of the Suzuki–Miyaura cross-coupling of alkyl iodides, especially if they can be generalized outside the class of benzylic substrates.

11.4   Conclusion

The field of stereospecific and stereoconvergent Suzuki–Miyaura coupling reactions has developed into a powerful method for several important classes of substrates. In the years to come, then, the challenge will be to expand this scope. Considering that, since its development, the Suzuki–Miyaura reaction has revolutionized the manner in which sp^2–sp^2 C–C bonds are prepared, and the further development of this reaction to include reactions in which stereochemistry is generated or preserved will undoubtedly have a huge impact on organic synthesis.

References

1. N. Miyaura, K. Yamada and A. Suzuki, *Tetrahedron Lett.*, 1979, 3437–3440.
2. N. Miyaura, T. Ishiyama, M. Ishikawa and A. Suzuki, *Tetrahedron Lett.*, 1986, **27**, 6369–6372.
3. N. Miyaura and A. Suzuki, *Chem. Rev.*, 1995, **95**, 2457–2483.
4. J. S. Carey, D. Laffan, C. Thomson and M. T. Williams, *Org. Biomol. Chem.*, 2006, **4**, 2337–2347.
5. R. K. Arvela, N. E. Leadbeater, M. S. Sangi, V. A. Williams, P. Granados and R. D. Singer, *J. Org. Chem.*, 2005, **70**, 161–168.
6. X.-Z. Wang and M.-Z. Deng, *J. Chem. Soc., Perkin Trans. 1*, 1996, 2663–2664.
7. J. P. Hildebrand and S. P. Marsden, *Synlett*, 1996, 893–894.
8. R. Jana, T. P. Pathak and M. S. Sigman, *Chem. Rev.*, 2011, **111**, 1417–1492.
9. E. C. Swift and E. R. Jarvo, *Tetrahedron*, 2013, **69**, 5799–5817.
10. A. Rudolph and M. Lautens, *Angew. Chem. Int. Ed.*, 2009, **48**, 2656–2670.
11. H. Doucet, *Eur. J. Org. Chem.*, 2008, 2013–2030.
12. S. R. Chemler, D. Trauner and S. J. Danishefsky, *Angew. Chem. Int. Ed.*, 2001, **40**, 4544–4568.
13. N. Miyaura, T. Ishiyama, H. Sasaki, M. Ishikawa, M. Sato and A. Suzuki, *J. Am. Chem. Soc.*, 1989, **111**, 314–321.
14. M. Sato, N. Miyaura and A. Suzuki, *Chem. Lett.*, 1989, 1405–1408.
15. G. Zou, Y. K. Reddy and J. R. Falck, *Tetrahedron Lett.*, 2001, **42**, 7213–7215.
16. G. A. Molander and C.-S. Yun, *Tetrahedron*, 2002, **58**, 1465–1470.
17. G. Zou and J. R. Falck, *Tetrahedron Lett.*, 2001, **42**, 5817–5819.
18. C. Han and S. L. Buchwald, *J. Am. Chem. Soc.*, 2009, **131**, 7532–7533.

19. K. Tamao, Y. Kiso, K. Sumitani and M. Kumada, *J. Am. Chem. Soc.*, 1972, **94**, 9268.
20. Y. Kiso, K. Tamao and M. Kumada, *J. Organomet. Chem.*, 1973, **50**, C12–C14.
21. N. Kataoka, Q. Shelby, J. P. Stambuli and J. F. Hartwig, *J. Org. Chem.*, 2002, **67**, 5553–5566.
22. T. Hayashi, Y. Matsumoto and Y. Ito, *J. Am. Chem. Soc.*, 1989, **111**, 3426–3428.
23. C. M. Crudden, Y. B. Hleba and A. C. Chen, *J. Am. Chem. Soc.*, 2004, **125**, 9200–9201.
24. D. S. Matteson, *Tetrahedron*, 1998, **54**, 10555–10606.
25. N. Selander, A. Kipke, S. Sebelius and K. J. Szabo, *J. Am. Chem. Soc.*, 2007, **129**, 13723–13731.
26. H. K. Ito, C. Kawakami and M. Sawamura, *J. Am. Chem. Soc.*, 2005, **127**, 16034–16035.
27. A. H. Hoveyda and Y. Lee, *J. Am. Chem. Soc.*, 2009, **131**, 3160–3161.
28. S. M. Smith, N. C. Thacker and J. M. Takacs, *J. Am. Chem. Soc.*, 2008, **130**, 3734–3735.
29. S. Trudeau, J. B. Morgan, M. Shrestha and J. P. Morken, *J. Org. Chem.*, 2005, **70**, 9538–9544.
30. L. T. Kliman, S. N. Mlynarski and J. P. Morken, *J. Am. Chem. Soc.*, 2009, **131**, 13210–13211.
31. G. I. McGrew, J. Temaismithi, P. J. Carroll and P. J. Walsh, *Angew. Chem. Int. Ed.*, 2010, **49**, 5541–5544.
32. C. Sun, B. Potter and J. P. Morken, *J. Am. Chem. Soc.*, 2014, **136**, 6534–6537.
33. G. A. Molander and P. E. Gormisky, *J. Org. Chem.*, 2008, **73**, 7481–7485.
34. G.-H. Fang, Z.-J. Yan and M.-Z. Deng, *Org. Lett.*, 2004, **6**, 357–360.
35. J. E. A. Luithle and J. Pietruszka, *J. Org. Chem.*, 1999, **64**, 8287–8297.
36. A. B. Charette and R. P. Freitas-Gil, *Tetrahedron Lett.*, 1997, **38**, 2809.
37. M. Rubina, M. Rubin and V. Gevorgyan, *J. Am. Chem. Soc.*, 2003, **125**, 7198–7199.
38. A. F. Littke and G. C. Fu, *Angew. Chem. Int. Ed.*, 2002, **41**, 4176–4211.
39. A. F. Littke, C. Y. Dai and G. C. Fu, *J. Am. Chem. Soc.*, 2000, **122**, 4020–4028.
40. A. van den Hoogenband, J. H. M. Lange, J. W. Terpstra, M. Koch, G. M. Visser, M. Visser, T. J. Korstanje and J. T. B. H. Jastrzebski, *Tetrahedron Lett.*, 2008, **49**, 4122–4124.
41. S. D. Dreher, P. G. Dormer, D. L. Sandrock and G. A. Molander, *J. Am. Chem. Soc.*, 2008, **130**, 9257–9259.
42. D. Imao, B. W. Glasspoole, V. S. Laberge and C. M. Crudden, *J. Am. Chem. Soc.*, 2009, **131**, 5024–5025.
43. J. Uenishi, J. M. Beau, R. W. Armstrong and Y. Kishi, *J. Am. Chem. Soc.*, 1987, **109**, 4756–4758.
44. T. C. Fessard, S. P. Andrews, H. Motoyoshi and E. M. Carreira, *Angew. Chem. Int. Ed.*, 2007, **46**, 9331–9334.

45. K. Okamoto, Y. Nishibayashi, S. Uemura and A. Toshimitsu, *Angew. Chem. Int. Ed.*, 2005, **44**, 3588–3591.

46. L. Prat, G. Dupas, J. Duflos, G. Queguiner, J. Bourguignon and V. Levacher, *Tetrahedron Lett.*, 2001, **42**, 4515–4518.

47. S. M. Podhajsky, Y. Iwai, A. Cook-Sneathen and M. S. Sigman, *Tetrahedron*, 2011, **67**, 4435–4441.

48. S. Nave, R. P. Sonawane, T. G. Elford and V. K. Aggarwal, *J. Am. Chem. Soc.*, 2010, **132**, 17096–17098.

49. G. A. Molander and N. Ellis, *Acc. Chem. Res.*, 2007, **40**, 275–286.

50. M. Butters, J. N. Harvey, J. Jover, A. J. J. Lennox, G. C. Lloyd-Jones and P. M. Murray, *Angew. Chem. Int. Ed.*, 2010, **49**, 5156–5160.

51. S. Darses and J. P. Genet, *Chem. Rev.*, 2008, **108**, 288–325.

52. S. Darses, J. P. Genet, J. L. Brayer and J. P. Demoute, *Tetrahedron Lett.*, 1997, **38**, 4393–4396.

53. J. Li and M. D. Burke, *J. Am. Chem. Soc.*, 2011, **133**, 13774–13777.

54. B. W. Glasspoole, K. Ghozati, J. Moir and C. M. Crudden, *Chem. Commun.*, 2012, **48**, 1230–1232.

55. L. Chausset-Boissarie, K. Ghozati, E. LaBine, J. L.-Y. Chen, V. K. Aggarwal and C. M. Crudden, *Chem. Eur. J.*, 2013, **19**, 17698–17701.

56. B. M. Partridge, L. Chausset-Boissarie, M. Burns, A. P. Pulis and V. K. Aggarwal, *Angew. Chem. Int. Ed.*, 2012, **51**, 11795–11799.

57. S. C. Mathew, B. W. Glasspoole, P. Eisenberger and C. M. Crudden, *J. Am. Chem. Soc.*, 2014, **136**, 5828–5831.

58. V. K. Aggarwal, M. Althaus, A. Mahmood, J. R. Suarez and S. P. Thomas, *J. Am. Chem. Soc.*, 2010, **132**, 4025–4028.

59. H. Lange, R. Huenerbein, R. Frohlich, S. Grimme and D. Hoppe, *Chem. Asian J.*, 2008, **3**, 78–87.

60. D. Masaki and M. Suginome, *J. Am. Chem. Soc.*, 2011, **133**, 4758–4761.

61. G. A. Molander and S. R. Wisniewski, *J. Am. Chem. Soc.*, 2012, **134**, 16856–16868.

62. T. Ohmura, T. Awano and M. Suginome, *J. Am. Chem. Soc.*, 2010, **132**, 13191–13193.

63. T. Awano, T. Ohmura and M. Suginome, *J. Am. Chem. Soc.*, 2011, **133**, 20738–20741.

64. D. L. Sandrock, L. Jean-Gerard, C.-Y. Chen, S. D. Dreher and G. A. Molander, *J. Am. Chem. Soc.*, 2010, **132**, 17108–17110.

65. S. M. Smith, G. L. Hoang, R. Pal, M. O. B. Khaled, L. S. W. Pelter, X. C. Zeng and J. M. Takacs, *Chem. Commun.*, 2012, **48**, 12180–12182.

66. H. Wu, S. Radomkit, J. M. O'Brien and A. H. Hoveyda, *J. Am. Chem. Soc.*, 2012, **134**, 8277–8285.

67. V. Lillo, A. Prieto, A. Bonet, M. M. Diaz-Requejo, J. Ramirez, P. J. Perez and E. Fernandez, *Organometallics*, 2009, **28**, 659–662.

68. H. Noguchi, T. Shioda, C.-M. Chou and M. Suginome, *Org. Lett.*, 2008, **10**, 377–380.

69. J. C. H. Lee, R. MacDonald and D. G. Hall, *Nat. Chem.*, 2011, **3**, 894–899.

70. K. Endo, T. Ohkubo, M. Hirokami and T. Shibata, *J. Am. Chem. Soc.*, 2010, **132**, 11033–11035.
71. T.-Y. Luh, M.-K. Leung and K.-T. Wong, *Chem. Rev.*, 2000, **100**, 3187–3204.
72. T. Ishiyama, S. Abe, N. Miyaura and A. Suzuki, *Chem. Lett.*, 1992, 691–692.
73. A. C. Frisch and M. Beller, *Angew. Chem. Int. Ed.*, 2005, **44**, 674–688.
74. K. S. Y. Lau, R. W. Fries and J. K. Stille, *J. Am. Chem. Soc.*, 1974, **96**, 4983–4986.
75. Y. Becker and J. K. Stille, *J. Am. Chem. Soc.*, 1978, **100**, 838–844.
76. I. D. Hills, M. R. Netherton and G. C. Fu, *Angew. Chem. Int. Ed.*, 2003, **42**, 5749–5752.
77. M. R. Netherton and G. C. Fu, *Angew. Chem. Int. Ed.*, 2002, **41**, 3910–3912.
78. E. M. O'Brien, E. A. Bercot and T. Rovis, *J. Am. Chem. Soc.*, 2003, **125**, 10498–10499.
79. A. López-Pérez, J. Adrio and J. C. Carretero, *Org. Lett.*, 2009, **11**, 5514–5517.
80. N. Rodriguez, C. R. de Arellano, G. Asensio and M. Medio-Simon, *Chem. Eur. J.*, 2007, **13**, 4223–4229.
81. A. He and J. R. Falck, *J. Am. Chem. Soc.*, 2010, **132**, 2524–2525.
82. J. Y. Yu and R. Kuwano, *Org. Lett.*, 2008, **10**, 973–976.
83. B. L. H. Taylor, E. C. Swift, J. D. Waetzig and E. R. Jarvo, *J. Am. Chem. Soc.*, 2011, **133**, 389–391.
84. M. A. Greene, I. M. Yonova, F. J. Williams and E. R. Jarvo, *Org. Lett.*, 2012, **14**, 4293–4296.
85. B. L. H. Taylor, M. R. Harris and E. R. Jarvo, *Angew. Chem. Int. Ed.*, 2012, **51**, 7790–7793.
86. H. M. Wisniewska, E. C. Swift and E. R. Jarvo, *J. Am. Chem. Soc.*, 2013, **135**, 9083–9090.
87. M. R. Harris, L. E. Hanna, M. A. Greene, C. E. Moore and E. R. Jarvo, *J. Am. Chem. Soc.*, 2013, **135**, 3303–3306.
88. Q. Zhou, H. D. Srinivas, S. Dasgupta and M. P. Watson, *J. Am. Chem. Soc.*, 2013, **135**, 3307–3310.
89. P. Maity, D. M. Shacklady-McAtee, G. P. A. Yap, E. R. Sirianni and M. P. Watson, *J. Am. Chem. Soc.*, 2013, **135**, 280–285.
90. F.-S. Han, *Chem. Soc. Rev.*, 2013, **42**, 5270–5298.
91. M. R. Netherton and G. C. Fu, *Adv. Synth. Catal.*, 2004, **346**, 1525–1532.
92. R. Giovannini, T. Stüdemann, G. Dussin and P. Knochel, *Angew. Chem. Int. Ed.*, 1998, **37**, 2387–2390.
93. A. Devasagayaraj, T. Stüdemann and P. Knochel, *Angew. Chem. Int. Ed.*, 1995, **34**, 2723–2725.
94. J. R. Zhou and G. C. Fu, *J. Am. Chem. Soc.*, 2003, **125**, 14726–14727.
95. J. Zhou and G. C. Fu, *J. Am. Chem. Soc.*, 2004, **126**, 1340–1341.
96. F. Gonzalez-Bobes and G. C. Fu, *J. Am. Chem. Soc.*, 2006, **128**, 5360–5361.

97. A similar result was also observed in the Miyaura borylation of the same substrates.[115]

98. Substituting Ni for both zero- and divalent Pd precatalysts effectively shuts down the cross-coupling of secondary alkyl electrophiles with arylboronic acids.[95]

99. Recently, Gagné and co-workers have taken advantage of the anomeric effect to add zerovalent Pd oxidatively to acetobromo-α-D-glucose with inversion of stereochemistry: C. Munro-Leighton, L. L. Adduci, J. J. Becker and M. R. Gagné, *Organometallics*, 2011, **30**, 2646–2649.

100. B. Saito and G. C. Fu, *J. Am. Chem. Soc.*, 2007, **129**, 9602–9603.

101. Z. Lu and G. C. Fu, *Angew. Chem. Int. Ed.*, 2010, **49**, 6676–6678.

102. F. González-Bobes and G. C. Fu, *J. Am. Chem. Soc.*, 2006, **128**, 5360–5361.

103. B. Saito and G. C. Fu, *J. Am. Chem. Soc.*, 2008, **130**, 6694–6695.

104. P. M. Lundin and G. C. Fu, *J. Am. Chem. Soc.*, 2010, **132**, 11027–11029.

105. C.-P. Zhang, H. Wang, A. Klein, C. Biewer, K. Stirnat, Y. Yamaguchi, L. Xu, V. Gomez-Benitez, D. A. David and A. Vicic, *J. Am. Chem. Soc.*, 2013, **135**, 8141–8144.

106. G. D. Jones, J. L. Martin, C. McFarland, O. R. Allen, R. E. Hall, A. D. Haley, R. J. Brandon, T. Konovalova, P. J. Desrochers, P. Pulay and D. A. Vicic, *J. Am. Chem. Soc.*, 2006, **128**, 13175–13183.

107. X. Hu, *Chem. Sci.*, 2011, **2**, 1867–1886.

108. Z. Li, Y.-Y. Jiang and Y. Fu, *Chem. Eur. J.*, 2012, **18**, 4345–4357.

109. S. L. Zultanski and G. C. Fu, *J. Am. Chem. Soc.*, 2011, **133**, 15362–15364.

110. B. L. H. Taylor and E. R. Jarvo, *J. Org. Chem.*, 2011, **76**, 7573–7576.

111. N. A. Owston and G. C. Fu, *J. Am. Chem. Soc.*, 2010, **132**, 11908–11909.

112. Z. Lu, A. Wilsily and G. C. Fu, *J. Am. Chem. Soc.*, 2011, **133**, 8154–8157.

113. B. W. Glasspoole and C. M. Crudden, *Nat. Chem.*, 2011, **3**, 912–913.

114. A. Wilsily, F. Tramutola, N. A. Owston and G. C. Fu, *J. Am. Chem. Soc.*, 2012, **134**, 5794–5797.

115. A. S. Dudnik and G. C. Fu, *J. Am. Chem. Soc.*, 2012, **134**, 10693–10697.

116. S. L. Zultanski and G. C. Fu, *J. Am. Chem. Soc.*, 2013, **135**, 624–627.

117. H.-Q. Do, E. R. R. Chandrashekar and G. C. Fu, *J. Am. Chem. Soc.*, 2013, **135**, 16288–16291.

CHAPTER 12

Direct Arylation via *C–H Activation*

UPENDRA SHARMA, ATANU MODAK, SOHAM MAITY, ARUN MAJI AND DEBABRATA MAITI*

Department of Chemistry, IIT Bombay, Powai, Mumbai-400076, India
*Email: dmaiti@chem.iitb.ac.in

12.1 Introduction

The discovery of new strategies and protocols that can improve a transformation in terms of number of steps, atom economy and versatility in the construction of complex scaffolds is a continuing target for chemists. In this context, the expanding frontier of C–H bond activation is one of the most attractive research areas of recent decades. The development of catalytic functionalization of unactivated C–H bonds for the synthesis of carbon–carbon and carbon–heteroatom bonds has revolutionized the whole approach towards the synthesis of practically important compounds.[1–4] Progress in the direct regioselective conversion of C–H bonds to C–C bonds has challenged the classical catalytic cross-coupling reactions involving organohalides along with organometallic coupling partners.[5] Despite significant progress in this area, the challenges still persist in terms of selectivity of C–H bond activation, realization of truly efficient catalytic systems and generality of reaction conditions. In this regard, several metal catalysts have been examined based on conscious judgments over time as understanding has gradually increases. Broadly, these transformations involve (a) activation of a C–H bond selectively followed by metallation, (b) addition of a coupling partner to the metal center and (c) reductive elimination of the

RSC Catalysis Series No. 21
New Trends in Cross-Coupling: Theory and Applications
Edited by Thomas J Colacot
© The Royal Society of Chemistry 2015
Published by the Royal Society of Chemistry, www.rsc.org

target molecule.[6] One way to impose site selectivity is to exploit the differences in the electronic environment within the substrate. On the other hand, introducing directing groups has also helped greatly in solving the riddle of selective C–H bond activation. In most cases, *ortho* functionalization has been observed,[7] but in few instances *meta* functionalization has also been achieved.[7,8]

The present discussion is mainly focused on recent progress in direct arylation. Notably, direct alkylation, alkenylation and alkynylation are also achievable *via* C–H bond activation catalysts.[9–11] An overview of direct arylation through C–H bond activation under catalysis with various metals such as Pd, Rh, Ru, Ir and Cu is presented. The literature up to 2013 has been surveyed for topics related to direct arylation through C–H activation.

12.2   Palladium-Catalyzed Reactions

Transition metal-catalyzed cross-coupling reactions are among the most popular and convenient routes for making new carbon–carbon and carbon–heteroatom bonds. The majority of these coupling reactions, including Suzuki, Heck, Kumada, Stille, Negishi, Hiyama and Buchwald–Hartwig coupling have utilized Pd catalysts. Although the Kumada reaction was discovered with Ni[0], the introduction of Pd by Murahashi's group further expanded the scope of the reaction. However, these traditional cross-coupling reactions require prefunctionalized substrates. In this context, Pd-catalyzed direct arylation *via* C–H activation has recently become one of the most investigated research topics. The catalytic cycle typically involves the activation of C–H bonds by a Pd[II] species, *trans*-metallation and subsequent reductive elimination to generate the desired product through C–C bond formation (Figure 12.1).[12]

Pd-catalyzed ligand-assisted C–H functionalization reactions generally take place at Pd[II] centers. A palladacycle intermediate is generated, which can undergo functionalization by two distinct pathways. Functionalization can occur by reductive elimination or β-hydride elimination involving a Pd[II/0] cycle (Scheme 12.1). Another pathway involves a Pd[II/IV] cycle where this

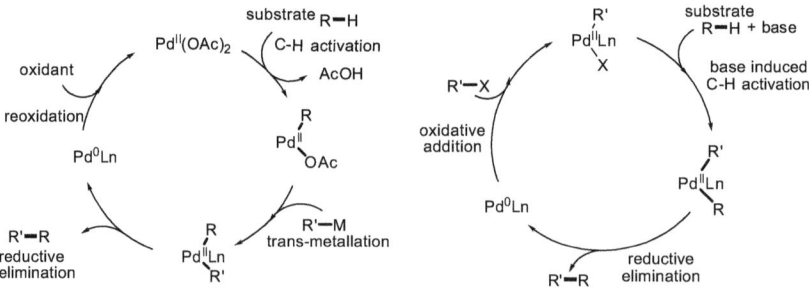

Figure 12.1   General mechanism of Pd-catalyzed C–H activation/C–C bond formation.

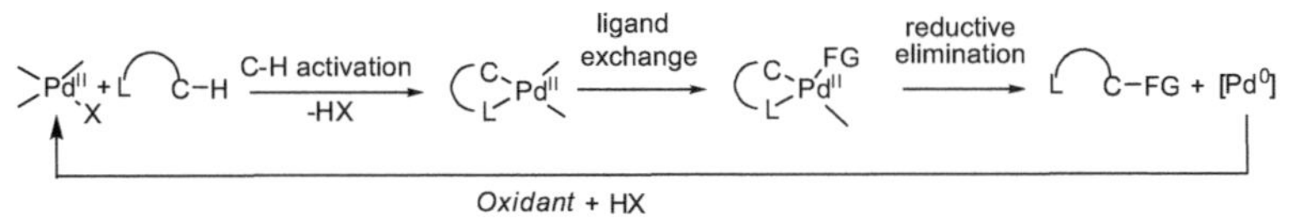

Scheme 12.1 Reductive functionalization pathway: $Pd^{II/0}$ catalytic cycle.

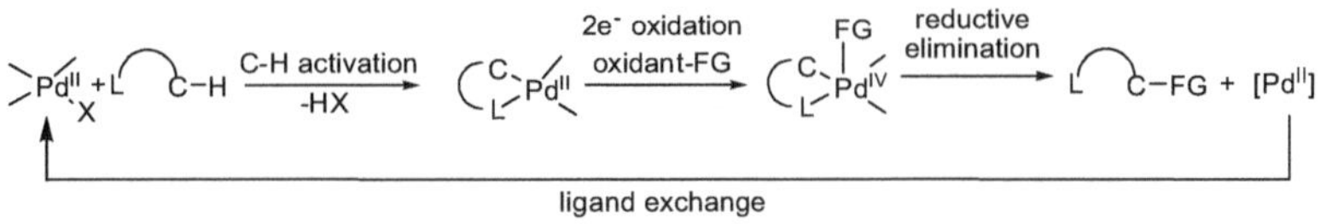

Scheme 12.2 Electrophilic functionalization pathway: two-electron oxidation of palladacycle.

transformation is achieved by using an electrophilic reagent (two-electron oxidation process) (Scheme 12.2).[4]

12.2.1 Intramolecular Direct Arylation

Pd-catalyzed intramolecular direct arylation can be traced back to 1982, when a product of such a reaction was obtained as a side product in the Heck coupling of ethyl acrylate with bromocinnolines (Scheme 12.3).[13]

Subsequently, related five- and six-membered ring compounds were synthesized through this approach (Scheme 12.4).[14,15] In the case of six-membered ring compounds, dehalogenation was also observed due to the lower nucleophilicity of simple arenes.

To overcome this problem, Rawal and co-workers used phenol-containing substrates to obtain the corresponding cyclized product with excellent yields and selectivity.[16] They proposed that the hydroxyl group undergoes deprotonation to generate a more reactive species (Scheme 12.5). Protection of the phenolic oxygen leads to lower yields.

In this context, Echavarren and co-workers proposed a mechanism for these palladium intramolecular-cyclization reactions,[17] which was further supported by the theoretical calculations (Figure 12.2).[18]

Fagnou's group carried out extensive investigations to understand intramolecular direct arylation reactions in detail.[19] Their improved reaction conditions tolerate different functional groups and are applicable to the formation of both six- and seven-membered rings (Scheme 12.6).

It was observed that ligand **A** (for a six-membered ring) and **B** (for a seven-membered ring) with $Pd(OAc)_2$ produced excellent results for aryl bromides, but these catalytic systems failed in the case of aryl chlorides and surprisingly for aryl iodides.[20] To overcome this, the authors applied more electron-rich NHC (**C**) and different alkylphosphine (**D** and **E**) ligands for the direct arylation of aryl chlorides.[21,22] Preformed catalysts with an NHC ligand (**C**) were found slightly less active than *in situ*-formed catalytic species.

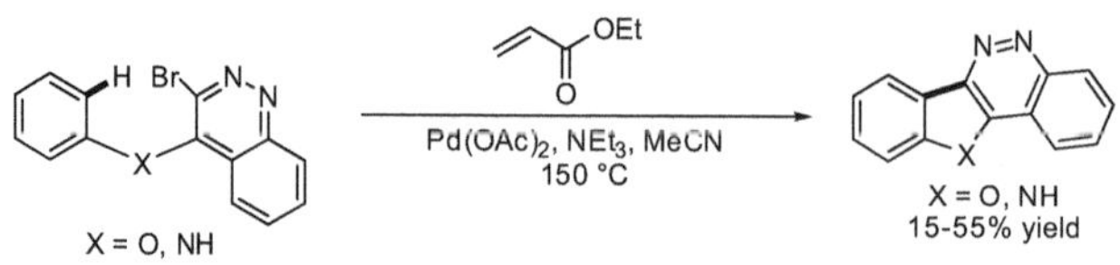

Scheme 12.3 First example of intramolecular direct arylation.

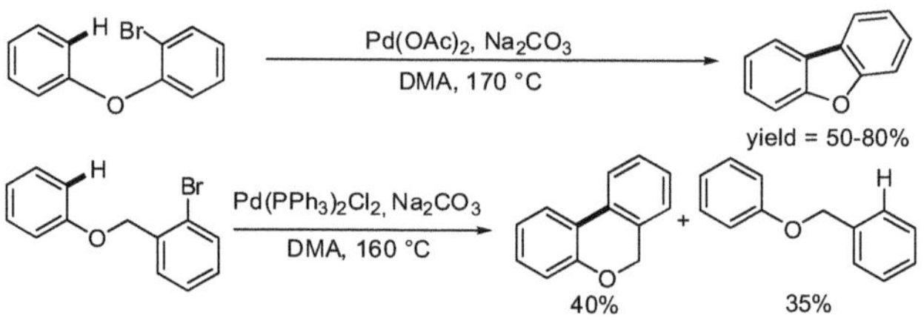

Scheme 12.4 Five- and six-membered ring formation through direct arylation.

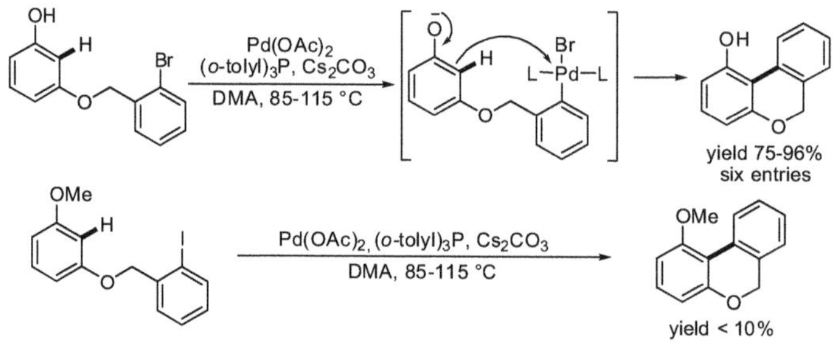

Scheme 12.5 Intramolecular *ortho*-arylation of phenols.

Generally, aryl iodides are the best substrates among aryl halides for transition metal-catalyzed cross-coupling reactions because of facile oxidative addition.[16,23] Surprisingly, when an aryl iodide was reacted in the presence of Pd(OAc)$_2$ with ligand **E** a lower TON (turnover number) was observed compared with an aryl bromide. Initially the reaction was very fast, but began to slow at 40% conversion and stopped at 60% conversion. Addition of silver salt resulted in an excellent improvement in the TON, which indicated that iodide may be inhibiting the reaction (Scheme 12.7).[22] It is well documented that halides can interact with Pd^{II} complexes to generate palladate species.[24] The stability of these complexes is greatest with iodide compared with bromide and chloride.[25] The binding of an iodide to form palladate species would block an additional coordination site and verify the arene–Pd interactions (Figure 12.3).

Later, Fagnou and co-workers discovered that intramolecular direct arylation of halides to form five- and six-membered hetero- and carbocyclic ring systems could be performed with Pd(OH)$_2$/C (Pearlman's catalyst) (Scheme 12.8).[26] To determine the nature of the active catalyst, the solid

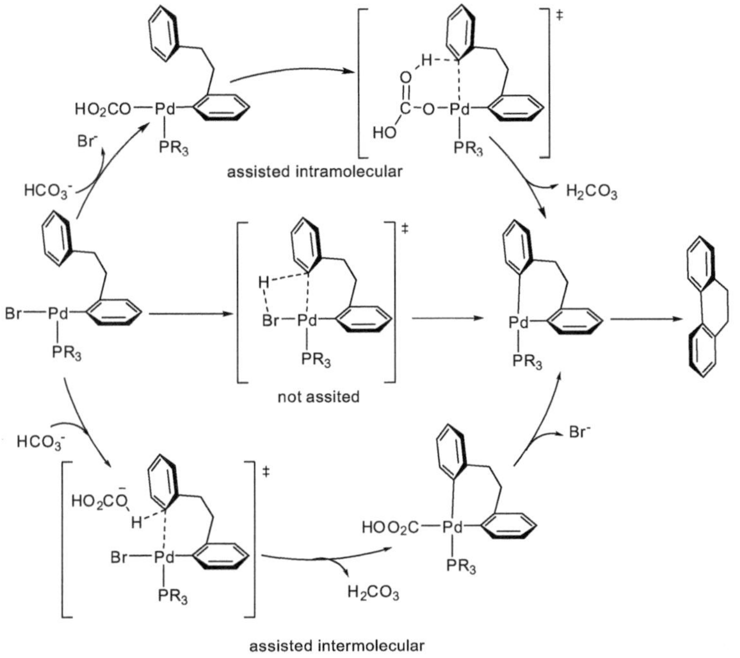

Figure 12.2 Proposed mechanism for the intramolecular *ortho*-arylation of phenols.

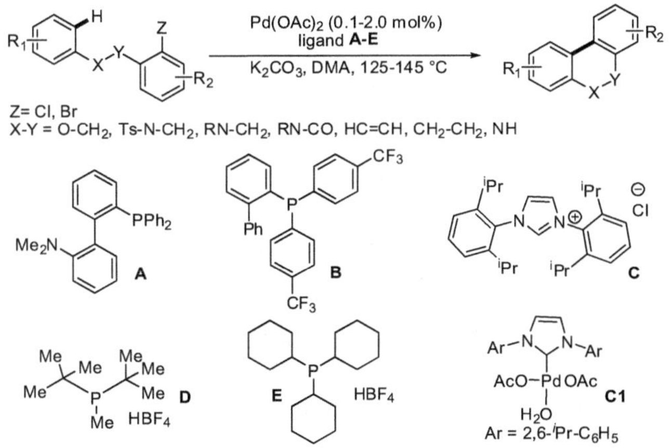

Scheme 12.6 Screening of various ligands for Pd-catalyzed intramolecular direct arylation.

support substrate was treated with $Pd(OH)_2/C$, followed by cleavage with TFA. Full conversion was observed, which indicated that a soluble catalyst species had leached into solution. The homogeneous nature of the active

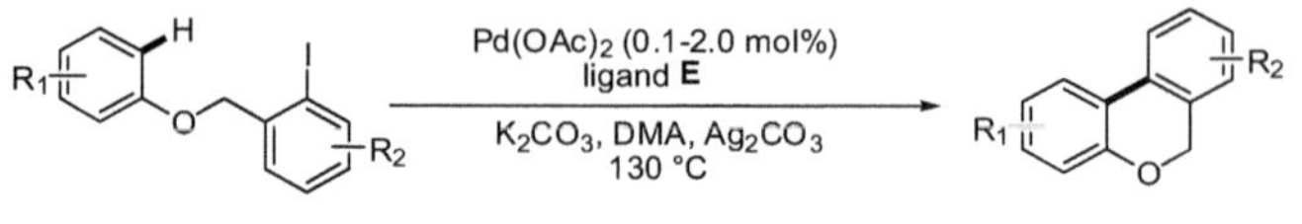

Scheme 12.7 Effect of silver salt in Pd-catalyzed intramolecular direct arylation.

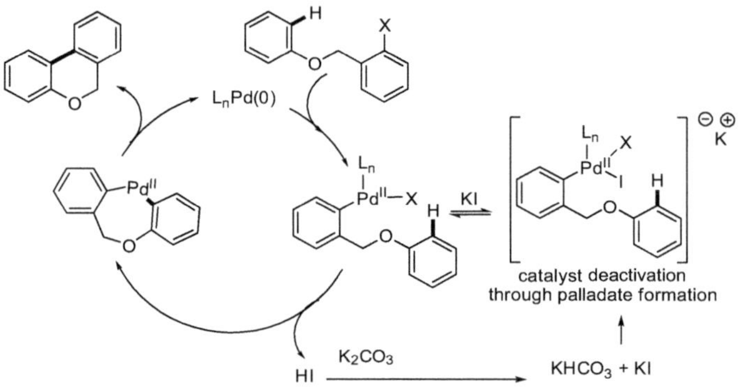

Figure 12.3 Proposed mechanism for Pd-catalyzed intramolecular arylation.

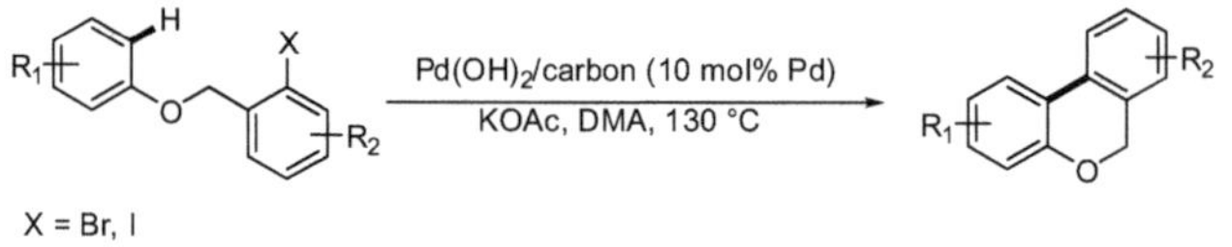

X = Br, I

Scheme 12.8 Application of Pearlman's catalyst for direct arylation.

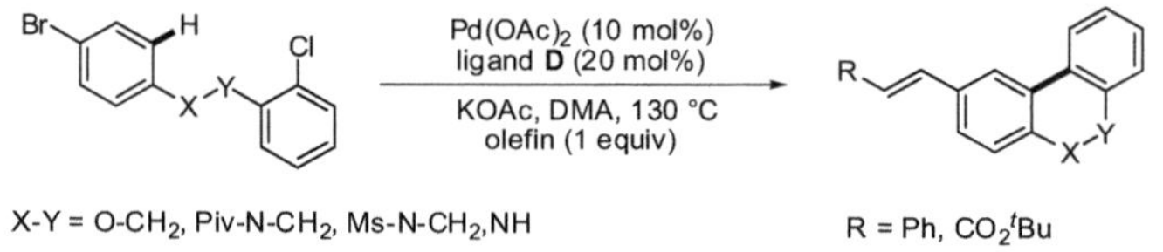

X-Y = O-CH$_2$, Piv-N-CH$_2$, Ms-N-CH$_2$, NH   R = Ph, CO$_2$^{t}Bu

Scheme 12.9 Tandem–sequential Heck–direct arylation.

species was further confirmed by the observation that no product was formed in the presence of solid-phase thiol scavenger resin.

Further study revealed that these methods can be applied to tandem Heck–direct arylation reaction sequences.[27] Substrates containing both bromide and chloride functional groups were reacted with a Heck acceptor (Scheme 12.9). An interesting tandem–sequential Heck–direct arylation–hydrogenation was observed on replacing the nitrogen atmosphere of the completed reaction with hydrogen (Scheme 12.10).

To gain an insight into the mechanism of Pd-catalyzed direct arylation reactions, Fagnou and co-workers conducted a series of experiments.[22] Competition experiments established a slight preference for electron-rich arenes. Direct arylation with a simple aryl bromide resulted in a primary

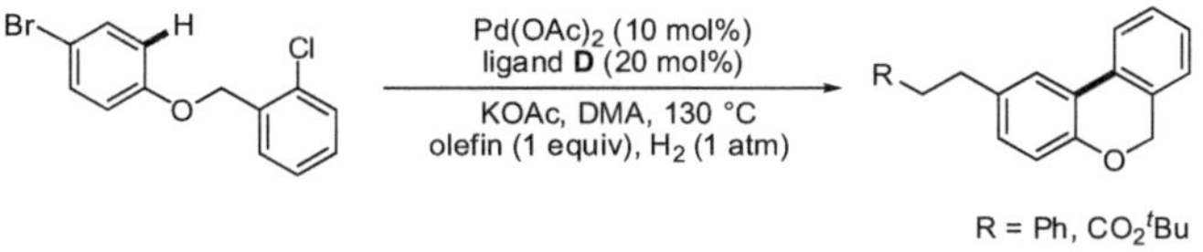

Scheme 12.10 Tandem–sequential Heck–direct arylation–hydrogenation.

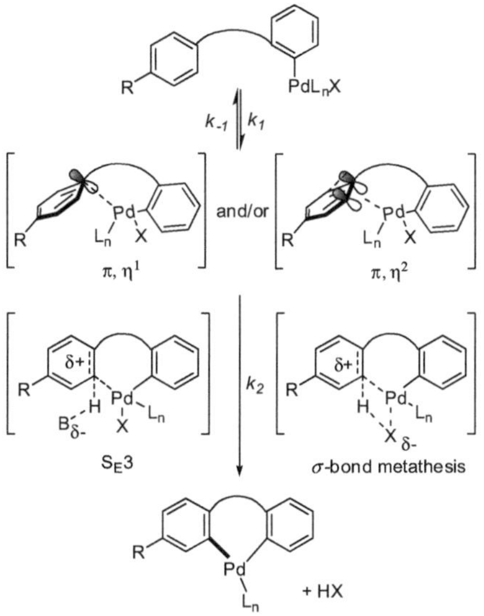

Figure 12.4 Proposed mechanism for Pd-catalyzed intramolecular direct arylation.

kinetic isotope effect of 4.25, excluding the possibility of electrophilic aromatic substitution. The presence of a primary kinetic isotope effect can be rationalized by considering the relative rates of coordination of the arene to give a π,η^2 or π,η^1 complex from the Pd^{II}–arene intermediate (k_1 and k_{-1}) and comparing them with the deprotonation step (k_2). The presence of a primary KIE implies that k_2 is kinetically significant. Deprotonation can occur in two ways: either external base deprotonates the arene(*via* the *substitution electrophilic trimolecular* or S_E3 pathway) while the Pd–C bond is forming or through σ-bond metathesis where an anionic ligand on the Pd removes the proton. It should be noted that σ-bond metathesis potentially includes the *concerted metallation–deprotonation* (CMD) pathway (Figure 12.4).

12.2.2 Intermolecular Direct Arylation

Fagnou and co-workers reported the direct arylation of electron-deficient perfluorinated arenes in 2006 (Scheme 12.11).[28] Mechanistic study revealed that the key C–H bond functionalization step occurs *via* a concerted metallation–C–H bond-cleaving process, which depends directly on the acidity of the C–H bond being cleaved. Two mechanistic pathways were

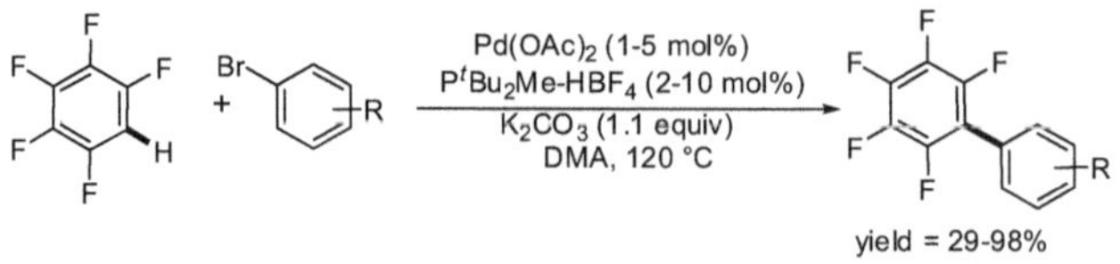

Scheme 12.11 Cross-coupling of bromoarenes with polyfluoroarenes.

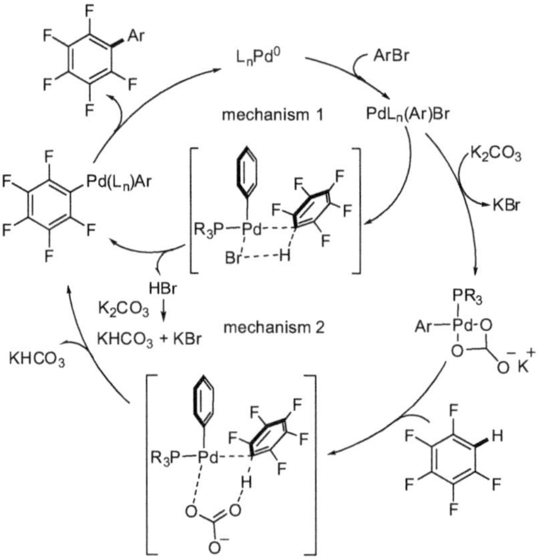

Figure 12.5 Proposed mechanism for cross-coupling of bromoarenes with polyfluoroarenes.

proposed: (a) concerted palladation and loss of HBr to afford a diarylpalladium(II) intermediate; and (b) exchange of the bromide ion with carbonate, which can allow for a related palladation–deprotonation process (Figure 12.5). In the case of substrates with more than one C–H bond available, products corresponding to di-and triarylation were also observed.

Yu's group has also contributed significantly to intermolecular direct arylation for C–C bond formation *via* C–H activation. They used the carboxylic acid group as a directing group for C–H activation/C–C coupling (Scheme 12.12).[29] Optimization studies revealed that carboxylates as substrates are much more effective than simple benzoic acids. A possible explanation for this is that the *O*-anion of the carboxylate has an impact on the C–H activation step where Pd^{II} inserts in the C–H bond.

This method was applicable to simple benzoic acids but bis-arylated compounds were also formed in significant amounts. Notably, *meta*-substituted benzoic acids led to regioselective arylation (Scheme 12.13).

Yu and co-workers also reported the direct arylation of benzoic and phenylacetic acids with potassium aryltrifluoroborates as the coupling partner (Schemes 12.14 and 15).[30] Coordination of the carboxylate moiety to K$^+$ was

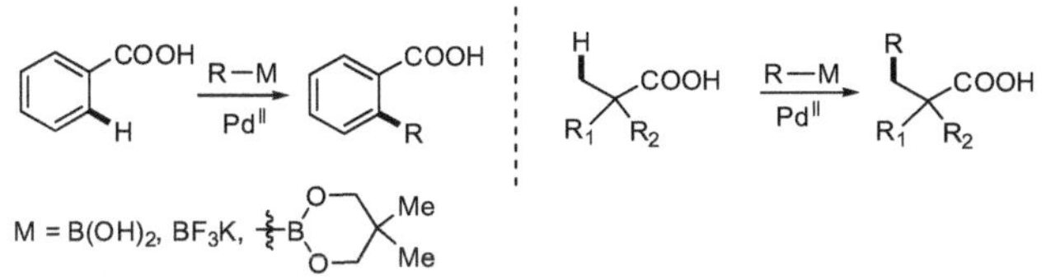

Scheme 12.12 Carboxyl functionalities as directing groups.

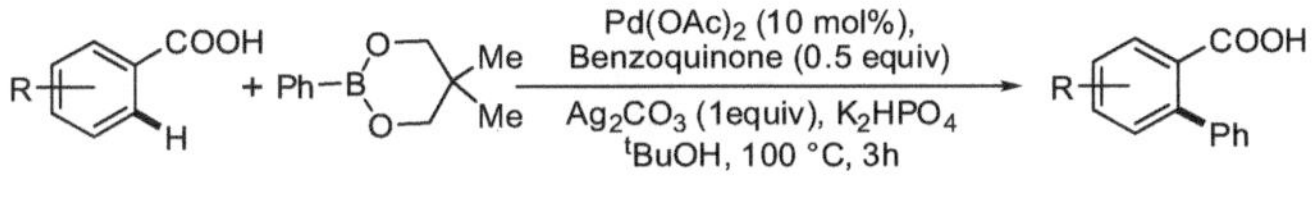

Scheme 12.13 *Ortho*-arylation of benzoic acids.

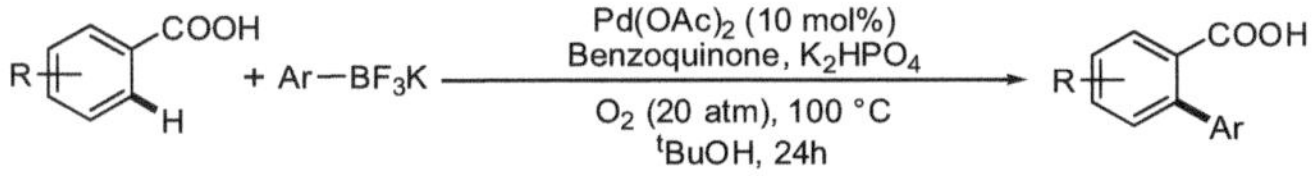

Scheme 12.14 *Ortho*-arylation of benzoic acids using Ar–BF$_3$K.

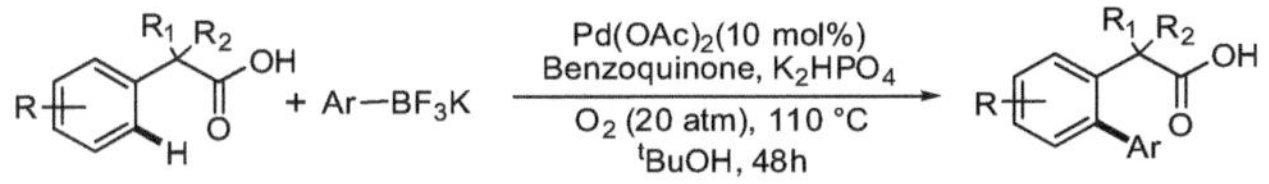

Scheme 12.15 Coupling of substituted arylacetic acids with–Ar-BF$_3$K.

found to be beneficial for C–H activation, but in the presence of Ag$^+$ the catalytic system was found to be deactivated through competitive binding with the carboxylate anion. This revised protocol was applicable to different substituted phenylacetic acids (Scheme 12.15). α-Substituted phenylacetic acids resulted in reasonable mono-selectivity due to steric crowding. In addition, strongly electron-donating and -withdrawing substituents were found to be compatible with these conditions, although the scope of this protocol needed to be improved.

In general, direct arylation reactions have several elementary steps such as C–H activation, transmetallation, reductive elimination and reoxidation, which can be improved by using appropriate ligands.[31] Yu's group improved the catalytic efficiency with the help of specific ligands, which made the reaction faster, afforded an improved scope of the reaction and gave higher yields compared with the previous methods (Scheme 12.16).

According to proposed catalytic cycle,[31] the substrate first coordinates to the Pd center, followed by C–H cleavage, which is facilitated due to the bound amino acid ligand (Figure 12.6). Transmetallation and reductive elimination give the desired product and generate Pd^0, which is oxidized back to Pd^{II}(active species) in the presence of Ag^I or O$_2$/BQ.

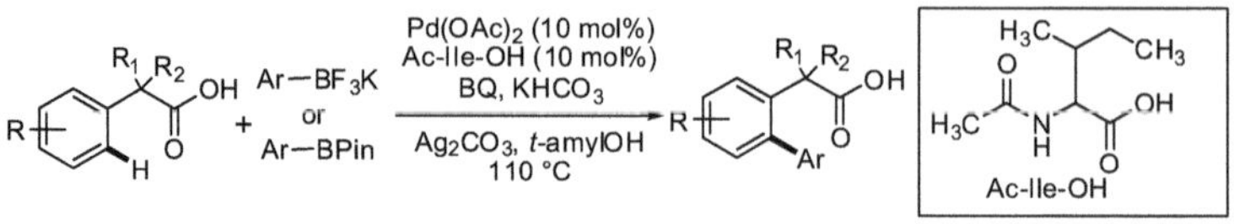

Scheme 12.16 Coupling of substituted phenylacetic acids with Ar–BF$_3$K/Ar–BX$_2$.

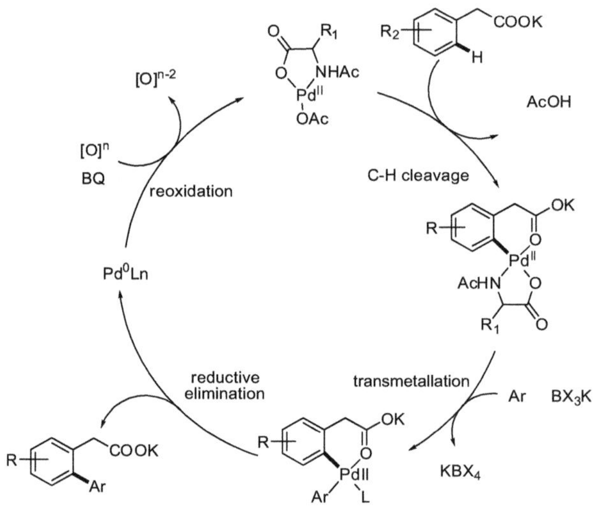

Figure 12.6 Proposed mechanism for Pd-catalyzed carboxylic acid-directed arylation.

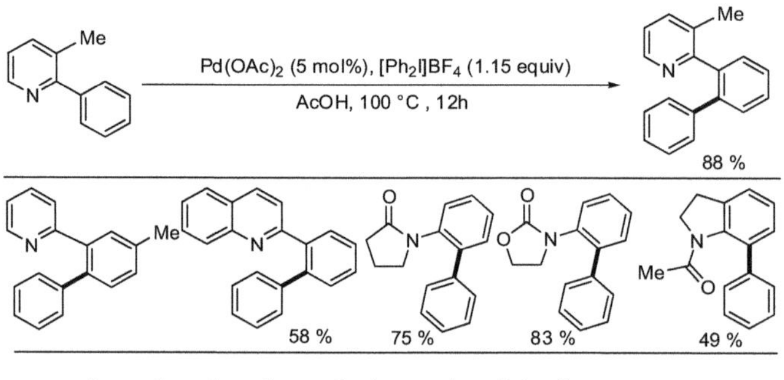

Scheme 12.17 Pd-catalyzed *ortho*-arylation using [Ph$_2$I]BF$_4$.

In 2005, Sanford and co-workers reported the Pd-catalyzed *ortho*-arylation of 2-arylpyridines using an electrophilic arylating agent, [Ar–I–Ar]BF$_4$ (Scheme 12.17).[32] Different types of *N*-heterocycles such as quinolines, oxazolidinones, pyrrolidinones and benzodiazepines were successfully arylated by this method. To show the functional group compatibility, [Mes–I–Ar]BF$_4$ was used as arylating reagent, where the mesityl group was found to facilitate the aryl transfer process due to steric reasons.

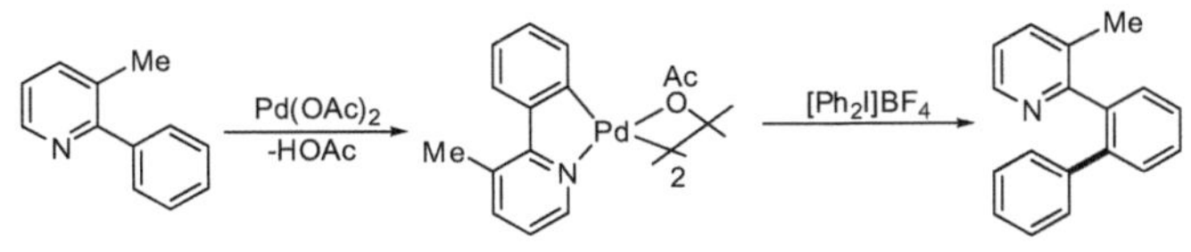

Scheme 12.18 Proposed mechanism of Pd-catalyzed *ortho*-arylation using [Ph$_2$I]BF$_4$.

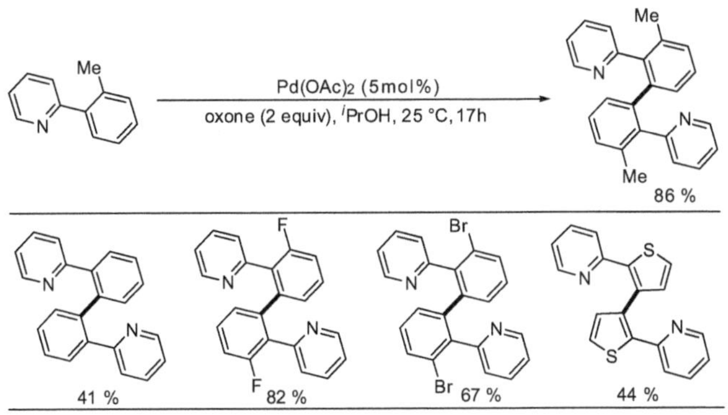

Scheme 12.19 Pd-catalyzed cross-dehydrogenative dimerization.

A Pd^{II}/Pd^{IV} cycle has been proposed (Scheme 12.18). Several experiments were performed which determined the first-order dependence on the hypervalent iodine reagent [I^{III}], a second-order dependence on [Pd] and an inverse third-order dependence on [arylpyridine]. Hammett studies and labeling experiments showed that the C–H activation process occurs after the turnover-limiting step, which is the oxidation of the cyclopalladated intermediate with the I^{III} reagent.

In 2006, Sanford and co-workers reported a Pd(OAc)$_2$-catalyzed cross-dehydrogenative dimerization of 2-arylpyridines (Scheme 12.19).[33] This ligand-free oxidative coupling was highly regioselective for 2-arylpyridine derivatives.

The proposed mechanism included initial cyclopalladation with Pd^{II} and subsequent oxidation to Pd^{IV}. Following another cyclopalladation at Pd^{IV}, reductive elimination gave the C–C coupled product, where oxone was presumed to play the role of a two-electron oxidant for the Pd^{II}/Pd^{IV} cycle (Figure 12.7).

Amide-directed arylation, alkylation, vinylation and carbonylation reactions were also achieved in simple aromatic ring systems (Scheme 12.20).[12] Different boron reagents were found to be suitable coupling partners in this case.

Interestingly, *para*-selective C–H arylation of monosubstituted arenes can also be achieved by using an amide as the directing group (Scheme 12.21).[34] No prefunctionalization was required and *para*-selectivity over *ortho*- and *meta*-arylation was fairly high.[35]

Buchwald and co-workers reported the direct arylation of anisole derivatives but the selectivity was not satisfactory as a mixture of regioisomers was

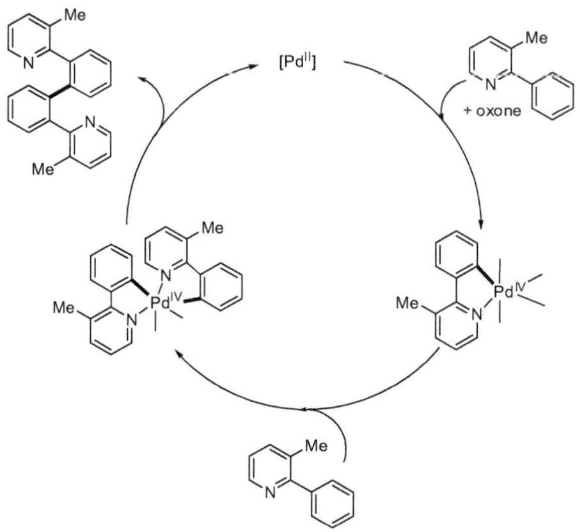

Figure 12.7 Plausible mechanism of Pd-catalyzed cross-dehydrogenative dimerization.

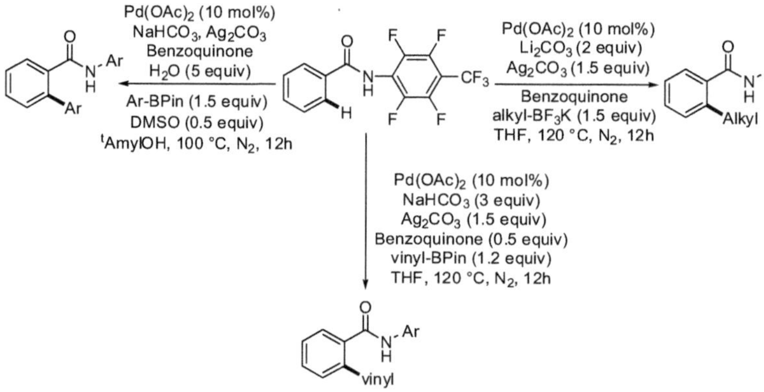

Scheme 12.20 Amide-directed C–H functionalization reactions.

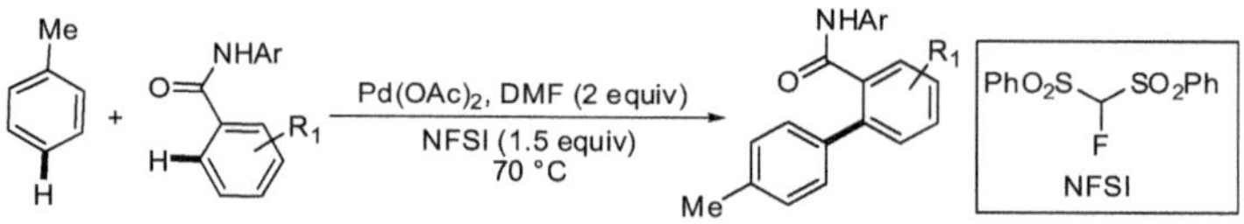

Scheme 12.21 Amide group-directed *para*-selective C–H arylation.

typically obtained.[35] The excellent selectivity achieved by Yu and co-workers can be attributed primarily to the acidic nature of the amide group, promoting stronger coordination to the metal center after deprotonation.

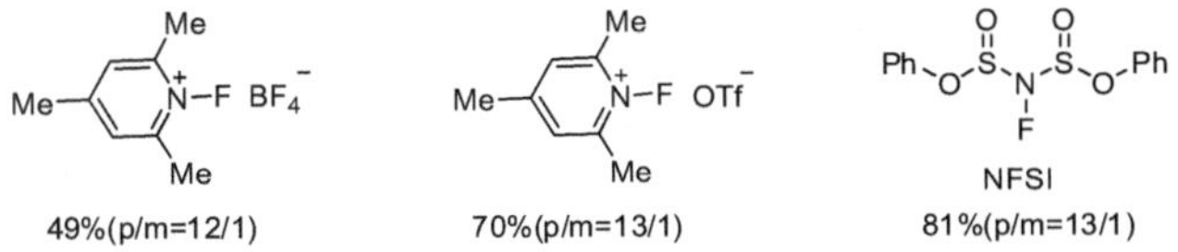

49% (p/m=12/1)   70% (p/m=13/1)   NFSI   81% (p/m=13/1)

Figure 12.8   Effect of different F$^+$ reagents.

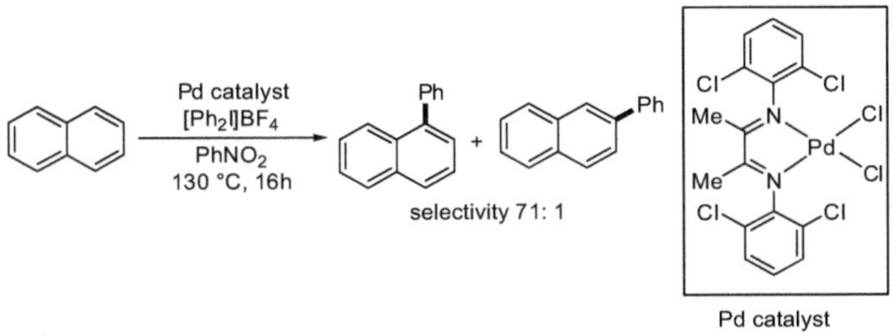

selectivity 71 : 1

Pd catalyst

Scheme 12.22   Pd-catalyzed site-selective arylation of naphthalene.

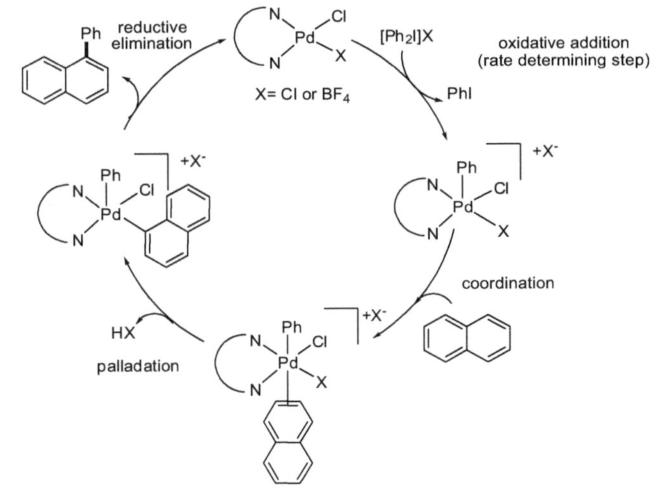

Figure 12.9   Proposed mechanism of the Pd-catalyzed direct arylation of naphthalene.

Also, F$^+$ reagents played the role of suitable oxidants, which was crucial for the success of the reaction (Figure 12.8).

In the absence of any directing group, Hickman and Sanford also reported a site-selective arylation of naphthalenes in 2011 (Scheme 12.22).[36] The highly selective (50 : 1) α-arylation with [Ph$_2$I]BF$_4$ as the arylating agent was realized by tuning the structure of the diimine ligand. The proposed mechanism consists of π-coordination of the naphthalene ring to the Pd^{IV} center as the fast step, after the rate-determining oxidation of Pd^{II} to Pd^{IV} *via* oxidative addition (Figure 12.9).

12.2.3   Direct Arylation of Heterocycles

The first example of a direct heterocycle arylation appeared in 1989, when Ohta and co-workers reported the C2 arylation of *N*-methylindole with a pyrazine halide (Scheme 12.23).[37] Further studies demonstrated that various π-electron-rich heterocycles can be arylated using similar reaction conditions.[38-42] Mechanistically, it was generally accepted that in direct arylation reactions, π-electron-rich substrates can react *via* an electrophilic palladation step and that the arylation is facilitated by the highly nucleophilic nature of these arenes.[43]

Sames and co-workers reported a Pd-catalyzed direct C2 arylation of *N*-substituted indoles with aryl iodides (Scheme 12.24).[43,44] They observed that the selectivity can be altered from the C2 to the C3 position on proper selection of the base and counter ion. Mechanistically, it was proposed that the selectivity arises from the migration of Pd during the metallation step.[43] Kinetic studies and isotope effects suggest an initial C3 arylation followed by deprotonation. On the other hand, the C2 regioisomer is obtained if migration of the Pd center to C2 takes place prior to deprotonation.

Using electrophilic arylating agents such as [Ar–I–Ar]BF$_4$, Sanford and co-workers reported a Pd-catalyzed C2 arylation of indoles in 2006 (Scheme 12.25).[45] The reaction was found to be highly C2 selective and C3 arylation was observed only when the C2 position was blocked.

A plausible mechanism involving a Pd^{II}/Pd^{IV} cycle was proposed. Previously, for arylation of indole, Pd^{II}/Pd^0 was believed to be involved in the turnover-limiting step along with a highly electron-rich phosphine ligand. The rate of the electrophilic palladation step was proposed to increase if a

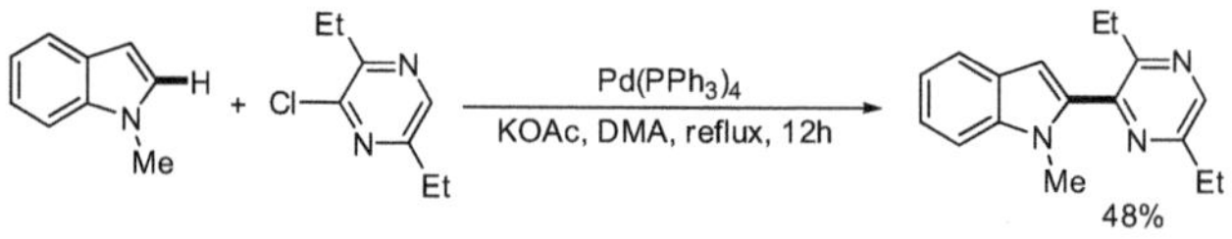

Scheme 12.23   First example of direct heterocycle arylation.

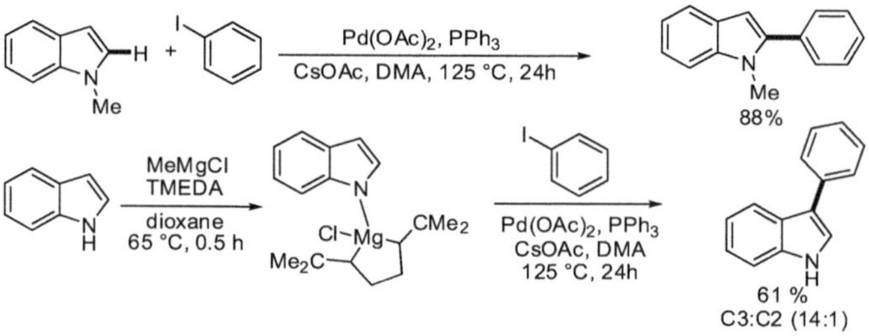

Scheme 12.24   Pd-catalyzed direct C2 arylation of an *N*-substituted indole with aryl iodide.

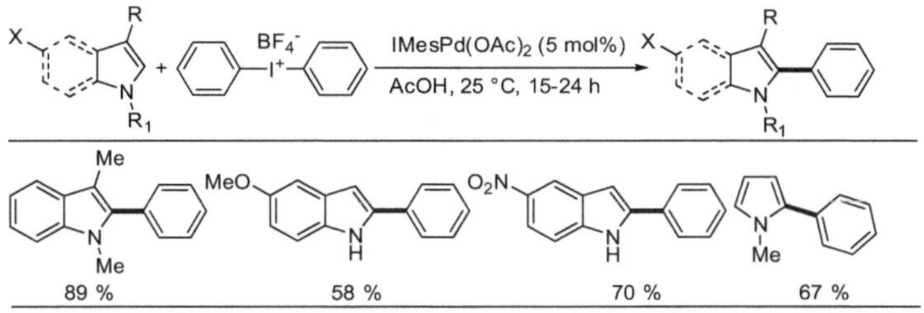

Scheme 12.25 Pd-catalyzed 2-arylation of indoles using [Ph$_2$I]BF$_4$.

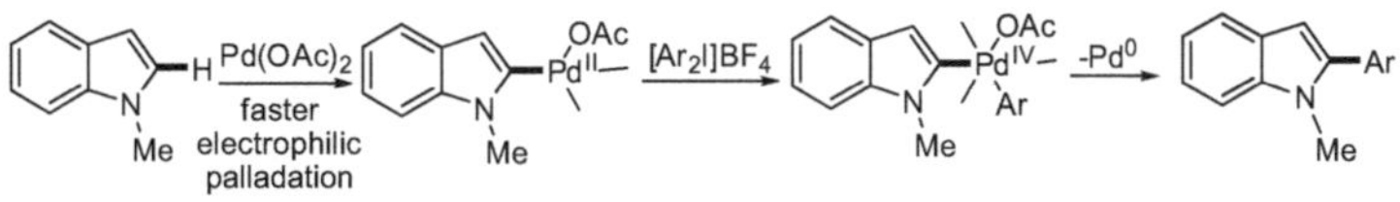

Scheme 12.26 Proposed mechanism of Pd-catalyzed C-arylation of indoles using [Ar$_2$I]BF$_4$.

more electron-deficient Pd^{II} catalyst was used. In such a situation, the formation of a σ-indole-Pd^{II} complex becomes faster and subsequent arylation with [Ar–I–Ar]BF$_4$ occurs *via* the alternative Pd^{II}/Pd^{IV} cycle (Scheme 12.26).[45]

Fagnou and co-workers used π-electron-deficient heterocycles to provide a synthetically useful method for the preparation of various 2-arylpyridines and 2-aryldiazines through a Pd-catalyzed coupling of corresponding *N*-oxides and aryl halides.[46–50] Both electron-rich and electron-deficient pyridine *N*-oxides and diazine *N*-oxides react exclusively at the 2-position. In addition, the aryl partner tolerated both sterically challenging and electronically diverse functional groups (Scheme 12.27). The method was also extended to other azine *N*-oxides such as quinoline and isoquinoline.[51] In this context, the use of *N*-iminopyridinium ylides as an alternative to *N*-oxides for analogous C–H arylation should be noted.[52]

They carried out experimental and theoretical studies to establish the mechanism of the direct arylation of different *N*-oxides.[53] The zero order with respect to aryl halide suggested a fast oxidative addition of the aryl halide to Pd^0, after which the catalyst is saturated as a Pd^{II} species. A 0.5 order with respect to catalyst indicated that the Pd^{II} species could possibly exist as an inactive dimer that is in equilibrium with the less favored active monomeric form of the catalyst. Studies with various Pd complexes indicated the involvement of acetate in the transition state as the deprotonating agent. The reaction likely proceeds *via* η^2-bound acetate on the Pd from which the pyridine *N*-oxide can displace one of the acetate oxygens and coordinate to the metal center *via* ligand substitution. The reaction is then proposed to proceed through a six-membered CMD transition state, leading to an Ar$_2$Pdbiaryl species that undergoes reductive elimination to provide the final product and regenerate the Pd^0 catalyst. The carbonate base used possibly deprotonates the acetic acid generated in this process, providing the acetate required to continue the catalytic cycle (Figure 12.10). The key

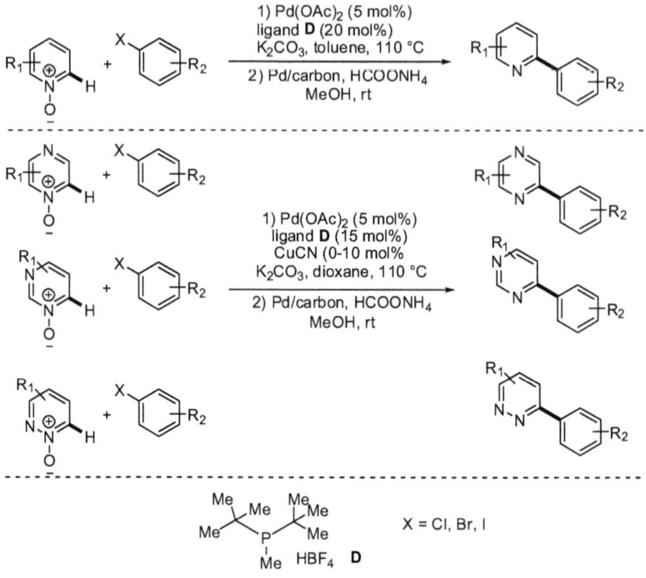

Scheme 12.27 Pd-catalyzed direct arylation of different π-electron-deficient heterocycles.

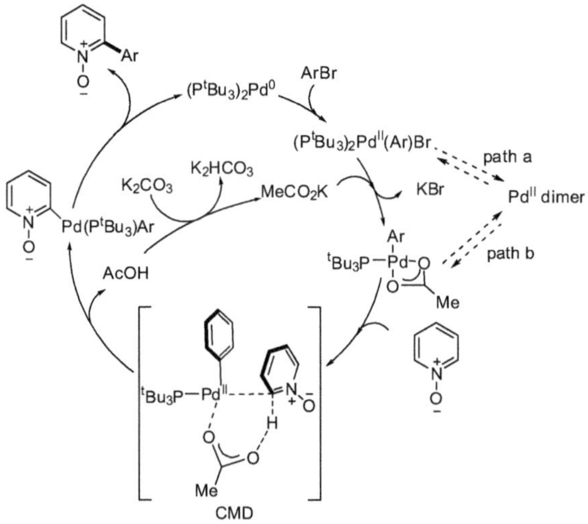

Figure 12.10 Proposed mechanism for direct arylation of different π-electron-deficient heterocycles.

step in this process, CMD, suggested by Fagnou and co-workers has attracted significant attention in the succeeding literature.

The direct C–H functionalization of the pyridine ring is problematic because of two distinct factors: pyridines are electronically deactivated and the pyridine nitrogen competes with the directing group for binding to the metal

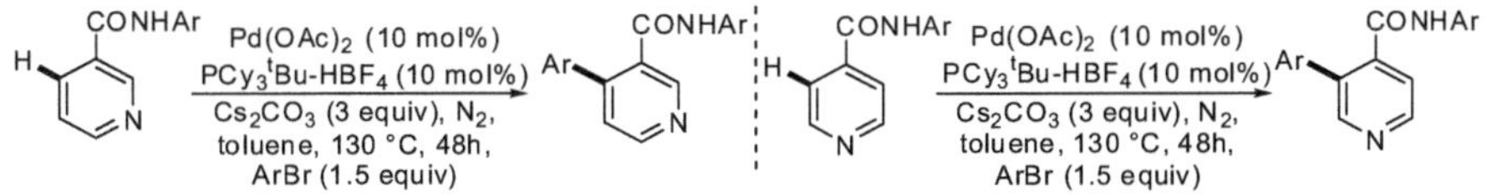

Scheme 12.28 Direct arylation reactions of pyridine derivatives with amide directing groups.

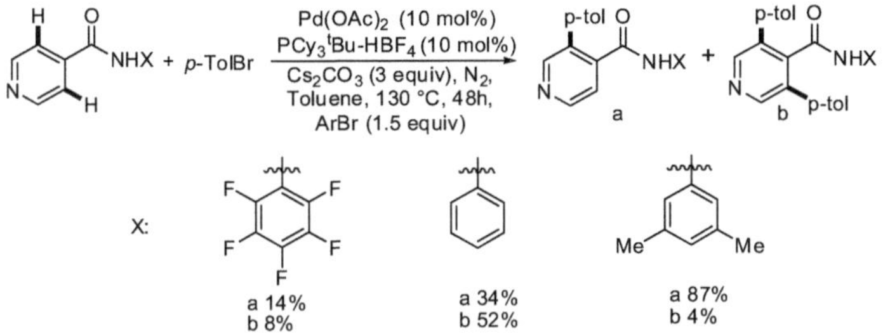

Scheme 12.29 C3 arylation reactions of pyridine derivatives and effect of directing group.

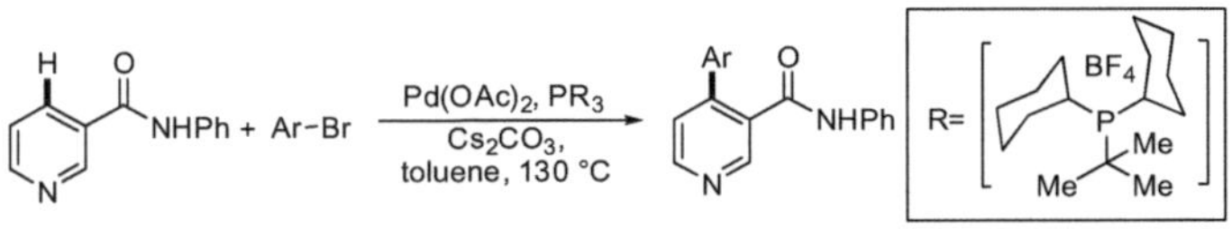

Scheme 12.30 C4 arylation reactions of a nicotinic acid derivative.

center. Yu and co-workers approached this problem by introducing amide as a directing group by which they achieved a complementary regioselectivity at C3 and C4 positions of pyridine while previously reported C2 arylation used *N*-oxides to determine the site of arylation (Schemes 12.28–12.30).[54–56]

The C2 arylation in the pyridine ring is a challenging task owing to the poor electron density on this carbon center and the counterproductive coordination between the metal and nitrogen atom. One way to enforce selectivity is to use *N*-oxides in place of pyridines or by protecting/blocking other positions that are prone to arylation/activation (Scheme 12.31).[57–59] Yu and co-workers described the selective C3 and C4 arylation of pyridines using amide as a directing group.[56] However, direct C3 arylation of pyridines[60] would be more interesting in terms of synthetic applicability if it could be carried out without any directing group. In this context, Yu and co-workers reported a directing group-free C3 arylation of unprotected pyridines using a simple Pd(OAc)$_2$/1,10-phenanthroline catalyst system (Scheme 12.31).

A plausible mechanism[60] is depicted in Figure 12.11. Phenanthroline is a strongly coordinating ligand and therefore metal–pyridine-*N* coordination can be destabilized. The dissociation rate increased the local concentration

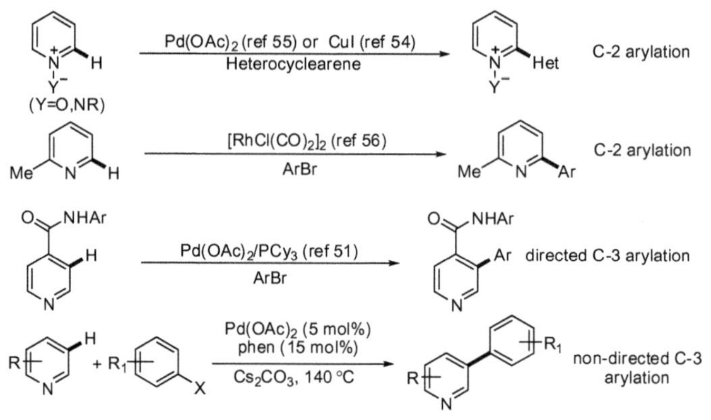

Scheme 12.31 Selective arylation in the pyridine ring.

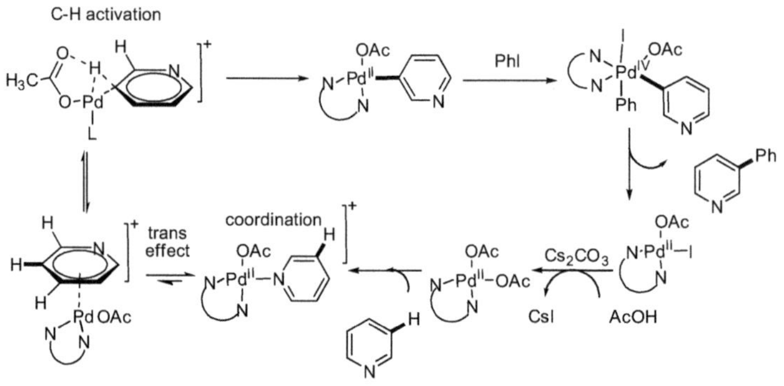

Figure 12.11 Mechanism of non-directed C3 arylation.

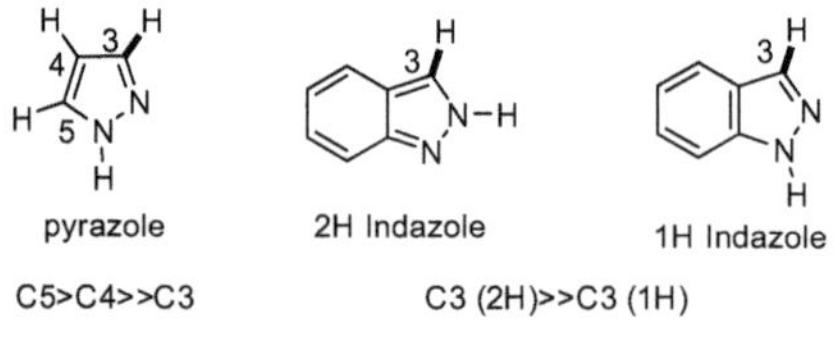

Figure 12.12 Reactivity pattern of pyrazole and indazole.

of Pd around different positions of pyridine. Notably, the C3 position is statistically favored as two equivalent carbons are available. Thus, proper orientation between the π-system of the pyridine ring and Pd leads to C–H bond cleavage.

Other than pyridines, site-selective direct arylations of indazoles, 1,2,3-triazole (C5),[61] thiophene (C2),[62] benzothiophene (C5),[62] pyrrole (C5),[62,63] thiazole (C2, C4 and C5),[62,64] oxazole, benzoxazole (C2),[65] imidazole (C5)[62] and pyrazoles have also been reported (Figure 12.12).[66] In particular,

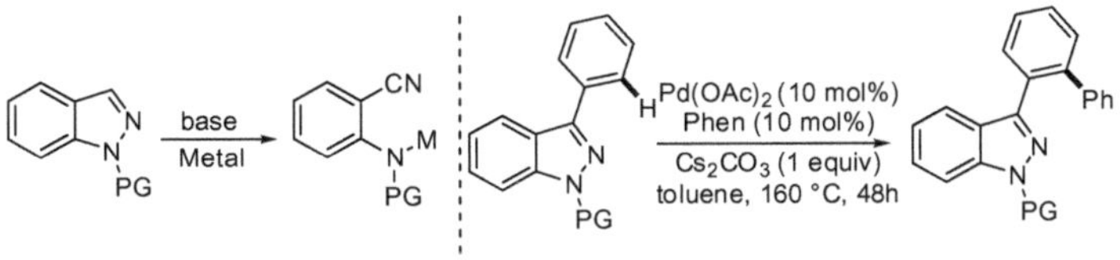

Scheme 12.32   Challenges in C3 arylation of indazole.

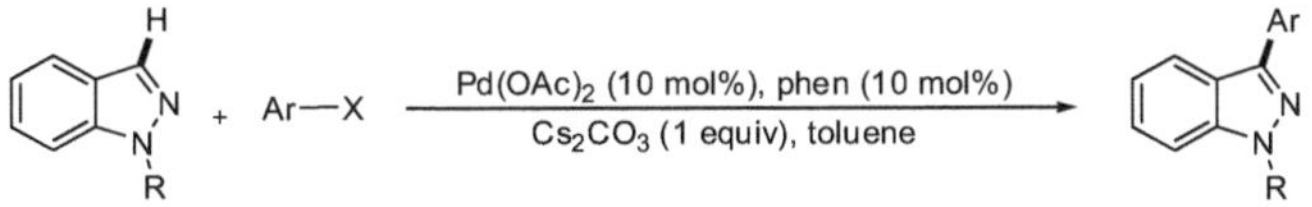

Scheme 12.33   Arylation of indazoles at the C3 position.

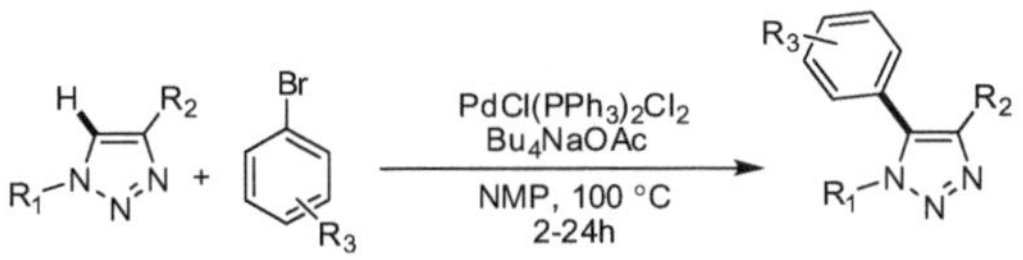

Scheme 12.34   Arylation of triazoles at the C5 position.

developing a new method for C3 arylation of 1*H*-indazole is difficult because the N–N bond has a tendency to hydrolyze in the presence of metal and base. Another problem is second C–H activation when the nitrogen at the 2-position of indazole ring can direct further arylation on the newly installed aryl group (Scheme 12.32).

To solve these problems, Yu and co-workers initially applied the previous conditions of C3 arylation of pyridines[60] on *N*-methylindazole using PhI as the coupling partner. Unfortunately, only a 24% yield of desired product was obtained along with 3% of undesired product from further directed arylation. Subsequent optimization produced the desired compound in 93% yield (Scheme 12.33).

1,2,3-Triazole, a multiple heteroatomic heterocycle, was also arylated by Gevorgyan and co-workers using palladium.[61] Under the reaction conditions, aryl bromides were used as the coupling partner (Scheme 12.34).

Further, in case of thiazoles, C2, C4 and C5 arylations were achieved at ambient temperature. In this case, azole *N*-oxides were used as the substrate and the selectivity obtained depended on the substrate design (Scheme 12.35).[64]

In case of simple heteroaromatics such as pyrrole, furan and thiophene, regioselective palladium-catalyzed arylations are also known. In 2009, Fagnou and co-workers reported the coupling reaction between heterocyclic compounds and an aryl bromide coupling partner (Scheme 12.36).[62]

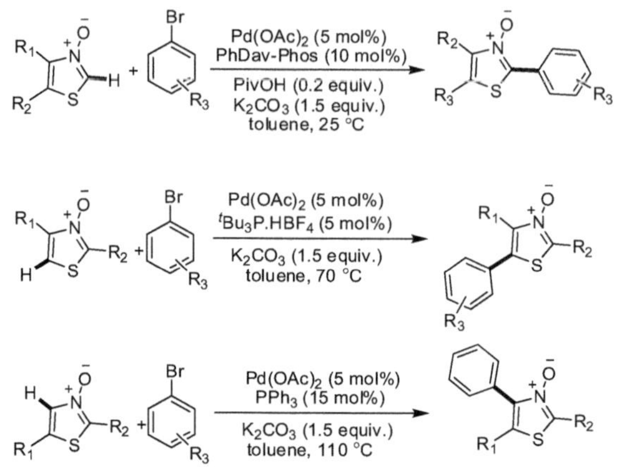

Scheme 12.35   C2, C4 and C5 arylations of azole *N*-oxides.

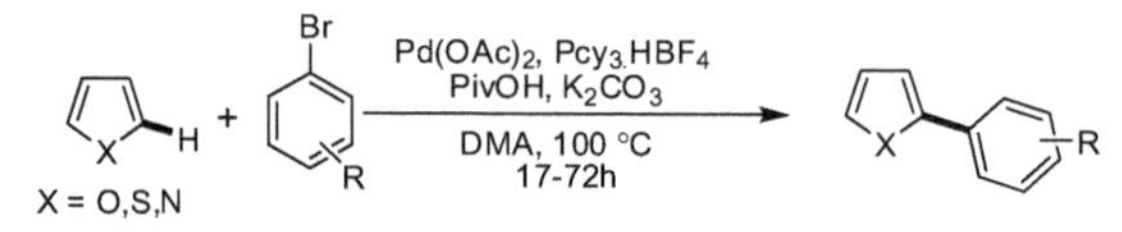

Scheme 12.36   C2 arylation of pyrrole, furan and thiophene.

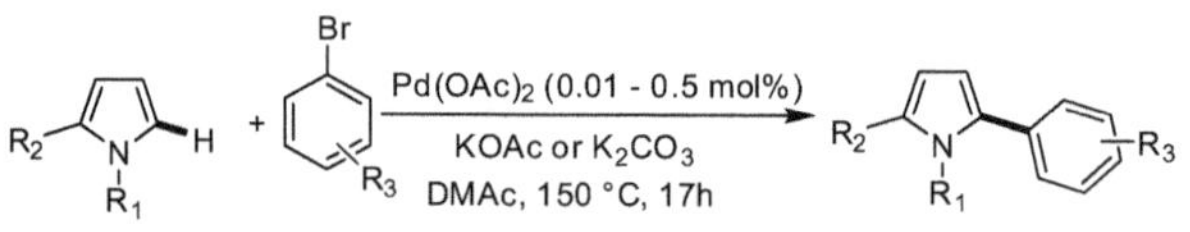

Scheme 12.37   C2 arylations of pyrroles.

Also in 2009, Roger and Doucet reported another protocol for the site-selective arylation of *N*-substituted pyrroles without using any external ligand.[67] In comparison with the previous work, this reaction consisted a bimolecular pathway without involving any pyrrolyl salt and minimized metal salt waste. The reaction was effective for a variety of *N*-substituted pyrrole substrates and functionalized aryl bromide coupling partner (Scheme 12.37). The reaction furnished a C2 arylated product whereas in case of substitution at C2, C5 arylation occurred. Notably, the reaction required high-boiling DMAc as a solvent, which is relatively non-toxic.

Fenner and co-workers implemented cheap and readily available aryl tosylates and mesylates, instead of aryl bromide, as the coupling partner for heterocyclic arylation (Scheme 12.38).[65]

Doucet and co-workers later developed a green protocol for heterocycle arylation using diethyl carbonate as solvent. The protocol was compatible with a number of benzoxazole, benzothiazole furan, thiophene and pyrrole derivatives (Scheme 12.39).[63]

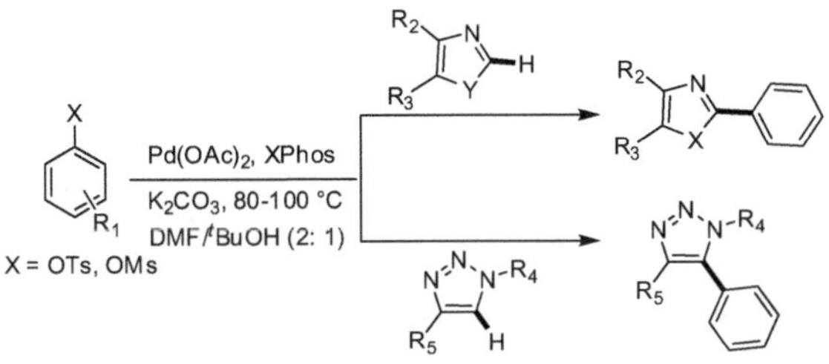

Scheme 12.38 Arylations of heterocycles.

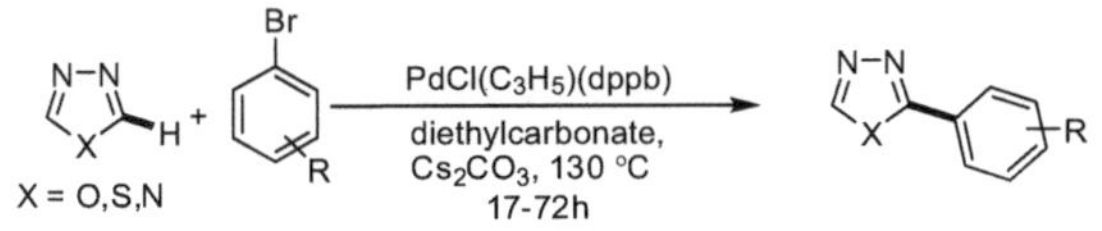

Scheme 12.39 Direct arylation of heterocycles in green media.

12.2.4 Direct C_{sp^3}–H Arylation

Despite the great advances in C–H activation,[4,10,68–71] C_{sp^3}–H bond functionalization remains challenging, mainly due to some inherent limitations. Direct arylation or functionalization of the C_{sp^2}–H bond is primarily facilitated by the interaction of stabilizing orbitals or specifically π-electrons with the metal center that allows metal–substrate binding and C–H bond cleavage. Similarly the functionalization of benzylic or C_{sp^3}–H bonds in the α-position to the heteroatom is facile because they are more acidic than alkyl groups at distant positions. However, owing to the absence of these types of orbitals or stabilizing effects in simple aliphatic systems, direct arylation of C_{sp^3}–H bonds is difficult and remains underdeveloped. Also, competitive β-hydride elimination gives rise to undesired side products and a decrease in the catalytic efficiency.

In 2005, Shabashov and Daugulis reported the use of a bidentate 8-aminoquinoline as a directing group for Pd-catalyzed C_{sp^3}–H direct arylation (Scheme 12.40).[72] They demonstrated that this directing group facilitates the transformation by coordinating to the metal center in a bidentate fashion, which provides stability to transient Pd^{IV} species. Picolinic acid was also used as the directing group for this transformation. This approach thus offers an excellent method for the functionalization of various carboxylic acids at the β-position by sequential amidation–C–H activation.

To probe the mechanism, they synthesized various palladacycle complexes as possible reaction intermediates (Scheme 12.41). From the structure of complex **1**, it was evident that bi-coordination of the directing group is necessary to stabilize the metal complex. Reaction of **1** with bromine leads to a stable Pd^{IV} complex (**2**). Reaction of **1** with aryl iodide leads to arylation of

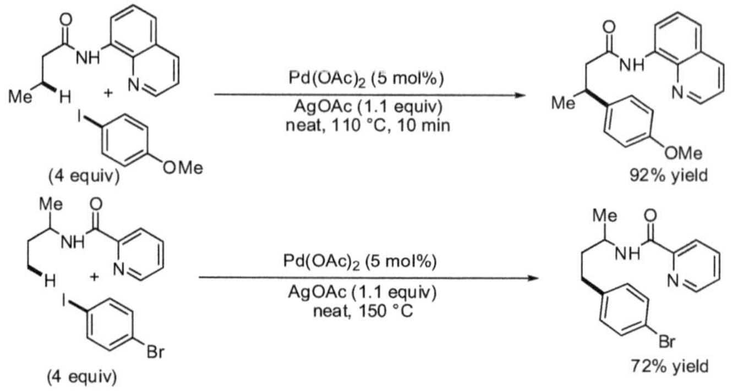

Scheme 12.40   Directing group-assisted Pd-catalyzed C_{sp^3}–H direct arylation.

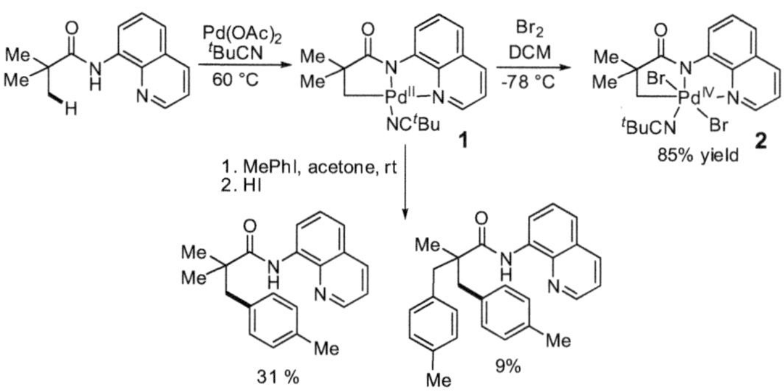

Scheme 12.41   Synthesis and reactivity of palladacycle complexes.

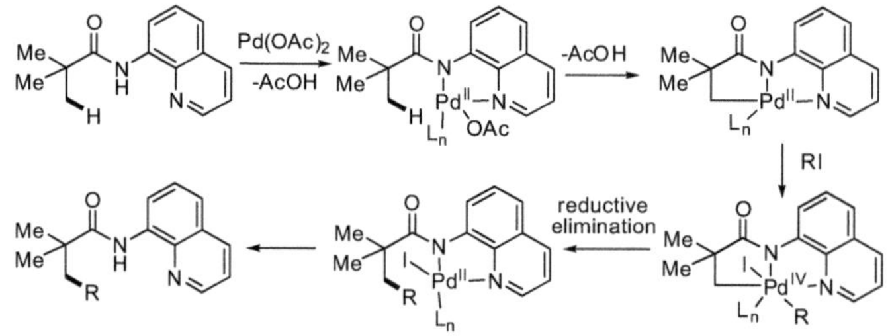

Scheme 12.42   Proposed mechanism for Pd-catalyzed C_{sp^3}–H direct arylation.

the C_{sp^3}–H bond where a Pd^{IV} species was not observed, indicating that probably **1** may be the reactive intermediate in the reaction. Based on these studies, they proposed a Pd^{II}/Pd^{IV} cycle to be operative (Scheme 12.42).

He and Chen extended this methodology for the arylation of the γ-C_{sp^3}–H bond in various aliphatic amines using picolinamide as the directing group.[73] They demonstrated that the regioselective outcome of the reaction depends on the relative conformation of the C–H bond with respect to the

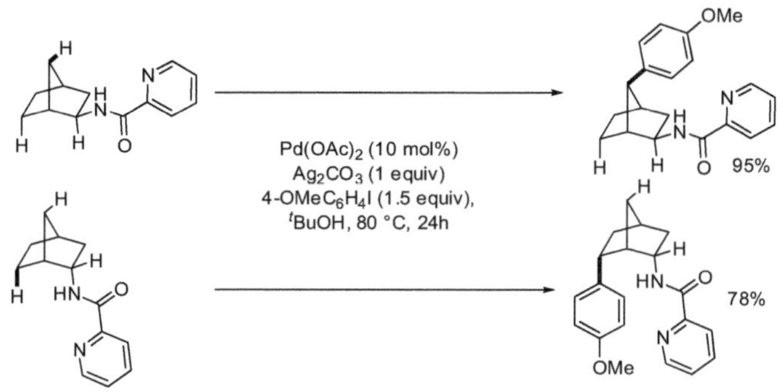

Scheme 12.43 Arylation of γ-C_{sp^3}–H aliphatic amines.

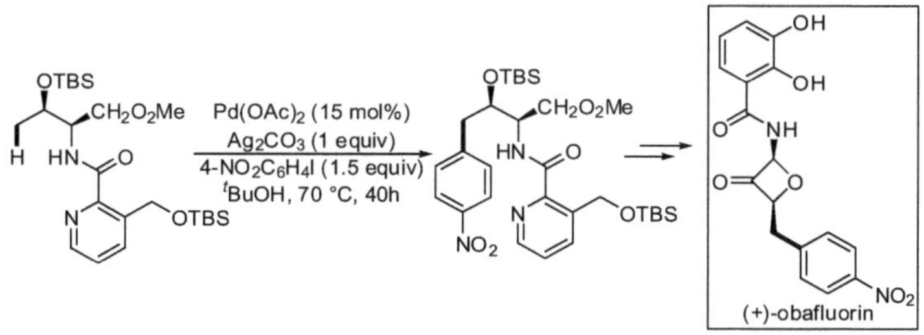

Scheme 12.44 Synthesis of (+)-obafluorin from a threonine derivative.

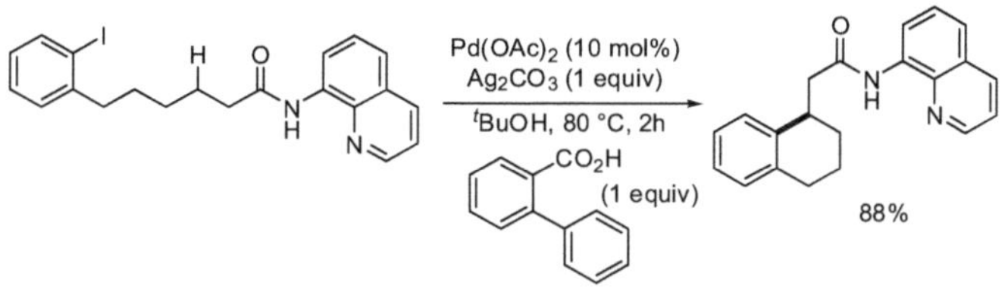

Scheme 12.45 Direct arylation of the C_{sp^3}–H bond in an intramolecular transformation.

picolinamide (Scheme 12.43). The synthetic utility of this method was demonstrated by synthesizing (+)-obafluorin from a threonine derivative (Scheme 12.44).

Chen and co-workers also utilized 8-aminoquinoline as a directing group for the Pd-catalyzed direct arylation of the C_{sp^3}–H bond in an intramolecular transformation depicted in Scheme 12.45.[74]

In this context, in addition to 8-aminoquinoline, carboxylic acid moieties were used as the directing group for the activation of aliphatic C_{sp^3}–H bond.[27] Aliphatic carboxylic acids were arylated by using Ar–B(OR)$_2$ (Scheme 12.46) and a Pd^0/Pd^{II} cycle was proposed. However, with an aryl iodide as coupling partner, a Pd^{II}/Pd^{IV} cycle was suggested (Scheme 12.47).

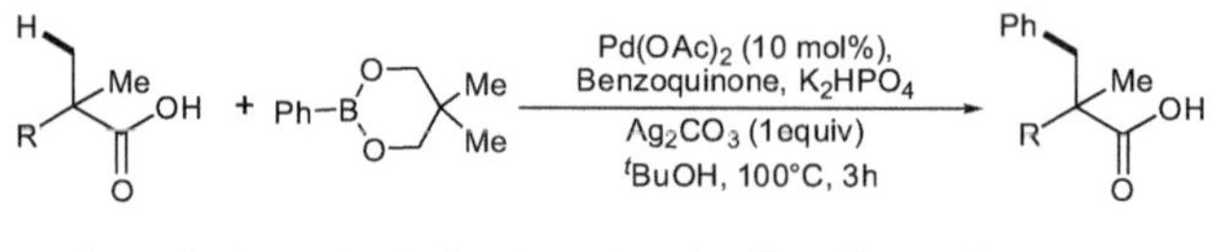

Scheme 12.46 β-Arylation of aliphatic acids (Pd^{II}/Pd^0 cycle).

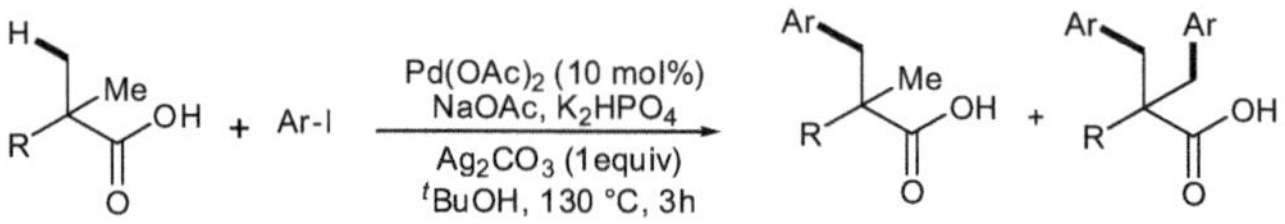

Scheme 12.47 β-Arylation by using Ar–I (Pd^{II}/Pd^{IV} cycle).

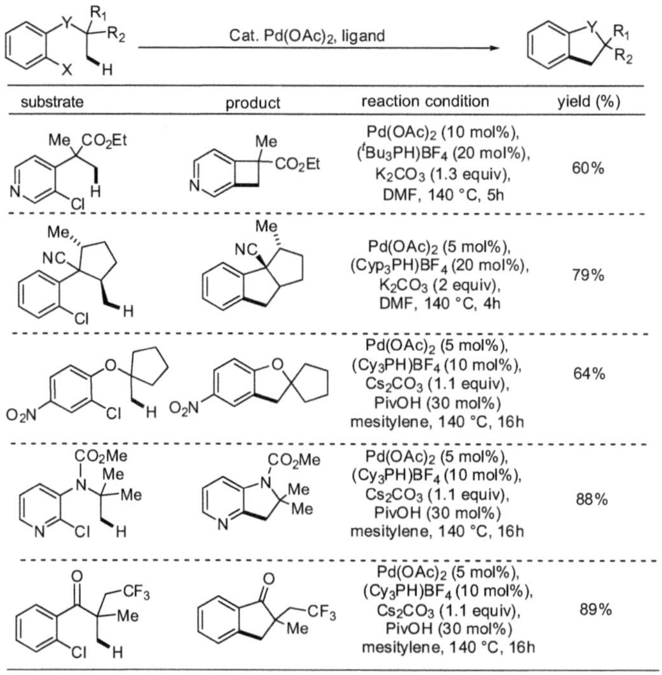

substrate	product	reaction condition	yield (%)
Me CO₂Et (pyridine, Cl)	Me CO₂Et	Pd(OAc)₂ (10 mol%), (^{t}Bu₃PH)BF₄ (20 mol%), K₂CO₃ (1.3 equiv), DMF, 140 °C, 5h	60%
Me, NC, Cl	Me, NC,	Pd(OAc)₂ (5 mol%), (Cyp₃PH)BF₄ (20 mol%), K₂CO₃ (2 equiv), DMF, 140 °C, 4h	79%
O₂N, O, Cl	O₂N, O	Pd(OAc)₂ (5 mol%), (Cy₃PH)BF₄ (10 mol%), Cs₂CO₃ (1.1 equiv), PivOH (30 mol%), mesitylene, 140 °C, 16h	64%
CO₂Me, Me, Me, Cl	CO₂Me, Me, Me	Pd(OAc)₂ (5 mol%), (Cy₃PH)BF₄ (10 mol%), Cs₂CO₃ (1.1 equiv), PivOH (30 mol%), mesitylene, 140 °C, 16h	88%
O, CF₃, Me, Cl	O, CF₃, Me	Pd(OAc)₂ (5 mol%), (Cy₃PH)BF₄ (10 mol%), Cs₂CO₃ (1.1 equiv), PivOH (30 mol%), mesitylene, 140 °C, 16h	89%

Scheme 12.48 Synthesis of four- and five-membered ring through C$_{sp^3}$–H arylation.

Fagnou and co-workers reported Pd-catalyzed intramolecular direct –H arylation with an aryl or heteroaryl halide, leading to the formation of valuable four- or five-membered ring compounds such as cyclobutarenes, indanes, indolines, dihydrobenzofuran and indanones (Scheme 12.48).[75,76] In cases where more than one C$_{sp^3}$–H bond can react, the arylation occurred regioselectively on the primary C–H bond. On substrates with more than one primary C–H bond, the regioselectivity trends correlated with the size of palladacycle intermediate, with five-membered rings favored over six- and

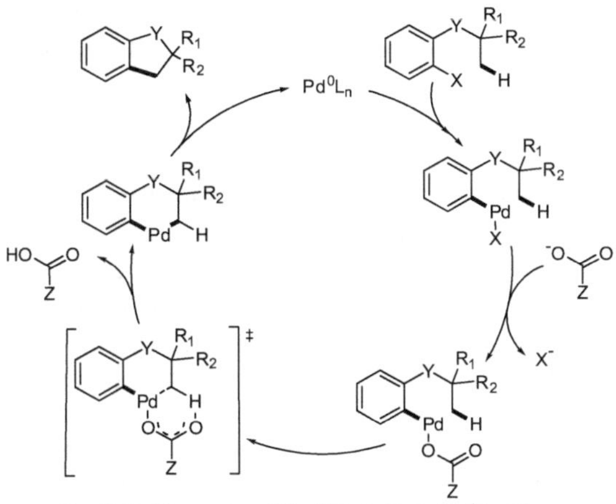

X = Cl, Br; Y = no atom, CR$_2$, N-R, O, C=O; Z = ^{t}Bu, O$^-$

Figure 12.13 Proposed mechanism.

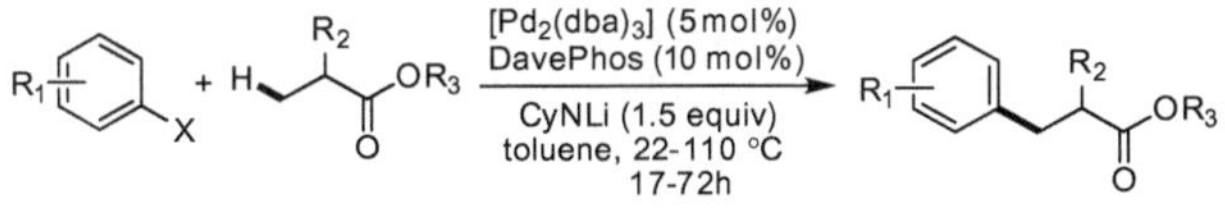

Scheme 12.49 β-Arylation of a carboxylic ester.

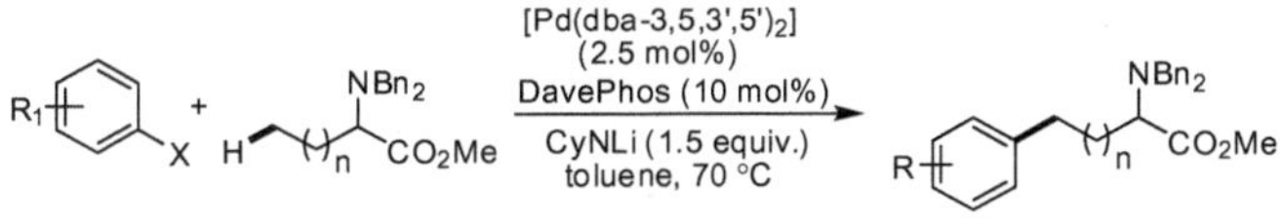

Scheme 12.50 Amine-directed long-range aliphatic arylation.

seven-membered rings. Based on computational and experimental studies, they proposed a general mechanism (Figure 12.13).

On the other hand, a related intermolecular C–C bond reaction through Pd-catalyzed C$_{sp^3}$–H activation was reported by Baudoin's group. In their first report in 2010,[77] β-arylation of a carboxylic ester was achieved with the aid of a strong base (Scheme 12.49). In a subsequent report in 2012,[78] they described an amine-directed long-range aliphatic arylation with aryl bromides (Scheme 12.50).

12.3 Rhodium-Catalyzed Reactions

Despite many recent developments in the field of Rh-catalyzed C–H activations, this area is far less developed compared with corresponding Pd-catalyzed reactions. Rh catalysts have a unique mode of reactivity for C–C bond formation *via* C–H activation and are typically highly regioselective.

owing to their broad range of functional group compatibility, both chelation-assisted and heteroatom-directed C–H bond arylation have been developed.

12.3.1 N-Atom-Directed C–H Arylation

Rh-catalyzed direct C–H arylation was carried out for the first time by Oi *et al.* with the well-known Wilkinson's catalyst in 1998.[79] In the presence of tetraarylstannanes, *ortho*-arylated arenes were obtained in good to moderate yields (Scheme 12.51). Mechanistically, it was proposed that the pyridine nitrogen plays a pivotal role to bring Rh in close proximity to the C–H bond of interest, subsequently facilitating *ortho* C–H activation.

In 2005, Miura and co-workers reported the N-atom directed *ortho*-arylation of benzophenone imine using sodium tetraphenylborate.[80,81] The scope of the protocol was limited owing to the formation of a hydrogenated by-product of imine in a considerable amount (Scheme 12.52).

They subsequently resolved the problem in 2008 using ethyl chloroacetate as hydrogen acceptor.[82] The new method was found to be more effective and was successfully implemented for various phenylazoles and tetraphenylborates. It was also observed that the sterically hindered azole ring promoted monoarylation over diarylation (Scheme 12.53).

In addition to the azole functionality as directing group, Miura and co-workers applied these conditions for *ortho*-arylation of azobenzenes.[82] In this case, [Rh(OMe)(COD)]₂ was found to be a more effective catalyst and phenylboronic was the arylating agent of choice (Scheme 12.54).

A transmetallation process between Rh^IX and arylboron species was proposed as the first step of the cycle to generate active arylated Rh^I complex, which interacts with the N-atom of the substrate. This anchoring of the Rh center facilitates the activation of the proximal *ortho*-C–H *via* an oxidative addition. Subsequent reductive elimination generates the arylated product

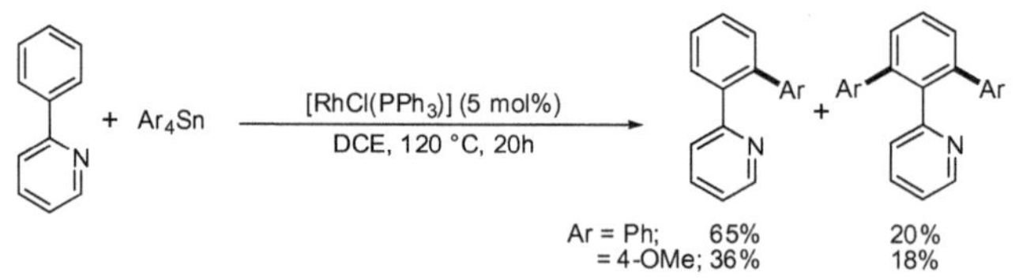

$$[RhCl(PPh_3)] (5 \text{ mol\%})$$
$$DCE, 120 \,°C, 20h$$

Ar = Ph;   65%   20%
= 4-OMe; 36%   18%

Scheme 12.51   Direct *ortho*-arylation of 2-phenylpyridine with tetraarylstannanes.

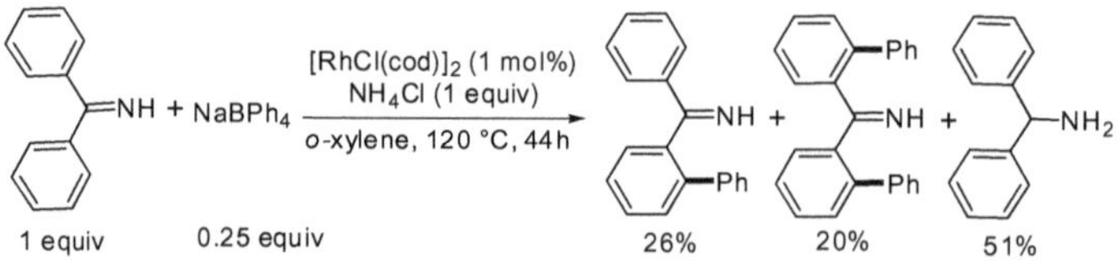

$$[RhCl(cod)]_2 (1 \text{ mol\%})$$
$$NH_4Cl (1 \text{ equiv})$$
$$o\text{-xylene, } 120 \,°C, 44h$$

1 equiv   0.25 equiv   26%   20%   51%

Scheme 12.52   Rh-catalyzed C–H arylation of benzophenone imine.

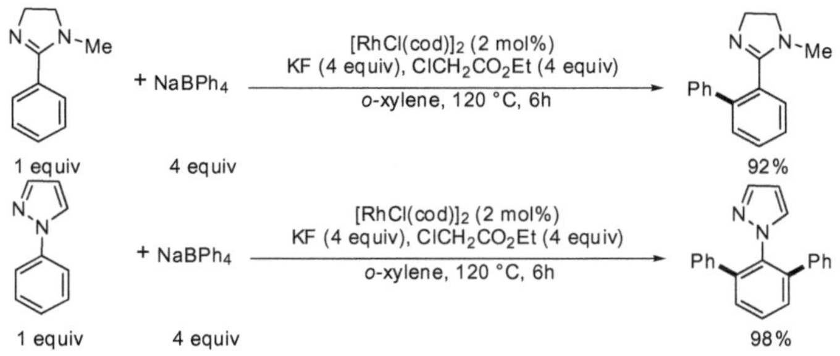

Scheme 12.53 Rh-catalyzed C–H arylation of phenylazoles.

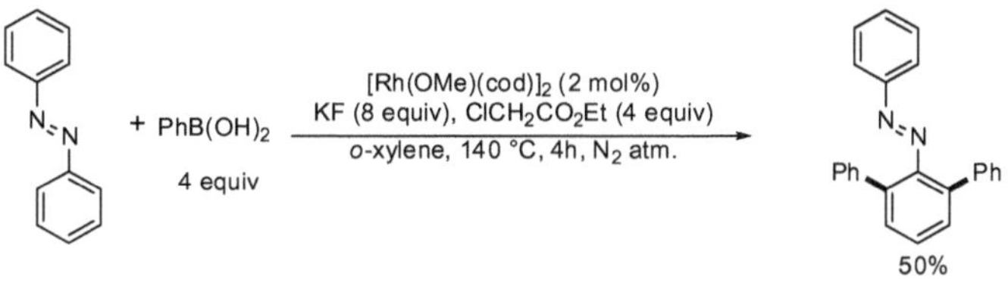

Scheme 12.54 Rh-catalyzed C–H arylation of azobenzene.

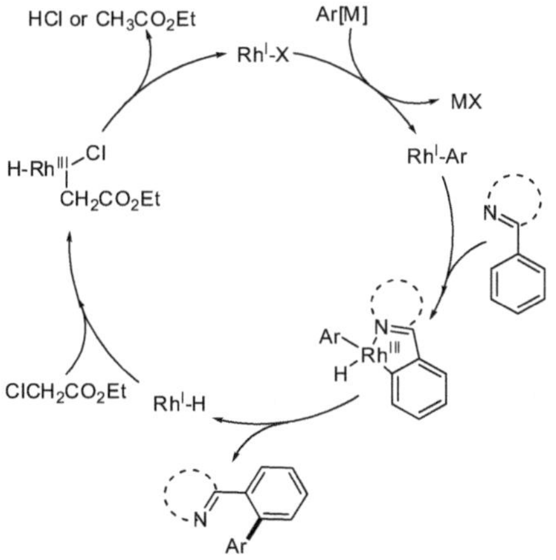

Figure 12.14 Proposed mechanism of Rh-catalyzed N-atom-directed C–H arylation.

and an $Rh^I H$ species. In this pathway, ethyl chloroacetate reacts with the $Rh^I H$ species to release HCl or CH_3CO_2Et with the regeneration of $Rh^I X$ species (Figure 12.14).[82] Notably, KF was used as a promoter in this reaction, although the exact role remained unclear.

Vogler and Studer described the direct C–H arylation of 2-pyridyl(hetero)-arenes using stoichiometric 2,2,6,6-tetramethylpiperidine *N*-oxide (TEMPO) as oxidant and $[RhCl(C_2H_4)_2]_2$ as catalyst.[83] In the case of 2-(thiophen-2-yl)pyridine, C–H arylation was achieved selectively at the C3 position of the

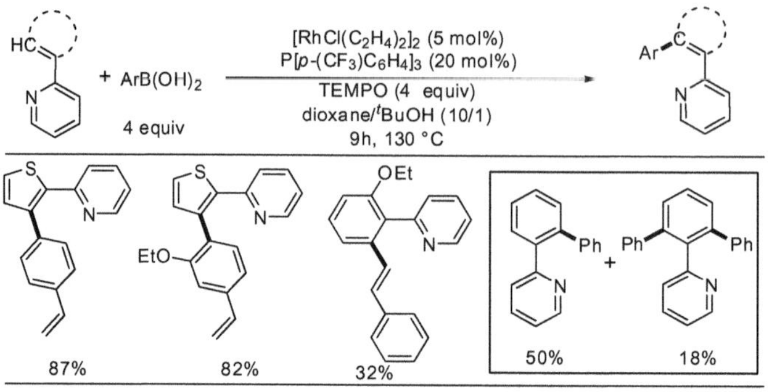

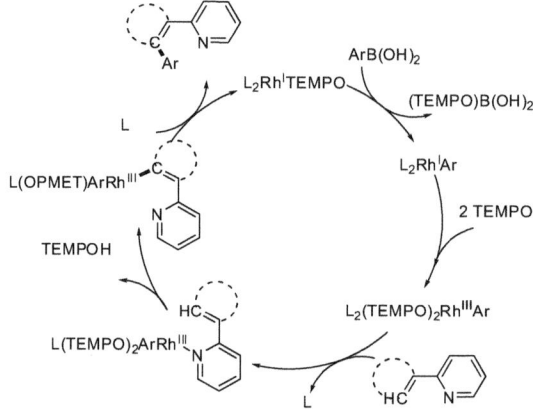

Scheme 12.55 Rh-catalyzed N-atom-directed C–H arylation using TEMPO as oxidant.

Figure 12.15 Rh-catalyzed direct C–H arylation using TEMPO as oxidant.

thiophene ring. For 2-phenylpyridine, again mono- and diarylation were observed (Scheme 12.55).

It was proposed that TEMPO oxidizes the L_2Rh^IAr species to $L_2(TEMPO)_2Rh^{III}Ar$ with the generation of TEMPO anion. This TEMPO anion act as a base and promotes base-assisted C–H activation. A successive reductive elimination was proposed to account for the formation of the desired arylated product and regeneration of Rh species (Figure 12.15).[83]

Furthermore, in the case of Rh-catalyzed direct arylation, acid chlorides[84] and acid anhydrides[85] were also used as coupling partner. In 2008, Zhao and Yu reported the C–H arylation of benzo[*h*]quinoline with a variety of acid chlorides in the presence of [Rh(COD)Cl]$_2$ as catalyst (Scheme 12.56).[84]

It was suggested that acid chlorides, after a successive oxidative addition and decarbonylation, form an arylRh^{III} chloride species, which is responsible for the direct C–H arylation of benzo[*h*]quinoline. The same reaction

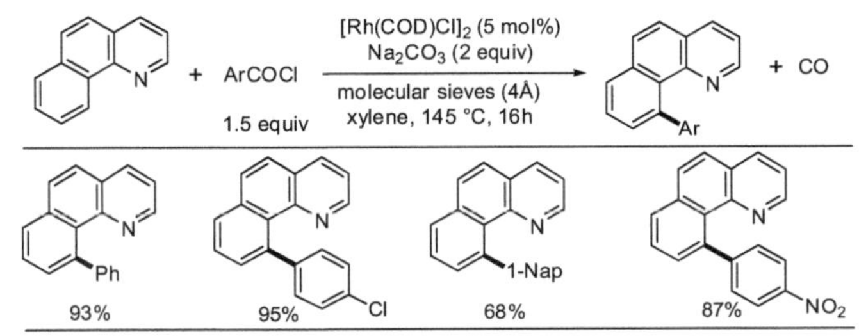

Scheme 12.56 Rh-catalyzed C–H functionalization of benzo[*h*]quinoline using an acid chloride.

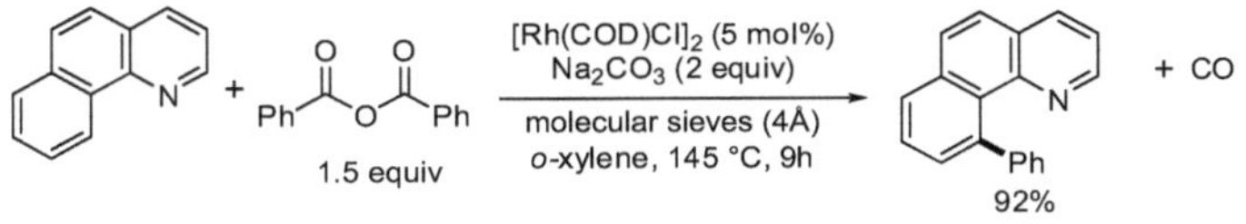

Scheme 12.57 Rh-catalyzed C–H functionalization of benzo[*h*]quinoline using an acid anhydride.

conditions were found also to be applicable for acid anhydrides instead of acid chlorides (Scheme 12.57).[85]

12.3.2 *Ortho*-Arylation of Phenols

In the case of *ortho*-metallation of phenols, the challenge lies in the formation of a four-membered metallacycle with the phenol itself. This challenge was solved by Bedford *et al.* in 2003, when they accomplished the *ortho*-arylation of phenols with aryl halides using a catalytic amount of Wilkinson's catalyst and phosphinite ligand (Scheme 12.58).[86]

According to their proposal, the aryl halide first interacts with $Rh^I L_n$, forming $ArRh^{III}XL_n$. In the next step, the phosphinite coordinates to the metal center and a base-mediated orthometallation occurs at the ligand's phenolate part, forming a five-membered chelating complex. Subsequent reductive elimination regenerates the $Rh^I L_n$ species with the release of an arylated phosphinite. Further phosphinyl group transfer between the phenol and arylated phosphinite ligand results in an arylated phenol along with regeneration of phosphinite co-catalyst (Figure 12.16).[86]

Notably, this protocol was confined to phenols having one *ortho* substitution. Bedford and Limmert reported an improved protocol for the arylation of phenol and its derivatives using $[Rh(COD)_2Cl]_2$ as catalyst and $P(NMe_2)_3$ as ligand, where the diarylated phenol was formed as the major product (Scheme 12.59).[87]

Oi *et al.* independently reported a similar catalytic system for the *ortho*-arylation of both substituted and unsubstituted phenols (Scheme 12.60).[88] A related Pd-catalyzed reaction in an intramolecular set-up was previously

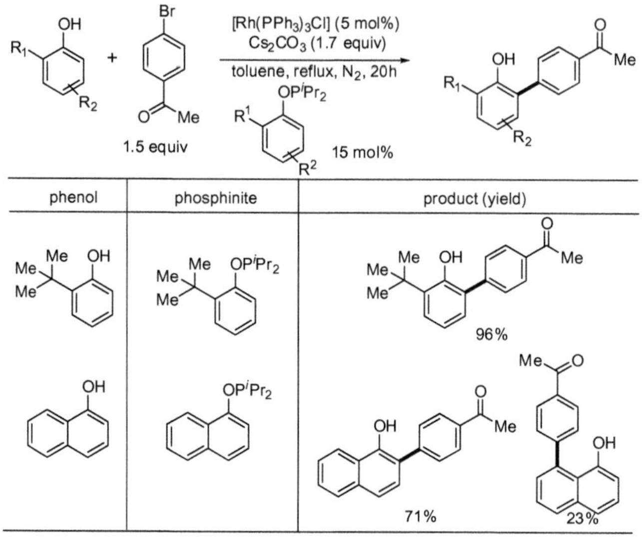

Scheme 12.58 Rh-catalyzed *ortho*-arylation of phenols.

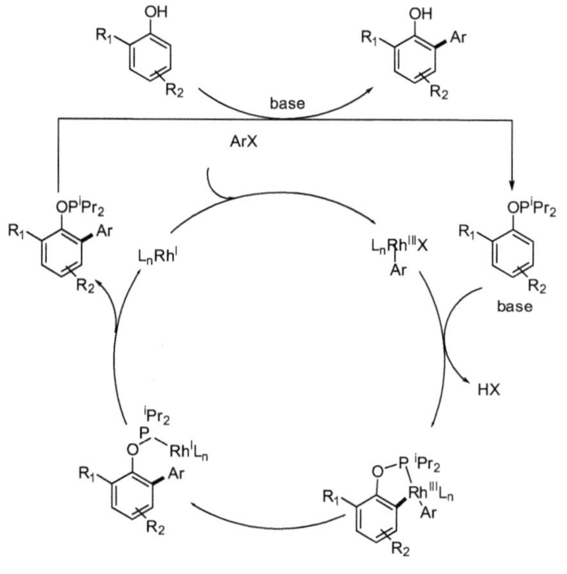

Figure 12.16 Proposed mechanism of Rh-catalyzed *ortho*-arylation of phenols.

described by Rawal and co-workers,[16] where electrophilic palladation dictated the regioselective outcome of the reaction. Oi *et al.*'s method involves a five-membered metallacycle intermediate for which only *ortho*-arylated products could be obtained.

Chlorophosphines were also found to be efficient ligands for this reaction (Scheme 12.61).[89]

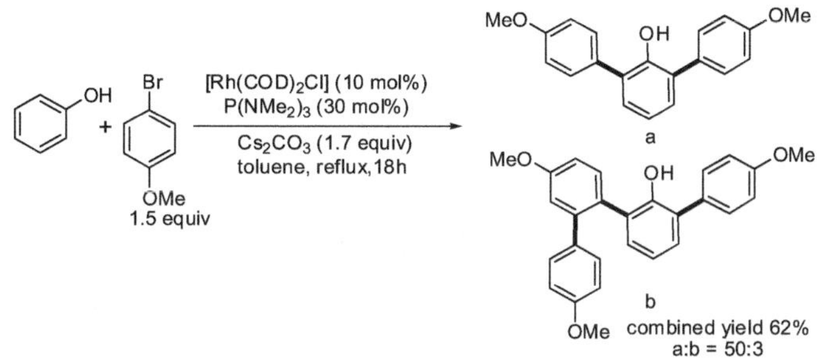

Scheme 12.59 Rh-catalyzed *ortho*-arylation of unsubstituted phenol using P(NMe₂)₃ as ligand.

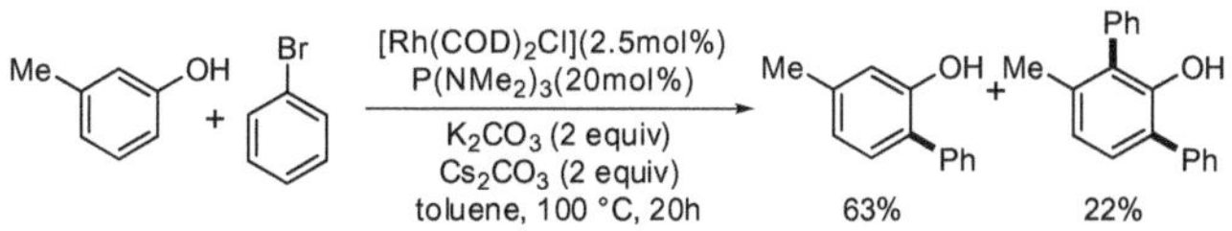

Scheme 12.60 Rh-catalyzed *ortho*-arylation of substituted phenol using P(NMe₂)₃ as ligand.

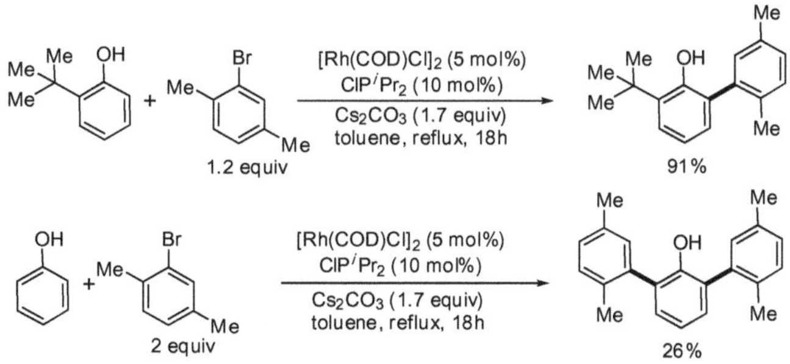

Scheme 12.61 Rh-catalyzed *ortho*-arylation of phenols with a chlorophosphine ligand.

12.3.3 C–H Functionalization Without Chelating Mechanism

Heteroarenes without a pendent directing group were arylated using Rh in 2004 by Bergman and co-workers.[90] Aryl iodides were used as the coupling partner in the presence of a catalytic amount of [Rh(coe)₂Cl]₂ and PCy₃ ligand. A variety of nitrogen- and oxygen-containing heteroarenes underwent arylation with good to moderate yields (Scheme 12.62).[90]

An *N*-heterocyclic carbene (NHC) resulting from the tautomerization of nitrogen-containing heteroarenes was proposed as a plausible intermediate. In the catalytic cycle, the benzimidazole substrate coordinates to the metal center, forming an Rh^{I}–NHC complex, which oxidatively combines with the aryl iodide. Finally, arylated heteroarene and HI are released *via* reductive elimination (Figure 12.17).[90]

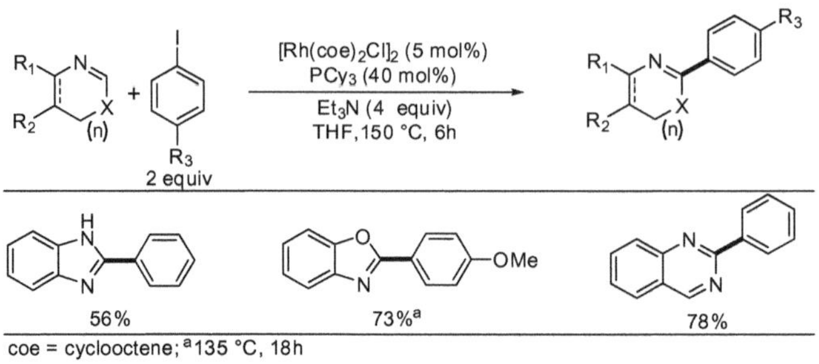

Scheme 12.62   Rh-catalyzed arylation of heteroarenes.

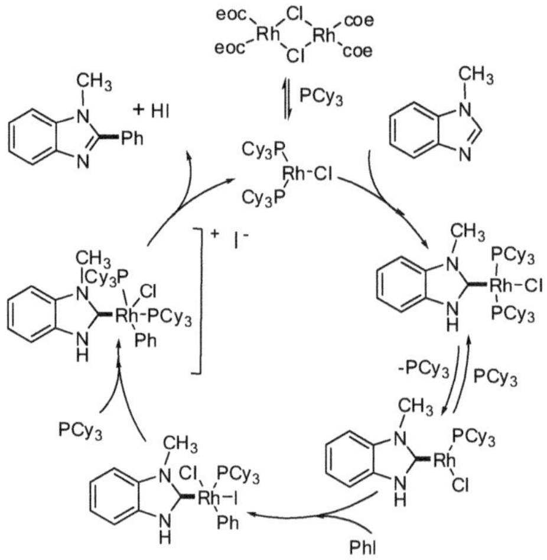

Figure 12.17   Proposed mechanism of Rh-catalyzed arylation of heteroarenes.

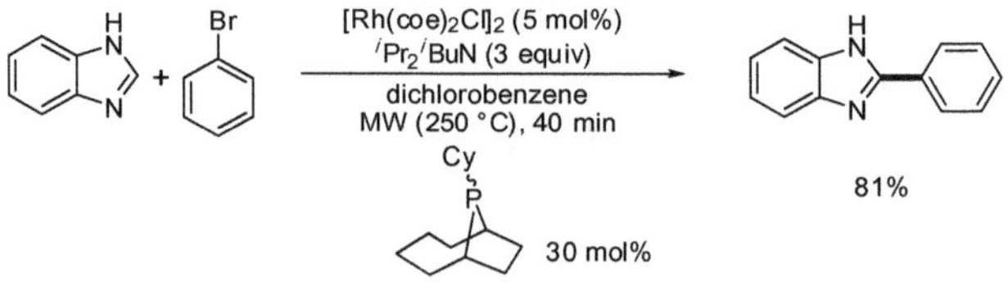

Scheme 12.63   Microwave-assisted Rh-catalyzed arylation of heteroarenes.

Bergman and co-workers also developed a microwave-assisted protocol with wider substrate scope and where by-product formation could be suppressed (Scheme 12.63).[91]

Sames and co-workers reported an Rh-catalyzed C2 arylation of indoles having a free NH using [Rh(coe)$_2$Cl$_2$]$_2$ in the presence of electron-deficient phosphine ligands.[92] Although indoles with anilide groups were compatible, no reaction was observed with 7-azaindole (Scheme 12.64).

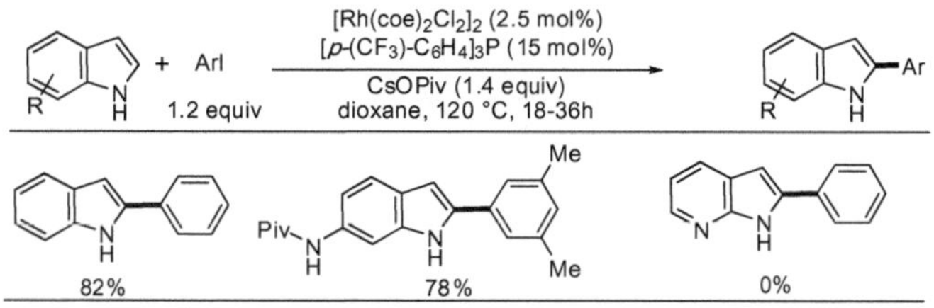

Scheme 12.64   Rh-catalyzed arylation of free NH-indole.

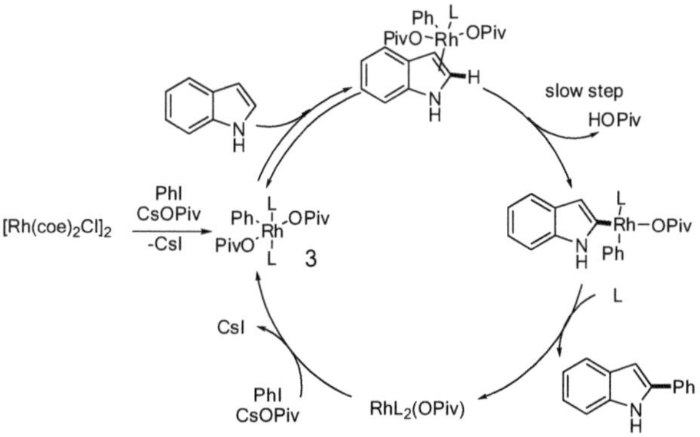

Figure 12.18   Proposed mechanism of the Rh-catalyzed arylation of free NH-indole.

According to the proposed mechanism, [Rh(coe)$_2$Cl$_2$]$_2$, in the presence of aryl iodide and cesium pivalate, generates complex **3**, which forms a π-complex with indole. Pivalate assisted the C–H bond dissociation of indole and the arylated indole product was formed after a subsequent reductive elimination (Figure 12.18).[92] Kinetic studies revealed a first-order dependence on complex **3** and indole and an inverse first-order dependence with respect to the ligand. A significant kinetic isotope effect was observed ($k_H/k_D = 3.0$), which suggests that the C2 metallation is likely to be the rate-limiting step (Figure 12.18).[92]

Nyori and co-workers reported the first Rh-catalyzed bis-heteroarene formation through direct C–H activation.[93,94] They developed an Rh catalyst having π-accepting ligands, which reduces the electron density on the metal, facilitating electrophilic metallation of electron-rich heteroarenes. This protocol involves (hetero)aryl iodides with simple heterocycles as coupling partners (Scheme 12.65). Based on some preliminary observations, an Rh^{I}/Rh^{III} catalytic cycle was proposed.

Rh-catalyzed C–H arylation through a radical pathway was reported by Proch and Kempe in 2007.[95] The active catalyst was generated *in situ* from [bis(2-pyridyl)amino]diphenylphosphane and 0.5 equiv. of [Rh(COD)Cl]$_2$ precursor (Scheme 12.66).[95] This bimetallic Rh catalyst was found to be effective in arylating unactivated simple arenes with aryl iodides, bromides and even chlorides.

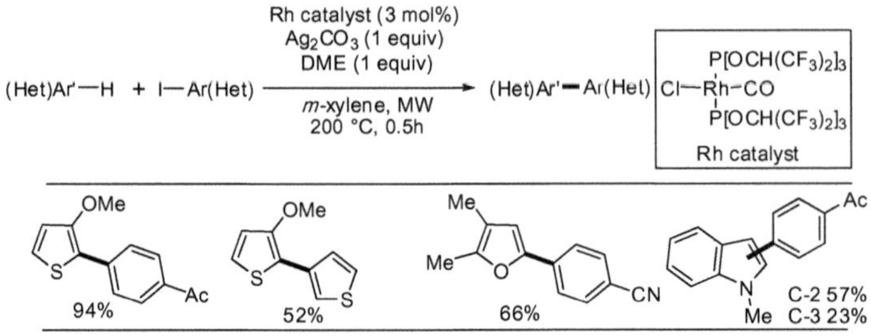

Scheme 12.65 Rh-catalyzed direct C–H arylation of (hetero)arenes.

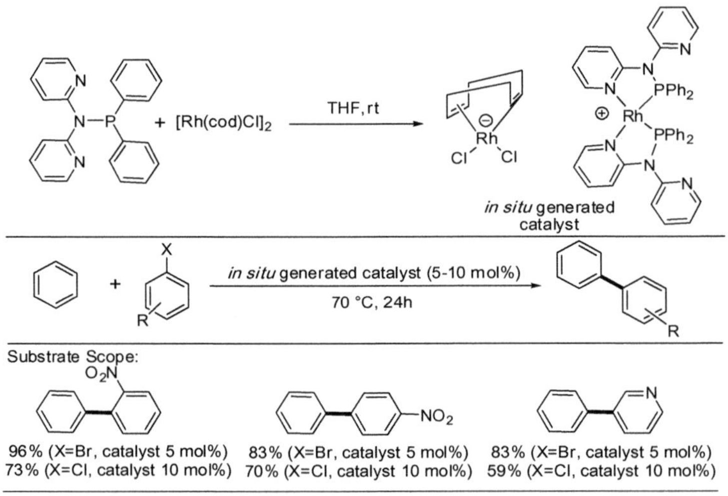

Scheme 12.66 Bimetallic Rh catalyst for C–H arylation.

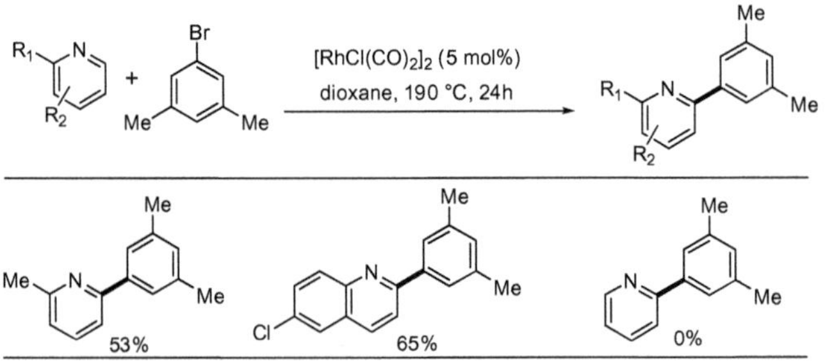

Scheme 12.67 Direct C–H arylation of pyridines and quinolines.

Pyridine and quinoline moieties were also arylated at the C2 position with aryl bromides using electron-poor $[RhCl(CO)_2]_2$ catalyst. Although pyridines and quinolines were arylated exclusively at the C2 position with a number of electron-donating and -withdrawing aryl bromides, unsubstituted pyridine remained unreacted under the standard conditions (Scheme 12.67).[59]

12.4 Ruthenium-Catalyzed Reactions

The initial work on transition metal-catalyzed C–H bond activation involved Pd and Rh under different reaction condition.[3,4,10,69,96–100] Over time, a variety of metal catalysts[101–106] have been evaluated to discover more general methods. An Ru catalyst proved to be a promising alternative as it is relatively cheaper than the late transition metals and more tolerant towards air and moisture.[6,9,107–111]

The use of Ru[0] catalyst precursors in the field of arylation *via* C–H activation was introduced by Murai *et al.* in 1993 (Scheme 12.68).[112] This discovery motivated further research in this area. In their pioneering work, Oi *et al.*[113,114] used an easy to prepare and stable Ru[II] catalyst. Ackermann *et al.*[115,116] and Dixneuf and co-workers[5,117,118] have also made significant contributions in this field.

12.4.1 Ru(0) Catalysts

In the case of Ru-catalyzed C–C bond formation reactions, the presence of a directing/chelating group in the molecular scaffold is essential to promote C–H bond activation. Initial work on C–H activation using Ru[0] catalyst was focused primarily on *ortho*-alkylation (Scheme 12.68).[112]

This particular alkylation, involving C_{sp^2}–H activation, was named the Murai reaction. In terms of the mechanism,[119–121] it has been proposed that *in situ*-generated Ru[0] cleaves the C–H bond *via* the assistance of the directing group to form a metallacycle. This metallacycle interacts with the R–X/alkene through an oxidative addition and the desired product is obtained upon reductive elimination, regenerating the active catalyst (Figure 12.19).

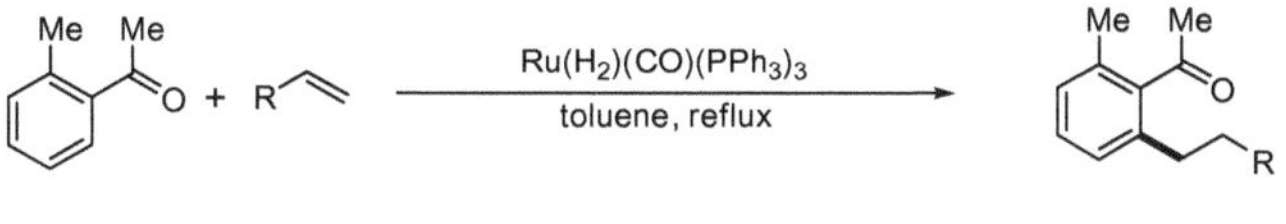

Scheme 12.68   Ru-catalyzed *ortho*-alkylation of an aryl ketone.

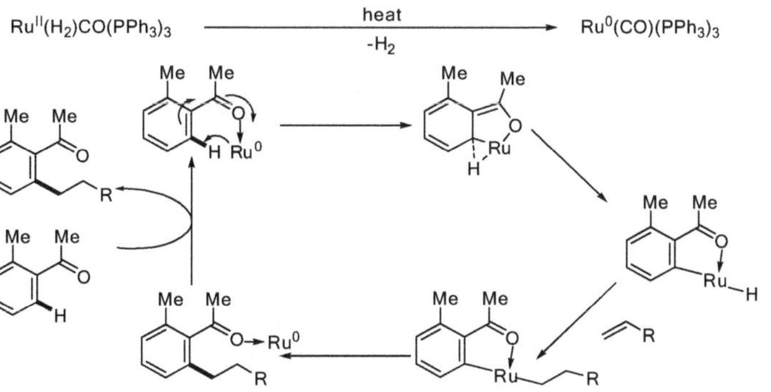

Figure 12.19   Catalytic cycle of alkylation.

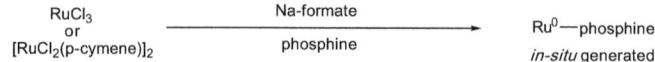

Scheme 12.69 *In situ* generation of Ru^0.

The reaction has been extended to various heterocycles such as thiophene, pyrrole and furan skeletons.[6,9,107–111] Upon testing a number of Ru^0 catalyst precursors, it was found that Ru(H$_2$)(CO)(PPh$_3$) showed the highest efficiency for oxygen-based directing groups (such as ketone, ester, aldehyde) whereas Ru$_3$(CO)$_{12}$ was more effective with nitrogen-based directing groups (such as imine, hydrazine, pyridine, oxazoline).[122] Other reported Ru^0 catalyst precursors include Ru(CO)$_2$(PPh$_3$)$_3$, RuH$_2$(H$_2$)(CO)(PCy$_3$) and Ru(H$_2$)(PPh$_3$)$_4$.[5]

In all of the above cases, the release of ligands or elimination of H$_2$ gas generated the active Ru^0 catalyst. Recently, Ru^0 species have been generated *in situ* following the reduction of Ru^{II} precursors[123–127] or Ru^{III} precursors[128,129] by formate in the presence of phosphines (Scheme 12.69).

12.4.2 Ru(II) Catalysts

Compared with the Ru^0 catalysts, Ru^{II} and Ru^{III} species are generally easy to prepare and more stable in air and water, hence it would be profitable if these complexes could be used as catalysts.[109] The first example of direct arylation in the presence of an Ru^{II} catalyst was reported in 2001 by Oi *et al.* with 2-phenylpyridine (Scheme 12.70).[113]

Other catalysts such as RuCl$_2$(PPh$_3$)$_3$ and [RuCl$_2$(COD)$_n$-4PPh$_3$ also showed comparable reactivity. This is indicative of the fact that all three catalyst systems are likely to involve analogous active species. For the [RuCl$_2$(C$_6$H$_6$)]$_2$ system, the reactivity pattern of the aryl halides was found to be PhBr > PhI > PhOTf >> PhCl.[113]

12.4.3 Directing Scaffolds

12.4.3.1 Pyridines

2-Arylpyridine was the first substrate to be used in Ru-catalyzed *ortho*-arylation through C–H bond activation. After their pioneering work on the *ortho*-arylation of 2-phenylpyridine using an aryl bromide,[113] Oi *et al.* extended this work to introduce a heteroaromatic group at the *ortho* position.[130] Changing the proportion of the aryl bromide allowed control of the relative formation of mono- and bis-arylated products. In the presence of 1–1.2 equiv. of aryl bromide, the monosubstituted products were obtained, whereas bis-arylated products were formed in good yields with 2.2 equiv. of coupling partner (Scheme 12.71). The mono-arylated product was obtained exclusively for 2'- and 3'-substituted arylpyridines. In the case of 3'-substituted arylpyridines, monoarylation occurred at the less hindered site.

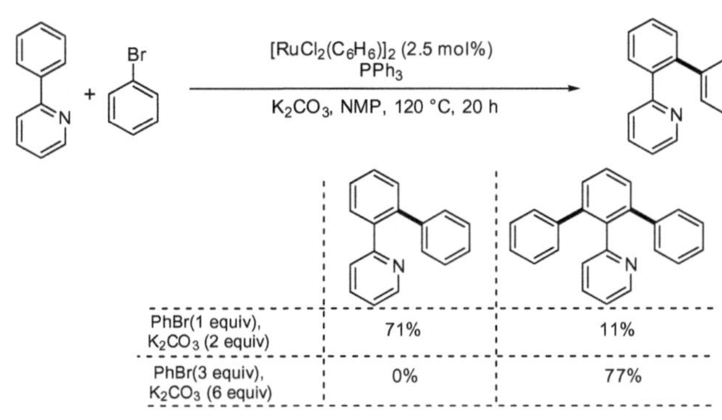

Scheme 12.70 *Ortho*-arylation using an Ru^{II}-based catalyst.

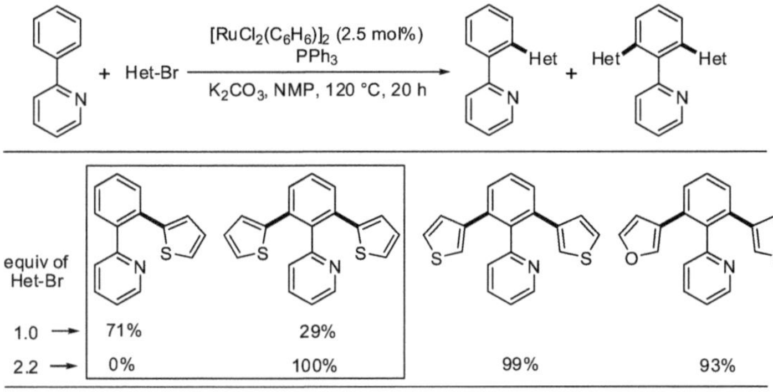

Scheme 12.71 *Ortho*-arylation of 2-phenylpyridine.

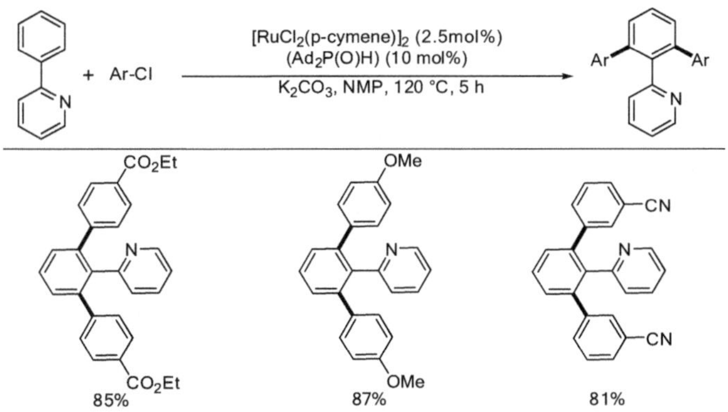

Scheme 12.72 *Ortho*-arylation of 2-phenylpyridine using a secondary phosphine oxide.

Ackermann *et al.* made a significant development in Ru-catalyzed arylation when they introduced phosphine oxides [R$_2$P(O)H] as a more efficient additive than phosphine ligands (Scheme 12.72).[115] Under these conditions,

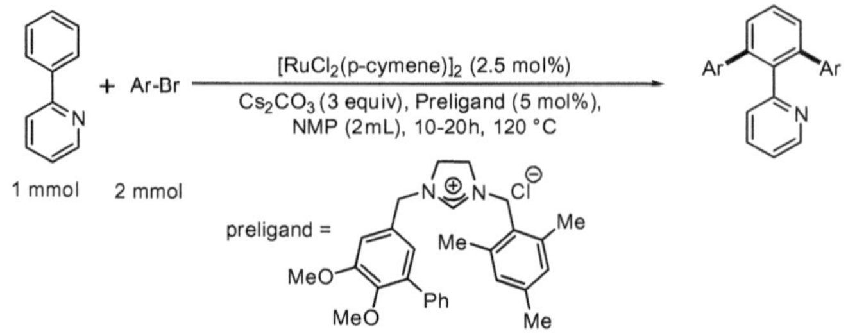

Scheme 12.73   *Ortho*-arylation of 2-phenylpyridine using NHC ligand.

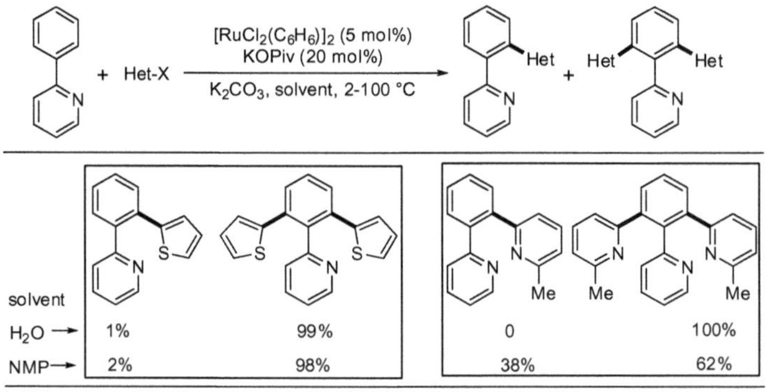

solvent				
H$_2$O →	1%	99%	0	100%
NMP→	2%	98%	38%	62%

Scheme 12.74   *Ortho*-arylation of 2-phenylpyridine in water.

the reaction system was found to be compatible with both aryl chlorides and aryl tosylates.[115,116]

Changing the halide partner to tosylate also allowed for better control of product ratios. Aryl chlorides formed bis-arylated product, whereas tosylates gave mono-arylated products preferentially. Notably, despite their increased accessibility and economic advantages, tosylates have not been used much as coupling partners owing to their decreased reactivity.[131] Among transition metals, palladium with some unique electron-rich ligand system is known to react with tosylates. For instance, DPPF ligand has been used for C–N bond formation even with heterocyclic tosylates.[132,133] In this respect, a few other well-known efficient ligands are XPhos, SPhos, *etc.*[134]

Özdemir and co-workers discovered the NHC complex of Ru^{II} for the arylation of 2-phenylpyridine in the presence of carbonate base (Scheme 12.73).[117]

Dixneuf and co-workers later showed that the same reaction can be carried out in water and better selectivity is shown for diarylation in water than the organic solvent NMP.[135] Under these conditions, pivalate was found to be a better ligand than acetate. The reaction proceeds even in the absence of KOPiv, but less efficiently (Scheme 12.74).

12.4.3.2 Imines

The imine functionality was successfully used as a directing group for Ru-catalyzed direct *ortho*-arylation.[114] Both ketimines and aldimines[136,137] were found to be reactive. In this case also, the mono-/bis-arylated product composition was found to depend on the amount of aryl bromides. An excess of aryl bromide enhances the formation of bis-arylated product, whereas the mono-arylated product was formed in a major amount when only 1 equiv. of the aryl bromide was used. The preference comes from the steric crowding between the alkyl group on the imine and the newly introduced aryl group, which disfavors a planar conformation of the imine, required for the introduction of a second aryl group (Scheme 12.75). For the same reason, bis-arylated product was obtained preferentially in good yield in the case of aldimines.

Dixneuf and co-workers reported the Ru[II] catalyzed diarylation of imines and ketamines.[138,139] In their first report, it was observed that PPh_3 inhibits the diarylation but *in situ*-generated $[Ru(OAc)_2(p\text{-cymene})]_2$ catalyst produce the diarylated product in a longer reaction time (Scheme 12.76). Subsequently they reported that the use of water in place of NMP favors the diarylation of ketamines (Scheme 12.77).

12.4.3.3 Oxazolines, Imidazolines and Pyrazoles

Maintaining structural similarity with pyridines and imines, various five-membered rings have been utilized as directing group for Ru-catalyzed *ortho*-arylation. Oi *et al.* reported the Ru-catalyzed *ortho*-arylation of

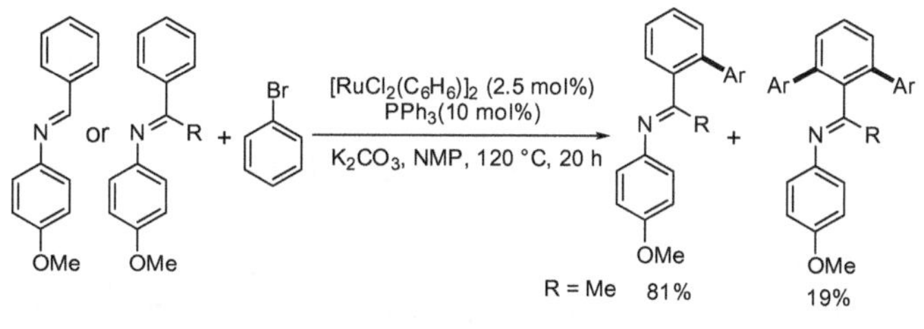

Scheme 12.75 *Ortho*-arylation of imine derivatives.

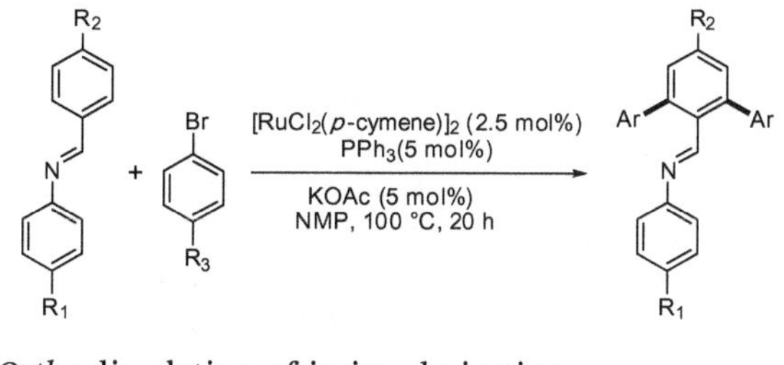

Scheme 12.76 *Ortho*-diarylation of imine derivatives.

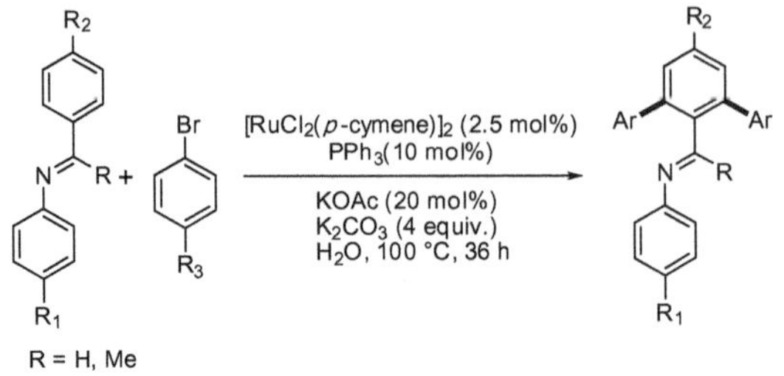

Scheme 12.77 *Ortho*-diarylation of imine derivatives with the use of water in place of NMP.

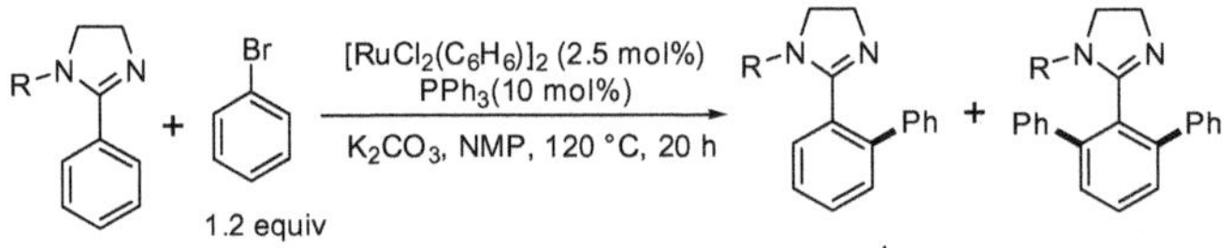

	combined yields	mono : diphenylated
R = H	64%	31:69
H (2.5 equiv)	90%	0:100
C(O)Me	76%	89:11
C(O)*t*-Bu	84%	86:14
C(O)Ph	88%	77:23
Ts	0%	

Scheme 12.78 *Ortho*-arylation of imidazoline derivatives.

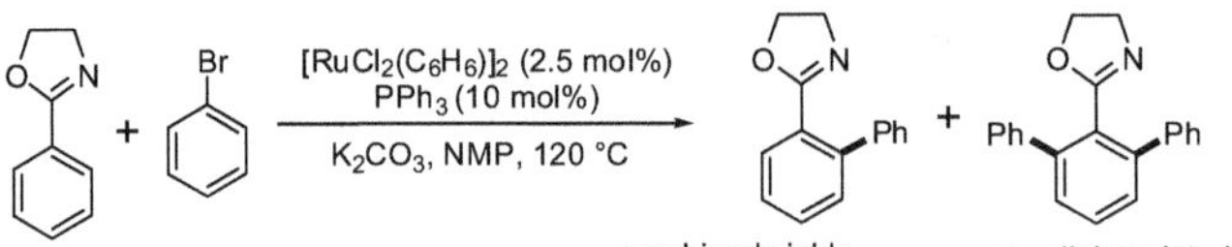

	combined yields	mono : diphenylated
1.2 equiv Ph-Br	60%	25:75
2.5 equiv Ph-Br	100%	0:100

Scheme 12.79 *Ortho*-arylation of an oxazoline derivative.

2-arylimidazolines with an aryl bromide and $[RuCl_2(C_6H_6)]_2$ as catalyst.[140] A mixture of mono- and bis-arylated products was obtained with a 31:69 ratio in the presence of 1.2 equiv. of phenyl bromide, whereas bis-arylated product was obtained predominantly with 2 equiv. In order to synthesize the mono-arylated product with appreciable yield, various *N*-substituted derivatives were tested, where the introduction of bulky substituents on nitrogen proved critical. Notably, *N*-tosyl derivatives did not form any product owing to the electron-withdrawing nature of the tosylate, which inhibits the binding of the nitrogen with the metal center (Scheme 12.78).

The same method was also applied with 2-phenyloxazolines as substrates.[140] A slight excess of the aryl bromide afforded a mixture of both mono- and bis-arylated products. In this case, the bis-aryl product was exclusively obtained with 2.5 equiv. of the aryl bromide (Scheme 12.79).

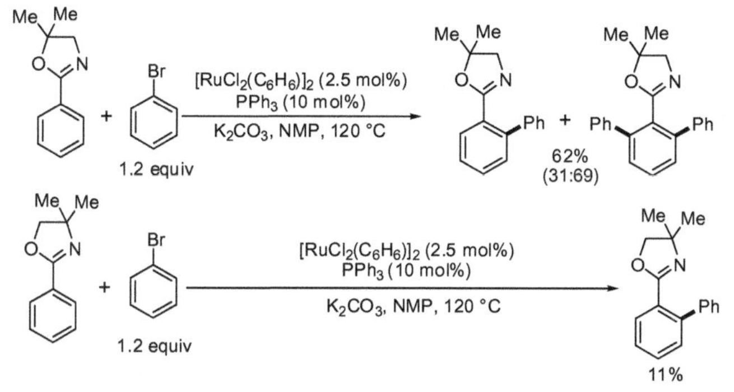

Scheme 12.80 Substituent effect on the *ortho*-arylation of an oxazoline derivative.

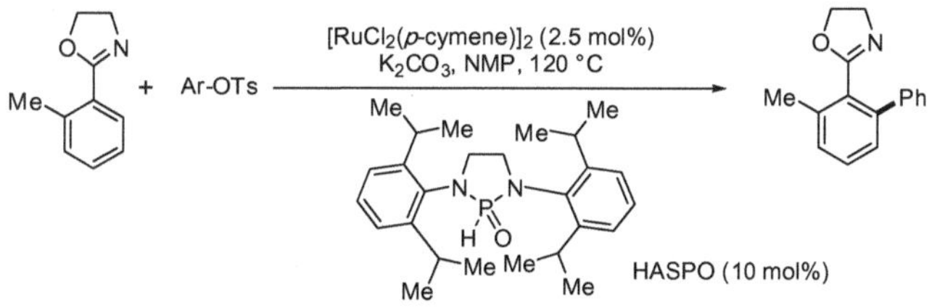

Scheme 12.81 *Ortho*-arylation of an oxazoline derivative using HASPO.

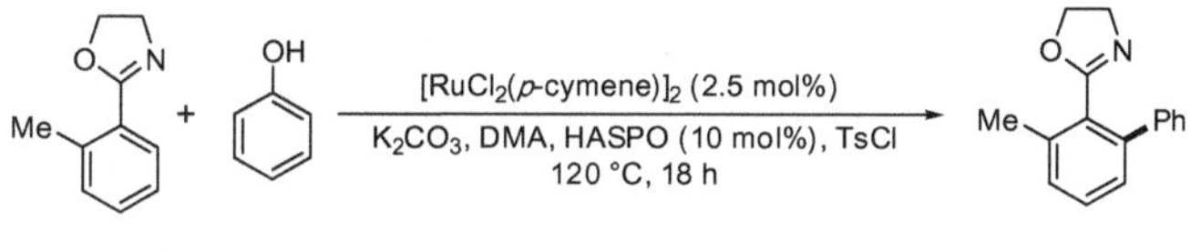

Scheme 12.82 *Ortho*-arylation of an oxazoline derivative with phenol partner.

The presence of substituents on the oxazoline ring at the 4-position dis-favored the product formation owing to increased steric crowding, hindering the coordination of the metal with the nitrogen (Scheme 12.80).

In the presence of a highly active heteroatom-assisted secondary phos-phine oxide (HASPO) ligand, both aryl chlorides and tosylates reacted with the aryloxazoline (Scheme 12.81).[116]

Ackermann and Mulzer reported phenol as a coupling partner for the arylation of oxazolines.[141] The efficiency of the ligand was found to be solvent dependent. In the presence of polar DMA (*N,N*-dimethylacetamide) solvent, HASPO preligand enabled the arylation, whereas in apolar toluene, MesCOOH ligand furnished the desired product with comparable yield. Under both conditions, sulfonyl chloride additive needs to be added. The reaction proceeds *via* the *in situ* activation of phenol to phenyl tosylate with TsCl (Scheme 12.82).

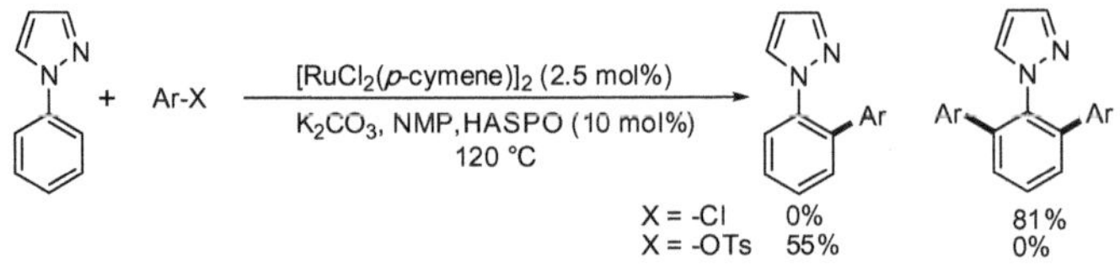

$$X = -Cl \quad 0\%$$
$$X = -OTs \quad 55\%$$
$$81\%$$
$$0\%$$

Scheme 12.83   *Ortho*-arylation of a pyrazole derivative.

The pyrazole moiety was also used as a directing group in Ru-catalyzed *ortho*-arylation reactions. Both aryl chlorides and tosylates were successfully employed in presence of HASPO.[116] Interestingly, aryl chloride partner gave bis-arylated product, whereas tosylate gave the mono-arylated product preferentially (Scheme 12.83).

The conditions developed by Oi *et al.* also worked successfully with other directing groups such as thiophenyl, furanyl, thiazolyl and pyridinyl bromides in the presence of $[RuCl_2(C_6H_6)]_2$ catalyst.[130]

12.4.3.4   *Ketones*

In the majority of cases, the metal-catalyzed arylation reaction involves aryl halides or pseudohalides as coupling partners. In contrast, Kakiuchi and co-workers reported an interesting transformation using arylboronic esters as coupling partners with $Ru(H_2)(CO)(PPh_3)_3$ as catalyst.[142,143] This protocol is efficient for arylating aromatic ketones along with fused homologs. The arylated product was accompanied by an equivalent amount of alcohol derived from the reduction of the ketone. A twofold increase in the amount of the ketone afforded the arylated ketone as major product. The formation of bis-arylated product was also observed in this case. The presence of a bulky α-group on the aryl ketone system preferentially leads to mono-arylated product (Scheme 12.84).

In the case of fused aromatic ketones, corresponding arylated products were obtained under the same conditions (Scheme 12.85).

Kakiuchi *et al.*[142] conducted several labeling experiments to gain a mechanistic insight into the reactions. They suggested that if a C–H bond cleaves prior to the metal–oxygen coordination, then the k_H/k_D values for the intra- and intermolecular competitive reaction should be nearly the same (Scheme 12.86).

In the intermolecular competition experiment, k_H/k_D remained almost same (1.06 and 1.09) after two different conversions (40% and 47%). A similar result was also observed in the intramolecular competition experiments: the k_H/k_D values remained almost constant (1.41 and 1.49) for two different conversions (50% and 75%). This observation indicates that the ruthenium coordinated with the substrate which led to the C–H bond activation. Considering the basicity of both the ketonic oxygen and π-electron cloud of the benzene ring and the selectivity for the newly formed C–C

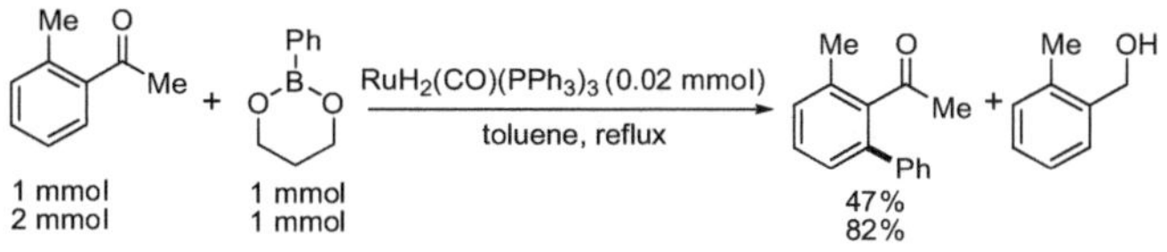

Scheme 12.84 *Ortho*-arylation of aromatic ketones.

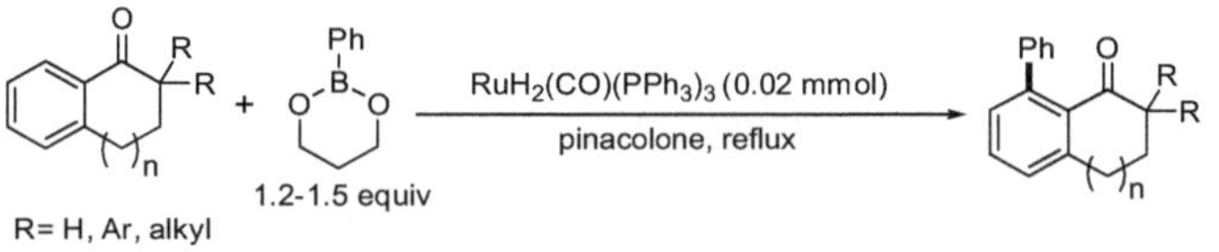

Scheme 12.85 *Ortho*-arylation of cyclic aromatic ketones.

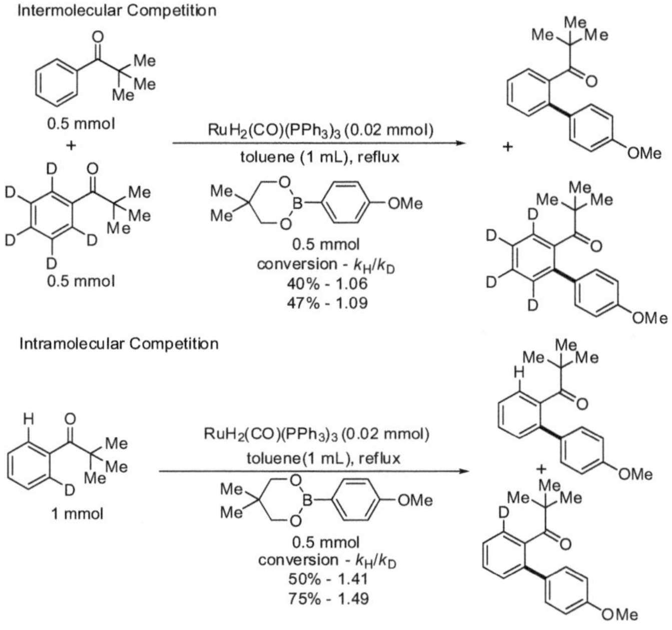

Scheme 12.86 Isotope labeling experiment.

bond, it is more likely that Ru coordinated to the ketonic oxygen prior to the C–H activation.

12.4.4 Green Protocol

Most Ru-catalyzed reactions lack operational simplicity owing to the involvement of hazardous or anaerobic conditions. In order to develop a more sustainable protocol, Ackermann and Potukuchi reported the direct arylation of 1,2,3-triazole[144] in non-toxic polyethylene glycol (PEG)[115] under a carboxylic acid-assisted pathway (Scheme 12.87).

Furthermore, non-toxic diethylcarbonate (DEC) has also been used as the reaction medium.[145–147] In 2005, Ackermann highlighted the use of water along with NMP.[115] Dixneuf and co-workers showed that the Ru-catalyzed arylation is feasible even in a completely aqueous medium in the presence of KOPiv.[148] Interestingly, under these conditions the reactivity order of the aryl halides is PhCl > PhBr > PhI, owing to their relative solubilities in water.

12.4.5   Mechanistic Illustration

The Ru-catalyzed arylation reaction can occur *via* two possible pathways, each of which involves two steps and differs in the order of these two steps. These steps are (a) oxidative addition of the aryl halide to form an aryl–Ru species and (b) ruthenation of the aromatic ring assisted by the directing group (Figure 12.20).[68]

Ackermann and Mulzer showed that complex **4** did not react with aryl chloride even at elevated temperature, whereas cyclometallation occurred readily (Figure 12.21). This phenomenon suggests facile cyclometallation prior to oxidative addition (pathway B).[141]

The reports of Davies *et al.* and Fernandez *et al.* on the beneficial effect of NaOAc as additive[149,150] and the DFT calculations by Dixneuf and co-

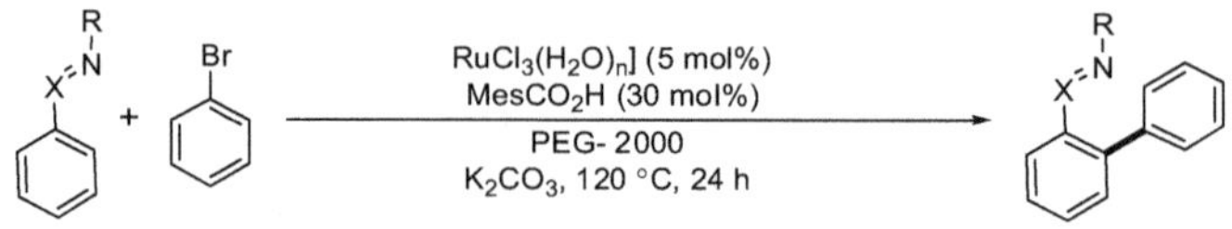

Scheme 12.87   Reaction in the presence of PEG-2000 as solvent.

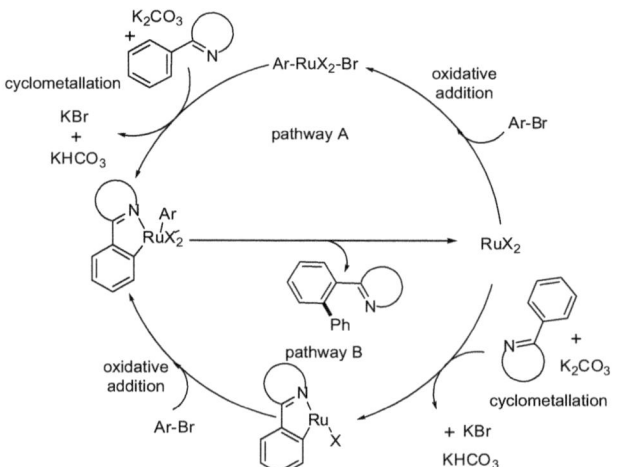

Figure 12.20   Probable catalytic cycles.

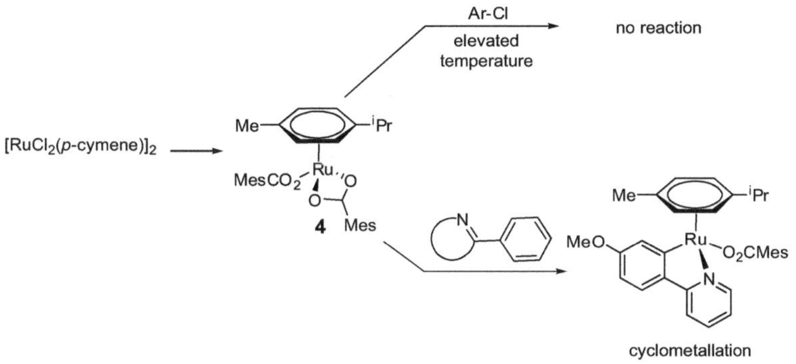

Figure 12.21 Mechanistic evidences towards pathway B.

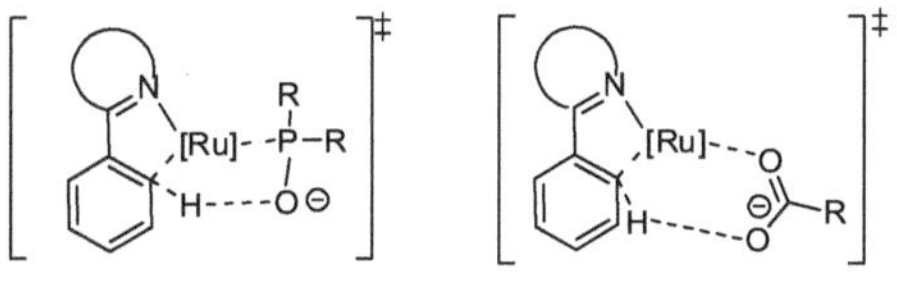

Figure 12.22 Probable transition states of ligand-assisted deprotonation.

workers[117] on a concerted metallation–deprotonation mechanism (CMD) provide strong support for this fact.[98,151–153] Based on other literature precedents, in most cases cyclometallation is likely to proceed *via* an oxidative addition. Unlike Ru⁰, cyclometallation with Ru^{II} does not proceed *via* an oxidative addition. The deprotonation of the *ortho* C–H bond facilitates the metallocycle formation without any variation of the oxidation state. The base-assisted C–H bond metallation proceeds *via* the formation of a five- or six-membered ring (Figure 12.22).

The previously used tertiary phosphine or NHC were not compatible with various scaffolds and also worked poorly.[154] In contrast, the Ru complex derived from a bifunctional sterically hindered secondary phosphine oxide (SPO)-based ligand was found to be more efficient. Recent reports by Ackermann and co-workers showcased carboxylates as potential co-catalysts under phosphine-free reaction conditions.[154,155]

12.5 Iridium-Catalyzed Reactions

Reports of Ir-catalyzed direct arylation *via* a C–H activation mechanism are scarce. For ease of discussion, they can be classified into two groups.

12.5.1 C–H Arylation Without Selectivity

In the first report of non-selective C–H arylation, an Ir–hydride complex was found suitable to catalyze the direct arylation of simple unsubstituted arenes.[156] Use of a strong base (KO^tBu) restricted the application of the method

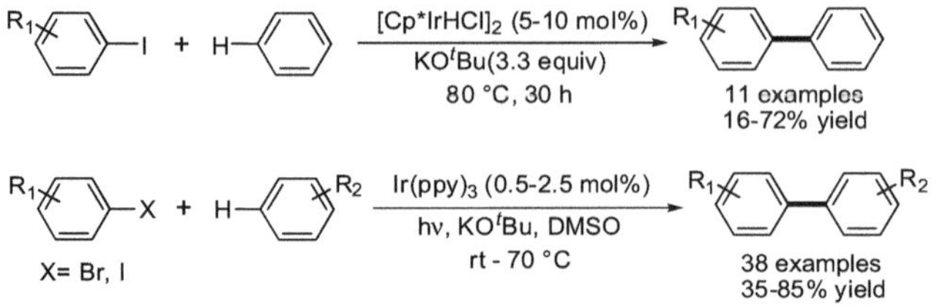

Scheme 12.88   Direct C–H arylation of aryl halides with benzene.

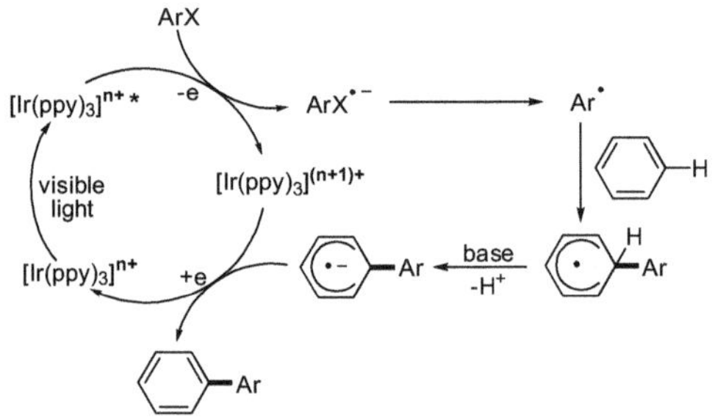

Figure 12.23   Proposed mechanism of photoredox homolytic aromatic substitution (HAS).

to aryl iodides only, and the reaction was only poorly regioselective when substituted arenes were used (Scheme 12.88, top). Recently, a similar type of transformation was reported using a photoredox iridium–polybipyridyl complex as the active catalyst (Scheme 12.88, bottom). In this protocol, aryl bromides were also applied as successful substrates for C–H arylation and the catalyst loading could be reduced to 0.5 mol% without any significant decrease in yield.[157]

Poor regioselectivity for substituted arenes with a mixture of *ortho*, *meta* and *para* isomers suggests that a radical-based reaction mechanism is likely to be operative. A recent breakthrough in the construction of biphenyl derivatives mediated by strong base without any involvement of a transition metal, popularly known as homolytic aromatic substitution (HAS), is likely to be involved in these two cases.[158] Thermal or photolytic excitation of the catalyst leads to the generation of an aryl halide radical anion through single-electron transfer (SET), followed by the formation of an aryl radical, which attacks the arene, resulting in a biaryl radical. Finally, after deprotonation in the presence of a strong base, the aryl radical anion transfers one electron to the oxidized metal catalyst to afford the biaryl product (Figure 12.23).[157]

12.5.2   C–H Arylation With Regioselectivity

It was shown later that the regioselectivity problem could be overcome and arylation of electron-rich heterocyclic arenes was performed selectively at the

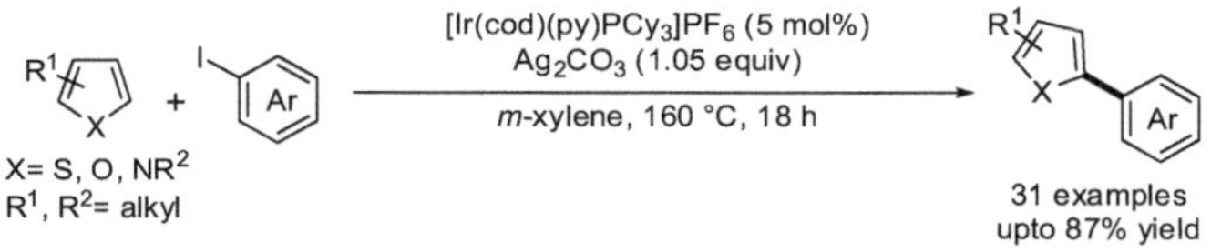

Scheme 12.89 Direct C–H arylation of heteroarenes at the 2-position.

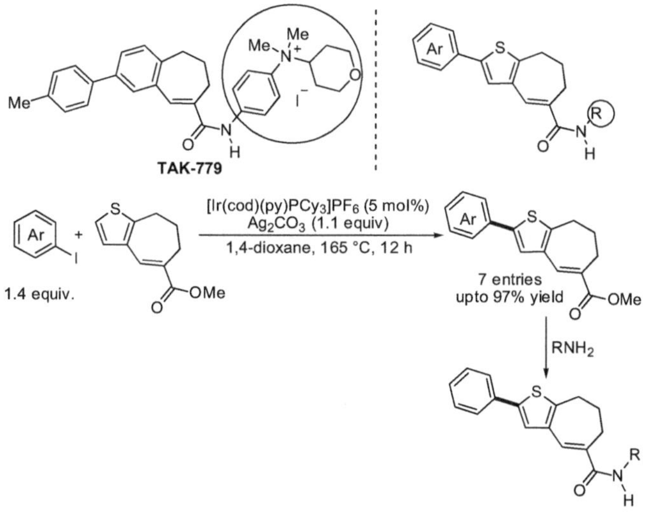

Scheme 12.90 Design of thiophene bioisosteric CCR5 receptor antagonists derived from TAK-779.

2-position of various heterocycles with aryl iodides using Crabtree's catalyst, [Ir(COD)(py)PCy$_3$]PF$_6$ (Scheme 12.89).[159] Highly substituted arenes could be synthesized by twofold functionalization of the heteroarenes (C2 and C5).

Recently, thiophene-based TAK-779 analogs have been prepared using the same strategy.[160] TAK-779 is a highly potent chemokine receptor 5 (CCR5) antagonist with a potential application as an HIV entry inhibitor (Scheme 12.90).

When anisole was used instead of thiophene, low-yielding direct arylation was found with an *orttho:meta:para* ratio of 37:17:46. This type of site selectivity indicates an electrophilic metallation process.[159] Hence, first, oxidative addition of aryl iodide leads to an Ir^{III} complex and Ag$_2$CO$_3$ serves to abstract the iodide, to generate am Ir^{III}–carbonate complex. The coordination of thiophene to electron-deficient Ir^{III} was followed by electrophilic concerted metallation–deprotonation (CMD) at the α-position of the thiophene ring. Finally, reductive elimination leads to the heterocyclic biaryl product with regeneration of the Ir^{I} species (Figure 12.24).[160,161]

Following the directing group approach, an Ir/Pd co-catalyzed directed C–H arylation with diaryliodonium regents was reported by Neufeldt and Sanford (Scheme 12.91).[162] Pd-catalyzed directing group-assisted C–H

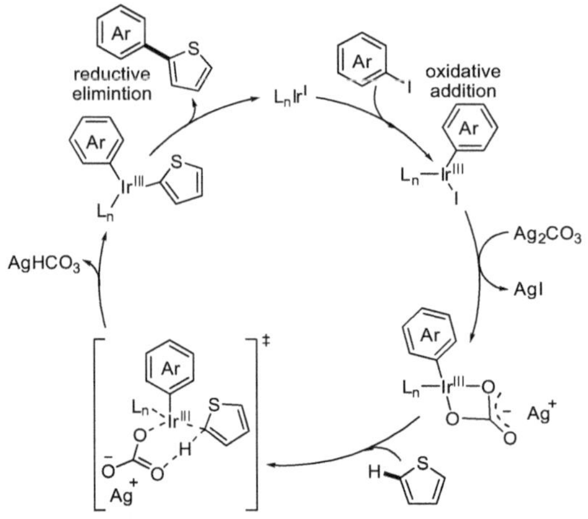

Figure 12.24 Proposed CMD mechanism for C–H arylation.

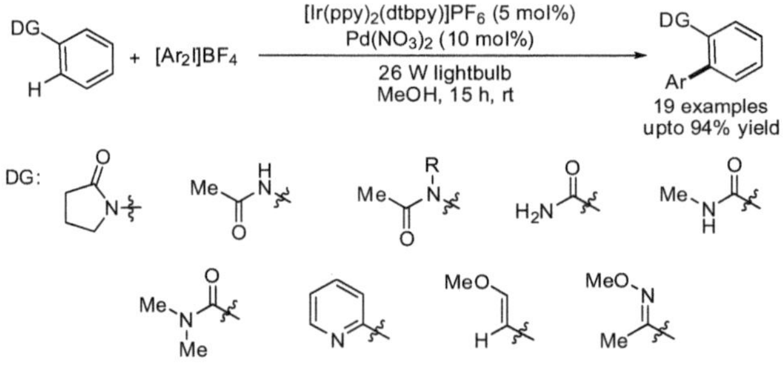

Scheme 12.91 Different directing group-assisted Ir/Pd co-catalyzed direct C–H arylation.

arylation reactions were sluggish, typically requiring high temperatures (80–110 °C) over extended periods of time (8–24 h), likely due to an ionic mechanism involving $2e^-$ oxidation of a palladacycle intermediate by Ar_2I^+. In this new approach, the Ir photocatalyst reverses the mechanism into a radical pathway, leading to a high-yielding transformation under mild conditions by generating $Ar^•$ from Ar_2I^+ *in situ*. The major advantage of this method is that the same conditions can be applied to a wide variety of directing groups at room temperature.

Through a series of experiments with radical scavengers, the radical nature of this reaction has been established with the involvement of $Pd^{II/IV}$ and $Ir^{III/IV}$ cycles. Initially, Ir^{3+} undergoes photoexcitation by visible light and the resultant $Ir^{3+}*$ can generate $Ar^•$ from Ar_2I^+. The aryl radical ($Ar^•$) can participate in the catalytic cycle by oxidizing the cyclopalladated complex. Then Ir^{4+} further oxidizes the Pd^{3+} center to Pd^{4+}, from where reductive elimination affords the biaryl products with regeneration of Pd^{2+} (Figure 12.25).[162]

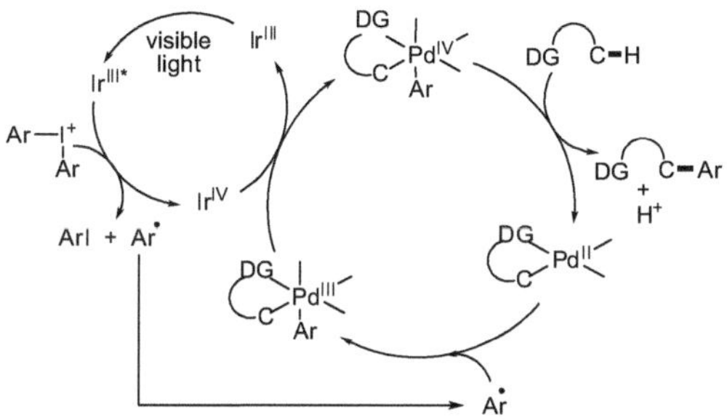

Figure 12.25 Mechanistic pathway for the Ir/Pd co-catalyzed direct C–H arylation.

12.6 Copper-Catalyzed Reactions

Transition metal-mediated direct arylation to construct biaryl motifs have been of significant interest, as by-products arising from prefunctionalized substrates can be avoided. Among these metals, Pd and Rh have been extensively used for their superior reactivity, where even relatively inactive chlorides have been employed as the coupling partners. However, from the viewpoint of catalyst loading, it would be highly advantageous if these catalysts could be replaced by first-row transition metals such as Cu. Owing to its relatively lower toxicity, low cost and ease of handling, significant attention has recently been focused on the development of Cu-mediated or -catalyzed C–H activation processes.

To address the regioselectivity issues of C–H activation processes with Cu, several approaches have been devised utilizing (a) the acidity of C–H bonds to impose site selectivity, (b) the inherent electronic nature of the substrates under consideration and (c) directing groups for chelation-assisted C–H activation.

In 2007, Do and Daugulis first developed a Cu-catalyzed direct C–H arylation of heterocycles with aryl iodides.[163] A strong base (LiO^{t}Bu) was employed for the deprotonation of benzoxazoles whereas KO^{t}Bu was sufficient to achieve high yields with more acidic imidazoles and triazoles. Similar reactivity was exhibited with polyfluoroarenes as coupling partners (Scheme 12.92).[164]

Do and Daugulis proposed a deprotonation–metallation–transmetallation sequence to explain the observed reactivity of these C–H bonds with pK_a values below 35 (in DMSO). From preliminary investigations, it was suggested that base-promoted deprotonation generated an aryllithium intermediate, which underwent transmetallation with Cu. This aryl–Cu intermediate then reacted with the iodoarenes to generate the C–C coupled product (Scheme 12.93).

The acidity of C–H bonds and the role of base were extremely important, as fluorobenzene remained unreacted and significant H–D crossover was observed under basic conditions with or without Cu. C–H bonds flanked by two electron-withdrawing groups were preferentially arylated. Notably, Fagnou and co-workers reported such an observation in the Pd-catalyzed arylation of polyfluoroarenes and aryl halides in 2006.[28] Moreover, an

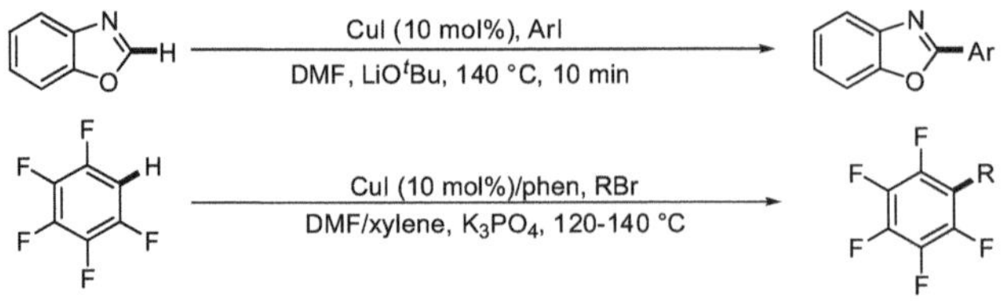

Scheme 12.92   Base-mediated arylation of heteroarenes and polyfluoroarenes.

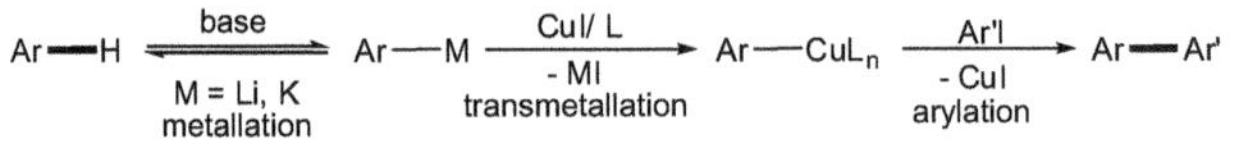

Scheme 12.93   Arylation sequence proposed by Do and Daugulis.

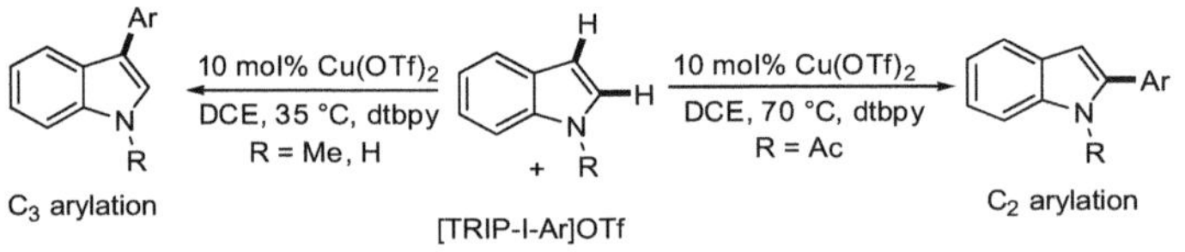

Scheme 12.94   Site-selective arylation of indoles with diaryliodine reagent.

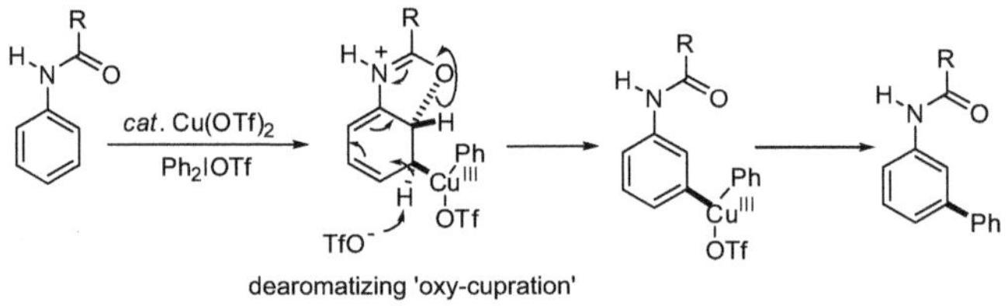

Scheme 12.95   Proposed mechanism for the observed *meta*-arylation of pivalinides.

aryl–copper–phenanthroline complex was synthesized and exhibited comparable reactivity to the catalytic reaction.[57,165]

In 2008, Gaunt and co-workers reported site-selective C2 and C3 arylation of indoles with diaryliodonium salts (Scheme 12.94).[166] The initial hypothesis on the mechanism was based on the high electrophilicity of Cu^{III} species as it has a d^8 configuration like Pd^{II} but a higher oxidation state. With free indole or *N*-methylindole, they were able to achieve selective arylation at the C3 position, thereby providing a complementary method to the Pd-catalyzed indole C2 arylation reported earlier by Sanford and co-workers.[45] Furthermore, exclusive transfer of the desired aryl group was made possible using a large "spectator" group, 2,4,6-triisopropylphenyl (TRIP), in the diaryliodine(III) reagent. However, with *N*-acetylindole, the C2 position was arylated preferentially.

With the same catalytic system, Phipps and Gaunt discovered an unprecedented *meta*-selective arylation of pivanilides in 2009 (Scheme 12.95).[167]

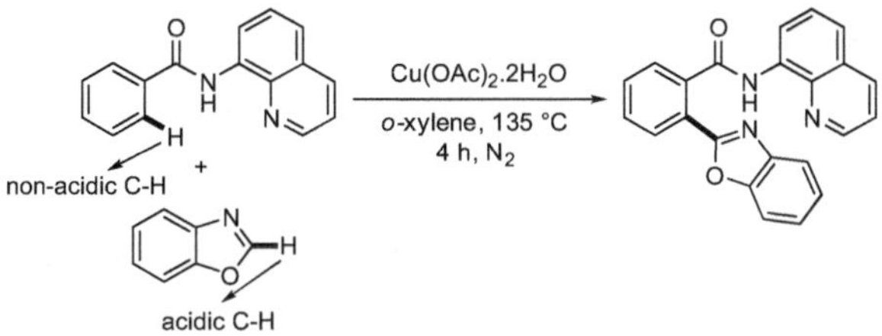

Scheme 12.96 Site-selective arylation of benzoic acid derivatives with 1,3-azoles.

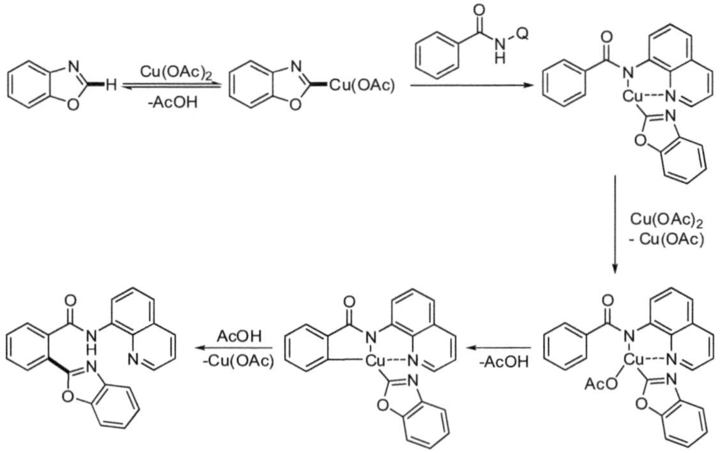

Scheme 12.97 Plausible reaction mechanism for the arylation of benzoic acid derivatives with 1,3-azoles.

A dearomatizing "oxy-cupration" aided by the coordination of anilide oxygen with the highly electrophilic Cu^{III} was held responsible for such an unexpected outcome. Later, the scope of this transformation was significantly expanded as Weinreb amides and even simple α-aryl ketones were selectively *meta*-arylated.[168]

Surprisingly, these transformations can also be achieved even without Cu at a slightly elevated temperature. A detailed understanding of the mechanism requires further investigation as the role of Cu and the diaryliodine reagent are not clear at present.[169]

Miura and co-workers exploited the inherent C–H acidity of electron-rich heteroarenes on one side and directing groups on other coupling partners to devise a C–C coupling through double C–H activation (Scheme 12.96). In an elegant study with a bidentate directing group, even non-acidic arene C–H bonds were selectively arylated with synthetically useful yields. This rigid coordination with the 8-aminoquinoline scaffold broadened the scope with respect to 1,3-azoles.[170] For instance, previously unreacted thiazole, imidazole and oxadiazole derivatives were found to be useful substrates under the present conditions.[171,172]

A preliminary understanding suggested acetate-mediated cupration of azoles followed by oxidative activation of relatively non-acidic arene C–H bonds to be the key steps in the desired transformation (Scheme 12.97).

12.7   Conclusion

The technique of C–H bond activation as a synthetic tool for the generation of C–C and C–X bonds has been found to be among the most efficient, mild and environmentally benign in comparison with more traditional approaches to the same goals. Transition metals such as Pd, Rh, Ru, Ir and Cu have been successfully implemented in catalysis. Different ligand scaffolds have been designed to enhance the catalytic activity. Various directing groups have also been recognized. However, a few fundamental challenges remain to be addressed. The presence of directing groups facilitates the incorporation of the –R group in specific locations, but remains intact in the product. The perfect approach would be the introduction of –R even in the absence of the directing group. Alternatively, a traceless moiety incorporated in the starting material leads to a final product without unwanted appendages. A few developments have already been made regarding this issue,[1] but these approaches are mostly either substrate specific and/or metal specific. In such an approach, the present challenge relies in developing a truly efficient system wherein a slight variation in the catalyst or ligand environment would result selectively in different regioisomers. In the absence of a directing group, the electronic and steric parameters dominate the selectivity. Hence the selectivity found in these cases is mostly complementary to the directing group-assisted pathways. One particularly attractive option for improved selectivity will be to incorporate the design elements needed for a certain molecular recognition into the catalyst scaffolding. This is especially important for activation of C–H bonds in molecules with no functional handles. The essential problem illustrated by methane C–H activation is such a case.[173,174] Despite all these shortcomings, the direct transformation of C–H to C–C has already enriched the synthetic methodology and further rapid advances in this area can be expected.

References

1. N. Kuhl, M. N. Hopkinson, J. Wencel-Delord and F. Glorius, *Angew. Chem. Int. Ed*, 2012, **51**, 10236–10254.
2. J. Wencel-Delord, T. Droge, F. Liu and F. Glorius, *Chem. Soc. Rev.*, 2011, **40**, 4740–4761.
3. P. Sehnal, R. J. K. Taylor and I. J. S. Fairlamb, *Chem. Rev.*, 2010, **110**, 824–889.
4. T. W. Lyons and M. S. Sanford, *Chem. Rev.*, 2010, **110**, 1147–1169.
5. P. B. Arockiam, C. Bruneau and P. H. Dixneuf, *Chem. Rev.*, 2012, **112**, 5879–5918.
6. L. Ackermann, *Chem. Rev.*, 2011, **111**, 1315–1345.
7. F. Juliá-Hernández, M. Simonetti and I. Larrosa, *Angew. Chem. Int. Ed.*, 2013, **52**, 11458–11460.
8. T. Truong and O. Daugulis, *Angew. Chem. Int. Ed.*, 2013, **51**, 11677–11679.

9. L. Ackermann, R. Vicente and A. R. Kapdi, *Angew. Chem. Int. Ed.*, 2009, **48**, 9792–9826.

10. D. A. Colby, R. G. Bergman and J. A. Ellman, *Chem. Rev.*, 2009, **110**, 624–655.

11. F. Monnier and M. Taillefer, *Angew. Chem. Int. Ed.*, 2009, **48**, 6954–6971.

12. M. Wasa, K. S. L. Chan and J.-Q. Yu, *Chem. Lett.*, 2011, **40**, 1004–1006.

13. D. E. Ames and D. Bull, *Tetrahedron*, 1982, **38**, 383–387.

14. D. E. Ames and A. Opalko, *Synthesis*, 1983, 234–235.

15. D. E. Ames and A. Opalko, *Tetrahedron*, 1984, **40**, 1919–1925.

16. D. D. Hennings, S. Iwasa and V. H. Rawal, *J. Org. Chem.*, 1997, **62**, 2–3.

17. D. Garcia-Cuadrado, A. A. C. Braga, F. Maseras and A. M. Echavarren, *J. Am. Chem. Soc.*, 2006, **128**, 1066–1067.

18. D. Garcia-Cuadrado, P. de Mendoza, A. A. C. Braga, F. Maseras and A. M. Echavarren, *J. Am. Chem. Soc.*, 2007, **129**, 6880–6886.

19. L.-C. Campeau and K. Fagnou, *Chem. Commun.*, 2006, 1253–1264.

20. L.-C. Campeau, M. Parisien, M. Leblanc and K. Fagnou, *J. Am. Chem. Soc.*, 2004, **126**, 9186–9187.

21. L.-C. Campeau, P. Thansandote and K. Fagnou, *Org. Lett.*, 2005, 7, 1857–1860.

22. L.-C. Campeau, M. Parisien, A. Jean and K. Fagnou, *J. Am. Chem. Soc.*, 2006, **128**, 581–590.

23. M. A. Campo, Q. Huang, T. Yao, Q. Tian and R. C. Larock, *J. Am. Chem. Soc.*, 2003, **125**, 11506–11507.

24. C. Amatore and A. Jutand, *Acc. Chem. Res.*, 2000, **33**, 314–321.

25. K. Fagnou and M. Lautens, *Angew. Chem. Int. Ed.*, 2002, **41**, 27–47.

26. M. Parisien, D. Valette and K. Fagnou, *J. Org. Chem.*, 2005, **70**, 7578–7584.

27. J. P. Leclerc, M. Andre and K. Fagnou, *J. Org. Chem.*, 2006, **71**, 1711–1714.

28. M. Lafrance, C. N. Rowley, T. K. Woo and K. Fagnou, *J. Am. Chem. Soc.*, 2006, **128**, 8754–8756.

29. R. Giri, N. Maugel, J.-J. Li, D.-H. Wang, S. P. Breazzano, L. B. Saunders and J.-Q. Yu, *J. Am. Chem. Soc.*, 2007, **129**, 3510–3511.

30. D.-H. Wang, T.-S. Mei and J.-Q. Yu, *J. Am. Chem. Soc.*, 2008, **130**, 17676–17677.

31. K. M. Engle, P. S. Thuy-Boun, M. Dang and J.-Q. Yu, *J. Am. Chem. Soc.*, 2011, **133**, 18183–18193.

32. D. Kalyani, N. R. Deprez, L. V. Desai and M. S. Sanford, *J. Am. Chem. Soc.*, 2005, **127**, 7330–7331.

33. K. L. Hull, E. L. Lanni and M. S. Sanford, *J. Am. Chem. Soc.*, 2006, **128**, 14047–14049.

34. X. Wang, D. Leow and J.-Q. Yu, *J. Am. Chem. Soc.*, 2011, **133**, 13864–13867.

35. G. Brasche, J. Garcila-Fortanet and S. L. Buchwald, *Org. Lett.*, 2008, **10**, 2207–2210.

36. A. J. Hickman and M. S. Sanford, *ACS Catal.*, 2011, **1**, 170–174.

37. Y. Akita, Y. Itagaki, S. Takizawa and A. Ohta, *Chem. Pharm. Bull.*, 1989, **37**, 1477–1480.
38. D. Alagille, R. M. Baldwin and G. D. Tamagnan, *Tetrahedron Lett.*, 2005, **46**, 1349–1351.
39. C. Hoarau, A. Du Fou de Kerdaniel, N. Bracq, P. Grandclaudon, A. Couture and F. Marsais, *Tetrahedron Lett.*, 2005, **46**, 8573–8577.
40. K. Kobayashi, A. Sugie, M. Takahashi, K. Masui and A. Mori, *Org. Lett.*, 2005, **7**, 5083–5085.
41. E. David, J. Perrin, S. Pellet-Rostaing, J. Fournier dit Chabert and M. Lemaire, *J. Org. Chem.*, 2005, **70**, 3569–3573.
42. M. Miura and M. Nomura, *Top. Curr. Chem.*, 2002, **219**, 211–241.
43. B. S. Lane, M. A. Brown and D. Sames, *J. Am. Chem. Soc.*, 2005, **127**, 8050–8057.
44. B. B. Toure, B. S. Lane and D. Sames, *Org. Lett.*, 2006, **8**, 1979–1982.
45. N. R. Deprez, D. Kalyani, A. Krause and M. S. Sanford, *J. Am. Chem. Soc.*, 2006, **128**, 4972–4973.
46. L.-C. Campeau, S. Rousseaux and K. Fagnou, *J. Am. Chem. Soc.*, 2005, **127**, 18020–18021.
47. J. P. Leclerc and K. Fagnou, *Angew. Chem. Int. Ed.*, 2006, **45**, 7781–7786.
48. L. C. Campeau, D. J. Schipper and K. Fagnou, *J. Am. Chem. Soc.*, 2008, **130**, 3266–3267.
49. D. J. Schipper, L. C. Campeau and K. Fagnou, *Tetrahedron*, 2009, **65**, 3155–3164.
50. D. J. Schipper, M. El-Salfiti, C. J. Whipp and K. Fagnou, *Tetrahedron*, 2009, **65**, 4977–4983.
51. L. C. Campeau, D. R. Stuart, J. P. Leclerc, M. Bertrand-Laperle, E. Villemure, H. Y. Sun, S. Lasserre, N. Guimond, M. Lecavallier and K. Fagnou, *J. Am. Chem. Soc.*, 2009, **131**, 3291–3306.
52. J. J. Mousseau and A. B. Charette, *Acc. Chem. Res.*, 2012, **46**, 412–424.
53. H. Y. Sun, S. I. Gorelsky, D. R. Stuart, L. C. Campeau and K. Fagnou, *J. Org. Chem.*, 2010, **75**, 8180–8189.
54. M. Wasa, K. M. Engle and J.-Q. Yu, *J. Am. Chem. Soc.*, 2009, **131**, 9886–9887.
55. Y. Kametani, T. Satoh, M. Miura and M. Nomura, *Tetrahedron Lett.*, 2000, **41**, 2655–2658.
56. M. Wasa, B. T. Worrell and J.-Q. Yu, *Angew. Chem. Int. Ed.*, 2010, **49**, 1275–1277.
57. H. O. Do, R. M. K. Khan and O. Daugulis, *J. Am. Chem. Soc.*, 2008, **130**, 15185–15192.
58. P. Xi, F. Yang, S. Qin, D. Zhao, J. Lan, G. Gao, C. Hu and J. You, *J. Am. Chem. Soc.*, 2010, **132**, 1822–1824.
59. A. M. Berman, J. C. Lewis, R. G. Bergman and J. A. Ellman, *J. Am. Chem. Soc.*, 2008, **130**, 14926–14927.
60. M. Ye, G.-L. Gao, A. J. F. Edmunds, P. A. Worthington, J. A. Morris and J.-Q. Yu, *J. Am. Chem. Soc.*, 2011, **133**, 19090–19093.

61. S. Chuprakov, N. Chernyak, A. S. Dudnik and V. Gevorgyan, *Org. Lett.*, 2007, **9**, 2333–2336.
62. B. Liegault, D. Lapointe, L. Caron, A. Vlassova and K. Fagnou, *J. Org. Chem.*, 2009, **74**, 1826–1834.
63. J. J. Dong, J. Roger, C. Verrier, T. Martin, R. Le Goff, C. Hoarau and H. Doucet, *Green Chem.*, 2010, **12**, 2053–2063.
64. L. C. Campeau, M. Bertrand-Laperle, J. P. Leclerc, E. Villemure, S. Gorelsky and K. Fagnou, *J. Am. Chem. Soc.*, 2008, **130**, 3276–3277.
65. L. Ackermann, A. Althammer and S. Fenner, *Angew. Chem. Int. Ed.*, 2009, **48**, 201–204.
66. M. Ye, A. J. F. Edmunds, J. A. Morris, D. Sale, Y. Zhang and J.-Q. Yu, *Chem. Sci.*, 2013, **4**, 2374–2379.
67. J. Roger and H. Doucet, *Adv. Synth. Catal.*, 2009, **351**, 1977–1990.
68. D. Alberico, M. E. Scott and M. Lautens, *Chem. Rev.*, 2007, **107**, 174–238.
69. K. Fagnou and M. Lautens, *Chem. Rev.*, 2002, **103**, 169–196.
70. U. Sharma, T. Naveen, A. Maji, S. Manna and D. Maiti, *Angew. Chem. Int. Ed.*, 2013, **52**, 12669–12673.
71. G. Rouquet and N. Chatani, *Angew. Chem. Int. Ed.*, 2013, **52**, 11726–11743.
72. D. Shabashov and O. Daugulis, *Org. Lett.*, 2005, 7, 3657–3659.
73. G. He and G. Chen, *Angew. Chem. Int. Ed.*, 2011, **50**, 5192–5196.
74. Y. Feng, Y. Wang, B. Landgraf, S. Liu and G. Chen, *Org. Lett.*, 2010, **12**, 3414–3417.
75. M. Lafrance, S. I. Gorelsky and K. Fagnou, *J. Am. Chem. Soc.*, 2007, **129**, 14570–14571.
76. S. Rousseaux, M. Davi, J. Sofack-Kreutzer, C. Pierre, C. E. Kefalidis, E. Clot, K. Fagnou and O. Baudoin, *J. Am. Chem. Soc.*, 2010, **132**, 10706–10716.
77. A. Renaudat, L. Jean-Gérard, R. Jazzar, C. E. Kefalidis, E. Clot and O. Baudoin, *Angew. Chem. Int. Ed.*, 2010, **49**, 7261–7265.
78. S. Aspin, A.-S. Goutierre, P. Larini, R. Jazzar and O. Baudoin, *Angew. Chem. Int. Ed.*, 2012, **51**, 10808–10811.
79. S. Oi, S. Fukita and Y. Inoue, *Chem. Commun.*, 1998, 2439–2440.
80. K. Ueura, T. Satoh and M. Miura, *Org. Lett.*, 2005, 7, 2229–2231.
81. K. Ueura, S. Miyamura, T. Satoh and M. Miura, *J. Organomet. Chem.*, 2006, **691**, 2821–2826.
82. S. Miyamura, H. Tsurugi, T. Satoh and M. Miura, *J. Organomet. Chem.*, 2008, **693**, 2438–2442.
83. T. Vogler and A. Studer, *Org. Lett.*, 2008, **10**, 129–131.
84. X. Zhao and Z. Yu, *J. Am. Chem. Soc.*, 2008, **130**, 8136–8137.
85. W. Jin, Z. Yu, W. He, W. Ye and W.-J. Xiao, *Org. Lett.*, 2009, **11**, 1317–1320.
86. R. B. Bedford, S. J. Coles, M. B. Hursthouse and M. E. Limmert, *Angew. Chem. Int. Ed.*, 2003, **42**, 112–114.
87. R. B. Bedford and M. E. Limmert, *J. Org. Chem.*, 2003, **68**, 8669–8682.

88. S. Oi, S.-i. Watanabe, S. Fukita and Y. Inoue, *Tetrahedron Lett.*, 2003, **44**, 8665–8668.

89. R. B. Bedford, M. Betham, A. J. M. Caffyn, J. P. H. Charmant, L. C. Lewis-Alleyne, P. D. Long, D. Polo-Ceron and S. Prashar, *Chem. Commun.*, 2008, 990–992.

90. J. C. Lewis, S. H. Wiedemann, R. G. Bergman and J. A. Ellman, *Org. Lett.*, 2003, **6**, 35–38.

91. J. C. Lewis, J. Y. Wu, R. G. Bergman and J. A. Ellman, *Angew. Chem. Int. Ed.*, 2006, **45**, 1589–1591.

92. X. Wang, B. S. Lane and D. Sames, *J. Am. Chem. Soc.*, 2005, **127**, 4996–4997.

93. S. Yanagisawa, T. Sudo, R. Noyori and K. Itami, *J. Am. Chem. Soc.*, 2006, **128**, 11748–11749.

94. S. Yanagisawa, T. Sudo, R. Noyori and K. Itami, *Tetrahedron*, 2008, **64**, 6073–6081.

95. S. Proch and R. Kempe, *Angew. Chem. Int. Ed.*, 2007, **46**, 3135–3138.

96. L.-C. Campeau and K. Fagnou, *Chem. Commun.*, 2006, 1253–1264.

97. X. Chen, K. M. Engle, D.-H. Wang and J.-Q. Yu, *Angew. Chem. Int. Ed.*, 2009, **48**, 5094–5115.

98. D. L. Davies, S. M. A. Donald and S. A. Macgregor, *J. Am. Chem. Soc.*, 2005, **127**, 13754–13755.

99. J. C. Lewis, R. G. Bergman and J. A. Ellman, *Acc. Chem. Res.*, 2008, **41**, 1013–1025.

100. E. M. Beck and M. J. Gaunt, in *C–H Activation*, ed. J.-Q. Yu and Z. Shi, Springer, Berlin, 2009, pp. 85–121.

101. A. Gunay and K. H. Theopold, *Chem. Rev.*, 2010, **110**, 1060–1081.

102. J. A. Labinger and J. E. Bercaw, *Nature*, 2002, **417**, 507–514.

103. I. A. I. Mkhalid, J. H. Barnard, T. B. Marder, J. M. Murphy and J. F. Hartwig, *Chem. Rev.*, 2009, **110**, 890–931.

104. I. V. Seregin and V. Gevorgyan, *Chem. Soc. Rev.*, 2007, **36**, 1173–1193.

105. C.-L. Sun, B.-J. Li and Z.-J. Shi, *Chem. Rev.*, 2011, **111**, 1293–1314.

106. A. E. Wendlandt, A. M. Suess and S. S. Stahl, *Angew. Chem. Int. Ed.*, 2011, **50**, 11062–11087.

107. L. Ackermann, *Chem. Commun.*, 2010, **46**, 4866–4877.

108. F. Kakiuchi and N. Chatani, in *Ruthenium Catalysts and Fine Chemistry*, ed. C. Bruneau and P. Dixneuf, Springer, Berlin, 2004, pp. 45–79.

109. F. Kakiuchi and N. Chatani, in *Ruthenium in Organic Synthesis*, ed. S.-I. Murahashi, Wiley-VCH Verlag GmbH, Weinheim, 2004, pp. 219–255.

110. F. Kakiuchi and N. Chatani, *Adv. Synth. Catal.*, 2003, **345**, 1077–1101.

111. F. Kakiuchi, T. Uetsuhara, Y. Tanaka, N. Chatani and S. Murai, *J. Mol. Catal. A: Chem.*, 2002, **182–183**, 511–514.

112. S. Murai, F. Kakiuchi, S. Sekine, Y. Tanaka, A. Kamatani, M. Sonoda and N. Chatani, *Nature*, 1993, **366**, 529–531.

113. S. Oi, S. Fukita, N. Hirata, N. Watanuki, S. Miyano and Y. Inoue, *Org. Lett.*, 2001, **3**, 2579–2581.

114. S. Oi, Y. Ogino, S. Fukita and Y. Inoue, *Org. Lett.*, 2002, **4**, 1783–1785.

115. L. Ackermann, *Org. Lett.*, 2005, 7, 3123–3125.
116. L. Ackermann, A. Althammer and R. Born, *Angew. Chem. Int. Ed.*, 2006, **45**, 2619–2622.
117. I. Ozdemir, S. Demir, B. Cetinkaya, C. Gourlaouen, F. Maseras, C. Bruneau and P. H. Dixneuf, *J. Am. Chem. Soc.*, 2008, **130**, 1156–1157.
118. F. Požgan and P. H. Dixneuf, *Adv. Synth. Catal.*, 2009, **351**, 1737–1743.
119. F. Kakiuchi, T. Kochi, E. Mizushima and S. Murai, *J. Am. Chem. Soc.*, 2010, **132**, 17741–17750.
120. T. Matsubara, N. Koga, D. G. Musaev and K. Morokuma, *J. Am. Chem. Soc.*, 1998, **120**, 12692–12693.
121. T. Matsubara, N. Koga, D. G. Musaev and K. Morokuma, *Organometallics*, 2000, **19**, 2318–2329.
122. F. Kakiuchi, T. Sato, T. Tsujimoto, M. Yamauchi, N. Chatani and S. Murai, *Chem. Lett.*, 1998, 1053–1054.
123. R. Martinez, R. Chevalier, S. Darses and J.-P. Genet, *Angew. Chem. Int. Ed.*, 2006, **45**, 8232–8235.
124. R. Martinez, J.-P. Genet and S. Darses, *Chem. Commun.*, 2008, 3855–3857.
125. R. Martinez, M.-O. Simon, R. Chevalier, C. Pautigny, J.-P. Genet and S. Darses, *J. Am. Chem. Soc.*, 2009, **131**, 7887–7895.
126. M.-O. Simon, R. Martinez, J.-P. Genêt and S. Darses, *Adv. Synth. Catal.*, 2009, **351**, 153–157.
127. M.-O. Simon, R. Martinez, J.-P. Genet and S. Darses, *J. Org. Chem.*, 2009, **75**, 208–210.
128. M.-O. Simon, J.-P. Genet and S. Darses, *Org. Lett.*, 2010, **12**, 3038–3041.
129. M.-O. Simon, G. Ung and S. Darses, *Adv. Synth. Catal.*, 2011, **353**, 1045–1048.
130. S. Oi, R. Funayama, T. Hattori and Y. Inoue, *Tetrahedron*, 2008, **64**, 6051–6059.
131. A. Littke, in *Modern Arylation Methods*, ed. L. Ackermann, Wiley-VCH, Weinheim, 2009, pp. 25–67.
132. W. K. Chow, C. M. So, C. P. Lau and F. Y. Kwong, *J. Org. Chem.*, 2010, **75**, 5109–5112.
133. M. L. H. Mantel, A. T. Lindhardt, D. Lupp and T. Skrydstrup, *Chem. Eur. J.*, 2010, **16**, 5437–5442.
134. L. Zhang and J. Wu, *J. Am. Chem. Soc.*, 2008, **130**, 12250–12251.
135. P. B. Arockiam, C. Fischmeister, C. Bruneau and P. H. Dixneuf, *Angew. Chem. Int. Ed.*, 2010, **49**, 6629–6632.
136. F. Kakiuchi, T. Tsujimoto, M. Sonoda, N. Chatani and S. Murai, *Synlett*, 2001, **2001**, 948–951.
137. F. Kakiuchi, M. Yamauchi, N. Chatani and S. Murai, *Chem. Lett.*, 1996, **25**, 111–112.
138. B. Li, C. B. Bheeter, C. Darcel and P. H. Dixneuf, *ACS Catal.*, 2011, **1**, 1221–1224.

139. B. Li, K. Devaraj, C. Darcel and P. H. Dixneuf, *Tetrahedron*, 2012, **68**, 5179–5184.
140. S. Oi, E. Aizawa, Y. Ogino and Y. Inoue, *J. Org. Chem.*, 2005, **70**, 3113–3119.
141. L. Ackermann and M. Mulzer, *Org. Lett.*, 2008, **10**, 5043–5045.
142. F. Kakiuchi, S. Kan, K. Igi, N. Chatani and S. Murai, *J. Am. Chem. Soc.*, 2003, **125**, 1698–1699.
143. F. Kakiuchi, Y. Matsuura, S. Kan and N. Chatani, *J. Am. Chem. Soc.*, 2005, **127**, 5936–5945.
144. L. Ackermann and H. K. Potukuchi, *Org. Biomol. Chem.*, 2010, **8**, 4503–4513.
145. P. Arockiam, V. Poirier, C. Fischmeister, C. Bruneau and P. H. Dixneuf, *Green Chem.*, 2009, **11**, 1871–1875.
146. B. Schaffner, F. Schaffner, S. P. Verevkin and A. Borner, *Chem. Rev.*, 2010, **110**, 4554–4581.
147. P. Tundo and M. Selva, *Acc. Chem. Res.*, 2002, **35**, 706–716.
148. P. B. Arockiam, C. Fischmeister, C. Bruneau and P. H. Dixneuf, *Angew. Chem. Int. Ed.*, 2010, **49**, 6629–6632.
149. D. L. Davies, O. Al-Duaij, J. Fawcett, M. Giardiello, S. T. Hilton and D. R. Russell, *Dalton Trans.*, 2003, 4132–4138.
150. S. Fernandez, M. Pfeffer, V. Ritleng and C. Sirlin, *Organometallics*, 1999, **18**, 2390–2394.
151. Y. Feng, M. Lail, K. A. Barakat, T. R. Cundari, T. B. Gunnoe and J. L. Petersen, *J. Am. Chem. Soc.*, 2005, **127**, 14174–14175.
152. A. D. Ryabov, *Chem. Rev.*, 1990, **90**, 403–424.
153. V. I. Sokolov, L. L. Troitskaya and O. A. Reutov, *J. Organomet. Chem.*, 1979, **182**, 537–546.
154. L. Ackermann, R. Vicente and A. Althammer, *Org. Lett.*, 2008, **10**, 2299–2302.
155. L. Ackermann, R. Vicente, H. K. Potukuchi and V. Pirovano, *Org. Lett.*, 2010, **12**, 5032–5035.
156. K.-i. Fujita, M. Nonogawa and R. Yamaguchi, *Chem. Commun.*, 2004, 1926–1927.
157. Y. Cheng, X. Gu and P. Li, *Org. Lett.*, 2013, **15**, 2664–2667.
158. E. Shirakawa, X. Zhang and T. Hayashi, *Angew. Chem. Int. Ed.*, 2011, **50**, 4671–4674.
159. B. Join, T. Yamamoto and K. Itami, *Angew. Chem. Int. Ed.*, 2009, **48**, 3644–3647.
160. A. Junker, J. Yamaguchi, K. Itami and B. Wunsch, *J. Org. Chem.*, 2013, **78**, 5579–5586.
161. M. García-Melchor, S. I. Gorelsky and T. K. Woo, *Chem. Eur. J.*, 2011, **17**, 13847–13853.
162. S. R. Neufeldt and M. S. Sanford, *Adv. Synth. Catal.*, 2012, **354**, 3517–3522.
163. H. Q. Do and O. Daugulis, *J. Am. Chem. Soc.*, 2007, **129**, 12404–12405.
164. H. Q. Do and O. Daugulis, *J. Am. Chem. Soc.*, 2008, **130**, 1128–1129.
165. H. Q. Do and O. Daugulis, *Chem. Commun.*, 2009, 6433–6435.

166. R. J. Phipps, N. P. Grimster and M. J. Gaunt, *J. Am. Chem. Soc.*, 2008, **130**, 8172–8174.

167. R. J. Phipps and M. J. Gaunt, *Science*, 2009, **323**, 1593–1597.

168. H. A. Duong, R. E. Gilligan, M. L. Cooke, R. J. Phipps and M. J. Gaunt, *Angew. Chem. Int. Ed.*, 2011, **50**, 463–466.

169. C. L. Ciana, R. J. Phipps, J. R. Brandt, F. M. Meyer and M. J. Gaunt, *Angew. Chem. Int. Ed.*, 2011, **50**, 458–462.

170. M. Nishino, K. Hirano, T. Satoh and M. Miura, *Angew. Chem. Int. Ed.*, 2012, **52**, 4457–4461.

171. M. Kitahara, N. Umeda, K. Hirano, T. Satoh and M. Miura, *J. Am. Chem. Soc.*, 2011, **133**, 2160–2162.

172. M. Nishino, K. Hirano, T. Satoh and M. Miura, *Angew. Chem. Int. Ed.*, 2012, **51**, 6993–6997.

173. S. I. Chan and S. S. F. Yu, *Acc. Chem. Res.*, 2008, **41**, 969–979.

174. O. Perraud, A. B. Sorokin, J.-P. Dutasta and A. Martinez, *Chem. Commun.*, 2013, **49**, 1288–1290.

CHAPTER 13

Cross-Coupling Chemistry in Continuous Flow

TIMOTHY NOËL* AND VOLKER HESSEL

Micro Flow Chemistry and Process Technology, Department of Chemistry and Chemical Engineering, Eindhoven University of Technology, Den Dolech 2 (STW 1.48), 5612 AZ Eindhoven, The Netherlands
*Email: t.noel@tue.nl

13.1 Introduction

Cross-coupling chemistry has attracted a lot of attention from researchers in academia and industry since it allows the reliable formation of carbon–carbon and carbon–heteroatom bonds.[1] This methodology has found widespread use in the synthesis of active pharmaceutical ingredients, natural products and other biologically active molecules and, additionally, it has found substantial application in material science. Since the initial breakthroughs in the early 1970s, much research effort has been devoted to bringing this chemistry to maturity, and substantial emphasis has been placed on the development of more active catalytic systems, allowing the reduction of the catalyst loading and the conversion of a wide array of substrate classes.[2] For economic and environmental reasons, a recent trend in cross-coupling involves the use of more abundant first-row transition metal-based catalysts.[2] More recently, continuous flow microreactors have been utilized to give another boost to cross-coupling chemistry. Noël and Buchwald recently reviewed the successful merger of continuous flow technology and cross-coupling chemistry.[3] In this chapter, we discuss the merits

RSC Catalysis Series No. 21
New Trends in Cross-Coupling: Theory and Applications
Edited by Thomas J Colacot
© The Royal Society of Chemistry 2015
Published by the Royal Society of Chemistry, www.rsc.org

of continuous flow technology for cross-coupling and report on recent evolutions seen in this field.

13.2 Advantages of Utilizing Continuous Flow Microreactor Technology

Owing to the advantages associated with continuous processing, microreactor technology has attracted growing interest from the chemical and processing engineering community (Table 13.1, Advantages).[4] The use of continuous flow microreactors provides a high degree of control over reaction/residence time and other process parameters (*e.g.* gas–liquid interfacial area). This feature results in enhanced reproducibility compared with traditional batch techniques. Owing to the small dimensions (typically 100–1000 μm internal diameter and several hundred microliters in volume), hazardous and explosive reactions can be carried out safely without risk of reactor failure (Figure 13.1). Toxic compounds can be generated *in situ* and immediately consumed in a follow-up reaction. Hence the storage of large quantities of hazardous compounds can be avoided. Moreover, the high surface-to-volume ratios permit rapid heat and mass transfer, which significantly reduces reaction times and allows fast exothermic reactions to be carried out by avoiding runaway conditions. Consequently, harsh and unusual process conditions (*e.g.* high temperatures and pressures) can be utilized in microreactors; these harsh conditions are called *novel process windows*.[5] The potential to integrate spectroscopic tools and inline analytical technology within the microreactor design allows for online reaction monitoring and optimization.[6] In addition, quick scalability without extensive optimization of the reaction parameters can be achieved in these devices by employing prolonged operating times, by a numbering-up strategy or smart scale-out.[7]

Despite the apparent advantages, moving from the traditional batch to continuous processing faces some challenges and witnessed initial skepticism (Table 13.1, Disadvantages). Many chemists have little to no experience with the engineering skills, such as knowledge about transport phenomena and process control, that are required to design a continuous flow protocol successfully. This will cost the researcher a certain time investment and requires a different way of thinking. In addition to developing a new chemical transformation, different process parameters, such as mixing, heat dissipation, multiphase flow characteristics and precipitate formation, have to be taken into account for a successful continuous flow reaction. Another potential issue can be the initial investment required to equip a laboratory with flow equipment. This involves buying suitable pumps, microreactors and other microfluidic equipment. However, owing to the increasing popularity of continuous flow chemistry, affordable microreactor solutions have been launched on to the market. Microcapillary tubing with concomitant microfluidic fittings are available in a wide variety of materials, diameters and lengths, and offer cheap alternatives to the more

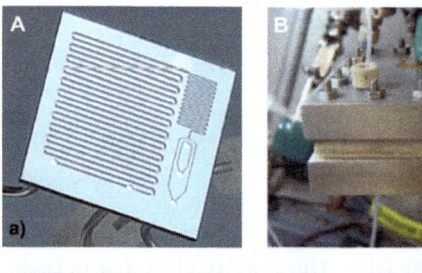
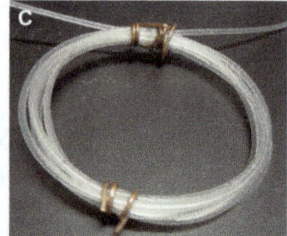
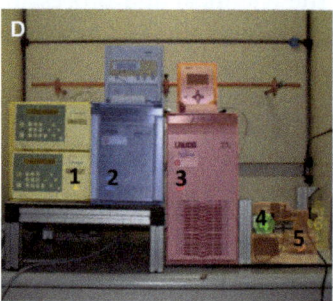
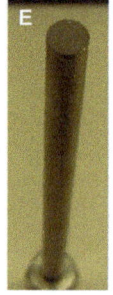
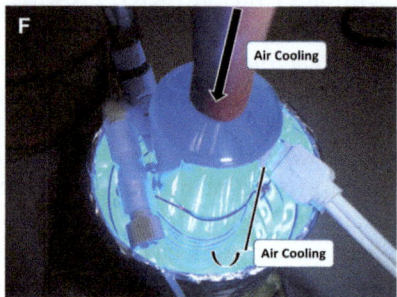

Figure 13.1 Selected examples of microreactor devices. (A) Silicon microreactor.[8a] Reproduced by permission of the Royal Society of Chemistry. (B) Teflon stack microreactor with integrated ultrasonication and heating for solids handling reactions.[8b] (C) PFA capillary microreactor.[8c] (D) Microcapillary assembly for high-temperature and -pressure applications with two HPLC pumps (1), capillary microreactor immersed in a heating bath (2), cooling zone (3), sample collection (4) and back-pressure regulator (5).[8d] (E) micro packed-bed reactor for biphasic reaction conditions.[8e] (F) Capillary microreactor for photochemical reactions.[8f]

Table 13.1 Advantages and disadvantages of continuous flow microreactors.

Advantages	Disadvantages
Excellent mixing (high mass transfer)	Handling of solids
Fast heating (high heat transfer)	Lack of engineering skills
Safety of operation	Initial investment cost
Novel process windows	Multidisciplinary
Precise control over reaction parameters	
Scale-up potential	
Reaction automation (self-optimization)	
Use of *in situ* spectroscopic tools	
Integrated multistep syntheses	

expensive microreactors made of silicon or stainless steel. One of the greatest issues in micro flow chemistry is dealing with solid-forming reactions, which usually lead to irreversible blockage of the microchannels. Often flow chemists will rethink the reaction conditions to avoid the formation of such solids. However, it should be noted that much effort from the flow chemistry community has resulted in the development of several solutions to deal with solid-forming reactions.[9]

13.3 Catalytic Systems for Cross-Coupling in Continuous Flow

13.3.1 Heterogeneous Supported Catalysts

A large number of the cross-coupling reactions that have been performed under continuous flow conditions have utilized immobilized palladium sources.[3,10] It has been generally stated that the use of immobilized palladium allows for facile recuperation and reuse of the catalyst source. Such immobilized catalysts are often loaded in a micro packed-bed reactor over which the reagents are directed. Owing to the large quantities of catalyst available in the packed bed, short reaction times are observed for a wide array of cross-coupling reactions. These apparent advantages have led to the development of solid supports (Table 13.2) on which catalysts can be immobilized *via* covalent bonding of the ligand or physical absorption of the metal.[11–14] One of the main drawbacks of a micro packed-bed reactor is the rather high pressure drop over the bed. This has been overcome by engineering novel polymer supports with high porosity, *i.e.* monolithic supports, which allow for excellent mass-transfer characteristics while minimizing the associated pressure drop.[15] Another possibility for avoiding large pressure drops consists in immobilizing the palladium catalyst on the reactor walls.[16]

Despite the low leaching behavior of Pd for many supports (Table 13.2, Entries 2–4), it is generally accepted that during cross-coupling reactions the catalytically active species is leached into solution.[17] After reaction, the leached metal can redeposit on the carrier material (so-called boomerang mechanism). Consequently, the actual amount of residual Pd in solution can be fairly low (<1 ppm) for batch reactions, which can mask the leaching behavior of the immobilized catalyst. Leaching can be promoted by using more polar solvents [*e.g. N,N*-dimethylformamide (DMF), *N*-methyl-2-pyrrolidinone (NMP)], by the oxidation state of the immobilized palladium [Pd(II) leaches more rapidly] and by the presence of soluble ligands. The true nature of the active catalyst can be revealed *via* a three-phase test in which an immobilized substrate is utilized to verify the homogeneous or heterogeneous character of the reaction.[18] In continuous flow, the packed bed functions as a huge palladium reservoir that gradually leaches the catalytically active species into a continuous liquid phase. As such, the palladium is dragged with the solution further downstream and, after reaction, the metal is deposited on the reactor walls. Such a leaching–deposition cycle can be compared with a chromatography effect (Figure 13.2). Eventually, all of the catalyst will leach from the microreactor, which results in complete deactivation of the catalyst bed.

Notwithstanding the homogeneous nature of cross-coupling reactions, the use of immobilized catalysts in micro packed-bed reactors can still be advantageous owing to their operational flexibility. Such microreactors have been made commercially available or, alternatively, can be rapidly made by packing open tubes (*e.g.* HPLC columns) with the immobilized catalyst.

Table 13.2 Selected examples of solid supports for immobilization of palladium.

Entry	Solid support	Cross-coupling reaction	Advantages	Leaching degree	Ref.
1	Carbon	Pd/C, Et₃N, MeCN, 170 °C — 81% (+ 10% homocoupling + 9% reduction)	Cheap	High (89% Pd was leached after 12 reactions)	11
2	Silicon dioxide	Siliacat DPP-Pd, KOH, THF/H₂O, 60 °C, 5 min residence time — > 95%, 1.71 mmol/h during 8 h, TON > 100	Good chemical stability Good thermal stability Cheap and abundant No swelling	Low (10 ppb in organic phase, 20 ppb in aqueous phase)	12
3	Polymer supports	PdEnCat, Bu₄NOMe, scCO₂ / MeOH (10:1), 100 °C, 166 bar — 81%	Wide variety of different supports Allows fine tuning of the support (metal–support interaction, pressure drop, etc.)	Low (10 ppm)	13
4	Magnetic nanoparticles	Pd on magnetic NP, Bu₃N, DMF, 120 °C — 63%	Allows for simultaneous heating via magnetic induction	Low (100 ppm for Heck; 34 ppm for Suzuki–Miyaura)	14

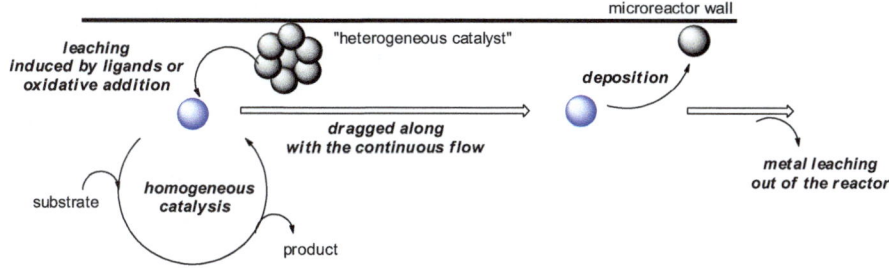

Figure 13.2 Leaching–deposition cycle leading to a chromatography effect in continuous flow reactors.

Since the degree of leaching can be fine-tuned by carefully selecting the appropriate carrier, contamination of the target product with trace amounts of metal can be minimized. This is especially important for the synthesis of active pharmaceutical ingredients (APIs), which require a very low metal content for clinical tests, and also any use of an API.

One interesting method to capture and reuse the leached palladium was developed for a sequential Heck alkenylation–hydrogenation.[19] For both steps, a micro packed-bed reactor filled with Pd/C was used to promote the reaction. It was observed that significant leaching occurred during the Heck alkenylation, after which the conversion decreased significantly for this transformation. However, the reactor used for the hydrogenation step could efficiently capture the leached palladium. By switching the order of the two reaction steps, the high activity of both reactor beds could be maintained and allowed the effective use of the immobilized catalyst to be prolonged.

13.3.2 Homogeneous Catalysts

Homogeneous catalytic systems allow for more flexibility with regard to fine-tuning of the catalytic properties than typically encountered with the immobilized versions. This fine-tuning gives access to more active catalysts, which are useful for more challenging substrate classes (*e.g.* heterocyclic substrates). A key parameter for achieving high activity is to ensure that the catalytically active Pd(0) species is efficiently formed. In this respect, several palladium precatalyst systems have been developed that ensure rapid and reliable formation of the $L_1Pd(0)$ species (Figure 13.3).[20,21] The use of continuous flow microreactors with intensified reaction conditions allows cross-coupling reactions to be greatly accelerated. It is therefore crucial that the active catalyst is formed rapidly, which can be achieved by using such precatalyst systems.

The Suzuki–Miyaura reaction of heteroaryl halides and heteroarylboronic acids is a challenging reaction that requires highly active catalyst systems. Moreover, it is known that 2-heteroarylboronic acids are prone to rapid decomposition due to protodeboronation in aqueous basic media. The use of an XPhos precatalyst (XPhos precat) allowed for the rapid carbon–carbon

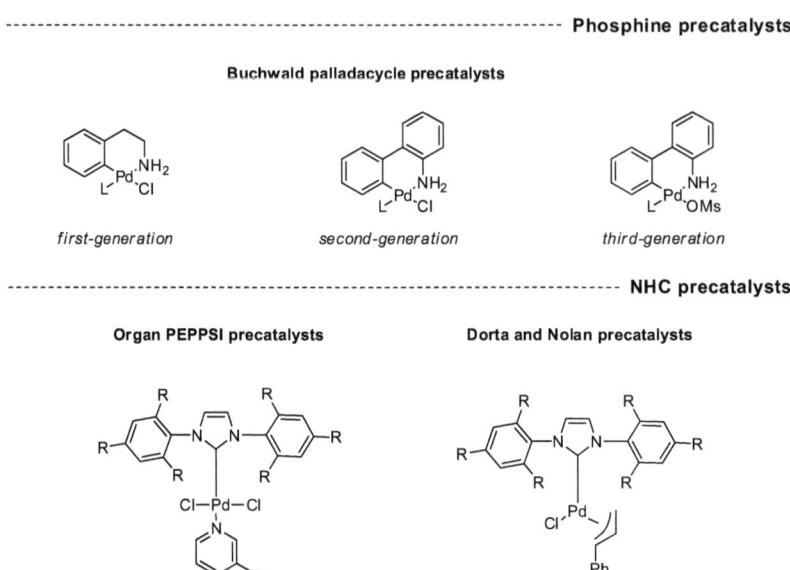

--- Phosphine precatalysts

Buchwald palladacycle precatalysts

first-generation   second-generation   third-generation

--- NHC precatalysts

Organ PEPPSI precatalysts   Dorta and Nolan precatalysts

Figure 13.3 Palladium precatalyst systems used in continuous flow microreactors.

bond formation of such challenging substrates in continuous flow.[22] The biphasic reaction system was directed over a packed-bed reactor, which was filled with stainless-steel spheres as an inert packing material and allowed for excellent mixing between the two immiscible phases. Most reactions could be completed within a 3 min residence time utilizing a low catalyst loading (0.5–1.5 mol% XPhos precat) at 90 °C. Even lower catalyst loadings (0.05–0.2 mol%) were feasible but required a lower reaction temperature owing to competitive protodeboronation and needed additional ligand to prevent catalyst deactivation owing to palladium black formation (Figure 13.4).

13.3.3   Catalyst Recycling of Homogeneous Catalysts

As mentioned above, the use of highly active homogeneous catalysts is required for more demanding cross-coupling reactions. However, its homogeneous nature makes it rather difficult to recover the expensive catalyst system. Several continuous recycling units have been developed for the recovery and reuse of such homogeneous catalyst systems.[23–25] Most examples make use of a biphasic system in which one phase selectively dissolves the catalyst, while the other dissolves the product. As such, the catalyst can be separated *via* a phase separation and recycled to the reactor inlet. One method involved the use of a fluorous-tagged palladium complex which could be contained in a fluorous solvent (Figure 13.5).[24] The reagents and products were soluble in an aqueous phase and merged with fluorous, catalyst-containing phase to establish a segmented flow regime. The polar

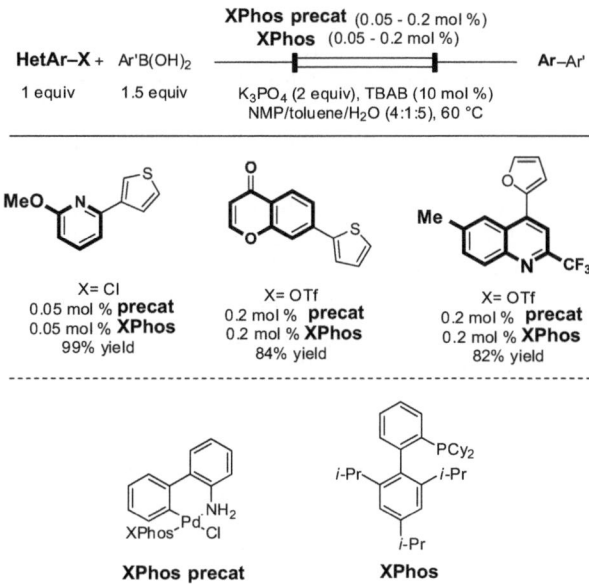

Figure 13.4 Biphasic Suzuki–Miyaura cross-coupling reactions in a micro packed-bed reactor.

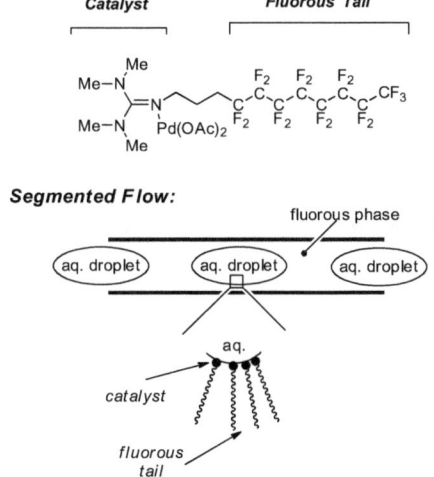

Figure 13.5 Fluorous-tagged Pd-catalyst for the continuous flow biphasic Suzuki–Miyaura reaction with catalyst recycling.

catalyst was oriented towards the aqueous droplet and allowed a Suzuki–Miyaura reaction to be catalyzed. The catalyst could be recycled four times without diminished activity. The extent of leaching of the catalyst into the aqueous phase was low, as indicated by inductively coupled plasma mass spectrometry (<0.5 ppm Pd). Another example involved the use of an ionic liquid-supported palladium catalyst for the Mizoroki–Heck reaction.[25] The reaction occurred in a single ionic liquid phase. The product was removed from the reaction stream *via* a hexane wash and impurities (*e.g.* ammonium salts) were removed *via* an aqueous wash, after which the catalyst-containing ionic liquid phase was redirected to the inlet. The catalyst could be recycled five times, after which the catalyst was still active and allowed 115.3 g of *trans*-butyl cinnamate to be produced (80% yield, 10 g h^{-1}).

Palladium-catalyzed hydroxylation is a useful transformation for synthesizing phenols starting from aryl halides. This challenging transformation utilizes sterically hindered dialkylbiarylphosphine ligands to facilitate the reaction. A continuous catalyst recycling unit was developed that allowed the expensive catalyst system to be recycled (Figure 13.6).[23] The conversion was set at 80%, which allowed the palladium species to be kept in its stable Pd^{2+} form and avoided the formation of palladium black upon recycling of the catalyst. The catalyst was soluble in the organic phase whereas the product was formed as a potassium phenolate which remained in the

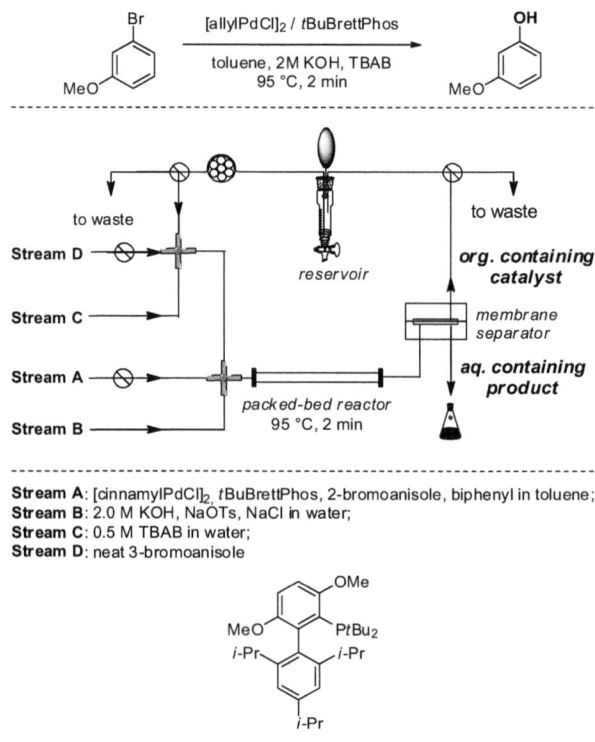

Figure 13.6   Continuous flow setup for catalyst recycling in the palladium-catalyzed hydroxylation of aryl halides.

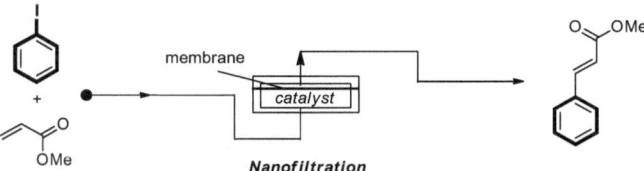

Figure 13.7 Continuous nanofiltration for catalyst retention in palladium-catalyzed Mizoroki–Heck coupling.

aqueous phase. A micro packed-bed reactor was utilized to ensure high mixing efficiency between the two phases. Upon exiting the microreactor, the two layers were separated in a micro extraction device. In this fashion, the catalyst was recycled a total of five times without a decrease in activity.

The examples given above involve biphasic reaction conditions in which the product and catalyst are separated based on a different solubility for one phase. The main drawback of this method is that the chemistry has to be compatible with the chosen biphasic reaction conditions. Recently, organic solvent nanofiltration has emerged as a new separation technique making use of a polymeric membranes to separate organic solvents from high molecular weight species, which is based on molecular weight cutoff, *i.e.* a membrane property. This technique allows the use of a single solvent phase and separates the catalyst from the solvent and reaction products. This technique was used in a continuous flow Heck-type coupling (Figure 13.7).[26] The experiment could be performed for 1000 h and allowed 1081 mol of product to be produced utilizing only 0.61 mmol of catalyst [turnover number (TON) = 1772]. Nevertheless, a significant amount of leaching was still observed (317 mg Pd per kg product). In comparison, there was 20-fold lower product contamination than was observed in a batch experiment without nanofiltration.

13.3.4 Homogeneous Catalysts in a Stationary Liquid Phase

A homogeneous catalyst can also be retained by dissolving it in a stationary liquid phase.[27] A porous support, *e.g.* mesoporous silica, is impregnated with a catalyst solution. Consequently, the catalyst is immobilized and can be retained after reaction, while the advantageous properties of high catalyst activity of homogeneous systems can be kept. The catalyst coating can be easily regenerated after deactivation by simply replacing the liquid film. As such, these catalyst systems are easier to produce and therefore more cost-efficient than methods that rely on a chemical bonding immobilization technique.

A supported aqueous phase catalyst was used for the Mizoroki–Heck reaction in flow.[28] A fused silica-coated capillary was impregnated with an aqueous solution of Pd(OAc)$_2$ and tris(2,4-dimethyl-5-sulfophenyl)phosphine (TXPTS). The reagents were introduced into the capillary microreactor and a

21% yield was obtained for the Mizoroki Heck reaction between phenyl iodide and *n*-butyl cinnamate within a 3 min residence time (the catalyst loading was below 0.001 mol%).

13.4 Carbon–Carbon Bond Formation in Continuous Flow

13.4.1 Suzuki–Miyaura Cross-Coupling in Flow

Suzuki–Miyaura cross-coupling (SMC) is one of the most utilized catalytic reactions in the pharmaceutical industry. It permits the reliable formation of carbon–carbon bonds by utilizing organoboron nucleophiles, which are stable and commercially available coupling partners. Owing to its broad utility in a variety of specialty applications, this reaction is one of the most studied cross-coupling examples in continuous flow microreactors.[3]

Typically, reaction times in continuous flow are much shorter than in the corresponding batch experiments. This is often due to the elevated amounts of catalyst available when utilizing immobilized palladium sources,[29] or to the rapid heat transfer and the elevated reaction temperatures that can be safely reached in microreactors. Efficient heat transfer allows for rapid activation of the molecules, which permits fast reactions. Microreactors are typically heated *via* conductive heating (*e.g.* hot-plates, oil-baths, GC ovens),[22] *via* resistive heating (*e.g.* cartridge heaters), *via* inductive heating[14] or *via* microwave irradiation.[30]

The ability to combine several reaction steps in one continuous and interrupted flow process constitutes a time- and cost-efficient alternative to the typical elaborate multistep procedures in synthetic organic chemistry.[31] However, utilizing a continuous flow strategy for multistep syntheses is not straightforward. First, reagents, solvents and impurities from the first re-action steps need to be compatible with the downstream reactions or need to be efficiently removed. Second, subsequent addition of dissolved reagents results in a diluted downstream reaction, which can make the reaction less efficient.

In the ideal situation, all reagents are compatible with the subsequent transformations and no intermediate purification is required. This consti-tutes an easily implemented flow strategy and resembles the one-pot batch strategies. An example involves a subsequent lithiation, borylation and SMC reaction sequence (Figure 13.8).[32] In a first PFA capillary microreactor, a lithium–halogen exchange was performed at room temperature. The reaction stream was subsequently merged with B(O^{i}Pr)$_3$ to produce the corresponding boronate coupling partner. Next, aryl halide, XPhos precat and aqueous KOH solution were added to enable the SMC reaction to pro-ceed. Owing to the partial insolubility of the boronate, acoustic irradiation was required to prevent microreactor clogging. The method was further ex-tended to the direct deprotonation of five-membered heterocycles. A similar protocol was developed to lithiate aryl halides with electrophilic functional

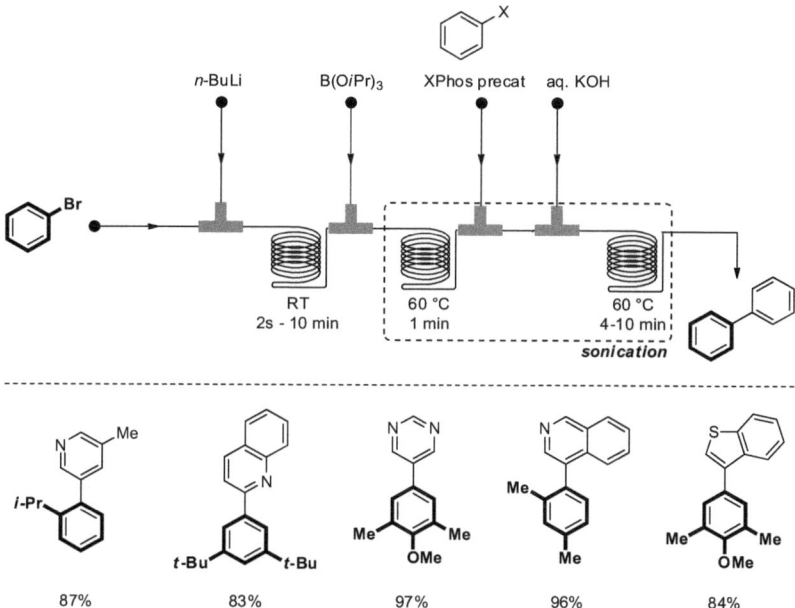

Figure 13.8 Continuous flow setup for a lithiation–borylation–SMC sequence: multistep synthesis without intermediate purification.

groups, *e.g.* esters.[33] Crucial for success was a very fast lithiation (a few milliseconds) followed by immediate quenching of the reaction stream with B(OMe)$_3$.

However, in most instances, impurities and excess of reagents are detrimental for follow-up catalytic reactions. In such cases, an in-line purification is mandatory for success of the multistep flow protocol. One approach involves the use of immobilized scavengers that are loaded into a micro packed-bed reactor. This strategy was used for the synthesis of a key intermediate of the fungicide boscalid (Figure 13.9).[34] In the first step, an SMC reaction was carried out with 0.25 mol% Pd(PPh$_3$)$_4$ and superheating of the reaction mixture. The presence of a homogeneous Pd catalyst in the hydrogenation step resulted in a significant amount of overreduced product. Therefore, an intermediate scavenging step was required; QuadraPure TU (a thiourea-based scavenger) was used to remove the remaining dissolved palladium. Next, the purified reaction stream was merged with hydrogen and directed over a Pt/C packed bed to permit the hydrogenation of the nitro group. The target compound was obtained in 77% overall yield.

Another way of purifying the reaction stream is to make use of unit operations. In contrast to the scavenging strategy, which requires replacement of the cartridges after saturation, no interruption of the continuous operation is required with unit operations to achieve the purification. An example of this principle is illustrated in Figure 13.10.[35] In the first step, phenols were reacted with triflic anhydride to yield the corresponding

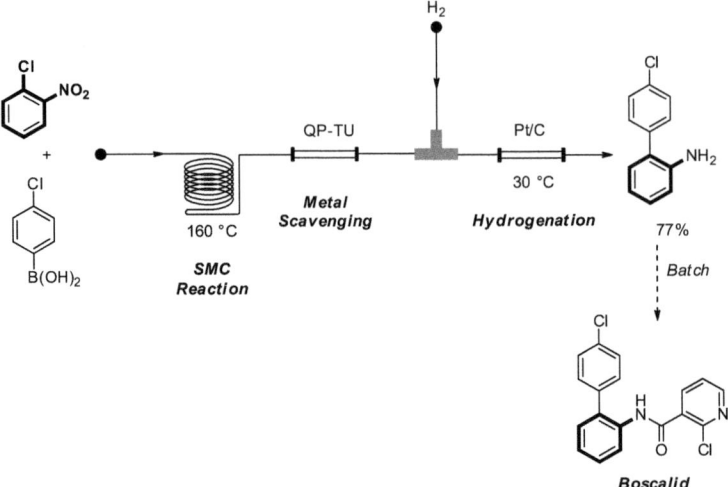

Figure 13.9 Continuous flow synthesis of a key intermediate in the synthesis of boscalid: multistep synthesis enabled by an intermediate purification by means of immobilized scavengers and catalysts.

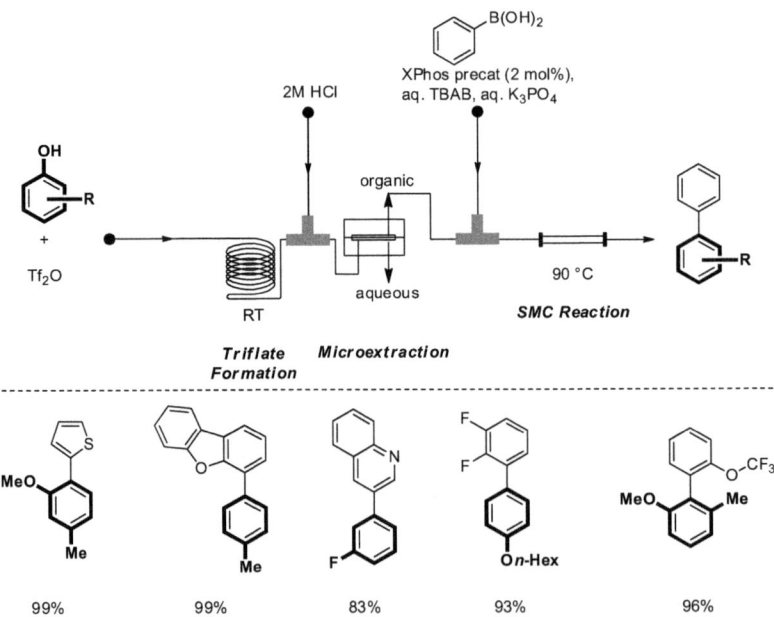

Figure 13.10 Continuous flow synthesis of biaryls *via* a triflate–SMC reaction sequence: multistep synthesis enabled by a microextraction.

triflates. Excess of triflic anhydride and ammonium salt by-products were removed *via* a microfluidic extraction. The organic phase was subsequently merged with XPhos precat (2 mol%), boronic acids, tetrabutylammonium

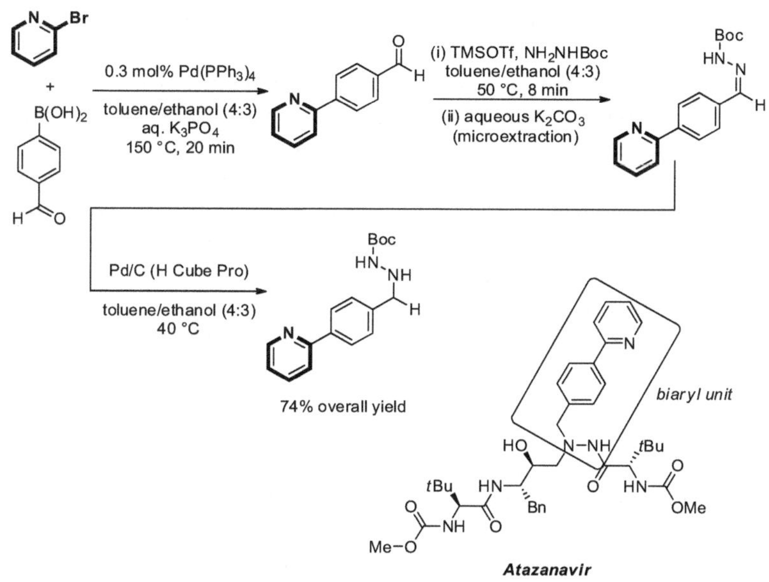

Figure 13.11 Continuous flow synthesis of the biaryl unit of atazanavir *via* a multistep continuous flow synthesis and microfluidic extraction.

bromide as a phase-transfer catalyst and aqueous potassium phosphate. The resulting biphasic mixture was delivered to a micro packed-bed reactor to ensure good mixing between the two phases. Several biaryls (14 examples) could be prepared in good to excellent overall yields (83–99%).

A similar strategy was used to prepare the biaryl unit of atazanavir, an HIV protease inhibitor.[36] A three-step continuous flow protocol was developed, including a Suzuki–Miyaura cross-coupling, hydrazine formation and hydrogenation step (Figure 13.11). Neutralization of the stream after hydrazine formation was achieved by quenching the mixture with aqueous potassium carbonate. The aqueous layer was separated from the product stream in an inline liquid–liquid extraction device. The overall yield for the continuous flow process was 74%.

13.4.2 Mizoroki–Heck Alkenylation in Flow

The Mizoroki–Heck alkenylation constitutes a powerful transformation to establish the coupling between an aryl halide and an olefin. Together with Suzuki–Miyaura coupling, this transformation has been very popular as a test reaction for the development of novel continuous flow microreactor techniques.

The combination of microreactors and spectroscopic detection techniques results in a so-called integrated microreactor environment, which allows for online reaction monitoring.[37] When such spectroscopic detection systems are further combined with logic and feedback control software, reactions can be optimized in a fully automated fashion. Such automated microreactor systems allow for rapid screening of reaction conditions, which is of great interest for the fine chemical and pharmaceutical industries. A self-optimizing system was used for the optimization and scale-up of a Mizoroki–Heck reaction between 4-chlorobenzotrifluoride and 2,3-dihydrofuran (Figure 13.12).[38] The reagents were introduced into a microreactor and samples were automatically taken by an inline HPLC system. HPLC was used to determine the conversion, yield and selectivity of the reaction and the data were analyzed *via* computer with DoE (Design of Experiments) software. Feedback was given with regard to the subsequent experiment and resulted in an adjustment of the concentrations and flow rates. After 19 automated experiments, the optimal yield and selectivity were found. Next, these optimized reaction conditions were transferred to a mesoscale flow reactor system, which provided a 50-fold scale-up without further reoptimization.

The vinylation of aryl halides *via* a Mizoroki–Heck reaction with ethylene is a challenging transformation in batch mode owing to poorly defined interfacial contact between the gas and liquid phases. Owing to the enhanced gas–liquid mass transfer, excellent mixing and the ability to process flammable gases in a safe and reliable fashion, the use of microreactors for gas-phase reactions has become popular.[39] More specifically, membrane reactors have been utilized to establish efficient gas–liquid contact. The gas

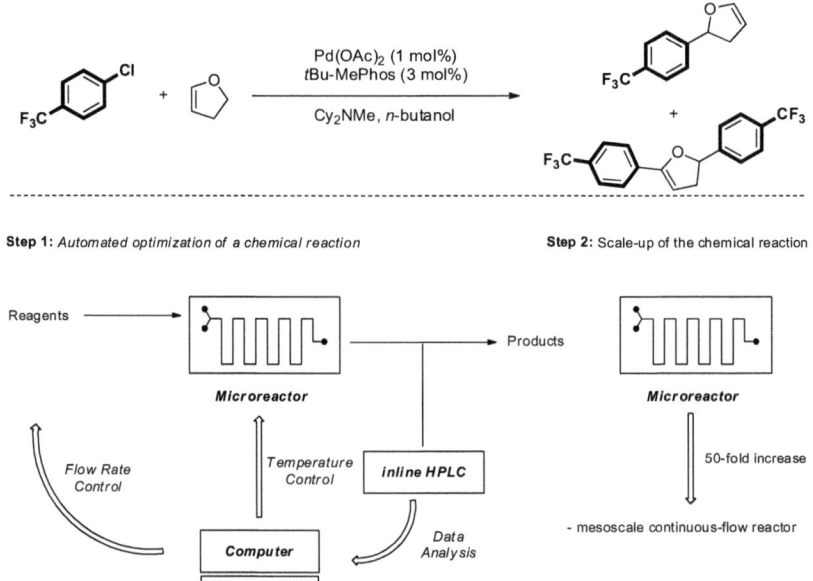

Figure 13.12   Self-optimization of the reaction conditions for a Mizoroki–Heck alkenylation by means of an automated "intelligent" microfluidic system and its subsequent scale-up in a mesoscale continuous flow reactor.

and liquid phases flow in two different compartments and are separated by a gas-permeable membrane. A membrane microreactor was used for the preparation of styrenes *via* the Mizoroki–Heck coupling between aryl iodides and ethylene (Figure 13.13).[40] The ethylene gas was brought into contact with the liquid reagents in a so-called tube-in-tube reactor, which consists of two concentric capillaries in which the central capillary is gas permeable. Next, the reaction was sent to a capillary microreactor heated to 120 °C to achieve full conversion. A back-pressure regulator was utilized to prevent out-gassing of the reaction mixture. The optimal catalyst system consisted of Pd(OAc)$_2$ with JohnPhos as a ligand and allowed a wide variety of styrenes to be prepared.

A multistep continuous flow system was developed to prepare enol ethers starting from *p-tert*-butylphenol (Figure 13.14).[41] The first reaction step involved triflate formation by reaction of the starting compound with triflic anhydride in dichloromethane. Impurities were removed *via* a liquid–liquid microextraction step and, subsequently, a solvent switch from dichloromethane to DMF was effected in a microfluidic distillation device. DMF was the optimal solvent to carry out the Mizoroki–Heck coupling step. Overall, 0.135 mmol product h^{-1} (79% yield) could be obtained with this system.

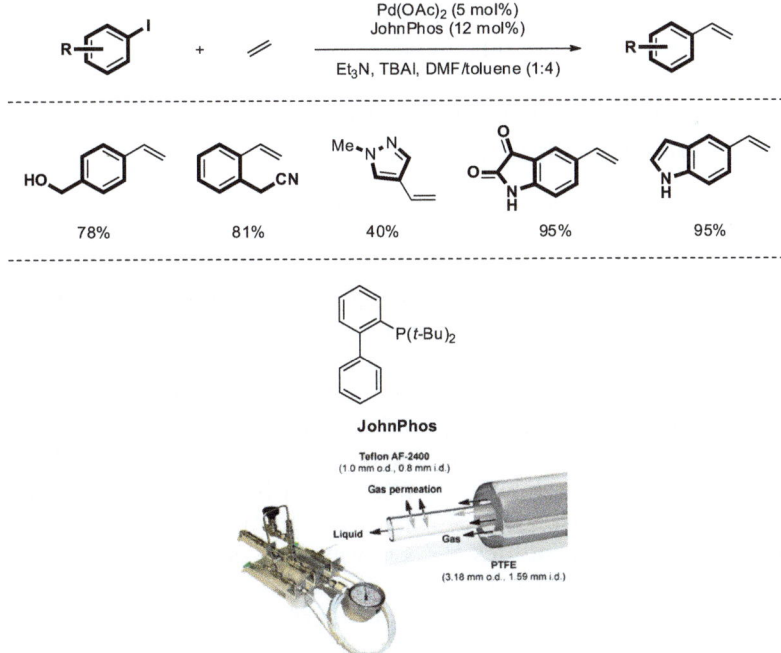

Figure 13.13 Gas–liquid Mizoroki–Heck vinylation in a tube-in-tube microreactor.[40] Copyright Wiley-VCH Verlag GmbH & Co. KGaA 2013. Reproduced with permission.

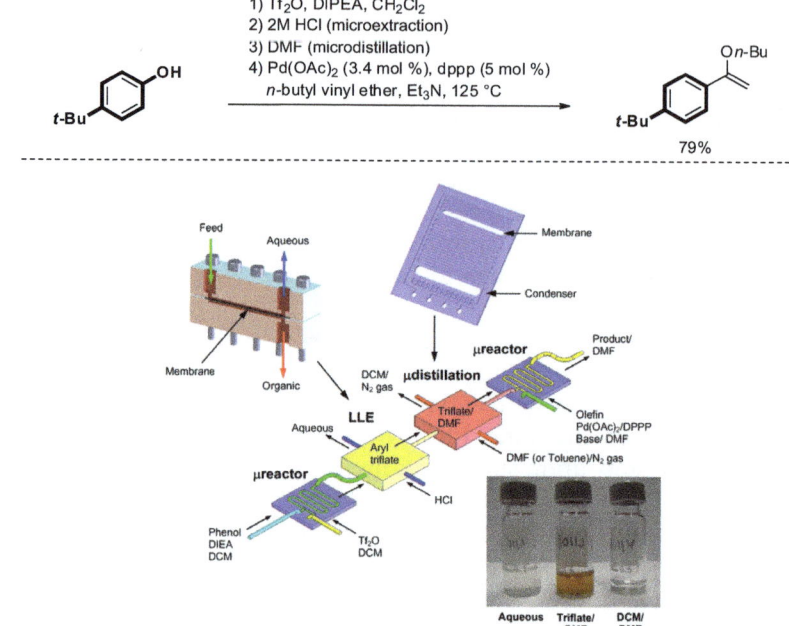

Figure 13.14 Microfluidic assembly for the synthesis of *n*-butyl vinyl ethers.[41] Copyright Wiley-VCH Verlag GmbH & Co. KGaA 2010. Reproduced with permission.

13.4.3 Palladium-Catalyzed Carbonylative Cross-Coupling in Flow

Despite the use of toxic carbon monoxide as a reagent, the palladium-catalyzed carbonylation reaction (Heck carbonylation) has become the method of choice for synthesizing carbonyl-containing compounds in a regioselective fashion. The large gas–liquid interfacial area encountered in microreactors allows for significant acceleration of this process. Such fast reactions are particularly advantageous for labeling of organic compounds with radioisotopes. Short reaction times are crucial to obtain high radio-chemical purities, which are required for molecular imaging applications. An example of a microreactor capable of performing fast carbonylations for radiolabeling is shown in Figure 13.15.[42] Carbon-11 ($t_{\frac{1}{2}}$ = 20.4 min) was utilized as the carbonyl source for the palladium-catalyzed aminocarbonylation reaction of aryl iodides.[43] ^{11}CO was first preconcentrated and subsequently brought in contact with the reagents in a microreactor. An annular flow regime was used to maximize the interfacial area, which allowed the reaction rate to be increased. A total of 15 min was required to finish the entire process, *i.e.* from generation of ^{11}CO to the collection of the product.

Microreactors have also been used to optimize rapidly and reliably reaction parameters such as reaction temperature, CO pressure and residence

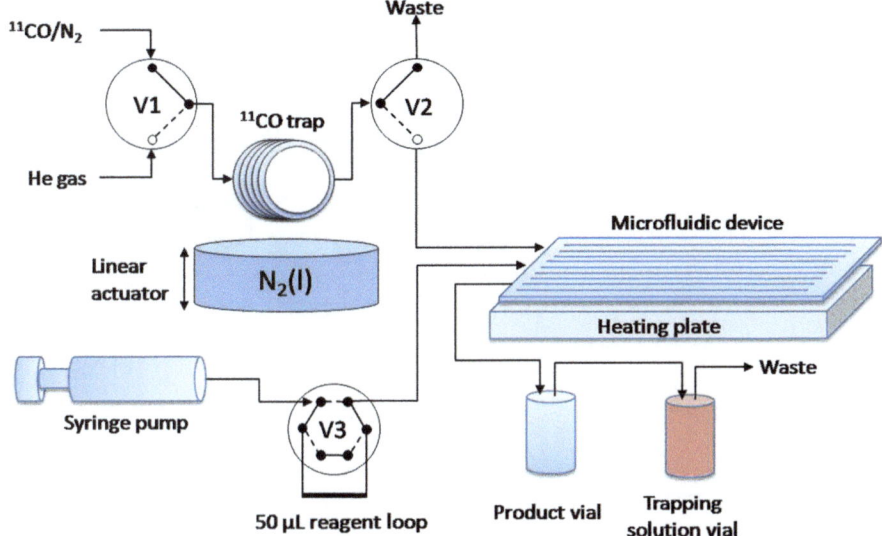

Figure 13.15 Microfluidic assembly for the synthesis of ^{11}C-labeled amide compounds *via* a palladium-catalyzed aminocarbonylation reaction.[43]
Copyright Wiley-VCH Verlag GmbH & Co. KGaA 2011.
Reproduced with permission.

time. This principle was demonstrated for the palladium-catalyzed aminocarbonylation of 4-bromobenzonitrile with morpholine as a nucleophile (Figure 13.16).[44] It was found that the ratio of amide to α-ketoamide could be fine-tuned by adjusting the temperature and pressure. High CO pressures and low reaction temperatures favored the formation of the α-ketoamide. This could be rationalized by the enhanced solubility of CO in the liquid phase, which allowed 2 equiv. of carbon monoxide to be inserted into the compound. On the other hand, low CO pressures and high reaction temperatures resulted in the selective formation of the amide product.

13.4.4 Alkyne Cross-Coupling in Flow

The cross-coupling between aryl, alkyl and vinyl halides and alkynes has been of great importance for the preparation of advanced materials. Several catalytic systems have been developed that make use of Cu catalysts (Castro–Stevens reaction), Pd catalysts (Heck–Cassar reaction) or a combination of both (Sonogashira reaction).

A high-temperature and high-pressure microreactor system allowed Heck–Cassar alkynylations to be carried out in water.[45] The substrates were merged neat with an aqueous solution of PdCl$_2$ and NaOH at very high pressure (25 MPa). Next, the combined reaction stream was mixed with water at 250 °C. Owing to the elevated temperature and pressure, the organic compounds are completely soluble in the supercritical water and no additional solvents are required. The palladium-catalyzed alkynylations could be

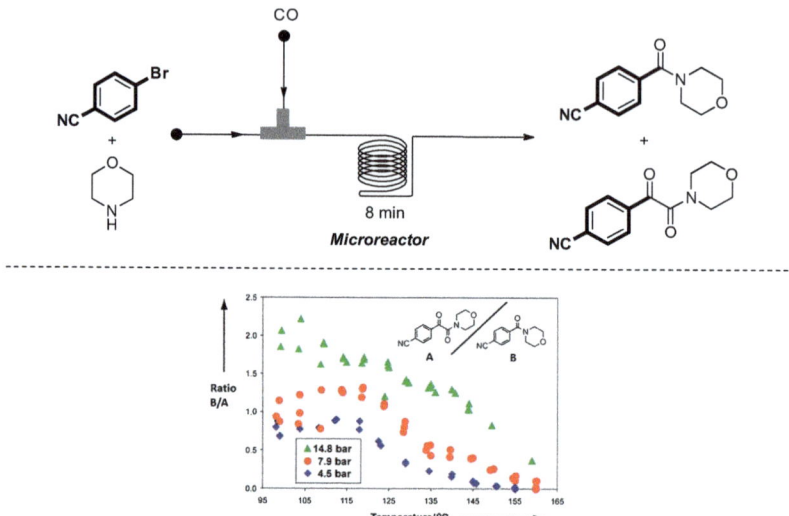

Figure 13.16 Aminocarbonylation of 4-bromobenzonitrile with morpholine using a Pd(OAc)$_2$/Xantphos catalyst system at high temperature and high pressure.[44] Copyright Wiley-VCH Verlag GmbH & Co. KGaA 2007. Reproduced with permission.

carried out with a residence time of less than 1 s. Upon exiting the microreactor, the reaction mixture was cooled and the products were readily separated from the aqueous mixture. In some cases, palladium black formation induced clogging of the microreactor. This was avoided by utilizing a capillary microreactor on which a layer of Pd–Cu alloy was deposited.[46] No additional homogeneous palladium catalyst was required and the formation of palladium black could be avoided.

Several Sonogashira reactions could be performed in a reactor sequence of a palladium-coated and copper capillary (Figure 13.17).[47] It was found that when the palladium-coated capillary was exclusively used, 3785 ppb of Pd (0.5 mol% Pd) was leached from the reactor under the reaction conditions. However, when a copper reactor was placed downstream, only 81.7 ppb of Pd was leached. This was attributed to a redeposition of Pd nanoparticles in the copper reactor. The remaining amount of palladium and copper was scavenged by placing QuadraPure TU (thiourea resin) after the copper capillary microreactor. By this means, leached Pd and Cu traces could be effectively removed from the product stream (<20 ppb). In total, 10 cycles were carried out with these reactors without a noticeable decrease in reactivity.

A two-step continuous flow protocol was developed to synthesize fluorinated alkynylarenes (Figure 13.18).[48] The first step involved the introduction of fluorine *via* nucleophilic displacement of a tosylate with tetrabutylammonium fluoride (TBAF). At 100 °C, the reaction could be finished within 2.5 min. The fluorinated alkylalkyne was subsequently introduced into a second reactor in which Heck–Cassar cross-coupling was performed at

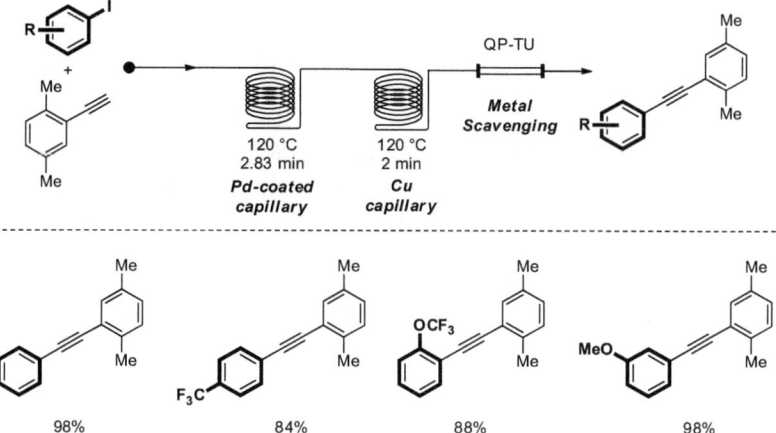

Figure 13.17 Sonogashira reactions in continuous flow utilizing a Pd-coated and Cu capillary microreactor sequence.

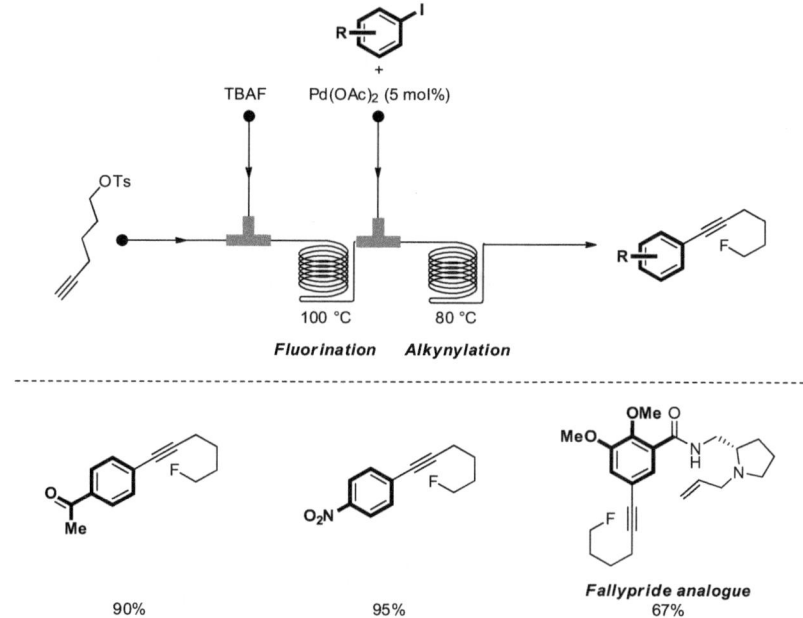

Figure 13.18 A two-step continuous flow protocol for the synthesis fluorinated alkynylarenes.

80 °C with Pd(OAc)$_2$ as a catalyst. The overall reaction sequence took less than 10 min and provides opportunities to introduce ^{18}F-labeled drug structures. Notably, this method could be used for the continuous flow synthesis of an analog of fallypride, a dopamine D2/3 receptor imaging agent (Figure 13.18).

13.4.5   Kumada–Corriu Cross-Coupling in Flow

The Kumada–Corriu cross-coupling reaction utilizes Grignard reagents as nucleophilic reagents in the transmetallation step. This transformation has been of historical importance since it was the first useful cross-coupling reaction. Its application in continuous flow has been limited; an example of this cross-coupling strategy in flow involved the use of a silicon-supported salen–nickel catalyst (Figure 13.19).[49] The catalyst was loaded in a micro packed-bed reactor and applied in the reaction between 4-bromoanisole and phenylmagnesium chloride. The coupled product was obtained in 62% yield. However, significant amounts of by-products were also observed (25% of reduced product and 7% of homocoupling). The yield decreased significantly after a few hours owing to deposition of magnesium salts on the catalyst support.

13.4.6   Murahashi Cross-Coupling in Flow

Similarly to Kumada–Corriu cross-coupling, the Murahashi cross-coupling reaction utilizes strong nucleophiles, in this case organolithium species, as coupling partners, which limits its practicality and functional group tolerability. An example of this methodology in flow is shown in Figure 13.20.[50] Aryllithium species are prepared *via* a halogen–lithium exchange in flow by treating aryl bromides with *n*-butyllithium. Owing to the excellent mixing and heat transfer capacities of microreactor technology, this reaction could be carried out at 0 °C and could thus be completed in a few seconds. Next, the aryllithium reagent was fed to a second reactor and merged with a stream of an aryl bromide and a PEPPSI-SIPr catalyst. The use of a homogeneous and very reactive catalyst was crucial to avoid side product formation.

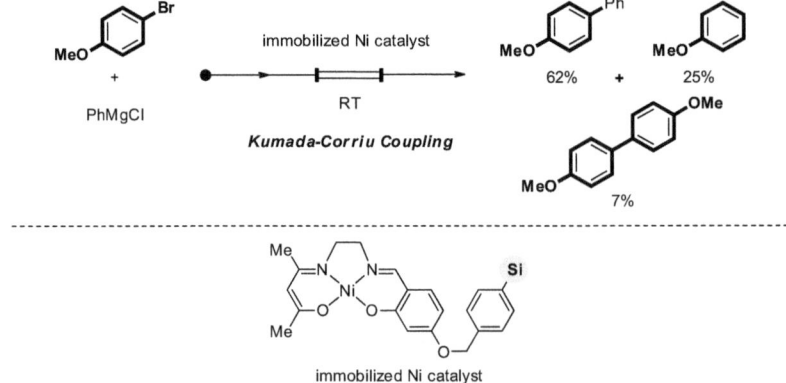

Figure 13.19   Kumada–Corriu coupling in flow using a packed-bed reactor filled with immobilized Ni catalyst.

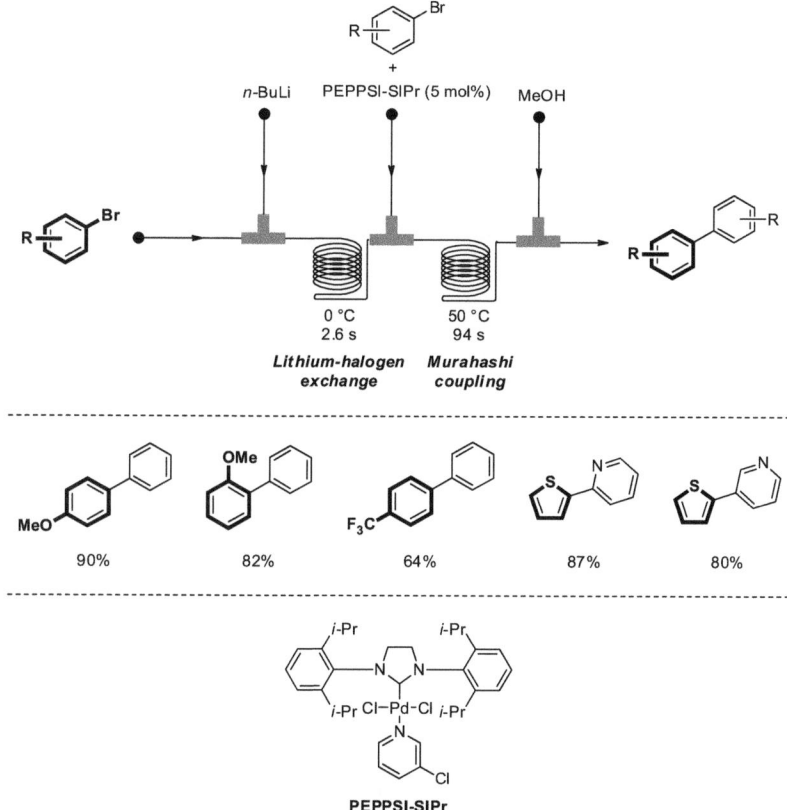

Figure 13.20 Microreactor assembly for the lithiation–Murahashi coupling sequence.

13.4.7 Hiyama Cross-Coupling in Flow

Hiyama coupling is a mild cross-coupling method that utilizes organosilicon nucleophiles. These coupling partners are inexpensive, easy to prepare and non-toxic. Despite these apparent advantages, only a single example can be found in which Hiyama coupling has been performed under continuous flow conditions (Figure 13.21).[51] Pd was immobilized on a diphenylphosphine-bonded silica gel, which was loaded in a micro packed-bed reactor. Several aryl bromides were coupled with trimethoxysilylbenzene in the presence of 2 equiv. of tetra-*n*-butylammonium fluoride, which activated silicon for transmetallation.

13.4.8 Negishi Cross-Coupling in Flow

Negishi cross-coupling utilizes organozinc coupling partners. Despite its popularity in batch, only a single example details its application in

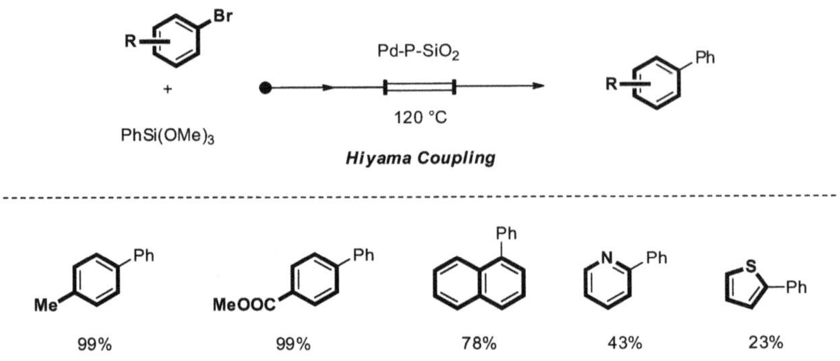

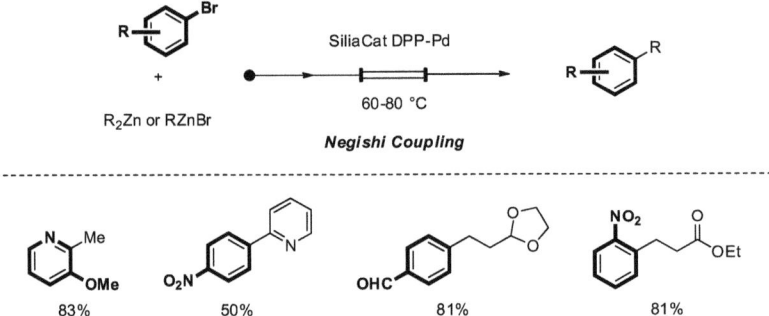

Figure 13.21 Continuous flow Hiyama coupling using a silica-supported catalyst.

Figure 13.22 Continuous flow Negishi coupling using a silica-supported catalyst.

continuous flow.[52] A silica-supported catalyst was loaded into a cartridge to form a micro packed-bed reactor (Figure 13.22). A wide variety of different substrates could be processed at 80 °C with a residence time of 5 min. Notably, a single cartridge allowed 20 consecutive reactions to be executed without a decrease in activity. Alternatively, a 4 h experiment could be performed without a noticeable decrease in efficiency. This resulted in a TON higher than 120. Contamination of the desired compounds with palladium was low (0.027 ppm Pd), suggesting a low leaching behavior of the immobilized catalyst.

13.4.9 Decarboxylative Cross-Coupling in Flow

Decarboxylative cross-coupling is a new strategy to construct carbon–carbon bonds. A copper co-catalyst is used to prepare the carbon nucleophile *via* a protodecarboxylation reaction of arylcarboxylic acids. A continuous flow strategy was developed to accelerate these reactions at 170 °C.[53] An important aspect of the study involved dealing with solubility issues, which otherwise led to clogging of the microreactor. First, NMP was used to ensure complete dissolution of all reagents. Second, aryl halides were replaced with aryl triflates because of the formation of potassium halide precipitates during the reaction. Third, in certain cases it was found that potassium carboxylates precipitated from the reaction mixture. This was solved by replacing the potassium *tert*-butoxide base with tetraethylammonium hydroxide. Good results were obtained for the preparation of a variety of biaryls (Figure 13.23). Compared with batch operation, better yields, fewer side reactions (less protodecarboxylation and homocoupling) and shorter residence times (1 h *versus* 16 h) were obtained.

Another example involves a combination of a decarboxylative coupling of propiolic acid followed by a Heck–Cassar coupling to produce unsymmetrical diarylalkynes (Figure 13.24).[54] In the first capillary, propiolic acid, aryl iodide and Pd(PPh$_3$)$_2$Cl$_2$ (5 mol%)–dppb (10 mol%) were combined and heated to

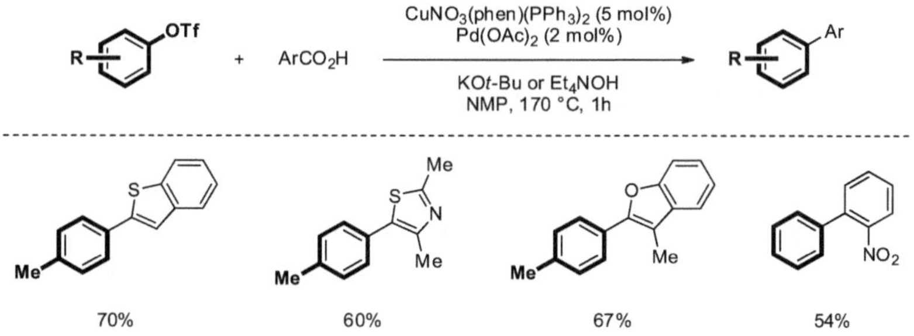

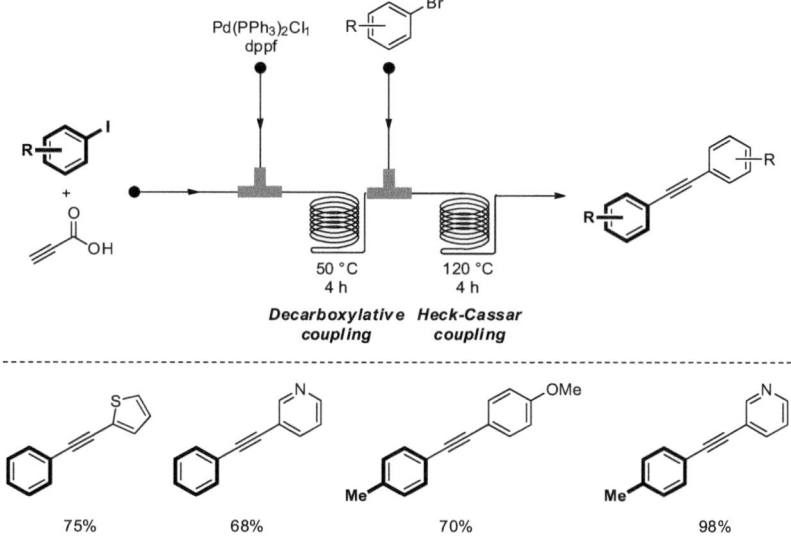

Figure 13.23 Decarboxylative cross-coupling of aryl triflates and arylcarboxylic acids in flow.

Figure 13.24 Synthesis of diarylalkynes *via* a decarboxylative coupling–Heck–Cassar coupling sequence in flow.

50 °C to produce the corresponding terminal alkyne. The product stream was subsequently brought into contact with an aryl bromide to perform the Heck–Cassar coupling at 120 °C. Although this procedure was used for the production of a variety of diarylalkynes, the total residence time was 8 h.

13.4.10 Palladium-Catalyzed α-Arylation/Alkylation in Flow

Palladium-catalyzed α-arylation represents a versatile reaction which allows the construction of carbon–carbon bonds between the α-position of a carbonyl and an aryl or vinyl moiety. A series of two micro packed-bed reactors was used for the preparation of 3,3-disubstituted oxindoles in continuous flow (Figure 13.25).[55] An XPhos second-generation precatalyst was used as a catalyst and allowed for rapid activation. Biphasic reaction conditions, with KOH as the base, were selected to avoid precipitate formation.

13.4.11 Direct Arylation in Flow

Inspired by selective biosynthetic pathways, C–H activation has emerged as a promising new area for the construction of carbon–carbon bonds. An intramolecular direct arylation to prepare biaryls has been transferred to continuous flow (Figure 13.26).[56] A binary solvent system, *N,N*-dimethylacetamide (DMA)–water (10 : 1), provided the highest conversion and selectivity and allowed optimal solubility of both organic and inorganic starting materials. Nevertheless, solubility issues were encountered during the course of the reaction and resulted in clogging of the microreactor. The use of ultrasonication could efficiently solve this problem, and several heterocyclic biaryls could be prepared in continuous flow. However, long reaction times (3–6 h) were still required to obtain high conversions.

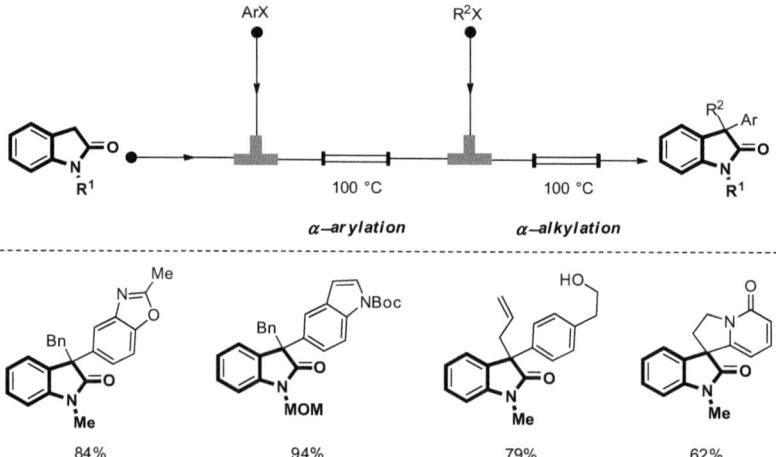

Figure 13.25 Continuous flow synthesis of 3,3 disubstituted oxindoles *via* a Pd-catalyzed α-arylation–alkylation.

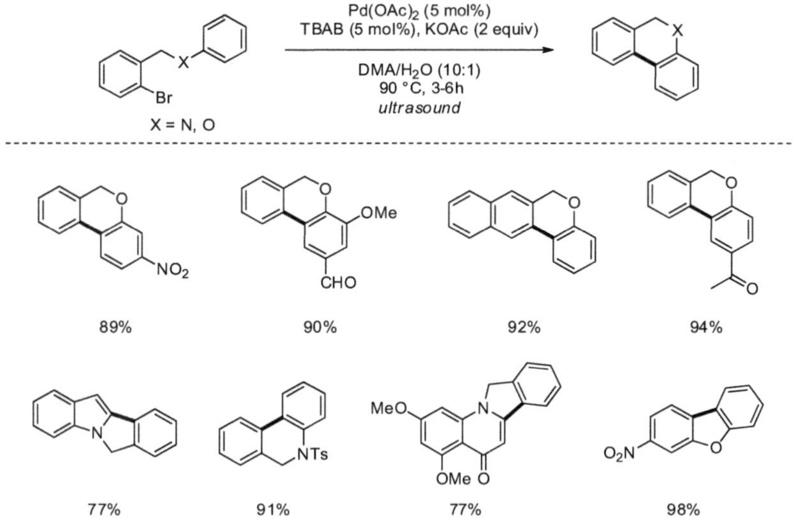

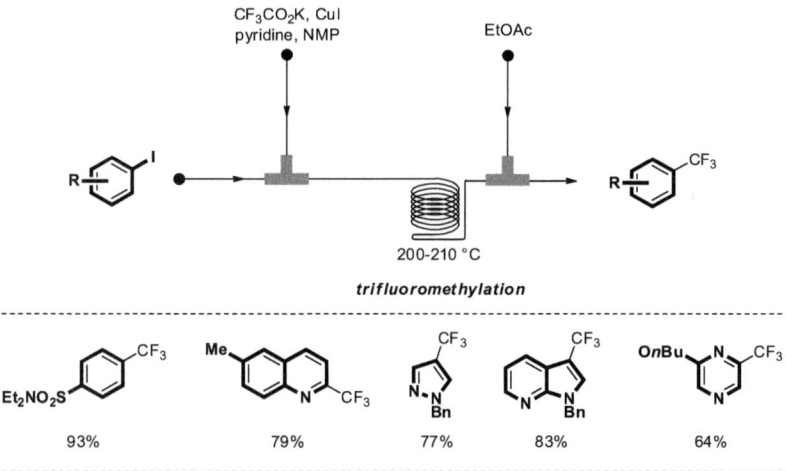

Figure 13.26 Continuous flow direct arylation to form heterocyclic biaryls.

Figure 13.27 Copper-mediated trifluoromethylation at elevated temperature in continuous flow.

13.4.12 Copper-Mediated Trifluoromethylation in Flow

The insertion of trifluoromethyl groups into drug compounds is an important strategy to improve their chemical and physical properties. A copper-mediated process was developed for the rapid trifluoromethylation in continuous flow using CF_3CO_2K as a CF_3 source (Figure 13.27).[57] The reagents were introduced into a stainless-steel capillary microreactor which was heated to 200 °C. At this temperature, copper-mediated trifluoromethylation

reactions, which normally take several hours or days in traditional batch reactors, could be accelerated significantly (16 min residence time).

13.5 Carbon–Heteroatom Bond Formation in Continuous Flow

13.5.1 C–N Bond Formation in Flow

The palladium-catalyzed amination reaction, also known as the Buchwald–
Hartwig reaction, constitutes one of the most important tools to establish a
C–N bond. However, formation of inorganic precipitates leads to irreversible
blockage of the microreactor channels. A first strategy involved the development of biphasic reaction conditions that allowed both organic and inorganic particles to be dissolved.[58] A micro packed-bed reactor, filled with
stainless-steel spheres as an inert packing material, was employed to establish vigorous mixing between the two phases and a phase-transfer catalyst
[tetrabutylammonium bromide (TBAB)] was used in combination with
aqueous potassium hydroxide as a base. Excellent yields of biarylamines
could be obtained in less than 10 min.

A second strategy involves the use of acoustic irradiation to break up
particle agglomerates.[59] This proved to be a powerful method to deal with
solid precipitates in continuous flow microreactors and allowed the C–N
coupling reaction to be performed under optimal reaction conditions.
A BrettPhos precatalyst (0.1–1 mol%) was used to effect the coupling of aryl
chlorides, bromides and triflates with primary amines without microreactor
clogging with short residence times (20 s–5 min) (Figure 13.28).
A microfluidic chip was developed by the same group to integrate both
ultrasonication and heating on the same device.[60]

This acoustic irradiation strategy was further used in multistep
continuous flow syntheses that involved an amination step.[61,62] The
synthesis of arylpyrazoles was achieved *via* a Pd-catalyzed amination
with hydrazines followed by a hydrolysis–cyclocondensation sequence with
β-diketones.[61] The first step was carried out in a capillary microreactor
which was immersed in an ultrasonic bath (Figure 13.29). Several arylpyrazoles could be prepared in good yield. Another example involves the
formation of 1-substituted benzotriazoles *via* an amination–hydrogenation–
diazotization–cyclization; sequence.[62] All reactions were performed consecu-
tively in an uninterrupted continuous flow setup and required ultrasonication
to avoid clogging in the amination step.

A third strategy involves the use of soluble organic bases and polar
solvents, which circumvents the formation of salt precipitation.[63–65] A
palladium-catalyzed amination strategy was used for the continuous flow
synthesis of imatinib base (Gleevec) (Figure 13.30).[66] The intermediate of
imatinib was synthesized in an automated fashion employing immobilized
reagents and a catch-and-release strategy.[67] The compound was subsequently released with DBU and merged with a stream of amine, BrettPhos

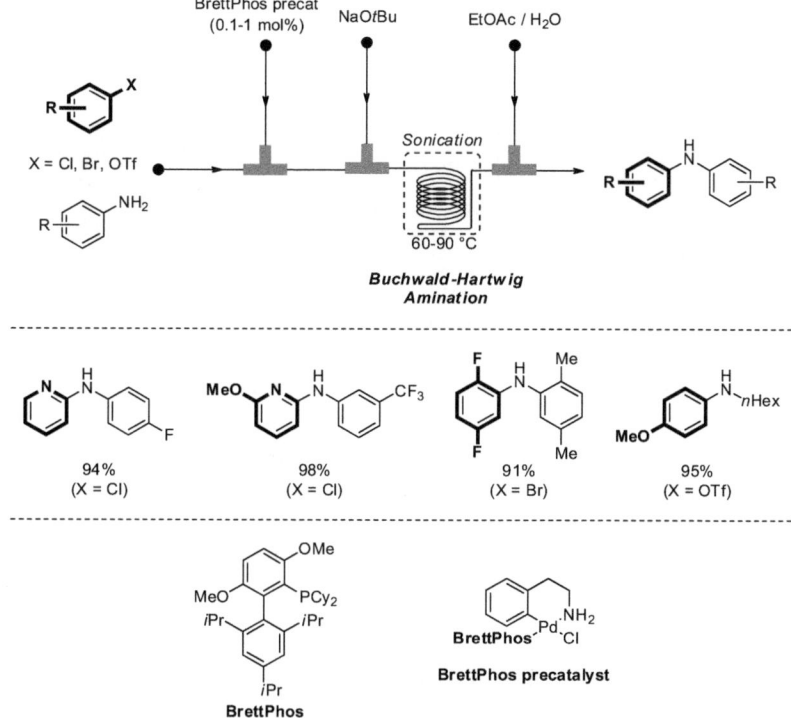

Figure 13.28 Continuous flow Buchwald–Hartwig amination of aryl halides/pseudohalides and primary amines assisted by acoustic irradiation.

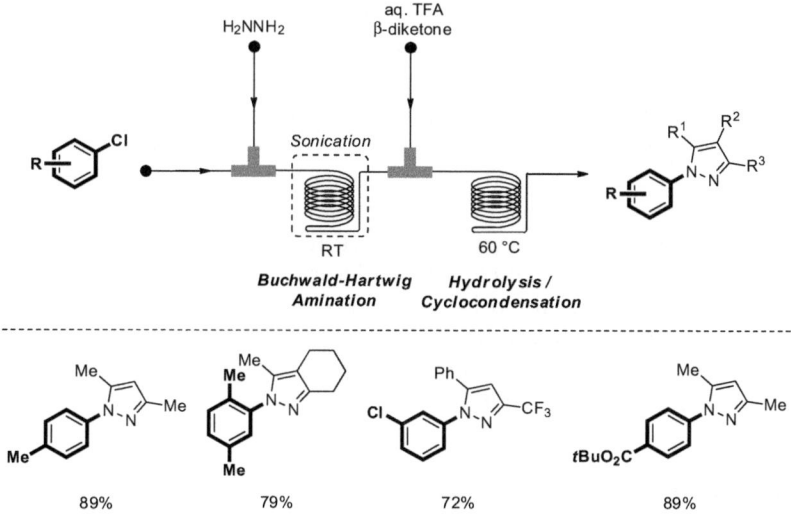

Figure 13.29 Continuous flow synthesis of arylpyrazoles *via* an amination–hydrolysis–cyclocondensation sequence.

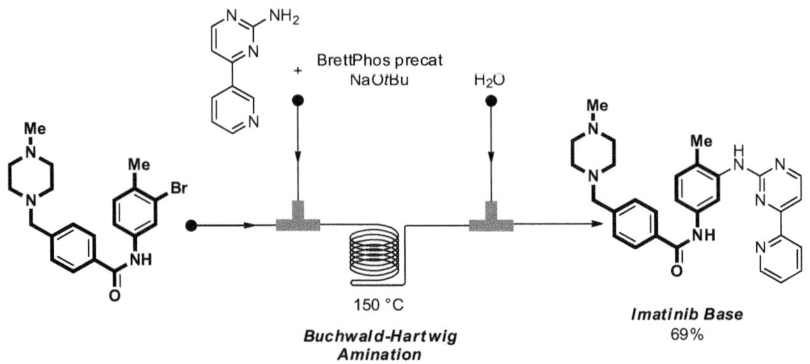

Figure 13.30   Continuous flow synthesis of imatinib base (Gleevec).

precatalyst (10 mol%) and NaO^{t}Bu in 1,4-dioxane–^{t}BuOH (2:1). The amination reaction was performed in a capillary microreactor heated at 150 °C. The polar solvent system minimized the extent of NaBr precipitation. However, the authors stated that an aqueous quench was required to avoid clogging of the back-pressure regulator.

13.5.2   C–O Bond Formation in Flow

The formation of carbon–oxygen bonds *via* a cross-dehydrogenative coupling is an interesting strategy that activates two C–H bonds and generates hydrogen as a sole by-product. A single example of this methodology in continuous flow has been reported.[68] The reaction involves copper-catalyzed C–O bond formation by direct α-C–H bond activation of ethers in the presence of *tert*-butyl hydroperoxide (TBHP) as the oxidant. Initial optimization studies were performed in a microwave batch reactor. Microwave equipment allows elevated reaction temperatures and pressures to be used; however, the chemistry is not scalable owing to the limited penetration capacity of the microwaves. Nevertheless, scale-up of the optimized reaction conditions can be achieved in continuous flow microreactors, which also show excellent heat transfer characteristics (so-called "microwave-to-flow" paradigm).[69] In flow, the reagents were mixed with TBHP in a micromixer and introduced into a 20 mL stainless-steel capillary microreactor at 130 °C (Figure 13.31). Similar results were obtained in continuous flow compared with microwave batch experiments. It should be noted that in batch mode multiple additions of TBHP were feasible, whereas in flow only a single addition was achieved, which explains the lower yields.

13.5.3   C–F Bond Formation in Flow

The introduction of fluorine in pharmaceuticals, agrochemicals and materials has attracted a great deal of attention since it allows tailoring of the

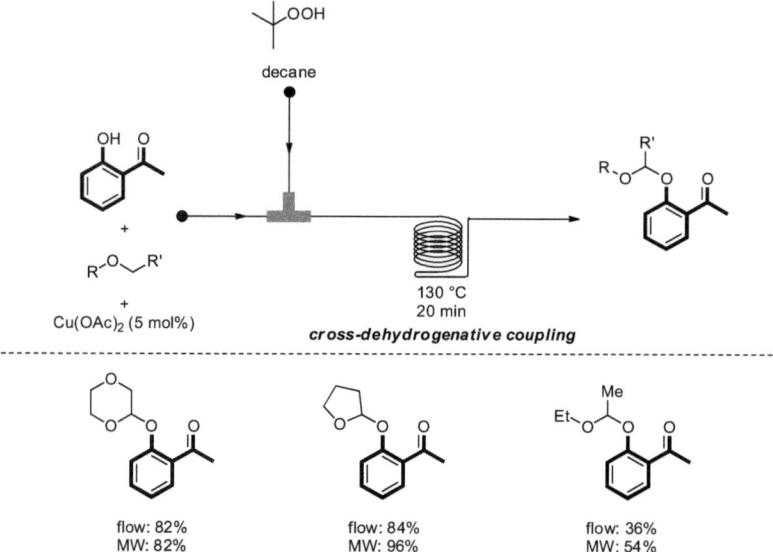

Figure 13.31 Continuous flow cross-dehydrogenative coupling *via* direct α-C–H bond activation of ethers.

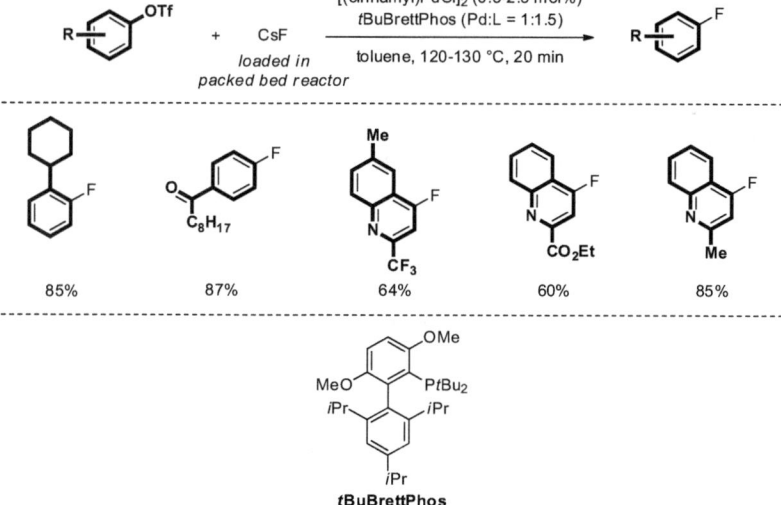

Figure 13.32 Pd-catalyzed fluorination of aryl triflates in a CsF-packed microreactor.

properties of the compounds. However, the development of synthetic methodologies to introduce such substituents has proven to be a challenging task. More specifically, transition metal-catalyzed processes to introduce fluorine have been hampered by a difficult reductive elimination step.[70] This was recently overcome by using a sterically demanding and electron-rich phosphine ligand, *t*BuBrettPhos, in combination with palladium metal to convert aryl triflates into aryl fluorides.[71] Cesium fluoride was used as the fluorine source and it was found that increasing amounts of CsF resulted in an accelerated fluorination protocol.[72] However, owing to the insolubility of CsF, large amounts of CsF were difficult to handle in standard batch vessels. A micro packed-bed reactor was filled with CsF and the reagents were directed over the fluoride bed. This allowed the process to be accelerated significantly; full conversion could be achieved for a broad variety of aryl triflates (Figure 13.32). Moreover, the use of this continuous flow protocol allowed the CsF bed to be reused several times without a decrease in reactivity. In addition, elevated temperatures and pressures provided additional acceleration compared with the batch protocol and allowed the catalyst loading to be reduced.

13.6   Conclusion and Outlook

The use of continuous flow microreactors for cross-coupling chemistry has provided a number of advantages with regard to practicality, time efficiency, safety, scale-up potential and the use of harsh reaction conditions. Further improvements in this field can be expected in the coming years.[73] Especially direct functionalization of C–H bonds remains underdeveloped and can probably benefit from the unusual reaction conditions provided by microreactors. As is evident from this review, gas–liquid cross-coupling reactions can significantly benefit from the intensified gas–liquid mass transfer characteristics, and many opportunities can still be found in this area. From a technology perspective, more efficient unit operations are required to facilitate multistep syntheses in flow. Moreover, increased automation of the entire process will be necessary to run the flow synthesis without intermediate handling. Success in these aspects will require an interdisciplinary approach in which chemists, process engineers, software developers and electrical engineers are involved. It should be noted that courses on flow chemistry and continuous manufacturing are still rare in the curriculum of many chemistry programs. If we want to embrace continuous flow technology, we should bring the chemists of the future into contact with its benefits early in their education.

References

1. (a) F. Diederich and A. de Meijere (eds), *Metal-Catalyzed Cross-Coupling Reactions*, Wiley-VCH, Weinheim, 2nd edn, 2004; (b) N. Miyaura, Cross-coupling reactions, a practical guide, *Top. Curr. Chem.*, 2002, **219**, 1–241.
2. (a) C. C. C. Johansson Seechum, M. O. Kitching, T. J. Colacot and V. Snieckus, *Angew. Chem. Int. Ed.*, 2012, **51**, 5062–5085; (b) H. Li, C. C. C. Johansson Seechum and T. J. Colacot, *ACS Catal.*, 2012, **2**, 1147–1164; (c) Z.-X. Wang and N. Liu, *Eur. J. Inorg. Chem.*, 2012, 901–911;

(d) D. S. Surry and S. L. Buchwald, *Chem. Sci.*, 2011, **2**, 27–50; (e) N. Marion and S. P. Nolan, *Acc. Chem. Res.*, 2008, **41**, 1440–1449; (f) A. F. Littke and G. C. Fu, *Angew. Chem. Int. Ed.*, 2002, **41**, 4176–4211.

3. T. Noël and S. L. Buchwald, *Chem. Soc. Rev.*, 2011, **40**, 5010–5029.

4. (a) I. Dencic, T. Noël, J. Meuldijk, M. de Croon and V. Hessel, *Eng. Life Sci.*, 2013, **13**, 326–343; (b) T. Noël, X. Wang and V. Hessel, *Chim. Oggi*, 2013, **31**(3, Suppl.), 10–14; (c) S. Sadler, A. R. Moeller and G. B. Jones, *Expert Opin. Drug Discov.*, 2012, **7**, 1107–1128; (d) L. Malet-Sanz and F. Susanne, *J. Med. Chem.*, 2012, **55**, 4062–4098; (e) R. L. Hartman, J. P. McMullen and K. F. Jensen, *Angew. Chem. Int. Ed.*, 2011, **50**, 7502–7519; (f) N. Kockmann, M. Gottsponer, B. Zimmermann and D. M. Roberge, *Chem. Eur. J.*, 2008, **14**, 7470–7477.

5. (a) V. Hessel, D. Kralisch, N. Kockmann, T. Noël and Q. Wang, *ChemSusChem*, 2013, **6**, 746–789; (b) S. C. Stouten, T. Noël, Q. Wang and V. Hessel, *Aust. J. Chem.*, 2013, **66**, 121–130; (c) V. Hessel, I. Vural-Gursel, Q. Wang, T. Noël and J. Lang, *Chem. Eng. Technol.*, 2012, **35**, 1184–1204; (d) V. Hessel, I. Vural-Gursel, Q. Wang, T. Noël and J. Lang, *Chem. Ing. Tech.*, 2012, **84**, 660–684.

6. J. P. McMullen and K. F. Jensen, *Annu. Rev. Anal. Chem.*, 2010, **3**, 19–42.

7. (a) G. Caygill, M. Zanfir and A. Gavriilidis, *Org. Process Res. Dev.*, 2006, **10**, 539–552; (b) N. Kockmann, M. Gottsponer and D. M. Roberge, *Chem. Eng. J.*, 2010, **167**, 718–726.

8. (a) R. L. Hartman and K. F Jensen, *Lab Chip*, 2009, **9**, 2495–2507; (b) see Ref. 57; (c) A. C. Varas, T. Noël, Q. Wang and V. Hessel, *ChemSusChem*, 2012, **5**, 1703–1707; (d) H. Kobayashi, B. Driessen, D. J. G. P. van Osch, A. Talla, S. Ookawara, T. Noël and V. Hessel, *Tetrahedron*, 2013, **69**, 2885–2890; (e) see Ref. 58; (f) X. Wang, G. D. Cuny and T. Noël, *Angew. Chem. Int. Ed.*, 2013, **52**, 7860–7864; (g) N. J. W. Straathof, D. J. G. P. van Osch, A. Schouten, X. Wang, J. C. Schouten, V. Hessel and T. Noël, *J. Flow Chem*, 2014, **4**, 12–17.

9. (a) B. S. Flowers and R. L. Hartman, *Challenges*, 2012, **3**, 194–211; (b) R. L. Hartman, *Org. Process Res. Dev.*, 2012, **16**, 870–887.

10. (a) C. G. Frost and L. Mutton, *Green Chem.*, 2010, **12**, 1687–1703; (b) W. R. Reynolds and C. G. Frost, in *Palladium-Catalyzed Coupling Reactions: Practical Aspects and Future Developments*, ed. A. Molnar, Wiley-VCH, Weinheim, 2013, pp. 409–443.

11. T. N. Glasnov, S. Findenig and C. O. Kappe, *Chem. Eur. J.*, 2009, **15**, 1001–1010.

12. J. M. de Munoz, J. Alcazar, A. de la Hoz and A. Diaz-Ortiz, *Adv. Synth. Catal.*, 2012, **354**, 3456–3460.

13. G. A. Leeke, R. C. D. Santos, B. Al-Duri, J. P. K. Seville, C. J. Smith, C. K. Y. Lee, A. B. Holmes and I. F. McConvey, *Org. Process Res. Dev.*, 2007, **11**, 144–148.

14. S. Ceylan, C. Friese, C. Lammel, K. Mazac and A. Kirschning, *Angew. Chem. Int. Ed.*, 2008, **47**, 8950–8953.

15. U. Kunz, H. Schonfeld, W. Solodenko, G. Jas and A. Kirschning, *Ind. Eng. Chem. Res.*, 2005, **44**, 8458–8467.

16. E. Comer and M. G. Organ, *J. Am. Chem. Soc.*, 2005, **127**, 8160–8167.

17. (a) A. Molnar, *Chem. Rev.*, 2011, **111**, 2251–2320; (b) L. Yin and J. Liebscher, *Chem. Rev.*, 2007, **107**, 133–173.

18. (a) S. J. Broadwater and D. T. McQuade, *J. Org. Chem.*, 2006, **71**, 2131–2134; (b) I. W. Davies, L. Matty, D. L. Hughes and P. J. Reider, *J. Am. Chem. Soc.*, 2001, **123**, 10139–10140.

19. X. Fan, M. G. Manchon, K. Wilson, S. Tennison, A. Kozynchenko, A. A. Lapkin and P. K. Plucinski, *J. Catal.*, 2009, **267**, 114–120.

20. For Buchwald palladacycle precatalysts, see:(a) D. S. Surry and S. L. Buchwald, *Chem. Sci.*, 2011, **2**, 27–50; (b) T. Noël, in *e-EROS Encyclopedia of Reagents for Organic Synthesis*, Wiuley, Hoboken, NJ, 2011, DOI: 10.1002/047084289X.rn01343; (c) N. C. Bruno and S. L. Buchwald, *Org. Lett.*, 2013, **15**, 2876–2879.

21. For NHC precatalysts, see:(a) C. Valente, S. Calimsiz, K. Hou Hoi, D. Mallik, M. Sayah and M. G. Organ, *Angew. Chem. Int. Ed.*, 2012, **51**, 3314–3332; (b) A. Chartoire, M. Lesieur, L. Falivene, A. M. Z. Slawin, L. Cavallo, C. S. J. Cazin and S. P. Nolan, *Chem. Eur. J.*, 2012, **18**, 4517–4521; (c) L. Wu, E. Drinkel, F. Gaggia, S. Capolicchio, A. Linden, L. Falivene, L. Cavallo and R. Dorta, *Chem. Eur. J.*, 2011, **17**, 12886–12890.

22. T. Noël and A. J. Musacchio, *Org. Lett.*, 2011, **13**, 5180–5183.

23. P. Li, J. S. Moore and K. F. Jensen, *ChemCatChem*, 2013, **5**, 1729–1733.

24. A. B. Theberge, G. Whyte, M. Frenzel, L. M. Fidalgo, R. C. R. Wootton and W. T. S. Huck, *Chem. Commun.*, 2009, 6225–6227.

25. S. Liu, T. Fukuyama, M. Sato and I. Ryu, *Org. Process Res. Dev.*, 2004, **8**, 477–481.

26. L. Peeva, J. Arbour and A. Livingston, *Org. Process. Res. Dev.*, 2013, **17**, 967–975.

27. For a review on supported ionic liquid phase catalysis, see: A. Riisager, R. Fehrmann, M. Haumann and P. Wasserscheid, *Eur. J. Inorg. Chem.*, 2006, 695–706.

28. S. C. Stouten, Q. Wang, T. Noël and V. Hessel, *Tetrahedron Lett.*, 2013, **54**, 2194–2198.

29. For some recent examples which use an immobilized palladium catalyst, see:(a) C. Pavia, E. Ballerini, L. A. Bivona, F. Giacalone, C. Aprile, L. Vaccaro and M. Gruttadauria, *Adv. Synth. Catal.*, 2013, **355**, 2007–2018; (b) G. O. D. Estrada, M. C. Flores, J. F. M. da Silva, R. O. M. A. de Souza and L. S. M. e Miranda, *Tetrahedron Lett.*, 2012, **53**, 4166–4168; (c) S. Wittmann, J.-P. Majoral, R. N. Grass, W. J. Stark and O. Reiser, *Green Process Synth.*, 2012, **1**, 275–279; (d) see Ref. 11.

30. P. Ohrngren, A. Fardost, F. Russo, J.-S. Schanche, M. Fagrell and M. Lahred, *Org. Process Res. Dev.*, 2012, **16**, 1053–1063.

31. D. Webb and T. F. Jamison, *Chem. Sci.*, 2010, **1**, 675–680.

32. W. Shu, L. Pellegatti, M. A. Oberli and S. L. Buchwald, *Angew. Chem. Int. Ed.*, 2011, **50**, 10665–10669.
33. A. Nagaki, Y. Moriwaki and J.-i. Yoshida, *Chem. Commun.*, 2012, **48**, 11211–11213.
34. T. N. Glasnov and C. O. Kappe, *Adv. Synth. Catal.*, 2010, **352**, 3089–3097.
35. T. Noël, S. Kuhn, A. J. Musacchio, K. F. Jensen and S. L. Buchwald, *Angew. Chem. Int. Ed.*, 2011, **50**, 10665–10669.
36. L. Dalla-Vechia, B. Reichart, T. Glasnov, L. S. M. Miranda, C. O. Kappe and R. O. M. A. de Souza, *Org. Biomol. Chem.*, 2013, **11**, 6806–6813.
37. J. P. McMullen and K. F. Jensen, *Annu. Rev. Anal. Chem.*, 2010, **3**, 19–42.
38. J. P. McMullen, M. T. Stone, S. L. Buchwald and K. F. Jensen, *Angew. Chem. Int. Ed.*, 2010, **49**, 7076–7080.
39. T. Noël and V. Hessel, *ChemSusChem*, 2013, **6**, 405–407.
40. S. L. Bourne, M. O'Brien, S. Kasinathan, P. Koos, P. Tolstoy, D. X. Hu, R. W. Bates, B. Martin, B. Schenkel and S. V. Ley, *ChemCatChem*, 2013, **5**, 159–172.
41. R. L. Hartman, J. R. Naber, S. L. Buchwald and K. F. Jensen, *Angew. Chem. Int. Ed.*, 2010, **49**, 899–903.
42. For a review on the use of microreactor technology for producing radioisotope-labeled compounds, see: G. Pascali, P. Watts and P. A. Salvadori, *Nucl. Med. Biol.*, 2013, **40**, 776–787.
43. P. W. Miller, H. Audrain, D. Bender, A. J. deMello, A. D. Gee, N. J. Long and R. Vilar, *Chem. Eur. J.*, 2011, **17**, 460–463.
44. E. R. Murphy, J. R. Martinelli, N. Zaborenko, S. L. Buchwald and K. F. Jensen, *Angew. Chem. Int. Ed.*, 2007, **46**, 1734–1737.
45. H. Kawanami, K. Matsushima, M. Sato and Y. Ikushima, *Angew. Chem. Int. Ed.*, 2007, **46**, 5129–5132.
46. R. Javaid, H. Kawanami, M. Chatterjee, T. Ishizaka, A. Suzuki and T. M. Suzuki, *Chem. Eng. J.*, 2011, **167**, 431–435.
47. L.-M. Tan, Z.-Y. Sem, W.-Y. Chong, X.-Q. Liu, Hendra, W. L. Kwan and C.-L. Ken Lee, *Org. Lett.*, 2013, **15**, 65–67.
48. M. S. Placzek, J. M. Chmielecki, C. Houghton, A. Calder, C. Wiles and G. B. Jones, *J. Flow Chem.*, 2013, **3**, 46–50.
49. N. T. S. Phan, D. H. Brown and P. Styring, *Green Chem.*, 2004, **6**, 526–532.
50. A. Nagaki, A. Kenmoku, Y. Moriwaki, A. Hayashi and J.-i. Yoshida, *Angew. Chem. Int. Ed.*, 2010, **49**, 7543–7547.
51. G. R. Yang, G. Bae, J. Choe, S. Lee and K. H. Song, *Bull. Korean Chem. Soc.*, 2010, **31**, 250–252.
52. B. Egle, J. de, M. Munoz, N. Alonso, W. M. De Borggraeve, A. de la Hoz, A. Diaz-Ortiz and J. Alcazar, *J. Flow Chem.*, 2014, **4**, 22–25.
53. P. P. Lange, L. J. Goossen, P. Podmore, T. Underwood and N. Sciammetta, *Chem. Commun.*, 2011, **47**, 3628–3630.
54. H. J. Lee, K. Park, G. Bae, J. Choe, K. H. Song and S. Lee, *Tetrahedron Lett.*, 2011, **52**, 5064–5067.
55. P. Li and S. L. Buchwald, *Angew. Chem. Int. Ed.*, 2011, **50**, 6396–6400.

56. L. Zhang, M. Geng, P. Teng, D. Zhao, X. Lu and J.-X. Li, *Ultrason. Sonochem.*, 2012, **19**, 250–256.
57. M. Chen and S. L. Buchwald, *Angew. Chem. Int. Ed.*, 2013, **52**, 11628–11631.
58. J. R. Naber and S. L. Buchwald, *Angew. Chem. Int. Ed.*, 2010, **49**, 9469–9474.
59. T. Noël, J. R. Naber, R. L. Hartman, J. P. McMullen, K. F. Jensen and S. L. Buchwald, *Chem. Sci.*, 2011, **2**, 287–290.
60. S. Kuhn, T. Noël, L. Gu, P. L. Heider and K. F. Jensen, *Lab Chip*, 2011, **11**, 2488–2492.
61. A. DeAngelis, D.-H. Wang and S. L. Buchwald, *Angew. Chem. Int. Ed.*, 2013, **52**, 3434–3437.
62. M. Chen and S. L. Buchwald, *Angew. Chem. Int. Ed.*, 2013, **52**, 4247–4250.
63. J. Hartwig, S. Ceylan, L. Kupracz, L. Coutable and A. Kirschning, *Angew. Chem. Int. Ed.*, 2013, **52**, 9813–9817.
64. Y. Zhang, T. F. Jamison, S. Patel and N. Mainolfi, *Org. Lett.*, 2011, **13**, 280–283.
65. A. Pommella, G. Tomaiuolo, A. Chartoire, S. Caserta, G. Toscano, S. P. Nolan and S. Guido, *Chem. Eng. J.*, 2013, **223**, 578–583.
66. M. D. Hopkin, I. R. Baxendale and S. V. Ley, *Chem. Commun.*, 2010, **46**, 2450–2452.
67. For a review on the use of immobilized reagents for the synthesis of natural products, see: M. Baumann, I. R. Baxendale and S. V. Ley, *Mol. Divers.*, 2011, **15**, 613–630.
68. G. S. Kumar, B. Pieber, K. R. Reddy and C. O. Kappe, *Chem. Eur. J.*, 2012, **18**, 6124–6128.
69. T. N. Glasnov and C. O. Kappe, *Chem. Eur. J.*, 2011, **17**, 11956–11968.
70. T. Liang, C. N. Neumann and T. Ritter, *Angew. Chem. Int. Ed.*, 2013, **52**, 8214–8264.
71. D. A. Watson, M. Su, G. Teverovskiy, Y. Zhang, J. Garcia-Fortanet, T. Kinzel and S. L. Buchwald, *Science*, 2009, **325**, 1661–1664.
72. T. Noël, T. J. Maimone and S. L. Buchwald, *Angew. Chem. Int. Ed.*, 2011, **50**, 8900–8903.
73. K. S. Elvira, X. Casadevall i Solvas, R. C. R. Wootton and A. J. deMello, *Nat. Chem.*, 2013, **5**, 905–915.

CHAPTER 14

Greener Approaches to Cross-Coupling

KEVIN H. SHAUGHNESSY

Department of Chemistry, University of Alabama, Box 870336, Tuscaloosa, AL 35487, USA
Email: kshaughn@as.ua.edu

14.1   Introduction

As society has become more aware of the environmental impact of chemical manufacturing, government regulation and public perception have caused the chemical industry to lessen its impact on our surroundings. The initial waves of environmental regulation focused on controlling chemical releases and cleaning hazardous waste sites. In 1990, the United States passed the Pollution Prevention Act, which changed the focus of environmental regulation from release control to source reduction. This change in philosophy caused chemists to consider how they design chemical processes to minimize the waste produced and the hazards of the final products, rather than focusing solely on how to deal with the hazardous materials produced. The concept of green chemistry, which is guided by a set of principles originally set out in 1998 by Anastas and Warner,[1] grew out of these efforts.

For the synthetic chemist, these principles provide a number of aspirational goals to strive to meet in the design of new processes. Atom economy should be maximized in reactions, while the use of unnecessary protection/deprotection steps should be minimized. Side products that cannot be avoided should be of limited toxicity. The use of solvents, which make up the major mass in most chemical processes, should be minimized. Alternatively,

RSC Catalysis Series No. 21
New Trends in Cross-Coupling: Theory and Applications
Edited by Thomas J Colacot
Published by the Royal Society of Chemistry, www.rsc.org

safe and renewable solvents should be used when required. Starting materials derived from renewable sources should be used if possible. Catalysts should be used to promote highly selective reactions under conditions that are not energy intensive.

As increasing attention is paid to the environmental impact of chemical processes, it is necessary to have metrics to evaluate these impacts.[2] Lifecycle assessment (LCA) is the most rigorous approach to evaluating fully the overall impact of a process.[3] LCA is a cradle-to-grave assessment that considers all facets of a process from the gathering of raw materials through the products' end of life when all materials are returned to the Earth. An LCA is an expensive and time-consuming task, which has led to the development of simpler approaches to assess the environmental impact of a process. The E-factor is often used to evaluate chemical processes as it can be easily calculated.[4] The E-factor is the ratio of the mass of waste to the mass of desired product produced in a process. The amount of waste is defined as the mass of all materials used in the process with the exception of water, which is usually not considered. Typical E-factors increase for smaller scale producers carrying out more complex processes, such as the pharmaceutical industry. Whereas bulk chemical producers have typical E-factors of <5 kg waste kg^{-1} product, the E-factor for the pharmaceutical industry is typically 25–100 kg waste kg^{-1} product. More recently the process mass intensity (PMI) has been proposed as an alternative to the E-factor.[5] The PMI is the total mass of materials used to produce a given mass of product. Unfortunately, these metrics are rarely applied in reports of more sustainable synthetic methods, so it is often difficult to evaluate the increased efficiency of these processes.[6]

As described throughout this book, palladium-catalyzed cross-coupling reactions are powerful methods for the formation of carbon–carbon and carbon–heteroatom bonds. As generally highly selective catalytic processes that can be run under relatively mild conditions, cross-coupling reactions adhere to many of the principles of green chemistry. Many opportunities to reduce the environmental impact of cross-coupling reactions remain. As a substitution reaction, cross-couplings have an inherent lack of atom economy (Scheme 14.1). Although the substitution by-product cannot be completely avoided, its environmental impact can be minimized. Chloride produces less waste on a per mass basis (lower E-factor) than iodide or bromide. Direct C–H activation avoids the use of halides altogether. In cross-coupling of organometallic nucleophiles, the metal by-product should be chosen to be of low toxicity. For example, Stille couplings use toxic organostannanes. This reaction has largely been supplanted by Suzuki coupling, which uses organoboron compounds that are much less toxic. Again,

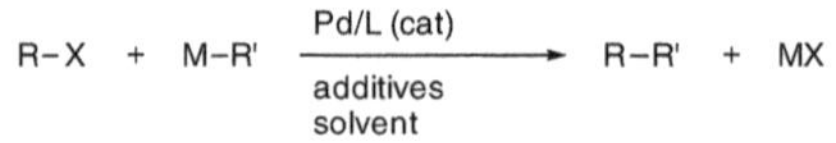

$$R-X \quad + \quad M-R' \quad \xrightarrow[\substack{\text{additives} \\ \text{solvent}}]{\text{Pd/L (cat)}} \quad R-R' \quad + \quad MX$$

Scheme 14.1

avoiding the metallic reagent altogether through the use of a C–H activation process would be most desirable.

Lifecycle analyses of pharmaceutical processes have shown that volatile organic solvents make up the bulk of waste that is produced.[7] Much of the effort to decrease the environmental impact of cross-coupling reactions has focused on using more benign solvents, such as water. Systems that avoid the use of stoichiometric additives, such as the base required in many cross-coupling reactions, can also lower the E-factor of these processes. The catalyst also provides an opportunity to enhance the sustainability of these processes. Palladium is one of the least abundant elements in the Earth's crust. Strategies to minimize the use of palladium by lowering the required catalyst loading or efficiently recycling the catalyst would improve the sustainability of its use as a catalyst. Alternatively, replacing palladium with more Earth-abundant metals, such as copper, nickel or iron, would be desirable.

In this chapter, various approaches to improving the "green-ness" of cross-coupling reactions are highlighted. The majority of this work has been focused on the use of alternative solvents, particularly water. Efforts to use new types of substrates in order to decrease the amount or toxicity of the reaction by-products are also highlighted.

14.2 Metal-Catalyzed Cross-Couplings in Aqueous Media

Water as an alternative solvent for organic chemistry has attracted significant attention.[8] Water is a non-toxic, non-flammable, relatively cheap and renewable solvent, in contrast to typical organic solvents that are potentially toxic, flammable and non-renewable. Caution should be taken when considering water as a green solvent, however. Once water has come into contact with organic compounds, it is classified as hazardous waste and cannot be returned to the environment without treatment. Aqueous waste can be more problematic than organic waste, because it often cannot be incinerated. Rather, it must be treated to remove contaminants or stored. Although water may not be a perfect green solvent, it does offer significant advantages over organic solvents in terms of safety and sustainability.

The use of water as a solvent can also simplify product purification, which can have a significant impact on waste production for a process. An aqueous-biphasic solvent system can allow the simple separation of organic products from inorganic by-products. If a water-soluble catalyst is used, that can also be retained in the aqueous layer. This simplified catalyst separation can overcome the significant challenge of separating homogeneous metal catalysts from organic products.[9] Furthermore, this approach provides the opportunity to recycle the aqueous catalyst solution, thus minimizing the amount of water and catalyst that must be used in the process. This concept was commercialized in the Rhône-Poulenc aqueous-phase hydroformylation of propene in the 1970s.[10] Significant effort has been devoted to the development of aqueous-phase metal-catalyzed cross-coupling reactions.

14.2.1   Hydrophilic Metal–Ligand Complexes for Aqueous-Phase Cross-Coupling

In order to use water to separate homogeneous catalysts from organic product streams in an aqueous-biphasic solvent system, it is necessary to design a water-soluble catalyst. In many cases, this is accomplished by using a hydrophilic supporting ligand for the metal catalyst.[11] A large variety of hydrophilic ligands have been applied to palladium-catalyzed cross-coupling reactions. The majority of examples involve modification of known privileged ligand structures by incorporation of a water-solubilizing group. Key examples and their catalytic applications are highlighted here.

14.2.1.1   *Pd-Catalyzed Cross-Coupling with Hydrophilic Phosphines*

The first example of palladium-catalyzed cross-coupling using a water-soluble ligand was reported by Casalnuovo and Calabrese.[12] They showed that Pd(TPPMS)$_3$ [TPPMS = sodium diphenyl(3-sulfonatophenyl)phosphine] provided an effective catalyst for Suzuki, Heck and Sonogashira coupling of hydrophilic and hydrophobic aryl iodides and bromides in a water–acetonitrile solvent system. Building on this seminal report, palladium-catalyzed cross-coupling using triphenylphosphine derivatives modified with sulfonate,[13] carboxylate,[14] guanidinium[15] and carbohydrate[16] water-solubilizing groups have been reported (Scheme 14.2). These ligands provide effective catalysts for Suzuki, Heck and Sonogashira couplings of aryl iodides and in some cases aryl bromides at elevated temperatures. Deactivated aryl bromides and chlorides are largely unreactive with these catalyst systems. Typically high catalyst loadings and high reaction temperatures are necessary, which make these systems impractical for large-scale application.

Increased ligand steric demand can improve catalyst performance in cross-coupling reactions. Sulfonation of tri(2,4-dimethylphenyl)phosphine provides trisodium tri(2,4-dimethyl-5-sulfonatophenyl)phosphine (TXPTS, Scheme 14.3) as a more hindered analog of TPPTS.[17] The TXPTS ligand provides more active catalysts than TPPTS for Suzuki, Heck and Sonogashira couplings.[18] Hydrophilic bidentate ligands have attracted limited attention in palladium-catalyzed cross-coupling. A sulfonated BINAS/Pd catalyst system was the first to catalyze the coupling of an aryl halide with amines in an aqueous solvent system (Scheme 14.4).[19]

Polyionic dendritic triphenylphosphine derivatives have been shown to provide high-activity palladium catalysts for cross-coupling reactions. Carboxylate-terminated dendritic phosphine **4** was shown to provide higher yields in Suzuki coupling of 4-iodotoluene with higher generation number with water as the only solvent (Scheme 14.5).[20] The hydrophobic dendrimer core may serve like a micelle, increasing interaction between the hydrophobic substrate and the catalyst. Hexacationic quaternary ammonium-substituted triphenylphosphine **5** provided a highly active catalyst for Suzuki

Scheme 14.2

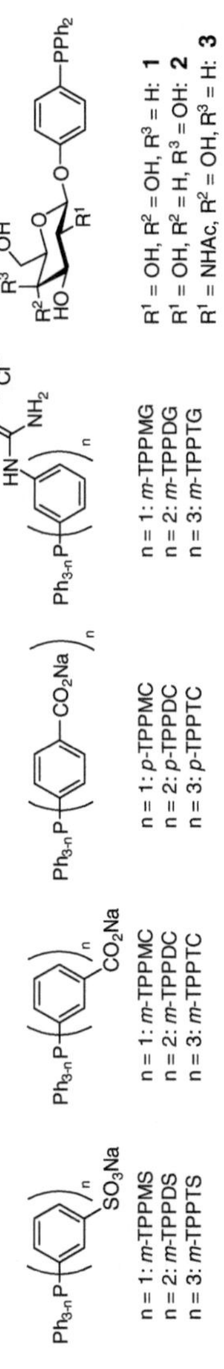

Scheme 14.3

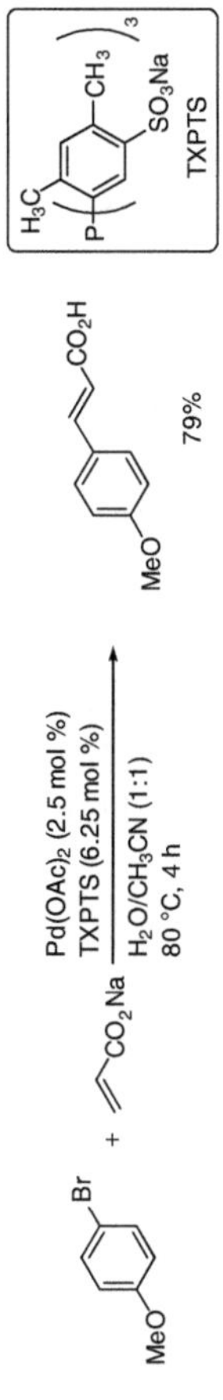

Scheme 14.4

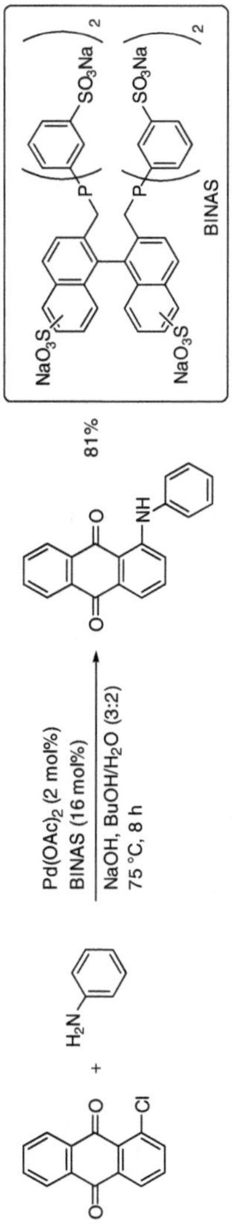

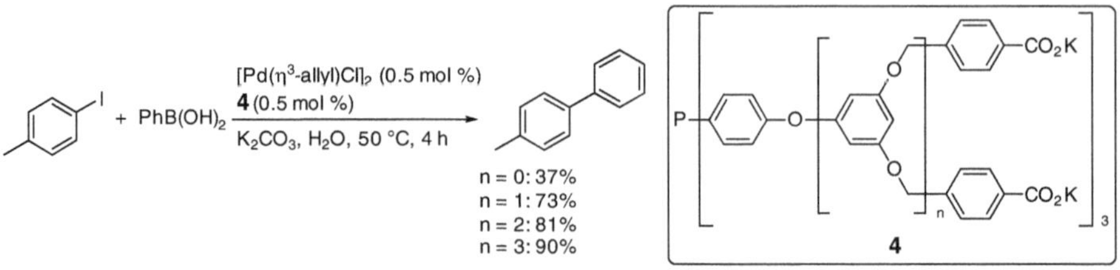

Scheme 14.5

couplings of aryl bromides in 10% aqueous methanol (Scheme 14.6). Complete conversion could be achieved in 1 h using 0.01 mol% palladium at 65 °C.[21] With this system, higher generation numbers gave higher rates with 4-chloronitrobenzene, but no generation effect was seen with 4-bromotoluene.[22]

Poly(ethylene glycol) (PEG) is an interesting water-solubilizing group. The resulting non-ionic phosphines are often soluble in moderately polar organic solvents, which facilitates synthesis and purification of the ligands. They are still hydrophilic enough to partition strongly into water in an aqueous-biphasic system. In addition, PEG shows temperature-dependent partitioning between water and organic phases. At low temperature, PEG-modified phosphines partition into the water layer, whereas the PEG-modified ligand partitions into the organic phase above its cloud point. Using a diphenylphosphine-terminated PEG oligomer $(n = 22)$ with a cloud point of 74 °C, little activity is seen below 72 °C.[23] At 74 °C, the catalyst activity increases dramatically. This ligand also provides recyclable catalysts for Suzuki couplings in water.[24]

Starting in the mid-1990s, a number of researchers found that sterically demanding and strongly electron-donating alkylphosphine ligands, such as tri-*tert*-butylphosphine, provide much more active and general catalysts for cross-coupling of aryl halides than triarylphosphines. Building on these advances, hydrophilic, sterically demanding alkylphosphine ligands have been prepared to give more active catalysts than TPPTS and related triarylphosphines. The first example of a ligand with these properties was *t*-Bu-Amphos, which provided the initial report of a room temperature Suzuki coupling of aryl halides in an aqueous solvent.[25] Sonogashira and Heck couplings can also be performed under milder conditions than are possible with TPPTS-derived catalysts.[26] The Pd/*t*-Bu-Amphos catalyst can be used for three reaction cycles using Pd(OAc)$_2$ as the palladium source, but up to 10 cycles are possible with a hydrophilic palladacycle precursor.[27] The *t*-Bu-Amphos/Pd catalyst has been applied to the Suzuki coupling of 7-bromo- and -iodocyclopenta[*d*][1,2]oxazines (**6**, Scheme 14.7).[28]

Although *t*-Bu-Amphos provides highly active catalysts for coupling of aryl bromides, aryl chlorides were largely unreactive. The cationic ammonium group decreases the electron-donating ability of *t*-Bu-Amphos compared with tri-*tert*-butylphosphine.[29] The DTBPPS ligand incorporates an anionic

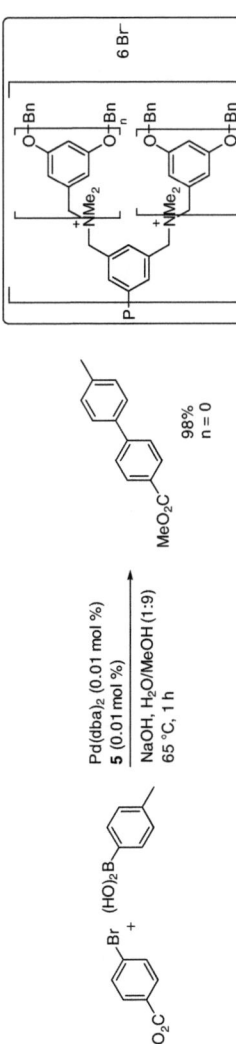

Scheme 14.6

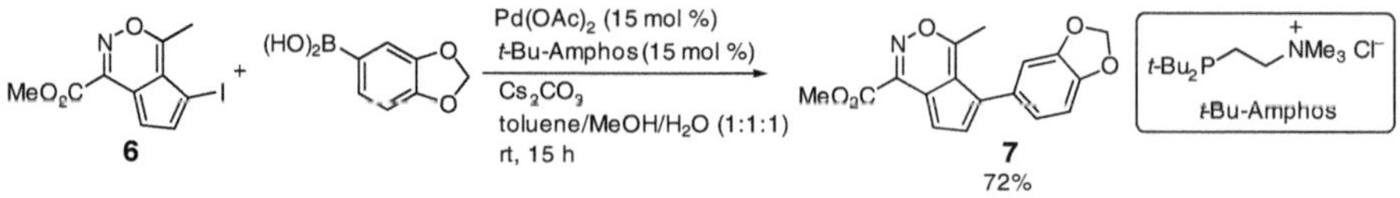

Scheme 14.7

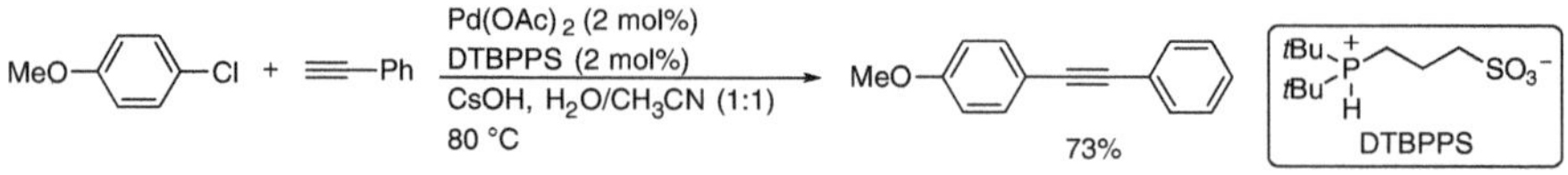

Scheme 14.8

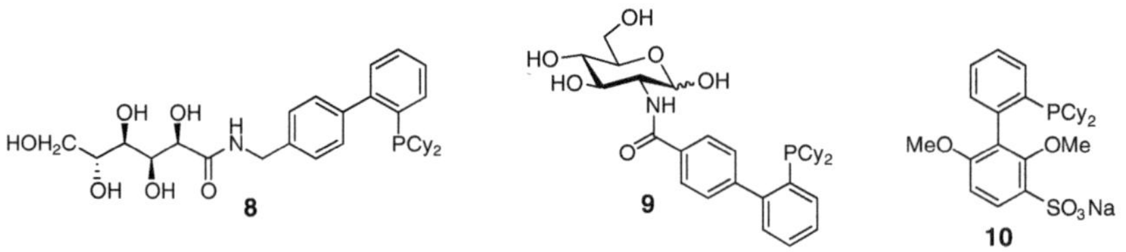

Scheme 14.9

sulfonate group to provide water solubility.[30] Computational studies showed that the ionization potential (IP) of Pd(DTBPPS) was lower than that of Pd(P*t*-Bu$_3$), whereas Pd(*t*-Bu-Amphos) had a higher IP. The anionic DTBPPS provides effective catalysts for the Suzuki and Sonogashira coupling of aryl bromides and chlorides (Scheme 14.8).

The dialkyl(2-biphenyl)phosphines developed by Buchwald and co-workers are highly effective ligands for a variety of cross-coupling reactions. Carbohydrate-modified 2-(dicyclohexylphosphino)biphenyl analogs with gluconamide (8)[31] and glucosamine (9)[32] moieties were synthesized and applied to Suzuki couplings of aryl bromides and chlorides in water (Scheme 14.9). Anderson and Buchwald reported the synthesis of a water-soluble ligand by sulfonation of SPhos.[33] The sulfonated SPhos (10) provides an effective catalyst for the Sonogashira coupling of aryl chlorides in water–acetonitrile. Plenio's group developed a highly water-soluble 9-dicyclohexyl-fluorenyl ligand (11) that provides active palladium catalysts for Suzuki and Sonogashira couplings of aryl and heteroaryl bromides and chlorides (Scheme 14.10).[34]

14.2.1.2 Palladium-Catalyzed Cross-Coupling with Hydrophilic NHC Ligands

Stable carbenes, such as 2-imidazolylidenes and 2-imidazolinylidenes are highly effective ligands for palladium-catalyzed cross-coupling that provide similar steric and electronic properties to electron-rich trialkylphosphines.[35]

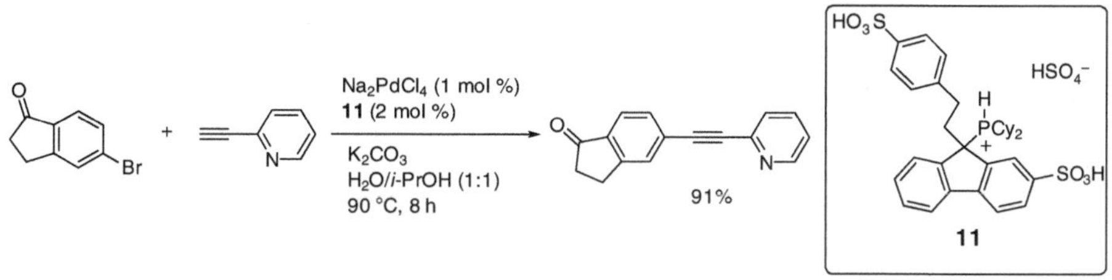

Scheme 14.10

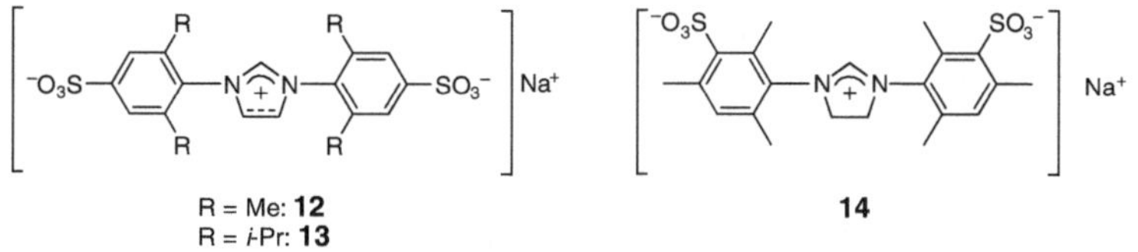

R = Me: **12**
R = *i*-Pr: **13**

14

Scheme 14.11

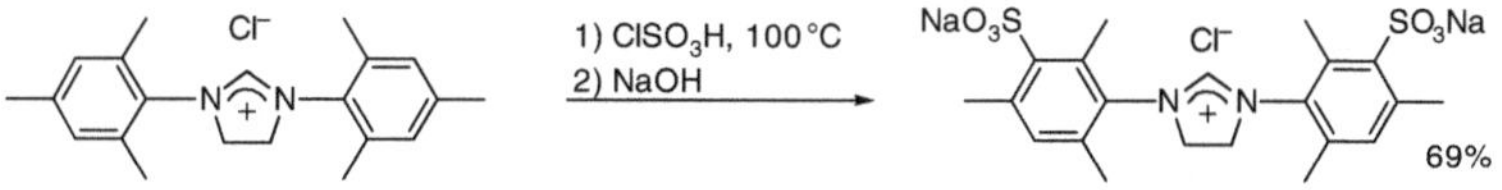

1) ClSO₃H, 100 °C
2) NaOH

69%

Scheme 14.12

15

16

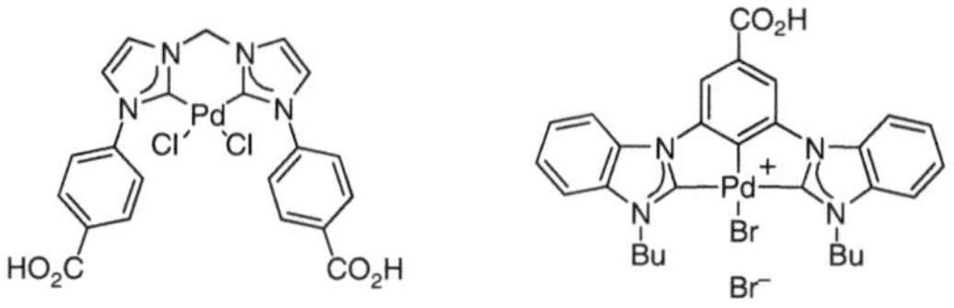

Scheme 14.13

There are fewer examples of water-soluble *N*-heterocyclic carbene (NHC) derivatives and their use in aqueous solvents. In part, this is due to the high basicity of the free carbene. Metal–NHC complexes are typically stable in water, which allows the use of Pd–NHC complexes in aqueous-phase catalysis. Sulfonated analogs of Mes and IPr imidazol(in)ium (**12–14**) were prepared starting from sulfonated aniline derivatives (Scheme 14.11).[36] The protonated NHC salts in combination with Na₂PdCl₂ gave effective catalysts for the Suzuki coupling of aryl chlorides in water at 100 °C. Imidazol(in)ium precursors of NHCs are sufficiently stable to tolerate electrophilic sulfonation, which provides an alternative method to the synthesis of sulfonated NHC precursors (Scheme 14.12).[37]

The carboxylate-substituted bis(NHC)–palladium complex **15** (Scheme 14.13) provides an active catalyst for the Suzuki coupling of aryl bromides in water at 100 °C.[38] The catalyst can be precipitated by addition of acid and reused in four cycles with a slow degradation of the reaction yield. Reusing the catalyst-containing aqueous layer gave improved catalyst recycling, particularly when TBAB was used as an additive. Very high catalyst productivity in the Suzuki coupling of aryl bromides was obtained with a palladium–pyridyl-bis(NHC) pincer complex (**16**).[39] Catalyst loadings as low as 8 ppm provided quantitative conversion of 4-bromoacetophenone to product. The aqueous catalyst solution could be used for three reaction cycles with no loss in activity. Attempted poisoning with mercury and polyvinylpyridine showed no effect on catalyst activity, which suggests that the active species is a molecular catalyst, rather than heterogeneous palladium particles. This catalyst effectively catalyzed the Heck coupling of aryl iodides (0.005 mol% Pd).[40]

NHC ligands supported on neutral hydrophilic polyether and polyol derivatives provide active catalysts for cross-coupling reactions in water. An *N*-PEG-substituted imidazolium salt in combination with Pd(OAc)$_2$ (0.5 mol%) afforded high yields in the Suzuki coupling of aryl bromides in 5–15 min.[41] Good yields were obtained in Suzuki couplings of aryl bromides using a bis(NHC)–palladium catalyst supported on hyperbranched polyglycerol (**17**, Scheme 14.14).[42] The catalyst could be separated by dialysis with approximately 80% recovery. The catalyst was successfully used for four reaction cycles with >97% yield, but further recycling was not possible owing to the loss of catalyst in each dialysis cycle.

14.2.1.3 Palladium-Catalyzed Cross-Coupling with Hydrophilic Nitrogen Ligands

Nitrogen-based ligands have received less attention than phosphines or NHCs in palladium-catalyzed cross-coupling reactions. The lower cost of nitrogen-based ligands compared with phosphines makes them attractive alternatives. Hydrophilic versions of nitrogen ligands have been reported to be effective in aqueous-phase cross-coupling reactions. 4,4'-Trimethylammonium-2,2'-bipyrdine (**18**) in combination with palladium provides active catalysts for Suzuki,[43] Hiyama[44] and Heck[45] couplings of aryl bromides and iodides (Scheme 14.15). Recycling of the aqueous catalyst solution was demonstrated for the Suzuki coupling of aryl bromides at 80 °C. Good yields were obtained for five reaction cycles, but the reaction time increased with each cycle, indicating a decrease in catalytic activity. Alkylammonium-substituted 2-(2-pyridyl)imidazole–palladium complex **19** (Scheme 14.16) catalyzes the Suzuki coupling of aryl bromides and chlorides in TBAB–water at low catalyst loading (0.1 mol%), but high temperature (120 °C).[46] TEM analysis showed that palladium nanoparticles form during the reaction, which the authors suggested are stabilized by the hydrophilic ligand.

Sulfonated aniline derivatives are readily available and provide easy access to sulfonated imine-based ligand systems. Excellent yields for Suzuki couplings were obtained using sulfonated β-ketoimine and β-diimine

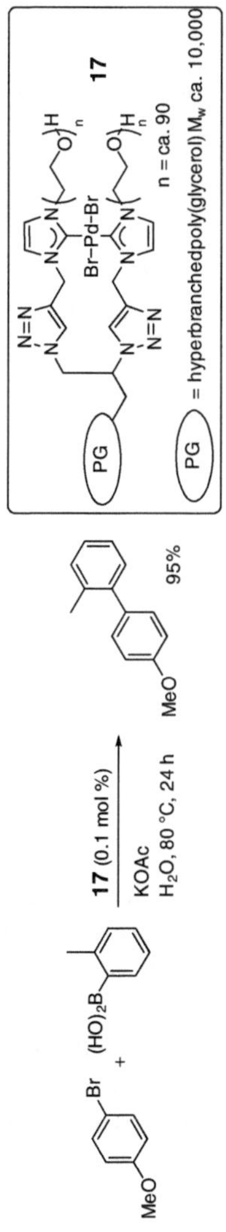

Scheme 14.14

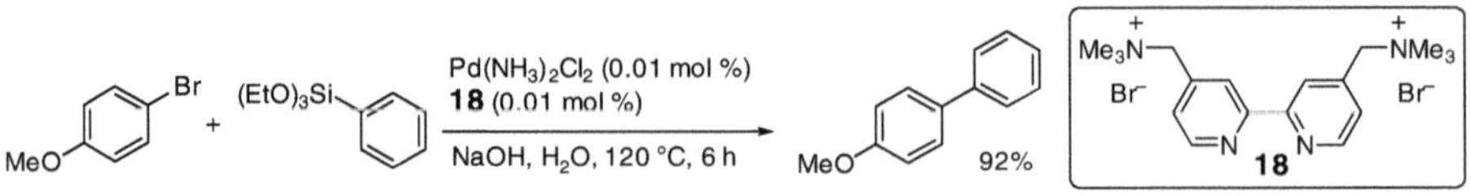

Scheme 14.15

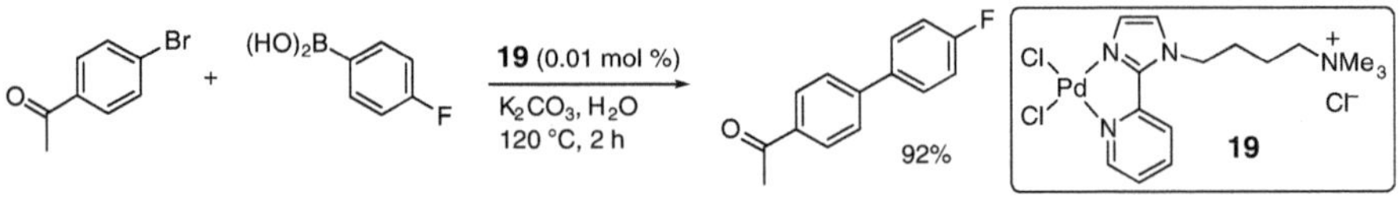

Scheme 14.16

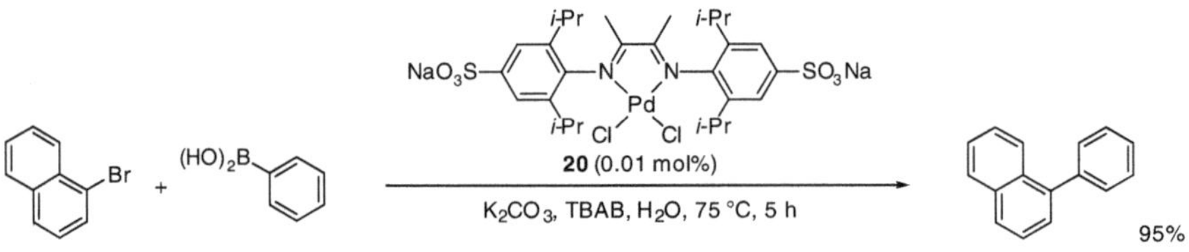

Scheme 14.17

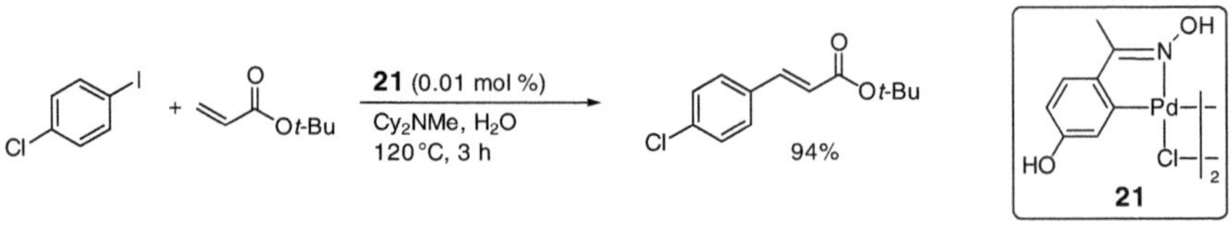

Scheme 14.18

ligands in combination with $PdCl_2$ (0.01 mol%) at 80 °C in water (Scheme 14.17).[47] Recycling of the catalyst was explored, but the conversion to product degraded rapidly after the second cycle.

EDTA is an inexpensive water-soluble ligand. Palladium (0.1 mol%) in combination with EDTA provides an effective catalyst for the Suzuki coupling of aryl iodides and bromides in water.[48] Tetrakis(2-hydroxyethyl)ethylenediamine complexed to palladium affords an active catalyst for aqueous-phase Suzuki coupling of aryl bromides.[49] Up to 100 000 turnovers were achieved with activated aryl bromides at 100 °C.

Oxime palladacycles were the first systems that allowed cross-coupling of aryl chlorides in aqueous solvent systems.[50] The palladium complex derived from 4'-hydroxyacetophenone oxime (**21**, Scheme 14.18) catalyzes the Suzuki coupling of aryl bromides and chlorides at 100 °C in water–TBAB in good yields.[51] This catalyst system is also effective in the Heck[52] and Hiyama coupling of aryl halides.[53] A related benzothiazole oxime palladacycle (**22**, Scheme 14.19) gave a highly active catalyst for the Suzuki and Heck

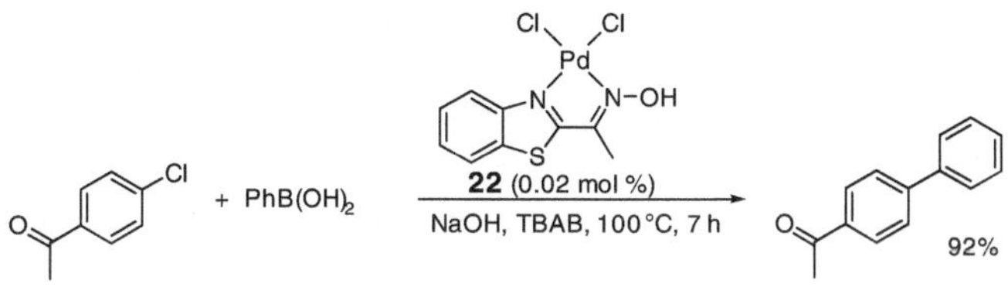

Scheme 14.19

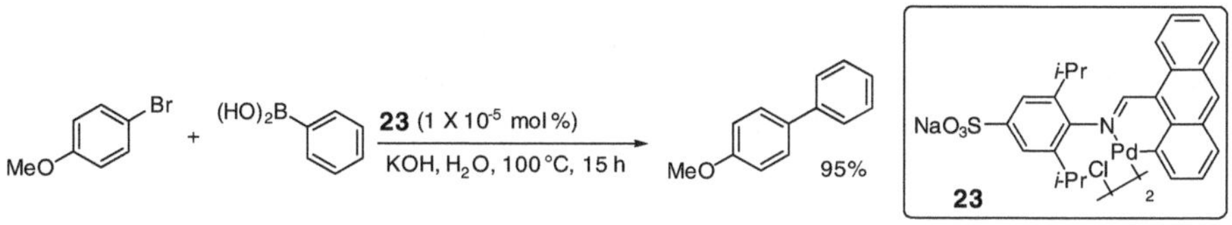

Scheme 14.20

couplings of aryl bromides and chlorides.[54] With 4-bromoacetophenone, over 100 000 turnovers with a turnover frequency (TOF) of 50 000 h^{-1} were achieved at 100 °C in water–TBAB, but the palladacycle could not be effectively recycled. In contrast, a polymer-supported version attached through an oxime ether could be used for four reaction cycles before seeing a decrease in conversion to product.

A recyclable catalyst was obtained from the cyclopalladated complex of anthracene-9-carbaldehyde imine containing a sulfonated aryl group on the imine nitrogen (**23**, Scheme 14.20). Up to 9 000 000 turnovers were achieved in the Suzuki coupling of 4-bromoanisole at 100 °C in 15 h.[55] The catalyst system could be used for up to six cycles, but the catalyst performance degraded, requiring longer reaction cycles. Analysis of the reaction mixture by TEM showed the formation of nanoparticles, which may be the true active species.

14.2.1.4 Copper- and Iron-Catalyzed Coupling in Aqueous Media

Although palladium has been the dominant metal used in cross-coupling reactions for the past several decades, there is a resurgence in interest in the use of the more Earth-abundant, and less costly, first-row transition metals, such as copper and iron.[56] Copper-catalyzed cross-coupling has a long history, going back to the early work of Ullmann and Goldberg.[57] The development of more active ligand-supported copper catalysts has promoted a resurgence of interest in copper.[58] With this renewed interest in alternative cross-coupling catalysts, examples of their application in aqueous solvents have been reported. Because the ligands used in these reactions are often more hydrophilic than typical ligands used in palladium-catalyzed reactions, it is often not necessary to introduce additional hydrophilic groups.

Diamine ligands are effective promoters of copper-catalyzed C–N bond formation in organic solvents and water. CuI in combination with *trans-N,N'*-dimethyl-1,2-diaminocyclohexane (DMCDA) catalyzes the arylation of azoles, amides and amines in water under air at 80 °C. The catalyst can be recovered with the aqueous-phase and reused four times with minimal decrease in reaction yield, although the system is limited to aryl iodides.[59] Intra-molecular cyclization of bromobenzofuran **24** in water catalyzed by CuI in the presence of *N,N'*-1,2-dimethyl-1,2-diaminoethane (DMEDA), which was used as base and ligand, affords benzo[4,5]furo[3,2-*b*]indole **25** (Scheme 14.21).[60] The aqueous, copper-catalyzed system gave superior re-sults to copper-catalyzed coupling in toluene or palladium-catalyzed ring closure. Copper(I) iodide catalyzes the cyclization of 2-bromobenzyl ketone derivatives to give benzofurans in the presence of TMEDA as the base and ligand.[61] Thiolation of aryl iodides in water at 120 °C is catalyzed by CuCl in combination with *trans*-1,2-diaminocyclohexane (CDA).[62] The aqueous catalyst solution could be used for three reaction cycles before a decrease in catalyst activity was observed.

A series of substituted phenanthroline derivatives were explored as lig-ands for the copper-catalyzed coupling of azoles, amines and amides with aryl iodides in water with PEG-400 as an additive.[63] 4,7-Dipyrrolylphenan-throline (**26**) was identified as the optimal ligand (Scheme 14.22). Copper(II) chloride/phenanthroline (phen) catalyzes the diarylation of thiocyanate to give diaryl sulfides in water from aryl iodides and bromides at 130 °C (Scheme 14.23).[64] Cu$_2$O in combination with pyridine-2-aldoxime catalyzed the hydroxylation of aryl halides in water with TBAB as promoter.[65] Glyoxal

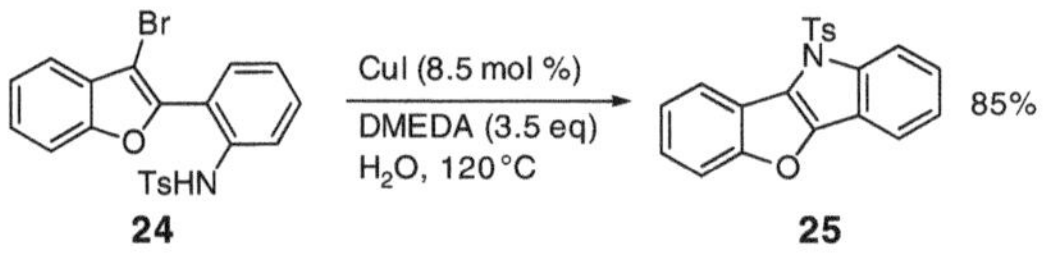

Scheme 14.21

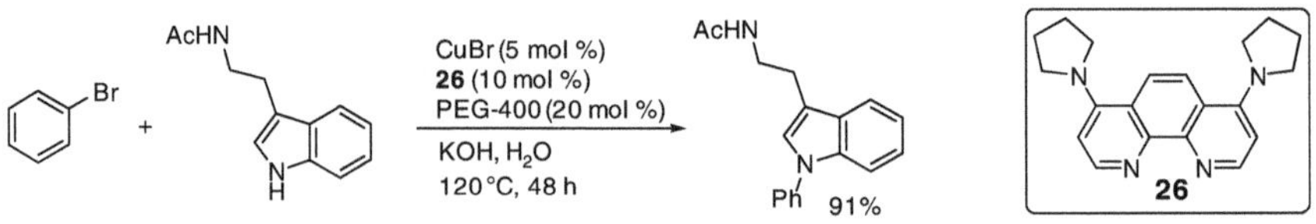

Scheme 14.22

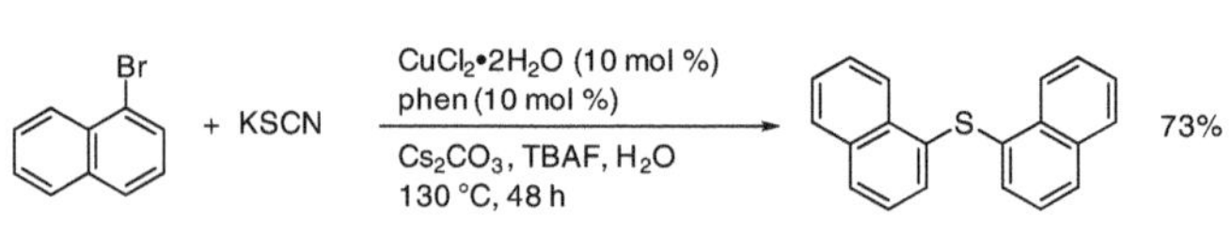

Scheme 14.23

bis(oxime) in combination with CuCl provides an effective catalyst for the arylation of azoles with aryl iodides and bromides in water at 100 °C.[66]

Amides and amino acids are a third class of privileged ligands for copper-catalyzed coupling reactions. Copper(II) oxide in combination with an oxalodihydrazide gives an effective catalyst for the selective monoarylation of ammonia in water–TBAB with a variety of aryl iodides and bromides.[67] Under microwave heating, this system catalyzes the coupling of aryl bromides and iodides with primary amines and anilines at 140 °C. Pyrrole-2-carbohydrazide is another effective ligand for the aqueous-phase copper-catalyzed arylation of amines.[68] The reaction can be accomplished within 5 min at 130 °C under microwave heating. Copper(I) iodide in combination with lithium picolinate has been applied to the hydroxylation of aryl iodides and bromides in water in the presence of TBAF at 130 °C.[69]

The water-soluble copper–salen complex 27 is a recyclable catalyst for the arylation of nitrogen nucleophiles in water. The catalyst promotes the arylation of azoles in water with TBAB at 100 °C (Scheme 14.24).[70] The catalyst solution was used for four reaction cycles with nearly identical yields for each cycle. Complex 27 is also effective for the arylation of primary aliphatic amines[71] and the selective monoarylation of ammonia in water.[72]

D-Glucosamine is an effective and environmentally benign ligand for the copper-catalyzed monoarylation of ammonia in water.[73] D-Glucosamine provides a more active catalyst than a range of other carbohydrates and more traditional ligands, such as phenanthroline or ethylenediamine. The Cu/D-glucosamine catalyst was also applied to the monoarylation of hydroxides to give phenols.[74]

Given the low cost and relatively low toxicity of iron, it is a desirable alternative to palladium or even copper as a catalyst. Recent research has shown that iron can be an effective catalyst,[75] although in some cases metal impurities appear to be the true active species.[76] Iron in combination with DMCDA[77] or DMEDA[78] has been shown to be an effective catalyst for the arylation of azoles and other nitrogen nucleophiles. By using the more hydrophilic ammonium-substituted bipyridine ligand 18, recycling of an iron catalyst was achieved in the arylation of thiols (Scheme 14.25).[79] Using

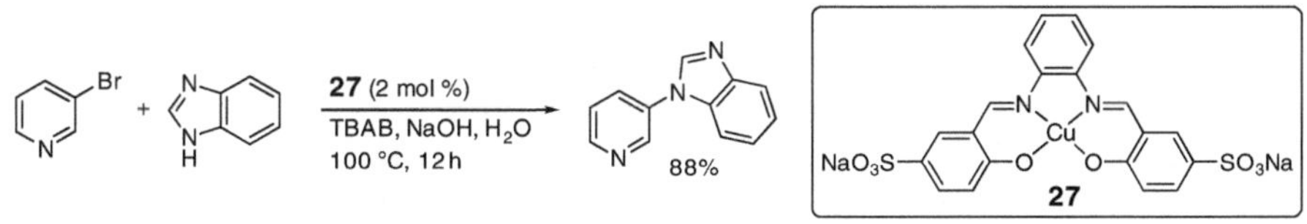

Scheme 14.24

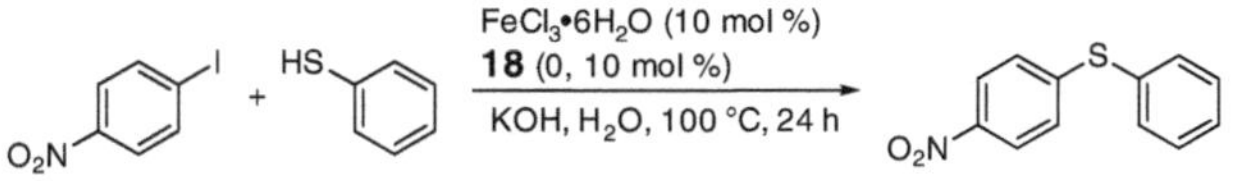

18	Cycle (% yield)					
	1	2	3	4	5	6
0 mol %	55	41	24	11	0	0
10 mol %	92	90	93	86	80	74

Scheme 14.25

the iron/**18** system, the reaction yield decreased from 92% to 74% over six reaction cycles, whereas in the absence of ligand a complete loss of activity occurred within five cycles.

Although the copper- and iron-catalyzed systems appear to offer the advantage of using a cheaper and more abundant metal than palladium, it should be noted that optimized palladium catalysts are typically much more active. Catalyst loadings for copper- and iron-catalyzed reactions are usually 5–10 mol%, whereas palladium loadings are often ≤1 mol%. Even with several orders of magnitude higher loadings, the cost of iron or copper catalysts can be less than that of palladium catalysts, however. From an environmental standpoint, the additional heavy metal may cause problems. In particular, the high loadings may make meeting residual metal specifications for pharmaceuticals challenging.

14.2.2 Hydrophobic Catalyst Systems

Much of the early effort in developing aqueous-phase cross-coupling reactions was focused on the design of hydrophilic catalysts to allow reactions to be carried out under biphasic conditions. Less attention has been devoted to the role that water as a unique reaction medium can play in reactions. In 1980, Rideout and Breslow noted that water accelerated Diels-Alder reactions of hydrophobic substrates.[80] They called this a hydrophobic effect and hypothesized that reactions with negative volumes of activation would be favored when carried out in water because the hydrophobic phase would want to minimize its interaction with water. This observation initially received limited attention, but over the past decade the ability of water to accelerate reactions, even when all reagents are insoluble in water, has attracted increasing attention.[81] The so-called on-water reactions often provide higher rates than homogeneous-phase reactions. The use of water can decrease the environmental impact of reactions compared with traditional organic-phase reactions. A disadvantage of on-water reactions is that the precious metal catalyst typically cannot be recycled.

14.2.2.1 Cross-Coupling Reactions on Water

Truly on-water cross-coupling reactions in which no surfactant or phase-transfer catalyst is used have been reported for a number of systems. Precatalysts $Pd(PPh_3)_4$ and $Pd(PPh_3)_2Cl_2$ effectively catalyze the Sonogashira coupling of aryl iodides and bromides on water using amine bases.[82] High-yielding, on-water Suzuki couplings of aryl and heteroaryl bromides can be achieved with low catalyst loadings (0.01–0.1 mol%) of $Pd(DPPF)Cl_2$.[83] Interestingly, with PEG as an additive, the catalyst could be retained in the aqueous phase and recycled, although the yield decreased from 91% to 80% over three cycles. A P_2N_2 ligand (**28**) in combination with $Pd(OAc)_2$ catalyzed the Suzuki coupling of aryl bromides on water with catalyst loadings as low as 0.001 mol% (Scheme 14.26).[84] Good yields were also obtained with

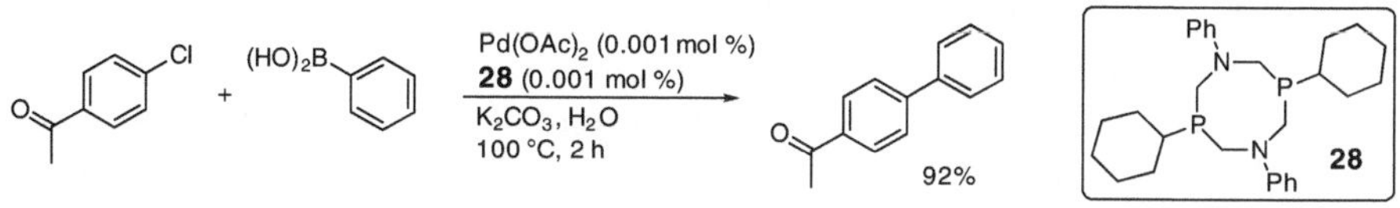

Scheme 14.26

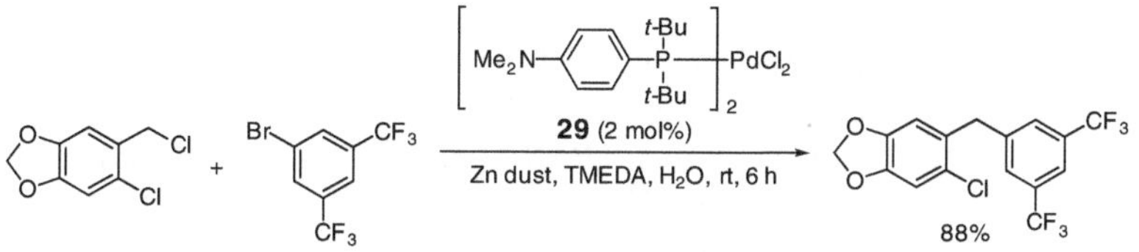

Scheme 14.27

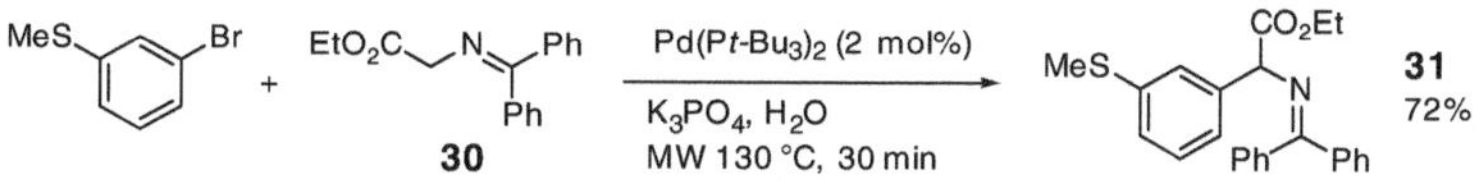

Scheme 14.28

activated aryl chlorides. Suzuki coupling of aryl chlorides was achieved on water using a Pd–IPr [IPr = 1,3-bis(2,6-diisopropylphenyl)imidazol-2-ylidine] precatalyst at 80 °C.[85]

The on-water protocol has been demonstrated for reactions that one would not normally expect to be tolerant of moisture. For example, the Negishi coupling of aryl halides with air- and moisture-sensitive organozinc compounds would not be expected to proceed with water as the reaction medium. Lipshutz and co-workers showed that under Barbier-type conditions, benzylic halides can be coupled with aryl, heteroaryl and vinyl halides using Pd/Amphos catalyst **29** on water at room temperature (Scheme 14.27).[86] The intermediacy of organozinc reagents was demonstrated by electrospray ionization mass spectrometric analysis of the reaction mixture.[87] Arylation of protected glycine ester **30**, with pK_a values too high for it to be significantly deprotonated in water, occurred efficiently on water with a Pd(Pt-Bu$_3$)$_2$ precatalyst (Scheme 14.28).[88] On-water Buchwald–Hartwig amination was achieved with a Pd$_2$(dba)$_3$/XPhos [XPhos = dicyclohexyl(2′,4′,6′-triisopropylbiphen-2-yl)phosphine] precatalyst system at 110 °C.[89]

14.2.2.2 Surfactant-Promoted Cross-Coupling Reactions

Although on-water reactions can occur efficiently in some cases, the poor interfacial mixing of water-suspended organic materials often results in poor yields and selectivities. A variety of phase-transfer agents have been explored to improve the interaction of hydrophobic materials in water. Quaternary ammonium ions, such as TBAB, are commonly used to promote reactions in water. The beneficial effect of quaternary ammonium salts was first noted in 1994 by Jeffery in studies of Heck couplings on water with hydrophobic catalyst systems.[90] Jefferey found that a range of tetrabutylammonium salts were effective promoters, whereas metal halide salts had no promoting effect.

TBAB is an effective promoter of ligand-free cross-couplings in water. The tetraalkylammonium ion can act as a phase-transfer catalyst (PTC), whereas the halide counterion is thought to stabilize palladium nanoparticles. By avoiding the use of a ligand, the cost of the process is decreased and waste is reduced. Ligand-free Pd(OAc)$_2$-catalyzed Suzuki coupling of aryl bromides occurs more effectively in water–TBAB than in organic solvents in which the reaction is homogeneous.[91] By preforming the trihydroxyborate, a stoichiometric base is not required, further improving the efficiency of the reaction.[92] The Pd(OAc)$_2$ catalyst in water–TBAB has been applied to the synthesis of tropolone–catechol natural product precursors (**32**, Scheme 14.29).[93] This method provided more consistent reactions and easier purification than the traditional Pd/PPh$_3$ organic solvent protocol.

Adding a long alkyl chain to the ammonium salt appears to improve the performance of the PTC. Cetyltrimethylammonium bromide (CTAB) provides higher yields in the Pd/C-catalyzed Suzuki coupling of aryl iodides in water than does TBAB.[94] Similar results were seen in the Heck coupling of aryl iodides catalyzed by PdCl$_2$.[95] N,N-Didodecyl-N',N',N''N''-tetramethylguanidinium bromide affords higher yields for the Suzuki coupling of aryl bromides than TBAB or guanidinium ions with shorter alkyl chains.[96] The aqueous-phase catalyst could be used for five reaction runs in the Suzuki coupling of 4-bromoacetophenone at 60 °C for 2 h with no decrease in reaction yield.

TBAB also promotes cross-coupling on water using catalysts supported by hydrophobic ligands, such as triphenylphosphine.[97] Water–TBAB was used as the solvent for Sonogashira couplings of aryl iodides and bromides catalyzed by palladacycle **33** under microwave heating (Scheme 14.30).[98] Suzuki coupling of heteroaryl chlorides was performed using the

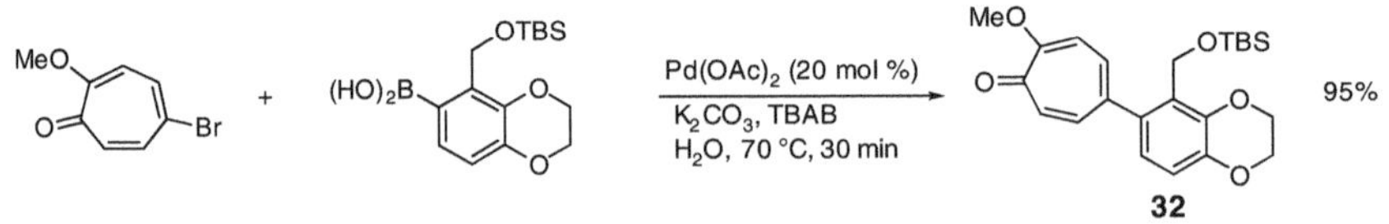

Scheme 14.29

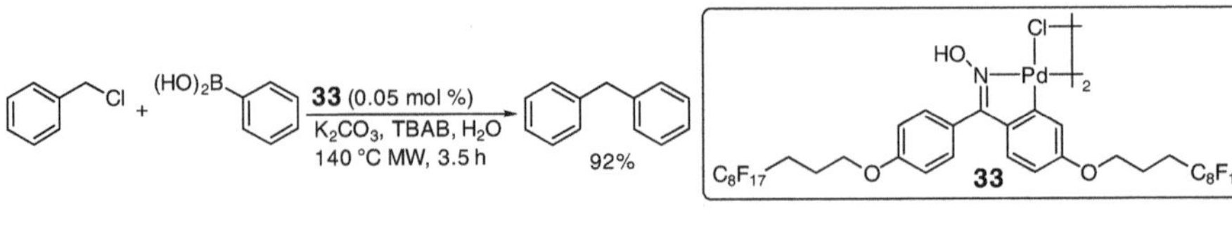

Scheme 14.30

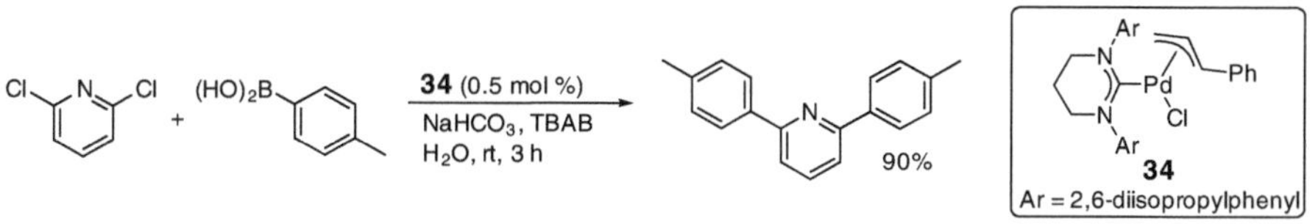

Scheme 14.31

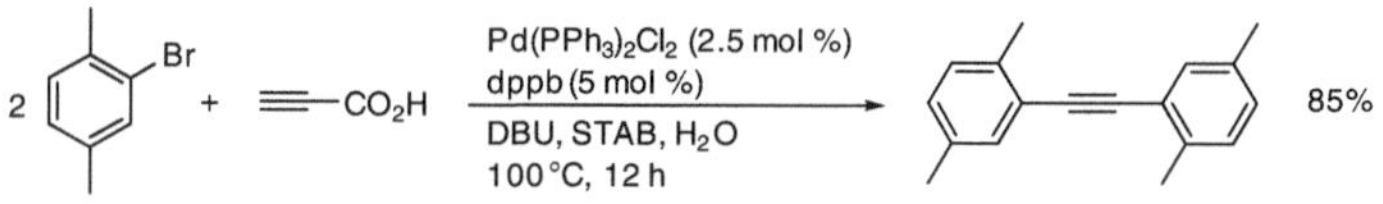

Scheme 14.32

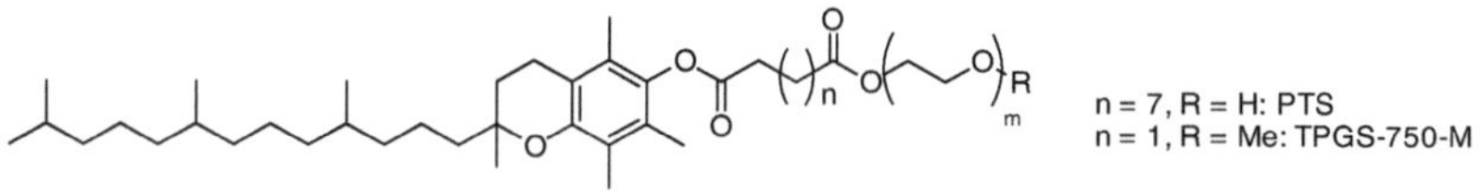

Scheme 14.33

palladium–NHC complex **34** in refluxing water with TBAB as promoter (Scheme 14.31).[99]

Stearyltrimethylammonium bromide (STAB) promotes the decarboxylative double Sonogashira coupling of aryl bromides and propiolic acid catalyzed by Pd(PPh$_3$)$_2$Cl$_2$/dppb on water (Scheme 14.32).[100] The hydrophobic catalyst gave better results than when the hydrophilic TPPTS ligand was used in place of dppb.

Anionic surfactants have also been applied to cross-coupling reactions on water, although less commonly than cationic surfactants. Sodium dodecyl sulfate (SDS) promotes the Hiyama coupling of aryl bromides and chlorides catalyzed by nanoparticles derived *in situ* from Na$_2$PdCl$_2$.[101] The same system was applied to the Suzuki coupling of aryl bromides[102] and the condensation of 2-iodophenols and alkynes to give benzofuran derivatives.[103]

Lipshutz and co-workers developed the nonionic surfactant PEG-α-tocopheryl sebacate (PTS, Scheme 14.33), which is a highly efficient surfactant for aqueous-phase palladium-catalyzed cross-coupling reactions. Suzuki,[104] Sonogashira,[105] Heck[106] and Miyaura borylation[107] couplings were all demonstrated in water–PTS (2 wt%) at room temperature using palladium/hydrophobic phosphine catalyst systems. Improved results for all reactions were achieved with the less expensive PEG-α-tocopheryl succinate (TPGS-750-M) surfactant.[108] The PTS and TPGS-750-M surfactants efficiently support the palladium complex and substrate in micelles to provide high reaction rates. The catalyst is constrained to the water–micelle phase, so catalyst separation and recycling are possible.

The micelle environment provides a hydrophobic reaction environment that allows reactions that normally are not favorable in water to occur

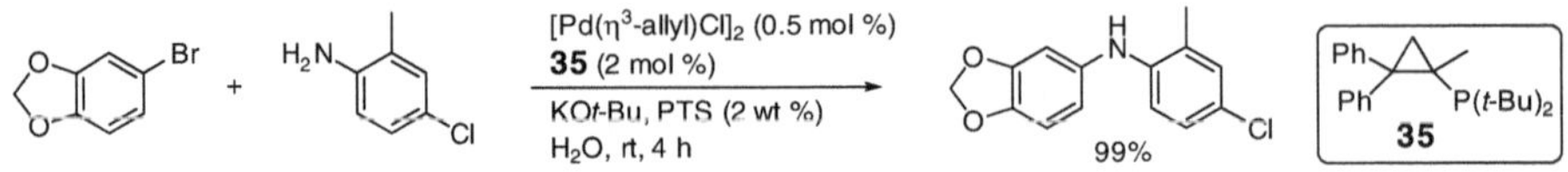

Scheme 14.34

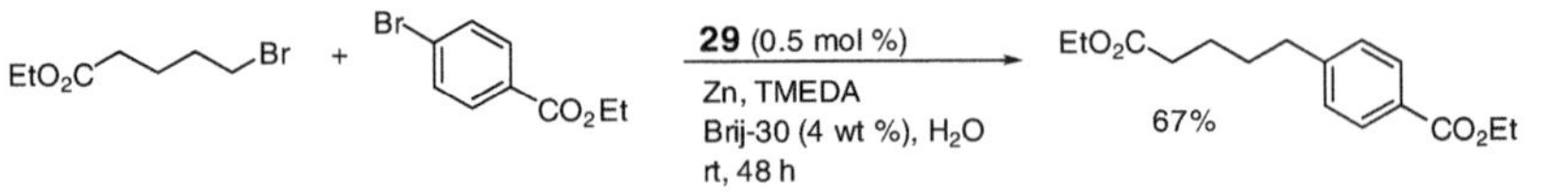

Scheme 14.35

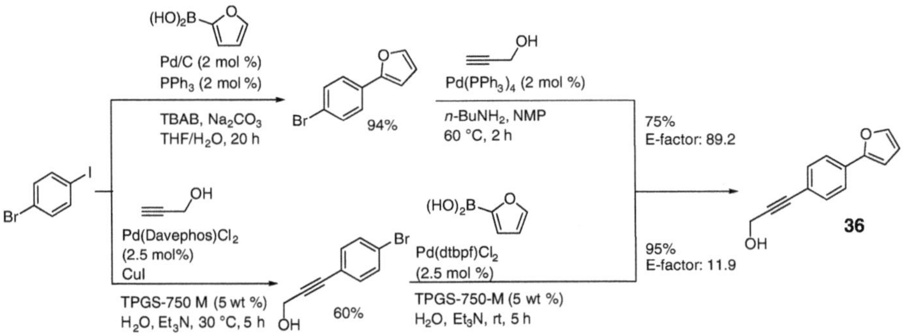

Scheme 14.36

efficiently. A few examples of Buchwald–Hartwig aminations of amines or enolates in aqueous reaction media have been reported. Amine arylation occurs efficiently using a Pd/cBRIDP catalyst system in water–PTS (Scheme 14.34).[109] Ketone α-arylation catalyzed by Pd/Pt-Bu₃ also proceeds efficiently in water–PTS.[110] More notably, Negishi couplings of aryl bromides and alkyl iodides, which react with zinc *in situ* to give organozinc reagents, are efficiently coupled using PTS on water.[111] Although benzyl halides can be coupled on water, less reactive alkyl iodides gave low yields in the absence of a surfactant. For less reactive alkyl bromides, Brij-30 (dodecyl tetraglyme) gave better results than PTS or TPGS-750-M (Scheme 14.35).[112] Brij-30 provides larger micelles than other nonionic surfactants. Larger micelles are thought to provide a larger payload of alkyl halide upon collision with heterogeneous zinc particles, which increases the rate of organozinc formation.

Comparison of aqueous-micellar catalysis with traditional organic-phase reactions show significant improvement in reaction efficiency for Suzuki, Heck and Sonogashira couplings.[6a] Researchers at Johnson and Johnson reported the synthesis of intermediate **36** by a sequential palladium-catalyzed cross-coupling approach under traditional organic solvent conditions (Scheme 14.36).[113] The reported process had an E-factor of 89.2 kg waste kg^{-1} product when including the aqueous workup. In contrast, an

E-factor of 11.9 kg waste kg^{-1} product was achieved by Lipshutz *et al.*[6a] using micellar catalysis conditions.

14.2.3 Heterogeneous Recyclable Catalysts

Heterogeneous catalysts are preferred for industrial-scale processes owing to their ease of separation from the product stream and the potential to carry out reactions in flow. The hydrophilic catalysts described above represent one approach to making a homogeneous catalyst heterogeneous as part of a liquid–liquid biphasic solvent system. Aqueous-phase reactions using solid heterogeneous catalyst systems have also received significant attention.[114] These catalyst systems can be broken down into two classes: nanoparticle palladium catalysts and molecular palladium species attached to insoluble support materials.

14.2.3.1 Nanoparticle Catalysts in Aqueous-Phase Cross-Coupling

Although much effort has been devoted to the design of optimal ligands to provide high-activity cross-coupling catalysts, the ligand adds to the cost of the system and increases the amount of waste material produced. Catalysts that do not rely on dative ligands, such as phosphines or NHCs, are cheaper and often more robust systems. Ligand-free catalyst systems can offer high productivity, although most such systems are less reactive than ligand-supported catalysts towards less reactive aryl electrophiles, such as aryl chlorides. Leadbeater and Marco demonstrated that Pd(OAc)$_2$ without other promoters is an effective catalyst for the Suzuki coupling of aryl bromides on water with low catalyst loading (0.4 mol%) at 150 °C with microwave heating.[115] Catalyst loadings as low as ppm levels gave high yields under these conditions.[116] By using acetone as a water-miscible organic co-solvent, the reaction temperature required could be lowered to 35 °C.[117]

Unsupported palladium nanoparticles can be highly effective catalysts, but often lose activity as the nanoparticles agglomerate to larger particles that precipitate from solution. As a result, unsupported nanoparticles often cannot be used for multiple reaction cycles. To increase the stability of the nanoparticles, they can be supported on polymeric or inorganic supports that serve to stabilize the nanoparticles and isolate them to avoid agglomeration.

A common approach to preparing supported nanoparticles is to use a polymer with coordinating sites where the palladium ions can coordinate and that can support the palladium nanoparticles. El-Sayed and co-workers reported an early example of palladium nanoparticles supported on poly(4-vinylpyridine) (PVP).[118] The Pd–PVP system effectively catalyzed the Suzuki coupling of aryl iodides in 40% aqueous ethanol. A highly effective catalyst system is obtained from palladium nanoparticles supported on polyaniline

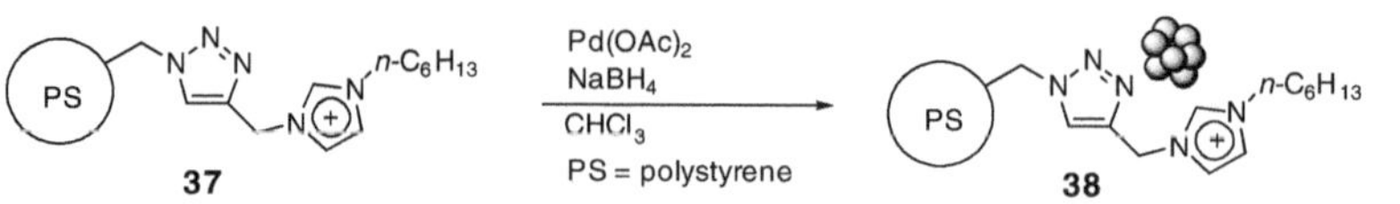

Scheme 14.37

(PANI).[119] The Pd–PANI catalyst system catalyzes the Suzuki coupling of aryl chlorides in water at 100 °C with 0.05 mol% palladium. Even aryl fluorides can be coupled under these conditions. The catalyst is recyclable for 10 reaction cycles with palladium leaching at the ppm level. The PANI material seems to be highly effective in preventing agglomeration of the active palladium particles. Palladium nanoparticle catalyst **38** supported on polystyrene modified with an imidazolium-substituted triazine (**37**) was effective in the Suzuki coupling of aryl bromides and showed no decrease in activity over five reaction cycles (Scheme 14.37).[120]

Polymers with anionic substituents can also effectively host palladium nanoparticles. Palladium nanoparticles supported on poly(4-styrenesulfonic acid-co-maleic anhydride) (PSSA-co-MA) were applied to the Suzuki coupling of aryl bromides in water.[121] The Pd-PSSA-co-MA catalyst system was used for six reaction cycles with the reaction yield decreasing from 96% to 86%. Palladium nanoparticles supported on poly(N-isopropylacrylamide-co-potassium methacrylate) hydrogel are effective catalysts for the Suzuki coupling of aryl iodides and bromides.[122] The hydrogel has a large water uptake capacity. When palladium(II) ions are added, they act as cross-linking agents and shrink the gel. Reduction of the palladium to give nanoparticles returns the water-swelling ability of the gel. The insoluble polymer could be recovered and reused for seven cycles with minimal decrease in activity. A core–shell nanosphere comprising a polystyrene core and a polyacrylate shell serves as a pH-responsive support for palladium nanoparticles.[123] Between pH 3 and 11, the shell is highly dispersed in water, allowing for efficient interaction between the catalyst and substrates. At low pH, the shell contracts and the particles can be separated by filtration. The catalyst was applied to the Suzuki coupling of aryl bromides and iodides and could be used for four reaction cycles, all of which gave quantitative yields.

Natural polymers are attractive and sustainable supports for catalysts as alternatives to synthetic polymers. In addition, biopolymers are water-compatible materials. Cellulose is a cheap and widely available plant-derived material. Palladium nanoparticles supported on bacterial cellulose prepared by fermentation of *Acetobacter xylinum* were highly active for Suzuki couplings of aryl halides, including chlorides, in water.[124] With a catalyst loading of 0.05 mol%, the catalyst gave high yields for four reaction cycles. The cellulose–palladium catalyst could also be used in the Heck coupling of aryl iodides in DMF. Chitosan [poly(β-D-glucosamine)]-supported nanoparticles were applied to the Suzuki coupling of aryl iodides and bromides in water at 150 °C under microwave heating.[125] The catalyst could be used for five reaction cycles with no decrease in product yield. Wool-supported palladium nanoparticles catalyzed the Heck coupling of aryl bromides in water with PEG-400 at 80 °C.[126] The catalyst was used for 10 reaction cycles, with a decrease in yield observed only in the tenth cycle. The palladium loading of the wool decreased from 12 to 11 wt% over the 10 cycles.

DNA binding protein from starved cells (DPS) from the thermophilic bacterium *Thermosynechococcus elongatus* (Te-DPS) serves as a support for palladium nanoparticles.[127] The DPS protein belongs to the ferritin family of metal chaperone proteins. The Pd–Te-DPS catalyst system (0.05 mol%) provides good yields in the Suzuki coupling of aryl iodides and bromides at 100 °C. The catalyst was used in a one-pot tandem reaction with an alcohol dehydrogenase enzyme to couple aryl iodides with 4-acetylphenylboronic acid followed by enantioselective enzymatic reduction of ketone **39** (Scheme 14.38). The biaryl alcohol product (**40**) was obtained in good yields (73–91%) and excellent enantioselectivity (>99%).

Palladium-on-carbon (Pd/C) is one of the earliest forms of heterogeneous catalysts developed. Although it has a long history in hydrogenation and hydrogenolysis catalysis, only recently has it been explored as a catalyst for cross-coupling reactions. Pd/C (0.05 mol% Pd) is an effective catalyst for the Suzuki coupling of aryl bromides with tetraarylborates in water (Scheme 14.39).[128] The Pd/C catalyst system can be recovered by filtration and used for up to five cycles with no decrease in reaction yield.[129] The filtrate from the Pd/C reactions shows no catalyst activity, which suggests that the active palladium remains on the carbon support. Pd/C gave better results in the aqueous-phase Suzuki coupling of aryl bromides and activated chlorides than palladium supported on metal oxide supports.[130] The hydrophobic carbon surface was proposed to attract the hydrophobic substrates. Palladium nanoparticles supported on amphiphilic carbon nanospheres prepared by hydrothermal treatment of glucose afforded good yields in the Suzuki coupling of aryl iodides, bromides and even unactivated chlorides.[131] The catalyst was used for four reaction cycles in the Suzuki

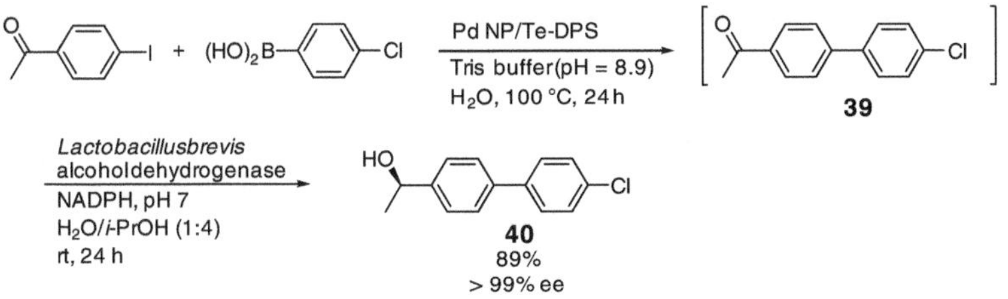

Scheme 14.38

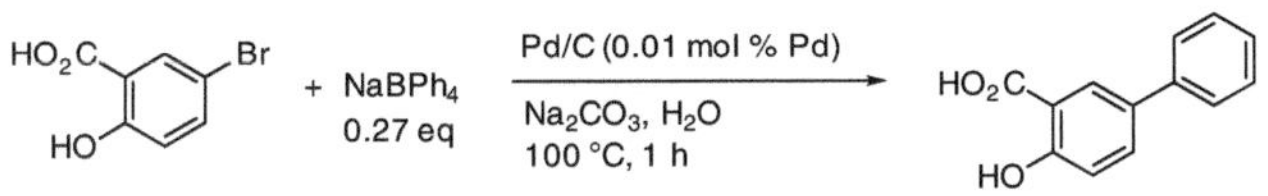

Scheme 14.39

arylation of 4-bromobenzaldehyde with the yield decreasing from 90% to 82%. The E-factors of Heck couplings of aryl iodides catalyzed by palladium on amphiphilic carbon nanospheres under a range of conditions were found to range between 4 and 6 kg waste kg^{-1} product.[6b] These E-factors were similar to those obtained for other heterogeneous palladium catalyst systems.

Most industrial heterogeneous catalyst systems are supported on inorganic supports, such as silica, alumina or zeolites. Palladium-exchanged NaY zeolite provides good activity for the Suzuki coupling of aryl iodides and bromides, but the catalyst activity rapidly degrades upon recycling.[132] Palladium nanoparticles obtained by reducing the Keggin-type polyoxometallate $K_5PPdW_{11}O_{39}$ were active for the Suzuki, Stille and Heck coupling of aryl bromides and activated aryl chlorides in water.[133] This system provides one of the few examples of a true aqueous-phase Buchwald–Hartwig amination reactions in the literature. Palladium nanoparticles supported on an MCM-41 surface modified with an alkylcyanosiloxane (43, Scheme 14.40) are effective catalysts for the Sonogashira coupling of aryl iodides in water.[134] The catalyst can be used for five reaction cycles with no decrease in reaction yield.

Metal–organic frameworks (MOF) are attractive supports because of their large surface area. The chromium terephthalate-based MOF MIL-101 was used as a support for palladium nanoparticles.[135] The MIL-101-supported palladium (0.9 mol%) catalyzed the Suzuki coupling of deactivated aryl chlorides in excellent yield at 80 °C in water–TBAB. The Pd–MIL-101 catalyst was also applied to the condensation of 2-iodoanilines with alkynes to give indoles in excellent yields (Scheme 14.41).[136] The MIL-101-supported catalyst could be used for 10 reaction cycles with minimal decrease in reaction yield, whereas an MCM-41-supported catalyst gave lower yields and showed a significant decrease in activity after six cycles.

Simple metal oxides can also serve as supports for palladium nanoparticles. Magnetic Fe_3O_4-supported palladium particles were applied to the Hiyama coupling of aryl bromides in water.[137] The catalyst could be

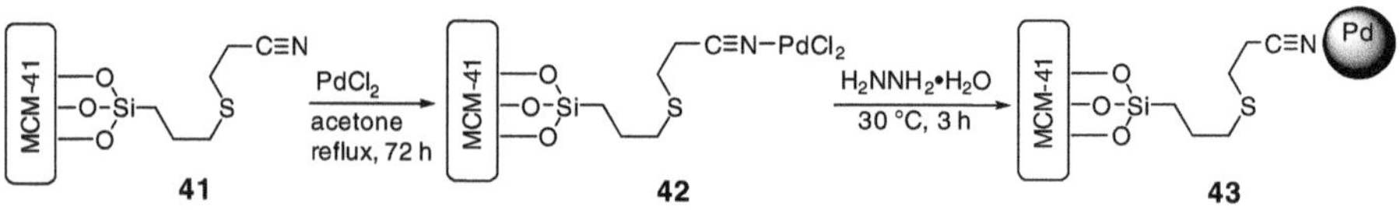

Scheme 14.40

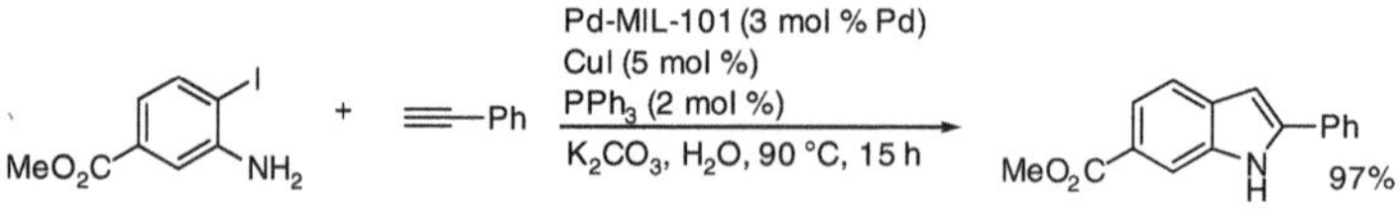

Scheme 14.41

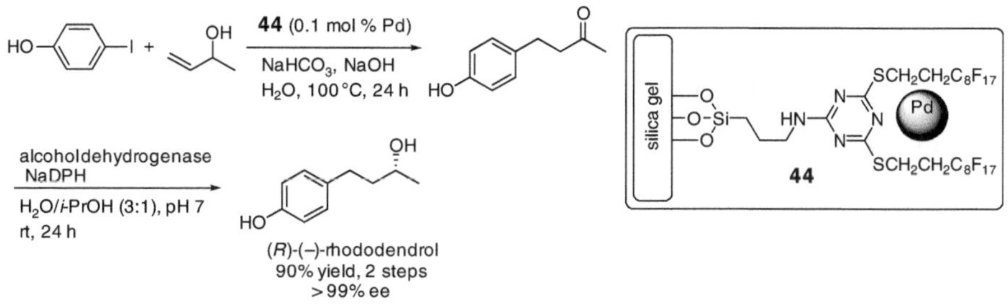

Scheme 14.42

recovered magnetically and used for five reaction cycles during which the yield decreased from 91% to 80%. Analysis of reaction filtrates by ICP-MS showed no palladium. Palladium nanoparticles supported on CeO_2 catalyze the Suzuki coupling of aryl bromides in water at room temperature.[138] The catalyst was very stable. After the reaction, it could be filtered, washed and dried in air and then reused. Over 11 reaction cycles, the yields ranged from 99 to 97% using 1 mol% palladium.

Surface-modified silica gel has been used to support palladium nano-particles in aqueous-phase cross-coupling reactions. Palladium nano-particles supported on a silica–starch composite are active catalysts for aqueous-phase Heck and Sonogashira couplings of aryl iodides, bromides and chlorides at 100 °C.[139] The supported catalyst could be used for six reaction cycles without a decrease in catalyst performance. Analysis of the recovered support by ICP-MS showed a decrease in palladium loading from 23.8 ppm prior to use to 22.1 ppm after the Sonogashira coupling. A triazine modified with fluorous alkyl tails was attached to silica gel co-valently through a trialkoxysilane substituent (**44**, Scheme 14.42). Palladium nanoparticles supported on the resulting materials were effective catalysts for the Sonogashira, Suzuki and Heck couplings of aryl iodides and bromides.[140] Up to 15 reaction cycles could be achieved using the supported catalysts (0.1 mol% Pd) with minimal decrease in reaction yield for both catalyst supports. The catalyst was applied to the tandem Heck coupling of 4-iodophenol and 3-buten-2-ol followed by enzymatic reduction of the resulting ketone to give (*R*)-(–)-rhododendrol in 90% yield and >99% *ee*.[141]

14.2.3.2 Supported Palladium Complexes

Supported palladium nanoparticles often provide highly recyclable catalysts for aqueous-phase cross-coupling reactions. There are few examples of catalysts that show high activity for deactivated aryl bromides and aryl chlorides, however. In order to combine the advantages of ligand-supported catalysts and heterogeneous systems, there has been significant interest in supported molecular catalysts in which the ligand is attached to an in-organic or polymeric support. The challenge in these systems is to maintain the high catalyst activity of the homogeneous system in the heterogeneous

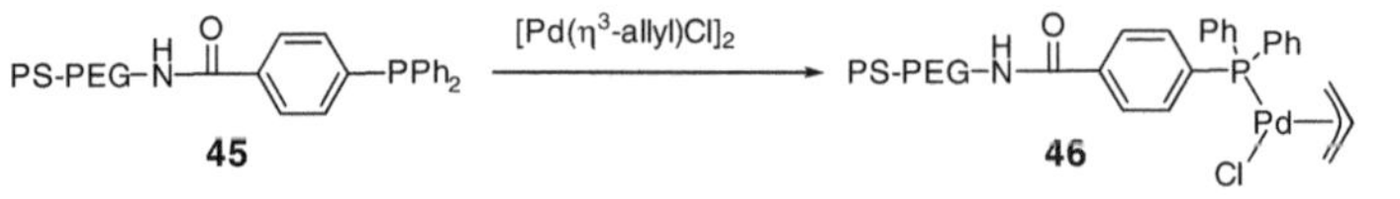

Scheme 14.43

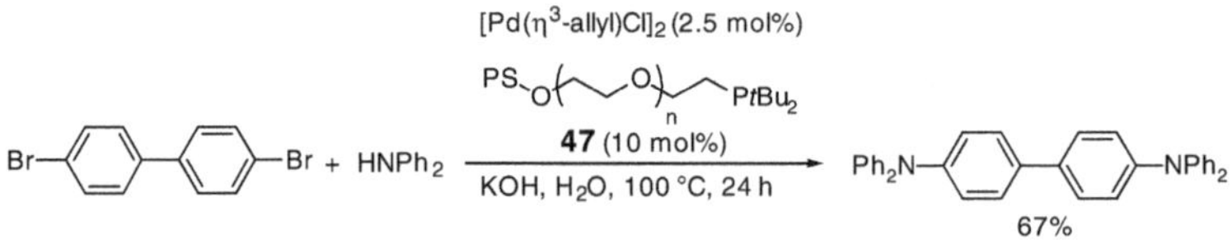

Scheme 14.44

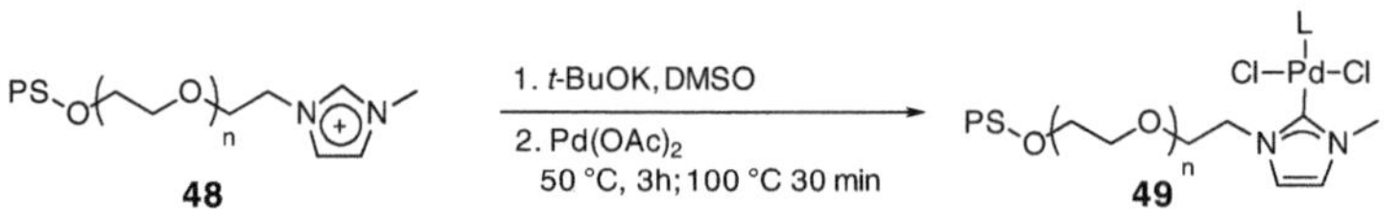

Scheme 14.45

analog. In addition, catalyst leaching due to metal decomplexation is often a significant problem in these systems.

Polymer-supported catalysts have attracted significant interest, with many of these systems using cross-linked polystyrene as the support. Because of the poor water-swelling properties of cross-linked polystyrene, it is often a poor support for aqueous-phase chemistry. Grafted polystyrene–PEG (PS–PEG) supports with a polystyrene core and a hydrophilic PEG shell provide good interaction of supported materials with water. PS–PEG-supported triphenylphosphine ligand **45** provides effective catalysts for the Suzuki coupling of aryl iodides and bromides and allylic acetates (Scheme 14.43).[142] The palladium is coordinated by mixing the heterogeneous ligand with [Pd(allyl)Cl]$_2$ to give supported complex **46**. Catalyst **46** is also effective for the Sonogashira coupling of aryl iodides.[143] Over five reaction cycles, the catalyst gave yields ranging from 99 to 87%.[144]

More effective ligands have also been attached to the PS–PEG support. A supported version of tri-*tert*-butylphosphine was prepared by appending a P*t*-Bu$_2$ unit to the end of the PEG chains (**47**, Scheme 14.44). The palladium complex of this supported ligand effectively catalyzed the coupling of aryl bromides with amines and thiols in water.[145] The reaction was effective in the synthesis of highly hydrophobic 4,4′-bis(diphenylamino)biphenyl. The PEG phase of the heterogeneous support likely provides a hydrophobic phase in which the reaction can occur efficiently. PS–PEG-supported Pd–NHC complex **49** has been applied to the Suzuki coupling of aryl iodides and bromides (Scheme 14.45).[146] The catalyst showed good recyclability with yields over five cycles ranging from 91 to 82%. Covalent attachment of the ligand does not appear to be required to achieve good recycling. Sulfonic acid-functionalized terpyridine **50** was attached to an ammonium-functionalized PS–PEG support through an ionic interaction and then complexed with palladium to give supported complex **51** (Scheme 14.46).[147] The catalyst was used for five reaction cycles in the Suzuki coupling of phenyl iodide, with quantitative conversion in each cycle. It should be noted that a high catalyst loading (5 mol%) and high temperature (100 °C) were used, although analysis of the aqueous phase by ICP-MS showed barely detectable levels of leached palladium.

Complexation of palladium(II) to a poly[*N*-isopropylacrylamide-co-diphenyl(4-styrenyl)phosphine] gave a cross-linked material that was highly active in the Suzuki coupling of aryl iodides and bromides.[148] The catalyst (0.005 mol%) was used for 10 cycles with an average yield of 95% per cycle in the Suzuki coupling of phenyl iodide and phenylboronic acid at 100 °C. The catalyst material can be generated as a membrane in a microfluidic channel by flowing an ethyl acetate solution of the polymeric ligand and an aqueous palladium salt solution separately into a Y-junction.[149] The networked membrane forms at the interface between the two phases throughout the channel. The same group reported a similar polymer in which the phosphine ligand had been replaced with an imidazole.[150] The resulting material catalyzed the Suzuki coupling of aryl chlorides at 100 °C using 0.1 mol% palladium in water. In the Suzuki coupling of 4-chloroanisole, no decrease in yield was seen over five reaction cycles. The catalyst was also applied to the coupling of allylic carbonates and organoboron nucleophiles in water or alcohol solvents.

Although cross-linked polystyrene is not appreciably swollen by water, it has been successfully used as a support in aqueous-phase cross-coupling reactions. Merrifield resin-modified palladium–salen complex **52** was an effective catalyst for the Suzuki coupling of aryl bromides in water–toluene (30 : 1) at 90 °C (Scheme 14.47).[151] Unlike the silica-supported analog of this catalyst, benzyltributylammonium chloride is not required as a PTC with the polystyrene support. A related Merrifield resin-supported palladium–salen complex (**53**) was applied to the Sonogashira coupling of aryl iodides and bromides in water at 100 °C.[152]

Glass–Merrifield resin composite Raschig rings were used to support a benzothiazole-based oxime palladacycle (Scheme 14.48).[54] The resulting catalyst (**55**) was effective for the Suzuki and Heck couplings of aryl bromides in water–TBAB at 100 °C. Activated aryl chlorides could also be coupled. The supported catalyst could be used for over 15 reaction cycles with 100% conversion to product in the Suzuki coupling using 0.5 mol% palladium. In contrast, a water-soluble version of this complex began to show lower yields after five cycles.

Johnson-Matthey has developed fiber-supported palladium complexes (FibreCats) for a range of cross-coupling reactions.[153] The support is a polyethylene or polypropylene fiber functionalized with vinyl monomers through electron beam irradiation. The initially reported system included a triphenylphosphine-modified fiber to which various palladium precursors

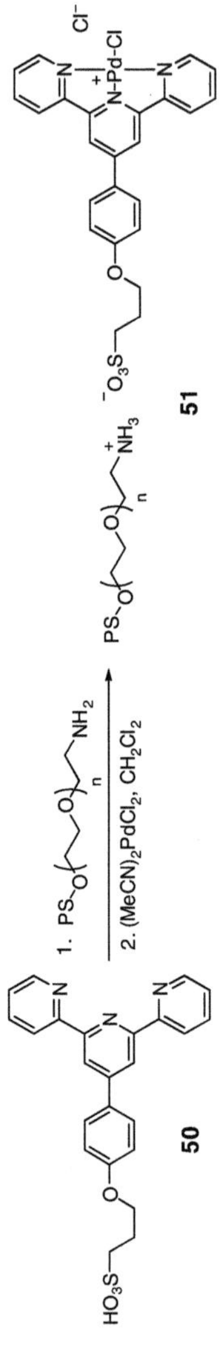

Scheme 14.46

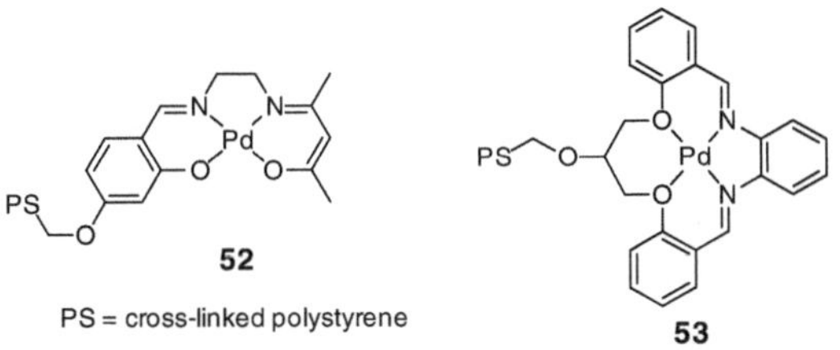
PS
52
PS = cross-linked polystyrene
53

Scheme 14.47

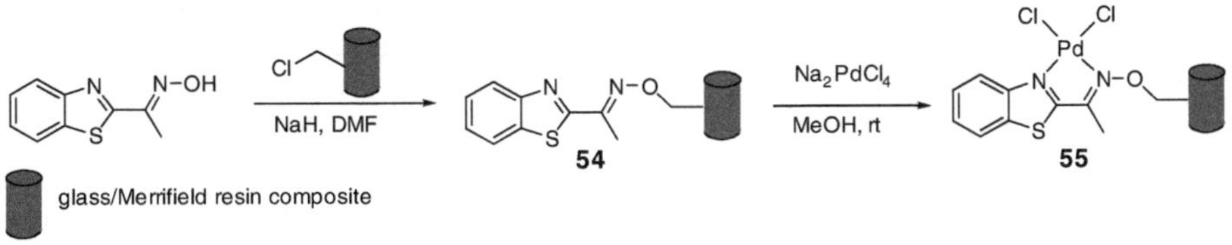
NaH, DMF
Na_2PdCl_4
MeOH, rt
54
55
glass/Merrifield resin composite

Scheme 14.48

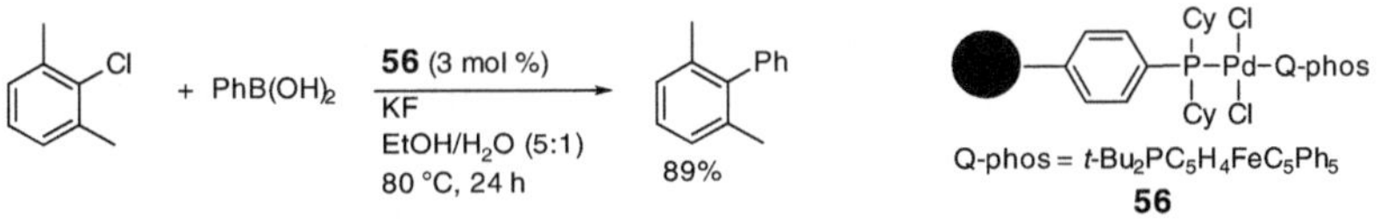
+ $PhB(OH)_2$
56 (3 mol %)
KF
$EtOH/H_2O$ (5:1)
80 °C, 24 h
89%
Q-phos = t-$Bu_2PC_5H_4FeC_5Ph_5$
56

Scheme 14.49

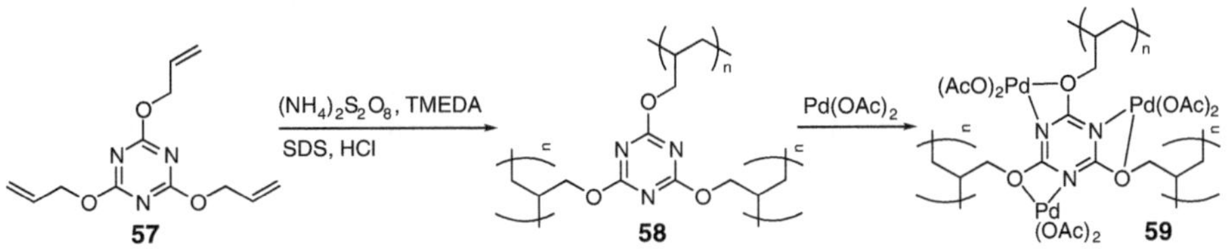
$(NH_4)_2S_2O_8$, TMEDA
SDS, HCl
$Pd(OAc)_2$
57
58
59

Scheme 14.50

could be coordinated. The resulting material was effective for Suzuki couplings of aryl bromides in water–ethanol (1 : 1). No palladium leaching was observed in water–ethanol or a range of organic solvents. Tunable catalyst systems can be generated by coordinating ligands to fiber-supported palladium complexes.[154] FibreCats with a supported dicyclohexylphenylphosphine ligand modified by sterically demanding, electron-rich ligands, such as tri-*tert*-butylphosphine, provided high activity for Suzuki coupling of aryl chlorides in water–ethanol (5 : 1) at 80 °C (Scheme 14.49). No palladium leaching (<0.01 ppm) was observed in these reactions.

Palladium supported on network polymer **58** formed by free radical polymerization of 2,4,6-triallyloxytriazine (**57**) catalyzes the Heck, Sonogashira and Suzuki couplings of aryl iodides and bromides in aqueous solvents (Scheme 14.50).[155] The recyclability of polymeric catalyst **59** was studied for the Suzuki coupling of iodobenzene and phenyl trihydroxyborate. A continuous decrease in yield was seen for successive cycles, from 95% in cycle 1 to 75% in cycle 6. Analysis of the catalyst system using hot filtration testing and a silica gel-supported catalyst poison suggested that the reaction occurs on the heterogeneous catalyst, rather than through soluble palladium species. A related polymer prepared by radical polymerization of triallylamine can also serve as a support for palladium.[156] The resulting material catalyzes Heck couplings of aryl iodides and bromides and shows modest activity for aryl chlorides in water.

Silica and other metal oxides also provide the potential to serve as supports for metal–ligand complexes. Whereas polymeric supports allow for a wide range of attachment chemistries, they are often physically fragile. In contrast, metal oxide materials can be more robust, but the attachment strategies are limited. A silica-supported acetophenone oxime palladacycle (**62**) was prepared by reacting alkene-functionalized palladacycle **61** with thiol-modified silica (**60**) (Scheme 14.51).[157] The silica gel-supported catalyst gave superior results in the Suzuki coupling of 4-chloroacetophenone compared with the same palladacycle attached to polystyrene, poly(ethyl acrylate) or MCM-41. The heterogeneous catalyst could be recovered by filtration, washed, dried and reused up to eight times with no decrease in performance. Silica-supported palladium ketoiminate complex **63** catalyzes the Suzuki, Stille and Sonogashira coupling of heteroaryl chlorides, including furans, using 0.5 mol% palladium at 60 °C in water–TBAB (Scheme 14.52).[158] The catalyst was used for six reaction cycles with no decrease in product yield. Palladium leaching to the reaction phase ranged from 0.003 to 0.15 ppm per cycle.

Magnetic iron oxide particles are interesting catalyst supports because they can be magnetically separated from reaction solutions, rather than using lower throughput processes, such as filtration or centrifugation. A palladium complex of bis(di-*tert*-butylphosphinomethyl)amine (**64,**

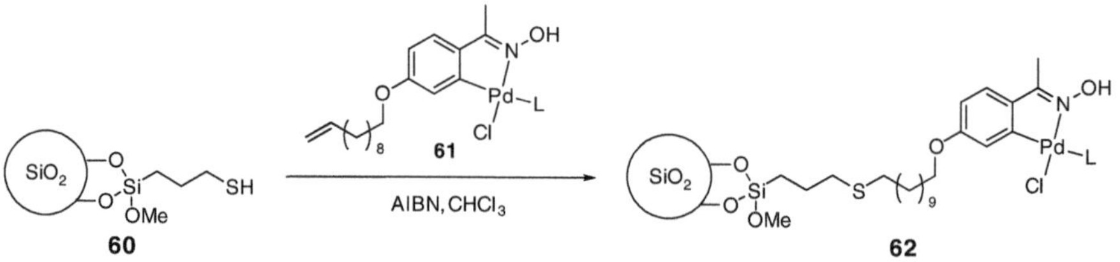

Scheme 14.51

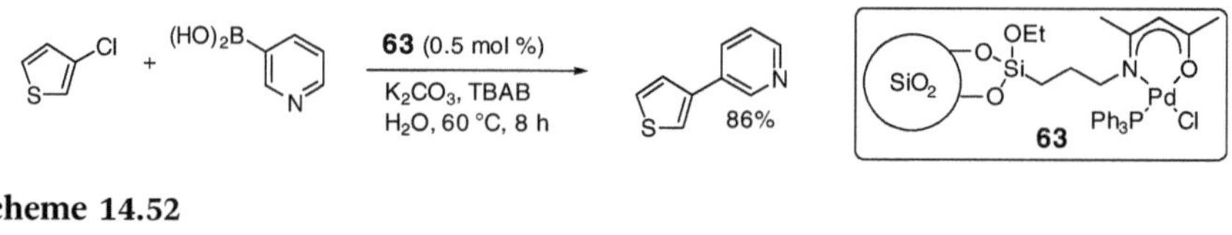

Scheme 14.52

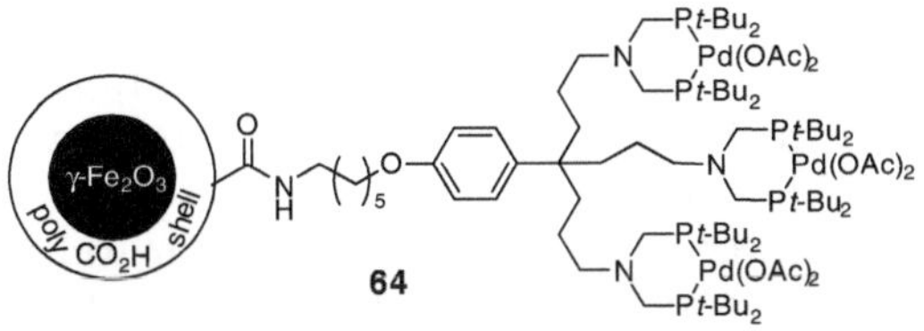

64

Scheme 14.53

Scheme 14.53) attached to a polymer-coated γ-Fe_2O_3 particle was applied to the aqueous-phase Sonogashira coupling of aryl bromides.[159] A surfactant was necessary to achieve high yields in water with Triton-X405 (2.1 wt%) providing the best results. The catalyst could be recycled for 10 reaction cycles with a slow decrease in reaction yield. In contrast, in methanol the product yields decreased precipitously after cycle six. Fe_3O_4 particles surface functionalized with imidazolium salts were used to immobilize a palladium bis(NHC) complex.[160] The catalyst was applied to the Suzuki coupling of aryl bromides in water–TBAB at 75 °C. The catalyst could be separated magnetically and reused for five reaction cycles with minimal decrease in reaction yield. Palladium leaching was 10 ppm in the first cycle, but decreased to $\sim$1 ppm for subsequent cycles. The relatively high leaching rates may be due to the fact that the catalyst was not covalently attached to the magnetic particle.

14.3 Other Potentially Green Alternatives to Traditional Organic Solvents

Replacement of the traditional volatile organic solvents used in organic synthetic methods represents one of the significant improvements that can be made to these methods. Water as an alternative to traditional organic solvents has attracted significant attention because of the beneficial properties that it offers as a solvent. Although water has garnered the most attention, other potentially environmentally benign alternative solvents can be considered. Ideally, these alternative solvents would be of low toxicity, non-volatile and derived from renewable resources.

14.3.1 Poly(Ethylene Glycol)

Poly(ethylene glycol) (PEG) oligomers are non-toxic, low-volatility liquids that are approved as food additives. PEG is miscible with water and polar organics, but immiscible with non-polar organic solvents. PEG is commonly used as an additive in aqueous-phase cross-coupling reactions to improve the interaction between water-soluble catalysts and organic substrates.

Mixed PEG-400–water solvents with PEG:water ratios ranging from 2:1 to 1:2 give optimal results in the $PdCl_2$-catalyzed Suzuki coupling of aryl bromides at room temperature.[161] Water alone or PEG-400 alone as solvent gave lower yields. The catalyst was poorly recyclable since the palladium nanoparticles were not well stabilized by the PEG. Similar results were seen in a Pd(OAc)$_2$-catalyzed Suzuki coupling using PEG as a co-solvent.[162] The best results were obtained with water as a co-solvent, whereas organic solvents gave much lower yields. Higher molecular weight PEG-2000 gave better results than PEG-600. In the Suzuki coupling of aryl bromides catalyzed by palladium supported on BaSO$_4$, good yields could be obtained in PEG-300 at 80 °C, but the reaction could be carried out at room temperature with 40% aqueous PEG 300.[163] The catalyst was used for three reaction cycles with no decrease in product yield. Good activity using PEG-300 alone as solvent was achieved in the Suzuki coupling of aryl iodides using 0.1 mol% Pd$_2$(dba)$_3$, but aryl bromides and chlorides gave low conversion.[164]

14.3.2   Glycerol

Glycerol is a widely used starting material in the drug, food, beverage and chemical industries. It is also a by-product produced in biodiesel production, which has led to an oversupply on world markets. As a result, there is interest in finding alternative uses for glycerol. Like water, glycerol has many attractive properties as a green solvent.[165] Glycerol is essentially non-toxic, non-flammable and non-volatile. It is also a relatively cheap and renewable resource as a by-product of biodiesel production. It has similar polarity and hydrogen bonding properties to water, which allows it to be used in biphasic reaction media. The high viscosity of glycerol along with the potential reactivity of the alcohol groups are challenges to be overcome in its use.

Wolfson and Dlugy reported the first study of glycerol as a solvent for palladium-catalyzed cross-coupling reactions.[166] Suzuki and Heck couplings of iodobenzene occurred in high yield using Pd(TPPTS)$_2$Cl$_2$ (2 mol%) as the precatalyst at 80 °C. Although higher initial yields were obtained with Pd(TPPTS)$_2$Cl$_2$, the resulting catalyst solution rapidly lost activity when recycled. In contrast, Pd(DPPF)Cl$_2$, consistently gave 80% yields in three cycles of the coupling of iodobenzene and phenylboronic acid. Palladium nanoparticles supported on aminopolysaccharide (AP) is an effective catalyst for the Heck β,β-diarylation of acrylates in glycerol (Scheme 14.54).[167] The organic products could be extracted from the glycerol phase using continuous extraction with supercritical CO$_2$. The system showed poor recyclability owing to the accumulation of by-products in the glycerol catalyst phase.

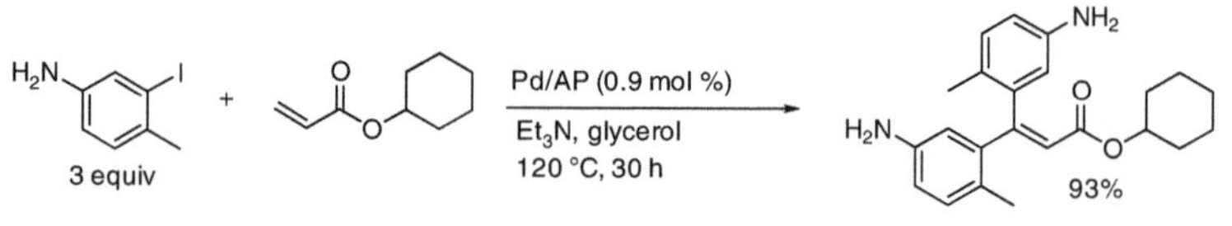

Scheme 14.54

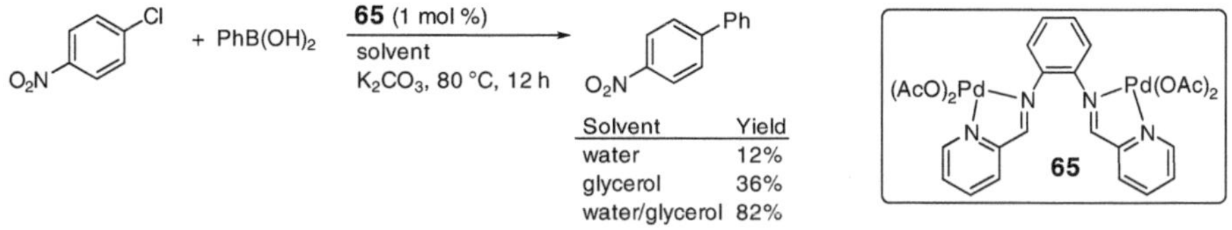

Scheme 14.55

Glycerol in combination with water improves the performance of complex **65** in Suzuki couplings of aryl chlorides (Scheme 14.55).[168] Complex **65** afforded good yields in water with aryl bromides, but only a 12% yield with 1-chloro-4-nitrobenzene. A 36% yield was obtained with glycerol as the solvent, but the yield increased to 82% with 1 : 1 water–glycerol. Good yields (62–86%) were obtained with unactivated and deactivated aryl chlorides under these conditions.

14.3.3 Compressed CO_2

Supercritical or liquid carbon dioxide (compressed CO_2) has attracted significant attention as an alternative solvent in a variety of applications, including organic synthesis.[169] Carbon dioxide is a widely available and inexpensive material. With increasing concern about its role in climate change, there is strong interest in finding uses for captured carbon dioxide. As a large-scale industrial material, it is relatively safe as it is non-flammable and fairly environmentally benign upon release. Compressed CO_2 is already widely used on an industrial scale in processes such as decaffeination of coffee. Compressed CO_2 has been explored as a replacement for volatile organic solvents in palladium-catalyzed coupling, although examples remain limited.

Early work showed that palladium-catalyzed cross-coupling reactions, including Heck, Suzuki and Stille couplings, could be performed in compressed CO_2.[170] The systems were generally modestly active and limited to aryl iodide substrates. Fluorinated ligands often provided the best results because of the superior solubility of fluorinated hydrocarbons in compressed CO_2 compared with non-fluorinated ligands. The high pressures required for these reactions (1000–3000 psi) has likely limited the study of these systems. These types of pressures are not a challenge for industrial-scale processes, however.

Nucleophilic reagents, such as amines, are a challenge for CO_2-phase reactions, owing to the electrophilicity of carbon dioxide. Amines and carbon dioxide exist in equilibrium with carbamic acids. Palladium-catalyzed carbon–nitrogen bond formation was achieved in compressed CO_2 using *N*-silylamines as the nitrogen source (Scheme 14.56).[171] The *N*-silylamines undergo transmetallation with palladium faster than they react with CO_2, resulting in efficient *N*-arylation. Using $Pd(OAc)_2$ in combination with XPhos (2 mol% catalyst), aryl bromides and chlorides could be coupled with

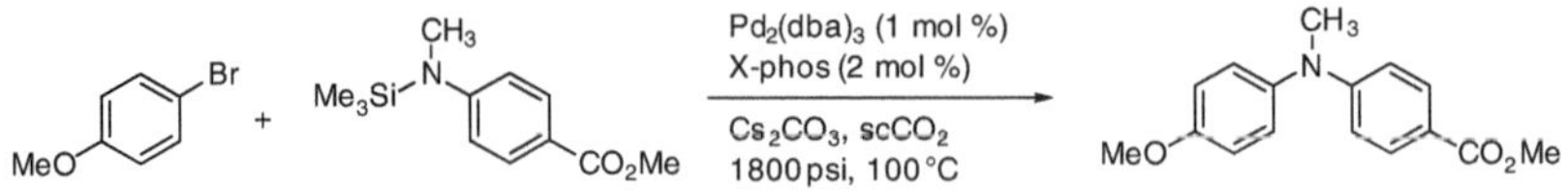

Scheme 14.56

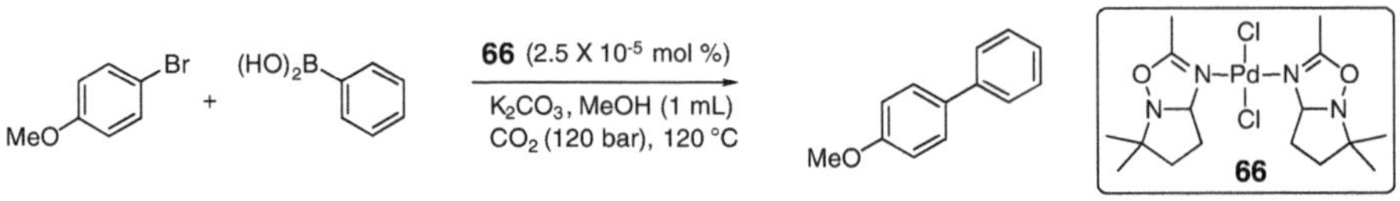

Scheme 14.57

N-silylamines at 100 °C and 1800 psi. Palladium–oxazoline complex **66** was highly active for the Suzuki coupling of aryl bromides in compressed CO_2 when methanol was used as a co-solvent (Scheme 14.57).[172] High yields were obtained with 2.5×10^{-5} mol% palladium at 120 °C and 120 bar. Triphenylphosphine-modified mesoporous SBA-15 silica complexed to palladium is effective for the Suzuki coupling of aryl bromides at 90 °C and 20 MPa.[173] The heterogeneous catalyst could be used for seven reaction cycles with no decrease in product yield.

14.3.4   Ionic Liquids

Ionic liquids (ILs) attracted significant attention as potential replacements for volatile organic solvents for organic reactions, including palladium-catalyzed cross-coupling, beginning in the early 2000s.[174] ILs are low-volatility materials, so there is no concern with their evaporation into the atmosphere. With proper choice of ions, they can be made to be immiscible with both water and non-polar organic solvents. As a result, catalyst and inorganic by-products can be removed through phase separation. Although ILs have been touted as green solvents, such a blanket assessment cannot be universally accepted. ILs cover a wide range of materials, some of which may be environmentally benign, whereas others are probably fairly harmful. If suitably designed, ILs could serve as safer and more sustainable replacements for volatile organic solvents.

Early examples of palladium-catalyzed coupling reactions were typically performed in 1-methyl-3-butylimidazolium tetrafluoroborate ([bmim][BF$_4$]). Using Pd(OAc)$_2$ as precatalyst, Pd–NHC complexes were observed in reaction mixtures during Heck coupling reactions.[175] Like water, IL phases can be used to constrain palladium catalysts, but it is not necessary to modify the catalyst species specially. Palladium in combination with P(*o*-tol)$_3$ catalyzed the Heck coupling of aryl bromides in [bmim][PF$_6$] at 180 °C under microwave heating.[176] The catalyst-containing IL solution could be used for five reaction cycles with no decrease in product yield. Similar results have been obtained with other systems.[177]

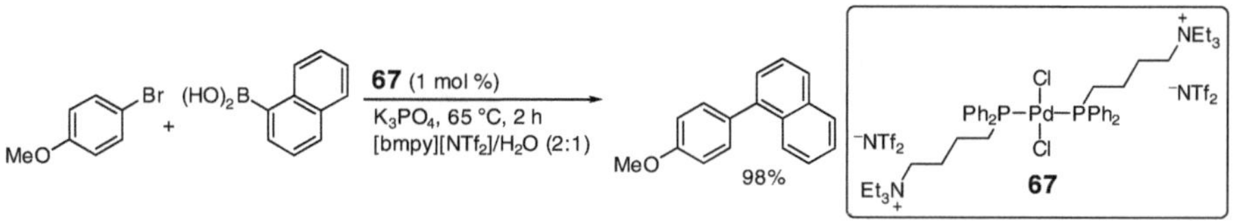

Scheme 14.58

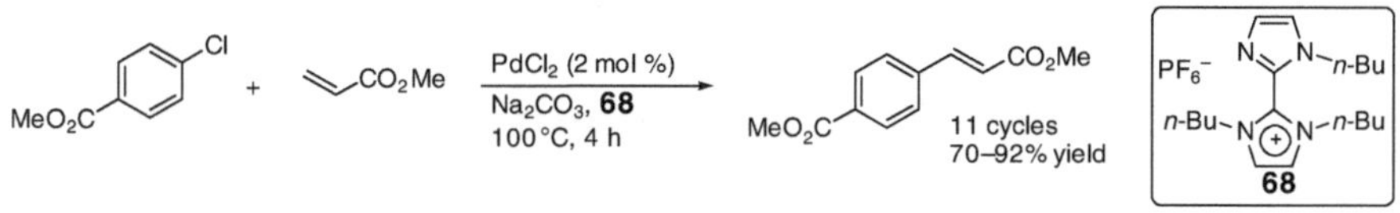

Scheme 14.59

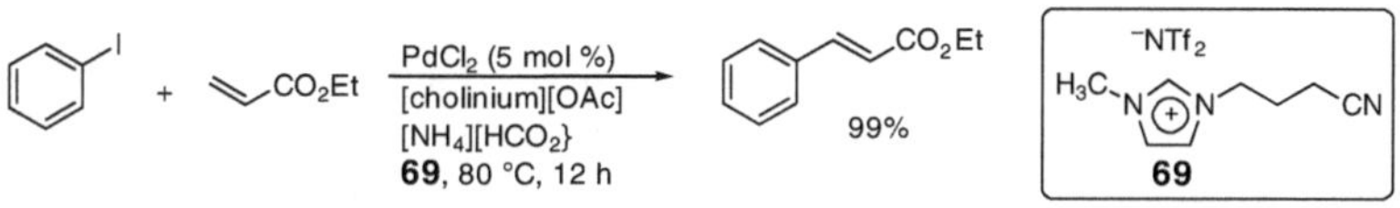

Scheme 14.60

Although recycling has been demonstrated without modification of palladium catalysts, these systems can potentially be leached away when extracted with organic solvents. Ligands tagged with ammonium ions have a strong affinity for IL phases, ensuring that they will not be extracted into non-polar organic solvents. Suzuki coupling of aryl bromides catalyzed by the IL-philic palladium complex **67** was performed in [bmpy][NTf₂]–water at 65 °C [bmpy = 1-butyl-1-methylpyrrolidinium, ⁻NTf₂ = bis(trifyl)amide] (Scheme 14.58).[178] The catalyst-containing IL phase could be recovered and reused for six reaction cycles with no decrease in product yield.

Task-specific ionic liquids (TSILs) containing coordinating groups have also been explored as potential reaction media. By including a ligand in the solvent, the palladium catalyst can be stabilized and constrained in the catalyst phase. 1,3-Dibutyl-2-(1′-butylimidazolyl)imidazolium PF₆⁻ (**68**) was used as solvent in the PdCl₂-catalyzed Heck coupling of aryl iodides and chlorides at 100 °C (Scheme 14.59).[179] The IL solution gave high yields over 11 reaction cycles using a range of different aryl halides. Nitrile-modified imidazolium TSIL **69** provides high yields in the Heck coupling of iodo-benzene, but the catalyst–IL solution rapidly lost activity when recycled (Scheme 14.60).[180] Nanoparticles generated from palladium acetate in the nitrile-modified TSIL were effective for the Hiyama coupling of aryl iodides, bromides and chlorides.[181] Again, the catalyst–IL solution rapidly lost activity when used in subsequent runs after recycling. It appears that the nitrile TSIL does not sufficiently stabilize the palladium nanoparticles against agglomeration.

N-Butyl-*N,N,N′,N′*-tetramethylguanidinium acetate is an IL, but is in equilibrium with acetic acid and the neutral guanidine. When used as a solvent for Heck coupling, the free guanidine can serve as a ligand for palladium.[182] Good yields were obtained with aryl iodides and bromides and even activated aryl chlorides at 140 °C. As the reaction proceeds, the acetate anion is replaced by bromide and acetic acid is formed. The IL–Pd solution could be reused until all of the acetate was consumed. Analysis of the IL–Pd solution after one reaction cycle showed that about 10% of the original palladium had been lost after removal of the product by extraction.

14.4 Alternative Reaction Conditions to Improve Sustainability

14.4.1 Microwave Heating

Microwave irradiation allows for rapid and even heating of reaction mixtures, unlike traditional thermal heating. As a result, microwave heating can provide more efficient thermal promotion of reactions with lower energy demand. The speed of heating can dramatically accelerate reaction rates compared with thermal heating. For microwave heating to be effective, the solvent must have a strong dipole. Water is an ideal solvent for microwave-promoted reactions because of its strong interaction with microwave radiation. Microwave heating is often used with catalysts that require high reaction temperatures (>100 °C) to provide high reaction rates.

Microwave heating was first applied to aqueous-phase palladium-catalyzed Suzuki coupling of PEG-supported aryl bromides as part of a rapid-throughput synthesis approach.[183] Using microwave heating, high yields were obtained with a ligand-free catalyst in PEG–water with a reaction time of 2–4 min. Suzuki coupling of aryl chlorides catalyzed by Pd/C was carried out at 120 °C using microwave heating.[115,184] Microwave irradiation gave rates that were 30 times higher than those with conventional heating and three times higher than those with ultrasound irradiation in aqueous-phase Suzuki coupling catalyzed by palladium nanoparticles supported on PVP.[185] Catalyst loadings as low as 2.5 ppm have been used for the Suzuki coupling of aryl bromides under microwave irradiation.[186] Microwave irradiation has also been applied to Hiyama,[53] Sonogashira[187] and Heck[188] couplings. α-Arylation of the protected glycine derivative **30** in water gave amino acid **31** (Scheme 14.28).[88] Copper-catalyzed Sonogashira couplings of aryl iodides can be carried out in <1 h under microwave irradiation compared with 24 h for conventional heating.[189]

14.4.2 Ultrasound

Ultrasound irradiation has been shown to accelerate many reactions, allowing them to be completed with shorter reaction times than with thermal activation. As a result of the shorter reaction times and lower reaction

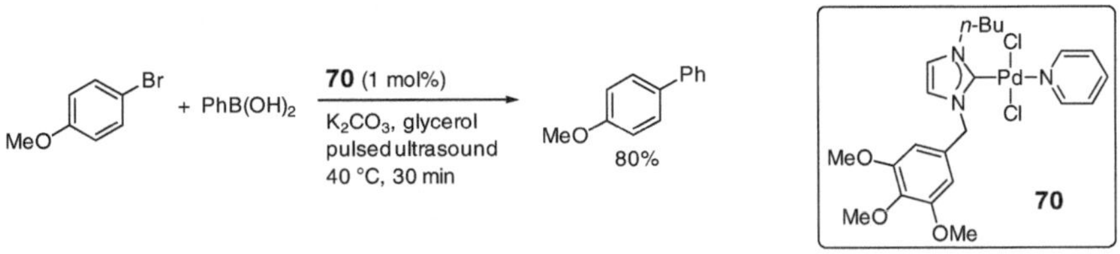

Scheme 14.61

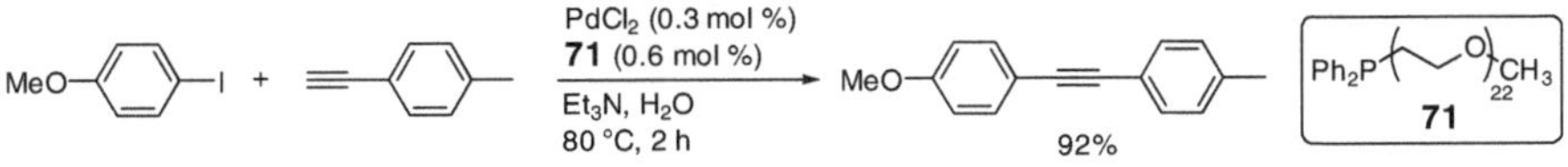

Scheme 14.62

temperatures, ultrasound-promoted reactions often give fewer side products than reactions promoted by heating. In the aqueous-phase Suzuki coupling of aryl iodides catalyzed by palladium supported on PVP, ultrasound-promoted reactions occurred 10 times faster than thermally promoted reactions (80 °C).[185] Ultrasound has also been used to promote the Pd(OAc)$_2$-catalyzed Heck coupling of aryl iodides in water–ethanol using SDS as a surfactant[190] Pulsed ultrasound allowed rapid Suzuki couplings of aryl iodides and bromides catalyzed by palladium complex **70** in glycerol at 40 °C (Scheme 14.61).[191] In the coupling of 4-bromoanisole with phenylboronic acid, conventional heating gave a 21% yield after 30 min, whereas pulsed ultrasound provided an 80% yield. The improved results under ultrasound irradiation were ascribed to the low solubility of the organic substrates in glycerol and the high viscosity of the solvent, which limited mass transport.

14.4.3 Thermomorphic Reaction Control

PEG or polyol-substituted compounds often display thermoregulated solvation. At low temperature, extensive hydrogen bonding with water allows them to partition primarily to the water layer of a water–organic biphasic system. As the temperature is increased, the hydrogen bonding is disrupted and the molecule becomes more hydrophobic and partitions into the organic phase. Thus, a catalyst with thermoreversible solvation properties can be partitioned into the organic phase at high temperature, then the catalyst partitions back into the aqueous phase when the reaction temperature is lowered. In this way, the catalysis can be carried out under nearly homogeneous conditions in the organic phase, while allowing the water-soluble catalyst to be recovered at ambient temperature.

Diphenylphosphino-substituted PEG **71** provides effective catalysts for thermoregulated Suzuki[24] and Sonogashira[23] couplings of aryl bromides and iodides (Scheme 14.62). The ligand has a cloud point of 93 °C. In Suzuki coupling, a significant increase in yield was seen above the cloud point, from

44% at 90 °C to 99% at 100 °C. The catalyst derived from this ligand is completely soluble in water at room temperature, but partitions into the substrate phase upon heating to 100 °C, leaving a colorless water layer. Upon cooling to room temperature, the catalyst partitions back into the water phase and the water-insoluble product precipitates. In Sonogashira coupling, the temperature effect was even more dramatic: at 72 °C, a 30% yield was obtained in the coupling of 4-iodonitrobenzene and phenylacetylene, whereas at 74 °C, the yield was 70% under otherwise identical conditions. The catalyst system could be used for four cycles before a significant decrease in activity occurred in both the Suzuki and Sonogashira reactions.

14.5 Greener Substrates for Cross-Coupling Reactions

As a substitution reaction, the traditional cross-coupling of an organic halide with a nucleophile is inherently atom inefficient. The resulting halide salt by-product represents a stoichiometric waste stream that must be treated. One approach would be to use leaving groups that leave innocuous by-products. Even more attractive would be if the leaving group were to act as a base in the reaction, thus eliminating the need for a stoichiometric base. The use of no leaving group would be the most atom economical approach.

14.5.1 Carboxylic Acid Derivatives as Coupling Partners

Decarboxylative or decarbonylative coupling of benzoic acid derivatives would release either CO_2 or CO as gaseous by-products, thus avoiding the production of metal halide waste. Benzoic acid derivatives can be used as either the nucleophile or electrophile in cross-coupling reactions. In a few cases, cross-coupling of different benzoic acid derivatives has been reported. Work on the development of this class of cross-coupling reaction has been reviewed previously,[192] so just key examples are highlighted here.

Stephan and co-workers first reported a cross-coupling reaction without salt by-product formation.[193] Palladium(II) chloride (0.25 mol%) catalyzed the Heck coupling of benzoic anhydride with alkenes in NMP at 140 °C to afford the arylated alkene (Scheme 14.63). Palladium inserts into the anhydride to give benzoate and a benzoyl–palladium intermediate that eliminates CO to give the aryl–palladium intermediate in the catalytic cycle. The benzoate released upon oxidative addition serves as the base in the reaction. In principle, the benzoic acid could be recovered and converted back to the

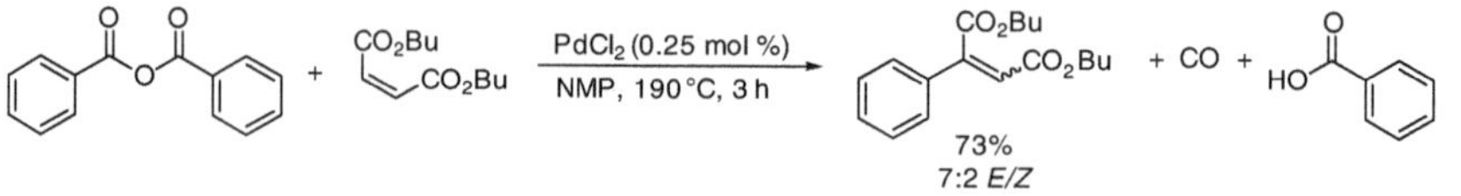

Scheme 14.63

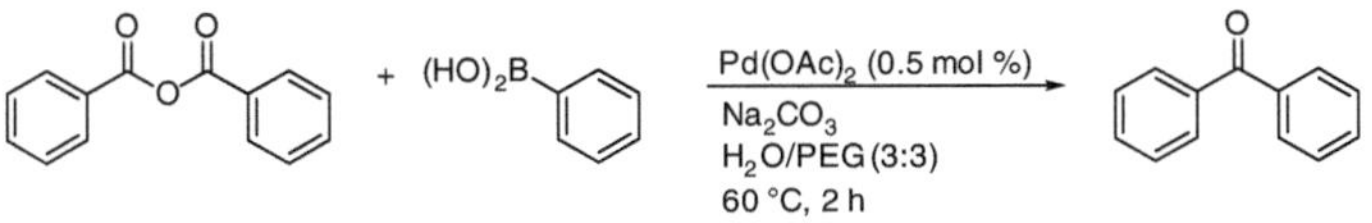

Scheme 14.64

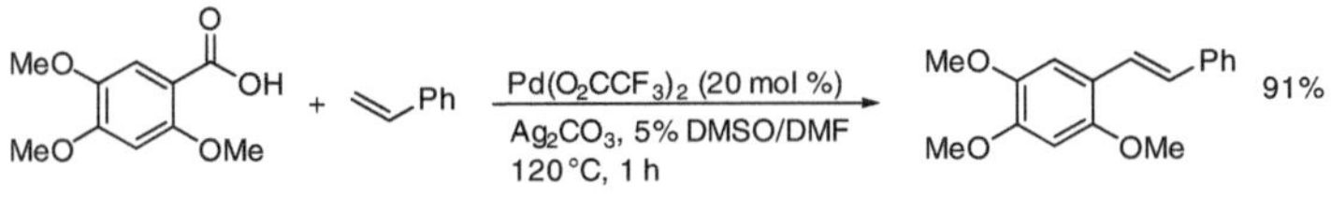

Scheme 14.65

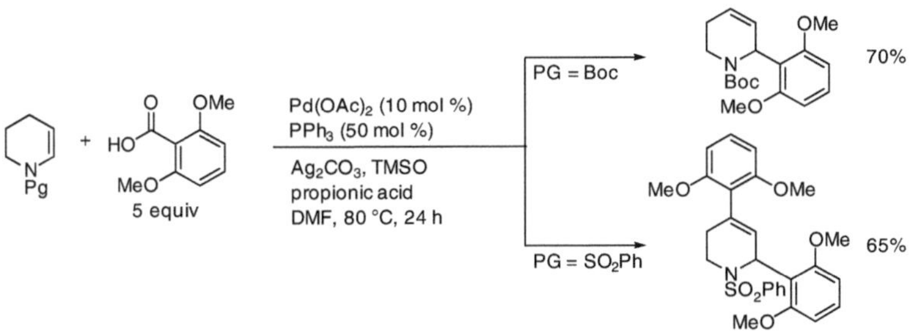

Scheme 14.66

anhydride. Benzoic anhydrides are coupled with arylboronic acids in water–PEG-2000 to give benzophenone derivatives using $PdCl_2$ as the precatalyst (Scheme 14.64).[194] In this case, transmetallation and reductive elimination occur prior to decarbonylation. The catalyst system lost activity after three reaction cycles. By changing to a water–[bmim][PF_6] solvent system, up to eight cycles could be achieved. Activated 4-nitrophenyl esters can also be used as substrates under these conditions.[195]

Direct decarboxylation of benzoic acid derivatives with palladium can be achieved at high temperature. This approach was first applied to cross-coupling by Myers and co-workers as part of a Heck-type coupling of benzoic acids and alkenes catalyzed by $Pd(O_2CCF_3)_2$ (20 mol%) at 120 °C using Ag_2CO_3 as an oxidant (Scheme 14.65).[196] At least one *ortho* substituent on the benzoic acid was required to prevent *ortho*-palladation. The decarboxylation reaction was proposed to involve coordination of the benzoate to palladium followed by β-aryl elimination to release CO_2.[197] Cyclic enamines and enamides can be mono- or diarylated *via* decarboxylative coupling with the selectivity determined by the protecting group (Scheme 14.66).[198] With a Boc protecting group, selective monoarylation occurs; whereas selective diarylation occurs with sulfonamide, amide and benzyl protecting groups. Decarboxylation of α-oxocarboxylic acids provides palladium–acyl species that can be coupled with aryltrifluoroborates to give benzophenone

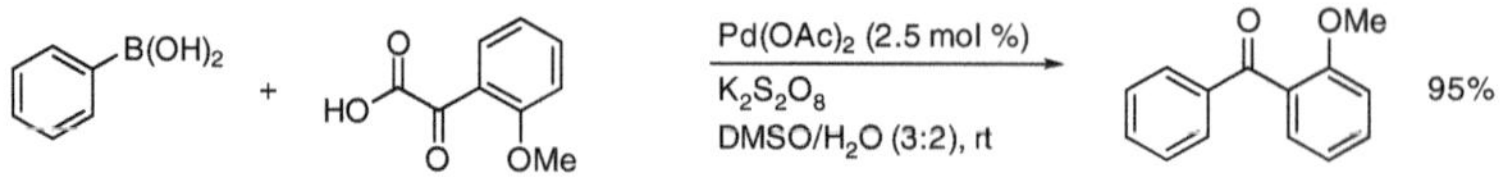

Scheme 14.67

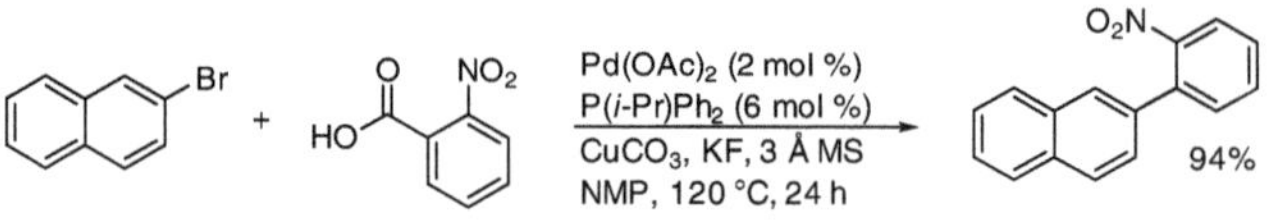

Scheme 14.68

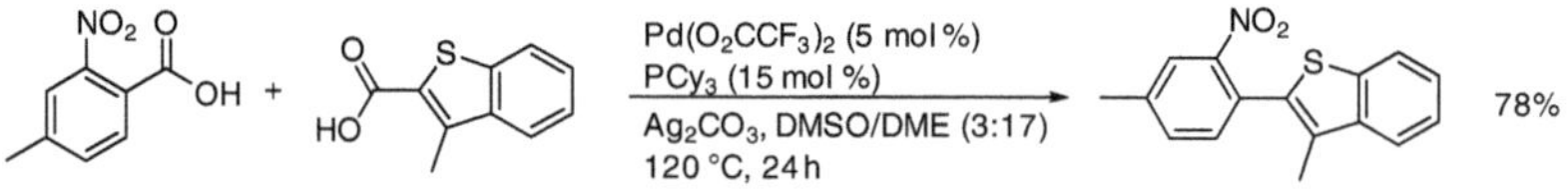

Scheme 14.69

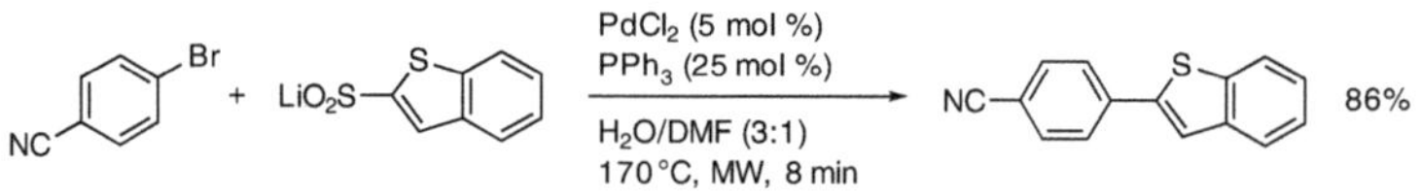

Scheme 14.70

derivatives at room temperature in DMSO–water using potassium persulfate as the oxidant (Scheme 14.67).[199] These decarboxylative reactions are less attractive than the decarbonylative couplings because they require a stoichiometric oxidant, such as Ag_2CO_3 or $K_2S_2O_8$.

Benzoic acids can also be used in place of typical organometallic partners in cross-coupling reactions with aryl halides. Palladium-catalyzed coupling of aryl halides with benzoic acids to give biaryls was reported independently by the groups of Goossen and Wagner (Scheme 14.68).[200] High temperatures and catalyst loadings are required along with a stoichiometric oxidant. Heterocyclic acids are also suitable substrates.[201] A combined palladium and copper catalyst system extended the reaction to aryl chlorides[202] and aryl sulfonates.[203]

Cross-coupling of two different arylcarboxylic acid derivatives is possible if they are electronically different (Scheme 14.69).[204] Electron-deficient benzoic acids can be coupled with electron-rich aromatic acid derivatives using $Pd(O_2CCF_3)_2$ as the catalyst (5 mol%) and Ag_2CO_3 as the oxidant in DMSO–DME (3 : 17) at 120 °C. Good yields of the cross-coupled products are obtained. Desulfinylative coupling of heteroarylsulfinates and aryl bromides catalyzed by palladium occurs effectively in water with a small amount of DMF (Scheme 14.70).[205]

14.5.2 Direct Coupling by C–H Activation Under Green Conditions

The direct arylation of arenes has been reviewed in Chapters 9 and 13. By directly activating C–H bonds, halide and metal by-products can be avoided. The current generation of direct C–H functionalization reactions often require stoichiometric oxidants or other additives that add to the waste output, which may offset the advantage of using a C–H bond as the reaction site. The use of water as a solvent in direct arylation reactions has attracted attention recently.[206] Greaney and co-workers reported the first examples of direct arylation using water as a solvent.[207] Aryl halides were coupled with thiazoles using $Pd(dppf)Cl_2$ as the catalyst and Ag_2CO_3 on water at 60 °C (Scheme 14.71). The on-water reaction proceeded faster than reactions performed under homogeneous conditions in acetonitrile. The authors noted that reactions run either on water or neat had similar reaction rates, suggesting the acceleration in water was really due to high local concentrations of the water-insoluble substrates and catalyst. This methodology was applicable to other heterocyclic ring systems.[208] The proper choice of base allowed for the selective 2- or 3-arylation of indole on water using $Pd(OAc)_2$/dppm (5 mol%) as the catalyst at 110 °C.[209] The 2-arylated product was formed using potassium acetate, whereas lithium hydroxide preferentially gave the 3-arylated product.

Palladium-catalyzed directed *ortho*-arylation of *N*-arylureas occurs efficiently in water at room temperature with Brij 35 (2 wt%) as surfactant (Scheme 14.72).[210] The Brij surfactant gave superior results to PTS and other nonionic surfactants. Addition of acetonitrile as a co-solvent shut down the catalytic reaction. Water accelerated the coupling of aryl iodides with pentafluorobenzene catalyzed by $Pd(OAc)_2$/MePhos [MePhos = dicyclohexyl(2′,6′-dimethylbiphen-2-yl)phosphine] at room temperature (Scheme 14.73).[211] The reaction required 16 h to reach completion in ethyl acetate or DMF. In contrast, the reaction gave higher yields after only 2 h in ethyl acetate–water (2.5 : 1). Interestingly, increasing the water ratio lowered the reaction yield.

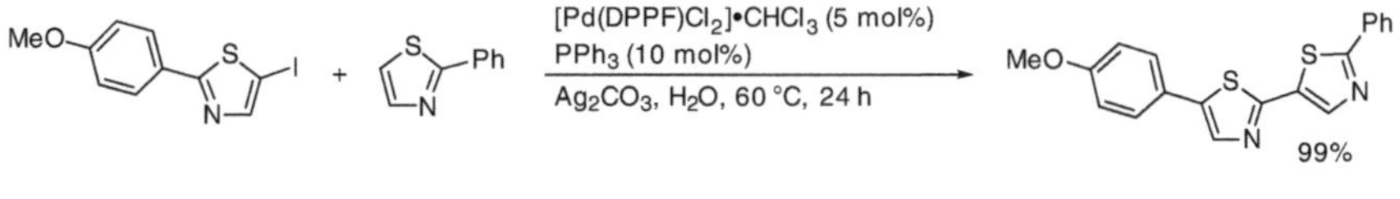

Scheme 14.71

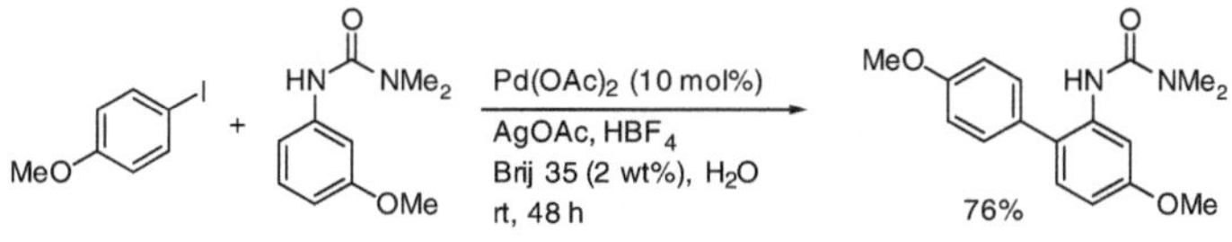

Scheme 14.72

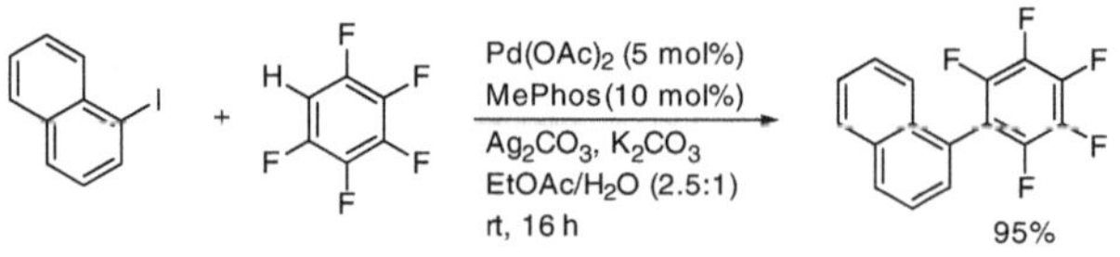

Scheme 14.73

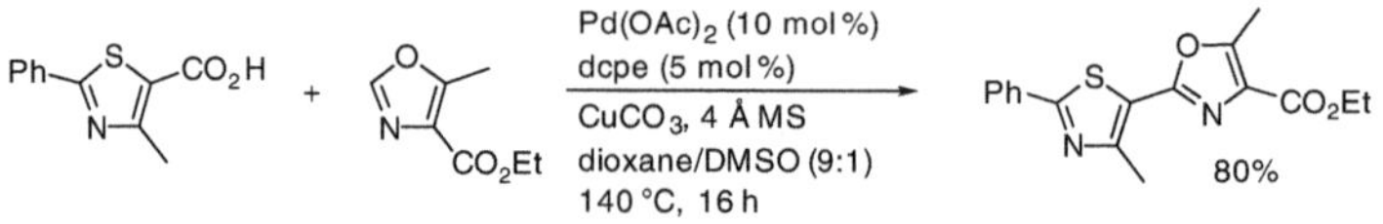

Scheme 14.74

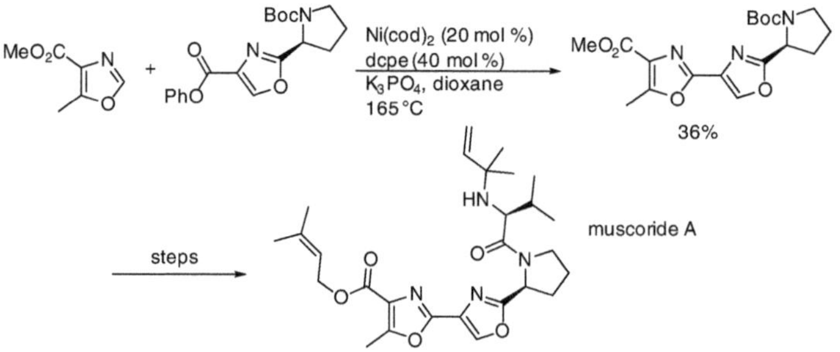

Scheme 14.75

The accelerating effect of water was attributed to dissolution of the inorganic reactants (Ag_2CO_3 and K_2CO_3).

Benzoic acid derivatives have also been used as coupling partners in the direct arylation of arenes. 2-Substituted oxazole- and thiazolecarboxylic acids undergo decarboxylative coupling with oxazoles to give biazoles in good yields using $Pd(OAc)_2$/dcpe [dcpe = 1,2-bis(dicyclohexylphosphino)ethane] (Scheme 14.74).[212] Under optimized conditions, a 10:1 ratio of cross-coupled to homo-coupled product was obtained. 2-Substituted benzoic acid derivatives were used to arylate tetra- and pentafluorobenzene derivatives using $Pd(O_2CCF_3)_2$/PCy_3 in the presence of Ag_2CO_3.[213] Phenyl esters were effective substrates for the 2-arylation of oxazoles catalyzed by $Ni(COD)_2$/dcpe.[214] This transformation was used as a key step in the synthesis of the natural product muscoride A (Scheme 14.75).

14.6 Conclusion and Future Prospects

The development of more environmentally sustainable approaches to cross-coupling catalysis represents an important area of future development in these incredibly powerful reactions. To date, significant effort has been devoted to developing solvent replacements, particularly the use of water in place of volatile organic solvents. Although water has many attractive features, the simple act of substituting water for an organic solvent does not inherently make a process green. In the future, efforts to consider the environmental impact of all aspects of a methodology must be applied in a holistic fashion. In addition to a safe and environmentally acceptable solvent, researchers should consider the substrates and the by-products that they will produce, work to minimize the amounts of precious metal catalysts and ligands used and work to develop reactions that require a minimum of additional stoichiometric reagents. In addition, methodologies should be developed that require minimal work-up and purification steps to provide isolated product and allow the recovery and reuse of the catalyst species. Finally, the energy demand should be minimized by the development of catalysts with sufficient activity to promote reactions at or near ambient temperature.

Recent advances in the activation of benzoic acid derivatives and direct C–H activation offer promising ways to decrease the environmental impact of cross-coupling reactions. In particular, replacing organometallic coupling partners, such as organoboron or organotin reagents, would effect significant decreases in waste production. To date, however, these systems often require stoichiometric oxidants, such as silver salts, and high temperatures to be efficient. The development of more active systems that can be reoxidized with O_2 would allow these systems to become highly attractive alternatives to traditional cross-coupling reactions.

As efforts are made to develop more sustainable cross-coupling methodologies, it will be critical that these methods be evaluated using metrics such as LCA or E-factors. The majority of current research reports involve new catalyst systems with a focus on catalyst performance metrics, such as yield, turnover number and turnover frequency. The environmental impacts of these methods cannot be easily compared, however; since few reports provide E-factors or other assessments of the sustainability of their processes. It is hoped that more researchers will begin to incorporate these measures in their work so that the true green potential of new methods can be evaluated.

References

1. P. T. Anastas and J. C. Warner, *Green Chemistry: Theory and Practice*, Oxford University Press, New York, 1998.
2. C. Jiménez-González, D. J. C. Constable and C. S. Ponder, *Chem. Soc. Rev.*, 2012, **41**, 1485–1498.
3. Scientific Applications International Corporation, *Life Cycle Assessment: Principles and Practice*, EPA/600/R-06/060, National Risk Management Research Laboratory, Environmental Protection Agency, Cincinnati, OH, 2006.
4. R. A. Sheldon, *Green Chem.*, 2007, **9**, 1273–1283.
5. (a) C. Jiménez-González, C. S. Ponder, Q. B. Broxterman and J. B. Manley, *Org. Process Res. Dev.*, 2011, **15**, 912–917; (b) C. Jiménez-González,

C. Ollech, W. Pyrz, D. L. Hughes, Q. B. Broxterman and N. Bhathela, *Org. Process Res. Dev.*, 2013, **17**, 239–246.

6. (a) B. H. Lipshutz, N. A. Isley, J. C. Fennewald and E. D. Slack, *Angew. Chem. Int. Ed.*, 2013, **52**, 10952–10958; (b) A. Kamal, V. Srinivasulu, B. N. Seshadri, N. Markandeya, A. Alarifi and N. Shankaraiah, *Green Chem.*, 2012, **14**, 2513–2522; (c) C. Pavia, E. Ballerini, L. A. Bivona, F. Giacalone, C. Aprile, L. Vaccaro and M. Gruttadauria, *Adv. Synth. Catal.*, 2013, **355**, 2007–2018.

7. C. Jiménez-González, A. D. Curzons, D. J. C. Constable and V. L. Cunningham, *Int. J. Life Cycle Assess.*, 2004, **9**, 114–121.

8. (a) C.-J. Li, *Chem. Rev.*, 2005, **105**, 3095–3165; (b) C.-J. Li and T.-H. Chan, *Comprehensive Organic Reactions in Aqueous Media*, Wiley, Hoboken, NJ, 2nd edn, 2007; (c) P. H. Dixneuf and V. Cadierno (eds), *Metal-Catalyzed Reactions in Water*, Wiley-VCH, Weinheim, 2013.

9. C. E. Garrett and K. Prasad, *Adv. Synth. Catal.*, 2004, **346**, 889–900.

10. E. G. Kuntz, *CHEMTECH*, 1987, **17**, 570–575.

11. K. H. Shaughnessy, *Chem. Rev.*, 2009, **109**, 643–710.

12. A. L. Casalnuovo and J. C. Calabrese, *J. Am. Chem. Soc.*, 1990, **112**, 4324–4330.

13. (a) J. P. Genêt, E. Blart and M. Savignac, *Synlett*, 1992, 715–717; (b) J. P. Genêt, A. Linquist, E. Blard, V. Mouries, M. Savignac and M. Vaultier, *Tetrahedron Lett.*, 1995, **36**, 1443–1446; (c) C. Dupuis, K. Adiey, L. Charruault, V. Michelet, M. Savignac and J.-P. Genêt, *Tetrahedron Lett.*, 2001, **42**, 6523–6526.

14. (a) R. Amengual, E. Genin, V. Michelet, M. Savignac and J.-P. Genêt, *Adv. Synth. Catal.*, 2002, **344**, 393–398; (b) E. Genin, R. Amengual, V. Michelet, M. Savignac, A. Jutand, L. Neuville and J. P. Genêt, *Adv. Synth. Catal.*, 2004, **346**, 1733–1741.

15. (a) H. Dibowski and F. P. Schmidtchen, *Tetrahedron*, 1995, **51**, 2325–2330; (b) A. Hessler, O. Stelzer, H. Dibowski, K. Worm and F. P. Schmidtchen, *J. Org. Chem.*, 1997, **62**, 2362–2369; (c) H. Dibowski and F. P. Schmidtchen, *Tetrahedron Lett.*, 1998, **39**, 525–528.

16. (a) M. Beller, J. G. E. Krauter and A. Zapf, *Angew. Chem. Int. Ed.*, 1997, **36**, 772–774; (b) S. Parisot, R. Kolodziuk, C. Goux-Henry, A. Iourtchenko and D. Sinou, *Tetrahedron Lett.*, 2002, **43**, 7397–7400.

17. H. Gulyás, Á. Szöllõsy, B. E. Hanson and J. Bakos, *Tetrahedron Lett.*, 2002, **43**, 2543–2546.

18. (a) L. R. Moore and K. H. Shaughnessy, *Org. Lett.*, 2004, **6**, 225–228; (b) L. R. Moore, E. C. Western, R. Craciun, J. M. Spruell, D. A. Dixon, K. P. O'Halloran and K. H. Shaughnessy, *Organometallics*, 2008, **27**, 576–593.

19. G. Wüllner, H. Jänsch, S. Kannenberg, F. Schubert and G. Boche, *Chem. Commun.*, 1998, 1509–1510.

20. H. Hattori, K.-i. Fujita, T. Muraki and A. Sakaba, *Tetrahedron Lett.*, 2007, **48**, 6817–6820.

21. D. J. M. Snelders, R. Kreiter, J. Firet, G. van Koten and R. J. M. Klein Gebbink, *Adv. Synth. Catal.*, 2008, **350**, 262–266.

22. D. J. M. Snelders, G. van Koten and R. J. M. Klein Gebbink, *J. Am. Chem. Soc.*, 2009, **131**, 11407–11416.
23. N. Liu, C. Liu, Q. Xu and Z. Jin, *Eur. J. Org. Chem.*, 2011, 4422–4428.
24. (a) N. Liu, C. Liu and Z. Jin, *J. Organomet. Chem.*, 2011, **696**, 2641–2647; (b) N. Liu, C. Liu, B. Yan and Z. Jin, *Appl. Organomet. Chem.*, 2011, **25**, 168–172.
25. K. H. Shaughnessy and R. S. Booth, *Org. Lett.*, 2001, **3**, 2757–2759.
26. R. B. DeVasher, L. R. Moore and K. H. Shaughnessy, *J. Org. Chem.*, 2004, **69**, 7919–7927.
27. R. Huang and K. H. Shaughnessy, *Organometallics*, 2006, **25**, 4105–4112.
28. S. Y. Cho, S. K. Kang, J. H. Ahn, J. D. Ha and J.-K. Choi, *Tetrahedron Lett.*, 2006, **47**, 5237–5240.
29. R. B. DeVasher, J. M. Spruell, D. A. Dixon, G. A. Broker, S. T. Griffin, R. D. Rogers and K. H. Shaughnessy, *Organometallics*, 2005, **24**, 962–971.
30. W. S. Brown, D. D. Boykin, M. Q. Sonnier Jr, W. D. Clark, F. V. Brown and K. H. Shaughnessy, *Synthesis*, 2008, 1965–1970.
31. M. Nishimura, M. Ueda and N. Miyaura, *Tetrahedron*, 2002, **58**, 5779–5787.
32. A. Konovets, A. Penciu, E. Framery, N. Percina, C. Goux-Henry and D. Sinou, *Tetrahedron Lett.*, 2005, **46**, 3205–3208.
33. K. W. Anderson and S. L. Buchwald, *Angew. Chem. Int. Ed.*, 2005, **44**, 6175–6177.
34. (a) C. Fleckenstein and H. Plenio, *Green Chem.*, 2007, **9**, 1287–1291; (b) C. Fleckenstein and H. Plenio, *J. Org. Chem.*, 2008, **73**, 3236–3244; (c) C. Fleckenstein and H. Plenio, *Chem. Eur. J.*, 2008, **14**, 4267–4279; (d) C. Fleckenstein and H. Plenio, *Green Chem.*, 2008, **10**, 563–570; (e) C. A. Fleckenstein and H. Plenio, *Chem. Eur. J.*, 2007, **13**, 2701–2716.
35. G. C. Fortman and S. P. Nolan, *Chem. Soc. Rev.*, 2011, **40**, 5151–5169.
36. C. Fleckenstein, S. Roy, S. Leuthäusser and H. Plenio, *Chem. Commun.*, 2007, 2870–2872.
37. S. Roy and H. Plenio, *Adv. Synth. Catal.*, 2010, **352**, 1014–1022.
38. L. Li, J. Wang, C. Zhou, R. Wang and M. Hong, *Green Chem.*, 2011, **13**, 2071–2077.
39. T. Tu, X. Feng, Z. Wang and X. Liu, *Dalton Trans.*, 2010, **39**, 10598–10600.
40. Z. Wang, X. Feng, W. Fang and T Tu, *Synlett*, 2011, 951–954.
41. N. Liu, C. Liu and Z. Jin, *Green Chem.*, 2012, **14**, 592–597.
42. M. Meise and R. Haag, *ChemSusChem*, 2008, **1**, 637–642.
43. W.-Y. Wu, S.-N. Chen and F.-Y. Tsai, *Tetrahedron Lett.*, 2006, **47**, 9267–9270.
44. S.-N. Chen, W.-Y. Wu and F.-Y. Tsai, *Tetrahedron*, 2008, **64**, 8164–8168.
45. S.-H. Huang, J.-R. Chen and F.-Y. Tsai, *Molecules*, 2010, **15**, 315–330.
46. C. Zhou, J. Wang, L. Li, R. Wang and M. Hong, *Green Chem.*, 2011, **13**, 2100–2106.
47. (a) X. Guo, J. Zhou, X. Li and H. Sun, *J. Organomet. Chem.*, 2008, **693**, 3692–3696; (b) J. Zhou, X. Guo, C. Tu, X. Li and H. Sun, *J. Organomet. Chem.*, 2009, **694**, 697–702.

48. D. N. Korolev and N. A. Bumagin, *Tetrahedron Lett.*, 2005, **46**, 5751–5754.
49. S. Gülcemal, İ. Kani, F. Yilmaz and B. Çetinkaya, *Tetrahedron*, 2010, **66**, 5602–5606.
50. D. A. Alonso and C. Nájera, *Chem. Soc. Rev.*, 2010, **39**, 2891–2902.
51. (a) L. Botella and C. Nájera, *Angew. Chem. Int. Ed.*, 2002, **41**, 179–181; (b) E. Alacid and C. Nájera, *Org. Lett.*, 2008, **10**, 5011–5014; (c) E. Alacid and C. Nájera, *J. Org. Chem.*, 2009, **74**, 2321–2327.
52. L. Botella and C. Nájera, *J. Org. Chem.*, 2005, **70**, 4360–4369.
53. E. Alacid and C. Nájera, *J. Org. Chem.*, 2008, **73**, 2315–2322.
54. K. M. Dawood, *Tetrahedron*, 2007, **63**, 9642–9651.
55. J. Zhou, X. Li and H. Sun, *J. Organomet. Chem.*, 2010, **695**, 297–303.
56. M. S. Holzwarth and B. Plietker, *ChemCatChem*, 2013, **5**, 1650–1679.
57. (a) F. Ullmann, *Ber. Dtsch. Chem. Ges.*, 1903, **36**, 2382–2384; (b) I. Goldberg, *Ber. Dtsch. Chem. Ges.*, 1906, **39**, 1691–1696.
58. (a) F. Monnier and M. Taillefer, *Angew. Chem. Int. Ed.*, 2009, **48**, 6954–6971; (b) D. S. Surry and S. L. Buchwald, *Chem. Sci.*, 2010, **1**, 13–21; (c) I. P. Beletskaya and A. V. Cheprakov, *Organometallics*, 2012, **31**, 7753–7808.
59. K. Swapna, S. N. Murthy and Y. V. D. Nageswar, *Eur. J. Org. Chem.*, 2010, 6678–6884.
60. M. Carril, R. SanMartin, E. Domínguez and I. Tellitu, *Green Chem.*, 2007, **9**, 219–220.
61. M. Carril, R. SanMartin, I. Tellitu and E. Dominguez, *Org. Lett.*, 2006, **8**, 1467–1470.
62. M. Carril, R. SanMartin, E. Domínguez and I. Tellitu, *Chem. Eur. J.*, 2007, **13**, 5100–5105.
63. J. Engel-Andreasen, B. Shimpukade and T. Ulven, *Green Chem.*, 2013, **15**, 336–340.
64. F. Ke, Y.-Y. Qu, Z.-Q. Jiang, Z. Li, D. Wu and X. Zhou, *Org. Lett.*, 2011, **13**, 454–457.
65. D. Yang and H. Fu, *Chem. Eur. J.*, 2010, **16**, 2366–2370.
66. X. Li, D. Yang, Y. Jiang and H. Fu, *Green Chem.*, 2010, **12**, 1097–1105.
67. F. Meng, X. Zhu, Y. Li, J. Xie, B. Wang, J. Yao and Y. Wan, *Eur. J. Org. Chem.*, 2010, 6149–6152.
68. J. Xie, X. Zhu, M. Huang, F. Meng, W. Chen and Y. Wan, *Eur. J. Org. Chem.*, 2010, 3219–3223.
69. L. Jing, J. Wei, L. Zhou, Z. Huang, Z. Li and X. Zhou, *Chem. Commun.*, 2010, **46**, 4767–4769.
70. Y. Wang, Z. Wu, L. Wang, Z. Li and X. Zhou, *Chem. Eur. J.*, 2009, **15**, 8971–8974.
71. Z. Wu, L. Zhou, Z. Jiang, D. Wu, Z. Li and X. Zhou, *Eur. J. Org. Chem.*, 2010, 4971–4975.
72. Z. Wu, Z. Jiang, D. Wu, H. Xiang and X. Zhou, *Eur. J. Org. Chem.*, 2010, 1854–1857.
73. K. G. Thakur, D. Ganapath and G. Sekar, *Chem. Commun.*, 2011, **47**, 5076–5078.

74. K. G. Thakur and G. Sekar, *Chem. Commun.*, 2011, **47**, 6692–6694.
75. A. Correa, G. Mancheño and C. Bolm, *Chem. Soc. Rev.*, 2008, **37**, 1108–1117.
76. S. L. Buchwald and C. Bolm, *Angew. Chem. Int. Ed.*, 2009, **48**, 5586–5587.
77. H. W. Lee, A. S. C. Chan and F. Y. Kwong, *Tetrahedron Lett.*, 2009, **50**, 5868–5871.
78. Y.-C. Teo, *Adv. Synth. Catal.*, 2009, **351**, 720–724.
79. W.-Y. Wu, J.-C. Wang and F.-Y. Tsai, *Green Chem.*, 2009, **11**, 326–329.
80. D. C. Rideout and R. Breslow, *J. Am. Chem. Soc.*, 1980, **102**, 7816–7817.
81. (a) A. Chanda and V. V. Fokin, *Chem. Rev.*, 2009, **109**, 725–748; (b) R. N. Butler and A. G. Coyne, *Chem. Rev.*, 2010, **110**, 6302–6337.
82. (a) S. Bhattacharya and S. Sengupta, *Tetrahedron Lett.*, 2004, **45**, 8733–8736; (b) J. T. Guan, T. Q. Weng, G.-A. Yu and S. H. Liu, *Tetrahedron Lett.*, 2007, **48**, 7129–7133.
83. N. Jiang and A. J. Ragauskas, *Tetrahedron Lett.*, 2006, **47**, 197–200.
84. A. Fihri, D. Luart, C. Len, A. Solhy, C. Chevrin and V. Polshettiwar, *Dalton Trans.*, 2011, **40**, 3116–3121.
85. (a) X.-X. Zhou and L.-X. Shao, *Synthesis*, 2011, 3138–3142; (b) Y. Zhang, M.-T. Feng and J.-M. Lu, *Org. Biomol. Chem.*, 2013, **11**, 2266–2272.
86. (a) C. Duplais, A.; Krasovskiy, A. Wattenberg and B. H. Lipshutz, *Chem. Commun.*, 2010, **46**, 562–564; (b) V. Krasovskaya, A. Krasovskiy and B. H. Lipshutz, *Chem. Asian J.*, 2011, **6**, 1974–1976; (c) V. Krasovskaya, A. Krasovskiy, A. Bhattacharjya and B. H. Lipshutz, *Chem. Commun.*, 2011, **47**, 5717–5719.
87. A. J. Ross, F. Dreiocker, M. Schäfer, J. Oomens, A. J. H. M. Meijer, B. T. Pickup and R. F. W. Jackson, *J. Org. Chem.*, 2011, **76**, 1727–1734.
88. O. Lagerlund, L. R. Odell, S. L. Mowbray, M. T. Nilsson, W. W. Krajewski, A. Nordqvist, A. Karlén and M. Larhed, *Comb. Chem. High Throughput Screening*, 2007, **10**, 783–789.
89. X. Huang, K. W. Anderson, D. Zim, L. Jiang, A. Klapars and S. L. Buchwald, *J. Am. Chem. Soc.*, 2003, **125**, 6653–6655.
90. T. Jeffery, *Tetrahedron Lett.*, 1994, **35**, 3051–3054.
91. (a) D. Badone, M. Baroni, R. Cardamone, A. Ielmini and U. Guzzi, *J. Org. Chem.*, 1997, **62**, 7170–7173; (b) J. C. Bussolari and D. C. Rehborn, *Org. Lett.*, 1999, **1**, 965–967.
92. B. Basu, K. Biswas, S. Kundu and S. Ghosh, *Green Chem.*, 2010, **12**, 1734–1738.
93. A. M. Deveau and T. L. Macdonald, *Tetrahedron Lett.*, 2004, **45**, 803–807.
94. A. Arcadi, G. Cerichelli, M. Chiarini, M. Correa and D. Zorzan, *Eur. J. Org. Chem.*, 2003, 4080–4086.
95. S. Bhattacharya, A. Srivastava and S. Sengugpta, *Tetrahedron Lett.*, 2005, **46**, 3557–3560.
96. L. Lin, Y. Li, S. Zhang and S. Li, *Synlett*, 2011, 1779–1783.
97. L. Bai, J.-X. Wang and Y. Zhang, *Green Chem.*, 2003, **5**, 615–617.
98. W. Susanto, C.-Y. Chu, W. J. Ang, T.-C. Chou, L.-C. Lo and Y. Lam, *Green Chem.*, 2012, **14**, 77–80.

99. E. L. Kolychev, A. F. Asachenko, P. B. Dzhevakov, A. A. Bush, V. V. Shuntikov, V. N. Khrustalev and M. S. Nechaev, *Dalton Trans.*, 2013, **42**, 6859–6866.

100. K. Park, G. Bae, A. Park, Y. Kim, J. Choe, K. H. Song and S. Lee, *Tetrahedron Lett.*, 2011, **52**, 576–580.

101. B. C. Ranu, R. Dey and K. Chattopadhyay, *Tetrahedron Lett.*, 2008, **49**, 3430–3432.

102. D. Saha, K. Chattopadhyay and B. Ranu, *Tetrahedron Lett.*, 2009, **50**, 1003–1006.

103. D. Saha, R. Dey and B. C. Ranu, *Eur. J. Org. Chem.*, 2010, 6067–6071.

104. (a) B. H. Lipshutz, T. B. Petersen and A. R. Abela, *Org. Lett.*, 2008, **10**, 1333–1336; (b) B. H. Lipshutz and A. R. Abela, *Org. Lett.*, 2008, **10**, 5329–5332.

105. B. H. Lipshutz, D. W. Chung and B. Rich, *Org. Lett.*, 2008, **10**, 3793–3796.

106. B. H. Lipshutz and B. R. Taft, *Org. Lett.*, 2008, **10**, 1329–1332.

107. B. H. Lipshutz, R. Moser and K. R. Voigtritter, *Isr. J. Chem.*, 2010, **50**, 691–695.

108. B. H. Lipshutz, S. Ghorai, A. R. Abela, R. Moser, T. Nishikata, C. Duplais, A. Krasovskiy, R. D. Gaston and R. C. Gadwood, *J. Org. Chem.*, 2011, **76**, 4379–4391.

109. B. H. Lipshutz, D. W. Chung and B. Rich, *Adv. Synth. Catal.*, 2009, **351**, 1717–1721.

110. M. Lessi, T. Masini, L. Nucara, F. Bellina and R. Rossi, *Adv. Synth. Catal.*, 2011, **353**, 501–507.

111. (a) A. Krasovskiy, C. Duplais and B. H. Lipschutz, *Org. Lett.*, 2010, **12**, 4742–4744; (b) A. Krasovskiy, C. Duplais and B. H. Lipshutz, *J. Am. Chem. Soc.*, 2009, **131**, 15592–15593; (c) A. Krasovskiy, I. Thomé, J. Graff, V. Krasovskaya, P. Konopelski, C. Duplais and B. H. Lipshutz, *Tetrahedron Lett.*, 2011, **52**, 2203–2205.

112. C. Duplais, A. Krasovskiy and B. H. Lipshutz, *Organometallics*, 2011, **30**, 6090–6097.

113. I. N. Houpis, D. Shields, U. Nettekoven, A. Schnyder, E. Bappert, K. Weerts, M. Canters and W. Vermuelen, *Org. Process Res. Dev.*, 2009, **13**, 596–606.

114. M. Lamblin, L. Nassar-Hardy, J.-C. Hierso, E. Fouquet and F.-X. Felpin, *Adv. Synth. Catal.*, 2010, **352**, 33–79.

115. (a) N. E. Leadbeater and M. Marco, *Org. Lett.*, 2002, **4**, 2973–2976; (b) N. E. Leadbeater and M. Marco, *J. Org. Chem.*, 2003, **68**, 888–892.

116. R. K. Arvela, N. E. Leadbeater, M. S. Sangi, V. A. Williams, P. Granados and R. D. Singer, *J. Org. Chem.*, 2005, **70**, 161–168.

117. L. Liu, Y. Zhang and B. Xin, *J. Org. Chem.*, 2006, **71**, 3994–3997.

118. Y. Li, X. M. Hong, D. M. Collard and M. A. El-Sayed, *Org. Lett.*, 2000, **2**, 2385–2388.

119. B. J. Gallon, R. W. Kojima, R. B. Kaner and P. L. Diaconescu, *Angew. Chem. Int. Ed.*, 2007, **46**, 7251–7254.

120. D. Zhang, C. Zhou and R. Wang, *Catal. Commun.*, 2012, **22**, 83–88.
121. Ö. Metin, F. Durap, M. Aydemir and S. Özkar, *J. Mol. Catal. A: Chem.*, 2011, **337**, 39–44.
122. K. S. Sivudu, N. M. Reddy, M. N. Prasad, K. M. Raju, Y. M. Mohan, J. S. Yadav, G. Sabitha and D. Shailaja, *J. Mol. Catal. A: Chem.*, 2008, **295**, 10–17.
123. M. Zhang and W. Zhang, *J. Phys. Chem. C*, 2008, **112**, 6245–6252.
124. P. Zhou, H. Wang, J. Yang, J. Tang, D. Sun and W. Tang, *RSC Adv.*, 2012, **2**, 1759–1761.
125. S.-S. Yi, D.-H. Lee, E. Sin and Y.-S. Lee, *Tetrahedron Lett.*, 2007, **48**, 6771–6775.
126. S. Wu, H. Ma, X. Jia, Y. Zhong and Z. Lei, *Tetrahedron*, 2011, **67**, 250–256.
127. A. Prastaro, P. Ceci, E. Chiancone, A. Boffi, R. Cirilli, M. Colone, G. Fabrizi, A. Stringaro and S. Cacchi, *Green Chem.*, 2009, **11**, 1929–1932.
128. G. Lu, R. Franzén, Q. Zhang and Y. Xu, *Tetrahedron Lett.*, 2005, **46**, 4255–4259.
129. T. Maegawa, Y. Kitamura, S. Sako, T. Udzu, A. Sakurai, A. Tanaka, Y. Kobayashi, K. Endo, U. Bora, T. Kurita, A. Kozaki, Y. Monguchi and H. Sajiki, *Chem. Eur. J.*, 2007, **13**, 5937–5943.
130. S. S. Soomro, C. Röhlich and K. Köhler, *Adv. Synth. Catal.*, 2011, **353**, 767–775.
131. C. B. Putta and S. Ghosh, *Adv. Synth. Catal.*, 2011, **353**, 1889–1896.
132. H. Bulut, L. Artok and S. Yilmaz, *Tetrahedron Lett.*, 2002, **44**, 289–291.
133. V. Kogan, Z. Aizenshtat, R. Popovitz-Biro and R. Neumann, *Org. Lett.*, 2002, **4**, 3529–3532.
134. M. Cai, Q. Xu and J. Sha, *J. Mol. Catal. A: Chem.*, 2007, **272**, 293–297.
135. B. Yuan, Y. Pan, Y. Li, B. Yin and H. Jiang, *Angew. Chem. Int. Ed.*, 2010, **49**, 4054–4058.
136. H. Li, Z.-H. Zhu, F. Zhang, S.-H. Xie, H.-X. Li, P. Li and X.-G. Zhou, *ACS Catal.*, 2011, **1**, 1604–1612.
137. B. Sreedhar, A. S. Kumar and D. Yada, *Synlett*, 2011, 1081–1084.
138. F. Amoroso, S. Colussi, A. Del Zotto, J. Llorca and A. Trovarelli, *J. Mol. Catal. A: Chem.*, 2010, **315**, 197–204.
139. A. Khalafi-Hezhad and F. Panahi, *Green Chem.*, 2011, **13**, 2408–2415.
140. R. Bernini, S. Cacchi, G. Fabrizi, G. Forte, F. Petrucci, A. Prastaro, S. Niembro, A. Shafir and A. Vallribera, *Green Chem.*, 2010, **12**, 150–158.
141. A. Boffi, S. Cacchi, P. Ceci, R. Cirilli, G. Fabrizi, A. Prastaro, S. Niembro, A. Shafir and A. Vallribera, *ChemCatChem*, 2011, **3**, 347–353.
142. Y. Uozumi, H. Danjo and T. Hayashi, *J. Org. Chem.*, 1999, **64**, 3384–3388.
143. Y. Uozumi and Y. Kobayashi, *Heterocycles*, 2003, **59**, 71–74.
144. T. Suzuka, Y. Okada, K. Ooshiro and Y. Uozumi, *Tetrahedron*, 2010, **66**, 1064–1069.
145. (a) Y. Hirai and Y. Uozumi, *Chem. Commun.*, 2010, **46**, 1103–1105; (b) Y. Hirai and Y. Uozumi, *Chem. Lett.*, 2011, **40**, 934–935.

694 Chapter 14

146. J.-W. Kim, J.-H. Kim, D.-H. Lee and Y.-S. Lee, *Tetrahedron Lett.*, 2006, **47**, 4745–4748.
147. T. Suzuka, T. Nagamine, K. Ogihara and M. Higa, *Catal. Lett.*, 2010, **139**, 85–89.
148. Y. M. A. Yamada, K. Takeda, H. Takahashi and S. Ikegami, *Org. Lett.*, 2002, **4**, 3371–3374.
149. Y. M. A. Yamada, T. Watanabe, T. Beppu, N. Fukuyama, K. Torii and Y. Uozumi, *Chem. Eur. J.*, 2010, **16**, 11311–11319.
150. Y. M. A. Yamada, S. M. Sarkar and Y. Uozumi, *J. Am. Chem. Soc.*, 2012, **134**, 3190–3198.
151. N. T. S. Phan and P. Styring, *Green Chem.*, 2008, **10**, 1055–1060.
152. Y. He and C. Cai, *J. Organomet. Chem.*, 2011, **696**, 2689–2692.
153. T. J Colacot., E. S. Gore and A. Kuber, *Organometallics*, 2002, **21**, 3301–3304.
154. T. J. Colacot, W. A. Carole, B. A. Neide and A. Harad, *Organometallics*, 2008, **27**, 5605–5611.
155. A. Modak, J. Mondal, M. Sasidharan and A. Bhaumik, *Green Chem.*, 2011, **13**, 1317–1331.
156. J. Mondal, A. Modak and A. Bhaumik, *J. Mol. Catal. A: Chem.*, 2011, **350**, 40–48.
157. C. Baleizão, A. Corma, H. García and A. Leyva, *J. Org. Chem.*, 2004, **69**, 439–446.
158. D.-H. Lee, J.-Y. Jung and M.-J. Jin, *Green Chem.*, 2010, **12**, 2024–2029.
159. D. Rosario-Amorin, M. Gaboyard, R. Clérac, S. Nlate and K. Heuzé, *Dalton Trans.*, 2011, **40**, 44–46.
160. A. Taher, J.-B. Kim, J.-Y. Jung, W.-S. Ahn and M.-J. Jin, *Synlett*, 2009, 2477–2482.
161. Z. Du, W. Zhou, F. Wang and J.-X. Wang, *Tetrahedron*, 2011, **67**, 4914–4918.
162. L. Liu, Y. Zhang and Y. Wang, *J. Org. Chem.*, 2005, **70**, 6122–6125.
163. A. L. F. de Souza, A. C. da Silva and O. A. C. Antunes, *Appl. Organomet. Chem.*, 2009, **23**, 5–8.
164. A. C. Silva, J. D. Senra, L. C. S. Aguiar, A. B. C. Simas, A. L. F. de Souza, L. F. B. Malta and O. A. C. Antunes, *Tetrahedron Lett.*, 2010, **51**, 3883–3885.
165. A. E. Díaz-Álvarez, J. Francos, B. Lastra-Barreira, P. Crochet and V. Cadierno, *Chem. Commun.*, 2011, **47**, 6208–6227.
166. A. Wolfson and C. Dlugy, *Chem. Pap.*, 2007, **61**, 228–232.
167. M. Delample, N. Villandier, J.-P. Douliez, S. Camy, J.-S. Condoret, Y. Pouilloux, J. Barrault and F. Jérôme, *Green Chem.*, 2010, **12**, 804–808.
168. B. Banik, A. Tairai, N. Shahnaz and P. Das, *Tetrahedron Lett.*, 2012, **53**, 5627–5630.
169. C. M. Rayner, *Org. Process Res. Dev.*, 2007, **11**, 121–132.
170. (a) N. Shezad, R. S. Oakes, A. A. Clifford and C. M. Rayner, *Tetrahedron Lett.*, 1999, **40**, 2221–2224; (b) D. K. Morita, D. R. Pesiri, S. A. David, W. H. Glaze and W. Tumas, *Chem. Commun.*, 1998, 1397–1398; (c) M. A. Carroll and A. B. Holmes, *Chem. Commun.*, 1998, 1395–1396;

(d) T. R. Early, R. S. Gordon, M. A. Carroll, A. B. Holmes, R. E. Shute and I. F. McConvey, *Chem. Commun.*, 2001, 1966–1967.

171. C. J. Smith, M. W. S. Tsang, A. B. Holmes, R. L. Danheiser and J. W. Tester, *Org. Biomol. Chem.*, 2005, **3**, 3767–3781.

172. R. R. Fernandes, J. Lasri, M. F. C. Guedes da Silva, A. M. F. Palavra, J. A. L. da Silva, J. J. R. Fráusto da Silva and A. J. L. Pombeiro, *Adv. Synth. Catal.*, 2011, **353**, 1153–1160.

173. X. Feng, M. Yan, T. Zhang, Y. Liu and M. Bao, *Green Chem.*, 2010, **12**, 1758–1766.

174. (a) V. Calò, A. Nacci and A. Monopoli, *Eur. J. Org. Chem.*, 2006, 3791–3802; (b) J. P. Hallett and T. Welton, *Chem. Rev.*, 2011, **111**, 3508–3576.

175. L. Xu, W. Chen and J. Xiao, *Organometallics*, 2000, **19**, 1123–1127.

176. K. S. A. Vallin, P. Emilsson, M. Larhed and A. Hallberg, *J. Org. Chem.*, 2002, **67**, 6243–6246.

177. (a) R. Wang, B. Twamley and J. M. Shreeve, *J. Org. Chem.*, 2006, **71**, 426–429; (b) A. J. Carmichael, M. J. Earle, J. D. Holbrey, P. B. McCormac and K. R. Seddon, *Org. Lett.*, 1999, **1**, 997–1000; (c) H. Hagiwara, Y. Shimizu, T. Hoshi, T. Suzuki, M. Ando, K. Ohkubo and C. Yokoyama, *Tetrahedron Lett.*, 2001, **42**, 4349–4351.

178. M. Lombardo, M. Chiarucci and C. Trombini, *Green Chem.*, 2009, **11**, 574–579.

179. J.-C. Xiao, B. Twamley and J. M. Shreeve, *Org. Lett.*, 2004, **6**, 3845–3847.

180. Z. Fei, D. Zhao, D. Pieraccini, W. H. Ang, T. J. Geldbach, R. Scopelliti, C. Chiappe and P. J. Dyson, *Organometallics*, 2007, **26**, 1588–1598.

181. C. Premi and N. Jain, *Eur. J. Org. Chem.*, 2013, 5493–5499.

182. S. Li, Y. Lin, H. Xie, S. Zhang and J. Xu, *Org. Lett.*, 2006, **8**, 391–394.

183. C. G. Blettner, W. A. König, W. Stenzel and T. Schotten, *J. Org. Chem.*, 1999, **64**, 3885–3890.

184. R. K. Arvela and N. E. Leadbeater, *Org. Lett.*, 2005, 7, 2101–2104.

185. A. L. F. de Souza, L. C. da Silva, B. L. Oliveira and O. A. C. Antunes, *Tetrahedron Lett.*, 2008, **49**, 3895–3898.

186. R. K. Arvela, N. E. Leadbeater, T. L. Mack and C. M. Kormos, *Tetrahedron Lett.*, 2006, **47**, 217–220.

187. G. Chen, X. Zhu, J. Cai and Y. Wan, *Synth. Commun.*, 2007, **37**, 1355–1361.

188. L. Botella and C. Nájera, *Tetrahedron Lett.*, 2004, **45**, 1833–1836.

189. G. Chen, J. Xie, J. Weng, X. Zhu, Z. Zheng, J. Cai and Y. Wan, *Synth. Commun.*, 2011, **41**, 3123–3133.

190. G. An, X. Ji, J. Han and Y. Pan, *Synth. Commun.*, 2011, **41**, 1464–1471.

191. A. Azua, J. A. Mata, P. Heymes, E. Peris, F. Lamaty, J. Martinez and E. Colacino, *Adv. Synth. Catal.*, 2013, **355**, 1107–1116.

192. (a) L. J. Goossen, K. Goossen and C. Stanciu, *Angew. Chem. Int. Ed.*, 2009, **48**, 3569–3571; (b) N. Rodríguez and L. J. Goossen, *Chem. Soc. Rev.*, 2011, **40**, 5030–5048; (c) W. I. Dzik, P. P. Lange and L. J. Goossen, *Chem. Sci.*, 2012, **3**, 2671–2678.

193. M. S. Stephan, A. J. J. M. Teunissen, G. K. M. Verzijl and J. G. de Vries, *Angew. Chem. Int. Ed.*, 1998, **37**, 662–664.

194. (a) B. Xin, Y. Zhang and K. Cheng, *J. Org. Chem.*, 2006, **71**, 5725–5731; (b) B.-W. Xin, *Synth. Commun.*, 2008, **38**, 2826–2837.
195. (a) L. J. Goossen and J. Paetzold, *Angew. Chem. Int. Ed.*, 2002, **41**, 1237–1241; (b) L. J. Goossen and J. Paetzold, *Angew. Chem. Int. Ed.*, 2004, **43**, 1095–1098.
196. (a) A. G. Myers, D. Tanaka and M. R. Mannion, *J. Am. Chem. Soc.*, 2002, **124**, 11250–11251; (b) D. Tanaka and A. G. Myers, *Org. Lett.*, 2003, **6**, 433–436.
197. D. Tanaka, S. P. Romeril and A. G. Myers, *J. Am. Chem. Soc.*, 2005, **127**, 10323–10333.
198. N. Gigant, L. Chausset-Boissarie and I. Gillaizeau, *Org. Lett.*, 2013, **15**, 816–819.
199. M. Li, C. Wang and H. Ge, *Org. Lett.*, 2011, **13**, 2062–2064.
200. (a) L. J. Goossen, N. Rogriguez, B. Melzer, C. Linder, G. Deng and L. M. Levy, *J. Am. Chem. Soc.*, 2007, **129**, 4824–4833; (b) J.-M. Becht, C. Catala, C. Le Drian and A. Wagner, *Org. Lett.*, 2007, **9**, 1781–1783.
201. (a) F. Zhang and M. F. Greaney, *Org. Lett.*, 2010, **12**, 4745–4747; (b) J.-M. Becht and C. Le Drian, *J. Org. Chem.*, 2011, **76**, 6327–6330; (c) C. K. Haley, C. D. Gilmore and B. M. Stoltz, *Tetrahedron*, 2013, **69**, 5732–5736.
202. L. J. Goossen, B. Zimmermann and T. Knauber, *Angew. Chem. Int. Ed.*, 2008, **47**, 7103–7106.
203. (a) L. J. Goossen, N. Rodríguez, P. P. Lange and C. Linder, *Angew. Chem. Int. Ed.*, 2010, **49**, 1111–1114; (b) B. Song, T. Knauber and L. J. Goossen, *Angew. Chem. Int. Ed.*, 2013, **52**, 2954–2958.
204. P. Hu, Y. Shang and W. Su, *Angew. Chem. Int. Ed.*, 2012, **51**, 5945–5949.
205. S. Sevigny and P. Forgione, *New J. Chem.*, 2013, **37**, 589–592.
206. B. Li and P. H. Dixneuf, *Chem. Soc. Rev.*, 2013, **42**, 5744–5767.
207. G. L. Turner, J. A. Morris and M. F. Greaney, *Angew. Chem. Int. Ed.*, 2007, **46**, 7996–8000.
208. (a) S. A. Ohnmacht, P. Mamone, A. J. Culshaw and M. F. Greaney, *Chem. Commun.*, 2008, 1241–1243; (b) S. A. Ohnmacht, A. J. Culshaw and M. F. Greaney, *Org. Lett.*, 2010, **12**, 224–226.
209. L. Joucla, N. Batail and L. Djakovitch, *Adv. Synth. Catal.*, 2010, **352**, 2929–2936.
210. T. Nishikata, A. R. Abela and B. H. Lipshutz, *Angew. Chem. Int. Ed.*, 2010, **49**, 781–784.
211. O. René and K. Fagnou, *Org. Lett.*, 2010, **12**, 2116–2119.
212. F. Zhang and M. F. Greaney, *Angew. Chem. Int. Ed.*, 2010, **49**, 2768–2771.
213. H. Zhao, Y. Wei, J. Xu, J. Kan, W. Su and M. Hong, *J. Org. Chem.*, 2011, **76**, 882–893.
214. K. Amaike, K. Muto, J. Yamaguchi and K. Itami, *J. Am. Chem. Soc.*, 2012, **134**, 13573–13576.

CHAPTER 15

Recent Large-Scale Applications of Transition Metal-Catalyzed Couplings for the Synthesis of Pharmaceuticals

JAVIER MAGANO*[a] AND JOSHUA R. DUNETZ[b]

[a] Chemical Research and Development, Pfizer Global Research and Development, Eastern Point Road, Groton, CT 06340, USA; [b] Process Chemistry, Gilead Sciences, 333 Lakeside Drive, Foster City, CA 94404, USA
*Email: javier.magano@pfizer.com

15.1 Introduction

Transition metal-catalyzed couplings[1] are well-established technologies that have found widespread application in both academic and industrial settings. Intense research in this area for the past four decades has resulted in an outstanding variety of catalytic systems that allow for the formation of C–C and C–heteroatom bonds under very mild conditions and that are applicable to a very broad range of substrates. Since the 1990s, the number of patents and publications in this area has grown exponentially and couplings such as Suzuki–Miyaura, Heck, Sonogashira and Negishi have been reported by the thousands.[2] Palladium is overwhelmingly the most

RSC Catalysis Series No. 21
New Trends in Cross-Coupling: Theory and Applications
Edited by Thomas J Colacot
© The Royal Society of Chemistry 2015
Published by the Royal Society of Chemistry, www.rsc.org

commonly used transition metal in large-scale couplings, followed by copper (C–N and C–O bond formation) and nickel (Kumada cross-coupling). Some of the reasons for this are: (a) the very thorough understanding of reaction mechanisms with Pd; (b) the very large number of commercially available Pd precatalysts;[3] and (c) the extensive list of ligands (mostly phosphines) that have been developed to perform in combination with this metal. The Nobel Prize in Chemistry awarded in 2010 to Professors Heck, Negishi and Suzuki for their contributions to "palladium-catalyzed cross-couplings in organic synthesis" has also contributed to the increasing popularity of these chemistries.[4]

Medicinal chemists have enthusiastically embraced these technologies as a way of accessing new molecules that would previously have been very difficult to prepare. At the same time, successful project progression in pharmaceutical companies requires the involvement of process groups to satisfy the increasing demand for active pharmaceutical ingredients (APIs) destined for pharmaceutical development and clinical trials. And it is at this stage of development in these groups where chemistries are pushed to the limit in terms of robustness and efficiency. As a result, process chemists are understandably conservative regarding the incorporation of new transformations into their synthetic arsenal due to the need for processes that are reliable, predictable and mechanistically well understood. However, since the late 1990s, the adoption of transition metal-catalyzed coupling reactions in process chemistry has shown an impressive growth, as demonstrated by the statistics in Table 15.1,[5] and we can arguably say that this trend can only continue to increase in the future. An excellent example of this is ring-closing metathesis, a technology that has been around for decades but that only very recently has received particular

Table 15.1   Statistics on the number of transition metal-catalyzed large-scale couplings (>100 mmol) published in mainstream literature by date.[5]

Type of coupling	Number of large-scale couplings (>100 mmol) by date		
	Before 2000	2001–2010	2011–December 2013
Suzuki–Miyaura	5	53	16
Heck	10	20	2
Sonogashira	4	17	5
Kumada–Corriu	3	3	3
Negishi	0	10	1
Stille	1	1	0
Hayashi-Miyaura	0	3	0
Tsuji–Trost	1	3	3
Carbonylation	5	8	1
Cyanation	1	14	0
Nozaki–Hiyama–Kishi	0	1	1
Ring-closing metathesis	0	6	3
C–N bond formation	0	13	9
Migita	0	2	3

attention and been implemented on a large scale in several instances (see Section 15.2.11) for the manufacture of drug candidates containing large rings, due in part to the increased availability of commercial Ru catalysts with higher reactivity and selectivity.

The Suzuki–Miyaura cross-coupling is by far the most widely used reaction in process chemistry. The commercial availability of a large number of boronic acids and esters, their stability in general and the possibility of coupling advanced intermediates due to outstanding functional group compatibility have contributed to the success of this reaction. A second group of reactions encompassing Heck, Sonogashira, Negishi, cyanation and C–N bond formation follow by number of examples. The latter is seeing increasing application on a large scale due to the prevalence of aromatic amines in today's APIs. Lastly, several transformations have seen very limited use in large scale applications, such as Stille coupling, because of the toxicity of organotin reagents, and Migita coupling, owing to the scarcity of sulfides, sulfoxides and sulfones in drugs and drug candidates. Interestingly, we have not found process applications of Hiyama coupling, despite the advantages that organosilicon reagents possess, such as low toxicity and excellent chemical stability.[6]

A major factor in the widespread application of these chemistries in both medicinal and process chemistry is the application of high-throughput experimentation (HTE) and design of experiments (DoE), which allows the screening of hundreds of metal source, ligand, base and solvent combinations in as quickly as a few days or weeks to solve a specific problem. Thus, it is not unusual to find unprecedented and totally unexpected reaction conditions which are highly substrate dependent and that otherwise would have never been identified under more traditional laboratory approaches. Further optimization of the metal and ligand loadings, metal-to-ligand ratio, concentration, temperature and reaction time can then provide the key to a clean, high-yielding and scalable process.

The best proof of the relevance of transition metal-catalyzed coupling reactions comes from the numerous applications to the synthesis of commercial drugs, as exemplified in Table 15.2.

This chapter describes case studies that have been reported recently by process groups in pharmaceutical companies and that, in most cases, have not been reviewed previously.[5] In addition to describing bond-forming processes, the examples contained in this chapter also pay special attention to work-ups and purifications that purge metal catalysts to provide intermediates or APIs of sufficient purity. These examples were selected based on two criteria: (a) the transformation has been implemented on a large-scale (>100 mmol) and (b) the article contains a detailed experimental procedure. Representative examples for each type of coupling are highlighted in the synthetic schemes. Finally, examples found exclusively in the patent literature have not been covered in this review because specific experimental details within these legal documents are often difficult to access and interpret.

Table 15.2 List of commercial drugs synthesized through transition metal-catalyzed couplings, with the arrows pointing to the bond(s) created.

Active ingredient name	Commercial name(s)	Structure	Type of coupling(s)	Ref.	Applicant	Indication	FDA approval date	Exclusivity expiration date (United States)
Atazanavir sulfate	Reyataz		Suzuki–Miyaura	7	Bristol-Myers Squibb	Protease inhibitor for the treatment of HIV	June 20, 2003	April 2017
Pitavastatin calcium	Livalo		Suzuki–Miyaura	8	Kowa/Sankyo	HMG-CoA reductase inhibitor for the treatment of high cholesterol	August 3, 2009	May 2015
Garenoxacin mesylate hydrate	Geninax		Suzuki–Miyaura	9	Astellas Pharma/Taisho Toyama Pharmaceutical Co./Toyama Chemical Co.	Quinolone antibiotic for the treatment of Gram-positive and Gram-negative bacterial infections	Never approved in the US or EU. Approved in Japan (2007).	Not applicable

#	Drug	Brand	Structure	Coupling	Company	Indication	Approval	Status
10	Losartan potassium	Cozaar		Suzuki–Miyaura (same coupling can be employed for structurally similar drugs such as candesartan, irbesartan and valsartan)	Merck	Angiotensin II receptor antagonist for the treatment of high blood pressure	April 28, 1995	Generic (August 2009)
11	Eltrombopag olamine	Promacta (USA), Revolade (EU)		Suzuki–Miyaura	GlaxoSmithKline	Treatment of idiopathic thrombo-cytopenia (abnormally low platelet count)	November 20, 2008	December 2021
12	Rofecoxib	Vioxx Ceoxx Ceeoxx		Suzuki–Miyaura	Merck	Non-steroidal anti-inflammatory	May 20, 1999	Withdrawn
13	Etoricoxib	Arcoxia		Suzuki–Miyaura (two instances)	Merck	COX-II inhibitor for the treatment of pain and inflammation	Never approved in the US. Approved in over 70 countries worldwide	Not applicable

Table 15.2 *(Continued)*

Active ingredient name	Commercial name(s)	Structure	Ref.	Type of coupling(s)	Applicant	Indication	FDA approval date	Exclusivity expiration date (United States)
Abiraterone acetate	Zytiga		14	Suzuki–Miyaura	Jansenn Biotech	Inhibitor of human cytochrome P450$_{17\alpha}$ for the treatment of castration-resistant prostate cancer	April 28, 2011	February 2014
Crizotinib	Xalkori		15	Suzuki–Miyaura	Pfizer	Treatment of metastatic non-small cell lung cancer	August 26, 2011	March 2025
Ruxolitinib phosphate	Jakafi Jakavi		16	Suzuki–Miyaura	Incyte/Novartis	Tyrosine-protein kinase JAK 1/2 inhibitor for treatment of myelofibrosis	November 16, 2011	December 2027
Vemurafenib	Zelboraf		17	Suzuki–Miyaura	Plexxikon/Genentech/F. Hoffmann La Roche	B-Raf enzyme inhibitor for the treatment of unresectable or metastatic melanoma	August 17, 2011	October 2026

Generic	Brand	No.	Coupling	Structure	Company	Indication		
Valsartan	Angiotan Diovan	18	Suzuki–Miyaura		Novartis	Non-peptidic angiotensin-II-receptor antagonist for the treatment of hypertension	December 23, 1996	September 2012
Perampanel	Fycompa	19	Suzuki–Miyaura, Stille, N-arylation		Eisai	AMPA receptor antagonist for the treatment of epilepsy	October 22, 2012	June 2021
Lapatinib ditosylate hydrate	Tykerb	20	Stille		SmithKline Beecham (now part of GlaxoSmith-Kline)	ErB-1 and ErB-2 dual kinase inhibitor for the treatment of breast cancer	March 13, 2007	July 2017
Rilpivirine hydro-chloride	Edurant	21	Heck		Janssen Products (originally developed by Tibotec)	Reverse transcriptase inhibitor for the treatment of HIV-1 infection	May 20, 2011	February 2021
Nabumetone	Relafen Relifex Gambaran (Belgium, Luxem-bourg)	22	Heck		Hoechst–Celanese (originally developed by Beecham)	Non-steroidal anti-inflammatory	December 1991	Generic

Table 15.2 (*Continued*)

Active ingredient name	Commercial name(s)	Structure	Type of coupling(s)	Ref.	Applicant	Indication	FDA approval date	Exclusivity expiration date (United States)
Naratriptan hydrochloride	Amerge Naramig		Heck	23	GlaxoSmithKline	Treatment of migraine	February 10, 1998	Generic (July 2010)
Almotriptan malate	Axert Almogran Almotrex Amignul		Heck (intramolecular)	24	Almirall	5-Hydroxytryptamine receptor subtype agonist for the treatment of migraine	May 7, 2001	May 2015
Eletriptan hydrobromide	Relpax		Heck	25	Pfizer	5-Hydroxytryptamine 1B/1D receptor agonist for the treatment of migraine	December 26, 2002	December 2016
Ondansetron	Zofran		Heck (intramolecular)	26	GlaxoSmithKline	Serotonin 5-HT$_3$ receptor antagonist for the treatment of nausea and vomiting after chemotherapy	January 1991	Generic (December 2006)

Drug	Brand names	Structure	Coupling	No.	Company	Indication	Date	Generic status
Naproxen	Aleve	Heck → / Carbonylation → ; CO_2H; MeO	Heck, carbonylation	27	Albermarle	Non-steroidal anti-inflammatory	1976	Generic
Ketoprofen	Orudis Oruvail	Heck → / Carbonylation → ; CO_2H	Heck, carbonylation	28	Originally developed by Rhône-Poulenc	Non-steroidal anti-inflammatory	1967	Generic
Axitinib	Inlyta	Migita → ; Heck → ; O; NHMe; N	Heck, Migita	29	Pfizer	Vascular endothelial growth factor (VEGF) antagonist for the treatment of solid tumors, including metastatic renal cell carcinoma	January 27, 2012	June 2020
Fexofenadine hydrochloride	Allegra Fexidine Telfast Fastofen Tilfur Vifas Telfexo Allerfexo	CO_2H; OH; N; •HCl; Ph Ph OH	Sonogashira	30 (McGill University)	Sanofi-Aventis (originally developed by Sepracor in 1993)	Antihistamine for the treatment of hay fever, allergy symptoms and urticaria	July 1996	Generic (February 2001)

Table 15.2 (*Continued*)

Active ingredient name	Commercial name(s)	Structure	Ref.	Type of coupling(s)	Applicant	Indication	FDA approval date	Exclusivity expiration date (United States)
Ponatinib hydrochloride	Iclusig		31	Sonogashira (two instances)	Ariad Pharmaceuticals	BCR-ABL inhibitor for the treatment of chronic myeloid leukemia and Ph+ acute lymphoblastic leukemia	December 14, 2012	December 2026
Terbinafine hydrochloride	Lamisil		32	Sonogashira	Novartis	Antifungal	March 17, 2000	Generic (June 2007)
Rizatriptan benzoate	Maxalt		33	Sonogashira followed by Larock indolization	Merck	5-HT$_{1D}$ receptor agonist for the treatment of migraine	June 29, 1998	Generic (December 2012)
Ezetimibe	Zetia Ezetrol		34	Negishi	Merck/Schering-Plough	Cholesterol absorption inhibitor	October 25, 2002	September 2013

Generic name	Brand name	Structure	No.	Coupling	Company	Indication	Approval date	Patent expiration
Vismodegib	Erivedge		35	Negishi	Genentech	SMO receptor antagonist for the treatment of basal cell carcinoma	January 30, 2012	November 2028
Tazarotene	Tazorac Avage Zorac		36	Negishi	Allergan	Topical retinoid for the treatment of acne, psoriasis and sun damaged skin	June 13, 1997	June 2011 (no generic version available)
Oseltamivir phosphate	Tamiflu		37	Carbonylation	Gilead Sciences/ F. Hoffmann-La Roche	Neuraminidase inhibitor for the treatment of influenza	October 27, 1999	December 2016
Ibuprofen	Motrin Nurofen Advil Nuprin		38	Carbonylation	Hoechst-Celanese	Non-steroidal anti-inflammatory	1961	Generic
Diflunisal	Dolobid		39	Kumada-Corriu	Merck	Non-steroidal anti-inflammatory	April, 1982	April 1992
Varenicline	Chantix Champix		40	Carbonyl α-arylation	Pfizer	Smoking cessation	May 11, 2006	November 2018

Table 15.2 (*Continued*)

Active ingredient name	Commercial name(s)	Structure	Type of coupling(s)	Ref.	Applicant	Indication	FDA approval date	Exclusivity expiration date (United States)
Abacavir sulfate	Ziagen		Tsuji–Trost	41	Glaxo Wellcome	Reverse transcriptase inhibitor for the treatment of HIV	December 18, 1998	Generic (December 2009)
Aripiprazole	Abilify Aripiprex		C–N bond formation	42	Otsuka (marketed with Bristol-Myers Squibb)	Partial dopamine agonist for the treatment of schizophrenia, bipolar disorder and clinical depression	November 15, 2002	October 2014
Eribulin mesylate	Halaven		Nozaki–Hiyama–Kishi (asymmetric)	43	Eisai	Mitotic inhibitor for the treatment of cancer	November, 2010	June 2019

Generic name	Brand name	Structure	Coupling type	Entry	Company	Description	Date 1	Date 2
Nilotinib hydrochloride monohydrate	Tasigna	(structure)	C–N bond formation	44	Novartis	Signal transduction inhibitor for the treatment of chronic myeloid leukemia	October 29, 2007	July 2023
Lumiracoxib	Prexige	(structure)	C–N bond formation	45	Novartis	Cyclooxygenase-2 selective inhibitor for the treatment of inflammation	Never approved in the US. Withdrawn from the EU and Australia	Not applicable
Imatinib mesylate	Gleevec Glivec	(structure)	C–N bond formation	46	Novartis	Treatment of chronic myeloid leukemia	May 10, 2001	January 2015
Raloxifene hydrochloride	Evista	(structure)	C–N bond formation (followed by amine displacement by aryl Grignard)	47	Eli Lilly	Estrogen receptor modulator for the treatment of osteoporosis	December 9, 1997	March 2014
Apixaban	Eliquis	(structure)	C–N bond formation (2 instances)	48	Bristol-Myers Squibb/Pfizer	Factor Xa inhibitor for the prevention of venous thrombo-embolic events after hip or knee replacement surgery	December 28, 2012	December 2019

Table 15.2 (*Continued*)

Active ingredient name	Commercial name(s)	Structure	Type of coupling(s)	Ref.	Applicant	Indication	FDA approval date	Exclusivity expiration date (United States)
Avanafil	Stendra Spedra		C–N bond formation	49	Vivus	PDE5 inhibitor for the treatment of erectile dysfunction	April 27, 2012	September 2020
Irbesartan	Aprovel Karvea Avapro		Ullmann biaryl synthesis	50	Sanofi-Aventis (marketed with Bristol-Myers Squibb)	Angiotensin II receptor antagonist for the treatment of hypertension	September 30, 1997	Generic (March 2012)
Carvedilol	Carvil Coreg Dilatrend Eucardic Carloc		Ullmann homo-coupling	51	Wyeth	Non-selective β-blocker/α-1 blocker of adrenergic receptor for the treatment of congestive heart failure	1995	Generic (March 5, 2007)

15.2 Carbon–Carbon Bond Formation

15.2.1 Suzuki–Miyaura Coupling

The Suzuki–Miyaura reaction is the transition metal-catalyzed coupling of an organoboron nucleophile with an electrophilic partner such as an organohalide or activated alcohol (*e.g.*, sulfonate). Initially reported in 1979,[52] the Suzuki reaction has become the most prevalent tool for the formation of aryl–aryl bonds. The process group at Merck published one of the earliest pharmaceutical examples of Suzuki coupling for the synthesis of losartan in 1994.[10] Since that time, this named reaction has been the most frequently reported Pd-catalyzed transformation for the large-scale synthesis of drug candidates.[53] The popularity of this coupling for the synthesis of highly functionalized compounds stems from its broad functional group compatibility and the commercial availability of many boronic acids and esters. Like most transition metal-catalyzed couplings, the Suzuki reaction can be sensitive to oxygen, but thorough inertion can minimize issues from oxygen contamination. Nickel can be an alternative catalyst to Pd for this transformation, although this is much less common.[54] Environmental considerations are paramount when scaling any process and the green aspects of the Suzuki reaction have been reviewed previously.[55] The recent applications of the Suzuki reaction in total synthesis have also been reviewed.[56]

The process group at Amgen implemented the Suzuki coupling of bromide **1** and boronic acid **2** for the synthesis of p38 MAP kinase inhibitor **4** for the treatment of conditions such as rheumatoid arthritis, Crohn's disease and psoriasis (Scheme 15.1).[53ar] A screen of various parameters including catalyst, base and solvent revealed PdCl$_2$(A-Phos)$_2$, K$_3$PO$_4$ and IPA/H$_2$O as the optimal conditions for this cross-coupling. In particular, the process team was interested in avoiding class 2 solvents (hence IPA and H$_2$O) and noted K$_3$PO$_4$ as a privileged base for large-scale Suzuki couplings that performs well for a wide range of substrates and does not off-gas upon neutralization.

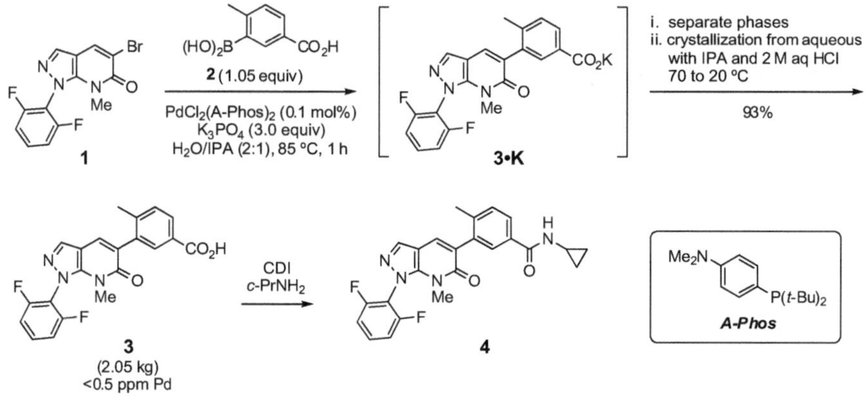

Scheme 15.1 Suzuki coupling for the synthesis of p38 MAP kinase inhibitor **4**.

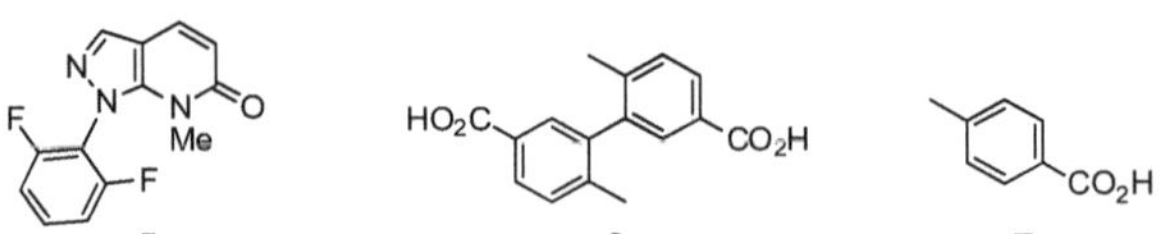

Figure 15.1   Impurities from the Suzuki coupling of bromide **1** and boronic acid **2**.

The A-Phos ligand proved to be the most effective at suppressing the formation of by-products desbromo **5**, homocoupled **6** and deboronated **7** (Figure 15.1).

On the kilogram scale, bromide **1** and boronic acid **2** in IPA were treated with aqueous K_3PO_4 to form a biphasic mixture (whereas IPA and H_2O typically form one phase, the presence of salts in the aqueous layer made the two phases immiscible). The biphasic mixture was charged with $PdCl_2(A\text{-}Phos)_2$ and the atmosphere was made inert before heating at 85 °C for 1 h. The Suzuki product **3** was formed as the potassium salt and resided predominantly in the aqueous phase. Washing the aqueous layer with toluene purged the Pd–ligand complex and subsequent acidification with aqueous HCl crystallized the Suzuki product as the acid. This pH adjustment was carried out at 70 °C with slow cooling to crystallize **3** with large particle size for easy filtration. Without acidification at elevated temperature, the crystallization of **3** was nucleation dominated and the resulting small particle size resulted in a very slow filtration. Ultimately, this process provided over 2 kg of Suzuki product **3** containing <0.5 ppm of Pd.

Fiorelli and co-workers at the Istituto Italiano di Tecnologia in Genova, Italy, developed the Suzuki coupling of boronic ester **10** and bromide **11** for their multi-gram synthesis of fatty-acid amide hydrolase inhibitor **13** for the treatment of nociceptive and inflammatory pain (Scheme 15.2).[53be] This cross-coupling was complicated by the hydrolytic lability of the carbamate found in both the bromide and Suzuki product **12**. For example, **11** hydrolyzed quickly in very polar solvents (DMF, MeCN) using acetate bases common to Suzuki couplings, but the carbamate proved much more stable in less polar solvents such as THF, dioxane and toluene. Initial couplings of boronic acid **8** and **11** in dioxane were sluggish at 80 °C and it was suspected that the low solubility of **8** in dioxane contributed to its poor conversion. The addition of water could improve the solubility of **8**; however, this benefit was offset by increased carbamate hydrolysis. As a result, the team investigated boronic esters such as **10**, which have better solubility in dioxane and better conversion in the Suzuki reaction. On the multi-gram scale, the boronic acid was converted to **10** by heating with ethylene glycol in dioxane and the boronic ester was treated with a mixture of bromide, CsOAc and $PdCl_2(dppf)$ at 80 °C to generate the Suzuki product **12** with only 7% combined by-product from carbamate hydrolysis of **11** and **12**. The reaction mixture was washed with aqueous citric acid solution to neutralize the base and sequester Pd and ultimately the Suzuki product was recrystallized from MeCN to afford 39 g of material in 66% yield. No information was provided on the amount of residual Pd in **12**.

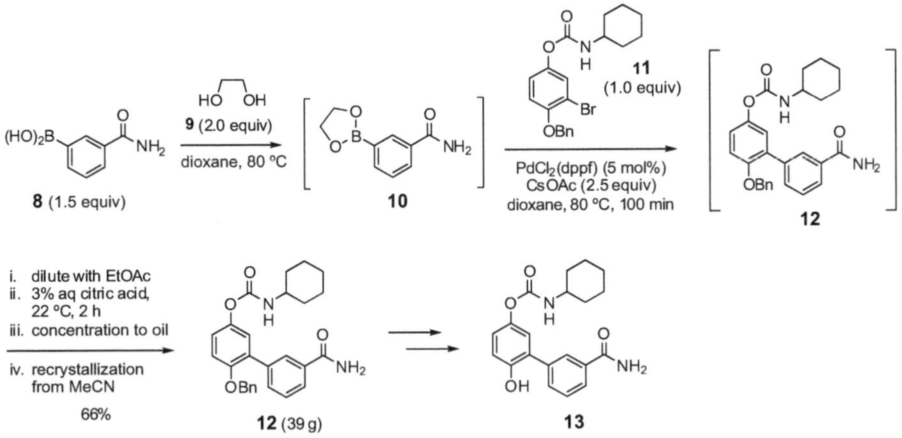

Scheme 15.2 Suzuki coupling of boronic ester **10** and bromide **11**.

15.2.2 Negishi Coupling

The Negishi cross-coupling[57] has seen more limited use by process chemistry groups in the pharmaceutical industry[58] in comparison with other types of couplings. Organozinc reagents display ionic properties that are intermediate between those of organoboron reagents and organomagnesium or organolithium compounds. As a result, they are more reactive than the former and, at the same time, more chemoselective than the latter, which makes them compatible with functional groups such as esters, nitriles, alcohols and amines. However, from the process chemist's point of view, Negishi couplings are operationally more complex since they involve (a) metallation of an organozinc precursor (ArX, ArH) to the corresponding arylmetal (ArLi, ArMgX), (b) zinc–metal exchange with typically $ZnCl_2$ or $ZnBr_2$ to provide the ArZnX species and (c) Pd(0)- or Ni(0)-catalyzed coupling of ArZnX with an electrophilic partner such as an aryl halide, triflate or acid chloride. Other practical factors to consider on a large scale are the substantial amount of zinc-contaminated waste (strictly regulated in most countries)[59] derived from the use of stoichiometric zinc and the need to purge this metal to satisfactory levels from intermediates and APIs, which can be very good metal chelators due to the presence of heterocycles. All these issues combined may have contributed to industry favoring Suzuki–Miyaura over Negishi cross-couplings.

The process group at Hoffmann-La Roche has described the synthesis of naphthylacetic acid CRTH2 receptor antagonist **19** as a candidate for the treatment of inflammatory diseases (Scheme 15.3).[58j] The introduction of the *p*-sulfonylbenzyl portion of the molecule was carried out *via* the Negishi coupling of triflate **15** and organozinc **17**. The original protocol for organozinc formation called for predrying Zn dust and LiCl at 170 °C and then activating the metal with 1,2-dibromoethane (three times) in THF at reflux

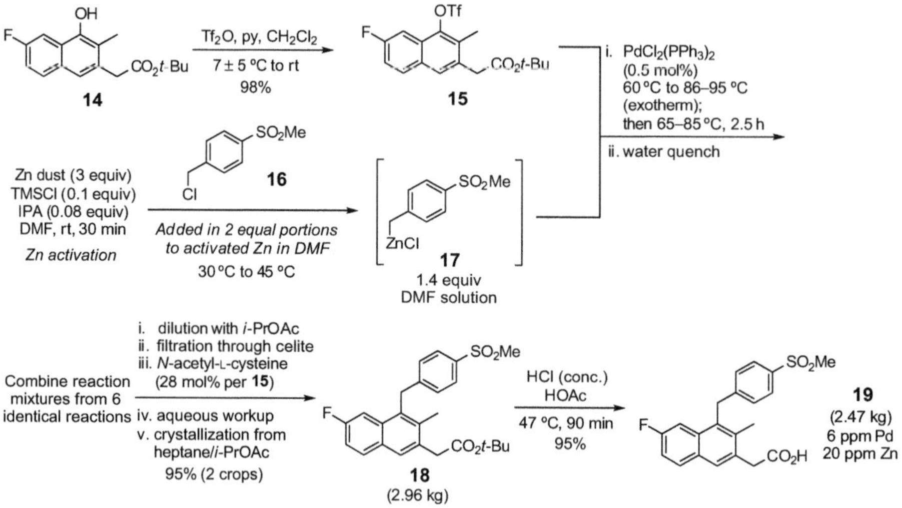

OH

Tf₂O, py, CH₂Cl₂
7 ± 5 °C to rt
98%

F...CO₂t-Bu

14

OTf

F...CO₂t-Bu

15

i. PdCl₂(PPh₃)₂ (0.5 mol%) 60 °C to 86–95 °C (exotherm); then 65–85 °C, 2.5 h
ii. water quench

SO₂Me

Cl

16

Zn dust (3 equiv)
TMSCl (0.1 equiv)
IPA (0.08 equiv)
DMF, rt, 30 min

Zn activation

Added in 2 equal portions to activated Zn in DMF
30 °C to 45 °C

SO₂Me

ZnCl

17
1.4 equiv
DMF solution

Combine reaction mixtures from 6 identical reactions

i. dilution with *i*-PrOAc
ii. filtration through celite
iii. *N*-acetyl-L-cysteine (28 mol% per **15**)
iv. aqueous workup
v. crystallization from heptane/*i*-PrOAc
95% (2 crops)

SO₂Me

F...CO₂t-Bu

18
(2.96 kg)

HCl (conc.)
HOAc
47 °C, 90 min
95%

SO₂Me

F...CO₂H

19
(2.47 kg)
6 ppm Pd
20 ppm Zn

Scheme 15.3 Negishi coupling for the preparation of CRTH2 receptor antagonist **19**.

followed by further activation with TMSCl.[60] This process was highly exothermic and caused considerable foaming, which made it not amenable to scale-up, and therefore an alternative was sought. In the absence of LiCl, the Negishi coupling was not reproducible. However, a simple treatment of Zn dust with TMSCl[61] in DMF generated hydrogen chloride after reaction with the water present in the solvent, which was the actual activating agent *via* Zn surface cleaning.

During the scale-up runs, TMSCl was added to a suspension of Zn dust in DMF and a small amount of IPA (0.06 equiv., to ensure enough hydrogen chloride production) and the resulting mixture was stirred at room temperature for 30 min. Following Zn activation, a solution of benzyl chloride **16** in DMF (containing half of the **16** used in the reaction) was added and the mixture was warmed to 30 °C to start organozinc formation. Zinc insertion was exothermic and, when the internal temperature reached 45 °C, the remainder of **16** was added as a solution in DMF over 25 min. PdCl₂(PPh₃)₂ (0.5 mol%), an inexpensive precatalyst that replaced the original Pd(OAc)₂ (0.5 mol%)/SPhos (1 mol%) combination, and triflate **15** were then added and the reaction mixture was heated to 60 °C. A second exotherm was observed that raised the temperature of the reaction to 86–95 °C over 10–15 min, which then subsided. After 2.5 h at 65–85 °C, the reaction mixture was quenched with water. The same experimental protocol was repeated five more times and all the reaction mixtures were combined and subjected to treatment with *N*-acetyl-L-cysteine (a metal scavenger for Pd and Zn; 28 mol% per triflate **15**). After an aqueous work-up, crystallization from heptane/*i*-PrOAc afforded 2.96 kg of Negishi product **18** in 95% yield (two crops). *tert*-Butyl ester cleavage with concentrated HCl provided API **19** with 6 and 20 ppm Pd

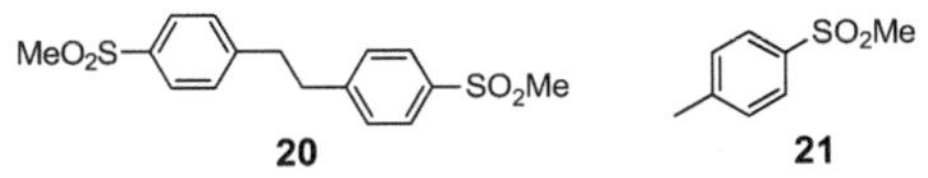

Figure 15.2 By-products from Negishi coupling *en route* to **19**.

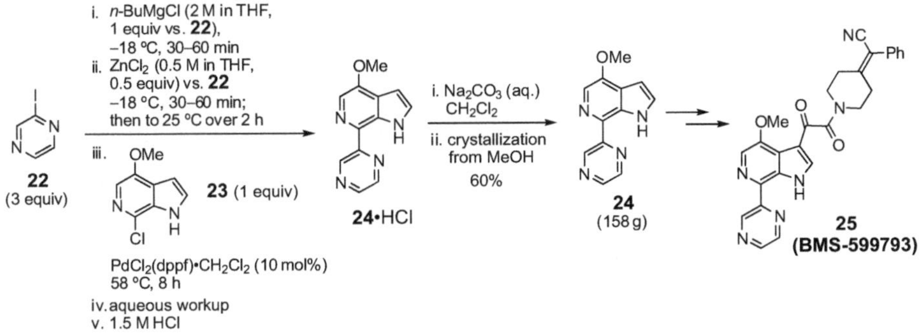

Scheme 15.4 Synthesis of **24** *via* Negishi coupling *en route* to BMS-599793.

and Zn, respectively. The two main by-products from the Negishi coupling were **20** from homocoupling and deschloro-**21** (Figure 15.2). Excess Zn (3 equiv.) minimized the formation of **20** and both were removed in the filtrates after the crystallization of Negishi product **18**.

A collaboration between Princeton API Services, J-Star Research, Scino-Pharm Taiwan and the International Partnership for Microbicides (IPM) resulted in the publication of the large-scale synthesis of BMS-599793 (**25**), a small-molecule HIV entry inhibitor developed by Bristol-Myers Squibb which was licensed to IPM to develop a topical microbicide for use in poor countries (Scheme 15.4).[58k] Medicinal chemistry employed a Stille coupling between 2-(tributylstannyl)pyrazine and a 7-chloro-6-azaindole to assemble the biaryl moiety of the molecule as the last step of the synthesis. However, the organotin reagents were expensive, toxic and difficulty to purge, which made this approach impractical. As alternatives, Suzuki and Negishi couplings were investigated and, based on impurity profiles and reaction kinetics, the researchers opted for developing the Negishi approach. Thus, 2-iodo-pyrazine (**22**) was treated with *n*-BuMgCl at −18 °C to generate the corresponding Grignard reagent. Low temperature was required to prevent the alkylation of the pyrazine *via* reaction of the Grignard formed with *n*-butyl halide by-product. Also, 1 equiv. of *n*-BuMgCl was employed, since excess Grignard reagent competed with the piperazine in the subsequent coupling with azaindole **23**. After pyrazine-Grignard formation, the addition of a THF solution of ZnCl$_2$ followed by warming to 25 °C produced the desired organozinc reagent. The stoichiometry of the ZnCl$_2$ was also important: 1 equiv. per **22** provided an organozinc that performed well in the Negishi coupling, but better results were obtained when only 0.5 equiv. was employed, which afforded complete consumption of chloroazaindole 23 and provided an adequate reaction rate (8 h reaction time). The Negishi coupling between the organozinc reagent (1.5 equiv. of diarylzinc species) and 23 was carried out in the presence of PdCl$_2$(dppf)·CH$_2$Cl$_2$ (10 mol%) at 58 °C for 8 h. The use of an excess of organozinc ensured complete consumption of 23. Upon completion of reaction, an aqueous work-up to remove inorganic salts followed by the addition of 1.5 M aqueous HCl afforded 24·HCl. Freebasing with aqueous Na$_2$CO$_3$, extraction into CH$_2$Cl$_2$ and crystallization from MeOH provided 158 g of 24 in 60% overall yield. No information was provided regarding the need for the very high Pd precatalyst loading or the residual Pd content in 24. This protocol was repeated several times to produce over 2 kg of Negishi product 24.

15.2.3 Kumada–Corriu Coupling

Owing to the very high reactivity and basicity of Grignard reagents, the Kumada–Corriu cross-coupling[62,63] has seen very limited use in process chemistry and has only been applied to relatively simple substrates, in both its Pd- and Ni-catalyzed versions.[64] It is worth pointing out that the Kumada reaction is, at least on a large scale, the only type of cross-coupling where a non-precious metal, Ni, has been employed in more instances than Pd. In addition to these two metals, Fe,[65] Co[66] and Mn[67] can catalyze this transformation but no applications in process chemistry have been found for Co or Mn. Some industrial applications of Kumada coupling for fine chemical production have been reviewed.[63h]

Zacuto and co-workers at Merck reported the preparation of 4-allylisoindoline (28) as a key component of a drug candidate (Scheme 15.5).[64f] The researchers implemented a Pd-catalyzed Kumada coupling to install the allyl moiety using commercially available allyl-Grignard reagents. High-throughput screening was applied to find the optimal Pd(OAc)$_2$–ligand combination and several ligand hits were identified: XPhos, PCy$_2$[(o-tol)indole], P(neopentyl)(t-Bu)$_2$·HBF$_4$, dippf, P(t-Bu)$_2$(Ph)·HBF$_4$, PCy$_2$(Mes) and P(t-Bu)$_3$·HBF$_4$. Based on cost and availability on scale, P(neopentyl)(t-Bu)$_2$·HBF$_4$ was selected for further development. Additional screening showed that better conversions were observed with a 1:2 metal:ligand ratio and that higher processing temperatures (>45 °C) and THF:toluene ratios

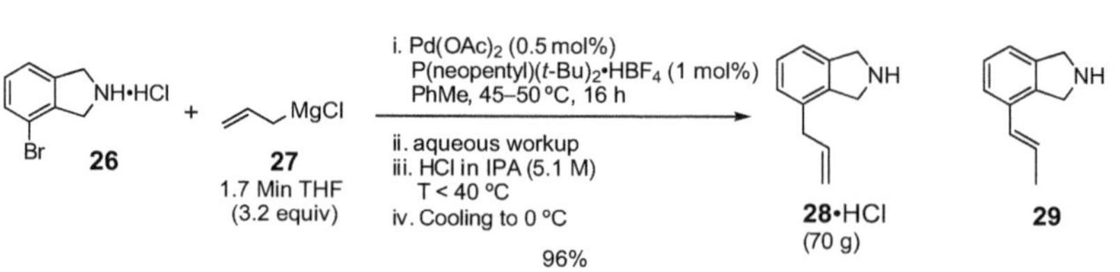

Scheme 15.5 Synthesis of 4-allylisoindoline (28) *via* Pd-catalyzed Kumada–Corriu cross-coupling.

tended to give more of isomer **29**. On scale, a mixture of **26**, Pd(OAc)$_2$ and ligand in toluene was treated with a 1.7 M solution of allylmagnesium chloride (**27**) in THF over 1 h at ≤25 °C. The resulting mixture was then heated at 45–50 °C for 16 h and, after cooling to room temperature, quenched into aqueous 15 wt% citric acid (to dissolve Mg salts and metal chelator). Following an aqueous work-up and concentration, 5.1 M HCl in IPA was added to afford 70 g of **28** · HCl in 96% yield. No information was provided on the level of residual metals in **28** · HCl.

Tewari and co-workers at Ranbaxy Research Laboratories in India described the preparation of olefin **33** *via* an Fe-catalyzed Kumada cross-coupling *en route* to cinacalcet HCl (**34**), a compound for the treatment of secondary hyperparathyroidism in patients with chronic kidney disease and hypocalcaemia in patients with parathyroid carcinoma (Scheme 15.6).[64h] The synthesis of **33** started with the preparation of Grignard **31** from 3-trifluoromethylbromobenzene (**30**) and Mg turnings in THF at reflux. Catalytic I$_2$ was employed to promote Grignard formation. The coupling of **31** with vinyl chloride **32** was carried out by first mixing a THF solution of **32** with an NMP solution of Fe(acac)$_3$ (1.7 mol%). The resulting mixture was added to the Grignard **31** solution at −50 °C while the internal T was held below 0 °C. After an additional 20–30 min, the mixture was quenched with dilute HCl to dissolve Mg salts and extracted into toluene. After a solvent switch and crystallization from *i*-PrOAc, 6.92 kg of **33** were obtained in 61% yield with less than 1% of the *cis* isomer. Before crystallization, the *trans:cis* ratio was 10:1, but both isomers afforded the final product cinacalcet after olefin hydrogenation. No information was given on the amount of residual Pd in Kumada product **33**.

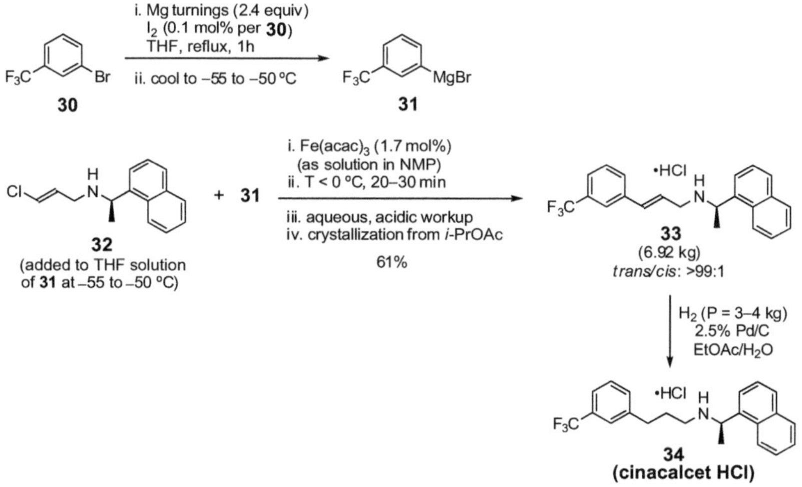

Scheme 15.6   Preparation of cinacalcet HCl *via* Fe-catalyzed Kumada–Corriu cross-coupling.

15.2.4   Enolate Arylation

The transition metal-catalyzed α-arylation of carbonyl groups had been scarcely studied[68] before the groups of Hartwig,[69] Buchwald[70] and Miura[71] published their work in the late 1990s on the coupling of ketones with aryl bromides. Since then, the scope of this technology has been expanded to include esters, β-keto esters, malononitriles, malonates, amides, aldehydes, sulfones and nitroalkanes. Also, asymmetric variants of this important transformation have been developed and applied to aldehydes, ketones, esters and amides.[72] Several reviews have been published in this area.[72,73] Pd is the metal of choice, but Ni and Cu have also been employed. Only recently has this technology begun to be applied by process groups in the pharmaceutical industry for the synthesis of drug candidates[74] and examples with ketones, esters, β-cyano esters, β-ketone esters and lactams have been reported.

Abele and co-workers at Actelion Pharmaceuticals in Switzerland described the large-scale preparation of drug intermediate 39 (Scheme 15.7).[74e] The phenyl group on the molecule was introduced *via* Pd-catalyzed α-arylation of enolate 36, generated by treating ester 35 with freshly prepared LDA at 0–10 °C, with bromobenzene (37). Screening of the reaction conditions showed that either $Pd(OAc)_2$ or $Pd_2(dba)_3$ in combination with $P(t\text{-}Bu)_3 \cdot HBF_4$ gave comparable results and that with chlorobenzene only $Pd_2(dba)_3$ performed satisfactorily. A delayed exotherm was observed for this reaction (the internal temperature rose to 25–29 °C but was easily controlled with a constant jacket temperature of 20 °C), but this exotherm remained constant up to the 4 kg scale. In the plant, neat 35 was added to cold (0–10 °C) LDA in a hexane–toluene mixture. After 10 min at 5–10 °C, $Pd_2(dba)_3$ (1 mol%) and $P(t\text{-}Bu)_3 \cdot HBF_4$ (1 mol%) were added, followed by thorough degassing. Bromobenzene was then added neat over 15 min and, after warming to 20 °C, the mixture was stirred at this temperature for 2 h 45 min.

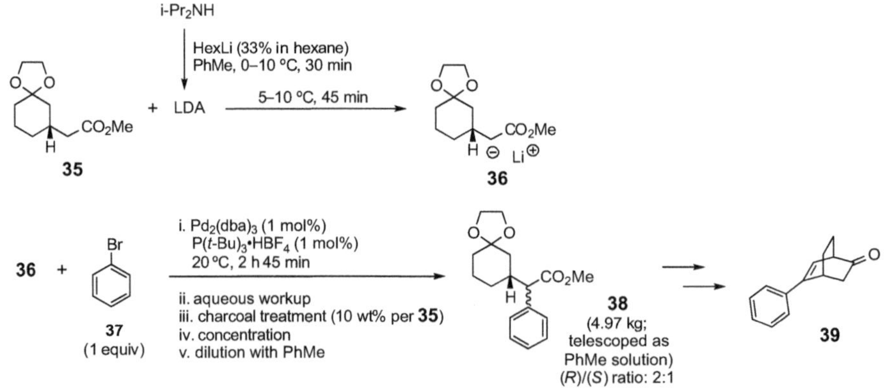

Scheme 15.7 Synthesis of intermediate 39 *via* Pd-catalyzed enolate arylation.

Upon completion of reaction, the mixture was quenched with aqueous citric acid (metal chelator), which precipitated Pd black and by-products from the ligand. Following phase separation and water washes, the product-rich organic layer was treated with charcoal (type not specified; 10 wt% per **35**) at 20–30 °C to remove further Pd residues and color. After filtration through Celite, concentration to remove water *via* azeotropic distillation and further dilution with toluene, 4.97 kg of **38** were isolated as a 2:1 diastereomeric mixture in a toluene solution that was used in the next step without further purification (96% purity by GC-MS). Both diastereomers could be employed in subsequent chemistry without the need to separate them at this stage. It was also mentioned that charcoal performed better than other metal scavenger treatments such as Silicycle's Si-Diamine to remove colored impurities. The combined yield for this step and the subsequent ester reduction to the corresponding alcohol with LiAlH$_4$ was 84%. No information was provided about the level of residual Pd in **38**.

Researchers at AMRI described the preparation of the triple reuptake inhibitor ALB 109780 (**43**), a candidate for the treatment of depressive disorders (Scheme 15.8).[74f] α-Arylation of lactam **40** was implemented in the presence Pd(OAc)$_2$ (5 mol%), (±)-BINAP (5 mol%) and NaO*t*-Bu as base in dioxane at 80 °C. Even though dioxane is carcinogenic and therefore generally avoided by process chemists, this solvent was chosen for further development since it provided better conversion and product purity than toluene. With the goal of taking this chemistry into the plant, calorimetric studies revealed that the addition of NaO*t*-Bu was exothermic but that heat evolution could be managed *via* slow dosing of the base. A second exotherm occurred when the reaction temperature reached 60 °C, which generated enough heat potentially to give rise to a runaway reaction if appropriate cooling was not applied. This second exotherm could also be controlled by slow addition of the aryl bromide at high temperature, but this approach led to the formation of more impurities. Based on these results, it was decided to carry out multiple small-scale reactions in which heat evolution could be more easily managed than in a larger vessel. In the laboratory, to a mixture of **40**, **41**, Pd(OAc)$_2$ and (±)-BINAP in 1,4-dioxane was added NaO*t*-Bu portionwise over 30 min. The resulting mixture was then heated to 60 °C (at this point the reaction became self-heating and external heating was stopped)

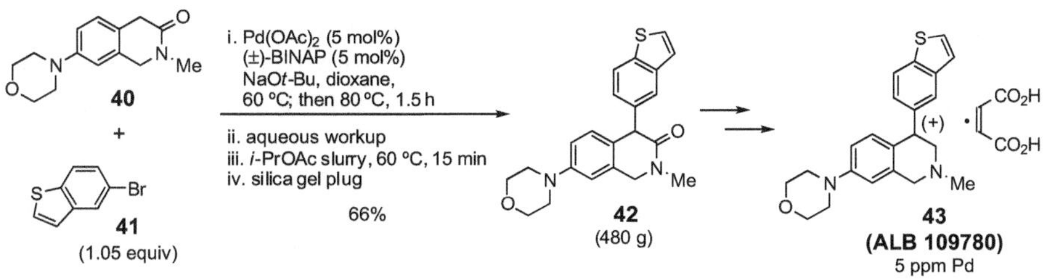

Scheme 15.8 Pd-mediated α-arylation *en route* to ALB 109780.

followed by further heating to 80 °C once the exotherm had subsided. After 1.5 h, the reaction was subjected to an aqueous work-up followed by a hot slurry in *i*-PrOAc and a silica plug using EtOAc as eluent to remove inorganic impurities. Racemic α-arylation product **42** was obtained in 66% yield as a brown solid (480 g). No information was provided on the level of residual Pd in **42**, but API **43** was isolated with 5 ppm Pd.

15.2.5   Sonogashira Coupling

The Cu and Pd co-catalyzed Sonogashira reaction (also known as the Sonogashira–Hagihara reaction)[75] has found an important niche in the toolbox of process chemists,[76] since it is the most useful method for preparing conjugated arenynes and enynes *via* coupling of terminal alkynes and sp[2]-hybridized carbons (aryl, heteroaryl and alkenyl halides). In addition to pharmaceuticals, the products derived from this reaction find applications in the synthesis of natural products, agrochemicals and molecular materials. Some of the advantages of the Sonogashira reaction are its technical simplicity, high functional group compatibility and the typically high yields of products. In addition to the Cu–Pd combination, this coupling can also be carried out by other metal–ligand complexes derived from Fe, Ru, Co, Ni, Ag, Au and In.[77]

The applications of the Sonogashira reaction have been recently reviewed.[78] Variations using only substoichiometric Pd[78c] or Cu (catalytic Stephens–Castro reaction[79]) have also been reported. The Sonogashira reaction has been employed in the synthesis of heterocycles containing nitrogen, oxygen and sulfur.[79a] In general, milder conditions can be employed with electron-rich acetylenes in comparison with less reactive electron-poor substrates. A common side reaction in Sonogashira couplings arises from alkyne homocoupling, but the levels of homocoupling by-products can be minimized by thorough degassing of the reaction mixture and through the use of high concentrations of a secondary or tertiary amine base to promote reduction of Pd(II) to Pd(0).

The process group at Wyeth Research described an application of a Sonogashira coupling for the large-scale preparation of GRN-529 (**47**), an mGluR5 negative allosteric modulator for the treatment of central nervous system disorders (Scheme 15.9).[76s] The original medicinal chemistry conditions for the coupling of aryl iodide **44** with 2-ethynylpyridine (**45**) necessitated very high PdCl$_2$(PPh$_3$)$_2$ (20 mol%) and CuI (20 mol%) loadings, employed Et$_3$N as base and required chromatography to purify alkyne **46**. All these factors led to reaction optimization for developing a scalable and more economical process. Thus, a solvent screen revealed that when the reaction was carried out in 2-MeTHF, a difficult phase cut was encountered owing to the very dark color of both the organic and aqueous phases. In addition, the crystallization of **46** from heptane–2-MeTHF gave material contaminated with black tar and very high levels of Pd (>1900 ppm). Switching to NMP as solvent maintained the dark impurities in solution while **46** could be

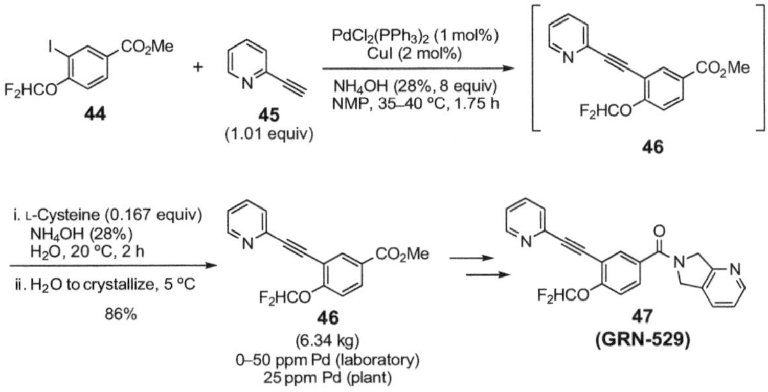

Scheme 15.9 Synthesis of intermediate **46** *via* Sonogashira coupling.

crystallized from water–NMP. Also, replacing Et_3N with NH_4OH as base allowed the reaction temperature to be decreased from 100 to 30–40 °C.

Further optimization showed that the amount of Pd precatalyst could be reduced to only 0.4 mol% provided that the reaction mixture was sparged subsurface with nitrogen gas, an operation that, however, could not be implemented in the kilogram-scale laboratory facility. Although vacuum/refill cycles were effective for removing oxygen, this less rigorous method required a slightly higher, but still acceptable at this stage of development, catalyst loading (0.8 mol%). To ensure robustness in the plant, the reaction was carried out with 1 mol% Pd precatalyst. At the same time, the amount of CuI was reduced from 20 to 2 mol% without adverse effects. Finally, the amount of 2-ethynylpyridine was also decreased from 1.5 to only 1.01 equiv., which still provided complete reaction and almost eliminated the formation of the dimer from alkyne homocoupling.

In the plant, to a solution of aryl iodide **44** and 2-ethynylpyridine (**45**) in NMP was added CuI and the contents of the reactor were degassed *via* three vacuum/refilling cycles (Scheme 15.9). $PdCl_2(PPh_3)_2$ was then charged, the mixture was heated to 35–40 °C and NH_4OH (28% solution) was added. Upon completion of reaction (1 h), a solution of L-cysteine (0.167 equiv. per **44**) in water and NH_4OH (28% solution; needed to fully dissolve the L-cysteine) was added to scavenge Pd and induce crystallization at 35–40 °C. Upon cooling to 20 °C, addition of water and further cooling to 5 °C, Sonogashira product **46** was isolated *via* filtration. Washing the cake with water (1×), NH_4OH (28%, 2×), H_2O (1×) and $MeOH$–H_2O (2:1; this final wash was to reduce the Pd content further) afforded 6.34 kg of **46** in 86% yield with only 25 ppm Pd.

The process group at Boehringer Ingelheim in the United States reported the preparation of BI 653048 BS (**52**), a glucocorticoid agonist candidate for the treatment of rheumatoid arthritis (Scheme 15.10).[76t] The introduction of the 1*H*-pyrrolo[2,3-*c*]pyridine moiety on the molecule was accomplished by the medicinal chemistry group in a two-step sequence that involved a Sonogashira coupling [$PdCl_2(PPh_3)_2$, CuI] followed by a DBU-promoted

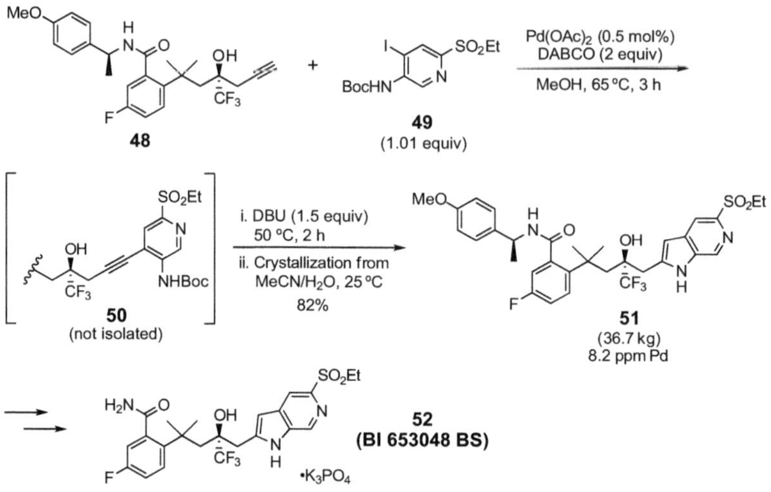

Scheme 15.10 Synthesis of intermediate **51** *via* Sonogashira coupling and cycliza-tion *en route* to BI 653048 BS.

cyclization. On a large scale, the researchers investigated a one-pot Sonogashira–cyclization sequence to couple alkyne **48** with aryl iodide **49** and thus avoid the isolation of intermediate **50**. Reaction optimization for the Sonogashira step showed that Pd(OAc)$_2$ gave the best results compared with other Pd sources and that, by performing the coupling in the absence of Cu salts, complete conversion and minimization of alkyne homocoupling were obtained. After screening bases (NaOMe, *N*-methylpyrrolidine, Et$_3$N, DBU, quinuclidine, *i*-PrNEt, piperidine, tetramethylguanidine, DABCO) and solvents (MeOH, EtOH, *n*-PrOH, IPA), DABCO and MeOH were selected for scale-up work. Unfortunately, no Sonogashira coupling was observed with DBU, which would have eliminated the need for two bases. In the plant, a mixture of **48**, **49** (1.01 equiv.), Pd(OAc)$_2$ (0.5 mol%) and DABCO (2 equiv.) in MeOH was heated at 65 °C for 3 h. Upon completion of coupling, DBU (1.5 equiv.) and further stirring at 50–55 °C effected the cyclization of **50** with loss of the Boc group to afford **51**. The isolation of **51** involved the addition of MeCN followed by water at 45–55 °C, seeding with crystals of **51** (0.005 wt% per **48**) and cooling to 25 °C to produce 36.7 kg of **51** in 82% yield with only 8.2 ppm Pd. These crystallization conditions afforded a fast-filtering solid in contrast to the original protocol using MTBE–water to precipitate **51**, which gave a very slow filtration rate and also higher residual Pd (30–70 ppm).

15.2.6 Heck Coupling

The Heck reaction has become an indispensable technology for C–C bond formation and it is currently second only to the Suzuki cross-coupling in number of applications.[2] In agreement with this statement, numerous inter-[80] and intramolecular[81] examples can be found in the field of process chemistry for the synthesis of pharmaceuticals.[82] Natural product synthesis[83] and other industrial applications[84] have also benefited from this reaction. Many types of substrates can now undergo Heck coupling (triflates, carbonyl and sulfonyl chlorides, diazonium salts, iodonium salts),[85] but only aryl halides (chlorides, bromides and iodides) have been used on a large scale for the synthesis of drug candidates. The asymmetric version[86] has also been developed but no examples on scale have been found. Numerous reviews on the Heck reaction[87] and its mechanistic aspects[88] have been published.

The process group at Merck reported the synthesis of hepatitis C virus (HCV) NS3/4a protease inhibitor vaniprevir (57, MK-7009). Two approaches to this macrocyclic structure were devised. The first involves a Heck reaction that brings together two advanced intermediates to give a substrate that then undergoes macrolactamization to afford the 20-membered macrocycle (Scheme 15.11).[80v] The second approach generates the macrocycle *via* ring-closing metathesis (see Section 15.2.11).

The Heck approach to vaniprevir (57) is shown in Scheme 15.11 and involves the coupling of aryl bromide 53 and unactivated, terminal olefin 54. A Pd source and ligand screen revealed that this transformation could be carried out under a variety of conditions (>85% assay yield) but, at the same time, 10–12% of undesired *exo* regioisomer 56 was produced. The fact that a wide variety of Pd–ligand combinations gave similar yields of 56 suggested that the ligand does not intervene in the step where the regioselectivity is determined. As a result, the Heck coupling was tested under ligandless conditions and comparable yield and regioselectivity were then observed. Further optimization showed that the Pd(OAc)$_2$ loading could be reduced to

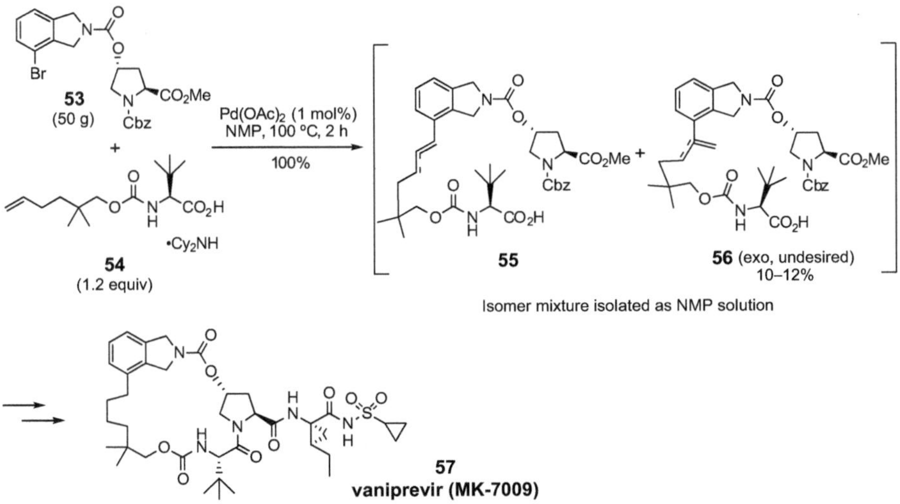

Scheme 15.11 Pd-catalyzed Heck approach to vaniprevir (**57**).

only 0.75 mol% in NMP as solvent without adverse effects. In the laboratory, to a degassed solution of **53** (50 g) and **54** at 45 °C (to help dissolve the two reactants and prevent foaming during nitrogen–vacuum purge cycles) was added Pd(OAc)$_2$ (1 mol%) followed by further degassing and heating to 100 °C. Upon completion of reaction (2 h; formation of a black precipitate, presumably Pd black, coincided with the end of the reaction), the dark mixture was cooled to room temperature and treated with BHT to prevent air oxidation. This NMP solution of intermediate **55** was telescoped directly into the next step (olefin hydrogenation). It was mentioned that the Pd from the Heck coupling was sufficient to catalyze the subsequent olefin reduction but, in order to ensure full conversion ($\geq$99.8%) to the saturated intermediate, 5 wt% of 10% Pd/C was added. No information was provided about the level of residual Pd after the Heck or hydrogenation steps or in the final API.

The preparation of drug candidate **61**, a neuropeptide calcitonin gene-related peptide receptor inhibitor for the treatment of migraine, was described by the process group at Bristol-Myers Squibb (Scheme 15.12).[80x] A Heck coupling was implemented between iodoaniline **58** and aminoacrylate **59** in the presence of only 0.005 mol% Pd(OAc)$_2$, TBAC (0.5 equiv.; catalyst stabilizer) and K$_2$CO$_3$ in a 1:1 THF–H$_2$O mixture at 65–70 °C with no additional ligand. Upon completion of reaction (5 h), the mixture was diluted with THF–H$_2$O and filtered through Celite and a 10 µm Cuno cartridge in-line filter to remove Pd by-products. After an aqueous work-up, the organic phase was concentrated and, following a solvent swap to MeOH, water was added to crystallize Heck product **60** in 66% yield and 94.1% purity. Intermediate **60** was then subjected to an extensive purification protocol that consisted of conversion to its mesylate salt followed by freebasing with aqueous, methanolic K$_2$CO$_3$ and an additional recrystallization from EtOAc to afford 14.7 kg of **60** (45% overall yield) with 98% purity. No information was provided on the level of residual Pd in **60**.

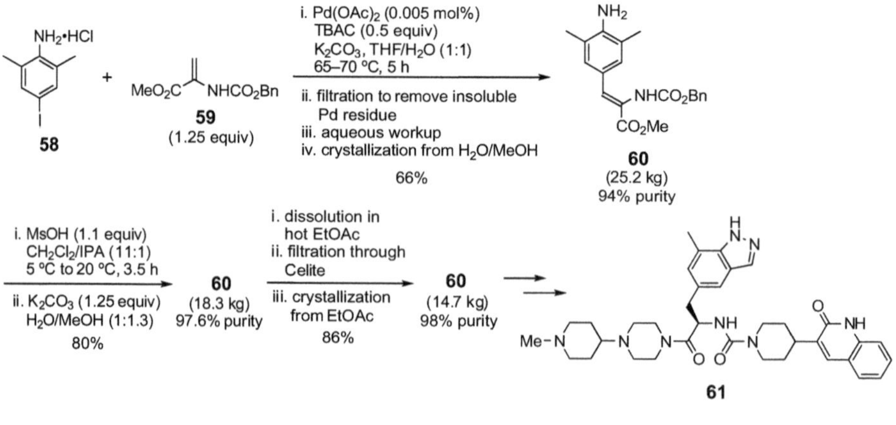

Scheme 15.12   Heck coupling for the preparation of aminoacrylate **60**.

15.2.7   Hayashi–Miyaura Coupling

The 1,4-addition (Michael addition) of alkenyl- and arylboronic acids to α,β-unsaturated ketones catalyzed by Rh(acac)(CO)$_2$ and dppb was first reported by Miyaura and co-workers in 1997.[89] Shortly after, Miyaura, Hayashi and co-workers went on to describe the first asymmetric version with the Rh(acac)(C$_2$H$_4$)$_2$/BINAP catalytic system.[90] Compared with other 1,4-additions, the Hayashi–Miyaura reaction has the following advantages: (a) boronic acids display better oxygen and moisture stability than other organometallic reagents; (b) lack of 1,2-addition by-products; (c) both alkenyl and arylboronic acids can be employed as nucleophiles; (d) α,β-unsaturated ketones, esters, amides, phosphonates and nitro compounds can perform as electrophiles; and (e) asymmetric induction can be accomplished through the use of chiral ligands.[91] Nevertheless, owing to the relatively recent disclosure of this technology and the relatively high cost of Rh, very few examples on a large scale have been reported to date.[92] Numerous reviews on the Rh-catalyzed Michael addition of organoboron reagents to activated olefins have been published in recent years.[91,93]

The medicinal chemistry group at Merck reported a Rh-catalyzed, asymmetric Hayashi–Miyaura coupling *en route* to telcagepant (MK-0974; 65), a calcitonin gene-related peptide receptor antagonist for the treatment of migraine (Scheme 15.13).[92b] Initially, when the researchers applied typical reaction conditions for the asymmetric Michael addition of 2,3-difluorophenylboronic acid (63) to nitroolefin 62 [Rh(acac)(C$_2$H$_4$)$_2$/(S)-BINAP (3 mol%) in dioxane–water (10:1) at 100 °C],[91,94] incomplete reaction was observed due to the extensive decomposition of 63 (normally observed for fluorophenylboronic acids). The decomposition of 63 could be minimized by running the reaction at 45 °C but at the expense of further charges of catalyst (10–20 mol%), which provided 64 in 80% yield and 93:7 *dr*. Further optimization was focused on the use of additives to increase the reaction kinetics.[95] It was found that the reaction could be carried out with 3 mol%

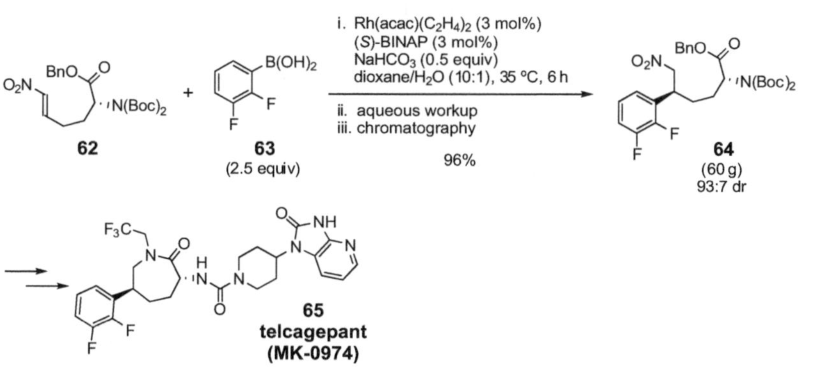

Scheme 15.13   Asymmetric Hayashi–Miyaura boronic acid addition *en route* to telcagepant.

catalyst and 2.5 equiv. boronic acid **63** at 35 °C in the presence of 0.5 equiv. NaHCO$_3$ (no list of additives was provided), which afforded Michael addition product **64** in 96% yield with the same 93:7 *dr* on a 60 g scale. This technology was further implemented on a 2 kg scale (no detailed experimental protocol was provided). No information was given on the level of residual Pd in **64**.

15.2.8   Tsuji–Trost Allylation

The Tsuji–Trost reaction was originally discovered by Tsuji *et al.*[96] in the mid-1960s and represented the first example of a Pd-mediated C–C bond formation.[97] Following this seminal discovery that employed stoichiometric Pd, conditions were developed that were substoichiometric in metal.[98] Trost and co-workers further developed this reaction to convert it into an asymmetric process.[99] In addition to Pd, examples of Ir-[100] and Fe-catalyzed[101] Tsuji–Trost reactions have been reported. The application of the Tsuji–Trost reaction in process chemistry is limited, but several examples have been published, mostly since the early 2000s.[102]

The large-scale preparation of the naturally occurring quinolizidine alkaloid (–)-huperzine A (**69**), a potent acetylcholinesterase inhibitor for the potential treatment of Alzheimer's disease, has been reported thanks to a collaboration between Shasun Pharma Solutions, Rhodia Pharma Solutions and Debiopharm SA Forum "aprés-demain" (Scheme 15.14).[102f] The bicyclic core on the molecule was installed *via* an asymmetric Pd-catalyzed annulation originally developed by Terashima's group in Japan between β-keto ester **66** and diacetate **67**.[103] Since Terashima's asymmetric synthesis employing Pd(OAc)$_2$ (20 mol%) and chiral ferrocenyl ligand **70** (40 mol%, Figure 15.3) gave **68** with modest enantioselectivity (64% *ee*) under a variety of reaction conditions, a screen of 33 mono- and diphosphine ligands was carried out in collaboration with Solvias. As a result, Taniaphos ligand SL-T002-1 (**71**, Figure 15.3) was selected as the ligand of choice. Further optimization of the reaction conditions (catalyst loading, solvent, concentration and temperature) allowed for a 20-fold reduction in the amount of Pd source and ligand. In the laboratory, the large-scale preparation of **68** involved stirring a mixture of ligand **71** and [PdCl(π-allyl)]$_2$ in acetone at 25 °C for 1 h to form the active catalyst. Diacetate **67** was then added and the mixture was

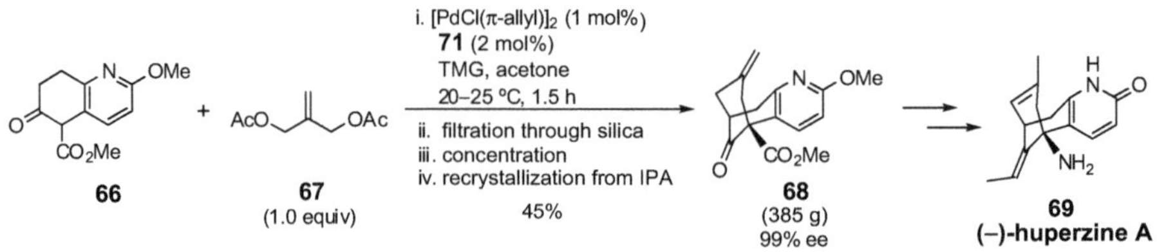

Scheme 15.14   Large-scale preparation of the bicyclic core of (–)-huperzine A *via* Tsuji–Trost coupling.

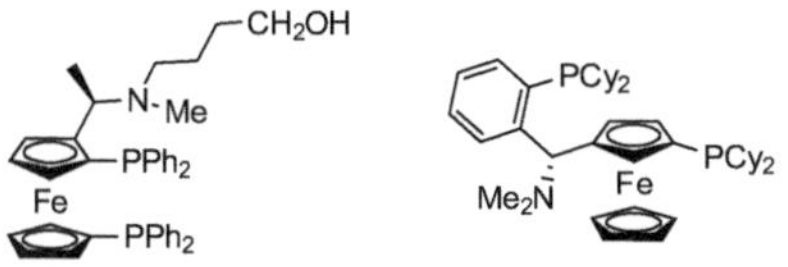

Figure 15.3 Structures of chiral ligands employed in the asymmetric synthesis of (–)-huperzine A.

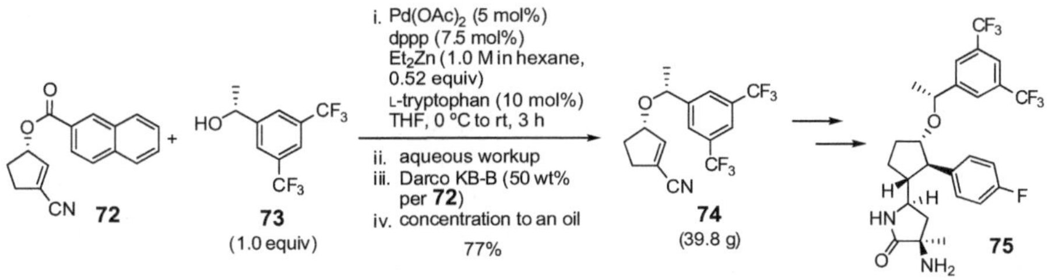

Scheme 15.15 Ether formation *via* Pd-catalyzed Tsuji–Trost allylation.

further stirred for a further 1 h. A solution of keto ester **66** and tetra-methylguanidine (TMG) in acetone was added over 30 min and the resulting mixture was stirred at 20–25 °C for 1 h. Upon completion of reaction, the acetone was removed *via* vacuum distillation and the residue was passed through a silica bed, eluting with hexane–EtOAc to remove Pd salts and ligand. The eluate was concentrated and the residue (84% *ee*) was re-crystallized from IPA to afford 385 g of bicyclic intermediate **68** in 45% yield and 99% *ee*. It was mentioned that the level of residual Pd in **68** was not determined but after converting (–)-huperzine to an imine derivative, the Pd content was below 20 ppm.

The process group at Merck described the synthesis of hNK-1 receptor antagonist **75**, a drug candidate for the treatment of chemotherapy-induced and postoperative nausea and vomiting (Scheme 15.15).[102e] Intermediate **72** was obtained *via* asymmetric enzymatic reduction of 3-cyanocyclopentenone followed by treatment with 2-naphthoyl chloride. The next step involved a Tsuji–Trost coupling between **72** and chiral benzylic alcohol **73** with Pd(OAc)$_2$ and 1,3-diphenylphosphinopropane (dppp) as ligand in the presence of catalytic amounts of Et$_2$Zn (reaction activator) and L-tryptophan (*N,O*-chelator to increase the reaction rate)[104] to generate ether **74** *via* an η^3-allylmetal species. The reaction proceeds through the formation of a Zn alkoxide that adds to the η^3-allylpalladium complex.[105] After completion of reaction, the mixture was filtered and the filtrates were subjected to an aqueous work-up. The product-rich organic phase was treated with Darco KB-B (50 wt% per **72**) at room temperature for 18 h, filtered and concentrated to afford 39.8 g of **74** as an oil that was used in the next step without further purification. It was mentioned that the acetate ester of **72** also afforded **74** but the naphthyl ester was chosen since it is a crystalline solid that facilitated the isolation of **72**. No information was provided on the levels of residual Pd in **74**.

15.2.9   Carbonylation

The first report on the Pd-catalyzed carbonylation of aryl and vinyl halides was disclosed by Heck's group 40 years ago.[106] Since then, the transition metal-catalyzed carbonylation reaction has been used extensively in both academia and industry for the synthesis of natural products and pharmaceuticals. Aryl halides, sulfonates and diazonium salts can be used as substrates. Depending on the reaction conditions, aldehydes, ketones and carboxylic acid derivatives can be readily accessed under very mild conditions that are compatible with many other functional groups. Some of the drawbacks of this reaction are the need for specialized equipment and the toxicity of CO. As an alternative to carbonylations run in batch mode, flow reactors represent an attractive approach particularly well suited for large-scale manufacturing.[107] Transition metal-catalyzed carbonylations have been reviewed.[108] Many examples of large-scale Pd-catalyzed carbonylations have been reported starting in the late 1990s, which show the interest in this technology by process chemistry groups.[109]

Wei and co-workers at Boehringer Ingelheim in the United States have reported the synthesis of sodium–hydrogen exchange type 1 inhibitor **78**, a drug candidate for the treatment of several heart diseases (Scheme 15.16).[109j] The introduction of the ester functionality was accomplished via a Pd-catalyzed carbonylation reaction on aryl bromide **76**. A ligand screen (dppp, dppf and BINAP) using Pd(OAc)$_2$ as metal source with K$_2$CO$_3$ as base in MeOH at 70 °C and 100 psi of CO showed the highest conversion with dppp. Additional experimentation with the Pd(OAc)$_2$/dppp catalytic system revealed that at higher temperature (110 °C), a faster reaction rate was observed but at the expense of generating up to 27% of carboxylic acid **79** (Figure 15.4). However, the use of Et$_3$N at a slightly lower temperature (90 °C) minimized

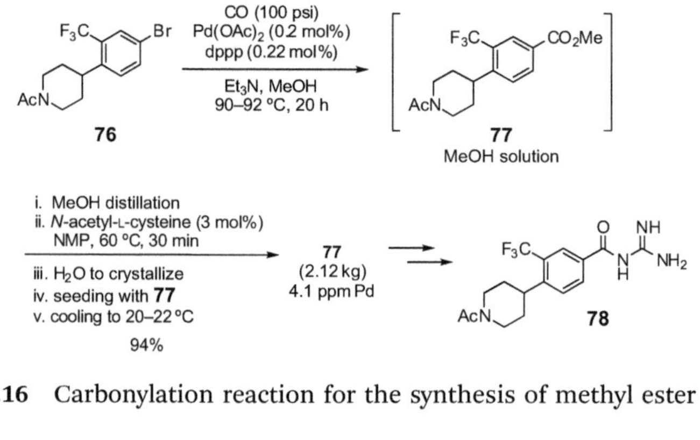

Scheme 15.16   Carbonylation reaction for the synthesis of methyl ester **77**.

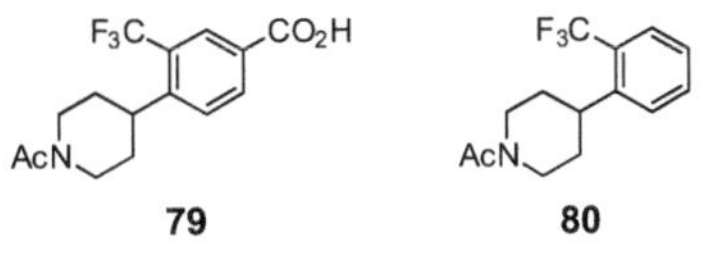

Figure 15.4   Major impurities from the carbonylation reaction of aryl bromide **76**.

the amount of **79** to only 5% and still provided an acceptable reaction rate. Desbromo impurity **80** (Figure 15.4) was also obtained in less than 2% yield. In the plant, a 20 L autoclave was charged with **76**, Et$_3$N and MeOH followed by purging with nitrogen gas. Pd(OAc)$_2$ and ligand were then added in degassed MeOH and the reactor was pressurized with CO to 100 psi. After 20 h at 90–92 °C, the reaction mixture was filtered through Celite and concentrated. After a solvent switch to NMP, the resulting mixture was treated with 3 mol% *N*-acetyl-L-cysteine, a Pd scavenger that forms a water-soluble complex with Pd, at 60 °C for 3 h. Dilution with water, seeding with **77** to promote crystallization and cooling afforded 2.12 kg of **77** in 94% yield and with only 4.1 ppm Pd. In addition to *N*-acetyl-L-cysteine, Cuno ZetaCarbon R53 and R55 filters were also tested but were less efficient at scavenging the Pd.

15.2.10   Cyanation

Transition metal-catalyzed cyanations have attracted considerable attention because the cyano group is a common motif in pharmaceuticals, agrochemicals, dyes and natural products. In addition, nitriles are useful handles that can be converted to a variety of other functional groups. The Pd-catalyzed cyanation of aryl halides[110,111] was introduced by Takagi *et al.* in 1973[112] and represents an attractive alternative to more traditional methods, such as the Sandmeyer[113] or Rosenmund–von Braun (run at high temperature)[114] reactions, which employ stoichiometric amounts of copper. In addition to mild conditions, Pd-catalyzed cyanations employ inexpensive cyanide sources such as sodium, potassium or zinc cyanide. Most examples of this transformation involve Pd as the metal source, but Ni[115] and Cu[116] can also be employed. Process chemistry groups have also taken advantage of this technology for the synthesis of aromatic nitriles, as evidenced by the large number of reports.[117] Several reviews on transition metal-catalyzed cyanations can be found in the literature.[110a,118]

Ryberg at AstraZeneca in Sweden published two very detailed articles on the implementation of a scalable cyanation protocol as the last step of the multi-kilogram synthesis of cyanoindole **82**, a drug candidate for the treatment of disorders related to glycogen synthase kinase 3 (Scheme 15.17).[117h,l] Conditions that employed Zn(CN)$_2$, Pd(dba)$_2$, dppf and Zn dust in DMF at 120 °C worked well on small scale, but on scale-up stalled reactions and decomposition of **81** and **82** were observed. With the goal of finding a more active catalyst that required milder conditions, a screen of Pd sources and ligands was undertaken. Pd[P(*t*-Bu)$_3$]$_2$, Pd(dba)$_2$/P(*t*-Bu)$_3$, [PdBrP(*t*-Bu)$_3$]$_2$,

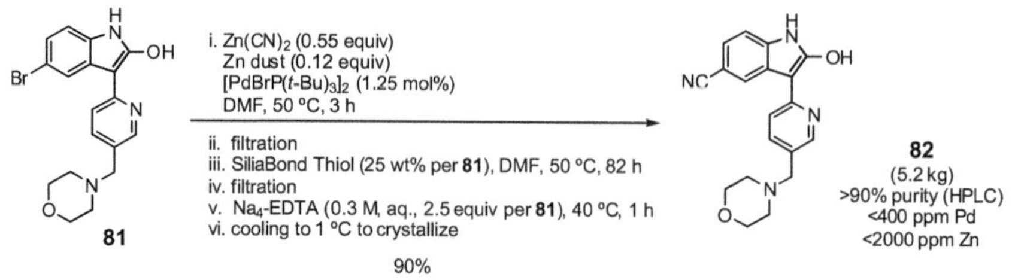

Scheme 15.17   Pd-catalyzed cyanation for the preparation of drug candidate **82**.

Pd(dba)$_2$/P(*t*-Bu$_3$)$_2$biphenyl (JohnPhos), Pd(dba)$_2$/PCy$_2$biphenyl (cyclohexyl JohnPhos) and Pd(dba)$_2$/QPhos [1,2,3,4,5-pentaphenyl-1'-(di-*t*-butylphosphino)ferrocene] were tested in DMF at 50 °C and the best conversion was observed with P(*t*-Bu)$_3$-based catalytic systems. Therefore, the Pd(dba)$_2$/P(*t*-Bu)$_3$ combination was chosen for further development. With this optimized catalytic system, control experiments showed that increasing the scale from 1 to 5 g caused the reactions to stall, which was ascribed to catalyst poisoning by excess cyanide in solution. For example, reactions to which the precatalyst Pd(dba)$_2$ was not added immediately after the rest of components or reactions that were not heated immediately after Pd(dba)$_2$ addition gave lower conversions. Since it was believed that this problem would only get worse on plant scale due to slower operations, the focus turned to testing different orders of addition of the reagents. Further experimentation showed that complete conversion was consistently obtained when a solution of Zn(CN)$_2$ in DMF was added to a mixture of aryl bromide, catalyst {Pd[P(*t*-Bu)$_3$]$_2$, Pd(dba)$_2$/ P(*t*-Bu)$_3$, [PdBrP(*t*-Bu)$_3$]$_2$, Pd(dba)$_2$/JohnPhos, Pd(dba)$_2$/cyclohexyl JohnPhos, Pd(dba)$_2$/QPhos, Pd(dba)$_2$/XPhos or Pd(dba)$_2$/P(*o*-tol)$_3$} and Zn dust in degassed DMF at 50 °C in 1 h. In the plant, a mixture of aryl bromide **81** and Zn dust (0.12 equiv.) in DMF was degassed with nitrogen until the oxygen level was <0.02 mg/L. [PdBrP(*t*-Bu)$_3$]$_2$ (1.25 mol%; the precatalyst of choice since it provided the cleanest reaction and is reasonably stable in air) was added and the reactor was made inert by applying vacuum and refilling with nitrogen. After the mixture had been heated to 40 °C, Zn(CN)$_2$ (0.55 equiv.) was added in one portion and the vessel was made inert again. The reaction mixture was heated at 50 °C for 3 h and filtered. The filtrates were then treated with SiliaBond Thiol (25 wt% per **81**) at 50 °C for 82 h to remove Pd. The metal scavenger was filtered off and, after concentration of the filtrates, Na$_4$-EDTA (0.3 M in water; Zn scavenger) was added and the mixture was stirred at 40 °C for 1 h. Upon cooling to 1 °C, 5.2 kg of nitrile **82** precipitated from solution and was collected by filtration in 90% yield and with <400 and <2000 ppm Pd and Zn, respectively. It was mentioned[117*l*] that subsequent citric acid salt formation further reduced the Pd and Zn levels to 1–2 and <1 ppm, respectively.

The process group at Actelion Pharmaceuticals in Switzerland described the large-scale preparation of ACT-209905 (**86**), an S1P$_1$ receptor agonist for the treatment of autoimmune diseases (Scheme 15.18).[117*m*] The synthesis of

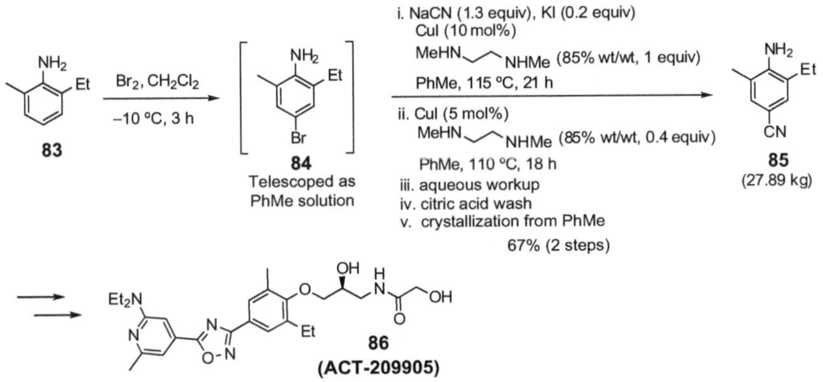

Scheme 15.18 Cu-catalyzed cyanation for the preparation of intermediate **85**.

the oxadiazole moiety on the molecule required the previous installation of a cyano group. Thus, aniline **83** was first converted to *p*-bromoaniline **84** with Br$_2$, which was telescoped as a toluene solution. A Cu-catalyzed cyanation protocol developed by Buchwald[116a] was then applied, in which **84** was treated with stoichiometric NaCN in the presence of catalytic KI and *N,N'*-dimethylethylenediamine as ligand in toluene at reflux to provide cyanoaniline **85**. The addition of a second charge of CuI/ligand was required to push the reaction to completion. Toluene was reported by Buchwald to perform better than polar, aprotic solvents such as DMF or sulfolane, in which the reactions tend to stall due to the higher solubility of cyanide. The role of KI is to generate small amounts of transient aryl iodide, which is then capable of reacting with NaCN to produce the target aryl cyanide. Upon completion of reaction, the mixture was subjected to an aqueous work-up that included a citric acid wash (10% aqueous) to sequester Cu. After concentration and cooling, 27.89 kg of cyanide **85** crystallized from toluene in 67% yield over two steps with 99% purity. No information on the residual Cu level in **85** was provided.

15.2.11 Ring-Closing Metathesis

The adoption of ring-closing metathesis (RCM)[119] by academic and medicinal chemistry groups for the preparation of medium and large rings *via* intramolecular olefin coupling has been widespread and a key contributing factor is that many of the Ru and Mo catalysts display remarkable functional group tolerance. The Nobel Prize in Chemistry awarded to Yves Chauvin, Richard Schrock and Robert Grubbs in 2005[120] is a clear testimony to the importance of this transformation in the chemical armamentarium. However, large-scale examples of RCM applied to the synthesis of pharmaceuticals have been scarce until recently,[121] for two main reasons: (a) the need for high dilution to minimize intermolecular cross-metathesis as a competing side reaction, which takes a heavy toll on the throughput of the process; and (b) the high cost and limited commercial availability of catalysts in large quantities. An additional consideration is that some of these catalysts are patent protected and would require substantial licensing fees for use in a commercial route. The situation changed when, in 2005, the process group at Boehringer Ingelheim reported the first pilot-plant implementation of an RCM reaction for the manufacture of hundreds of kilograms of API.[121a] After this seminal publication, other process groups in pharmaceutical companies embarked on similar approaches to large-sized rings, especially in the active area of HCV protease inhibitors, such as the examples shown below. Olefin metathesis[122] and also the green aspects[123] of this transformation have been reviewed.

Kong and co-workers at Merck recently reported the synthesis of vaniprevir (57, MK-7009), a potent hepatitis C virus NS3/4a protease inhibitor (Scheme 15.19), which employs RCM as the key step to generate the macrocyclic portion of the molecule.[121g] An earlier report also by the process group at Merck described the preparation of the macrocycle that employed a macrolactamization approach.[80v] The researchers investigated the three possible options to assemble the macrocycle *via* RCM (Scheme 15.20). Thus, the conversion of diene **91** into RCM product **90** had been investigated previously[80v] but it was abandoned owing to the requirements for a high catalyst loading (Neolyst M1, 0.3 equiv.; Zhan 1B, 0.1 equiv.) and high dilution (<3 mM) to obtain an acceptable yield. Since the preparation of diene **87** was considered to be more straightforward than that of **93**, the RCM of the allyl–allyl substrate **87** was further investigated.

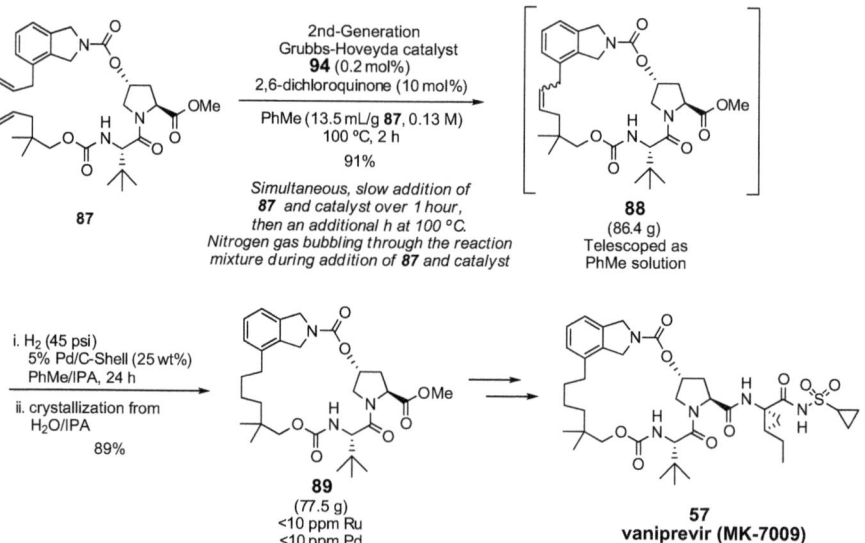

Scheme 15.19 Synthesis of saturated intermediate **89** *via* RCM *en route* to vaniprevir.

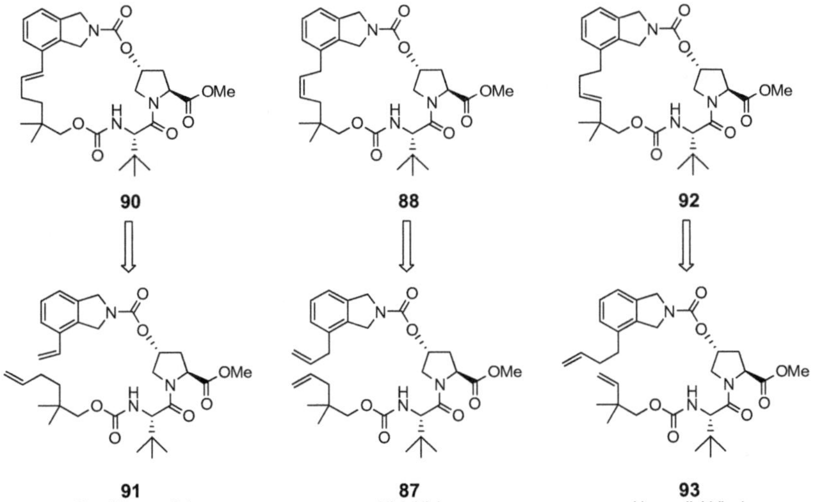

90   **88**   **92**

91   **87**   **93**
Styryl-Homoallyl   Allyl-Allyl   Homoallyl-Vinyl

Scheme 15.20 Three possible strategies for macrocyclization *via* RCM *en route* to vaniprevir (**57**).

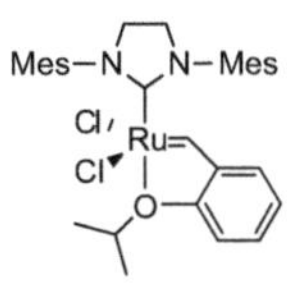

94

Figure 15.5 Structure of second-generation Grubbs–Hoveyda catalyst.

When the reaction was first attempted with 1 or 5 mol% of the second-generation Grubbs–Hoveyda catalyst **94** (Figure 15.5) in toluene at 60 °C, modest yields were obtained (57 and 67%, respectively) after the subsequent hydrogenation step. A closer look at the reaction kinetics revealed that the catalyst was very active at the initial stages of the reaction but its activity decreased as the reaction proceeded. In addition, during the final stages of the reaction, the formation of oligomers and other impurities increased. However, this problem was easily solved *via* slow addition of the catalyst (1 mol%) to the reaction mixture (50 mL of toluene per gram of diene **87**) at 60 °C over 1 h, which afforded cyclized product **89** in 82% yield for the RCM and hydrogenation steps combined.

The high dilution employed during the process development described above was then addressed with the goal of increasing the throughput of the process. Thus, when the reaction was carried out in 30 vol. of solvent (0.058 M), the yield decreased to 61% and 19-membered by-product **95** was obtained, presumably *via* double-bond isomerization caused by the Ru–H complex which is formed after Ru catalyst decomposition.[124] Double-bond isomerization can be limited through the use of quinone additives, which are known to prevent catalyst decomposition.[125] As a result, several quinones (1,4-benzoquinone, 2,6-dimethylbenzoquinone, 2,6-dichlorobenzoquinone, 2,6-dimethoxybenzoquinone and 2,3,5,6-tetrafluorobenzoquinone) were tested and 2,6-dichlorobenzoquinone (10 mol%) afforded desired 20-membered ring **88** in 77% conversion and less than 1% of **95** from olefin migration prior to RCM (Figure 15.6) in 20 vol. of solvent (0.086 M). In addition, the catalyst loading could be reduced to only 0.2 mol%. Further optimization showed that the yield increased to 84% at 100 °C and to 88% when the reaction was sparged subsurface with nitrogen gas to displace the generated ethylene gas. A higher concentration (13.5 vol., 0.13 M) decreased the yield to 78% due to the formation of large amounts of oligomers (>15%). However, this high concentration could be maintained through the slow addition of both diene and Ru catalyst while increasing the yield to 91%.

The reaction was run in the laboratory by dissolving 2,6-dichlorobenzoquinone in degassed toluene and adding 10 vol.% of a solution of diene **87** in toluene. The mixture was heated to 100 °C and, while bubbling nitrogen gas through the resulting solution, a solution of the catalyst in toluene and the remaining diene solution were added over 1 h. After an additional 1 h at 100 °C, the mixture was cooled to room temperature and, following dilution with IPA, was subjected to catalytic hydrogenation to reduce the olefin. Intermediate **89** (77.5 g) was obtained in 81% yield for the two steps combined after crystallization from water–IPA and with <10 ppm each of Pd and Ru.

RCM was employed by Wei and co-workers in the process chemistry group at Boehringer Ingelheim in the United States for the synthesis of 15-membered macrocyclic hepatitis C virus protease inhibitor BI 201302 (**100**, Scheme 15.21).[121h] The same group had previously published the synthesis of a closely related desbromo analog[121a–d] and some of the learnings from that project were applied to the synthesis of this new candidate. Thus, using first-generation Grubbs catalyst **102**, the effect of substituents on the amide (R group, Scheme 15.22) was investigated. It was found that when R = H, carbene transfer took place at the vinylcyclopropane to give **103** (presumably due to chelation of the Ru to the carbonyl group of the ester) leading to epimerization and, as a consequence, a mixture of RCM products **104** and **105**. On the other hand, when R = Boc (**96**), a preference for carbene transfer to the other olefin was observed perhaps due to subtle conformational

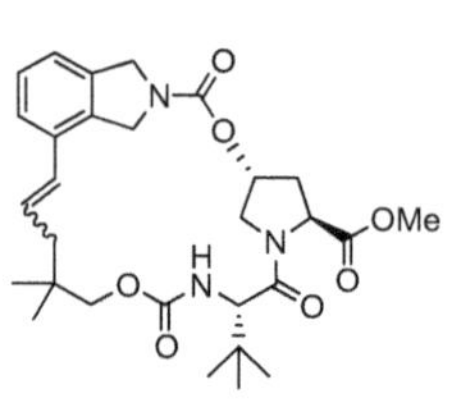

95

Figure 15.6   Structure of 19-membered ring by-product *via* olefin isomerization.

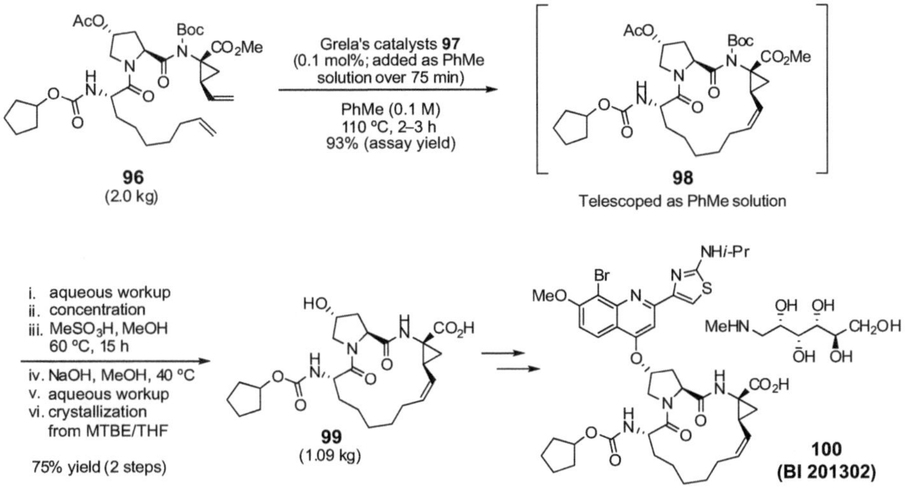

Scheme 15.21 Synthesis of BI 201302 (**100**) *via* RCM.

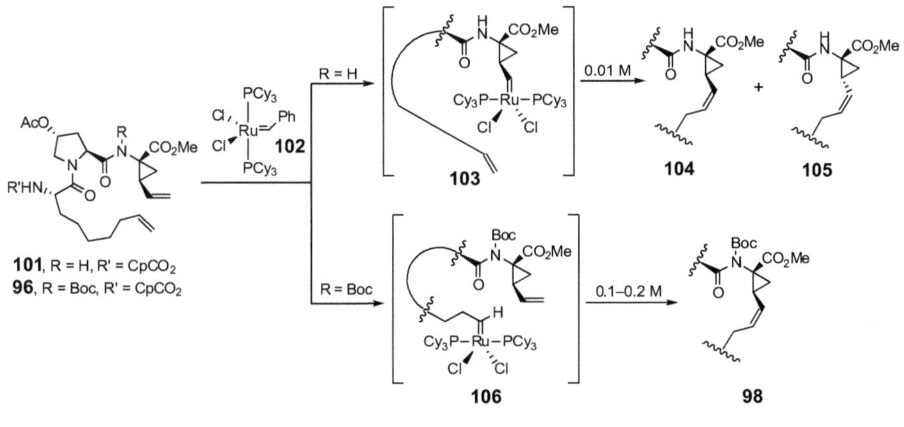

Scheme 15.22 Preliminary studies to investigate the effect of *N*-substituents on the RCM step.

effects and the reduced ability of the ester group to stabilize the Ru species due to its electron-withdrawing effect. This simple change increased the reaction rate by 3–4-fold, the effective molarity (k_{intra}/k_{inter}) of the reaction by one order of magnitude (0.1–0.2 M) and avoided epimerization on the cyclopropyl ring. The Boc substituent is therefore believed to have both a kinetic and a thermodynamic effect, the latter due to reduced ring strain in **98** through conformational changes.

In the plant, the RCM step was carried out by heating a solution of diene **96** in toluene (2 kg of **96** in 30 L of solvent) to 110 °C. A toluene solution of Grela's catalysts **97**[126] (Figure 15.7; no mention was made about why this catalyst is preferred over first-generation Grubbs catalyst, but it may be due

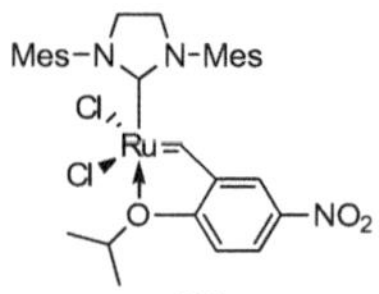
Mes-N
N-Mes
Cl
Ru
Cl
O
NO_2

97

Figure 15.7 Structure of Grela's catalyst for the synthesis of BI 201302.

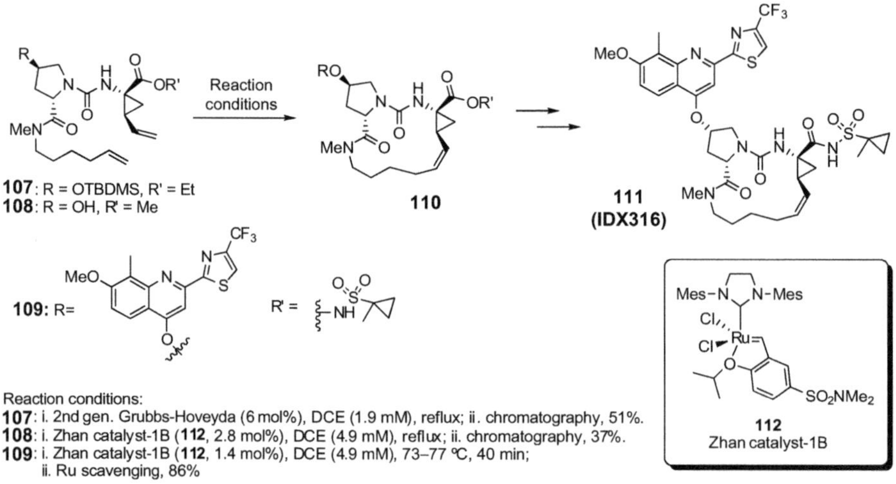
Reaction conditions
107: R = OTBDMS, R' = Et
108: R = OH, R' = Me
110
111
(IDX316)
109: R=
R' =
112
Zhan catalyst-1B
SO_2NMe_2
Reaction conditions:
107: i. 2nd gen. Grubbs-Hoveyda (6 mol%), DCE (1.9 mM), reflux; ii. chromatography, 51%.
108: i. Zhan catalyst-1B (112, 2.8 mol%), DCE (4.9 mM), reflux; ii. chromatography, 37%.
109: i. Zhan catalyst-1B (112, 1.4 mol%), DCE (4.9 mM), 73–77 °C, 40 min;
ii. Ru scavenging, 86%

Scheme 15.23 Medicinal chemistry approaches to IDX316 (**111**).

to intellectual property reasons) was then slowly added over ~75 min, which resulted in ethylene gas evolution. After 2–3 h, the mixture was subjected to an aqueous work-up and the organic phase (93% assay yield for olefin **98**) was first treated with methanolic $MeSO_3H$ to cleave the Boc group followed by aqueous NaOH (one-pot operations) to hydrolyze the two ester groups. Acidification to pH 2–3 with 12 M HCl and crystallization from MTBE–THF afforded 1.09 kg of carboxylic acid intermediate **99** in 75% yield (two steps). No information was provided on the levels of residual Ru after RCM or in the final API.

Mayes and co-workers at Idenix Pharmaceuticals recently described the large-scale syntheses of IDX316 (**111**; Scheme 15.23) and IDX320 (**114**; Scheme 15.24), two macrocyclic candidates that rely on RCM for ring formation.[121i] Compared with the macrolide from Boehringer Ingelheim described above, **111** and **114** have similar structures and the same indication. The medicinal chemistry route to IDX316, the first candidate selected for development, tested the RCM step on three different substrates (Scheme 15.23). The initial conditions using TBDMS-protected intermediate **107** required 1,2-dichloroethane (DCE) as solvent, very high dilution (1.24 kg of DCE per gram of diene; 1.9 mM) and a very high catalyst loading (92 mol% of second-generation Grubbs–Hoveyda catalyst **94**; Figure 15.5). This process

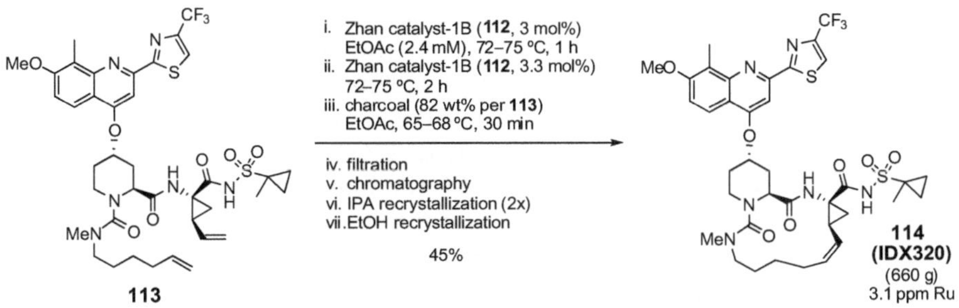

Scheme 15.24 Synthesis of more potent candidate IDX320 (**114**).

led to the formation of large amounts of oligomers and isomers, which needed extensive chromatographic purification. Alternatively, RCM on unprotected substrate **108** (which eliminated the two steps for hydroxy protection and deprotection) with Zhan catalyst-1B (**112**) and higher concentration (4.9 mM) resulted in a lower yield after chromatographic purification due to the formation of macrocyclic oligomers. Better results were obtained for advanced intermediate **109** with only 1.4 mol% Zhan catalyst-1B (**112**) at 4.9 mM concentration, which afforded IDX316 in 86% yield on a <1 g scale. Several Ru removal methods were investigated to lower the metal content to acceptable levels. Treatment with 2-mercaptonicotinic acid followed by activated charcoal, filtration through silica and MeOH trituration afforded material with only 20 ppm Ru but at the expense of a considerable decrease in yield (15–20%). Crystallization/recrystallization from a number of solvents (MeOH–DCE, heptane–EtOAc) lowered the Ru level to ≤20 ppm but a decrease in yield was also observed. In addition, reactions that had been carried out with lower catalyst loading but at higher concentration provided material with 300 ppm Ru, even after MeOH trituration. Finally, a combination of Ph$_3$PO and functionalized silica was chosen as the preferred method since it provided material in high purity and with very low levels of Ru (<5 ppm) regardless of reaction concentration. Thus, once the RCM reaction was complete, Ph$_3$PO (100 equiv. per Ru) was added at 75 °C and the mixture was then concentrated (the decrease in yield was 17% and the Ru content at this point was 180–390 ppm). After trituration in MeOH to remove the Ph$_3$PO-Ru by-products and concentration, the resulting solid was then treated with Siliabond-DMT (silica-bound dimercaptotriazine) in 2-MeTHF at reflux for 16 h. Filtration and concentration afforded **110** with only 3.6 ppm Ru and 98% purity with a 2–6% decrease in yield.

With the discovery of more potent analog IDX320 (**114**), the research efforts on IDX316 were discontinued. However, a similar RCM synthetic strategy for macrocycle generation in **114** was implemented that built on the learnings from **111** (Scheme 15.24). Preliminary experiments with Zhan catalyst-1B (**112**) at 5.0 mM concentration in DCE gave up to 10% of a cyclic dimer and, after chromatographic purification, **114** was obtained in only 34% yield. When the reaction was run at concentrations of 2.5 and 1.5 mM, the isolated yields were 56 and 60–65% respectively. Since higher dilution in DCE did not have a substantial effect on the yield, it was decided to investigate alternative and greener solvents. Thus, at 2.4 mM in MTBE at reflux, slower reaction kinetics were observed, most likely due to the lower boiling point of this solvent. However, in EtOAc at 5.0 mM, **114** was produced in 63% yield, and at 2.6–3.0 mM, the isolated yield increased to 69–74%. In the plant, the RCM step was performed in 500 mL of EtOAc per gram of diene (2.4 mM) with 3.3 mol% Zhan catalyst-1B (added in two portions) at reflux with continuous nitrogen gas degassing to displace ethylene gas. After 3 h, charcoal was added to deactivate the catalyst and the mixture was stirred at 65–68 °C for 30 min. The reaction mixture was cooled to 20–25 °C, filtered and concentrated. The residue was chromatographed on silica and the purified material was then subjected to two recrystallizations from IPA and one from EtOH to afford 660 g of **114** in 45% yield with only 3.1 ppm Ru.

15.2.12   Nozaki–Hiyama–Kishi Coupling

The synthesis of homoallylic alcohols *via* addition of allylchromium(III) salts to aldehydes and ketones was reported by Nozaki, Hiyama and co-workers in 1977.[127] The organochromium reagents were synthesized by treating allyl halides with $CrCl_2$ and reacted with the carbonyl electrophile. An interesting observation by both Nozaki's[128] and Kishi's[129] groups was that the addition of Ni salts facilitated the reaction and led to a more reliable and reproducible process that also expanded the nucleophile scope to the less reactive aryl and alkenyl halides.[130] In addition to allyl, aryl and alkenyl halides, this transformation can now be applied to alkynyl and propargyl halides and also triflates, sulfonates and phosphates. This Barbier-type addition of organohalides or pseudohalides to carbonyl compounds is nowadays called the Nozaki–Hiyama–Kishi (NHK) reaction.[131] Some interesting features of this transformation include (a) chemoselectivity of the organochromium reagents for aldehydes even in the presence of ketones; (b) high functional group compatibility on both nucleophile and electrophile; (c) γ-monosubstituted allylchromium reagents usually afford homoallyl alcohols with excellent *anti* selectivity, regardless of the stereochemistry of the starting halide (*E* or *Z* configuration); and (d) complete retention of double-bond geometry is observed when alkenyl halides are employed as coupling partners.[131a]

   Despite the clear advantages of this transformation, especially for chiral alcohol generation in its asymmetric version, the NHK coupling has received little attention in process chemistry.[43,132] Notable exceptions are the Herculean syntheses of anti-cancer marine natural products discodermolide at Novartis[132] and INN erubulin mesylate (Halaven), a synthetic analog of halichondrin B, at Eisai.[43]

   A collaboration between the process groups at Novartis in Switzerland and the United States and Paterson at the University of Cambridge in the UK resulted in the publication of the multi-gram synthesis of discodermolide (**120**).[132] The introduction of one of the 13 stereogenic centers and the terminal diene on the molecule was accomplished following Paterson's one-pot, two-step NHK–Peterson elimination strategy (Scheme 15.25).[133] The Ni-free NHK reaction involved substrate-controlled, stereoselective coupling between aldehyde **115** and allyl bromide **116**. The protocol that was implemented in the plant called for the sequential addition of solutions of aldehyde **115** and allyl bromide **116** (5.6 equiv.) in THF to a cold (0 °C) suspension of CrCl$_2$ (4.3 equiv.) in the same solvent. After stirring at 0 °C for 15 min, the reaction mixture was warmed to 15 °C and further stirred at this temperature for 1 h to afford a mixture of *anti*-adducts **117** (major) and **118** (minor; the actual diastereomeric ratio was not disclosed in the report). MeOH and 6 M aqueous KOH were then added to effect Peterson *syn*-elimination.[134] Originally, KH was employed as base, but for safety reasons KOH was chosen on scale.[135] Even though both **117** and **118** provided the required (*Z*)-olefin, they underwent *syn*-elimination at different rates. Following an aqueous work-up and concentration, the residue was purified by chromatography on silica to afford 1.27 kg of intermediate **119** as an oil with a combined yield of 81%. No information was provided about the level of residual Pd in intermediate **119**.

Researchers at Eisai in the United States and Japan described the synthesis of INN eribulin mesylate (Halaven, **126**), a very complex molecular structure that has recently been approved by the US Food and Drug Administration (FDA) for the treatment of metastatic breast cancer.[43] One of the 19 stereogenic centers on the molecule was generated *via* an asymmetric NHK reaction[43a] between aldehyde **122** and alkenyl triflate **123** in the presence of chiral ligand (*S*)-**121** (Scheme 15.26).[136] Prior to performing this particular coupling, the researchers had investigated a similar transformation with different coupling partners employing ReactIR to monitor the reaction. Thus, it was determined that (a) the complexation between ligand (*R*)-**121** (the *R* enantiomer was used in that case) and CrCl$_2$ required a temperature >30 °C; (b) the oxygen level had to be closely monitored; and (c) better results were obtained when the aldehyde and coupling partners were added immediately after the addition of NiCl$_2$ while keeping the reaction

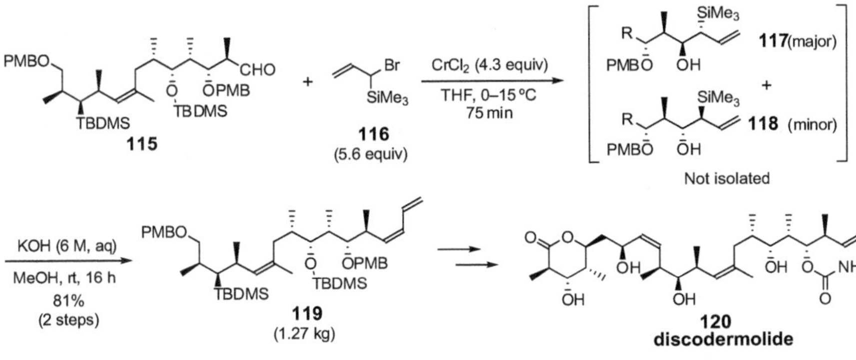

Scheme 15.25   NHK coupling during the synthesis of discodermolide (**120**).

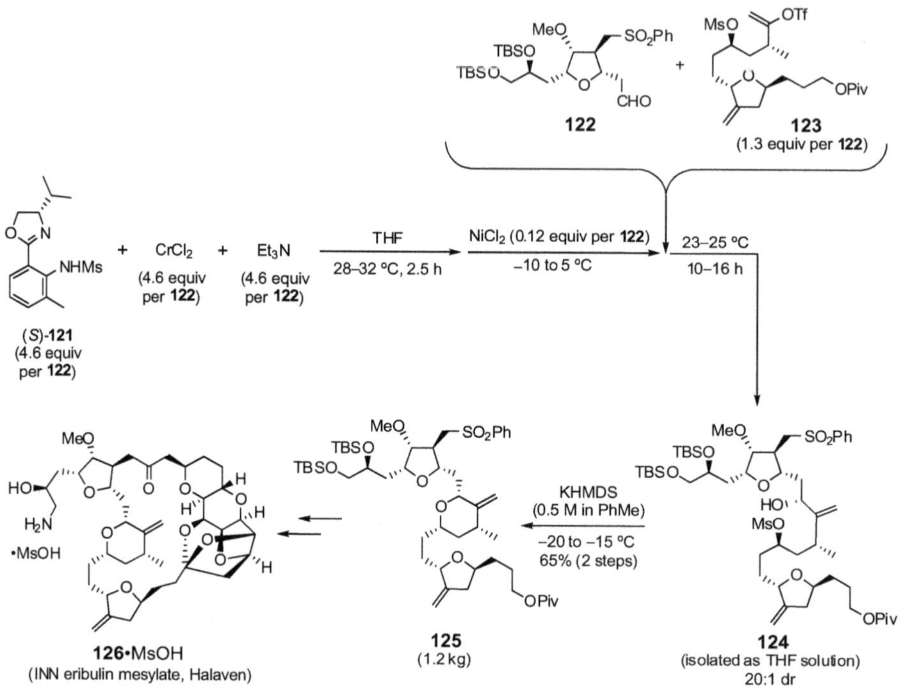

Scheme 15.26 Synthesis of Halaven mesylate (**126 · MsOH**) *via* Nozaki–Hiyama-Kishi coupling.

temperature in the 0–5 °C range. Based on these findings, the process that was implemented in the plant called for dissolving ligand (*S*)-**121** (4.6 equiv. per aldehyde **122**) in THF and purging the reaction with nitrogen gas until the oxygen level fell below 200 ppm. CrCl$_2$ (4.6 equiv. per **122**) was then added and the mixture was warmed to 28–32 °C. After the addition of Et$_3$N (4.6 equiv. per **122**) at 30–35 °C, the mixture was stirred at 28–32 °C for at least 2.5 h. The reaction mixture was then cooled to −10 to 5 °C and NiCl$_2$ (0.12 equiv. per **122**) was added followed by more nitrogen purging. A solution of aldehyde **122** and triflate **123** in THF was then added and the resulting mixture was stirred at 23–25 °C for 10–16 h. Following an aqueous work-up, chiral alcohol **124** was obtained in 20:1 diastereomeric ratio and isolated as a THF solution that was used in the next step (intramolecular Williamson ether formation) to afford 1.2 kg of intermediate **125** with a combined yield of 65%. No information was provided on the level of residual metal in **124** or final API.

15.2.13   C–H Activation

The design of new methods to activate C–H bonds and transform them into C–C and C–heteroatom bonds has seen impressive developments in recent years[137] and they have the potential to change and streamline the way in which chemical syntheses are planned and executed since no additional functionalization needs to be introduced in the molecule. However, two major challenges exist to turn C–H activation into a more practical synthetic method: (a) the lack of reactivity of the C–H bond; and (b) the need to discriminate between several of these bonds within the same molecule to obtain the desired regioselectivity. As with many other transition metal-catalyzed reactions, Pd is the metal of choice, but other metals such as Pt, Au and Cu can also promote this transformation.[138]

The novelty of the C–H activation protocols has prevented them from being customarily incorporated into process chemistry. Only four large-scale examples were found in the literature, two of them for oxindole synthesis,[139] one for biaryl formation[140] and one for direct benzylation of a heteroaryl chloride.[141]

Magano and co-workers at Pfizer described the kilogram-scale preparation of oxindole **129**, a key intermediate in the synthesis of serine palmitoyl transferase enzyme inhibitor **130**, a candidate for the treatment of cardio-vascular diseases (Scheme 15.27).[139b,c] The medicinal chemistry preparation of **129** afforded this material in 41% yield after five steps from commercially available but expensive 3-fluoro-4-nitrobenzoic acid.[142] With the goal of reducing the cost and number of steps, the researchers focused on an alternative approach that relied on C–H activation for oxindole synthesis from α-chloroacetanilides, as described by Hennessy and Buchwald a few years earlier.[143] The required precursor, intermediate **127** was readily synthesized in two steps from inexpensive methyl *p*-aminobenzoate. With **127** at hand, the cyclization step to **129** was attempted in toluene, following the conditions reported by Hennessy and Buchwald, but sticky mixtures were obtained owing to the low solubility of **127** in this solvent, which also led to considerable impurity formation. A solvent screen [DMF, MeCN, 2-MeTHF, 2-MeTHF–IPA (1:1, 4:1)] revealed that the latter combination gave homogeneous mixtures and complete conversion in 1 h was observed to afford **129** in 70% yield with good purity on a 60 g scale. However, high Pd(OAc)$_2$

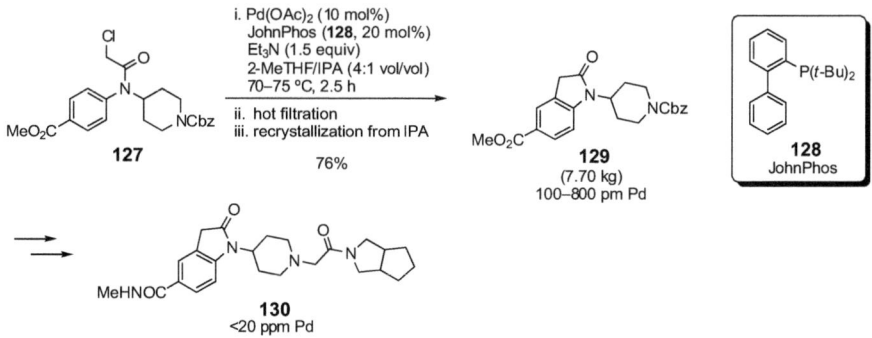

Scheme 15.27 Pd-catalyzed C–H activation for synthesis of oxindole **129**.

(10 mol%) and JohnPhos ligand **128** (20 mol%) loadings were required to achieve full consumption of **127**, much higher than in the original publication (1–3 and 2–6 mol%, respectively; the reason for this is not clear). In the kilogram-laboratory facility, **127** was treated with Pd(OAc)$_2$, JohnPhos ligand (**128**) and Et$_3$N (1.5 equiv.) in 2-MeTHF–IPA (4:1 v/v) at 70–75 °C for 2.5 h. Upon completion of reaction, the hot mixture (to avoid precipitation of **129**) was filtered through Celite and concentrated at reduced pressure. To the residue was added IPA and, after heating the resulting mixture to reflux to redissolve the solids, **129** crystallized upon cooling to 25 °C. Filtration and IPA washing of the cake afforded 7.7 kg of **129** in 76% yield and 98.4% purity. This material contained 100–800 ppm Pd but no efforts were made to effect metal removal at this stage since downstream chemistry lowered the metal content to acceptable levels (<20 ppm).

Several mechanisms were proposed by Hennessy and Buchwald for this transformation based on kinetic isotope effect data (Scheme 15.28).[143] After oxidative addition into the C–Cl bond of **I** (rate-determining step based on observed isotope effect), Pd-enolate **II** can undergo electrophilic aromatic substitution to provide palladacycle **III** followed by loss of HCl and reductive elimination to afford oxindole **V** (Path A). A second possibility (Path B) is that **II** undergoes carbopalladation to give species **VI** followed by *anti*-elimination of HPdCl (or, alternatively, isomerization followed by *syn*-β-hydride elimination) to generate **V**. Hennessy and Buchwald suggested that both Paths A and B can take place simultaneously. Finally, a true C–H activation mechanism is possible (Path C) that can occur *via* either σ-bond metathesis or a

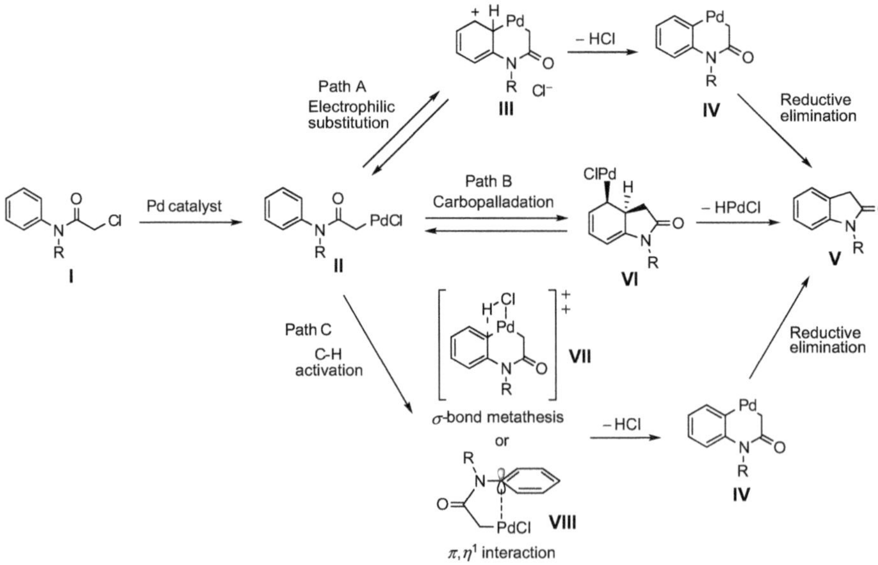

Scheme 15.28 Proposed mechanisms for oxindole synthesis *via* C–H activation.

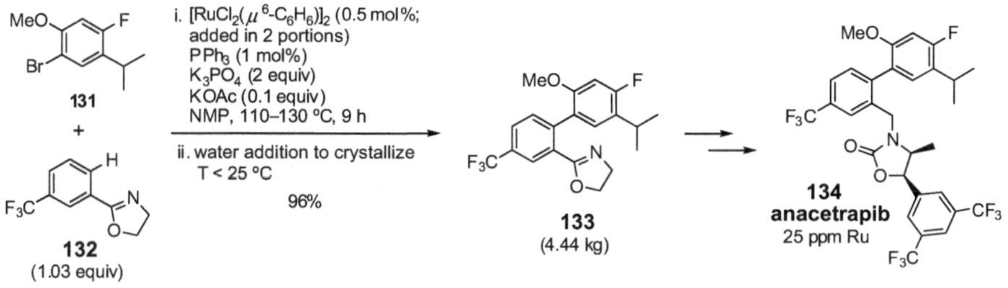

Scheme 15.29   Ru-catalyzed biaryl formation *via* C–H activation.

π,η^1 interaction in which the Pd enolate behaves as a π-acid that weakens the C–H bond and facilitates its cleavage.

Ouellet and co-workers at Merck Frosst in Canada described the synthesis of anacetrapib (**134**), a potent and selective inhibitor of cholesteryl ester transfer protein for the prevention and treatment of hypercholesterolemia (Scheme 15.29).[140] The biaryl core of the molecule was assembled *via* Ru-catalyzed cross-coupling between aryl bromide **131** and aryloxazoline **132**.[144] Initial experiments with 2.5 mol% [RuCl$_2$(μ^6-C$_6$H$_6$)]$_2$, 10 mol% PPh$_3$ and K$_3$PO$_4$ (2 equiv.) in NMP at 120 °C for 2 h afforded clean **133** with good conversion. Further experimentation revealed that alternative ligands such as PCy$_3$ (electron rich), P(4-F-C$_6$H$_4$)$_3$ (electron poor) or dppf (bidentate) gave lower yields of **133** compared with PPh$_3$. In addition, inorganic bases (K$_2$CO$_3$, Cs$_2$CO$_3$) performed better than organic bases (*i*-Pr$_2$NEt, Et$_3$N) and polar aprotic solvents (DMF, DMAc, NMP) were preferred over toluene or xylenes. After NMP had been selected as the preferred solvent, it was observed that whereas the reactions carried out with 5 mol% Ru gave consistent results, the reactions with lower Ru loadings (0.5–1 mol%) provided variable conversions (30–99%). The reason for this variability was ascribed to the level of γ-butyrolactone present in the NMP, as lots of solvent that contained higher levels of this impurity afforded better conversions. This assumption was corroborated experimentally when a failed reaction that employed γ-butyrolactone-free NMP was spiked with this impurity, which then proceeded to high conversion. It was speculated that the reason for this enhanced catalytic activity in the presence of this impurity could be the hydrolysis of the lactone to the corresponding hydroxy acid by the K$_3$PO$_4$ in the reaction medium.[145] A similar activating effect was observed when additives such as KOAc, TFA, pivalic acid and 4-methylbenzoic acid were tested.

The final experimental conditions called for the addition of a degassed NMP solution containing [RuCl$_2$(μ^6-C$_6$H$_6$)]$_2$ (0.5 mol%) and PPh$_3$ (1 mol%, added as two equal portions) to a degassed solution of **131**, **132**, K$_3$PO$_4$ and KOAc in NMP at 130 °C. After the first catalyst portion had been added, the mixture was stirred at 130 °C for 2 h, followed by the addition of the second portion of catalyst and an additional 7 h at 110 °C. Upon completion of reaction, the mixture was cooled to room temperature and water was added

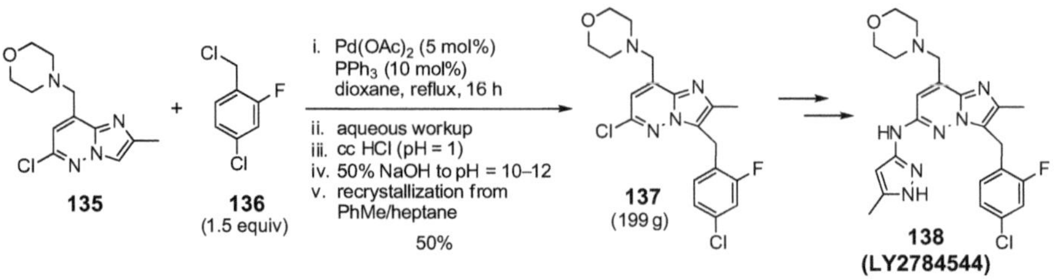

Scheme 15.30 Benzylation *via* C–H activation during the synthesis of LY2784544.

to crystallize **133**, which provided 4.44 kg of intermediate in 96% yield. No information was provided on the Ru content in **133**, but downstream chemistry afforded anacetrapib with only 25 ppm Ru.

An unusual direct benzylation *via* Pd-catalyzed C–H activation on a large scale was implemented at Lilly *en route* to LY2784544 (**138**), a highly selective JAK2-V617F inhibitor for the treatment of chronic disorders related to splenomegaly and the development of leukemia (Scheme 15.30).[141] Based on some previous literature reports that employed this type of transformation on heterocyclic substrates,[146] the preparation of intermediate **137** was investigated through the coupling of aryl chloride **135** and benzyl chloride **136**. Since initial experiments showed that coupling at the desired 3-position of the heterocycle took place, a thorough screen of the conditions was undertaken. Thus, Pd(OAc)$_2$ in combination with several ligands [PPh$_3$, 2-Ph$_2$P-2'-(Me$_2$N)biphenyl, P(4-FPh)$_3$, P(4-CF$_3$Ph)$_3$, XPhos, P(mesityl)$_3$, DPEphos, Xantphos, JackiePhos, P(C$_6$F$_5$)$_3$] and bases (Na$_2$CO$_3$, K$_2$CO$_3$, Cs$_2$CO$_3$) in 1,4-dioxane at reflux revealed that PPh$_3$ and K$_2$CO$_3$ provided the best results in terms of *in situ* yield (70%). Further screening of Pd sources [Pd(TFA)$_2$, Pd(OPiv)$_2$, PdCl$_2$, Pd$_2$(dba)$_3$, Pd(PPh$_3$)$_4$] showed that better *in situ* yields (HPLC wt% of **137**) were obtained as the basicity of the Pd counterion increased [Pd(OAc)$_2$: 70 wt%; Pd(OTFA)$_2$: 33 wt%], whereas Pd(OPiv)$_2$ (51 wt%) gave a slightly lower *in situ* yield than Pd(OAc)$_2$. PdCl$_2$, Pd(PPh$_3$)$_4$ and Pd$_2$(dba)$_3$, which lack a basic ligand, gave poor results unless a basic additive such as NaOAc (10%) was added, which increased the *in situ* yield to 56–70%. With this information at hand, the catalytic cycle shown in Scheme 15.31, which proceeds *via* a concerted metallation–deprotonation mechanism, was proposed.[147] Thus, after Pd(0) generation, oxidative addition on benzyl chloride **136** leads to intermediate **II**, which then undergoes concerted metallation–deprotonation facilitated by acetate ion in the presence of **135** to give rise to intermediate **IV** *via* **III**. Reductive elimination generates the desired product **137** and restarts the catalytic cycle. It was also mentioned that employing NaOAc as base instead of K$_2$CO$_3$ shut down the reaction, perhaps due to Pd poisoning by acetate ion.

On a laboratory scale, a mixture of Pd(OAc)$_2$ (5 mol%), PPh$_3$ (10 mol%), K$_2$CO$_3$, **135** and **136** in dioxane [other solvents were also tested (PhMe, CPME, *t*-AmOH, THF, 2-MeTHF, anisole, MTBE, 2-BuOH, DME, diglyme),

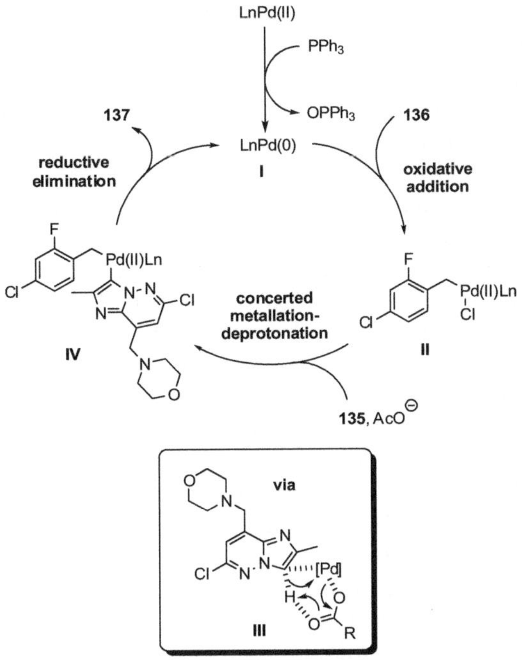

Scheme 15.31 Proposed mechanism for the coupling of **135** and **136**.

but they provided lower *in situ* yields] under a nitrogen atmosphere was heated at reflux for 16 h (85% conversion; longer reaction times did not increase the conversion, presumably due to catalyst decomposition). Toluene, water and concentrated HCl were added to extract the HCl salt of **137** into the aqueous phase. After two toluene washes to remove non-basic organic impurities, pH adjustment of the aqueous layer to 1 precipitated the HCl salt of **137**, which was collected by filtration; **137** · HCl was then taken up in a 2-MeTHF–water mixture and the pH was increased to 10–12 with 50% NaOH to extract **137** into the organic layer. After further extractions of the aqueous layer with 2-MeTHF, the combined organic extracts were dried and concentrated. Purified **137** was then isolated after recrystallization from toluene–heptane in 50% yield. No information was provided about the level of residual Pd in **137** or API **138**.

15.3 Carbon–Heteroatom Bond Formation

15.3.1 Carbon–Nitrogen Bond Formation

15.3.1.1 *Copper-Catalyzed C–N Bond Formation*

The traditional conditions for the copper-catalyzed Ullmann condensation required stoichiometric copper and harsh conditions that would be in-compatible with today's complex molecular structures found in pharmaceuticals. However, thanks to great advances in the design of new catalytic systems through the use of ligands such as aliphatic diamines, aromatic amines, amino acids, amino alcohols, imines and 1,3-dicarbonyl compounds, the Ullmann coupling employing substoichiometric amounts of metal has seen a resurgence in recent years that has greatly expanded the application of this technology to make C–N bonds.[148] Since CuI and many of the ligands currently used are inexpensive, this technology has found several applications in the field of process chemistry.[149]

The process group at GlaxoSmithKline described the implementation of a Cu-catalyzed Ullmann coupling during the synthesis of GSK2137305 (142), a spirocyclic glycine transporter inhibitor for the treatment of neurological and neuropsychiatric disorders (Scheme 15.32).[149d] The coupling of aryl bromide 139 and 4-methylimidazole (140) was initially carried out in the presence of CuI (0.6 equiv.) and L-histidine (1.2 equiv.), but the high ligand loading gave rise to substantial by-product formation that complicated the isolation of 141. In addition, a 4:1 regioisomeric mixture of N-1 and N-3 coupling products was obtained and the undesired regioisomer carried through subsequent chemistry to generate the corresponding regioisomer of API 142. With the goal of minimizing the catalyst loading and improving the regioselectivity of the reaction, screening studies were carried out with a number of commercially available ligands [L-proline, ethylene glycol, 1,10-phenanthroline, ninhydrin, (±)-*trans*-1,2-diaminocyclohexane, 2-dimethyl-aminoethanol, *N,N'*-dimethylethylenediamine, *trans*-*N,N'*-dimethylcyclo-hexane-1,2-diamine and 8-hydroxyquinoline] in several solvents (DMSO, *o*-xylene, toluene, 1,4-dioxane, propionitrile, NMP, *n*-butyl acetate, DMPU, 1-butanol, 1-pentanol). Alcohols in combination with amines were the most promising hits, but the isolation of 141 still proved difficult, especially with low Cu levels (target <100 ppm). The best results were obtained when the reaction was carried out in the absence of ligand, since it is known that heterocycles can perform this role in this type of coupling. In the plant, a mixture of 139, 140, CuI (15 mol%) and K_2CO_3 in DMSO (a preferred solvent on scale) was heated at 130 °C for 36 h. Upon completion of reaction, the mixture was cooled to 60–70 °C, diluted with *i*-PrOAc and treated with 3 wt% aqueous L-cysteine (metal chelator). Following an aqueous work-up and

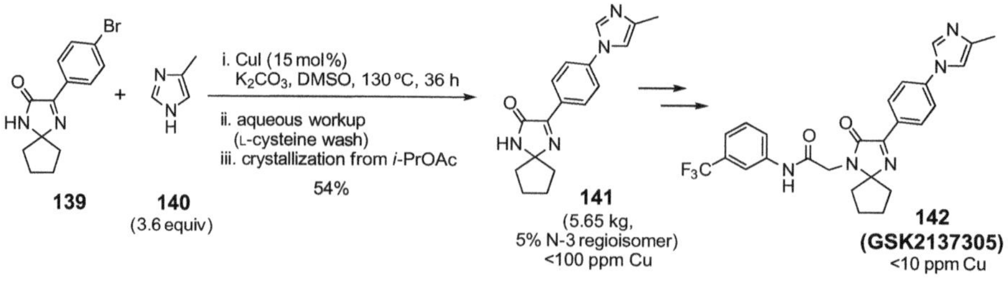

Scheme 15.32   Ullmann coupling of aryl bromide **139** and imidazole **140** *en route* to GSK2137305.

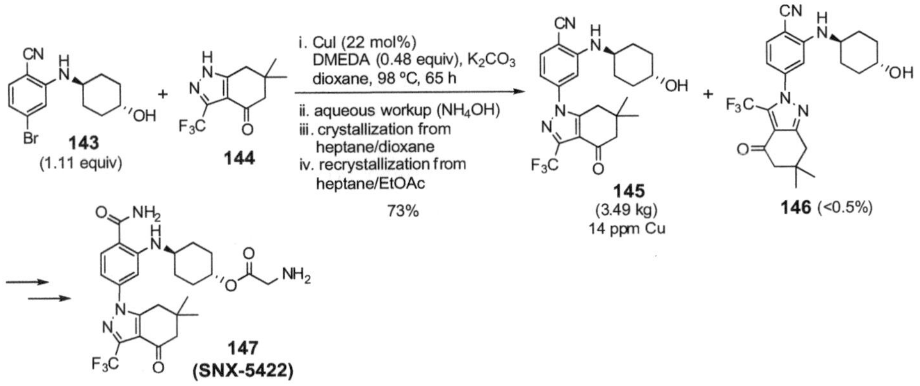

Scheme 15.33 Cu-catalyzed C–N coupling for the preparation of SNX-5422 (**147**).

concentration of the organic phase, 5.65 kg of coupling product **141** crystallized in 54% yield from *i*-PrOAc with ~5% of the undesired N-3 regioisomer and less than 100 ppm Cu. The final target for residual Cu in the API **142** was <10 ppm.

A collaboration between AMRI, Serenex and Pfizer reported the process development and scale-up of the synthesis of SNX-5422 (**147**), a heat shock protein 90 inhibitor for the treatment of cancer (Scheme 15.33).[149e] The Cu-catalyzed Ullmann coupling between aryl bromide **133** and pyrazole **144** was investigated based on some precedent for the coupling of azoles with aryl halides.[150] For cost and commercial availability reasons, the focus was placed on *N,N*-dimethylethylenediamine (DMEDA) and L-proline as ligands. Several solvents were screened [toluene, 1,4-dioxane (degassed and not), 10% aqueous 1,4-dioxane and DMSO] at 100 °C with K_2CO_3 as base for 24 h. Degassed dioxane (to reduce catalyst deactivation) gave the best results in terms of low production of the N-2 isomer, which had to be limited to <1% in order to provide regioisomer-free API. Aqueous dioxane afforded better conversion but at the expense of generating 15% of the N-2 isomer **146**, and similar results were obtained when other bases such as KH_2PO_4 and Cs_2CO_3 were tested. On scale, a nitrogen-sparged mixture of **143**, **144** and K_2CO_3 in dioxane was treated with a degassed solution of CuI (22 mol%) and DMEDA (0.48 equiv.) in dioxane and the resulting mixture was heated at 98 °C for 65 h. Upon cooling, aqueous NH_4OH (copper chelator) and saturated aqueous NH_4Cl were added. After phase separation, the organic layer was concentrated and diluted at 60 ± 5 °C with heptane. Further cooling to 20 °C crystallized **145**, which was subjected to recrystallization from heptane–EtOAc to afford 3.49 kg of purified **145** in 73% yield with <0.5% of the N-2 regioisomer **146** and only 14 ppm Cu.

15.3.1.2   *Palladium-Catalyzed C–N Bond Formation*

Since the first report of a Pd-catalyzed amination of an aryl halide described by Migita and co-workers in 1983 using a stoichiometric amount of

n-Bu$_3$SnNEt$_2$,[151] numerous improvements have been implemented to streamline this type of reaction, especially through the work of Buchwald's and Hartwig's groups.[152] The harsh reaction conditions of the original Ullmann coupling that employed stoichiometric copper and high temperatures[153] have now been replaced by milder, catalytic methods that can be applied to more complex substrates. These advances in aromatic amine preparation have found widespread use in the areas of natural product synthesis[154] and for the preparation of other materials of interest.[155] In process chemistry, the Buchwald–Hartwig cross-coupling has gained increasing importance in the synthesis of pharmaceutical candidates due to the preeminence of aromatic amines in current drug substances.[156] Aryl bromides and iodides have commonly been used, but more inexpensive and readily available aryl chlorides can also serve as substrates. Recent advances in Pd-catalyzed amination facilitated by sterically demanding ligands have been reviewed.[157]

Researchers at Eli Lilly described the preparation of JAK2 inhibitor LY2784544 (**151**), a drug candidate for the treatment of several myeloproliferative disorders (Scheme 15.34).[156i] During the first-generation synthesis of **151**, conditions developed for the coupling of **148** and an analog of **149** were also applied to the coupling of **148** and **149** without extensive optimization. Since the coupling with **149** was slower than with the *N*-PMB analog (most likely due to steric reasons), xylenes were chosen over toluene to run the reaction at higher temperature (xylene–water azeotrope at 99–103 °C). In the plant, to a mixture of **148**, **149** and aqueous NaOH in xylenes was added Xantphos (0.8 mol%) and Pd$_2$(dba)$_3$ (0.4 mol%). The resulting mixture was then heated at reflux for 40 h, followed by cooling to below 60 °C, dilution with THF (to solubilize **150**), filtration through Celite to remove insoluble material, phase separation and an aqueous wash. The resulting organic phase was treated with SiliaBond Thiol from SiliCycle (Pd scavenger, 8.2 wt% per **148**) at 50–60 °C for 10 h. Filtration, concentration of the filtrates and addition of heptane crystallized 82.4 kg of **150** in 79% yield. It was mentioned that during scale-up runs, the slower kinetics of the reaction allowed the formation of larger quantities of a by-product that was characterized as the bis-addition product of **149** to both chloride groups on the molecule.

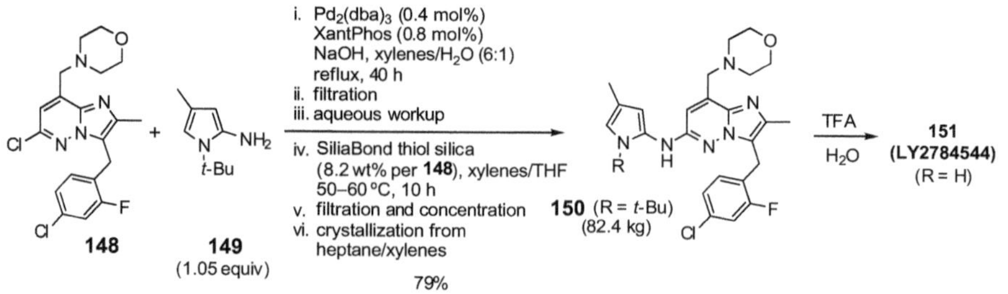

Scheme 15.34   Pd-catalyzed amination *en route* to LY2784544.

A possible explanation for the slower rate of cross-coupling is that poorer mixing of the biphasic mixture resulted in a base-starved organic phase. However, recrystallization from H_2O–IPA was able to reduce the level of this impurity to <0.5% (even though this protocol was not implemented in the plant based on the experimental section of the article). No information was provided on the level of residual Pd in **150** after the functionalized silica treatment.

Goodyear and co-workers at Merck reported the large-scale synthesis of non-nucleoside reverse transcriptase inhibitor MK-6186 (**155**), a drug candidate for the treatment of HIV infection (Scheme 15.35).[156*l*] The formation of intermediate **154** was accomplished through the Pd-catalyzed coupling of aryl bromide **152** and benzophenone hydrazone (**153**), which is a convenient and easier to handle substitute for hydrazine hydrate on a large scale ($NH_2NH_2 \cdot H_2O$ had been employed by the medicinal chemistry group in the original synthesis of **154**). The use of strong bases promoted the formation of amidine **156**, whereas weaker bases (K_3PO_4, Cs_2CO_3) in combination with very active ligands such as Xantphos, DPEPhos and BINAP afforded by-product **157** in addition to **154**. A ligand screen revealed that the best combination to minimize impurity formation was $Pd_2(dba)_3$/dppf together with saturated, aqueous K_3PO_4 in CPME. Further optimization led to the use of $PdCl_2(dppf) \cdot CH_2Cl_2$, 2 M K_3PO_4 and toluene as solvent, which afforded complete conversion after 19 h at reflux. In the plant, to a degassed solution of **152** and **153** in toluene was added the Pd precatalyst (2 mol%) followed by further degassing. A degassed aqueous 2 M K_3PO_4 solution was then added and the resulting mixture was refluxed for 19 h. After cooling and an aqueous work-up, the toluene layer was treated with macroporous poly-styrene-2,4,6-trimercaptotriazine (MP-TMT; polymer-bound 2,4,6-tri-mercaptotriazine) at 20 °C for 17 h to remove residual Pd and Fe. Following

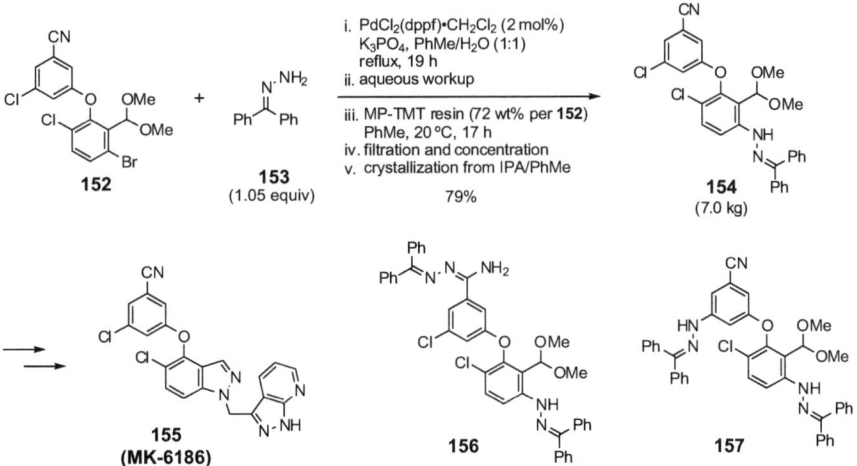

Scheme 15.35 Buchwald–Hartwig amination with benzophenone hydrazone.

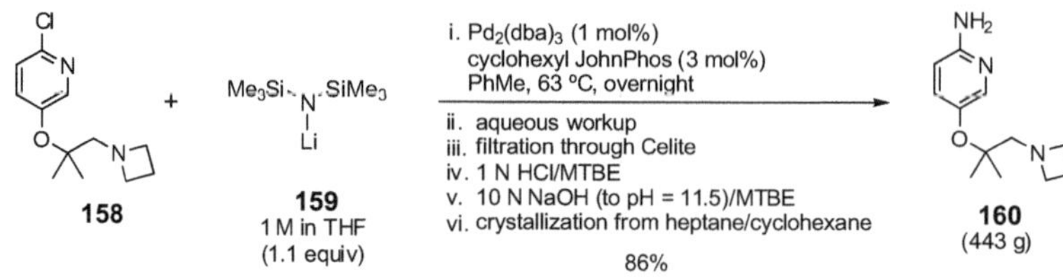

Scheme 15.36 Buchwald–Hartwig amination using LHMDS as an ammonia surrogate.

filtration and concentration, the addition of IPA and cooling to 0 °C crystallized 7.0 kg of **154** in 79% yield. No information was provided on the level of residual Pd in **154** or final API.

The process group at Hoffmann-La Roche reported the preparation of aniline **160**, a key component in several Bruton's tyrosine kinase (BTK) inhibitors under development for the treatment of rheumatoid arthritis (Scheme 15.36).[156o] The conversion of 2-chloropyridine **158** into amine **160** required the use of ammonia or an appropriate surrogate, such as LHMDS,[158] or tert-butyl carbamate.[159] Since the latter approach failed, the researchers opted for optimizing the introduction of the amino group via LHMDS (**159**). Initial experimentation showed that Pd$_2$(dba)$_3$ and 2-(dicyclohexylphosphino)biphenyl (cyclohexyl JohnPhos) in THF worked well on a small scale but, on scale-up, multiple Pd and ligand charges were necessary to achieve complete conversion. The use of 2-(di-tert-butylphosphino)biphenyl (JohnPhos) slowed the cross-coupling reaction whereas NaHMDS provided only trace amounts of amine **160**. More reproducible results on scale were obtained when THF was replaced with toluene and when a THF solution of LHMDS was slowly added to a mixture of **158** and Pd$_2$(dba)$_3$/ligand. On a large scale, a degassed solution of **158**, Pd$_2$(dba)$_3$ (1 mol%) and cyclohexyl JohnPhos (3 mol%) in toluene at 45 °C was treated with a 1 M solution of LHMDS in THF (one-quarter of the total volume of solution) over 5 min. Following an exotherm that increased the internal temperature to 48 °C, the mixture was heated to 62 °C and the remainder of the LHMDS solution was added over 90 min. The reaction mixture was then stirred at 63 °C overnight. After an aqueous quench and phase separation, the toluene layer was filtered through Celite to remove Pd by-products and concentrated. The residue was redissolved in MTBE and extracted with 1 N HCl to form the water-soluble HCl salt while leaving the non-basic impurities in the organic layer. After neutralization with 10 N NaOH to regenerate the free base of **160** (pH adjustment to 11.5) and extraction into MTBE, a solvent switch to 2-MeTHF followed by a second solvent switch to cyclohexane and addition of heptane crystallized 443 g of **160** in 86% yield. The level of residual Pd in **160** was not disclosed, but it was mentioned that several subsequent steps also employed Pd catalysts, which made its removal at this stage unnecessary.

15.3.2 Carbon–Sulfur Bond Formation (Migita Thioether Synthesis)

The Pd-catalyzed arylation of sulfides, also known as the Migita reaction,[160] has attracted less attention in industrial and academic settings than similar transformations with amines, amides and alcohols, most likely due to the limited presence of the sulfide, sulfoxide and sulfone functionalities in pharmaceuticals and natural products. Even more scarce are reports from process groups in the pharmaceutical industry.[161] However, a very detailed mechanistic study of this reaction using palladium complexes of Josiphos ligand CyPF-*t*-Bu (Josiphos SL-J009-1) was reported by Hartwig's group.[162] Thiols are stronger nucleophiles than alcohols and amines, which should favor the transmetallation step. However, they are prone to oxidation and avoiding disulfide formation still remains a challenge in this reaction.

Since the original reports by Migita and co-workers,[160] numerous improvements have been described, such as the work by Murata and Buchwald[163] and Hartwig and co-workers[164] on the use of phosphine ligands (dippf and CyPF-*t*-Bu, respectively) in combination with Pd(OAc)$_2$ which now allows the use of aryl chlorides or bromides and also aliphatic and aromatic thiols. Many other ligands have been tested in this transformation.[163,165] Copper has been shown to be an effective substitute for Pd for C–S bond formation.[166] CuBr in the presence of a phosphazene base[167] or CuI with 2,9-dimethyl-1,10-phenanthroline (neocuproine) has been shown to promote diaryl thioether formation.[168] Transition metal-catalyzed C–S bond formation was previously reviewed in 2000.[169]

A recent example by Yang and co-workers at AMRI involved the preparation of **162**, a potent tubulin inhibitor for the treatment of leukemia and lymphoma (Scheme 15.37).[161e] Compound **162** is a derivative of vinblastine, a natural product belonging to the vinca alkaloids family. The first two members of this family, vinblastine and vincristine, were isolated from *Catharanthus roseus* (Madagascar periwinkle) in the 1950s and, since then, several members of this family have been prepared through total synthesis. However, owing to the molecular complexity of the carbon skeleton in this series of compounds, structural modification of the natural products is a more straightforward approach to synthesize vinca derivatives as potential therapeutic agents. Thus, starting from vinblastine sulfate, iodination with NIS afforded iodoarene **161**, which was telescoped as an NMP solution into the thiomethylation step (Migita coupling). When this transformation was performed in a sealed tube at 65 °C with 15 equiv. of MeSH, the reaction proceeded to completion to generate methyl thioether **162**. However, when the reaction was carried out in an open vessel under a nitrogen sweep, incomplete conversion was observed due to the loss of MeSH. On a large scale, the NMP solution of **161** was sparged with nitrogen gas for 30 min, which allowed for a lower Pd precatalyst loading (7.5 mol% *versus* the original 10 mol%) to accomplish complete consumption of **161** with 1.5–1.7 equiv. of

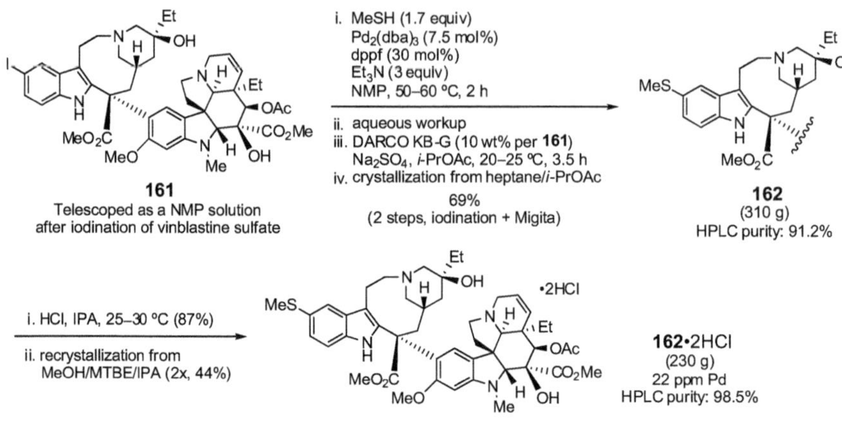

Scheme 15.37 Introduction of a thiomethyl group *via* Migita coupling during the synthesis of **162 · 2HCl**.

MeSH (less than 1.2 equiv. gave incomplete reaction). In addition, decreasing the Pd$_2$(dba)$_3$:dppf ratio from 1:2.5 to 1:4 gave lower levels of unreacted **161** (<0.6%). Further optimization also reduced the amount of Et$_3$N from 10 to only 3 equiv. Upon completion of reaction, the mixture was partitioned between *i*-PrOAc and brine to remove NMP. The *i*-PrOAc extracts were treated with 0.5 M H$_2$SO$_4$ and water to form the sulfate salt of **162**. The aqueous layer was then treated with 20% Na$_2$CO$_3$ and, after extraction of the free base of **162** into *i*-PrOAc, the organic extracts were washed with brine and then treated with Na$_2$SO$_4$ (drying agent) and DARCO KB-G (palladium scavenger; 10 wt% per vinblastine sulfate) for 3.5 h (longer processing times increased the impurity level) at 20–25 °C. After filtration and concentration, the resulting solution was added to heptane to crystallize **162** in 69% yield with 91.2% purity. Salt formation (2 M HCl in IPA) and two recrystallizations from MeOH–MTBE–IPA afforded 230 g of **162 · 2HCl** in 38% combined yield with 22 ppm Pd.

It was mentioned that crude **162** contained 500–1000 ppm Pd, which could only be reduced to 100–200 ppm after HCl salt formation and two recrystallizations. As a result, several Pd scavengers were tested [Darco-G60, trithiocyanuric acid (trimercaptotriazine), polymer-bound ethylenediamine, Smopex-111pp, Smopex-112pp, Smopex-234pp, PL-TMT MP resin (polymer-supported trimercaptotriazine) and DARCO KB-G] and DARCO KB-G was selected as it provided material with only 20–40 ppm Pd at 10–20 wt% loading. Another favorable advantage of this scavenger is the low cost compared with more expensive alternatives such as Smopex and PL-TMT MP resin.

Norris and Leeman at Pfizer published a very thorough study on a Migita reaction during the synthesis of **168**, a drug candidate for the treatment of asthma (Scheme 15.38).[161b] The Migita coupling of aryl bromide **163** and 4-fluorothiophenol (**164**) was initially carried out in the presence of

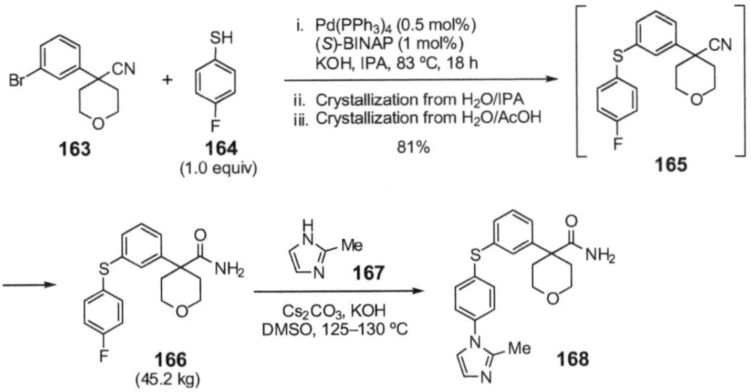

Scheme 15.38 Migita coupling *en route* to diaryl ether **168**.

Pd(PPh₃)₄ to afford diaryl thioether **165** in 62% yield. However, additional development work revealed that the addition of catalytic amounts of bidentate phosphorus ligands accelerated the rate of reaction and increased the yield. Thus, additional PPh₃ provided **165** in 74% yield, whereas with (*S*)-BINAP the yield increased to 82%. Interestingly, while (*R*)- and (*S*)-BINAP performed similarly well in the reaction, (±)-BINAP displayed a kinetic profile faster than without additional ligand but slower than with enantio-merically pure BINAP. The reason for this rate difference was unclear, but it may be related to the different solubilities of the optically pure and racemic ligand in the reaction medium. Another interesting observation was that, in the absence of additional ligand, nitrile hydrolysis to the amide preceded the cross-coupling step as the dominant pathway, whereas with added (*S*)-BINAP, diaryl thioether formation took place preferentially before nitrile hydrolysis.

The large-scale campaign was carried out by charging the two reactants, Pd precatalyst, ligand, base and solvent and heating the resulting mixture to a steady reflux (internal temperature of 83 °C). Upon completion of reaction, water was added to crystallize the product, which, after heating to 50–55 °C and cooling to 20 °C, was isolated by filtration. The wet cake was then dissolved in AcOH at 95–100 °C and treated with activated charcoal (5 wt% per **163**) and filter aid (6.7 wt% per **163**) for 30 min. After cooling to 65 °C, the mixture was filtered and to the filtrates was added water to crystallize 45.2 kg of Migita product **166** in 81% yield. No information was provided on the level of residual Pd in purified **166**.

The catalytic cycle shown in Scheme 15.39 was proposed. Initially, the four PPh₃ groups in Pd(PPh₃)₄ (**I**) would be displaced by BINAP to afford an 18-electron species **II** that can exist in equilibrium with a 16-electron species **III**. Dissociation of one of the BINAP ligands can provide a 14-electron species **IV** (believed to be the active catalyst) that can then undergo oxidative addition with ArBr to provide Fitton Rick intermediate **V** (usually a *trans* complex with monodentate ligands but *cis* with bidentate ligands).[170] Replacement of the

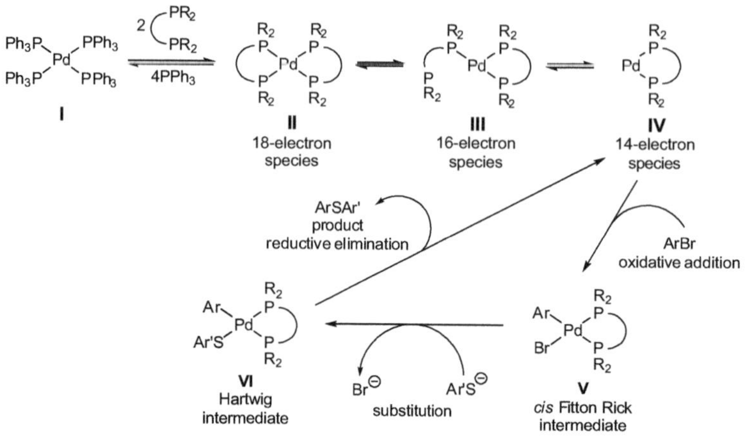

Scheme 15.39 Proposed catalytic cycle for Migita coupling employing bidentate ligands.

bromide ligand with thiolate would generate Hartwig complex **VI**, which can then reductively eliminate desired product ArSAr′ and **IV**, which restarts the catalytic cycle.

15.4 Conclusion

As the examples throughout this chapter have shown, process chemists can now leverage a wide range of transition metal-catalyzed couplings for C–C and C–heteroatom bond formation. The efficiency of these technologies has reached a point where cost is not a deterring factor for their application on scale due to low catalyst loadings in many cases.

Residual metal removal must be addressed if the coupling is positioned near the end of the synthetic route and many possibilities for this exist.[171] Activated carbon is the preferred choice owing to cost and availability as prepacked cartridges, but early process groups can resort to more sophisticated metal scavengers such as functionalized silicas to remove a variety of metals.

In addition to Pd, the preferred metal in most couplings, the application of first-row metals such as Fe, Ni and Cu in coupling chemistry represents an attractive and promising alternative to precious metals. Copper has been the most widely employed non-precious metal on the process scale with recent advances in C–N bond formation. Nickel follows as the preferred metal for Kumada couplings. Iron has seen much more limited use and only one example was found applied to Kumada coupling on a large scale. We expect to see more of this chemistry in the future. With the increasing cost of Pd, the investigation of non-precious metals such as Fe, Ni and Cu should result in new technologies that may replace Pd and provide more sustainable processes.

List of Abbreviations

acac	acetylacetone
Am	amyl
API	active pharmaceutical ingredient
BHT	2,6-di-*tert*-butyl-4-methylphenol
BINAP	(1,1'-binaphthalene-2,2'diyl)bis(diphenylphosphine)
Bu	butyl
Cp	cyclopentyl
CPME	cyclopentyl methyl ether
CyPF-*t*-Bu	(R)-1-[(S_P)-2-(dicyclohexylphosphino)ferrocenyl]ethyldi-*tert*-butylphosphine (Josiphos SL-J009-1)
DABCO	1,4-diazabicyclo[2.2.2]octane
(R,R)-DACH-Ph	(1R,2R)-(+)-1,2-diaminocyclohexane-N,N'-bis(2-diphenylphosphinobenzoyl)
dba	dibenzylideneacetone
DBU	1,8-diazabicyclo[5.4.0]undec-7-ene
DCE	1,2-dichloroethane
dippf	diisopropylphosphinoferrocene
DMAc	N,N-dimethylacetamide
DME	1,2-dimethoxyethane
DMEDA	N,N'-dimethylethylenediamine
DMF	N,N-dimethylformamide
DMT	dimercaptotriazine
DPEphos	bis[(2-diphenylphosphino)phenyl] ether
dppf	diphenylphosphinoferrocene
dppp	diphenylphosphinopropane
EDTA	ethylenediaminetetraacetic acid
equiv.	equivalents
HCV	hepatitis C virus
IPA	2-propanol
KHMDS	potassium bis(trimethylsilyl)amide
LDA	lithium diisopropylamide
LHMDS	lithium bis(trimethylsilyl)amide
Mes	mesityl
2-MeTHF	2-methyltetrahydrofuran
MP-TMT	macroporous polystyrene trimercaptotriazine
MTBE	*tert*-butyl methyl ether
NaHMDS	sodium bis(trimethylsilyl)amide
NIS	N-iodosuccinimide
NMP	1-methyl-2-pyrrolidone
P	pressure
PMB	p-methoxybenzyl
Pr	propyl
rt	room temperature
SPhos	2-dicyclohexylphosphino-2',6'-dimethoxy-1,1'-biphenyl

T	temperature
TBAC	tetra-*n*-butylammonium chloride
TBDMS	*tert*-butyldimethylsilyl
THF	tetrahydrofuran
TMG	1,1,3,3-tetramethylguanidine
TMS	trimethylsilyl
TMT	trimercaptotriazine
tol	tolyl
vol.	volumes $(mL\ g^{-1}\ or\ L\ kg^{-1})$
Xantphos	4,5-bis(diphenylphosphino)-9,9-dimethylxanthene
XPhos	2-dicyclohexylphosphino-2′,4′,6′-triisopropylbiphenyl

Acknowledgements

The authors thank Drs Joseph R. Martinelli and Jason Mulder for reviewing the manuscript and providing numerous comments and suggestions to improve it.

References

1. A. de Meijere and F. Diederich, eds., *Metal-Catalyzed Cross-Coupling Reactions*, 2nd. edn, Wiley-VCH, Weinheim, 2004.
2. C. C. C. Johansson, M. O. Kitching, T. J. Colacot and V. Snieckus, *Angew. Chem. Int. Ed.*, 2012, **51**, 5062.
3. For a recent review on preformed Pd precatalysts for cross-coupling reactions, see: H. Li, C. C. C. Johansson and T. J. Colacot, *ACS Catal.*, 2012, **2**, 1147
4. For Nobel Lectures by Professors Suzuki and Negishi, see: (a) A. Suzuki, *Angew. Chem. Int. Ed.*, 2011, **50**, 6722; (b) E.-i. Negishi, *Angew. Chem. Int. Ed.*, 2011, **50**, 6738. For Nobel Lecture by Professor Heck, see: http://www.nobelprize.org/nobel_prizes/chemistry/laureates/2010/heck-lecture.html (accessed 28 September 2013).
5. The statistics shown in Table 15.1 were obtained from J. Magano and J. R. Dunetz, *Chem. Rev.* 2011, **111**, 2177 and updated to include examples from the mainstream literature up to December 2013. For additional large-scale applications, see: J. Magano and J. R. Dunetz, eds. *Transition Metal-Catalyzed Couplings in Process Chemistry: Case Studies from the Pharmaceutical Industry*, Wiley-VCH, Weinheim, 2013.
6. (a) S. E. Denmark and M. H. Ober, *Aldrichim. Acta*, 2003, **36**, 75; (b) S. E. Denmark and C. S. Regens, *Acc. Chem. Res.*, 2008, **41**, 1486; (c) S. E. Denmark, *J. Org. Chem.*, 2009, **74**, 2915.
7. Z. Xu, J. Singh, M. D. Schwinden, B. Zheng, T. P. Kissick, B. Patel, M. J. Humora, F. Quiroz, L. Dong, D.-M. Hsieh, J. E. Heikes, M. Pudipeddi, M. D. Lindrud, S. K. Srivastava, D. R. Kronenthal and R. H. Mueller, *Org. Process Res. Dev.*, 2002, **6**, 323.
8. (a) K. Harada, S. Nishino, K. Hirotsu, H. Shima, N. Okada, T. Harada, A. Nakamura and H. Oda, *PCT Int. Appl.*, WO 2002064569 A1 20020822,

2002; (b) Y. Fujikawa, M. Suzuki, H. Iwasaki, M. Sakashita and M. Kitahara, *Eur. Pat. Appl.*, EP 304063 A2 19890222, 1989; (c) N. Miyachi, Y. Yanagawa, H. Iwasaki, Y. Ohara and T. Hiyama, *Tetrahedron Lett.*, 1993, **34**, 8267.

9. (a) Y. Todo, K. Hayashi, M. Takahata, Y. Watanabe and H. Narita, *PCT Int. Appl.*, WO 9729102 A1 19970814, 1997; (b) M. Yamada, S. Hamamoto, K. Hayashi, K. Takaoka, H. Matsukura, M. Yotsuji, K. Yonezawa, K. Ojima, T. Takamatsu, K. Taya, H. Yamamoto, T. Kiyoto and H. Kotsubo, *PCT Int. Appl.*, WO 9921849 A1 19990506, 1999.

10. R. D. Larsen, A. O. King, C. Y. Chen, E. G. Corley, B. S. Foster, F. E. Roberts, C. Yang, D. R. Lieberman, R. A. Reamer, D. M. Tschaen, T. R. Verhoeven, P. J. Reider, Y. S. Lo, L. T. Rossano, A. S. Brookes, D. Meloni, J. R. Moore and J. F. Arnett, *J. Org. Chem.*, 1994, **59**, 6391.

11. E. O. Delorme, K. J. Duffy, P. I. Lamb, J. I. Luengo and S.-S. C. Tian, *PCT Int. Appl.*, WO 2002057300 A1 20020725, 2002.

12. (a) R. Desmond, U. Dolling, B. Marcune, R. Tillyer and D. Tschaen, *PCT Int. Appl.*, WO 9608482 A1 19960321, 1996; (b) B. Hancock, C. Winters, B. Gertz and E. Ehrich, *PCT Int. Appl.*, WO 9744028 A1 19971127, 1997.

13. R. W. Friesen, C. Brideau, C. C. Chan, S. Charleson, D. Deschênes, D. Dubé, D. Ethier, R. Fortin, J. Y. Gauthier, Y. Girard, R. Gordon, G. M. Greig, D. Riendeau, C. Savoie, Z. Wang, E. Wong, D. Visco, L. J. Xu and R. N. Young, *Bioorg. Med. Chem. Lett.*, 1998, **8**, 2777.

14. G. A. Potter, I. R. Hardcastle and M. Jarman, *Org. Prep. Proced. Int.*, 1997, **29**, 123.

15. (a) J. J. Cui, L. A. Funk, L. Jia, P.-P. Kung, J. J. Meng, M. D. Nambu, M. A. Pairish, H. Shen and M. B. Tran-Dube, *PCT Int. Appl.*, WO 2006021881 A2 20060302, 2006; (b) J. J. Cui, L. A. Funk, L. Jia, P.-P. Kung, J. J. Meng, M. D. Nambu, M. A. Pairish, H. Shen and M. Tran-Dube, *US Pat. Appl. Publ.*, US 20060046991 A1 20060302, 2006.

16. (a) C. Campas-Moya, *Drugs Future*, 2010, **35**, 457; (b) J. D. Rodgers, S. Shepard, T. P. Maduskuie Jr, H. Wang, N. Falahatpisheh, M. Rafalski, A. G. Arvanitis, L. Storace, R. K. Jalluri, J. S. Fridman and K. Vaddi, *US Pat. Appl. Publ.*, US 20090181959 A1 20090716, 2009.

17. (a) P. N. Ibrahim, D. R. Artis, R. Bremer, S. Mamo, M. Nespi, C. Zhang, J. Zhang, Y.-L. Zhu, J. Tsai, K.-P. Hirth, G. Bollag, W. Spevak, H. Cho, S. J. Gillette, G. Wu, H. Zhu and S. Shi, *PCT Int. Appl.*, WO 2007002325 A1 20070104, 2007; (b) P. N. Ibrahim, D. R. Artis, R. Bremer, G. Habets, S. Mamo, M. Nespi, C. Zhang, J. Zhang, Y.-L. Zhu, R. Zuckerman, B. West, Y. Suzuki, J. Tsai, K.-P. Hirth, G. Bollag, W. Spevak, H. Cho, S. J. Gillette, G. Wu, H. Zhu and S. Shi, *PCT Int. Appl.*, WO 2007002433 A1 20070104, 2007.

18. (a) P. Bühlmayer, P. Furet, L. Criscione, M. de Gasparo, S. Whitebread, T. Schmidlin, R. Lattmann and J. Wood, *Bioorg. Med. Chem. Lett.*, 1994, **4**, 29; (b) P. Bühlmayer, F. Ostermayer and T. Schmidlin, *Eur. Pat. Appl.*, EP 443983 A1 19910828, 1991.

19. (a) S. Nagato, K. Ueno, K. Kawano, Y. Norimine, K. Ito, T. Hanada, M. Ueno, H. Amino, M. Ogo, S. Hatakeyama, Y. Urawa, H. Naka,

A. J. Groom, L. Rivers and T. Smith, *PCT Int. Appl.*, WO 2001096308 A1 20011220, 2001; (b) K. Koyakumaru, Y. Matsuo and Y. Satake, *PCT Int. Appl.*, WO 2004009553 A1 20040129, 2004; (c) A. Kayano and K. Nishiura, *PCT Int. Appl.*, WO 2006004100 A1 20060112, 2006.

20. B. F. Whitehead, P. T. C. Ho, A. B. Suttle and A. N. Pandite, *PCT Int. Appl.*, WO 2007143483 A2 20071213, 2007.

21. (a) J. E. G. Guillemont, P. Palandjian, M. R. De Jonge, L. M. H. Koymans, H. M. Vinkers, F. F. D. Daeyaert, J. Heeres, K. J. A. Van Aken, P. J. Lewi and P. A. J. Janssen, *PCT Int. Appl.*, WO 2003016306 A1 20030227, 2003; (b) J. Guillemont, E. Pasquier, P. Palandjian, D. Vernier, S. Gaurrand, P. J. Lewi, J. Heeres, M. R. de Jonge, L. M. H. Koymans, F. F. D. Daeyaert, M. H. Vinkers, E. Arnold, K. Das, R. Pauwels, K. Andries, M.-P. de Béthune, E. Bettens, K. Hertogs, P. Wigerinck, P. Timmerman and P. A. J. Janssen, *J. Med. Chem.*, 2005, **48**, 2072; (c) D. P. R. Schils, J. J. M. Willems, B. P. A. M. J. Medaer, E. T. J. Pasquier, P. A. J. Janssen, J. Heeres and R. G. G. Leenders, *PCT Int. Appl.*, WO 2004016581 A1 20040226, 2004.

22. M. Aslam and V. Elango, *US Pat.*, US 5225603 A 19930706, 1993.

23. (a) A. W. Oxford, D. Butina and M. R. Owen, *Eur. Pat. Appl.*, EP 303507 A2 19890215, 1989; (b) P. Blatcher, M. Carter, R. Hornby and M. R. Owen, *PCT Int. Appl.*, WO 9509166 A1 19950406, 1995.

24. J. Bosch, T. Roca, M. Armengol and D. Fernández-Forner, *Tetrahedron*, 2001, **57**, 1041.

25. (a) R. J. Ogilvie, *PCT Int. Appl.*, WO 2002050063 A1 20020627, 2002; (b) J. E. Macor and M. J. Wythes, *US Pat.*, US 5545644 A 19960813, 1996.

26. (a) N. Godfrey, I. H. Coates, J. A. Bell, D. C. Humber and G. B. Ewan, *US Pat., 4957609*, 1990; (b) H. Iida, Y. Yuasa and C. Kibayashi, *J. Org. Chem.*, 1980, **45**, 2938.

27. T.-C. Wu, *US Pat.*, US 5536870 A 19960716, 1996.

28. C. Ramminger, D. Zim, V. R. Lando, V. Fassina and A. L. Monteiro, *J. Braz. Chem. Soc.*, 2000, **11**, 105.

29. B. P. Chekal, S. M. Guinness, B. M. Lillie, R. W. McLaughlin, C. W. Palmer, R. J. Post, J. E. Sieser, R. A. Singer, G. W. Sluggett, R. Vaidyanathan and G. J. Withbroe, *Org. Process Res. Dev.*, 2014, **18**, 266.

30. S. H. Kawai, R. J. Hambalek and G. Just, *J. Org. Chem.*, 1994, **59**, 2620.

31. (a) W.-S. Huang, C. A. Metcalf, R. Sundaramoorthi, Y. Wang, D. Zou, R. M. Thomas, X. Zhu, L. Cai, D. Wen, S. Liu, J. Romero, J. Qi, I. Chen, G. Banda, S. P. Lentini, S. Das, Q. Xu, J. Keats, F. Wang, S. Wardwell, Y. Ning, J. T. Snodgrass, M. I. Broudy, K. Russian, T. Zhou, L. Commodore, N. I. Narasimhan, Q. K. Mohemmad, J. Iuliucci, V. M. Rivera, D. C. Dalgarno, T. K. Sawyer, T. Clackson and W. C. Shakespeare, *J. Med. Chem.*, 2010, **53**, 4701; (b) D. Zou, W.-S. Huang, R. M. Thomas, J. A. C. Romero, J. Qi, Y. Wang, X. Zhu, W. C. Shakespeare, R. Sundaramoorthi, C. A. Metcalf III, D. C. Dalgarno and T. K. Sawyer, *PCT Int. Appl.*, WO 2007075869 A2 20070705, 2007.

32. U. Beutler, J. Mazacek, G. Penn, B. Schenkel and D. Wasmuth, *Chimia*, 1996, **50**, 154.

33. (a) C.-y. Chen and R. D. Larsen, *PCT Int. Appl.*, WO 9806725 A1 19980219, 1998; (b) C.-y. Chen, R. D. Larsen and T. R. Verhoeven, *PCT Int. Appl.*, WO 9532197 A1 19951130, 1995; (c) C.-y. Chen, D. R. Lieberman, R. D. Larsen, R. A. Reamer, T. R. Verhoeven, P. J. Reider, I. F. Cottrell and P. G. Houghton, *Tetrahedron Lett.*, 1994, **35**, 6981.

34. W. D. Vaccaro, R. Sher and H. R. Davis Jr, *Bioorg. Med. Chem.*, 1998, **6**, 1429.

35. (a) K. D. Robarge, S. A. Brunton, G. M. Castanedo, Y. Cui, M. S. Dina, R. Goldsmith, S. E. Gould, O. Guichert, J. L. Gunzner, J. Halladay, W. Jia, C. Khojasteh, M. F. T. Koehler, K. Kotkow, H. La, R. L. La Londe, K. Lau, L. Lee, D. Marshall, J. C. Marsters, L. J. Murray, C. Qian, L. L. Rubin, L. Salphati, M. S. Stanley, J. H. A. Stibbard, D. P Sutherlin, S. Ubhayaker, S. Wang, S. Wong and M. Xie, *Bioorg. Med. Chem. Lett.*, 2009, **19**, 5576; (b) J. Gunzner, D. Sutherlin, M. Stanley, L. Bao, G. Castanedo, R. Lalonde, S. Wang, M. Reynolds, S. Savage, K. Malesky and M. Dina, *PCT Int. Appl.*, WO 2006028958 A2 20060316, 2006.

36. (a) R. A. S. Chandraratna, *Eur. Pat. Appl.*, EP 284288 A1 19880928, 1988; (b) R. A. S. Chandraratna, *US Pat.*, US 5602130 A 19970211, 1997.

37. S. Abrecht, P. Harrington, H. Iding, M. Karpf, R. Trussardi, B. Wirz and U. Zutter, *Chimia*, 2004, **58**, 621.

38. (a) M. Beller, in *Applied Homogeneous Catalysis with Organometallic Compounds*, Vol. 1, ed. B. Cornils and W. A. Herrmann, VCH, Weinheim, 1996, p. 148; (b) E. J. Jang, K. H. Lee, J. S. Lee and Y. G. Kim, *J. Mol. Catal. A: Chem.*, 1999, **138**, 25.

39. C. Giodano, L. Coppi and F. Minsci, *Eur. Pat. Appl.*, EP 494419 A2 19920715, 1992.

40. R. A. Singer, J. D. McKinley, G. Barbe and R. A. Farlow, *Org. Lett.*, 2004, **6**, 2357.

41. (a) H. Kapeller, C. Marschner, M. Weissenbacher and H. Griengl, *Tetrahedron*, 1998, **54**, 1439; (b) H. F. Olivo and J. Yu, *J. Chem. Soc., Perkin Trans. 1*, 1998, 391.

42. (a) S. Morita, K. Kitano, J. Matsubara, T. Ohtani, Y. Kawano, K. Otsubo and M. Uchida, *Tetrahedron*, 1998, **54**, 4811; (b) Y. Oshiro, S. Sato, N. Kurahashi, T. Tanaka, T. Kikuchi, K. Tottori, Y. Uwahodo and T. Nishi, *J. Med. Chem.*, 1998, **41**, 658; (c) Y. Oshiro, S. Sato and N. Kurahashi, *Eur. Pat. Appl.*, EP 367141 A2 19900509, 1990.

43. (a) B. C. Austad, F. Benayoud, T. L. Calkins, S. Campagna, C. E. Chase, H.-w. Choi, W. Christ, R. Costanzo, J. Cutter, A. Endo, F. G. Fang, Y. Hu, B. M. Lewis, M. D. Lewis, S. McKenna, T. A. Noland, J. D. Orr, M. Pesant, M. J. Schnaderbeck, G. D. Wilkie, T. Abe, N. Asai, Y. Asai, A. Kayano, Y. Kimoto, Y. Komatsu, M. Kubota, H. Kuroda, M. Mizuno, T. Nakamura, T. Omae, N. Ozeki, T. Suzuki, T. Takigawa, T. Watanabe and K. Yoshizawa, *Synlett.*, 2013, **24**, 327; (b) B. C. Austad, T. L. Calkins, C. E. Chase, F. G. Fang, T. E. Horstmann, Y. Hu, B. M. Lewis, X. Niu,

T. A. Noland, J. D. Orr, M. J. Schnaderbeck, H. Zhang, N. Asakawa, N. Asai, H. Chiba, T. Hasebe, Y. Hoshino, H. Ishizuka, T. Kajima, A. Kayano, Y. Komatsu, M. Kubota, H. Kuroda, M. Miyazawa, K. Tagami and T. Watanabe, *Synlett*, 2013, **24**, 333.

44. W. S. Huang and W. C. Shakespeare, *Synthesis*, 2007, 2121.

45. (a) R. A. Fujimoto, L. W. Mcquire, B. B. Mugrage, J. H. Van Duzer and D. Xu, *PCT Int. Appl.*, WO 9911605 A1 19990311, 1999; (b) M. Acemoglu, T. Allmendinger, J. V. Calienni, J. Cercus, O. Loiseleur, G. Sedelmeier and D. Xu, *PCT Int. Appl.*, WO 2001023346 A2 20010405, 2001.

46. O. Loiseleur, D. Kaufmann, S. Abel, H. M. Buerger, M. Meisenbach, B. Schmitz and G. Sedelmeier, *PCT Int. Appl.*, WO 2003066613 A1 20030814, 2003.

47. D. A. Bradley, A. G. Godfrey and C. R. Schmid, *Tetrahedron Lett.*, 1999, **40**, 5155.

48. (a) D. Pinto, M. Quan, M. Orwat, Y.-L. Li, W. Han, J. Qiao, P. Lam and S. Koch, *PCT Int. Appl.*, WO 2003026652 A1 20030403, 2003; (b) D. J. P. Pinto, M. J. Orwat, S. Koch, K. A. Rossi, R. S. Alexander, A. Smallwood, P. C. Wong, A. R. Rendina, J. M. Luettgen, R. M. Knabb, K. He, B. Xin, R. R. Wexler and P. Y. S. Lam, *J. Med. Chem.*, 2007, **50**, 5339.

49. K. Yamada, K. Matsuki, K. Omori and K. Kikkawa, *US Pat. Appl. Publ.*, US 20040142930 A1 20040722, 2004.

50. (a) C. A. Bernhart, P. M. Perreaut, B. P. Ferrari, Y. A. Muneaux, J.-L. A. Assens, J. Clément, F. Haudricourt, C. F. Muneaux, J. E. Taillades, M.-A. Vignal, J. Gougat, P. R. Guiraudou, C. A. Lacour, A. Roccon, C. F. Cazaubon, J.-C. Brelière, G. Le Fur and D. Nisato, *J. Med. Chem.*, 1993, **36**, 3371; (b) B. Ferrari, J. Taillades, P. Perreaut, C. Bernhart, J. Gougat, P. Guiraudou, C. Cazaubon, A. Roccon, D. Nisato, G. Le Fur and J. C. Brelière, *Bioorg. Med. Chem. Lett.*, 1994, **4**, 45.

51. (a) K. Lauer and E. Kiegel, *Ger. Pat.*, DE 2928483 B1 19800417, 1980; (b) S. Huang, H. Yu, X. Sun and W. Gao, *Zhongguo Yaowu Huaxue Zazhi*, 2000, **10**, 293.

52. (a) N. Miyaura, K. Yamada and A. Suzuki, *Tetrahedron Lett.*, 1979, **20**, 3437; (b) N. Miyaura and A. Suzuki, *J. Chem. Soc., Chem. Commun.*, 1979, 866.

53. (a) D. S. Ennis, J. McManus, W. Wood-Kaczmar, J. Richardson, G. E. Smith and A. Carstairs, *Org. Process Res. Dev.*, 1999, **3**, 248; (b) S. Caron, S. S. Massett, D. E. Bogle, M. J. Castaldi and T. F. Braish, *Org. Process Res. Dev.*, 2001, **5**, 254; (c) D. D. Winkle and K. M. Schaab, *Org. Process Res. Dev.*, 2001, **5**, 450; (d) Z. Xu, J. Singh, M. D. Schwinden, B. Zheng, T. P. Kissick, B. Patel, M. J. Humora, F. Quiroz, L. Dong, D.-M. Hsieh, J. E. Heikes, M. Pudipeddi, M. D. Lindrud, S. K. Srivastava, D. R. Kronenthal and R. H. Mueller, *Org. Process Res. Dev.*, 2002, **6**, 323; (e) C.-y. Chen, P. Dagneau, E. J. J. Grabowski, R. Oballa, P. O'Shea, P. Prasit, J. Robichaud, R. Tillyer and X. Wang, *J. Org. Chem.*, 2003, **68**, 2633; (f) V. Derdau, R. Oekonomopulos and G. Schubert, *J. Org. Chem.*, 2003, **68**, 5168; (g) D. A. Conlon, B. Pipik, S. Ferdinand,

C. R. LeBlond, J. R. Sowa Jr, B. Izzo, P. Collins, G.-J. Ho, J. M. Williams, Y.-J. Shi and Y. Sun, *Adv. Synth. Catal.*, 2003, **345**, 931; (h) Y. Urawa, M. Miyazawa, N. Ozeki and K. Ogura, *Org. Process Res. Dev.*, 2003, 7, 191; (i) C. P. Ashcroft, S. Challenger, A. M. Derrick, R. Storey and N. M. Thomson, *Org. Process Res. Dev.*, 2003, 7, 362; (j) M. F. Lipton, M. A. Mauragis, M. T. Maloney, M. F. Veley, D. W. VanderBor, J. J. Newby, R. B. Appell and E. D. Daugs, *Org. Process Res. Dev.*, 2003, 7, 385; (k) S. J. Mickel, G. H. Sedelmeier, D. Niederer, F. Schuerch, M. Seger, K. Schreiner, R. Daeffler, A. Osmani, D. Bixel, O. Loiseleur, J. Cercus, H. Stettler, K. Schaer, R. Gamboni, A. Bach, G.-P. Chen, W. Chen, P. Geng, G. T. Lee, E. Loeser, J. McKenna, F. R. Kinder, K. Konigsberger, K. Prasad, T. M. Ramsey, N. Reel, O. Repič, L. Rogers, W.-C. Shieh, R.-M. Wang, L. Waykole, S. Xue, G. Florence and I. Paterson, *Org. Process Res. Dev.*, 2004, **8**, 113; (l) T. E. Jacks, D. T. Belmont, C. A. Briggs, N. M. Horne, G. D. Kanter, G. L. Karrick, J. J. Krikke, R. J. McCabe, J. G. Mustakis, T. N. Nanninga, G. S. Risedorph, R. E. Seamans, R. Skeean, D. D. Winkle and T. M. Zennie, *Org. Process Res. Dev.*, 2004, **8**, 201; (m) J. F. Payack, E. Vazquez, L. Matty, M. H. Kress and J. McNamara, *J. Org. Chem.*, 2005, **70**, 175; (n) S. P. Keen, C. J. Cowden, B. C. Bishop, K. M. J. Brands, A. J. Davies, U. H. Dolling, D. R. Lieberman and G. W. Stewart, *J. Org. Chem.*, 2005, **70**, 1771; (o) D. R. Gauthier Jr, J. Limanto, P. N. Devine, R. A. Desmond, R. H. Szumigala Jr, B. S. Foster and R. P. Volante, *J. Org. Chem.*, 2005, **70**, 5938; (p) M. S. Jensen, R. S. Hoerrner, W. Li, D. P. Nelson, G. J. Javadi, P. G. Dormer, D. Cai and R. D. Larsen, *J. Org. Chem.*, 2005, **70**, 6034; (q) N. A. Magnus, J. A. Aikins, J. S. Cronin, W. D. Diseroad, A. D. Hargis, M. E. LeTourneau, B. E. Parker, S. M. Reutzel-Edens, J. P. Schafer, M. A. Staszak, G. A. Stephenson, S. L. Tameze and L. M. H. Zollars, *Org. Process Res. Dev.*, 2005, **9**, 621; (r) J. Y. L. Chung, R. J. Cvetovich, M. McLaughlin, J. Amato, F.-R. Tsay, M. Jensen, S. Weissman and D. Zewge, *J. Org. Chem.*, 2006, **71**, 8602; (s) D. A. Conlon, A. Drahus-Paone, G.-J. Ho, B. Pipik, R. Helmy, J. M. McNamara, Y.-J. Shi, J. M. Williams, D. Macdonald, D. Deschênes, M. Gallant, A. Mastracchio, B. Roy and J. Scheigetz, *Org. Process Res. Dev.*, 2006, **10**, 36; (t) M. Cameron, B. S. Foster, J. E. Lynch, Y.-J. Shi and U.-H. Dolling, *Org. Process Res. Dev.*, 2006, **10**, 398; (u) F. A. J. Kerdesky, M. R. Leanna, J. Zhang, W. Li, J. E. Lallaman, J. Ji and H. E. Morton, *Org. Process Res. Dev.*, 2006, **10**, 512; (v) T. Itoh, S. Kato, N. Nonoyama, T. Wada, K. Maeda, T. Mase, M. M. Zhao, J. Z. Song, D. M. Tschaen and J. M. McNamara, *Org. Process Res. Dev.*, 2006, **10**, 822; (w) D. J. Ager, K. Anderson, E. Oblinger, Y. Shi and J. VanderRoest, *Org. Process Res. Dev.*, 2007, **11**, 44; (x) W. D. Miller, A. H. Fray, J. T. Quatroche and C. D. Sturgill, *Org. Process Res. Dev.*, 2007, **11**, 359; (y) B. Li, R. A. Buzon and Z. Zhang, *Org. Process Res. Dev.*, 2007, **11**, 951; (z) L. A. Hobson, W. A. Nugent, S. R. Anderson, S. S. Deshmukh, J. J. Haley III, P. Liu, N. A. Magnus, P. Sheeran, J. P. Sherbine, B. R. P. Stone and J. Zhu, *Org.*

Process Res. Dev., 2007, **11**, 985; (aa) K. Daïri, Y. Yao, M. Faley, S. Tripathy, E. Rioux, X. Billot, D. Rabouin, G. Gonzalez, J.-F. Lavallée and G. Attardo, *Org. Process Res. Dev.*, 2007, **11**, 1051; (ab) C. A. Busacca, M. Cerreta, Y. Dong, M. C. Eriksson, V. Farina, X. Feng, J.-Y. Kim, J. C. Lorenz, M. Sarvestani, R. Simpson, R. Varsolona, J. Vitous, S. J. Campbell, M. S. Davis, P.-J. Jones, D. Norwood, F. Qiu, P. L. Beaulieu, J.-S. Duceppe, B. Haché, J. Brong, F.-T. Chiu, T. Curtis, J. Kelley, Y. S. Lo and T. H. Powner, *Org. Process Res. Dev.*, 2008, **12**, 603; (ac) K. M. Bullock, M. B. Mitchell and J. F. Toczko, *Org. Process Res. Dev.*, 2008, **12**, 896; (ad) J. Limanto, B. T. Dorner, F. W. Hartner and L. Tan, *Org. Process Res. Dev.*, 2008, **12**, 1269; (ae) D. M. Barnes, J. Barkalow, Y. Chen, A. Gupta, A. R. Haight, J. E. Hengeveld, F. A. J. Kerdesky, B. J. Kotecki, B. Macri and A. Pal, *Org. Process Res. Dev.*, 2009, **13**, 225; (af) O. R. Thiel, M. Achmatowicz, C. Bernard, P. Wheeler, C. Savarin, T. L. Correll, A. Kasparian, A. Allgeier, M. D. Bartberger, H. Tan and R. D. Larsen, *Org. Process Res. Dev.*, 2009, **13**, 230; (ag) M. H. Yates, T. M. Koenig, N. J. Kallman, C. P. Ley and D. Mitchell, *Org. Process Res. Dev.*, 2009, **13**, 268; (ah) R. N. Richey and H. Yu, *Org. Process Res. Dev.*, 2009, **13**, 315; (ai) K. Menzel, F. Machrouhi, M. Bodenstein, A. Alorati, C. Cowden, A. W. Gibson, B. Bishop, N. Ikemoto, T. D. Nelson, M. H. Kress and D. E. Frantz, *Org. Process Res. Dev.*, 2009, **13**, 519; (aj) I. N. Houpis, D. Shilds, U. Nettekoven, A. Schnyder, E. Bappert, K. Weerts, M. Canters and W. Vermuelen, *Org. Process Res. Dev.*, 2009, **13**, 598; (ak) G. L. Allsop, A. J. Cole, M. E. Giles, E. Merifield, A. J. Noble, M. A. Pritchett, L. A. Purdie and J. T. Singleton, *Org. Process Res. Dev.*, 2009, **13**, 751; (al) R. Hanselmann, G. E. Job, G. Johnson, R. Lou, J. G. Martynow and M. M. Reeve, *Org. Process Res. Dev.*, 2010, **14**, 152; (am) X. Jiang, G. T. Lee, E. B. Villhauer, K. Prasad and M. Prashad, *Org. Process Res. Dev.*, 2010, **14**, 883; (an) M. J. Fray, A. T. Gillmore, M. S. Glossop, D. J. McManus, I. B. Moses, C. F. B. Praquin, K. A. Reeves and L. R. Thompson, *Org. Process Res. Dev.*, 2010, **14**, 263; (ao) M. Whiting, K. Harwood, F. Hossner, P. G. Turner and M. C. Wilkinson, *Org. Process Res. Dev.*, 2010, **14**, 820; (ap) G. W. Stewart, K. M. J. Brands, S. E. Brewer, C. J. Cowden, A. J. Davies, J. S. Edwards, A. W. Gibson, S. E. Hamilton, J. D. Katz, S. P. Keen, P. R. Mullens, J. P. Scott, D. J. Wallace and C. S. Wise, *Org. Process Res. Dev.*, 2010, **14**, 849; (aq) V. I. Elitzin, K. A Harvey, H. Kim, M. Salmons, M. J. Sharp, E. A. Tabet and M. A. Toczko, *Org. Process Res. Dev.*, 2010, **14**, 912; (ar) R. R. Milburn, O. R. Thiel, M. Achmatowicz, X. Wang, J. Zigterman, C. Bernard, J. T. Colyer, E. DiVirgilio, R. Crockett, T. L. Correll, K. Nagapudi, K. Ranganathan, S. J. Hedley, A. Allgeier and R. D. Larsen, *Org. Process Res. Dev.*, 2011, **15**, 31; (as) C. Molinaro, S. Shultz, A. Roy, S. Lau, T. Trinh, R. Angelaud, P. D. O'Shea, S. Abele, M. Cameron, E. Corley, J.-A. Funel, D. Steinhuebel, M. Weisel and S. Krska, *J. Org. Chem.*, 2011, **76**, 1062; (at) D. M. Bowles, D. C. Boyles, C. Choi, J. A. Pfefferkorn and

S. Schuyler, *Org. Process Res. Dev.*, 2011, **15**, 148; (au) H. Malmgren, H. Cotton, B. Frøstup, D. S. Jones, M.-L. Loke, D. Peters, S. Schultz, E. Sölver, T. Thomsen and J. Wennerberg, *Org. Process Res. Dev.*, 2011, **15**, 408; (av) Q. Huang, P. F. Richardson, N. W. Sach, J. Zhu, K. K.-C. Liu and G. L. Smith, *Org. Process Res. Dev.*, 2011, **15**, 556; (aw) S. D. Walker, C. J. Borths, E. DiVirgilio, L. Huang, P. Liu, H. Morrison, K. Sugi, M. Tanaka, J. C. S. Woo and M. M. Faul, *Org. Process Res. Dev.*, 2011, **15**, 570; (ax) D. J. Wallace, C. A. Baxter, K. J. M. Brands, N. Bremeyer, S. E. Brewer, R. Desmond, K. M. Emerson, J. Foley, P. Fernandez, W. Hu, S. P. Keen, P. Mullens, D. Muzzio, P. Sajonz, L. Tan, R. D. Wilson, G. Zhou and G. Zhou, *Org. Process Res. Dev.*, 2011, **15**, 831; (ay) P. D. de Koning, D. McAndrew, R. Moore, I. B. Moses, D. C. Boyles, K. Kissick, C. L. Stanchina, T. Cuthbertson, A. Kamatani, L. Rahman, R. Rodriguez, A. Urbina, A. Sandoval and P. R. Rose, *Org. Process Res. Dev.*, 2011, **15**, 1018; (az) M. Berwe, W. Jöntgen, J. Krüger, Y. Cancho-Grande, T. Lampe, M. Michels, H. Paulsen, S. Raddatz and S. Weigand, *Org. Process Res. Dev.*, 2011, **15**, 1348; (ba) C. Boos, D. M. Bowles, C. Cai, A. Casimiro-Garcia, X. Chen, C. A. Hulford, S. M. Jennings, E. J. Kiser, D. W. Piotrowski, M. Sammons and R. A. Wade, *Tetrahedron Lett.*, 2011, **52**, 7025; (bb) M. A. Graham, S. A. Raw, D. M. Andrews, C. J. Good, Z. S. Matusiak, M. Maybury, E. S. E Stokes and A. T. Turner, *Org. Process Res. Dev.*, 2012, **16**, 1283; (bc) A. T. Gillmore, M. Badland, C. L. Crook, N. M. Castro, D. J. Critcher, S. J. Fussell, K. J. Jones, M. C. Jones, E. Kougoulos, J. S. Mathew, L. McMillan, J. E. Pearce, F. L. Rawlinson, A. E. Sherlock and R. Walton, *Org. Process Res. Dev.*, 2012, **16**, 1897; (bd) S. R. Breining, J. F. Genus, J. P. Mitchener, T. J. Cuthbertson, R. Heemstra, M. S. Melvin, G. M. Dull and D. Yohannes, *Org. Process Res. Dev.*, 2013, **17**, 413; (be) C. Fiorelli, R. Scarpelli, D. Piomelli and T. Bandiera, *Org. Process Res. Dev.*, 2013, **17**, 359; (bf) F. Hicks, Y. Hou, M. Langston, A. McCarron, E. O'Brien, T. Ito, C. Ma, C. Matthews, C. O'Bryan, D. Provencal, Y. Zhao, J. Huang, Q. Yang, L. Heyang, M. Johnson, Y. Sitang and L. Yuqiang, *Org. Process Res. Dev.*, 2013, **17**, 829; (bg) Y. Wang, K. Przyuski, R. C. Roemmele, R. L. Hudkins and R. P. Bakale, *Org. Process Res. Dev.*, 2013, **17**, 846.

54. (a) N. A. Owston and G. C. Fu, *J. Am. Chem. Soc.*, 2010, **132**, 11908; (b) P. M. Lundin and G. C. Fu, *J. Am. Chem. Soc.*, 2010, **132**, 11027; (c) B. Saito and G. C. Fu, *J. Am. Chem. Soc.*, 2008, **130**, 6694; (d) V. Gracias and R. Iyengar, *Chemtracts*, 2005, **18**, 339; (e) D. Zim and A. L. Monteiro, *Tetrahedron Lett.*, 2002, **43**, 4009; (f) B. H. Lipshutz, *Adv. Synth. Catal.*, 2001, **343**, 313; (g) D. Zim, V. R. Lando, J. Dupont and A. L. Monteiro, *Org. Lett.*, 2001, **3**, 3049; (h) B. H. Lipshutz, J. A. Sclafani and P. A. Blomgren, *Tetrahedron*, 2000, **56**, 2139; (i) S. Saito, S. Oh-tani and N. Miyaura, *J. Org. Chem.*, 1997, **62**, 8024.

55. (a) V. Polshettiwar, A. Decottignies, C. Len and A. Fihri, *ChemSusChem*, 2010, **3**, 502; (b) R. Franzen and Y. Xu, *Can. J. Chem.*, 2005, **83**, 266; (c) L. Bai and J.-X. Wang, *Curr. Org. Chem.*, 2005, **9**, 535.

56. M. M. Heravi and E. Hashemi, *Tetrahedron*, 2012, **68**, 9145.

57. (a) E-i. Negishi, in *Handbook of Organopalladium Chemistry for Organic Synthesis*, Vol. 1, ed. E-i. Negishi, Wiley, New York, 2002, pp. 229–248; (b) D. Chen, S. Kotti and G. Li, *Chemtracts*, 2005, **18**, 193; (c) E.-i. Negishi, Q. Hu, Z. Huang, M Qian and G. Wang, *Aldrichim. Acta*, 2005, **38**, 71.

58. (a) M. Baenziger, C. P Mak, H. Muehle, F. Nobs, W. Prokoszovich, J. L. Reber and U. Sunay, *Org. Process Res. Dev.*, 1997, **1**, 395; (b) Y.-Y. Ku, T. Grieme, P. Raje, P. Sharma, H. E Morton, M. Rozema and S. A. King, *J. Org. Chem.*, 2003, **68**, 3238; (c) J. A. Ragan, J. W. Raggon, P. D. Hill, B. P. Jones, R. E. McDermott, M. J. Munchhof, M. A. Marx, J. M. Casavant, B. A. Cooper, J. L. Doty and Y. Lu, *Org. Process Res. Dev.*, 2003, 7, 676; (d) S. Cai, M. Dimitroff, T. McKennon, M. Reider, L. Robarge, D. Ryckman, X. Shang and J. Therrien, *Org. Process Res. Dev.*, 2004, **8**, 353; (e) D. Denni-Dischert, W. Marterer, M. Bänziger, N. Yusuff, D. Batt, T. Ramsey, P. Geng, W. Michael, R.-M. B. Wang, F. Taplin Jr, R. Versace, D. Cesarz and L. B. Perez, *Org. Process Res. Dev.*, 2006, **10**, 70; (f) Z. Liu and J. Xiang, *Org. Process Res. Dev.*, 2006, **10**, 285; (g) R. W. Scott, S. N. Neville, A. Urbina, D. Camp and N. Stankovic, *Org. Process Res. Dev.*, 2006, **10**, 296; (h) C. Pérez-Balado, A. Willemsens, D. Ormerod, W. Aelterman and N. Mertens, *Org. Process Res. Dev.*, 2007, **11**, 237; (i) C. H. V. A. Sasikala, P. R. Padi, V. Sunkara, P. Ramayya, P. K. Dubey, V. B. R. Uppala and C. Praveen, *Org. Process Res. Dev.*, 2009, **13**, 907; (j) L. Shu, P. Wang, C. Gu, W. Liu, L. M. Alabanza and Y. Zhang, *Org. Process Res. Dev.*, 2013, **17**, 651; (k) S. Pikul, H. Cheng, A. Cheng, C. D. Huang, A. Ke, L. Kuo, A. Thompson and S. Wilder, *Org. Process Res. Dev.*, 2013, **17**, 907.

59. US Environmental Protection Agency, zinc waste regulation search, http://nlquery.epa.gov/epasearch/epasearch?querytext= zinc+waste+regulation&fld=&areaname=&typeofsearch=epa& areacontacts=http%3A%2F%2Fwww.epa. gov%2Fepahome%2Fcomments.htm&areasearchurl=&result_template= epafiles_default.xsl&filter=sample4filt.hts (accessed 26 December 2013).

60. G. Manolikakes, C. M. Hernandez, M. A. Schade, A. Metzger and P. Knochel, *J. Org. Chem.*, 2008, **73**, 8422.

61. (a) E. Erdik, *Tetrahedron*, 1987, **43**, 2203; (b) H. J. C. Deboves, C. A. G. N. Montalbetti and R. F. W. Jackson, *J. Chem. Soc., Perkin Trans. 1*, 2001, 1876.

62. (a) R. J. P. Corriu and J. P. Masse, *J. Chem. Soc., Chem. Commun.*, 1972, 144; (b) K. Tamao, K. Sumitani and M. Kumada, *J. Am. Chem. Soc.*, 1972, **94**, 4374; (c) K. Tamao, K. Sumitani, M. Zembayashi, A. Fijioka, S.-I. Kodama, I. Nakajima, A. Minato and M. Kumada, *Bull. Chem. Soc. Jpn.*, 1976, **49**, 1958; (d) M. Kumada, K. Tamao and K. Sumitani, *Org. Synth.*, 1978, **58**, 127; (e) M. Kumada, *Pure Appl. Chem.*, 1980, **52**, 669; (f) T. Hayashi and M. Kumada, *Acc. Chem. Res.*, 1982, **15**, 395.

63. For recent developments, see: (a) J. Terao and N. Kambe, *Acc. Chem. Res.*, 2008, **41**, 1545; (b) O. Vechorkin, V. Proust and X. Hu, *J. Am. Chem.*

Soc., 2009, **131**, 9756; (c) N. Yoshikai, H. Matsuda and E. Nakamura, *J. Am. Chem. Soc.*, 2009, **131**, 9590; (d) N. Yoshikai, H. Matsuda and E. Nakamura, *J. Am. Chem. Soc.*, 2008, **130**, 15258; (e) L. Ackermann, R. Born, J. H. Spatz and D. Meyer, *Angew. Chem. Int. Ed.*, 2005, **44**, 7216; (f) V. P. W. Böhm, C. W. K. Gstöttmayr, T. Weskamp and W. A. Herrmann, *Angew. Chem. Int. Ed.*, 2001, **40**, 3387; (g) G. Y. Li, *Angew. Chem. Int. Ed.*, 2001, **40**, 1513; (h) T. Banno, Y. Hayakawa and M. Umeno, *J. Organomet. Chem.*, 2002, **653**, 288.

64. (a) G. Marzoni and M. D. Varney, *Org. Process Res. Dev.*, 1997, **1**, 81; (b) G. Bold, A. Fässler, H.-G. Capraro, R. Cozens, T. Klimkait, J. Lazdins, J. Mestan, B. Poncioni, J. Rösel, D. Stover, M. Tintelnot-Blomley, F. Acemoglu, W. Beck, E. Boss, M. Eschbach, T. Hürlimann, E. Masso, S. Roussel, K. Ucci-Stoll, D. Wyss and M. Lang, *J. Med. Chem.*, 1998, **41**, 3387; (c) P. W. Manley, M. Acemoglu, W. Marterer and W. Pachinger, *Org. Process Res. Dev.*, 2003, **7**, 436; (d) X.-j. Wang, L. Zhang, L. L. Smith-Keenan, I. N. Houpis and V. Farina, *Org. Process Res. Dev.*, 2007, **11**, 60; (e) X. Fan, Y.-L. Song and Y. Q. Long, *Org. Process Res. Dev.*, 2008, **12**, 69; (f) M. J. Zacuto, C. S. Shultz and M. Journet, *Org. Process Res. Dev.*, 2011, **15**, 158; (g) J.-A. Funel, G. Schmidt and S. Abele, *Org. Process Res. Dev.*, 2011, **15**, 1420; (h) N. Tewari, N. Maheshwari, R. Medhane, H. Nizar and M. Prasad, *Org. Process Res. Dev.*, 2012, **16**, 1566.

65. (a) B. D. Sherry and A. Fürstner, *Acc. Chem. Res.*, 2008, **41**, 1500; (b) B. Plietker, *Iron Catalysis in Organic Chemistry*, Wiley-VCH, Weinheim, 2008.

66. (a) W. Hess, J. Treutwein and G. Hilt, *Synthesis*, 2008, 3537; (b) C. Gosmini, J.-M. Begouin and A. Moncomble, *Chem. Commun.*, 2008, 3221; (c) H. Yorimitsu and K. Oshima, *Pure Appl. Chem.*, 2006, **78**, 441.

67. G. Cahiez, C. Duplais and J. Buendia, *Chem. Rev.*, 2009, **109**, 1434.

68. For early reports on the Ni- and Pd-catalyzed arylation of K, Li, Zn and Sn enolates, see: (a) M. F. Semmelhack, R. D. Stauffer and T. D. Rogerson, *Tetrahedron Lett.*, 1973, **14**, 4519; (b) A. A. Millard and M. W. Rathke, *J. Am. Chem. Soc.*, 1977, **99**, 4833; (c) J. F. Fauvarque and A. Jutand, *J. Organomet. Chem.*, 1979, **177**, 273; (d) R. Galarini, A. Musco, R. Pontellini and R. Santi, *J. Mol. Catal.*, 1992, **72**, L11; (e) I. Kuwajima and E.-i. Nakamura, *Acc. Chem. Res.*, 1985, **18**, 181; (f) M. Kosugi, I. Hagiwara, T. Sumiya and T. Migita, *Bull. Chem. Soc. Jpn.*, 1984, **57**, 242; (g) M. Kosugi, Y. Negishi, M. Kameyama and T. Migita, *Bull. Chem. Soc. Jpn.*, 1985, **58**, 3383.

69. B. C. Hamann and J. F. Hartwig, *J. Am. Chem. Soc.*, 1997, **119**, 12382.

70. M. Palucki and S. L. Buchwald, *J. Am. Chem. Soc.*, 1997, **119**, 11108.

71. T. Satoh, Y. Kawamura, M. Miura and M. Nomura, *Angew. Chem. Int. Ed. Engl.*, 1997, **36**, 1740.

72. C. C. C. Johansson and T. J. Colacot, *Angew. Chem. Int. Ed.*, 2010, **49**, 676.

73. For reviews, see: (a) G. C. Lloyd-Jones, *Angew. Chem. Int. Ed.*, 2002, **41**, 953; (b) D. A. Culkin and J. F. Hartwig, *Acc. Chem. Res.*, 2003, **36**, 234; (c) F. Bellina and R. Rossi, *Chem. Rev.*, 2010, **110**, 1082.

74. (a) O. Dirat, J. M. Elliott, R. A. Jelley, B. Jones and M. Reader, *Tetrahedron Lett.*, 2006, **47**, 1295; (b) C. A. Busacca, M. Cerreta, Y. Dong, M. C. Eriksson, V. Farina, X. Feng, J.-Y. Kim, J. C. Lorenz, M. Sarvestani, R. Simpson, R. Varsolona, J. Vitous, S. J. Campbell, M. S. Davis, P.-J. Jones, D. Norwood, F. Qiu, P. L. Beaulieu, J.-S. Duceppe, B. Haché, J. Brong, F.-T. Chiu, T. Curtis, J. Kelley, Y. S. Lo and T. H. Powner, *Org. Process Res. Dev.*, 2008, **12**, 603; (c) X. Jiang, B. Gong, K. Prasad and O. Repič, *Org. Process Res. Dev.*, 2008, **12**, 1164; (d) C. Ohta, S.-i. Kuwabe, T. Shiraishi, I. Shinohara, H. Araki, S. Sakuyama, T. Makihara, Y. Kawanaka, S. Ohuchida and T. Seko, *J. Org. Chem.*, 2009, **74**, 8298; (e) S. Abele, R. Inauen, J.-A. Funel and T. Weller, *Org. Process Res. Dev.*, 2012, **16**, 129; (f) Q. Yang, L. G. Ulysse, M. D. McLaws, D. K. Keefe, B. P. Haney, C. Zha, P. R. Guzzo and S. Liu, *Org. Process Res. Dev.*, 2012, **16**, 499.

75. (a) H. Doucet and J.-C. Hierso, *Angew. Chem. Int. Ed.*, 2007, **46**, 834; (b) K. Sonogashira, in *Metal-Catalyzed Cross-Coupling Reactions*, Vol. 1, ed. F. Diederich and A. de Meijere, Wiley-VCH, Weinheim, 2004, p. 319; (c) K. Sonogashira, in *Handbook of Organopalladium Chemistry for Organic Synthesis*, ed. E.-i. Negishi, Wiley, New York, 2002, pp. 493–530; (d) K. Sonogashira, *J. Organomet. Chem.*, 2002, **653**, 46; (e) K. Sonogashira, in *Comprehensive Organic Synthesis*, Vol. 3, ed. B. M. Trost and I. Fleming, Pergamon Press, Oxford, 1991, p. 521.

76. (a) A. V. Thomas, H. H. Patel, L. A. Reif, S. R. Chemburkar, D. P. Sawick, B. Shelat, M. K. Balmer and R. R. Patel, *Org. Process Res. Dev.*, 1997, **1**, 294; (b) L. S. Bleicher, N. D. P. Cosford, A. Herbaut, J. S. McCallum and I. A. McDonald, *J. Org. Chem.*, 1998, **63**, 1109; (c) C. J. Barnett, T. M. Wilson and M. E. Kobierski, *Org. Process Res. Dev.*, 1999, **3**, 184; (d) J. W. B. Cooke, R. Bright, M. J. Coleman and K. P. Jenkins, *Org. Process Res. Dev.*, 2001, **5**, 383; (e) K. Königsberger, G.-P. Chen, R. R. Wu, M. J. Girgis, K. Prasad, O. Repič and T. J. Blacklock, *Org. Process Res. Dev.*, 2003, 7, 733; (f) B. M. Andresen, M. Couturier, B. Cronin, M. D'Occhio, M. D. Ewing, M. Guinn, J. M. Hawkins, J. Jasys, S. D. LaGreca, J. P. Lyssikatos, G. Moraski, K. Ng, J. W. Raggon, A. M. Stewart, D. L. Tickner, J. L. Tucker, F. J. Urban, E. Vazquez and L. Wei, *Org. Process Res. Dev.*, 2004, **8**, 643; (g) F. W. Hartner, Y. Hsiao, K. K. Eng, N. R. Rivera, M. Palucki, L. Tan, N. Yashua, D. L. Hughes, S. Weissman, D. Zewge, T. King, D. Tschaen and R. P. Volante, *J. Org. Chem.*, 2004, **69**, 8723; (h) D. H. B. Ripin, D. E. Bourassa, T. Brandt, M. J. Castaldi, H. N. Frost, J. Hawkins, P. J. Johnson, S. S. Massett, K. Neumann, J. Phillips, J. W. Raggon, P. R. Rose, J. L. Rutherford, B. Sitter, A. M. Stewart III, M. G. Vetelino and L. Wei, *Org. Process Res. Dev.*, 2005, **9**, 440; (i) R. L. Dorow, P. M. Herrinton, R. A. Hohler, M. T. Maloney, M. A. Mauragis, W. E. McGhee, J. A. Moeslein,

J. W. Strohbach and M. F. Veley, *Org. Process Res. Dev.*, 2006, **10**, 493; (j) N. A. Magnus, W. D. Diseroad, R. Nevill Jr and J. P. Wepsiec, *Org. Process Res. Dev.*, 2006, **10**, 556; (k) H. Li, Z. Xia, S. Chen, K. Koya, M. Ono and L. Sun, *Org. Process Res. Dev.*, 2007, **11**, 246; (l) R. N. Richey and H. Yu, *Org. Process Res. Dev.*, 2009, **13**, 315; (m) I. N. Houpis, D. Shilds, U. Nettekoven, A. Schnyder, E. Bappert, K. Weerts, M. Canters and W. Vermuelen, *Org. Process Res. Dev.*, 2009, **13**, 598; (n) S. Yu, A. Haight, B. Kotecki, L. Wang, K. Lukin and D. R. Hill, *J. Org. Chem.*, 2009, **74**, 9539; (o) X. Deng, J. T. Liang, M. Peterson, R. Rynberg, E. Cheung and N. S. Mani, *J. Org. Chem.*, 2010, **75**, 1940; (p) M. A. Berliner, E. M. Cordi, J. R. Dunetz and K. E. Price, *Org. Process Res. Dev.*, 2010, **14**, 180; (q) N. M. Deschamps, V. I. Elitzin, B. Liu, M. B. Mitchell, M. J. Sharp and E. A. Tabet, *J. Org. Chem.*, 2011, **76**, 712; (r) J. Y. L. Chung, D. Steinhuebel, S. W. Krska, F. W. Hartner, C. Cai, J. Rosen, D. E. Mancheno, T. Pei, L. DiMichele, R. G. Ball, C.-y. Chen, L. Tan, A. D. Alorati, S. E. Brewer and J. P. Scott, *Org. Process Res. Dev.*, 2012, **16**, 1832; (s) J. B. Sperry, R. M. Farr, M. Levent, M. Ghosh, S. M. Hoagland, R. J. Varsolona and K. Sutherland, *Org. Process Res. Dev.*, 2012, **16**, 1854; (t) J. T. Reeves, D. R. Fandrick, Z. Tan, J. J. Song, S. Rodriguez, B. Qu, S. Kim, O. Niemeier, Z. Li, D. Byrne, S. Campbell, A. Chitroda, P. DeCroos, T. Fachinger, V. Fuchs, N. C. Gonnella, N. Grinberg, N. Haddad, B. Jäger, H. Lee, J. C. Lorenz, S. Ma, B. A. Narayanan, L. J. Nummy, A. Premasiri, F. Roschangar, M. Sarvestani, S. Shen, E. Spinelli, X. Sun, R. J. Varsolona, N. Yee, M. Brenner and C. H. Senanayake, *J. Org. Chem.*, 2013, **78**, 3616; (u) J. Kuethe, Y.-L. Zhong, N. Yasuda, G. Beutner, K. Linn, M. Kim, B. Marcune, S. D. Dreher, G. Humphrey and T. Pei, *Org. Lett.*, 2013, **15**, 4174.

77. H. Plenio, *Angew. Chem. Int. Ed.*, 2008, **47**, 6954.
78. (a) M. M. Heravi and S. Sadjadi, *Tetrahedron*, 2009, **65**, 7761; (b) R. Chinchilla and C. Nájera, *Chem. Rev.*, 2007, **107**, 874; (c) R. Chinchilla and C. Nájera, *Chem. Soc. Rev.*, 2011, **40**, 5084.
79. R. D. Stephens and C. E. Castro, *J. Org. Chem.*, 1963, **28**, 3313.
80. (a) A. O. King, E. G. Corley, R. K. Anderson, R. D. Larsen, T. R. Verhoeven, P. J. Reider, Y. B. Xiang, M. Belley, Y. Leblanc, M. Labelle, P. Prasit and R. J. Zamboni, *J. Org. Chem.*, 1993, **58**, 3731; (b) D. R. Sidler, J. W. Sager, J. J. Bergan, K. M. Wells, M. Bhupathy and R. P. Volante, *Tetrahedron: Asymmetry*, 1997, **8**, 161; (c) G. Bold, A. Fässler, H.-G. Capraro, R. Cozens, T. Klimkait, J. Lazdins, J. Mestan, B. Poncioni, J. Rösel, D. Stover, M. Tintelnot-Blomley, F. Acemoglu, W. Beck, E. Boss, M. Eschbach, T. Hürlimann, E. Masso, S. Roussel, K. Ucci-Stoll, D. Wyss and M. Lang, *J. Med. Chem.*, 1998, **41**, 3387; (d) D. C. Waite and C. P. Mason, *Org. Process Res. Dev.*, 1998, **2**, 116; (e) Z. J. Song, M. Zhao, R. Desmond, P. Devine, D. M. Tscaen, R. Tillyer, L. Frey, R. Heid, F. Xu, B. Foster, J. Li, R. Reamer, R. Volante, E. J. J. Grabowski, U. H. Dolling and P. Reider, *J. Org. Chem.*, 1999, **64**, 9658; (f) J. Singh, O. K. Kim, T. P. Kissick, K. J. Natalie, B. Zhang,

G. A. Crispino, D. M. Springer, J. A. Wichtowski, Y. Zhang, J. Goodrich, Y. Ueda, B. Y. Luh, B. D. Burke, M. Brown, A. P. Dutka, B. Zheng, D.-M. Hsieh, M. J. Humora, J. T. North, A. J. Pullockaran, J. Livshits, S. Swaminathan, Z. Gao, P. Schierling, P. Ermann, R. K. Perrone, M. C. Lai, J. Z. Gougoutas, J. D. DiMarco, J. J. Bronson, J. E. Heikes, J. A. Grosso, D. R. Kronenthal, T. W Denzel and R. H. Mueller, *Org. Process Res. Dev.*, 2000, **4**, 488; (g) S. Katayama, N. Ae and R. Nagata, *J. Org. Chem.*, 2001, **66**, 3474; (h) J. W. Raggon and W. M. Snyder, *Org. Process Res. Dev.*, 2002, **6**, 67; (i) S. Caron, E. Vazquez, R. W. Stevens, K. Nakao, H. Koike and Y. Murata, *J. Org. Chem.*, 2003, **68**, 4104; (j) N. Yasuda, Y. Hsiao, M. S. Jensen, N. R. Rivera, C. Yang, K. M. Wells, J. Yau, M. Palucki, L. Tan, P. G. Dormer, R. P. Volante, D. L. Hughes and P. J. Reider, *J. Org. Chem.*, 2004, **69**, 1959; (k) D. J. Plata, M. R. Leanna, M. Rasmussen, M. A. McLaughlin, S. L. Condon, F. A. Kerdesky, S. A. King, M. J. Peterson, E. J. Stoner and S. J. Wittenberg, *Tetrahedron*, 2004, **60**, 10171; (l) M. D. Wallace, M. A. McGuire, M. S. Yu, L. Goldfinger, L. Liu, W. Dai and S. Shilcrat, *Org. Process Res. Dev.*, 2004, **8**, 738; (m) T. J. Kwok and J. A. Virgilio, *Org. Process Res. Dev.*, 2005, **9**, 694; (n) D. R. Gauthier Jr, J. Limanto, P. N. Devine, R. A. Desmond, R. H. Szumigala Jr, B. S. Foster and R. P. Volante, *J. Org. Chem.*, 2005, **70**, 5938; (o) D. H. B. Ripin, D. E. Bourassa, T. Brandt, M. J. Castaldi, H. N. Frost, J. Hawkins, P. J. Johnson, S. S. Massett, K. Neumann, J. Phillips, J. W. Raggon, P. R. Rose, J. L. Rutherford, B. Sitter, A. M. Stewart III, M. G. Vetelino and L. Wei, *Org. Process Res. Dev.*, 2005, **9**, 440; (p) D. Camp, C. F. Matthews, S. T. Neville, M. Rouns, R. W. Scott and Y. Truong, *Org. Process Res. Dev.*, 2006, **10**, 814; (q) S. Abbas, L. Ferris, A. K. Norton, L. Powell, G. E. Robinson, P. Siedlecki, R. J. Southworth, A. Stark and E. G. Williams, *Org. Process Res. Dev.*, 2008, **12**, 202; (r) D. Schils, F. Stappers, G. Solberghe, R. van Heck, M. Coppens, D. Van der Heuvel, P. Van der Donck, T. Callewaert, F. Meeussen, E. De Bie, K. Eersels and E. Schouteden, *Org. Process Res. Dev.*, 2008, **12**, 530; (s) E. J. Flahive, B. L. Ewanicki, N. W. Sach, S. A. O'Neill-Slawecki, N. S. Stankovic, S. Yu, S. M. Guinness and J. Dunn, *Org. Process Res. Dev.*, 2008, **12**, 637; (t) X. Jiang, G. T. Lee, K. Prasad and O. Repič, *Org. Process Res. Dev.*, 2008, **12**, 1137; (u) E. A. Voight, H. Yin, S. V. Downing, S. A. Calad, H. Matsuhashi, I. Giordano, A. J. Hennessy, R. M. Goodman and J. L. Wood, *Org. Lett.*, 2010, **12**, 3422; (v) Z. J. Song, D. M. Tellers, M. Journet, J. T. Kuethe, D. Lieberman, G. Humphrey, F. Zhang, Z. Peng, M. S. Waters, D. Zewge, A. Nolting, D. Zhao, R. A. Reamer, P. G. Dormer, K. M. Belyk, I. W. Davies, P. N. Devine and D. M. Tschaen, *J. Org. Chem.*, 2011, **76**, 7804; (w) U. S. Kumar, V. R. Sankar, M. M. Rao, T. S. Jaganathan and R. B. Reddy, *Org. Process Res. Dev.*, 2012, **16**, 1917; (x) R. O. Cann, C.-P. H. Chen, Q. Gao, R. L. Hanson, D. Hsieh, J. Li, D. Lin, R. L. Parsons, Y. Pendri, R. B. Nielsen, W. A. Nugent, W. L. Parker, S. Quinlan, N. P. Reising, B. Remy, J. Sausker and X. Wang, *Org. Process Res. Dev.*, 2012, **16**, 1953.

81. (a) D. Bankston, F. Fang, E. Huie and S. Xie, *J. Org. Chem.*, 1999, **64**, 3461; (b) A. Banks, G. F. Breen, D. Caine, J. S. Carey, C. Drake, M. A. Forth, A. Gladwin, S. Guelfi, J. F. Hayes, P. Maragni, D. O. Morgan, P. Oxley, A. Perboni, M. E. Popkin, F. Rawlinson and G. Roux, *Org. Process Res. Dev.*, 2009, **13**, 1130.
82. M. Prashad, *Top. Organomet. Chem.*, 2004, **6**, 181.
83. (a) J. T. Link, L. E. Overman, in *Metal-Catalyzed Cross-Coupling Reactions*, ed. F. Diederich and P. J. Stang, Wiley-VCH, Weinheim, 1998, Chapter 6; (b) S. Bräse and A. de Meijere, in *Metal-Catalyzed Cross-Coupling Reactions*, ed. F. Diederich and P. J. Stang, Wiley-VCH, Weinheim, 1998, Chapter 3; (c) K. C. Nicolaou and E. J. Sorensen, *Classics in Total Synthesis*; VCH, Weinheim, 1996; Chapter 31.
84. (a) M. Beller, A. Zapf and W. Mägerlein, *Chem. Eng. Technol.*, 2001, **24**, 575; (b) J. G. de Vries, *Can. J. Chem.*, 2001, **79**, 1086.
85. (a) N. J. Whitecombe, K. Hii and S. Gibson, *Tetrahedron*, 2001, **57**, 7449; (b) V. Farina, *Adv. Synth. Catal.*, 2004, **346**, 1553; (c) A. F. Littke and G. C. Fu, *Org. Synth.*, 2005, **81**, 63; (d) A. F. Littke and G. C. Fu, *J. Am. Chem. Soc.*, 2001, **123**, 6989; (e) B. Basu and T. Frejd, *Acta Chem. Scand.*, 1996, **50**, 316.
86. (a) M. Shibasaki, E. M. Vogl and T. Ohshima, *Adv. Synth. Catal.*, 2004, **346**, 1533; (b) M. Shibasaki, C. D. J. Boden and A. Kojima, *Tetrahedron*, 1997, **53**, 7371.
87. (a) D. A. Alonso and C. Najera, *Sci Synth.*, 2010, **47a**, 439; (b) F. Alonso, I. P. Beletskaya and M. Yus, *Tetrahedron*, 2005, **61**, 11771; (c) N. J. Whitcombe, K. K. Hii and S. E. Gibson, *Tetrahedron*, 2001, **57**, 7449; (d) I. P. Beletskaya and A. V. Cheprakov, *Chem. Rev.*, 2000, **100**, 3009; (e) A. De Meijere and S. Bräse, in *Transition Metal Catalyzed Reactions*, ed. S. G. Davies and S.-I. Murahashi, Blackwell Science, Oxford, 1999; (f) R. F. Heck, in *Comprehensive Organic Synthesis*, Vol. 4, ed. B. M. Trost, Pergamon Press, New York, 1991, Chapter 4.3; (g) G. T. Crisp, *Chem. Soc. Rev.*, 1998, **27**, 427; (h) J. T. Link and L. E. Overman, *Chemtech*, 1998, **28**, 19; (i) W. A. Herrmann, in *Applied Homogeneous Catalysis with Organometallic Compounds*, ed. B. Cornils and W. A. Herrmann, VCH, Weinheim, 1996; (j) S. E. Gibson and R. J. Middleton, *Contemp. Org. Synth.*, 1996, **3**, 447; (k) T. Jeffery, in *Advances in Metal–Organic Chemistry*, Vol. 5, ed. L. S. Liebeskind, JAI Press, Greenwich, CT, 1996, pp. 153–260; (l) W. Cabri and I. Candiani, *Acc. Chem. Res.*, 1995, **28**, 2; (m) A. de Meijere and F. E. Meyer, *Angew. Chem. Int. Ed.*, 1995, **33**, 2379; (n) L. E. Overman, *Pure Appl. Chem.*, 1994, **66**, 1423; (o) R. F. Heck, *Palladium Reagents in Organic Synthesis*, Academic Press, London, 1985; (p) G. D. Daves and A. Hallberg, *Chem. Rev.*, 1989, **89**, 1433; (q) M. T. Reetz, in *Transition Metal Catalyzed Reactions*, ed. S. G. Davies and S.-I. Murahashi, Blackwell Science: Oxford, 1999; (r) R. F. Heck, *Org. React.*, 1982, **27**, 345; (s) R. F. Heck, *Acc. Chem. Res.*, 1979, **12**, 146; (t) R. F. Heck, *Pure Appl. Chem.*, 1978, **50**, 691.

88. (a) J. P. Knowles and A. Whiting, *Org. Biomol. Chem.*, 2007, **5**, 31; (b) N. T. S. Phan, M. Van Der Sluys and C. W. Jones, *Adv. Synth. Catal.*, 2006, **348**, 609.

89. M. Sakai, H. Hayashi and N. Miyaura, *Organometallics*, 1997, **16**, 4229.

90. (a) Y. Takaya, M. Ogasawara, T. Hayashi, M. Sakai and N. Miyaura, *J. Am. Chem. Soc.*, 1998, **120**, 5579; (b) Y. Takaya, T. Senda, H. Kurushima, M. Ogasawara and T. Hayashi, *Tetrahedron: Asymmetry*, 1999, **10**, 4047.

91. T. Hayashi and K. Yamasaki, *Chem. Rev.*, 2003, **103**, 2829.

92. (a) S. Brock, D. R. J. Hose, J. D. Moseley, A. J. Parker, I. Patel and A. J. Williams, *Org. Process Res. Dev.*, 2008, **12**, 496; (b) C. S. Burgey, D. V. Paone, A. W. Shaw, J. Z. Deng, D. N. Nguyen, C. M. Potteiger, S. L. Graham, J. P. Vacca and T. M. Williams, *Org. Lett.*, 2008, **10**, 3235.

93. (a) T. Hayashi, *Synlett*, 2001, 879; (b) K. Fagnou and M. Lautens, *Chem. Rev.*, 2003, **103**, 169; (c) T. Hayashi, *Pure Appl. Chem.*, 2004, **76**, 465; (d) K. Yoshida and T. Hayashi, Rhodium-catalyzed additions of boronic acids to alkenes and carbonyl compounds, in *Boronic Acids: Preparation and Applications in Organic Synthesis and Medicine*, ed. D. G. Hall, Wiley-VCH, Weinheim, 2005, pp. 171–203; (e) N. Miyaura, *Bull. Chem. Soc. Jpn.*, 2008, **81**, 1535.

94. T. Hayashi, T. Senda and M. Ogasawara, *J. Am. Chem. Soc.*, 2000, **122**, 10716.

95. (a) R. Itooka, Y. Iguchi and N. Miyaura, *J. Org. Chem.*, 2003, **68**, 6000; (b) T. Hayashi, M. Takahashi, Y. Takaya and M. Ogasawara, *J. Am. Chem. Soc.*, 2002, **124**, 5052.

96. J. Tsuji, H. Takahashi and M. Morikawa, *Tetrahedron Lett.*, 1965, **6**, 4387.

97. (a) J. Tsuji, Palladium-catalyzed nucleophilic substitution involving allylpalladium, propargylpalladium and related derivatives: the Tsuji–Trost reaction and related carbon–carbon bond formation reactions: overview of the palladium-catalyzed carbon–carbon bond formation via π-allylpalladium and propargylpalladium intermediates, in *Handbook of Organopalladium Chemistry for Organic Synthesis*, ed. E.-i. Negishi, Vol. 2, Wiley, New York, 2002, pp. 1669–1687; (b) L. Acemoglu and J. M. J. Williams, Synthetic scope of the Tsuji–Trost reaction with allylic halides, carboxylates, ethers and related oxygen nucleophiles as starting compounds, in *Handbook of Organopalladium Chemistry for Organic Synthesis*, ed. E.-i. Negishi, Vol. 2, Wiley, New York, 2002, pp. 1689–1705; (c) M. Moreno-Manas and R. Pleixats, Palladium-catalyzed allylation with allyl carbonates, in *Handbook of Organopalladium Chemistry for Organic Synthesis*, ed. E.-i. Negishi Vol. 2, Wiley, New York, 2002, pp. 1707–1767; (d) C. Courillon, S. Thorimbert and M. Malacria, Palladium-catalyzed substitution reactions of alkenyl epoxides, in *Handbook of Organopalladium Chemistry for Organic Synthesis*, ed. E.-i. Negishi, Vol. 2, Wiley, New York, 2002, pp. 1795–1810; (e) B. M. Trost and T. R. Verhoeven, in *Comprehensive Organometallic Chemistry*, Vol. 8, ed.

G. Wilkinson, F. G. A. Stone and E. W. Abel, Pergamon Press, Oxford, 1982, pp. 799–938; (f) B. M. Trost, *Acc. Chem. Res.*, 1980, **13**, 385; (g) B. M. Trost, *Pure Appl. Chem.*, 1981, **53**, 2357; (h) J. Tsuji, *Pure Appl. Chem.*, 1982, **54**, 197; (i) J. A. Davies, in *Comprehensive Organometallic Chemistry II*, Vol. 9, ed. E. W Abel., F. G. A. Stone and G. Wilkinson, Pergamon Press, Oxford, 1995, pp. 291–390; (j) J. Tsuji, *Tetrahedron*, 1986, **42**, 4361.

98. (a) G. Hata, K. Takahashi and A. Miyake, *Chem. Commun.*, 1970, 1392; (b) K. E. Atkins, W. E. Walker and R. M. Manyik, *Tetrahedron Lett.*, 1970, **11**, 3821.

99. (a) B. M. Trost and D. L. Van Vranken, *Chem. Rev.*, 1996, **96**, 395; (b) A. Pfaltz and M. Lautens, in *Comprehensive Asymmetric Catalysis*, ed. E. N. Jacobsen, A. Pfaltz and H. Yamamoto, Springer, Heidelberg, 2000, p. 833; (c) B. M. Trost and C. Lee, in *Catalytic Asymmetric Synthesis*, ed. I. Ojima, Wiley-VCH, New York, 2nd edn, 2000, p. 593; (d) B. M. Trost and M. L. Crawley, *Chem. Rev.*, 2003, **103**, 2921; (e) B. M. Trost, *J. Org. Chem.*, 2004, **69**, 5813; (f) B. M. Trost and C. Jiang, *Synthesis*, 2006, 369.

100. R. Takeuchi and S. Kezuka, *Synthesis*, 2006, 3349.

101. Z.-Y. Jiang, C.-H. Zhang, F.-L. Gu, K.-F. Yang, G.-Q. Lai, L.-W. Xu and C.-G. Xia, *Synlett*, 2010, 1251.

102. (a) G. R. Humphrey, R. A. Miller, P. J. Pye, K. Rossen, R. A. Reamer, A. Maliakal, S. S. Ceglia, E. J. J. Grabowski, R. P. Volante and P. J. Reider, *J. Am. Chem. Soc.*, 1999, **121**, 11261; (b) M. Palucki, J. M. Um, N. Yasuda, D. A. Conlon, F.-R. Tsay, F. W. Hartner, Y. Hsiao, B. Marcune, S. Karady, D. L. Hughes, P. G. Dormer and P. J. Reider, *J. Org. Chem.*, 2002, **67**, 5508; (c) D. J. Plata, M. R. Leanna, M. Rasmussen, M. A. McLaughlin, S. L. Condon, F. A. Kerdesky, S. A. King, M. J. Peterson, E. J. Stoner and S. J. Wittenberg, *Tetrahedron*, 2004, **60**, 10171; (d) G. Xu, D. Tang, Y. Gai, G. Wang, H. Kim, Z. Chen, L. T. Phan, Y. S. Or and Z. Wang, *Org. Process Res. Dev.*, 2010, **14**, 504; (e) K. R. Campos, A. Klapars, Y. Kohmura, D. Pollard, H. Ishibashi, S. Kato, A. Takezawa, J. H. Waldman, D. J. Wallace, C.-y. Chen and N. Yasuda, *Org. Lett.*, 2011, **13**, 1004; (f) S. R. Tudhope, J. A. Bellamy, A. Ball, D. Rajasekar, M. Azadi-Ardakani, H. S. Meera, J. M. Gnanadeepam, R. Saiganesh, F. Gibson, L. He, C. H. Behrens, G. Underiner, J. Marfurt and N. Favre, *Org. Process Res. Dev.*, 2012, **16**, 635; (g) F. González-Bobes, N. Kopp, L. Li, J. Deerberg, P. Sharma, S. Leung, M. Davies, J. Bush, J. Hamm and M. Hrytsak, *Org. Process Res. Dev.*, 2012, **16**, 2051.

103. S. Kaneko, T. Yoshino, T. Katoh and S. Terashima, *Tetrahedron: Asymmetry*, 1997, **8**, 829.

104. J. P. Roberts and C. Lee, *Org. Lett.*, 2005, **7**, 2679.

105. H. Kim and C. Lee, *Org. Lett.*, 2002, **4**, 4369.

106. A. Schoenberg, I. Bartoletti and R. F. Heck, *J. Org. Chem.*, 1974, **39**, 3318.

107. C. B. Kelly, C. Lee, M. A. Mercadante and N. E. Leadbeater, *Org. Process Res. Dev.*, 2011, **15**, 717.

108. (a) B. Gabriele, G. Salerno and M. Costa, *Top. Organomet. Chem.*, 2006, **18**, 239; (b) L. Ashfield and C. F. J. Barnard, *Org. Process Res. Dev.*, 2007, **11**, 39; (c) C. F. J. Barnard, *Organometallics*, 2008, **27**, 5402; (d) C. F. J. Barnard, *Org. Process Res. Dev.*, 2008, **12**, 566; (e) A. Brennführer, H. Neumann and M. Beller, *Angew. Chem. Int. Ed.*, 2009, **48**, 4114; (f) R. Grigg and S. Mutton, *Tetrahedron*, 2010, **66**, 5515.

109. (a) K. E. Henegar, S. W. Ashford, T. A. Baughman, J. C. Sih and R.-L. Gu, *J. Org. Chem.*, 1997, **62**, 6588; (b) M. Journet, D. Cai, L. M. DiMichele, D. L. Hughes, R. D. Larsen, T. R. Verhoeven and P. J. Reider, *J. Org. Chem.*, 1999, **64**, 2411; (c) G. G. Wu, Y. Wong and M. Poirier, *Org. Lett.*, 1999, **1**, 745; (d) S. K. Etridge, J. F. Hayes, T. C. Walsgrove and A. S. Wells, *Org. Process Res. Dev.*, 1999, **3**, 60; (e) A. R. Daniewski, W. Liu, K. Püntener and M. Scalone, *Org. Process Res. Dev.*, 2002, **6**, 220; (f) R. J. Herr, D. J. Fairfax, H. Meckler and J. D. Wilson, *Org. Process Res. Dev.*, 2002, **6**, 677; (g) R. J. Atkins, A. Banks, R. K. Bellingham, G. F. Breen, J. S. Carey, S. K. Etridge, J. F. Hayes, N. Hussain, D. O. Morgan, P. Oxley, S. C. Passey, T. C. Walsgrove and A. S. Wells, *Org. Process Res. Dev.*, 2003, 7, 663; (h) E. E. Boros, S. A. Burova, G. A. Erickson, B. A. Johns, C. S. Koble, N. Kurose, M. J. Sharp, E. A. Tabet, J. B. Thompson and M. A. Toczko, *Org. Process Res. Dev.*, 2007, **11**, 899; (i) F. Gosselin, R. A. Britton, I. W. Davies, S. J. Dolman, D. Gauvreau, R. S. Hoerrner, G. Hughes, J. Janey, S. Lau, C. Molinaro, C. Nadeau, P. D. O'Shea, M. Palucki and R. Sidler, *J. Org. Chem.*, 2010, 75, 4154; (j) W. Tang, N. D. Patel, X. Wei, D. Byrne, A. Chitroda, B. Narayanan, A. Sienkiewicz, L. J. Nummy, M. Sarvestani, S. Ma, N. Grinberg, H. Lee, S. Kim, Z. Li, E. Spinelli, B.-S. Yang, N. Yee and C. H. Senanayake, *Org. Process Res. Dev.*, 2013, **17**, 382.

110. (a) M. Sundermeier, A. Zapf and M. Beller, *Eur. J. Inorg. Chem.*, 2003, 3513; (b) M. Sundermeier, A. Zapf and M. Beller, *Angew. Chem. Int. Ed.*, 2003, **42**, 1661; (c) M. Sundermeier, S. Mutyala, A. Zapf, A. Spannenberg and M. Beller, *J. Organomet. Chem.*, 2003, **684**, 50; (d) T. Schareina, A. Zapf and M. Beller, *Chem. Commun.*, 2004, 1388; (e) T. Schareina, A. Zapf, W. Mägerlein, N. Müller and M. Beller, *Tetrahedron Lett.*, 2007, **48**, 1087; (f) S. Velmathi and N. E. Leadbeater, *Tetrahedron Lett.*, 2008, **49**, 4693; (g) P. Anbarasan, T. Schareina and M. Beller, *Chem. Soc. Rev.*, 2011, **40**, 5049.

111. For transition metal-catalyzed cyanations of aryl chlorides, see: M. Sundermeier, A. Zapf, S. Mutyala, W. Baumann, J. Sans, S. Weiss and M. Beller, *Chem. Eur. J.*, 2003, **9**, 1828

112. K. Takagi, T. Okamoto, Y. Sakakibara and S. Oka, *Chem. Lett.*, 1973, 471.

113. For a review on Sandmeyer cyanations, see: I. P. Beletskaya, A. S. Sigeev, A. S. Peregudov and P. V. Petrovskii, *J. Organomet. Chem.*, 2004, **689**, 3810

114. (a) K. W. Rosenmund and E. Struck, *Chem. Ber.*, 1919, **52**, 1749; (b) J. von Braun and G. Manz, *Liebigs Ann. Chem.*, 1931, **488**, 111

Reviews:; (c) J. Lindley, *Tetrahedron*, 1984, **40**, 1433; (d) T. Ito and K. Watanabe, *Bull. Chem. Soc. Jpn.*, 1968, **41**, 419.

115. R. K. Arvela and N. E. Leadbeater, *J. Org. Chem.*, 2003, **68**, 9122.

116. (a) J. Zanon, A. Klapars and S. L. Buchwald, *J. Am. Chem. Soc.*, 2003, **125**, 2890; (b) T. Schareina, A. Zapf and M. Beller, *Tetrahedron Lett.*, 2005, **46**, 2585; (c) H.-J. Cristau, A. Ouali, J.-F. Spindler and M. Taillefer, *Chem. Eur. J.*, 2005, **11**, 2483; (d) T. Schareina, A. Zapf, W. Mägerlein, N. Müller and M. Beller, *Chem. Eur. J.*, 2007, **13**, 6249; (e) Y. Ren, Z. Liu, S. Zhao, X. Tian, J. Wang, W. Yin and S. He, *Catal. Commun.*, 2009, **10**, 768.

117. (a) P. E. Maligres, M. S. Waters, F. Fleitz and D. Askin, *Tetrahedron Lett.*, 1999, **40**, 8193; (b) J. Beaudin, D. E. Bourassa, P. Bowles, M. J. Castaldi, R. Clay, M. A. Couturier, G. Karrick, T. W. Makowski, R. E. McDermott, C. N. Meltz, M. Meltz, J. E Phillips, J. A. Ragan, D. H. Brown Ripin, R. A. Singer, J. L. Tucker and L. Wei, *Org. Process Res. Dev.*, 2003, **7**, 873; (c) T. D. Nelson, C. R. LeBlond, D. E. Frantz, L. Matty, J. V. Mitten, D. G. Weaver, J. C. Moore, J. M. Kim, R. Boyd, P.-Y. Kim, K. Gbewonyo, M. Brower, M. Sturr, K. McLaughlin, D. R. McMasters, M. H. Kress, J. M. McNamara and U. H. Dolling, *J. Org. Chem.*, 2004, **69**, 3620; (d) M. R. Pitts, P. McCormack and J. Whittall, *Tetrahedron*, 2006, **62**, 4705; (e) B. C. Kim, S. Y. Hwang, T. H. Lee, J. H. Chang, H.-w. Choi, K. W. Lee, B. S. Choi, Y. K. Kim, J. H. Lee, W. S. Kim, Y. S. Oh, H. B. Lee, K. Y. Kim and H. Shin, *Org. Process Res. Dev.*, 2006, **10**, 881; (f) X. Wang, B. Zhi, J. Baum, Y. Chen, R. Crockett, L. Huang, S. Eisenberg, J. Ng, R. Larsen, M. Martinelli and P. Reider, *J. Org. Chem.*, 2006, **71**, 4021; (g) C.-y. Chen, L. F. Frey, S. Shultz, D. J. Wallace, K. M. Marcantonio, J. F Payack, E. Vazquez, S. A. Springfield, G. Zhou, P. Liu, G. R. Kieczykowski, A. M. Chen, B. D. Phenix, U. Singh, J. Strine, B. Izzo and S. W. Krska, *Org. Process Res. Dev.*, 2007, **11**, 616; (h) P. Ryberg, *Org. Process Res. Dev.*, 2008, **12**, 540; (i) S. Challenger, Y. Dessi, D. E. Fox, L. C. Hesmondhalgh, P. Pascal, A. J. Pettman and J. D. Smith, *Org. Process Res. Dev.*, 2008, **12**, 575; (j) D. J. Wallace, K. R. Campos, C. S. Schultz, A. Klapars, D. Zewge, B. R. Crump, B. D. Phenix, J. C. McWilliams, S. Krska, Y. Sun, C.-y. Chen and F. Spindler, *Org. Process Res. Dev.*, 2009, **13**, 84; (k) Y. Ren, Z. Liu, S. He, S. Zhao, J. Wang, R. Niu and W. Yin, *Org. Process Res Dev.*, 2009, **13**, 764; (l) P. Ryberg, *Top. Organomet. Chem.*, 2012, **42**, 125; (m) G. Schmidt, S. Reber, M. H. Bolli and S. Abele, *Org. Process Res. Dev.*, 2012, **16**, 595.

118. G. P. Ellis and T. M. Romney-Alexander, *Chem. Rev.*, 1987, **87**, 779.

119. (a) R. H. Grubbs, *Handbook of Metathesis*, Wiley-VCH, Weinheim, 2003; (b) D. Astruc, *New J. Chem.*, 2005, **29**, 42.

120. (a) Y. Chauvin, *Angew. Chem. Int. Ed.*, 2006, **45**, 3741; (b) R. R. Schrock, *Angew. Chem. Int. Ed.*, 2006, **45**, 3748; (c) R. H. Grubbs, *Angew. Chem. Int. Ed.*, 2006, **45**, 3760.

121. (a) T. Nicola, M. Brenner, K. Donsbach and P. Kreye, *Org. Process Res. Dev.*, 2005, **9**, 513; (b) N. K. Yee, V. Farina, I. N. Houpis, N. Haddad,

R. P. Frutos, F. Gallou, X.-j. Wang, X. Wei, R. D. Simpson, X. Feng, V. Fuchs, Y. Xu, J. Tan, L. Zhang, J. Xu, L. L. Smith-Keenan, J. Vitous, M. D. Ridges, E. M. Spinelli, M. Johnson, K. Donsbach, T. Nicola, M. Brenner, E. Winter, P. Kreye and W. Samstag, *J. Org. Chem.*, 2006, **71**, 7133; (c) C. Shu, X. Zeng, M.-H. Hao, X. Wei, N. K. Yee, C. A. Busacca, Z. Han, V. Farina and C. H. Senanayake, *Org. Lett.*, 2008, **10**, 1303; (d) H. Wang, S. N. Goodman, Q. Dai, G. W. Stockdale and W. M. Clark Jr, *Org. Process Res. Dev.*, 2008, **12**, 226; (e) V. Farina, C. Shu, X. Zeng, X. Wei, Z. Han, N. K. Yee and C. H. Senanayake, *Org. Process Res. Dev.*, 2009, **13**, 250; (f) H. Wang, H. Matsuhashi, B. D. Doan, S. N. Goodman, X. Ouyang and W. M. Clark Jr, *Tetrahedron*, 2009, **65**, 6291; (g) J. Kong, C. Chen, J. Balsells-Padros, Y. Cao, R. F. Dunn, S. J. Dolman, J. Janey, H. Li and M. J. Zacuto, *J. Org. Chem.*, 2012, **77**, 3820; (h) X. Wei, C. Shu, N. Haddad, X. Zeng, N. D. Patel, Z. Tan, J. Liu, H. Lee, S. Shen, S. Campbell, R. J. Varsolona, C. A. Busacca, A. Hossain, N. Y. Yee and C. H. Senanayake, *Org. Lett.*, 2013, **15**, 1016; (i) J. Arumugasamy, K. Arunachalam, D. Bauer, A. Becker, C. A. Caillet, R. Glynn, G. M. Latham, J. Lim, J. Liu, B. A. Mayes, A. Moussa, E. Rosinovsky, A. E. Salanson, A. F. Soret, A. Stewart, J. Wang and X. Wu, *Org. Process Res. Dev.*, 2013, **17**, 811.

122. (a) M. Schuster and S. Blechert, *Angew. Chem. Int. Ed.*, 1997, **36**, 2037; (b) R. H. Grubbs and S. Chang, *Tetrahedron*, 1998, **54**, 4413; (c) S. K. Armstrong, *J. Chem. Soc., Perkin Trans. 1*, 1998, 371; (d) A. Fürstner, M. Picquet, C. Bruneau and P. H. Dixneuf, *Chem. Commun.*, 1998, 1315; (e) T. M. Trnka and R. H. Grubbs, *Acc. Chem. Res.*, 2001, **34**, 18; (f) R. H. Grubbs, *Tetrahedron*, 2004, **60**, 7117; (g) K. C. Nicolau, P. G. Bulger and D. Sarlah, *Angew. Chem. Int. Ed.*, 2005, **44**, 4490; (h) T. J. Donohoe, A. J. Orr and M. Bingham, *Angew. Chem. Int. Ed.*, 2006, **45**, 2664; (i) A. H. Hoveyda and A. R. Zhugralin, *Nature*, 2007, **450**, 243.

123. (a) D. Burtscher and K. Grela, *Angew. Chem. Int. Ed.*, 2009, **48**, 442; (b) H. Clavier, K. Grela, A. Kirschning, M. Mauduit and S. P. Nolan, *Angew. Chem. Int. Ed.*, 2007, **46**, 6786.

124. S. H. Hong, M. W. Day and R. H. Grubbs, *J. Am. Chem. Soc.*, 2004, **126**, 7414.

125. S. H. Hong, D. P. Sanders, C. W. Lee and R. H. Grubbs, *J. Am. Chem. Soc.*, 2005, **127**, 17160.

126. A. Michrowska, R. Bujok, S. Harutyunyan, V. Sashuk, G. Dolgonos and K. Grela, *J. Am. Chem. Soc.*, 2004, **126**, 9318.

127. Y. Okude, S. Hirano, T. Hiyama and H. Nozaki, *J. Am. Chem. Soc.*, 1977, **99**, 3179.

128. K. Takai, M. Tagashira, T. Kuroda, K. Oshima, K. Utimoto and H. Nozaki, *J. Am. Chem. Soc.*, 1986, **108**, 6048.

129. H. Jin, J.-i. Uenishi, W. J. Christ and Y. Kishi, *J. Am. Chem. Soc.*, 1986, **108**, 5644.

130. K. Takai, K. Kimura, T. Kuroda, T. Hiyama and H. Nozaki, *Tetrahedron Lett.*, 1983, **24**, 5281.

131. For selected reviews, see: (a) G. C. Hargaden and P. J. Guiry, *Adv. Synth. Catal.*, 2007, **349**, 2407; (b) A. Fürstner, *Chem. Rev.*, 1999, **99**, 991; (c) L. A. Wessjohann and G. Scheid, *Synthesis*, 1999, 1; (d) P. Cintas, *Synthesis*, 1992, 248.

132. S. J. Mickel, G. H. Sedelmeier, D. Niederer, F. Schuerch, M. Seger, K. Schreiner, R. Daeffler, A. Osmani, D. Bixel, O. Loiseleur, J. Cercus, H. Stettler, K. Schaer, R. Gamboni, A. Bach, G.-P. Chen, W. Chen, P. Geng, G. T. Lee, E. Loeser, J. McKenna, F. R. Kinder, K. Konigsberger, K. Prasad, T. M. Ramsey, N. Reel, O. Repič, L. Rogers, W.-C. Shieh, R.-M. Wang, L. Waykole, S. Xue, G. Florence and I. Paterson, *Org. Process Res. Dev.*, 2004, **8**, 113.

133. I. Paterson and A. Schlapbach, *Synlett*, 1995, 498.

134. For a review of the Peterson olefination, see: D. J. Ager, *Org. React.*, 1990, **38**, 1

135. (a) J. Buckley, R. L. Webb, T. Laird and R. J. Ward, *Chem. Eng. News*, 1982, **60**, 5; (b) G. DeWall, *Chem. Eng. News*, 1982, **60**, 5, 43.

136. H.-W. Choi, D. Demeke, F. A. Kang, Y. Kishi, K. Nakajima, P. Nowak, Z.-K. Wan and C. Xie, *Pure Appl. Chem.*, 2003, **75**, 1.

137. (a) G. Dyker, *Angew. Chem. Int. Ed.*, 1999, **38**, 1698; (b) M. Miura and T. Satoh, *Top. Organomet. Chem.*, 2005, **14**, 55; (c) B.-J. Li, S.-D. Yang and Z.-J. Shi, *Synlett*, 2008, 949; (d) L. Ackermann, *Pure Appl. Chem.*, 2010, **82**, 1403; (e) L. Ackermann, *Chem. Commun.*, 2010, **46**, 4866; (f) T. W. Lyons and M. S. Sanford, *Chem. Rev.*, 2010, **110**, 1147; (g) J. Wencel-Delord, T. Dröge, F. Liu and F. Glorius, *Chem. Soc. Rev.*, 2011, **40**, 4740; (h) K. Hirano and M. Miura, *Synlett*, 2011, 294; (i) H. Li, B.-J. Li and Z.-J. Shi, *Catal. Sci. Technol.*, 2011, **1**, 191; (j) N. Kuhl, M. N. Hopkinson, J. Wencel-Delord and F. Glorius, *Angew. Chem. Int. Ed.*, 2012, **51**, 10236; (k) J. Yamaguchi, A. D. Yamaguchi and K. Itami, *Angew. Chem. Int. Ed.*, 2012, **51**, 8960.

138. (a) O. Daugulis, H.-Q. Do and D. Shabashov, *Acc. Chem. Res.*, 2009, **42**, 1074; (b) H. Li, C.-L. Sun, M. Yu, D.-G. Yu, B.-J. Li and Z.-J. Shi, *Chem. Eur. J.*, 2011, **17**, 3593 and references therein.

139. (a) A. Choy, N. Colbry, C. Huber, M. Pamment and J. Van Duine, *Org. Process Res. Dev.*, 2008, **12**, 884; (b) E. J. Kiser, J. Magano, R. J. Shine and M. H. Chen, *Org. Process Res. Dev.*, 2012, **16**, 255; (c) J. Magano, E. J. Kiser, R. J. Shine and M. H. Chen, *Org. Synth.*, 2013, **90**, 74.

140. S. G. Ouellet, A. Roy, C. Molinaro, R. Angelaud, J.-F. Marcoux, P. D. O'Shea and I. W. Davies, *J. Org. Chem.*, 2011, **76**, 1436.

141. A. N. Campbell, K. P. Cole, J. R. Martinelli, S. A. May, D. Mitchell, P. M. Pollock and K. A. Sullivan, *Org. Process Res. Dev.*, 2013, **17**, 273.

142. G. L. Bolton, R. H. Hutchings, J. T. Kohrt, W. K. C. Park and C. A. Van Huis, *PCT Int. Appl.*, WO 2008084300 A1 20080717, 2008.

143. E. J. Hennessy and S. L. Buchwald, *J. Am. Chem. Soc.*, 2003, **125**, 12084.

144. S. Oi, E. Aizawa, Y. Ogino and Y. Inoue, *J. Org. Chem.*, 2005, **70**, 3113.

145. (a) L. Ackermann, R. Vicente and A. Althammer, *Org. Lett.*, 2008, **10**, 2299; (b) L. Ackermann and R. Vicente, *Org. Lett.*, 2009, **11**, 4922;

(c) L. Ackermann and P. Novák, *Org. Lett.*, 2009, **11**, 4966; (d) L. Ackermann, P. Novák, R. Vicente, V. Pirovano and H. K. Potukuchi, *Synthesis*, 2010, 2245.

146. (a) D. Lapointe and K. Fagnou, *Org. Lett.*, 2009, **11**, 4160; (b) C. Verrier, C. Hoarau and F. Marsais, *Org. Biomol. Chem.*, 2009, 7, 647; (c) L. Ackermann, *Chem. Commun.*, 2010, **46**, 4866; (d) L. Ackermann, S. Barfuesser and J. Pospech, *Org. Lett.*, 2010, **12**, 724; (e) T. Mukai, K. Hirano, T. Satoh and M. Miura, *Org. Lett.*, 2010, **12**, 1360; (f) L. Ackermann, S. Barfuesser, C. Kornhaass and A. R. Kapdi, *Org. Lett.*, 2011, **13**, 3082; (g) S. Sahnoun, S. Messaoudi, J.-D. Brion and M. Alami, *ChemCatChem*, 2011, **3**, 893; (h) Y. Zhu and V. H. Rawal, *J. Am. Chem. Soc.*, 2011, **134**, 111; (i) W. Mai, J. Yuan, Z. Li, L. Yang, Y. Xiao, P. Mao and L. Qu, *Synlett*, 2012, **23**, 938.

147. (a) D. L. Davies, S. M. A. Donald and S. A. Macgregor, *J. Am. Chem. Soc.*, 2005, **127**, 13754; (b) D. Garcia-Cuadrado, A. A. C. Braga, F. Maseras and A. M. Echavarren, *J. Am. Chem. Soc.*, 2006, **128**, 1066; (c) S. I. Gorelsky, D. Lapointe and K. Fagnou, *J. Am. Chem. Soc.*, 2008, **130**, 10848; (d) D. Balcells, E. Clot and O. Eisenstein, *Chem. Rev.*, 2010, **110**, 749; (e) L. Ackermann, *Chem. Rev.*, 2011, **111**, 1315.

148. (a) J. P. Collman and M. Zhong, *Org. Lett.*, 2000, **2**, 1233; (b) A. Klapars, J. C. Antilla, X. Huang and S. L. Buchwald, *J. Am. Chem. Soc.*, 2001, **123**, 7727; (c) D. Ma and C. Xia, *Org. Lett.*, 2001, **3**, 2583; (d) A. Klapars, X. Huang and S. L. Buchwald, *J. Am. Chem. Soc.*, 2002, **124**, 7421; (e) F. Y. Kwong and S. L. Buchwald, *Org. Lett.*, 2003, **5**, 793; For reviews, see: (f) K. Kunz, U. Scholz and D. Ganzer, *Synlett*, 2003, 2428; (g) S. V. Ley and A. W. Thomas, *Angew. Chem. Int. Ed.*, 2003, **42**, 5400; (h) I. P. Beletskaya and A. V. Cheprakov, *Coord. Chem. Rev.*, 2004, **248**, 2337; (i) D. Ma and Q. Cai, *Acc. Chem. Res.*, 2008, **41**, 1450; (j) F. Monnier and M. Taillefer, *Angew. Chem. Int. Ed.*, 2008, **47**, 3096; (k) V. I. Sorokin, *Mini-Rev. Org. Chem.*, 2008, **5**, 323; (l) F. Monnier and M. Taillefer, *Angew. Chem. Int. Ed.*, 2009, **48**, 6954.

149. (a) A. Ghosh, J. E. Sieser, S. Caron, M. Couturier, K. Dupont-Gaudet and M. Girardin, *J. Org. Chem.*, 2006, **71**, 1258; (b) Y.-M. Pu, Y.-Y. Ku, T. Grieme, L. A. Black, A. V. Bhatia and M. Cowart, *Org. Process Res. Dev.*, 2007, **11**, 1004; (c) A. Ribecai, S. Bacchi, M. Delpogetto, S. Guelfi, A. M. Manzo, A. Perboni, P. Stabile, P. Westerduin, M. Hourdin, S. Rossi, S. Provera and L. Turco, *Org. Process Res. Dev.*, 2010, **14**, 895; (d) J. P. Graham, N. Langlade, J. M. Northall, A. J. Roberts and A. J. Whitehead, *Org. Process Res. Dev.*, 2011, **15**, 44; (e) S. Duan, S. Venkatraman, X. Hong, K. Huang, L. Ulysse, B. I. Mobele, A. Smith, L. Lawless, A. Locke and R. Garigipati, *Org. Process Res. Dev.*, 2012, **16**, 1787.

150. (a) D. Ma, Q. Cai and H. Zhang, *Org. Lett.*, 2003, **5**, 2453; (b) J. C. Antilla, J. M. Baskin, T. E. Barder and S. L. Buchwald, *J. Org. Chem.*, 2004, **69**, 5578.

151. M. Kosugi, M. Kameyama and T. Migita, *Chem. Lett.*, 1983, 927.

152. (a) D. S. Surry and S. L. Buchwald, *Angew. Chem. Int. Ed.*, 2008, **47**, 6338; (b) S. L. Buchwald, C. Mauger, G. Mignani and U. Scholz, *Adv. Synth.*

Catal., 2006, **348**, 23; (c) C. C. Mauger and G. A. Mignani, *Aldrichim. Acta*, 2006, **39**, 17; (d) A. R. Muci and S. L. Buchwald, *Top. Curr. Chem.*, 2002, **219**, 131; (e) J. F. Hartwig, in *Modern Arene Chemistry*, ed. D Astruc, Wiley-VCH, Weinheim, 2002, pp. 107–168; (f) A. S. Guram, R. A. Rennels and S. L. Buchwald, *Angew. Chem. Int. Ed.*, 1995, **34**, 1348; (g) S. Wagaw, B. H. Yang and S. L. Buchwald, *J. Am. Chem. Soc.*, 1998, **120**, 6621; (h) S. Wagaw, B. H. Yang and S. L. Buchwald, *J. Am. Chem. Soc.*, 1999, **121**, 10251; (i) J. F. Hartwig, *Angew. Chem. Int. Ed.*, 1998, **37**, 2090; (j) J. Louie and J. F. Hartwig, *Tetrahedron Lett.*, 1995, **36**, 3609.

153. (a) K. Kunz, U. Scholz and D. Ganzer, *Synlett*, 2003, 2428; (b) S. V. Ley and A. W. Thomas, *Angew. Chem. Int. Ed.*, 2003, **42**, 5400.

154. J. Buckingham, *Dictionary of Natural Products*, Chapman and Hall/CRC Press, New York, 1994.

155. (a) R. O. Loufty, C. K. Hsiao and P. Kazmaier, *J. Photogr. Sci. Eng.*, 1983, **27**, 5; (b) G. D'Aprano, M. Leclerc, G. Zotti and G. Schiavon, *Chem. Mater.*, 1995, 7, 33.

156. (a) T. Mase, I. N. Houpis, A. Akao, I. Dorziotis, K. Emerson, T. Hoang, T. Iida, T. Itoh, K. Kamei, S. Kato, Y. Kato, M. Kawasaki, F. Lang, J. Lee, J. Lynch, P. Maligres, A. Molina, T. Nemoto, S. Okada, R. Reamer, J. Z. Song, D. Tschaen, T. Wada, D. Zewge, R. P. Volante, P. J. Reider and K. Tomimoto, *J. Org. Chem.*, 2001, **66**, 6775; (b) C. C. Mauger and G. A. Mignani, *Org. Process Res. Dev.*, 2004, **8**, 1065; (c) G. E. Robinson, O. R. Cunningham, M. Dekhane, J. C. McManus, A. O'Kearney-McMullan, A. M. Mirajkar, V. Mishra, A. K. Norton, B. Venugopalan and E. G. Williams, *Org. Process Res. Dev.*, 2004, **8**, 925; (d) H.-J. Federsel, M. Hedberg, F. R. Qvarnström and W. Tian, *Org. Process Res. Dev.*, 2008, **12**, 512; (e) J. T. Kuethe, K. G. Childers, G. R. Humphrey, M. Journet and Z. Peng, *Org. Process Res. Dev.*, 2008, **12**, 1201; (f) J. Magano, A. Akin, M. H. Chen, K. Giza, J. Moon and J. Saenz, *Synth. Commun.*, 2008, **38**, 3631; (g) D. J. Wallace, K. R. Campos, C. S. Schultz, A. Klapars, D. Zewge, B. R. Crump, B. D. Phenix, J. C. McWilliams, S. Krska, Y. Sun, C.-y. Chen and F. Spindler, *Org. Process Res. Dev.*, 2009, **13**, 84; (h) J. Chan, B. J. Burke, K. Baucom, K. Hansen, M. M Bio, E. DiVirgilio, M. Faul and J. Murry, *J. Org. Chem.*, 2011, **76**, 1767; (i) D. Mitchell, K. P. Cole, P. M. Pollock, D. M. Coppert, T. P. Burkholder and J. R. Clayton, *Org. Process Res. Dev.*, 2012, **16**, 70; (j) J. Li, D. Smith, S. Krishnananthan, R. A. Hartz, B. Dasgupta, V. Ahuja, W. D. Schmitz, J. J. Bronson, A. Mathur, J. C. Barrish and B.-C. Chen, *Org. Process Res. Dev.*, 2012, **16**, 156; (k) D. K. Leahy, L. V. Desai, R. P. Deshpande, A. V. Mariadass, S. Rangaswamy, S. K. Rajagopal, L. Madhavan and S. Illendula, *Org. Process Res. Dev.*, 2012, **16**, 244; (l) A. Goodyear, X. Linghu, B. Bishop, C. Chen, E. Cleator, M. McLaughlin, F. J. Sheen, G. W. Stewart, Y. Xu and J. Yin, *Org. Process Res. Dev.*, 2012, **16**, 605; (m) M. Betti, G. Castagnoli, A. Panico, S. S. Coccone and P. Wiedenau, *Org. Process Res. Dev.*, 2012, **16**, 1739; (n) A. Witt, P. Teodorovic, M. Linderberg, P. Johanson and A. Minidis, *Org. Process Res. Dev.*, 2013,

17, 672; (o) L. M. Alabanza, Y. Dong, P. Wang, J. A. Wright, Y. Zhang and A. J. Briggs, *Org. Process Res. Dev.*, 2013, **17**, 876.

157. R. J. Lundgren and M. Stradiotto, *Aldrichim. Acta*, 2012, **45**, 59.

158. (a) S. Lee, M. Jørgensen and J. F. Hartwig, *Org. Lett.*, 2001, **3**, 2729; (b) X. Huang and S. L. Buchwald, *Org. Lett.*, 2001, **3**, 3417.

159. S. Bhagwanth, A. G. Waterson, G. M. Adjabeng and K. R. Hornberger, *J. Org. Chem.*, 2009, **74**, 4634.

160. (a) M. Kosugi, T. Shimizu and T. Migita, *Chem. Lett.*, 1978, 13; (b) T. Migita, T. Shimizu, Y. Asami, J. Shiobara, Y. Kato and M. Kosugi, *Bull. Chem. Soc. Jpn.*, 1980, **53**, 1385; (c) M. Kosugi, T. Ogata, M. Terada, H. Sano and T. Migita, *Bull. Chem. Soc. Jpn.*, 1985, **58**, 3657.

161. (a) J. Magano, A. Acciacca, V. Beylin, J. Spence, P. Dunn and M. Hughes, *Synth. Commun.*, 2007, **37**, 3569; (b) T. Norris and K. Leeman, *Org. Process Res. Dev.*, 2008, **12**, 869; (c) J. Malmgren, B. Bäckström, E. Sølver and J. Wennerberg, *Org. Process Res. Dev.*, 2008, **12**, 1195; (d) P. D. de Koening, L. Murtagh, J. P. Lawson, R. A. Vonder Embse, S. A. Kunda and W. Kong, *Org. Process Res. Dev.*, 2011, **15**, 1046; (e) M. A. Mortensen, C. Guo, N. T. Reynolds, L. Wang, M. A. Helle, D. K. Keefe, B. P. Haney, B. J. Paul, P. R. Bruzinski, M. A. Wolf, N. L. Malinowski and Q. Yang, *Org. Process Res. Dev.*, 2012, **16**, 1811; (f) D. J. Cowan, J. L. Collins, M. B. Mitchell, J. A. Ray, P. W. Sutton, A. A. Sarjeant and E. E. Boros, *J. Org. Chem.*, 2013, **78**, 12726.

162. E. Alvaro and J. F. Hartwig, *J. Am. Chem. Soc.*, 2009, **131**, 7858.

163. M. Murata and S. L. Buchwald, *Tetrahedron*, 2004, **60**, 7397.

164. M. A. Fernández-Rodríguez, Q. Shen and J. F. Hartwig, *J. Am. Chem. Soc.*, 2006, **128**, 2180.

165. (a) N. Zheng, J. C. McWilliams, F. J. Fleitz, J. D. Armstrong III and R. P. Volante, *J. Org. Chem.*, 1998, **63**, 9606; (b) U. Schopfer and A. Schlapbach, *Tetrahedron*, 2001, **57**, 3069; (c) G. Y. Li, G. Zheng and A. F. Noonan, *J. Org. Chem.*, 2001, **66**, 8677; (d) G. Y. Li, *Angew. Chem. Int. Ed.*, 2001, **40**, 1513; (e) T. Itoh and T. Mase, *Org. Lett.*, 2004, **6**, 4587; (f) C. Mispelaere-Canivet, J.-F. Spindler, S. Perrio and P. Beslin, *Tetrahedron*, 2005, **61**, 5253.

166. (a) K. Kunz, U. Scholz and D. Ganzer, *Synlett*, 2003, 2428; (b) S. V. Ley and A. W. Thomas, *Angew. Chem. Int. Ed.*, 2003, **42**, 5400.

167. C. Palomo, M. Oiarbide, R. López and E. Gómez-Bengoa, *Tetrahedron Lett.*, 2000, **41**, 1283.

168. C. G. Bates, R. K. Gujadhur and D. Venkataraman, *Org. Lett.*, 2002, **4**, 2803.

169. T. Kondo and T.-a. Mitsudo, *Chem. Rev.*, 2000, **100**, 3205.

170. P. Fitton and E. A. Rick, *J. Organomet. Chem.*, 1971, **28**, 287.

171. J. Magano, Large-scale applications of transition metal removal techniques in the manufacture of pharmaceuticals, in *Transition Metal-Catalyzed Couplings in Process Chemistry: Case Studies from the Pharmaceutical Industry*, ed. J. Magano and J. R. Dunetz, Wiley-VCH, Weinheim, 2013, pp. 313–355.

CHAPTER 16

Palladium Detection Techniques for Active Pharmaceutical Ingredients Prepared via Cross-Couplings

KAZUNORI KOIDE

Department of Chemistry, University of Pittsburgh, 219 Parkman Avenue, Pittsburgh, PA 15260, USA
Email: koide@pitt.edu

16.1 Palladium Catalysis in the Pharmaceutical Industry

The generality of palladium catalysis in forming carbon–carbon and carbon–heteroatom bonds has made it an indispensable tool in organic synthesis. Most of the known palladium-catalyzed cross-coupling reactions and closely related reactions were discovered in the 1960s and 1970s.[1-11] However, the pharmaceutical industry did not adopt these reactions until the 1980s and early 1990s.[12] The adoption was delayed partly because residual palladium in synthetic compounds could not be removed to meet safety standards. Removing palladium from synthetic materials is not yet a solved problem, but dozens of palladium-scavenging techniques have become available,[13-24] making palladium-catalyzed reactions employable in pharmaceutical production.[12,25] In the syntheses of 128 drugs carried out at AstraZeneca,

RSC Catalysis Series No. 21
New Trends in Cross-Coupling: Theory and Applications
Edited by Thomas J Colacot
© The Royal Society of Chemistry 2015
Published by the Royal Society of Chemistry, www.rsc.org

GlaxoSmithKline and Pfizer prior to 2006, 22% of carbon–carbon bond-forming reactions were palladium-catalyzed.[26]

Federal agencies regulate the standards for palladium contamination in active pharmaceutical ingredients (APIs), hence toxicity is worth briefly reviewing. In 1975, the absorption, distribution, metabolism and excretion of palladium species was reported.[27] Although the effects of palladium on the isolated rat heart were studied,[28] their relevance in live rats was unclear.[29] In addition, palladium allergy is well documented,[30] and its underlying molecular mechanism has been elucidated,[31] although likewise it has not been relevantly applied to toxicity. Nonetheless, as a precaution, the permitted daily exposure to palladium based on a 50 kg person is limited to 100 µg per day, according to the United States Pharmacopeia (USP).[32] This limit is stricter in other parts of the world and may become even more stringent in the near future.[33]

Generally, trace palladium is acceptable at levels up to 10 ppm in APIs (*e.g.*, 2 µg of palladium in a 200 mg ibuprofen tablet). Preparing APIs with such a low concentration of palladium is not trivial; Yang and co-workers at AMRI found 500–1500 ppm of palladium in a synthetic product after a Buchwald–Hartwig coupling and it could be decreased only to 100–200 ppm even after salt formation and two recrystallizations.[23] Subsequently, they evaluated 13 scavenging methods to lower palladium optimally to 12 ppm after a 24-h incubation.[23] In another example, Wang and co-workers at Merck performed a palladium-catalyzed cross-coupling and screened 19 methods to develop an efficient method for removing palladium.[22] These two examples show that screenings for efficient metal scavengers are a necessary part of the process to produce safe and pure APIs.

Most academic researchers in synthetic organic chemistry do not measure trace palladium because they are not subjected to regulations concerning trace metals or are not trained to analyze trace metals. However, observations in Koide's laboratory indicated unreported, but commonplace, incidents in which residual palladium catalyzed unintended reactions. The following exemplifies these unreported problems: upon completion of a Mizoroki–Heck cross-coupling to form ketone 3 (Figure 16.1), King *et al.* attempted to reduce the carbonyl group to form alcohol 4 by means of a catalytic CBS reduction method.[34] However, trace palladium from the previous reaction catalyzed the hydrogenation of the olefin, forming the undesired alcohol 5 in 3–10% yield. This problem was solved by a less desirable, stoichiometric enantioselective reduction to form the desired alcohol 4.

As evidenced by the previous example, palladium contamination can force chemists to change a synthetic route. In another example, GR127935, originally produced *via* a Suzuki–Miyaura coupling between 6 and 7 (Figure 16.2), remained unacceptably contaminated by palladium after recrystallization of the drug. This contamination prompted the chemists to

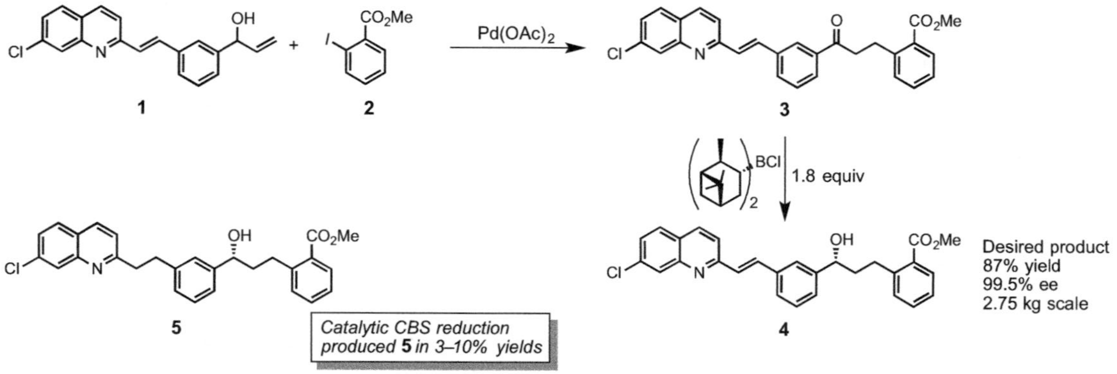

Figure 16.1 CBS reduction of ketone **3** that was contaminated with trace palladium produced an inseparable mixture of the desired product **4** and by-product **5**. This problem was solved by using a stoichiometric amount of a chiral boron reagent.

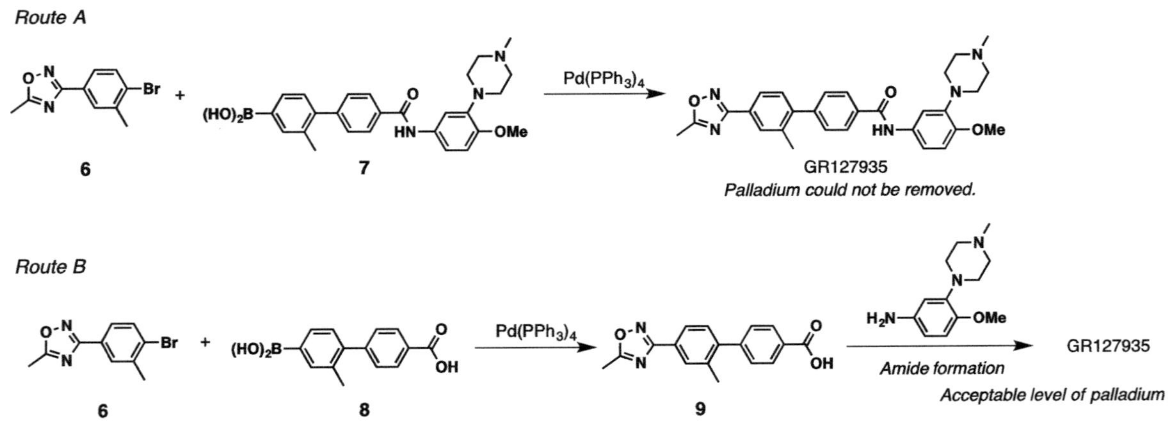

Figure 16.2 Route A was the original synthetic route to GR127935. Owing to the palladium contamination in the drug, Route B was used to carry out a palladium-catalyzed reaction at an earlier stage.

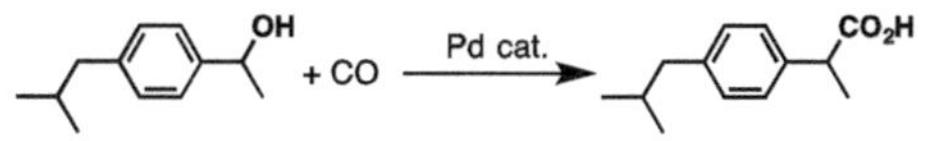

Figure 16.3 Commercial synthesis of ibuprofen.
Adapted from US Patent 4 981 995.

carry out the key carbon–carbon bond-forming reaction at an earlier stage of the synthesis, as shown in Route B.[35]

Ibuprofen, a staple in everyday medicine cabinets, is produced by palladium catalysis (Figure 16.3). The commercial production of ibuprofen by Hoechst Celanese ended with the conversion of alcohol **10** to ibuprofen, as described in one patent.[36] The removal of trace palladium required an extensive purification process, warranting two additional patents.[16,37]

Some of the papers dedicated solely to palladium removal were authored by 8–13 synthetic and analytical chemists, indicating that great manpower is required in the pharmaceutical industry to remove palladium from APIs.[22,23,38] It is difficult to separate palladium from APIs because it forms complexes with functionalized organic molecules, many of which are surprisingly stable during purification processes. Among the palladium–organic molecule complexes is a well-characterized framework between palladium and pyridine investigated by Fujita's group: a $Pd_{12}(pyridine)_{24}$ spherical complex underwent a ligand exchange with a half-life of ~20 days.[39] In another study, cinchona alkaloid–palladium complexes were so stable that they could be used as catalysts.[40] These examples highlight strong bindings between palladium and drug-like scaffolds, accounting for the difficulty in removing residual palladium. Larger pyridine derivatives prevent palladium black from forming.[41] One must consider whether palladium black (insoluble solid) is formed when filtration techniques are used as part of palladium removal.

Different palladium complexes require different scavenging protocols to reduce the palladium content effectively to <10 ppm. A combinatorial approach for metal removal with various solvents, scavengers and salts produces many dozens of samples for metal analysis;[19] it should be noted that current analytical techniques are not amenable to large sample numbers. Furthermore, palladium often binds tightly to organic compounds or precipitates as palladium black, hindering metal detection. Analytical samples for spectroscopy are therefore prepared by treating metal-containing organic materials with aqueous HCl, HNO_3 or aqua regia. Standard techniques for metal analysis include inductively coupled plasma mass spectrometry (ICP-MS), inductively coupled plasma optical emission spectrometry (ICP-OES) and atomic absorption spectrometry (AAS). Additionally, ICP-MS, ICP-OES and AAS instruments must be operated by well-trained analytical chemists.

Instruments are first calibrated using standard metal solutions and need regular cleaning, delaying the analysis for hours. Upon completion of these preparatory steps, samples are injected one at a time, with extensive washing with acids between samples to minimize memory effects. Although ICP-MS, ICP-OES and AAS are well established, it should be noted that the accuracy continues to be improved.[42]

With these analytical techniques, the whole operation at a pharmaceutical company may be as follows: a synthetic chemist carries out a palladium-catalyzed reaction, adds an aqueous solution containing metal scavengers, extracts the product by adding an organic solvent and removes the aqueous layer, evaporates the organic solvent, purifies the resulting crude material by recrystallization or column chromatography, isolates the product and verifies the chemical structure. After isolation or during the extraction step, the synthetic chemist employs as many methods as are available to remove palladium. This synthetic chemist fills many vials with these samples and submits them to an analytical laboratory. Some companies have the facility for trace metal analysis on-site, whereas other, less well equipped companies may ship the samples to an external analytical laboratory. In both cases, the synthetic process is on hold until the data for metal analysis become available. If the analysis shows sufficient sample purity, the drug production will resume. If the analysis reveals the palladium content to be too high, a second round of metal scavenging and analysis must be carried out, further delaying the synthetic process. In order to supply APIs in a timely manner to clinics, the metal removal and analytical processes must be expedited.

16.2 Expediting Palladium Analysis by New Detection Methods

One approach for expediting the purification process during drug production is to change the paradigm of metal analysis. If a synthetic chemist can quantify trace metals without extensive training, delays in pharmaceutical production will be minimal. This should sound familiar to synthetic organic chemists, as we all appreciate the accessibility of NMR, IR and LC-MS instruments to gain insights quickly into the reaction events without any direct help from others (of course, instruments must be maintained by experts). HPLC techniques allow synthetic organic chemists to measure purity and enantiomeric excess values routinely within the same facility.

Colorimetric methods have had an impact on the practice of synthetic organic chemistry. For example, the presence of primary and secondary amines and formyl groups can readily be confirmed by a ninhydrin test and Fehling's test, respectively. Likewise, it will be beneficial if synthetic chemists can measure trace palladium in their laboratories so that the pharmaceutical production can proceed without cumbersome analytical techniques. Many groups have developed fluorimetric and colorimetric methods to try to accomplish such a change.

16.3 Criteria for New Fluorimetric and Colorimetric Methods to Quantify Trace Amounts of Palladium

The following criteria are important for the development of fluorimetric or colorimetric methods to quantify trace palladium in synthetic APIs:

1. *The method must be selective for palladium* – If a palladium-catalyzed reaction has been performed, the remaining trace metal is most likely palladium, although other metals may be present. Therefore, spectacular selectivity for palladium is not necessary, but other commonly found metals such as sodium, calcium, iron, copper and magnesium should not give false data.

2. *The method must be able to quantify 10 parts per billion (ppb) (94 nM) palladium in solution* – As an example, if an API contains 10 ppm of palladium in the solid state (10 μg of Pd per gram of API) and 10 mg of such an API sample is dissolved in 1 mL of water, the resulting solution will be 100 ppb (940 nM) Pd. This API solution may be added to a sensor solution by, for example, 10-fold dilution (*e.g.*, 20 μL of the API solution into 180 μL of sensor solution). Therefore, when palladium ions are exposed to a chemodosimeter or chemosensor, the palladium concentration may be about 10 ppb (94 nM). Even when this high sensitivity cannot be achieved, fluorimetric methods may still be valuable if they can provide information about the relative amounts of palladium in API samples.

3. *The method must be independent of the structures of palladium complexes to quantify total palladium* – The pharmaceutical industry is concerned with the total amount of palladium. In order for the method to be comparable to ICP-MS, ICP-OES or AAS, the method must be independent of the palladium species in APIs.

4. *The method must not be subject to interference from APIs* – An API often binds to palladium with high affinity. Even in the presence of such an API, the fluorimetric or colorimetric method must detect all palladium species. In order to detect 10 ppm of palladium in an API, the method must be effective in the presence of 100 000 equivalents of API relative to palladium by weight.

5. *The method must be cheaper, faster and more user friendly than currently used technology* – Currently accepted analytical methods such as ICP-MS continue to be the methods of choice for the quality control of final pharmaceutical products. Fluorimetric and colorimetric methods for palladium detection will be used by synthetic organic chemists to identify the most effective protocol for palladium scavenging during, not after, synthetic processes. Learning and mastering these techniques should take less than a day.

6. *The method must be rigorously validated by comparison with the common standard technology* – Fluorimetric and colorimetric methods are not

yet commonly used for trace metal analysis. Therefore, prior to the implementation of either of these methods, the new method must be rigorously compared with a proven technique for validation.

7. *The linear range for palladium quantification must be as broad as possible –* An API may contain palladium in the 2–2000 ppm range. Therefore, the lower and upper limits of quantification need to span at least three orders of magnitude.

16.4   Fluorimetric and Colorimetric Methods for Quantifying Trace Amounts of Palladium

16.4.1   Chemosensors for Palladium

Chemosensors are molecules that reversibly bind to an analyte and change the optical properties. The generally quick binding event is beneficial, allowing rapid determination of palladium concentration. A major limitation is that one molecule of palladium can only generate one molecule of a fluorescent complex at best, providing only moderately sensitive methods. In other words, chemosensors do not utilize catalysis to generate a fluorescent signal. Furthermore, APIs may interfere with the resulting signal, as these methods generally rely on metal–heteroatom bindings.

Nielsen's group exploited the affinity of sulfur atoms towards palladium, as depicted in Figure 16.4(a); the thiourea derivative **11** bound to palladium to form the complex **12** that exhibited a unique UV–visible spectrum.[43] With this method, the product of a Sonogashira coupling **15** [Figure 16.4(b)] was analyzed for trace palladium. Their colorimetric method and inductively coupled plasma sector field mass spectrometry indicated 376 and 610 ppm of palladium, respectively. Notably, Pd, Pt, Cu and Ni showed different absorption spectra when bound to **11**.

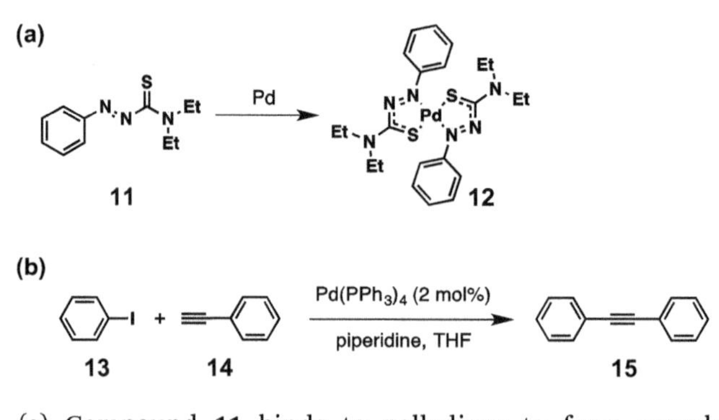

Figure 16.4   (a) Compound **11** binds to palladium to form complex **12**. (b) A Sonogashira coupling was performed between **13** and **14** and product **15** was analyzed for its trace palladium content using **11**.
Reproduced with permission from Ref. 43. Copyright 2006 Georg Thieme Verlag KG.

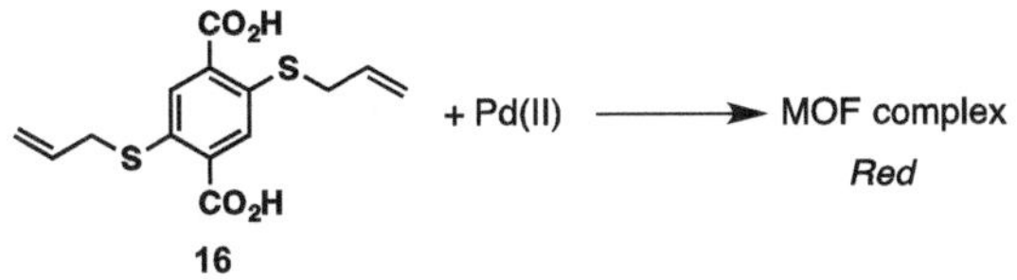

Figure 16.5   Pd(II) binds to the sulfide, olefin and the carboxylic acid, forming a large, red MOF complex.
Reproduced with permission from Ref. 44. Copyright 2013 American Chemical Society.

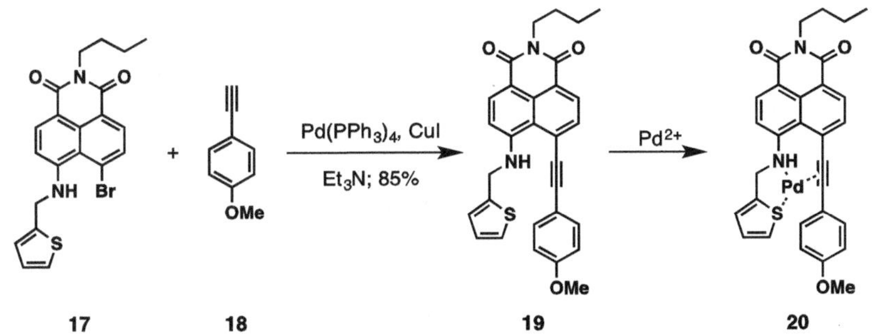

Figure 16.6   A Sonogashira coupling between **17** and **18** afforded palladium sensor **19**. This sensor bound to Pd^{2+} to form complex **20**, which showed a distinct spectrum from **19**.
Adapted from Ref. 45.

Also taking advantage of the strong Pd–S binding, Xu and co-workers used metal–organic framework (MOF) chemistry to design compound **16** (Figure 16.5).[44] This compound formed a palladium-containing red MOF complex, allowing the colorimetric detection of palladium species. The presence of >1 ppm of Pd(MeCN)$_2$Cl$_2$ could be visualized by this method.

Qian's group prepared palladium sensor **19** by the Sonogashira coupling of **17** and **18** (Figure 16.6).[45] This sensor might have been contaminated with palladium, although determination of the metal was not reported. Sensor **19** formed complex **20** with palladium(II), showing a weaker yellow fluorescence than the parent compound **19**, with the same λ_{max} value. This on–off switch triggered by the target analyte is not as desirable as an off–on switch.

Peng and co-workers developed a rhodamine-based detection method for palladium species. Coordination of palladium species opened the spirocyclic structure of **21** [Figure 16.7(a)], producing the fluorescent product **22**.[46] Ferric ions also increased the fluorescence to a minor extent, but this method was mostly selective for palladium. Real-world samples were not used in this work. Later, the same group improved the selectivity toward palladium species using **23** [Figure 16.7(b)] and proposed, based on time-of-flight mass spectrometry (TOF-MS) data, that the fluorescent product might be the dimeric compound **24**.[47] They developed the third-generation sensor **25** [Figure 16.7(c)]. This sensor bound to Pd^{2+} reversibly, selectively and

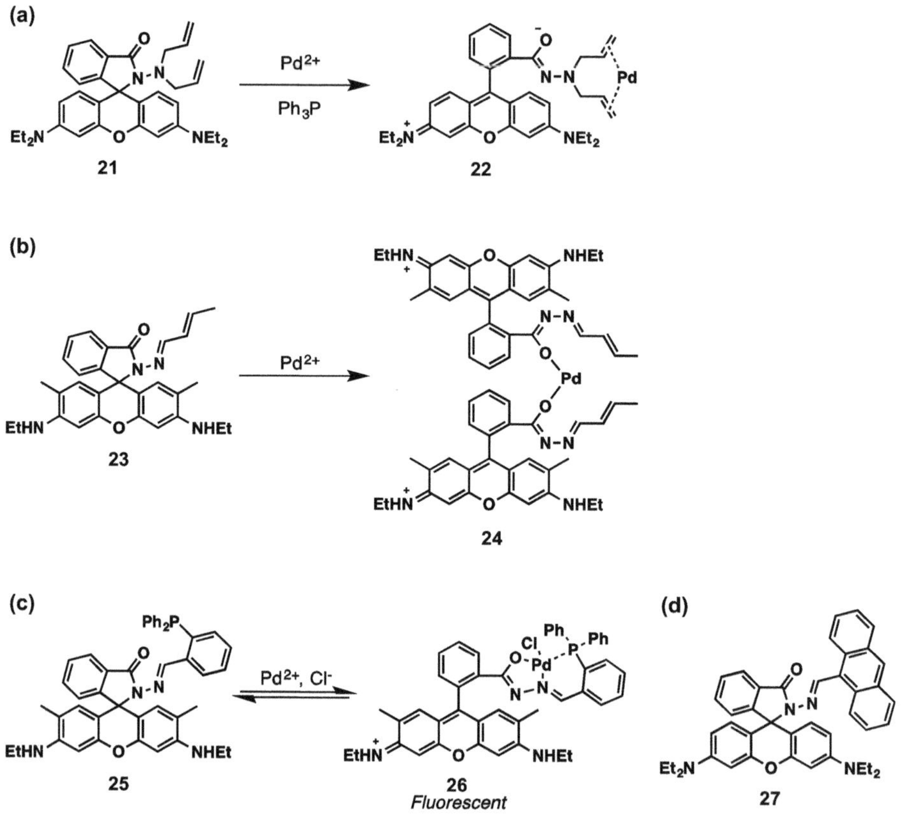

Figure 16.7 (a)–(c) Palladium sensors or chemodosimeters developed by Peng's group. Adapted from Refs 46, 47 and 48, respectively. (d) Palladium indicator developed by Niu's group. Reproduced with permission from Ref. 49. Copyright 2013 Springer-Verlag.

rapidly to form the green fluorescent complex **26**. They applied this fluorimetric method to the quantification of palladium in environmental samples and compared the results with those obtained by ICP-AES, which gave satisfactory agreement.[48] Compound **27** [Figure 16.7(d)], developed by Niu's group, was used to detect Pd^{2+} and may also react similarly to **23**.[49]

Zeng and co-workers developed a closely related approach to the detection of Pd^{2+}.[50] The non-fluorescent compound **28** (Figure 16.8) formed the palladium complex **29**, generating the fully conjugated fluorescent structure. This complex then rearranged to the more stable **30**. The detection limit was stated to be 1.49 nM, although it is not known how this limit was calculated. Both **25** and **28** exploited Pd–P binding as an additional interaction. It has not yet been determined how the phosphine tolerates real-world samples containing other transition metals or oxidants.

Bai's group developed two on–off chemosensors for Pd^{2+} (Figure 16.9).[51,52] The first, polymer **31**, bound to Pd^{2+} and the resulting complex was less

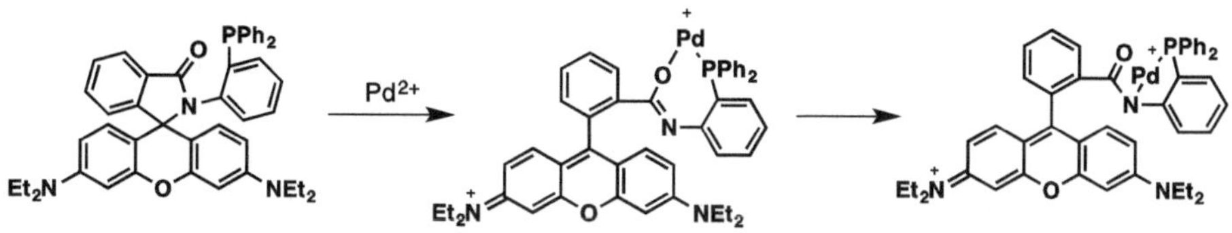

28 29 *Fluorescent* 30 *Fluorescent*

Figure 16.8 The non-fluorescent compound **28** reacted with Pd^{2+} to form the fully conjugated, fluorescent intermediate **29**. This O–Pd–P complex rearranged to the more stable N–Pd–P complex **30**.
Adapted from Ref. 50.

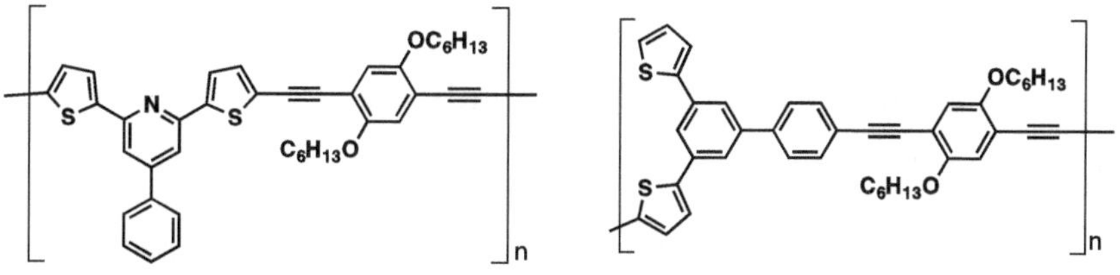

31 32

Figure 16.9 Structures of polymer-based palladium sensors developed by Bai's group.
Adapted from Refs 51, 52 and 53.

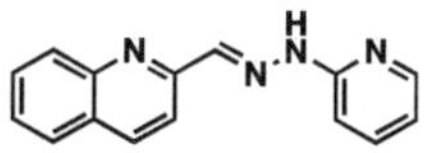

33

Figure 16.10 Structure of the palladium sensor developed by Mukherjee *et al.*
Adapted from Ref. 55.

fluorescent than the starting material. The same group reported the second-generation sensor **32**.[53] As the final step in the preparation of these sensors was a palladium-catalyzed Sonogashira coupling, the sensors were presumably contaminated with palladium. Another polymer-based fluorescent sensor for palladium was reported by Wang and Liao.[54] Although the chemical structures of the palladium complexes were not well defined, it was reported that poly(*o*-phenylenediamine) bound selectively to Pd^{2+} and emitted signals at 304 nm as the λ_{max}.

Mukherjee's group developed another on–off fluorescent chemosensor (Figure 16.10).[55] Blue fluorescent hydrazone **33** became less fluorescent upon binding to Pd^{2+}. Using this hydrazine as a palladium chelator, Pd^{2+} in water could be selectively extracted into an organic layer and visualized by the blue color.

Another fluorescent chemosensor for palladium(II) was developed by Tae and co-workers (Figure 16.11).[56] The rhodamine derivative **34** was shown to bind to 2 equiv. of Pd^{2+} in water to form the fluorescent complex **35**. This complex formation was selective for Pd^{2+}.

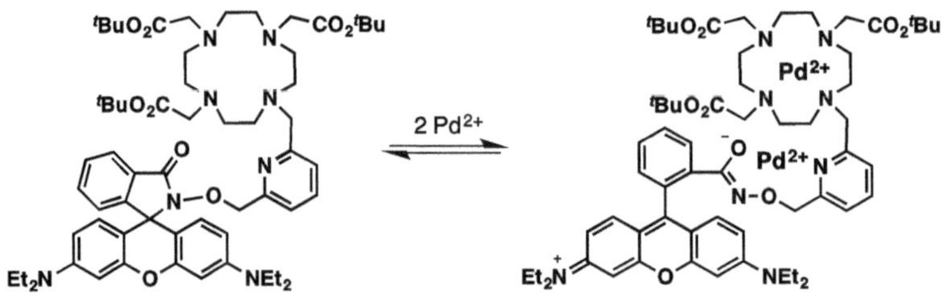

Figure 16.11 Chemosensor **34** binds to two palladium atoms to form the fluorescent complex **35**.
Adapted from Ref. 56.

16.4.2 Chemodosimeters for Palladium

Chemodosimeters are molecules that undergo irreversible reactions and optical changes. Generally, the resulting product is more fluorescent than the corresponding chemodosimeter, providing an off–on switch. The irreversible transformations are catalyzed by palladium(0) or palladium(II). Because the reactions are generally catalytic with respect to palladium, the resulting fluorescence signals are amplified and thus the methods are more sensitive than those with chemosensors. Chemodosimeter-based methods are inherently time sensitive, which can be both advantageous and disadvantageous. Advantageously, a minute amount of palladium can be detected by signal amplification over time. Disadvantageously, if the fluorogenic reactions rapidly consume the chemodosimeter, the method will not be quantitative.

16.4.2.1 Chemodosimeters Based on Oxidation

Tang and co-workers were the first to report a fluorimetric method for palladium.[57] In their study, palladium(II) catalyzed the oxidation of salicylaldehyde furfuralhydrazone (SAFH, Figure 16.12) into an unspecified fluorescent product in the presence of $KBrO_3$. This method was successfully applied to the quantification of palladium in a geological sample.[57] The metal selectivity was discussed, although the data were not presented.

16.4.2.2 Chemodosimeters Based on Desulfurization

Anslyn's group reported a method based on the chemistry depicted in Figure 16.13; the sulfur atom of SQ1:SEt coordinated with palladium(II) to form the fully conjugated, optically distinct product SQ1.[58] This method was evaluated by measuring the palladium content in **38**, prepared by Suzuki–Miyaura coupling between **36** and **37**. Since the coupling product contained no heteroatoms, the robustness in more complex samples was undetermined. Importantly, Anslyn and co-workers were the first to

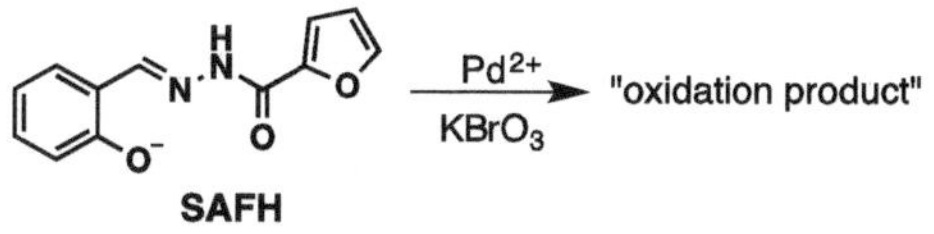

Figure 16.12   SAFH was oxidized to a fluorescent product by Pd^{2+} and KBrO$_3$. Reproduced with permission from Ref. 57. Copyright 2004 Taylor & Francis.

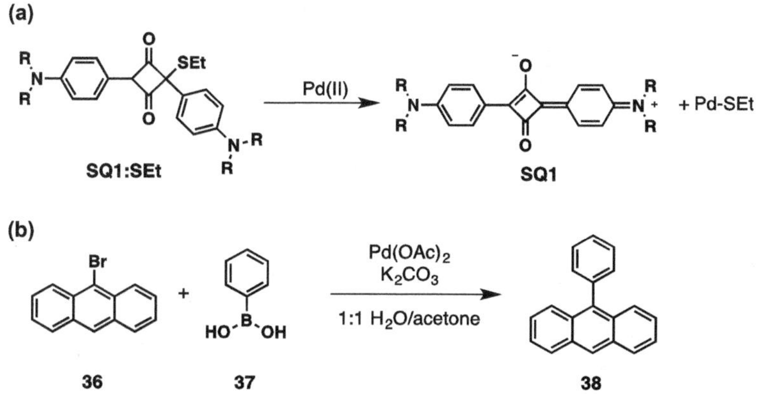

Figure 16.13   (a) SQ1:SEt reacted with Pd(II) to form the fully conjugated product SQ1, showing a distinct absorption spectrum from that of the starting material. (b) Compound **38** was prepared by a Suzuki–Miyaura cross-coupling and analyzed for its palladium content using the method shown in (a). Adapted from Ref. 58.

recognize the importance of sample preparation when trace palladium was quantified in synthetic organic molecules. Namely, acid digestion of synthetic samples played a critical role in the quantification of palladium species.

16.4.2.3   Chemodosimeters Based on Deallylation

16.4.2.3.1   2007 Version from Koide's Group.   In 2007, Koide's group discovered that a hydroxymethyl variant of O-alkylated dichlorofluorescein, **A**, (Figure 16.14a) was nearly non-fluorescent.[59] In contrast, the dealkylated product of **A**, Pittsburgh Green, was about 400 times more fluorescent than **A**, making its green fluorescence signal as intense as that of fluorescein. Pittsburgh Green was strongly fluorescent in the pH range 5.5–10 (Figure 16.14b);[60] therefore, the fluorescence-based assays were developed within this range.

Koide and co-workers hypothesized that the palladium-catalyzed deallylation of allyl Pittsburgh Green ether (APE, Figure 16.15) could be used for the fluorimetric detection of palladium. This ether was synthesized according to the synthetic scheme depicted in Figure 16.15.[61] They were able to

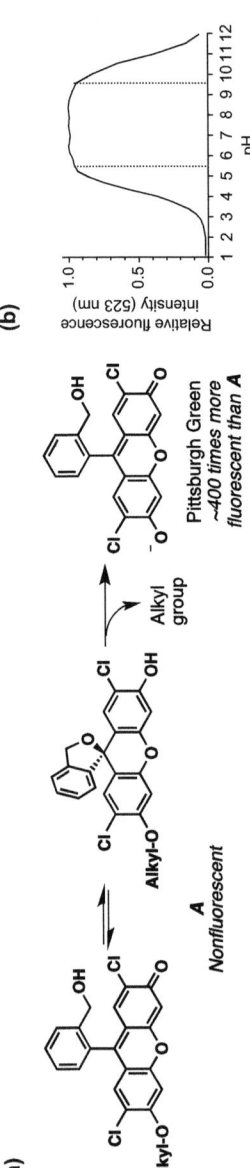

Figure 16.14 (a) General scheme for the fluorogenic conversion of ether **A** to Pittsburgh Green. (b) Fluorescence intensity of Pittsburgh Green is pH dependent. Adapted from Ref. 60.

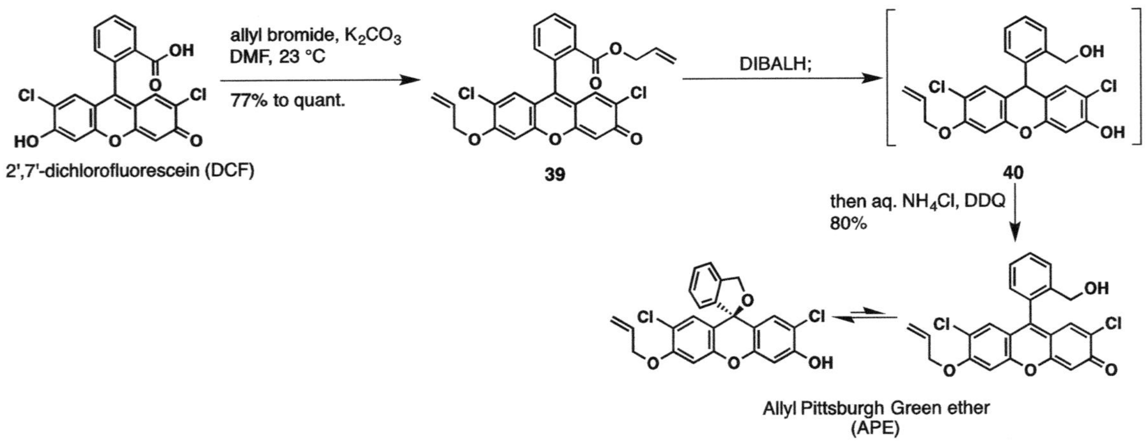

Figure 16.15 Synthesis of APE. Reproduced with permission from Refs 59 and 61. Copyright 2007, 2004 American Chemical Society.

synthesize 4–8 g (9.4–18.7 mmol) of APE, as needed. These are substantial amounts, as each palladium analysis requires 2.5–10 nmol of APE in 0.2 mL of solution.

A potential pitfall was that different palladium species in complex samples could react with APE at different rates, not meeting criterion 3 in section 16.3. To overcome this potential problem, excess $NaBH_4$ and Ph_3P were added to the deallylation reaction solutions to ensure that all the palladium species with various oxidation states would be converted to $Pd(PPh_3)_4$ *in situ*. Under these reaction conditions ($NaBH_4$, Ph_3P, DMSO, pH 7 phosphate buffer), the fluorescence intensity, as a measure of the conversion of APE to Pittsburgh Green, was linearly correlated with palladium concentration. As shown in Figure 16.16(a), palladium was the most efficient metal for catalyzing the deallylation reaction. Platinum, with slightly under half the efficiency, was the second most reactive metal for this transformation. A better selectivity for palladium over platinum could be achieved by carrying out the assays in a pH 4 buffer.[62] Although a slight signal could be observed with $CuCl_2$, it was found later that the particular batch of $CuCl_2$ was contaminated with traces of palladium.[63] As expected, 2 μM $PdCl_2$ and 2 μM $Pd(PPh_3)_4$ produced the same fluorescence intensity due to the presence of excess PPh_3. This fluorimetric method was used by Semagina and co-workers to elucidate the mechanism for the deallylation of APE catalyzed by palladium nanoparticles and their soluble derivatives.[64]

When the deallylation was carried out in THF, the reaction mixture turned red in a palladium concentration-dependent manner [Figure 16.16(b)]. This result indicated that the conversion of APE to Pittsburgh Green could be used as a colorimetric method to detect palladium in the micromolar range. The origin of the red color is currently being investigated.

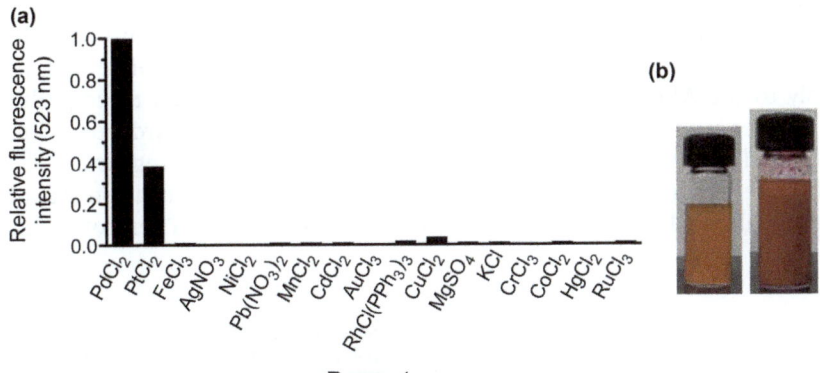

Reagent
(All the reagents were at the same concentration)

Figure 16.16 (a) Metal selectivity for the deallylation of APE in the presence of Ph_3P. (b) Palladium-catalyzed deallylation of APE in the presence of $NaBH_4$ in THF. [$PdCl_2$] = 10 μM (left) and 1.6 mM (right).
Reproduced with permission from Ref. 59. Copyright 2007 American Chemical Society.

16.4.2.3.2   2009 Version from Koide's Group.   Removal of an allyl group from allylic ethers has routinely been performed using $Pd(PPh_3)_4$ in organic solvents. In aqueous media, the deallylation might be more efficient with different phosphine ligands. Therefore, to optimize further the 2007 method, Koide's group screened commercial phosphine ligands in the presence of $NaBH_4$ in 1:9 DMSO–pH 7 buffer and found that John-Phos and tri(2-furyl)phosphine (TFP) were optimal.[65] Because TFP is more economical and more soluble in water, this ligand was chosen for further studies. This 2009 version also showed linearity between the palladium concentration and fluorescence signals.

They evaluated this 2009 version to determine how quantitative it can be in the presence of drugs or drug-like compounds. Drug or drug-like samples were prepared to contain 8.5 ppm of palladium (8.5 μg of palladium per gram of drug). The expected amount of palladium could be observed in 17 out of 18 samples in either a pH 7 or a pH 10 buffer.

The same group also measured trace palladium after **41** was prepared by a Suzuki-Miyaura coupling (Figure 16.17). The APE method showed that the palladium content was 1884 ± 169 ppm. Analyses by ICP-MS and ICP-OES methods, conducted by two independent institutions, indicated that the palladium content in the indole derivative was 1813 ± 316 and 2097 ± 108 ppm, respectively.

16.4.2.3.3   2010 Version from Koide's Group.   At this point, Koide's group had several inquiries about the APE method from pharmaceutical companies. Although their actual data could not be made available to us, it became clear that the APE method did not meet their expectations regarding robustness. Simply put, the success rate was only modest ($\sim 50\%$ at best), not meeting criterion 6 in section 16.3. Users might have taken too much of each sample for analysis; conventionally, the more sample one uses, the more accurate the analysis is. However, this concept does not apply to the APE method; as Figure 16.18(a) shows, when trace palladium in amines was quantified, the deviation from the standard curve became increasingly noticeable for amine concentrations >2 mg mL^{-1}.[66] This could be explained by eqn (16.1); reducing the concentration of an API should allow the equilibrium to shift toward the left. Therefore, the APE

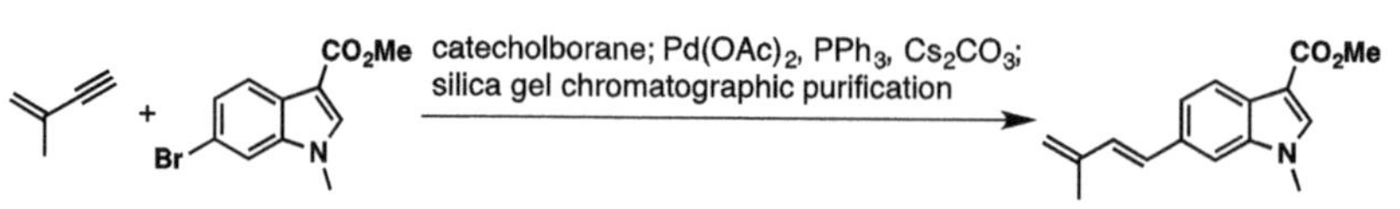

Figure 16.17   Indole derivative **41** was prepared by Suzuki–Miyaura coupling and subjected to palladium analysis.
Reproduced with permission from Ref. 65. Copyright 2009 American Chemical Society.

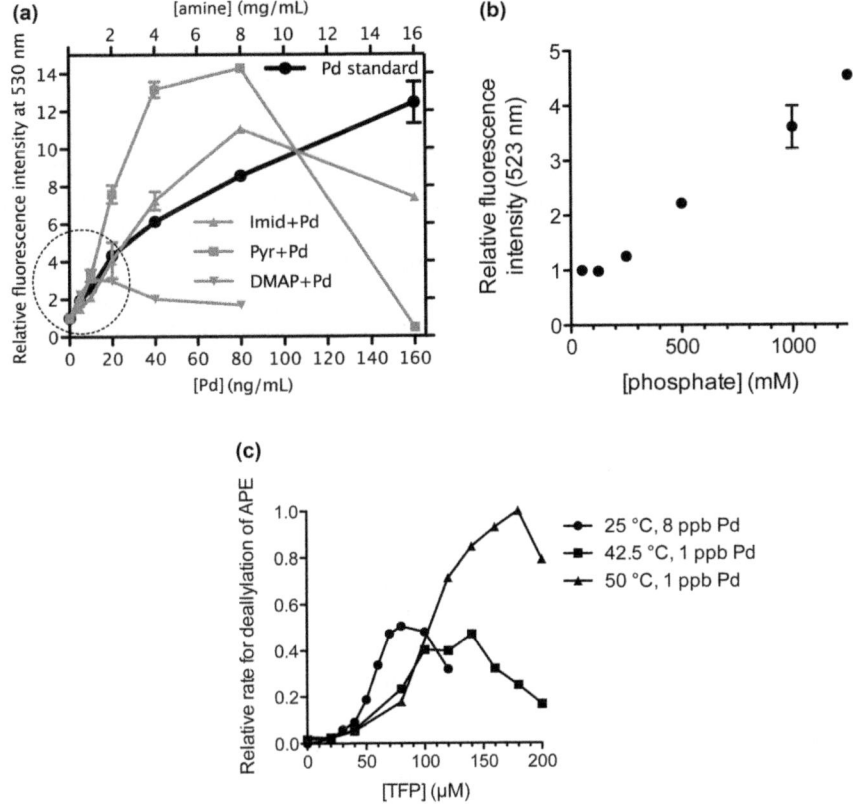

Figure 16.18 (a) Apparent concentrations of palladium are different from actual concentrations when samples are too concentrated. In this experiment, all operations were conducted at room temperature with palladium:API mimic ratios of 1 : 100 000 w/w (*i.e.*, 10 ppm palladium in API mimic). (b) Phosphate ions accelerate the deallylation of APE in a concentration-dependent manner. (c) The concentrations of TFP strongly influence the rates for the deallylation of APE at various temperatures. Adapted from Ref. [66].

method performs best when the concentration of palladium-contaminated API is below 2 mg mL^{-1} to promote the dissociation of palladium from the API.

$$Pd(0/II) + API \rightleftharpoons Pd(0/II) \cdot API \qquad (16.1)$$

Interestingly, the rate of the palladium-catalyzed deallylation of APE is linearly correlated with the concentration of phosphate ions [Figure 16.18(b)]. The reaction rate was not influenced by Na_2SO_4 or $LiClO_4$ concentrations in the range 0.05–50 mM or by Tris buffer (pH 7) in the range 0.1–1 M salt concentration, excluding the possibility for ionic strength-accelerated kinetics. Because the 1.23 M phosphate pH 7 buffer is commercially available, the remaining experiments were performed using this buffer.

One of the most important discoveries in this study was that the rate of deallylation was highly dependent upon TFP concentrations; optimal TFP concentrations were in the range 70–100, 100–140 and 160–180 μM at 25, 43 and 50 °C, respectively [Figure 16.18(c)].[66] The apparent second-order kinetics with respect to the concentration of TFP at lower concentrations suggested that two TFP molecules were involved in the rate-determining step.

It should be noted that these optimal TFP concentrations are atypical reaction conditions in synthetic organic chemistry; in most palladium-catalyzed organic reactions, the stoichiometry of the phosphine ligands is less than 10 equivalents relative to palladium. In this aqueous Tsuji–Trost-type reaction, the optimal TFP stoichiometry was in the range 13 000–19 000 equivalents, presumably because at such low palladium concentrations excessive TFP was necessary to shift the binding equilibrium towards a palladium–TFP complex.

The rate of deallylation of APE showed a first-order dependence at low concentrations of $NaBH_4$ and a zero-order dependence for $[NaBH_4]$ ≥30 μM.[66] Hence the reaction exhibited saturation kinetics; the rate-determining step was the reduction of palladium(II) to palladium(0) at low concentrations of $NaBH_4$. At higher concentrations of $NaBH_4$, the rate-determining step was presumably the reaction between a nucleophile and a π-allylpalladium complex. The reaction proceeded even in the absence of $NaBH_4$, because TFP could, albeit slowly, reduce palladium(II) to palladium(0). However, collaboration between Koide's group and Merck Research Laboratories revealed that the addition of $NaBH_4$ could improve the percentage signal recovery when palladium species were quantified in the presence of drug-like compounds.[67]

The molecularity of the deallylation was first order in APE,[66] indicating that one molecule of APE was involved in the transition state of the rate-determining step. From this and aforementioned studies, the rate-determining step was likely the nucleophilic attack of the π-allylpalladium complex.

Under the conditions developed above ([TFP] = 120 μM, [NaBH₄] = 1 mM, [APE] = 12.5 μM, 45 °C, 30 min, 1 : 9 v/v DMSO–1.25 M phosphate pH 7 buffer), the metals shown in Figure 16.19(a) were tested at a concentration of 100 nM, with the exception of palladium, which was tested at a concentration of 10 nM. The method was most responsive to palladium despite the lower concentration. Interestingly, unlike the previous method with PPh_3 that showed a moderate response with platinum, this method did so with rhodium. To examine further the generality of the method, palladium and each of the metals listed in Figure 16.19(b) were mixed in a 1 : 10 ratio and used for the deallylation reaction. Except for the mixture of palladium and rhodium, the combination of metals did not produce false data, indicating that metal contaminants in palladium samples do not interfere with the APE method.

Figure 16.20 shows how the APE method evolved over time. The 2010 version was 19 times more sensitive than the 2009 version. The rate

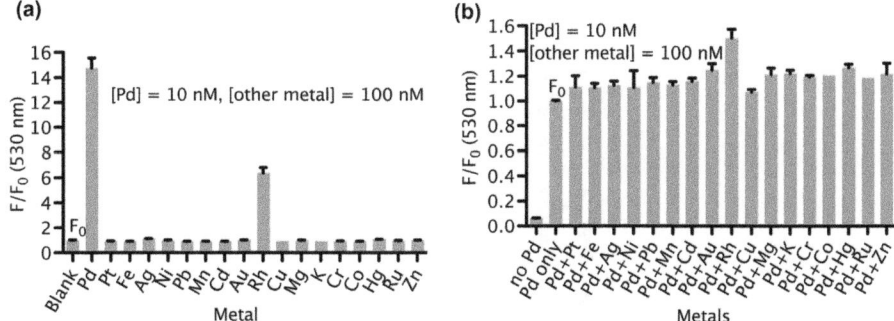

Figure 16.19 (a) Palladium was the by far the most reactive metal for the deallylation of APE. (b) In the presence of other metals, palladium-catalyzed deallylation of APE did not suffer interference by those metals.
Adapted from Ref. 66.

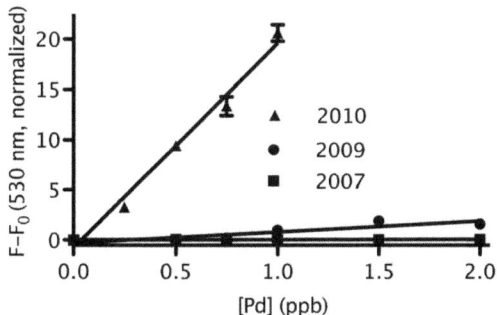

Figure 16.20 Comparison of the 2007, 2009 and 2010 methods with respect to sensitivity.
Adapted from Ref. 66.

enhancement can be explained by the increased phosphate concentration, elevated temperature and optimized TFP concentration. The 2010 version could quantify 0.1 ppb of palladium in solution.

16.4.2.3.4 Scope of the Fluorimetric Method with Various APIs. The utility of the new assay was evaluated with functionalized organic compounds (Figure 16.21) spiked with palladium (palladium:organic compound = 1 : 100 000 w/w). The average signal recovery was 99 ± 25%. Out of 26 compounds arbitrarily chosen based on drug-like functional groups, two posed a problem: apparent and actual palladium concentrations were different by more than 30%. This problem was solved by diluting the samples [*cf.*, Figure 16.18(a)]. Features worthy of note are that (1) this method was compatible with potential fluorescence quenchers by the photoinduced electron-transfer mechanism, (2) even strong metal chelators did not interfere with the method and (3) both acidic and basic functional groups were compatible with the fluorimetric method.

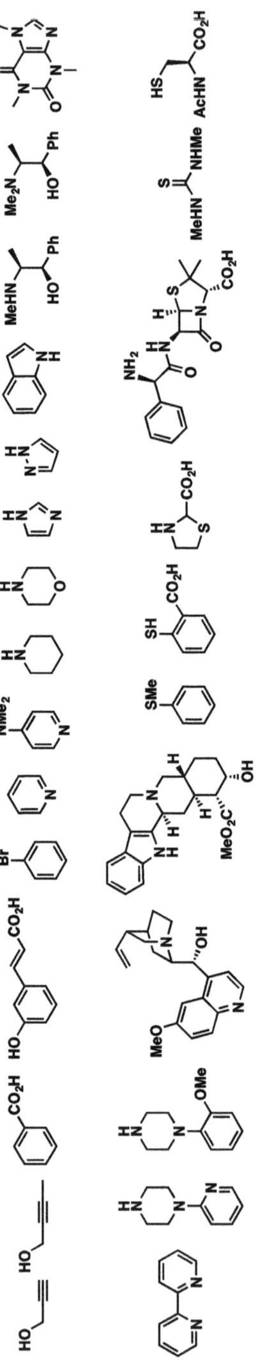

Figure 16.21 Palladium was added to each of these compounds in a 1 : 100 000 w/w ratio. Adapted from Ref. 66.

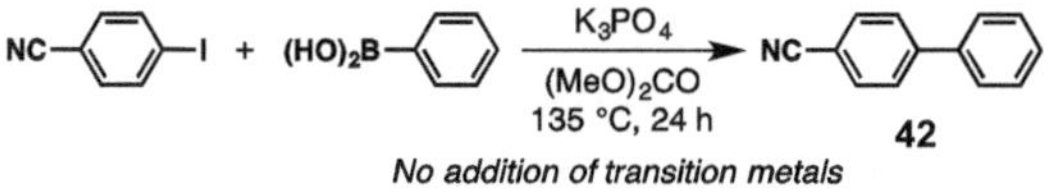

Figure 16.22 Compound **42** was produced by a Suzuki–Miyaura-type reaction without the addition of transition metals.
Adapted from Ref. 71.

16.4.2.3.5 Applications. Whether "palladium-free" reactions that normally require palladium are truly devoid of palladium remains controversial.[68,69] A well-known case was a Suzuki–Miyaura coupling that proved to be catalyzed by trace amounts of palladium present in Na_2CO_3.[70] This report by Leadbeater and Marco prompted others to analyze cross-coupling reagents more rigorously before claiming "palladium-free". Inamoto's group reported a cross-coupling between 4-iodobenzonitrile and phenylboronic acid in the presence of K_3PO_4 in dimethyl carbonate.[71] Without the addition of palladium reagents, the cross-coupling proceeded to form the biphenyl product **42** (Figure 16.22). The Inamoto–Koide team analyzed the crude mixture of the reaction, treated it with nitric acid and subjected the resulting sample to the deallylation-based fluorimetric method to detect palladium. It was concluded that 0.78 ppb of palladium was present in the original reaction solution, which was in reasonable agreement with an independent ICP-MS analysis of a closely related sample with 0.42 ppb of palladium.[72]

As the previous cases illustrate,[70,72,73] commercial reagents may contain trace amounts of palladium. Koide's group tested chemicals that might be relevant to cross-coupling reactions and found that particular batches of CuI and Cs_2CO_3 contained palladium. In some instances, the fluorimetric method could be directly applied to solid samples. For example, round-bottomed flasks in the Koide laboratory were subjected to the fluorimetric method and some were found to be contaminated with palladium.[63] Williams and Koide were also able to quantify palladium in ore samples without metal extraction with acids.[74]

Although the APE method was sufficiently robust and accurate in the Koide laboratory, a Merck team wished to make it more accurate and broadly applicable with their API samples. Pretreatment of samples with aqua regia proved to be beneficial for accuracy, although HCl sufficed in many cases.[67] The deallylation can proceed without $NaBH_4$ but its use improved the percentage signal recovery with complex samples produced at Merck.[67] By increasing the APE concentration and shortening the time for deallylation to 10–15 min, the linear range was extended to 1–30 ppb.[67] Therefore, by measuring fluorescence signals first at 15 and then at 60 min, 0.1–30 ppb of palladium can be quantified, meeting criterion 7 above. Importantly, the APE-based fluorogenic method was rigorously compared with ICP-OES with 12 API samples and proved to be accurate.[67]

Following these studies, the APE method now meets criteria 1–6 above, with comments warranted for criterion 5. An ICP-MS instrument costs $100 000–250 000 and requires frequent maintenance and repair;

a fluorescence plate reader costs $6000–15 000 and the plate reader in the Koide laboratory has not required repairs during the past 5 years. Therefore, as far as palladium analysis is concerned, a fluorescence method is more consistently available at a lower cost. ICP-MS can analyze most metals; a fluorescence plate reader can be used for a limited number of metals and many enzyme assays. An ICP-MS analysis of one sample takes only a few minutes; the APE method takes 15–60 min. However, the APE method can analyze dozens to hundreds of samples in parallel in the 15–60 min span – fractions of a minute are necessary per sample when many samples are analyzed. The APE method can be employed by a first-year undergraduate student within a few hours of training, whereas many ICP-MS operators are PhD graduates. Although not ideal, users must prepare a solution devoid of $NaBH_4$ and add a freshly made $NaBH_4$ solution in aqueous NaOH as needed because an $NaBH_4$ solution in DMSO–pH 7 buffer cannot be stored for more than half a day. Commercially available solutions of $NaBH_4$ in aqueous NaOH were found to be ineffective in the Koide laboratory (unpublished results). Nonetheless, this method may be the most practical to date, as it has been implemented at Merck.

If given a specific concentration of palladium ions as a threshold, the method developed by Baker and Phillips[75] may prove useful (Figure 16.23); the Alloc-protected aniline derivative **43** underwent a palladium-catalyzed deallylation–decarboxylation reaction to form **44** *in situ*. This compound was then hydrolyzed to generate 2 equiv. of fluoride ions. Each fluoride removed a TBS group from **45** to form 4-aminobenzaldehyde and 2 equiv. of fluoride ions. Thus, the whole system was designed to propagate the color intensity originating from 4-aminobenzaldehyde.[75] The same group developed another method in which the palladium-catalyzed deallylation and well-established Fmoc chemistry were combined to amplify absorption signals autocatalytically.[76]

The deallylation process was applied to another chromophore by Liu's group (Figure 16.24);[77] the Alloc derivative **47** was subjected to palladium-catalyzed deallylation conditions similar to those in previous work[59] to form amine **48**. The starting material was blue fluorescent and the product was green fluorescent, allowing for a ratiometric detection of palladium.[77] The metal selectivity favored palladium. They also screened anions to discover that this deallylation chemistry was not subject to interference by F^-, Cl^-, Br^-, I^-, CNS^-, HSO_4^-, $H_2PO_4^-$, NO_3^-, ClO_4^- and AcO^-.[77]

Qian's group developed a colorimetric method based on the palladium-catalyzed conversion of the blue-fluorescent allyl ether **49** to the yellow-fluorescent product **50** [Figure 16.25(a)].[78] To demonstrate its utility, they formed biaryl **53** [Figure 16.25(b)], subjected it to three rounds of purification and the palladium content was monitored after each purification step using paper strips. The color change diminished after each purification step due to the removal of palladium from the material. Their successful demonstration indicated that it may be possible to develop paper strips for palladium detection in the API production process. Their method has not yet

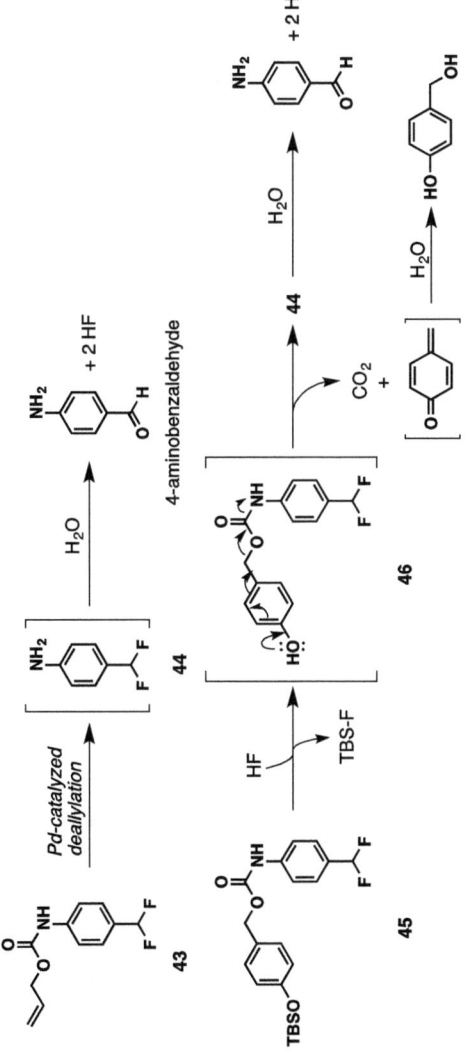

Figure 16.23   Pd-catalyzed Alloc removal was coupled with fluoride-mediated desilylation. Reproduced with permission from Ref. 75. Copyright 2011 American Chemical Society.

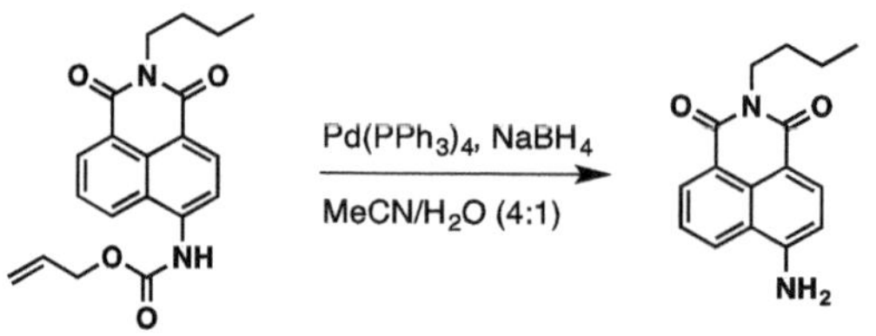

47
blue fluorescent

48
green fluorescent

Figure 16.24   Pd-catalyzed Alloc removal generates the fluorescent product **48**. Reproduced with permission from Ref. 77. Copyright 2011 American Chemical Society.

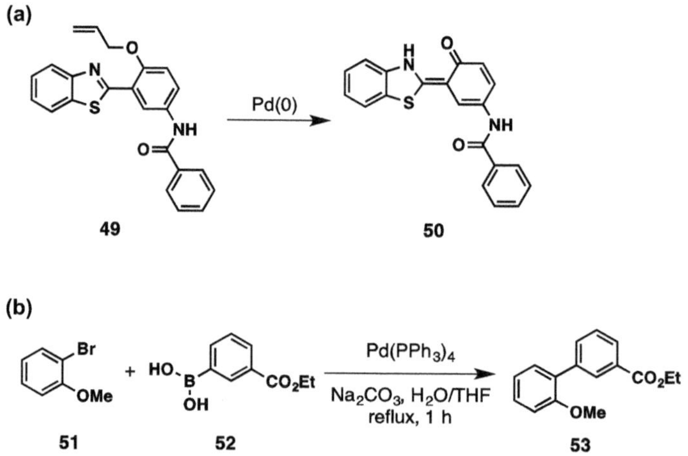

(a)

49   **50**

(b)

51   **52**   **53**

Figure 16.25   (a) Pd-catalyzed deallylation converts **49** (blue fluorescent) to **50** (yellow fluorescent). (b) Compound **53** was subjected to palladium quantification.
Adapted from Ref. 78.

been validated by comparison against ICP-MS or ICP-OES with real-world drug-like samples.

16.4.2.4   Chemodosimeters Based on Depropargylation

Ahn and co-workers synthesized and used chemodosimeter **54** for various palladium species (Figure 16.26).[79] Their attempted mechanistic studies could not elucidate the mechanism of the depropargylation. Nonetheless, they successfully demonstrated the utility of the method in zebrafish exposed to palladium ions. Similar depropargylations were used by the Du and Bai groups later with a different fluorophore.[80,81] Du's group applied their method to live RAW 264.7 macrophage cells.[80] The depropargylation chemistry was coupled with a lanthanide-based luminescent complex, which could quantify palladium species in micromolar concentrations.[82]

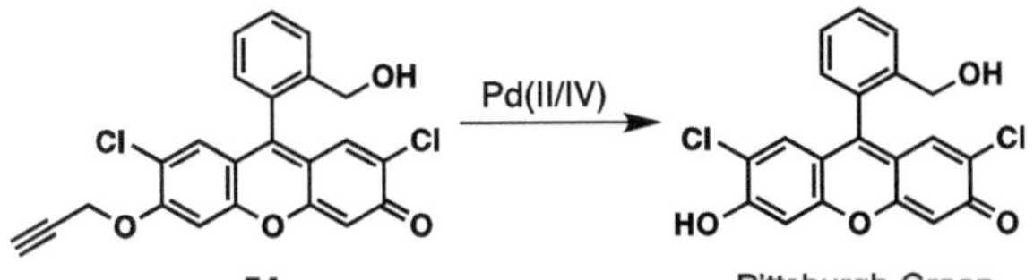

Figure 16.26 Propargylic ether **54** was converted to Pittsburgh Green by palladium catalysis.
Adapted from Ref. 79.

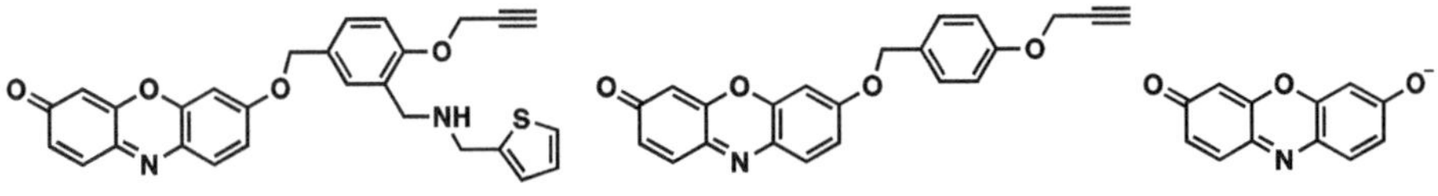

Figure 16.27 Compound **55** undergoes Pd-catalyzed depropargylation to form fluorescent resorufin eight times faster than **56**.
Adapted from Ref. 83.

Kim and co-workers indicated that the depropargylation of **56** catalyzed by palladium was too slow, but could be accelerated by installing an aminomethylthiophene unit, as shown in **55** (Figure 16.27). The palladium-catalyzed depropargylation to form the fluorescent product resorufin was eight times faster for **55** because the thiophene moiety bound to palladium(II).[83]

Using chemodosimeters POF and AOF for palladium,[81] Bai's group developed a method for palladium speciation (Figure 16.28). They were able to distinguish among Pd(dppf)$_2$Cl$_4$, K$_2$PdCl$_6$, Pd(PPh$_3$)$_4$ and Pd(PPh$_3$)$_2$Cl$_2$ as the species designated HF may or may not be generated from POF and AOF and the resulting complexation with palladium showed different emission spectra.[84] As Figure 28 indicates, each palladium species showed a unique combination of the fluorescence colors, although PdCl$_2$ and K$_2$PdCl$_6$ could not be distinguished from each other.

Liu's group combined depropargylation chemistry and spirocycle-opening chemistry (Figure 16.29); the non-fluorescent propargylic amine **58** was converted to the fluorescent rhodamine derivative **59** by palladium catalysis.[85] They attempted to elucidate the mechanism for the depropargylation, but the NMR spectrum proved to be too ambiguous to draw a conclusion.

As demonstrated by the above studies, palladium-catalyzed depropargylation provides a useful platform for palladium-selective fluorogenic reactions. Intramolecular acceleration of this transformation, shown by Kim and co-workers,[83] may lead to an even faster depropargylation method in the future. However, rational design of faster depropargylation will require better mechanistic insights.

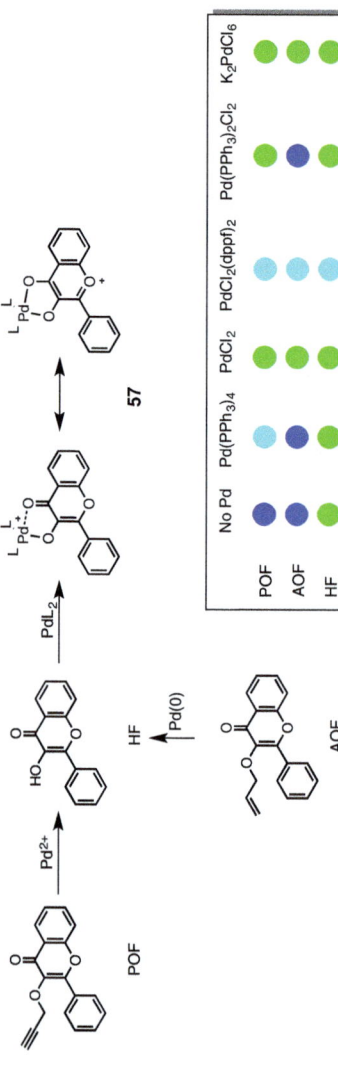

Figure 16.28 Pd-catalyzed depropargylation of POF and deallylation of AOF provide compound HF. Pd–HF complexation affords a colorimetric means to read the signals. Adapted from Ref. 84.

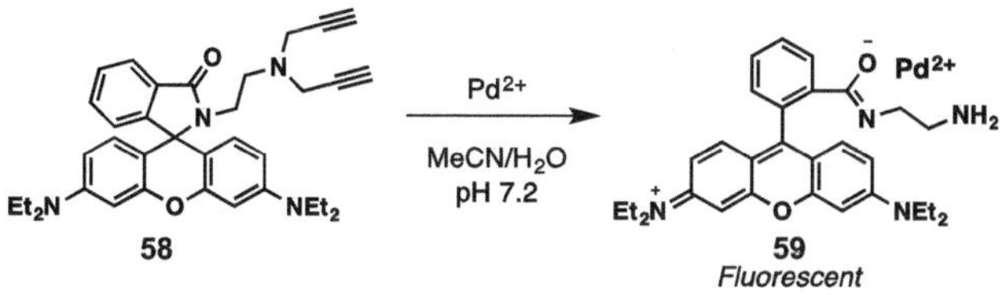

Figure 16.29 Depropargylation of **58** forms fluorescent complex **59**. Adapted from Ref. 85.

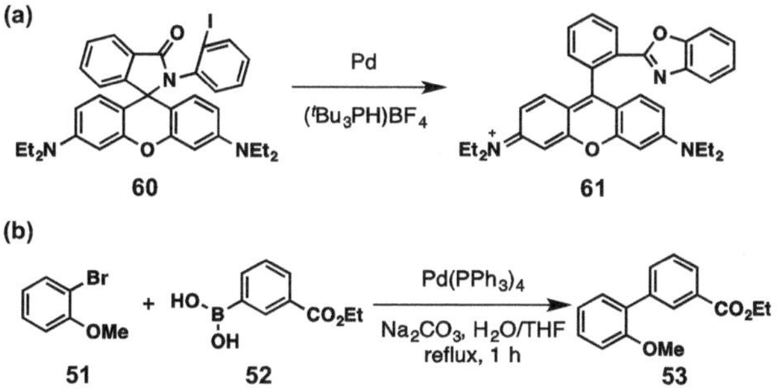

Figure 16.30 (a) Pd-catalyzed cyclization of non-fluorescent aryl iodide **60** generates fluorescent product **61**. (b) Compound **53** was subjected to palladium analysis.
Reproduced with permission from Ref. 86. Copyright 2010 American Chemical Society.

16.4.2.5 Chemodosimeters Based on Cross-Coupling Reactions

Ahn's group used an intramolecular Buchwald–Hartwig-type cyclization [Figure 16.30(a)] to convert the non-fluorescent rhodamine derivative **60** to the fluorescent rhodamine derivative **61**.[86] Although platinum and iron catalyzed this fluorimetric conversion to some extent, palladium was by far the most efficient. They applied this method to quantify palladium in **53**, prepared by a Suzuki–Miyaura coupling [Figure 16.30(b)].[86] After each of the three rounds of purification, the palladium concentration was 10.2 ppm, 960 ppb and 110 ppb, respectively.[86] These values are unusually low compared with typical palladium contents reported by pharmaceutical companies (see above). Unfortunately, their data were not compared with those produced by other analytical methods, such as ICP-MS, to validate their method. In addition, their reported lower concentrations appear to be below the limit of quantification. Compound **60** was colorless, whereas **61** was red, making this method chromogenic.

Although the Heck reaction has potential to be used in the development of chemodosimeters for palladium, thus far there is only one report in the literature. Hong and co-workers used *N*-methylpyridinium unit **62** and aryl bromide **63** to form the green fluorescent product **64** catalyzed by Pd(OAc)$_2$

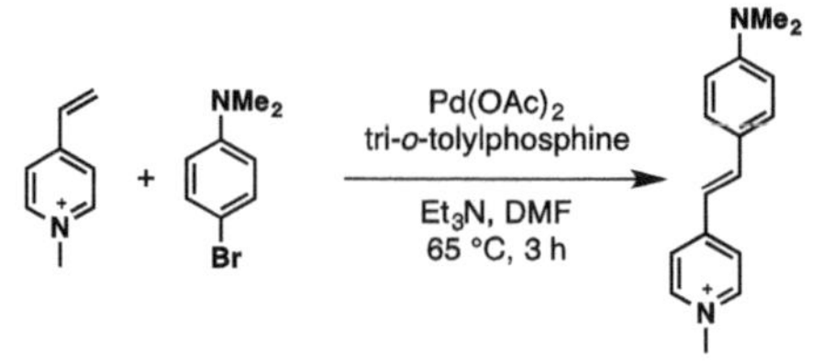

Figure 16.31 The combination of **62** and **63** was used as a means to detect the presence of palladium as the Heck reaction product **64** emits fluorescence.
Adapted from Ref. 87.

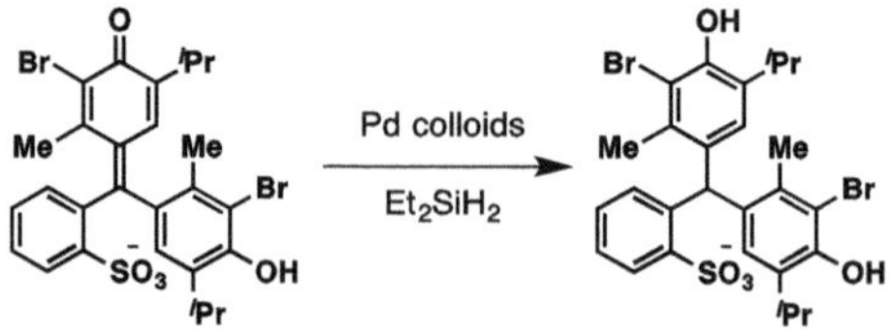

Figure 16.32 Palladium colloids catalyze the reduction of **65** to **66**, as manifested by a color change.
Reproduced with permission from Ref. 88. Copyright 2012 American Chemical Society.

in the presence of tri-*o*-tolylphosphine in DMF (Figure 16.31).[87] Its utility with real-world samples remains to be seen.

16.4.2.6 Chemodosimeters Based on Reduction

A conceptually interesting system was developed by Mohapatra and Phillips (Figure 16.32).[88] Soluble palladium species were first converted to palladium colloids. In the presence of Et_2SiH_2, the palladium colloids catalyzed the reduction of **65** to **66**. The change could be used as a colorimetric method. Pyridine interfered in this method, implying that further development may be necessary for practical use.

16.5 Conclusion

Many fluorimetric and colorimetric methods have been developed, and several of them have been tested with real-world samples and validated. The iteration of development and evaluation with real-world samples continues to enable researchers to improve the accuracy, reproducibility and user-friendliness of methods. It was stated in 2011, "An ideal metal detection method would quantitatively analyze metals in solution as it exits the scavenger setup, similarly to several process analytical technologies already integrated for large-scale manufactures, in the pharmaceutical industry. Unfortunately, such quantitative analytical methods are to date not yet available ..."[21] There may be some on the horizon.

Acknowledgements

The author thanks Matthew Tracey, Dianne Pham and Jessica Williams in his research group for critical reading of the manuscript. Part of our research was supported by the US National Science Foundation (CHE-0911092).

References

1. K. Tamao, K. Sumitani and M. Kumada, *J. Am. Chem. Soc.*, 1972, **94**, 4374.
2. J. P. Corriu and J. P. Masse, *J. Chem. Soc., Chem. Commun.*, 1972, 144.
3. R. F. Heck and J. P. Nolley, *J. Org. Chem.*, 1972, **37**, 2320.
4. T. Mizoroki, K. Mori and A. Ozaki, *Bull. Chem. Soc. Jpn.*, 1971, **44**, 581.
5. A. O. King, N. Okukado and E. I. Negishi, *J. Chem. Soc., Chem. Commun.*, 1977, 683.
6. N. Miyaura, K. Yamada and A. Suzuki, *Tetrahedron Lett.*, 1979, **20**, 3437.
7. K. Sonogashira, Y. Tohda and N. Hagihara, *Tetrahedron Lett.*, 1975, 4467.
8. J. Tsuji, H. Takahashi and M. Morikawa, *Tetrahedron Lett.*, 1965, 4387.
9. B. M. Trost and T. J. Fullerton, *J. Am. Chem. Soc.*, 1973, **95**, 292.
10. M. Kosugi, K. Sasazawa, Y. Shimizu and T. Migita, *Chem. Lett.*, 1977, 301.
11. D. Milstein and J. K. Stille, *J. Am. Chem. Soc.*, 1978, **100**, 3636.
12. J. Magano and J. R. Dunetz, *Chem. Rev.*, 2011, **111**, 2177.
13. Y. Urawa, M. Miyazawa, N. Ozeki and K. Ogura, *Org. Process Res. Dev.*, 2003, 7, 191.
14. J. T. Bien, G. C. Lane and M. R. Oberholzer, *Top. Organomet. Chem.*, 2004, **6**, 263.
15. C. E. Garrett and K. Prasad, *Adv. Synth. Catal.*, 2004, **346**, 889.
16. X. Sava, M. Roeper, C. Orgill and J. Cooper, *WO Pat.*, WO2004096747X, 2006.
17. E. Flahive, B. Ewanicki, S. Yu, P. D. Higginson, N. W. Sach and I. Morao, *QSAR Comb. Sci.*, 2007, **26**, 679.
18. N. Galaffu, S. P. Man, R. D. Wilkes and J. R. H. Wilson, *Org. Process Res. Dev.*, 2007, **11**, 406.
19. K. M. Bullock, M. B. Mitchell and J. F. Toczko, *Org. Process Res. Dev.*, 2008, **12**, 896.
20. J. P. Huang, X. X. Chen, S. X. Gu, L. Zhao, W. X. Chen and F. E. Chen, *Org. Process Res. Dev.*, 2010, **14**, 939.
21. G. Reginato, P. Sadler and R. D. Wilkes, *Org. Process Res. Dev.*, 2011, **15**, 1396.
22. L. Wang, L. Green, Z. Li, J. McCabe Dunn, X. Bu, C. J. Welch, C. Li, T. Wang, Q. Tu, E. Bekos, D. Richardson, J. Eckert and J. Cui, *Org. Process Res. Dev.*, 2011, **15**, 1371.

23. M. A. Mortensen, C. Guo, N. T. Reynolds, L. L. Wang, M. A. Helle, D. K. Keefe, B. P. Haney, B. J. Paul, P. R. Bruzinski, M. A. Wolf, N. L. Malinowski and Q. Yang, *Org. Process Res. Dev.*, 2012, **16**, 1811.
24. C. J. Welch, J. Albaneze-Walker, W. R. Leonard, M. Biba, J. DaSilva, D. Henderson, B. Laing, D. J. Mathre, S. Spencer, X. D. Bu and T. B. Wang, *Org. Process Res. Dev.*, 2005, **9**, 198.
25. C. A. Busacca, D. R. Fandrick, J. J. Song and C. H. Senanayake, *Adv. Synth. Catal.*, 2011, **353**, 1825.
26. J. S. Carey, D. Laffan, C. Thomson and M. T. Williams, *Org. Biomol. Chem.*, 2006, **4**, 2337.
27. W. Moore, D. Hysell, L. Hall, K. Campbell and J. Stara, *Environ. Health Perspect.*, 1975, **10**, 63.
28. T. Peric, V. L. Jakovljevic, V. Zivkovic, J. Krkeljic, Z. D. Petrovic, D. Simijonovic, S. Novokmet, D. M. Djuric and S. M. Jankovic, *Med. Chem.*, 2012, **8**, 9.
29. T. S. Peric and S. M. Jankovic, *J. Med. Biochem.*, 2013, **32**, 20.
30. A. Faurschou, T. Menne, J. D. Johansen and J. P. Thyssen, *Contact Dermatitis*, 2011, **64**, 185.
31. D. Rachmawati, H. J. Bontkes, M. I. Verstege, J. Muris, B. M. E. von Blomberg, R. J. Scheper and I. M. W. van Hoogstraten, *Contact Dermatitis*, 2013, **68**, 331.
32. United States Pharmacopeial Convention, *<232> Elemental Impurities – Limits*, 2013, http://www.usp.org/sites/default/files/usp_pdf/EN/USPNF/key-issues/c232_final.pdf (accessed 19 May 2014).
33. A. M. Thayer, *Chem. Eng. News*, 2013 August 19, 10.
34. A. O. King, E. G. Corley, R. K. Anderson, R. D. Larsen, T. R. Verhoeven, P. J. Reider, Y. B. Xiang, M. Belley and Y. Leblanc, *J. Org. Chem.*, 1993, **58**, 3731.
35. M. Butters, D. Catterick, A. Craig, A. Curzons, D. Dale, A. Gillmore, S. P. Green, I. Marziano, J. P. Sherlock and W. White, *Chem. Rev.*, 2006, **106**, 3002.
36. V. Elango, M. A. Murphy, K. G. Davenport, G. N. Mott, E. G. Zey and G. L. Moss, *US Pat.*, US 4981995, 1991.
37. E. G. Zey, T. H. Shockley, D. A. Ryan and G. L. Moss, *US Pat.*, US 5155551, 1992.
38. E. J. Flahive, B. L. Ewanicki, N. W. Sach, S. A. O'Neill-Slawecki, N. S. Stankovic, S. Yu, S. M. Guinness and J. Dunn, *Org. Process Res. Dev.*, 2008, **12**, 637.
39. S. Sato, Y. Ishido and M. Fujita, *J. Am. Chem. Soc.*, 2009, **131**, 6064.
40. L. D. Pachón, I. Yosef, T. Z. Markus, R. Naaman, D. Avnir and G. Rothenberg, *Nat. Chem.*, 2009, **1**, 160.
41. T. Iwasawa, M. Tokunaga, Y. Obora and Y. Tsuji, *J. Am. Chem. Soc.*, 2004, **126**, 6554.
42. A. S. Al-Ammar and J. Northington, *J. Anal. At. Spectrom.*, 2011, **26**, 1531.
43. K. T. Nielsen, K. Bechgaard and F. C. Krebs, *Synthesis*, 2006, 1639.

44. J. He, M. Q. Zha, J. S. Cui, M. Zeller, A. D. Hunter, S. M. Yiu, S. T. Lee and Z. T. Xu, *J. Am. Chem. Soc.*, 2013, **135**, 7807.
45. L. P. Duan, Y. F. Xu and X. H. Qian, *Chem. Commun.*, 2008, 6339.
46. H. L. Li, J. L. Fan, J. J. Du, K. X. Guo, S. G. Sun, X. J. Liu and X. J. Peng, *Chem. Commun.*, 2010, **46**, 1079.
47. H. L. Li, J. L. Fan, F. L. Song, H. Zhu, J. J. Du, S. G. Sun and X. J. Peng, *Chem. Eur. J.*, 2010, **16**, 12349.
48. H. L. Li, J. L. Fan, M. M. Hu, G. H. Cheng, D. H. Zhou, T. Wu, F. L. Song, S. G. Sun, C. Y. Duan and X. J. Peng, *Chem. Eur. J.*, 2012, **18**, 12242.
49. J. L. Zhang, L. Zhang, Y. M. Zhou, T. S. Ma and J. Y. Niu, *Microchim. Acta*, 2013, **180**, 211.
50. S. T. Cai, Y. Lu, S. He, F. F. Wei, L. C. Zhao and X. S. Zeng, *Chem. Commun.*, 2013, **49**, 822.
51. B. Liu, Y. Bao, F. Du, H. Wang, J. Tian and R. Bai, *Chem. Commun.*, 2011, **47**, 1731.
52. B. Liu, H. G. Dai, Y. Y. Bao, F. F. Du, J. Tian and R. K. Bai, *Polym. Chem.*, 2011, **2**, 1699.
53. B. Liu, Y. Y. Bao, H. Wang, F. F. Du, J. Tian, Q. B. Li, T. S. Wang and R. K. Bai, *J. Mater. Chem.*, 2012, **22**, 3555.
54. Z. Wang and F. Liao, *Synth. Met.*, 2012, **162**, 444.
55. S. Mukherjee, S. Chowdhury, A. K. Paul and R. Banerjee, *J. Lumin.*, 2011, **131**, 2342.
56. H. Kim, K. S. Moon, S. Shim and J. Tae, *Chem. Asian J.*, 2011, **6**, 1987.
57. B. Tang, H. Zhang and Y. Wang, *Anal. Lett.*, 2004, **37**, 1219.
58. R. J. T. Houk, K. J. Wallace, H. S. Hewage and E. V. Anslyn, *Tetrahedron*, 2008, **64**, 8271.
59. F. L. Song, A. L. Garner and K. Koide, *J. Am. Chem. Soc.*, 2007, **129**, 12354.
60. K. Koide, F. L. Song, E. D. de Groh, A. L. Garner, V. D. Mitchell, L. A. Davidson and N. A. Hukriede, *ChemBioChem*, 2008, **9**, 214.
61. B. A. Sparano, S. P. Shahi and K. Koide, *Org. Lett.*, 2004, **6**, 1947.
62. A. L. Garner and K. Koide, *Chem. Commun.*, 2009, 86.
63. D. Li, L. D. Campbell, B. A. Austin and K. Koide, *ChemPlusChem*, 2012, 77, 281.
64. Y. W. Yang, L. D. Unsworth and N. Semagina, *J. Catal.*, 2011, **281**, 137.
65. A. L. Garner, F. L. Song and K. Koide, *J. Am. Chem. Soc.*, 2009, **131**, 5163.
66. F. L. Song, E. J. Carder, C. C. Kohler and K. Koide, *Chem. Eur. J.*, 2010, **16**, 13500.
67. X. Bu, K. Koide, E. J. Carder and C. J. Welch, *Org. Process Res. Dev.*, 2013, **17**, 108.
68. N. E. Leadbeater, *Nat. Chem.*, 2010, **2**, 1007.
69. I. Thome, A. Nijs and C. Bolm, *Chem. Soc. Rev.*, 2012, **41**, 979.
70. N. E. Leadbeater and M. Marco, *J. Org. Chem.*, 2003, **68**, 5660.
71. K. Inamoto, C. Hasegawa, K. Hiroya, Y. Kondo, T. Osako, Y. Uozumi and T. Doi, *Chem. Commun.*, 2012, **48**, 2912.
72. K. Inamoto, L. D. Campbell, T. Doi and K. Koide, *Tetrahedron Lett.*, 2012, **53**, 3147.

73. R. K. Arvela, N. E. Leadbeater, M. S. Sangi, V. A. Williams, P. Granados and R. D. Singer, *J. Org. Chem.*, 2005, **70**, 161.
74. J. M. Williams and K. Koide, *Ind. Eng. Chem. Res.*, 2013, **52**, 8612.
75. M. S. Baker and S. T. Phillips, *J. Am. Chem. Soc.*, 2011, **133**, 5170.
76. H. Mohapatra, K. M. Schmid and S. T. Phillips, *Chem. Commun.*, 2012, **48**, 3018.
77. J. Jiang, H. E. Jiang, W. Liu, X. L. Tang, X. Zhou, R. T. Liu and W. S. Liu, *Org. Lett.*, 2011, **13**, 4922.
78. L. Cui, W. P. Zhu, Y. F. Xu and X. H. Qian, *Anal. Chim. Acta*, 2013, **786**, 139.
79. M. Santra, S. K. Ko, I. Shin and K. H. Ahn, *Chem. Commun.*, 2010, **46**, 3964.
80. B. Zhu, C. Gao, Y. Zhao, C. Liu, Y. Li, Q. Wei, Z. Ma, B. Du and X. Zhang, *Chem. Commun.*, 2011, **47**, 8656.
81. B. Liu, H. Wang, T. S. Wang, Y. Y. Bao, F. F. Du, J. Tian, Q. B. A. Li and R. K. Bai, *Chem. Commun.*, 2012, **48**, 2867.
82. E. Pershagen, J. Nordholm and K. E. Borbas, *J. Am. Chem. Soc.*, 2012, **134**, 9832.
83. W. X. Ren, T. Pradhan, Z. Yang, Q. Y. Cao and J. S. Kim, *Sens. Actuators B*, 2012, **171**, 1277.
84. Y. X. Wang, B. Liu, J. Tian, Q. B. A. Li, F. F. Du, T. S. Wang and R. K. Bai, *Analyst*, 2013, **138**, 779.
85. R. Balamurugan, C. C. Chien, K. M. Wu, Y. H. Chiu and J. H. Liu, *Analyst*, 2013, **138**, 1564.
86. M. E. Jun and K. H. Ahn, *Org. Lett.*, 2010, **12**, 2790.
87. S. Y. Yu, H.-W. Rhee and J.-I. Hong, *Tetrahedron Lett.*, 2011, **52**, 1512.
88. H. Mohapatra and S. T. Phillips, *Anal. Chem.*, 2012, **84**, 8927.

Subject Index

References to figures are given in *italic* type. References to tables are given in **bold** type.